The Most Accepted
Crash Course Programme

JEE Main in 40 Days

PHYSICS

The Most Accepted
Crash Course Programme

JEE Main in 40 Days

PHYSICS

arihant
ARIHANT PRAKASHAN (Series), MEERUT

ARIHANT PRAKASHAN (Series), MEERUT

卐 **Administrative & Production Offices**

Regd. Office
'Ramchhaya' 4577/15, Agarwal Road, Darya Ganj, New Delhi -110002
Tele: 011- 47630600, 43518550

卐 **Head Office**

Kalindi, TP Nagar, Meerut (UP) - 250002
Tel: 0121-7156203, 7156204

卐 **Sales & Support Offices**

Agra, Ahmedabad, Bengaluru, Bareilly, Chennai, Delhi, Guwahati, Hyderabad, Jaipur, Jhansi, Kolkata, Lucknow, Nagpur & Pune.

卐 **ISBN** 978-93-25292-03-1

卐 **PRICE** ₹385.00

PO No : TXT-XX-XXXXXXX-X-XX

Published by Arihant Publications (India) Ltd.

For further information about the books published by Arihant, log on to www.arihantbooks.com or e-mail at info@arihantbooks.com

Follow us on    

PREFACE

It is a fact that nearly 10 lacs students would be in the race with you in JEE Main, the gateway to some of the prestigious engineering and technology institutions in the country, requires that you take it seriously and head-on. A slight underestimation or wrong guidance will ruin all your prospects. You have to earmark the topics in the syllabus and have to master them in concept-driven-problem-solving ways, considering the thrust of the questions being asked in JEE Main.

The book **40 Days JEE Main Physics** serves the above cited purpose in perfect manner. At whatever level of preparation you are before the exam, this book gives you an accelerated way to master the whole JEE Main Physics Syllabus. It has been conceived keeping in mind the latest trend of questions, and the level of different types of students.

The whole syllabus of Physics has been divided into day-wise-learning modules with clear groundings into concepts and sufficient practice with solved and unsolved questions on that day. After every few days you get a Unit Test based upon the topics covered before that day. On last three days you get three full-length Mock Tests, making you ready to face the test. It is not necessary that you start working with this book in 40 days just before the exam. You may start and finish your preparation of JEE Main much in advance before the exam date. This will only keep you in good frame of mind and relaxed, vital for success at this level.

Salient Features

- Concepts discussed clearly and directly without being superfluous. **Only the required material** for JEE Main being described comprehensively to keep the students focussed.
- Exercises for each day give you the collection of **only the Best Questions** of the concept, giving you the perfect practice in less time.
- Each day has two Exercises; **Foundation Questions Exercise** having Topically Arranged Questions & **Progressive Question Exercise** having higher Difficulty Level Questions.
- **All types of Objective Questions** included in Daily Exercises (Single Option Correct, Assertion & Reason, etc).
- Along with Daywise Exercises, there above also the **Unit Tests** & **Full Length Mock Tests.**
- At the end, there are all Online Solved Papers of JEE Main 2020; January & September attempts.

We are sure that 40 Days Physics for JEE Main will give you a fast way to prepare for Physics without any other support or guidance.

Publisher

CONTENTS

Preparing JEE Main 2021 Physics in 40 Days !

SYLLABUS

PHYSICS

NOTE The syllabus contains two Sections - A & B. Section A pertains to the Theory Part, having 80% weightage, while Section B contains Practical Component (Experimental Skills) having 20% weightage.

SECTION- A

UNIT 1 Physics and Measurement

Physics, technology and society, SI units, Fundamental and derived units. Least count, accuracy and precision of measuring instruments, Errors in measurement, Significant figures. Dimensions of Physical quantities, dimensional analysis and its applications.

UNIT 2 Kinematics

Frame of reference. Motion in a straight line: Position-time graph, speed and velocity. Uniform and non-uniform motion, average speed and instantaneous velocity.

Uniformly accelerated motion, velocity-time, position time graphs, relations for uniformly accelerated motion.

Scalars and Vectors, Vector addition and Subtraction, Zero Vector, Scalar and Vector products, Unit Vector, Resolution of a Vector. Relative Velocity, Motion in a plane, Projectile Motion, Uniform Circular Motion.

UNIT 3 Laws of Motion

Force and Inertia, Newton's First Law of motion; Momentum, Newton's Second Law of motion; Impulse; Newton's Third Law of motion. Law of conservation of linear momentum and its applications, Equilibrium of concurrent forces.

Static & Kinetic friction, laws of friction, rolling friction.

Dynamics of uniform circular motion: Centripetal force and its applications.

UNIT 4 Work, Energy and Power

Work done by a constant force and a variable force; kinetic and potential energies, work-energy theorem, power.

Potential energy of a spring, conservation of mechanical energy, conservative and nonconservative forces; Elastic and inelastic collisions in one and two dimensions.

UNIT 5 Rotational Motion

Centre of mass of a two-particle system, Centre of mass of a rigid body; Basic concepts of rotational motion; moment of a force, torque, angular momentum, conservation of angular momentum and its applications; moment of inertia, radius of gyration. Values of moments of inertia for simple geometrical objects, parallel and perpendicular axes theorems and their applications.

Rigid body rotation, equations of rotational motion.

UNIT 6 Gravitation

The universal law of gravitation.

Acceleration due to gravity and its variation with altitude and depth.

Kepler's laws of planetary motion.

Gravitational potential energy; gravitational potential.

Escape velocity. Orbital velocity of a satellite.

Geo-stationary satellites.

UNIT 7 Properties of Solids & Liquids

Elastic behaviour, Stress-strain relationship, Hooke's. Law, Young's modulus, bulk modulus, modulus of rigidity.

Pressure due to a fluid column; Pascal's law and its applications.

Viscosity, Stokes' law, terminal velocity, streamline and turbulent flow, Reynolds number. Bernoulli's principle and its applications.

Surface energy and surface tension, angle of contact, application of surface tension - drops, bubbles and capillary rise.

Heat, temperature, thermal expansion; specific heat capacity, calorimetry; change of state, latent heat.

Heat transfer-conduction, convection and radiation, Newton's law of cooling.

UNIT 8 Thermodynamics

Thermal equilibrium, zeroth law of thermo-dynamics, concept of temperature. Heat, work and internal energy. First law of thermodynamics.
Second law of thermodynamics: reversible and irreversible processes. Camot engine and its efficiency.

UNIT 9 Kinetic Theory of Gases

Equation of state of a perfect gas, work done on compressing a gas.

Kinetic theory of gases - assumptions, concept of pressure. Kinetic energy and temperature: rms speed of gas molecules; Degrees of freedom, Law of equipartition of energy, applications to specific heat capacities of gases; Mean free path, Avogadro's number.

UNIT 10 Oscillations and Waves

Periodic motion - period, frequency, displacement as a function of time. Periodic functions. Simple harmonic motion (S.H.M.) and its equation; phase; oscillations of a spring - restoring force and force constant; energy in S.H.M. - kinetic and potential energies; Simple pendulum - derivation of expression for its time period; Free, forced and damped oscillations, resonance.

Wave motion Longitudinal and transverse waves, speed of a wave. Displacement relation for a progressive wave. Principle of superposition of waves, reflection of waves, Standing waves in strings and organ pipes, fundamental mode and harmonics, Beats, Doppler effect in sound.

UNIT 11 Electrostatics

Electric charges Conservation of charge, Coulomb's law-forces between two point charges, forces between multiple charges; superposition principle and continuous charge distribution.

Electric field Electric field due to a point charge, Electric field lines, Electric dipole, Electric field due to a dipole, Torque on a dipole in a uniform electric field.

Electric flux, Gauss's law and its applications to find field due to infinitely long, uniformly charged straight wire, uniformly charged infinite plane sheet and uniformly charged thin spherical shell.
Electric potential and its calculation for a point charge, electric dipole and system of charges; Equipotential surfaces, Electrical potential energy of a system of two point charges in an electrostatic field.

Conductors and insulators, Dielectrics and electric polarization, capacitor, combination of capacitors in series and in parallel, capacitance of a parallel plate capacitor with and without dielectric medium between the plates, Energy stored in a capacitor.

UNIT 12 Current Electricity

Electric current, Drift velocity, Ohm's law, Electrical resistance, Resistances of different materials, V-I characteristics of Ohmic and nonohmic conductors, Electrical energy and power, Electrical resistivity, Colour code for resistors; Series and parallel combinations of resistors; Temperature dependence of resistance.

Electric Cell and its Internal resistance, potential difference and emf of a cell, combination of cells in series and in parallel.

Kirchhoff's laws and their applications. Wheatstone bridge, Metre bridge.

Potentiometer - principle and its applications.

UNIT 13 Magnetic Effects of Current and Magnetism

Biot-Savart law and its application to current carrying circular loop. Ampere's law and its applications to infinitely long current carrying straight wire and solenoid. Force on a moving charge in uniform magnetic and electric fields Cyclotron.

Force on a current-carrying conductor in a uniform magnetic field. Force between two parallel current-carrying conductors-definition of ampere. Torque experienced by a current loop in uniform magnetic field, Moving coil galvanometer, its current sensitivity and conversion to ammeter and voltmeter.

Current loop as a magnetic dipole and its magnetic dipole moment. Bar magnet as an equivalent solenoid, magnetic field lines; Earth's magnetic field and magnetic elements. Para, dia and ferro-magnetic substances

Magnetic susceptibility and permeability, Hysteresis, Electromagnets and permanent magnets.

UNIT 14 Electromagnetic Induction and Alternating Currents

Electromagnetic induction; Faraday's law, induced emf and current; Lenz's Law, Eddy currents. Self and mutual inductance.

Alternating currents, peak and rms value of alternating current/ voltage; reactance and impedance; LCR series circuit, resonance; Quality factor, power in AC circuits, wattless current. AC generator and transformer.

UNIT 15 Electromagnetic Waves

Electromagnetic waves and their characteristics. Transverse nature of electromagnetic waves.

Electromagnetic spectrum (radio waves, microwaves, infrared, visible, ultraviolet, X-rays, gamma rays). Applications of e.m. waves.

UNIT 16 Optics

Reflection and refraction of light at plane and spherical surfaces, mirror formula, Total internal reflection and its applications, Deviation and Dispersion of light by a prism, Lens Formula, Magnification, Power of a Lens, Combination of thin lenses in contact, Microscope and Astronomical Telescope (reflecting and refracting) and their magnifying powers.

Wave optics wave front and Huygens' principle, Laws of reflection and refraction using Huygen's principle. Interference, Young's double slit experiment and expression for fringe width, coherent sources and sustained interference of light. Diffraction due to a single slit, width of central maximum. Resolving power of microscopes and astronomical telescopes, Polarisation, plane polarized light; Brewster's law, uses of plane polarized light and Polaroids.

UNIT 17 Dual Nature of Matter and Radiation

Dual nature of radiation. Photoelectric effect, Hertz and Lenard's observations; Einstein's photoelectric equation; particle nature of light. Matter waves-wave nature of particle, de Broglie relation. Davisson-Germer experiment.

UNIT 18 Atoms and Nuclei

Alpha-particle scattering experiment; Rutherford's model of atom; Bohr model, energy levels, hydrogen spectrum.

Composition and size of nucleus, atomic masses, isotopes, isobars; isotones. Radioactivity-alpha, beta and gamma particles/rays and their properties; radioactive decay law. Mass-energy relation, mass defect; binding energy per nucleon and its variation with mass number, nuclear fission and fusion.

UNIT 19 Electronic Devices

Semiconductors; semiconductor diode: I-V characteristics in forward and reverse bias; diode as a rectifier; I-V characteristics of LED, photodiode, solar cell, and Zener diode; Zener diode as a voltage regulator. Junction transistor, transistor action, characteristics of a transistor transistor as an amplifier (common emitter configuration) and oscillator. Logic gates (OR, AND, NOT, NAND & NOR). Transistor as a switch.

UNIT 20 Communication Systems

Propagation of electromagnetic waves in the atmosphere; Sky and space wave propagation, Need for modulation, Amplitude and Frequency Modulation, Bandwidth of signals, Bandwidth of Transmission medium, Basic Elements of a Communication System (Block Diagram only)

SECTION- B

UNIT 21 Experimental Skills

Familiarity with the basic approach and observations of the experiments and activities

1. Vernier callipers - its use to measure internal and external diameter and depth of a vessel.
2. Screw gauge - its use to determine thickness/ diameter of thin sheet/wire.
3. Simple Pendulum - dissipation of energy by plotting a graph between square of amplitude and time.
4. Metre Scale - mass of a given object by principle of moments.
5. Young's modulus of elasticity of the material of a metallic wire.
6. Surface tension of water by capillary rise and effect of detergents.
7. Coefficient of Viscosity of a given viscous liquid by measuring terminal velocity of a given spherical body.
8. Plotting a cooling curve for the relationship between the temperature of a hot body and time.
9. Speed of sound in air at room temperature using a resonance tube.
10. Specific heat capacity of a given (i) solid and (ii) liquid by method of mixtures.
11. Resistivity of the material of a given wire using metre bridge.
12. Resistance of a given wire using Ohm's law.
13. Potentiometer
    (i) Comparison of emf of two primary cells.
    (ii) Internal resistance of a cell.
14. Resistance and figure of merit of a galvanometer by half deflection method.
15. Focal length of
    (i) Convex mirror
    (ii) Concave mirror
    (iii) Convex lens
16. Using parallax method. Plot of angle of deviation vs angle of incidence for a triangular prism.
17. Refractive index of a glass slab using a travelling microscope.
18. Characteristic curves of a p-n junction diode in forward and reverse bias.
19. Characteristic curves of a Zener diode and finding reverse break down voltage.
20. Characteristic curves of a transistor and finding current gain and voltage gain.
21. Identification of Diode, LED, Transistor, IC, Resistor, Capacitor from mixed collection of such items.
22. Using multimeter to
    (i) Identify base of a transistor.
    (ii) Distinguish between npn and pnp type transistor.
    (iii) See the unidirectional flow of current in case of a diode and an LED.
    (iv) Check the correctness or otherwise of a given electronic component (diode, transistor or IC).

HOW THIS BOOK IS
USEFUL FOR YOU ?

As the name suggest, this is the perfect book for your recapitulation of the whole syllabus, as it provides you a capsule course on the subject covering the syllabi of JEE Main, with the smartest possible tactics as outlined below:

1. REVISION PLAN

The book provides you with a practical and sound revision plan.

The chapters of the book have been designed day-wise to guide the students in a planned manner through day-by-day, during those precious 35-40 days. Every day you complete a chapter/a topic, also take an exercise on the chapter. So that you can check & correct your mistakes, answers with hints & solutions also have been provided. By 37th day from the date you start using this book, entire syllabus gets revisited.

Again, as per your convenience/preparation strategy, you can also divide the available 30-35 days into two time frames, first time slot of 3 weeks and last slot of 1 & 1/2 week. Utilize first time slot for studies and last one for revising the formulas and important points. Now fill the time slots with subjects/topics and set key milestones. Keep all the formulas, key points on a couple of A4 size sheets as ready-reckner on your table and go over them time and again. If you are done with notes, prepare more detailed inside notes and go over them once again. Study all the 3 subjects every day. Concentrate on the topics that have more weightage in the exam that you are targeting.

2. MOCK TESTS

Once you finish your revision on 37th day, the book provides you with full length mock tests for day 38th, 39th, & 40th, thereby ensures your total & full proof preparation for the final show.

The importance of solving previous years' papers and 10-15 mock tests cannot be overemphasized. Identify your weaknesses and strengths. Work towards your strengths i.e., devote more time to your strengths to be 100% sure and confident. In the last time frame of 1 & 1/2 week, don't take-up anything new, just revise what you have studied before. Be exam-ready with quality mock tests in between to implement your winning strategy.

3. FOCUS TOPICS

Based on past years question paper trends, there are few topics in each subject which have more questions in exam than other. So far Physics is concerned it may be summed up as below:

Electricity, Magnetism, Modern Physics, Mechanics, Radioactivity, Wave Options and Heat Transfer. More than 80% of questions are normally asked from these topics.

However, be prepared to find a completely changed pattern for the exam than noted above as examiners keep trying to weed out 'learn by rot practice'. One should not panic by witnessing a new pattern , rather should be tension free as no one will have any upper hand in the exam.

4. IMPROVES STRIKE RATE AND ACCURACY

The book even helps to improve your strike rate & accuracy. When solving practice tests or mock tests, try to analyze where you are making mistakes-where are you wasting your time; which section you are doing best. Whatever mistakes you make in the first mock test, try to improve that in second. In this way, you can make the optimum use of the book for giving perfection to your preparation.

What most students do is that they revise whole of the syllabus but never attempt a mock and thus they always make mistake in main exam and lose the track.

5. LOG OF LESSONS

During your preparations, make a log of Lesson's Learnt. It is specific to each individual as to where the person is being most efficient and least efficient. Three things are important - what is working, what's not working and how would you like to do in your next mock test.

6. TIME MANAGEMENT

Most candidates who don't make it to good medical colleges are not good in one area- Time Management. And, probably here lies the most important value addition that's the book provides in an aspirant's preparation. Once the students go through the content of the book precisely as given/directed, he/she learns the tactics of time management in the exam.

Realization and strengthening of what you are good at is very helpful, rather than what one doesn't know. Your greatest motto in the exam should be, how to maximize your scoring with the given level of preparation. You have to get about 200 plus marks out of a total of about 400 marks for admission to a good NIT (though for a good branch one needs to do much better than that). Remember that one would be doomed if s/he tries to score 400 in about 3 hours.

7. ART OF PROBLEM SOLVING

The book also let you to master the art of problem solving. The key to problem solving does not lie in understanding the solution to the problem but to find out what clues in the problem leads you to the right solution. And, that's the reason Hints & Solutions are provided with the exercises after each chapter of the book. Try to find out the reason by analyzing the level of problem & practice similar kind of problems so that you can master the tricks involved. Remember that directly going though the solutions is not going to help you at all.

8. POSITIVE PERCEPTION

The book put forth for its readers a 'Simple and Straightforward' concept of studies, which is the best possible, time-tested perception for 11th hour revision / preparation.

The content of the book has been presented in such a lucid way so that you can enjoy what you are reading, keeping a note of your already stressed mind & time span.

Cracking JEE Main is not a matter of life and death. Do not allow panic and pressure to create confusion. Do some yoga and prayers. Enjoy this time with studies as it will never come back.

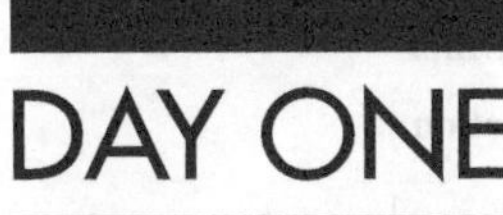

DAY ONE

Units and Measurement

Learning & Revision for the Day

- Physics
- Units
- Significant Figures
- Accuracy and Precision
- Errors in Measurement
- Dimensions of Physical Quantities

Physics

Physics is the study of matter and its motion, as well as space and time using concepts such as energy, force, mass and charge. It is an experimental science, creating theories that are tested against observation.

Scope and Excitement

Scope of Physics is very vast, as it deals with a wide variety of disciplines such as mechanics, heat, light, etc.

It also deals with very large magnitude of astronomical phenomenon as well as very small magnitude involving electrons, protons, etc.

Nature of Physical Laws

Physics is the study of nature and natural phenomena. All observations and experiments in physics lead to certain facts. These facts can be explained on the basis of certain laws.

Physics, Technology and Society

Connection between physics, technology and society can be seen in many examples like working of heat engines gave rise to thermodynamics. Wireless communication technology arose from basic laws of electricity and magnetism. Lately discovery of silicon chip triggered the computer revolution.

Units

Measurement of any physical quantity involves comparison with a certain basic, widely accepted reference standard called **unit**.

PREP MIRROR

Your Personal Preparation Indicator

- No. of Questions in Exercises (x)—
- No. of Questions Attempted (y)—
- No. of Correct Questions (z)—

(Without referring Explanations)

- Accuracy Level ($z/y \times 100$)—
- Prep Level ($z/x \times 100$)—

In order to expect good rank in JEE, your Accuracy Level should be above 85 & Prep Level should be above 75.

Fundamental and Derived Units

Fundamental units are the units which can neither be derived from one another, nor they can be further resolved into more simpler units.

These are the units of fundamental quantity. However, **derived units** are the units of measurement of all physical quantities which can be obtained from fundamental units.

System of Units

A complete set of these units, both fundamental and derived unit is known as the **system of units**. The common systems are given below:

1. **CGS System** (Centimetre, Gram, Second) are often used in scientific work. This system measures, Length in centimetre (cm), Mass in gram (g), Time in second (s).
2. **FPS System** (Foot, Pound, Second) It is also called the British Unit System. This unit measures, Length in foot (foot), Mass in gram (pound), Time in second (s).
3. **MKS System** In this system also length, mass and time have been taken as fundamental quantities and corresponding fundamental units are metre, kilogram and second.
4. **International System (SI) of Units** It is an extended version of the MKS (Metre, Kilogram, Second) system. It has seven base units and two supplementary units. Seven base quantities and two supplementary quantities, their units along with definitions are tabulated below.

Base Quantity	Basic Units	
	Name and Symbol	**Definintion**
Length	metre (m)	The metre is the length of path travelled by light in vacuum during a time interval of 1/299,792,458 part of a second.
Mass	kilogram (kg)	It is the mass of the international prototype of the kilogram (a platinum iridium alloy cylinder) kept at International Bureau of Weights and Measures, at Sevres (France).
Time	second (s)	The second is the duration of 9,192, 631,770 periods of the radiation corresponding to the transition between the two hyperfine levels of the ground state of cesium-133 atom.
Electric current	Ampere (A)	The ampere is that constant current, which if maintained in two straight, parallel conductors of infinite length placed 1 m apart in vacuum would produce a force equal to 2×10^{-7} Nm^{-1} on either conductor.
Thermodyn -amic temperature	Kelvin (K)	The kelvin is $\frac{1}{273.16}$th fraction of the thermodynamic temperature of the triple point of water.
Amount of substance	mole (mol)	The mole is the amount of substance of a system, which contains as many elementary entities as there are atoms in 0.012 kg of carbon-12.
Luminous intensity	candela (cd)	The candela is the luminous intensity in a given direction of a source emitting monochromatic radiation of frequency 540×10^{12} Hz and having a radiant intensity of $\frac{1}{683}$ W sr^{-1} in that direction.

Supplementary Units

Supplementary Quantity	Name and Symbol	Definition
Plane angle	radian (rad)	It is angle subtended at the centre by an arc of a circle having a length equal to the radius of the circle.
Solid angle	steradian (sr)	It is the solid angle which is having its vertex at the centre of the sphere, it cuts-off an area of the surface of sphere equal to that of a square with the length of each side equal to the radius of the sphere.

NOTE
- Angle subtended by a closed curve at an inside points is 2π rad.
- Solid angle subtended by a closed surface at an inside point is 4π steradian.

Significant Figures

In the measured value of a physical quantity, the digits about the correction of which we are sure, plus the last digits which is doubtful, are called the significant figures.

Larger the number of significant figures obtained in a measurement, greater is the accuracy of the measurement.

Rules for Counting Significant Figures

- All the non-zero digits are significant. In 2.738 the number of significant figures is 4.
- All the zeros between two non-zero digits are significant, no matter where the decimal point is. As examples 209 and 3.002 have 3 and 4 significant figures, respectively.
- If the measured number is less than 1, the zero (s) on the right of decimal point and to the left of the first non-zero digit are non-significant. In $\underline{0.00}807$, first three underlined zeros are non-significant and the number of significant figures is only 3.

- The terminal or trailing zero (s) in a number without a decimal point are not significant. Thus, 12.3 m = 1230 cm = 12300 mm has only 3 significant figures.
- The trailing zero(s) in number with a decimal point are significant. Thus, 3.800 kg has 4 significant figures.
- A choice of change of units does not change the number of significant digits or figures in a measurement.
- To remove ambiguities in determining number of significant figures, a measurement is usually expressed as '$a \times 10^b$', where $1 \le a \le 10$ and b is the order of magnitude.

Rules for Arithmetic Operations with Significant Figures

- In addition or subtraction, the final results should retain as many decimal places as there are in the number with the least decimal place. As an example sum of 423.5 g, 164.92 g and 24.381 g is 612.801 g, but it should be expressed as 612.8 g only because the least precise measurement (423.5 g) is correct to only one decimal place.
- In multiplication or division, the final result should retain as many significant figures as there are in the original number with the least significant figures.
  For example Suppose an expression is performed like

$$(243 \times 1243)/(44.65) = 676.481522$$

Rounding the above result upto three significant figures result would become 676.

Rules for Rounding off the Uncertain Digits

- The preceding digit is raised by 1 if the insignificant digit to be dropped is more than 5 and is left unchanged if the latter is less than 5. e.g. 18.764 will be rounded off to 18.8 and 18.74 to 18.7.
- If the insignificant figure is 5 and the preceding digit is even, then the insignificant digit is simply dropped. However, if the preceding digit is odd, then it is raised by one so as to make it even. e.g. 17.845 will be rounded off to 17.84 and 17.875 to 17.88.

Accuracy and Precision

The accuracy of a measurement is a measure of how close the measured value is to the true value of the quantity. However, precision tells us to what resolution or limit, the quantity is measured by a measuring instrument.

Least Count

The least count of a measuring instrument is the least value, that can be measured using the instrument. It is denoted as LC.

(i) **Least count of vernier callipers**

$$\text{LC} = \frac{\text{Value of 1 main scale division}}{\text{Total number of vernier scale division}}$$

(ii) **Least count of screw gauge**

$$\text{LC} = \frac{\text{Value of 1 pitch scale reading}}{\text{Total number of head scale division}}$$

Errors in Measurement

The difference in the true value (mean value) and measured value of a quantity is called error of measurement. Different types of error are given below:

(i) Absolute error,

$$a_{\text{mean}} = a_0 = \frac{a_1 + a_2 + a_3 + \ldots a_n}{n} = \frac{1}{n}\sum_{i=1}^{i=n} a_i$$

$$\Delta a_1 = \text{mean value} - \text{observed value}$$

$$\Delta a_1 = a_0 - a_1$$

$$\Delta a_2 = a_0 - a_2$$

$$\vdots \quad \vdots \quad \vdots$$

$$\Delta a_n = a_0 - a_n$$

(ii) Mean absolute error,

$$\Delta a_{\text{mean}} = \frac{[|\Delta a_1| + |\Delta a_2| + |\Delta a_3| + \ldots + |\Delta a_n|]}{n} = \frac{\sum_{i=1}^{n} |\Delta a_i|}{n}$$

(iii) Relative or fractional error $= \dfrac{\Delta a_{\text{mean}}}{a_{\text{mean}}}$

(v) Percentage error,

$$\delta_a = \text{Relative error} \times 100\,\% = \frac{\Delta a_{\text{mean}}}{a_{\text{mean}}} \times 100\%$$

Combination of Errors

(i) If $X = A + B$, then $(\Delta X) = \pm(\Delta A + \Delta B)$

(ii) If $X = ABC$, then $\left(\dfrac{\Delta X}{X}\right)_{\text{max}} = \pm\left[\dfrac{\Delta A}{A} + \dfrac{\Delta B}{B} + \dfrac{\Delta C}{C}\right]$

(iii) If $X = A^k B^l C^n$, then $\left(\dfrac{\Delta X}{X}\right) = \pm\left[k\dfrac{\Delta A}{A} + l\dfrac{\Delta B}{B} + n\dfrac{\Delta C}{C}\right]$

Dimensions of Physical Quantities

The dimensions of a physical quantity are the powers to which the fundamental (base) quantities are raised, to represent that quantity.

Consider the physical quantity force.

'Force = mass × acceleration = mass × length × $(\text{time})^{-2}$'

Thus, the dimension of force are 1 in mass [M]

1 in length [L] and −2 in time $[T^{-2}]$, that is $[MLT^{-2}]$.

- Dimensions of a physical quantity do not depend on its magnitude or the units in which it is measured.

Principle of Homogeneity of Dimensions and Applications

According to this principle, a correct dimensional equation must be homogeneous, i.e. dimensions of all the terms in a physical expression must be same.

LHS (dimension) = RHS (dimension)

Uses of Dimensions

(i) To check the correctness of a given physical equation.

(ii) Derivation of formula.

(iii) Dimensional formula is useful to convert the value of a physical quantity from one system to the other. Physical quantity is expressed as a product of numerical value and unit. In any system of measurement, this product remains constant.

Let dimensional formula of a given physical quantity be $[M^a L^b T^c]$. If in a system having base units $[M_1, L_1, T_1]$ the numerical value of given quantity be n_1 and numerical value n_2 in another unit system having the base units $[M_2, L_2, T_2]$, then

$$Q = n_1 u_1 = n_2 u_2$$

$$n_1 [M_1^a L_1^b T_1^c] = n_2 [M_2^b L_2^b T_2^c]$$

$$n_2 = n_1 \left[\frac{M_1}{M_2}\right]^a \left[\frac{L_1}{L_2}\right]^b \left[\frac{T_1}{T_2}\right]^c$$

Dimensions of Important Physical Quantities

Physical Quantity	SI Unit	Dimensional Formula
Power	Watt (W)	$[ML^2T^{-3}]$
Pressure, stress, coefficient of elasticity (ρ, σ, η)	Pascal (Pa) or Nm^{-2}	$[ML^{-1}T^{-2}]$
Frequency, angular frequency	Hz or s^{-1}	$[T^{-1}]$
Angular momentum	$kg\,m^2 s^{-1}$	$[ML^2T^{-1}]$
Torque	Nm	$[ML^2T^{-2}]$
Gravitational constant (G)	$N\,m^2\,kg^{-2}$	$[M^{-1}L^3T^{-2}]$
Moment of inertia	$kg\,m^2$	$[ML^2]$
Acceleration, acceleration due to gravity	ms^{-2}	$[LT^{-2}]$
Force, thrust, tension, weight	Newton (N)	$[MLT^{-2}]$
Linear momentum, impulse	$kg\,ms^{-1}$ or Ns	$[MLT^{-1}]$
Work, energy, KE, PE, thermal energy, internal energy, etc.	Joule (J)	$[ML^2T^{-2}]$
Surface area, area of cross-section	m^2	$[L^2]$
Electric conductivity	Sm^{-1}	$[M^{-1}L^{-3}T^3A^2]$
Young's modulus, Bulk modulus	Pa	$[ML^{-1}T^{-2}]$
Compressibility	$m^2\,N^{-1}$	$[M^{-1}LT^2]$
Magnetic Flux	Wb	$[ML^2T^{-2}A^{-1}]$
Magnetic Flux density (σ)	Wb/m^2	$[MT^{-2}A^{-1}]$
Intensity of a wave	Wm^{-2}	$[MT^{-3}]$
Photon flux density	$m^{-2}\,s^{-1}$	$[L^{-2}T^{-1}]$
Luminous energy	Lm s	$[ML^2T^{-2}]$
Luminance	Lux	$[MT^{-3}]$
Specific heat capacity	$Jkg^{-1}K^{-1}$	$[L^2T^{-2}K^{-1}]$
Latent heat of vaporisation	Jkg^{-1}	$[L^2T^{-2}]$
Coefficient of Thermal conductivity	$Wm^{-1}K^{-1}$	$[MLT^{-3}K^{-1}]$
Electric voltage	JC^{-1}	$[ML^2T^{-3}A^{-1}]$
Magnetisation	Am^{-1}	$[L^{-1}A]$
Magnetic induction	T	$[MT^{-2}A^{-1}]$
Planck's constant	J-s	$[ML^2T^{-1}]$
Radioactive decay constant	Bq	$[T^{-1}]$
Binding energy	MeV	$[ML^2T^{-2}]$

DAY PRACTICE SESSION 1

FOUNDATION QUESTIONS EXERCISE

1 In which of the following systems of units, a Weber is the unit of magnetic flux?

(a) CGS (b) MKS (c) SI (d) FPS

2 In an experiment, the angles are required to be measured using an instrument. 29 divisions of the main scale exactly coincide with the 30 divisions of the vernier scale. If the smallest division of the main scale is half-a-degree (0.5°), then the least count of the instrument is → AIEEE 2009

(a) one minute (b) half minute
(c) one degree (d) half degree

3 A student measured the length of a rod and wrote it as 3.50 cm.Which instrument did he use to measure it? → JEE Main 2014

(a) A meter scale
(b) A vernier calliper where the 10 divisions in vernier scale matches with 9 divisions in main scale and main scale has 10 divisions in 1 cm
(c) A screw gauge having 100 divisions in the circular scale and pitch as 1 mm
(d) A screw gauge having 50 divisions in the circular scale and pitch as 1 mm

4 N division on main scale of a vernier callipers coincide with $(N+1)$ division of the vernier scale if each division on main scale is of "a" units, least count of instrument is

(a) $\frac{N+1}{a}$ (b) $\frac{a}{N+1}$ (c) $\frac{N-1}{a}$ (d) $\frac{a}{N-1}$

5 One 8 centimetre on the main scale of a vernier calliper is divided into 10 equal parts. If 10 of the divisions of the vernier coincide with small divisions on the main scale, the least count of the callipers is

(a) 0.005 cm (b) 0.02 cm
(c) 0.01 cm (d) 0.05 cm

6 The respective number of significant figures for the numbers 23.023, 0.0003 and 2.1×10^{-3} are

(a) 5, 1, 2 (b) 5, 1, 5 (c) 5, 5, 2 (d) 4, 4, 2

7 A bee of mass 0.000087 kg sits on a flower of mass 0.0123 kg. What is the total mass supported by the stem of the flower up to appropriate significant figures?

(a) 0.012387 kg (b) 0.01239 kg
(c) 0.0124 kg (d) 0.012 kg

8 The radius of a uniform wire is $r = 0.021$ cm. The value of π is given to be 3.142. What is the area of cross-section of the wire up to appropriate significant figures?

(a) 0.0014 cm^2 (b) 0.00139 cm^2
(c) 0.001386 cm^2 (d) 0.0013856 cm^2

9 A man runs 100.5 m in 10.3 s. Find his average speed up to appropriate significant figures.

(a) 9.71 ms^{-1} (b) 9.708 ms^{-1}
(c) 9.7087 ms^{-1} (d) 9.70874 ms^{-1}

10 If the length of rod A is 3.25 ± 0.01 cm and that of B is 4.19 ± 0.01 cm, then the rod B is longer than rod A by

(a) (0.94 ± 0.00) cm (b) (0.94 ± 0.01) cm
(c) (0.94 ± 0.02) cm (d) (0.094 ± 0.005) cm

11 You measure two quantities as $A = 1.0$ m $\pm$ 0.2 m, $B = 2.0$ m $\pm$ 0.2 m. We should report correct value for $\sqrt{AB}$ as

(a) 1.4 m $\pm$ 0.4 m (b) 1.41 m $\pm$ 0.15 m
(c) 1.4 m $\pm$ 0.3 m (d) 1.4 m $\pm$ 0.2 m

12 A student measured the length of the pendulum 1.21 m using a metre scale and time for 25 vibrations as 2 min 20 sec using his wrist watch, absolute error in g is

(a) 0.11 ms^{-2} (b) 0.88 ms^{-2}
(c) 0.44 ms^{-2} (d) 0.22 ms^{-2}

13 The absolute error in density of a sphere of radius 10.01 cm and mass 4.692 kg is

(a) 3.59 kgm^{-3} (b) 4.692 kgm^{-3}
(c) 0 (d) 1.12 kgm^{-3}

14 A sphere has a mass of 12.2 kg $\pm$ 0.1 kg and radius 10 cm $\pm$ 0.1 cm, the maximum % error in density is

(a) 10% (b) 2.4% (c) 3.83% (d) 4.2%

15 If error in measurement of radius of sphere is 1%, what will be the error in measurement of volume?

(a) 1% (b) $\frac{1}{3}$% (c) 3% (d) 10%

16 What is the percentage error in the measurement of the time period T of a pendulum, if the maximum errors in the measurements of l and g are 2% and 4%, respectively?

(a) 6% (b) 4% (c) 3% (d) 5%

17 The density of a material in the shape of a cube is determined by measuring three sides of the cube and its mass. If the relative errors in measuring the mass and length are respectively 1.5% and 1% , the maximum error in determining the density is → JEE Main 2018

(a) 2.5% (b) 3.5%
(c) 4.5% (d) 6%

18 Which one of the following represents the correct dimensions of the coefficient of viscosity?

(a) $[ML^{-1}T^{-2}]$ (b) $[MLT^{-1}]$
(c) $[ML^{-1}T^{-1}]$ (d) $[ML^{-2}T^{-2}]$

19 Which of the following sets share different dimensions?

(a) Pressure, Young's modulus, stress
(b) Emf, potential difference, electric potential
(c) Heat, work done, energy
(d) Dipole moment, electric flux, electric field

20 Out of the following pairs, which one does not have identical dimensions?

(a) Angular momentum and Planck's constant
(b) Impulse and momentum
(c) Moment of inertia and moment of a force
(d) Work and torque

21 The dimensions of $\frac{e^2}{4\pi\varepsilon_0 hc}$, where e, ε_0, h and c are the electronic charge, electric permittivity, Planck's constant and velocity of light in vacuum respectively, are

(a) $[M^0L^0T^0]$ (b) $[ML^0T^0]$ (c) $[M^0LT^0]$ (d) $[M^0L^0T]$

22 In the relation $X = 3YZ^2$, X and Z represent the dimensions of capacitance and magnetic induction respectively, dimensions of Y are

(a) $[MT^{-1}Q^{-1}]$ (b) $[M^{-3}T^4L^{-2}Q^4]$
(c) $[M^{-3}T^{-1}L^{-1}Q^4]$ (d) $[ML^2T^{-2}A^{-2}]$

23 The dimensions of magnetic field in M,L,T and C (coulomb) is given as

(a) $[MLT^{-1}C^{-1}]$ (b) $[MT^2C^{-2}]$ (c) $[MT^{-1}C^{-1}]$ (d) $[MT^{-2}C^{-1}]$

24 Dimensions of $\frac{1}{\mu_0\varepsilon_0}$, where symbols have their usual meaning, are

(a) $[L^{-1}T]$ (b) $[L^2T^2]$ (c) $[L^2T^{-2}]$ (d) $[LT^{-1}]$

25 If the acceleration due to gravity is 10 ms^{-2} and units of length and time are changed to kilometre and hours respectively, the numerical value of acceleration is

(a) 360000 (b) 72000 (c) 36000 (d) 129600

26 If E = energy, G = gravitational constant, I = impulse and M = mass, then dimensions of $\frac{GIM^2}{E^2}$ are same as that of

(a) time (b) mass (c) length (d) force

27 Let $[\varepsilon_0]$ denotes the dimensional formula of the permittivity of vacuum. If M = mass, L = length, T = time and A = electric current, then → JEE Main 2013

(a) $[\varepsilon_0] = [M^{-1}L^{-3}T^2A]$ (b) $[\varepsilon_0] = [M^{-1}L^{-3}T^4A^2]$
(c) $[\varepsilon_0] = [M^{-2}L^2T^{-1}A^{-2}]$ (d) $[\varepsilon_0] = [M^{-1}L^2T^{-1}A^2]$

28 With the usual notations, the following equation

$$s = u + \frac{1}{2}a(2t - 1) \text{ is}$$

(a) only numerically correct
(b) only dimensionally correct
(c) Both numerically and dimensionally correct
(d) Neither numerically nor dimensionally correct

29 The dimensions of $\frac{a}{b}$ in the equation $p = \frac{a - t^2}{bx}$ where, p is pressure, x is distance and t is time, are

(a) $[M^2LT^{-3}]$ (b) $[MT^{-2}]$ (c) $[ML^3T^{-2}]$ (d) $[LT^{-3}]$

30 The velocity of a particle is given as $v = a + bt + ct^2$. If the velocity is measured in ms^{-1}, then units of a and c are

(a) ms^{-1} and ms^{-3} (b) ms^{-2} and ms
(c) m^2s^{-1} and ms^2 (d) ms and ms^{-1}

31 In the following dimensionally consistent equation, we have, $F = \frac{X}{\text{Linear density}} + Y$, where F = force.

The dimensional formula for X and Y are

(a) $[M^2L^0T^{-2}]$; $[MLT^{-2}]$ (b) $[M^2L^{-2}T^{-2}]$; $[MLT^{-2}]$
(c) $[MLT^{-2}]$; $[ML^2T^{-2}]$ (d) $[M^0L^0T^0]$; $[ML^0T^0]$

32 The dimensions of self-inductance are

(a) $[ML^{-2}T^{-2}A^{-2}]$ (b) $[ML^2T^{-2}A^{-2}]$
(c) $[ML^2T^{-2}A^{-1}]$ (d) $[ML^{-2}T^{-2}A^{-1}]$

33 Match List I with List II and select the correct answer using the codes given below the lists. → 2013 Main

	Column I		Column II
A.	Boltzmann constant	p.	$[ML^2T^{-1}]$
B.	Coefficient of viscosity	q.	$[ML^{-1}T^{-1}]$
C.	Planck constant	r.	$[MLT^{-3}K^{-1}]$
D.	Thermal conductivity	s.	$[ML^2T^{-2}K^{-1}]$

Codes

	A	B	C	D		A	B	C	D
(a)	p	q	r	s	(b)	s	q	p	r
(c)	s	r	p	q	(d)	r	s	q	p

Direction (Q. Nos. 34-37) *Each of these questions contains two statements : Statement* I *(Assertion) and Statement* II *(Reason). Each of these questions also has four alternative choices, only one of which is the correct answer. You have to select one of the codes (a), (b), (c), (d) given below*

(a) Statement I is true, Statement II is true; Statement II is the correct explanation for Statement I
(b) Statement I is true, Statement II is true; Statement II is not the correct explanation for Statement I
(c) Statement I is true; Statement II is false
(d) Statement I is false; Statement II is true

34 **Statement I** The order of accuracy of measurement depends on the least count of the measuring instrument.

Statement II The smaller the least count, the greater is the number of significant figures in the measured value.

35 **Statement I** The dimensional method cannot be used to obtain the dependence of the work done by a force F on the angle θ between force F and displacement x.
Statement II Angle can be measured in radians but it has no dimensions.

36 **Statement I** The mass of an object is 13.2 kg in the measurement there are 3 significant figures.
Statement II The same mass when expressed in grams as 13200 g has five significant figures.

37 **Statement I** Method of dimensions cannot be used for deriving formula containing trigonometrical ratios.
Statement II This is because trigonometrical ratios have no dimensions.

DAY PRACTICE SESSION 2

PROGRESSIVE QUESTIONS EXERCISE

1 The dimensions of angular momentum, latent heat and capacitance are, respectively. → JEE Main (Online) 2013

(a) $[ML^2T^1A^2]$, $[L^2T^{-2}]$, $[M^{-1}L^{-2}T^2]$
(b) $[ML^2T^{-2}]$, $[L^2T^2]$, $[M^{-1}L^{-2}T^4A^2]$
(c) $[ML^2T^{-1}]$, $[L^2T^{-2}]$, $[ML^2TA^2]$
(d) $[ML^2T^{-1}]$, $[L^2T^{-2}]$, $[M^{-1}L^{-2}T^4A^2]$

2 The speed (v) of ripples on the surface of water depends on surface tension (σ), density (ρ) and wavelength (λ). The square of speed (v) is proportional to

(a) $\frac{\sigma}{\rho\lambda}$ (b) $\frac{\rho}{\sigma\lambda}$ (c) $\frac{\lambda}{\sigma\rho}$ (d) $\rho\lambda\sigma$

3 Dimensions of resistance in an electrical circuit, in terms of dimensions of mass M, length L, time T and current I, would be

(a) $[ML^2T^{-3}I^{-1}]$ (b) $[ML^2T^{-2}]$
(c) $[ML^2T^{-1}I^{-1}]$ (d) $[ML^2T^{-3}I^{-2}]$

4 In the relation $p = \frac{\alpha}{\beta} e^{-\frac{\alpha z}{k\theta}}$, p is pressure, z is distance, k is Boltzmann constant and θ is the temperature. The dimensional formula of β will be

(a) $[M^0L^2T^0]$ (b) $[ML^2T]$
(c) $[ML^0T^{-1}]$ (d) $[M^0L^2T^{-1}]$

5 The dimensions of σb^4 (σ = Stefan's constant and b = Wien's constant) are

(a) $[M^0L^0T^0]$ (b) $[ML^4T^{-3}]$ (c) $[ML^{-2}T]$ (d) $[ML^6T^{-3}]$

6 If Planck's constant (h) and speed of light in vacuum (c) are taken as two fundamental quantities, which one of the following can, in addition,be taken to express length, mass and time in terms of the three chosen fundamental quantities?

(i) Mass of electron (m_e)
(ii) Universal gravitational constant (G)
(iii) Charge of electron (e)
(iv) Mass of proton (m_p)

(a) (i),(ii) and (iii) (b) (i) and (iii)
(c) (i), (ii) and (iv) (d) (i) only

7 The length and breadth of a rectangular sheet are 16.2 cm and 10.1 cm , respectively. The area of the sheet in appropriate significant figures and error is

(a) 164 ± 3 cm^2 (b) 163.62 ± 2.6 cm^2
(c) 163.6 ± 2.6 cm^2 (d) 163.62 ± 3 cm^2

8 Resistance of a given wire is obtained by measuring the current flowing in it and the voltage difference applied across it. If the percentage errors in the measurement of the current and the voltage difference are 3% each, then error in the value of resistance of the wire is

(a) 6% (b) zero (c) 1% (d) 3%

9 A screw gauge gives the following reading when used to measure the diameter of a wire.
Main scale reading : 0 mm
Circular scale reading : 52 divisions
Given that 1 mm on main scale corresponds to 100 divisions of the circular scale.
The diameter of wire from the above data is

(a) 0.052 cm (b) 0.026 cm
(c) 0.005 cm (d) 0.52 cm

10 A screw gauge with a pitch of 0.5 mm and a circular scale with 50 divisions is used to measure the thickness of a thin sheet of aluminium. Before starting the measurement, it is found that when the two jaws of the screw gauge are brought in contact, the 45th division coincides with the main scale line and that the zero of the main scale is barely visible. What is the thickness of the sheet, if the main scale reading is 0.5 mm and the 25th division coincides with the main scale line?

(a) 0.75 mm (b) 0.80 mm
(c) 0.70 mm (d) 0.50 mm

11 The following observations were taken for determining surface tension T of water by capillary method. Diameter of capillary, $d = 1.25 \times 10^{-2}$ m rise of water, $h = 1.45 \times 10^{-2}$m. Using $g = 9.80$ m/s^2 and the simplified relation $T = \frac{dhg}{4} \times 10^3$N/m, the possible error in surface tension is closest to → JEE Main 2017 (Offline)

(a) 1.5% (b) 2.4% (c) 10% (d) 0.15%

12 A student measures the time period of 100 oscillations of a simple pendulum four times. The data set is 90s, 91s, 92s and 95s. If the minimum division in the measuring clock is 1s, then the reported mean time should be
→ JEE Main 2016 (Offline)

(a) (92 ± 2)s (b) (92 ± 5)s (c) (92 ± 1.8)s (d) (92 ± 3)s

13 The period of oscillation of a simple pendulum is $T = 2\pi\sqrt{L/g}$. Measured value of L is 20.0 cm known to 1mm accuracy and time for 100 oscillations of the pendulum is found to be 90 s using a wrist watch of resolution. The accuracy in the determination of g is
→ JEE Main 2015

(a) 2% (b) 3% (c) 1% (d) 5%

14 The current voltage relation of diode is given by $I = (e^{1000V/T} - 1)$ mA, where the applied voltage V is in volt and the temperature T is in kelvin. If a student makes an error measuring $\pm$ 0.01V while measuring the current of 5 mA at 300K, what will be the error in the value of current in mA?
→ JEE Main 2013

(a) 0.2 mA (b) 0.02 mA (c) 0.5 mA (d) 0.05 mA

Direction (Q. Nos. 15-16) *Each of these questions contains two statements : Statement I (Assertion) and Statement II (Reason). Each of these questions also has four alternative choices, only one of which is the correct answer. You have to select one of the codes (a), (b), (c), (d) given below*

(a) Statement I is true, Statement II is true; Statement II is the correct explanation for Statement I
(b) Statement I is true, Statement II is true; Statement II is not the correct explanation for Statement I
(c) Statement I is true; Statement II is false
(d) Statement I is false; Statement II is true

15 **Statement I** The value of velocity of light is 3×10^8 ms^{-1} and acceleration due to gravity is 10 ms^{-2} and the mass of proton is 1.67×10^{-27} kg.

Statement II The value of time in such a system is 3×10^7s.

16 **Statement I** The distance covered by a body is given by $s = u + \frac{1}{2}\frac{a}{t}$, where the symbols have usual meaning.

Statement II We can add, substract or equate quantities which have same dimensions.

ANSWERS

SESSION 1	1 (c)	2 (a)	3 (b)	4 (b)	5 (b)	6 (a)	7 (d)	8 (a)	9 (a)	10 (c)
	11 (d)	12 (d)	13 (a)	14 (c)	15 (c)	16 (c)	17 (c)	18 (c)	19 (d)	20 (c)
	21 (a)	22 (b)	23 (c)	24 (c)	25 (d)	26 (a)	27 (b)	28 (d)	29 (b)	30 (a)
	31 (a)	32 (b)	33 (b)	34 (b)	35 (a)	36 (c)	37 (a)			
SESSION 2	1 (d)	2 (a)	3 (d)	4 (a)	5 (b)	6 (a)	7 (a)	8 (a)	9 (a)	10 (b)
	11 (a)	12 (a)	13 (b)	14 (a)	15 (b)	16 (d)				

Hints and Explanations

SESSION 1

1 A weber is the unit of magnetic flux in SI system.

2 Least count

$= \dfrac{\text{Value of main scale division}}{\text{Number of divisions on vernier scale}}$

$= \dfrac{1}{30}\text{ MSD} = \dfrac{1}{30} \times \dfrac{1^\circ}{2} = \dfrac{1^\circ}{60} = 1\text{ min}$

3 If student measures 3.50cm, it means that there is an uncertainly of $\dfrac{1}{100}$ cm.

For vernier scale with 1 MSD = 1 mm and 9MSD = 10 VSD

LC of vernier calliper = 1 MSD – 1VSD

$= \dfrac{1}{10}\left(1 - \dfrac{9}{10}\right) = \dfrac{1}{100}\text{cm}$

4 $(N + 1)$ VSD = N MSD

$\therefore \quad 1\text{ VSD} = \dfrac{N}{N+1}\text{ MSD}$

Least count

= (1 MSD – 1VSD) (value of MSD)

$= \left(1 - \dfrac{N}{N+1}\right) \times a = \dfrac{a}{N+1}$

5 $1\text{ MSD} = \dfrac{1}{10}\text{ cm} = 0.1\text{ cm}$,

10 VSD = 8 MSD

Hence, we get

$1\text{ VSD} = \dfrac{8}{10}\text{ MSD} = \dfrac{8}{10} \times (0.1) = 0.08\text{ cm}$

Thus, the least count = 1 MSD – 1VSD

= 0.1 – 0.08 = 0.02 cm

6 Number of significant figures in 23.023 = 5

Number of significant figures in 0.0003 = 1

Number of significant figures in $2.1 \times 10^{-3} = 2$

7 The mass of the bee has 2 significant figures in kg, whereas the mass of the flower has three significant figures. Hence, the sum (0.012387) must be rounded off to the third decimal place. Therefore, the correct significant figure is 0.012.

8 $A = \pi r^2 = 3.142 \times (0.021)^2$

$= 0.00138562\text{ cm}^2$.

Now, there are only two significant figures in 0.021 cm. Hence, the result must be rounded off to two significant figures as $A = 0.0014\text{ cm}^2$.

9 Average speed $= \dfrac{100.5\text{ m}}{10.3\text{ s}}$

$= 9.75728\text{ ms}^{-1}$

The distance has four significant figures but the time has only three. Hence, the result must be rounded off to three significant figure to 9.71 ms^{-1}.

10 As, $A = 3.25 \pm 0.01$ cm

and $B = 4.19 \pm 0.01$ cm

$\therefore \quad Y = B - A$

$= 4.19 - 3.25 = 0.94$ cm

and $\Delta Y = \Delta B + \Delta A$

= 0.01 cm + 0.01 cm = 0.02 cm

$\therefore \quad Y = (0.94 \pm 0.02)$ cm

11 Here, $A = 1.0\text{ m} \pm 0.2\text{ m}$,

$B = 2.0\text{ m} \pm 0.2\text{ m}$

$x = \sqrt{AB} = \sqrt{(1.0)(2.0)} = 1.414\text{ m}$

Rounding off to two significant digits,

$x = \sqrt{AB} = 1.4\text{ m}$

Now, $\dfrac{\Delta x}{x} = \dfrac{1}{2}\left[\dfrac{\Delta A}{A} + \dfrac{\Delta B}{B}\right]$

$= \dfrac{1}{2}\left[\dfrac{0.2}{1.0} + \dfrac{0.2}{2.0}\right] = \dfrac{0.6}{2 \times 2.0}$

$\Delta x = \dfrac{0.6\,x}{2 \times 2.0} = 0.15 \times 1.414$

$= 0.2121$

Rounding off to one significant digit,

$\Delta x = 0.2$ m

Hence, $\sqrt{AB} = x \pm \Delta x = (1.4 \pm 0.2)$ m

12 As, $T = 2\pi\sqrt{\dfrac{l}{g}}$ or $g = \dfrac{4\pi^2 l}{T^2}$

$\Rightarrow \quad \left(\dfrac{\Delta g}{g} = \dfrac{\Delta L}{L} + \dfrac{2\Delta T}{T}\right)$

So, absolute error in g is

$\Delta g = \left(\dfrac{\Delta L}{L} + \dfrac{2\Delta T}{T}\right) g$

$= \left(\dfrac{0.01}{1.21} + \dfrac{2 \times 1}{140}\right) \times 9.8$

$= (0.0227 \times 9.8) = 0.22\text{ ms}^{-2}$

13 $\therefore \rho = \dfrac{M}{\frac{4}{3}\pi r^3} = \dfrac{4.692 \times 3}{4 \times 3.14 \times (10.01)^3 \times 10^{-6}}$

$\rho = 1.12 \times 10^3\text{ kg - m}^{-3}$

$\dfrac{\Delta\rho}{\rho} = \dfrac{\Delta M}{M} + \dfrac{3\Delta r}{r}$

$\therefore \Delta\rho = \left(\dfrac{0.001}{4.692} + \dfrac{3 \times 0.01}{10.01}\right) \times 1.12 \times 10^3$

$= 3.59\text{ kgm}^{-3}$

14 $\therefore$ Density, $\rho = \dfrac{M}{V} = \dfrac{M}{\frac{4}{3}\pi r^3}$,

$\dfrac{d\rho}{\rho} \times 100 = \left(\dfrac{\Delta M}{M} + \dfrac{3\Delta r}{r}\right) \times 100$

$= \left(\dfrac{0.1}{12.2} + 3 \times \dfrac{0.1}{10}\right) \times 100$

$= 3.83\%$

15 As, $V = \dfrac{4}{3}\pi r^3$

Hence, $\dfrac{\Delta V}{V} \times 100 = 3\dfrac{\Delta r}{r} \times 100$

$= 3 \times 1\% = 3\%$

16 Since, the time period,

$T = \dfrac{1}{2\pi}\sqrt{\dfrac{l}{g}}$

Thus, for calculating the error, we get

$\dfrac{\Delta T}{T} = \pm\left[\dfrac{1}{2}\dfrac{\Delta l}{l} + \dfrac{1}{2}\dfrac{\Delta g}{g}\right]$

$= \pm\left[\dfrac{1}{2} \times 2\% + \dfrac{1}{2} \times 4\%\right] = \pm 3\%$

17 $\because$ Density, $\rho = \dfrac{\text{Mass}}{\text{Volume}} = \dfrac{M}{L^3}$ or $\rho = \dfrac{M}{L^3}$

$\Rightarrow$ Error in density $\dfrac{\Delta\rho}{\rho} = \dfrac{\Delta M}{M} + \dfrac{3\Delta L}{L}$

So, maximum % error in measurement of ρ is

$\dfrac{\Delta\rho}{\rho} \times 100 = \dfrac{\Delta M}{M} \times 100 + \dfrac{3\Delta L}{L} \times 100$

or % error in density = 1.5 + 3 × 1

% error = 4.5%

18 By Newton's formula $\eta = \dfrac{F}{A\,(\Delta V / \Delta Z)}$

$\therefore$ Dimensions of η

$= \dfrac{\text{Dimensions of force}}{(\text{Dimensions of area} \times \text{Dimensions of velocity gradient})}$

$= \dfrac{[\text{MLT}^{-2}]}{[\text{L}^2][\text{T}^{-1}]} = [\text{ML}^{-1}\text{T}^{-1}]$

19 Dipole moment = charge × distance

Electric flux = electric field × area

20 $I = mr^2$

$\therefore \quad [I] = [\text{ML}^2]$

τ moment of force $= r \times F$

$\therefore \quad [\tau] = [\text{L}][\text{MLT}^{-2}] = [\text{ML}^2\text{T}^{-2}]$

21 $\left[\dfrac{e^2}{4\pi\varepsilon_0 hc}\right]$

$= \dfrac{[\text{AT}]^2}{[\text{M}^{-1}\text{L}^{-3}\text{T}^4\text{A}^2]\cdot[\text{ML}^2\text{T}^{-1}]\cdot[\text{LT}^{-1}]}$

$= [\text{M}^0\text{L}^0\text{T}^0]$

22 $X = [C] = [M^{-1} L^{-2}T^2Q^2]$,
$Z = [B] = [MT^{-1}Q^{-1}]$
$Y = \frac{X}{Z^2} = \frac{[M^{-1}T^2L^{-2}Q^2]}{[MT^{-1}Q^{-1}]^2} = [M^{-3}T^4L^{-2}Q^4]$

23 From the relation $F = qvB$
$\Rightarrow [MLT^{-2}] = [C][LT^{-1}][B]$
$\Rightarrow [B] = [MC^{-1}T^{-1}]$

24 As we know that, formula of velocity is
$v = \frac{1}{\sqrt{\mu_0\varepsilon_0}} \Rightarrow v^2 = \frac{1}{\mu_0\varepsilon_0} = [LT^{-1}]^2$ $\therefore$
$\frac{1}{\mu_0\varepsilon_0} = [L^2T^{-2}]$

25 $n_2 = n_1\left[\frac{L_1}{L_2}\right]\left[\frac{T_1}{T_2}\right]^{-2}$
$= 10\left[\frac{\text{metre}}{\text{km}}\right]\left[\frac{\text{sec}}{\text{h}}\right]^{-2}$
$n_2 = 10\left[\frac{\text{m}}{10^3\text{ m}}\right]\left[\frac{\text{sec}}{3600\text{sec}}\right]^{-2}$
$=129600$

26 $\left[\frac{GIM^2}{E^2}\right] = \frac{[M^{-1}L^3T^{-2}]\times[MLT^{-1}]\times[M]^2}{[ML^2T^{-2}]^2}$
$= [M^0L^0T]$

So, dimensions of $\frac{GIM^2}{E^2}$ are same as that of time.

27 Electrostatic force between two charges,
$F = \frac{1}{4\pi\varepsilon_0}\frac{q_1 q_2}{R^2}$
$\Rightarrow \varepsilon_0 = \frac{q_1 q_2}{4\pi FR^2}$
Substituting the units.
Hence, $\varepsilon_0 = \frac{C^2}{N\text{-}m^2} = \frac{[AT]^2}{[MLT^{-2}][L^2]}$
$= [M^{-1} L^{-3} T^4 A^2]$

28 s = distance travelled, u = velocity.
So, dimensionally it is not a correct equation.

29 $p = \frac{a - t^2}{bx}$
$\Rightarrow pbx = a - t^2$
$\Rightarrow [pbx] = [a] = [T^2]$
or $[b] = \frac{[T^2]}{[p][x]} = \frac{[T^2]}{[ML^{-1}T^{-2}][L]}$
$= [M^{-1}T^4]$
$\therefore \left[\frac{a}{b}\right] = \frac{[T^2]}{[M^{-1}T^4]} = [MT^{-2}]$

30 Unit of a = unit of v = m/s = ms^{-1}
and unit of c = unit of $\frac{v}{t^2}$
$= \frac{m/s}{s^2} = m/s^3 = ms^{-3}$

31 $[F] = \left[\frac{X}{\text{Linear density}}\right] + [Y]$.
So, the dimensions of Y are the same as that of F, i.e. $[Y] = [F] = [MLT^{-2}]$
Now, $[MLT^{-2}] = \left[\frac{X}{ML^{-1}}\right]$
$\Rightarrow X = [M^2 L^0T^{-2}]$

32 The self-inductance L of a coil in which the current varies at a rate $\frac{dI}{dt}$ and is given by $e = -L\frac{dI}{dt}$, where e is the electromotive force (emf) induced in the coil.
Now, the dimensions of emf are the same as that of the potential difference, i.e. $[ML^2T^{-3}I^{-1}]$
Now, $L = \frac{-e}{dI/dt}$.
Hence, the dimensions of L are
$[L] = \frac{\text{dimensions of } e}{\text{dimensions of } I/\text{dimensions of } t}$
$= \frac{[ML^2 T^{-3} I^{-1}]}{[I/T]}$
$= [ML^2T^{-2}I^{-2}]$

33 (A) $U = \frac{1}{2}kT$
$\Rightarrow [ML^2T^{-2}] = [k]\,K$
$\Rightarrow [K] = [ML^2T^{-2}K^{-1}]$
(B) $F = \eta A\frac{dv}{dx}$
$\Rightarrow [\eta] = \frac{[MLT^{-2}]}{[L^2LT^{-1}L^{-1}]} = [ML^{-1} T^{-1}]$
(C) $E = h\nu \Rightarrow [ML^2T^2] = [h]\,[T^{-1}]$
$\Rightarrow [h] = [ML^2T^{-1}]$
(D) $\frac{dQ}{dt} = \frac{kA\Delta\theta}{l}$
$\Rightarrow [k] = \frac{[ML^2T^{-3}L]}{[L^2K]} = [MLT^{-3}K^{-1}]$

34 The least count of a measuring device is the least distance (resolution/accuracy), that can be measured using the device.
Greater is the number of significant figures obtained in a measurement, greater is its precision and for this the least count of the measuring instrument should be smaller.

35 Work done is $W = Fx\cos\theta$. Since, θ is dimensionless, the dependence of W on θ cannot be determined by the dimensional method.

36 The degree of accuracy (and hence the number of significant figures) of a measurement cannot be increased by changing the unit.

37 It is true that trigonometrical ratios do not have dimensions. Therefore, method of finding dimensions cannot be utilized for deriving formula involving trigonometrical ratio.

SESSION 2

1 Angular momentum $= r\times P = [LM\,LT^{-1}]$
$= [ML^2T^{-1}]$
Latent heat, $L = \frac{Q}{M} = [L^2T^{-2}]$
Capacitance, $C = \frac{q}{V} = \frac{(AT)^2}{W}$ $\left(\text{as, } V = \frac{W}{q}\right)$
$= [A^2T^2M^{-1}L^{-2}T^{+2}]$
$= [M^{-1}L^{-2}T^4A^2]$

2 Let $v \propto \sigma^a\rho^b\lambda^c$. Equating dimensions on both sides
$[M^0 L^1 T^{-1}] = k\,[MT^{-2}]^a[ML^{-3}]^b[L]^c$
$[M^0LT^{-1}] = k\,[M]^{a+b}[L]^{-3b+c}\,[T]^{-2a}$
Equating the powers of M, L, T on both sides, we get
$a + b = 0$ and $-3b + c = 1; -2a = -1$
Solving, we get $a = \frac{1}{2}, b = -\frac{1}{2}, c = -\frac{1}{2}$
$\therefore v \propto \sigma^{1/2}\rho^{-1/2}\lambda^{-1/2}$ $\therefore v^2 \propto \frac{\sigma}{\rho\lambda}$

3 Resistance, $R = \frac{\text{Potential difference}}{\text{Current}}$
$= \frac{V}{I} = \frac{W}{qI}$
($\because$ Potential difference is equal to work done per unit charge)
So, dimensions of R
$= \frac{[\text{Dimensions of work}]}{[\text{Dimensions of charge}][\text{Dimensions of current}]}$
$= \frac{[ML^2T^{-2}]}{[IT]\,[I]} = [ML^2T^{-3}I^{-2}]$

4 In the given equation, $\frac{\alpha z}{k\theta}$ should be dimensionless
$\therefore [\alpha] = \left[\frac{k\theta}{z}\right]$
$\Rightarrow [\alpha] = \frac{[ML^2\,T^{-2}\,K^{-1}\times K]}{[L]} = [MLT^{-2}]$
and $[p] = \left[\frac{\alpha}{\beta}\right]$
$\Rightarrow [\beta] = \left[\frac{\alpha}{p}\right] = \frac{[MLT^{-2}]}{[ML^{-1}T^{-2}]} = [M^0L^2T^0]$

5 $\lambda_m T = b$ or $b^4 = \lambda_m^4 T^4$
and $\frac{\text{energy}}{\text{area}\times\text{time}} = \sigma T^4$
or $\sigma = \frac{\text{energy}}{(\text{area}\times\text{time})\,T^4}$

$$\sigma b^4 = \left(\frac{\text{energy}}{\text{area}\times\text{time}}\right)\lambda_m^4$$

or $[\sigma b^4] = \frac{[ML^2T^{-2}]}{[L^2][T]}[L^4] = [ML^4T^{-3}]$.

6 $h = [ML^2T^{-1}]$; $c = [LT^{-1}]$, $m_e = [M]$, $G = [M^{-1}L^3T^{-2}]$, $e = AT$; $m_p = M$,

$$\frac{hc}{G} = \frac{[M^1 L^2T^{-1}][LT^{-1}]}{[M^{-1}L^3T^{-2}]} = [M^2]$$

$$\Rightarrow \quad M = \sqrt{\frac{hc}{G}}$$

$$\frac{h}{c} = \frac{[ML^2 T^{-1}]}{[LT^{-1}]} = [ML]$$

$$\Rightarrow \quad L = \frac{h}{cM} = \frac{h}{c}\sqrt{\frac{G}{hc}} = \frac{\sqrt{Gh}}{c^{3/2}}$$

From $c = [LT^{-1}]$,

$$T = \frac{L}{c} = \frac{\sqrt{Gh}}{c^{3/2}c} = \frac{\sqrt{Gh}}{c^{5/2}}$$

Hence, out of (i), (ii) and (iii) any one can be taken to express L, M, T in terms of three chosen fundamental quantities.

7 Here, $l = (16.2 \pm 0.1)$ cm ;

$b = (10.1 \pm 0.1)$ cm

$A = l \times b = 16.2\times 10.1 = 163.62$

Rounding off to three significant digits, $A = 164\ \text{cm}^2$

$$\frac{\Delta A}{A} = \left(\frac{\Delta l}{l} + \frac{\Delta b}{b}\right) = \frac{0.1}{16.2} + \frac{0.1}{10.1}$$

$$= \frac{1.01 + 1.62}{16.2\times 10.1} = 2.63\ \text{cm}^2$$

Rounding off to one significant figure, $\Delta A = 3\text{cm}^2$

$\therefore \quad A = (164 \pm 3)\ \text{cm}^2$

8 From Ohm's law, $R = \frac{V}{I}$

$\Rightarrow \ln R = \ln V - \ln I$

$$\Rightarrow \quad \frac{\Delta R}{R} = \frac{\Delta V}{V} + \frac{\Delta I}{I} = 3\% + 3\% = 6\%$$

9 Diameter of wire, d = MSR + CSR $\times$ LC

$$= 0 + 52 \times \frac{1}{100} = 0.52\ \text{mm} = 0.052\ \text{cm}$$

10 Least count $= \frac{\text{pitch}}{\text{(number of division on circular scale)}}$

$$= \frac{0.5}{50}\ \text{mm}$$

$\therefore \quad$ LC = 0.01

Negative zero error $= -5\times \text{LC} = -0.005$ mm

Measured value = main scale reading + screw guage reading − zero error

$= 0.5\ \text{mm} + \{25\times 0.01 - (-0.05)\}$ mm

$= 0.80$ mm

11 By given formula, we have surface tension,

$$T = \frac{dhg}{4} \times 10^3\ \frac{N}{m} \qquad \left(\because r = \frac{d}{2}\right)$$

$$\Rightarrow \frac{\Delta T}{T} = \frac{\Delta d}{d} + \frac{\Delta h}{h} \quad \text{[given, g is constant]}$$

So, percentage error is

$$= \frac{\Delta T}{T} \times 100$$

$$= \left(\frac{\Delta d}{d} + \frac{\Delta h}{h}\right) \times 100$$

$$= \left(\frac{0.01 \times 10^{-2}}{1.25\times 10^{-2}} + \frac{0.01 \times 10^{-2}}{1.45\times 10^{-2}}\right)\times 100$$

$= 1.5\%$

$$\therefore \frac{\Delta T}{T} \times 100 = 1.5\%$$

12 Arithmetic mean time of a oscillating simple pendulum $= \frac{\Sigma x_i}{N}$

$$= \frac{90 + 91 + 92 + 95}{4} = 92\ \text{s}$$

Mean error is

$$= \frac{\Sigma|\bar{x} - x_i|}{N} = \frac{2 + 1 + 3 + 0}{4} = 1.5$$

Given, minimum division in the measuring clock, is 1 s. Thus, the reported mean time of a oscillating simple pendulum = (92 ± 2) s.

13 Given, time period, $T = 2\pi\sqrt{\frac{L}{g}}$

Thus,changes can be expressed as

$$= \frac{2T}{T} = \pm\frac{\Delta L}{L} \pm \frac{\Delta g}{g}$$

According to the question, we can write

$$\frac{\Delta L}{L} = \frac{0.1\text{cm}}{20.0\text{cm}} = \frac{1}{200}$$

Again time period $T = \frac{90}{100}$ s

and $\Delta T = \frac{1}{200}\text{s} \Rightarrow \frac{\Delta T}{T} = \frac{1}{90}$

Now, $\quad T = 2\pi\sqrt{\frac{L}{g}}$

$$\therefore \frac{\Delta g}{g} = \frac{\Delta L}{L} + \frac{2\Delta T}{T} \text{ or } \frac{\Delta g}{g}\times 100\%$$

$$= \left(\frac{\Delta L}{L}\right)\times 100\% + \left(\frac{2\Delta T}{T}\right)\times 100\%$$

$$= \left(\frac{1}{200}\times 100\right)\% + 2\times\frac{1}{90}\times 100\%$$

$= 2.72\% = 3\%$

Thus, accuracy in the determination of g is approx 3 %.

14 Given, $I = (e^{1000V/T} - 1)$ mA

$dV = \pm\ 0.01\text{V}, T = 300\ \text{K},\ I = 5\text{mA},$

$I = e^{1000V/T} - 1,\ I + 1 = e^{1000V/T}$

Taking log on both sides, we get

$$\log (I + 1) = \frac{1000V}{T}$$

$$\Rightarrow \quad \frac{d(I + 1)}{I + 1} = \frac{1000}{T}dV$$

$$\frac{dI}{I + 1} = \frac{1000}{T}dV$$

$$\Rightarrow \quad dI = \frac{1000}{T}\times (I + 1)\, dV$$

$$dI = \frac{1000}{300}\times (5 + 1)\times 0.01$$

$= 0.2$ mA

So, error in the value of current is 0.2 mA.

15 $[c] = [LT^{-1}] = 3 \times 10^8\ \text{ms}^{-1}$

and $[g] = [LT^{-2}] = 10\ \text{ms}^{-2}$

So, $\quad \frac{c}{g} = \frac{[LT^{-1}]}{[LT^{-2}]} = T$

$$\therefore \quad T = \frac{3 \times 10^8}{10} = 3 \times 10^7\ \text{s}$$

16 The physical quantities can be equated, added or subtracted only when they have same dimensions. The distance covered by a body is $s = u + \frac{1}{2}\frac{a}{t}$

$$[L] = [LT^{-1}] + \frac{[LT^{-2}]}{[T]}$$

$$[L] = [LT^{-1}] + [LT^{-3}]$$

As every term of equation is not is not having same dimensions, so it is a wrong expression for distance.

DAY TWO

Kinematics

Learning & Revision for the Day

- Frame of Reference
- Motion in a Straight Line
- Uniform and Non-uniform Motion
- Uniformly Accelerated Motion
- Elementary Concept of Differentiation and Integration for Describing Motion
- Graphs

Frame of Reference

The frame of reference is a suitable coordinate system involving space and time used as a reference to study the motion of different bodies. The most common reference frame is the cartesian frame of reference involving (x, y, z and t).

(i) **Inertial Frame of Reference** A frame of reference which is either at rest or moving with constant velocity is known as inertial frame of reference. Inertial frame of reference is one in which Newton's first law of motion holds good.

(ii) **Non-Inertial Frame of Reference** A frame of reference moving with some acceleration is known as non-inertial frame of reference. Non-inertial frame of reference in one which Newton's law of motion does not hold good.

Motion in a Straight Line

The motion of a point object in a straight line is one dimensional motion. During such a motion the point object occupies definite position on the path at each instant of time. Different terms used to described motion are defined below:

Distance and Displacement

- **Distance** is the total length of the path travelled by a particle in a given interval of time. It is a scalar quantity and its SI unit is metre (m).
- **Displacement** is shortest distance between initial and final positions of a moving object. It is a vector quantity and its SI unit is metre.

  From the given figure, mathematically it is expressed as,

  $$\Delta r = r_2 - r_1$$

- Displacement of motion may be zero or negative but path length or distance can never be negative.
- For motion between two points displacement is single valued while distance depends on actual path and so can have many values.
- Magnitude of displacement can never be greater than distance. However, it can be equal, if the motion is along a straight line without any change in direction.

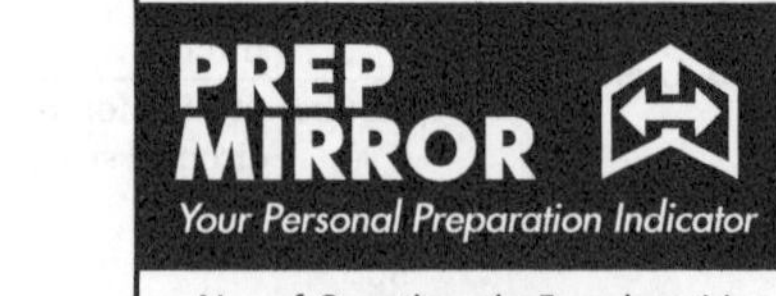

- No. of Questions in Exercises (x)—
- No. of Questions Attempted (y)—
- No. of Correct Questions (z)—
  (Without referring Explanations)

- Accuracy Level ($z/y \times 100$)—
- Prep Level ($z/x \times 100$)—

In order to expect good rank in JEE, your Accuracy Level should be above 85 & Prep Level should be above 75.

Speed and Velocity

- **Speed** is defined as the total path length (or actual distance covered) by time taken by object.

$$\text{Speed} = \frac{\text{Distance}}{\text{Time taken}}$$

It is scalar quantity. Its SI unit is m/s.

- Average Speed, $v_{av} = \frac{\text{Total distance travelled}}{\Delta t}$
- When a body travels equal distance with speeds v_1 and v_2, the average speed (v) is the harmonic mean of the two speeds.

$$\frac{2}{v} = \frac{1}{v_1} + \frac{1}{v_2}$$

- When a body travels for equal time with speeds v_1 and v_2, the average speed v is the arithmetic mean of the two speeds.

$$v_{av} = \frac{v_1 + v_2}{2}$$

- **Velocity** is defined as ratio of displacement and corresponding time interval taken by an object.

i.e. velocity $= \frac{\text{Displacement}}{\text{time interval}}$

- Average velocity $= \frac{\text{Total displacement}}{\text{Total time taken}} = \frac{x_2 - x_1}{t_2 - t_1} = \frac{\Delta x}{\Delta t}$

Here, x_2 and x_1 are the positions of a particle at the time t_2 and t_1 respectively, with respect to a given frame of reference.

- For a moving body speed can never be negative or zero while velocity can be negative and zero.
- The **instantaneous speed** is average speed for infinitesimal small time interval (i.e. $\Delta t \to 0$)

i.e. Instantaneous speed, $v = \lim_{\Delta t \to 0} \frac{\Delta s}{\Delta t} = \frac{ds}{dt}$

- The instantaneous velocity (or simply velocity) v of a moving particle is $v = \lim_{\Delta t \to 0} \frac{\Delta x}{\Delta t} = \frac{dx}{dt}$

It (at a particular time) can be calculated as the slope (at that particular time) of the graph of x *versus* t.

Uniform and Non-uniform Motion

- An object is said to be in uniform motion if its velocity is uniform i.e. it undergoes equal displacement in equal may be intervals of time, howsoever small these interval.
- An object is said to be in non-uniform motion if its undergoes equal displacement in unequal intervals of time., howsoever small these intervals may be.

Acceleration

Acceleration of an object is defined as rate of change of velocity. It is a vector quantity having unit m/s^2 or ms^{-2}. It can be positive, zero or negative.

Average and Instantaneous Acceleration If velocity of a particle at instant t is v_1 and at instant t_2 is v_2, then

- Average acceleration, $a_{av} = \frac{v_2 - v_1}{t_2 - t_1} = \frac{\Delta v}{\Delta t}$
- Instantaneous acceleration, $a = \lim_{\Delta t \to 0} \frac{\Delta v}{\Delta t} = \frac{dv}{dt}$

Uniformly Accelerated Motion

- A motion, in which change in velocity in each unit of time is constant, is called an uniformly accelerated motion. So, for an uniformly accelerated motion, acceleration is constant.
- For uniformly accelerated motion are given below

Equations of motion, $v = u + at$ ...(i)

$$s = ut + \frac{1}{2}at^2 \quad \text{...(ii)}$$

and $$v^2 = u^2 + 2as \quad \text{...(iii)}$$

where, u = initial velocity, v = velocity at time t and s = displacement of particle at time t.

- Equation of uniformaly accelerated motion under gravity are

(i) $v = u - gt$ (ii) $h = ut - \frac{1}{2}gt^2$ (iii) $v^2 = u^2 - 2gh$

Elementary Concept of Differentiation and Integration for Describing Motion

- At an instant t, the body is at point $P(x, y, z)$.

Thus, velocity along X-axis, $v_x = \frac{dx}{dt}$

Acceleration along X-axis is $a_x = \frac{dv_x}{dt}$

Velocity along Y-axis is $v_y = \frac{dy}{dt}$

Acceleration along Y-axis is $a_y = \frac{dv_y}{dt}$

Similarly, $v_z = \frac{dz}{dt}$ and $a_z = \frac{dv_z}{dt}$

- **For a accelerating body**

(i) If a_x variable, $x = \int v_x dt$, $\int dv_x = \int a_x dt$

(ii) If a_y is variable, $y = \int v_y\, dt$, $\int dv_y = \int a_y dt$

(iii) If a_z is variable, $z = \int v_z\, dt$, $\int dv_z = \int a_z dt$

Also, distance travelled by a particle is $s = \int |v|\, dt$

(i) x-component of displacement is $\Delta x = \int v_x\, dt$

(ii) y-component of displacement is $\Delta y = \int v_y\, dt$

(iii) z-component of displacement is $\Delta z = \int v_z\, dt$

Graphs

During motion of the particle, its parameters of kinematical analysis changes with time. These can be represented on the graph, which are given as follows:

Position-Time Graph

(i) Position-time graph gives instantaneous value of displacement at any instant.

(ii) The slope of tangent drawn to the graph at any instant of time gives the instantaneous velocity at that instant.

(iii) The *s-t* graph cannot make sharp turns.

Different Cases of Position-Time Graph

Different Cases	s- t Graph	The main Features of Graph
At rest		Slope = $v = 0$
Uniform motion	$s = vt$	Slope = constant, v = constant $a = 0$
Uniformly accelerated motion with $u = 0, s = 0$ at $t = 0$	$s = \frac{1}{2}at^2$	$u = 0$, i.e. Slope of *s-t* graph at $t = 0$, should be zero.
Uniformly accelerated motion with $u \neq 0$ but $s = 0$ at $t = 0$	$s = ut + \frac{1}{2}at^2$	Slope of *s-t* graph gradually goes on increasing
Uniformly retarded motion	t_0	θ is decreasing so, v is decreasing, a is negative

Velocity-Time Graph

(i) Velocity-time graph gives the instantaneous value of velocity at any instant.

(ii) The slope of tangent drawn on graph gives instantaneous acceleration.

(iii) Area under *v-t* graph with time axis gives the value of displacement covered in given time.

(iv) The *v-t* curve cannot take sharp turns.

Different Cases in Velocity-Time Graph

Different Cases	v- t Graph	The main Features of Graph
Uniform motion	v = constant	(i) $\theta = 0^\circ$ (ii) v = constant (iii) Slope of *v-t* graph = $a = 0$
Uniformly accelerated motion with $u = 0$ and $s = 0$ at $t = 0$	$v = at$	So slope of *v-t* graph is constant $u = 0$ i.e. so, a = constant $u = 0$ i.e. $v = 0$ at $t = 0$
Uniformly accelerated motion with $u \neq 0$ but $s = 0$ at $t = 0$	$v = u + at$	Positive constant acceleration because θ is constant and $<90^\circ$ but the initial velocity of the particle is positive
Uniformly decelerated motion	$v = u - at$, t_0	Slope of *v-t* graphs = $-a$ (retardation)
Non-uniformly accelerated motion		Slope of *v-t* graph increases with time. θ is increasing, so, acceleration is increasing
Non-uniformly decelerating motion		θ is decreasing, so acceleration decreasing

DAY PRACTICE SESSION 1

FOUNDATION QUESTIONS EXERCISE

1 An aeroplane flies 400 m from North and then flies 300 m South and then flies 1200 m upwards, then net displacement is

(a) 1200 m (b) 1300 m (c) 1400 m (d) 1500 m

2 The correct statement from the following is

(a) A body having zero velocity will not necessarily have zero acceleration
(b) A body having zero velocity will necessarily have zero acceleration
(c) A body having uniform speed can have only uniform acceleration
(d) A body having non-uniform velocity will have zero acceleration

3 A vehicle travels half the distance L with speed v_1 and the other half with speed v_2, then its average speed is

(a) $\frac{v_1+v_2}{2}$ (b) $\frac{2v_1+v_2}{v_1+v_2}$ (c) $\frac{2v_1v_2}{v_1+v_2}$ (d) $\frac{2(v_1+v_2)}{v_1v_2}$

4 A particle moves along the sides AB, BC, CD of a square of side 25 m with a velocity of 15 m/s. Its average velocity is

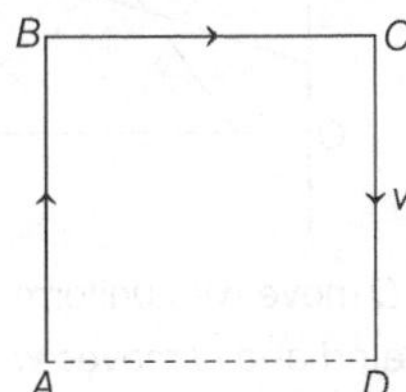

(a) 5 m/s (b) 7.5 m/s (c) 10 m/s (d) 15 m/s

5 A body sliding down on a smooth inclined plane slides down 1/4th of plane's length in 2 s. It will slide down the complete plane in

(a) 4 s (b) 5 s (c) 2 s (d) 3 s

6 Three particles P, Q and R are situated at corners of an equilateral triangle of side length (d). At $t = 0$, they started to move such that P is moving towards Q, Q is moving towards R and R is moving towards P at every instant. After how much time (in second) will they meet each other?

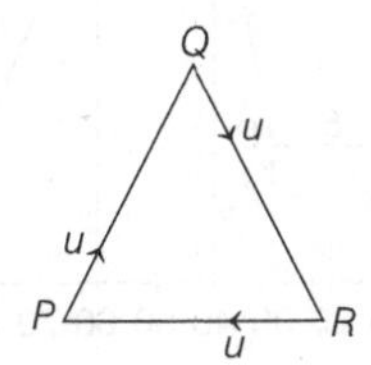

(a) $\frac{d}{u}$ (b) $\frac{2d}{3u}$ (c) $\frac{2d}{\sqrt{(3)}u}$ (d) $\frac{d}{\sqrt{(3)}\,u}$

7 A body is thrown vertically upwards in air, when air resistance is taken into account, the time of ascent is t_1 and time of descent is t_2, then which of the following is true?

(a) $t_1 = t_2$ (b) $t_1 < t_2$
(c) $t_1 > t_2$ (d) $t_1 \geq t_2$

8 A stone falls freely from rest and the total distance covered by it in the last second of its motion equals the distance covered by it in the first three seconds of its motion. The stone remains in the air for

(a) 6 s (b) 5 s (c) 7 s (d) 4 s

9 The motor of an electric train can give it an acceleration of 1 ms^{-2} and brakes can give a negative acceleration of 3 ms^{-2}. The shortest time in which the train can make a trip between the two stations 1215 m apart is

(a) 113.6 s (b) 56.9 s
(c) 60 s (d) 55 s

10 A train is moving along a straight path with a uniform acceleration. Its engine passes a pole with a velocity of 60 kmh^{-1} and the end (guard's van) passes across the same pole with a velocity of 80 kmh^{-1}. The middle point of the train will pass the same pole with a velocity

(a) 70 kmh^{-1} (b) 70.7 kmh^{-1}
(c) 65 kmh^{-1} (d) 75 kmh^{-1}

11 The acceleration experienced by a moving boat after its engine is cut-off, is given by $a = -kv^3$, where k is a constant. If v_0 is the magnitude of velocity at cut-off, then the magnitude of the velocity at time t after the cut-off is

(a) $\frac{v_0}{2ktv_0^2}$ (b) $\frac{v_0}{1+2ktv_0^2}$
(c) $\frac{v_0}{\sqrt{1-2kv_0^2}}$ (d) $\frac{v_0}{\sqrt{1+2ktv_0^2}}$

12 A body moving with an uniform acceleration describes 12 m in the 3rd second of its motion and 20 m in the 5th second. Find the velocity after the 10th second.

(a) 40 ms^{-1} (b) 42 ms^{-1}
(c) 52 ms^{-1} (d) 4 ms^{-1}

13 A train accelerating uniformly from rest attains a maximum speed of 40 ms^{-1} in 20 s. It travels at this speed for 20 s and is brought to rest with an uniform retardation in the next 40 s. What is the average velocity during this period?

(a) $\frac{80}{3}$ ms^{-1} (b) 25 ms^{-1}
(c) 40 ms^{-1} (d) 30 ms^{-1}

14 A frictionless wire AB is fixed on a sphere of radius R. A very small spherical ball slips on this wire. The time taken by this ball to slip from A to B is

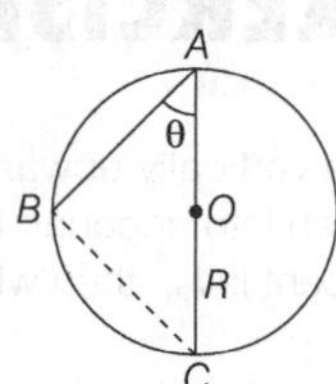

(a) $\frac{2\sqrt{gR}}{g\cos\theta}$ (b) $2\sqrt{gR}\,\frac{\cos\theta}{g}$ (c) $2\sqrt{\frac{R}{g}}$ (d) $\frac{gR}{\sqrt{g}\cos\theta}$

15 A balloon is going upwards with velocity $12\ \text{ms}^{-1}$. It releases a packet when it is at a height of 65 m from the ground. How much time the packet will take to reach the ground if $g = 10\ \text{ms}^{-2}$?

(a) 5 s (b) 6 s (c) 7 s (d) 8 s

16 A ball is dropped from the top of a building. The ball takes 0.5 s to fall past the 3 m length of window some distance below from the top of building. With what speed does the ball pass the top of window?

(a) $6\ \text{ms}^{-1}$ (b) $12\ \text{ms}^{-1}$ (c) $7\ \text{ms}^{-1}$ (d) $3.5\ \text{ms}^{-1}$

17 A body starts from the origin and moves along the axis such that the velocity at any instant is given by $v = 4t^3 - 2t$ where, t is in second and the velocity in ms^{-1}. Find the acceleration of the particle when it is at a distance of 2 m from the origin.

(a) $28\ \text{ms}^{-2}$ (b) $22\ \text{ms}^{-2}$ (c) $12\ \text{ms}^{-2}$ (d) $10\ \text{ms}^{-2}$

18 A point initially at rest moves along the x-axis. Its acceleration varies with time as $a = (5t + 6)\ \text{ms}^{-2}$. If it starts from the origin, the distance covered by it in 2s is

(a) 18.66 m (b) 14.33 m (c) 12.18 m (d) 6.66 m

19 A rod of length l leans by its upper end against a smooth vertical wall, while its other end leans against the floor. The end that leans against the wall moves uniformly downwards. Then,

(a) the other end also move uniformly
(b) the speed of other end goes on increasing
(c) the speed of other end goes on decreasing
(d) the speed of other end first decreases and then increases

20 Which of the following distance-time graphs is not possible?

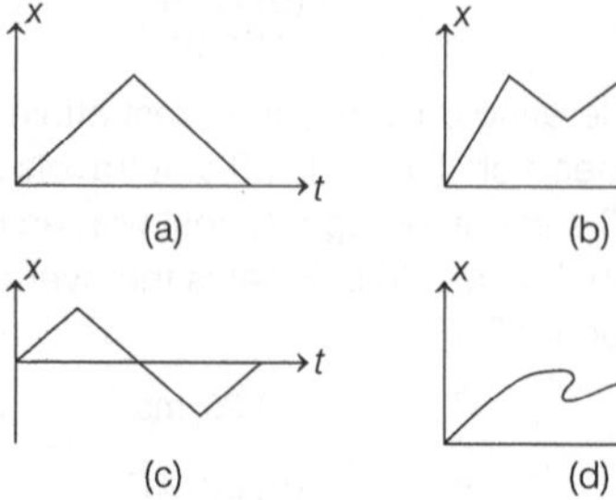

21 Look at the graphs (i) to (iv) in figure carefully and choose, which of these can possibly represent one-dimensional motion of particle?

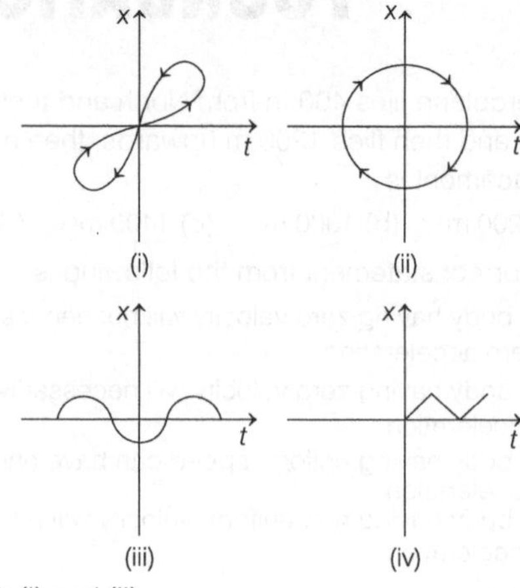

(a) Both (i) and (ii)
(b) Only (iv)
(c) Only (iii)
(d) Both (iii) and (iv)

22 Figure shows the time-displacement curve of the particles P and Q. Which of the following statement is correct?

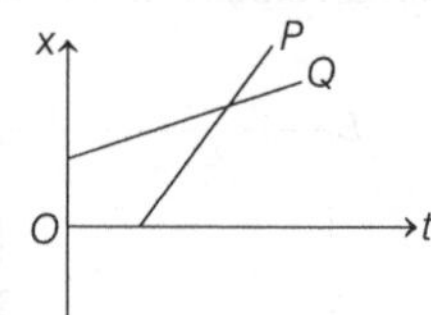

(a) Both P and Q move with uniform equal speed
(b) P is accelerated and Q moves with uniform speed but the speed of P is more than the speed of Q
(c) Both P and Q moves with uniform speeds but the speed of P is more than the speed of Q
(d) Both P and Q moves with uniform speeds but the speed of Q is more than the speed of P

23 The velocity *versus* time curve of a moving point is shown in the figure below. The maximum acceleration is

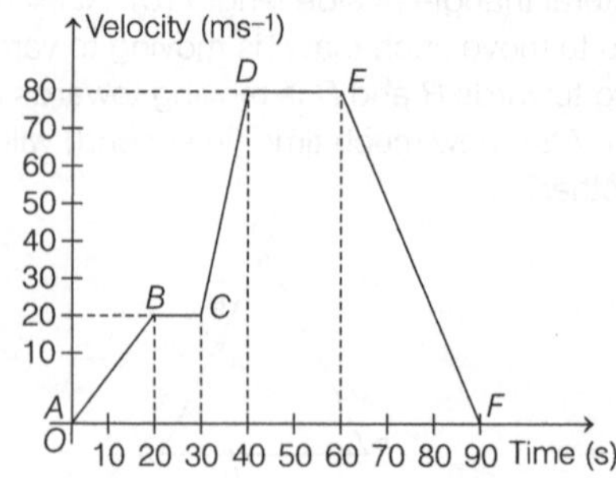

(a) $1\ \text{ms}^{-2}$ (b) $6\ \text{ms}^{-2}$
(c) $2\ \text{ms}^{-2}$ (d) $1.5\ \text{ms}^{-2}$

24 The velocity-time graph of a body in a straight line is as shown in figure.

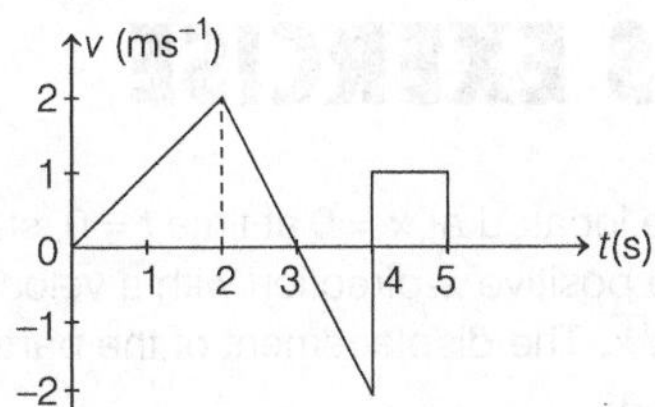

The displacement of the body in five seconds is

(a) 2 m (b) 3 m (c) 4 m (d) 5 m

25 All the graphs below are intended to represent the same motion. One of them does it incorrectly. Pick it up.

→ JEE Main 2018

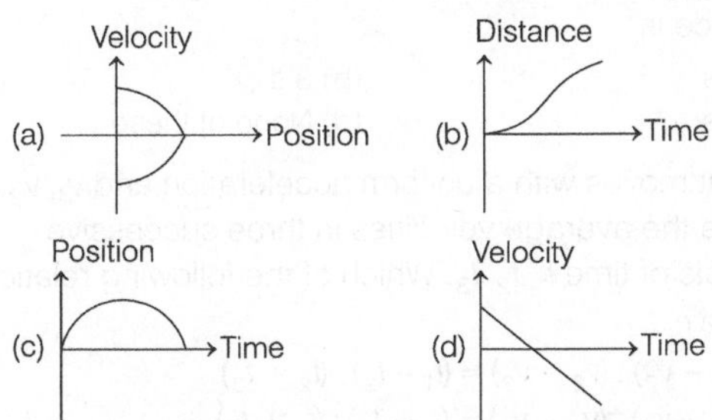

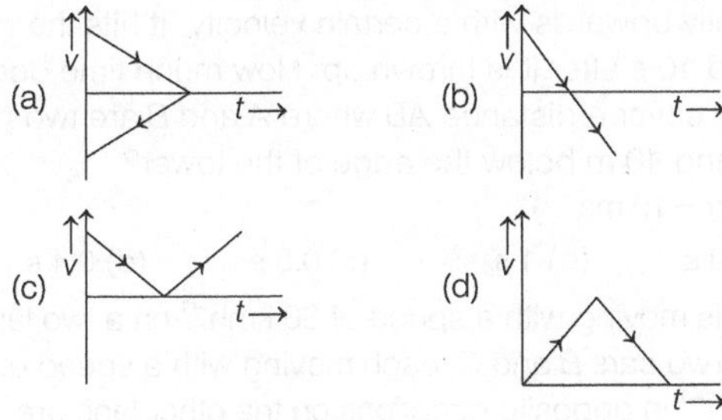

26 A body is thrown vertically upwards. Which one of the following graphs correctly represent the velocity *versus* time?

→ JEE Main 2017 (Offline)

(a) (b) (c) (d)

27 When two bodies move uniformly towards each other the distance between them decreases by 8 ms^{-1}. If both bodies move in the same direction with different speeds, the distance between them increases by 2 ms^{-1}. The speeds of two bodies will be

(a) 4 ms^{-1} and 3 ms^{-1}
(b) 4 ms^{-1} and 2 ms^{-1}
(c) 5 ms^{-1} and 3 ms^{-1}
(d) 7 ms^{-1} and 3 ms^{-1}

Direction (Q. Nos. 28-30) *Each of these questions contains two statements : Statement* I *(Assertion) and Statement* II *(Reason). Each of these questions also has four alternative choices, only one of which is the correct answer. You have to select one of the codes (a), (b), (c), (d) given below.*

(a) Statement I is true, Statement II is true, Statement II is the correct explanation for Statement I
(b) Statement I is true, Statement II is true, Statement II is not the correct explanation for Statement I
(c) Statement I is true, Statement II is false
(d) Statement I is false, Statement II is true

28 **Statement I** A particle moving with a constant velocity, changes its direction uniformly.

Statement II In a uniform motion, the acceleration is zero.

29 **Statement I** Two objects moving with velocities $\mathbf{v}_1$ and $\mathbf{v}_2$ in the opposite directions, have their relative velocity along the direction of the one with a larger velocity.

Statement II The relative velocity between two bodies moving with velocity $\mathbf{v}_1$ and $\mathbf{v}_2$ in same direction is given by $\mathbf{v} = \mathbf{v}_1 - \mathbf{v}_2$

30 **Statement I** Acceleration of a moving particle can be change without changing direction of velocity.

Statement II If the direction of velocity changes, so the direction of acceleration also changes.

DAY PRACTICE SESSION 2

PROGRESSIVE QUESTIONS EXERCISE

1 The velocity-time plot for a particle moving on a straight line is as shown in figure, then

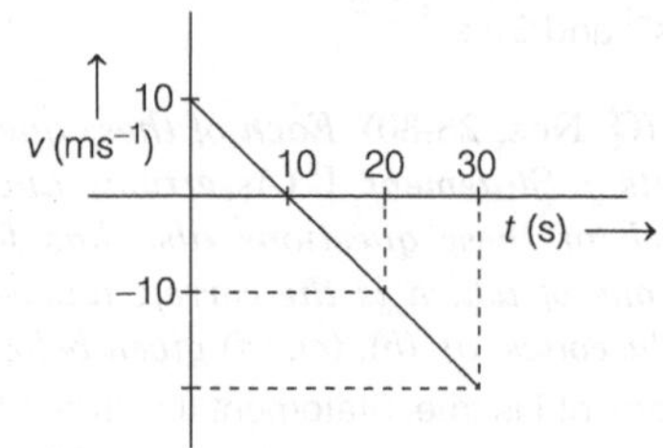

(a) the particle has a constant acceleration
(b) the particle has never turned around
(c) the average speed in the interval 0 to 10 s is the same as the average speed in the interval 10 s to 20 s
(d) Both (a) and (c) are correct

2 A body is at rest at $x = 0$. At $t = 0$, it starts moving in the positive x-direction with a constant acceleration. At the same instant another body passes through $x = 0$ moving in the positive x-direction with a constant speed. The position of the first body is given by $x_1(t)$ after time t and that of the second body by $x_2(t)$ after the same time interval. Which of the following graphs correctly describes $(x_1 - x_2)$ as a function of time t?

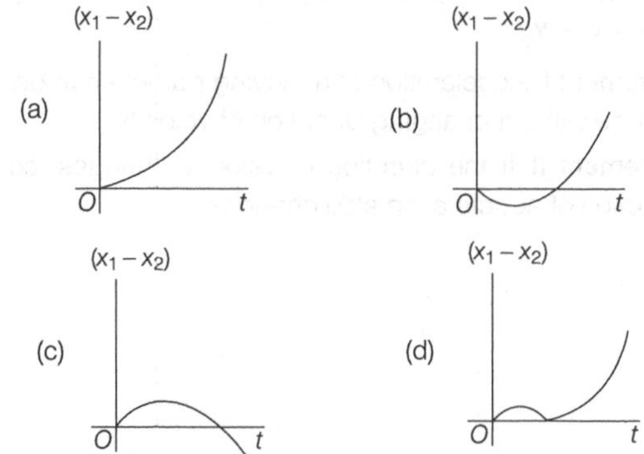

3 The velocity of a particle is $v = v_0 + gt + ft^2$. If its position is $x = 0$ at $t = 0$, then its displacement after unit time $(t = 1)$ is

(a) $v_0 + 2g + 3f$
(b) $v_0 + \frac{g}{2} + \frac{f}{3}$
(c) $v_0 + g + f$
(d) $v_0 + \frac{g}{2} + f$

4 A particle located at $x = 0$ at time $t = 0$, starts moving along the positive x-direction with a velocity v that varies as $v = \alpha\sqrt{x}$. The displacement of the particle varies with time as

(a) t^2 (b) t (c) $t^{1/2}$ (d) t^3

5 A stone is dropped from a certain height and reaches the ground in 5 s. If the stone is stopped after 3 s of its fall and then allowed to fall again, then the time taken by the stone to reach the ground after covering the remaining distance is

(a) 2 s (b) 3 s
(c) 4 s (d) None of these

6 A point moves with a uniform acceleration and v_1, v_2, v_3 denote the average velocities in three successive intervals of time t_1, t_2, t_3. Which of the following relations is correct?

(a) $(v_1 - v_2) : (v_2 - v_3) = (t_1 - t_2) : (t_2 + t_3)$
(b) $(v_1 - v_2) : (v_2 - v_3) = (t_1 + t_2) : (t_2 + t_3)$
(c) $(v_1 - v_2) : (v_2 - v_3) = (t_1 - t_2) : (t_1 - t_3)$
(d) $(v_1 - v_2) : (v_2 - v_3) = (t_1 - t_2) : (t_2 - t_3)$

7 From the top of a tower of height 50 m, a ball is thrown vertically upwards with a certain velocity. It hits the ground 10 s after it is thrown up. How much time does it take to cover a distance AB where A and B are two points 20 m and 40 m below the edge of the tower? (take, $g = 10\,\text{ms}^{-2}$)

(a) 2.0 s (b) 1 s (c) 0.5 s (d) 0.4 s

8 Car A is moving with a speed of 36 kmh^{-1} on a two lane road. Two cars B and C, each moving with a speed of 54 kmh^{-1} in opposite directions on the other lane are approaching car A. At certain instant of time, when the distance $AB = AC = 1$ km, the driver of car B decides to overtake A before C does. What must be the minimum acceleration of car B, so as to avoid an accident ?

(a) 1ms^{-2} (b) 4 ms^{-2} (c) 2 ms^{-2} (d) 3 ms^{-2}

9 The displacement x of a particle varies with time, according to the relation $x = \frac{a}{b}(1 - e^{-bt})$. Then

(a) the particle can not reach a point at a distance x from its starting position, if $x > a/b$
(b) at $t = 1/b$, the displacement of the particle is nearly $(2/3)(a/b)$
(c) the velocity and acceleration of the particle at $t = 0$ are a and $-ab$ respectively
(d) the particle will come back its starting point as $t \to \infty$

10 From the top of a tower, a stone is thrown up which reaches the ground in time t_1. A second stone thrown down, with the same speed, reaches the ground in time t_2. A third stone released from rest, from the same location, reaches the ground in a time t_3. Then,

(a) $\frac{1}{t_3} = \frac{1}{t_2} - \frac{1}{t_1}$ (b) $t_3^2 = t_1^2 - t_2^2$
(c) $t_3 = \frac{t_1 + t_2}{2}$ (d) $t_3 = \sqrt{t_1 t_2}$

11 A bullet moving with a velocity of 100 ms^{-1} can just penetrate two plancks of equal thickness. The number of such plancks penetrated by the same bullet, when the velocity is doubled, will be

(a) 4 (b) 6 (c) 8 (d) 10

12 The acceleration in ms^{-2} of a particle is given by, $a = 3t^2 + 2t + 2$ where, t is time. If the particle starts out with a velocity $v = 2\ ms^{-1}$ at $t = 0$, then the velocity at the end of 2 s is

(a) 36 ms^{-1} (b) 18 ms^{-1} (c) 12 ms^{-1} (d) 27 ms^{-1}

13 A car, starting from rest, accelerates at the rate f through a distance s, then continues at constant speed for time t and then decelerates at the rate $\frac{f}{2}$ to come to rest. If the total distance travelled is 15 s, then

(a) $s = ft$ (b) $s = \frac{1}{6}ft^2$
(c) $s = \frac{1}{72}ft^2$ (d) $s = \frac{1}{4}ft^2$

14 The displacement of a particle is given by $x = (t - 2)^2$ where, x is in metres and t in seconds. The distance covered by the particle in first 4 seconds is

(a) 4 m (b) 8 m (c) 12 m (d) 16 m

15 A metro train starts from rest and in five seconds achieves 108 kmh^{-1}. After that it moves with constant velocity and comes to rest after travelling 45 m with uniform retardation. If total distance travelled is 395 m, find total time of travelling.

(a) 12.2 s (b) 15.3 s
(c) 9 s (d) 17.2 s

16 From a tower of height H, a particle is thrown vertically upwards with a speed u. The time taken by the particle to hit the ground, is n times that taken by it to reach the highest point of its path. The relation between H, u and n is → 2014 JEE Main

(a) $2gH = n^2u^2$ (b) $gH = (n-2)^2u^2$
(c) $2gH = nu^2(n-2)$ (d) $gH = (n-2)^2u^2$

17 An object, moving with a speed of 6.25 ms^{-1}, is decelerated at a rate given by $\frac{dv}{dt} = -2.5\sqrt{v}$, where, v is the instantaneous speed. The time taken by the object, to come to rest, would be

(a) 2 s (b) 4 s (c) 8 s (d) 1 s

18 A ball is released from the top of a tower of height h metre. It takes T second to reach the ground. What is the position of the ball in $\frac{T}{3}$ s?

(a) $\frac{h}{9}$ m from the ground (b) $\frac{7h}{9}$ m from the ground
(c) $\frac{8h}{9}$ m from the ground (d) $\frac{17h}{18}$ m from the ground

19 Two stones are thrown up simultaneously from the edge of a cliff 240 m high with initial speed at 10 ms^{-1} and 40 ms^{-1}, respectively. Which of the following graph best represents the time variation of relative position of the second stone with respect to the first? (Assume stones do not rebound after hitting the ground and neglect air resistance, take $g = 10\ ms^{-2}$. The figures are schematic and not drawn to scale → JEE Main 2015

(a)
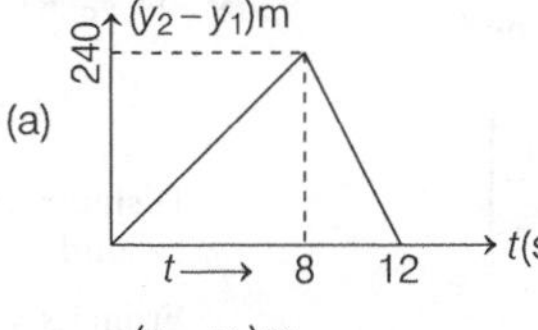

(b)
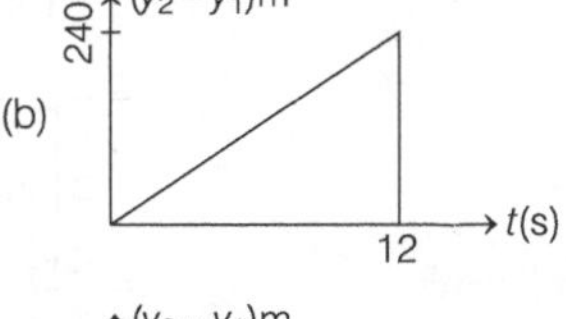

(c)
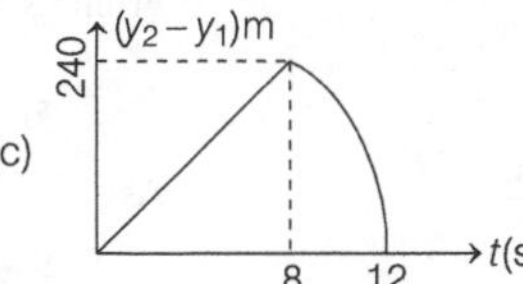

(d)
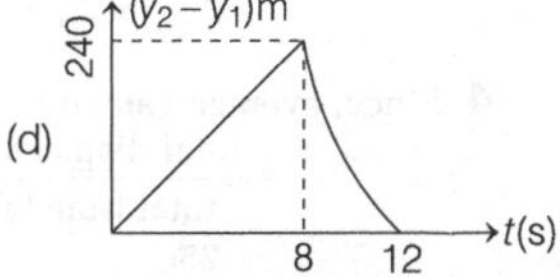

ANSWERS

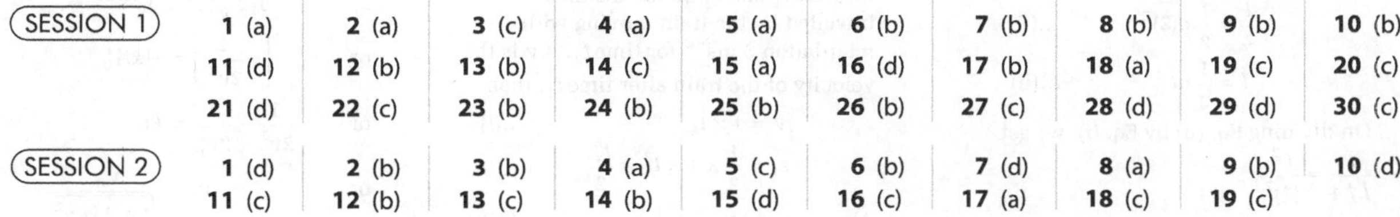

SESSION 1	1 (a)	2 (a)	3 (c)	4 (a)	5 (a)	6 (b)	7 (b)	8 (b)	9 (b)	10 (b)
	11 (d)	12 (b)	13 (b)	14 (c)	15 (a)	16 (d)	17 (b)	18 (a)	19 (c)	20 (c)
	21 (d)	22 (c)	23 (b)	24 (b)	25 (b)	26 (b)	27 (c)	28 (d)	29 (d)	30 (c)
SESSION 2	1 (d)	2 (b)	3 (b)	4 (a)	5 (c)	6 (b)	7 (d)	8 (a)	9 (b)	10 (d)
	11 (c)	12 (b)	13 (c)	14 (b)	15 (d)	16 (c)	17 (a)	18 (c)	19 (c)	

Hints and Explanations

SESSION 1

1 Displacement along North
$= 400 - 300 = 100\,\text{m}$
Upward displacement = 1200 m
$\therefore$ Net displacement
$= \sqrt{(100)^2 + (1200)^2}$
$= 1204.15\,\text{m} \approx 1200\,\text{m}$

2 When a body is projected vertically upwards, at the highest point of its motion, the velocity of the body becomes zero but acceleration is not zero.

3 Time taken to travel first half distance,

$$t_1 = \frac{L/2}{v_1} = \frac{L}{2v_1}$$

Time taken to travel second half distance,

$$t_2 = \frac{L/2}{v_2} = \frac{L}{2v_2}$$

Total time $= t_1 + t_2$

$$= \frac{L}{2v_1} + \frac{L}{2v_2}$$

$$= \frac{L}{2}\left[\frac{1}{v_1} + \frac{1}{v_2}\right]$$

$$\therefore \text{ Average speed} = \frac{\text{Total distance}}{\text{Total time}}$$

$$= \frac{L}{\frac{L}{2}\left[\frac{1}{v_1} + \frac{1}{v_2}\right]}$$

$$= \frac{2v_1 v_2}{v_1 + v_2}$$

4 Since, average velocity

$$= \frac{\text{total displacement}}{\text{total time taken}}$$

$$= \frac{25}{\left(\frac{75}{15}\right)}$$

$$= \frac{25 \times 15}{75} = 5\,\text{m/s}$$

5 As, $u = 0$ and a is a constant

$$\frac{l}{4} = \frac{1}{2}a(2)^2 \quad \ldots\text{(i)}$$

$$l = \frac{1}{2}at^2 \quad \ldots\text{(ii)}$$

On dividing Eq. (ii) by Eq. (i), we get

$$\frac{l}{l/4} = \frac{t^2}{(2)^2},$$

$$t = 4\,\text{s}$$

6 The person at P will travel a distance PO, with velocity along $PO = u\cos 30°$
Here,

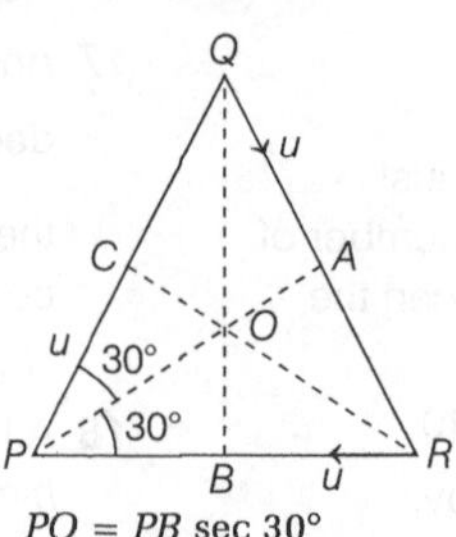

$$PO = PB \sec 30°$$

$$= \frac{d}{2} \times \frac{2}{\sqrt{3}} = \frac{d}{\sqrt{3}}$$

$\therefore$ Time of meeting,

$$t = \frac{\text{distance}}{\text{velocity}} = \frac{d/\sqrt{3}}{u\cos 30°}$$

$$= \frac{d/\sqrt{3}}{u\sqrt{3}/2}$$

$$= \frac{2d}{3u}\,\text{s}$$

7 When a body is thrown up, its velocity goes on decreasing as air resistance is small. When a body falls down, its velocity goes on increasing as air resistance is large, t_2 increases.

8 As, $s_n = u + \frac{g}{2}(2n - 1)$

$$= 0 + \frac{g}{2}(2n - 1)$$

Distance travelled in the first three second

From $s = ut + \frac{1}{2}at^2$

$$s_3 = 0 \times 3 + \frac{1}{2} \times g \times 3^2 = \frac{9}{2}g$$

As, $S_n = s_3$

$$\frac{g}{2}(2n - 1) = \frac{9}{2}g$$

$$2n - 1 = 9$$

$$n = 5\text{s}$$

9 Let s_1 be the distance travelled by the train moving with acceleration 1ms^{-2} for time t_1 and s_2 be the distance travelled by the train moving with retardation $3\ \text{ms}^{-2}$ for time t_2. If v is the velocity of the train after time t_1, then

$$v = 1 \times t_1 \quad \ldots\text{(i)}$$

$$s_1 = \frac{1}{2} \times 1 \times t_1^2 = \frac{t_1^2}{2} \quad \ldots\text{(ii)}$$

Also, $v = 3t_2 \quad \ldots\text{(iii)}$

and $s_2 = vt_2 - \frac{1}{2} \times 3 \times t_2^2$

$$= t_1 t_2 - \frac{3}{2}t_2^2$$

From Eqs. (i) and (iii), we get

$$t_1 = 3t_2 \text{ or } t_2 = \frac{t_1}{3}$$

$$s_1 + s_2 = \frac{t_1^2}{2} + t_1 \times \frac{t_1}{3} - \frac{3}{2} \times \frac{t_1^2}{9} = \frac{2}{3}t_1^2$$

$$1215 = \frac{2}{3}t_1^2$$

$$\Rightarrow \quad t_1 = \sqrt{\frac{3 \times 1215}{2}}$$

$$= 42.69\,\text{s}$$

Total time $= t_1 + t_2 = t_1 + \frac{t_1}{3} = 56.9\,\text{s}$

10 From $v^2 - u^2 = 2as$

$$\Rightarrow \quad \frac{80^2 - 60^2}{2a} = s$$

$$\Rightarrow \quad s = \frac{6400 - 3600}{2a}$$

$$= \frac{1400}{a}$$

The middle point of the train has to cover a distance

$$\frac{s}{2} = \frac{700}{a}$$

From $v^2 - u^2 = 2as$

$$v^2 - 60^2 = 2a \times \frac{700}{a} = 1400$$

$$v^2 = 1400 + 3600$$

$$v = \sqrt{5000}$$

$$= 70.7\ \text{kmh}^{-1}$$

11 Given, acceleration $a = -kv^3$
Initial velocity at cut-off, $v_1 = v_0$
Initial time of cut-off, $t = 0$ and final time after cut-off, $t_2 = t$

Again, $a = \frac{dv}{dt} = -kv^3$ or $\frac{dv}{v^3} = -kdt$

Integrating both sides, with in the condition of motion.

$$\int_{v_0}^{v} \frac{dv}{v^3} = -\int_0^t k\,dt$$

or $$\left[-\frac{1}{2v^2}\right] = -[kt]_0^t$$

or $$\frac{1}{2v^2} - \frac{1}{2v_0^2} = kt$$

or $$v = \frac{v_0}{\sqrt{1 + 2kt\,v_0^2}}$$

12 Using, $s_n = u + \frac{a}{2}(2n-1)$

$$12 = u + \frac{a}{2}(2\times 3 - 1) \quad ...(i)$$

$$20 = u + \frac{a}{2}(2\times 5 - 1) \quad ...(ii)$$

On subtracting Eq. (i) from Eq. (ii), we get

$$8 = \frac{a}{2}(10-6) = 2a$$

$$a = 4\text{ ms}^{-2}$$

From Eq. (i), $12 = u + \frac{4}{2}\times 5$

$$u = 2\text{ ms}^{-1}$$

From $v = u + at = 2 + 4\times 10$

$$= 42\text{ ms}^{-1}$$

13 As, $v = u + at_1 \quad ...(i)$

$$40 = 0 + a\times 20$$

$$a = 2\text{ ms}^{-2}$$

Now, $v^2 - u^2 = 2as$

$$40^2 - 0 = 2\times 2\times s_1$$

$$s_1 = 400\text{ m}$$

$$s_2 = v\times t_2 \quad ...(ii)$$

$$= 40\times 20 = 800\text{ m}$$

and $v = u + at \quad ...(iii)$

$$0 = 40 + a\times 40,$$

$$a = -1\text{ ms}^{-2}$$

Also, $v^2 - u^2 = 2as$

$$0^2 - 40^2 = 2(-1)s_3$$

$$s_3 = 800\text{ m}$$

∴ Total distance travelled

$$= s_1 + s_2 + s_3$$

$$= 400 + 800 + 800$$

$$= 2000\text{ m}$$

and total time taken = 20 + 20 + 40 = 80 s

∴ Average velocity $= \frac{2000}{80} = 25\text{ ms}^{-1}$

14 Acceleration of the body down the plane = $g\cos\theta$

Distance travelled by ball in time t second is

$$AB = \frac{1}{2}(g\cos\theta)t^2 \quad ...(i)$$

From Δ ABC,

$$AB = 2R\cos\theta \quad ...(ii)$$

From Eqs. (i) and (ii), we get

$$2R\cos\theta = \frac{1}{2}g\cos\theta t^2$$

$$t^2 = \frac{4R}{g}$$

or $t = 2\sqrt{\frac{R}{g}}$

15 $a = +g = 10\text{ ms}^{-2}$,

$s = 65\text{ m}, t = ?$

As, $s = ut + \frac{1}{2}at^2$

$$\Rightarrow 65 = -12t + 5t^2$$

$$5t^2 - 12t - 65 = 0$$

This gives,

$$t = \frac{12 \pm \sqrt{144 + 1300}}{10} = \frac{12\pm 38}{10} = 5\text{ s}$$

16 From $s = ut + \frac{1}{2}at^2$,

$$x = 0 + \frac{1}{2}\times 10t^2 = 5t^2 \quad ...(i)$$

Also, $x + 3 = 0 + \frac{1}{2}\times 10(t+0.5)^2$

$$= 5\left(t^2 + \frac{1}{4} + t\right) \quad ...(ii)$$

Subtract Eq. (i) from Eq. (ii), we get

$$3 = 5\left(\frac{1}{4} + t\right) = \frac{5}{4} + 5t$$

$$3 - \frac{5}{4} = 5t$$

$$\frac{7}{4} = 5t \text{ or } t = \frac{7}{20}\text{ s}$$

From $v = u + at$,

$$v = 0 + 10\times\frac{7}{20} = 3.5\text{ ms}^{-1}$$

17 $v = 4t^3 - 2t \quad ...(i)$

$$\frac{dx}{dt} = 4t^3 - 2t$$

On integration, we get,

$$x = 2 = t^4 - t^2$$

Let $t^2 = \alpha$

∴ $2 = \alpha^2 - \alpha \quad ...(ii)$

Let $t^2 = \alpha$

$$\alpha^2 - \alpha - 2 = 0$$

$$(\alpha - 2)(\alpha + 1) = 0$$

∴ $\alpha = 2, \alpha = -1$,

which is not possible

$$t^2 = \alpha = 2 \text{ or } t = \sqrt{2},$$

Differentiating Eq. (i) w.r.t. t,

$$\frac{dv}{dt} = 12t^2 - 2$$

$$a = 12\times 2 - 2 = 22\text{ ms}^{-2}$$

18 Acceleration, $a = \frac{dv}{dt} = 5t + 6$

On integrating, we get

$$v = \frac{5}{2}t^2 + 6t = \frac{dx}{dt}$$

Integrating again,

$$x = \frac{5}{6}t^3 + \frac{6}{2}t^2$$

At $t = 2\text{s}$,

$$x = \frac{5}{6}\times 8 + 3\times 4 = 18.66\text{ m}$$

19 If $(x, 0)$ and $(y, 0)$ are the coordinates of the end points of the rod at a given location, then $x^2 + y^2 = l^2$

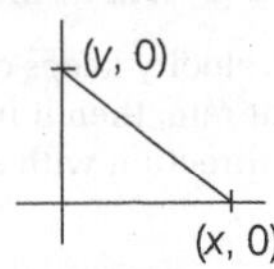

Differentiating it w.r.t. t, we get

$$2x\frac{dx}{dt} + 2y\frac{dy}{dt} = 0$$

$$\frac{dx}{dt} = -y\frac{dy/dt}{x}$$

and $v_x = -\frac{y}{x}v_y$

As, y decreases, x increases, so v_x decreases.

v_x becomes zero when y is zero.

20 The distance travelled can never be negative in one dimensional motion.

21 In one dimensional motion, there is a single value of displacement at one particular time.

22 As x-t graph is a straight line in either case, velocity of both is uniform. As the slope of x-t graph for P is greater, therefore, velocity of P is greater than that of Q.

23 Maximum acceleration is represented by the maximum slope of the velocity-time graph. Thus, it is the portion CD of the graph, which has a slope $= \frac{80-20}{40-30}$

$$= 6\text{ ms}^{-2}.$$

24 Displacement is the algebraic sum of area under velocity-time graph.

As, displacement = area of triangles + area of rectangle

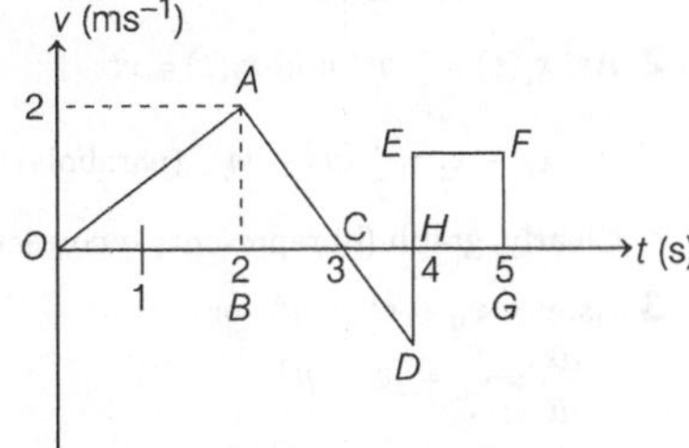

$$\Delta OAB + \Delta ABC + \Delta CDH + HEFG$$

$$= \frac{1}{2}\times 2\times 2 + \frac{1}{2}\times 1\times 2 + \frac{1}{2}\times 1\times(-2) + 1\times 1$$

$$= 2 + 1 - 1 + 1 = 3\text{ m}$$

25 If velocity *versus* time graph is a straight line with negative slope, then acceleration is constant and negative. With a negative slope distance-time graph will be parabolic $\left(s = ut - \frac{1}{2}at^2\right)$.

So, option (b) will be incorrect.

26 Initially velocity keeps on decreasing at a constant rate, then it increases in negative direction with same rate.

27 **Case I**

Relative velocity is $v_1 + v_2 = 8$

Case II

Relative velocity is $v_1 - v_2 = 2$

On solving,

$v_1 = 5\,\text{ms}^{-1}, v_2 = 3\,\text{ms}^{-1}$

28 When a particle moves with constant velocity, then acceleration of particle is zero and hence particle is not able to change the direction. Hence, statement I is false while statement II is true. Hence, correct answer is (d).

29 When two objects moving in opposite direction, then their relative velocity becomes $(v_1 + v_2)$, hence statement I is false. When moves in same direction, then relative velocity $v = (v_1 - v_2)$, hence statement II is true. Hence, correct answer is (d).

30 Without changing direction of velocity, it is possible to change the acceleration of a moving particle, hence statement I is true, while statement II is false. Hence, correct answer is (c).

SESSION 2

1 The slope of velocity-time graph gives acceleration. Since, the given graph is a straight line and slope of graph is constant. Hence acceleration is constant. Thus, (a) is correct. The area of *v-t* graph between 0 to 10 s is same as between 10 s to 20 s.

2 As, $x_1(t) = \frac{1}{2}at^2$ and $x_2(t) = vt$

$\therefore\ x_1 - x_2 = \frac{1}{2}at^2 - vt$ (parabola)

Clearly, graph (b) represents it correctly.

3 As, $v = v_0 + gt + ft^2$ or

$\frac{dx}{dt} = v_0 + gt + ft^2$

$\Rightarrow dx = (v_0 + gt + ft^2)\,dt$

So, $\int_0^x dx = \int_0^1 (v_0 + gt + ft^2)\,dt$

$\Rightarrow\quad x = v_0 + \frac{g}{2} + \frac{f}{3}$

4 Given, $v = \alpha\sqrt{x}$

or $\frac{dx}{dt} = \alpha\sqrt{x}$ $\quad\left(\because v = \frac{dx}{dt}\right)$

or $\frac{dx}{\sqrt{x}} = \alpha\,dt$

On integration,

$$\int_0^x \frac{dx}{\sqrt{x}} = \int_0^t \alpha\,dt$$

[$\because$ at $t = 0$, $x = 0$ and let at any time t, particle is at x]

$\Rightarrow\quad \left.\frac{x^{1/2}}{1/2}\right|_0^x = \alpha t$

or $x^{1/2} = \frac{\alpha}{2}t$

or $x = \frac{\alpha^2}{4} \times t^2$

$\therefore\quad x \propto t^2$

5 From $s = ut + \frac{1}{2}at^2$,

$s = 0 + \frac{1}{2} \times 10 \times 5^2 = 125\,\text{m}$

Distance covered in 3 s,

$= 0 + \frac{1}{2} \times 10 \times 3^2 = 45\,\text{m}$

Distance to be covered $= 125 - 45 = 80\,\text{m}$

From $s = ut + \frac{1}{2}at^2$

$80 = 0 + \frac{1}{2} \times 10t^2$

$\Rightarrow\quad t^2 = \frac{80}{5} = 16$

$\therefore\quad t = 4\,\text{s}$

6 Suppose velocity at O = zero

As average velocity in interval t_1 is v_1,

$\therefore$ Velocity at $A = v_1$

As average velocity in interval t_2 is v_2,

$\therefore$ Velocity at $B = (v_2 - v_1)$

As average velocity in interval t_3 is v_3,

Velocity at $C = (v_3 - v_2 + v_1)$

Using $v = u + at$

$v_1 = 0 + at_1$ ...(i)

$(v_2 - v_1) = 0 + a(t_1 + t_2)$ ...(ii)

$(v_3 - v_2 + v_1) = 0 + a(t_1 + t_2 + t_3)$ ...(iii)

Subtract Eq. (i) from Eq. (iii), we get

$(v_3 - v_2) = a(t_2 + t_3)$ ...(iv)

Divide Eq. (ii) by Eq. (iv), we get

$$\frac{(v_2 - v_1)}{(v_3 - v_2)} = \frac{a(t_1 + t_2)}{a(t_2 + t_3)}$$

$$\frac{(v_1 - v_2)}{(v_2 - v_3)} = \frac{t_1 + t_2}{t_2 + t_3}$$

7 Given, $v = -u, a = g = 10\,\text{ms}^{-2}$,

$s = 50\,\text{m}, t = 10\,\text{s}$

As, $s = ut + \frac{1}{2}at^2$,

$\Rightarrow\ 50 = -u \times 10 + \frac{1}{2} \times 10 \times 10^2$

On solving,

$u = 45\,\text{ms}^{-1}$

If t_1 and t_2 are the timings taken by the ball to reach the points A and B respectively, then

$20 = -45t_1 + \frac{1}{2} \times 10 \times t_1^2$

$40 = -45t_2 + \frac{1}{2} \times 10 \times t_2^2$

On solving, we get

$t_1 = 9.4\,\text{s}$ and $t_2 = 9.8\,\text{s}$

Time taken to cover the distance AB,

$= (t_2 - t_1)$

$= 9.8 - 9.4 = 0.4\,\text{s}$

8 Let us suppose that the cars A and B are moving in the positive x-direction. Then, car C is moving in the negative x-direction.

Therefore,

$v_A = 36\,\text{kmh}^{-1} = 10\,\text{ms}^{-1}$

$v_B = 54\,\text{kmh}^{-1} = 15\,\text{ms}^{-1}$

and $v_C = -54\,\text{kmh}^{-1} = -15\,\text{ms}^{-1}$

Thus, the relative speed of B with respect to A is,

$v_{BA} = v_B - v_A$

$= 15 - 10 = 5\,\text{ms}^{-1}$

and the relative speed of C with respect to A is,

$v_{CA} = v_C - v_A = -15 - 10$

$= -25\,\text{ms}^{-1}$

At time $t = 0$, the distance between A and B = distance between A and $C = 1\,\text{km} = 1000\,\text{m}$.

The car C covers a distance $AC = 1000\,\text{m}$ and reaches car A at a time t given by

$$t = \frac{AC}{|v_{CA}|}$$

$$= \frac{1000\,\text{m}}{25\,\text{ms}^{-1}} = 40\,\text{s}$$

Car B will overtake car A just before car C does and the accident can be avoided if it acquires a minimum acceleration a such that it covers a distance, $s = AB = 1000$ m in time $t = 40$ s travelling with a relative speed of

$u = v_{BA} = 5\,\text{ms}^{-1}$.

This gives, from

$s = ut + \frac{1}{2}at^2, a = 1\,\text{ms}^{-2}$

9 Velocity of the particle is given by

$$v = \frac{dx}{dt} = \frac{d}{dt}\left\{\frac{a}{b}(1-e^{-bt})\right\} = ae^{-bt}$$

Acceleration of the particle is given by

$$\alpha = \frac{dv}{dt} = \frac{d}{dt}(ae^{-bt}) = -abe^{-bt}$$

At $t = 1/b$, the displacement of the particle is

$$x = \frac{a}{b}(1-e^{-1}) \approx \frac{a}{b}\left(1-\frac{1}{3}\right) = \frac{2}{3}\frac{a}{b}$$

$$\left(\because e^{-1} \approx \frac{1}{3}\right)$$

Thus, choice (b) is correct. At $t = 0$, the value v and α are $v = ae^{-0} = a$

and $\alpha = -ab\,e^{-0} = ab$

The displacement x is maximum, when

$$t \to \infty,$$

i.e. $x_{max} = \frac{a}{b}(1-e^{-\infty}) = \frac{a}{b}$

10 We know that, $h = ut + \frac{1}{2}gt^2$

$\Rightarrow$ $h = -ut + \frac{1}{2}gt^2$

and $t = \frac{2u}{g} = t_1 - t_2$ ...(i)

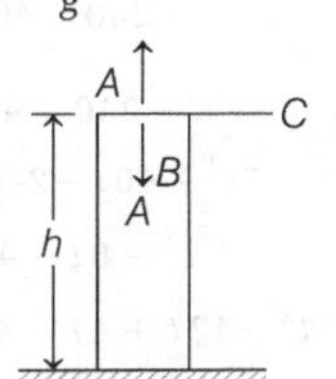

Case I $h = -ut_1 + \frac{1}{2}gt_1^2$ ...(ii)

Case II $h = +ut_2 + \frac{1}{2}gt_2^2$ ...(iii)

Case III $h = \frac{1}{2}gt_3^2$ ...(iv)

This gives,

$$\frac{2h}{g} = \frac{2u}{g}t_2 + t_2^2 \quad ...(v)$$

Solving these, give us

$$t_3^2 = (t_1 - t_2)t_2 + t_2^2$$

$\Rightarrow$ $t_3 = \sqrt{t_1 t_2}$

11 Given that the initial velocity of the bullet in the first case is $u_1 = 100\text{ ms}^{-1}$.

Initial number of plancks, $n_1 = 2$

Initial stopping distance

$$= s_1 = n_1 x = 2x,$$

with x as the thickness of one planck.

Similarly, Initial velocity of the bullet in second case,

$$u_2 = 2 \times 100 = 200\text{ ms}^{-1}$$

We know that the relation for the stopping distance s is

$$v^2 = u^2 + 2as$$

Since, $v = 0$,

So, $2as = -u^2$

As, $s \propto u^2$

Hence, $\frac{s_1}{s_2} = \left(\frac{u_1}{u_2}\right)^2 = \left(\frac{100}{200}\right)^2 = \frac{1}{4}$

Thus, $s_2 = 4s_1 = 8x$

Hence, the number of plancks

$$= n_2 = \frac{s_2}{x} = 8$$

12 Given, $a = \frac{dv}{dt} = 3t^2 + 2t + 2$

$\Rightarrow$ $dv = (3t^2 + 2t + 2)\,dt$

On integrating, this gives

$$\int_u^v dv = \int_0^2 (3t^2 + 2t + 2)\,dt$$

$\Rightarrow$ $v - u = \left[\frac{3t^3}{3} + \frac{2t^2}{2} + 2t\right]_0^2$

$\Rightarrow$ $v = u + [t^3 + t^2 + 2t]_0^2$

$$v = 2 + [2^3 + 2^2 + 2 \times 2]$$

$$= 2 + 16$$

$$= 18\text{ ms}^{-1}$$

13 The velocity-time graph for the given situation can be drawn as below.

Magnitudes of slope of $OA = f$

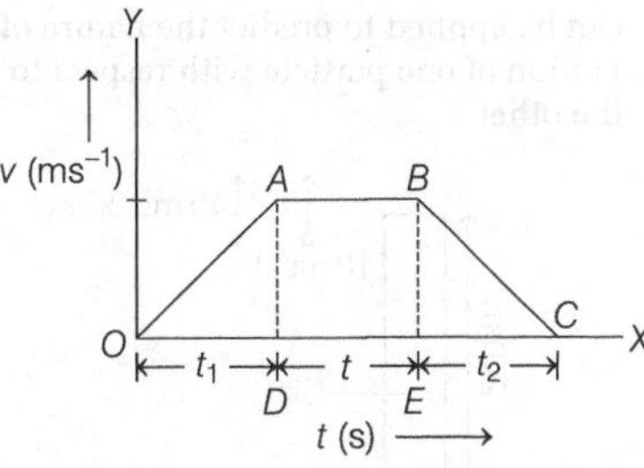

and slope of $BC = \frac{f}{2}$

$$v = ft_1 = \frac{f}{2}t_2$$

$$t_2 = 2t_1$$

In graph area of Δ OAD gives distance,

$$s = \frac{1}{2}ft_1^2 \quad ...(i)$$

Area of rectangle $ABED$ gives distance travelled in time t

$$s_2 = (ft_1)t$$

Distance travelled in time t_2,

$$s_3 = \frac{1}{2}\frac{f}{2}(2t_1)^2$$

Thus, $s_1 + s_2 + s_3 = 15s$

$\Rightarrow$ $s + (ft_1)t + ft_1^2 = 15s$

or $s + (ft_1)t + 2s = 15s$

$$\left(s = \frac{1}{2}ft_1^2\right)$$

or $(ft_1) = 12s$ ...(ii)

From Eqs. (i) and (ii), we have

$$\frac{12s}{s} = \frac{(ft_1)t}{\frac{1}{2}(ft_1)t_1}$$

or $t_1 = \frac{t}{6}$

From Eq. (i), we get

$$s = \frac{1}{2}f(t_1)^2$$

$$= \frac{1}{2}f\left(\frac{t}{6}\right)^2 = \frac{1}{72}ft^2$$

14 Here, $x = (t-2)^2$

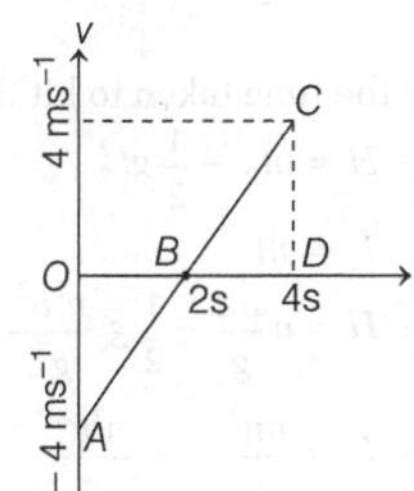

Velocity, $v = \frac{dx}{dt} = 2(t-2)\text{ ms}^{-1}$

Acceleration, $a = \frac{dv}{dt} = 2\text{ ms}^{-2}$

(i.e. uniform)

When $t = 0, v = -4\text{ ms}^{-1}$,

$t = 2\text{ s},\ v = 0,\ t = 4\text{s},\ v = 4\text{ ms}^{-1}$

Velocity (v)- time (t) graph of this motion is as shown in figure.

Distance travelled

= Area AOB + Area BCD

$$= \frac{4 \times 2}{2} + \frac{4 \times 2}{2} = 8\text{ m}$$

15 Given, $v = 108\text{ kmh}^{-1} = 30\text{ ms}^{-1}$

From first equation of motion

$$v = u + at$$

$\therefore$ $30 = 0 + a \times 5$ $(\because u = 0)$

or $a = 6\text{ ms}^{-2}$

So, distance travelled by metro train in 5 s

$$s_1 = \frac{1}{2}at^2 = \frac{1}{2} \times (6) \times (5)^2 = 75\text{ m}$$

Distance travelled before coming to rest = 45 m

So, from third equation of motion

$$0^2 = (30)^2 - 2a' \times 45$$

or $a' = \frac{30 \times 30}{2 \times 45} = 10\text{ ms}^{-2}$

Time taken in travelling 45 m is

$t_3 = \frac{30}{10} = 3$ s

Now, total distance = 395 m

i.e. $75 + s' + 45 = 395$ m

or $s' = 395 - (75 + 45) = 275$ m

$\therefore \quad t_2 = \frac{275}{30} = 9.2$ s

Hence, total time taken in whole journey $= t_1 + t_2 + t_3$

$= 5 + 9.2 + 3 = 17.2$ s

16 Time taken to reach the maximum height, $t_1 = \frac{u}{g}$

If t_2 is the time taken to hit the ground,

i.e. $-H = ut_2 - \frac{1}{2} gt_2^2$

But $t_2 = nt_1$ [Given]

So, $-H = u\frac{nu}{g} - \frac{1}{2} g \frac{n^2u^2}{g^2}$

$-H = \frac{nu^2}{g} - \frac{1}{2}\frac{n^2u^2}{g}$

$H = \frac{1}{2}\frac{n^2u^2}{g} - \frac{nu^2}{g} = \frac{n^2u^2 - 2nu^2}{2g}$

$2gH = n^2u^2 - 2nu^2$

$2gH = nu^2(n-2)$

17 Given, $\frac{dv}{dt} = -2.5\sqrt{v}$

$\Rightarrow \quad \frac{dv}{\sqrt{v}} = -2.5dt$

$\Rightarrow \quad \int_{6.25}^{0} v^{-1/2}\, dv = -2.5 \int_0^t dt$

$\Rightarrow \quad -2.5[t]_0^t = [2v^{1/2}]_{6.25}^{0}$

$= 2(-\sqrt{6.25}) = 2 \times 2.5$

$\Rightarrow \quad t = 2$ s

18 From equation law of motion gives,

$s = ut + \frac{1}{2} gT^2$

or $h = 0 + \frac{1}{2} gT^2 \quad (\because u = 0)$

$\Rightarrow \quad T = \sqrt{\left(\frac{2h}{g}\right)}$

At, $t = \frac{T}{3}$ s,

$s = 0 + \frac{1}{2} g\left(\frac{T}{3}\right)^2$

or $s = \frac{1}{2} g \cdot \frac{T^2}{9}$

$\Rightarrow \quad s = \frac{g}{18} \times \frac{2h}{g} \quad \left(\therefore T = \sqrt{\frac{2h}{g}}\right)$

or $s = \frac{h}{9}$ m

Hence, the position of ball from the ground $= h - \frac{h}{9} = \frac{8h}{9}$ m

19 Central idea concept of relative motion can be applied to predict the nature of motion of one particle with respect to the other.

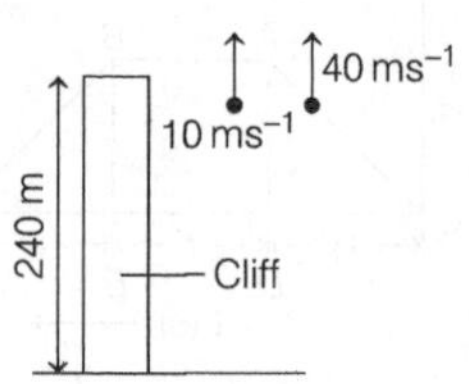

Consider the stones thrown up simultaneously as shown in the diagram below. As motion of the second particle with respect to the first we have relative acceleration

$|a_{21}| = |a_2 - a_1| = g - g = 0.$

Thus, motion of first particle is straight line with respect to second particle till the first particle strikes ground at a time is given by

$-240 = 10t - \frac{1}{2} \times 10 \times t^2$

or $t^2 - 2t - 48 = 0$

or $t^2 - 8t + 6t - 48 = 0$

or $t = 8, -6$

[As, $t = -6$s is not possible]

i.e., $t = 8$ s

Thus, distance covered by second particle with respect to first particle in 8 s is

$s_{12} = (v_{21})t = (40 - 10)$ (8s)

$= 30 \times 8$

$= 240$ m

Similarly, time taken by second particle to strike the ground is given by

$-240 = 40t - \frac{1}{2} \times 10 \times t$

or $-240 = 40t - 5t^2$

or $5t^2 - 40t - 240 = 0$

or $t^2 - 8t - 48 = 0$

$t^2 - 12t + 4t - 48 = 0$

or $t(t - 12) + 4(t - 12) = 0$

or $t = 12, -4$

(As, $t = -4$ s is not possible)

i.e. $t = 12$ s

Thus, after 8 s, magnitude of relative velocity will increase upto 12 s when second particle strikes the ground. Hence, graph (c) is the correct description.

DAY THREE

Scalar and Vector

Learning & Revision for the Day

- Scalar and Vector Quantities
- Laws of Vector Addition
- Subtraction of Vectors
- Multiplication or Division of a Vector by a Scalar
- Product of Vectors
- Resolution of a vector
- Relative velocity
- Motion in a Plane
- Projectile Motion

Scalar and Vector Quantities

A **scalar quantity** is one whose specification is completed with its magnitude only. e.g. mass, distance, speed, energy, etc.

A **vector quantity** is a quantity that has magnitude as well as direction. e.g. Velocity, displacement, force, etc.

Position and Displacement Vectors

A vector which gives position of an object with reference to the origin of a coordinate system is called position vector.

The vector which tells how much and in which direction on object has changed its position in a given interval of time is called displacement vector.

General Vectors and Notation

- **Zero Vector** The vector having zero magnitude is called **zero vector** or **null vector**. It is written as 0. The initial and final points of a zero vector overlap, so its direction is arbitrary (not known to us).
- **Unit Vector** A vector of unit magnitude is known as an **unit vector**. Unit vector for **A** is $\hat{\mathbf{A}}$ (read as A cap).

$$\mathbf{A} = A\,\hat{\mathbf{A}}$$

(Direction: $\hat{\mathbf{A}}$; Magnitude: A)

- **Orthogonal Unit Vectors** The unit vectors along X-axis,s, Y-axis and Z-axis are denoted by $\hat{\mathbf{i}}, \hat{\mathbf{j}}$ and $\hat{\mathbf{k}}$. These are the orthogonal unit vectors.

$$\hat{\mathbf{i}} = \frac{\mathbf{x}}{x}, \hat{\mathbf{j}} = \frac{\mathbf{y}}{y}, \hat{\mathbf{k}} = \frac{\mathbf{z}}{z}$$

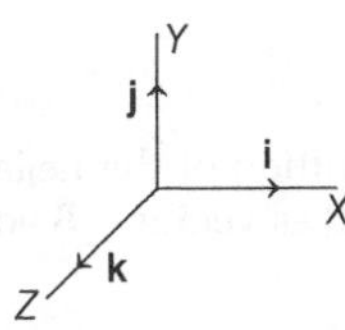

- No. of Questions in Exercises (x)—
- No. of Questions Attempted (y)—
- No. of Correct Questions (z)—
  (Without referring Explanations)

- Accuracy Level ($z/y \times 100$)—
- Prep Level ($z/x \times 100$)—

In order to expect good rank in JEE, your Accuracy Level should be above 85 & Prep Level should be above 75.

- **Parallel Vector** Two vectors are said to be parallel, if they have same direction but their magnitudes may or may not be equal.
- **Antiparallel Vector** Two vectors are said to be anti-parallel when
  (i) both have opposite direction
  (ii) one vectors is scalar non zero negative multiple of another vector.
- **Collinear Vector** Collinear vector are those which act along same line.
- **Coplanar Vector** Vector which lies on the same plane are called coplanar vector.
- **Equal Vectors** Two vectors **A** and **B** are equal, if they have the same magnitude and the same direction.

Laws of Vector Addition

1. Triangle Law

If two non-zero vectors are represented by the two sides of a triangle taken in same order than the resultant is given by the closing side of triangle in opposite order, i.e.

$$\mathbf{R} = \mathbf{A} + \mathbf{B}$$

The resultant R can be calculated as

$$|\mathbf{A} + \mathbf{B}| = R = \sqrt{A^2 + B^2 + 2AB\cos\theta}$$

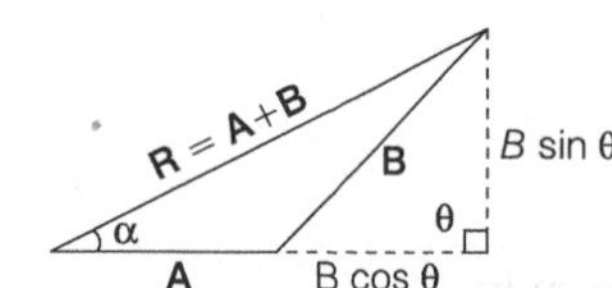

If resultant R makes an angle α with vector A, then

$$\tan\alpha = \frac{B\sin\theta}{A + B\cos\theta}$$

2. Parallelogram Law

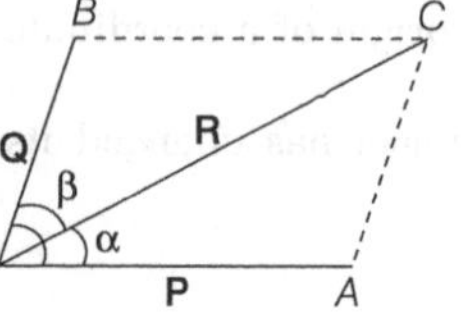

According to parallelogram law of vector addition, if two vector acting on a particle are represented in magnitude and direction by two adjacent side of a parallelogram, then the diagonal of the parallelogram represents the magnitude and direction of the resultant of the two vector acting as the particle.

i.e. $\mathbf{R} = \mathbf{P} + \mathbf{Q}$

Magnitude of the resultant **R** is given by

$$|\mathbf{R}| = \sqrt{P^2 + Q^2 + 2PQ\cos\theta}$$

$$\tan\alpha = \frac{Q\sin\theta}{P + Q\cos\theta} \Rightarrow \tan\beta = \frac{P\sin\theta}{Q + P\cos\theta}$$

Subtraction of Vectors

Vector subtraction makes use of the definition of the negative of a vector. We define the operation **A** − **B** as vector − **B** added to vector **A**. $\mathbf{A} - \mathbf{B} = \mathbf{A} + (-\mathbf{B})$

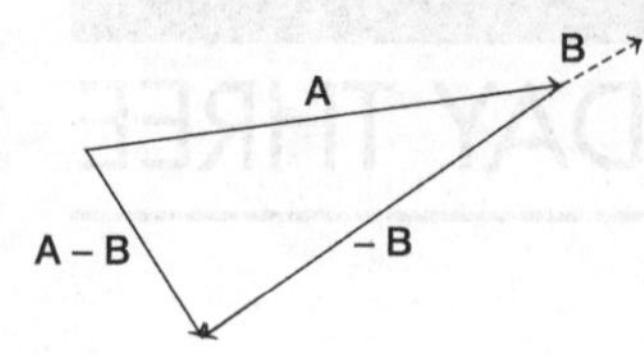

Thus, vector subtraction is really a special case of vector addition. The geometric construction for subtracting two vectors is shown in the above figure.

If θ be the angle between A and B,

then $|\mathbf{A} - \mathbf{B}| = \sqrt{A^2 + B^2 - 2AB\cos\theta}$

If the vectors form a closed n sided polygon with all the sides in the same order, then the resultant is zero.

Multiplication or Division of a Vector by a Scalar

The multiplication or division of a vector by a scalar gives a vector. For example, if vector **A** is multiplied by the scalar number 3, the result, written as 3**A**, is a vector with a magnitude three times that of **A**, pointing in the same direction as **A**. If we multiply vector **A** by the scalar − 3, the result is −3**A**, a vector with a magnitude three times that of **A**, pointing in the direction opposite to **A** (because of the negative sign).

Products of Vectors

The two types of products of vectors are given below

Scalar or Dot Product

The scalar product of two vectors A and B is defined as the product of magnitudes of A and B multiplied by the cosine of smaller angle between them. i.e. $\mathbf{A} \cdot \mathbf{B} = AB\cos\theta$

Properties of Dot Product

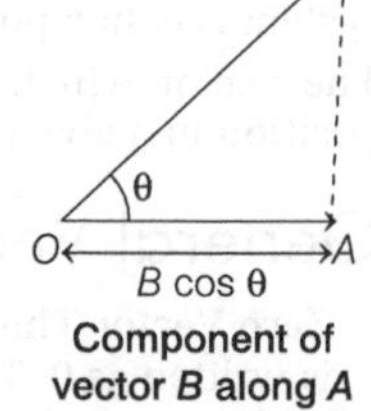

Component of vector B along A

- Dot product or scalar product of two vectors gives the scalar two vectors given the scalar quantity.
- It is commutative in nature. i.e. $\mathbf{A} \cdot \mathbf{B} = \mathbf{B} \cdot \mathbf{A}$.
- Dot product is distributive over the addition of vectors. i.e. $\mathbf{A} \cdot (\mathbf{B} + \mathbf{C}) = \mathbf{A} \cdot \mathbf{B} + \mathbf{A} \cdot \mathbf{C}$
- $\hat{\mathbf{i}} \cdot \hat{\mathbf{i}} = \hat{\mathbf{j}} \cdot \hat{\mathbf{j}} = \hat{\mathbf{k}} \cdot \hat{\mathbf{k}} = 1$, because angle between two equal vectors is zero.
- If two vectors A and B are perpendicular vectors, then $\mathbf{A} \cdot \mathbf{B} = AB\cos 90^\circ = 0$ and $\hat{\mathbf{i}} \cdot \hat{\mathbf{j}} = \hat{\mathbf{j}} \cdot \hat{\mathbf{k}} = \hat{\mathbf{k}} \cdot \hat{\mathbf{i}} = 0$

The Vector Product

The vector product of **A** and **B**, written as $\mathbf{A} \times \mathbf{B}$, produces a third vector **C** whose magnitude is $C = AB\sin\theta$. where, θ is the smaller of the two angles between **A** and **B**.

Because of the notation, $\mathbf{A} \times \mathbf{B}$ is also known as the **cross product**, and it is spelled as '**A** cross **B**'.

Properties of Cross Product

- Vector or cross product of two vectors gives the vector quantity.
- Cross product of two vectors does not obey the commutative law. i.e. $\mathbf{A} \times \mathbf{B} \neq \mathbf{B} \times \mathbf{A}$;
  Here, $\mathbf{A} \times \mathbf{B} = -\mathbf{B} \times \mathbf{A}$
- Cross product of two vectors is distributive over the addition of vectors.
  $$\mathbf{A} \times (\mathbf{B} + \mathbf{C}) = \mathbf{A} \times \mathbf{B} + \mathbf{A} \times \mathbf{C}$$
- Cross product of two equal vectors is given by $\mathbf{A} \times \mathbf{A} = 0$
  Similarly, $\hat{\mathbf{i}} \times \hat{\mathbf{i}} = (1 \times 1 \times \sin 0°)\, \hat{\mathbf{n}} = 0$
  $\hat{\mathbf{j}} \times \hat{\mathbf{j}} = (1 \times 1 \times \sin 0°)\, \hat{\mathbf{n}} = 0$
  $\hat{\mathbf{k}} \times \hat{\mathbf{k}} = (1 \times 1 \times \sin 0°)\, \hat{\mathbf{n}} = 0$
- Cross product of two perpendicular vectors is given as $\mathbf{A} \times \mathbf{B} = (AB \sin 90°)\, \hat{\mathbf{n}} = (AB)\, \hat{\mathbf{n}}$
- For two vectors $\mathbf{A} = a_x \hat{\mathbf{i}} + a_y \hat{\mathbf{j}} + a_z \hat{\mathbf{k}}$ and $\mathbf{B} = b_x \hat{\mathbf{i}} + b_y \hat{\mathbf{j}} + b_z \hat{\mathbf{k}}$.
  $$\mathbf{A} \times \mathbf{B} = \begin{vmatrix} \hat{\mathbf{i}} & \hat{\mathbf{j}} & \hat{\mathbf{k}} \\ a_x & a_y & a_z \\ b_x & b_y & b_z \end{vmatrix}$$
- Cross product of vectors $\hat{\mathbf{i}}, \hat{\mathbf{j}}$ and $\hat{\mathbf{k}}$ are following cyclic rules as follows $\hat{\mathbf{i}} \times \hat{\mathbf{j}} = \hat{\mathbf{k}}$, $\hat{\mathbf{j}} \times \hat{\mathbf{k}} = \hat{\mathbf{i}}$ and $\hat{\mathbf{k}} \times \hat{\mathbf{i}} = \hat{\mathbf{j}}$

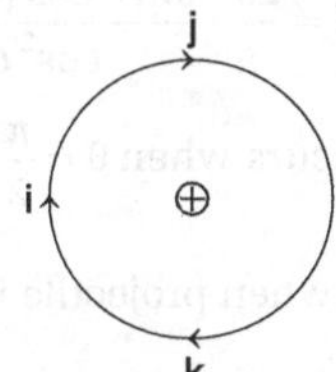

Cyclic representation for unit vectors $\hat{\mathbf{i}}$, $\hat{\mathbf{j}}$ and $\hat{\mathbf{k}}$

NOTE
- Vector triple product is given by
  $$\mathbf{A} \times (\mathbf{B} \times \mathbf{C}) = \mathbf{B}\,(\mathbf{A} \cdot \mathbf{C}) - \mathbf{C}\,(\mathbf{A} \cdot \mathbf{B})$$

Resolution of a Vector

The process of splitting of a single vector into two or more vectors in different direction is called resolution of a vector. Consider a vector A in the X-Y plane making an angle θ with the X-axis. The X and Y components of A are A_x and A_y respectively.

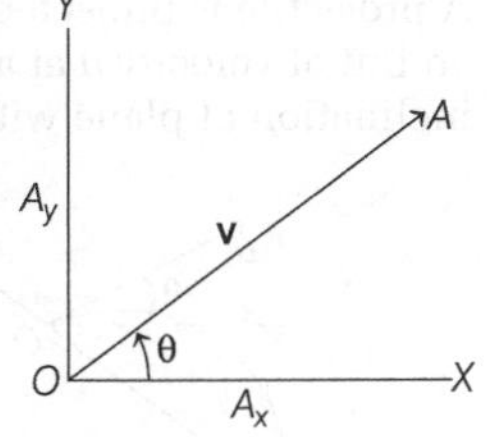

Thus $\mathbf{A}_x = \mathbf{A}_{xi} = (A \cos \theta)\hat{\mathbf{i}}$ along X-direction

$\mathbf{A}_y = \mathbf{A}_{yj} = (A \sin \theta)\hat{\mathbf{j}}$ along Y-direction

From triangle law of vector addition

$$|\mathbf{A}| = |\mathbf{A}_{xi} + \mathbf{A}_{yj}| = \sqrt{A_x^2 + A_y^2}$$

and $$\tan \theta = \frac{A_y}{A_x} = \theta = \tan^{-1}\left(\frac{A_y}{A_x}\right)$$

Relative Velocity

The time rate of change of relative position of one object with respect to another is called relative velocity.

Different Cases

Case I If both objects A and B move along parallel straight lines in the opposite direction, then relative velocity of B w.r.t. A is given as,

$$\mathbf{v}_{BA} = \mathbf{v}_B - (-\mathbf{v}_A) = \mathbf{v}_B + \mathbf{v}_A$$

If both objects A and B move along parallel staight lines in the same direction, then

$$\mathbf{v}_{AB} = \mathbf{v}_B - \mathbf{v}_A$$

Case II Crossing the River To cross the river over shortest distance, i.e. to cross the river straight, the man should swim upstream making an angle θ with **OB** such that, **OB** gives the direction of resultant velocity ($\mathbf{v}_{mR}$) of velocity of swimmer $\mathbf{v}_M$ and velocity of river water $\mathbf{v}_R$ as shown in figure. Let us consider

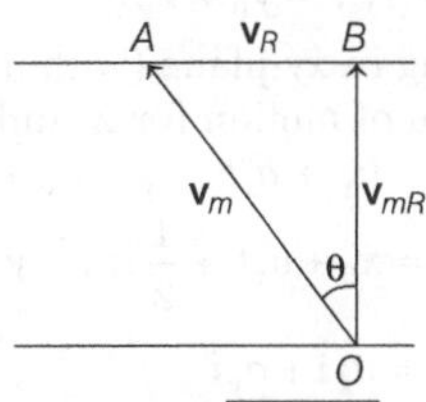

In ΔOAB, $\sin \theta = \dfrac{v_R}{v_m}$ and $v_{mR} = \sqrt{v_m^2 - v_R^2}$

The time taken to cross the river given by

$$t_1 = \frac{d}{v_{mR}} = \frac{d}{\sqrt{v_m^2 - v_R^2}}$$

Case III To cross the river in possible shortest time The man should go along OA. Now, the swimmer will be going along OB, which is the direction of resultant velocity $\mathbf{v}_{mR}$ of v_m and v_R.

In ΔOAB, $\tan \theta = \dfrac{AB}{OA} = \dfrac{v_R}{v_m}$

and $$v_{mR} = \sqrt{v_m^2 + v_R^2}$$

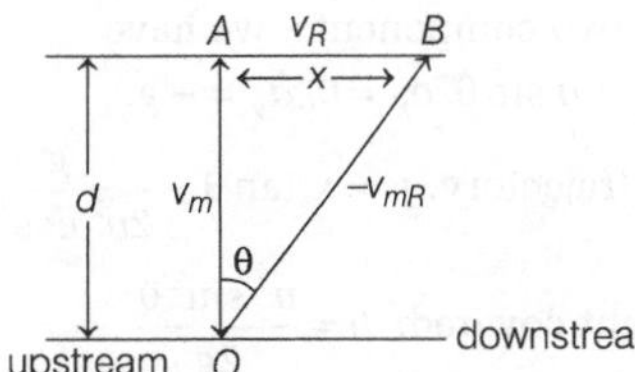

Time of crossing the river,

$$t = \frac{d}{v_m} = \frac{OB}{v_{mR}} = \frac{\sqrt{x^2 + d^2}}{\sqrt{v_m^2 + v_R^2}}$$

The boat will be reaching the point B instead of point A. If $AB = x$,

then, $\tan \theta = \dfrac{v_R}{v_m} = \dfrac{x}{d} \Rightarrow x = \dfrac{dv_R}{v_m}$

Motion in a Plane

Let the object be at position A and B at timing t_1 and t_2, where $OA = \mathbf{r}_1$, and $OB = \mathbf{r}_2$

Suppose O be the origin for measuring time and position of the object (see figure).

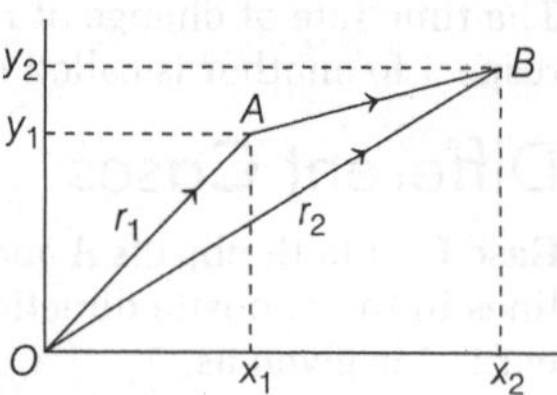

- Displacement of an object form position A to B is
$$AB = \mathbf{r} = \mathbf{r}_2 - \mathbf{r}_1 = (x_2 - x_1)\hat{\mathbf{i}} - (y_2 - y_1)\hat{\mathbf{j}}$$
- Velocity, $\mathbf{v} = \dfrac{\mathbf{r}_2 - \mathbf{r}_1}{t_2 - t_1}$
- A particle moving in X-Y plane (with uniform velocity) then, its equation of motion for X and Y axes are
$$\mathbf{v} = v_x\hat{\mathbf{i}} + v_y\hat{\mathbf{j}},\ \mathbf{r}_0 = x_0\hat{\mathbf{i}} + y_0\hat{\mathbf{j}} \text{ and } \mathbf{r} = x\hat{\mathbf{i}} + y\hat{\mathbf{j}}$$
$$x = x_0 + v_x t,\ y = y_0 + v_y t$$
- A particle moving in xy-plane (with uniform acceleration), then its equation of motion for X and Y-axes are
$$v_x = u_x + a_x t,\ v_y = u_y + a_y t$$
$$x = x_0 + u_x t + \frac{1}{2}a_x t^2,\ y = y_0 + u_y t + \frac{1}{2}a_y t^2$$
$$\mathbf{a} = a_x\hat{\mathbf{i}} + a_y\hat{\mathbf{i}}$$

Projectile Motion

Projectile is an object which once projected in a given direction with given velocity and is then free to move under gravity alone. The path described by the projectile is called its trajectory.

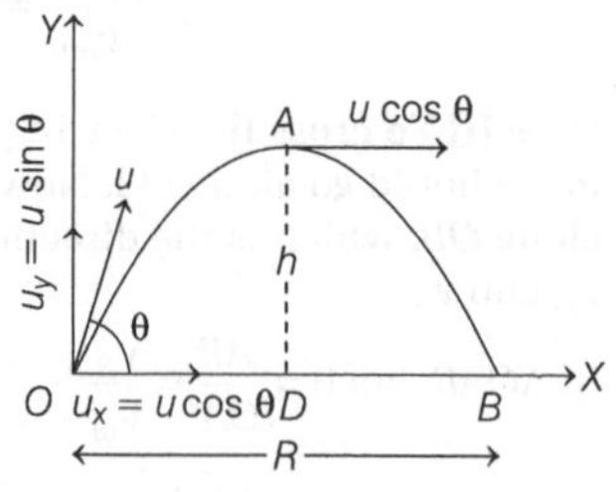

Let a particle is projected at an angle θ from the ground with initial velocity u.

Resolving u in two components, we have

$u_x = u\cos\theta,\ u_y = u\sin\theta,\ a_x = 0,\ a_y = -g.$

- Equation of trajectory, $y = x\tan\theta - \dfrac{g}{2u^2\cos^2\theta}x^2$
- Vertical height covered, $h = \dfrac{u^2\sin^2\theta}{2g}$
- Horizontal range, $R = OB = u_x T,\ R = \dfrac{u^2\sin 2\theta}{g}$

NOTE Maximum range occurs when $\theta = 45°$

Projectile Motion in Horizontal Direction From Height (h)

Let a particle be projected in horizontal direction with speed u from height h.

- Equation of trajectory, $y = \dfrac{gx^2}{2u^2}$
- Time of flight, $T = \dfrac{\sqrt{2h}}{g}$
- Horizontal range, $R = u\sqrt{\dfrac{2h}{g}}$
- Velocity of projectile at any time, $v = \sqrt{u^2 + g^2t^2}$

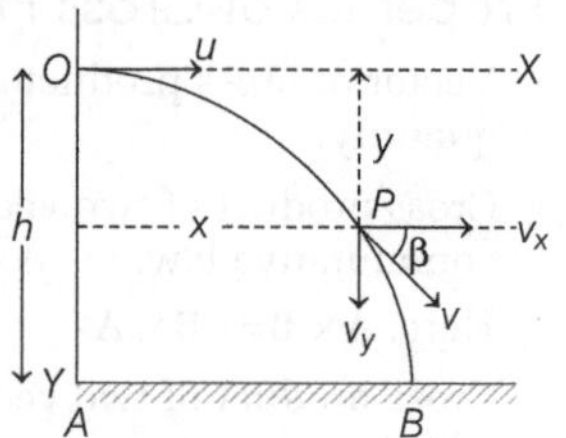

Projectile Motion Up an Inclined Plane

Let a particle be projected up with speed u from an inclined plane which makes an angle α with the horizontal and velocity of projection makes an angle θ with the inclined plane.

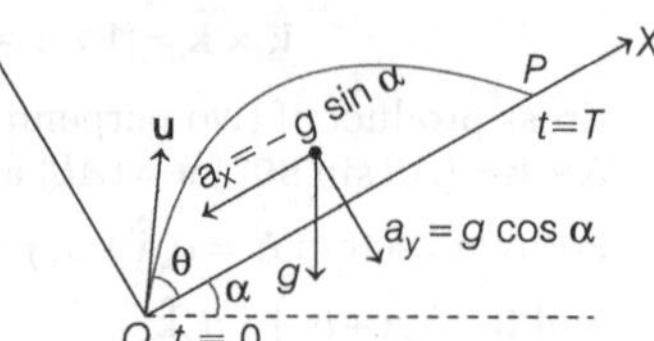

- Time of flight on an inclined plane, $T = \dfrac{2u\sin\theta}{g\sin\alpha}$
- Maximum height, $h = \dfrac{u^2\sin^2\theta}{2g\cos\alpha}$
- Horizontal range, $R = \dfrac{2u^2}{g}\dfrac{\sin\theta\cos(\theta+\alpha)}{\cos^2\alpha}$
- Maximum range occurs when $\theta = \dfrac{\pi}{2} - \dfrac{\alpha}{2}$
- $R_{max} = \dfrac{u^2}{g(1+\sin\alpha)}$ when projectile is thrown upwards.
- $R_{max} = \dfrac{u^2}{g(1-\sin\alpha)}$ when projectile is thrown downwards.

Projectile Motion Down an Inclined Plane

A projectile is projected down the plane from the point O with an initial velocity u at an angle θ with horizontal. The angle of inclination of plane with horizontal α. Then,

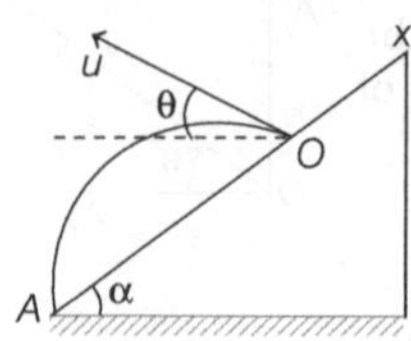

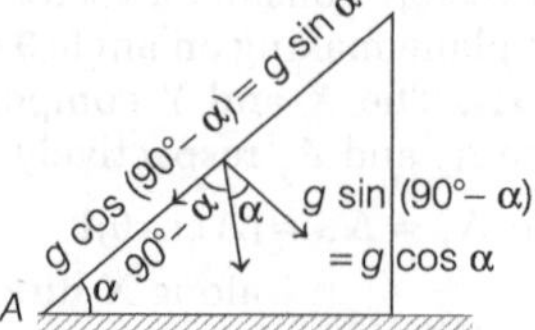

- Time of flight down an inclined plane, $T = \dfrac{2u\sin(\theta+\alpha)}{g\cos\alpha}$
- Horizontal range, $R = \dfrac{u^2}{g\cos^2\alpha}[\sin(2\theta+\alpha) + \sin\alpha]$

(DAY PRACTICE SESSION 1)

FOUNDATION QUESTIONS EXERCISE

1 If **A** and **B** are two non-zero vectors having equal magnitude, the angle between the vectors **A** and **A** − **B** is

(a) 0° (b) 90° (c) 180°
(d) dependent on the orientation of **A** and **B**

2 A vector having magnitude of 30 unit, makes equal angles with each of the X, Y and Z-axes. The components of the vector along each of X, Y and Z-axes are

(a) $10\sqrt{3}$ unit (b) $\frac{10}{\sqrt{3}}$ unit (c) $15\sqrt{3}$ unit (d) 10 unit

3 A particle has an initial velocity $3\hat{i} + 4\hat{j}$ and an acceleration of $0.4\hat{i} + 0.3\hat{j}$. Its speed after 10 s is

(a) 10 units (b) $7\sqrt{2}$ units (c) 7 units (d) 8.5 units

4 Unit vector perpendicular to vector $\mathbf{A} = 3\hat{i} + \hat{j}$ and $\mathbf{B} = 2\hat{i} - \hat{j} - 5\hat{k}$ both, is

(a) $\pm\frac{3\hat{j} - 2\hat{k}}{\sqrt{11}}$ (b) $\pm\frac{(\hat{i} - 3\hat{j} + \hat{k})}{\sqrt{11}}$
(c) $\pm\frac{-\hat{j} + 2\hat{k}}{\sqrt{13}}$ (d) $\pm\frac{\hat{i} + 3\hat{j} - \hat{k}}{\sqrt{13}}$

5 A force $\mathbf{F} = (5\hat{i} + 3\hat{j} + 2\hat{k})$ N is applied over a particle which displaces it from its origin to the point $\mathbf{r} = (2\hat{i} - \hat{j})$ m. The work done on the particle (in joule) is

(a) −7 (b) +7 (c) +10 (d) +13

6 If $\mathbf{A} \times \mathbf{B} = \mathbf{B} \times \mathbf{A}$, then the angle between **A** and **B** is

(a) π (b) $\frac{\pi}{3}$ (c) $\frac{\pi}{2}$ (d) $\frac{\pi}{4}$

7 A ball rolls off the top of a stair way with a horizontal velocity of u ms^{-1}. If the steps are h metre high and b metre wide, the ball will hit the edge of the nth step, where n is

(a) $\frac{2hu}{gb^2}$ (b) $\frac{2hu^2}{gb^2}$ (c) $\frac{2hu^2}{gb}$ (d) $\frac{hu^2}{gb^2}$

8 Two paper screens A and B are separated by a distance of 200 m. A bullet pierces A and B. The hole in B is 40 cm below the hole in A. If the bullet is travelling horizontally at the time of hitting A, then the velocity of the bullet at A is

(a) 200 ms^{-1} (b) 400 ms^{-1} (c) 600 ms^{-1} (d) 700 ms^{-1}

9 A projectile is fired at an angle of 30° with the horizontal such that the vertical component of its initial velocity is 80 ms^{-1}. Its time of flight is T. Its velocity at $t = \frac{T}{4}$ has a magnitude of nearly (take, g = 10 ms^{-2})

(a) 180 ms^{-1} (b) 155 ms^{-1} (c) 145 ms^{-1} (d) 140 ms^{-1}

10 A ball is thrown from the ground with a velocity of $20\sqrt{3}$ ms^{-1} making an angle of 60° with the horizontal. The ball will be at a height of 40 m from the ground after a time t equal to (take, g = 10 ms^{-2})

(a) $\sqrt{2}$ s (b) $\sqrt{3}$ s (c) 2 s (d) 3 s

11 A body is projected with a velocity v_1 from the point A as shown in the figure. At the same time, another body is projected vertically upwards from B with velocity v_2. The point B lies vertically below the highest point. For both the bodies to collide, $\frac{v_2}{v_1}$ should be

(a) 2 (b) 0.5 (c) $\sqrt{\frac{3}{2}}$ (d) 1

12 A person aims a gun at a bird from a point, at a horizontal distance of 100 m. If the gun can induce a speed of 500 ms^{-1} to the bullet, at what height above the bird must he aim his gun in order to hit it? (take, g = 10 ms^{-2})

(a) 10 cm (b) 20 cm (c) 50 cm (d) 100 cm

13 A cannon ball has the same range R on a horizontal plane for two angles of projection. If h_1 and h_2 are the greatest heights in the two paths for which this is possible, then

(a) $R = (h_1h_2)^{1/4}$ (b) $R = 3\sqrt{h_1h_2}$
(c) $R = 4\sqrt{h_1h_2}$ (d) $R = \sqrt{h_1h_2}$

14 A projectile is thrown in the upward direction making an angle of 60° with the horizontal direction with a velocity of 147 ms^{-1}. Then, the time after which its inclination with the horizontal is 45°, is

(a) 15 s (b) 10.98 s (c) 5.49 s (d) 2.745 s

15 A projectile projected with a velocity u at an angle θ passes through a given height h two times at t_1 and t_2. Then,

(a) $t_1 + t_2 = T$ (time of flight) (b) $t_1 + t_2 = \frac{T}{2}$
(c) $t_1 + t_2 = 2T$ (d) $\sqrt{t_1t_2} = T$

16. A particle is projected at angle of 60° with the horizontal having a kinetic energy K. The kinetic energy at the highest point is

(a) K (b) zero
(c) $K/4$ (d) $K/2$

17 A boy playing on the roof of a 10 m high building throws a ball with a speed of 10 ms^{-1} at an angle of 30° with the horizontal. How far from the throwing point, will the ball be at the height of 10 m from the ground?

(take, $g = 10\ ms^{-2}$, $\sin 30° = \frac{1}{2}$, $\cos 30° = \frac{\sqrt{3}}{2}$)

(a) 5.20 m (b) 4.33 m (c) 2.60 m (d) 8.66 m

18 A ball projected from ground at an angle of 45° just clears a wall in front. If point of projection is 4 m from the foot of wall and ball strikes the ground at a distance of 6 m on the other side of the wall, the height of the wall is → JEE Main (Online) 2013

(a) 4.4 m (b) 2.4 m (c) 3.6 m (d) 1.6 m

19 Neglecting the air resistance, the time of flight of a projectile is determined by

(a) $U_{vertical}$ (b) $U_{horizontal}$
(c) $U = U^2_{vertical} + U^2_{horizontal}$ (d) $U = (U^2_{vertical} + U^2_{horizontal})^{1/2}$

20 The horizontal range of a projectile is $4\sqrt{3}$ times its maximum height. Its angle of projection will be

(a) 45° (b) 60° (c) 90° (d) 30°

21 A projectile is fired at an angle of 45° with the horizontal. Elevation angle of the projectile at its highest point as seen from the point of projection is

(a) 60° (b) $\tan^{-1}(\sqrt{3}/2)$
(c) $\tan^{-1}(1/2)$ (d) 45°

22 A man can swim with a speed of 4 kmh^{-1} in still water.How long does he take to cross a river 1km wide, if the river flows steadily 3 kmh^{-1} and he makes his strokes normal to the river current.How far down the river does he go, when he reaches the other bank?

(a) 800 m (b) 900 m
(c) 400 m (d) 750 m

23 A swimmer crosses a flowing stream of width d to and fro in time t_1. The time taken to cover the same distance up and down the stream is t_2. Then, the time the swimmer would take to swim across a distance 2 d in still water is

(a) $\frac{t_1^2}{t_2}$ (b) $\frac{t_2^2}{t_1}$ (c) $\sqrt{t_1 t_2}$ (d) $(t_1 + t_2)$

24 A man standing on a road has to hold his umbrella at 30° with the vertical to keep the rain away. He throws the umbrella and starts running at 10 kmh^{-1}. He finds that the raindrops are hitting his head vertically. The actual speed of raindrops is

(a) 20 kmh^{-1} (b) $10\sqrt{3}$ kmh^{-1}
(c) $20\sqrt{3}$ kmh^{-1} (d) 10 kmh^{-1}

25 A passenger train is moving at 5 ms^{-1} . An express train is travelling at 30 ms^{-1}, on the same track and rear side of the passenger train at some distance. The driver in express train applied brakes to avoid collision. If the retardation due to brakes is 4 ms^{-2}, the time in which the accident is avoided after the application of brakes is

(a) 4.25 s (b) 5.25 s (c) 6.25 s (d) 7.25 s

26 A boat takes 2 h to travel 8 km and back in a still water lake. If the velocity of water is 4 kmh^{-1}, the time taken for going upstream of 8 km and coming back is

(a) 2 h (b) 2 h and 40 min
(c) 1 h and 20 min
(d) Cannot be estimated from the given information

27 A car is travelling with a velocity of 10 kmh^{-1} on a straight road. The driver of the car throws a parcel with a velocity of $10\sqrt{2}$ kmh^{-1} when the car is passing by a man standing on the side of the road. If the parcel is to reach the man, the direction of throw makes the following angle with the direction of the car.

(a) 135° (b) 45° (c) $\tan^{-1}(\sqrt{2})$ (d) $\tan\left(\frac{1}{2}\right)$

28 A point P moves in counter- clockwise direction on a circular path as shown in the figure.

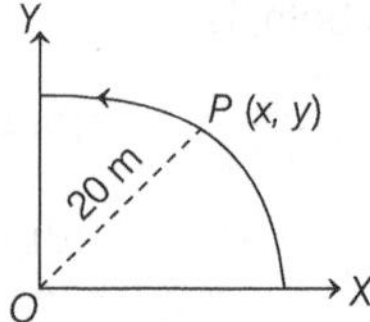

The movement of p is such that it sweeps out a length $s = t^3 + 5$, where, s is in metre and t is in second.The radius of the path is 20 m. The acceleration of P when t = 2s is nearly

(a) 13 ms^{-2} (b) 12 ms^{-2} (c) 7.2 ms^{-2} (d) 14 ms^{-2}

29 For a particle in uniform circular motion the acceleration **a** at a point $P(R, \theta)$ on the circle of radius R is (here, θ is measured from the X-axis)

(a) $-\frac{v^2}{R}\cos\theta\,\hat{i} + \frac{v^2}{R}\sin\theta\,\hat{j}$ (b) $-\frac{v^2}{R}\sin\theta\,\hat{i} + \frac{v^2}{R}\cos\theta\,\hat{j}$
(c) $-\frac{v^2}{R}\cos\theta\,\hat{i} - \frac{v^2}{R}\sin\theta\,\hat{j}$ (d) $\frac{v^2}{R}\hat{i} + \frac{v^2}{R}\hat{j}$

Direction (Q. Nos. 30-34) *Each of these questions contains two statements : Statement* I *(Assertion) and Statement* II *(Reason). Each of these questions also has four alternative choices, only one of which is the correct answer. You have to select one of the codes (a), (b), (c), (d) given below :*

(a) Statement I is true, Statement II is true; Statement II is the correct explanation for Statement I
(b) Statement I is true, Statement II is true; Statement II is not the correct explanation for Statement I
(c) Statement I is true; Statement II is false
(d) Statement I is false; Statement II is true

30 Statement I Rain is falling vertically downwards with a velocity of 3 kmh^{-1}. A man walks with a velocity of 4 kmh^{-1}. Relative velocity of rain w.r.t. man is 5 kmh^{-1}.

Statement II Relative velocity of rain w.r.t. man is given by $\mathbf{v}_{rm} = \mathbf{v}_r - \mathbf{v}_m$

31 Statement I For the projection angle $\tan^{-1}(4)$, the horizontal and maximum height of a projectile are equal.

Statement II The maximum range of a projectile is directly proportional to the square of velocity and inversely proportional to the acceleration due to gravity.

32 Statement I In order to hit a target, a man should point his rifle in the same direction as the target.

Statement II The horizontal range of bullet is dependent on the angle of projection with the horizontal.

33 Statement I The resultant of three vectors **OA**, **OB** and **OC** as shown in the figure is $R(1+\sqrt{2})$. R is the radius of the circle.

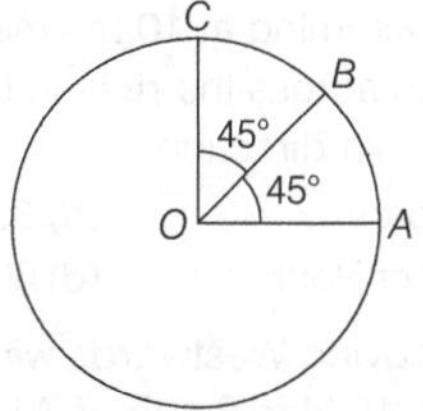

Statement II **OA**+ **OC** is acting along **OB** and (**OA**+ **OC**)+**OB** is acting along **OB**.

34 Statement I Angle between $\hat{i} + \hat{j}$ and $\hat{i}$ is 45°.

Statement II $\hat{i} + \hat{j}$ is equally include to both $\hat{i}$ and $\hat{j}$ and the angle between $\hat{i}$ and $\hat{j}$ is 90°.

DAY PRACTICE SESSION 2

PROGRESSIVE QUESTIONS EXERCISE

1 If $\mathbf{F}_1$ and $\mathbf{F}_2$ are two vectors of equal magnitude F such that $|\mathbf{F}_1 \cdot \mathbf{F}_2| = |\mathbf{F}_1 \times \mathbf{F}_2|$, then $|\mathbf{F}_1 + \mathbf{F}_2|$ is equal to

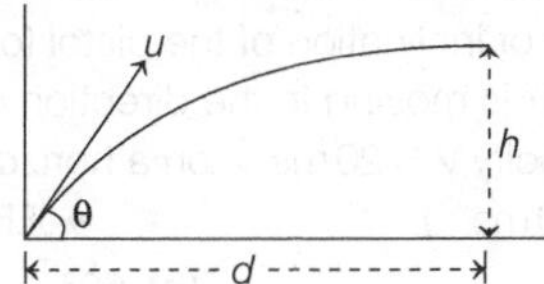

(a) $\sqrt{(2+\sqrt{2})}F$ (b) $2F$
(c) $F\sqrt{2}$ (d) None of these

2 If a stone is to hit at a point which is at a distance d away and at a height h above the point from where the stone starts, then what is the value of initial speed u if stone is launched at an angle θ?

(a) $\dfrac{g}{\cos\theta}\sqrt{\dfrac{d}{2(d\tan\theta - h)}}$ (b) $\dfrac{d}{\cos\theta}\sqrt{\dfrac{g}{2(d\tan\theta - h)}}$
(c) $\sqrt{\dfrac{gd^2}{h\cos^2\theta}}$ (d) $\sqrt{\dfrac{gd^2}{(d-h)}}$

3 A projectile can have the same range R for two angles of projection. If T_1 and T_2 be the time of flight in the two cases, then the product of the two time of flight is directly proportional to

(a) R (b) $\dfrac{1}{R}$
(c) $\dfrac{1}{R^2}$ (d) R^2

4 A projectile is given an initial velocity of $(\mathbf{i} + 2\mathbf{j})\ ms^{-1}$ where, **i** is along the ground and **j** is along the vertical. If $g = 10\ ms^{-2}$, the equation of its trajectory is
→ JEE Main 2013

(a) $Y = X - 5X^2$ (b) $Y = 2X - 5X^2$
(c) $4Y = 2X - 5X^2$ (d) $4Y = 2X - 25X^2$

5 A particle of mass m is projected with a velocity v making an angle of 30° with the horizontal. The magnitude of angular momentum of the projectile about the point of projection when the particle is at its maximum height h is

(a) $\dfrac{\sqrt{3}mv^2}{2g}$ (b) zero
(c) $\dfrac{mv^3}{\sqrt{2}g}$ (d) $\dfrac{\sqrt{3}}{16}\dfrac{mv^3}{g}$

6 The coordinates of a moving particle at any time t are given by $x = \alpha t^3$ and $y = \beta t^3$. The speed of the particle at time t is given by

(a) $t^2\sqrt{\alpha^2+\beta^2}$ (b) $\sqrt{\alpha^2+\beta^2}$
(c) $3t\sqrt{\alpha^2+\beta^2}$ (d) $3t^2\sqrt{\alpha^2+\beta^2}$

7 A ball whose kinetic energy is E, is projected at an angle of 45° with respect to the horizontal. The kinetic energy of the ball at the highest point of its flight will be

(a) E (b) $\dfrac{E}{\sqrt{2}}$
(c) $\dfrac{E}{2}$ (d) zero

8 A river is flowing from West to East with a speed of 5 m - min^{-1}. A man on the South bank of the river, is capable of swimming at 10 m - min^{-1} in still water, he wants to swim across the river in the shortest time. He should swim in a direction

(a) due to North (b) 30° East of North
(c) 30° West of North (d) 60° East of North

9 A ship A is moving Westwards with a speed of 10 kmh^{-1} and a ship B, 100 km South of A is moving Northwards with a speed of 10 kmh^{-1}. The time after which the distance between them is shortest and the the shortest distance between them are

(a) 0 h, 100 km (b) 5h, $50\sqrt{2}$ km
(c) $5\sqrt{2}$ h, 50 km (d) $10\sqrt{2}$ h, $50\sqrt{2}$ km

10 A particle is moving Eastwards with a velocity of 5 ms^{-1}. In 10s, the velocity changes to 5 ms^{-1} Northwards. The average acceleration in this time is

(a) $\frac{1}{\sqrt{2}}$ms^{-2} towards North-East
(b) $\frac{1}{2}$ ms^{-2} towards North
(c) zero
(d) $\frac{1}{\sqrt{2}}$ ms^{-2} towards North-West

11 A boy can throw a stone upto a maximum height of 10 m. The maximum horizontal distance that the boy can throw the same stone upto will be

(a) $20\sqrt{2}$ m (b) 10 m
(c) $10\sqrt{2}$ m (d) 20 m

12 A water fountain on the ground sprinkles water all around it. If the speed of water coming out of the fountain is v, the total area around the fountain that gets wet is

(a) $\pi\frac{v^4}{g^2}$ (b) $\frac{\pi}{2}\frac{v^4}{g^2}$
(c) $\pi\frac{v^2}{g^2}$ (d) $\pi\frac{v^2}{g}$

13 Two fixed frictionless inclined plane making an angle 30° and 60° with the vertical are as shown in the figure. Two blocks A and B are placed on the two planes. What is the relative vertical acceleration of A with respect to B?

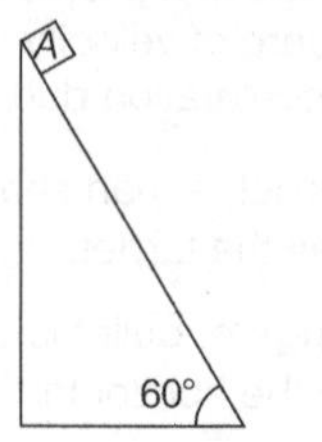

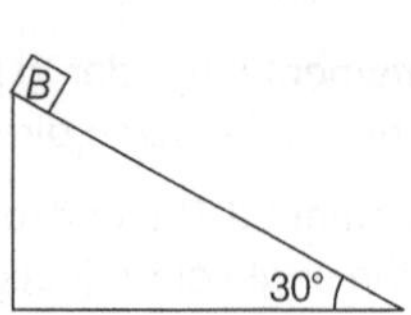

(a) 4.9 ms^{-2} in horizontal direction
(b) 9.8 ms^{-2} in vertical direction
(c) zero
(d) 4.9 ms^{-2} in vertical direction

14 A particle is moving with velocity $v = k(Y\,\mathbf{i} + X\,\mathbf{j})$, where k is a constant. The general equation for its path is

(a) $Y = X^2$ + constant (b) $Y^2 = X$ + constant
(c) XY = constant (d) $Y^2 = X^2$ + constant

15 The maximum range of a bullet fired from a toy pistol, mounted on a car at rest is $R_0 = 40$ m. What will be the acute angle of inclination of the pistol for maximum range when the car is moving in the direction of firing with uniform velocity $v = 20$ ms^{-1}, on a horizontal surface? (take, $g = 10$ ms^{-2}) → JEE Main (Online) 2013

(a) 30° (b) 60°
(c) 75° (d) 45°

ANSWERS

SESSION 1	**1** (d)	**2** (a)	**3** (b)	**4** (b)	**5** (b)	**6** (a)	**7** (b)	**8** (d)	**9** (c)	**10** (c)
	11 (b)	**12** (b)	**13** (c)	**14** (c)	**15** (a)	**16** (c)	**17** (d)	**18** (c)	**19** (a)	**20** (d)
	21 (c)	**22** (d)	**23** (a)	**24** (a)	**25** (c)	**26** (b)	**27** (b)	**28** (d)	**29** (c)	**30** (a)
	31 (b)	**32** (d)	**33** (a)	**34** (a)						
SESSION 2	**1** (a)	**2** (b)	**3** (a)	**4** (b)	**5** (d)	**6** (d)	**7** (c)	**8** (a)	**9** (b)	**10** (d)
	11 (d)	**12** (a)	**13** (d)	**14** (d)	**15** (b)					

Hints and Explanations

SESSION 1

1 Suppose angle between two vectors **A** and **B** of equal magnitude is θ. Then, angle between **A** and **A** − **B** will be $\frac{180^\circ - \theta}{2}$ or $90^\circ - \frac{\theta}{2}$.

Hence, this angle will depend on the angle between **A** and **B** or θ.

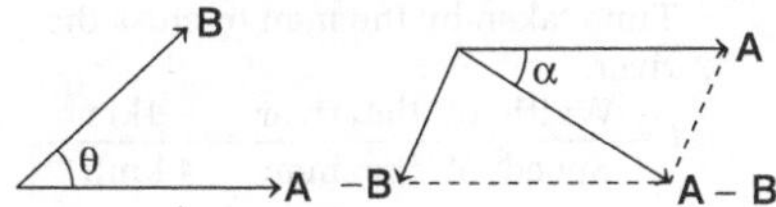

2 $A_x = A_y = A_z$

Now, $A = \sqrt{A_x^2 + A_y^2 + A_z^2} = \sqrt{3}\, A_x$

$\therefore \quad A_x = \frac{A}{\sqrt{3}} = \frac{30}{\sqrt{3}} = 10\sqrt{3}$

Similarly $A_y = A_x = 10\sqrt{3}$ unit

3 $\mathbf{u} = 3\mathbf{i} + 4\mathbf{j};\ \mathbf{a} = 0.4\,\hat{\mathbf{i}} + 0.3\,\hat{\mathbf{j}}$

$\mathbf{v} = \mathbf{u} + \mathbf{a}t$

$= 3\hat{\mathbf{i}} + 4\hat{\mathbf{j}} + (0.4\hat{\mathbf{i}} + 0.3\,\hat{\mathbf{j}})\,10$

$= 3\,\hat{\mathbf{i}} + 4\,\hat{\mathbf{j}} + 4\,\hat{\mathbf{i}} + 3\,\hat{\mathbf{j}}$

$= 7\,\hat{\mathbf{i}} + 7\,\hat{\mathbf{j}}$

Speed $= \sqrt{7^2 + 7^2} = 7\sqrt{2}$ units

4 Given $\mathbf{A} = 3\hat{\mathbf{i}} + \hat{\mathbf{j}},\ \mathbf{B} = 2\hat{\mathbf{i}} - \hat{\mathbf{j}} - 5\hat{\mathbf{k}}$

The unit vector in the normal direction is

$$\hat{\mathbf{n}} = \pm \frac{\mathbf{A} \times \mathbf{B}}{|\mathbf{A}|\ |\mathbf{B}| \sin\theta}$$

Here, $\mathbf{A} \times \mathbf{B} = \begin{vmatrix} \hat{\mathbf{i}} & \hat{\mathbf{j}} & \hat{\mathbf{k}} \\ 3 & 1 & 0 \\ 2 & -1 & -5 \end{vmatrix}$

$= -5\hat{\mathbf{i}} + 15\hat{\mathbf{j}} - 5\hat{\mathbf{k}}$

$|\mathbf{A}| = \sqrt{3^2 + 1^2} = \sqrt{10}$

$|\mathbf{B}| = \sqrt{(2)^2 + (-1)^2 + (-5)^2} = \sqrt{30}$

$\cos\theta = \frac{\mathbf{A} \cdot \mathbf{B}}{|\mathbf{A}|\,|\mathbf{B}|}$

$= \frac{5}{10\sqrt{3}} - \frac{1}{2\sqrt{3}}$

$\therefore \quad \sin\theta = \sqrt{1 - \cos^2\theta} = \frac{\sqrt{11}}{2\sqrt{3}}$

$\therefore \quad \hat{\mathbf{n}} = \pm \frac{-5\hat{\mathbf{i}} + 15\hat{\mathbf{j}} - 5\hat{\mathbf{k}}}{\sqrt{10} \cdot \sqrt{30} \cdot \frac{\sqrt{11}}{2\sqrt{3}}}$

$= \pm \frac{(\hat{\mathbf{i}} - 3\hat{\mathbf{j}} + \hat{\mathbf{k}})}{\sqrt{11}}$

5 Work done in displacing the particle

$W = \mathbf{F} \cdot \mathbf{r} = (5\hat{\mathbf{i}} + 3\hat{\mathbf{j}} + 2\hat{\mathbf{k}}) \cdot (2\hat{\mathbf{i}} - \hat{\mathbf{j}})$

$= 5 \times 2 + 3 \times (-1) + 2 \times 0 = 10 - 3$

$= 7$ J

6 $(\mathbf{A} \times \mathbf{B}) = (\mathbf{B} \times \mathbf{A})$ (given)

$\Rightarrow \quad (\mathbf{A} \times \mathbf{B}) - (\mathbf{B} \times \mathbf{A}) = \mathbf{0}$

or $\quad (\mathbf{A} \times \mathbf{B}) + (\mathbf{A} \times \mathbf{B}) = \mathbf{0}$

$[\because (\mathbf{B} \times \mathbf{A}) = -(\mathbf{A} \times \mathbf{B})]$

or $\quad 2(\mathbf{A} \times \mathbf{B}) = \mathbf{0}$

$\Rightarrow \quad 2AB \sin\theta = 0$

or $\quad \sin\theta = 0$

$[\because |\mathbf{A}| = A \neq 0, |\mathbf{B}| = B \neq 0]$

$\Rightarrow \quad \theta = 0$ or π

7 Let the ball strike the, nth step after t second. Vertical distance travelled by the ball $= nh = \frac{1}{2} g t^2$ ...(i)

Horizontal distance travelled by the ball $= nb = ut$ or $t = \frac{nb}{u}$

So from Eq (i), we get

$nh = \frac{1}{2} g \left(\frac{nb}{u}\right)^2$ or $n = \frac{2u^2 h}{g b^2}$

8 Refering to the figure,

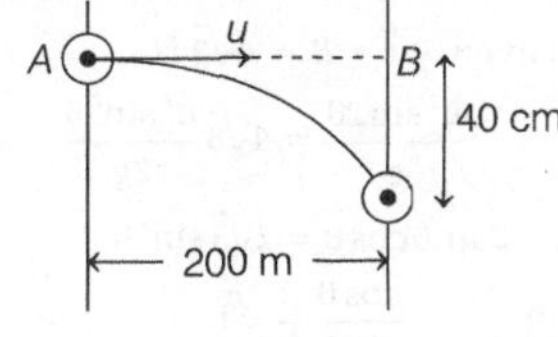

$200 = ut$ or $t = \frac{200}{u}$

Also, $\quad \frac{40}{100} = \frac{1}{2} \times 9.8 \left(\frac{200}{u}\right)^2$

On solving, $\quad u = 700\ \text{ms}^{-1}$

9 Vertical component of the initial velocity,

$u_y = u \sin 30^\circ$

or $u = \frac{u_y}{\sin 30^\circ} = \frac{80}{1/2} = 160\ \text{ms}^{-1}$

$t = \frac{T}{4} = \frac{2u \sin 30^\circ}{4 \times g} = \frac{2 \times 80}{4 \times 10} = 4$ s

$v_x = u \cos 30^\circ = 160 \times \frac{\sqrt{3}}{2}$

$= 80\sqrt{3}\ \text{ms}^{-1}$

$\therefore \quad v = \sqrt{v_x^2 + v_y^2} = \sqrt{u_x^2 + (u_y - gt)^2}$

$= \sqrt{(80\sqrt{3})^2 + (80 - 10 \times 4)^2}$

$= 144.3\ \text{ms}^{-1}$

$= 145\ \text{ms}^{-1}$

10 As, $s = u \sin\theta t - \frac{1}{2} g t^2$

So, $40 = 20\sqrt{3} \times \frac{\sqrt{3}t}{2} - \frac{1}{2} \times 10 \times t^2$

or $\quad 5t^2 - 30t + 40 = 0$

or $\quad t^2 - 6t + 8 = 0$ or $t = 2$ or 4

The minimum time $t = 2$ s

11 The two bodies will collide at the highest point if both cover the same vertical height in the same time. So,

$\frac{v_1^2 \sin^2 30^\circ}{2g} = \frac{v_2^2}{2g}$

or $\quad \frac{v_2}{v_1} = \sin 30^\circ = \frac{1}{2} = 0.5$

12 Let the gun be fired with a velocity u from the point O on the bird at B, making an angle θ with the horizontal direction. Therefore, the height of the aim of the person be at height BA (h) above the bird.

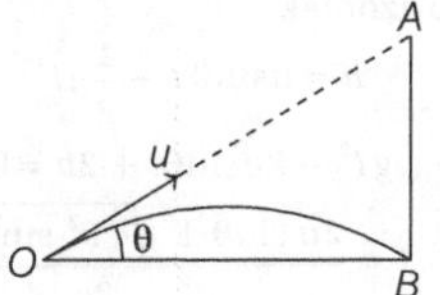

Here, horizontal range

$= \frac{u^2 \sin 2\theta}{g} = 100$

or $\quad \frac{(500)^2 \sin 2\theta}{10} = 100$

or $\quad \sin 2\theta = \frac{100 \times 10}{(500)^2} = \frac{1}{250} = \sin 14'$

or $\quad 2\theta = 14'$ or $\theta = 7'$

$= \frac{7}{60} \times \frac{\pi}{180}$ rad

As, angle $= \frac{\text{arc}}{\text{radius}}$

$\therefore \quad \theta = \frac{AB}{OB}$

or $\quad AB = \theta \times OB$

$= \frac{7}{60} \times \frac{\pi}{180} \times (100 \times 100)$ cm

$= 20$ cm

13 The cannon ball will have the same horizontal range for the angle of projection θ and $(90^\circ - \theta)$. So,

$h_1 = \frac{u^2 \sin^2\theta}{2g}$ and $h_2 = \frac{u^2 \cos^2\theta}{2g}$

$h_1 h_2 = \frac{1}{4}\left(\frac{u^2 \sin\theta \cos\theta}{g}\right)^2 = \frac{1}{4} \times \frac{R^2}{4}$

or $R = 4\sqrt{h_1 h_2}$

14 Horizontal component of the velocity at an angle 60°
= Horizontal component of the velocity at an angle 45°.

i.e. $u \cos 60° = v \cos 45°$

$$147 \times \frac{1}{2} = v \times \frac{1}{\sqrt{2}}$$

or $v = \frac{147}{\sqrt{2}}\ \text{ms}^{-1}$

Vertical component of $u = u \sin 60°$

$$= \frac{147\sqrt{3}}{2}\ \text{m}$$

Vertical component of $v = v \sin 45°$

$$= \frac{147}{\sqrt{2}} \times \frac{1}{\sqrt{2}} = \frac{147}{2}\ \text{m}$$

But $v_y = u_y + at$

$\therefore \frac{147}{2} = \frac{147\sqrt{3}}{2} - 9.8\,t$

or $9.8\,t = \frac{147}{2}(\sqrt{3} - 1)$

$\therefore t = 5.49$ s

15 For a projectile fired with a velocity u inclined at an angle θ with the horizontal.

$$h = u \sin\theta\, t - \frac{1}{2} g t^2$$

$\Rightarrow g t^2 - 2u \sin\theta t + 2h = 0$

$$\therefore t = \frac{2u \sin\theta \pm \sqrt{4u^2 \sin^2\theta - 8gh}}{2g}$$

$$\Rightarrow t_1 = \frac{2u \sin\theta + \sqrt{4u^2 \sin^2\theta - 8gh}}{2g}$$

and $t_2 = \frac{2u \sin\theta - \sqrt{4u^2 \sin^2\theta - 8gh}}{2g}$

$$\therefore t_1 + t_2 = \frac{2u \sin\theta}{g} = T$$

16 Kinetic energy at highest point,

$$(\text{KE})_H = \frac{1}{2} m v^2 \cos^2\theta = K \cos^2\theta$$

$$= K(\cos 60°)^2 = \frac{K}{4}$$

17

$$OP = R = \frac{u^2 \sin 2\theta}{g}$$

$$= \frac{10^2 \times \sin(2 \times 30°)}{10}$$

$$= \frac{10\sqrt{3}}{2} = 5\sqrt{3} = 8.66\ \text{m}$$

18 As, range $= 10 = \frac{u^2 \sin 2\theta}{g}$

$\Rightarrow u^2 = 10\,g$

$\therefore u = 10\text{ms}^{-1}$ (as, $g = 10\text{ms}^{-2}$)

$$Y = x \tan\theta - \frac{1}{2} \frac{gx^2}{2v_0^2 \cos^2\theta}$$

$$= 4 \tan 45° - \frac{1}{2} \frac{g \times 16}{2 v_0^2 \cos^2 45°}$$

$$= 4 \times 1 - \frac{1}{2} \frac{10 \times 16}{2 \times 10 \times 10 \times \frac{1}{2}}$$

$= 4 - 0.8 = 3.2 \approx 3.6$ m

19 Time of flight (T) is $2t$

$$\therefore T = 2t = \frac{2u \sin\theta}{g} = \frac{2}{g} \times U_{\text{vertical}}$$

20 Let u be initial velocity of projection at angle θ with the horizontal.
Then horizontal range,

$$R = \frac{u^2 \sin 2\theta}{g}$$

and maximum height $H = \frac{u^2 \sin^2\theta}{2g}$

Given, $R = 4\sqrt{3}\,H$

$$\therefore \frac{u^2 \sin 2\theta}{g} = 4\sqrt{3} \frac{u^2 \sin^2\theta}{2g}$$

$\therefore 2\sin\theta\cos\theta = 2\sqrt{3}\sin^2\theta$

or $\frac{\cos\theta}{\sin\theta} = \sqrt{3}$

or $\cot\theta = \sqrt{3} = \cot 30°$

$\theta = 30°$

21 Maximum height,

$$H = \frac{u^2 \sin^2 45°}{2g} = \frac{u^2}{4g} = AC$$

Horizontal range,

$$R = \frac{u^2 \sin 2 \times 45°}{g} = \frac{u^2}{g}$$

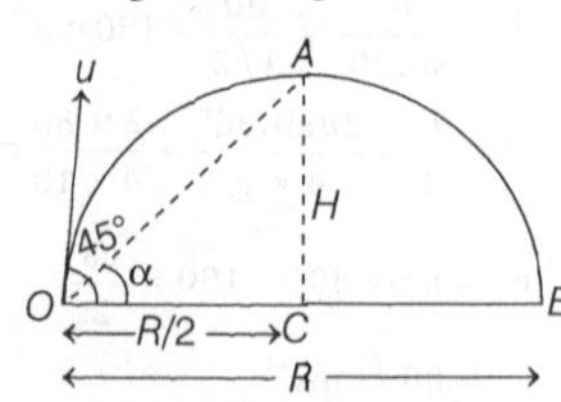

$\therefore OC = R/2 = u^2/(2g)$

$$\tan\alpha = \frac{AC}{OC} = \frac{u^2/4g}{u^2/2g} = \frac{1}{2}$$

$\therefore \alpha = \tan^{-1}(1/2)$

22 Given, speed of man $v_m = 4$ km/h

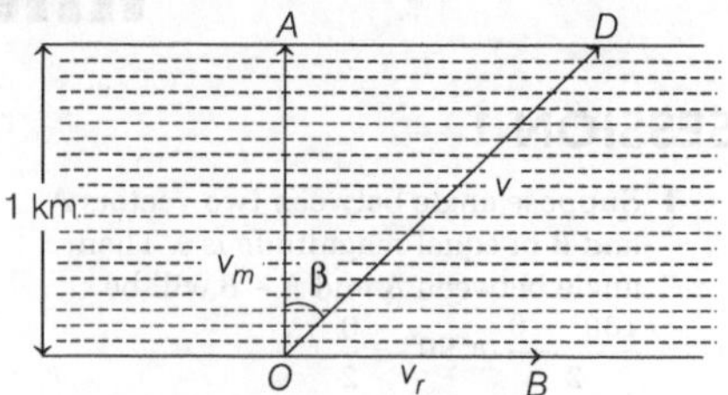

Speed of river, $v_r = 3\ \text{kmh}^{-1}$
Width of the river $d = 1$ km
Time taken by the man to cross the river,

$$t = \frac{\text{Width of the river}}{\text{Speed of the man}} = \frac{1\text{km}}{4\ \text{kmh}^{-1}}$$

$= \frac{1}{4}\text{h} = \frac{1}{4} \times 60 = 15$ min

Distance travelled along the river

$= v_r \times t$

$= 3 \times \frac{1}{4} = \frac{3}{4}\text{km} = \frac{3000}{4} = 750$ m

23 Let u be the velocity of the swimmer and v be the velocity of the river flow. Then

$$t_1 = \frac{2d}{\sqrt{u^2 - v^2}}$$

and $t_2 = \frac{d}{u - v} + \frac{d}{u + v} = \frac{2ud}{u^2 - v^2}$

If t is the time taken by the swimmer to swim a distance $2d$ in still water, then

$$t = \frac{2d}{u}$$

$$t_2 \times t = \frac{2ud}{u^2 - v^2} \times \frac{2d}{u} = \frac{4d^2}{u^2 - v^2}$$

From above, $t_2 \times t = t_1^2$ or $t = \frac{t_1^2}{t_2}$

24 When the man is at rest w.r.t. to the ground, the rain comes to him at an angle 30° with the vertical. This is the direction of the velocity of raindrops w.r.t. to the ground.

Here, $\mathbf{v}_{rg}$ = velocity of the rain w.r.t. the ground.

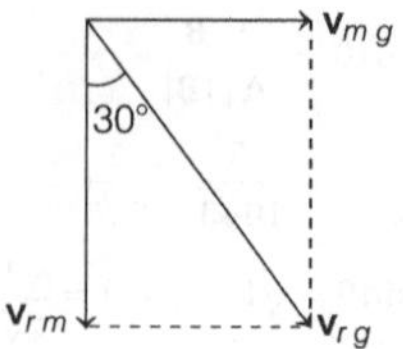

$\mathbf{v}_{rm} = \mathbf{v}_{rg} + \mathbf{v}_{gm} = \mathbf{v}_{rg} - \mathbf{v}_{mg} = \mathbf{v}_{rg} + (-\mathbf{v}_{mg})$

$\mathbf{v}_{mg}$ = velocity of the man w.r.t the ground,

and $\mathbf{v}_{rm}$ = velocity of the rain w.r.t. the man.

Here, $v_{rg} \sin 30° = v_{mg} = 10 \text{ kmh}^{-1}$

$\Rightarrow \quad v_{rg} = \frac{10}{\sin 30°} = 20 \text{ kmh}^{-1}$

25 Initial relative velocity of the express train w.r.t. the passenger train,

$u_{ep} = u_e - u_p = 30 - 5 = 25 \text{ms}^{-1}$

Final relative velocity of the express train w.r.t. the passenger train, $v_{ep} = 0$ (because express train comes to rest relative to passenger train)

From first equation of motion,

$v_{ep} = u_{ep} - at \Rightarrow 0 = 25 - 4t$

$\Rightarrow \quad 4t = 25 \quad \Rightarrow t = 6.25 \text{ s}$

26 Total distance travelled by the boat in 2 h in still water = 8 + 8 = 16 km

Therefore, speed of boat in still water,

$v_b = \frac{16}{2} = 8 \text{ km h}^{-1}$

Effective velocity when boat moves upstream $= v_b - v_w = 8 - 4$

$= 4 \text{ km h}^{-1}$,

Therefore, time taken to travel from one end to other $= \frac{8}{4} = 2 \text{ h}$

Effective velocity when boat moves downstream $= v_b + v_w = 8 + 4$

$= 12 \text{ km h}^{-1}$

The time taken to travel 8 km distance

$= \frac{8}{12} = \frac{2}{3} \text{h} = 40 \text{ min}$

Total time taken = 2 h + 40 min

= 2 h and 40 min

27 Let v_1 be the velocity of the car and v_2 be the velocity of the parcel. The parcel is thrown at an angle θ from Q, it reaches the man at M.

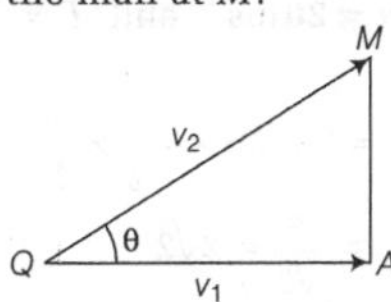

$\therefore \cos\theta = \frac{v_1}{v_2} = \frac{10}{10\sqrt{2}} = \frac{1}{\sqrt{2}} = \cos 45°$

So, $\theta = 45°$

28 Given, $s = t^3 + 5$

$\therefore$ Speed $v = \frac{ds}{dt} = 6t$

and rate of change of speed

$a_t = \frac{dv}{dt} = 6t$

$\therefore$ Tangential acceleration at $t = 2\text{s}$

$a_t = 6 \times 2 = 12 \text{ ms}^{-2}$

and at $t = 2\text{s}, v = 3(2)^2 = 12 \text{ ms}^{-1}$

$\therefore$ Centripetal acceleration,

$a_c = \frac{v^2}{R} = \frac{144}{20} \text{ ms}^{-2}$

Net acceleration $= \sqrt{a_c^2 + a_t^2} \approx 14 \text{ ms}^{-2}$

29 For a particle in uniform circular motion

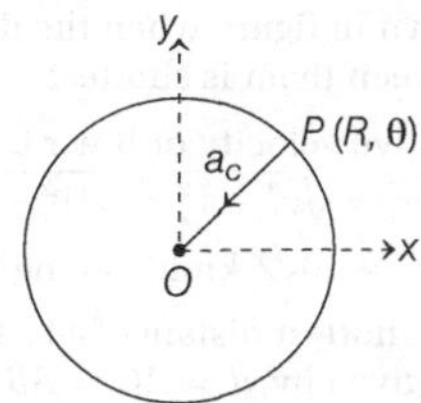

$\mathbf{a} = \frac{v^2}{R}$ towards centre of circle

$\therefore \mathbf{a} = \frac{v^2}{R}(-\cos\theta\,\hat{\mathbf{i}} - \sin\theta\,\hat{\mathbf{j}})$

or $\mathbf{a} = -\frac{v^2}{R}\cos\theta\,\hat{\mathbf{i}} - \frac{v^2}{R}\sin\theta\,\hat{\mathbf{j}}$

30 Velocity of rain

$\mathbf{v}_r = -3\hat{\mathbf{j}}$ (vertically downward)

Velocity of man

$\mathbf{v}_m = 4\hat{\mathbf{i}}$

$\therefore$ Relative velocity of rain w.r.t. man

$\mathbf{v}_{rm} = \mathbf{v}_r - \mathbf{v}_m = -3\hat{\mathbf{j}} - 4\hat{\mathbf{i}}$

$\therefore |\mathbf{v}_{rm}| = \sqrt{(-3)^2 + (-4)^2}$

$= 5 \text{kmh}^{-1}$

31 Horizontal range of the projectile,

$$R = \frac{u^2 \sin 2\theta}{g}$$

$$R_{max} = \frac{u^2}{g} \qquad (\theta = 45°)$$

The maximum height attained by the projectile

$$H = \frac{u^2 \sin^2\theta}{2g}$$

If $H = R$, then

$$\frac{u^2 \sin^2\theta}{2g} = \frac{u^2 \sin 2\theta}{g}$$

$$\frac{\sin^2\theta}{2} = 2\sin\theta\cos\theta$$

So, $\tan\theta = 4$

$\therefore \quad \theta = \tan^{-1}(4)$

32 To hit a target, the man should aim his rifle at a point higher than the target as the bullet suffers a vertical deflection $(y = \frac{1}{2}gt^2)$ due to acceleration due to gravity.

33 $OA = OC$

OA + **OC** is along **OB** (bisector) and its magnitude is

$$2R \cos 45° = R\sqrt{2}$$

(**OA** + **OC**) + **OB** is along **OB** and its magnitude is

$$R\sqrt{2} + R = R(1 + \sqrt{2})$$

34 $\cos\theta = \frac{(\hat{\mathbf{i}} + \hat{\mathbf{j}}) \cdot \hat{\mathbf{i}}}{|(\hat{\mathbf{i}} + \hat{\mathbf{j}})| \cdot |\hat{\mathbf{i}}|} = \frac{1}{\sqrt{2}} = \cos 45°$

So, $\theta = 45°$

Hence, angle between $\hat{\mathbf{i}}$ and $\hat{\mathbf{j}}$ is 90°.

SESSION 2

1 $F F \cos\theta = F F \sin\theta$

or $\tan\theta = 1$ or $\theta = 45°$

$\therefore |\mathbf{F}_1 + \mathbf{F}_2|$

$= \sqrt{F^2 + F^2 + 2FF \cos 45°}$

$= \sqrt{(2 + \sqrt{2})}\, F$

2 Vertical distance covered by a projectile is given by

$$h = (u\sin\theta)t - \frac{1}{2}gt^2 \qquad \text{...(i)}$$

$$d = u\cos\theta \times t$$

or $t = \frac{d}{u\cos\theta}$

From Eq. (i), we get

$$h = u\sin\theta \times \frac{d}{u\cos\theta} - \frac{1}{2}g\frac{d^2}{u^2\cos^2\theta}$$

or $h = \frac{d}{\cos\theta}\sqrt{\frac{g}{2(d\tan\theta - h)}}$

3 We know that the range is

$$T = \frac{2u \sin\theta}{g}$$

According to the question, the range of the projectile is the same for complementary angles,

So, $T_1 = \frac{2u \sin\theta}{g}$

$\Rightarrow \quad T_2 = \frac{2u \sin(90° - \theta)}{g}$

Again, the range of the projectile is

$$R = \frac{u^2 \sin 2\theta}{g}$$

$\therefore T_1 T_2 = \frac{2u \sin\theta}{g} \times \frac{2u \sin(90° - \theta)}{g}$

$= \frac{2u^2 \sin 2\theta}{g^2} = \frac{2R}{g}$

$\therefore \quad T_1 T_2 \propto R$

4 Initial velocity, $\mathbf{u} = \mathbf{i} + 2\mathbf{j} \text{ ms}^{-1}$

Magnitude of velocity

$u = \sqrt{(1)^2 + (2)^2} = \sqrt{5} \text{ms}^{-1}$

Equation of trajectory of projectile

$$Y = X\tan\theta - \frac{gX^2}{2u^2}(1 + \tan^2\theta)$$

$$\left[\tan\theta \frac{Y}{X} = \frac{2}{1} = 2\right]$$

$\therefore Y = X \times 2 - \frac{10(X)^2}{2(\sqrt{5})^2}[1 + (2)^2]$

$= 2X - \frac{10(X^2)}{2 \times 5}(1 + 4) = 2X - 5X^2$

5 Angular momentum of the projectile, $L = mv_h r_\perp = m(v\cos\theta)h$

where, h is the maximum height

$$= m(v\cos\theta)\left(\frac{v^2\sin^2\theta}{2g}\right);$$

$$L = \frac{mv^3\sin^2\theta\cos\theta}{2g} = \frac{\sqrt{3}\,mv^3}{16g} \quad [\because \theta = 30°]$$

6 Since, $x = \alpha t^3$ and $y = \beta t^3$

$$\therefore \quad \mathbf{r} = x\hat{\mathbf{i}} + y\hat{\mathbf{j}} = \alpha t^3\hat{\mathbf{i}} + \beta t^3\hat{\mathbf{j}}$$

Now, $\mathbf{v} = \frac{d\mathbf{r}}{dt} = \alpha t^2 \times 3\hat{\mathbf{i}} + \beta t^2 \times 3\hat{\mathbf{j}}$

Thus, $|\mathbf{v}| = \sqrt{(3\alpha t^2)^2 + (3\beta t^2)^2}$

$$= \sqrt{9\alpha^2t^4 + 9\beta^2t^4}$$

$$= 3t^2\sqrt{\alpha^2 + \beta^2}$$

7 At the highest point of its flight, vertical component of velocity is zero and only horizontal component is left which is

$$u_x = u\cos\theta$$

Given, $\theta = 45°$

$$\therefore \quad u_x = u\cos 45° = \frac{u}{\sqrt{2}}$$

Hence, at the highest point kinetic energy

$$E' = \frac{1}{2}mu_x^2 = \frac{1}{2}m\left(\frac{u}{\sqrt{2}}\right)^2$$

$$= \frac{1}{2}m\left(\frac{u^2}{2}\right) = \frac{E}{2} \quad \left(\because \frac{1}{2}mv^2 = E\right)$$

8 Let the swimmer starts swimming with velocity v along AC in a direction making an angle θ with AB as shown in the figure. If d is the width of the river, time taken by the swimmer to cross the river will be

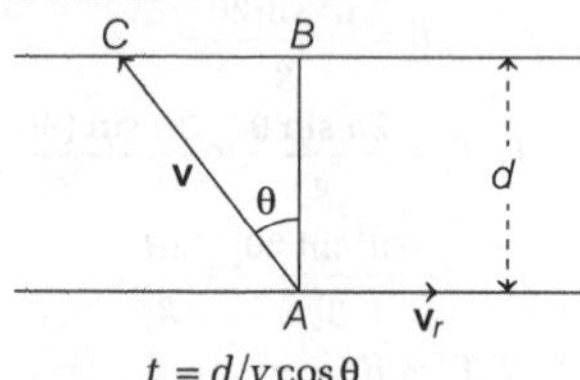

$$t = d/v\cos\theta$$

As component of AB will be $v\cos\theta$, This time will be minimum, when $\cos\theta = \max = 1$, i.e. $\theta = 0°$.

So, the swimmer should swim in North direction.

9 Let the ships A and B be at positions as shown in figure when the distance between them is shortest.

Relative velocity of B w.r.t. A is

$$v_r = \sqrt{v_1^2 + v_2^2} = \sqrt{10^2 + 10^2}$$

$$= 10\sqrt{2}\ \text{kmh}^{-1} \text{ along } BC$$

The shortest distance between A and C is d given by, $d = AC = AB\sin 45°$

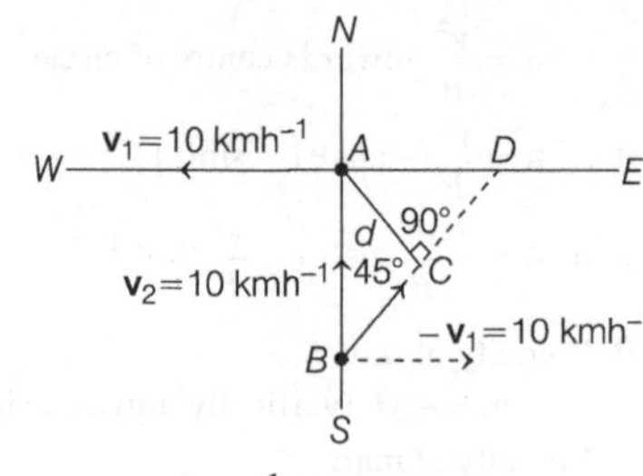

$$= 100 \times \frac{1}{\sqrt{2}} = 50\sqrt{2}\ \text{km}$$

Shortest time, $t = \frac{d}{v_r} = \frac{50\sqrt{2}}{10\sqrt{2}} = 5\,\text{h}$

10 $\mathbf{v}_1 = +5\hat{\mathbf{i}}$, $\mathbf{v}_2 = +5\hat{\mathbf{j}}$

$$\Delta\mathbf{v} = \mathbf{v}_2 - \mathbf{v}_1 = 5\hat{\mathbf{j}} - 5\hat{\mathbf{i}}$$

$$|\Delta\mathbf{v}| = 5\sqrt{2}$$

$$\therefore \quad a = \frac{|\Delta\mathbf{v}|}{t} = \frac{5\sqrt{2}}{10} = \frac{1}{\sqrt{2}}\text{ms}^{-2}$$

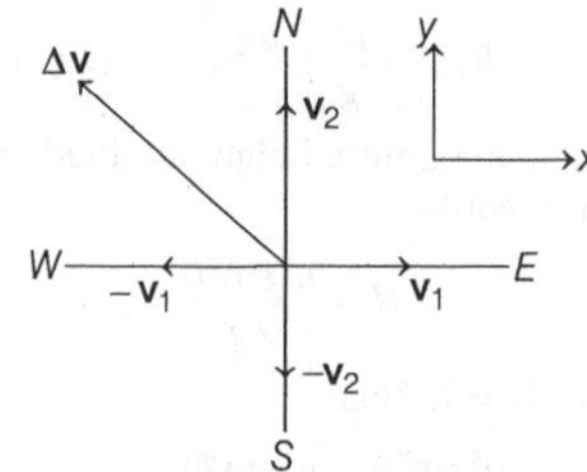

For direction,

$$\tan\alpha = -\frac{5}{5} = -1$$

Average acceleration is $\frac{1}{\sqrt{2}}$ ms^{-2} towards North-West.

11 Maximum speed with which the boy can throw stone is

$$u = \sqrt{2gh} = \sqrt{2\times 10\times 10}$$

$$= 10\sqrt{2}\ \text{ms}^{-1}.$$

Range is maximum when projectile is thrown at an angle of 45°.

Thus, $R_{\max} = \frac{u^2}{g} = \frac{(10\sqrt{2})^2}{10} = 20\,\text{m}$

12 Maximum range of water coming out of the fountain,

$$R_m = \frac{v^2}{g}$$

$\therefore$ Total area around fountain,

$$A = \pi R_m^2 = \pi\frac{v^4}{g^2}$$

13 $mg\sin\theta = ma$

$$\therefore \quad a = g\sin\theta$$

where, a is along the inclined plane.

$\therefore$ Vertical component of acceleration is $g\sin^2\theta$.

$\therefore$ Relative vertical acceleration of A with respect to B is

$$g(\sin^2 60° - \sin^2 30°) = \frac{g}{2} = 4.9\ \text{ms}^{-2}$$

(in vertical direction)

14 Given, velocity $v = kY\mathbf{i} + kX\mathbf{j}$

$$\frac{dX}{dt} = kY, \frac{dY}{dt} = kX$$

$$\frac{dY}{dX} = \frac{dY}{dt}\times\frac{dt}{dX} = \frac{kX}{kY}$$

$$YdY = XdX;$$

$$Y^2 = X^2 + C$$

where, c = constant.

15 According to question, $\frac{u^2}{g} = 40$

$$\therefore \quad u = 20\,\text{ms}^{-1} \text{ and } T = \frac{2u\sin 45°}{g}$$

$$= 2\times 20\times\frac{1}{\sqrt{2}}\times\frac{1}{10}$$

$$= \frac{4}{\sqrt{2}} = 2\sqrt{2}$$

When car is moving with speed, $v = 20\ \text{ms}^{-1}$, then

$$(v\cos\theta + 20)\times t = 40$$

$$(20\cos\theta + 20)\times 2\sqrt{2} = 40$$

$$\Rightarrow \quad \cos^2\frac{\theta}{2} = \frac{1}{2\sqrt{2}} \approx 60°$$

DAY FOUR

Laws of Motion

Learning & Revision for the Day

- Concept of Forces
- Inertia
- Newton's Laws of Motion
- Principle of Conservation of Linear Momentum
- Free Body Diagram
- Connected Motion
- Equilibrium of concurrent Forces
- Friction

Concept of Forces

A push or a pull exerted on any object, is defined to be a force. It is a vector quantity. Force can be grouped into two types:

- **Contact forces** are the forces that act between two bodies in contact, e.g. tension, normal reaction, friction etc.
- **Non-contact forces** are the forces that act between two bodies separated by a distance without any actual contact. e.g. gravational force between two bodies and electrostatic form between two charges etc.

Inertia

The inability of a body to change by itself its state of rest or state of uniform motion along a straight line is called inertia of the body.

As inertia of a body is measured by the mass of the body. Heavier the body, greater the force required to change its state and hence greater is its inertia. There are three type of inertia (i) inertia of rest (ii) inertia of motion (iii) inertia of direction.

Newton's Laws of Motion

First Law of Motion (Law of Inertia)

It states that a body continues to be in a state of rest or of uniform motion along a straight line, unless it is acted upon by some external force the change the state. This is also called law of inertia.

If $F = 0, \Rightarrow v = \text{constant} \Rightarrow a = 0$

- This law defines force.
- The body opposes any external change in its state of rest or of uniform motion.
- It is also known as the **law of inertia** given by Galileo.

PREP MIRROR

Your Personal Preparation Indicator

- No. of Questions in Exercises (x)—
- No. of Questions Attempted (y)—
- No. of Correct Questions (z)—
  (Without referring Explanations)

- Accuracy Level ($z/y \times 100$)—
- Prep Level ($z/x \times 100$)—

In order to expect good rank in JEE, your Accuracy Level should be above 85 & Prep Level should be above 75.

Linear Momentum

It is defined as the total amount of motion of a body and is measured as the product of the mass of the body and its velocity. The momentum of a body of mass m moving with a velocity $\mathbf{v}$ is given by $\mathbf{p} = m\mathbf{v}$.

Its unit is kg-ms^{-1} and dimensional formula is [ML T^{-1}]

Second Law of Motion

The rate of change of momentum of a body is directly proportional to the applied force and takes place in the direction in which the force acts.

According to second law, $F \propto \frac{dp}{dt}$ or $F = k\frac{dp}{dt}$

where, k is constant.

as, $\frac{dp}{dt} = \frac{d}{dt}(mv) = ma$ or $\frac{m\,dv}{dt} = ma$

i.e., second law can be written as

$$F = \frac{dp}{dt} = ma$$

The SI unit of force is newton (N) and in CGS system is dyne.

$$1\,\text{N} = 10^5 \text{ dyne}$$

Impulse

Impulse received during an impact is defined as the product of the average force and the time for which the force acts.

Impulse, $\mathbf{I} = \mathbf{F}_{av}\, t$

Impulse is also equal to the total change in momentum of the body during the impact.

Impulse, $\mathbf{I} = \mathbf{p}_2 - \mathbf{p}_1$

Impulse = Change in momentum

Third Law of Motion

To every action, there is an equal and opposite reaction.

$$\mathbf{F}_{12} = -\mathbf{F}_{21}$$

- Action and reaction are mutually opposite and act on two different bodies.
- The force acting on a body is known as action.
- When a forc acts on a body, then the reaction acts normally to the surface of the body.

Principle of Conservation of Linear Momentum

It states that if no external force is acting on a system, the momentum of the system remains constant.

According to second law of motion, $\mathbf{F} = \frac{d\mathbf{p}}{dt}$

If no force is acting, then $\mathbf{F} = 0$

$$\therefore \quad \frac{d\mathbf{p}}{dt} = 0 \Rightarrow \mathbf{p} = \text{constant}$$

or $$m_1\mathbf{v}_1 = m_2\mathbf{v}_2 = \text{constant}$$

Applications of Conservation of Linear Momentum

The propulsion of rockets and jet planes is based on the principle of conservation of linear momentum.

- Upward thrust on the rocket, $F = -\frac{u\,dm}{dt} - mg$ and if effect of gravity is neglected, then $F = -\frac{u\,dm}{dt}$.
- Instantaneous upward velocity of the rocket

$$v = u \ln\left(\frac{m_0}{m}\right) - gt$$

and neglecting the effect of gravity

$$v = u \ln\left(\frac{m_0}{m}\right) = 2.303\, u \log_{10}\left(\frac{m_0}{m}\right)$$

where, m_0 = initial mass of the rocket including that of the fuel,

u = initial velocity of the rocket at any time t,

m = mass of the rocket left,

v = velocity acquired by the rocket,

$\frac{dm}{dt}$ = rate of combination of fuel.

- **Burnt out speed** of the rocket is the speed attained by the rocket when the whole of fuel of the rocket has been burnt. Burnt out speed of the rocket

$$v_b = u \log_e\left(\frac{m_0}{m_r}\right) = 3.303\, u \log_{10}\left(\frac{m_0}{m_r}\right)$$

Apparent Weight of a Boy in a Lift

Actual weight of the body is mg. Here we consider the apparent weight of a man standing in a moving lift.

- If lift is accelerating upward with acceleration a, then apparent weight of the body is $R = m(g + a)$.
- If lift is accelerating downward at the rate of acceleration a, then apparent weight of the body is $R = m(g - a)$
- If lift is moving upward or downward with constant velocity, then apparent weight of the body is equal to actual weight.
- If the lift is falling freely under the effect of gravity, then it is called weightlessness condition.

Free Body Diagram

A free body diagram (FBD) consists of a diagramatize representation of a single body or sub-system of bodies isolated from its surroundings showing all forces acting on it.

While sketching a free body diagram the following points should be kept in mind.

- Normal reaction (N) always acts normal to the surface on which the body is kept.
- When two objects A and B are connected by a string, the tension for object A is towards B and for object B, it is towards A.

Connected Motion

- If two blocks of masses m_1 and m_2 are placed on a perfectly smooth surface and are in contact, then

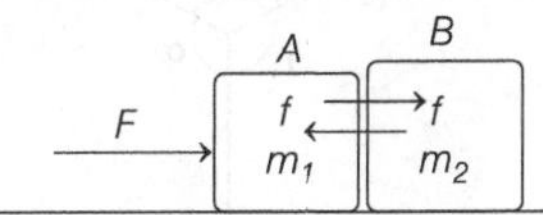

Acceleration of the blocks, $a = \dfrac{F}{m_1 + m_2}$

and the contact force (acting normally) between the two blocks is $f = m_2 a = \dfrac{F\, m_2}{(m_1 + m_2)}$.

- A block system is shown in the figure

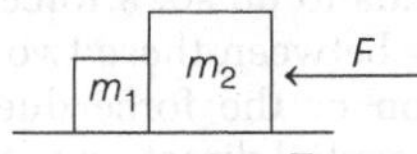

Acceleration of the blocks $a = \dfrac{F}{m_1 + m_2}$

Contact force between two blocks

$$f = m_1 a = \frac{m_1 F}{(m_1 + m_2)}$$

- For a block of mass m placed on a fixed, perfectly smooth inclined plane of angle θ, the forces acting on the block are as shown in the figure. Obviously, here $a = g \sin\theta$

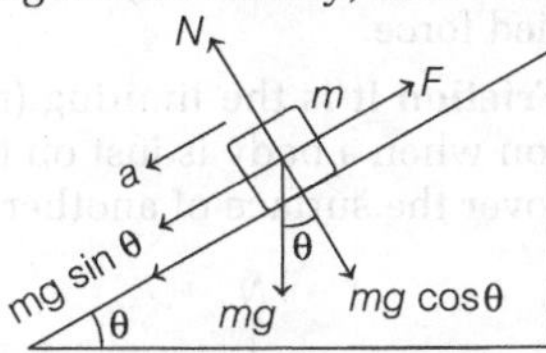

- If a block of mass m is placed on a smooth movable wedge of mass M, which in turn is placed on smooth surface, then a force **F** is applied on the wedge, horizontally.

The acceleration of the wedge and the block is

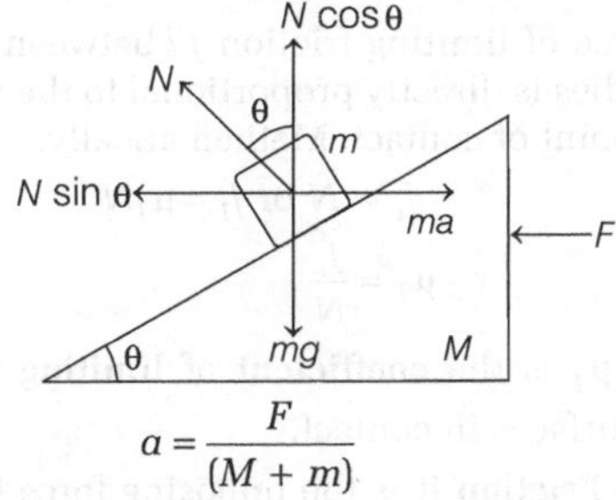

$$a = \frac{F}{(M + m)}$$

Force on the block, $F = (M + m)a = (M + m)g \tan\theta$

- For a block system shown in the figure, acceleration of the system

$$a = \frac{F}{m_1 + m_2 + m_3}$$

Tension in the string,

$$T_1 = (m_2 + m_3)a = \frac{F(m_3 + m_2)}{(m_1 + m_2 + m_3)}$$

and tension $T_2 = m_3 a = \dfrac{F\, m_3}{m_1 + m_2 + m_3}$

- For a block system shown in the figure, acceleration of the system

$$a = \frac{F}{m_1 + m_2 + m_3}$$

Tension in the string,

$$T_1 = \frac{m_1 F}{m_1 + m_2 + m_3}$$

and $T_2 = \dfrac{(m_1 + m_2)F}{m_1 + m_2 + m_3}$

- For a block system suspended freely from a rigid support as shown in the figure, the acceleration of the system $a = 0$.

String tension,

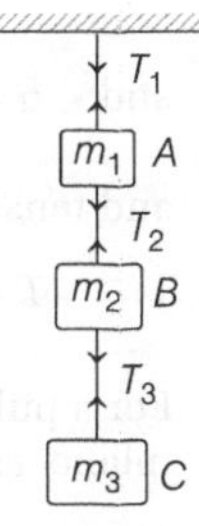

$$T_1 = (m_1 + m_2 + m_3)g$$
$$T_2 = (m_2 + m_3)g$$

and $T_3 = m_3 g$

- For a block system and a pulley as shown in the figure, value of the acceleration of the system

$$a = \frac{(m_1 + m_2 - m_3)g}{(m_1 + m_2 + m_3)}$$

Tension, $T_1 = \dfrac{2m_1 m_3 g}{(m_1 + m_2 + m_3)}$

Tension, $T_2 = \dfrac{2m_3(m_1 + m_2)g}{(m_1 + m_2 + m_3)}$

and tension T

$$T = 2T_2 = \frac{4m_3(m_1 + m_2)g}{(m_1 + m_2 + m_3)}$$

- For the pulley and block arrangement as shown in the figure, we have

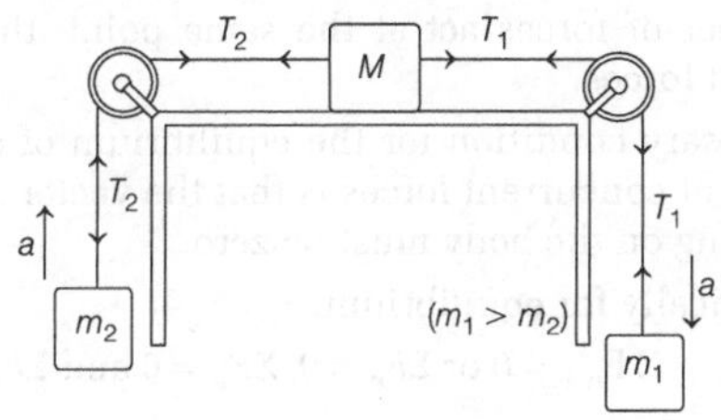

Net acceleration,

$$a = \frac{\text{Net accelerating force}}{\text{Total mass}} = \frac{(m_1 - m_2)g}{(m_1 + m_2 + M)}$$

Tension, $T_1 = m_1(g - a) = \frac{(M + 2m_2)m_1 g}{(M + m_1 + m_2)}$

and Tension, $T_2 = m_2(g + a) = \frac{(M + 2m_1)m_2 g}{(M + m_1 + m_2)}$

- For the system of block and pulley, with a smooth inclined plane as shown in the figure, we have

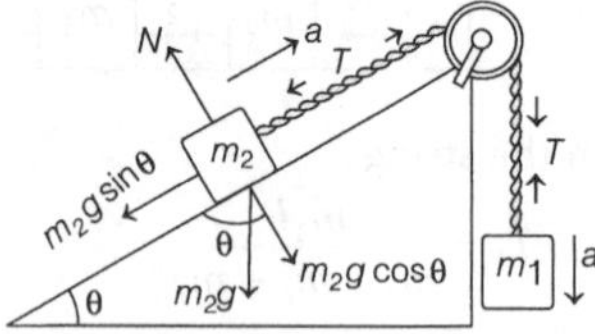

Net acceleration,

$$a = \frac{(m_1 - m_2 \sin\theta)g}{(m_1 + m_2)}, \text{ if } m_1 g > m_2 g \sin\theta$$

and $a = \frac{(m_2 \sin\theta - m_1)g}{m_1 + m_2}$, if $m_1 g < m_2 g \sin\theta$

and tension in the string

$$T = m_1(g - a) = \frac{m_1 m_2 (1 + \sin\theta)}{(m_1 + m_2)}$$

- For a pulley and block system on a smooth double inclined plane as shown in the figure, we have

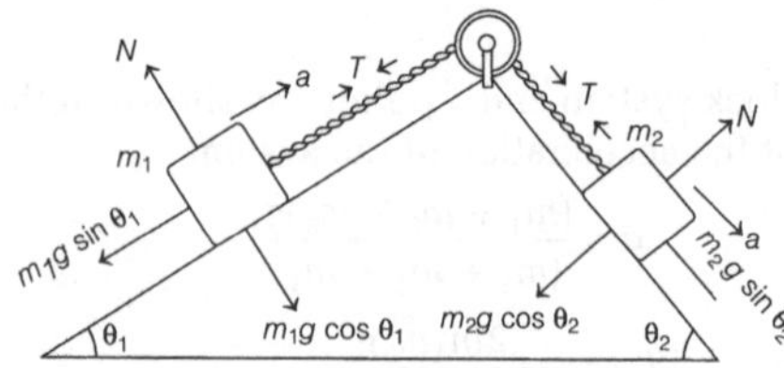

Net acceleration, $a = \frac{(m_2 \sin\theta_2 - m_1 \sin\theta_1)g}{(m_1 + m_2)}$,

for $\theta_2 > \theta_1, m_2 > m_1$

and tension in the string,

$$T = \frac{m_1 m_2 (\sin\theta_1 + \sin\theta_2)g}{(m_1 + m_2)}$$

Equilibrium of Concurrent Forces

If a number of forces act at the same point, they are called concurrent forces.

The necessary condition for the equilibrium of a body under the action of concurrent forces is that the vector sum of all the forces acting on the body must be zero.

Mathematically for equilibrium,

$$\Sigma \mathbf{F}_{\text{net}} = 0 \text{ or } \Sigma F_x = 0, \Sigma F_y = 0 \text{ and } \Sigma F_z = 0$$

Lami Theorem For three concurrent forces in equilibrium position.

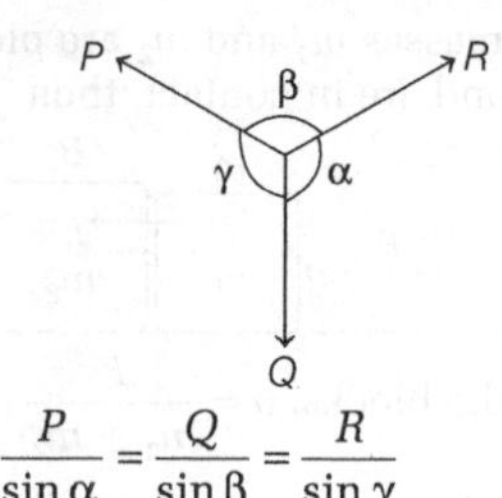

$$\frac{P}{\sin\alpha} = \frac{Q}{\sin\beta} = \frac{R}{\sin\gamma}$$

Friction

Whenever an object actually slides or rolls over the surface of another body or tends to do so, a force opposing the relative motion starts acting between these two surfaces in contact. It is known as **friction** or the force due to friction. Force of friction acts in a tangential direction to the surfaces in contact.

Types of Friction

The four types of friction are given below

1. **Static Friction** It is the opposing force that comes into play when one body is at rest and a force acts to move it over the surface of another body.

   It is a self adjusting force and is always equal and opposite to the applied force.

2. **Limiting Friction** It is the limiting (maximum) value of static friction when a body is just on the verge of starting its motion over the surface of another body.

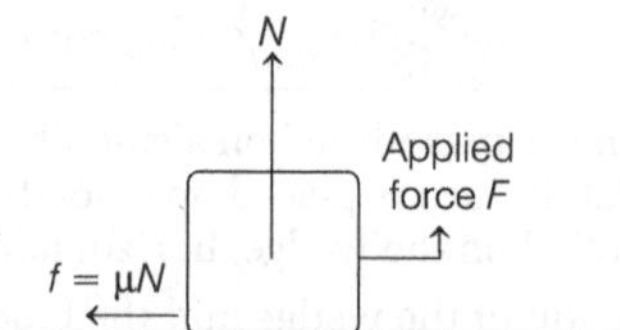

   The force of limiting friction fl between the surfaces of two bodies is directly proportional to the normal reaction at the point of contact. Mathematically,

$$f_l \propto N \text{ or } f_l = \mu_l N$$

$$\Rightarrow \quad \mu_l = \frac{f_l}{N}$$

   where, μ_l is the **coefficient of limiting friction** for the given surfaces in contact.

3. **Kinetic Friction** It is the opposing force that comes into play when one body is actually slides over the surface of another body. Force of kinetic friction f_k is directly proportional to the normal reaction N and the ratio $\frac{f_k}{N}$ is called **coefficient of kinetic friction** μ_k, value of μ_k is slightly less than μ_e $(\mu_k < \mu_l)$.

   Whenever limiting friction is converted into kinetic friction, body started motion with a lurch.

4. **Rolling Friction** It is the opposing force that comes into play when a body of symmetric shape (wheel or cylinder or disc, etc.) rolls over the surface of another body. Force of rolling friction f_r is directly proportional to the normal reaction N and inversely proportional to the radius (r) of the wheel.

Thus, $f_r \propto \frac{N}{r}$ or $f_r = \mu_r \frac{N}{r}$

The constant μ_r is known as the **coefficient of rolling friction** μ_r has the unit and dimensions of length. Magnitudewise $\mu_r << \mu_k$ or μ_l.

- The value of rolling friction is much smaller than the value of sliding friction.
- Ball bearings are used to reduce the wear and tear and energy loss against friction.

Angle of Friction

Angle of friction is defined as the angle θ which the resultant R of the force of limiting friction f_l and normal reaction N, subtends with the normal reaction.

The tangent of the angle of friction is equal to the coefficient of friction. i.e. $\mu = \tan\theta$

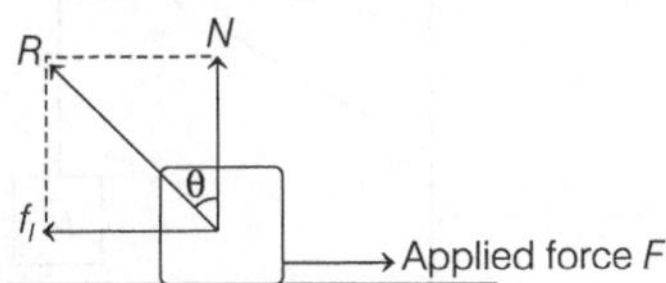

Angle of Repose

Angle of repose is the least angle of the inclined plane (of given surface) with the horizontal such that the given body placed over the plane, just begins to slide down, without getting accelerated.

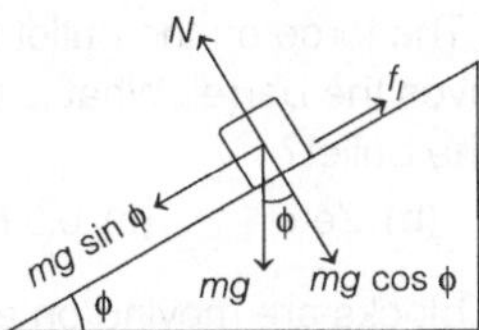

The tangent of the angle of repose is equal to the coefficient of friction.

Hence, we conclude that angle of friction (θ) is equal to the angle of repose (ϕ).

In limiting condition, $f_1 = mg \sin\phi$

$\Rightarrow$ $N = mg\cos\phi$

$$\frac{f_1}{N} = \tan\phi$$

$$\therefore \quad \frac{f_1}{N} = \mu_s = \tan\phi$$

Acceleration of a Block on Applying a Force on a Rough Surface

- Acceleration of a block on a horizontal surface is as shown in the figure.

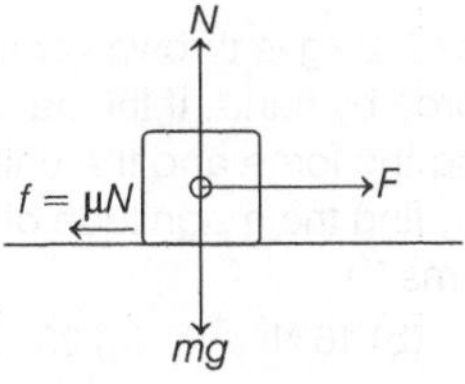

$$a = \frac{F - f}{m} = \frac{F - \mu mg}{m}$$

or $$a = \frac{F}{m} - \mu g$$

where, μ = coefficient of kinetic friction between the two surfaces in contact.

- Acceleration of block sliding down a rough inclined plane as shown in the figure is given by

$$a = g(\sin\alpha - \mu\cos\alpha)$$

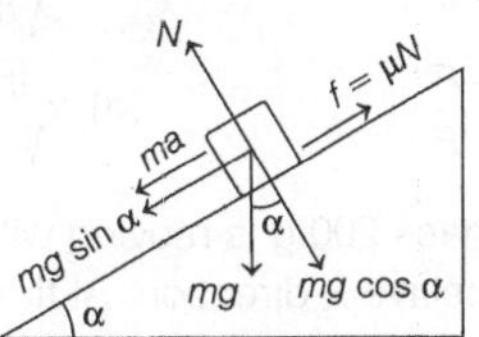

- Retardation of a block sliding up a rough inclined plane as shown in the figure is $a = g(\sin\alpha + \mu\cos\alpha)$

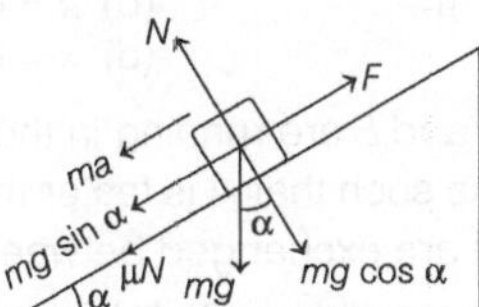

DAY PRACTICE SESSION 1

FOUNDATION QUESTIONS EXERCISE

1 Five forces inclined at an angle of 72° w.r.t. each other act on a particle of mass m placed at the origin. Four forces are of magnitude F_1 and one has a magnitude F_2. Find the resultant acceleration of the particle.

(a) $\frac{F_2 - F_1}{m}$ (b) Zero

(c) $\frac{F_2 + F_1}{m}$ (d) $\frac{F_2 - 4F_1}{m}$

2 A ball of mass 0.2 kg is thrown vertically upwards by applying a force by hand. If the hand moves by 0.2 m while applying the force and the ball goes upto 2 m height further, find the magnitude of the force. (take, $g = 10\ ms^{-2}$)

(a) 4 N (b) 16 N (c) 20 N (d) 22 N

3 A player catches a cricket ball of mass 150 g, moving at a rate of $20\ ms^{-1}$. If the catching process is completed in 0.1 s, the force of the blow exerted by the ball on the hand of the player is equal to

(a) 150 N (b) 3 N (c) 30 N (d) 300 N

4 A ball of mass m is thrown vertically upwards with a velocity v. If air exerts an average resisting force F, the velocity with which the ball returns to the thrower is

(a) $v\sqrt{\frac{mg}{mg + F}}$ (b) $v\sqrt{\frac{F}{mg + F}}$

(c) $v\sqrt{\frac{mg - F}{mg + F}}$ (d) $v\sqrt{\frac{mg + F}{mg}}$

5 A block of mass 200 g is moving with a velocity of $5\ ms^{-1}$ along the positive x-direction. At time $t = 0$, when the body is at $x = 0$, a constant force 0.4 N is directed along the negative x-direction , is applied on the body for 10 s. What is the position x of the body at $t = 2.5$ s?

(a) $x = 6.75$ m (b) $x = 6.25$ m

(c) $x = 6$ m (d) $x = 6.50$ m

6 Two trains A and B are running in the same direction on parallel tracks such that A is faster than B. If packets of equal weight are exchanged between the two, then

(a) A will be retarded but B will be accelerated

(b) A will be accelerated but B will be retarded

(c) there will not be any change in the velocity of A but B will be accelerated

(d) there will not be any change in the velocity of B, but A will be accelerated

7 Which of the four arrangements in the figure correctly shows the vector addition of two forces F_1 and F_2 to yield the third force F_3?

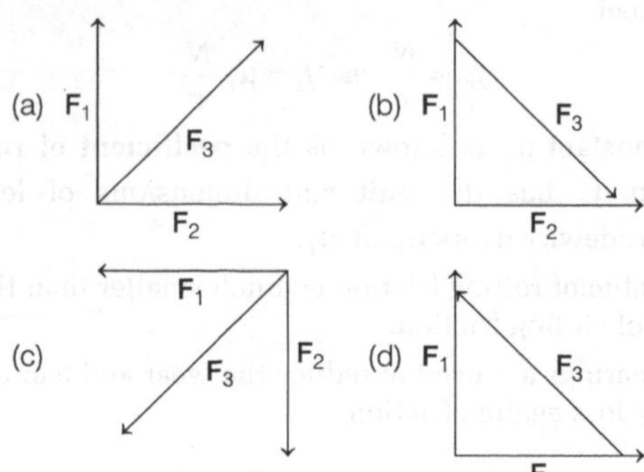

8 A string of negligible mass, going over a clamped pulley of mass m supports a block of mass M as shown in the figure. The force on the pulley by the clamp is given by

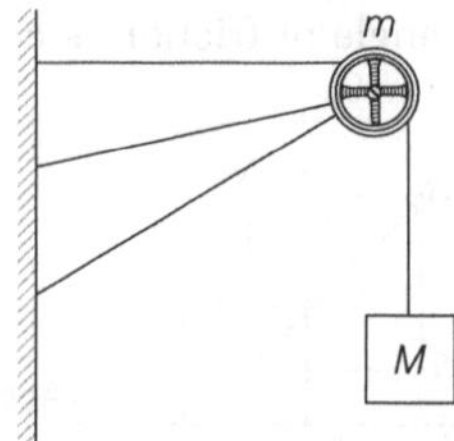

(a) $\sqrt{2}Mg$ (b) $\sqrt{[(M + m)^2 + m^2]g}$

(c) $2Mg$ (d) $\sqrt{[(M + m) + m]^2 g}$

9 A bullet is fired from a gun. The force on the bullet is given by $F = (600 - 2 \times 10^5 t)$, where F is in newton and t is in second. The force on the bullet becomes zero as soon as it leaves the barrel. What is the average impulse imparted to the bullet?

(a) 9 N-s (b) Zero (c) 0.9 N-s (d) 1.8 N-s

10 Two wooden blocks are moving on a smooth horizontal surface, such that the mass m remains stationary with respect to the block of mass M as shown in the figure. The magnitude of force P is

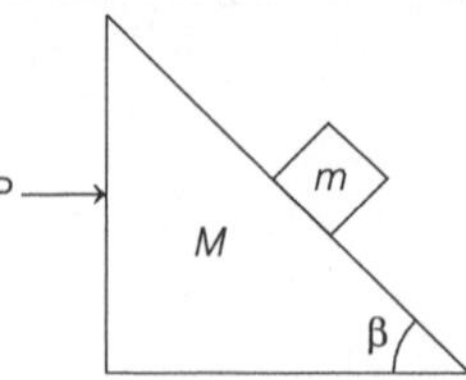

(a) $g \tan\beta$ (b) $mg\cos\beta$

(c) $(M + m)\ \text{cosec}\ \beta$ (d) $(M + m)\ g\tan\beta$

11 The figure below shows a uniform rod of length 30 cm having a mass of 3.0 kg. The strings as shown in the figure are pulled by constant forces of 20 N and 32 N. Find the force exerted by the 20 cm part of the rod on the 10 cm part. All the surfaces are smooth and the strings are light

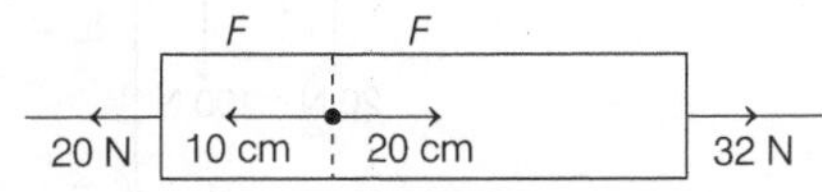

(a) 36 N (b) 12 N (c) 64 N (d) 24 N

12 Two masses $m_1 = 5$ kg and $m_2 = 4.8$ kg, tied to a string, are hanging over a light frictionless pulley. What is the acceleration of the masses produced when system is free to move? (take, $g = 9.8\ \text{ms}^{-2}$)

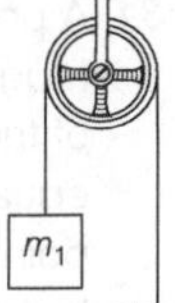

(a) $0.2\ \text{ms}^{-2}$ (b) $9.8\ \text{ms}^{-2}$
(c) $5\ \text{ms}^{-2}$ (d) $4.8\ \text{ms}^{-2}$

13 A light string passing over a smooth light pulley, connects two blocks of masses m_1 and m_2 (vertically). If the acceleration of the system is $(g/8)$, then the ratio of masses is

(a) 8 : 1 (b) 9 : 7 (c) 4 : 3 (d) 5 : 3

14 Two bodies of equal masses are connected by a light inextensible string passing over a smooth frictionless pulley. The amount of mass that should be transferred from one to another, so that both the masses move 50 m in 5 s is

(a) 30% (b) 40% (c) 70% (d) 50%

15 A man slides down a light rope, whose breaking strength is η times his weight. What should be his maximum acceleration, so that the rope does not break?

(a) $g(1-\eta)$ (b) ηg (c) $\frac{g}{1+\eta}$ (d) $\frac{g}{1-\eta}$

16 Two blocks of mass $M_1 = 20$kg and $M_2 = 12$kg are connected by a metal rod of mass 8 kg. The system is pulled vertically up by applying a force of 480 N as shown in figure. The tension at the mid-point of the rod is

→ JEE Main (Online) 2013

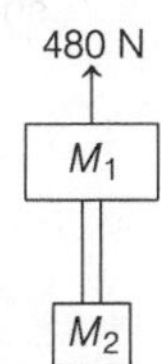

(a) 144 N (b) 96 N (c) 240 N (d) 190 N

17 Two blocks of masses m and M are connected by means of a metal wire of cross-sectional area A passing over a frictionless fixed pulley as shown in the figure. The system is then released. If $M = 2\,m$, then the stress produced in the wire is → JEE Main (Online) 2013

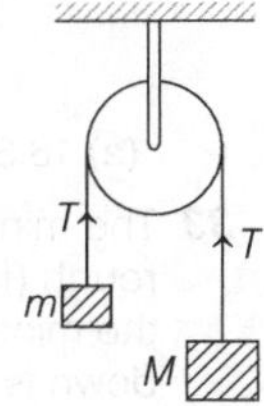

(a) $\frac{2\,mg}{3A}$ (b) $\frac{4\,gm}{3A}$ (c) $\frac{gm}{A}$ (d) $\frac{3\,mg}{4A}$

18 A satellite in a force-free space sweeps out stationary interplanetary dust at a rate $\frac{dM}{dt} = \alpha v$, where M is the mass and v is the velocity of the satellites and α is a constant. The deceleration of the satellite is

(a) $\frac{2\alpha v^2}{M}$ (b) $-\frac{\alpha v^2}{M}$ (c) $-\frac{\alpha v^2}{2M}$ (d) $-\alpha v^2$

19 A lift is moving down with an acceleration a. A man in the lift drops a ball inside the lift. The acceleration of the ball as observed by the man in the lift and a man standing stationary on the ground are respectively

(a) g, g (b) a, a (c) $(g-a), g$ (d) a, g

20 A spring balance is attached to the ceiling of a lift. A man hangs his bag on the string and the balance reads 49 N, when the lift is stationary. If the lift moves downwards with an acceleration of $5\ \text{ms}^{-2}$, the reading of the spring balance would be

(a) 24 N (b) 74 N (c) 15 N (d) 49 N

21 A block A is able to slide on the frictionless incline of angle θ and length l, kept inside an elevator going up with uniform velocity v. Time taken by the block to slide down the length of the incline, if released from rest is

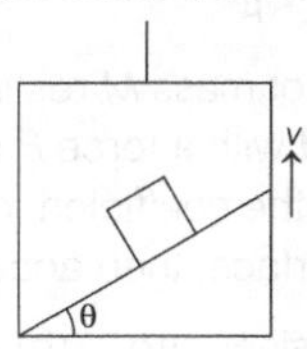

(a) $\sqrt{\frac{2l}{(g+a)\sin\theta}}$ (b) $\sqrt{\frac{2l}{g}}$

(c) $\sqrt{\frac{2l}{g\sin\theta}}$ (d) $\sqrt{\frac{2l}{\sin\theta}}$

22 A plane is inclined at an angle θ with the horizontal. A body of mass m rests on it. If the coefficient of friction is μ, then the minimum force that has to be applied parallel to the inclined plane, so as to make the body to just move up the inclined plane, is

(a) $mg\sin\theta$ (b) $\mu mg\cos\theta$
(c) $\mu mg\cos\theta - mg\sin\theta$ (d) $\mu mg\cos\theta + mg\sin\theta$

23 A 40 kg slab rests on a frictionless floor. A 10 kg block rests on the top of the slab. The static coefficients of friction between the block and the slab is 0.60, while the kinetic coefficient is 0.40.

The 10 kg block is acted upon by a horizontal force of 100 N. If $g = 9.8\ \text{ms}^{-2}$, the resultant acceleration of the slab will be

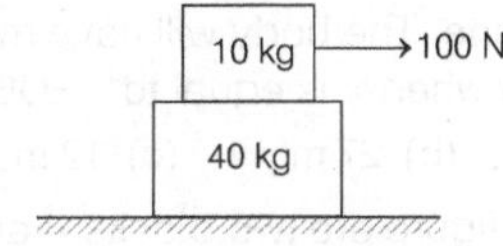

(a) $0.98\ \text{ms}^{-2}$ (b) $1.47\ \text{ms}^{-2}$
(c) $1.52\ \text{ms}^{-2}$ (d) $6.1\ \text{ms}^{-2}$

24 A horizontal force of 10 N is necessary to just hold a block stationary against a wall. The coefficient of friction between the block and the wall is 0.2. The weight of the block is

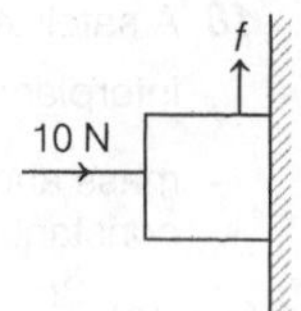

(a) 20 N (b) 50 N
(c) 100 N (d) 2 N

25 What is the maximum value of the force F, such that the block as shown in the arrangement, does not move?

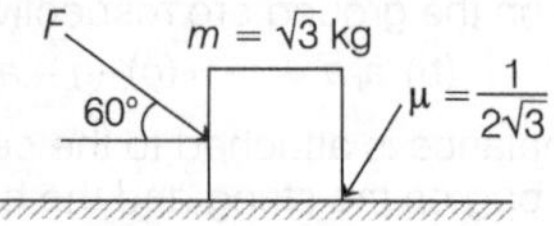

(a) 20 N (b) 10 N (c) 12 N (d) 15 N

26 A block of mass M is held against a rough vertical wall by pressing it with a finger. If the coefficient of friction between the block and the wall is μ and the acceleration due to gravity is g, then minimum force required to be applied by the finger to hold the block against the wall?

(a) $\frac{Mg}{2\mu}$ (b) $\frac{Mg}{\mu}$ (c) $\frac{2M}{\mu g}$ (d) $\frac{2Mg}{\mu}$

27 A wooden block of mass M resting on a rough horizontal surface, is pulled with a force F at an angle with the horizontal. If μ is the coefficient of kinetic friction between block and the surface, then acceleration of the block is

(a) $\frac{F}{M}(\cos\phi + \mu \sin\phi) - \mu g$ (b) $\frac{F}{M}\sin\phi$
(c) $\mu F \cos\phi$ (d) $\mu F \sin\phi$

28 A block of mass 10 kg is placed at a distance of 5 m from the rear end of a long trolley as shown in the figure. The coefficient of friction between the block and the surface below is 0.2. Starting from rest, the trolley is given an uniform acceleration of 3 ms^{-2}. At what distance from the starting point will the block fall off the trolley? (take, $g = 10\ ms^{-2}$)

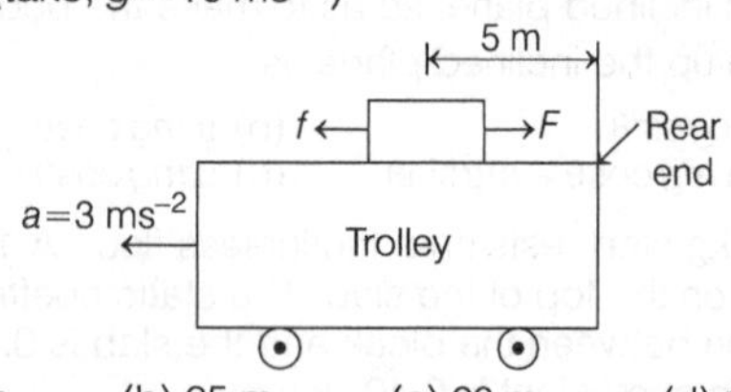

(a) 15 m (b) 25 m (c) 20 m (d) 10 m

29 A body starts from rest on a long inclined plane of slope 45°. The coefficient of friction between the body and the plane varies as $\mu = 0.3\ x$, where x is distance travelled down the plane. The body will have maximum speed (for $g = 10\ m s^{-2}$) when x is equal to → JEE Main (Online) 2013

(a) 9.8 m (b) 27 m (c) 12 m (d) 3.33 m

30 Given in the figure are two blocks A and B of weight 20 N and 100 N, respectively. These are being pressed against a wall by a force F as shown in figure.If the coefficient of friction between the blocks is 0.1 and between block B and the wall is 0.15, the frictional force applied by the wall in block B is → JEE Main 2015

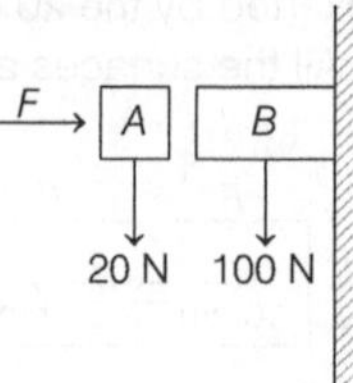

(a) 100 N (b) 80 N (c) 120 N (d) 150 N

31 A point particle of mass m, moves along the uniformly rough track PQR as shown in the figure. The coefficient of friction between the particle and the rough track equals μ. The particle is released, from rest , from the point P and it comes to rest at a point R. The energies, lost by the ball, over the parts, PQ and QR, of the track, are equal to each other, and no energy is lost when particle changes direction from PQ to QR. The values of the coefficient of friction μ and the distance $x(= QR)$, are respectively close to

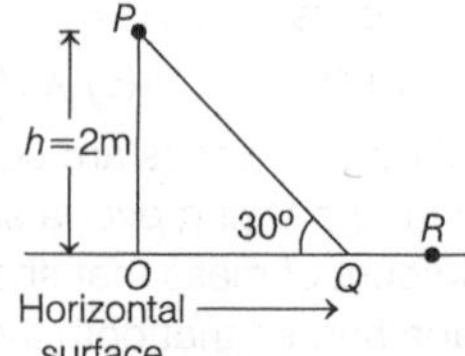

→ JEE Main 2016 (Offline)

(a) 0.2 and 6.5 m (b) 0.2 and 3.5 m
(c) 0.29 and 3.5 m (d) 0.29 and 6.5 m

32 Two masses $m_1 = 5$ kg and $m_2 = 10$ kg connected by an inextensible string over a frictionless pulley, are moving as shown in the figure. The coefficient of friction of horizontal surface is 0.15. The minimum weight m that should be put on top of m_2 to stop the motion is → JEE Main 2018

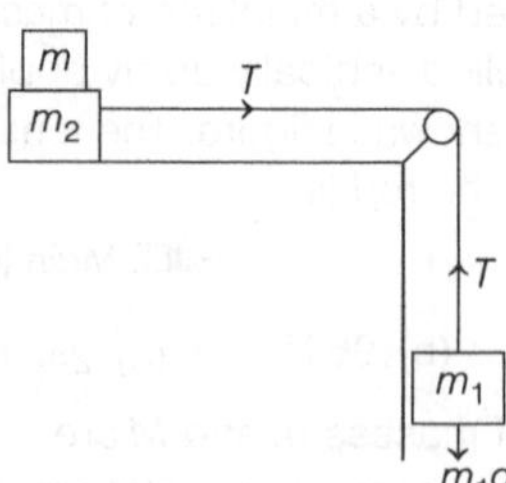

(a) 18.3 kg (b) 27.3 kg (c) 43.3 kg (d) 10.3 kg

33 The minimum force required to start pushing a body up a rough (frictional coefficient μ) inclined plane is F_1 while the minimum force needed to prevent it from sliding down is F_2. If the inclined plane makes an angle θ from the horizontal such that $\tan\theta = 2\mu$, then the ratio $\frac{F_1}{F_2}$ is

(a) 4 (b) 1 (c) 2 (d) 3

34 A box of mass 80 kg kept on a horizontal weighing machine of negligible mass, attached to a massless platform P that slides down at 37° incline. The weighing machine read 72 kg. Box is always at rest w.r.t. weighing machine. Then, coefficient of friction between the platform and incline is

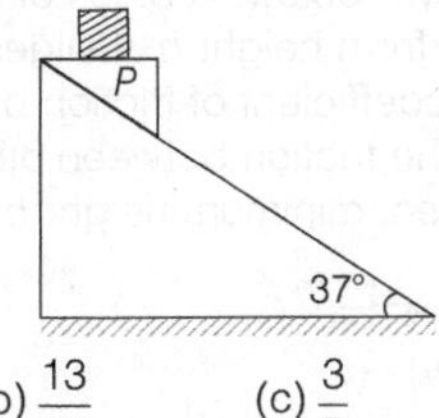

(a) $\frac{12}{13}$ (b) $\frac{13}{24}$ (c) $\frac{3}{5}$ (d) $\frac{11}{24}$

35 A block of mass m is placed against a vertical surface by a spring of unstretched length l. If the coefficient of friction between the block and the surface is μ, then choose the correct statement.

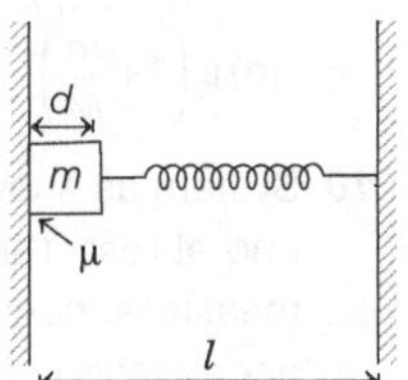

(a) If spring constant $k = \frac{2mg}{\mu d}$, block will not be in equilibrium.

(b) Minimum spring constant k_{min} to keep the block of mass m in equilibrium is $\frac{mg}{\mu d}$.

(c) If spring constant is $k = \frac{2mg}{\mu d}$, the normal reaction is $\frac{mg}{\mu}$.

(d) In the part (c), force of friction is $2mg$.

Direction (Q. Nos. 36-40) *Each of these questions contains two statements : Statement* I *(Assertion) and Statement* II *(Reason). Each of these questions also has four alternative choices, only one of which is the correct answer. You have to select one of the codes (a), (b), (c), (d) given below*

(a) Statement I is true, Statement II is true, Statement II is the correct explanation for Statement I

(b) Statement I is true, Statement II is true, Statement II is not the correct explanation for Statement I

(c) Statement I is true, Statement II is false

(d) Statement I is false, Statement II is true

36 Statement I When the car accelerates horizontally along a straight road, the accelerating force is given by the push of the rear axle on the wheels.

Statement II When the car accelerates, the rear axle rotates with a greater frequency.

37 Statement I It is easier to pull a heavy object than to push it on a level ground.

Statement II The magnitude of frictional force depends on the nature of the two surfaces in contact.

38 Statement I A cloth covers a table. Some dishes are kept on it. The cloth can be pulled out without dislodging the dishes from the table.

Statement II For every action there is an equal and opposite reaction.

39 Statement I A bullet is fired from a rifle. If the rifle recoils freely, the kinetic energy of the rifle is less than that of the bullet.

Statement II In the case of a rifle-bullet system, the law of conservation of momentum is violated.

40 Statement I Newton's second law is applicable on a body with respect to an inertial frame of reference.

Statement I In order to apply Newton's second law on a body observed from a non-inertial frame of reference. We apply line pseudo force an imaginary force.

DAY PRACTICE SESSION 2

PROGRESSIVE QUESTIONS EXERCISE

1 A flexible uniform chain of mass m and length l suspended vertically, so that its lower end just touches the surface of a table. When the upper end of the chain is released, it falls with each link coming to rest the instant it strikes the table. The force exerted by the chain on the table at the moment when y part of the chain has already rested on the table is

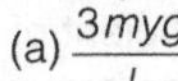

(a) $\frac{3myg}{l}$ (b) $\frac{3mg}{l}$ (c) $\frac{2}{3}\frac{mg}{l}$ (d) $\frac{1}{3}\frac{mg}{l}$

2 Two fixed frictionless inclined plane making the angles 30° and 60° with the vertical are shown in the figure. Two blocks A and B are placed on the two planes. What is the relative vertical acceleration of A with respect to B?

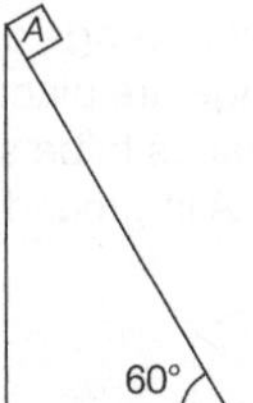

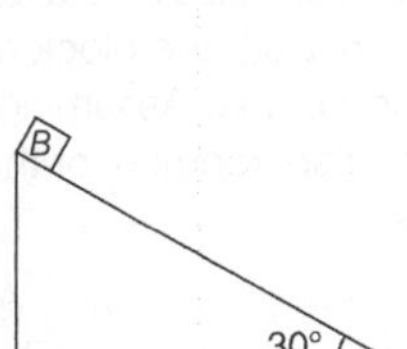

(a) 4.9 ms^{-2} in horizontal direction

(b) 9.8 ms^{-2} in vertical direction

(c) zero

(d) 4.9 ms^{-2} in vertical direction

3 If M is the mass of a rocket, r is the rate of ejection of gases with respect to the rocket, then acceleration of the rocket, $\frac{dv}{dt}$ is equal to

(a) $\frac{ru}{(M-rt)}$ (b) $\frac{(M-rt)}{ru}$ (c) $\frac{ru}{(M+rt)}$ (d) $\frac{ru}{M}$

4 A person 40 kg is managing to be at rest between two vertical walls by pressing one wall A by his hands and feet and B with his back. The coefficient of friction is 0.8 between his body and the wall. The force with which the person pushes the wall is

(a) 100 N (b) 50 N (c) 150 N (d) 200 N

5 An insect crawls up a hemispherical surface very slowly. The coefficient of friction between the insect and the surface is 1/3. If the line joining the centre of the hemispherical surface to the insect makes an angle α with the vertical, the maximum possible value of α is given by

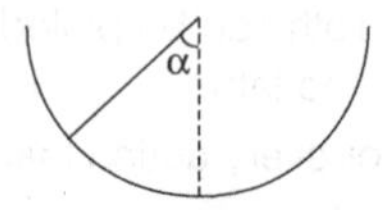

(a) $\cot\alpha = 3$ (b) $\sec\alpha = 3$
(c) $\operatorname{cosec}\alpha = 3$ (d) None of these

6 A given object takes n times more time to slide down a 45° rough inclined plane in comparison to slides down a perfectly smooth 45° incline. The coefficient of kinetic friction between the object and the incline is

(a) $\frac{1}{1-n^2}$ (b) $1-\frac{1}{n^2}$ (c) $\sqrt{1-\frac{1}{n^2}}$ (d) $\sqrt{\frac{1}{1-n^2}}$

7 When a body slides down from rest along a smooth inclined plane making an angle of 45° with the horizontal,it takes time T. When the same body slides down from rest along a rough inclined plane making the same angle and through the same distance,it is seen to take time pT, where p is some number greater than 1. The coefficient of friction between the body and the rough plane will be

(a) $1-p^2$ (b) $1-\frac{1}{p^2}$ (c) p^2-1 (d) p^2

8 Block A of mass m is placed over a wedge of the same mass m. Both the block and wedge are placed on a fixed inclined plane. Assuming all surfaces to be smooth. Then, displacement of the block A in ground frame in 1 s is

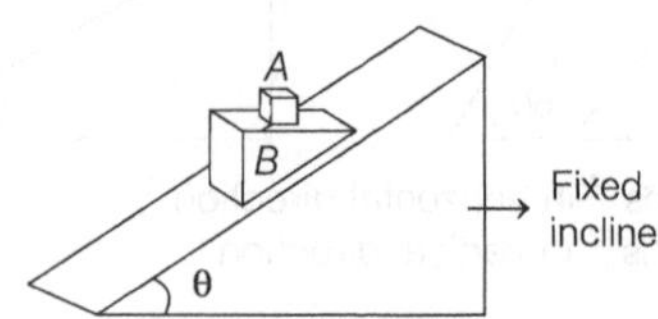

(a) $\left[\frac{g\sin^2\theta}{1+\sin^2\theta}\right]$ (b) $\left[\frac{g\cos^2\theta}{1+\cos^2\theta}\right]$
(c) $\left[\frac{g\tan^2\theta}{1+\tan^2\theta}\right]$ (d) $\left[\frac{g\sin^2\theta}{1+\sin\theta}\right]$

9 A carriage of mass M and length l is joined to the end of a slope as shown below. A block of mass m is released from the slope from height h. It slides till end of the carriage. The coefficient of friction between block and carriage is μ (the friction between other surfaces are negligible). Then, minimum height h is

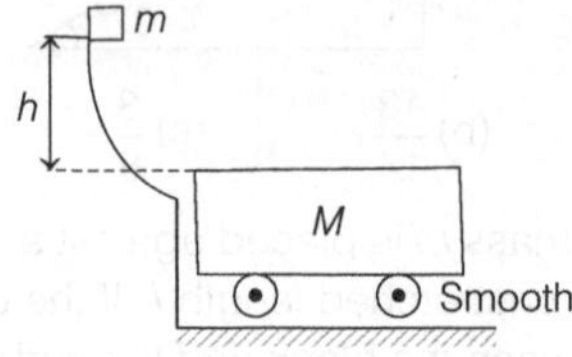

(a) $\mu\left(1+\frac{M}{m}\right)l$ (b) $\mu\left(2+\frac{m}{M}\right)l$
(c) $\mu\left(1+\frac{m}{M}\right)l$ (d) $2\mu\left(1+\frac{m}{M}\right)l$

10 System as shown in figure is in equilibrium and at rest. The spring and string are massless, now the string is cut. The acceleration of the masses $2m$ and m just after the string is cut, will be

(a) $\frac{g}{2}$ upwards, g downwards
(b) g upwards, $\frac{g}{2}$ downwards
(c) g upwards, $2g$ downwards
(d) $2g$ upwards, g downwards

11 A block of mass m is at rest under the action of a force F, acting against a wall, as shown in the figure. Which of the following statement is incorrect?

(a) $f = mg$ (where, f is the frictional force)
(b) $F = N$ (where, N is the normal force)
(c) F will not produce torque
(d) N will not produce torque

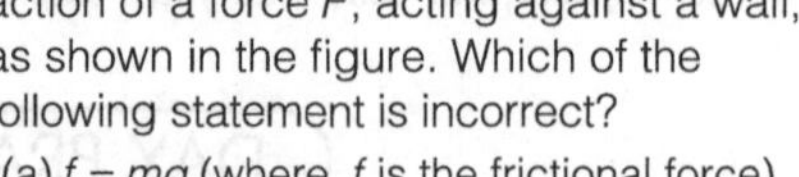

12 The upper-half of an inclined plane with an inclination ϕ, is perfectly smooth, while the lower half is rough. A body starting from rest at the top will again come to rest at the bottom, if the coefficient of friction for the lower half is given by

(a) $2\sin\phi$ (b) $2\cos\phi$ (c) $2\tan\phi$ (d) $\tan\phi$

13 A block of mass m is placed on a surface with a vertical cross-section given by $y = x^3/6$. If the coefficient of friction is 0.5, the maximum height above the ground at which the block can be placed without slipping is
→ JEE Main 2014

(a) 1/6 m (b) 2/3 m (c) 1/3 m (d) 1/2 m

14 A particle of mass m is at rest at the origin at time $t = 0$. It is subjected to a force $F(t) = F_0 e^{-bt}$ in the x-direction. Its speed $v(t)$ is depicted by which of the following curves?
→ JEE Main 2013

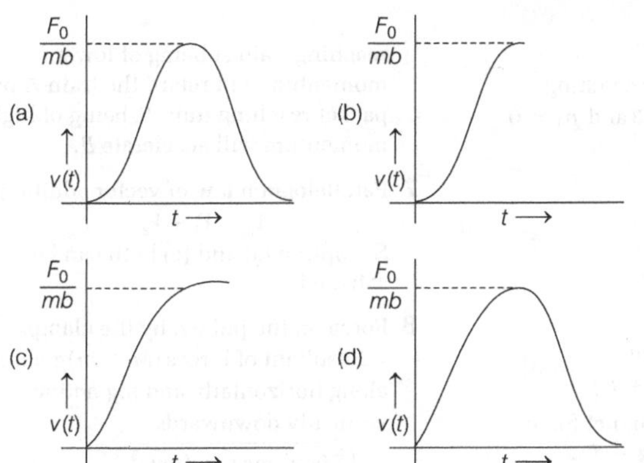

15 The figure shows the position-time (x-t) graph of one-dimensional motion of a body of mass 0.4 kg. The magnitude of each impulse is

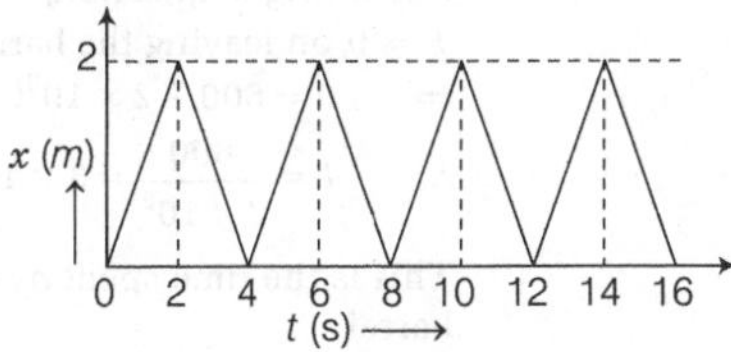

(a) 0.4N - s (b) 0.8 N - s
(c) 1.6N - s (d) 0.2 N - s

16 A worker is raising himself and the crate on which he stands with an acceleration of 5 ms^{-2} by a massless rope and pulley arrangement. Mass of the worker is 100 kg and that of the crate is 50 kg. If T is the tension in the rope and F be the force of contact between the worker and the floor and if $g = 10\,\text{ms}^{-2}$, then

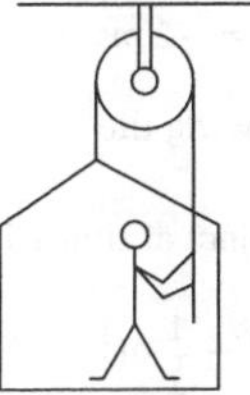

(a) $T = 2250$N, $F = 1125$N
(b) $T = 1125$N, $F = 2250$N
(c) $T = 1125$N, $F = 375$N
(d) $T = 1125$N, $F = 750$N

17 In the system shown in figure, masses of the blocks are such that when system is released, acceleration of pulley P_1 is a upwards and acceleration of block 1 is a_1, upwards. It is found that acceleration of block 3 is same as that of 1 both in magnitude and direction.

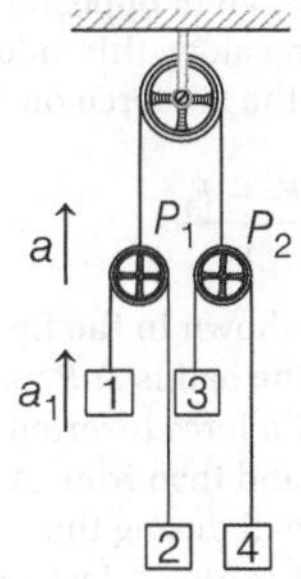

Given that, $a_1 > a > \frac{a_1}{2}$.

Match the following.

	Column I		Column II
A.	Acceleration of 2	1.	$2a + a_1$
B.	Acceleration of 4	2.	$2a - a_1$
C.	Acceleration of 2 w.r.t.3	3.	Upwards
D.	Acceleration of 2 w.r.t. 4	4.	Downwards

Codes

	A	B	C	D
(a)	2,3	1,4	4	3
(b)	2,4	1	4	1,3
(c)	4	1,3	2	1,2

(d) No above matching is correct

ANSWERS

SESSION 1	**1** (a)	**2** (d)	**3** (c)	**4** (c)	**5** (b)	**6** (a)	**7** (a,c)	**8** (b)	**9** (c)	**10** (d)
	11 (d)	**12** (a)	**13** (b)	**14** (b)	**15** (a)	**16** (d)	**17** (b)	**18** (b)	**19** (c)	**20** (a)
	21 (c)	**22** (d)	**23** (a)	**24** (d)	**25** (a)	**26** (b)	**27** (a)	**28** (a)	**29** (d)	**30** (c)
	31 (c)	**32** (b)	**33** (d)	**34** (b)	**35** (b)	**36** (a)	**37** (b)	**38** (b)	**39** (c)	**40** (a)
SESSION 2	**1** (a)	**2** (d)	**3** (a)	**4** (d)	**5** (a)	**6** (b)	**7** (b)	**8** (a)	**9** (c)	**10** (a)
	11 (d)	**12** (c)	**13** (a)	**14** (c)	**15** (b)	**16** (c)	**17** (a)			

Hints and Explanations

SESSION 1

1 According to polygon law, resultant of four forces, each of magnitude F_1 acting at an angle of 72°, is along the fifth side of the polygon taken an in opposite order. As F_2 is acting along this side of polygon, therefore the net force on the particle $= F_2 - F_1$

Acceleration $(a) = \dfrac{F_2 - F_1}{m}$

2 The situation is as shown in the figure. At an initial time, the ball is at P, then under the action of a force (exerted by hand) from P to A and then from A to B, let acceleration of ball during the motion from P to A is $a\,\text{ms}^{-2}$ [assumed to be constant] in an upward direction and velocity of ball at A is $v\ \text{ms}^{-1}$.

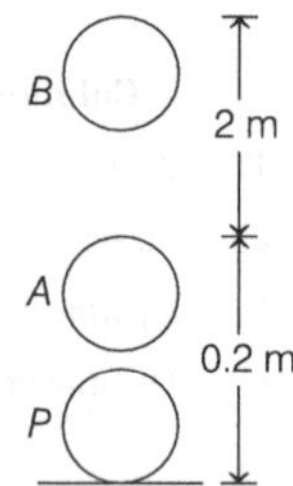

Then for PA, $v^2 = 0^2 + 2a \times 0.2$

For AB, $0 = v^2 - 2 \times g \times 2$

$\Rightarrow v^2 = 2g \times 2$

From above equations,

$$a = 10g = 100\ \text{ms}^{-2}$$

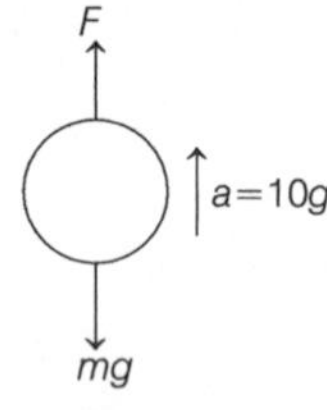

Then for PA, from FBD of ball is

$$F - mg = ma$$

[F is the force exerted by hand on the ball]

$\Rightarrow F = m(g + a) = 0.2(11g) = 22\text{N}$

3 This is the question based on impulse-momentum theorem.

$$|F\Delta t| = |\text{Change in momentum}|$$

$\Rightarrow F \times 0.1 = |p_f - p_i|$

As the ball will stop after catching,

$p_i = mv_i = 0.15 \times 20 = 3$ and $p_f = 0$

$\Rightarrow F \times 0.1 = 3$

$\Rightarrow F = 30\text{N}$

4 For an upward motion

Retarding force $= mg + F$

Retardation $(a) = \dfrac{mg + F}{m}$

Distance, $s = \dfrac{v^2}{2a} = \dfrac{v^2 m}{2(mg + F)}$ ... (i)

For the downward motion, net force $= mg - F$

$\therefore$ Acceleration $(a') = \dfrac{mg - F}{m}$

Distance $(s') = \dfrac{v'^2}{2a'} = \dfrac{v'^2 m}{2(mg - F)}$ As, $s = s'$

$\therefore v' = v\sqrt{\dfrac{mg - F}{mg + F}}$

5 Given, $u = 5\text{ms}^{-1}$, along positive x-direction

$F = -0.4\text{ N}$, along negative x-direction

$M = 200\text{g} = 0.2\text{ kg}$

Thus, the acceleration

$$a = \frac{F}{M} = -\frac{0.4}{0.2} = -2\ \text{ms}^{-2}$$

The negative sign showing the retardation.

The position of the object at time t is given by

$$x = x_0 + ut + \frac{1}{2}at^2$$

At $t = 0$, the body is at $x = 0$, therefore $x_0 = 0$.

Hence, $x = ut + \dfrac{1}{2}at^2$

Since, the force acts during the time interval from $t = 0$ to $t = 10$ s, the motion is accelerated only within this time interval. The position of the body at $t = 2.5$ s is given by

$$x = 5 \times 2.5 + \frac{1}{2} \times (-2) \times (2.5)^2 = 6.25\text{m}$$

6 Initially, the momentum of the packet in train A is more than in train B. When packets are changed, the packet reaching train A being of lower momentum will retard the train A but packet reaching train B, being of higher momentum will accelerate B.

7 Parallelogram law of vector addition

$$\mathbf{F}_3 = \mathbf{F}_1 + \mathbf{F}_2$$

So, option (a) and (c) both can be satisfied.

8 Force on the pulley, by the clamp = Resultant of forces $(M + m)g$ acting along horizontally and mg acting vertically downwards

$= \sqrt{(Mg + mg)^2 + (mg)^2}$

$= \sqrt{[(M + m)^2 + m^2]}g$

9 As, $F = 600 - 2 \times 10^5 t$

At, $t = 0$, $F = 600\text{N}$

According to question,

$F = 0$, on leaving the barrel,

$\Rightarrow 0 = 600 - 2 \times 10^5 t$

$\therefore t = \dfrac{600}{2 \times 10^5} = 3 \times 10^{-3}\text{s}$

This is the time spent by the bullet in the barrel

Average force $= \dfrac{600 + 0}{2} = 300\text{N}$

Average impulse imparted $= F \times t$

$= 300 \times 3 \times 10^{-3} = 0.9$ N-s

10 Different forces involved are as shown in the figure.

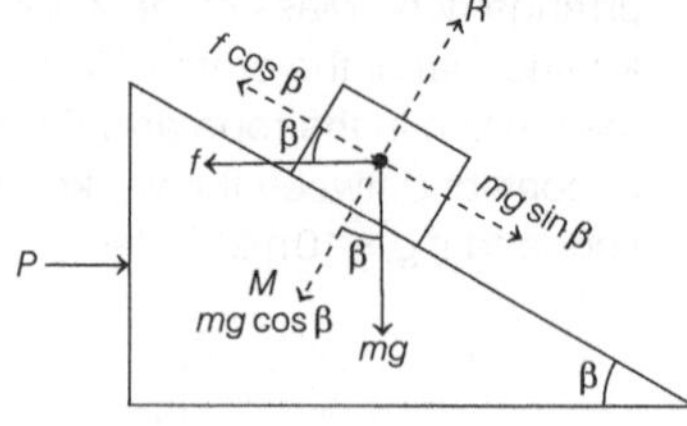

Observing the figure, we have acceleration of the system,

$$a = \frac{P}{M + m}$$

Force on block of mass $(m) = \dfrac{Pm}{M + m}$

If f is pseudo force on m in the direction opposite to force P, then

$$f = \frac{Pm}{M + m}$$

As it is clear from the figure,

$$f\cos\beta = mg\sin\beta$$

$$\frac{Pm}{(M+m)}\cos\beta = mg\sin\beta$$

$$P = g(M+m)\frac{\sin\beta}{\cos\beta}$$

or, $$P = (M+m)\,g\tan\beta$$

11 Net force on the rod,

$$f = 32 - 20 = 12\text{N}$$

Acceleration of the rod

$$= \frac{f}{m} = \frac{12}{3} = 4\,\text{ms}^{-2}$$

Equation of motion of the 10 cm part is

$$F - 20 = m \times a = 1 \times 4,$$
$$F = 4 + 20 = 24\text{ N}$$

Similarly, equation of motion of 20 cm part is

$$32 - F = m \times a = 2 \times 4,$$
$$F = 32 - 8 = 24\text{ N}$$

12 On releasing, the motion of the system will be according to the figure

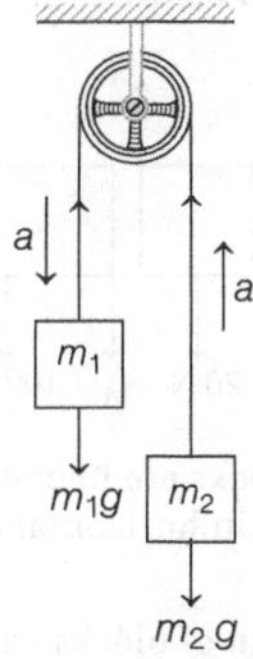

The equations of motion of blocks are,

$$m_1g - T = m_1a \quad \text{... (i)}$$

and $$T - m_2g = m_2a \quad \text{... (ii)}$$

On solving,

$$a = \left(\frac{m_1 - m_2}{m_1 + m_2}\right)g \quad \text{... (iii)}$$

Here, $m_1 = 5\text{ kg}$,

$m_2 = 4.8\text{ kg}$

$g = 9.8\text{ ms}^{-2}$

$$\therefore \quad a = \left(\frac{5 - 4.8}{5 + 4.8}\right) \times 9.8$$

$$= \frac{0.2}{9.8} \times 9.8$$

$$= 0.2\,\text{ms}^{-2}$$

13 As, $a = \frac{(m_1 - m_2)g}{m_1 + m_2} = \frac{g}{8}$,

$$\frac{m_1 - m_2}{m_1 + m_2} = \frac{1}{8}$$

or $8m_1 - 8m_2 = m_1 + m_2$

or $7m_1 = 9m_2$

$$\frac{m_1}{m_2} = \frac{9}{7}$$

14 As, $s = ut + \frac{1}{2}at^2$

$$\Rightarrow 50 = 0 \times 5 + \frac{1}{2} \times a \times (5)^2$$

$$\therefore \quad a = \frac{100}{25} = 4\text{ ms}^{-2}$$

Let, mass of one become m_1 and that of other m_2, where $m_1 > m_2$. As m_1 moves downwards with acceleration $a = 4\text{ ms}^{-2}$

$$a = \left(\frac{m_1 - m_2}{m_1 + m_2}\right)g$$

So, $$4 = \left(\frac{m_1 - m_2}{m_1 + m_2}\right)10$$

$$\left(\frac{m_1 - m_2}{m_1 + m_2}\right) = \frac{a}{g} = \frac{4}{10} = \frac{2}{5}$$

$\therefore$ Percentage of mass transferred

$$= \left(\frac{m_1 - m_2}{m_1 + m_2}\right) \times 100$$

$$= \frac{2}{5} \times 100 = 40\%$$

15 As, $mg - R = ma$

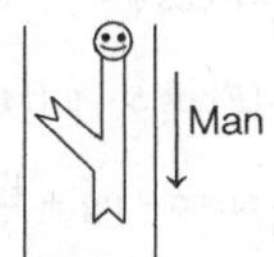

$$mg - \eta mg = ma$$

$$mg(1 - \eta) = ma \Rightarrow a = g(1 - \eta)$$

16 For block of mass M_1,

$$\frac{480 - T_1 - 20\,g}{20} = a$$

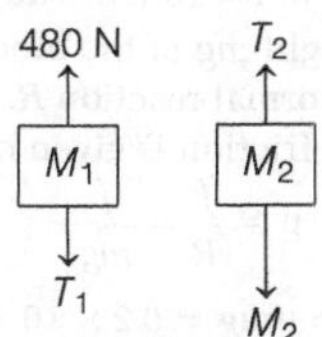

Also, for block of mass M_2,

$$\frac{T_2 - 12\,g}{12} = a$$

Since, a is common for all the individuals of the system

$$\Rightarrow \quad \frac{480 - T_1 - 20g}{20} = \frac{T_2 - 12g}{12}$$

After taking $g = 10\text{ ms}^{-2}$ this gives

$$5T_2 + 3T_1 = 1440 \quad \text{...(i)}$$

Now, for the metal rod, tension at both of its end are dissimilar and

$$T_1 - T_2 = 80 \quad \text{...(ii)}$$

$(\because g = 10\text{ ms}^{-2})$

Now, from Eqs. (i) and (ii), we get

$$T_1 = 230\text{N} \text{ and } T_2 = 150\text{N}$$

$\therefore$ Tension at mid-point

$$= T_1 - 4g = 190\text{ N}$$

17 Tension, $T = \left(\frac{2m_1m_2}{m_1 + m_2}\right)g$

$$= \left(\frac{2m \times 2m}{m + 2m}\right)g$$

(where, $m_1 = m$ and $m_2 = 2m$)

$$= \frac{4}{3}mg$$

$$\therefore \quad \text{Stress} = \frac{\text{Force (Tension)}}{\text{Area}}$$

$$= \frac{\frac{4}{3}mg}{A} = \frac{4\,mg}{3A}$$

18 It is known that the thrust

$$= -v\left(\frac{dM}{dt}\right) = -v(\alpha v)$$

Hence, the retardation produced

$$= \frac{\text{thrust}}{\text{mass}} = -\frac{\alpha v^2}{M}$$

19 When ball dropped, acceleration of the ball is g as will be observed by a man standing stationary on the ground. The man inside the lift is having its own downward acceleration, a. Therefore, relative acceleration of the ball as observed by the man in the lift will must be $= (g - a)$.

20 When the lift is stationary, then

$$R = mg$$
$$49 = m \times 9.8$$
$$m = \frac{49}{9.8}\text{ kg} = 5\text{ kg}$$

If a is the downward acceleration of the lift then, $R = m(g - a)$

$$= 5(9.8 - 5) = 24\text{ N}$$

21 The situation is given in figure below

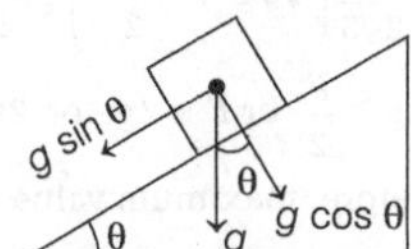

From equation of motion,

$$\frac{1}{2}g\sin\theta\, t^2 = l$$

$$\Rightarrow \quad t = \sqrt{\frac{2l}{g\sin\theta}}$$

22 To move the body up the inclined plane, the force required,

$$= mg\sin\theta + \mu R$$
$$= mg\sin\theta + \mu mg\cos\theta$$

23 Limiting force of friction of block on slab $\mu\, m_1 g = 0.6 \times 10 \times 9.8 = 58.8\text{N}$
Since, the applied force = 100N on block, which is greater than the force of limiting friction, the block will accelerate on the slab, due to which, the force acting on the slab will be that due to the kinetic friction ($\mu_k m_1 g$).
Hence, acceleration of the slab,

$$a = \frac{\mu_k m_1 g}{m_2} = \frac{0.4 \times 10 \times 9.8}{40} = 0.98\ \text{ms}^{-2}$$

24 Let, R be the normal contact force by wall on the block.

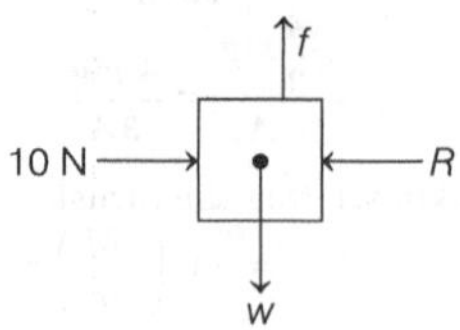

$R = 10\text{N},\ f_L = w$ and $f = \mu R$

$\therefore\quad \mu R = w$ or $w = 0.2 \times 10 = 2\text{N}$

25 Free body diagram (FBD) of the block (shown by a dot) is as shown in the figure

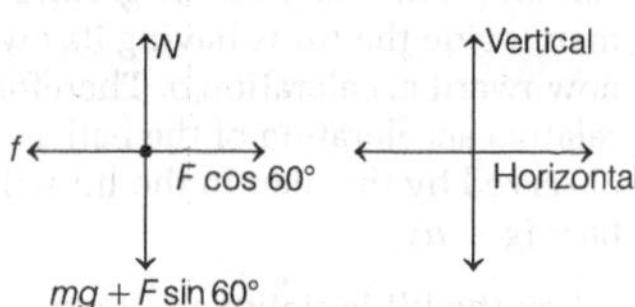

For vertical equilibrium of the block

$$N = mg + F\sin 60^\circ$$
$$= \sqrt{3}g + \sqrt{3}\frac{F}{2} \quad \ldots\text{(i)}$$

For no motion, force of friction

$$f \geq F\cos 60^\circ$$

or $\quad \mu N \geq F\cos 60^\circ$

or $\quad \dfrac{1}{2\sqrt{3}}\left(\sqrt{3}g + \dfrac{\sqrt{3}F}{2}\right) \geq \dfrac{F}{2}$

or $\quad g \geq \dfrac{F}{2}$ or $F \leq 2g$ or 20N

Therefore, maximum value of F is 20 N.

26 Given, mass of the block = M
Coefficient of friction between the block and the wall = μ
Let, a force F be applied on the block to hold the block against the wall. The normal reaction of mass be N and force of friction acting upward be f. In equilibrium, vertical and horizontal forces should be balanced separately.

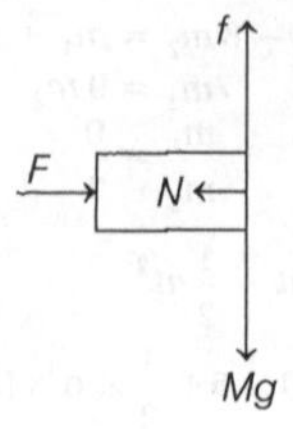

$$f = Mg \quad \ldots\text{(i)}$$

and $\quad F = N \quad \ldots\text{(ii)}$

But force of friction $(f) = \mu N$

$$= \mu F \quad \ldots\text{(iii)}$$

[using Eq. (ii)]

From Eqs. (i) and (iii), we get

$$\mu F = Mg \quad \text{or} \quad F = \frac{Mg}{\mu}$$

27 Here, $R = Mg - F\sin\phi$

$\therefore\quad f = \mu R \approx \mu(Mg - F\sin\phi)$

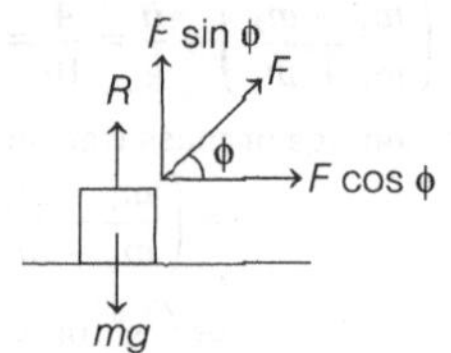

Net force, $F\cos\phi - f = Ma$

$$\therefore\quad a = \frac{1}{M}[F\cos\phi - f]$$
$$\Rightarrow a = \frac{1}{M}[F\cos\phi - \mu(Mg - F\sin\phi)]$$
$$= \frac{F}{M}\cos\phi - \mu g + \frac{\mu F}{M}\sin\phi$$
$$= \frac{F}{M}(\cos\phi + \mu\sin\phi) - \mu g$$

28 Given, acceleration of the trolley $(a) = 3\ \text{ms}^{-2}$.
Therefore, the force acting on the block is $F = ma = 10 \times 3 = 30\,\text{N}$.
The weight mg of the block is balanced by the normal reaction R. The force of limiting friction is given by

$$\mu = \frac{f}{R} = \frac{f}{mg}$$
$$f = \mu mg = 0.2 \times 10 \times 10 = 20\,\text{N}$$

The net force on the block is towards right and is given by

$$F' = F - f = 30 - 20 = 10\,\text{N}$$

So, $\quad a' = \dfrac{F'}{m} = \dfrac{10}{10} = 1\ \text{ms}^{-2}$.

Let, t be the time taken for the block to fall off from the rear end for the trolley. Then, the block has to travel a distance $s' = 5\text{m}$ to fall off. Now, since the trolley starts from rest. So, $u = 0$ and using $s = ut + \dfrac{1}{2}at^2$, we can determine t as $\sqrt{10}$ s.

The distance covered by the trolley in this time,

$$s' = ut + \frac{1}{2}at^2 = 0 + \frac{1}{2} \times 3 \times 10 = 15\text{m}$$

29 From Newton's second law,

$$\frac{mg\sin\theta - \mu\, mg\cos\theta}{m} = a$$

Now, distance covered by the particle,

$$v^2 = u^2 + 2as$$
$$\Rightarrow v = \sqrt{2\left(\frac{mg\sin\theta - \mu\, mg\cos\theta}{m}\right)x}$$
$$= \sqrt{2\,gx\sin\theta - 0.6\,x^2\,g\cos\theta}$$

v should be maximum when $\dfrac{dv}{dx} = 0$

$$\Rightarrow \frac{d\sqrt{2gx\sin\theta - 0.6\,x^2\,g\cos\theta)}}{dx} = 0$$

By differentiating, we get $x = 3.33$ m

30 In vertical direction, weights are balanced by frictional forces.
Consider FBD of block A and B as shown in diagram below.

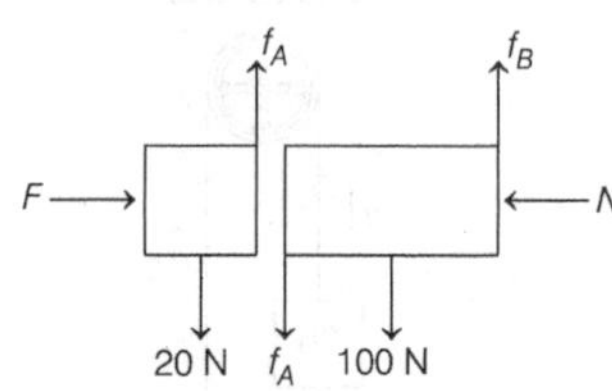

As the blocks are in equilibrium, balance forces are in horizontal and vertical direction.

The system of blocks $(A + B)$

$$F = N$$

For block A, $f_A = 20\text{N}$

and for block B,

$$f_B = f_A + 100 = 20 + 100 = 120\text{ N}$$

31 Energy lost over path $PQ = \mu\, mg\cos\theta \times 4$

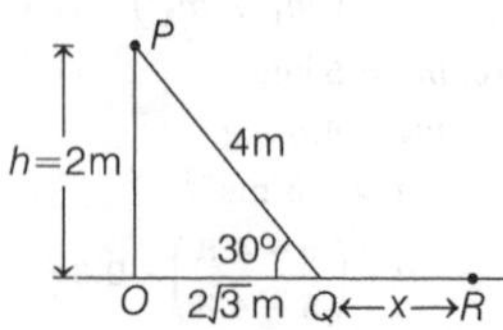

Energy lost over path $QR = \mu\, mg\, x$
i.e. $\mu\, mg\cos 30^\circ \times 4 = \mu\, mg\, x$

$(\because \theta = 30^\circ)$

$$x = 2\sqrt{3} = 3.45\text{m}$$

From Q to R energy loss is half of the total energy loss.

i.e. $\mu\, mgx = \dfrac{1}{2} \times mgh \Rightarrow \mu = 0.29$

The values of the coefficient of friction μ and the distance $x(= QR)$ are 0.29 and 3.5.

32 Motion stops when pull due to $m_1 \le$ force of friction between m and m_2 and surface.

$\Rightarrow \quad m_1 g \le \mu(m_2 + m)g$

$\Rightarrow \quad 5 \times 10 \le 0.15(10 + m) \times 10$

$\Rightarrow \quad m \ge 23.33$ kg

Here, nearest value is 27.3 kg

So, $m_{min} = 27.3$ kg

33 $F_1 = mg\,(\sin\theta + \mu\cos\theta)$ [as body just in position to move up, friction force downward]

$F_2 = mg(\sin\theta - \mu\cos\theta)$ [as body just in position to slide down, friction upward]

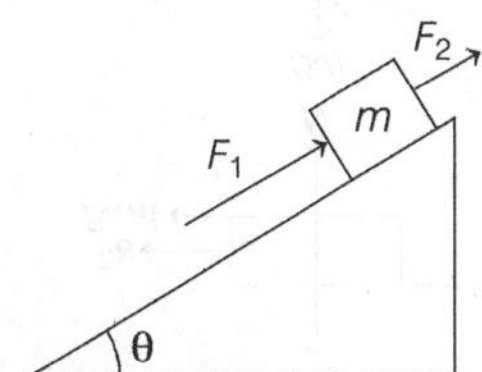

$$\therefore \frac{F_1}{F_2} = \frac{\sin\theta + \mu\cos\theta}{\sin\theta - \mu\cos\theta}$$

$$= \frac{\tan\theta + \mu}{\tan\theta - \mu} = \frac{2\mu + \mu}{2\mu - \mu} = 3$$

34 Here, $(80 - 72)g = ma_y$

or $\quad a_y = 1\ \text{m/s}^2$

$\therefore a\sin 37° = 1\ \text{m/s}^2$ or $a = \frac{5}{3}\ \text{m/s}^2$

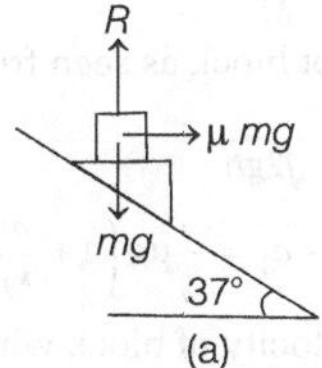

(a)

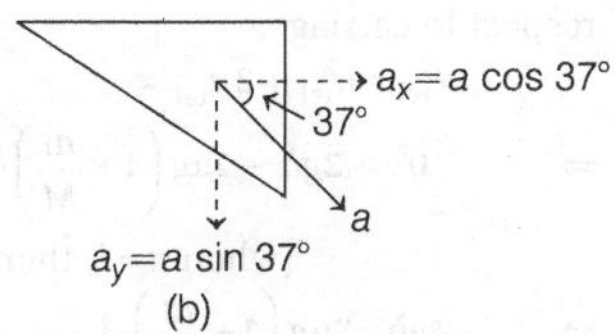

(b)

Now, we apply Newton's second law of motion on the box in the direction of acceleration,

$$mg\sin 37° - \mu\, mg\cos 37° = m \times \frac{5}{3}$$

or $\mu = \frac{13}{24}$

Hence, (b) is the correct option.

35 Free body diagram of the block is

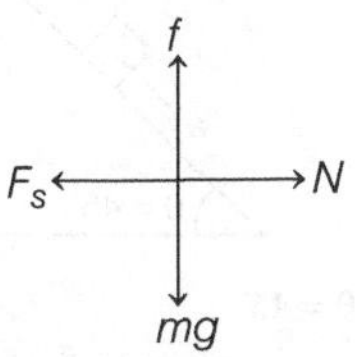

Here, $N = F_s = kd$

and $mg = f \le \mu N = \mu kd$

or $k \ge \frac{mg}{\mu d}$

Hence, (b) is the correct option.

36 When a car accelerates, the engine rotates the rear axle which exerts a push on the wheels to move.

37 Both Statements are correct. But Statement II does not explain correctly, Statement I.

Correct explanation is there is increase in normal reaction when the object is pushed and there is decrease in normal reaction when the object is pulled (but strictly, not horizontally).

38 The cloth can be pulled out without dislodging the dishes from the table due to law of inertia, which is Newton's first law. While, Statement II is true, but it is Newton's third law.

39 If the bullet is fired from the rifle, the momentum of bullet-rifle system is conserved.

It means, $\quad M_b v_b = M_r v_r \quad$ ... (i)

and $$\frac{E_{k(b)}}{E_{k(r)}} = \frac{\frac{1}{2}M_b v_b^2}{\frac{1}{2}M_r v_r^2}$$

$$= \frac{M_r}{M_b}$$

As, $M_r > M_b$ (mass of rifle is greater than the mass of bullet).

Hence, $E_{k(b)} > E_{k(r)}$. So, the kinetic energy of bullet is greater than the kinetic energy of rifle.

40 In order to apply Newton's second law on a body observed from a non-inertial frame of reference pseudo force is considered in a direction opposite to real acceleration.

SESSION 2

1 Suppose, F = force on the table due to the weight of the chain on the table + momentum of the chain transmitted on the table

$\Rightarrow \quad F = F_1 + F_2$ (Let)

Now, $\quad F_1 = \frac{m}{l} yg,$

$$dp = dmv = \frac{m}{l} dy\,\sqrt{2gy}$$

$$\frac{dp}{dt} = F_2 = \frac{m}{l}\frac{dy}{dt}\sqrt{2gy}$$

$$= \frac{m}{l}(2gy)\left[\frac{dy}{dt} = \sqrt{2gy}\right]$$

$$\therefore \quad F = \frac{m}{l} yg + \frac{2myg}{l}$$

$$= \frac{3myg}{l}$$

2 Force applying on the block

$$F = mg\sin\theta$$

or $\quad mg\sin\theta = ma$

$\therefore \quad a = g\sin\theta$

where, a is along the inclined plane.

$\therefore$ Vertical component of acceleration is $g\sin^2\theta$.

$\therefore$ Relative vertical acceleration of A with respect to B is

$$g\,[\sin^2 60° - \sin^2 30°]$$

$$= \frac{g}{2} = 4.9\ \text{ms}^{-2}$$

[in vertical direction]

3 Here, initial mass of the rocket = M

$$\frac{dm}{dt} = r$$

Relative velocity of gases w.r.t. rocket $= v$

then, acceleration of the rocket

$$a = \frac{F}{m} = \frac{u(dm/dt)}{\left(M - \frac{dm}{dt} \times t\right)}$$

$$= \frac{ur}{(M - rt)}$$

4 Balanced horizontal force, $N_1 = N_2$

Balanced vertical force, $2\mu N_1 = mg$

$$\mu N_1 = \frac{mg}{2} = \frac{40 \times 10}{2}$$

$$= 200\ \text{N}$$

$\therefore$ Man pushes the wall with 200 N

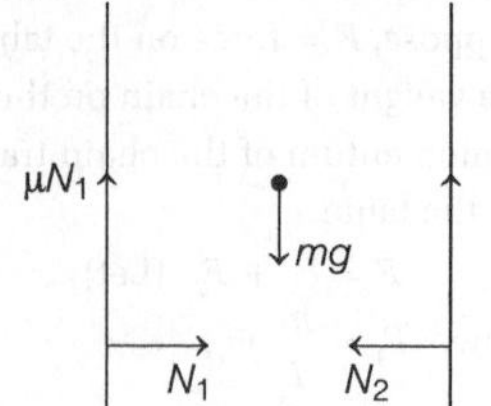

5 As, it is clear from the figure

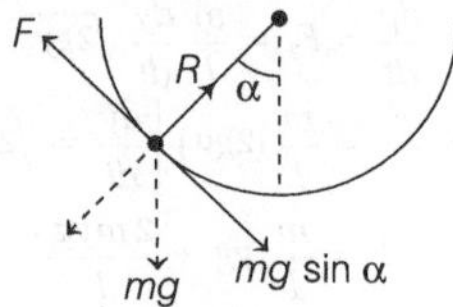

$$F = mg\sin\alpha$$

and $R = mg\cos\alpha$

$\Rightarrow \quad \frac{F}{R} = \tan\alpha$

i.e. $\mu = \tan\alpha = \frac{1}{3}$

$\Rightarrow \cot\alpha = 3$

6 On smooth inclined plane, the acceleration of body,

$$a = g\sin\theta$$

If s be the distance travelled, then

$$s = \frac{1}{2}g\sin\theta t_1^2 \quad \ldots(i)$$

On rough inclined plane, the acceleration is given by

$$a = g\sin\theta - \mu g\cos\theta$$

$$\therefore \quad s = \frac{1}{2}(g\sin\theta - \mu g\cos\theta)t_2^2 \quad \ldots(ii)$$

From Eqs. (i) and (ii), we get

$$\frac{t_2^2}{t_1^2} = \frac{\sin\theta}{g\sin\theta - g\cos\theta}$$

But $t_1 = nt_1$ or $\frac{t_2^2}{t_1^2} = n^2$

$$\therefore \quad n^2 = \frac{\sin\theta}{\sin\theta - \mu\cos\theta}$$

Solving, we get

$$\mu = \frac{n^2-1}{n^2} \times \frac{\sin\theta}{\cos\theta}$$

$$= \frac{n^2-1}{n^2} \times \frac{\sin 45^\circ}{\cos 45^\circ} = \frac{n^2-1}{n^2} \times \left(\frac{\frac{1}{\sqrt{2}}}{\frac{1}{\sqrt{2}}}\right)$$

$$= \frac{n^2-1}{n^2} = 1 - \frac{1}{n^2}$$

$$\mu = 1 - \frac{1}{n^2}$$

7 **On smooth inclined plane** Acceleration of a body sliding down a smooth inclined plane, $a = g\sin\theta$

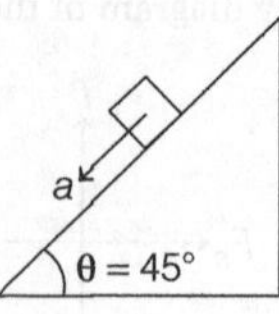

Here, $\theta = 45^\circ$

$$\therefore \quad a = g\sin 45^\circ = \frac{g}{\sqrt{2}}$$

Let the travelled distance be s.
Using equation of motion,

$$s = ut + \frac{1}{2}at^2, \text{ we get}$$

$$s = 0.t + \frac{1}{2}\frac{g}{\sqrt{2}}T^2 \text{ or } s = \frac{gT^2}{2\sqrt{2}} \quad \ldots(i)$$

On rough inclined plane Acceleration of the body

$$a = g(\sin\theta - \mu\cos\theta)$$

$$= g(\sin 45^\circ - \mu\cos 45^\circ) = \frac{g(1-\mu)}{\sqrt{2}}$$

$$\left(\text{As}, \sin 45^\circ = \cos 45^\circ = \frac{1}{\sqrt{2}}\right)$$

Again using equation of motion

$$s = ut + \frac{1}{2}at^2, \text{ we get}$$

$$s = 0(pT) + \frac{1}{2}\frac{g(1-\mu)}{\sqrt{2}}(pT)^2$$

$$\text{or} \quad s = \frac{g(1-\mu)p^2T^2}{2\sqrt{2}} \quad \ldots(ii)$$

From Eqs. (i) and (ii), we get

$$\frac{gT^2}{2\sqrt{2}} = \frac{g(1-\mu)p^2T^2}{2\sqrt{2}}$$

or $(1-\mu)p^2 = 1$

or $1 - \mu = \frac{1}{p^2}$ or $\mu = \left(1 - \frac{1}{p^2}\right)$

8 Free body diagram for A and B is

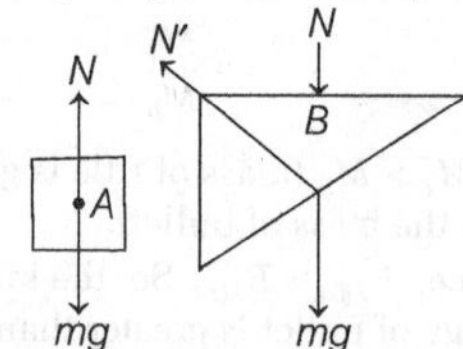

For A

$$mg - N = m(a\sin\theta) \quad \ldots(i)$$

[as block has only vertically downward acceleration]

For B

$$(N + mg)\sin\theta = ma \quad \ldots(ii)$$

On solving Eqs. (i) and (ii), we get

$$a = \left(\frac{2g\sin\theta}{1+\sin^2\theta}\right)$$

The acceleration of block A is

$$a_A = a\sin\theta = \frac{2g\sin^2\theta}{1+\sin^2\theta}$$

$\therefore$ Displacement of block A in 1 s is

$$s = 0.1 + \frac{1}{2}\left(\frac{2g\sin^2\theta}{1+\sin^2\theta}\right) \times (1)^2$$

$$\text{or} \quad s = \left[\frac{g\sin^2\theta}{1+\sin^2\theta}\right]$$

Hence, (a) is the correct option.

9

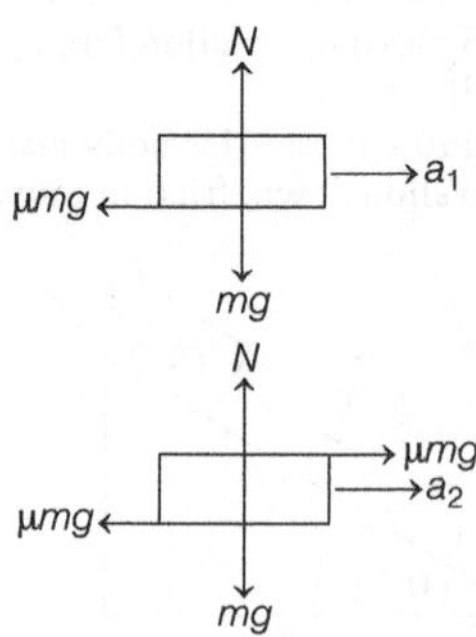

Velocity of block, just before reaching the carriage is

$$v = \sqrt{2gh}$$

Now, acceleration of block,

$$a_1 = \frac{-\mu mg}{m} = -\mu g$$

Acceleration of carriage,

$$a_2 = \frac{\mu mg}{M}$$

At $t = 0$, motion of block as seen from the carriage is

$$u_{rel} = v = \sqrt{2gh}$$

and $a_{rel} = a_1 - a_2 = -\mu g\left(1 + \frac{m}{M}\right)$

Now, relative velocity of block when block moves through distance x with respect to carriage,

$$v_{rel}^2 = u_{rel}^2 + 2a_{rel}x$$

$$\Rightarrow \quad 0^2 = 2gh - 2\mu g\left(1 + \frac{m}{M}\right) \cdot l$$

(when $x = l$, then $v_{rel} = 0$)

$$\Rightarrow \quad 2gh = 2\mu g\left(1 + \frac{m}{M}\right) \cdot l$$

$$\Rightarrow \quad h = \mu\left(1 + \frac{m}{M}\right) \cdot l$$

Hence, (c) is the correct option.

10 Initially under the equilibrium of mass m, $T = mg$

Now, the string is cut. Therefore, $T = mg$ force is decreased on the mass m upwards and downwards on the mass $2m$.

$$\therefore \quad a_m = \frac{mg}{m} = g \quad \text{(downwards)}$$

$$\text{and} \quad a_{2m} = \frac{mg}{2m} = \frac{g}{2} \quad \text{(upwards)}$$

11 This is the equilibrium of coplanar forces.

Hence, $\Sigma F_x = 0$

$\therefore \quad F = N$

$\therefore \quad \Sigma F_y = 0, \; f = mg, \; \Sigma \tau_c = 0$

$\therefore \quad \tau_N + \tau_f = 0$

Since, $\quad \tau_f \neq 0$

$\therefore \quad \tau_N \neq 0$

Thus, N will produce torque.

12 According to the work-energy theorem,

$$\Sigma W = \Delta K = 0$$

$\Rightarrow$ Work done by friction + Work done by gravity = 0

$$\Rightarrow -(\mu \, mg\cos\phi)\frac{l}{2} + mgl\sin\phi = 0$$

or $\frac{\mu}{2}\cos\phi = \sin\phi$ or $\mu = 2\tan\phi$

13 A block of mass m is placed on a surface with a vertical cross-section, then

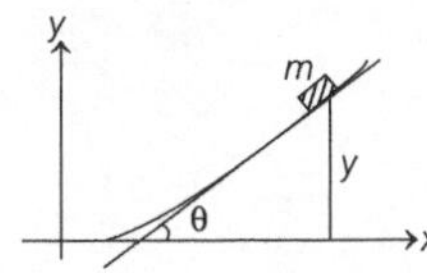

$$\tan\theta = \frac{dy}{dx} = \frac{d\left(\frac{x^3}{6}\right)}{dx} = \frac{x^2}{2}$$

At limiting equilibrium, we get

$$\mu = \tan\theta$$

$$0.5 = \frac{x^2}{2} \Rightarrow x^2 = 1 \Rightarrow x = \pm 1$$

Now, putting the value of x in $y = \frac{x^3}{6}$, we get

When $x = 1$ $\quad y = \frac{(1)^3}{6} = \frac{1}{6}$

When $x = -1$ $\quad y = \frac{(-1)^3}{6} = \frac{-1}{6}$

So, the maximum height above the ground at which the block can be placed without slipping is 1/6 m.

14 As the force is exponentially decreasing, so it's acceleration, i.e. rate of increase of velocity will decrease with time. Thus, the graph of velocity will be an increasing curve with decreasing slope with time.

$$a = \frac{F}{m} = \frac{F_0}{m}e^{-bt} = \frac{dv}{dt}$$

$$\Rightarrow \int_0^v dv = \int_0^t \frac{F_0}{m}e^{-bt}$$

$$\Rightarrow \quad v = \frac{F_0}{m}\left(\frac{1}{-b}\right)e^{-bt}\Bigg|_0^t = \frac{F_0}{mb}e^{-bt}\Bigg|_t^0$$

$$= \frac{F_0}{mb}(e^0 - e^{-bt})$$

$$= \frac{F_0}{mb}(1 - e^{-bt})$$

with $v_{\max} = \frac{F_0}{mb}$ $[a + t = \infty]$

15 From the graph, it is a straight line so, motion is uniform because of impulse direction of velocity changes as can be seen from the slope of the graph.

Initial velocity, $v_1 = \frac{2}{2} = 1 \text{ ms}^{-1}$

Final velocity, $v_2 = -2/2 = -1 \text{ ms}^{-1}$

$$p_i = mv_1 = 0.4 \text{ N-s}$$

and $\quad p_f = mv_2 = -0.4 \text{ N-s}$

Now, impulse,

$$J = p_f - p_i = -0.4 - 0.4 = -0.8 \text{ N-s}$$

$$\Rightarrow \quad |J| = 0.8 \text{ N-s}$$

16 Free body diagram for crate

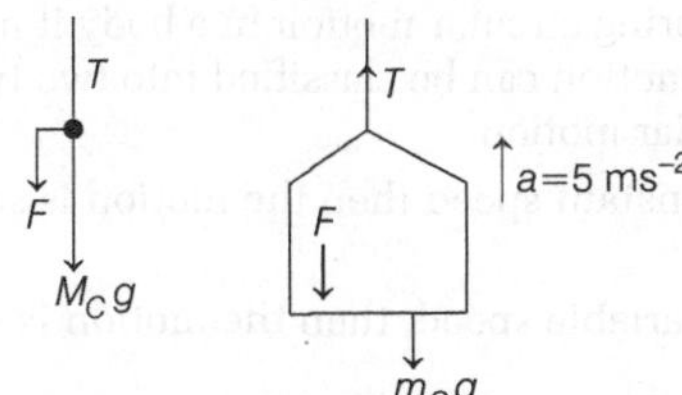

So, for vertical equilibrium of the crate.

$$T - F - M_c g = M_c a$$

[$\because M_c$ = mass of crate = 50 kg]

$$\Rightarrow \quad T - F - 500 = 250$$

$$\Rightarrow \quad T - F = 750 \quad \ldots\text{(i)}$$

Free body diagram for worker

$$T + F - m_w g = m_w a$$

[$\because m_w$ = mass of worker]

$$\Rightarrow \quad T + F - 1000 = 500$$

$$\Rightarrow \quad T + F = 1500 \quad \ldots\text{(ii)}$$

Solving Eqs. (i) and (ii), we get

$$T = 1125\text{N} \text{ and } F = 375\text{N}$$

17 Let the accelerations of various blocks are as shown in figure. Pulley P_2 will have downward acceleration a.

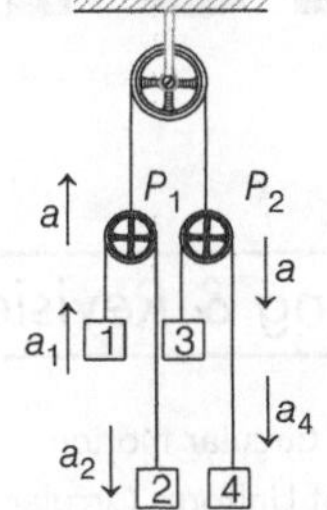

Analysing the diagram,

$$a = \frac{a_1 + a_2}{2} \Rightarrow a_2 = 2a - a_1 > 0$$

So, acceleration of block 2 is upward

Hence, (A)→ (2,3)

and $\quad a = \frac{a_1 + a_4}{2}$

$$\Rightarrow \quad a_4 = 2a + a_1 > 0$$

So, acceleration of block 4 is downward.

Hence, (B)→ (1,4).

This is downwards.

Hence, (c)→ (4)

Acceleration of block 2 with respect to block 4.

$$a_2/4 = a_2 - (-a_4) = 4a > 0$$

This is upward.

Hence (D)→(3) (C)→ (4)

Hence, A→ (2,3), B→ (1,4), C→ (4), D→ (3)

DAY FIVE

Circular Motion

Learning & Revision for the Day

- Concept of Circular Motion
- Dynamics of Uniform Circular Motion
- Forces in Circular Motion
- Applications of Centripetal and Centrifugal Forces

Concept of Circular Motion

Circular motion is a two dimension motion. To bring circular motion in a body it must be given some initial velocity and a force. Circular motion can be classified into two types- Uniform circular motion and Non uniform circular motion.

When an object moves in a circular path at a constant speed then the motion is said to be a **uniform circular motion.**

When an object moves in a circular path with variable speed, then the motion is said to be **non-uniform circular motion.**

Terms Related to Circular Motion

1. Angular Displacement

It is defined as the angle turned by the particle from some reference line. Angular displacement $\Delta\theta$ is usually measured in radians.

Finite angular displacement $\Delta\theta$ is a scalar but an infinitesimally small displacement is a vector.

2. Angular Velocity

It is defined as the rate of change of the angular displacement of the body.

From figure a particle moving on circular track of radius r is showing angular displacement $\Delta\theta$ in Δt time and in this time period, it covers a distance Δs along the circular track, then

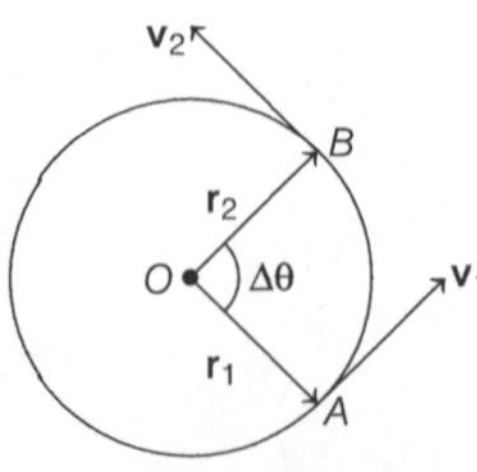

$\therefore$ Angular velocity, $\omega = \lim_{\Delta t \to 0} \left(\dfrac{\Delta\theta}{\Delta t}\right) = \dfrac{d\theta}{dt}$

It is an axial vector whose direction is given by the right hand rule. Its unit is rad/s.

- No. of Questions in Exercises (x)—
- No. of Questions Attempted (y)—
- No. of Correct Questions (z)—
  (Without referring Explanations)
- Accuracy Level ($z/y \times 100$)—
- Prep Level ($z/x \times 100$)—

In order to expect good rank in JEE, your Accuracy Level should be above 85 & Prep Level should be above 75.

3. Angular Acceleration

It is the rate of change of angular velocity.

Thus, $$\alpha = \frac{d\omega}{dt} = \frac{d^2\theta}{dt^2}$$

Its unit is rad/s^2.

Dynamics of Uniform Circular Motion

If a particle, is performing circular motion with a uniform speed, then motion of the particle is called uniform circular motion. In such a case,

$$\frac{dv}{dt} = 0 \text{ and } a_c = \omega^2 r = \frac{v^2}{r} \qquad [\because v = r\omega]$$

Thus, if a particle moves in a circle of radius r with a uniform speed v, then its acceleration is $\frac{v^2}{r}$, towards the centre. This acceleration is termed as **centripetal acceleration.**

NOTE • In non-uniform circular motion Resultant acceleration of the body is $a = \sqrt{a_r^2 + a_T^2} = \sqrt{\frac{v^4}{r^2} + r^2\alpha^2}$

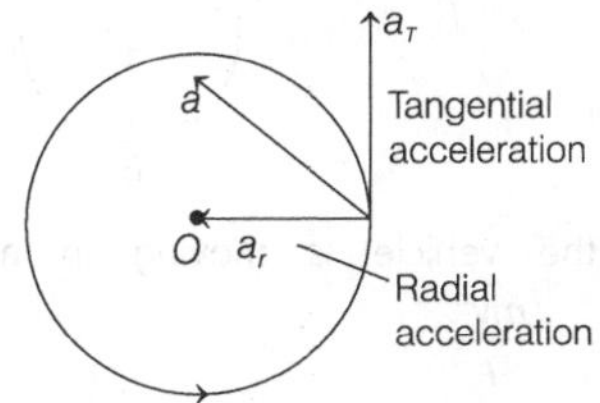

Forces in Circular Motion

In circular motion of an object two kinds of forces occur which are described below

Centripetal Force

The centripetal force is required to move a body along a circular path with a constant speed. The direction of the centripetal force is along the radius, acting towards the centre of the circle, on which the given body is moving.

Centripetal force,

$$F = \frac{mv^2}{r} = mr\omega^2 = mr\,4\pi^2 v^2 = mr\frac{4\pi^2}{T^2} \qquad [\because v = r\omega]$$

Work done by centripetal force is always zero as it is perpendicular to velocity and hence instantaneous displacement.

Centrifugal Force

'Certrifugal force can be defined as the radially directed outward force acting on a body in circular motion, as observed by a person moving with the body.'

Mathematically, centrifugal force $= \frac{mv^2}{r} = mr\omega^2 \quad [\because v = r\omega]$

Applications of Centripetal and Centrifugal Forces

Some of the most important applications of centripetal and centrifugal forces are given below

Motion of a Vehicle on a Level Circular Road

When a vehicle negotiates a circular path, it requires a centripetal force.

In such cases the lateral force of friction may provide the required centripetal force. Thus, for maintaining its circular path required centripetal force.

$$\left(\frac{mv^2}{r}\right) \le \text{frictional force } (\mu\, mg)$$

Maximum speed $v_{\max} = \sqrt{\mu rg}$

where, μ = coefficient of friction between road and vehicle tyres and r = radius of circular path.

Bending of a Cyclist

When a cyclist goes round turns in a circular track, then angle made by cyclist with vertical level is given by

$$\tan\theta = \frac{v^2}{rg} \approx \theta$$

$$= \tan^{-1}\left(\frac{v^2}{rg}\right)$$

Banking of a Curved Road

For the safe journey of a vehicle on a curved (circular) road, without any risk of skidding, the road is slightly raised towards its outer end.

Let the road be banked at an angle θ from the horizontal, as shown in the figure.

If b is width of the road and h is height of the outer edge of the road as compared to the inner edge, then

$$\tan\theta = \frac{v^2}{rg} = \frac{h}{b}$$

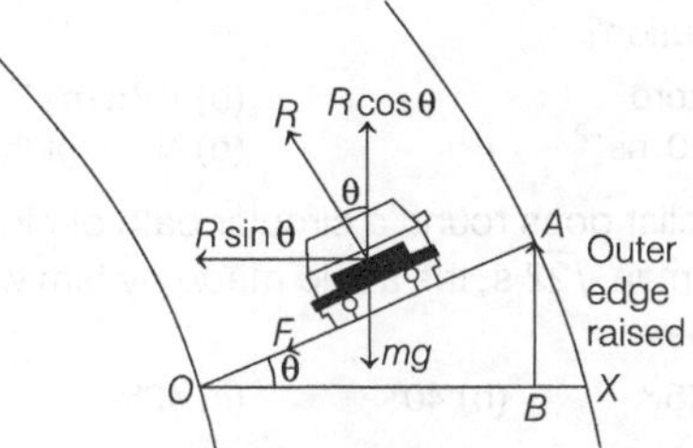

In case of friction is present between road and tyre, then Maximum speed,

$$v_{\max} = \sqrt{\frac{rg(\mu_s + \tan\theta)}{1 - \mu_s\tan\theta}},$$

where μ_s = coefficient of static friction.

Motion of a Cyclist in a Death Well

For equilibrium of cyclist in a death well, the normal reaction N provides the centripetal force needed and the force of friction balances his weight mg.

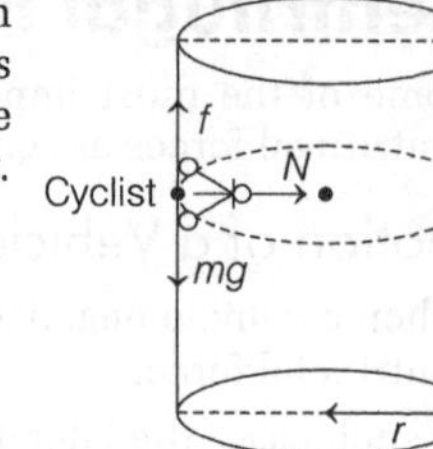

Thus, $N = \dfrac{mv^2}{r}$

and $f = \mu N = mg$

$\Rightarrow \quad v_{max} = \sqrt{\dfrac{rg}{\mu}}$

Motion along a Vertical Circle

In non-uniform circular motion speed of object decreases due to effect of gravity as the object goes from its lowest position A to highest position B.

- At the lowest point A, the tension T_L and the weight mg are in mutually opposite directions and their resultant provides the necessary centripetal force,

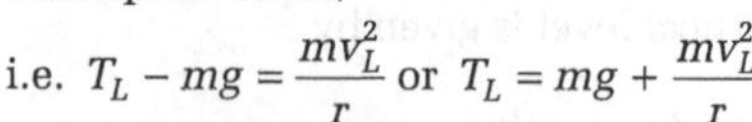

i.e. $T_L - mg = \dfrac{mv_L^2}{r}$ or $T_L = mg + \dfrac{mv_L^2}{r}$

- At the highest point B, tension T_H and the weight mg are in the same direction and hence,

$$T_H + mg = \frac{mv_H^2}{r} \text{ or } T_H = \frac{mv_H^2}{r} - mg$$

Moreover, v_L and v_H are correlated as $v_H^2 = v_L^2 - 2gr$.

- In general, if the revolving particle, at any instant of time, is at position C, inclined at an angle θ from the vertical, then

$$v^2 = v_L^2 - 2gr(1 - \cos\theta)$$

and $$T = mg\cos\theta + \frac{mv^2}{r}$$

- In the critical condition of just looping the vertical loop, (i.e. when the tension just becomes zero at the highest point B), we obtain the following results

$$T_H = 0, T_L = 6\,mg, v_L = \sqrt{5\,rg}$$

and $$v_H = \sqrt{rg}$$

In general, $T_L - T_H = 6\,mg$

NOTE
- When a vehicle is moving over a convex bridge, the maximum velocity $v = \sqrt{rg}$, where r is the radius of the road.
- When the vehicle is at the maximum height, the reaction of the road, is $N_1 = mg - \dfrac{mv^2}{r}$

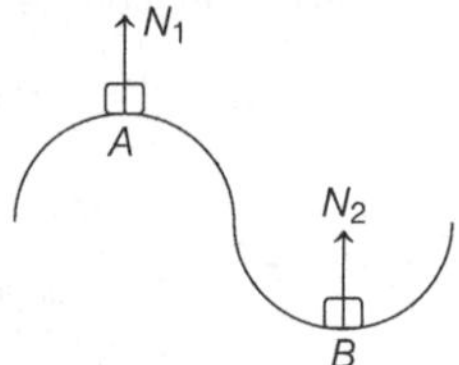

- When the vehicle is moving in a dip B, then $N_2 = mg + \dfrac{mv^2}{r}$

DAY PRACTICE SESSION 1

FOUNDATION QUESTIONS EXERCISE

1 A particle is moving along a circular path of radius 5 m, moving with a uniform speed of 5 m s^{-1}. What will be the average acceleration, when the particle completes half revolution?

(a) zero (b) $10/\pi$ m s^{-2}
(c) 10 m s^{-2} (d) None of these

2 A cyclist goes round a circular path of circumference 34.3 m in $\sqrt{22}$ s, the angle made by him with the vertical will be

(a) 45° (b) 40° (c) 42° (d) 48°

3 A particle undergoes a uniform circular motion. About which point on the plane of the circle, will the angular momentum of the particle, remain conserved?

(a) About centre of the circle
(b) On the circumference of the circle
(c) Inside the circle
(d) Outside the circle

4 A wheel is rotating at 900 rpm about its axis. When the power is cut-off, it comes to rest in 1 min. The angular retardation in rad s^{-2} is

(a) $\pi/2$ (b) $\pi/4$ (c) $\pi/6$ (d) $\pi/8$

5 A roller is made by joining together two corners at their vertices O. It is kept on two rails AB and CD which are placed a symmetrically (see the figure), with its axis perpendicular to CD and its centre O at the centre of line joining AB and CD (see the figure). It is given a light path, so that it starts rolling with its centre O moving parallel to CD in the direction shown. As it moves, the roller will tend to → **JEE Main 2016 (Offline)**

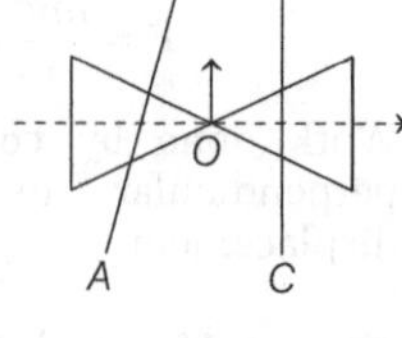

(a) turn left (b) turn right
(c) go straight
(d) turn left and right alternately

6 A particle moves in a circular path with decreasing speed. Choose the correct statement.
(a) Angular momentum remains constant
(b) Acceleration **a** is acting towards the centre
(c) Particle moves in a spiral path with decreasing radius
(d) The direction of angular momentum remains constant

7 A mass of 2 kg is whirled in a horizontal circle with the help of a string, at an initial speed of 5 rev/min. Keeping the radius constant, the tension in the string is doubled. The new speed is nearly
(a) 14 rpm (b) 10 rpm
(c) 2.25 rpm (d) 7 rpm

8 A cyclist is riding with a speed of 27 kmh^{-1}. As he approaches a circular turn on the road of radius 80 m, he applies brakes and reduces his speed at the constant rate of 0.5 ms^{-1} every second. What is the magnitude and direction of the net acceleration of the cyclist on the circular turn?
(a) 0.86 ms^{-2} at 54° to the velocity
(b) 0.6 ms^{-2} at 54° to the velocity
(c) 0.3 ms^{-2} at 75° to the velocity
(d) 0.7 ms^{-2} at 68° to the velocity

9 Two cars of masses m_1 and m_2 are moving in circles of radii r_1 and r_2, respectively. Their speeds are such that they make complete circles in the same time t. The ratio of their centripetal acceleration is → AIEEE 2012
(a) $m_1r_1 : m_2r_2$ (b) $m_1 : m_2$
(c) $r_1 : r_2$ (d) 1 : 1

10 A ball of mass 0.25 kg attached to the end of a string of length 1.96 m is moving in a horizontal circle. The string will break if the tension is more than 25 N. What is the maximum speed with which the ball can be moved?
(a) 14 ms^{-1} (b) 3 ms^{-1} (c) 3.92 ms^{-1} (d) 5 ms^{-1}

11 A car is moving along a straight horizontal road with a speed v_0. If the coefficient of friction between the tyres and the road is μ, the shortest distance in which the car can be stopped is
(a) $\frac{v_0^2}{2\mu g}$ (b) $\frac{v_0}{\mu g}$
(c) $\left(\frac{v_0}{\mu g}\right)^2$ (d) $\frac{v_0}{\mu}$

12 A racing car travel on a track (without banking) $ABCDEFA$. ABC is a circular arc of radius $2R$. CD and FA are straight paths of length R and DEF is a circular arc of radius $R = 100$ m. The coefficient of friction on the road is $\mu = 0.1$. The maximum speed of the car is 50 ms^{-1}. The minimum time for completing one round is

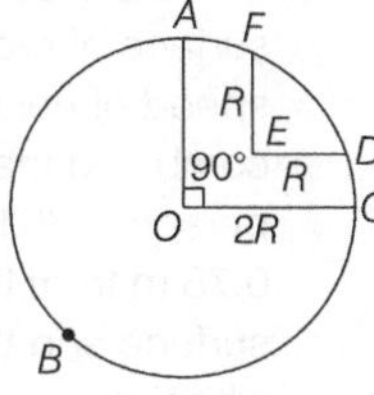

(a) 89.5 s (b) 86.3 s (c) 91.2 s (d) 41.3 s

13 A cyclist starts from centre O of a circular park of radius 1 km and moves along the path $OPRQO$ as shown in figure. If he maintains constant speed of 10 ms^{-1}, what is his acceleration at point R in magnitude and direction?

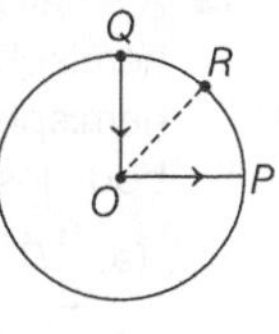

(a) 0.1 ms^{-2} along RO (b) 0.01 ms^{-2} along OR
(c) 1 ms^{-2} along RO (d) 0.1 rad s^{-2} along RO

14 A car is moving in a circular horizontal track of radius 10 m with a constant speed of 10 ms^{-1}. A plump bob is suspended from the roof of the car by a light rigid rod of length 1.00 m. The angle made by the rod with the track is
(a) 60° (b) 30°
(c) 45° (d) zero

15 A frictionless track $ABCD$ ends in a semi-circular loop of radius R. A body slides down the track from point A which is at a height $h = 5$ cm. Maximum value of R for the body to successfully complete the loop is

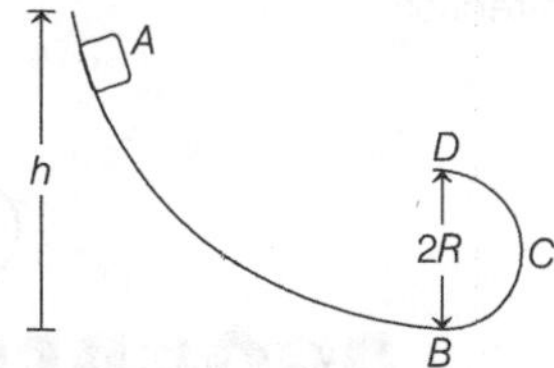

(a) 5 cm (b) $\frac{15}{4}$ cm (c) $\frac{10}{3}$ cm (d) 2 cm

16 A weightless thread can bear a tension upto 3.7 kg-wt. A stone of mass 500 g is tied to it and revolved in a circular path of radius 4 m in a vertical plane. If $g = 10\ ms^{-2}$, then the maximum angular velocity of the stone will be
(a) 4 rad s^{-1} (b) 16 rad s^{-1}
(c) $\sqrt{21}$ rad s^{-1} (d) 2 rad s^{-1}

17 A bob of mass m suspended by a light string of length L is whirled into a vertical circle as shown in figure. What will be the trajectory of the particle, if the string is cut at B.

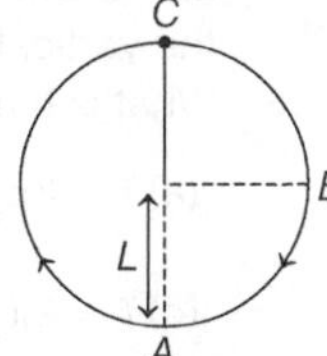

(a) Vertically upward
(b) Vertically downward
(c) Horizontally towards left
(d) Horizontally towards right

18 A particle is moving in a vertical circle. The tensions in the string when passing through two positions at angles 30° and 60° from vertical (lowest positions) are T_1 and T_2, respectively. Then
(a) $T_1 = T_2$
(b) $T_2 > T_1$
(c) $T_1 > T_2$
(d) tension in the string always remains the same

19 A small body of mass m slides down from the top of a hemisphere of radius R. The surface of block and hemisphere are frictionless. The height at which the body lose contact with the surface of the sphere is

(a) $\frac{3}{2}R$ (b) $\frac{2}{3}R$ (c) $\frac{1}{2}R$ (d) $\frac{1}{3}R$

Direction (Q. Nos. 20-24) *Each of these questions contains two statements : Statement* I *(Assertion) and Statement* II *(Reason). Each of these questions also has four alternative choices, only one of which is the correct answer. You have to select one of the codes (a), (b), (c), (d) given below :*

(a) Statement I is true, Statement II is true; Statement II is the correct explanation for Statement I
(b) Statement I is true, Statement II is true; Statement II is not the correct explanation for Statement I
(c) Statement I is true; Statement II is false
(d) Statement I is false; Statement II is true

20 **Statement I** A car is moving in a horizontal circular plane with varying speed, then the frictional force is neither pointing towards the radial direction nor along the tangential direction.

Statement II Components of the frictional force are providing the necessary tangential and centripetal acceleration, in the above situation.

21 **Statement I** A particle moving in a vertical circle, has a maximum kinetic energy at the highest point of its motion.

Statement II The magnitude of the velocity remains constant for a particle moving in a horizontal plane.

22 **Statement I** The centripetal force and the centrifugal force never cancel out.

Statement II They do not act at the same time.

23 **Statement I** Improper banking of roads causes wear and tear of tyres.

Statement II The necessary centripetal force in that event is provided by friction between the tyres and roads.

24 **Statement I** When a particle moves in a circle with a uniform speed, there is a change in both its velocity and acceleration.

Statement II The centripetal acceleration in circular motion is dependent on the angular velocity of the body.

DAY PRACTICE SESSION 2

PROGRESSIVE QUESTIONS EXERCISE

1 A heavy sphere of mass m is suspended by a string of length l. The sphere is made to revolve about a vertical line passing through the point of suspension, in a horizontal circle such that the string always remains inclined to the vertical making an angle θ. What is the period of revolution?

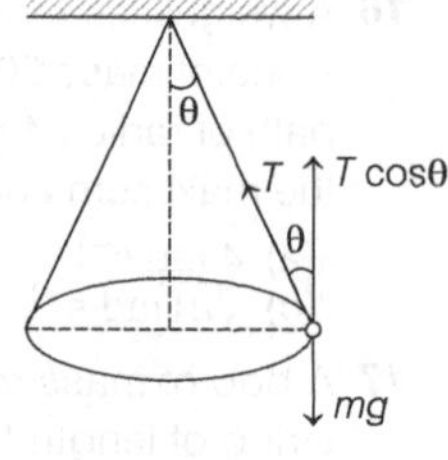

(a) $T = 2\pi\sqrt{\frac{l}{g}}$ (b) $T = 2\pi\sqrt{\frac{l\cos\theta}{g}}$

(c) $T = 2\pi\sqrt{\frac{l\sin\theta}{g}}$ (d) $T = 2\pi\sqrt{\frac{l\tan\theta}{g}}$

2 Two wires AC and BC are tied at C of small sphere of mass 5 kg, which revolves at a constant speed v in the horizontal circle of radius 1.6 m. The minimum value of v is

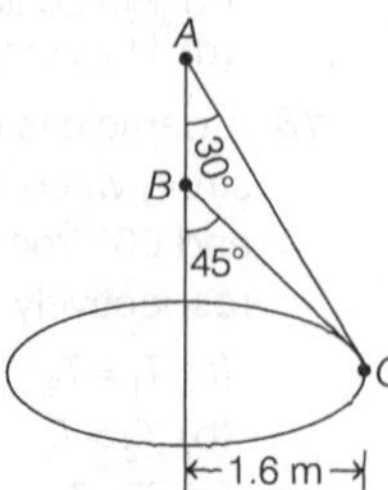

(a) 8.01 ms^{-1}
(b) 1.6 ms^{-1}
(c) 0
(d) 3.96 ms^{-1}

3 Two small spherical balls are free to move on the inner surface of the rotating spherical chamber of radius $R = 0.2$ m. If the balls reach a steady state at angular position $\theta = 45°$, the angular speed ω of device is

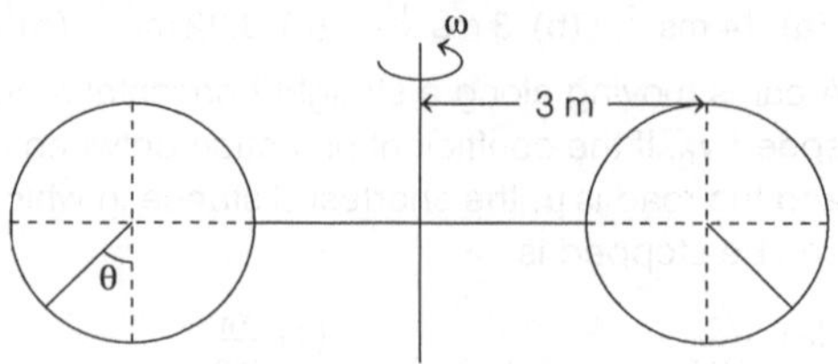

(a) 8 rad s^{-1} (b) 2 rad s^{-1}
(c) 3.64 rad s^{-1} (d) 9.34 rad s^{-1}

4 The skate board negotiates the circular surface of radius 4.5 m. At $\theta = 45°$, its speed of centre of mass is 6 ms^{-1}. The combined mass of skate board and the person is 70 kg and his centre of mass is 0.75 m from the surface. The normal reaction between the surface and the skate board wheel is

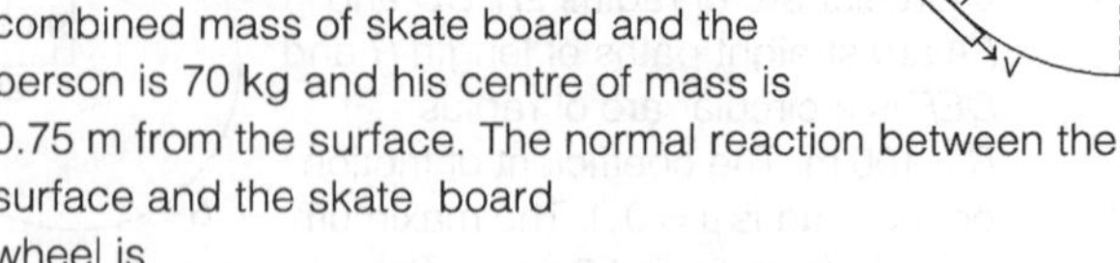

(a) 500 N (b) 2040 N
(c) 1157 N (d) zero

5 A particle is moving with a uniform speed in a circular orbit of radius R in a central force inversely proportional to the nth power of R. If the period of rotation of the particle is T, then → JEE Main 2018

(a) $T \propto R^{3/2}$ for any n (b) $T \propto R^{\frac{n}{2}+1}$

(c) $T \propto R^{(n+1)/2}$ (d) $T \propto R^{n/2}$

6 A bob of mass M is suspended by a massless string of length L. The horizontal velocity v at position A is just sufficient to make it reach the point B. The angle θ at which the speed of the bob is half of that at A, satisfies

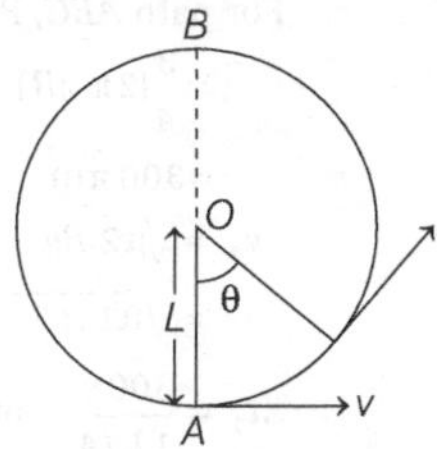

(a) $\theta = \frac{\pi}{4}$ (b) $\frac{\pi}{4} < \theta < \frac{\pi}{2}$

(c) $\frac{\pi}{2} < \theta < \frac{3\pi}{4}$ (d) $\frac{3\pi}{4} < \theta < \pi$

7 A roller coaster is designed such that riders experience 'weightlessness' as they go round the top of a hill whose radius of curvature is 20 m. The speed of the car at the top of the hill is between

(a) 14 ms^{-1} and 15 ms^{-1} (b) 15 ms^{-1} and 16 ms^{-1}

(c) 16 ms^{-1} and 17 ms^{-1} (d) 13 ms^{-1} and 14 ms^{-1}

8 Two particles revolve concentrically in a horizontal plane in the same direction. The time required to complete one revolution for particle A is 3 min, while for particle B is 1 min. The time required for A to complete one revolution relative to B is

(a) 2 min (b) 1.5 min (c) 1 min (d) 1.25 min

9 A skier plans to ski on smooth fixed hemisphere of radius R. He starts from rest from a curved smooth surface of height $\left(\frac{R}{4}\right)$. The angle θ at which he leaves the hemisphere is

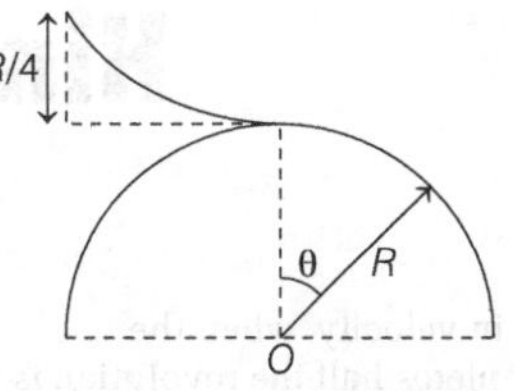

(a) $\cos^{-1}\left(\frac{2}{3}\right)$ (b) $\cos^{-1}\left(\frac{5}{\sqrt{3}}\right)$

(c) $\cos^{-1}\left(\frac{5}{6}\right)$ (d) $\cos^{-1}\left(\frac{5}{2\sqrt{3}}\right)$

10 A point moves along a circle with a speed $V = kt$, where $k = 0.5\ m/s^2$. Then total acceleration of the point at the moment when it has covered the nth fraction of the circle after the beginning of motion, where $n = \frac{1}{10}$

(a) 0.8 m/s^2 (b) 1. 2 m/s^2

(c) 1.6 m/s^2 (d) 2.0 m/s^2

11 A circular tube of mass M is placed vertically on a horizontal surface as shown in the figure. Two small spheres, each of mass m, just fit in the tube, are released from the top. If θ gives the angle between radius vector of either ball with the verticle, then for what value of the ratio $\frac{M}{m}$, tube breaks its contact with ground when $\theta = 60°$. (Neglect any friction).

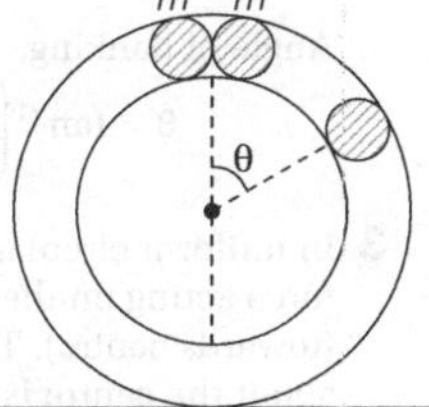

(a) $\frac{1}{2}$ (b) $\frac{2}{3}$ (c) $\frac{\sqrt{3}}{2}$ (d) None of these

12 A particle is moving in a circle of radius R in such a way that at any instant the normal and tangential component of its acceleration are equal. if its speed at $t = 0$ is v_o. The time taken to complete the first revolution is

(a) $\frac{R}{v_0}$ (b) $\frac{R}{v_0}e^{-2\pi}$

(c) $\frac{R}{v_0}(1+e^{-2\pi})$ (d) $\frac{R}{v_0}(1-e^{-2\pi})$

ANSWERS

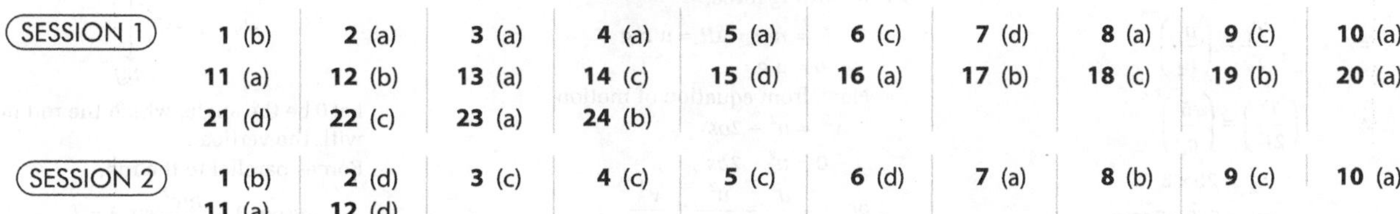

SESSION 1	**1** (b)	**2** (a)	**3** (a)	**4** (a)	**5** (a)	**6** (c)	**7** (d)	**8** (a)	**9** (c)	**10** (a)
	11 (a)	**12** (b)	**13** (a)	**14** (c)	**15** (d)	**16** (a)	**17** (b)	**18** (c)	**19** (b)	**20** (a)
	21 (d)	**22** (c)	**23** (a)	**24** (b)						
SESSION 2	**1** (b)	**2** (d)	**3** (c)	**4** (c)	**5** (c)	**6** (d)	**7** (a)	**8** (b)	**9** (c)	**10** (a)
	11 (a)	**12** (d)								

Hints and Explanations

SESSION 1

1 The change in velocity, when the particle completes half the revolution is given by
$\Delta v = 5\text{ms}^{-1} - (-5\text{ms}^{-1}) = 10\text{ms}^{-1}$

Now, the time taken to complete half the revolution is given by

$$t = \frac{\pi r}{v} = \frac{\pi \times 1}{1} = \pi \text{ s}$$

So, the average acceleration

$$= \frac{\Delta v}{t} = \frac{10}{\pi} \text{ms}^{-2}$$

2 Here, $2\pi r = 34.3 \Rightarrow r = \frac{34.3}{2\pi}$

and $v = \frac{2\pi r}{T} = \frac{2\pi r}{\sqrt{22}}$

Angle of banking,

$$\theta = \tan^{-1}\left(\frac{v^2}{rg}\right) = 45°$$

3 In uniform circular motion, the only force acting on the particle is centripetal (towards centre). Torque of this force about the centre is zero. Hence, angular momentum about centre remains conserved.

4 Use $\omega = \omega_0 + \alpha t$ ...(i)

Here, $\omega_0 = 900 \text{rpm}$
$= (2\pi \times 900)/60 \text{ rad s}^{-1}$
$\omega = 0$ and $t = 60 \text{ s}$

Then, Eq. (i) gives

$$\alpha = -\frac{\pi}{2} \text{rad s}^{-2}$$

5 As, the wheel rolls forward the radius of the wheel decreases along *AB*, hence for the same number of rotations it moves less distance along *AB*, hence it turns left.

6 A particle moves in a spiral path with decreasing radius.

7 The tension in the string will provide necessary centripetal force.

$\therefore \quad T = mr\omega^2$
$= mr4\pi^2 n^2$

$\therefore \quad T \propto n^2$

$\Rightarrow \quad \frac{T_1}{T_2} = \left(\frac{n_1}{n_2}\right)^2$

$\therefore \quad \left(\frac{T}{2T}\right) = \left(\frac{5}{n_2}\right)^2$

$n_2^2 = 25 \times 2$

$n_2 = 5\sqrt{2} \approx 7 \text{ rpm}$

8 Speed of the cyclist $(v) = 27 \text{ kmh}^{-1}$

$= 27 \times \frac{5}{18}$

$\left(\because 1\text{kmh}^{-1} = \frac{5}{18}\text{ms}^{-1}\right)$

$= \frac{15}{2}\text{ms}^{-1} = 7.5 \text{ ms}^{-1}$

Radius of the circular turn $(r) = 80\text{m}$

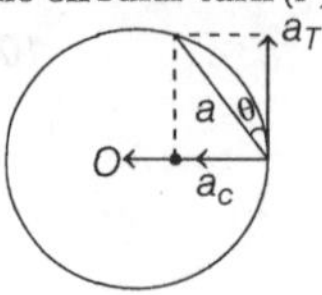

$\therefore$ Centripetal acceleration acting on the cyclist

$$a_c = \frac{v^2}{r} = \frac{(15/2)^2}{80} = \frac{225}{4 \times 80} \text{ms}^{-2}$$
$= 0.70 \text{ms}^{-2}$

Tangential acceleration applied by brakes

$$a_T = 0.5 \text{ms}^{-2}$$

Centripetal acceleration and tangential acceleration act perpendicular to each other.

$\therefore$ Resultant acceleration,

$a = \sqrt{a_c^2 + a_T^2}$
$= \sqrt{(0.7)^2 + (0.5)^2}$
$= \sqrt{0.49 + 0.25}$
$= \sqrt{0.74} = 0.86 \text{ ms}^{-2}$

If resultant acceleration makes an angle θ with the direction of velocity, then

$$\tan\theta = \frac{a_c}{a_T} = \frac{0.7}{0.5} = 1.4 = \tan 54°28'$$
$$\theta = 54°28'$$

9 As their period of revolution is same, so, their angular speed is same centripetal acceleration is $a = \omega^2 r$.

Thus, $\frac{a_1}{a_2} = \frac{\omega^2 r_1}{\omega^2 r_2} = \frac{r_1}{r_2}$

10 As, $T = mv^2/r$

Hence, $v = \sqrt{Tr/m}$
$= \sqrt{25 \times 1.96 / 0.25} = 14 \text{ ms}^{-1}$

11 Retarding force,

$F = ma = \mu R = \mu \, mg$
$a = \mu \, g$

Now, from equation of motion,

$v^2 = u^2 - 2as$

$\therefore \quad 0 = u^2 - 2as$

or $s = \frac{u^2}{2a} = \frac{u^2}{2\mu g} = \frac{v_0^2}{2\mu g}$

12 Balancing frictional force for centripetal force,

$$\frac{mv^2}{r} = f = \mu N = \mu mg$$

where, N is normal reaction

$\therefore \quad v = \sqrt{\mu rg}$

(where, r is radius of the circular track)

For path *ABC*, Path length

$= \frac{3}{4}(2\pi \, 2R) = 3\pi R = 3\pi \times 100$
$= 300 \, \pi\text{m}$

$v_1 = \sqrt{\mu 2 \, Rg}$
$= \sqrt{0.1 \times 2 \times 100 \times 10} = 14.14 \text{ ms}^{-1}$

$\therefore t_1 = \frac{300\pi}{14.14} = 66.6 \text{ s}$

For path *DEF*, Path length

$= \frac{1}{4}(2\pi R) = \frac{\pi \times 100}{2} = 50\pi$

$v_2 = \sqrt{\mu Rg} = \sqrt{0.1 \times 100 \times 10}$
$= 10 \text{ ms}^{-1}$

$t_2 = \frac{50\pi}{10} = 5 \, \pi\text{s} = 15.7 \text{ s}$

For paths *CD* and *FA*,
Path length $= R + R = 2R = 200\text{m}$

$t_3 = \frac{200}{50} = 4.0 \text{s}$

$\therefore$ Total time for completing one round

$t = t_1 + t_2 + t_3$
$= 66.6 + 15.7 + 4.0 = 86.3 \text{s}$

13 Acceleration of the cyclist at point *R* = centripetal acceleration (a_c)

$$a_c = \frac{v^2}{r} = \frac{(10)^2}{1000} = \frac{100}{1000}$$
$= 0.1 \text{ ms}^{-2}$, along *RO*

14 Centrifugal force on the rod,

$F = \frac{mv^2}{r}$ along *BF*.

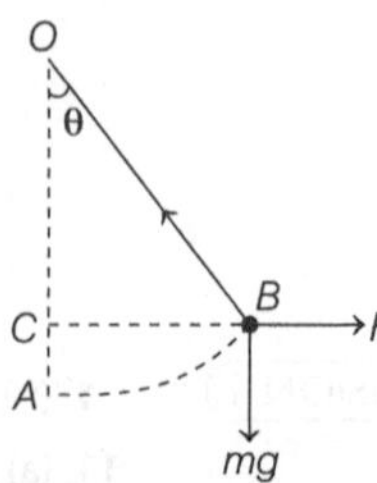

Let θ be the angle, which the rod makes with the vertical.

Forces parallel to the rod,

$$mg\cos\theta + \frac{mv^2}{r}\sin\theta = T$$

Force perpendicular to the rod

$$mg\sin\theta - \frac{mv^2}{r}\cos\theta$$

The rod would be balanced if

$$mg\sin\theta - \frac{mv^2}{r}\cos\theta = 0$$

$$mg\sin\theta = \frac{mv^2}{r}\cos\theta$$

This gives $\tan\theta = \frac{v^2}{rg} = \frac{(10)^2}{10\times 10}$

$= 1 = \tan 45°$

Here $\theta = 45°$

15 Velocity at the bottom is $\sqrt{2gh}$

For completing the loop,

$$\sqrt{2gh} = \sqrt{5gR}$$

Hence, $R = 2h/5$

$= (2\times 5)/5 = 2$ cm

16 As, $T_{max} = mr\omega^2 + mg$

$3.7\times 10 = 0.5\times 4\omega^2 + 0.5\times 10$

or $\omega^2 = \frac{32}{2} = 16$ or $\omega = 4$ rad s^{-1}

17 When bob is whirled into a vertical circle, the required centripetal force is obtained from the tension in the string. When string is cut,tension in string becomes zero and centripetal force is not provided, hence bob start to move in a straight line path along the direction of its velocity.

At point B, the velocity of B is vertically downward therefore, when string is cut at B, bob moves vertically downward.

18 $T = \frac{mv^2}{r} + mg\cos\theta$

$\therefore \quad T \propto \cos\theta$

$$\frac{T_1}{T_2} = \frac{\cos 30°}{\cos 60°} = \frac{\sqrt{3}/2}{1/2} = \frac{\sqrt{3}}{1}$$

or $T_1 > T_2$.

19 Suppose body slips at point B

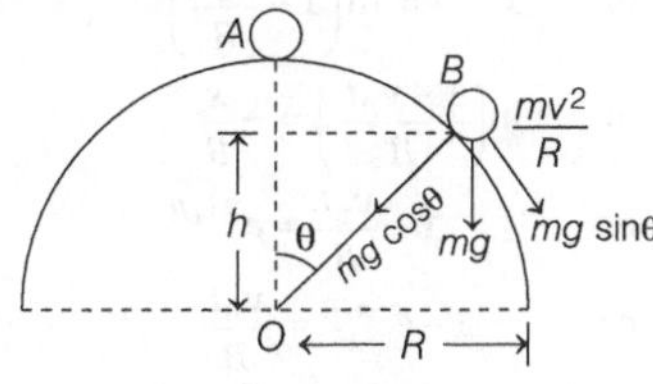

$$mg\cos\theta = \frac{mv^2}{R}$$

$[\because v = \sqrt{2g(R-h)}]$

$$g\cos\theta = \frac{2g(R-h)}{R}$$

$$\cos\theta = \frac{2(R-h)}{R}$$

$$\frac{h}{R} = \frac{2(R-h)}{R} \quad \left[\cos\theta = \frac{h}{R}\right]$$

$\therefore \quad h = \frac{2}{3}R$

20 In the present case, the tangential component of frictional force is responsible for changing the speed of car while component along the radial direction is providing necessary centripetal force, hence net friction force is neither towards radial nor along tangential direction.

21 As the kinetic energy at the highest point is zero.

22 We know that centripetal and centrifugal forces act at the same time on two different bodies. Thus, they never cancel out.

23 If the roads are not properly banked, the force of friction between tyres and road provides the necessary centripetal force, which causes the wear and tear of tyres.

24 A particle in a circular motion has the shown feature. The velocity of particle in circular motion

$$\mathbf{v} = r\omega\mathbf{e}_t \quad \text{... (i)}$$

Thus, we see that velocity of the particle is $r\omega$ along $\mathbf{e}_t$ or in tangent direction. So, it changes as the particle rotates the circle. Acceleration of the particle

$$\mathbf{a} = -\omega^2 r\mathbf{e}_r + \frac{dv}{dt}\mathbf{e}_t \quad \text{...(ii)}$$

Thus, acceleration of a particle moving in a circle has two components one along $\mathbf{e}_t$ (along tangent) and the other along $\mathbf{e}_r$ (or towards centre). Of these the first one is called the tangential $\mathbf{a}_t$ and other is called centripetal $\mathbf{a}_r$.

From Eq. (ii), it is obvious that acceleration depends on angular velocity (ω) of the body.

SESSION 2

1 Here, $\frac{mv^2}{r} = T\sin\theta$ and $mg = T\cos\theta$

Dividing these two, we get

$$\tan\theta = \frac{v^2}{rg} = \frac{r\omega^2}{g} = \frac{4\pi^2 r}{gT^2}$$

$$\frac{\sin\theta}{\cos\theta} = \frac{4\pi^2}{gT^2} l\sin\theta \quad [\because r = l\sin\theta]$$

$$\Rightarrow \quad T^2 = \frac{4\pi^2}{g} l\cos\theta$$

$$\Rightarrow \quad T = 2\pi\sqrt{\frac{l\cos\theta}{g}}$$

2

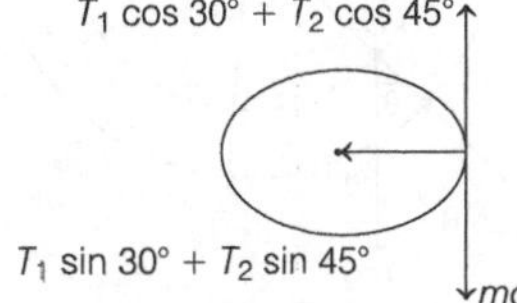

From the figure,

$$T_1\cos 30° + T_2\cos 45° = mg$$

$$T_1\sin 30° + T_2\sin 45° = \frac{mv^2}{r}$$

$$\Rightarrow \quad T_1 = \frac{mg - \frac{mv^2}{r}}{\frac{(\sqrt{3}-1)}{2}}$$

But $T_1 \geq 0$

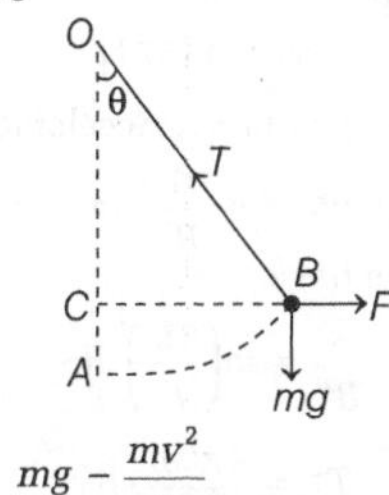

$$\frac{mg - \frac{mv^2}{r}}{\frac{\sqrt{3}-1}{2}} \geq 0$$

$$\Rightarrow \quad mg \geq \frac{mv^2}{r} \Rightarrow v \leq \sqrt{rg}$$

$$\therefore \quad v_{max} = \sqrt{rg}$$

$$= \sqrt{1.6\times 9.8} = 3.96 \text{ ms}^{-1}$$

3 Given, $R = 0.2$ m

From the figure,

$$r = 3R + R\sin 45°$$

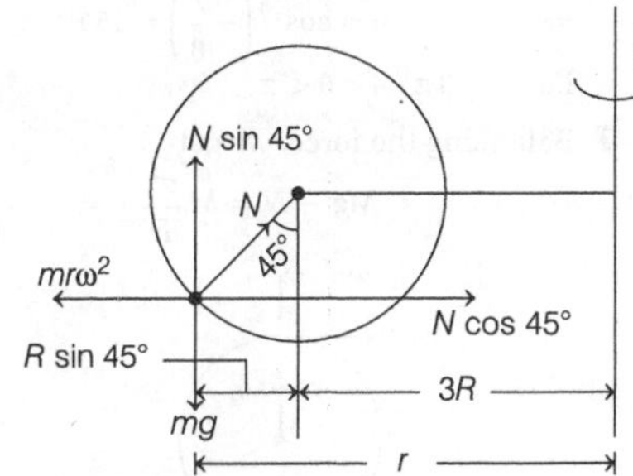

In the frame of rotating spherical chamber,

$$N\cos 45° = mr\omega^2$$

$$N\sin 45° = mg$$

$$\Rightarrow \tan 45° = \frac{mg}{mr\omega^2} = \frac{g}{r\omega^2}$$

$$\Rightarrow \omega = \sqrt{\frac{g}{r}} = \sqrt{\frac{g}{3R + \frac{R}{\sqrt{2}}}} = 3.64 \text{ rad s}^{-1}$$

4

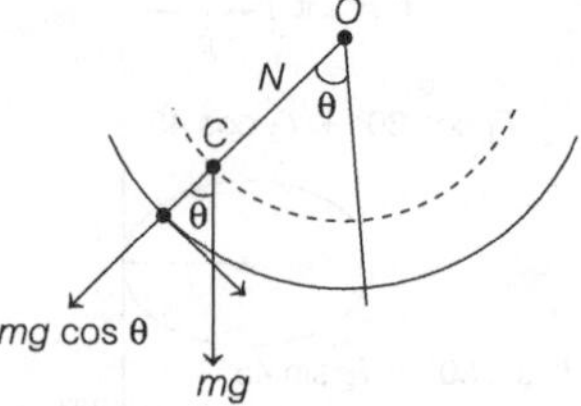

From the diagram,

$$N - mg\cos\theta = \frac{mv^2}{r}$$

$$\therefore \quad N = \frac{mv^2}{r} + mg\cos\theta$$

From the figure,

$\theta = 45°$, $m = 70$ kg, $g = 9.8 \text{ ms}^{-2}$,

$v = 6 \text{ ms}^{-2}$, $r = CO = 4.5 - 0.75 = 3.75$ m

Putting these values in above equation, we get

$$N = 1157 \text{ N}$$

5 $\because$ Force = Mass × Acceleration = $m\omega^2 R$

and given, $F \propto \frac{1}{R^n} \Rightarrow F = \frac{k}{R^n}$

So, we have

$$\frac{k}{R^n} = m\left(\frac{2\pi}{T}\right)^2 R$$

$$\Rightarrow \quad T^2 = \frac{4\pi^2 m}{k} \cdot R^{n+1} \Rightarrow T \propto R^{\frac{n+1}{2}}$$

6 Velocity of the bob at the point A

$$v = \sqrt{5gL} \quad \ldots(i)$$

From conservation of energy,

i.e. KE + PE = constant

$$\Rightarrow \quad \left(\frac{v}{2}\right)^2 = v^2 - 2gh \quad \ldots(ii)$$

$$h = L(1 - \cos\theta) \quad \ldots(iii)$$

Solving Eqs. (i), (ii) and (iii), we get

$$\cos\theta = -\frac{7}{8}$$

$$\Rightarrow \quad \theta = \cos^{-1}\left(-\frac{7}{8}\right) = 151°$$

i.e. $3\pi/4 < \theta < \pi$

7 Balancing the force, we get

$$Mg - N = M\frac{v^2}{R}$$

For weightlessness, $N = 0$

$$\therefore \quad \frac{Mv^2}{R} = Mg \quad \text{or } v = \sqrt{Rg}$$

Putting the values,

$R = 20$ m, $g = 10 \text{ ms}^{-2}$

So, $v = \sqrt{20 \times 10} = 14.14 \text{ ms}^{-1}$

Thus, the speed of the car at the top of the hill is between 14 ms^{-1} and 15 ms^{-1}.

8 Here, $(\omega_1 - \omega_2)t = 2\pi$ or $t = \frac{2\pi}{\omega_1 - \omega_2}$

where, $\omega_1 = \frac{2\pi}{T_1}$ and $\omega_2 = \frac{2\pi}{T_2}$

$$\therefore t = \frac{2\pi}{\left(\frac{2\pi}{T_1} - \frac{2\pi}{T_2}\right)} \text{ or } t = \frac{T_1 T_2}{T_2 - T_1} = \frac{3 \times 1}{3 - 1}$$

or $t = 1.5$ min

Hence, (b) is the correct option.

9 At the time of leaving contact, normal reaction must be zero.

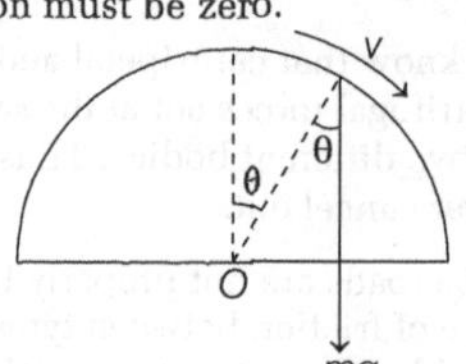

i.e. $N = 0$

$$\therefore mg\cos\theta = \frac{mv^2}{R} = \frac{m(2gh)}{R}$$

or $\cos\theta = \frac{2h}{R}$

or $\cos\theta = \frac{2}{R}\left[\frac{R}{4} + R(1 - \cos\theta)\right]$

or $3\cos\theta = \frac{5}{2}$ or $\cos\theta = \frac{5}{6}$

$$\therefore \quad \theta = \cos^{-1}\left(\frac{5}{6}\right)$$

Hence, (c) is the correct option.

10 $\because v = \frac{ds}{dt} = kt$

$$\therefore \int_0^s ds = k\int_0^t t\,dt \Rightarrow s = \frac{1}{2}kt^2 \quad \ldots(i)$$

After completion of nth fraction of circle

$$s = 2\pi rn \quad \ldots(ii)$$

From Eqs. (i) and (ii), we get

$$t^2 = \left[\frac{4\pi rn}{k}\right]$$

Now, tangential acceleration,

$$a_T = \frac{dv}{dt} = \frac{d}{dt}(kt) = k$$

and normal acceleration,

$$a_n = \frac{v^2}{r} = \frac{k^2 t^2}{r}$$

or $a_n = \frac{k^2}{r} \times \frac{4\pi rn}{\kappa} = 4\pi nk$

$$\therefore a = \sqrt{a_t^2 + a_n^2} = \sqrt{[k^2 + 16\pi^2 n^2 k^2]}$$

$$= k\sqrt{1 + 16\pi^2 n^2}$$

$$= 0.50\sqrt{1 + 16 \times (3.14)^2 \times (0.1)^2}$$

$$= 0.8 \text{ m/s}^2$$

Hence, (a) is the correct option.

11 Speed of each particle at angle θ is

$$v = \sqrt{2gh}$$

[After applying conservation of energy]

where, $h = R(1 - \cos\theta)$

$$\therefore \quad v = \sqrt{2gR(1 - \cos\theta)}$$

If N is normal reaction, then

$$N + mg\cos\theta = \frac{mv^2}{R}$$

or $N + mg\cos\theta = \frac{m}{R} \times 2gR(1 - \cos\theta)$

or $N + mg\cos\theta = 2mg(1 - \cos\theta)$

or $N = 2m - 3mg\cos\theta$ ...(i)

The tube will breaks its contact with ground when,

$$2N\cos\theta \geq Mg$$

where, we put the value of N from Eq. (i) is above relation, then we get

$4mg\cos\theta - 6mg\cos^2\theta = Mg$

Put $\theta = 60°$ (given)

$\therefore 4mg\cos 60° - 6mg\cos^2 60° = Mg$

or $2mg - \frac{3mg}{2} = Mg \Rightarrow \frac{M}{m} = \frac{1}{2}$

Hence, (a) is the correct option.

12 Given, $a_t = a_r$

or $\frac{dv}{dt} = \frac{v^2}{R}$ or $\frac{dv}{v^2} = \frac{1}{R}dt$

or $\int_{v_0}^{v} v^{-2}dv = \frac{1}{R}\int_0^t dt$

or $\left[\frac{1}{v_0} - \frac{1}{v}\right] = \frac{t}{R}$

or $\frac{1}{v} = \frac{1}{v_0} - \frac{t}{R} = \frac{R - v_0 t}{Rv_0}$

or $v = \frac{R.v_0}{R - v_0 t}$

As $v = \frac{dx}{dt}$

$\therefore \frac{dx}{dt} = \frac{Rv_0}{R - v_0 t}$

or $\int_0^x dx = \int_0^t \frac{Rv_0}{R - v_0 t}dt$

or $x = Rv_0\left(\frac{-1}{v_0}\right)[\ln(R - v_0 t)]_0^t$

or $x = -R\ln\left(1 - \frac{v_0 t}{R}\right)$

or $\ln\left(\frac{1 - v_0 t}{R}\right) = -\frac{x}{R}$

or $1 - \frac{v_0 t}{R} = e^{-x/R}$

or $1 - e^{-x/R} = \frac{v_0 t}{R}$

or $t = \frac{R}{v_0}(1 - e^{-x/R})$.

After completing one revolution, $x = 2\pi R$

$\therefore t = \frac{R}{v_0}(1 - e^{-2\pi})$

Hence, (d) is the correct option.

DAY SIX

Work, Energy and Power

Learning & Revision for the Day

- Work
- Conservative and Non-conservative Force
- Energy
- Work-Energy Theorem
- Law of Conservation of Energy
- Power
- Collision

Work

Work is said to be done, when a body a displaced through some distance in the direction of applied force. The SI unit work is joule (J) and in CGS it is erg.

$$1 \text{ joule (J)} = 10^7 \text{ erg}$$

The work done by the force **F** in displacing the body through a distance **s** is

$$W = (F\cos\theta)s = Fs\cos\theta = \mathbf{F}\cdot\mathbf{s}$$

where, $F\cos\theta$ is the component of the force, acting along the direction of the displacement produced. SI unit of work is joule (J).

$$1\text{ J} = 1\text{ N-m}$$

Work is a scalar quantity. Work can be of three types

(i) Positive work (ii) Negative work and (iii) Zero work.

- **Positive work** If value of the angle θ between the directions of **F** and **s** is either zero or an acute angle.
- **Negative work** If value of angle θ between the directions of **F** and **s** is either 180° or an obtuse angle.
- As work done $W = \mathbf{F}\cdot\mathbf{s} = Fs\cos\theta$, hence **work done can be zero,** if
  - (i) No force is being applied on the body, i.e. $F = 0$.
  - (ii) Although the force is being applied on a body but it is unable to cause any displacement in the body, i.e. $F \neq 0$ but $s = 0$.
  - (iii) Both F and s are finite but the angle θ between the directions of force and displacement is 90°. In such a case

$$W = \mathbf{F}\cdot\mathbf{s} = Fs\cos\theta = Fs\cos 90° = 0$$

- No. of Questions in Exercises (x)—
- No. of Questions Attempted (y)—
- No. of Correct Questions (z)—
  (Without referring Explanations)
- Accuracy Level ($z/y \times 100$)—
- Prep Level ($z/x \times 100$)—

In order to expect good rank in JEE, your Accuracy Level should be above 85 & Prep Level should be above 75.

Work Done by Variable Force

Work done by a variable force is given by $W = \int \mathbf{F} \cdot d\mathbf{s}$

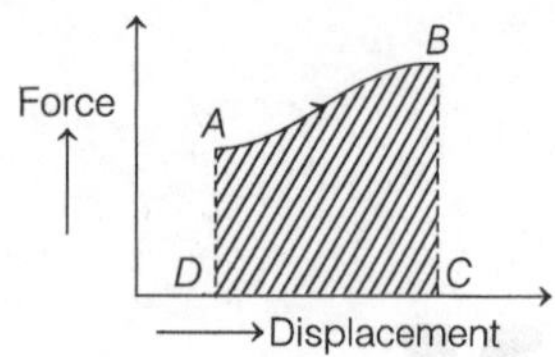

It is equal to the area under the force-displacement graph, along with proper sign.

Work done = Area of *ABCDA*

Conservative and Non-conservative Force

A force is said to be **conservative** if work done by or against the force in moving a body depends only on the initial and final positions of the body and not on the nature of path followed between the initial and the final position. Gravitational force, force of gravity, electrostatic force are some examples of conservative forces (fields).

A force is said to be **non-conservative** if work done by or against the force in moving a body from one positions to another, depends on the path followed between these two positions. Force of friction and viscous force are the examples of non-conservative forces.

Energy

Energy is defined as the capacity or ability of a body to do work. Energy is scalar and its units and dimensions are the same as that of work. Thus, SI unit of energy is J.

Some other commonly used units of energy are

$$1 \text{ erg} = 10^{-7} \text{ J},$$
$$1 \text{ cal} = 4.186 \text{ J} \cong 4.2 \text{ J},$$
$$1 \text{ kcal} = 4186 \text{ J},$$
$$1 \text{ kWh} = 3.6 \times 10^{6} \text{ J},$$

and $1 \text{ electron volt } (1 \text{ eV}) = 1.60 \times 10^{-19} \text{ J}$

Kinetic Energy

- Kinetic energy is the capacity of a body to do work by virtue of its motion. A body of mass m, moving with a velocity v, has a kinetic energy, $K = \frac{1}{2} mv^2$.
- Kinetic energy of a body is always positive irrespective of the sign of velocity v. Negative kinetic energy is impossible. Kinetic energy is correlated with momentum as,

$$K = \frac{p^2}{2m} \quad \text{or} \quad p = \sqrt{2mK}$$

- Kinetic energy for a system of particle will be

$$K = \frac{1}{2} \sum_i m_i v_i^2$$

- Relation between kinetic energy and force is

$$\frac{\text{KE}}{\text{Force}} = \frac{v \times t}{2}$$

where, v is velocity and t is time.

Potential Energy

Potential energy is the energy stored in a body or a system by virtue of its position in a field of force or due to its configuration. Potential energy is also called **mutual energy** or **energy of the configuration.**

Value of the potential energy in a given position can be defined only by assigning some arbitrary value to the reference point. Generally, reference point is taken at infinity and potential energy at infinity is taken as zero. In that case,

$$U = -W = -\int_{\infty}^{r} \mathbf{F} \cdot d\mathbf{r}$$

Potential energy is a scalar quantity. It may be positive as well as negative.

Different types of potential energy are given below.

Gravitational Potential Energy

It is the energy associated with the state of separation between two bodies which interact *via* the gravitational force.

- The gravitational potential energy of two particles of masses m_1 and m_2 separated by a distance r is

$$U = \frac{-Gm_1m_2}{r}.$$

- If a body of mass m is raised to a height h from the surface of the earth, the change in potential energy of the system (earth+body) comes out to be

$$\Delta U = \frac{mgh}{\left(1 + \frac{h}{R}\right)}$$

or $\Delta U \approx mgh$ if $h << R$

Thus, the potential energy of a body at height h, i.e. mgh is really the change in potential energy of the system for $h << R$.

- For the gravitational potential energy, the zero of the potential energy is chosen on the ground.

Electric Potential Energy

The electric potential energy of two point charges q_1 and q_2 separated by a distance r in vacuum is given by

$$U = \frac{1}{4\pi\varepsilon_0} \frac{q_1q_2}{r}$$

where, $\frac{1}{4\pi\varepsilon_0} = 9.1 \times 10^9 \frac{\text{N-m}^2}{\text{C}^2} = \text{constant}$

Potential Energy of a Spring

Whenever an elastic body (say a spring) is either stretched or compressed, work is being done against the elastic spring force. The work done is $W = \frac{1}{2}kx^2$,

where, k is spring constant and x is the displacement.

And elastic potential energy, $U = \frac{1}{2}kx^2$

If spring is stretched from initial position x_1 to final position x_2, then

work done = Increment in elastic potential energy

$$= \frac{1}{2}k(x_2^2 - x_1^2).$$

Work-Energy Theorem

Accordingly, work done by all the forces (conservative or non-conservative, external or internal) acting on a particle or an object is equal to the change in its kinetic energy of the particle. Thus, we can write

$$W = \Delta K = K_f - K_i$$

We can also write, $K_f = K_i + W$

Which says that $\begin{pmatrix}\text{Kinetic energy after}\\ \text{the net work is done}\end{pmatrix}$

$$= \begin{pmatrix}\text{Kinetic energy before}\\ \text{the net work done}\end{pmatrix} + \begin{pmatrix}\text{The net}\\ \text{work done}\end{pmatrix}$$

Law of Conservation of Energy

The mechanical energy E of a system is the sum of its kinetic energy K and its potential energy U.

$$E = K + U$$

When the forces acting on the system are conservative in nature, the mechanical energy of the system remains constant,

$$K + U = \text{constant} \Rightarrow \Delta K + \Delta U = 0$$

There are physical situations, where one or more non-conservative force act on the system but net work done by them is zero, then too the mechanical energy of the system remains constant.

If $\quad \Sigma W_{net} = 0$

Mechanical energy, E = constant.

Power

- It is a quantity that measures the rate at which work is done or energy is transformed.

  Average power $(P)_{av} = \frac{W}{t}$

- The shorter is the time taken by a person or a machine in performing a particular task, the larger is the power of that person or machine.
- Power is a scalar quantity and its SI unit is watt, where, $1\text{W} = 1\text{ J/s}$
- Instantaneous power,

$$\mathbf{P}_{inst} = \frac{dW}{dt} = \frac{\mathbf{F}\cdot d\mathbf{s}}{dt} = \mathbf{F}\cdot\mathbf{v}$$

- Some other commonly used units of power are

$$1\text{ kW} = 10^3\text{ W},$$
$$1\text{ MW} = 10^6\text{ W},$$
$$1\text{ HP} = 746\text{ W}$$

Collision

The physical interaction of two or more bodies in which each equal and opposite forces act upon each other causing the exchange of energy and momentum is called collision. Collisions are classified as

(i) elastic collisions and (ii) inelastic collisions.

Elastic Collision in One Dimension

In a perfectly elastic collision, total energy and total linear momentum of colliding particles remains conserved. Moreover, the forces involved in interaction are conservative in nature and the total kinetic energy before and after the collision, remains unchanged.

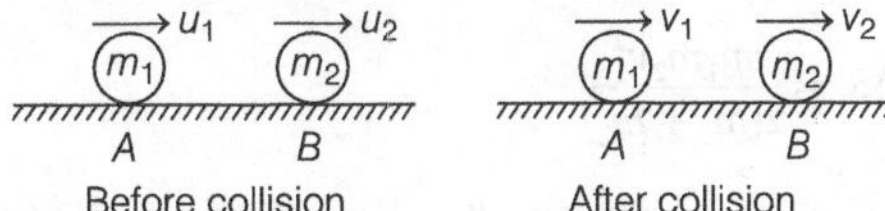

In above figure, two bodies A and B of masses m_1 and m_2 and having initial velocities u_1 and u_2 in one dimension, collide elastically and after collision move with velocities v_1 and v_2, then we find that

- Relative velocity of approach = Relative velocity of separation, i.e. $u_1 - u_2 = v_2 - v_1$
- $v_1 = \left(\frac{m_1 - m_2}{m_1 + m_2}\right)u_1 + \left(\frac{2m_2}{m_1 + m_2}\right)u_2$

  and $v_2 = \left(\frac{2m_1}{m_1 + m_2}\right)u_1 + \left(\frac{m_2 - m_1}{m_1 + m_2}\right)u_2$

Elastic Collision in Two Dimensions

- In this type of collision, the two particles or objects moving along different directions collide with each other.

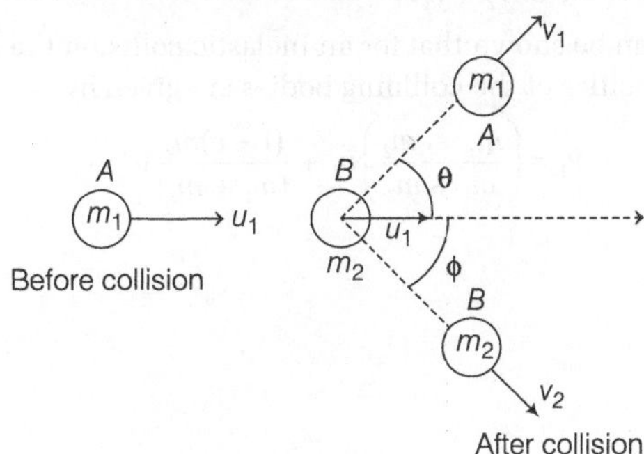

- As linear momentum is conserved.

  ∴ Along the x-axis

$$m_1u_1 + m_2u_2 = m_1v_1 \cos\theta + m_2v_2 \cos\phi \qquad \text{...(i)}$$

  and along the y-axis

$$0 = m_1v_1 \sin\theta - m_2v_2 \sin\phi \qquad \text{...(ii)}$$

- As the total kinetic energy remains unchanged.

  Hence, $$\frac{1}{2}m_1u_1^2 + \frac{1}{2}m_2u_2^2 = \frac{1}{2}m_1v_1^2 + \frac{1}{2}m_2v_2^2 \qquad \text{...(iii)}$$

  We can solve these equations provided that either the value of θ and ϕ is known to us.

Inelastic Collision in One Dimension

In an inelastic collision, the total linear momentum as well as total energy remain conserved but total kinetic energy after collision is not equal to kinetic energy before collision.

For inelastic collision,

Common speed, $v = \dfrac{m_1v_1}{m_1 + m_2}$

and loss of kinetic energy $\Delta K = \dfrac{m_1m_2(v_1 - v_2)^2}{2(m_1 + m_2)}$

Here $v_2 = 0$

∴ $$\Delta K = \frac{m_1m_2v_1^2}{2(m_1 + m_2)}$$

Medium ← m_1 $\xrightarrow{v_1}$ m_2 $\xrightarrow{v_2}$

Before collision v_2=0

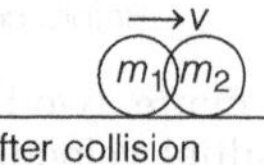

After collision

Coefficient of Restitution (e)

- It is defined as the ratio of relative velocity of separation to the relative velocity of approach.
- Coefficient of restitution, $e = \dfrac{v_2 - v_1}{u_1 - u_2}$
- For a perfectly elastic collision, $e = 1$.
- For a perfectly inelastic collision, $e = 0$.
- If $0 < e < 1$, the collision is said to be **partially elastic.**
- In a perfectly inelastic collision, $e = 0$ which means that $v_2 - v_1 = 0$ or $v_2 = v_1$.
- It can be shown that for an inelastic collision the final velocities of the colliding bodies are given by

$$v_1 = \left(\frac{m_1 - em_2}{m_1 + m_2}\right)u_1 + \frac{(1 + e)m_2}{(m_1 + m_2)}u_2$$

and $$v_2 = \frac{(1 + e)m_1}{(m_1 + m_2)}u_1 + \left(\frac{m_2 - em_1}{m_1 + m_2}\right)u_2$$

- If a particle of mass m, moving with velocity u, hits an identical stationary target inelastically, then final velocities of projectile and target are correlated as

  i.e. $$m_1 = m_2 = m \text{ and } u_2 = 0;$$
  $$\frac{v_1}{v_2} = \frac{1 - e}{1 + e}$$

- In case of extreme inelastic collision (in which colliding objects stick together after collision)

m_1 $\xrightarrow{v_1}$ m_2 rest — $m_1 m_2$ $\xrightarrow{v_2}$

Before collision — After collision

- Final velocity $(v_2) = \left[\dfrac{m_1}{m_1 + m_2}\right]v_1$
- Ratio of kinetic energy before and after collision is

$$\frac{KE_1}{KE_2} = \left[\frac{m_1}{m_1 + m_2}\right]$$

- Fraction of kinetic energy lost in the collision is

$$\frac{KE_1 - KE_2}{KE_1} = \left[\frac{m_2}{m_1 + m_2}\right]$$

Rebounding of a Ball on Collision with the Floor

- Speed of the ball after the nth rebound

$$v_n = e^n v_0 = e^n\sqrt{2gh_0}$$

- Height covered by the ball after the nth rebound

$$h_n = e^{2n}h_0$$

- Total distance (vertical) covered by the ball before it stops bouncing

$$H = h_0 + 2h_1 + 2h_2 + 2h_3 + \ldots = h_0\left(\frac{1 + e^2}{1 - e^2}\right)$$

- Total time taken by the ball before it stops bouncing

$$T = t_0 + t_1 + t_2 + t_3 + \ldots$$
$$= \sqrt{\frac{2h_0}{g}} + 2\sqrt{\frac{2h_1}{g}} + 2\sqrt{\frac{2h_2}{g}} + \ldots$$
$$= \sqrt{\frac{2h_0}{g}}\left(\frac{1 + e}{1 - e}\right)$$

DAY PRACTICE SESSION 1

FOUNDATION QUESTIONS EXERCISE

1 A body of mass 2 kg, initially at rest, is acted upon simultaneously by two forces, one of 4N and other of 3N, acting at right angles to each other. The work done by the body after 20 s is

(a) 500 J (b) 1250 J (c) 2500 J (d) 5000 J

2 A body of mass 500 g is taken up an inclined plane of length 10 m and height 5 m and then released to slide down to the bottom. The coefficient of friction between the body and the plane is 0.1. What is the amount of work done in the round trip?

(a) 5 J (b) 15 J (c) $5\sqrt{3}$ J (d) $\frac{5}{\sqrt{3}}$ J

3 A block of mass 5 kg is initially at rest on a horizontal frictionless surface. A horizontal force $\mathbf{F} = (9 - x^2)\hat{\mathbf{i}}$ Newton acts on it, when the block is at $x = 0$. The maximum work done by the block between $x = 0$ and $x = 3$ m in joule is

(a) 18 J (b) 15 J (c) 20 J (d) 24 J

4 An object is displaced from point A (2m, 3m, 4m) to a point B (1m, 2m, 3m) under a constant force $\mathbf{F} = (2\hat{\mathbf{i}} + 3\hat{\mathbf{j}} + 4\hat{\mathbf{k}})$N, then the work done by this force in this process is

(a) 9 J (b) –9 J (c) 18 J (d) –18 J

5 An open water tight railway wagon of mass 5×10^3 kg coasts with an initial velocity of 1.2 ms^{-1} on a railway track without friction. Rain falls vertically downwards on the wagon.

What change occurs in the kinetic energy of the wagon, after it has collected 10^3 kg of water?

(a) 900 J (b) 300 J (c) 600 J (d) 1200 J

6 A particle moves in a straight line with retardation proportional to its displacement. The loss in kinetic energy of the particle, for any displacement x, is proportional to

(a) x^2 (b) e^x (c) x (d) $\log_e x$

7 Which of the diagrams as shown in figure most closely shows the variation in kinetic energy of the earth as it moves once around the sun in its elliptical orbit?

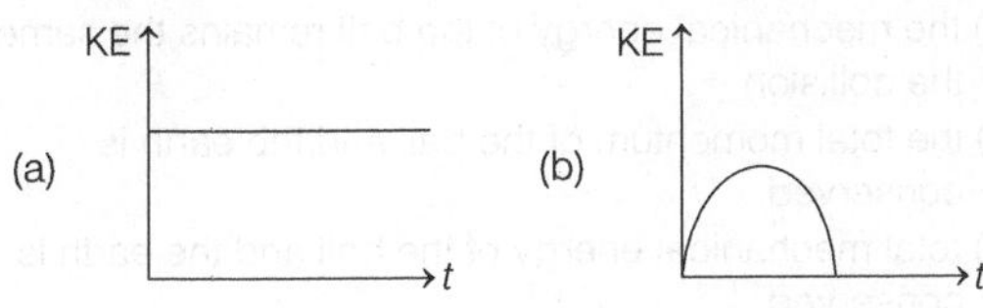

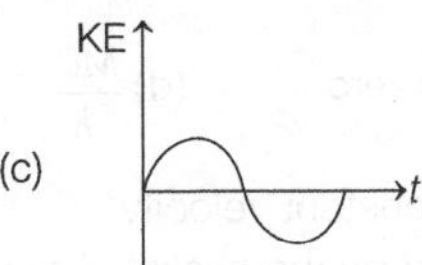

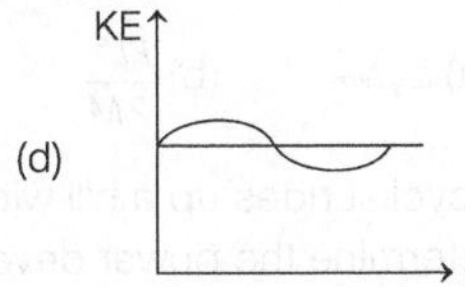

8 A particle is moving in a circular path of radius a under the action of an attractive potential $U = -\frac{k}{2r^2}$. Its total energy is

(a) $-\frac{k}{4a^2}$ (b) $\frac{k}{2a^2}$ (c) zero (d) $-\frac{3}{2}\frac{k}{a^2}$

9 The potential energy of a 1 kg particle free to move along the x-axis is given by $V(x) = \left(\frac{x^4}{4} - \frac{x^2}{2}\right)$ J.

The total mechanical energy of the particle is 2 J. Then, the maximum speed (in ms^{-1}) is

(a) $\frac{3}{\sqrt{2}}$ (b) $\sqrt{2}$ (c) $\frac{1}{\sqrt{2}}$ (d) 2

10 A particle is placed at the origin and a force $F = kx$ acts on it (where, k is a positive constant). If $U(0) = 0$, the graph of $U(x)$ *versus* x will be (where, U is the potential energy function)

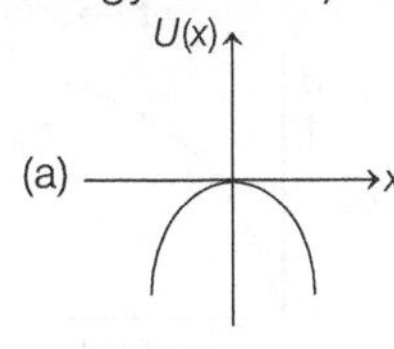

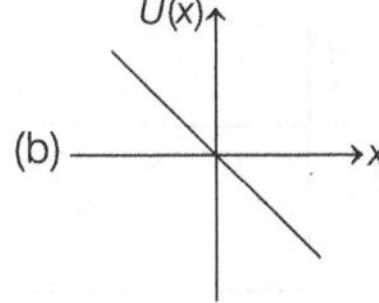

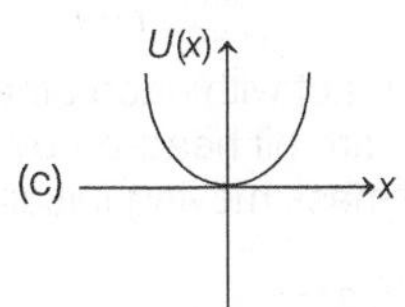

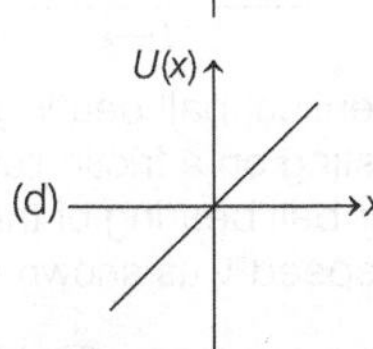

11 A bullet fired into a fixed target losses half of its velocity after penetrating distance of 3 cm. How much further it will penetrate before coming to rest, assuming that it faces constant resistance to its motion?

(a) 3.0 cm (b) 2.0 cm (c) 1.5 cm (d) 1.0 cm

12 A 1.5 kg block is initially at rest on a horizontal frictionless surface, when a horizontal force in the positive direction of x-axis, is applied to the block. The force is given by $\mathbf{F} = (4 - x^2)\hat{\mathbf{i}}$ Newton, where x is in metre and the initial position of the block is at $x = 0$. The maximum kinetic energy of the block between $x = 0$ and $x = 2.0$ m is

(a) 6.67 J (b) 5.33 J (c) 8.67 J (d) 2.44 J

13 A block of mass M moving on a frictionless horizontal surface, collides with a spring of spring constant k and compresses it by length L. The maximum momentum of the block, after collision is

(a) $L\sqrt{Mk}$ (b) $\frac{kL^2}{2M}$ (c) zero (d) $\frac{ML^2}{k}$

14 A cyclist rides up a hill with a constant velocity. Determine the power developed by the cyclist, if the length of the connecting rod of the pedal is $r = 25$ cm, the time of revolution of the rod is $t = 2$ s and the mean force exerted by his foot on the pedal is $F = 15$ kgf.

(a) 115.6 W (b) 215.6 W (c) 15.6 W (d) 11.56 W

15 Power supplied to a particle of mass 2 kg varies with time as $P = 3t^2/2$ W, where t is in second. If velocity of the particle at $t = 0$ is $v = 0$, the velocity of the particle at $t = 2$ s, will be

(a) 1 ms^{-1} (b) 4 ms^{-1} (c) 2 ms^{-1} (d) $2\sqrt{2}$ ms^{-1}

16 A body is moving unidirectionally under the influence of a source of constant power supplying energy. Which of the diagrams as shown in figure correctly shows the displacement-time curve for its motion?

(a)
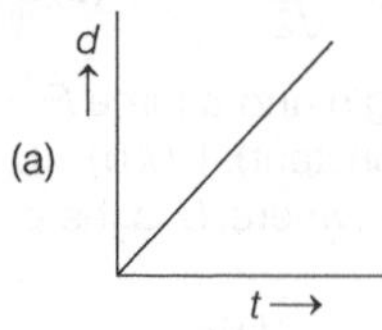

(b)
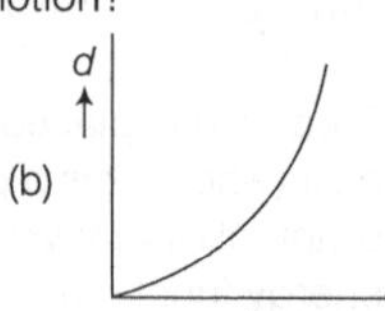

(c)
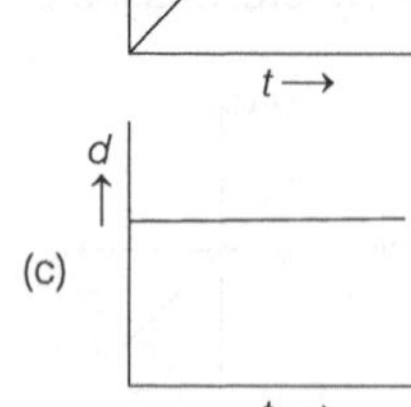

(d)
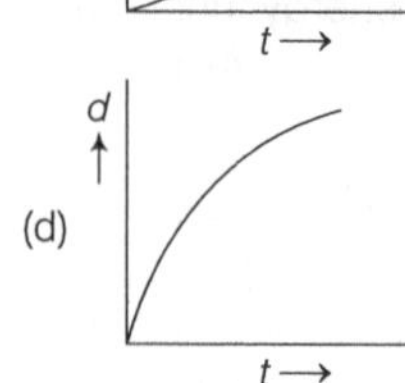

17 Two identical ball bearings in contact with each other and resting on a frictionless table are hit head-on by another ball bearing of the same mass moving initially with a speed v as shown in

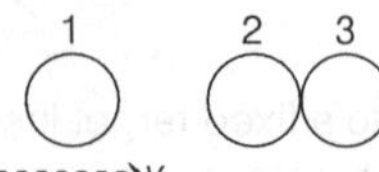

If the collision is elastic, which of the following is a possible result after collision?

(a)
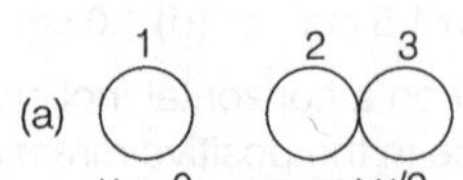

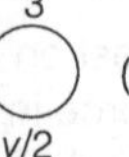

(b)
1 2 3
v = 0 → v

(c)
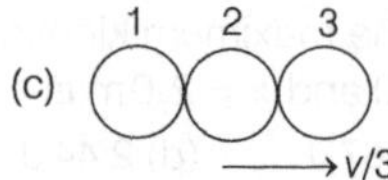

(d)
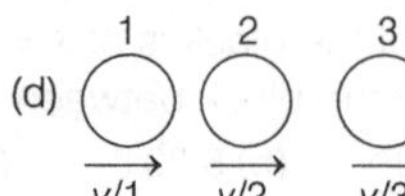

18 Two balls of masses m_1 and m_2 are separated from each other and a charge is placed between them. The whole system is at rest on the ground. Suddenly, the charge explodes and the masses are pushed apart. The mass m_1 travels a distance S_1 and then it stops. If the coefficient of friction between the balls and the ground are the same, mass m_2 stops after covering the distance

(a) $S_2 = \frac{m_1}{m_2} S_1$ (b) $S_2 = \frac{m_2}{m_1} S_1$

(c) $S_2 = \frac{m_1^2}{m_2^2} S_1$ (d) $S_2 = \frac{m_2^2}{m_1^2} S_1$

19 A shell is fired from a cannon with a velocity v ms^{-1} at an angle θ with the horizontal direction. At the highest point in its path, it explodes into 2 pieces of equal masses. One of the pieces retraces its path to the cannon. The speed in ms^{-1} of the other piece, immediately after the explosion is

(a) $3v\cos\theta$ (b) $2v\cos\theta$ (c) $v\cos\theta$ (d) $\sqrt{\frac{3}{2}}v\cos\theta$

20 A particle of mass m moving with a velocity u makes an elastic one dimensional collision with a stationary particle of mass m. They are in contact for a very short interval of time T.

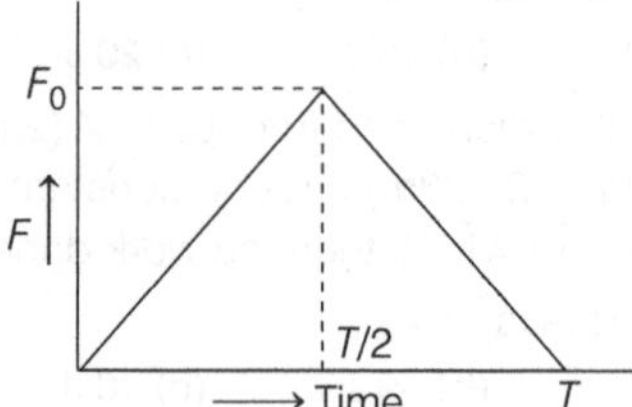

The force of interaction increases from zero to F_0 in $\frac{T}{2}$ and then decreases linearly to zero in further time interval $\frac{T}{2}$. The magnitude of F_0 is

(a) $\frac{mu}{T}$ (b) $\frac{2mu}{T}$ (c) $\frac{mu}{2T}$ (d) None of these

21 A block of mass 0.50 kg is moving with a speed of 2.00 ms^{-1} on a smooth surface. It strikes another mass of 1.00 kg and then they move together as a single body. The energy loss during the collision is

(a) 0.16 J (b) 1.00 J (c) 0.67 J (d) 0.34 J

22 A ball hits the floor and rebounds after an inelastic collision. In this case

(a) the momentum of the ball just after the collision is the same as that just before the collision

(b) the mechanical energy of the ball remains the same in the collision

(c) the total momentum of the ball and the earth is conserved

(d) total mechanical energy of the ball and the earth is conserved

23 Two particles of masses m_1 and m_2 in projectile motion have velocities $\mathbf{v}_1$ and $\mathbf{v}_2$ respectively at time $t = 0$. They collide at time t_0. Their velocities become $\mathbf{v}'_1$ and $\mathbf{v}'_2$ at time $2t_0$ while still moving in air. The value of $|(m_1 \mathbf{v}'_1 + m_2 \mathbf{v}'_2) - (m_1 \mathbf{v}_1 + m_2 \mathbf{v}_2)|$ is

(a) zero (b) $(m_1 + m_2)gt_0$
(c) $2(m_1 + m_2)gt_0$ (d) $\frac{1}{2}(m_1 + m_2)gt_0$

24 A ball is dropped on the ground from a height of 2m. If the coefficient of restitution is 0.6, the height to which the ball will rebound is

(a) 0.72 m (b) 1.72 m (c) 2.72 m (d) 1 m

25 A body of mass m is accelerated uniformly from rest to a speed v in a time interval T. The instantaneous power delivered to the body as a function of time (t), is given by

(a) $\frac{mv^2}{T^2}t$ (b) $\frac{mv^2}{T^2}t^2$ (c) $\frac{1}{2}\frac{mv^2}{T^2}t$ (d) $\frac{1}{2}\frac{mv^2}{T^2}t^2$

26 A person trying to lose weight by burning fat lifts a mass of 10 kg upto a height of 1 m, 1000 times. Assume that the potential energy lost each time he lowers the mass is dissipated. How much fat will he use up considering the work done only when the weight is lifted up? Fat supplies 3.8×10^7 J of energy per kg which is converted into mechanical energy with a 20% of efficiency rate. (Take, $g = 9.8\ \text{ms}^{-2}$) → JEE Main 2016 (Offline)

(a) 2.45×10^{-3} kg (b) 6.45×10^{-3} kg
(c) 9.89×10^{-3} kg (d) 12.89×10^{-3} kg

27 A time dependent force $F = 6t$ acts on a particle of mass 1 kg. If the particle starts from rest, the work done by the force during the first 1 s will be → JEE Main 2017 (Offline)

(a) 22 J (b) 9 J
(c) 18 J (d) 4.5 J

28 Match the categories of energy given in column I with their formula given in column II and select the correct option from the choices given below.

	Column I		Column II
A.	Spring energy	1.	$\frac{1}{2}mv^2$
B.	Kinetic energy	2.	mgh
C.	Potential energy	3.	$\frac{1}{2}kx^2$

	A	B	C
(a)	3	2	1
(b)	3	1	2
(c)	1	2	3
(d)	2	3	1

Direction (Q. Nos. 29-32) *These question consists of two statements each printed as Statement I and Statement II. While answering these questions you are required to choose any one of the following five responses.*

(a) If both Statement I and Statement II are correct and Statement II is the correct explanation of Statement I.
(b) If both Statement I and Statement II are correct but Statement II is not correct explanation of Statement I.
(c) If Statement I is correct but Statement II is incorrect.
(d) If Statement I is incorrect but Statement II is correct.
(e) If both Statement I and Statement II are incorrect

29 **Statement I** A body cannot have energy without having momentum but it can have momentum without having energy.

Statement II Momentum and energy have same dimensions.

30 **Statement I** When a machine-gun fires n bullets per second with kinetic energy K, then the power of the machine-gun is $P = nK$.

Statement II Power $= \frac{\text{Work}}{\text{Time}} = \frac{nk}{1}$

31 **Statement I** A quick collision between two bodies is more violent than a slow collision; even when the initial and the final velocities are identical.

Statement II The momentum is greater in the first case.

32 **Statement I** A point particle of mass m moving with speed v collides with stationary point particle of mass M. If the maximum energy loss possible is given as $f\left(\frac{1}{2}mv^2\right)$, then $f = \left(\frac{M}{M+m}\right)$.

Statement II Maximum energy loss occurs when the particles get stuck together as a result of the collision. → JEE Main 2013

DAY PRACTICE SESSION 2

PROGRESSIVE QUESTIONS EXERCISE

1 A block of mass 0.5 kg has an initial velocity of 10 ms^{-1} while moving down an inclined plane of angle 30°, the coefficient of friction between the block and the inclined surface is 0.2. The velocity of the block, after it covers a distance of 10 m, is

(a) 17 ms^{-1} (b) 13 ms^{-1}
(c) 24 ms^{-1} (d) 8 ms^{-1}

2 Two inclined frictionless tracks, one gradual and the other steep meet at *A* from where two stones are allowed to slide down from rest, one on each track as shown in figure.

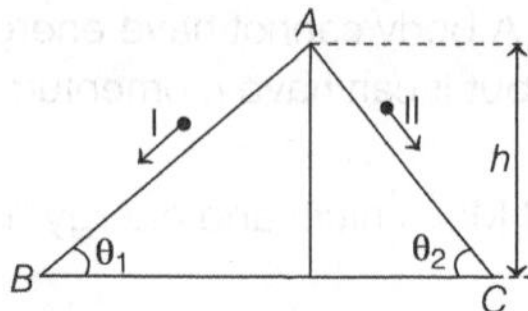

Which of the following statement is correct ?

(a) Both the stones reach the bottom at the same time but not with the same speed
(b) Both the stones reach the bottom with the same speed and stone I reaches the bottom earlier than stone II
(c) Both the stones reach the bottom with the same speed and stone II reaches the bottom earlier than stone I
(d) Both the stones reach the bottom at different times and with different speeds

3 Suppose the average mass of raindrops is 3×10^{-5} kg and their average terminal velocity 9 ms^{-1}. Then the energy transferred by rain to each square metre of the surface at a place which receives 100 cm of rain in a year is

(a) 302×10^2 J (b) 102×10^5 J
(c) 4.05×10^4 J (d) 9.2×10^3 J

4 A ball of mass *m* is dropped from a height *h* on a massless platform fixed at the top of a vertical spring as shown below. The platform is depressed by a distance *x*. What will be the value of the spring constant?

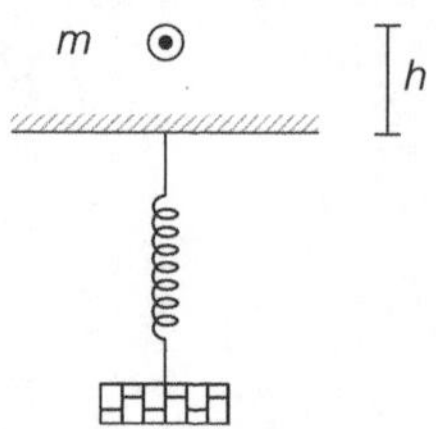

(a) $2mg/x$ (b) $2mgh/x$
(c) $2mg(h+x)/x^2$ (d) $2mg(h) + 2mghx/x^2$

5 A force $\mathbf{F} = -k(y\,\mathbf{i} + x\,\mathbf{j})$, acts on a particle moving in the *xy*-plane. Starting from the origin, the particle is taken along the positive *x*-axis to the point $(a, 0)$ and is then taken parallel to the *y*-axis to the point (a, a). The total work done by the force is

(a) $-2ka^2$ (b) $2ka^2$ (c) $-ka^2$ (d) ka^2

6 A uniform chain of length 2 m is kept on a table such that a length of 60 cm hangs freely from the edge of the table. The total mass of the chain is 4 kg. What is the work done in pulling the entire chain on the table?

(a) 7.2 J (b) 3.6 J (c) 120 J (d) 1200 J

7 A 70 kg man leaps vertically into the air from a crouching position. To take the leap the man pushes the ground with a constant force *F* to raise himself.

The centre of gravity rises by 0.5 m before he leaps. After the leap the centre of gravity rises by another 1 m. The maximum power delivered by the muscles is (take, $g = 10\ ms^{-2}$). → **JEE Main (Online) 2013**

(a) 6.26×10^3 W at the start (b) 6.26×10^3 W at take off
(c) 6.26×10^4 W at the start (d) 6.26×10^4 W at take off

8 If two springs S_1 and S_2 of force constants k_1 and k_2, respectively are stretched by the same force, it is found that more work is done on spring S_1 than on spring S_2. The correct option is

(a) $w_1 = w_2$ (b) $k_1 < k_2$
(c) $k_1 > k_2$ (d) None of these

9 At time $t = 0$ s particle starts moving along the *x*-axis. If its kinetic energy increases uniformly with time *t*, the net force acting on it must be proportional to

(a) $\sqrt{t}$ (b) constant (c) t (d) $\frac{1}{\sqrt{t}}$

10 It is found that, if a neutron suffers an elastic collinear collision with deuterium at rest, fractional loss of its energy is P_d; while for its similar collision with carbon nucleus at rest, fractional loss of energy is P_c. The values of P_d and P_c are respectively

(a) (.89, .28) (b) (.28, .89)
(c) (0, 0) (d) (0, 1)

11 A particle of mass *m* moving in the *x*-direction with speed 2*v* is hit by another particle of mass 2*m* moving in the *y*-direction with speed *v*. If the collision is perfectly inelastic, the percentage loss in the energy during the collision is close to → **JEE Main 2015**

(a) 44% (b) 50% (c) 56% (d) 62%

12 Consider a rubber ball freely falling from a height $h = 4.9$ m onto a horizontal elastic plate. Assume that the duration of collision is negligible and the collision with the plate is totally elastic. Then, the velocity as a function of time the height as function of time will be

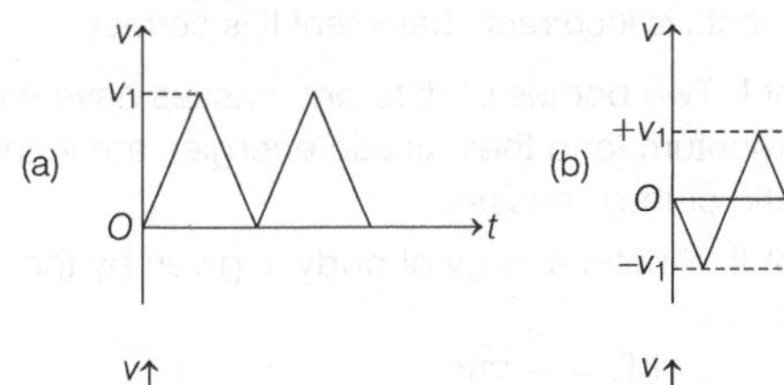

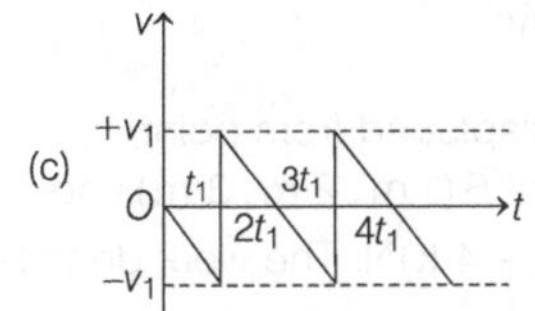

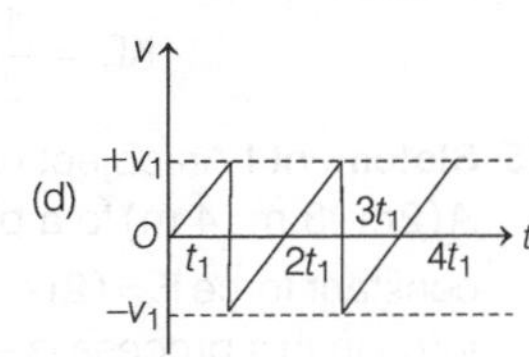

13 A body of mass $m = 10^{-2}$ kg is moving in a medium and experiences a frictional force $F = -kv^2$. Its initial speed is $v_0 = 10\,\text{ms}^{-1}$. If after 10 s, its energy is $\frac{1}{8}mv_0^2$, the value of k will be → JEE Main 2017 (Offline)

(a) $10^3\,\text{kgs}^{-1}$ (b) $10^{-4}\,\text{kgm}^{-1}$
(c) $10^{-1}\,\text{kgm}^{-1}\text{s}^{-1}$ (d) $10^{-3}\,\text{kgm}^{-1}$

14 A uniform chain of length l and weight w is hanging from its ends A and B which are close together. At a given instant end B is released. The tension at A when B has fallen a distance $x < \frac{l}{2}$ is

(a) $\frac{w}{2}\left[\frac{3x}{l} - 2\right]$ (b) $\frac{w}{2}\left[3 - \frac{3x}{4}\right]$
(c) $\frac{w}{2}\cdot\left[1 + \frac{3x}{l}\right]$ (d) $\frac{w}{2}\left[\frac{3x}{l} + 4\right]$

15 A body is moved along a straight line by a machine delivering a constant power. The distance moved by the body in time t is proportional to

(a) $t^{3/4}$ (b) $t^{3/2}$
(c) $t^{1/4}$ (d) $t^{1/2}$

16 In a collinear collision, a particle with an initial speed v_0 strikes a stationary particle of the same mass. If the final total kinetic energy is 50% greater than the original kinetic energy, the magnitude of the relative velocity between the two particles after collision, is → JEE Main 2018

(a) $\frac{v_0}{4}$ (b) $\sqrt{2}\,v_0$
(c) $\frac{v_0}{2}$ (d) $\frac{v_0}{\sqrt{2}}$

17 A mass m moves with a velocity v and collides inelastically with another identical mass. After collision, the 1st mass moves with velocity $\frac{v}{\sqrt{3}}$ in a direction perpendicular to the initial direction of motion. Find the speed of the second mass after collision.

(a) v (b) $\sqrt{3}\,v$ (c) $\frac{2}{\sqrt{3}}v$ (d) $\frac{v}{\sqrt{3}}$

18 The potential energy function for the force between two atoms in a diatomic molecule is approximately given by $U(x) = \frac{a}{x^{12}} - \frac{b}{x^6}$, where a and b are constants and x is the distance between the atoms. If the dissociation energy of the molecule is $D = [U(x = \infty) - U_{\text{at equilibrium}}]$, D is

(a) $\frac{b^2}{2a}$ (b) $\frac{b^2}{12a}$ (c) $\frac{b^2}{4a}$ (d) $\frac{b^2}{6a}$

19 Under the action of force, 3 kg body moves such that its position x as a function of time t is given by $x = \frac{t^3}{3}$, where x in meter and t in second. Then, work done by the force in the first 2 s is

(a) 8 J (b) 16 J (c) 24 J (d) 30 J

20 When a rubber band is stretched by a distance x, it exerts a restoring force of magnitude $F = ax + bx^2$, where, a and b are constants. The work done in stretching the unstretched rubber band by L is → JEE Main 2014

(a) $aL^2 + bL^3$ (b) $\frac{1}{2}(aL^2 + bL^3)$
(c) $\frac{aL^2}{2} + \frac{bL^3}{3}$ (d) $\frac{1}{2}\left(\frac{aL^2}{2} + \frac{bL^3}{3}\right)$

21 A spring gun having a spring of spring constant k is placed at a height h. A ball of mass m is placed in its barrel and compressed by a distance x. Where shall we place a box on the ground, so that the ball lands in the box?

(a) $\frac{kh}{mg}x$ (b) $\sqrt{\frac{2kh}{mg}}x$ (c) $\sqrt{\frac{kh}{2mg}}x$ (d) $\frac{kh}{2mg}$

22 A particle of mass m is projected from the ground with an initial speed u_0 at an angle α with the horizontal. At the highest point of its trajectory, it makes a completely inelastic collision with another identical particle, which was thrown vertically upward from the ground with the same initial speed u_0. The angle that the composite system makes with the horizontal immediately after the collision is

(a) $\frac{\pi}{4}$ (b) $\frac{\pi}{4} + \alpha$ (c) $\frac{\pi}{4} - \alpha$ (d) $\frac{\pi}{2}$

23 Two small particles of equal masses start moving in opposite directions from a point A in a horizontal circular orbit. Their tangential velocities are v and $2v$ respectively, as shown in the figure. Between collisions, the particles move with constant speeds. After making how many elastic collisions, other than that at A, these two particles will again reach the point A?

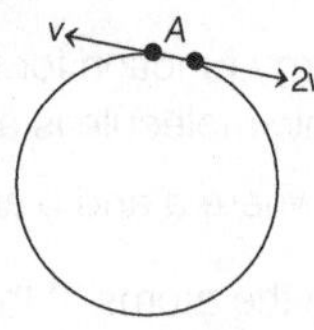

(a) 4 (b) 3 (c) 2 (d) 1

Direction (Q. Nos. 24-25) *Each of these questions contains two statements : Statement I (Assertion) and Statement II (Reason). Each of these questions also has four alternative choices, only one of which is the correct answer. You have to select one of the codes (a), (b), (c), (d) given below*

(a) Statement I is correct, Statement II is correct; Statement II is the correct explanation for Statement I
(b) Statement I is correct, Statement II is correct; Statement II is not the correct explanation for Statement I
(c) Statement I is correct; Statement II is incorrect
(d) Statement I is incorrect; Statement II is correct

24 Statement I Two bodies of different masses have the same momentum, and their kinetic energies are in the inverse ratio of their masses.

Statement II Kinetic energy of body is given by the relation.

$$\text{KE} = \frac{1}{2}mv^2$$

25 Statement I An object is displaced from point $A(2\text{m}, 3\text{ m}, 4\text{ m})$ to a point $B(1\text{ m}, 2\text{ m}, 3\text{ m})$ under a constant force $\mathbf{F} = (2\mathbf{i} + 3\mathbf{j} + 4\mathbf{k})$N. The work done by the force in this process is −9 J.

Statement II Work done by a force, an object can be given by the relation,

$$W = \int_{r_1}^{r_2} \mathbf{F} \cdot dr \quad \text{or} \quad W = \mathbf{F} \cdot \mathbf{s}$$

ANSWERS

SESSION 1									
1 (c)	**2** (c)	**3** (a)	**4** (b)	**5** (c)	**6** (a)	**7** (d)	**8** (c)	**9** (a)	**10** (a)
11 (d)	**12** (b)	**13** (a)	**14** (a)	**15** (c)	**16** (b)	**17** (b)	**18** (c)	**19** (a)	**20** (b)
21 (c)	**22** (c)	**23** (c)	**24** (a)	**25** (a)	**26** (d)	**27** (d)	**28** (b)	**29** (e)	**30** (a)
31 (a)	**32** (a)								

SESSION 2									
1 (b)	**2** (c)	**3** (c)	**4** (c)	**5** (c)	**6** (b)	**7** (b)	**8** (b)	**9** (d)	**10** (a)
11 (c)	**12** (c)	**13** (b)	**14** (c)	**15** (b)	**16** (b)	**17** (c)	**18** (c)	**19** (c)	**20** (c)
21 (b)	**22** (a)	**23** (c)	**24** (a)	**25** (a)					

Hints and Explanations

SESSION 1

1 Resultant force,

$$F = \sqrt{3^2 + 4^2} = 5\text{N}$$

$$a = \frac{F}{m} = \frac{5}{2} = 2.5\,\text{m/s}^2$$

$$u = 0, t = 20\text{ s}$$

$$\therefore\ s = ut + \frac{1}{2}at^2$$

$$= 0 \times 20 + \frac{1}{2} \times 2.5 \times (20)^2 = 500\text{ m}$$

Hence, work done

$$W = F \cdot s = 5 \times 500 = 2500\text{ J}$$

2 Work done in the round trip = total work done against friction while moving up and down the plane

$$= 2(\mu\, mg\cos\theta) \times s$$

$$= 2(0.1 \times 0.5 \times 10 \times \frac{\sqrt{3}}{2} \times 10)$$

$$= 5\sqrt{3}\text{ J} \quad \left[\because \cos\theta = \frac{5\sqrt{3}}{10} = \frac{\sqrt{3}}{2}\right]$$

3 Work done among horizontal is

$$dW = F \cdot dx$$

or $dW = (9 - x^2) \cdot dx$

For total work done, integrate both sides with proper limit

$$\int_0^W dW = \int_0^3 (9 - x^2)dx$$

or $$W = \left[9x - \frac{x^3}{3}\right]_0^3$$

$$= \left(27 - \frac{27}{3}\right) - (0 - 0) = 18\text{ J}$$

$\therefore$ Maximum work done is 18 J.

4 Since, $\mathbf{F}$ = constant, we can also use

$$W = \mathbf{F} \cdot \mathbf{s}$$

Here, $\mathbf{s} = \mathbf{r}_f - \mathbf{r}_i = (\hat{\mathbf{i}} + 2\hat{\mathbf{j}} + 3\hat{\mathbf{k}}) - (2\hat{\mathbf{i}} + 3\hat{\mathbf{j}} + 4\hat{\mathbf{k}})$

$$= (-\hat{\mathbf{i}} - \hat{\mathbf{j}} - \hat{\mathbf{k}})$$

$$\therefore\ W = (2\hat{\mathbf{i}} + 3\hat{\mathbf{j}} + 4\hat{\mathbf{k}}) \cdot (-\hat{\mathbf{i}} - \hat{\mathbf{j}} - \hat{\mathbf{k}})$$

$$= -2 - 3 - 4 = -9\text{ J}$$

5 If v' is the final velocity of the wagon, then applying the principle of conservation of linear momentum, we get

$$5 \times 10^3 \times 1.2 = (5 \times 10^3 + 10^3) \times v'$$

$$v' = 1\text{ ms}^{-1}$$

Change in KE

$$= \frac{1}{2}(6 \times 10^3) \times 1^2 - \frac{1}{2}(5 \times 10^3)(1.2)^2$$

$= -600$ J [minus sign for the loss in kinetic energy]

6 $a = -kx$ or $\dfrac{vdv}{dx} = -kx$

$$\Rightarrow\ vdv = -kxdx$$

Let the velocity change from v_0 to v.

$$\Rightarrow\ \int_{v_0}^{v} dv = -\int_0^x kx\,dx$$

$$\Rightarrow\ \frac{v^2 - v_0^2}{2} = -\frac{kx^2}{2}$$

$$\Rightarrow\ m\left(\frac{v^2 - v_0^2}{2}\right) = -\frac{mkx^2}{2}$$

$$\Rightarrow\ \Delta k \propto x^2$$

[Δk is loss in kinetic energy]

7 As, the earth moves ones around the sun in its elliptical orbit, its KE, is maximum when it is closest to the sun and minimum when it is farthest from the sun. Also, KE of the earth is never zero during its motion choice (d) is correct.

8 $\therefore$ Force $= -\dfrac{dU}{dr}$

$$\Rightarrow\ F = -\frac{d}{dr}\left(\frac{-k}{2r^2}\right) = -\frac{k}{r^3}$$

As particle is on circular path, this force must be centripetal force.

$$\Rightarrow\ |F| = \frac{mv^2}{r}$$

So, $$\frac{k}{r^3} = \frac{mv^2}{r} \Rightarrow \frac{1}{2}mv^2 = \frac{k}{2r^2}$$

$\therefore$ Total energy of particle = KE + PE

$$= \frac{k}{2r^2} - \frac{k}{2r^2}$$

$= 0$

Total energy = 0

9 $V(x) = \left(\dfrac{x^4}{4} - \dfrac{x^2}{2}\right)$ J

For minimum value of V, $\dfrac{dV}{dx} = 0$

$$\Rightarrow\ \frac{4x^3}{4} - \frac{2x}{2} = 0$$

$$\Rightarrow\ x = 0, x = \pm 1$$

So, $V_{\min}(x = \pm 1) = \dfrac{1}{4} - \dfrac{1}{2} = \dfrac{-1}{4}$ J

Now, $K_{\max} + V_{\min}$ = total mechanical energy

$$\Rightarrow\ K_{\max} = \left(\frac{1}{4}\right) + 2$$

or $$K_{\max} = \frac{9}{4}$$

or $$\frac{mv^2}{2} = \frac{9}{4}$$

or $$v = \frac{3}{\sqrt{2}}\text{ ms}^{-1}$$

10 From $F = -\dfrac{dU}{dx}$

$$\int_0^{U(x)} dU = -\int_0^x F\,dx = -\int_0^x (kx)dx$$

$$\therefore\ U(x) = -\frac{kx^2}{2} \text{ as } U(0) = 0$$

11 According to the work-energy theorem,

$$W = \Delta K$$

***Case* I**

$$-F \times 3 = \frac{1}{2}m\left(\frac{v_0}{2}\right)^2 - \frac{1}{2}mv_0^2$$

where, F is the resistive force and v_0 is the initial speed.

***Case* II** Let, the further distance travelled by the bullet before coming to rest is s.

$$\therefore\ -F(3 + s) = K_f - K_i = -\frac{1}{2}mv_0^2$$

$$\Rightarrow\ -\frac{1}{8}mv_0^2(3 + s) = -\frac{1}{2}mv_0^2$$

or $$\frac{1}{4}(3 + s) = 1$$

or $$\frac{3}{4} + \frac{s}{4} = 1$$

or $$s = 1\text{ cm}$$

12 From the work-energy theorem, kinetic energy of the block at a distance x is

$$K = \int_0^x F\,dx = \int_0^x (4 - x^2)dx = 4x - \frac{x^3}{3}$$

For kinetic energy to be maximum,

$$\frac{dK}{dx} = 0$$

$$\frac{d}{dx}\left(4x - \frac{x^3}{3}\right) = 0$$

$$4 - x^2 = 0 \text{ or } x = \pm 2\text{ m}$$

At $x = +2$m, $\dfrac{d^2K}{dx^2}$ = negative

i.e. Kinetic energy is maximum.

$$K_{\max} = 4x - \frac{x^3}{3}$$

$$= 4(2) - \frac{2^3}{3} = 5.33\text{ J}$$

13 According to the conservation of energy,

$$\frac{1}{2}kL^2 = \frac{1}{2}Mv^2$$

$$\Rightarrow\ kL^2 = \frac{(Mv)^2}{M}$$

or $$MkL^2 = p^2 \quad [\because p = Mv]$$

$$\Rightarrow\ p = L\sqrt{Mk}$$

14 $v = r\omega = r\frac{2\pi}{t} = \frac{1}{4} \times \frac{2\pi}{2} = \frac{\pi}{4}\ \text{ms}^{-1}$

$P = F \times v = (15 \times 9.8) \times \frac{\pi}{4} = 115.6\ \text{W}$

15 From the work-energy theorem,

$$\Delta KE = W_{net}$$
$$K_f - K_i = \int P\,dt$$
$$\frac{1}{2}mv^2 - 0 = \int_0^2 \left(\frac{3}{2}t^2\right)dt$$
$$\frac{1}{2}(2)v^2 = \frac{3}{2}\int_0^2 t^2 dt$$
$$= \frac{3}{2} \times \left[\frac{t^3}{3}\right]_0^2 = 4$$
$$\Rightarrow \quad v = 2\ \text{ms}^{-1}$$

16 Here, $P = [\text{ML}^2\ \text{T}^{-3}]$ = constant

As mass M of body is fixed

$[\text{L}^2\ \text{T}^{-3}]$ = constant

$\Rightarrow \quad \frac{[\text{L}^2]}{[\text{T}^3]}$ = constant

$\Rightarrow \quad [\text{L}] \propto [\text{T}^{3/2}]$

$\Rightarrow$ Displacement $\propto t^{3/2}$

17 As the ball bearings are identical, their masses are equal. In elastic collision, their velocities are interchanged. In collision between 1 and 2; velocity of 1 becomes zero and velocity of 2 becomes v. In collision between 2 and 3, velocity of 2 becomes zero and velocity of 3 becomes v.

18 From the conservation of momentum, we get

$$m_1 v_1 = m_2 v_2 \quad \text{...(i)}$$

also, $\frac{1}{2}m_1 v_1^2 = f_1 S_1 = \mu m_1 g S_1 \quad \text{...(ii)}$

and $\frac{1}{2}m_2 v_2^2 = f_2 S_2 = \mu m_2 g S_2 \quad \text{...(iii)}$

where, μ = coefficient of friction

On dividing Eq. (ii) by Eq. (iii), we get

$$\frac{m_1 v_1^2}{m_2 v_2^2} = \frac{m_1 S_1}{m_2 S_2} \quad \text{...(iv)}$$

Using Eqs. (i) and (iv), we get

$$\frac{v_1}{v_2} = \frac{m_1 S_1}{m_2 S_2}$$

or $\frac{m_2}{m_1} = \frac{m_1 S_1}{m_2 S_2} \Rightarrow S_2 = \frac{m_1^2}{m_2^2} S_1$

19 Velocity at the highest point $= v\cos\theta$

Applying the principle of conservation of linear momentum, we get

$2m(v\cos\theta) = m(-v\cos\theta) + mv'$

$v' = 3\,v\cos\theta$

20 The collision will cause an exchange of velocities. The change in momentum of any particle = mu, which is equal to the impulse = area under the force-time graph

$$mu = \frac{1}{2}F_0 \times T$$
$$\Rightarrow \quad F_0 = 2\frac{mu}{T}$$

21 $m_1 u_1 + m_2 u_2 = (m_1 + m_2)v$

$\Rightarrow \quad v = 2/3\ \text{ms}^{-1}$

Energy loss

$= \frac{1}{2}(0.5) \times (2)^2 - \frac{1}{2}(1.5) \times \left(\frac{2}{3}\right)^2$

$= 0.67\ \text{J}$

22 In an inelastic collision only momentum of the system may remain conserved. Some energy can be lost in the form of heat, sound, etc.

23 $|(m_1 \mathbf{v}_1' + m_2 \mathbf{v}_2') - (m_1 \mathbf{v}_1 + m_2 \mathbf{v}_2)|$

= |change in momentum of the two particles |

= |External force on the system | × time interval

$= (m_1 + m_2)\,g\,(2t_0) = 2\,(m_1 + m_2)\,g t_0$

24 Let v is the final velocity of ball on reaching the ground, then

$$v = \sqrt{2gh} = \sqrt{2 \times g \times 2}$$

or $\quad v = 2\sqrt{g}$

For upward motion

$$u = 2\sqrt{g} \times e \quad \text{and} \quad v = 0$$

$\therefore$ Height upto which the ball will rebound is

$$H = \frac{u^2}{2g} = \frac{(2\sqrt{g} \times e)^2}{2g}$$
$$= \frac{4g \times e^2}{2g} = 2 \times 0.6 \times 0.6 = 0.72\ \text{m}$$

25 $F = ma = \frac{mv}{T} \quad \left(\because a = \frac{v-0}{T}\right)$

Instantaneous power $= Fv = mav$

$$= \frac{mv}{T}at$$
$$= \frac{mv}{T}\frac{v}{T}t = \frac{mv^2}{T^2}t$$

26 Given, potential energy burnt by lifting weight $= mgh = 10 \times 9.8 \times 1 \times 1000$

$= 9.8 \times 10^4$

If mass lost by a person be m, then

energy dissipated $= m \times \frac{2}{10} \times 3.8 \times 10^7$ J

$\Rightarrow \quad 9.8 \times 10^4 = m \times \frac{1}{5} \times 3.8 \times 10^7$

$\Rightarrow \quad m = \frac{5}{3.8} \times 10^{-3} \times 9.8$

$= 12.89 \times 10^{-3}$ kg

27 From Newton's second law,

$$\frac{\Delta p}{\Delta t} = F \Rightarrow \Delta p = F\Delta t$$

$\therefore \quad p = \int dp = \int_0^1 F\,dt$

$\Rightarrow \quad p = \int_0^1 6t\,dt = 3\ \text{kg}\left(\frac{\text{m}}{\text{s}}\right)$

Also, change in kinetic energy

$$\Delta k = \frac{\Delta p^2}{2m} = \frac{3^2}{2 \times 1} = 4.5$$

From work-energy theorem,

Work done = change in kinetic energy

So, work done $= \Delta k = 4.5$ J

28 Kinetic energy $= \frac{1}{2}mv^2$

Potential energy $= mgh$

Spring energy $= \frac{1}{2}kx^2$

29 We know that a body may not have momentum but may have potential energy by virtue of its position as in case of a stretched or a compressed spring. But when the body does not contain energy then its kinetic energy is zero hence, its momentum is also zero.

Dimensions of momentum

$$(mv) = [\text{MLT}^{-1}]$$

Dimensions of energy

$$\left(\frac{1}{2}mv^2\right) = [\text{ML}^2\text{T}^{-2}]$$

30 $\therefore$ Power $= \frac{\text{Work}}{\text{Time}} = \frac{n \times K}{1} = \frac{nK}{1}$

31 Momentum $p = mv$ or $p \propto v$,

i.e. momentum is directly proportional to the velocity, so the momentum is greater in a quicker collision between two bodies than in a slower one. Hence, due to greater momentum, quicker collision between two bodies will be more violent even if the initial and the final velocities are identical.

32 Energy $E = \frac{p^2}{2m}$, where p is momentum,

m is the mass moving of the particle. Maximum energy loss occurs when the particles get stuck together as a result of the collision.

Maximum energy loss (ΔE)

$$= \frac{p^2}{2m} - \frac{p^2}{2(m+M)}$$

where, $(m + M)$ is the resultant mass when the particles get stuck.

$$\Delta E = \frac{p^2}{2m}\left[1 - \frac{m}{m+M}\right] = \frac{p^2}{2m}\left[\frac{M}{m+M}\right]$$

Also, $p = mv$

$$\therefore \quad \Delta E = \frac{m^2v^2}{2m}\left[\frac{M}{m+M}\right]$$
$$= \frac{mv^2}{2}\left[\frac{M}{m+M}\right]$$

Comparing the expression with

$$\Delta E = f\left(\frac{1}{2}mv^2\right), f = \frac{M}{m+M}$$

SESSION 2

1 Here, $m = 0.5$ kg, $u = 10$ ms^{-1}, $\theta = 30°$, $\mu = 0.2$, $s = 10$ m

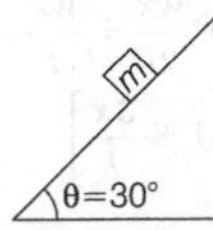

Acceleration down the plane,

$$a = g(\sin\theta - \mu\cos\theta)$$
$$= 10(\sin 30° - 0.2\cos 30°)$$
$$= 3.268 \text{ ms}^{-2}$$

From second equation of motion,

$$v^2 = u^2 + 2as$$
$$= 10^2 + 2(3.268) \times 10$$
$$= 165.36$$
$$v = \sqrt{165.36}$$
$$\approx 13 \text{ ms}^{-1}$$

2 As, both surfaces I and II are frictionless and two stones slide from rest from the same height, therefore, both the stones reach the bottom with same speed $\left(\frac{1}{2}mv^2 = mgh\right)$.

As acceleration down plane II is larger ($a_2 = g\sin\theta_2$ is greater than $a_1 = g\sin\theta_1$), therefore, stone II reaches the bottom earlier than stone I.

3 Given, average mass of rain drop $(m) = 3.0 \times 10^{-5}$ kg

Average terminal velocity $= (v)$
$= 9$ ms^{-1}

Height $(h) = 100$ cm $= 1$m

Density of water $(\rho) = 10^3$ kgm^{-3}

Area of the surface $(A) = 1\text{m}^2 = A \times h$
$= 1 \times 1$
$= 1\text{m}^3$

Mass of the water due to rain (M)
$=$ Volume $\times$ density
$= V \times \rho = 1 \times 10^3$
$= 10^3$ kg

$\therefore$ Energy transferred to the surface

$$= \frac{1}{2}mv^2 = \frac{1}{2} \times 10^3 \times (9)^2$$
$$= 40.5 \times 10^3 \text{ J}$$
$$= 4.05 \times 10^4 \text{ J}$$

4 Here, $mg(h + x) = \frac{1}{2}kx^2$

$$\Rightarrow \quad kx^2 = 2mg(h + x)$$
$$\Rightarrow \quad k = \frac{2mg(h + x)}{x^2}$$

5 $W_1 = \int_0^a \mathbf{F} dx = \int_0^a -k(y\hat{\mathbf{i}} + x\hat{\mathbf{j}})\hat{\mathbf{i}}\, dx$

$$= \int_0^a -k(0\hat{\mathbf{i}} + x\hat{\mathbf{j}})\hat{\mathbf{i}}\, dx = \text{zero}$$

$$W_2 = \int_0^a \mathbf{F} \cdot d\mathbf{y} = \int_0^a -k(y\hat{\mathbf{i}} + x\hat{\mathbf{j}})\,\hat{\mathbf{j}}\, dy$$
$$= \int_0^a -k(a\hat{\mathbf{i}} + a\hat{\mathbf{j}})\hat{\mathbf{j}}\, dy$$
$$= -ka\int_0^a dy = -ka^2$$

Total work done,

$$W = W_1 + W_2 = 0 - ka^2 = -ka^2$$

6 Mass per unit length $= \frac{M}{L}$

$$= \frac{4}{2} = 2 \text{ kg m}^{-1}$$

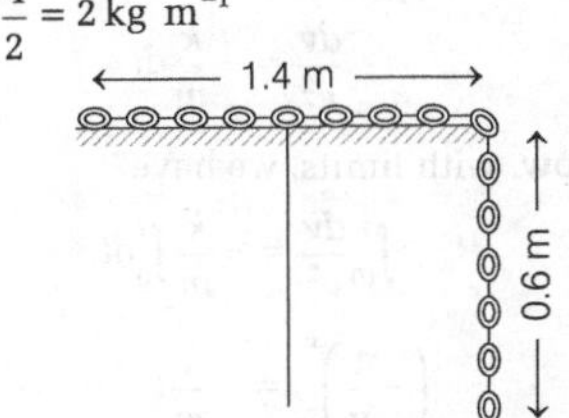

The mass of 0.6 m of chain $= 0.6 \times 2$
$= 1.2$ kg

The height of the centre of mass of the hanging part

$$h = \frac{0.6 + 0}{2} = 0.3 \text{ m}$$

Hence, work done in pulling the chain on the table = work done against the force of gravity

i.e. $W = mgh = 1.2 \times 10 \times 0.3 = 3.6$ J

7 As, $\quad P = F \cdot v$

So, $\quad \frac{dP}{dt} = F \cdot \frac{dv}{dt}$

To deliver the maximum power $\frac{dP}{dt} = 0$,

which gives

$$P_{\text{max}} = 6.26 \times 10^3 \text{ W}$$

8 As no relation between k_1 and k_2 is given in the question.

Then, for same force

$$W = Fx = F\frac{F}{k} = \frac{F^2}{k} \Rightarrow W \propto \frac{1}{k}$$

i.e. $\quad W_1 > W_2 \Rightarrow k_1 < k_2$

9 Given, $\frac{dk}{dt} = \text{constant} \Rightarrow k \propto t$

$$\Rightarrow \quad v \propto \sqrt{t}$$

Also, $\quad p = Fv = \frac{dk}{dt} = \text{constant}$

$$\Rightarrow \quad F \propto \frac{1}{v} \Rightarrow F \propto \frac{1}{\sqrt{t}}$$

10 Neutron-Deuterium collision;

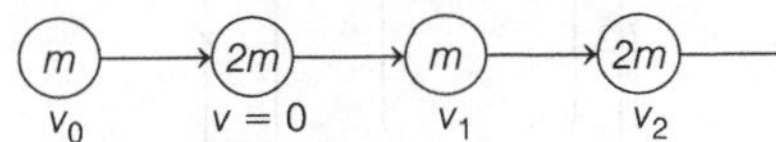

Momentum conservation gives;

$$mv_0 = mv_1 + 2mv_2$$
$$\Rightarrow \quad v_0 = v_1 + 2v_2 \quad \text{...(i)}$$

Collision given is elastic .

So, coefficient of restitution, $e = 1$

$$\therefore \quad e = 1 = \frac{\text{Velocity of separation}}{\text{Velocity of approach}}$$
$$\Rightarrow \quad 1 = \frac{v_2 - v_1}{v_0 - 0}$$
$$\Rightarrow v_0 = v_2 - v_1 \quad \text{...(ii)}$$

On adding Eqs. (i) and (ii), we get

$$2v_0 = 3v_2 \Rightarrow \frac{2v_0}{3} = v_2$$

So, from Eq. (i), we get

$$v_1 = v_0 - 2v_2 = v_0 - \frac{4v_0}{3}$$
$$\Rightarrow \quad v_1 = -\frac{v_0}{3}$$

Fractional loss of energy of neutron

$$= \left(\frac{-K_f + K_i}{K_i}\right)_{\text{for neutron}}$$
$$= \frac{-\frac{1}{2}mv_1^2 + \frac{1}{2}mv_0^2}{\frac{1}{2}mv_0^2} = \frac{-\frac{v_0^2}{9} + v_0^2}{v_0^2}$$
$$= \left(-\frac{1}{9} + 1\right) = \frac{8}{9} = 0.88$$
$$= 0.89$$

Similarly, for neutron-carbon atom collision;

Momentum conservation gives;

$$v_0 = v_1 + 12v_2 \text{ and } e = 1$$
$$\Rightarrow \quad v_0 = v_2 - v_1$$

So, $\quad v_1 = \frac{11}{13}v_0$

$$\therefore \text{ Loss of energy} = \left(-\frac{121}{169} + 1\right) = 0.28$$

So, $\quad P_d = 0.89$ and $P_c = 0.28$

11 Conservation of linear momentum can be applied but energy is not conserved. Consider the movement of two particles as shown below :

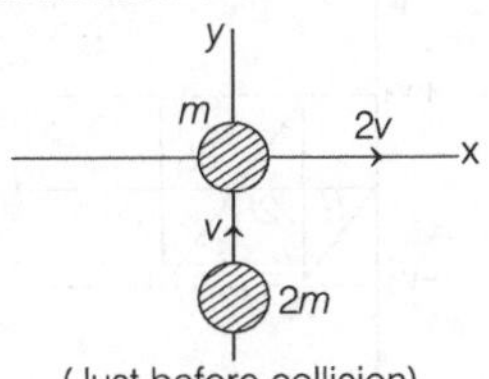

(Just before collision)

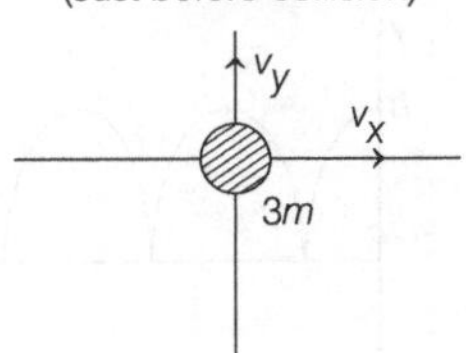

(Just after collision)

According to conservation of linear momentum in x-direction we have
$(p_1)x = (p_1)x$ or $2mv = (2m + m)v_x$

or $v_2 = \frac{2}{3}v$

As, conserving linear momentum in y-direction, we get

$$(p_i)y = (p_t)y$$

or $2\,mv = (2m + m)v_x$

or $v_y = \frac{2}{3}v$

Initial kinetic energy of the two particles system is

$$E = \frac{1}{2}m(2v)^2 + \frac{1}{2}(2m)(v)^2$$
$$= \frac{1}{2}\times 4mv^2 + \frac{1}{2}\times 2mv^2$$
$$= 2\,mv^2 + mv^2 = 3mv^2$$

Final energy of the combined two particles system is

$$E_t = \frac{1}{2}(3m)(v_x^2 + v_y^2)$$
$$= \frac{1}{2}(3m)\left[\frac{4v^2}{9} + \frac{4v^2}{9}\right]$$
$$= \frac{3m}{2}\left[\frac{8v^2}{9}\right] = \frac{4mv^2}{3}$$

Loss in the energy,

$$\Delta E = E_i - E_f$$
$$= mv^2\left[3 - \frac{4}{3}\right] = \frac{5}{3}mv^2$$

Percentage loss in the energy during the collision

$$\frac{\Delta E}{E_i}\times 100 = \frac{\frac{5}{3}mv^2}{3mv^2}\times 100$$
$$= \frac{5}{9}\times 100 = 56\%$$

12 As we know that for vertical motion,
$h = \frac{1}{2}gt^2$ [parabolic]

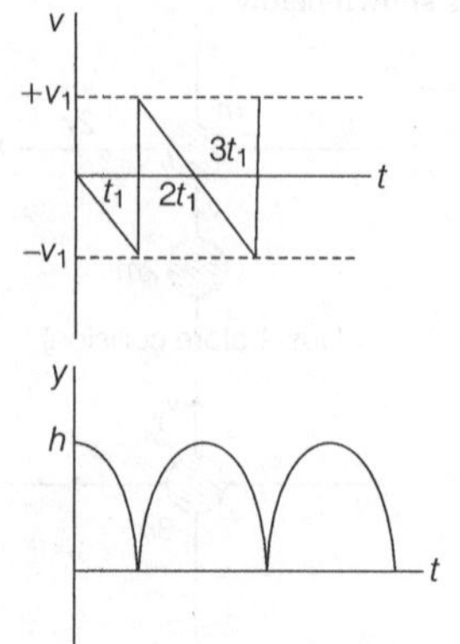

$v = -gt$ and after the collision,
$v = gt$ (straight line). [$\because v = u + gt$]

Collision is perfectly elastic, then ball reaches to same height again and again with same velocity.
Hence, option (c) is true.

13 Given force, $F = -kv^2$

$\therefore$ Acceleration, $a = \frac{-k}{m}v^2$

or $\frac{dv}{dt} = \frac{-k}{m}v^2$

$\Rightarrow$ $\frac{dv}{v^2} = -\frac{k}{m}dt$

Now, with limits, we have

$$\int_{10}^{v}\frac{dv}{v^2} = -\frac{k}{m}\int_0^t dt$$

$\Rightarrow$ $\left(-\frac{1}{v}\right)_{10}^{v} = -\frac{k}{m}t$

$\Rightarrow$ $\frac{1}{v} = 0.1 + \frac{kt}{m}$

$\Rightarrow$ $v = \frac{1}{0.1 + \frac{kt}{m}} = \frac{1}{0.1 + 1000k}$

$\Rightarrow$ $\frac{1}{2}\times m\times v^2 = \frac{1}{8}mv_0^2$

$\Rightarrow$ $v = \frac{v_0}{2} = 5$

$\Rightarrow$ $\frac{1}{0.1 + 1000k} = 5$

$\Rightarrow$ $1 = 0.5 + 5000k$

$\Rightarrow$ $k = \frac{0.5}{5000}$

$\Rightarrow$ $k = 10^{-4}$ kg/m

14 Mass per unit length,

$$\lambda = \frac{m}{l} = \frac{w}{lg}$$

Velocity, $v^2 = 2g\left(\frac{x}{2}\right)$

or $v = \sqrt{gx}$

Change in momentum when an element dx falls is $\frac{w}{lg}\sqrt{gx}\,dx$

$$F = \frac{dp}{dt} = \frac{w}{lg}\sqrt{gx}\frac{dx}{dt} = \frac{w}{l}x$$
$$\left(\because \frac{dx}{dt} = \sqrt{gx}\right)$$

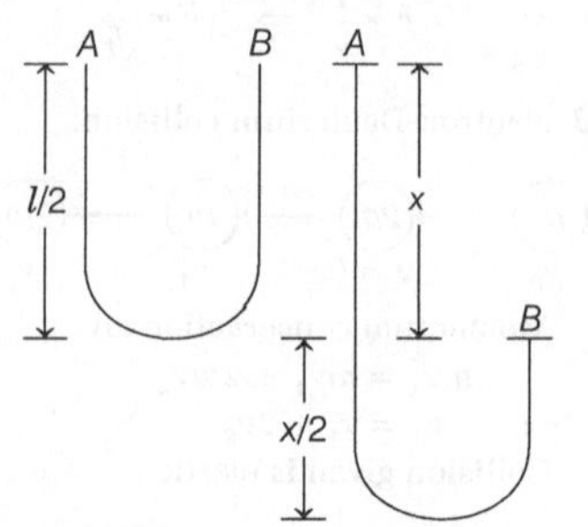

Tension at A

= Weight of half the chain
+ Weight of $\frac{x}{2}$ length + F

$$= \frac{w}{2} + \frac{wx}{2l} + \frac{wx}{l}$$
$$= \frac{w}{2}\left[1 + \frac{3x}{l}\right]$$

15 P = constant

$\Rightarrow$ $Fv = P$ [$\because P$ = force × velocity]

$\Rightarrow$ $Ma \times v = P$

$\Rightarrow$ $va = \frac{P}{M}$

$\Rightarrow v\times\left[\frac{vdv}{ds}\right] = \frac{P}{M}$ $\left[\because a = \frac{vdv}{ds}\right]$

$\Rightarrow$ $\int_0^v v^2dv = \int_0^s \frac{P}{M}ds$

[Assuming at $t = 0$ it starts from rest, i.e. from $s = 0$]

or $\frac{v^3}{3} = \frac{P}{M}s$

$\Rightarrow$ $v = \left(\frac{3P}{M}\right)^{1/3}s^{1/3}$

$\Rightarrow$ $\frac{ds}{dt} = ks^{1/3}$ $\left[k = \left(\frac{3P}{M}\right)^{1/3}\right]$

$\Rightarrow$ $\int_0^s \frac{ds}{s^{1/3}} = \int_0^t k\,dt$

or $\frac{s^{2/3}}{2/3} = kt$

$\Rightarrow$ $s^{2/3} = \frac{2}{3}kt$

or $s = \left(\frac{2}{3}k\right)^{3/2}\times t^{3/2}$

or $s \propto t^{3/2}$

16 Momentum is conserved in all type of collisions,
Final kinetic energy is 50% more than initial kinetic energy

$\Rightarrow$ $\frac{1}{2}mv_2^2 + \frac{1}{2}mv_1^2$

$$= \frac{150}{100}\times\frac{1}{2}mv_0^2 \quad \ldots\text{(i)}$$

m m
$\rightarrow v_0$
Before collision

m m
$\rightarrow v_1$ $\rightarrow v_2$
After collision

Conservation of momentum gives,

$$mv_0 = mv_1 + mv_2$$
$$v_0 = v_2 + v_1 \quad \ldots\text{(ii)}$$

From Eqs. (i) and (ii), we have

$$v_1^2 + v_2^2 + 2v_1v_2 = v_0^2$$

$$\Rightarrow \quad 2v_1v_2 = \frac{-v_0^2}{2}$$

$$\therefore \ (v_1 - v_2)^2 = (v_1 + v_2)^2 - 4v_1v_2 = 2v_0^2$$

or $\quad v_{\text{rel}} = \sqrt{2}v_0$

17 In x-direction,
Apply conservation of momentum, we get

$$mu_1 + 0 = mv_x$$

$$\Rightarrow \quad mv = mv_x$$

$$\Rightarrow \quad v_x = v$$

1 → 2

Before collision

$v/\sqrt{3}\,m$ ↑ 1   2 → v_x, ↓ v_y

After collision

In y-direction, apply conservation of momentum, we get

$$0 + 0 = m\left(\frac{v}{\sqrt{3}}\right) - mv_y \Rightarrow v_y = \frac{v}{\sqrt{3}}$$

Velocity of second mass after collision,

$$v' = \sqrt{\left(\frac{v}{\sqrt{3}}\right)^2 + v^2} = \sqrt{\frac{4}{3}v^2}$$

or $\quad v' = \frac{2}{\sqrt{3}}v$

18 $U(x) = \frac{a}{x^{12}} - \frac{b}{x^6}$

$U(x = \infty) = 0$

As, $\quad F = -\frac{dU}{dx} = -\left[\frac{12a}{x^{13}} + \frac{6b}{x^7}\right]$

At equilibrium, $F = 0$

$$\therefore \quad x^6 = \frac{2a}{b}$$

$$\therefore \ U_{\text{at equilibrium}} = \frac{a}{\left(\frac{2a}{b}\right)^2} - \frac{b}{\left(\frac{2a}{b}\right)} = \frac{-b^2}{4a}$$

$$\therefore \ D = [U(x = \infty) - U_{\text{at equilibrium}}] = \frac{b^2}{4a}$$

19 Here, $m = 3$ kg, $t = 2$ s,

$$x = \frac{t^3}{3} \quad \ldots\text{(i)}$$

Work done,

$$W = \int dW = \int F \cdot dx \quad \ldots\text{(ii)}$$

Differentiate Eq. (i) w.r.t. time, then we get

$$\frac{dx}{dt} = \frac{3t^2}{3} \text{ or } dx = t^2dt$$

or $\quad v = t^2 \quad \left[\text{as } \frac{dx}{dt} = v = \text{velocity}\right]$

or $\frac{dv}{dt} = 2t$ [on difference above relation]

or $\quad a = 2t \quad \left[\text{as } \frac{dv}{dt} = a = \text{acceleration}\right]$

$\therefore \quad F = ma = 6t$

Hence, work done, put the value of F and dx in Eq. (ii), we get

$$W = \int_0^2 6t \times t^2 dt = \left[6 \times \frac{t^4}{4}\right]_0^2 = \frac{6}{4}[2^4 - 0^4]$$

$$= \frac{3}{2} \times 16 = 24 \text{ J}$$

20 We know that change in potential energy of a system corresponding to a conservative internal force as,

$$U_f - U_i = -W = -\int \mathbf{F} \cdot d\mathbf{r}$$

Given, $\quad F = ax + bx^2$

We know that work done in stretching the rubber band by L is

$$|dW| = |Fdx|$$

$$|W| = \int_0^L (ax + bx^2)dx$$

$$= \left[\frac{ax^2}{2}\right]_0^L + \left[\frac{bx^3}{3}\right]_0^L$$

$$= \left[\frac{aL^2}{2} - \frac{a \times (0)^2}{2}\right] + \left[\frac{b \times L^3}{3} - \frac{b \times (0)^3}{3}\right]$$

$$|W| = \frac{aL^2}{2} + \frac{bL^3}{3}$$

21 From law of conservation of energy, we have

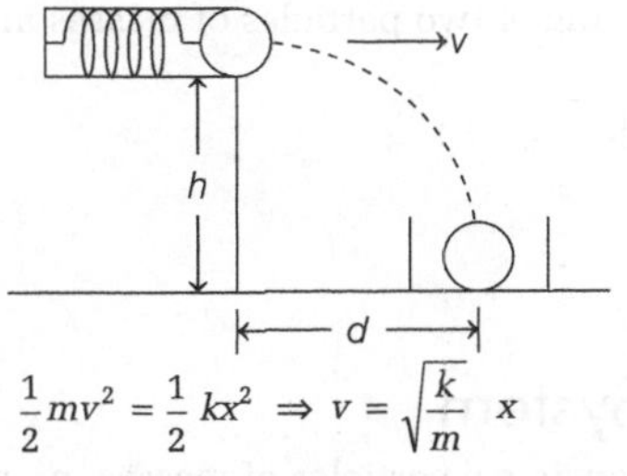

$$\frac{1}{2}mv^2 = \frac{1}{2}kx^2 \Rightarrow v = \sqrt{\frac{k}{m}}\,x$$

Time taken to fall, $t = \sqrt{\frac{2h}{g}}$

So, horizontal distance travelled is

$$d = vt = \sqrt{\frac{k}{m}}\,x\sqrt{\frac{2h}{g}}\,x = \sqrt{\frac{2kh}{mg}}\,x$$

22 From momentum conservation equation, we have,

m → $u_0 \cos \alpha$ ↑ $\sqrt{u_0^2 - 2gH}$, m $\hat{\mathbf{j}}$, $\hat{\mathbf{i}}$

$$\mathbf{p}_i = \mathbf{p}_f$$

$$\therefore \ m(u_0 \cos \alpha)\hat{\mathbf{i}} + m(\sqrt{u_0^2 - 2gH})\hat{\mathbf{j}} = (2m)\mathbf{v} \quad \ldots\text{(i)}$$

$$H = \frac{u_0^2 \sin^2 \alpha}{2g} \quad \ldots\text{(ii)}$$

From Eqs. (i) and (ii), we get

$$\mathbf{v} = \frac{u_0 \cos \alpha}{2}\hat{\mathbf{i}} + \frac{u_0 \cos \alpha}{2}\hat{\mathbf{j}}$$

Since, both components of $\mathbf{v}$ are equal. Therefore, it is making 45° with horizontal.

23 At first collision one particle having speed $2v$ will rotate 240° $\left(\text{or } \frac{4\pi}{3}\right)$ while other particle having speed v will rotate 120° $\left(\text{or } \frac{2\pi}{3}\right)$. At first collision they will exchange their velocities. Now as shown in figure, after two collisions they will again reach at point A.

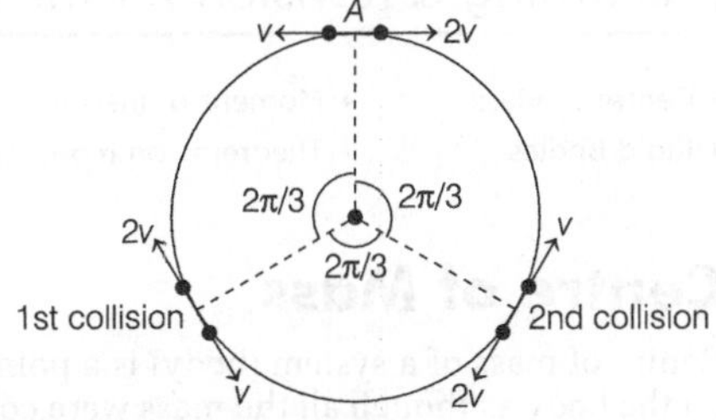

24 According to the principle of conservation of momentum

$$m_1v_1 = m_2v_2 \Rightarrow \frac{v_1}{v_2} = \frac{m_2}{m_1}$$

Again, $\quad \frac{\text{KE}_1}{\text{KE}_2} = \frac{\frac{1}{2}m_1v_1^2}{\frac{1}{2}m_2v_2^2}$

$$= \frac{m_1}{m_2} \times \left(\frac{m_2}{m_1}\right)^2 = \frac{m_2}{m_1}$$

$$\therefore \quad \text{KE} \propto \frac{1}{m}$$

25 $W = \int_{\mathbf{r}_1}^{\mathbf{r}_2} \mathbf{F} \cdot d\mathbf{r}$

$$\int_{(2\text{m}, 3\text{m}, 4\text{m})}^{(1\text{m}, 2\text{m}, 3\text{m})} (2\hat{\mathbf{i}} + 3\hat{\mathbf{j}} + 4\hat{\mathbf{k}}) \cdot (dx\hat{\mathbf{i}} + dy\hat{\mathbf{j}} + dz\hat{\mathbf{k}})$$

$$= [2x + 3y + 4z]_{(2\text{m}, 3\text{m}, 4\text{m})}^{(1\text{m}, 2\text{m}, 3\text{m})} = -9 \text{ J}$$

Alternate

Since, $\mathbf{F}$ = constant, we can also use

$$W = \mathbf{F} \cdot \mathbf{s}$$

Here, $\mathbf{s} = \mathbf{r}_f - \mathbf{r}_i$

$$= (\hat{\mathbf{i}} + 2\hat{\mathbf{j}} + 3\hat{\mathbf{k}}) - (2\hat{\mathbf{i}} + 3\hat{\mathbf{j}} + 4\hat{\mathbf{k}})$$

$$= (-\hat{\mathbf{i}} - \hat{\mathbf{j}} - \hat{\mathbf{k}})$$

$$\therefore \quad W = (2\hat{\mathbf{i}} + 3\hat{\mathbf{j}} + 4\hat{\mathbf{k}})(-\hat{\mathbf{i}} - \hat{\mathbf{j}} - \hat{\mathbf{k}})$$

$$= -2 - 3 - 4 = -9 \text{ J}$$

DAY SEVEN

System of Particles and Rigid Body

Learning & Revision for the Day

- Centre of Mass
- Rigid Bodies
- Moment of Inertia
- Theorems on Moment of Inertia
- Moment of Inertia of Some Geometrical Oojects

Centre of Mass

Centre of mass of a system (body) is a point that moves when external forces are applied on the body as though all the mass were concentrated at that point and all external forces were applied there.

Centre of Mass of Two Particle System

Centre of mass of a two particles system consisting of two particles of masses m_1, m_2 and respective position vectors $\mathbf{r}_1, \mathbf{r}_2$ is given by

$$\mathbf{r}_{CM} = \frac{m_1\mathbf{r}_1 + m_2\mathbf{r}_2}{m_1 + m_2}$$

If $m_1 = m_2 = m$ (say), then $\mathbf{r}_{CM} = \frac{\mathbf{r}_1 + \mathbf{r}_2}{2}$

Centre of Mass of *n*-Particle System

Centre of mass of $\mathbf{r}_{CM}$ particles system which consists n-particles of masses $m_1, m_2, \ldots, m_n$ with $\mathbf{r}_1, \mathbf{r}_2, \ldots, \mathbf{r}_n$ as their position vectors at a given instant of time is given by

$$\mathbf{r}_{CM} = \frac{m_1\mathbf{r}_1 + m_2\mathbf{r}_2 + \ldots + m_n\mathbf{r}_n}{m_1 + m_2 + \ldots + m_n} = \frac{\sum_{i=1}^{n} m_i\mathbf{r}_i}{M}$$

Cartesian Components of the Centre of Mass

The position vectors $\mathbf{r}_{CM}$ and $\mathbf{r}_i$ are related to their cartesian components by

$$\mathbf{r}_{CM} = x_{CM}\hat{i} + y_{CM}\hat{j} + z_{CM}\hat{k} \quad \text{and} \quad \mathbf{r}_i = x_i\hat{i} + y_i\hat{j} + z_i\hat{k}$$

The cartesian components of $\mathbf{r}_{CM}$ are given by

$$x_{CM} = \frac{\sum_{i=1}^{n} m_i x_i}{M}, \; y_{CM} = \frac{\sum_{i=1}^{n} m_i y_i}{M} \text{ and } z_{CM} = \frac{\sum_{i=1}^{n} m_i z_i}{M}$$

Your Personal Preparation Indicator

- No. of Questions in Exercises (x)—
- No. of Questions Attempted (y)—
- No. of Correct Questions (z)—
  (Without referring Explanations)
- Accuracy Level ($z/y \times 100$)—
- Prep Level ($z/x \times 100$)—

In order to expect good rank in JEE, your Accuracy Level should be above 85 & Prep Level should be above 75.

Motion of Centre of Mass

The position vector $\mathbf{r}_{CM}$ of the centre of mass of n particle system is defined by

$$\mathbf{r}_{CM} = \frac{m_1\mathbf{r}_1 + m_2\mathbf{r}_2 + m_3\mathbf{r}_3 + \ldots + m_n\mathbf{r}_n}{m_1 + m_2 + m_3 \ldots m_n}$$

$$= \frac{1}{M}(m_1\mathbf{r}_1 + m_2\mathbf{r}_2 + m_3\mathbf{r}_3 + \ldots + m_n\mathbf{r}_n)$$

$$\frac{d\mathbf{r}_{CM}}{dt} = \frac{1}{M}\left(m_1\frac{d\mathbf{r}_1}{dt} + m_2\frac{d\mathbf{r}_2}{dt} + \ldots + m_n\frac{d\mathbf{r}}{dt}\right)$$

Velocity of centre of mass $\mathbf{v}_{CM} = \frac{1}{M}(m_1\mathbf{v}_1 + m_2\mathbf{v}_2 + \ldots m_n\mathbf{v}_n)$

$$\mathbf{v}_{CM} = \frac{\sum_{i=1}^{n} m_i v_i}{M}$$

- Similarly, **acceleration of centre of mass** is given by

$$\mathbf{a}_{CM} = \frac{\sum_{i=1}^{n} m_i \mathbf{a}_i}{M}$$

- From Newton's second law of motion,

$$M\mathbf{a}_{CM} = \mathbf{F}_1 + \mathbf{F}_2 + \ldots \mathbf{F}_n \Rightarrow M\mathbf{a}_{CM} = \mathbf{F}_{Ext}$$

For an isolated system, if external force on the body is zero.

$$\mathbf{F} = M\mathbf{a}_{CM} = M\frac{d}{dt}(\mathbf{v}_{CM}) = 0 \Rightarrow \mathbf{v}_{CM} = \text{constant}$$

i.e. Centre of mass of an isolated system moves with uniform velocity along a straight line path and momentum remain conserved.

NOTE • If some mass or area is removed from a rigid body, then the position of centre of mass of the remaining portion is obtained from the following formula

$$\mathbf{r}_{CM} = \frac{m_1\mathbf{r}_1 - m_2\mathbf{r}_2}{m_1 - m_2} \text{ or } \mathbf{r}_{CM} = \frac{A_1\mathbf{r}_1 - A_2\mathbf{r}_2}{A_1 - A_2}$$

Momentum Conservation

Let us consider a system of particles of masses $m_1, m_2 \ldots m_n$ are respective velocities $\mathbf{v}_1, \mathbf{v}_2 \ldots \mathbf{v}_n$. The total linear momentum of the system would be the vector sum of the momentum of the individual particles.

i.e. $p = p_1 + p_2 + p_3 \ldots p_n = m_1\mathbf{v}_1 + m_2\mathbf{v}_2 + \ldots + m_n\mathbf{v}_n$

Velocity of centre of mass of a system

$$\mathbf{v}_{CM} = \frac{1}{M}(m_1\mathbf{v}_1 + m_2\mathbf{v}_2 + \ldots + m_n\mathbf{v}_n)$$

where, M is the total mass of the system, therefore

$$\mathbf{p} = M\mathbf{v}_{CM}$$

Thus, total linear momentum of a system of particles is equal to the product of the total mass of the system and the velocity of its centre of mass.

Again $\frac{dp}{dt} = M\frac{dv}{dt} = Ma = F_{ext}$

If $F_{ext} = 0$, then $\frac{dp}{dt} = 0$, i.e. $p = \text{constant}$

If external force of a system is zero, then momentum of system of particle remain constant.

Rigid Bodies

A rigid body is defined as that body which does not undergo any change in shape and volume when external forces are applied on it. When a force is applied on a rigid body, the distance between any two particles of the body will remain unchanged, however, larger the forces may be.

Coordinates of centre of mass of a rigid body are

$$X_{CM} = \frac{1}{M}\int x\,dm,$$

$$Y_{CM} = \frac{1}{M}\int y\,dm$$

and $$Z_{CM} = \frac{1}{M}\int z\,dm$$

Centre of Mass of Some Rigid Bodies

- The centre of mass of a uniform rod is located at its mid-point.

- Centre of mass of a uniform rectangular, square or circular plate lies at its centre.

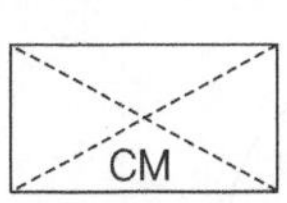

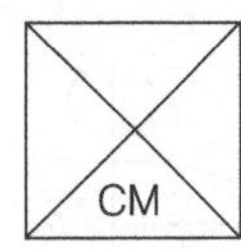

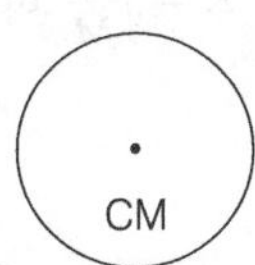

- Centre of mass of a uniform semi-circular ring lies at a distance of $h = \frac{2R}{\pi}$ from its centre, on the axis of symmetry, where R is the radius of the ring.

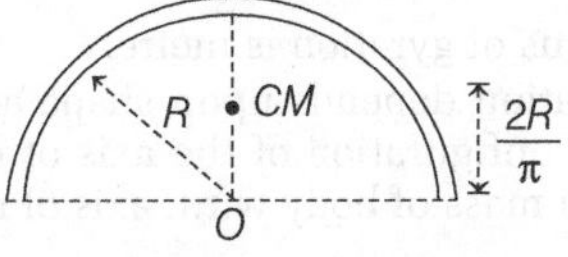

- Centre of mass of a uniform semi-circular disc of radius R lies at a distance of $h = \frac{4R}{3\pi}$ from the centre on the axis of symmetry as shown in figure.

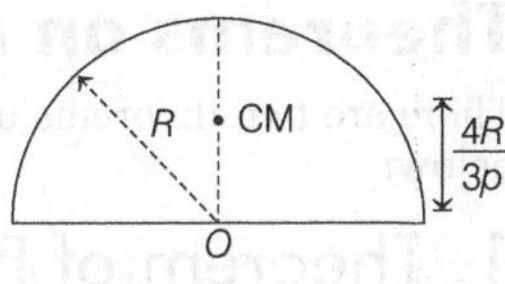

- Centre of mass of a hemispherical shell of radius R lies at a distance of $h = \frac{R}{2}$ from its centre on the axis of symmetry as shown in figure.

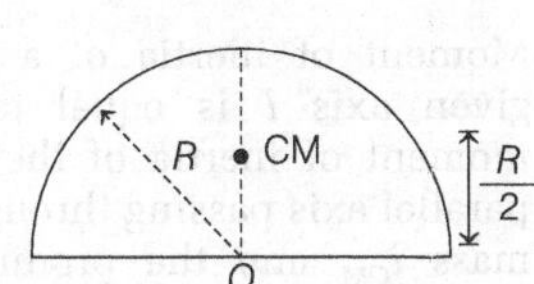

- Centre of mass of a solid hemisphere of radius R lies at a distance of $h = \frac{3R}{8}$ from its centre on the axis of symmetry.

R CM 3R/8 O

Moment of Inertia

Moment of inertia of a rotating body is its property to oppose any change in its state of uniform rotation.

If in a given rotational system particles of masses $m_1, m_2, m_3, \dots$ be situated at normal distances $r_1, r_2, r_3, \dots$ from the axis of rotation, then moment of inertia of the system about the axis of rotation is given by

$$I = m_1 r_1^2 + m_2 r_2^2 + m_3 r_3^2 + \dots = \Sigma mr^2$$

For a rigid body having continuous mass distribution

$$I = \int dm\, r^2$$

SI unit of moment of inertia is kg m^2. It is neither a scalar nor a vector i.e. it is a tensor.

Radius of Gyration

Radius of gyration of a given body about a given axis of rotation is the normal distance of a point from the axis, where if whole mass of the body is placed, then its moment of inertia will be exactly same as it has with its actual distribution of mass. Thus, radius of gyration

$$K = \sqrt{\frac{I}{M}} \text{ or } K = \left[\frac{r_1^2 + r_2^2 + r_3^2 + \dots + r_n^2}{n}\right]^{1/2}$$

Axis of rotation

SI unit of radius of gyration is metre.

Radius of gyration depends upon shape and size of the body, position and configuration of the axis of rotation and also on distribution of mass of body w.r.t. axis of rotation.

Theorems on Moment of Inertia

There are two theorems based on moment of inertia are given below:

1. Theorem of Parallel Axes

Moment of inertia of a body about a given axis I is equal to the sum of moment of inertia of the body about a parallel axis passing through its centre of mass I_{CM} and the product of mass of body (M) and square of normal distance between the two axes. Mathematically, $I = I_{CM} + Md^2$

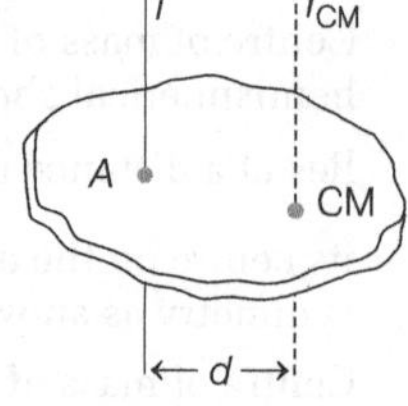

2. Theorem of Perpendicular Axes

The sum of moment of inertia of a plane laminar body about two mutually perpendicular axes lying in its plane is equal to its moment of inertia about an axis passing through the point of intersection of these two axes and perpendicular to the plane of laminar body.

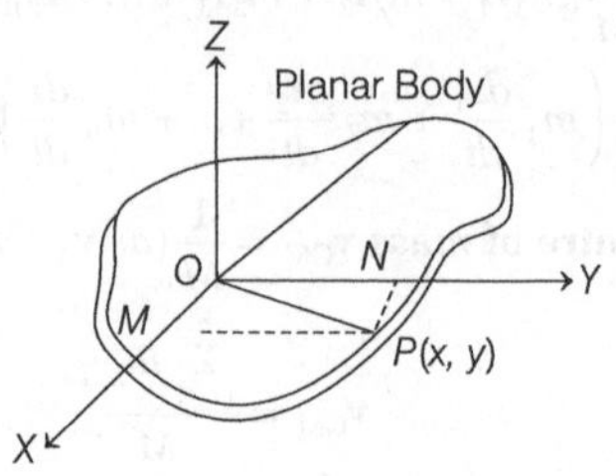

If I_x and I_y be moment of inertia of the body about two perpendicular axes in its own plane and I_z be the moment of inertia about an axis passing through point O and perpendicular to the plane of lamina, then

$$I_z = I_x + I_y$$

In theorem of perpendicular axes, the point of intersection of the three axes (x, y and z) may be any point on the plane.

Moment of Inertia of Some Geometrical Objects

- **Uniform Ring of Mass M and Radius R** About an axis passing through the centre and perpendicular to plane of ring $I = MR^2$. About a diameter $I = \frac{1}{2}MR^2$
- **Uniform Circular Disc of Mass M and Radius R** About an axis passing through the centre and perpendicular to plane of disc $I = \frac{1}{2}MR^2$. About a diameter $I = \frac{1}{4}MR^2$
- **Thin Uniform Rod of Mass M and Length l** About an axis passing through its centre and perpendicular to the rod, $I = \frac{1}{12}Ml^2$
- **Uniform Solid Cylinder of Mass M, Length l and Radius R** About its own axis, $I = \frac{1}{2}MR^2$. About an axis passing through its centre and perpendicular to its length $I = M\left[\frac{l^2}{12} + \frac{R^2}{4}\right]$
- **Uniform Solid Sphere** About its diameter $I = \frac{2}{5}MR^2$. About its tangent $I = \frac{7}{5}MR^2$

DAY PRACTICE SESSION 1

FOUNDATION QUESTIONS EXERCISE

1 A body A of mass M while falling vertically downwards under gravity breaks into two parts; a body B of mass $\frac{1}{3}M$ and, a body C of mass $\frac{2}{3}M$. The centre of mass of bodies B and C taken together shifts compared to that of body A towards

(a) depends on height of breaking
(b) does not shift
(c) body C
(d) body B

2 Distance of the centre of mass of a solid uniform cone from its vertex is z_0. If the radius of its base is R and its height is h, then z_0 is equal to → **JEE Main 2015**

(a) $\frac{h^2}{4R}$ (b) $\frac{3h}{4}$ (c) $\frac{5h}{8}$ (d) $\frac{3h^2}{8R}$

3 A circular disc of radius R is removed from a bigger circular disc of radius $2R$, such that the circumference of the discs coincide. The centre of mass of the new disc is αR from the centre of the bigger disc. The value of α is

(a) $\frac{1}{3}$ (b) $\frac{1}{2}$ (c) $\frac{1}{6}$ (d) $\frac{1}{4}$

4 Which of the following points is the likely position of the centre of mass of the system as shown in figure?

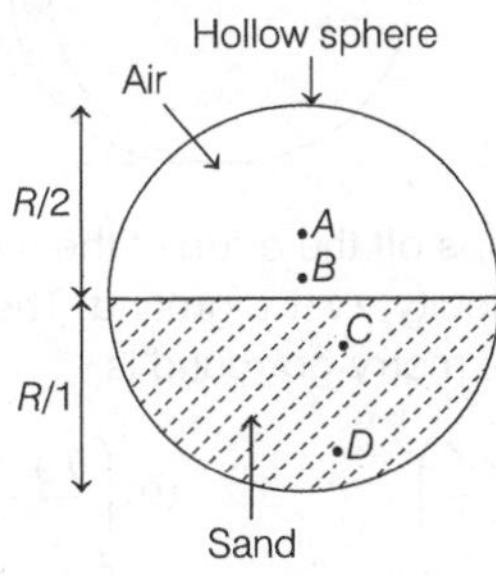

(a) A (b) B (c) C (d) D

5 Consider a two particles system with particles having masses m_1 and m_2. If the first particle is pushed towards the centre of mass through a distance d, by what distance should the second particle be moved, so as to keep the centre of mass at the same position?

(a) $\frac{m_2}{m_1}d$ (b) $\frac{m_1}{m_1+m_2}d$ (c) $-\frac{m_1}{m_2}d$ (d) d

6 A string of negligible thickness is wrapped several times around a cylinder kept on a rough horizontal surface. A man standing at a distance l_m from the cylinder holds one end of the string and pulls the cylinder towards him. There is no slipping anywhere. The length in (m) of the string passed through the hand of the man while the cylinder reached his hands is of

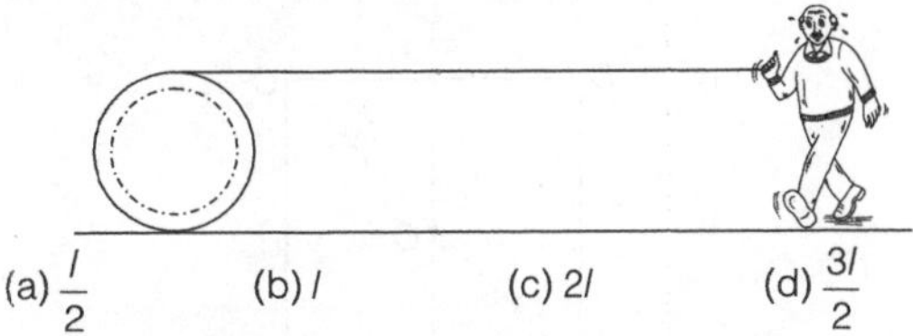

(a) $\frac{l}{2}$ (b) l (c) $2l$ (d) $\frac{3l}{2}$

7 A wheel has mass of the rim 1 kg, having 50 spokes each of mass 5 g. The radius of the wheel is 40 cm. The moment of inertia

(a) 0.273 kg-m^2 (b) 1.73 kg-m^2
(c) 0.173 kg-m^2 (d) 2.73 kg-m^2

8 The surface density of a circular disc of radius a depends on the distance as $\rho(r) = A + Br$. The moment of inertia about the line perpendicular to the plane of the disc is

(a) $\pi a^4\left(\frac{A}{2}+\frac{2a}{5}B\right)$ (b) $\pi a^4\left(\frac{A}{2}+\frac{2B}{5}\right)$
(c) $2\pi a^3\left(\frac{A}{2}+\frac{Ba}{5}\right)$ (d) None of these

9 Four point masses, each of value m, are placed at the corners of a square $ABCD$ of side l. The moment of inertia of this system about an axis passing through A and parallel to BD is

(a) $2ml^2$ (b) $\sqrt{3}ml^2$ (c) $3ml^2$ (d) ml^2

10 The ratio of the radii of gyration of a circular disc and a circular ring of the same radii about a tangential axis perpendicular to plane of disc or ring is

(a) 1 : 2 (b) $\sqrt{5} : \sqrt{6}$
(c) 2 : 3 (d) $\frac{\sqrt{3}}{2}$

11 A uniform square plate has a small piece Q of an irregular shape removed and glued to the centre of the plate leaving a hole behind. The moment of inertia about the z-axis, then

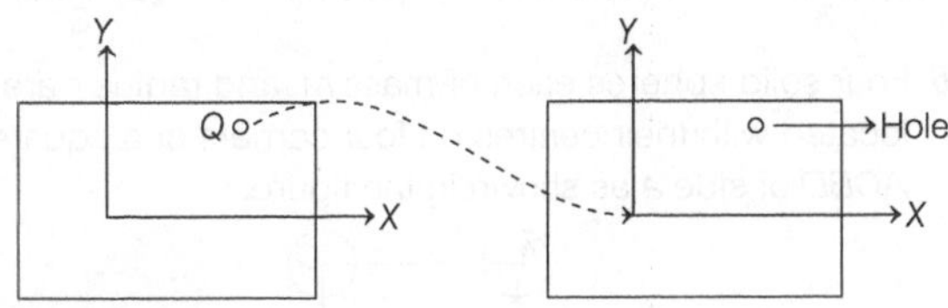

(a) increased
(b) decreased
(c) same
(d) changed in unpredicted manner

12 Consider a uniform square plate of side a and mass m. The moment of inertia of this plate about an axis perpendicular to its plane and passing through one of its corners is

(a) $\frac{5}{6}ma^2$ (b) $\frac{1}{12}ma^2$ (c) $\frac{7}{12}ma^2$ (d) $\frac{2}{3}ma^2$

13 For the given uniform square lamina $ABCD$, whose centre is O.

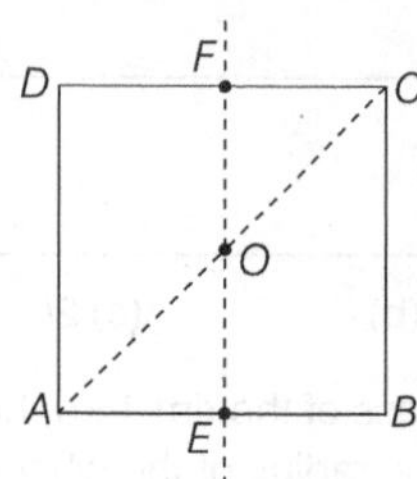

(a) $\sqrt{2}\, I_{AC} = I_{EF}$ (b) $I_{AD} = 3I_{EF}$
(c) $I_{AC} = I_{EF}$ (d) $I_{AC} = \sqrt{2}\, I_{EF}$

14 A solid sphere of radius R has moment of inertia I about its geometrical axis. It is melted into a disc of radius r and thickness t. If it's moment of inertia about the tangential axis (which is perpendicular to plane of the disc), is also equal to I, then the value of r is equal to

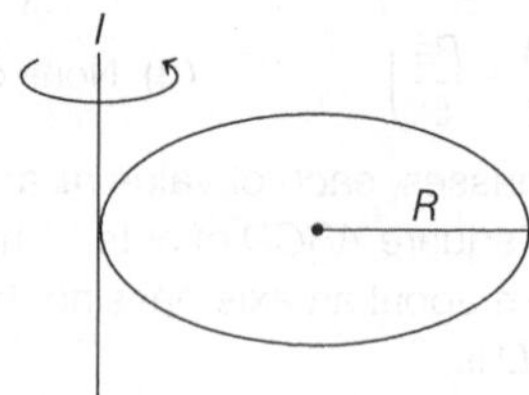

(a) $\frac{2}{\sqrt{15}}R$ (b) $\frac{2}{\sqrt{5}}R$ (c) $\frac{3}{\sqrt{15}}R$ (d) $\frac{\sqrt{3}}{\sqrt{15}}R$

15 The moment of inertia of a system of four rods each of length l and mass m about the axis shown is

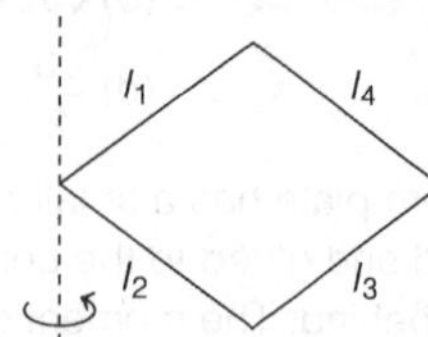

(a) $\frac{2}{3}ml^2$ (b) $2\,ml^2$ (c) $3\,ml^2$ (d) $\frac{8}{3}ml^2$

16 Four solid spheres each of mass m and radius r are located with their centres on four corners of a squares $ACBD$ of side a as shown in the figure.

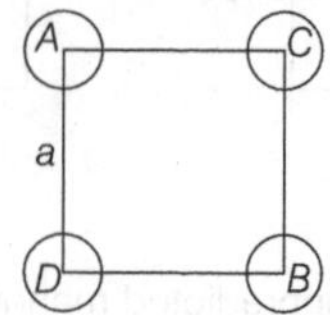

The moment of inertia of the system of four spheres about diagonal AB is

(a) $\frac{m}{5}(8r^2 + 5a^2)$ (b) $\frac{m}{5}(7r^2 + 4a^2)$
(c) $\frac{m}{5}(5r^2 + 8a^2)$ (d) $\frac{m}{5}(3r^2 + 5a^2)$

17 From a uniform circular disc of radius R and mass $9\,M$, a small disc of radius $\frac{R}{3}$ is removed as shown in the figure. The moment of inertia of the remaining disc about an axis perpendicular to the plane of the disc and passing through centre of disc is **→ JEE Main 2018**

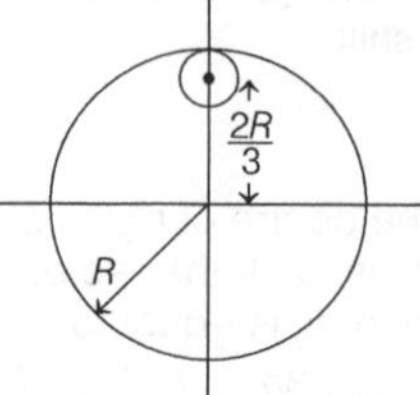

(a) $4MR^2$ (b) $\frac{40}{9}MR^2$ (c) $10MR^2$ (d) $\frac{37}{9}MR^2$

18 A child with mass m is standing at the edge of a merry-go-round having moment of inertia I, radius R and initial angular velocity ω_0 as shown in the figure.

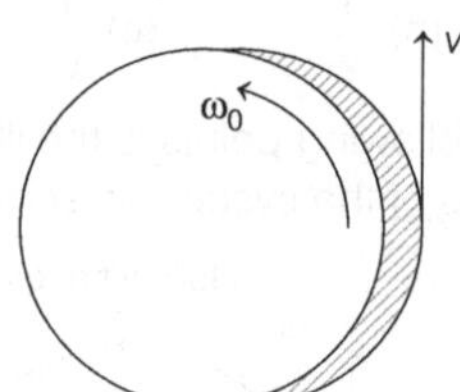

The child jumps off the edge of the merry-go-round with tangential velocity v w.r.t. ground. The new angular velocity of the merry-go-round is

(a) $\left(\frac{I\omega_0^2 - mv^2}{I}\right)^{\frac{1}{2}}$ (b) $\left(\frac{(I + mR^2)\omega_0^2 - mv^2}{I}\right)^{\frac{1}{2}}$

(c) $\left(\frac{I\omega_0 - mvR}{I}\right)$ (d) $\left[\frac{(I + mR^2)\,\omega_0 - mvR}{I}\right]$

Direction (Q. Nos. 19-20) *are the Assertion and Reason type. Each of the these contains two Statements; Assertion and Reason. Each of these question also has four alternative choice,only one of which is correct. You have to select the correct choices from the codes (a),(b), (c) and (d) given below:*

(a) If both Assertion and Reason are true and the Reason is correct explanation of the Assertion
(b) If both Assertion and Reason are true but Reason is not correct explanation of the Assertion
(c) If Asserion is true but Reason is false
(d) If both Assertion and Reason are false

19 **Assertion** (A) A constant force F is applied on the two blocks and one spring system as shown in the figure. Velocity of centre of mass increases linearly with time.

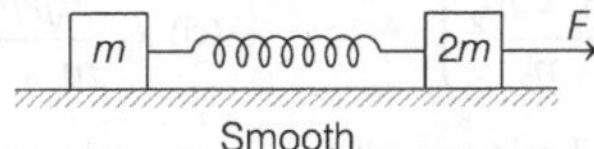

Reason (R) Acceleration of centre of mass is constant.

20 **Assertion** (A) There is a triangular plate as shown in the figure A dotted axis is lying in the plane of the slab. As the axis is moved downwards, moment of inertia of the slab will first decreases, then increases.

Reason (R) Axis is first moving towards its centre of mass, then it is receding from it.

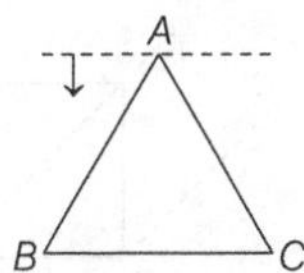

DAY PRACTICE SESSION 2

PROGRESSIVE QUESTIONS EXERCISE

1 Mass of bigger disc having radius $2R$ is M. A disc of radius R is cut from bigger disc as shown in figure. Moment of inertia of disc about an axis passing through periphery and perpendicular to plane is

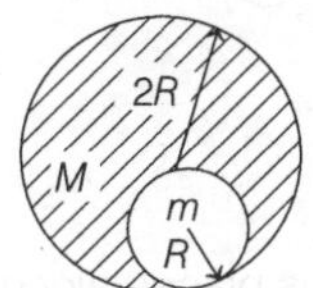

(a) $\frac{27\ MR^2}{8}$ (b) $\frac{29\ MR^2}{8}$

(c) $3.5\ MR$ (d) $2MR^2$

2 Four spheres of diameter $2a$ and mass M are placed with their centres on the four corners of a square of side b. Then the moment of inertia of the system about an axis along one of the sides the square is

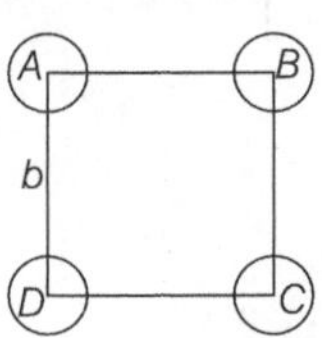

(a) $\frac{4}{5}Ma^2 + 2Mb^2$ (b) $\frac{8}{5}Ma^2 + 2Mb^2$

(c) $\frac{8}{5}Ma^2$ (d) $\frac{4}{5}Ma^2 + 4Mb^2$

3 From a solid sphere of mass M and radius R, a cube of maximum possible volume is cut. Moment of inertia of cube about an axis passing through its centre and perpendicular to one of its faces is → **JEE Main 2015**

(a) $\frac{MR^2}{32\sqrt{2}\pi}$ (b) $\frac{MR^2}{16\sqrt{2}\pi}$

(c) $\frac{4MR^2}{9\sqrt{3}\pi}$ (d) $\frac{4MR^2}{3\sqrt{3}\pi}$

4 Seven identical circular planar discs, each of mass M and radius R are welded symmetrically as shown in the figure. The moment of inertia of the arrangement about the axis normal to the plane and passing through the point P is

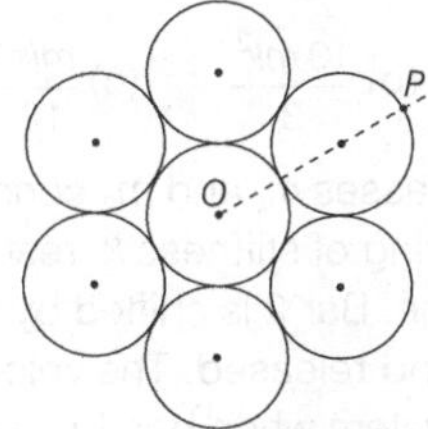

→ **JEE Main 2018**

(a) $\frac{19}{2}MR^2$ (b) $\frac{55}{2}MR^2$

(c) $\frac{73}{2}MR^2$ (d) $\frac{181}{2}MR^2$

5 The moment of inertia of a uniform cylinder of length l and radius R about its perpendicular bisector is I. What is the ratio l/R such that the moment of inertia is minimum?

→ **JEE Main 2017 (Offline)**

(a) $\frac{\sqrt{3}}{2}$ (b) 1

(c) $\frac{3}{\sqrt{2}}$ (d) $\sqrt{\frac{3}{2}}$

6 A particle moves parallel to X-axis with constant velocity v as shown in the figure. The angular velocity of the particle about the origin O is

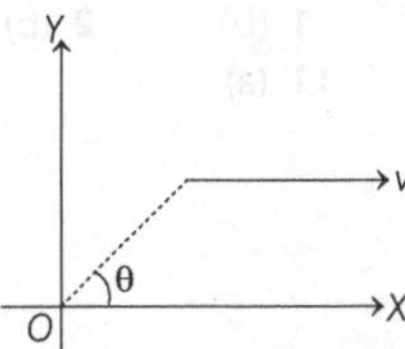

(a) remains constant (b) continuously increasing

(c) continuously decreasing (d) oscillates

7 Let I_1 and I_2 be the moment of inertia of a uniform square plate about an axis as shown in the figure. Then, the ratio $I_1 : I_2$ is

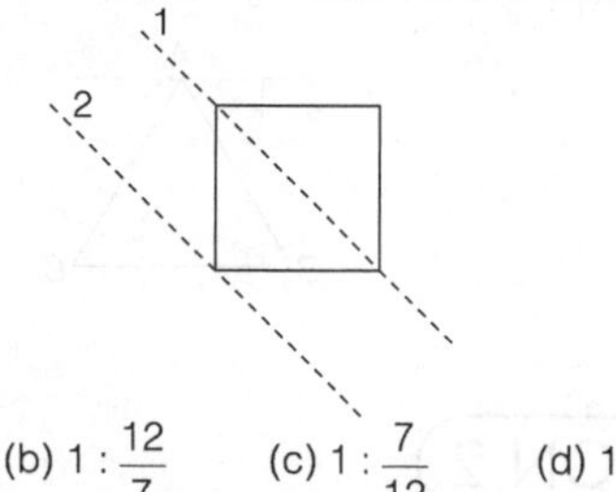

(a) $1 : \frac{1}{7}$ (b) $1 : \frac{12}{7}$ (c) $1 : \frac{7}{12}$ (d) 1 : 7

8 A thin rod of length $4l$, mass $4m$ is bent at the points as shown in the figure. What is the moment of inertia of the rod about the axis passing through O and perpendicular to the plane of the paper?

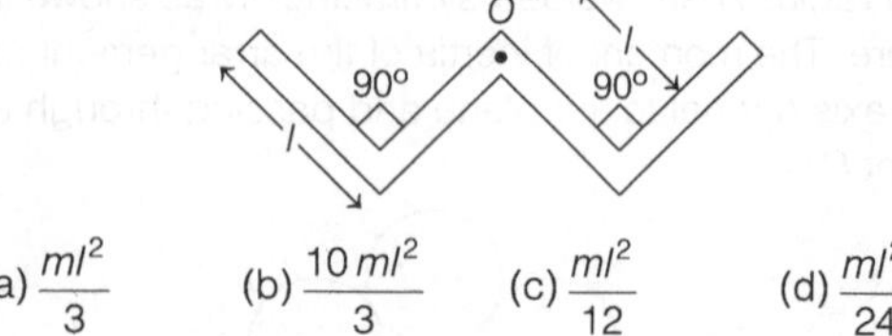

(a) $\frac{ml^2}{3}$ (b) $\frac{10\,ml^2}{3}$ (c) $\frac{ml^2}{12}$ (d) $\frac{ml^2}{24}$

9 Two bars of masses m_1 and m_2 connected by a weightless spring of stiffness k, rest on a smooth horizontal plane. Bar 2 is shifted by a small distance x_0 to the left and released. The velocity of the centre of mass of the system when bar 1 breaks off the wall is

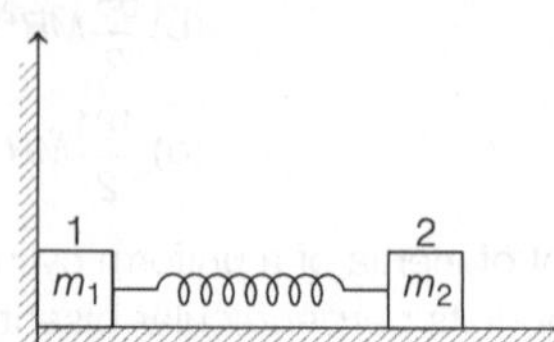

(a) $\frac{\sqrt{km_2}}{m_1 + m_2}$ (b) $\frac{x_0}{m_1 + m_2}\sqrt{km_2}$

(c) $x_0 k\left(\frac{m_1 + m_2}{m_2}\right)$ (d) $x_0 \frac{\sqrt{km_1}}{m_1 + m_2}$

10 A racing bike is travelling along a straight track at a constant velocity of 40 m/s. A fixed CCTV camera is recording the event as shown in the figure. In order to keep the bike in view, in the position shown, the angular velocity of the CCTV camera should be

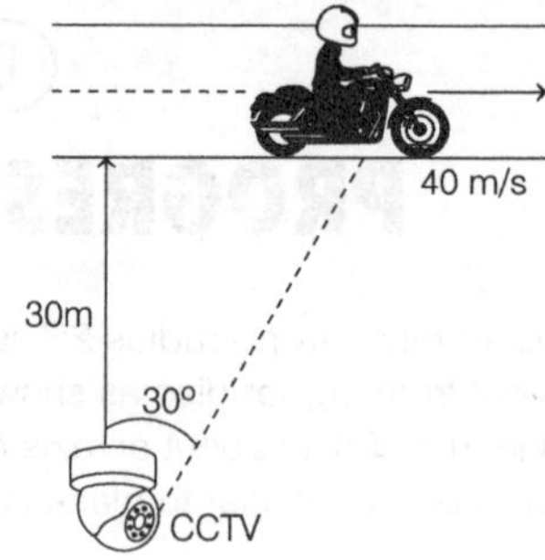

(a) 3 rad/s
(b) 2 rad/s
(c) 1 red/s
(d) 4 rad/s

11 A rod of length L is placed along the X-axis between $x = 0$ and $x = L$. The linear mass density (mass/length) ρ of the rod varies with the distance x from the origin as $\rho = a + bx$. Here, a and b are constants. The position of centre of mass of the rod is

(a) $\left[\frac{3aL + 2bL^2}{6a + 3bL}, 0, 0\right]$ (b) $\left[\frac{6aL + 3bL^2}{3a + 2bL}, 0, 0\right]$

(c) $\left[\frac{aL + bL^2}{2a + 3bL}, 0, 0\right]$ (d) None of these

ANSWERS

SESSION 1	**1** (b)	**2** (b)	**3** (a)	**4** (c)	**5** (c)	**6** (c)	**7** (c)	**8** (a)	**9** (c)	**10** (d)
	11 (b)	**12** (d)	**13** (c)	**14** (a)	**15** (d)	**16** (a)	**17** (a)	**18** (d)	**19** (a)	**20** (a)
SESSION 2	**1** (b)	**2** (b)	**3** (c)	**4** (d)	**5** (d)	**6** (c)	**7** (d)	**8** (b)	**9** (a)	**10** (c)
	11 (a)									

Hints and Explanations

SESSION 1

1 The position of centre of mass remains unaffected because breaking of mass into two parts is due to internal forces.

2 We know that centre of mass of a uniform solid cone of height (h) is at height $\frac{h}{4}$ from base, therefore

As, $h - z_0 = \frac{h}{4}$ or $z_0 = h - \frac{h}{4} = \frac{3h}{4}$

3 In this question, distance of centre of mass of new disc from the centre of mass of remaining disc is αR.

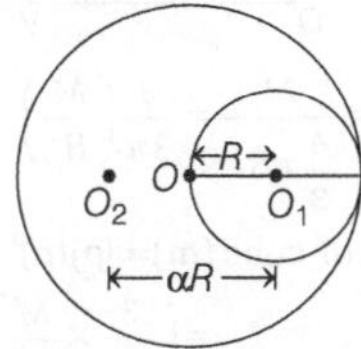

Mass of original disc $= M$

Mass of disc removed

$$= \frac{M}{\pi(2R)^2} \times \pi R^2 = \frac{M}{4}$$

Mass of remaining disc

$$= M - \frac{M}{4} = \frac{3M}{4}$$

$$\therefore \; -\frac{3M}{4}\alpha R + \frac{M}{4}R = 0$$

$$\Rightarrow \quad \alpha = \frac{1}{3}$$

Note *In this question, the given distance must be αR for real approach to the solution.*

4 Centre of mass of a system lies towards that part of the system, having bigger mass. In the given diagram, lower part is heavier, hence CM of the system lies below the horizontal diameter.

Hence, (c) is correct option.

5 To keep the centre of mass at the same position, velocity of centre of mass is zero, so

$$\frac{m_1\mathbf{v}_1 + m_2\mathbf{v}_2}{m_1 + m_2} = 0$$

where, $\mathbf{v}_1$ and $\mathbf{v}_2$ are velocities of particles 1 and 2 respectively.

$$\Rightarrow m_1\frac{d\mathbf{r}_1}{dt} + m_2\frac{d\mathbf{r}_2}{dt} = 0$$

$$\left[\because \mathbf{v}_1 = \frac{d\mathbf{r}_1}{dt} \text{ and } \mathbf{v}_2 = \frac{d\mathbf{r}_2}{dt}\right]$$

$$\Rightarrow \quad m_1 d\mathbf{r}_1 + m_2 d\mathbf{r}_2 = 0$$

$d\mathbf{r}_1$ and $d\mathbf{r}_2$ represent the change in displacement of particles.

Let second particle has been displaced by distance x.

$$\Rightarrow m_1(d) + m_2(x) = 0 \text{ or } x = -\frac{m_1 d}{m_2}$$

6 If velocity of centre of mass is v, then velocity of contact is 0 and that of the top is 2 v, hence when centre of mass covers a distance l, thread covers a distance $2l$.

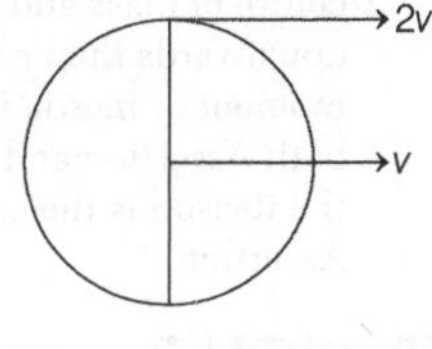

7 $I = mr^2 + 50\frac{ml^2}{3}$

$$= 1 \times (0.4)^2 + \frac{50\,(5 \times 10^{-3})\,(0.4)^2}{3}$$

$$= 0.16\,(1.083)$$

$$= 0.173 \text{ kg- m}^2$$

8 $dm = 2\pi r\, dr\, (\rho) = (A + Br)\,(2\pi r\, dr)$

$$I = \int_0 dmr^2 = \frac{\pi A a^4}{2} + \frac{2\pi B a^5}{5}$$

$$= \pi a^4\left(\frac{A}{2} + \frac{2a}{5}B\right)$$

9 The situation is as shown in figure.

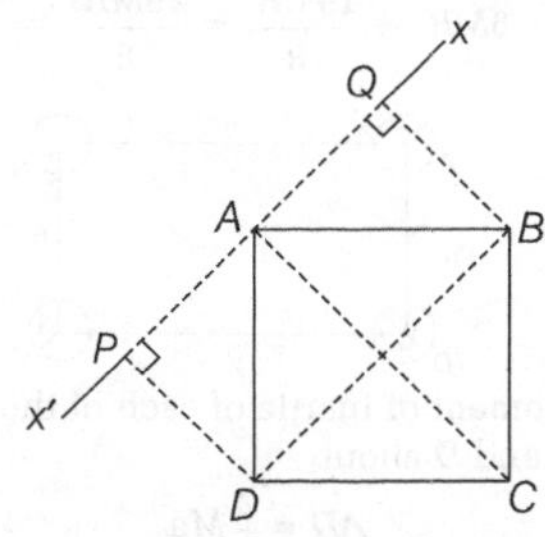

$$I_{xx'} = m \times DP^2 + m \times BQ^2 + m \times CA^2$$

$$= m \times 2 \times \left(\frac{\sqrt{2}\,l}{2}\right)^2 + m \times (\sqrt{2}\,l)^2$$

$$= 3\,ml^2$$

10 Radius of gyration,

$$K = \sqrt{\frac{I}{m}}$$

$$\Rightarrow \quad K_{\text{disc}} = \sqrt{\frac{\frac{1}{2}mR^2 + nR^2}{R}} = \sqrt{\frac{3}{2}}R$$

$$K_{\text{ring}} = \sqrt{\frac{mR^2 + mR^2}{m}} = \sqrt{2}R$$

$$\frac{K_{\text{disc}}}{K_{\text{ring}}} = \frac{\sqrt{\frac{3}{2}}}{\sqrt{2}} = \frac{\sqrt{3}}{2}$$

11 According to the theorem of perpendicular axes, $I_z = I_x + I_y$ with the hole, I_x and I_y, both decreases (gluing the removed piece at the centre of square plate does not affect I_z).

Hence, I_z decreases overall.

12 Moment of inertia of square plate about xy is $\frac{ma^2}{6}$. Moment of inertia about zz' can be computed using parallel axes theorem

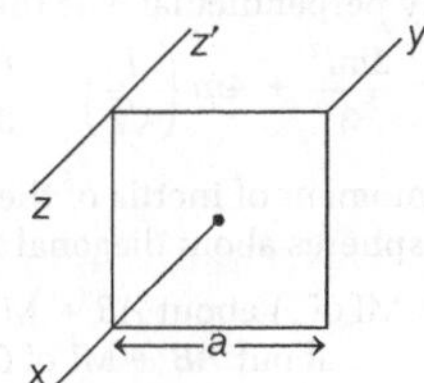

$$I_{zz'} = I_{xy} + m\left(\frac{a}{\sqrt{2}}\right)^2$$

$$= \frac{ma^2}{6} + \frac{ma^2}{2}$$

$$= \frac{2ma^2}{3}$$

13 Let the each side of square lamina is d.

So, $\quad I_{EF} = I_{GH}$ (due to symmetry)

and $\quad I_{AC} = I_{BD}$ (due to symmetry)

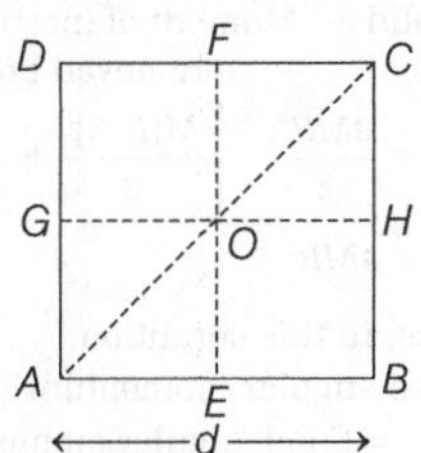

Now, according to theorem of perpendicular axes,

$$I_{AC} + I_{BD} = I_0$$

$$\Rightarrow \quad 2I_{AC} = I_0 \quad \text{...(i)}$$

$$\text{and} \quad I_{EF} + I_{GH} = I_0$$

$$\Rightarrow \quad 2I_{EF} = I_0 \quad \text{...(ii)}$$

From Eqs. (i) and (ii), we get

$$I_{AC} = I_{EF}$$

14 $\frac{2}{5}MR^2 = \frac{1}{2}Mr^2 + Mr^2$

or $\frac{2}{5}MR^2 = \frac{3}{2}Mr^2$

$\therefore \quad r = \frac{2}{\sqrt{15}}R$

15 Consider a square lamina, then

$4m\left(\frac{l^2}{12} + \frac{l^2}{12}\right)$ about

Centre of mass $= \frac{2ml^2}{3}$

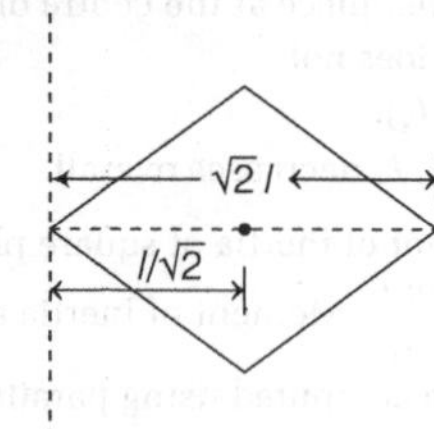

Apply perpendicular axis theorem,

$= \frac{2ml^2}{3} + 4m\left(\frac{l}{\sqrt{2}}\right)^2 = \frac{8}{3}ml^2$

16 The moment of inertia of the system of four spheres about diagonal *AB* is

I_{AB} = MI of *A* about *AB* + MI of *B* about *AB* + MI of *C* about *AB* + MI of *D* about *AB*

$= \frac{2}{5}mr^2 + \frac{2}{5}mr^2 + \left(\frac{2}{5}mr^2 + \frac{1}{2}ma^2\right) + \left(\frac{2}{5}mr^2 + \frac{1}{2}ma^2\right)$

$= \frac{8}{5}mr^2 + ma^2 = m\left(\frac{8r^2}{5} + a^2\right)$

$= \frac{m}{5}(8r^2 + 5a^2)$

17 Moment of inertia of remaining solid = Moment of inertia of complete solid − Moment of inertia of removed portion

$\therefore \quad I = \frac{9MR^2}{2} - \left[\frac{M(R/3)^2}{2} + M\left(\frac{2R}{3}\right)^2\right]$

$\Rightarrow I = 4MR^2$

18 Since, in this condition,

Initial angular momentum = Final angular momentum

$\therefore (I + mR^2)\omega_0 = (mvR) + I\omega'$

or $\quad \omega' = \frac{(I + mR^2)\omega_0 - mvR}{I}$

Hence, (d) is the correct option.

19 Total mass of the system

$= m + 2m = 3m$

Force applied on the system is *F*.

$\therefore \quad a_{CM} = \frac{F}{3m}$

= constant as *F* is constant

$\therefore \quad v_{CM} = a_{CM} \times t$

or $v_{CM} = \frac{F}{3m} \times t$ or $v_{CM} \propto t$

Hence, both Assertion and Reason are true and the Reason is the correct explanation of Assertion.

20 Moment of inertia $(I) = mr^2$, where *r* is distance from the axis of rotation to the centre of mass. When dotted axis moved downward (towards centre of mass), *r* decreases result moment of inertia decrease and when dotted axis cross the centre of mass and moved further downwards then *r* increases result moment of inertia increases. Hence, both Assertion and Reason are true and the Reason is the correct explanation of Assertion.

SESSION 2

1 Surface density of motional disc is

$\sigma = \frac{M}{\pi(2R)^2} = \frac{M}{4\pi R^2}$

Mass of cutting portion is

$m_1 = \sigma \times \pi R^2 = \frac{M}{4}$

$I = I_1 - I_2$

where,

I_1 = Moment of inertia of disc about given axis without cutting portion

I_2 = Moment of inertia due to cutting portion

$I = \frac{M(2R)^2}{2} + M(2R)^2 - \left[\frac{m_1 R^2}{2} + m_1(3R)^2\right]$

$= 6MR^2 - \frac{19MR^2}{8} = \frac{29MR^2}{8}$

2

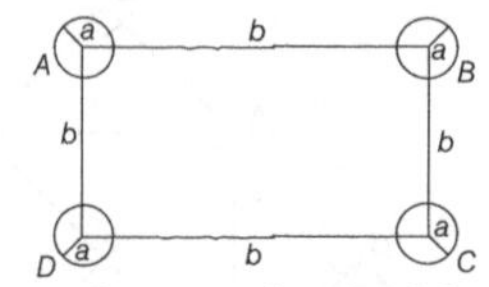

Moment of inertia of each of the spheres *A* and *D* about

$AD = \frac{2}{5}Ma^2$

Moment of inertia of each of the spheres *B* and *C* about *AD*

$= \left(\frac{2}{5}Ma^2 + Mb^2\right)$

Using theorem of parallel axis, we get

Total moment of inertia

$= I\left[\frac{2}{5}Ma^2\right] \times 2 + \left[\frac{2}{5}Ma^2 + Mb^2\right] \times 2$

$I = \frac{8}{5}Ma^2 + 2Mb^2$

3 Consider the cross-sectional view of a diametric plane as shown in the adjacent diagram.

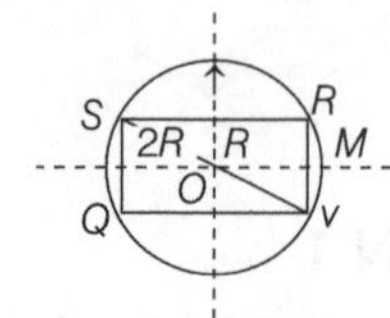

Cross-sectional view of the cube and sphere

Using geometry of the cube

$PQ = 2R = (\sqrt{3})\,a$ or $a = \frac{2R}{\sqrt{3}}$

Volume density of the solid sphere,

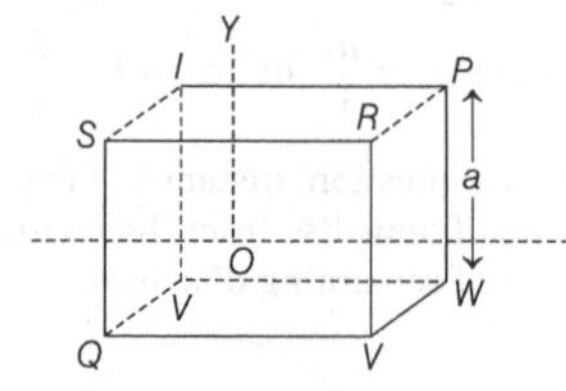

$\rho = \frac{M}{\frac{4}{3}\pi R^3} = \frac{3}{3\pi}\left(\frac{M}{R^3}\right)$

Mass of cube $(m) = (\rho)(a)^3$

$= \left(\frac{3}{4\pi} \times \frac{M}{R^3}\right)\left[\frac{2R}{\sqrt{3}}\right]^3$

$= \frac{3M}{4\pi R^3} \times \frac{8R^3}{3\sqrt{3}} = \frac{2M}{\sqrt{3}\pi}$

Moment of inertia of the cube about given axis is

$I_y = \frac{m}{12}(a^2 + a^2) = \frac{ma^2}{6}$

$\Rightarrow I_y = \frac{ma^2}{6} = \frac{2M}{\sqrt{3}\pi} \times \frac{1}{6} \times \frac{4R^2}{3} = \frac{4MR^2}{9\sqrt{3}\pi}$

4 Moment of inertia of an outer disc about the axis through centre is

$= \frac{MR^2}{2} + M(2R)^2 = MR^2\left(4 + \frac{1}{2}\right) = \frac{9}{2}MR^2$

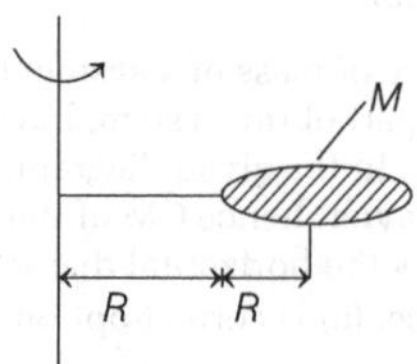

For 6 such discs,

Moment of inertia $= 6 \times \frac{9}{2}MR^2$

$= 27MR^2$

So, moment of inertia of system

$= \frac{MR^2}{2} + 27MR^2 = \frac{55}{2}MR^2$

Hence, $I_P = \frac{55}{2}MR^2 + (7M \times 9R^2)$

$\Rightarrow \quad I_P = \frac{181}{2}MR^2$

$I_{system} = \frac{181}{2}MR^2$

5 MI of a solid cylinder about its perpendicular bisector of length is

$$I = m\left(\frac{l^2}{12} + \frac{R^2}{4}\right)$$

$$\Rightarrow \quad I = \frac{mR^2}{4} + \frac{ml^2}{12}$$

$$= \frac{m^2}{4\pi\rho l} + \frac{ml^2}{12} \qquad [\because \rho\pi R^2 l = m]$$

For I to be maximum,

$$\frac{dI}{dl} = -\frac{m^2}{4\pi\rho}\left(\frac{1}{l^2}\right) + \frac{ml}{6} = 0$$

$$\Rightarrow \quad \frac{m^2}{4\pi\rho} = \frac{ml^3}{6}$$

Now, putting $m = \rho\pi R^2 l$

$$\therefore \quad l^3 = \frac{3}{2\pi\rho}\cdot\rho\pi R^2 l$$

$$\frac{l^2}{R^2} = \frac{3}{2}$$

$$\therefore \quad \frac{l}{R} = \sqrt{\frac{3}{2}}$$

6 Since, $\omega = \dfrac{v}{r}$

But, here

$$v = v\sin\theta$$

and $\quad r = b/\sin\theta$

$$\therefore \quad \omega = \frac{v\sin\theta}{b/\sin\theta} = \frac{v\sin^2\theta}{b}$$

or $\quad \omega \propto \sin^2\theta$ [as v and b are constants]

$\therefore$ When particle moves further parallel to the X-axis, θ decreases result angular velocity decreases.

Hence, (c) is the correct option.

7 Here, $I_1 = \dfrac{ma^2}{12}$ and $I_2 = I_1 + mr^2$

From figure, we get

$$r = \frac{a}{\sqrt{2}}$$

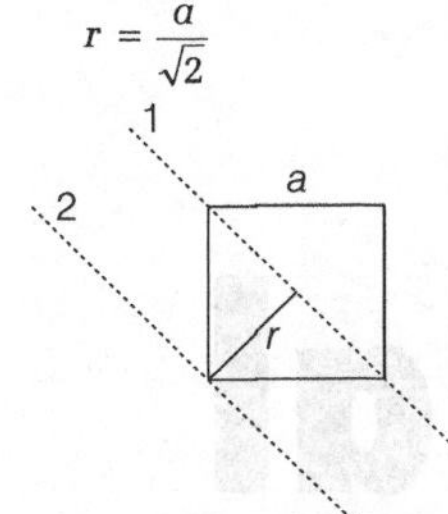

$$\therefore I_2 = \frac{ma^2}{12} + \frac{ma^2}{2} = \frac{7}{12}ma^2$$

Hence, $I_1 : I_2 = 1 : 7$

Thus, (d) is the correct option.

8 From figure,

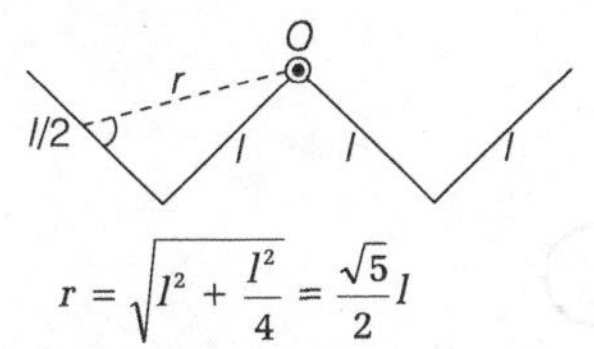

$$r = \sqrt{l^2 + \frac{l^2}{4}} = \frac{\sqrt{5}}{2}l$$

Due to symmetry,

$$I = 2[I_1 + I_2] = 2\left[\frac{ml^2}{3} + \left(\frac{ml^2}{12} + mr^2\right)\right]$$

$$= 2\left[\frac{ml^3}{3} + \frac{ml^2}{12} + \frac{5ml^2}{4}\right] = \frac{10}{3}ml^2$$

Hence, (b) is the correct option.

9 Here, $\frac{1}{2}kx_0^2 = \frac{1}{2}m_2v_2^2$

$$\therefore \quad v_2 = \sqrt{\frac{k}{m_2}}\,x_0$$

Since, $v_{CM} = \dfrac{m_1v_1 + m_2v_2}{m_1 + m_2}$

$$\therefore \quad v_{CM} = \frac{m_1 \times 0 + m_2\sqrt{\dfrac{k}{m_2}}\,x_0}{m_1 + m_2} = \frac{\sqrt{k\,m_2}}{m_1 + m_2}\cdot x_0$$

Hence, (a) is the correct option.

10 Here, $\tan\theta = \dfrac{x}{30}$ or $x = 30\tan\theta$

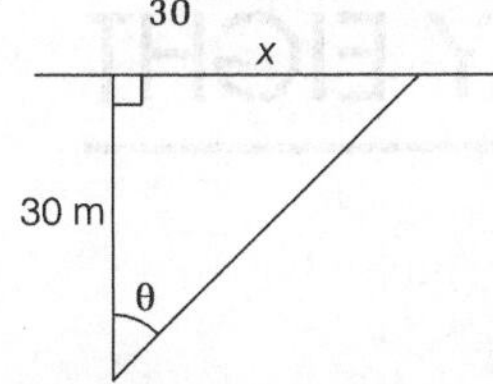

Now, differentiate both sides w.r.t. time,

$$\frac{dx}{dt} = 30 \times \sec^2\theta\cdot\frac{d\theta}{dt}$$

But $\dfrac{dx}{dt} = v_{\text{bike}}$ and $\dfrac{d\theta}{dt} = W$

$$\therefore \quad W = \frac{v_{\text{bike}}}{30\sec^2\theta} = \frac{40}{30\sec^2 30^\circ}$$

or $\quad W = 1$ rad/s

Hence, (c) is the correct option.

11 Let dm is the mass of element AB of length dx at distance x

$$\therefore \quad dm = \rho\cdot dx = (a + bx)\cdot dx$$

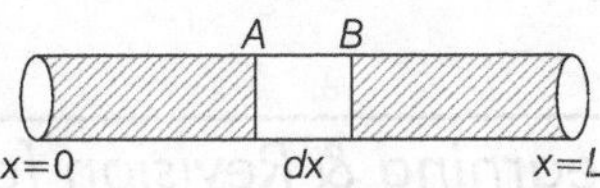

The centre of mass of the element has coordinates $(x, 0, 0)$.

Therefore, x-coordinate of centre of mass of the rod will be

$$x_{CM} = \frac{\int_0^L x\cdot dm}{\int_0^L dm} = \frac{\int_0^L x\,(a + bx)\,dx}{\int_0^L (a + bx)\,dx} = \frac{\left[\dfrac{ax^2}{2} + \dfrac{bx^3}{3}\right]_0^L}{\left[ax + \dfrac{bx^2}{2}\right]_0^L}$$

or $\quad x_{CM} = \dfrac{3aL + 2bL^2}{6a + 3bL}$

$$y_{CM} = 0,$$

$$z_{com} = 0$$

$\therefore$ Centre of mass of the rod will be

$$\left[\frac{3aL + 2bL^2}{6a + 3bL}, 0, 0\right]$$

Hence, (a) is the correct option.

DAY EIGHT

Rotational Motion

Learning & Revision for the Day

- Concept of Rotational Motion
- Equation of Rotational Motion
- Moment of Force or Torque
- Angular Momentum
- Law of Conservation of Angular Momentum
- Equilibrium of a Rigid Bodies
- Rigid Body Rotation

Concept of Rotational Motion

In rotation of a rigid body about a fixed axis, every particle of the body moves in a circle, which lies in a plane perpendicular to the axis and has its centre on the axis.

- Rotational motion is characterised by angular displacement $d\theta$ and angular velocity $\omega = \frac{d\theta}{dt}$.
- If angular velocity is not uniform, then rate of change of angular velocity is called the **angular acceleration**.

  Angular acceleration, $\alpha = \frac{d\omega}{dt}$.

  SI unit of angular acceleration is rad/s^2.
- Angular acceleration α and linear tangential acceleration $\mathbf{a}_t$ are correlated as $\mathbf{a}_t = \alpha \times \mathbf{r}$.

Equation of Rotational Motion

If angular acceleration α is uniform, then equations of rotational motion may be written as

(i) $\omega = \omega_0 + \alpha t$

(ii) $\theta = \omega_0 t + \frac{1}{2}\alpha t^2$

(iii) $\omega^2 - \omega_0^2 = 2\alpha\theta$

(iv) $\theta_{nth} = \omega_0 + \frac{\alpha}{2}(2n-1)$

- No. of Questions in Exercises (x)—
- No. of Questions Attempted (y)—
- No. of Correct Questions (z)—
  (Without referring Explanations)
- Accuracy Level ($z/y \times 100$)—
- Prep Level ($z/x \times 100$)—

In order to expect good rank in JEE, your Accuracy Level should be above 85 & Prep Level should be above 75.

Moment of Force or Torque

Torque (or moment of a force) is the turning effect of a force applied at a point on a rigid body about the axis of rotation. Mathematically, torque, $\tau = \mathbf{r} \times \mathbf{F} = |\mathbf{r} \times \mathbf{F}|\,\hat{\mathbf{n}} = rF\sin\theta\,\hat{\mathbf{n}}$

where, $\hat{\mathbf{n}}$ is a unit vector along the axis of rotation. Torque is an axial vector and its SI unit is newton-metre (N-m).

- The torque about axis of rotation is independent of choice of origin O, so long as it is chosen on the axis of rotation AB.
- Only normal component of force contributes towards the torque. Radial component of force does not contribute towards the torque.
- A torque produces angular acceleration in a rotating body. Thus, torque, $\tau = I\alpha$
- Moment of a couple (or torque) is given by product of position vector $\mathbf{r}$ between the two forces and either force $\mathbf{F}$. Thus, $\tau = \mathbf{r} \times \mathbf{F}$
- If under the influence of an external torque, τ the given body rotates by $d\theta$, then work done, $dW = \tau \cdot d\theta$.
- In rotational motion, power may be defined as the scalar product of torque and angular velocity, i.e. Power $P = \tau \cdot \omega$.

Angular Momentum

The moment of linear momentum of a given body about an axis of rotation is called its angular momentum. If $\mathbf{p}=m\mathbf{v}$ be the linear momentum of a particle and $\mathbf{r}$ is its position vector from the point of rotation, then

Angular momentum, $\mathbf{L} = \mathbf{r} \times \mathbf{p} = r\,p\sin\theta\,\hat{\mathbf{n}} = mvr\sin\theta\,\hat{\mathbf{n}}$

where, $\hat{\mathbf{n}}$ is a unit vector in the direction of rotation. Angular momentum is an axial vector and its SI unit is kg-m^{2}s^{-1} or J-s.

- For rotational motion of a rigid body, **angular momentum** is equal to the product of angular velocity and moment of inertia of the body about the axis of rotation. Mathematically, $L = I\omega$.
- According to the second law of rotational motion, the rate of change of angular momentum of a body is equal to the external torque applied on it and takes place in the direction of torque. Thus,
$$\tau = \frac{dL}{dt} = \frac{d}{dt}(I\omega) = I\frac{d\omega}{dt} = I\alpha \qquad \left[\because \alpha = \frac{d\omega}{dt}\right]$$
- Total effect of a torque applied on a rotating body in a given time is called angular impulse. **Angular impulse** is equal to total change in angular momentum of the system in given time. Thus, angular impulse,
$$J = \int_0^{\Delta t} \tau\, dt = \Delta L = L_f - L_i$$
- The angular momentum of a system of particles about the origin is $\quad L = \sum^{n} r_i \times p_i$

Law of Conservation of Angular Momentum

According to the law of conservation of angular momentum, if no external torque is acting on a system, then total vector sum of angular momentum of different particles of the system remains constant.

We know that, $\quad \dfrac{dL}{dt} = \tau_{ext}$

Hence, if $\tau_{ext} = 0$, then $\dfrac{dL}{dt} = 0 \Rightarrow L = $ constant

Therefore, in the absence of any external torque, total angular momentum of a system must remain conserved.

Comparison of Linear and Rotational Motion

	Linear Motion	Rotational Motion
1.	Linear momentum, $p = mv$	Angular momentum, $L = I\omega$, $L = \sqrt{2IE}$
2.	Force, $F = ma$	Torque, $T = I\alpha$
3.	Electric energy, $E = \frac{1}{2}mv^2$	Rotational energy, $E = \frac{1}{2}I\omega^2$

Equilibrium of a Rigid Bodies

For mechanical equilibrium of a rigid body, two condition need to be satisfied.

1. Translational Equilibrium

A rigid body is said to be in translational equilibrium, if it remains at rest or moving with a constant velocity in a particular direction. For this, the net external force or the vector sum of all the external forces acting on the body must be zero,

i.e. $\quad \mathbf{F} = 0$ or $F = \Sigma F_i = 0$

2. Rotational Equilibrium

A rigid body is said to be in rotational equilibrium, if the body does not rotate or rotates with constant angular velocity. For this, the net external torque or the vector sum of all the torques acting on the body is zero.

For the body to be in rotational equilibrium,
$$\tau_{ext} = 0,\ \frac{dL}{dt} = 0 \text{ or } \Sigma\tau_i = 0$$

Rigid Body Rotation

Spinning

When the body rotates in such a manner that its axis of rotation does not move, then its motion is called spinning motion.

In spinning rotational kinetic energy is given by, $K_R = \frac{1}{2}I\omega^2$.

Rotational kinetic energy is a scalar having SI unit joule (J). Rotational kinetic energy is related to angular momentum as per relation,
$$K_R = \frac{L^2}{2I} \text{ or } L = \sqrt{2IK_R}$$

Pure Rolling Motion

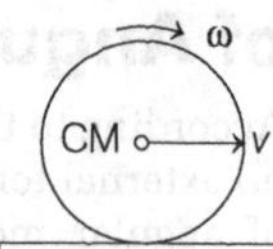

Let a rigid body, having symmetric surface about its centre of mass, is being spined at a certain angular speed and placed on a surface, so that plane of rotation is perpendicular to the surface. If the body is simultaneously given a translational motion too, then the net motion is called **rolling motion.**

The total kinetic energy in rolling motion,

$$K_N = K_R + K_T$$

$$= \frac{1}{2}mv^2 + \frac{1}{2}I\omega^2 = \frac{1}{2}mv^2 + \frac{1}{2}mv^2\left(\frac{K^2}{R^2}\right)$$

$$K_N = \frac{1}{2}mv^2\left(1+\frac{K^2}{R^2}\right)$$

Rolling Without Slipping

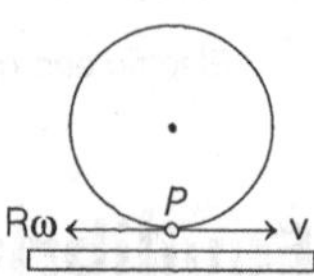

If the given body rolls over a surface such that there is no relative motion between the body and the surface at the point of contact, then the motion is called **rolling without slipping.**

Impure Rolling Motion

In impure rolling motion, the point of contact of the body with the platform is not relatively at rest w.r.t. platform on which, it is performing rolling motion, as a result sliding occurs at point of contact.

For impure rolling motion, $v_{AB} \neq 0$ i.e. $v - R\omega \neq v_0$

If platform is stationary, i.e. $v_0 = 0$, then $v \neq R\omega$

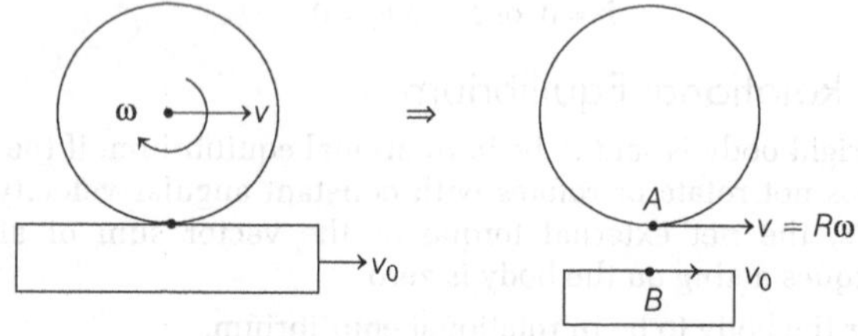

Rolling on an Inclined Plane

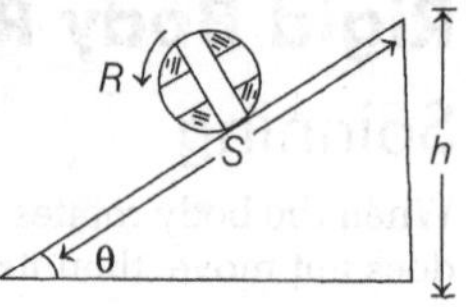

When a body of mass m and radius R rolls down on inclined plane of height h and angle of inclination θ, it loses potential energy. However, it acquires both linear and angular speeds and hence gain kinetic energy of translation and that of rotation.

By conservation of mechanical energy, $mgh = \frac{1}{2}mv^2\left(1+\frac{K^2}{R^2}\right)$

- **Velocity at the lowest point** $v = \sqrt{\dfrac{2gh}{1+\dfrac{K^2}{R^2}}}$
- **Acceleration in motion** From second equation of motion, $v^2 = u^2 + 2as$

  By substituting $u = 0$, $s = \dfrac{h}{\sin\theta}$ and $v = \sqrt{\dfrac{2gh}{1+\dfrac{K^2}{R^2}}}$, we get

  $$a = \frac{g\sin\theta}{1+\dfrac{K^2}{R^2}}$$
- **Time of descent** From first equation of motion, $v = u + at$

  By substituting $u = 0$ and value of v and a from above expressions $t = \dfrac{1}{\sin\theta}\sqrt{\dfrac{2h}{g}\left[1+\dfrac{K^2}{R^2}\right]}$

From the above expressions, it is clear that,

$$v \propto \frac{1}{\sqrt{1+\dfrac{K^2}{R^2}}};\ a \propto \frac{1}{1+\dfrac{K^2}{R^2}};\ t \propto \sqrt{1+\frac{K^2}{R^2}}$$

Important Terms Related to Inclined Plane

- Here, factor $\left(\dfrac{K^2}{R^2}\right)$ is a measure of moment of inertia of a body and its value is constant for given shape of the body and it does not depend on the mass and radius of a body.
- Velocity, acceleration and time of descent (for a given inclined plane) all depends on $\dfrac{K^2}{R^2}$. Lesser the moment of inertia of the rolling body lesser will be the value of $\dfrac{K^2}{R^2}$. So, greater will be its velocity and acceleration and lesser will be the time of descent.

Rotation about Axis through Centre of Mass (Centroidal Rotation)

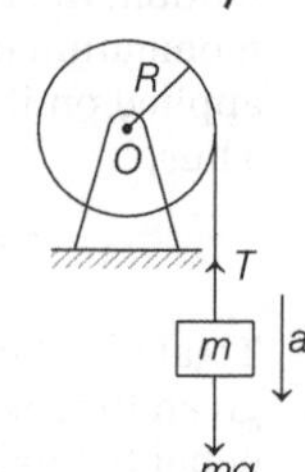

When an object of given mass is tied to a light string wound over a pulley whose moment of inertia is I and radius R as shown in the figure. The wheel bearing is frictionless and the string does not slip on the rim, then

Tension in the string is

$$T = \frac{I}{I + mR^2} \times mg$$

and acceleration, $a = \dfrac{mR^2}{I + mR^2} \cdot g$

DAY PRACTICE SESSION 1

FOUNDATION QUESTIONS EXERCISE

1 When a body is projected at an angle with the horizontal in the uniform gravitational field of the earth, the angular momentum of the body about the point of projection, as it proceeds along the path

(a) remains constant
(b) increases continuously
(c) decreases continuously
(d) initially decreases and after reaching highest point increases

2 A force of $-F\hat{\mathbf{k}}$ acts on O, the origin of the coordinate system. The torque about the point $(1, -1)$ is

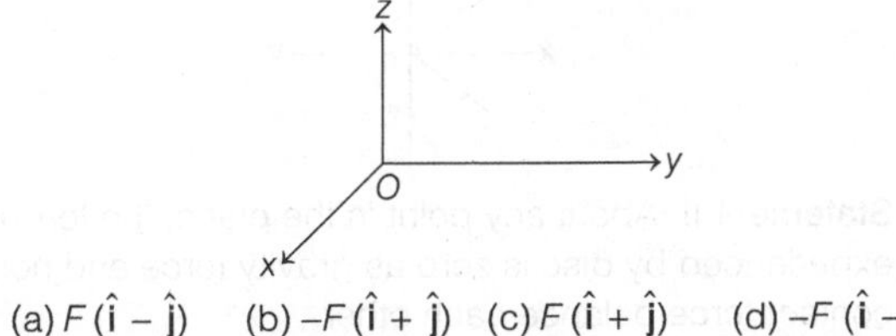

(a) $F(\hat{\mathbf{i}} - \hat{\mathbf{j}})$ (b) $-F(\hat{\mathbf{i}} + \hat{\mathbf{j}})$ (c) $F(\hat{\mathbf{i}} + \hat{\mathbf{j}})$ (d) $-F(\hat{\mathbf{i}} - \hat{\mathbf{j}})$

3 Angular momentum of the particle rotating with a central force is constant due to

(a) constant force (b) constant linear momentum
(c) zero torque (d) constant torque

4 A particle of mass $m = 3$ kg moves along a straight line $4y - 3x = 2$, where x and y are in metre, with constant velocity $v = 5\,\text{ms}^{-1}$. The magnitude of angular momentum about the origin is

(a) $12\,\text{kg-m}^2\,\text{s}^{-1}$ (b) $8.0\,\text{kg-m}^2\,\text{s}^{-1}$
(c) $6.0\,\text{kg-m}^2\,\text{s}^{-1}$ (d) $4.5\,\text{kg-m}^2\,\text{s}^{-1}$

5 A particle of mass m moves along line PC with velocity v as shown in the figure. What is the angular momentum of the particle about O?

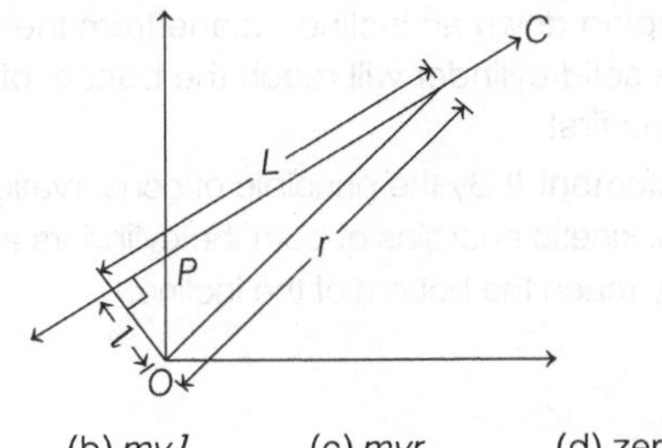

(a) mvL (b) mvl (c) mvr (d) zero

6 A particle of mass 5 g is moving with a uniform speed of $3\sqrt{2}\,\text{cms}^{-1}$ is in the xy-plane along the line $y = 2\sqrt{5}$ cm. The magnitude of its angular momentum about the origin in g-$\text{cm}^2\,\text{s}^{-1}$ is

(a) zero
(b) 30
(c) $30\sqrt{2}$
(d) $30\sqrt{10}$

7 A bob of mass m attached to an inextensible string of length l is suspended from a vertical support. The bob rotates in a horizontal circle with an angular speed ω rad/s about the vertical support. About the point of suspension → JEE Main 2014

(a) angular momentum is conserved
(b) angular momentum changes in magnitude but not in direction
(c) angular momentum changes in direction but not in magnitude
(d) angular momentum changes both in direction and magnitude

8 A particle of mass 2 kg is moving such that at time t, its position, in metre, is given by $\mathbf{r}(t) = 5\,\hat{\mathbf{i}} - 2t^2\,\hat{\mathbf{j}}$. The angular momentum of the particle at $t = 2$ s about the origin in $\text{kg m}^{-2}\,\text{s}^{-1}$ is → JEE Main (Online) 2013

(a) $-80\,\hat{\mathbf{k}}$
(b) $(10\hat{\mathbf{i}} - 16\hat{\mathbf{j}})$
(c) $-40\,\hat{\mathbf{k}}$
(d) $40\,\hat{\mathbf{k}}$

9 A uniform disc of radius a and mass m, is rotating freely with angular speed ω in a horizontal plane, about a smooth fixed vertical axis through its centre. A particle, also of mass m, is suddenly attached to the rim of the disc and rotates with it. The new angular speed is

(a) $\omega/6$
(b) $\omega/3$
(c) $\omega/2$
(d) $\omega/5$

10 If the radius of the earth contracts $\frac{1}{n}$ of its present day value, length of the day will be approximately.

(a) $\frac{24}{n}$ h (b) $\frac{24}{n^2}$ h
(c) $24\,n$h (d) $24\,n^2$h

11 A particle of mass m is moving along the side of a square of side a, with a uniform speed v in the X-Y plane as shown in the figure.

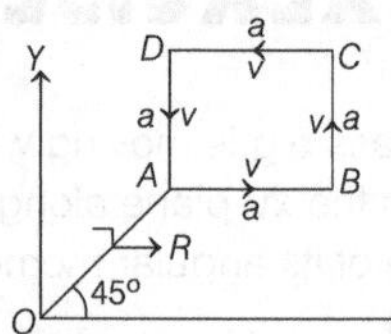

Which of the following statements is false for the angular momentum **L** about the origin? →JEE Main 2016 (Offline)

(a) $\mathbf{L} = -\frac{mv}{\sqrt{2}} R\,\hat{\mathbf{k}}$, when the particle is moving from A to B.

(b) $\mathbf{L} = mv\left(\frac{R}{\sqrt{2}} + a\right)\hat{\mathbf{k}}$, when the particle is moving from B to C.

(c) $\mathbf{L} = mv\left(\frac{R}{\sqrt{2}} - a\right)\hat{\mathbf{k}}$, when the particle is moving from C to D.

(d) $\mathbf{L} = \frac{mv}{\sqrt{2}} R\,\hat{\mathbf{k}}$, when the particle is moving from D to A.

12 A uniform rod AB of mass m and length l at rest on a smooth horizontal surface. An impulse P is applied to the end B. The time taken by the rod to turn through at right angle is

(a) $2\pi\frac{ml}{P}$ (b) $2\frac{\pi P}{ml}$ (c) $\frac{\pi}{12}\frac{ml}{P}$ (d) $\frac{\pi P}{ml}$

13 A solid cylinder of mass 20 kg rotates about its axis with angular speed 100 rad/s. The radius of the cylinder is 0.25 m. Then the kinetic energy associated with the rotation of the cylinder and the magnitude of angular momentum of the cylinder about its axis is respectively

(a) 3200 J, 62.5 J-s (b) 3125 J, 62.5 J-s
(c) 3500 J, 68 J-s (d) 3400 J, 63.5 J-s

14 A ring and a disc having the same mass, roll without slipping with the same linear velocity. If the kinetic energy of the ring is 8 J, that of the disc must be

(a) 2 J (b) 4 J (c) 6 J (d) 16 J

15 A round uniform body of radius R, mass M and moment of inertia I, rolls down (without slipping) an inclined plane making an angle θ with the horizontal. Then, its acceleration is

(a) $\frac{g\sin\theta}{1 + I/MR^2}$ (b) $\frac{g\sin\theta}{1 + MR^2/I}$ (c) $\frac{g\sin\theta}{1 - I/MR^2}$ (d) $\frac{g\sin\theta}{1 - MR^2/I}$

Direction (Q. Nos. 16-18) *Each of these questions contains two statements : Statement I and Statement II. Each of these questions also has four alternative choices, only one of which is the correct answer. You have to select one of the codes (a), (b), (c), (d) given here.*

(a) Statement I is true, Statement II is true; Statement II is the correct explanation for Statement I
(b) Statement I is true, Statement II is true; Statement II is not the correct explanation for Statement I
(c) Statement I is true; Statement II is false
(d) Statement I is false; Statement II is true

16 **Statement I** A disc A moves on a smooth horizontal plane and rebounds elastically from a smooth vertical wall (Top view is as shown in the figure) in this case about any point on line xy. The angular momentum of the disc remains conserved.

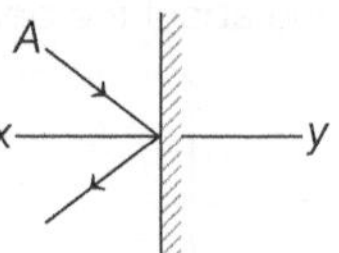

Statement II About any point in the plane, the torque experienced by disc is zero as gravity force and normal contact force balance each other.

17 **Statement I** A block is kept on a rough horizontal surface, under the action of a force F as shown in the figure. The torque of normal contact force about centre of mass is having zero value.

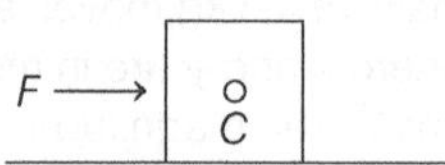

Statement II The point of application of normal contact force may pass through centre of mass.

18 **Statement I** Two cylinders, one hollow (metal) and the other solid (wood) with the same mass and identical dimensions are simultaneously allowed to roll without slipping down an inclined plane from the same height. The solid cylinder will reach the bottom of the inclined plane first.

Statement II By the principle of conservation of energy, the total kinetic energies of both the cylinders are equal, when they reach the bottom of the incline.

DAY PRACTICE SESSION 2

PROGRESSIVE QUESTIONS EXERCISE

1 A thin circular ring of mass m and radius R is rotating about its axis with a constant angular velocity ω. Two objects each of mass M are attached gently to the opposite ends of a diameter of the ring. The ring now rotates with an angular velocity ω'

(a) $\frac{\omega(m+2M)}{m}$ (b) $\frac{\omega(m-2M)}{(m+2M)}$

(c) $\frac{\omega m}{(m+M)}$ (d) $\frac{\omega m}{(m+2M)}$

2 A T shaped object with dimensions as shown in the figure, is lying on a smooth floor. A force is applied at the point P parallel to AB, such that the object has only the translational motion without rotation. Find the location of P with respect to C.

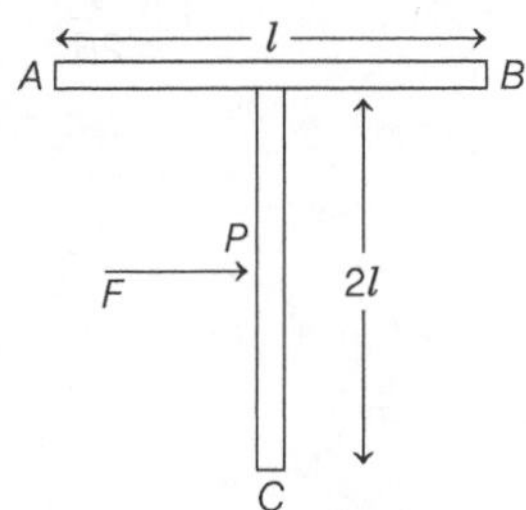

(a) $\frac{2}{3}l$ (b) $\frac{3}{2}l$ (c) $\frac{4}{3}l$ (d) l

3 A thin uniform rod of length l and mass m is swinging freely about horizontal axis passing through its end. Its maximum angular speed is ω. Its centre of mass rises to a maximum height of

(a) $\frac{1}{3}\frac{l^2\omega^2}{g}$ (b) $\frac{1}{6}\frac{l\omega}{g}$ (c) $\frac{1}{2}\frac{l^2\omega^2}{g}$ (d) $\frac{1}{6}\frac{l^2\omega^2}{g}$

4 A small object of uniform density rolls up a curved surface with an initial velocity v. It reached up to a maximum height of $\frac{3v^2}{4g}$ with respect to the initial position. The object is

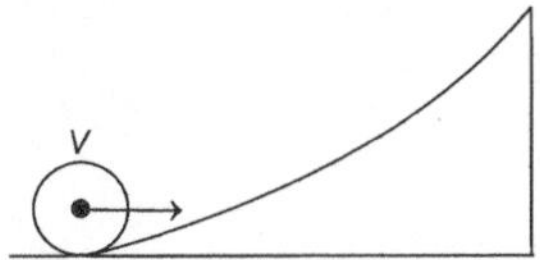

(a) ring (b) solid sphere
(c) hollow sphere (d) disc

5 A ring of mass M and radius R is rotating about its axis with angular velocity ω. Two identical bodies each of mass m are now gently attached at the two ends of a diameter of the ring. Because of this, the kinetic energy loss will be → JEE Main (Online) 2013

(a) $\frac{m(M+2m)}{M}\omega^2R^2$ (b) $\frac{Mm}{(M+2m)}\omega^2R^2$

(c) $\frac{Mm}{(M-2m)}\omega^2R^2$ (d) $\frac{(M+m)M}{(M+2m)}\omega^2R^2$

6 A hoop of radius r and mass m rotaing with an angular velocity ω_0 is placed on a rough horizontal surface. The initial velocity of the centre of the hoop is zero. What will be the velocity of the centre of the hoop, when it ceases to slip? → JEE Main 2013

(a) $\frac{r\omega_0}{4}$ (b) $\frac{r\omega_0}{3}$

(c) $\frac{r\omega_0}{2}$ (d) $r\omega_0$

7 An annular ring with inner and outer radii R_1 and R_2 is rolling without slipping with a uniform angular speed. The ratio of the forces experienced by the two particles situated on the inner and outer parts of the ring, $\frac{F_1}{F_2}$ is

(a) $\frac{R_2}{R_1}$ (b) $\left(\frac{R_1}{R_2}\right)^2$

(c) 1 (d) $\frac{R_1}{R_2}$

8 A slender uniform rod of mass M and length l is pivoted at one end so that it can rotate in a vertical plane (see the figure). There is negligible friction at the pivot. The free end is held vertically above the pivot and then released. The angular acceleration of the rod when it makes an angle θ with the vertical, is

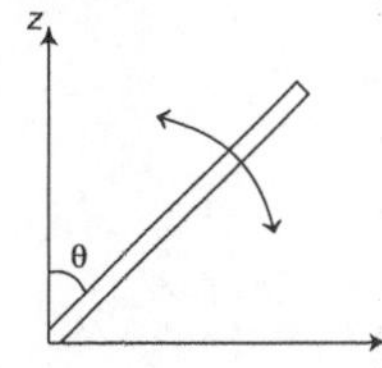

→ JEE Main 2017 (Offline)

(a) $\frac{2g}{3l}\sin\theta$ (b) $\frac{3g}{2l}\cos\theta$ (c) $\frac{2g}{3l}\cos\theta$ (d) $\frac{3g}{2l}\sin\theta$

9 A horizontal force F acts on the sphere at its centre as shown. Coefficient of friction between ground and sphere is μ. What is the maximum value of F, for which there is no slipping?

(a) $F \le \frac{5}{2}\mu mg$ (b) $F \le \frac{7}{2}\mu mg$

(c) $F \le \frac{9}{2}\mu mg$ (d) $F \le \frac{3}{2}\mu mg$

10 A solid uniform disc of mass m rolls without slipping down a fixed inclined plank with an acceleration a. The frictional force on the disc due to surface of the plank is

(a) $\frac{1}{4}ma$ (b) $\frac{3}{2}ma$ (c) ma (d) $\frac{1}{2}ma$

11 A cylinder having radius 0.4 m, initially rotating (at $t = 0$) with $\omega_0 = 54$ rad/s is placed on a rough inclined plane with $\theta = 37°$ having frictional coefficient $\mu = 0.5$. The time taken by the cylinder to start pure rolling is $[g = 10 \text{m} / \text{s}^2]$.

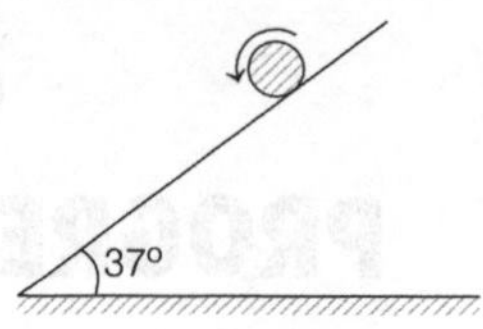

(a) 5.4 s (b) 2.4 s
(c) 1.4 s (d) 1.2 s

12 Angular acceleration of the Cylinder (B) shown in the figure is (all strings and pulley are ideal)

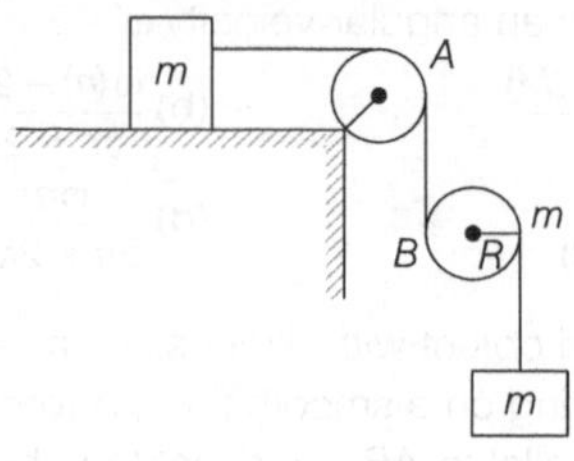

(a) $\frac{2g}{3R}$ (b) $\frac{2g}{5R}$

(c) $\frac{2g}{R}$ (d) $\frac{g}{2R}$

ANSWERS

SESSION 1	**1** (b)	**2** (c)	**3** (c)	**4** (c)	**5** (b)	**6** (d)	**7** (c)	**8** (a)	**9** (b)	**10** (b)
	11 (c,d)	**12** (c)	**13** (b)	**14** (c)	**15** (a)	**16** (a)	**17** (a)	**18** (b)		
SESSION 2	**1** (d)	**2** (c)	**3** (d)	**4** (d)	**5** (b)	**6** (c)	**7** (d)	**8** (d)	**9** (b)	**10** (d)
	11 (d)	**12** (b)								

Hints and Explanations

SESSION 1

1 $\tau_0 = mg \times r_\perp$

As, $r_\perp$ is continuously increasing or torque is continuously increasing on the particle. Hence, angular momentum is continuously increasing.

2 $\tau = \mathbf{r} \times \mathbf{F}$

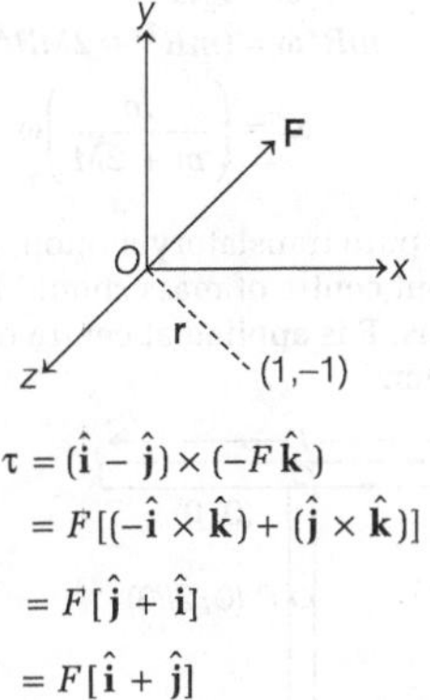

$$\tau = (\hat{\mathbf{i}} - \hat{\mathbf{j}}) \times (-F\hat{\mathbf{k}})$$
$$= F[(-\hat{\mathbf{i}} \times \hat{\mathbf{k}}) + (\hat{\mathbf{j}} \times \hat{\mathbf{k}})]$$
$$= F[\hat{\mathbf{j}} + \hat{\mathbf{i}}]$$
$$= F[\hat{\mathbf{i}} + \hat{\mathbf{j}}]$$

3 Torque due to central force is zero,

$$\tau = \frac{d}{dt}(L) = 0$$

$\therefore \quad L = \text{constant}$

4 Equation of straight line AB is

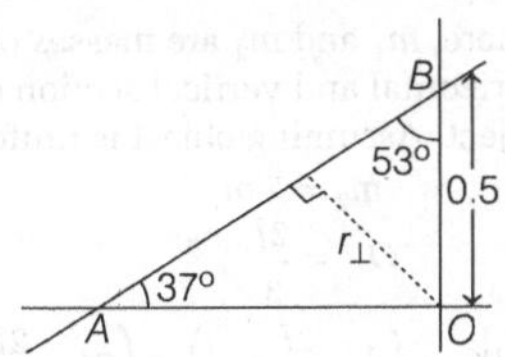

$$4y - 3x = 2$$

or $\quad y = \frac{3}{4}x + 0.5$

$\therefore$ Slope $= \tan\theta = \frac{3}{4}$

or $\quad \theta = 37°$

$$r_\perp = 0.5 \sin 53°$$
$$= 0.4 \text{ m}$$

$\therefore$ Angular momentum $(L) = mvr_\perp$

or $\quad L = 3 \times 5 \times 0.4$
$$= 6 \text{kg - m}^2\text{s}^{-1}$$

5 Angular momentum of particle about O,

$$\mathbf{L} = m(\mathbf{r} \times \mathbf{v})$$
$$|\mathbf{L}| = mr\,v\sin\theta$$
$$= mv(r\sin\theta)$$
$$= mvl$$

6 $L = mvr_1$

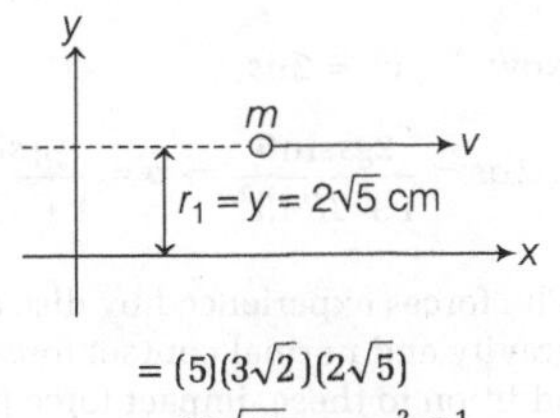

$$= (5)(3\sqrt{2})(2\sqrt{5})$$
$$= 30\sqrt{10} \text{ g- cm}^2\text{s}^{-1}$$

7 Angular momentum of the pendulum about the suspension point O is

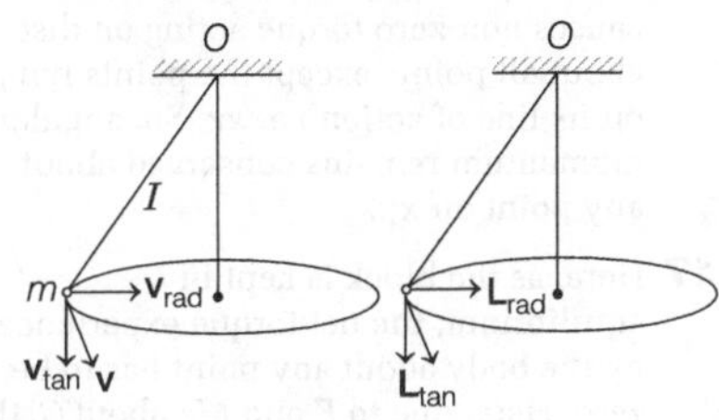

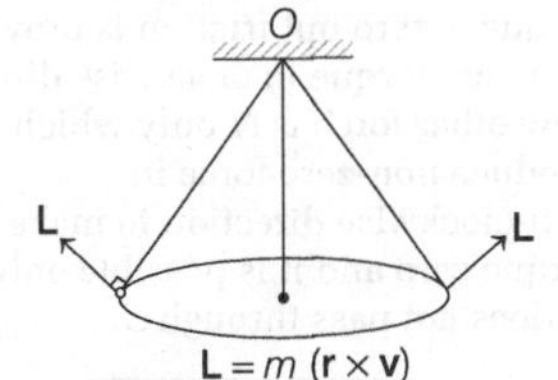

Then, $\mathbf{v}$ can be resolved into two components, radial component r_{rad} and tangential component $r_{\tan}$. Due to v_{rad}, $\mathbf{L}$ will be tangential and due to $v_{\tan}$, $\mathbf{L}$ will be radially outwards as shown. So, net angular momentum will be as shown in figure, whose magnitude will be constant ($|L| = mvl$). But its direction will change as shown in the figure.

$$\mathbf{L} = m(\mathbf{r} \times \mathbf{v})$$

where, r = radius of circle.

8 Given, $m = 2$ kg, $\mathbf{r}(t) = 5\hat{\mathbf{i}} - 2t^2\hat{\mathbf{j}}$

Angular momentum $(L) = \mathbf{r} \times \mathbf{p}$

$\therefore$ Velocity, $v = \frac{dr}{dt} = \frac{d}{dt}(5\hat{\mathbf{i}} - 2t^2\hat{\mathbf{j}})$

$$= -8\hat{\mathbf{j}} \quad (\text{at } t = 2\text{s})$$

$\therefore \quad p = mv$
$$= 2 \times (-8\hat{\mathbf{j}}) = -16\hat{\mathbf{j}}$$

Therefore,
$$L = \mathbf{r} \times \mathbf{p}$$
$$= (5\hat{\mathbf{i}} - 2t^2\hat{\mathbf{j}}) \times (-16)\hat{\mathbf{j}} \quad (\text{at } t = 2\text{s})$$
$$= -80\hat{\mathbf{k}}$$

9 Conserving angular momentum

$$I_1\omega_1 = I_2\omega_2$$

$$\therefore \quad \omega_2 = \frac{I_1}{I_2}\omega_1 = \frac{\left(\dfrac{ma^2}{2}\right)}{\left(\dfrac{ma^2}{2}\right) + ma^2}\omega = \frac{\omega}{3}$$

10 $\quad I\omega = \text{constant}$

or $\quad \frac{R^2}{T} = \text{constant}$

$$\left(\text{As, } I \propto R^2 \text{ and } \omega \propto \frac{1}{T}\right)$$

$\therefore \quad T \propto R^2$

As, $\quad R' = \frac{1}{n}R$

$\therefore \quad T' = \frac{T}{n^2} = \frac{24}{n^2}\text{h}$

11 For a particle of mass m is moving along the side of a square of side a. Such that Angular momentum $\mathbf{L}$ about the origin

$$= \mathbf{L} = \mathbf{r} \times \mathbf{p}$$
$$= rp\sin\theta\,\hat{\mathbf{n}}$$

or $\quad \mathbf{L} = \mathbf{r}\,(\mathbf{p})\,\hat{\mathbf{n}}$

When a particle is moving from D to A,

$$\mathbf{L} = \frac{R}{\sqrt{2}}\,mv(-\hat{\mathbf{k}})$$

A particle is moving from A to B,

$$\mathbf{L} = \frac{R}{\sqrt{2}}\,mv(-\hat{\mathbf{k}})$$

and it moves from C to D,

$$\mathbf{L} = \left(\frac{R}{\sqrt{2}} + a\right)mv\,(\hat{\mathbf{k}})$$

For B to C, we have

$$\mathbf{L} = \left(\frac{R}{\sqrt{2}} + a\right)mv\,(\hat{\mathbf{k}})$$

Hence, options (c) and (d) are incorrect.

12 Angular impulse $= P \times \frac{l}{2}$

= change in angular momentum

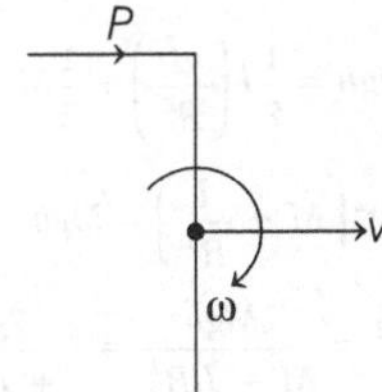

$\therefore \quad \frac{Pl}{2} = I\omega = \left(\frac{ml^2}{12}\right)\omega$

$\therefore \quad \omega = \frac{6P}{ml}$

Now, $t = \frac{\theta}{\omega} = \frac{\pi/2}{6P/ml}$

$= \frac{\pi ml}{12P}$

13 Moment of inertia of the solid cylinder about its axis of symmetry,

$I = \frac{1}{2}MR^2$

$= \frac{1}{2}\times 20\times(0.25)^2$

$= 10\times 0.0625$

$= 0.625\,\text{kg-m}^2 = 62.5\,\text{J-s}$

Kinetic energy associated with the rotation of the cylinder is given by

$K = \frac{1}{2}I\omega^2$

$= \frac{1}{2}\times 0.625\times(100)^2$

$= 0.3125\times 10000 = 3125\,\text{J}$

14 For ring, $\frac{K_R}{K_T} = 1$

For disc, $\frac{K_R}{K_T} = \frac{1}{2}$

For ring, total kinetic energy

$= 2(K_T) = 8\text{ J}$ (given)

Hence, $K_T = 4\text{ J}$

For disc, total kinetic energy $= \frac{3}{2}K_T$

$= \frac{3}{2}\times 4 = 6\text{ J}$

15 Assuming that no energy is used up against friction, the loss in potential energy is equal to the total gain in the kinetic energy.

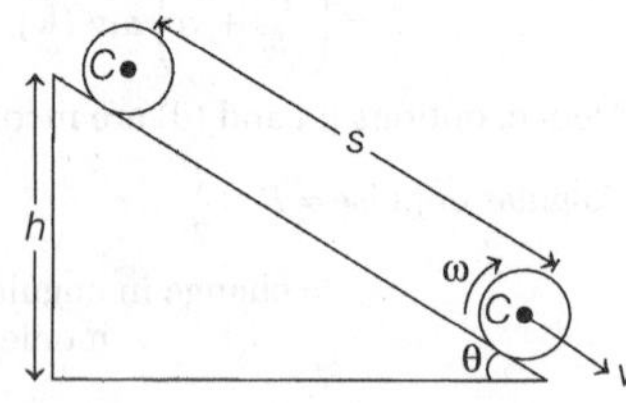

i.e. $Mgh = \frac{1}{2}I\left(\frac{v^2}{R^2}\right) + \frac{1}{2}Mv^2$

or $\frac{1}{2}v^2\left(M + \frac{I}{R^2}\right) = Mgh$

or $v^2 = \frac{2Mgh}{M + I/R^2} = \frac{2gh}{1 + I/MR^2}$

If s be the distance covered along the plane, then

$h = s\sin\theta$

$\therefore \quad v^2 = \frac{2gs\sin\theta}{1 + I/MR^2}$

Now, $v^2 = 2as$

$\therefore 2as = \frac{2gs\sin\theta}{1 + I/MR^2} \Rightarrow a = \frac{g\sin\theta}{1 + I/MR^2}$

16 The forces experienced by disc are gravity and normal contact force. In addition to these, impact force (during collision) will act on the disc along line xy. Gravity and normal contact force balance each other (in terms of force and torque both), but impact force causes non-zero torque acting on disc about all points except the points lying on its line of action i.e. xy. So, angular momentum remains conserved about any point on xy.

17 Here, as the block is kept in equilibrium, the net torque experienced by the body about any point has to be zero. Here, due to F and Mg about C, the torque is zero but friction is providing non-zero torque in clockwise direction, now other force is N only which can produce non-zero force in anti-clockwise direction to make net torque zero and it is possible only when N does not pass through C.

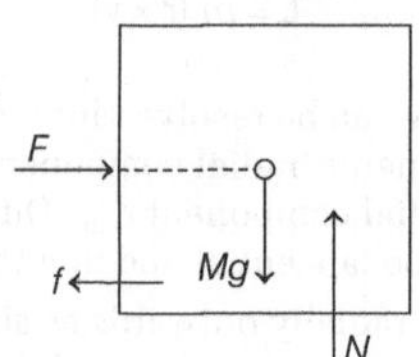

18 In case of pure rolling on inclined plane,

$a = \frac{g\sin\theta}{1 + I/mR^2}$

$I_{\text{Solid}} < I_{\text{Hollow}}$

$\therefore \quad a_{\text{Solid}} > a_{\text{Hollow}}$

$\therefore$ Solid cylinder will reach the bottom first. Further, in case pure rolling on stationary ground, work done by friction is zero. Therefore, mechanical energy of both the cylinders will remain constant.

$\therefore (\text{KE})_{\text{Hollow}} = (\text{KE})_{\text{Solid}}$

= decrease in PE = mgh

SESSION 2

1 As, no external torque is acting on the system, angular momentum of system remains conserved.

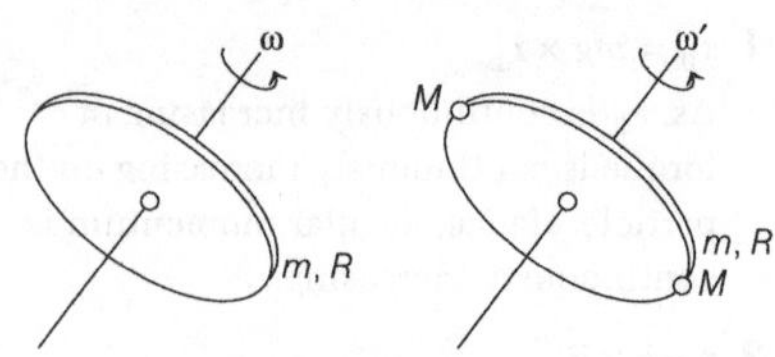

i.e. $I_1\omega = I_2\omega'$

$\Rightarrow \quad mR^2\omega = (mR^2 + 2MR^2)\omega'$

$\Rightarrow \quad \omega' = \left(\frac{m}{m + 2M}\right)\omega$

2 For pure translatory motion, net torque about centre of mass should be zero. Thus, **F** is applied at centre of mass of system.

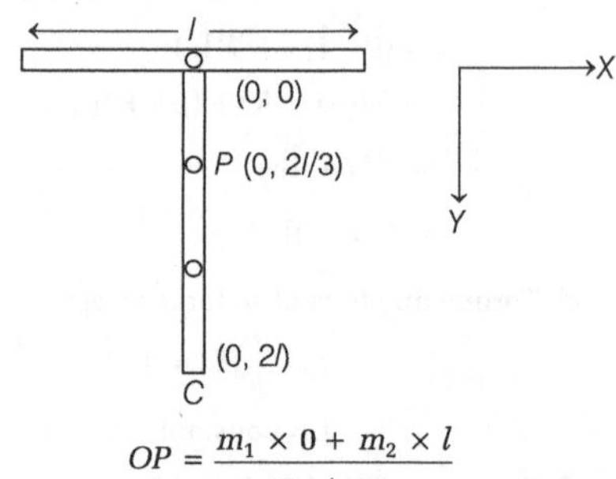

$OP = \frac{m_1\times 0 + m_2\times l}{m_1 + m_2}$

where, m_1 and m_2 are masses of horizontal and vertical section of the object. Assuming object is uniform,

$m_2 = 2m_1$

$\Rightarrow \quad OP = \frac{2l}{3}$

$\therefore \; PC = \left(l - \frac{2l}{3} + l\right) = \left(2l - \frac{2l}{3}\right) = \frac{4l}{3}$

3 If centre of mass rises to a maximum height, then loss in KE = Gain in PE, we get

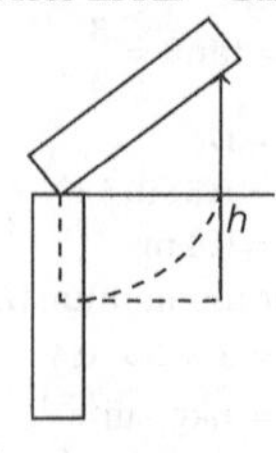

$\frac{1}{2}I\omega^2 = mgh$

$\frac{1}{2}\times\frac{1}{3}ml^2\omega^2 = mgh \Rightarrow h = \frac{1}{6}\frac{l^2\omega^2}{g}$

4 $\frac{1}{2}mv^2 + \frac{1}{2}I\left(\frac{v}{R}\right)^2 = mg\left(\frac{3v^2}{4g}\right)$

$\therefore \quad I = \frac{1}{2}mR^2$

Therefore, the body is a disc.

5 By conservation of angular momentum,

$I_1\omega_1 = I_2\omega_2$ ...(i)

where, $I_1 = mR^2$ ...(ii)

and $I_2 = 2mR^2 + MR^2$ ...(iii)

Now change in,

$\text{KE} = \frac{1}{2}L_1 I_1^2 - \frac{1}{2}L_2 I_2^2$ ...(iv)

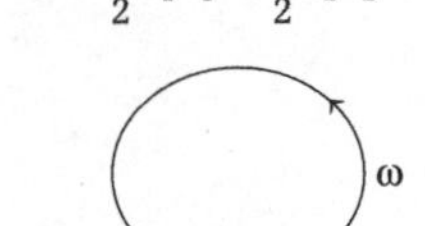

Substituting the values from Eqs. (i), (ii) and (iii) into Eq. (iv), we get,

Change in KE $= \left(\frac{Mm}{M+2m}\right)\omega^2 R^2$

6

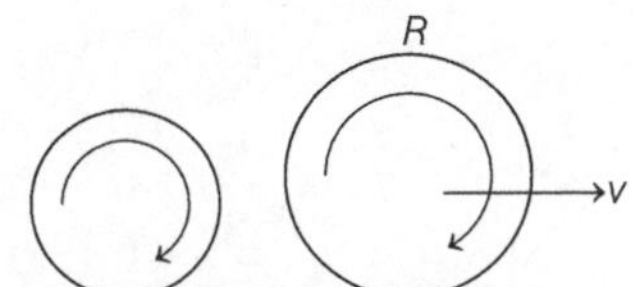

From conservation of angular momentum

$mr^2\omega_0 = mvr + mr^2 \times \frac{v}{r}$

$\Rightarrow \quad v = \frac{\omega_0 r}{2}$

7 Since ω is constant, v would also be constant. So, no net force or torque is acting on ring. The force experienced by any particle is only along radial direction, or we can say the centripetal force.

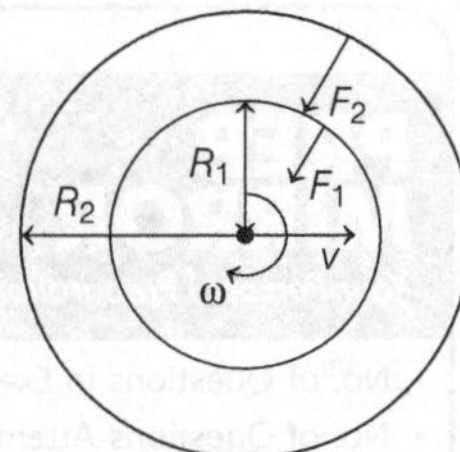

The force exprienced by inner part, $F_1 = m\omega^2 R_1$ and the force experienced by outer part,

$F_2 = m\omega^2 R_2$

$\frac{F_1}{F_2} = \frac{R_1}{R_2}$

8 As the rod rotates in vertical plane so a torque is acting on it, which is due to the vertical component of weight of rod.

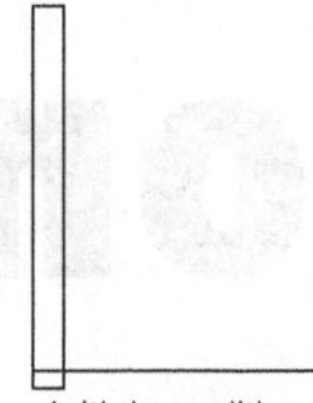

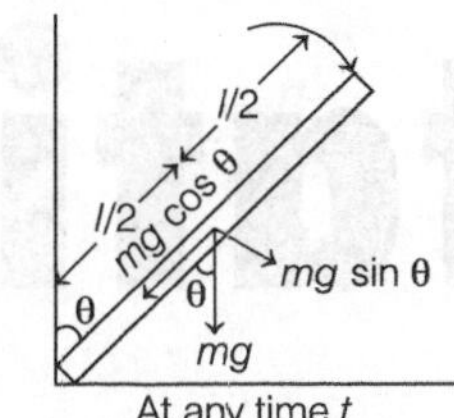

Now, torque τ = force × perpendicular distance of line of action of force from axis of rotation

$= mg \sin\theta \times \frac{l}{2}$

Again, torque $\tau = I\alpha$

where, I = moment of inertia

$= \frac{ml^2}{3}$

[Force and torque frequency along axis of rotation passing through in end]

α = angular acceleration

$\therefore \quad mg \sin\theta \times \frac{l}{2} = \frac{ml^2}{3}\alpha$

$\therefore \quad \alpha = \frac{3g \sin\theta}{2l}$

9 Here, $F - f = ma$ ... (i)

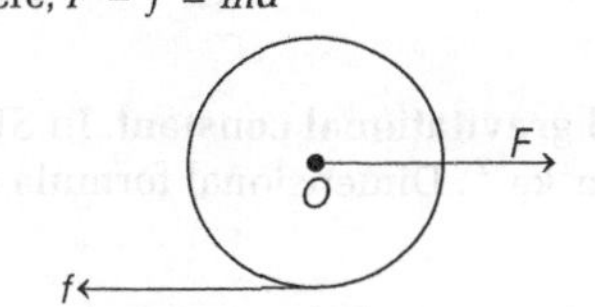

and $\tau = I\alpha = I\left(\frac{a}{R}\right)$

$\Rightarrow \quad f.R = \frac{2}{5}mR^2 \times \frac{a}{R}$

or $f = \frac{2}{5}ma$ ...(ii)

From Eqs. (i) and (ii), we get

$f = \frac{2}{7}F$

or $\frac{2}{7}F \le \mu mg$

or $F \le \frac{7}{2}\mu mg$

Hence, (b) is the correct option.

10 Since, $a = \frac{g\sin\theta}{1 + I/mR^2}$

or $a = \frac{g\sin\theta}{1+\frac{1}{2}}$;

[For uniform solid disc, $I = \frac{1}{2}mR^2$]

$= \frac{2}{3}g\sin\theta$

or $g\sin\theta = \frac{3}{2}a$

Now, force of friction $(f) = \frac{mg\sin\theta}{1+\frac{mR^2}{I}}$

or $f = \frac{m(3a/2)}{1+2} = \frac{1}{2}ma$

Hence, (d) is the correct option.

11 $a = (\mu g\cos\theta) + (g\sin\theta)$

$= 0.5 \times 10 \times 0.8 + 10 \times 0.6$

or $a = 10\,\text{m/s}^2$

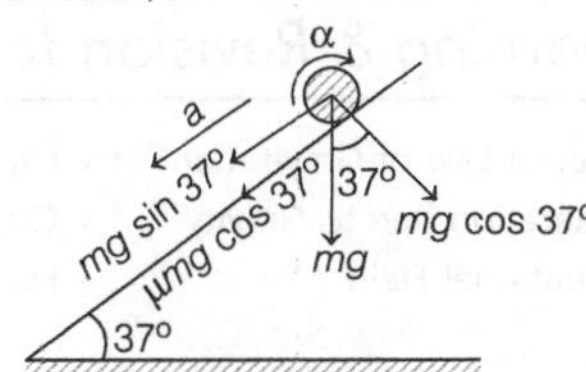

And angular acceleration (α)

$= \frac{(\mu\ mg\cos\theta)R}{\frac{1}{2}mR^2}$

or $\alpha = \frac{2\mu\, g\cos\theta}{R}$

or $\alpha = \frac{2 \times 0.5 \times 10 \times 0.8}{0.4} = 20\,\text{rad/s}^2$

Pure rolling will start, when

$v = R\omega$

where, $v = \alpha \times t$ and $\omega = (\omega_0 - \alpha t)$

$\therefore \quad at = R(\omega_0 - \alpha t)$

or $10t = 0.4(54 - 20t)$

or $t = 1.2$ s

Hence, (d) is the correct option.

12 Linear acceleration of cylinder (B) is given by the relation

$a = \frac{mR^2}{I + m'R^2} \cdot g$

where, $I = \frac{1}{2}mR^2$ (for solid cylinder) and

$m' = 2m$

$\therefore \quad a = \frac{mR^2}{\frac{1}{2}mR^2 + 2mR^2} \cdot g$

$= \frac{2mR^2}{5mR^2} \cdot g = \frac{2g}{5}$

∴ Angular acceleration (α) of the cylinder B is

$\alpha = \frac{a}{R}$ or $\alpha = \frac{2g}{5R}$

Hence, (b) is the correct option.

DAY NINE

Gravitation

Learning & Revision for the Day

- Universal Law of Gravitation
- Acceleration due to Gravity
- Gravitational Field
- Gravitational Potential
- Gravitational Potential Energy
- Escape Velocity
- Kepler's Laws of Planetary Motion
- Satellite
- Geostationary Satellite

Universal Law of Gravitation

In this universe, each body attracts other body with a force that is directly proportional to the product of their masses and inversely proportional to the square of the distance between them.

Let m_1 and m_2 be the masses of two bodies and r be the separation between them.

$$F = G\frac{m_1 m_2}{r^2}.$$

The proportionality constant G is called **universal gravitational constant**. In SI system, value of gravitational constant G is $6.67 \times 10^{-11} \text{Nm}^2 \text{kg}^{-2}$. Dimensional formula of G is $[\text{M}^{-1}\text{L}^3\text{T}^{-2}]$.

Acceleration due to Gravity

The acceleration of an object during its free fall towards the earth is called acceleration due to gravity.

If M is the mass of earth and R is the radius, the earth attracts a mass m on its surface with a force F given by

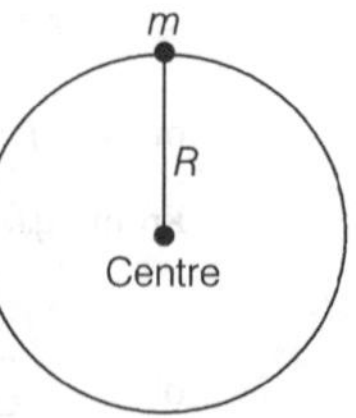

$$F = \frac{GMm}{R^2}$$

This force imparts an acceleration to the mass m which is known as acceleration due to gravity (g).

By Newton's law, we have

$$\text{Acceleration } (g) = \frac{F}{m} = \frac{\frac{GMm}{R^2}}{m} = \frac{GM}{R^2}$$

On the surface of earth, $g = \dfrac{GM}{R^2}$

Substituting the values of G, M, R, we get $g = 9.81 \text{ ms}^{-2}$.

Mass of the earth $m = 6 \times 10^{24}$ kg and radius of the earth $R = 6.4 \times 10^6$ m.

- No. of Questions in Exercises (x)—
- No. of Questions Attempted (y)—
- No. of Correct Questions (z)—
  (Without referring Explanations)
- Accuracy Level ($z/y \times 100$)—
- Prep Level ($z/x \times 100$)—

In order to expect good rank in JEE, your Accuracy Level should be above 85 & Prep Level should be above 75.

Variation in g with Altitude and Depth with respect to Earth

The value of g is variable and can vary in same cases as mentioned below

1. **Value of acceleration due to gravity (g) at a height (h)** from the surface of the earth is given by $g' = \frac{gR^2}{(R+h)^2}$

   If $h << R$, then $g' = g\left[1 - \frac{2h}{R}\right]$

2. **Value of acceleration due to gravity (g) at a depth (d)** from the surface of the earth is given by

   $$g' = g\left(1 - \frac{d}{R}\right)$$

   At the centre of the earth $d = R$ and hence, $g' = 0$.

Variation in the Value of g Due to Rotation of the Earth

Due to rotation of the earth, the value of g decreases as the speed of rotation of the earth increases. The value of acceleration due to gravity at a latitude is

$$g'_\lambda = g - R\omega^2 \cos^2 \lambda$$

Following conclusions can be drawn from the above discussion

(a) The effect of centrifugal force due to rotation of the earth is to reduce the effective value of g.

(b) The effective value of g is not truely in vertical direction.

(c) At the equators, $\lambda = 0°$

Therefore, $g' = g - R\omega^2$ (minimum value)

(d) At the poles, $\lambda = 90°$

Therefore, $g' = g$ (maximum value)

Gravitational Field

The space surrounding a material body in which its gravitational force of attraction can be experienced is called its gravitational field.

Gravitational Field Intensity (I)

Gravitational field intensity at any point is defined as the force experienced by any test mass devided by the magnitude of test mass when placed at the desired point.

Mathematically,

Gravitational field intensity, $\mathbf{I} = \frac{\mathbf{F}}{m_0}$

where, m_0 is a small test mass. The SI unit of gravitational intensity is N kg^{-1}.

- Gravitational intensity at a point situated at a distance r from a point mass M is given by $\mathbf{I} = \frac{GM}{r^2}$
- Gravitational field intensity due to a solid sphere (e.g. earth) of mass M and radius R at a point distant r from its centre ($r > R$) is $\mathbf{I} = \frac{GM}{r^2}$

  and at the surface of solid sphere, $\mathbf{I} = \frac{GM}{R^2}$.

  However, for a point $r < R$, we find that $\mathbf{I} = \frac{GMr}{R^3}$

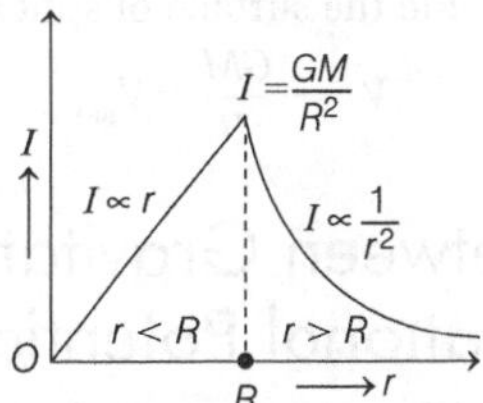

Variation of gravitational field intensity in solid sphere

- Due to a body in the form of uniform shell gravitational field intensity at a point outside the shell ($r > R$) is given by $\mathbf{I} = \frac{GM}{r^2}$

  But at any point inside the shell, gravitational intensity is zero.
- Gravitational intensity at a point due to the combined effect of different point masses is given by the vector sum of individual intensities.

  Thus, $\mathbf{I} = \mathbf{I}_1 + \mathbf{I}_2 + \mathbf{I}_3 + \ldots$

Gravitational Potential

Gravitational potential at any point in a gravitational field is defined as the work done in bringing a unit mass from infinity to that point.

Gravitational potential, $V = \lim_{m_0 \to 0} \frac{W}{m_0}$.

Gravitational potential due to a point mass is $V = -\frac{GM}{r}$.

Gravitational potential is always negative. It is a scalar term and its SI unit is J kg^{-1}.

For Solid Sphere

- At a point outside the solid sphere, (e.g. earth), i.e.

  $$r > R \quad V = -\frac{GM}{r}.$$
- At a point on the surface, $V = -\frac{GM}{R}$.
- At a point inside the sphere, ($r < R$).

  $$V = -\frac{GM}{2R^3}(3R^2 - r^2) = -\frac{GM}{2R}\left[3 - \frac{r^2}{R^2}\right]$$
- At the centre of solid sphere,

  $$V = -\frac{3GM}{2R} = \frac{3}{2} V_{\text{surface}}$$

For Spherical Shell

- At a point outside the shell,

$$V = -\frac{GM}{r} \text{ where, } r > R.$$

- At a point on the surface of spherical shell,

$$V = -\frac{GM}{R}$$

- At any point inside the surface of spherical shell

$$V = -\frac{GM}{R} = V_{\text{surface}}$$

Relation between Gravitational Field and Gravitational Potential

If $\mathbf{r}_1$ and $\mathbf{r}_2$ are position of two points in the gravitation field with intensity (I), then change in gravitational potential

$$V(\mathbf{r}_2) - V(\mathbf{r}_1) = -\int_{r_1}^{r_2} \mathbf{I} \cdot d\mathbf{r}$$

$$\Rightarrow \quad dV = -\mathbf{I} \cdot d\mathbf{r}$$

$$dr = dx\,\hat{\mathbf{i}} + dy\,\hat{\mathbf{j}} + dz\,\hat{\mathbf{k}}$$

and $\quad I = I_x\,\hat{\mathbf{i}} + I_y\,\hat{\mathbf{j}} + I_z\hat{\mathbf{k}}$, then

$$dV = -\mathbf{I}.dr = -I_x dx - I_y dy - I_z dz$$

Thus, $\quad \mathbf{I} = -\frac{\partial V}{\partial x}\hat{\mathbf{i}} - \frac{\partial V}{\partial y}\hat{\mathbf{j}} - \frac{\partial V}{\partial z}\hat{\mathbf{k}}$

Remember that partial differentiation indicates that variation of gravitational potential in counter along the variation of x-coordinate, then other coordinates (i.e. y and z) are assumed to be constant.

Kepler's Laws of Planetary Motion

Kepler discovered three empirical laws which accurately describe the motion of planets.
These laws are

1. **Law of Orbits** All the planets move around the sun in an elliptical orbit with sun at one of the focus of ellipse.

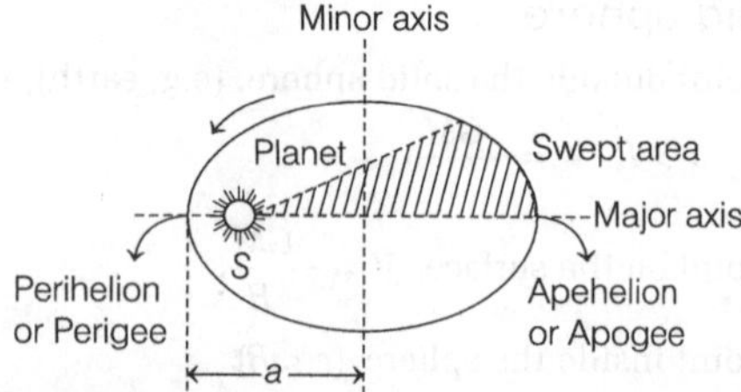

2. **Law of Areas** The line joining the sun to the planet sweeps out equal areas in equal intervals of time, i.e. areal velocity of the planet w.r.t. sun is constant. This is called the law of area, which indicates that a planet moves faster near the sun and slowly when away from the sun.

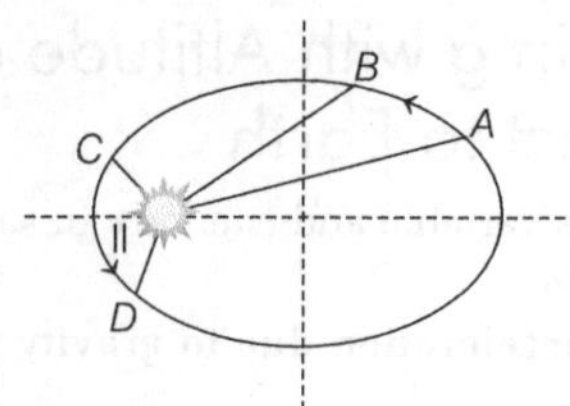

3. **Law of Periods** The square of the planet's time period of revolution is directly proportional to the cube of semi-major axis of its orbit.

$$T^2 \propto a^3$$

where a is the semi-major axis.

Gravitational Potential Energy

Gravitational potential energy of a body or system is negative of work done by the conservative gravitational forces F in bringing it from infinity to the present position.
Mathematically, gravitational potential energy

$$U = -W = -\int_{\infty}^{\mathbf{r}} \mathbf{F} \cdot d\mathbf{r}$$

- The gravitational potential energy of two particles of masses m_1 and m_2 separated by a distance r is given by

$$U = -\frac{Gm_1m_2}{r}$$

- The gravitational potential energy of mass m at the surface of the earth is

$$U = -\frac{GMm}{R}$$

- Difference in potential energy of mass m at a height h from the earth's surface and at the earth's surface is

$$U_{(R+h)} - U_R = \frac{mgh}{1 + \frac{h}{R}} \simeq mgh, \text{ if } h << R$$

- For three particles system,

$$U = -\left[\frac{Gm_1m_2}{r_{12}} + \frac{Gm_1m_3}{r_{13}} + \frac{Gm_2m_3}{r_{23}}\right]$$

- For n-particles system, $\frac{n(n-1)}{2}$ pairs form and total potential energy of the system is sum of potential energies of all such pairs.

Escape Velocity

It is the minimum velocity with which a body must be projected from the surface of the earth so that it escapes the gravitational field of the earth. We can also say that a body, projected with escape velocity, will be able to go to a point which is at infinite distance from the earth.

The value of escape velocity from the surface of a planet of mass M, radius R and acceleration due to gravity g is

$$v_{\text{escape}} = \sqrt{\frac{2GM}{R}} = \sqrt{2gR}$$

Escape velocity does not depend upon the mass or shape or size of the body as well as the direction of projection of the body. For earth value of escape velocity is 11.2 kms^{-1}.

Satellites

Anybody that revolves around earth or any planet is called satellite. These can be natural (e.g. Moon) or artificial. The artificial satellites are man made satellites launched from the earth. The path of these satellites are elliptical with the centre of earth at a foci of the ellipse. However, as a first approximation we may consider the orbit of satellite as circular.

Orbital Velocity of Satellite

Orbital velocity of a satellite is the velocity required to put the satellite into its orbit around the earth. The orbital velocity of satellite is given by

$$v_o = \sqrt{\frac{GM}{r}} = \sqrt{\frac{gR^2}{r}} = \sqrt{\frac{gR^2}{(R+h)}}$$

If $h << R$ or $r \simeq R$, then

$$v_o = \sqrt{\frac{GM}{R}} = \sqrt{gR} = 7.9 \text{ kms}^{-1}$$

$$v_{\text{escape}} = \sqrt{2} v_{\text{orbital}}$$

Period of Revolution

It is the time taken by a satellite to complete one revolution around the earth.

$$\text{Revolution period, } T = \frac{2\pi r}{v_o} = 2\pi\sqrt{\frac{r^3}{GM}} = 2\pi\sqrt{\frac{r^3}{gR^2}} = \sqrt{\frac{3\pi}{G.e}}$$

$$\text{If } r \simeq R, \text{ then } T = 2\pi\sqrt{\frac{R}{g}} = 84.6 \text{ min.}$$

Height of Satellite in Terms of Period

The height of the satellite (from the earth planet) can be determined by its time period and *vice-versa*.

As the height of the satellite in terms of time period,

$$h = r - R = \left[\frac{gR^2T^2}{4\pi^2}\right]^{1/3} - R.$$

Energy of Satellite

Kinetic energy of satellite, $K = \frac{1}{2} m v_0^2 = \frac{GMm}{2r}$.

Potential energy of satellite, $U = -\frac{GMm}{r}$

and total energy of satellite $E = K + U = -\frac{GMm}{2r} = -K$.

Binding Energy of Satellite

It is the energy required to remove the satellite from its orbit and take it to infinity.

Binding energy $= -E = +\frac{GMm}{2r}$

Angular Momentum of Satellite

Angular momentum of a satellite, $L = m v_0 r = \sqrt{m^2 GMr}$

Geostationary Satellite

If an artificial satellite revolves around the earth in an equatorial plane with a time period of 24 h in the same sense as that of the earth, then it will appear stationary to the observer on the earth. Such a satellite is known as a **geostationary satellite** or **parking satellite**.

DAY PRACTICE SESSION 1

FOUNDATION QUESTIONS EXERCISE

1 A mass M is split into two parts m and $(M - m)$, which are separated by a certain distance. The ratio m/M which maximizes the gravitational force between the parts is
(a) 1 : 4 (b) 1 : 3
(c) 1 : 2 (d) 1 : 1

2 Particles of masses $2M$, m and M are respectively at points A, B and C with $AB = \frac{1}{2}(BC)$, m is much-much smaller than M and at time $t = 0$, they are all at rest as given in figure. At subsequent times before any collision takes place.

A B C
2M m M

(a) M will remain at rest
(b) M will move towards M
(c) M will move towards $2M$
(d) M will have oscillatory motion

3 If one moves from the surface of the earth to the moon, what will be the effect on its weight?
(a) Weight of a person decreases continuously with height from the surface of the earth
(b) Weight of a person increases with height from the surface of the earth
(c) Weight of a person first decreases with height and then increases with height from the surface of the earth
(d) Weight of a person first increases with height and then decreases with height from the surface of the earth

4 At the surface of a certain planet acceleration due to gravity is one-quarter of that on the earth. If a brass ball is transported on this planet, then which one of the following statements is not correct?

(a) The brass ball has same mass on the other planet as on the earth
(b) The mass of the brass ball on this planet is a quarter of its mass as measured on the earth
(c) The weight of the brass ball on this planet is a quarter of the weight as measured on the earth
(d) The brass ball has the same volume on the other planet as on the earth

5 If both the mass and radius of the earth, each decreases by 50%, the acceleration due to gravity would

(a) remain same (b) decrease by 50%
(c) decrease by 100% (d) increase by 100%

6 A research satellite of mass 200 kg circles the earth in an orbit of average radius $\frac{3R}{2}$,where R is the radius of the earth. Assuming the gravitational pull on a mass of 1kg on the earth's surface to be 10 N, the pull on the satellite will be

(a) 880 N (b) 889 N
(c) 885 N (d) 892 N

7 The height at which the acceleration due to gravity becomes $\frac{g}{9}$ (where, g = the acceleration due to gravity on the surface of the earth) in terms of R, the radius of the earth is

(a) $2h$ (b) $\frac{h}{\sqrt{3}}$ (c) $\frac{h}{2}$ (d) $\sqrt{2}\,h$

8 The change in the value of g at a height h above the surface of the earth is the same as at a depth d below the surface of the earth. When both d and h are much smaller than the radius of the earth, then which one of the following is correct?

(a) $d=\frac{h}{2}$ (b) $d=\frac{3h}{2}$ (c) $d=2h$ (d) $d=h$

9 Average density of the earth

(a) does not depend on g
(b) is a complex function of g
(c) is directly proportional to g
(d) is inversely proportional to g

10 The variation of acceleration due to gravity g with distance d from centre of the Earth is best represented by (R = Earth's radius) → JEE Main 2017 (Offline)

(a) 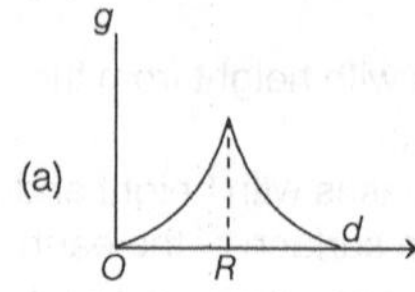

(b) 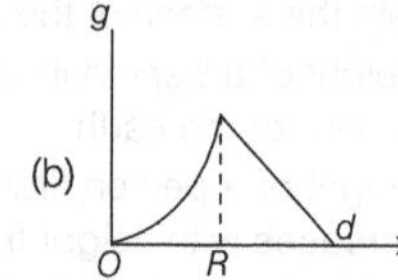

(c) 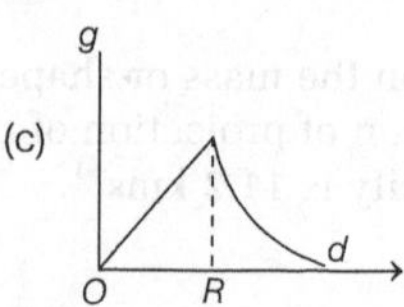

(d) 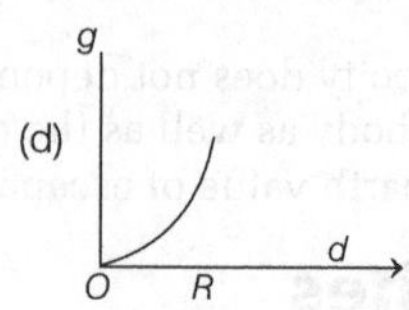

11 The gravitational field, due to the 'left over part' of a uniform sphere (from which a part as shown has been 'removed out') at a very far off point, P located as shown, would be (nearly) → JEE Main 2015

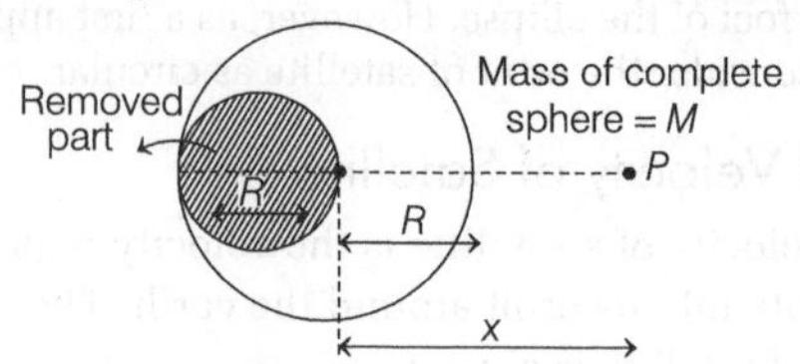

(a) $\frac{5}{6}\frac{GM}{R^2}$ (b) $\frac{8}{9}\frac{GM}{R^2}$ (c) $\frac{7}{8}\frac{GM}{R^2}$ (d) $\frac{6}{7}\frac{GM}{R^2}$

12 A solid sphere is of density ρ and radius R. The gravitational field at a distance r from the centre of the sphere, when $r < R$, is

(a) $\frac{\rho\pi GR^3}{r}$ (b) $\frac{4\pi G\rho r^2}{3}$ (c) $\frac{4\pi G\rho R^3}{3r^2}$ (d) $\frac{4\pi G\rho r}{3}$

13 The maximum vertical distance through which a full dressed astronaut can jump on the earth is 0.5 m. Estimate the maximum vertical distance through which he can jump on the moon, which has a mean density 2/3rd that of the earth and radius one quarter that of the earth.

(a) 1.5 m (b) 3 m (c) 6 m (d) 7.5 m

14 The mass of a spaceship is 1000 kg. It is to be launched from the earth's surface out into free space. The value of g and R (radius of earth) are 10 m/s^2 and 6400 km respectively. The required energy for this work will be

(a) 6.4×10^{11} J (b) 6.4×10^{8} J
(c) 6.4×10^{9} J (d) 6.4×10^{10} J

15 If g is the acceleration due to gravity on the earth's surface, the gain in the potential energy of an object of mass m raised from the surface of the earth to a height equal to the radius R of the earth, is

(a) $2mgR$ (b) $\frac{1}{2}mgR$ (c) $\frac{1}{4}mgR$ (d) mgR

16 Two bodies of masses m and $4\,m$ are placed at a distance r. The gravitational potential at a point on the line joining them where the gravitational field is zero is

(a) $-\frac{4Gm}{r}$ (b) $-\frac{6Gm}{r}$ (c) $-\frac{9Gm}{r}$ (d) zero

17 A planet in a distant solar system is 10 times more massive than the earth and its radius is 10 times smaller. Given that the escape velocity from the earth is 11 kms^{-1}, the escape velocity from the surface of the planet would be

(a) 1.1 kms^{-1} (b) 11 kms^{-1}
(c) 110 kms^{-1} (d) 0.11 kms^{-1}

18 A projectile is fired vertically upwards from the surface of the earth with a velocity kv_e, where v_e is the escape velocity and $k < 1$. If R is the radius of the earth, the maximum height to which it will rise measured from the centre of the earth will be

(a) $\frac{1-k^2}{R}$ (b) $\frac{R}{1-k^2}$

(c) $R(1-k^2)$ (d) $\frac{R}{1+k^2}$

19 Two cars of masses m_1 and m_2 are moving in circles of radii r_1 and r_2, respectively. Their speeds are such that they make complete circles in the same time t. The ratio of their centripetal acceleration is

(a) $m_1r_1 : m_2r_2$ (b) $m_1 : m_2$ (c) $r_1 : r_2$ (d) 1 : 1

20 What is the direction of areal velocity of the earth around the sun?

(a) Perpendicular to positon of the earth w.r.t. the sun at the focus
(b) Perpendicular to velocity of the earth revolving in the elliptical path
(c) Parallel to angular displacement
(d) Both (a) and (b)

21 The time period of a satellite of the earth is 5 h. If the separation between the earth and the satellite is increased to 4 times the previous value, the new time period will become

(a) 10 h (b) 80 h (c) 40 h (d) 20 h

22 A satellite goes along an elliptical path around earth. The rate of change of area swept by the line joining earth and the satellite is proportional to

(a) $r^{1/2}$ (b) r (c) $r^{3/2}$ (d) r^2

23 A satellite of mass m revolves around the earth of radius R at a height x from its surface. If g is the acceleration due to gravity on the surface of the earth, the orbital speed of the satellite is

(a) gx (b) $\frac{gR}{R-x}$ (c) $\frac{gR^2}{R+x}$ (d) $\left(\frac{gR^2}{R+x}\right)^{1/2}$

24 Two particles of equal mass m go around a circle of radius R under the action of their mutual gravitational attraction. The speed of each particle with respect to their centre of mass is → AIEEE 2011

(a) $\sqrt{\frac{Gm}{R}}$ (b) $\sqrt{\frac{Gm}{4R}}$ (c) $\sqrt{\frac{Gm}{3R}}$ (d) $\sqrt{\frac{Gm}{2R}}$

25 Suppose the gravitational force varies inversely as the nth power of distance. Then the time period of a planet in circular orbit of radius R around the sun will be proportional to

(a) $R^{\left(\frac{n+1}{2}\right)}$ (b) $R^{\left(\frac{n-1}{2}\right)}$ (c) R^n (d) $R^{\left(\frac{n-2}{2}\right)}$

26 The gravitational force exerted by the sun on the moon is about twice as great as the gravitational force exerted by the earth on the moon, but still the moon is not escaping from gravitational influence of the earth. Mark the option which correctly explains the above system.

(a) At some point of time the moon will escape from the earth
(b) Separation between the moon and sun is larger than the separation between the earth and moon
(c) The moon-earth system is bounded one and a minimum amount of energy is required to escape the moon from the earth
(d) None of the above

27 What is the minimum energy required to launch a satellite of mass m from the surface of a planet of mass M and radius R in a circular orbit at an altitude of $2R$?

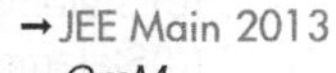

(a) $\frac{5\,GmM}{6R}$ (b) $\frac{2GmM}{3R}$ (c) $\frac{GmM}{2R}$ (d) $\frac{GmM}{3R}$

28 The curves for potential energy (E_P) and kinetic energy (E_K) of two particles system are as shown in the figure. At what points the system will be bound

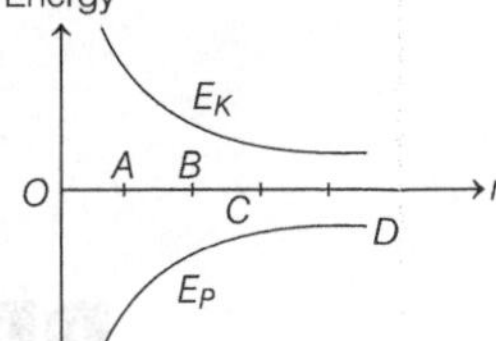

(a) Only at point A
(b) Only at point D
(c) At points A and D
(d) At points A, B and C

29 The earth moves around the sun in an elliptical orbit as shown in the figure. The ratio $\frac{OA}{OB} = x$. The ratio of the speeds of the earth at B and at A is

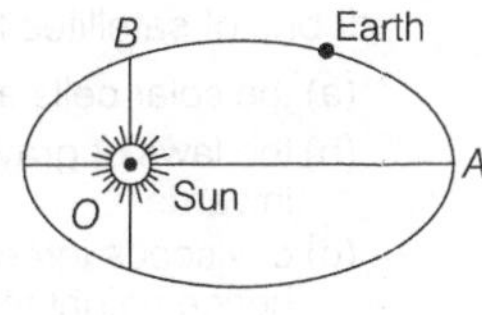

(a) $\sqrt{x}$ (b) x (c) x^2 (d) $x\sqrt{x}$

30 A satellite of mass m_s revolving in a circular orbit of radius r_s around the earth of mass M, has a total energy E. Then its angular momentum will be

(a) $(2Em_sr_s)^{1/2}$ (b) $(2Em_sr_s)$
(c) $(2Em_sr_s^2)^{1/2}$ (d) $(2Em_sr_s^2)$

31 Match the term related to gravitation given in Column I with their formula given in Column II and select the correct option from the choices given below :

	Column I		Column II
A.	Gravitational potential at a point outside the solid sphere	1.	$\frac{3GM}{2R}$
B.	Gravitational potential at a point on the surface of sphere	2.	$\frac{GM}{r}$
C.	Gravitational potential at the centre of solid sphere	3.	$\frac{GM}{R}$

	A	B	C
(a)	1	3	2
(b)	2	3	1
(c)	1	2	3
(d)	3	2	1

Direction (Q. Nos. 32-33) *Each of these questions contains two statements : Statement I and Statement II. Each of these questions also has four alternative choices, only one of which is the correct answer. You have to select one of the codes (a), (b), (c), (d) given below*

(a) Statement I is true, Statement II is true; Statement II is the correct explanation for Statement I

(b) Statement I is true, Statement II is true; Statement II is not the correct explanation for Statement I

(c) Statement I is true; Statement II is false

(d) Statement I is false; Statement II is true

32 Statement I An astronaut in an orbiting space station above the earth experiences weightlessness.

Statement II An object moving around the earth under the influence of the earth's gravitational force is in a state of 'free-fall'.

33 Statement I Kepler's laws for planetary motion are consequence of Newton's laws.

Statement II Kepler's laws can be derived by using Newton's laws.

34 Assertion (A) If the earth were a hollow sphere, gravitational field intensity at any point inside the earth would be zero.

Reason (R) Net force on a body inside the sphere is zero.

(a) If both Assertion and Reason are correct and Reason is the correct explanation of Assertion

(b) If both Assertion and Reason are true but Reason is not the correct explanation of Assertion

(c) If Assertion is true but Reason is false

(d) If Assertion is false but Reason is true

DAY PRACTICE SESSION 2

PROGRESSIVE QUESTIONS EXERCISE

1 Satellites orbitting the earth have finite life and sometimes debris of satellites fall to the earth. This is because

(a) the solar cells and batteries in satellites run out

(b) the laws of gravitation predict a trajectory spiralling inwards

(c) of viscous forces causing the speed of satellite and hence height to gradually decrease

(d) of collisions with other satellites

2 A body is released from a point distance r from the centre of earth. If R is the radius of the earth and $r > R$, then the velocity of the body at the time of striking the earth will be

(a) $\sqrt{gR}$ (b) $\sqrt{2gR}$

(c) $\sqrt{\frac{2gRr}{r-R}}$ (d) $\sqrt{\frac{2gR(r-R)}{r}}$

3 Two small satellites move in circular orbits around the earth, at distances r and $r + \Delta r$ from the centre of the earth. Their time periods of rotation are T and $T + \Delta T$, $(\Delta r << r, \Delta T << T)$. Then, ΔT is equal to

(a) $\frac{3}{2}T\frac{\Delta r}{r}$ (b) $\frac{2}{3}T\frac{\Delta r}{r}$ (c) $\frac{-3}{2}T\frac{\Delta r}{r}$ (d) $T\frac{\Delta r}{r}$

4 A straight rod of length L extends from $x = a$ to $x = L + a$. The gravitational force, it exerts on a point mass m at $x = 0$, if the mass per unit length is $A + Bx^2$, is

(a) $Gm\left[A\left(\frac{1}{a+L}-\frac{1}{a}\right)+BL\right]$ (b) $Gm\left[A\left(\frac{1}{a}-\frac{1}{a+L}\right)-BL\right]$

(c) $Gm\left[A\left(\frac{1}{(a+L)}-\frac{1}{a}\right)-BL\right]$ (d) $Gm\left[A\left(\frac{1}{a}-\frac{1}{a+L}\right)+BL\right]$

5 A ring having non-uniform distribution of mass having mass M and radius R is being considered. A point mass m_0 is taken slowly from A to B along the axis of the ring. In doing so, work done by the external force against the gravitational force exerted by ring is

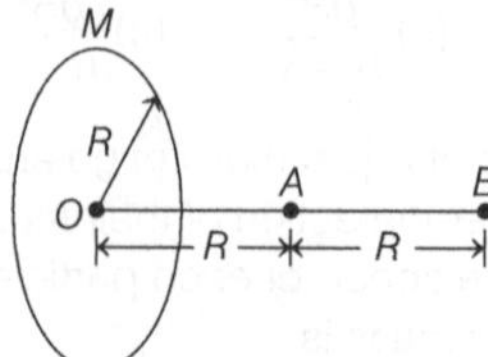

(a) $\frac{GMm_0}{\sqrt{2}R}$

(b) $\frac{GMm_0}{R}\left[\frac{1}{\sqrt{2}}-\frac{1}{\sqrt{5}}\right]$

(c) $\frac{GMm_0}{R}\left[\frac{1}{\sqrt{5}}-\frac{1}{\sqrt{2}}\right]$

(d) It is not possible to find the required work as the nature of distribution of mass is not known

6 With what minimum speed should m be projected from point C in the presence of two fixed masses M each at A and B as shown in the figure, such that mass m should escape the gravitational attraction of A and B?

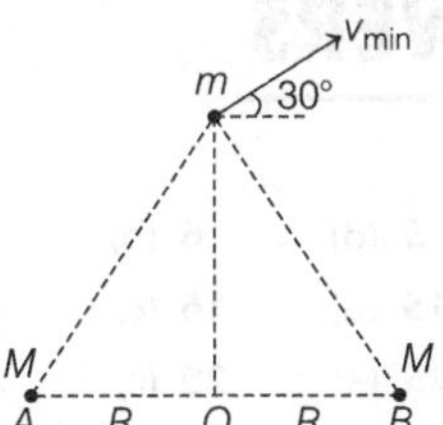

(a) $\sqrt{\frac{2GM}{R}}$ (b) $\sqrt{\frac{2\sqrt{2}GM}{R}}$

(c) $2\sqrt{\frac{GM}{R}}$ (d) $2\sqrt{2}\sqrt{\frac{GM}{R}}$

7 A particle of mass 10 g is kept on the surface of a uniform sphere of mass 100 kg and radius 10 cm. Find the work to be done against the gravitational force between them, to take the particle far away from the sphere, (you may take $G = 6.67 \times 10^{-11}$ Nm^2/kg^2)

(a) 13.34×10^{-10} J (b) 3.33×10^{-10} J
(c) 6.67×10^{-9} J (d) 6.67×10^{-10} J

8 Two metallic spheres each of mass M are suspended by two strings each of length L. The distance between the upper ends of strings is L. The angle which the strings will make with the vertical due to mutual attraction of the spheres is

(a) $\tan^{-1}\left[\frac{GM}{gL}\right]$ (b) $\tan^{-1}\left[\frac{GM}{2gL}\right]$

(c) $\tan^{-1}\left[\frac{GM}{gL^2}\right]$ (d) $\tan^{-1}\left[\frac{2GM}{gL^2}\right]$

9 A body of mass 2 kg is moving under the influence of a central force whose potential energy is given by $U = 2r^3$ joule. If the body is moving in a circular orbit of 5 m, its energy will be

(a) 250 J (b) 125 J
(c) 625 J (d) 650 J

10 A mass m is placed at P at a distance h along the normal through the centre O of a thin circular ring of mass M and radius r as shown in figure. If the mass is moved further away such that OP becomes $2h$, by what factor, the force of gravitation will decrease, if $h = r$?

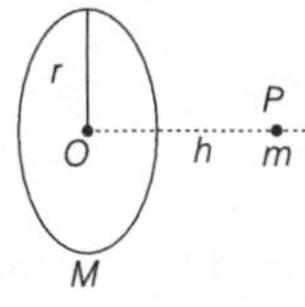

(a) $\frac{3\sqrt{2}}{4\sqrt{3}}$ (b) $\frac{5\sqrt{2}}{\sqrt{3}}$ (c) $\frac{4\sqrt{3}}{5}$ (d) $\frac{4\sqrt{2}}{5\sqrt{5}}$

11 An artificial satellite of the earth is launched in circular orbit in equatorial plane of the earth and satellite is moving from West to East. With respect to a person on the equator, the satellite is completing one round trip in 24 h. Mass of the earth is, $M = 6 \times 10^{24}$kg. For this situation, orbital radius of the satellite is

(a) 2.66×10^4km (b) 6400 km
(c) 36,000 km (d) 29,600 km

12 From a solid sphere of mass M and radius R, a spherical portion of radius $\left(\frac{R}{2}\right)$ is removed as shown in the figure.

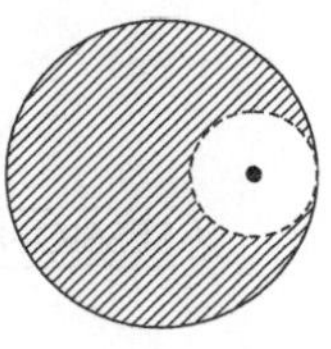

Taking gravitational potential $V = 0$ at $I = \infty$, the potential at the centre of the cavity thus formed is (I = gravitational constant). → JEE Main 2015

(a) $\frac{-GM}{2R}$ (b) $\frac{-GM}{R}$ (c) $\frac{-2GM}{3R}$ (d) $\frac{-2GM}{R}$

13 A satellite is revolving in a circular orbit at a height h from the Earth's surface (radius of Earth $R, h << R$). The minimum increase in its orbital velocity required, so that the satellite could escape from the Earth's gravitational field, is close to (Neglect the effect of atmosphere) → JEE Main 2017 (Offline)

(a) $\sqrt{2gR}$ (b) $\sqrt{gR}$ (c) $\sqrt{gR/2}$ (d) $\sqrt{gR}\,(\sqrt{2}-1)$

14 Four particles, each of mass M and equidistant from each other, move along a circle of radius R under the action of their mutual gravitational attraction, the speed of each particle is → JEE Main 2015

(a) $\sqrt{\frac{GM}{R}}$ (b) $\sqrt{2\sqrt{2}\frac{GM}{R}}$

(c) $\sqrt{\frac{GM}{R}(1+2\sqrt{2})}$ (d) $\frac{1}{2}\sqrt{\frac{GM}{R}(1+2\sqrt{2})}$

Direction (Q. Nos. 15-16) *Each of these questions contains two statements : Statement* I *and Statement* II. *Each of these questions also has four alternative choices, only one of which is the correct answer. You have to select one of the codes (a), (b), (c), (d) given below*

(a) Statement I is true, Statement II is true; Statement II is the correct explanation for Statement I
(b) Statement I is true, Statement II is true; Statement II is not the correct explanation for Statement I
(c) Statement I is true; Statement II is false
(d) Statement I is false; Statement II is true

15 **Statement I** Three orbits are marked as 1, 2 and 3. These three orbits have same semi-major axis although their shapes (eccentricities) are different. The three identical satellites are orbiting in these three orbits, respectively. These three satellites have the same binding energy.

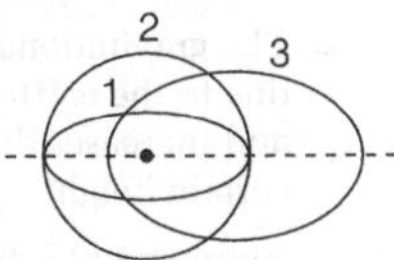

Statement II Total energy of a satellite depends on the semi-major axis of orbit according to the expression,

$$E = \frac{-GMm}{2r}$$

16 **Statement I** Two satellites are following one another in the same circular orbit. If one satellite tries to catch another (leading one) satellite, then it can be done by increasing its speed without changing the orbit.

Statement II The energy of the earth satellites system in circular orbits is given by $E = \frac{-GMm}{2r}$, where, r is the radius of the circular orbit.

ANSWERS

SESSION 1									
1 (c)	**2** (c)	**3** (c)	**4** (b)	**5** (d)	**6** (b)	**7** (c)	**8** (c)	**9** (c)	**10** (c)
11 (c)	**12** (d)	**13** (b)	**14** (d)	**15** (b)	**16** (c)	**17** (c)	**18** (b)	**19** (c)	**20** (d)
21 (c)	**22** (a)	**23** (d)	**24** (b)	**25** (a)	**26** (c)	**27** (a)	**28** (d)	**29** (b)	**30** (c)
31 (b)	**32** (a)	**33** (d)	**34** (a)						

SESSION 2									
1 (c)	**2** (d)	**3** (a)	**4** (d)	**5** (b)	**6** (b)	**7** (d)	**8** (c)	**9** (c)	**10** (d)
11 (a)	**12** (b)	**13** (d)	**14** (d)	**15** (a)	**16** (d)				

Hints and Explanations

SESSION 1

1 As, $F = \frac{Gm(M-m)}{x^2}$

For maximum gravitational force

$\frac{dF}{dm} = \frac{G}{x^2}(M - 2m) = 0$ or $\frac{m}{M} = \frac{1}{2}$

2 The particle B will move towards the greater force between forces by A and B.

Force on B due to $A = F_{BA} = \frac{G(2Mm)}{(AB)^2}$ towards BA

Force on B due to $C = F_{BC} = \frac{GMm}{(BC)^2}$ towards BC

As, $(BC) = 2AB$

$\Rightarrow F_{BC} = \frac{GMm}{(2AB)^2} = \frac{GMm}{4(AB)^2} < F_{BA}$

Hence, m will move towards BA (i.e. $2M$).

3 The gravitational attraction on a body due to the earth decreases with height and increases due to the moon at a certain height.

At one point it becomes zero and with further increase in height the gravitational attraction of the moon becomes more than that of the earth.

4 The mass of a body is always constant and does not change with position.

5 Here, $g = GM/R^2$

and $g' = \frac{G(M/2)}{(R/2)^2} = \frac{2GM}{R^2} = 2g$

$\therefore$ % increase in $g = \left(\frac{g' - g}{g}\right) \times 100$

$= \left(\frac{2g - g}{g}\right) \times 100$

$= 100\%$

6 Here, $mg = 10$ or $1 \times g = 10$

$\Rightarrow g = 10\,\text{ms}^{-2}$

Now, $g' = g\frac{R^2}{r^2} = 10 \times \frac{R^2}{(3R/2)^2} = \frac{40}{9}$

Pull on satellite $\approx m'g'$

$\approx 200 \times \frac{40}{9}$

$\approx 889\text{N}$

7 Acceleration due to gravity at height h is, $g' = \frac{GM}{(R+h)^2}$

$\Rightarrow \frac{g}{9} = \frac{GM}{R^2} \cdot \frac{R^2}{(R+h)^2} = g\left(\frac{R}{R+h}\right)^2$

$\Rightarrow \frac{1}{9} = \left(\frac{R}{R+h}\right)^2$

$\Rightarrow \frac{R}{R+h} = \frac{1}{3}$

$\Rightarrow 3R = R + h$

$\Rightarrow R = \frac{h}{2}$

8 $g_h = g\left(1 - \frac{2h}{R}\right)$ ...(i)

and $g_d = g\left(1 - \frac{d}{R}\right)$ ...(ii)

As per statement of the problem,

$g_h = g_d$

i.e. $g\left(1 - \frac{2h}{R}\right) = g\left(1 - \frac{d}{R}\right)$

$\Rightarrow 2h = d$

9 As, $g = \frac{GM}{R^2}$;

$M = \left(\frac{4}{3}\pi R^3\right)\rho$

$\therefore g = \frac{4G}{3}\frac{\pi R^3}{R^2}\rho$

$\Rightarrow \rho = \left(\frac{3}{4\pi GR}\right)g$;

$\therefore \rho \propto g$

(where, ρ = average density of the earth)

10 Inside the earth's surface,

$g = \frac{GM}{R^3}d$ i.e. $g \propto d$

Outside the earth's surface,

$g = \frac{Gm}{d^2}$ i.e. $g \propto \frac{1}{d^2}$

So, till earth's surface g increases linearly with distance r, shown only in graph (c).

11 We have,

$M' = \frac{M}{\frac{4}{3}\pi R^3} \times \frac{4}{3}\pi\left(\frac{R}{2}\right)^3 = \frac{M}{8}$

$\therefore$ Gravitational field at

$P = \frac{GM}{R^2} - \frac{GM}{8R^2}$

$= G \times \frac{M}{R^2}\left(1 - \frac{1}{8}\right) = \frac{7}{8}\frac{GM}{R^2}$

12 Intensity,

$I = \frac{GM}{R^3}r = \frac{Gr}{R^3}\left(\frac{4}{3}\pi R^3\rho\right) = \frac{4\pi G\rho r}{3}$

13 On the moon, $g_m = \frac{4}{3}\pi G(R/4)(2\rho/3)$

$= \frac{1}{6}\left(\frac{4}{3}\pi GR\rho\right) = \frac{1}{6}g$

Work done in jumping

$= m \times g \times 0.5 = m \times (g/6)h_1$

$h_1 = 0.5 \times 6 = 3.0$ m

14 Potential energy on the earth surface is $-mgR$ while in free space it is zero. So, to free the spaceship, minimum required energy is

$K = mgR = 10^3 \times 10 \times 6400 \times 10^3$J

$= 6.4 \times 10^{10}$ J

15 Gravitational potential energy of body on the earth's surface is

$$U = -\frac{GM_e m}{R}$$

At a height h from earth's surface, its value is

$$U_h = -\frac{GM_e m}{(R+h)} = -\frac{GM_e m}{2R} \quad [\text{as } h = R]$$

where, M_e = mass of earth,
m = mass of body
and R = radius of earth.

∴ Gain in potential energy

$$= U_h - U$$
$$= -\frac{GM_e m}{2R} - \left(-\frac{GM_e m}{R}\right)$$
$$= -\frac{GM_e m}{2R} + \frac{GM_e m}{R}$$
$$= \frac{GM_e m}{2R} = \frac{gR^2 m}{2R} \left[\text{as } g = \frac{GM_e}{R^2}\right]$$
$$= \frac{1}{2} mgR$$

16 Let gravitational field is zero at P as shown in figure.

$$\therefore \quad \frac{Gm}{x^2} = \frac{G(4m)}{(r-x)^2}$$
$$\Rightarrow \quad 4x^2 = (r-x)^2 \Rightarrow 2x = r - x$$
$$\Rightarrow \quad x = \frac{r}{3}$$
$$\therefore \quad V_P = -\frac{Gm}{x} - \frac{G(4m)}{r-x} = -\frac{9Gm}{r}$$
$$(\because x = r/3)$$

17 Mass of planet, $M_p = 10M_e$,
where, M_e is mass of earth.
Radius of planet, $R_p = \frac{R_e}{10}$,
where R_e is radius of earth.
Escape velocity is given by,

$$v_e = \sqrt{\frac{2GM}{R}}$$

For planet, $v_p = \sqrt{\frac{2G \times M_p}{R_p}}$

$$= \sqrt{\frac{100 \times 2GM_e}{R_e}}$$
$$= 10 \times v_e$$
$$= 10 \times 11$$
$$= 110 \text{ kms}^{-1}$$

18 From law of conservation of energy,

$$\frac{1}{2}mv^2 = \frac{mgh}{1 + \frac{h}{R}}$$

Here, $v = kv_e = k\sqrt{2gR}$
and $h = r - R$

$$\frac{1}{2}m\,k^2 \cdot 2gR = \frac{mg\,(r-R)}{1 + \frac{r-R}{R}},$$
$$k^2R\left[1 + \frac{r-R}{R}\right] = r - R$$
$$k^2 r = r - R$$
$$\Rightarrow \quad r = \frac{R}{1-k^2}$$

19 As their period of revolution is same, so their angular speed is also same. Centripetal acceleration is circular path, $a = \omega^2 r$.

Thus, $\frac{a_1}{a_2} = \frac{\omega^2 r_1}{\omega^2 r_2} = \frac{r_1}{r_2}$

20 Areal velocity of the earth around the sun is given by

$$\frac{d\mathbf{A}}{dt} = \frac{\mathbf{L}}{2m}$$

where, **L** is the angular momentum and m is the mass of the earth. But angular momentum

$$\mathbf{L} = \mathbf{r} \times \mathbf{p} = \mathbf{r} \times m\mathbf{v}$$

∴ Areal velocity

$$\left(\frac{d\mathbf{A}}{dt}\right) = \frac{1}{2m}(\mathbf{r} \times m\mathbf{v})$$
$$= \frac{1}{2}(\mathbf{r} \times \mathbf{v})$$

Therefore, the direction of areal velocity $\left(\frac{d\mathbf{A}}{dt}\right)$ is the direction of $(\mathbf{r} \times \mathbf{v})$, i.e. perpendicular to the plane containing **r** and **v**.

21 According to Kepler's law, $T^2 \propto r^3$

$$\therefore \quad \left(\frac{T}{T^1}\right)^2 = \left(\frac{r}{r^1}\right)^3 \Rightarrow \frac{25}{(T')^2} = \frac{r^3}{64r^3}$$

or $\quad T' = \sqrt{1600}$ or $T' = 40\,\text{h}$

22 Areal velocity $= \frac{dA}{dt} = \frac{L}{2m} = \frac{mvr}{2m} = \frac{vr}{2}$

$$= \frac{r}{2}\sqrt{\frac{GM}{r}} = \frac{1}{2}\sqrt{GMr}$$

So, $\quad \frac{dA}{dt} \propto \sqrt{r}$

23 The gravitational force exerted on satellite at a height x is

$$F_G = \frac{GM_e m}{(R+x)^2}$$

where, M_e = mass of the earth.
Since, gravitational force provides the necessary centripetal force, so

$$\frac{GM_e m}{(R+x)^2} = \frac{mv_o^2}{(R+x)}$$

where, v_o is orbital speed of satellite

$$\Rightarrow \quad \frac{GM_e m}{(R+x)} = mv_o^2$$

or $\quad \frac{gR^2 m}{(R+x)} = mv_o^2 \quad \left(\because g = \frac{GM_e}{R^2}\right)$

or $\quad v_o = \sqrt{\left[\frac{gR^2}{(R+x)}\right]} = \left[\frac{gR^2}{(R+x)}\right]^{1/2}$

24 As gravitational force provides necessary centripetal force.

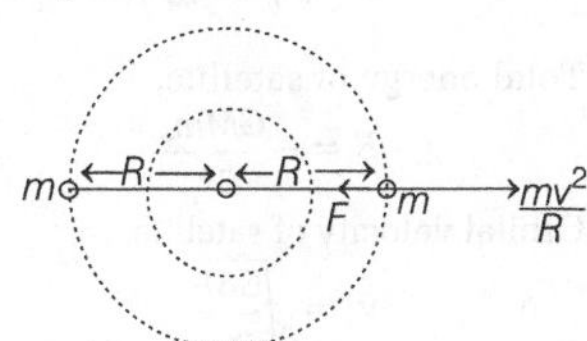

i.e. $F = \frac{Gm^2}{(2R)^2} = \frac{mv^2}{R}$

$$\Rightarrow \quad v = \sqrt{\frac{Gm}{4R}}$$

25 The necessary centripetal force required for a planet to move around the sun = gravitational force exerted on it

i.e. $\frac{mv^2}{R} = \frac{GM_e m}{R^n}$ or $v = \left(\frac{GM_e}{R^{n-1}}\right)^{1/2}$

Now, $\quad T = \frac{2\pi R}{v} = 2\pi R \times \left(\frac{R^{n-1}}{GM_e}\right)^{1/2}$

$$= 2\pi\left(\frac{R^2 \times R^{n-1}}{Gm_e}\right)^{1/2}$$
$$= 2\pi\left(\frac{R^{(n+1)/2}}{(Gm_e)^{1/2}}\right)$$

or $\quad T \propto R^{(n+1)/2}$

26 Option (c) is correct, a minimum amount of energy equal to |TE| of the moon-earth system has to be given to break (unbound) the system, the sun is exerting force on the moon but not providing any energy.

27 From conservation of energy
Total energy at the planet = Total energy at the altitude

$$\frac{-GMm}{R} + (\text{KE})_{\text{surface}}$$
$$= \frac{-GMm}{3R} + \frac{1}{2}mv_A^2 \quad \text{...(i)}$$

In its orbit, the necessary centripetal force is provided by gravitational force.

$$\therefore \quad \frac{mv_A^2}{(R+2R)} = \frac{GMm}{(R+2R)^2}$$
$$\Rightarrow \quad v_A^2 = \frac{GM}{3R} \quad \text{...(ii)}$$

From Eqs. (i) and (ii), we get

$$(\text{KE})_{\text{surface}} = \frac{5}{6}\frac{GMm}{R}$$

28 The system will be bound at all these points where the total energy $= (E_P + E_K)$ is negative.

In the given curve, at points A, B and C, the $E_P > E_K$.

29 According to law of conservation of angular momentum;

$$mv_A \times OA = mv_B \times OB,$$

$$\frac{v_B}{v_A} = \frac{OA}{OB} = x$$

30 Total energy of satellite,

$$E = -\frac{GMm_s}{2r_s} \quad \text{...(i)}$$

Orbital velocity of satellite,

$$v_s = \sqrt{\frac{GM}{r_s}}$$

Angular momentum of satellite is given by

$$L = m_s v_s r_s = m_s\left(\frac{GM}{r_s}\right)^{1/2} r_s = (GM\,m_s^2 r_s)^{1/2}$$

$$= (2Em_s r_s^2)^{1/2} \quad \text{[from Eq. (i)]}$$

31 Gravitational potential,

At outside point (solid sphere) $= \frac{GM}{r}$

At the surface (solid sphere) $= \frac{GM}{R}$

At the centre (solid sphere) $= \frac{3GM}{2R}$

Hence, (b) is correct.

32 Force acting on astronaut is utilised in providing necessary centripetal force, thus he feels weightlessness, as he is in the state of free fall.

33 Kepler's laws are based on observations, hit and trial method and already recorded data but later on Newton proved their correctness using his laws.

34 It is clear that the net force on the body inside the hollow sphere is zero hence, then net gravitational field intensity $\left(E = \frac{F}{m}\right)$ at any point inside the earth must also be zero.

SESSION 2

1 As the total energy of the earth satellite bounded system is negative $\left(\frac{-GM}{2a}\right)$.

Where, a is radius of the satellite and M is mass of the earth. Due to the viscous force acting on satellite, energy decreases continuously and radius of the orbit or height decreases gradually.

2 Using law of conservation of energy,

$$-\frac{GMm}{r} = \frac{1}{2}mv^2 - \frac{GMm}{R}$$

$$\Rightarrow \frac{v^2}{2} = \frac{GM}{R} - \frac{GM}{r} = GM\left(\frac{r-R}{rR}\right)$$

$$= gR\left(\frac{r-R}{r}\right) \quad \left(\because \frac{GM}{R^2} = g\right)$$

$$\therefore \quad v = \sqrt{2gR(r-R)/r}$$

3 According to Kepler's law, $T^2 \propto r^3$

$$T^2 = kr^3 \quad \text{... (i)}$$

Differentiating it, we have

$$2T\,\Delta T = 3kr^2\Delta r$$

Dividing it by Eq. (i), we get

$$\frac{2T\,\Delta T}{T^2} = \frac{3kr^2\Delta r}{kr^3} \Rightarrow \Delta T = \frac{3}{2}T\frac{\Delta r}{r}$$

4 Mass of the element of length dx at a distance x from the origin

$$= (A + Bx^2)\,dx$$

$$dF = \frac{Gm(A + Bx^2)dx}{x^2}$$

On integrating,

$$F = Gm\int_a^{a+L}\frac{(A+Bx^2)dx}{x^2}$$

$$= Gm\int_a^{a+L}\left(\frac{A}{x^2} + B\right)dx$$

$$= Gm\left[A\left(\frac{1}{a} - \frac{1}{a+L}\right) + BL\right]$$

5 Even though the distribution of mass is unknown we can find the potential due to ring on any axial point because from any axial point the entire mass is at the same distance (whatever would be the nature of distribution).

Potential at A due to ring is,

$$V_A = -\frac{GM}{\sqrt{2}R}$$

Potential at B due to ring is,

$$V_B = -\frac{GM}{\sqrt{5}R}$$

$$dU = U_f - U_i$$

$$= U_B - U_A = m_0(V_B - V_A)$$

$$= \frac{GMm_0}{R}\left[-\frac{1}{\sqrt{5}} + \frac{1}{\sqrt{2}}\right]$$

$$W_{gr} = -W_{ext}$$

$$\Rightarrow W_{gr} = -dU = -W_{ext}$$

$$\therefore W_{ext} = dU = \frac{GMm_0}{R}\left[\frac{1}{\sqrt{2}} - \frac{1}{\sqrt{5}}\right]$$

6 Here, $K_i + U_i = K_f + U_f$

$$\therefore \frac{1}{2}mv^2 - \frac{2GMm}{\sqrt{2}R} = 0 + 0$$

$$\text{or } \frac{1}{2}mv^2 = \frac{2GMm}{\sqrt{2}R} \text{ or } v = \sqrt{\frac{2\sqrt{2}GM}{R}}$$

Hence, (b) is the correct option.

7 The initial potential energy of the particle = Work done.

$$U_i = -\frac{GMm}{r}$$

$$U_i = -\frac{6.67 \times 10^{-11} \times 100 \times 10^{-2}}{0.1}$$

$$U_i = -\frac{6.67 \times 10^{-11}}{0.1}$$

$$= -6.67 \times 10^{-10}\ \text{J}$$

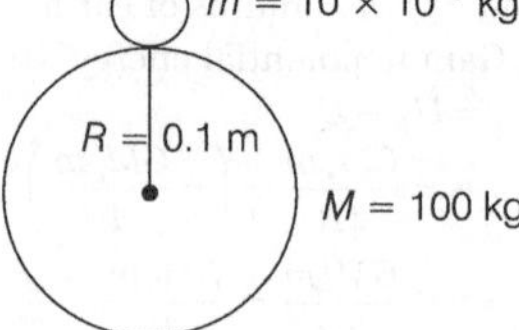

We know that, work done = Difference in potential energy

$$\therefore \quad W = \Delta U = U_f - U_i$$

$$\Rightarrow \quad W = -U_i \quad [\because U_f = 0]$$

$$= 6.67 \times 10^{-10}\ \text{J}$$

8 The metallic spheres will be at positions as shown in the figure.

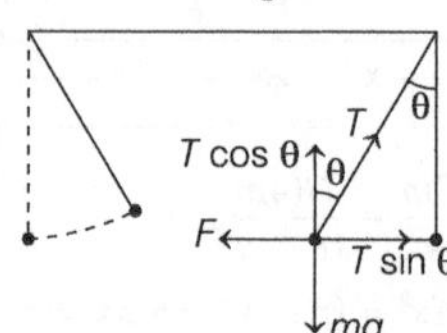

From the figure,

$$T\sin\theta = F = \frac{GM \times M}{L^2} = \frac{GM^2}{L^2}$$

and $T\cos\theta = Mg$

$$\Rightarrow \quad \tan\theta = \frac{GM}{gL^2}$$

$$\Rightarrow \quad \theta = \tan^{-1}\left(\frac{GM}{gL^2}\right)$$

9 Given, $U = 2r^3$

$$\therefore \quad F = \frac{-dU}{dr} = -6r^2$$

Now, $\frac{mv^2}{r} = |F| = 6r^2$

or $\frac{1}{2}mv^2 = \frac{1}{2}(6r^3) = 3r^3$

As, $\quad TE = K + U$

$$= 3r^3 + 2r^3$$

$$= 5r^3$$

$\therefore$ At $\quad r = 5\,\text{m}$,

$$TE = 625\,\text{J}$$

Hence, (c) is the correct option.

10 Gravitational force acting on an object of mass m, placed at point P at a distance h along the normal through the centre of a circular ring of mass M and radius r is given by

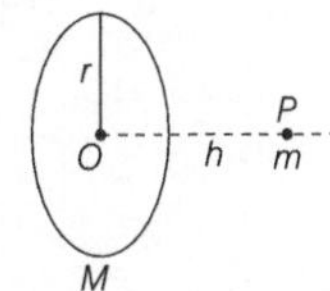

$$F = \frac{GMmh}{(r^2 + h^2)^{3/2}}$$

When mass is displaced upto distance $2h$, then

$$F' = \frac{GMm \times 2h}{(r^2 + (2h)^2)^{3/2}} = \frac{2GMmh}{(r^2 + 4h^2)^{3/2}}$$

When $h = r$, then

$$F = \frac{GMm \times r}{(r^2 \times r^2)^{3/2}} = \frac{GMm}{2\sqrt{2}r^2}$$

and $F' = \dfrac{2GMmr}{(r^2 + 4r^2)^{3/2}} = \dfrac{2GMm}{5\sqrt{5}r^2}$

$$\therefore \quad \frac{F'}{F} = \frac{4\sqrt{2}}{5\sqrt{5}} \text{ or } F' = \frac{4\sqrt{2}}{5\sqrt{5}}F$$

11 Here, time period of satellite w.r.t. observer on equator is 24 h and the satellite is moving from West to East, so angular velocity of satellite w.r.t. the earth's axis of rotation (considered as fixed) is,

$$\omega = \frac{2\pi}{T_s} + \frac{2\pi}{T_e},$$

where, T_s and T_e are time periods of satellite and the earth, respectively

$$\omega = \frac{\pi}{6} \text{ rad/h}$$
$$= 1.45 \times 10^{-4} \text{ rad s}^{-1}$$

From $v = \sqrt{\dfrac{GM}{r}}$

$$\Rightarrow \quad r\omega = \sqrt{\frac{GM}{r}}$$

$$\Rightarrow \quad r^{3/2} = \frac{\sqrt{GM}}{\omega}$$

$$= \frac{\sqrt{6.67 \times 10^{-11} \times 6 \times 10^{24}}}{1.45 \times 10^{-4}}$$

$$\Rightarrow \quad r = 2.66 \times 10^7 \text{ m}$$
$$= 2.66 \times 10^4 \text{ km}$$

12 Consider cavity as negative mass and apply superposition of gravitational potential. Consider the cavity formed in a solid sphere as shown in figure.

$V(\infty) = 0$

According to the equation, potential at an internal point P due to complete sphere.

$$V_3 = -\frac{GM}{2R^3}\left[3R^2 - \left(\frac{R}{2}\right)^2\right]$$

$$\frac{-GM}{2R^3}\left[3R^2 - \frac{R^2}{4}\right] = \frac{-GM}{2R^3}\left[\frac{11R^2}{4}\right]$$
$$= \frac{-11\,GM}{8R}$$

Mass of removed part

$$= \frac{M}{\frac{4}{3} \times \pi R^3} \times \frac{4}{3}\pi\left(\frac{R}{2}\right)^3 = \frac{M}{8}$$

Potential at point P due to removed part

$$V_2 = \frac{-3}{2} \times \frac{GM/8}{R/2} = \frac{-3GM}{8R}$$

Thus, potential due to remaining part at point P.

$$V_P = V_3 - V_2 = \frac{-11GM}{8R} - \left(-\frac{3GM}{8R}\right)$$
$$= \frac{(-11+3)\,GM}{8R} = \frac{-GM}{R}$$

13 Given, a satellite is revolving in a circular orbit at a height h from the Earth's surface having radius of Earth R, i.e. $h << R$.

Orbital velocity of a satellite,

$$v = \sqrt{\frac{GM}{R+h}} = \sqrt{\frac{GM}{R}} \quad (\text{as } h << R)$$

Velocity required to escape,

$$\frac{1}{2}mv'^2 = \frac{GMm}{R+h}$$

$$v' = \sqrt{\frac{2GM}{R+h}} = \sqrt{\frac{2GM}{R}}$$
$$(h << R)$$

$\therefore$ Minimum increase in its orbital velocity required to escape from the Earth's gravitational field.

$$v' - v = \sqrt{\frac{2GM}{R}} - \sqrt{\frac{GM}{R}}$$
$$= \sqrt{2gR} - \sqrt{gR}$$
$$= \sqrt{gR}\,(\sqrt{2} - 1) \qquad \left(\because g = \frac{GM}{R^2}\right)$$

14 Net force acting on any one particle M.

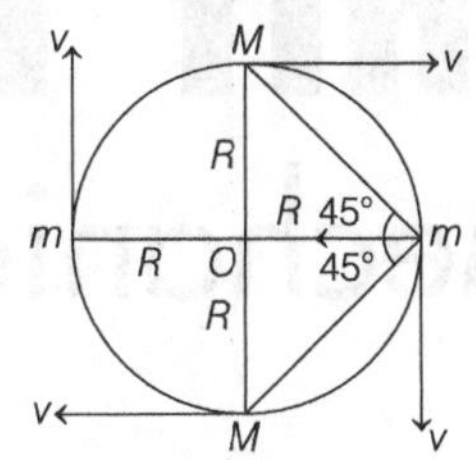

$$= \frac{GM^2}{(2R)^2} + \frac{GM^2}{(R\sqrt{2})^2}\cos 45^\circ + \frac{GM^2}{(R\sqrt{2})^2}\cos 45^\circ$$

$$= \frac{GM^2}{R^2}\left(\frac{1}{4} + \frac{1}{\sqrt{2}}\right)$$

This force will equal to centripetal force.

So, $\dfrac{Mv^2}{R} = \dfrac{GM^2}{R^2}\left(\dfrac{1}{4} + \dfrac{1}{\sqrt{2}}\right)$

$$v = \sqrt{\frac{GM}{4R}(1 + 2\sqrt{2})}$$
$$= \frac{1}{2}\sqrt{\frac{GM}{R}(2\sqrt{2} + 1)}$$

Hence, speed of each particle in a circular motion is

$$\frac{1}{2}\sqrt{\frac{GM}{R}(2\sqrt{2} + 1)}$$

15 Total energy of earth (planet)-satellite system is independent of eccentricity of orbit and it depends on semi-major axis and masses of the planet and satellite.

16 Here, Statement I is wrong because as speed of one satellite increases, its kinetic energy and hence total energy increases, i.e. total energy becomes less negative and hence r increases, i.e. orbit changes.

DAY TEN

Unit Test 1 (Mechanics)

1 Taking into account the significant figures, what should be the value of 9.99 + 0.0099 ?

(a) 9.9999 (b) 10.00 (c) 10.0 (d) 10

2 The maximum error in the measurement of mass and length of the cube are 3% and 2%, respectively. The maximum error in the measurement of density will be

(a) 5% (b) 6% (c) 7% (d) 9%

3 If C and L denote the capacitance and inductance, then the dimensional formula for C-L is same as that for

(a) frequency (b) time period
(c) (frequency)^2 (d) (time period)^2

4 The dimensions of (velocity)^2 ÷ radius are the same as that of

(a) Planck's constant (b) Gravitational constant
(c) Dielectric constant (d) None of these

5 A soap bubble oscillates with time period T, which in turn depends on the pressure (p), density (ρ) and surface tension (σ). Which of the following correctly represents the expression for T^2?

(a) $\dfrac{\rho\sigma^2}{p^3}$ (b) $\dfrac{\rho p^3}{\sigma}$
(c) $\dfrac{p^3\sigma}{\rho}$ (d) $\dfrac{\rho}{p^3\sigma}$

6 An automobile travels on a straight road for 40 km at 30 km h^{-1}. It then continues in the same direction for another 40 km at 60 km h^{-1}. What is the average velocity of the car during its 80 km trip?

(a) 30 km h^{-1} (b) 50 km h^{-1}
(c) 40 km h^{-1} (d) 60 km h^{-1}

7 A particle had a speed of 18 ms^{-1}. After 2.4 s, its speed was 30 ms^{-1} in the opposite direction. What were the magnitude and direction of the average acceleration of the particle during this 2.4 s interval?

(a) 10 ms^{-2} (b) 15 ms^{-2} (c) 20 ms^{-2} (d) 25 ms^{-2}

8 A rock is dropped from a 100 m high cliff. How long does it take to fall first 50 m and the second 50 m?

(a) 2 s, 3 s (b) 1.5 s, 2.5 s
(c) 1.2 s, 3.2 s (d) 3.2 s, 1.3 s

9 Two bodies of masses M_1 and M_2 are dropped from heights H_1 and H_2, respectively. They reach the ground after time T_1 and T_2, respectively. Which of the following relation is correct?

(a) $\dfrac{T_1}{T_2} = \left[\dfrac{H_1}{H_2}\right]^{1/2}$ (b) $\dfrac{T_1}{T_2} = \dfrac{H_1}{H_2}$

(c) $\dfrac{T_1}{T_2} = \left[\dfrac{M_1}{M_2}\right]^{1/2}$ (d) $\dfrac{T_1}{T_2} = \dfrac{M_1}{M_2}$

10 As a rocket is accelerating vertically upwards at 9.8 ms^{-2} near the earth's surface, it releases a projectile with zero speed relative to rocket. Immediately after release, the acceleration (in ms^{-2}) of the projectile is [take, $g = 9.8$ ms^{-2}]

(a) zero (b) 9.8 ms^{-2}, up
(c) 9.8 ms^{-2}, down (d) 19.6 ms^{-2}, down

11 A bullet is fired in a horizontal direction with a muzzle velocity of 300 ms^{-1}. In the absence of air resistance, how far will it drop in travelling a horizontal distance of 150 m? [take, $g = 10$ ms^{-2}]

(a) 1.25 cm (b) 12.5 cm (c) 1.25 m (d) 1.25 mm

12 A fixed mortar fires a bomb at an angle of 53° above the horizontal with a muzzle velocity of 80 ms^{-1}. A tank is advancing directly towards the mortar on level ground at a constant speed of 5 ms^{-1}. The initial separation (at the instant mortar is fired) between the mortar and tank, so that the tank would be hit is [take, $g = 10\ ms^{-2}$]

(a) 678.4 m (b) 614.4 m
(c) 64 m (d) None of these

13 A vector **a** is turned without a change in its length through a small angle $d\theta$. The value of $|\Delta \mathbf{a}|$ and Δa are respectively

(a) $0, a d\theta$ (b) $a d\theta, 0$
(c) 0, 0 (d) None of these

14 The vectors from origin to the points A and B are $\mathbf{A} = 3\hat{i} - 6\hat{j} + 2\hat{k}$ and $\mathbf{B} = 2\hat{i} + \hat{j} - 2\hat{k}$, respectively. The area of the ΔOAB be

(a) $\frac{5}{2}\sqrt{17}$ sq units (b) $\frac{2}{5}\sqrt{17}$ sq units
(c) $\frac{3}{5}\sqrt{17}$ sq units (d) $\frac{5}{3}\sqrt{17}$ sq units

15 The sum of the magnitudes of the two forces acting at a point is 18 and the magnitude of their resultant is 12. If the resultant is at 90° with the force of smaller magnitude, what are the magnitude of the forces?

(a) 12, 5 (b) 14, 4 (c) 5, 13 (d) 10, 8

16 A metal sphere is hung by a string fixed to a wall. The sphere is pushed away from the wall by a stick. The forces acting on the sphere are shown in the second diagram. Which of the following statements is wrong?

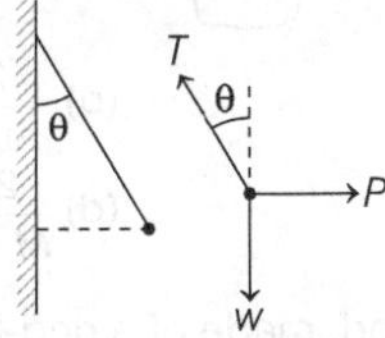

(a) $P = w\tan\theta$ (b) $\mathbf{T} + \mathbf{P} + \mathbf{w} = 0$
(c) $T^2 = P^2 + w^2$ (d) $T = P + w$

17 Three blocks A, B and C of masses 5 kg, 3 kg and 2 kg respectively are placed on a horizontal surface. The coefficient of friction between C and surface is 0.2 while between A and surface is zero and between B and surface is zero. If a force $F = 10$ N is first applied on A as shown and then in second case on C (shown dotted), then the ratio of normal contact force between B and C in first with respect to the second case is [Take, $g = 10\ ms^{-2}$]

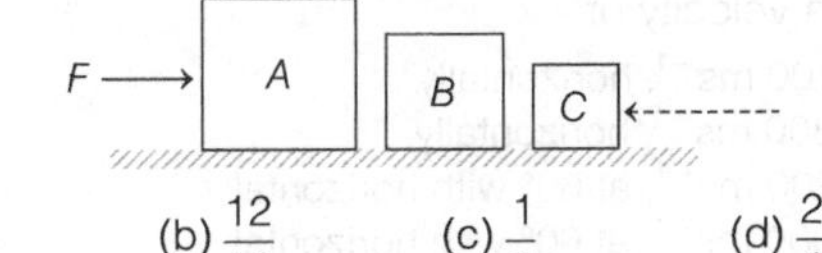

(a) $\frac{13}{12}$ (b) $\frac{12}{13}$ (c) $\frac{1}{2}$ (d) $\frac{2}{1}$

18 A 20 kg block is placed on top of a 50 kg block as shown. A horizontal force F acting on A causes an acceleration of 3 ms^{-2} to A and 2 ms^{-2} to B as shown. For this situation mark out the correct statement.

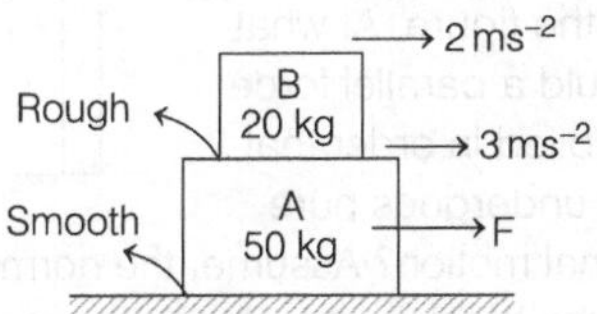

(a) The friction force between A and B is 40 N
(b) The net force acting on A is 150 N
(c) The value of F is 190 N
(d) All of the above

19 The figure below shows a man standing stationary w.r.t. a horizontal conveyor belt which is accelerating at 1 ms^{-2}. What is the net force on the man in this situation?

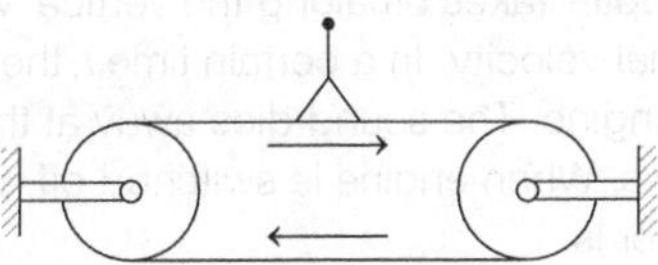

Take mass of the person to be as 70 kg. If the maximum acceleration of the belt, for which the man remains stationary w.r.t. the belt, is 3 ms^{-2}, then the coefficient of static friction between the man's shoes and the belt would be

(a) 70 N, 0.2 (b) 70 N, 0.3
(c) 700 N, 0.1 (d) 700 N, 0.3

20 A parachutist is in free fall before opening her parachute. The net force on her has a magnitude F and is directed downwards. This net force is somewhat less than, her weight w because of air resistance. Then, she opens her parachute. At the instant after her parachute fully inflates, the net force on her would be

(a) greater than F and still directed downwards
(b) less than F and still directed downwards
(c) zero
(d) directed upwards but could be more or less than F

21 The drive shaft of an automobile rotates at 3600 rpm and transmits 80 HP up from the engine to the rear wheels. The torque developed by the engine is

(a) 16.58 N-m (b) 0.022 N-m
(c) 158.31 N-m (d) 141.6 N-m

22 A disk starts rotating from rest about its axis with an angular acceleration equal to $\alpha = 10\ rads^{-2}$, where t is time in seconds. At $t = 0$, the disk is at rest. The time taken by the disk to make its, first complete revolution is

(a) $\left[\frac{6\pi}{5}\right]^{1/3}$ (b) $\left[\frac{3\pi}{10}\right]^{1/3}$ (c) $\left[\frac{2\pi}{5}\right]^{1/2}$ (d) $\left[\frac{6\pi}{13}\right]^{1/3}$

23 A rectangular block of mass M and height a is resting on a smooth level surface. A force F is applied to one corner as shown in the figure. At what point should a parallel force $3F$ be applied in order that the block undergoes pure translational motion? Assume, the normal contact force between the block and surface, passes through the centre of gravity of the block.

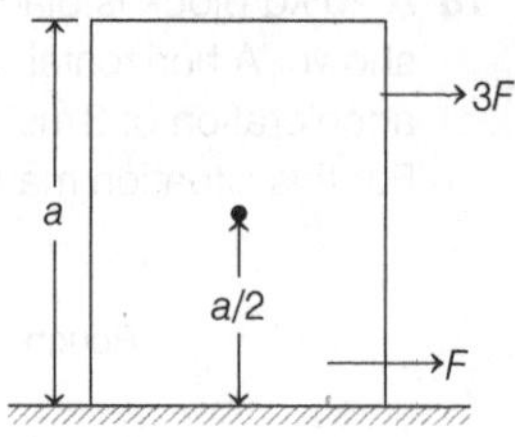

(a) $\frac{a}{3}$, vertically above centre of gravity
(b) $\frac{a}{6}$, vertically above centre of gravity
(c) No such point exists
(d) It is not possible

24 A helicopter takes off along the vertical with 3 ms^{-2} with zero initial velocity. In a certain time t, the pilot switches off the engine. The sound dies away at the point of take off in 30 s. When engine is switched off, velocity of the helicopter is

(a) 80 ms^{-1} (b) 30 ms^{-1} (c) 25 ms^{-1} (d) 100 ms^{-1}

25 To maintain a rotor at an uniform angular speed of 200 $rads^{-1}$, an engine needs to transmit a torque of 180 N-m. What is the power required by the engine? (Assume efficiency of the engine to be 80%)

(a) 36 kW (b) 18 kW (c) 45 kW (d) 54 kW

26 When a ball is whirled in a circle and the string supporting the ball is released, the ball flies off tangentially. This is due to

(a) the action of centrifugal force
(b) inertia for linear motion
(c) centripetal force
(d) some unknown cause

27 When a particle is moving in a vertical circle,

(a) its radial and tangential acceleration both are constant
(b) its radial and tangential acceleration both are varying
(c) its radial acceleration is constant but tangential acceleration is varying
(d) its radial acceleration is varying but tangential acceleration is constant

28 A particle of mass m slides on a quarter part of a smooth sphere of radius R as shown in the figure. It is released from rest at A, the normal contact force exerted by surface on the particle, when it reaches P is

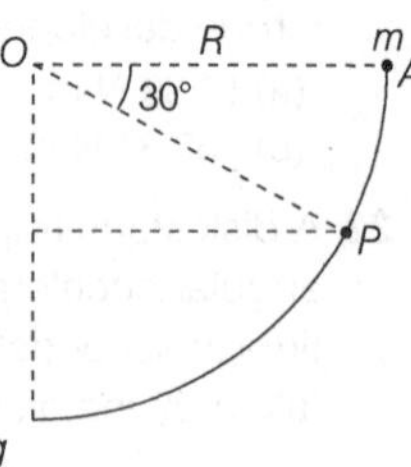

(a) $\frac{mg}{2}$ (b) $\frac{3mg}{2}$
(c) $mg \times \frac{\sqrt{3}}{2} + mg$ (d) $\frac{mg\sqrt{3}}{2}$

29 If an object weighs 270 N at the earth's surface, what will be the weight of the object at an altitude equal to twice the radius of the earth?

(a) 270 N (b) 90 N
(c) 30 N (d) 60 N

30 At its aphelion, the planet mercury is 6.99×10^{10} m from the sun, and at its perihelion it is 4.6×10^{10} m from the sun. If its orbital speed at aphelion is 3.88×10^4 ms^{-1}, then its perihelion orbital speed would be

(a) 3.88×10^4 ms^{-1} (b) 5.90×10^4 ms^{-1}
(c) 5.00×10^4 ms^{-1} (d) 5.5×10^4 ms^{-1}

31 If R is the radius of the orbit of a geosynchronous satellite and another satellite is orbiting around the earth in a circular orbit of radius $\frac{R}{2}$, then its time period would be

(a) $6\sqrt{2}$ h (b) 6 h
(c) 12 h (d) Cannot be determined

32 A cylinder of mass M and radius r is mounted on a frictionless axle over a well. A rope of negligible mass is wrapped around the cylinder and a bucket of mass m is suspended from the rope. The linear acceleration of the bucket will be

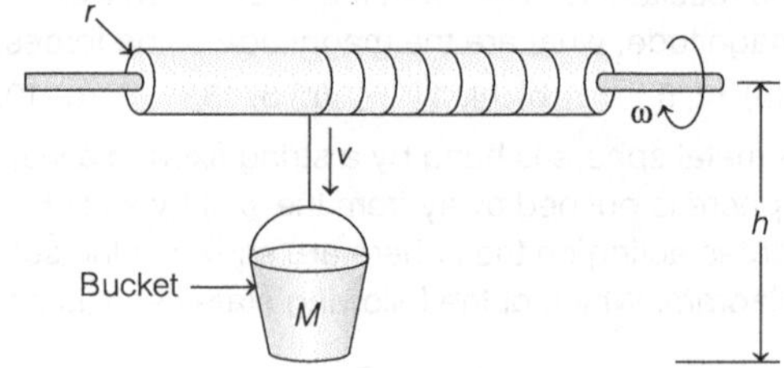

(a) $\frac{mg}{M+2m}$ (b) $\frac{2mg}{m+2M}$
(c) $\frac{mg}{2M+m}$ (d) $\frac{2mg}{M+2m}$

33 A merry-go-round, made of a ring-like platform of radius R and mass M, is revolving with angular speed ω. A person of mass M is standing on it. At one instant, the person jumps off the round, radially away from the centre of the round (as seen from the round). The speed of the round of afterwards is

(a) 2ω (b) ω (c) $\omega/2$ d) 0

34 A canon ball is fired with a velocity of 200 ms^{-1} at an angle of 60° with the horizontal. At the highest point, it explodes into three equal fragments. One goes vertically upwards with a velocity of 100 ms^{-1} and other goes vertically downwards with 100 ms^{-1}. The third one moves with a velocity of

(a) 100 ms^{-1}, horizontally
(b) 300 ms^{-1}, horizontally
(c) 200 ms^{-1}, at 60° with horizontal
(d) 300 ms^{-1}, at 60° with horizontal

35 A rocket of initial mass (including fuel) 15000 kg ejects mass at a constant rate of 25 kgs^{-1} with a constant relative speed of 15 kms^{-1}. The acceleration of the rocket, 5 min after the blast is [Neglect gravity effect]

(a) 40 ms^{-2} (b) 50 ms^{-2}
(c) 60 ms^{-2} (d) 45 ms^{-2}

36 An elevator of total mass (elevator + passenger) 1800 kg is moving up with a constant speed of 2 ms^{-1}. Frictional force of 2000 N is opposing its motion. The minimum power delivered by the motor to the elevator is [take, $g = 10\ ms^{-2}$]

(a) 36 kW (b) 4 kW
(c) 40 kW (d) – 40 kW

37 A bead of mass 1/2 kg starts from rest from a point A to B move in a vertical plane along a smooth fixed quarter ring of radius 5 m, under the action of a constant horizontal force $F = 5$ N as shown. The speed of the bead as it reaches the point B is [Take $g = 10\ ms^{-2}$]

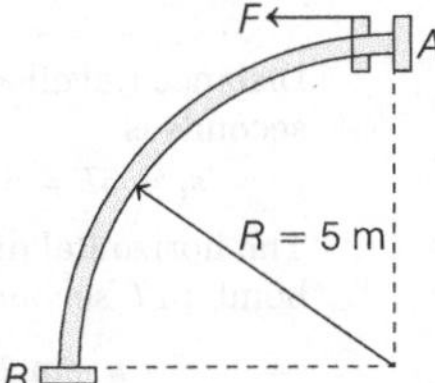

(a) 14.14 ms^{-1} (b) 7.07 ms^{-1} (c) 5 ms^{-1} (d) 25 ms^{-1}

38 A car (treat it as particle) of mass m is accelerating on a level smooth road under the action of a single force F. The power delivered to the car is constant and equal to P. If the velocity of the car at an instant is v, then after travelling how much distance, it becomes double?

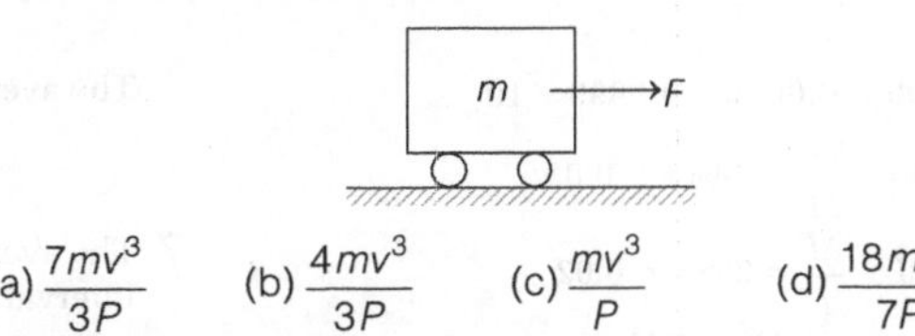

(a) $\dfrac{7mv^3}{3P}$ (b) $\dfrac{4mv^3}{3P}$ (c) $\dfrac{mv^3}{P}$ (d) $\dfrac{18mv^3}{7P}$

Direction (Q. Nos. 39-40) *Each of these questions contains two statements Statement I and Statement II. Each of these questions also has four alternative choices, only one of which is the correct answer. You have to select one of the codes (a), (b), (c) and (d) given below*

(a) Statement I is true, Statement II is true; Statement II is the correct explanation for Statement I
(b) Statement I is true, Statement II is true; Statement II is not the correct explanation for Statement I
(c) Statement I is true; Statement II is false
(d) Statement I is false; Statement II is true

39 By considering the earth to be non-spherical

Statement I As, one moves from equator to the pole of the earth, the value of accelaration due to gravity increases.

Statement II If the earth stops rotating about its own axis, the value of accelaration due to gravity will be same at pole and at equator.

40 **Statement I** Total torque on a system is independent of the origin if the total external force is zero.

Statement II Torque due to a couple is independent of the origin.

ANSWERS

1. (d)	**2.** (d)	**3.** (d)	**4.** (d)	**5.** (a)	**6.** (c)	**7.** (c)	**8.** (d)	**9.** (a)	**10.** (c)
11. (c)	**12.** (a)	**13.** (b)	**14.** (a)	**15.** (c)	**16.** (d)	**17.** (a)	**18.** (d)	**19.** (b)	**20.** (d)
21. (c)	**22.** (c)	**23.** (b)	**24.** (a)	**25.** (c)	**26.** (b)	**27.** (b)	**28.** (b)	**29.** (c)	**30.** (b)
31. (a)	**32.** (d)	**33.** (a)	**34.** (b)	**35.** (b)	**36.** (c)	**37.** (a)	**38.** (a)	**39.** (c)	**40.** (a)

Hints and Explanations

1 $9.99 + 0.0099 = 9.9999 = 10$

2 Here, $\frac{\Delta M}{M} = 3\% = \pm 0.03$

and $\frac{\Delta l}{l} = 2\% = \pm 0.02$

Hence, $\frac{\Delta V}{V} = \frac{3\Delta l}{l} = \pm 0.06$

Now, $\frac{\Delta \rho}{\rho} = \frac{\Delta M}{M} + \frac{\Delta V}{V}$ $\left[\because \text{Density} = \frac{\text{Mass}}{\text{Volume}}\right]$

$= \pm 0.09 = 9\%$

3 Time period of C-L oscillations is given by $2\pi\sqrt{LC}$.

Hence, $[LC] = [\text{time period}]^2$

4 Dimensional formula of $(\text{velocity})^2 \div$ radius $= \frac{[M^0LT^{-1}]^2}{[M^0LT^0]} = [M^0LT^{-2}]$

$= [\text{acceleration}]$

Note *In circular motion, centripetal acceleration is* $\frac{V^2}{R}$.

5 Here, $T^2 = p^a \rho^b \sigma^c$ ...(i)

Putting the dimensions of the quantities in RHS, we get

$= [ML^{-1}T^{-2}]^a [ML^{-3}]^b \; [MT^{-2}]^c$

$= [M^{a+b+c} L^{-a-3b} T^{-2a-2c}]$

Hence, $a + b + c = 0$ ...(ii)

$-a - 3b = 0$ ...(iii)

and $-2a - 2c = 2$ ...(iv)

On solving, Eqs. (ii) (iii) and (iv), we get

$a = -3b, b = 1$ and $c = 2$

So, after putting the values of a, b and c in Eq. (i), we get

$$T^2 = \frac{\rho\sigma^2}{p^3}$$

6 Average velocity $(v_{av}) = \frac{\Delta x}{\Delta t}$

where, Δx is the displacement in a given time interval. For first part of the journey, time taken by the car,

$$\Delta t_1 = \frac{40}{30} = 1.33 \text{ h}$$

For second part of the journey time taken by the car

$$\Delta t_2 = \frac{40}{60} = 0.67 \text{ h}$$

Hence, the total displacement

$\Delta x = \Delta x_1 + \Delta x_2$

$= 40 + 40 = 80$ km

and the total time interval

$\Delta t = \Delta t_1 + \Delta t_2 = 1.33 + 0.67 = 2$ h

The average velocity is

$$v_{av} = \frac{80}{2} = 40 \text{ kmh}^{-1}$$

7 The average acceleration in the given interval is

$$a_{av} = \frac{(v_2 - v_1)}{(t_2 - t_1)}$$

Take, $v_1 = 18 \text{ ms}^{-1}, v_2 = -30 \text{ ms}^{-1}$, $t_1 = 0$ and $t_2 = 2.4$s

$$a_{av} = \frac{-30 - 18}{2.4} = -20 \text{ ms}^{-2}$$

The $-$ ve sign indicates that the acceleration is opposite to the initial direction of motion.

8 Using second equation of motion,

$y = ut + \frac{1}{2}at^2$, at $v_0 = 0$

and solve $y = v_0 t + \frac{1}{2}gt^2$ for t

So, $t = \sqrt{2y/g}$

For $y = 50$ m

So, $t_1 = \sqrt{\frac{2(50)}{9.8}} = 3.2$ s

For $y = 100$ m

$t_2 = \sqrt{\frac{2\,(100)}{9.8}} = 4.5$ s

Hence, the difference is the time taken to fall the second 50 m

$= 4.5 - 3.2 = 1.3$ s

9 Distance of fall is independent of the mass of the bodies

$H_1 = \frac{1}{2}gT_1^2$ and $H_2 = \frac{1}{2}gT_2^2$

Hence, $\frac{T_1}{T_2} = \left[\frac{H_1}{H_2}\right]^{1/2}$

10 As nothing has been mentioned that w.r.t. which frame of reference is to be found, it means we have to compute w.r.t frame of reference of earth. As the object is released, its acceleration w.r.t ground is only due to the influence of gravity of the earth and hence is equal to 9.8 ms^{-2} in the downward direction.

11 Let the bullet, dropped by y metre while covering a horizontal distance of 150 m.

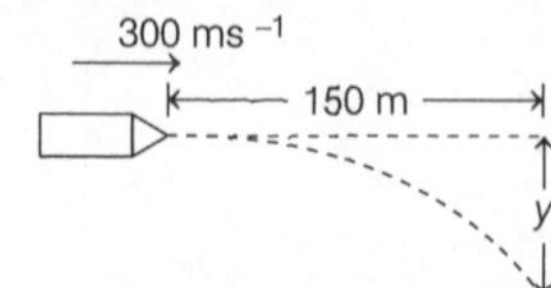

Let t be the time taken by the bullet to cover a horizontal distance of 150 m, then

$150 = 300t \Rightarrow t = \frac{1}{2}$s

$y = \frac{1}{2}gt^2 = \frac{1}{2} \times 10 \times \frac{1}{4} = 1.25$ m

12 The situation is shown clearly in figure.

Time of flight of bomb is $T = \frac{2u\sin\theta}{g}$

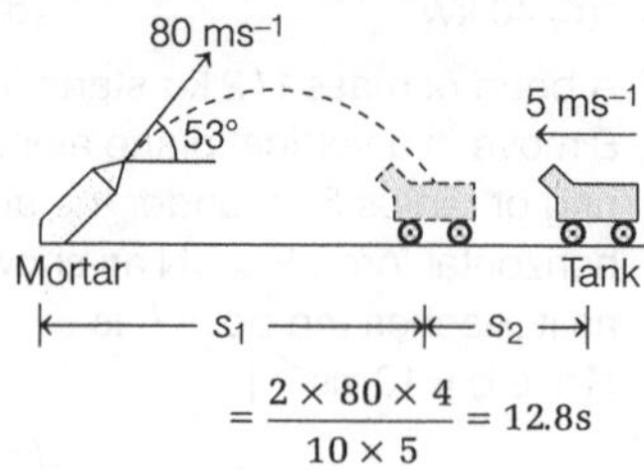

$= \frac{2 \times 80 \times 4}{10 \times 5} = 12.8$s

Distance travelled by the tank in T seconds is

$s_1 = 5T = 5 \times 12.8 = 64$ m

The horizontal distance travelled by bomb in T seconds is

$$s_2 = \frac{u^2 \sin 2\theta}{g}$$

$$= \frac{80^2 \times 2 \times \frac{3}{5} \times \frac{4}{5}}{10}$$

$= 614.4$ m

$[\because \sin 2\theta = 2\sin\theta \cdot \cos\theta]$

So, required separation

$s = s_1 + s_2 = 678.4$ m

13 From the figure $|\mathbf{OA}| = a$ and $|\mathbf{OB}| = a$.

Also, from the triangle rule

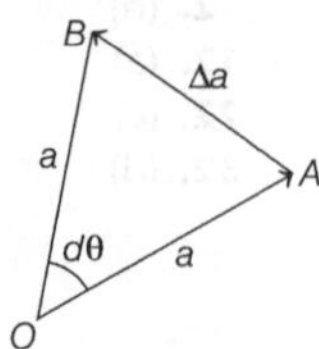

$\mathbf{OB} - \mathbf{OA} = \mathbf{AB} = \Delta\mathbf{a}$

$\Rightarrow |\Delta\mathbf{a}| = AB$

Using, Angle $= \frac{\text{Arc}}{\text{Radius}}$

$\Rightarrow AB = a \cdot d\theta$

So, $|\Delta\mathbf{a}| = a \cdot d\theta$

Δa means a change in magnitude of the vector, i.e.,

$|\mathbf{OB}| - |\mathbf{OA}| \Rightarrow a - a = 0$

So, $\Delta a = 0$

14 Given $OA = \mathbf{a} = 3\hat{i} - 6\hat{j} + 2\hat{k}$ and $OB = \mathbf{b} = 2\hat{i} + \hat{j} - 2\hat{k}$

$$\therefore (\mathbf{a} \times \mathbf{b}) = \begin{vmatrix} \hat{i} & \hat{j} & \hat{k} \\ 3 & -6 & 2 \\ 2 & 1 & -2 \end{vmatrix}$$

$$= (12 - 2)\hat{i} + (4 + 6)\hat{j} + (3 + 12)\hat{k}$$

$$= 10\hat{i} + 10\hat{j} + 15\hat{k}$$

$$\Rightarrow |\mathbf{a} \times \mathbf{b}| = \sqrt{10^2 + 10^2 + 15^2} = \sqrt{425} = 5\sqrt{17}$$

Area of $\Delta OAB = \frac{1}{2}|\mathbf{a} \times \mathbf{b}|$

$$= \frac{5\sqrt{17}}{2} \text{ sq units.}$$

15 Let P be the smaller force and Q be the greater force, then according to the problem

$$P + Q = 18 \quad \text{... (i)}$$

$$R = \sqrt{P^2 + Q^2 + 2PQ\cos\theta} = 12 \quad \text{... (ii)}$$

$$\tan\phi = \frac{Q\sin\theta}{P + Q\cos\theta} = \tan 90° = \infty$$

$$\therefore \quad P + Q\cos\theta = 0 \quad \text{... (iii)}$$

By solving Eqs. (i), (ii) and (iii), we get

$$P = 5 \text{ and } Q = 13$$

16

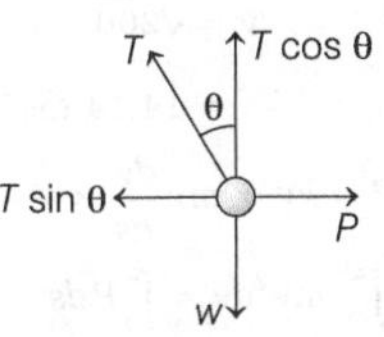

As the metal sphere is in equilibrium under the effect of the three forces therefore, $T + P + w = 0$

From the figure, $T\cos\theta = w$ ...(i)

$T\sin\theta = P$ ... (ii)

From Eqs. (i) and (ii), we get

$$P = W\tan\theta$$

and $T^2 = P^2 + w^2$

[as, $\sin^2\theta + \cos^2\theta = 1$]

17 Friction force between C and surface is $f = \mu \times 2g = 0.2 \times 2 \times 10 = 4\text{N}$

***Case* I** $(5 + 3 + 2)a = F - f$

$$\Rightarrow \quad a = \frac{10 - 4}{10} = \frac{6}{10}$$

For C, $N_1 - f = 2a$

$$\Rightarrow \quad N_1 = 2 \times \frac{6}{10} + 4 = \frac{26}{5}\text{N}$$

***Case* II** $F - f = (5 + 3 + 2)a$

$$\Rightarrow \quad a = \frac{10 - 4}{10} = \frac{6}{10}$$

For C, $F - N_2 - f = 2a$

$$\Rightarrow \quad N_2 = 10 - 4 - 2 \times \frac{6}{10} = \frac{24}{5}\text{N}$$

Required ratio is, $\frac{N_1}{N_2} = \frac{26}{24} = \frac{13}{12}$

18 As the acceleration of A and B are different, it means there is a relative motion between A and B. The free body diagram of A and B can be drawn as

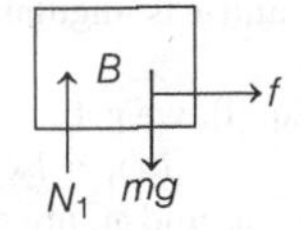

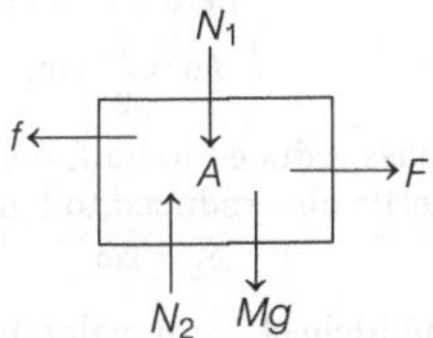

For A, $F - f = Ma_A = 50 \times 3 = 150$ N

For B, $f = ma_B = 20 \times 2 = 40$ N

So, $F = 150 + 40 = 190$ N

19 As person remains stationary w.r.t. belt, so acceleration of the person w.r.t. ground is the same as that of belt w.r.t. ground. So, net force acting on the person is, $f = ma = 70 \times 1 = 70\text{N}$. Let coefficient of static friction between the man's shoes and belt is μ_S, then

$$a_{max} = \mu_s g$$

$$\mu_S = \frac{3}{10} = 0.3$$

20 As the parachute inflates fully, the force of air friction increases by a large amount and the parachutist starts decelerating, i.e. net force acting on her is in upward direction but the magnitude of the net force cannot be determined from given information.

21 From $P = \tau\omega$

$$\tau = \frac{P}{\omega}$$

It is given,

$P = 80\,\text{HP} = 80 \times 746$ W

$= 59680$ N-ms^{-1}

$$\omega = 3600 \text{ rpm} = \frac{3600}{60} \times 2\pi \text{ rad s}^{-1}$$

$$= 120\pi \text{ rad s}^{-1}$$

So, $\tau = 158.31$ N-m

22 As, $\omega = \omega_0 \pm \alpha t$

$$\Rightarrow \quad \omega_0 = 10t$$

or $\frac{d\theta}{dt} = 10t$

On integrating, (for one complete revolution)

$$\int_0^{2\pi} d\theta = \int_0^T 10t\,dt$$

$$\Rightarrow \quad 2\pi = \frac{10T^2}{2}$$

$$\Rightarrow \quad T = \left(\frac{2\pi}{5}\right)^{\frac{1}{2}}$$

23 The free body diagram of the block can be drawn as shown. As body has to move in a pure translational motion, the torque about the centre of gravity must be zero.

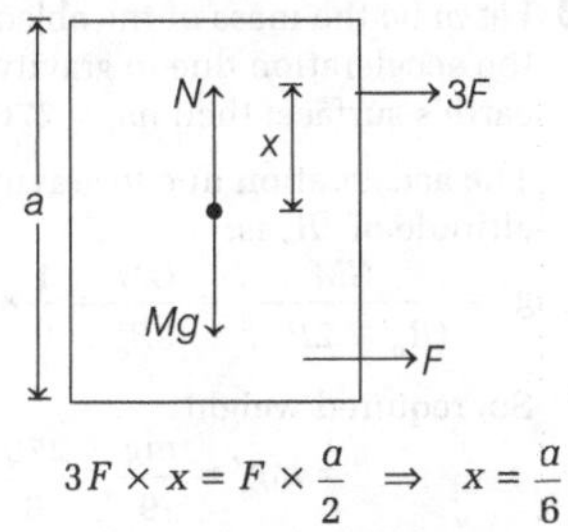

$$3F \times x = F \times \frac{a}{2} \Rightarrow x = \frac{a}{6}$$

24 The altitude of the helicopter when engine is switched off $h = \frac{at_1^2}{2}$. Sound is not heard after $t_2 = t_1 + \frac{at_1^2}{2c}$, where c = speed of sound.

$$at_1^2 + 2ct_1 - 2ct_2 = 0$$

$$\Rightarrow \quad t_1 = \frac{-2c \pm \sqrt{4c^2 + 8cat_2}}{2a}$$

$$t_1 = -\frac{c}{a} + \sqrt{\frac{c^2}{a^2} + \frac{2c}{a}t_2}$$

$$\therefore v = at_1 = -c + \sqrt{c^2 + 2cat_2}$$

$$= -320 + \sqrt{(320)^2 + 2 \times 320 \times 3 \times 30}$$

$$= \sqrt{1600 \times (10)^2} - 320$$

$$= 400 - 320 = 80 \text{ ms}^{-1}$$

25 Power required for rotor, $P = \tau \cdot \omega$

$= 180 \times 200 = 36$ kW

Power of engine, $P_0 = \frac{P}{0.8} = 45$ kW

[as efficiency is 80%]

26 This is due to Intertia for linear motion of ball.

27 In a vertical circle, both radial and tangential components of the acceleration change direction at every instant.

28 Apply Work-Energy theorem at A and P,

$$\frac{mv^2}{2} - 0 = mg \times \frac{R}{2}$$

Use dynamical equations at P,

$$N - mg\cos 60° = \frac{mv^2}{R}$$

$$\Rightarrow N = mg + \frac{mg}{2} = \frac{3\,mg}{2}$$

29 Let m be the mass of the object and g is the acceleration due to gravity at the earth's surface, then $mg = 270$ N.

The acceleration due to gravity at an altitude of $2R_e$ is,

$$g' = \frac{GM}{(R_e + 2R_e)^2} = \frac{GM}{9R_e^2} = \frac{1}{9} \times g$$

So, required weight

$$= mg' = \frac{mg}{9} = \frac{270}{9} = 30\text{ N}$$

30 From the conservation of angular momentum,

$$m\,v_A r_A = m v_p r_p$$

$$\Rightarrow v_p = \frac{3.88 \times 10^4 \times 6.99 \times 10^{10}}{4.6 \times 10^{10}}$$

$$= 5.90 \times 10^4 \text{ ms}^{-1}$$

31 From the Kepler's law, $T^2 \propto r^3$

$$\Rightarrow \frac{T_1}{T_2} = \left(\frac{r_1}{r_2}\right)^{3/2}$$

$$\Rightarrow \frac{24}{T_2} = \left(\frac{R}{R/2}\right)^{3/2} = 2^{3/2}$$

$$\Rightarrow T_2 = \frac{24}{2^{3/2}} = \frac{24}{2\sqrt{2}} = 6\sqrt{2}\text{ h}$$

32 Weight of bucket acts downwards while tension T in opposite direction

$$mg - T = ma$$

Also, $\tau = I\alpha = rT$

$$\Rightarrow \frac{1}{2}Mr^2\alpha = rT \Rightarrow \frac{1}{2}M(r\alpha) = T$$

or $$T = \frac{Ma}{2}$$

$\therefore$ $$a = \frac{mg - T}{m}$$

$$\Rightarrow a = \frac{mg - \frac{Ma}{2}}{m} = \frac{2mg - Ma}{2m}$$

$$\Rightarrow 2ma + Ma = 2mg$$

$$\Rightarrow a = \frac{2mg}{2m + M}$$

33 As no external torque acts on the system, angular momentum should be conserved.

Hence, $I\omega$ = constant ...(i)

where, I is moment of inertia of the system and ω is angular velocity of the system.

From Eq. (i), we get

$$I_1\omega_1 = I_2\omega_2$$

[where ω_1 and ω_2 are angular velocities before and after jumping]

$$\Rightarrow I\omega = \frac{I}{2} \times \omega_2$$

[as mass reduced to half, hence moment of inertia also reduced to half]

$$\Rightarrow \omega_2 = 2\omega$$

34 At the highest point, velocity before explosion is $v\cos 60°$ along X-axis.

By law of conservation of momentum,

$$(mv\cos 60°)\,\hat{\mathbf{i}} = \frac{m}{3}(100\hat{\mathbf{j}}) + \frac{m}{3}(-100\,\hat{\mathbf{j}}) + \frac{mv'}{3}$$

$$\Rightarrow v' = \frac{3v}{2} = \frac{3 \times 200}{2}\hat{\mathbf{i}}$$

$$= 300\,\hat{\mathbf{i}}\text{ ms}^{-1}$$

or 300 ms^{-1} along X-axis or horizontally.

35 Thrust force acting on the rocket is,

$$F = v_{\text{rel}}\frac{dm}{dt}$$

$$F = 15 \times 1000 \times 25\text{N}$$

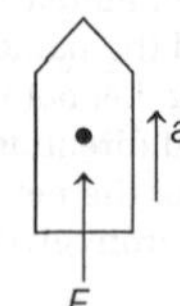

Mass of rocket at $t = 5$ min after the blasting starts, is

$$m = 15000 - 25 \times 5 \times 60 = 7500$$

So, $F = ma$

$$\Rightarrow a = \frac{F}{m}$$

$$= \frac{15 \times 25000}{7500}$$

$$= 50\text{ ms}^{-2}$$

Note *If gravity is not neglected, then $F - mg = ma$ check whether acceleration is constant here.*

36 The net downward force on the elevator is,

$$F_1 = mg + f = 18000 + 2000 = 20000\text{ N}$$

So, the motor has to work against this force.

To move the elevator with a constant speed, the minimum power delivered by the motor to the elevator must be,

$$P = \mathbf{F} \cdot \mathbf{v} = 20000 \times 2 = 40\text{ kW}$$

37 Applying the work-energy theorem,

$$\frac{1}{2} \times mv^2 - 0 = F \times R + mg \times R$$

$$\Rightarrow \frac{1}{2} \times \frac{1}{2} \times v^2 = 5 \times 5 + \frac{1}{2} \times 10 \times 5 = 50$$

$$v = \sqrt{200}$$

$$= 14.14\text{ ms}^{-1}$$

38 As, $P = Fv = mv\dfrac{dv}{ds} \times v$

$$\Rightarrow \int_v^{2v} mv^2 dv = \int_0^s P\,ds$$

$$\Rightarrow s = \frac{7mv^3}{3P}$$

39 As one go from equator to pole of the earth, the value of g increase due to decrease in latitude (λ). Also, the earth is non-spherical, this implies the value of g, at the poles and equitorial point on the earth's surface are unequal due to its different distances from earth's centre.

40 If net force on the system is zero, it can be resolved into two equal and opposite forces which can be considered to form a couple.

DAY ELEVEN

Oscillations

Learning & Revision for the Day

- Periodic Motion
- Simple Harmonic Motion
- Oscillations of a Spring
- Force and Energy in SHM
- Composition of Two SHMs
- Simple Pendulum
- Free, Damped, Forced and Resonant Vibrations

Periodic Motion

A motion which repeats itself over a regular interval of time is called a periodic motion. A periodic motion in which a body moves back and forth repeatedly about a fixed point (called mean position) is called **oscillatory** or **vibratory motion.**

- **Period** The regular interval of time after which periodic motion repeats itself is called period of the motion.
- **Frequency** The number of times of motion repeated in one second is called frequency of the periodic motion. Every oscillatory motion is periodic but every periodic motion is not an oscillatory motion.
- **Displacement as a Function of Time** In a periodic motion each displacement value is repeated after a regular interval of time, displacement can be represented as a function of time.

$$y = f(t)$$

- **Periodic Function** A function which repeats its value after a fix interval of time is called a periodic function.

$$y(t) = y(t + T)$$

where, T is the period of the function.

Trigonometric functions $\sin\theta$ and $\cos\theta$ are simplest periodic functions having period of 2π.

Simple Harmonic Motion

Simple Harmonic Motion (SHM) is that type of oscillatory motion in which the particle moves to and fro or back and forth about a fixed point under a restoring force, whose magnitude is directly proportional to its displacement

i.e. $$F \propto x \text{ or } F = -kx$$

where, k is a positive constant called the **force constant** or **spring factor** and x is displacement.

Differential equations of SHM, for linear SHM, $\dfrac{d^2 y}{dt^2} + \omega^2 y = 0$,

for angular SHM, $\dfrac{d^2\theta}{dt^2} + \omega^2\theta = 0$

- No. of Questions in Exercises (x)—
- No. of Questions Attempted (y)—
- No. of Correct Questions (z)—
  (Without referring Explanations)

- Accuracy Level ($z/y \times 100$)—
- Prep Level ($z/x \times 100$)—

In order to expect good rank in JEE, your Accuracy Level should be above 85 & Prep Level should be above 75.

Terms Related to SHM

The few important terms related to simple harmonic motion are given as

- **Displacement** The displacement of a particle executing SHM is, in general, expressed as $y = A \sin(\omega t - \phi)$.

  where, A is the amplitude of SHM, ω is the angular frequency $\left(\text{where } \omega = \frac{2\pi}{T} = 2\pi\nu\right)$ and ϕ is the initial phase of SHM. However, displacement may also be expressed as $x = A\cos(\omega t - \phi)$.

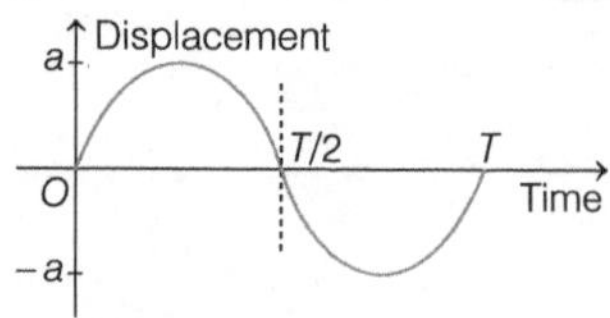

- **Amplitude** The maximum displacement on either side of mean position is called amplitude of SHM.
- **Velocity** The velocity of a particle executing SHM at an instant is defined as the time rate of change of its displacement at that instant.

  Velocity, $\frac{dy}{dt} = v = \omega\sqrt{A^2 - y^2}$

  At the mean position ($y = 0$), during its motion $v = A\omega = v_{max}$ and at the extreme positions ($y = \pm A$), $v = 0$.

  $\therefore$ Velocity amplitude, $v_{max} = A\omega$

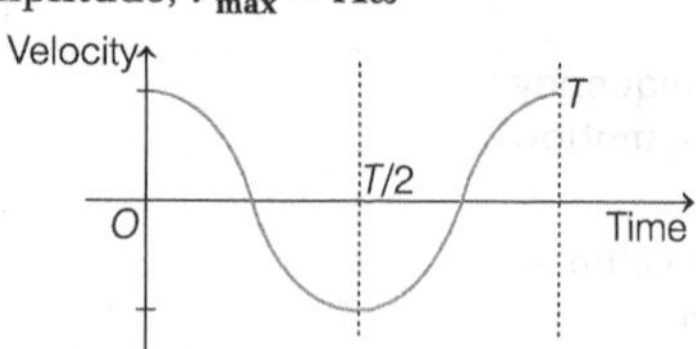

- **Acceleration** The acceleration of a particle executing SHM at an instant is defined as the time rate of change of velocity at that instant.

  Acceleration, $\frac{d^2y}{dt^2} = a = -\omega^2 y$

  The acceleration is also a variable.

  At the mean position ($y = 0$), acceleration $a = 0$ and at the extreme position ($y = \pm A$), the acceleration is $a_{max} = -A\omega^2$.

  $\therefore$ Acceleration amplitude, $a_{max} = A\omega^2$

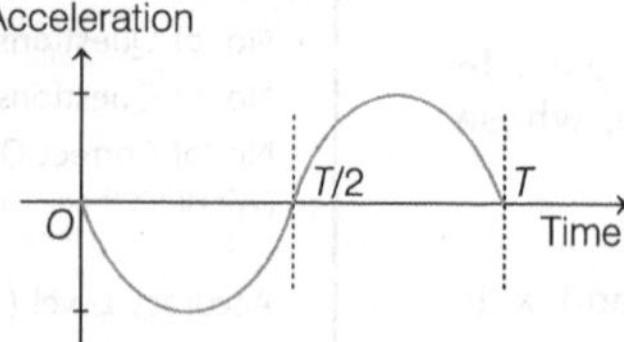

- **Phase** Phase is that physical quantity which tells about the position and direction of motion of any particle at any moment. It is denoted by ϕ.
- **Phase Difference** If two particles perform S.H.M and their equations are

  $$y_1 = a\sin(\omega t + \phi_1) \text{ and } y_2 = a\sin(\omega t + \phi_2)$$

  phase difference $\Delta\phi = (\omega t + \phi_2) - (\omega t + \phi_1) = \phi_2 - \phi_1$

- **Time Period** The time taken by a particle to complete one oscillation is called time period. It is denoted by T.

  $\therefore$ Time period of SHM,

  $$T = \frac{2\pi}{\omega} = 2\pi\sqrt{\frac{|y|}{|a|}} = 2\pi\sqrt{\frac{\text{Displacement}}{\text{Acceleration}}}$$

- **Frequency and Angular Frequency** It is defined as the number of oscillations executed by body per second. SI unit of frequency is hertz.

  Angular frequency of a body executing periodic motion is equal to product of frequency of the body with factor 2π. Angular frequency, $\omega = 2\pi n$.

Oscillations of a Spring

If the mass is once pulled, so as to stretch the spring and is then released, then a restoring force acts on it which continuously tries to restore its mean position.

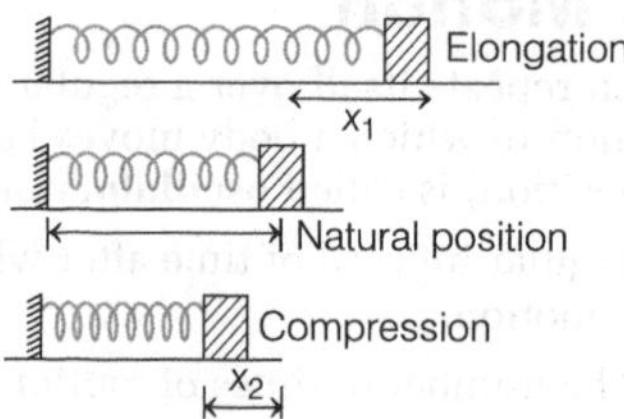

Restoring force $\quad F = -kl$,

where k is force constant and l is the change in length of the spring.

Here, $\quad x_1 = x_2 = l$

- The spring pendulum oscillates simple harmonically having time period and frequency given by

  $$T = 2\pi\sqrt{\frac{m}{k}}$$

  and $$\nu = \frac{1}{2\pi}\sqrt{\frac{k}{m}}$$

- If the spring is not light but has a mass m_s, then

  $$T = 2\pi\sqrt{\frac{m + 1/3\, m_s}{k}}$$

- If two masses m_1 and m_2, connected by a spring, are made to oscillate on a horizontal surface, then its period will be $T = 2\pi\sqrt{\frac{\mu}{k}}$

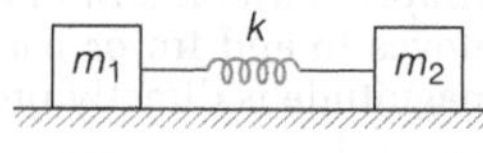

  where, $\mu = \frac{m_1 m_2}{m_1 + m_2}$ = reduced mass of the system.

Series Combination of Springs

If two springs of spring constants k_1 and k_2 are joined in series (horizontally or vertically), then their equivalent spring constant k_s is given by

$$\frac{1}{k_s} = \frac{1}{k_1} + \frac{1}{k_2} \Rightarrow k_s = \frac{k_1 k_2}{k_1 + k_2}$$

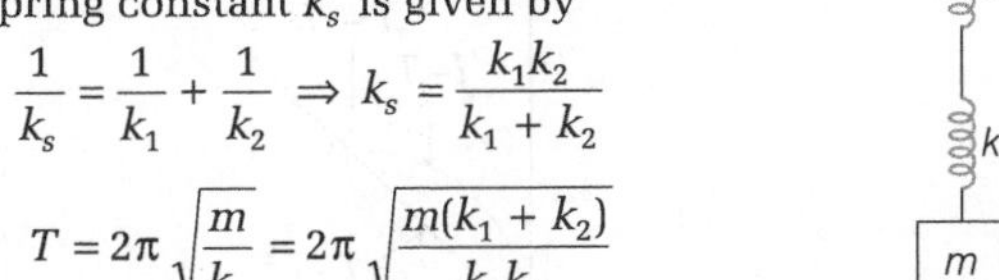

$$\therefore \quad T = 2\pi\sqrt{\frac{m}{k_s}} = 2\pi\sqrt{\frac{m(k_1 + k_2)}{k_1 k_2}}$$

Parallel Combination of Springs

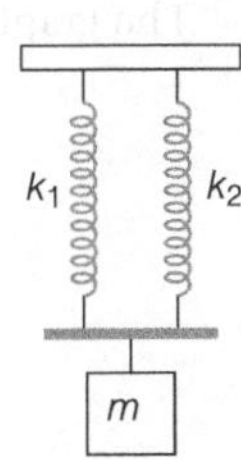

If two springs of spring constants k_1 and k_2 are joined in parallel as shown in figure, then their equivalent spring constant $k_p = k_1 + k_2$ hence,

$$T = 2\pi\sqrt{\frac{m}{k_p}} = 2\pi\sqrt{\frac{m}{(k_1 + k_2)}}$$

Force and Energy in SHM

- **Force** For an object executing SHM, a force always acts on it, which tries to bring it in mean position, i.e. it is always directed towards mean position.

  The equation of motion, $\mathbf{F} = m\mathbf{a}$,

  $$\therefore \quad F = -m\omega^2 x \quad [\because a = -\omega^2 x]$$

  $$= -kx \quad \left[\because \omega = \sqrt{\frac{k}{m}}\right]$$

  Here, negative sign shows that direction of force is always opposite to the direction of displacement.

- **Energy** If a particle of mass m is executing SHM, then at a displacement x from the mean position, the particle possesses potential and kinetic energy.

  At any displacement x,

  Potential energy, $U = \frac{1}{2} m\omega^2 x^2 = \frac{1}{2} k x^2$

  Kinetic energy, $K = \frac{1}{2} m\omega^2 (A^2 - x^2) = \frac{1}{2} k (A^2 - x^2)$

  Total energy, $E = U + K = \frac{1}{2} m\omega^2 A^2 = 2\pi^2 m\nu^2 A^2$

  If there is no friction, the total mechanical energy, $E = K + U$, of the system always remains constant even though K and U change.

Composition of Two SHMs

If a particle is acted upon two separate forces each of which can produce a simple harmonic motion. The resultant motion of the particle would be a combination of two SHMs.

For which $\quad \mathbf{F}_1 + \mathbf{F}_2 = m\frac{d^2\mathbf{r}}{dt}$

and $\mathbf{r}_1 + \mathbf{r}_2 = \mathbf{r}$ = resultant position of the particle
where, m = mass of the particle.
$\mathbf{r}_1, \mathbf{r}_2$ = positions of the particle under two forces.

There are two cases

- When two SHM are in same direction the resultant is given by

  $$x = x_1 + x_2 = A\sin(\omega t + \beta)$$

  where, $x_1 = A_1 \sin \omega t$,

  $$x_2 = A_2 \sin(\omega t + \phi)$$

  $$A = \sqrt{A_1^2 + 2A_1 A_2 \cos\phi + A_2^2}$$

  and $\tan\beta = \frac{A_2 \sin\phi}{A_1 + A_2\cos\phi}$

  For any value of ϕ other than 0 and π resultant amplitude is between $|A_1 - A_2|$ and $A_1 + A_2$.

- When two SHM are mutually perpendicular to each other.

  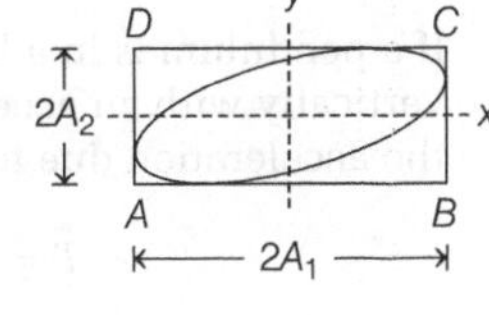

  The resultant SHM is given by

  $$\frac{x^2}{A_1^2} + \frac{y^2}{A_2^2} - \frac{2xy\cos\phi}{A_1 A_2} = \sin^2\phi \text{ (ellipse)}$$

  where, $x = A_1 \sin \omega t$ and $y = A_2 \sin(\omega t + \phi)$

  Here, x is always between $-A_1$ to $+A_1$ and y is always between $-A_2$ to $+A_2$.

NOTE **Special Cases in Composition of Two SHMs**

- When $\phi = 0$, $y = \frac{A_2}{A_1}x$

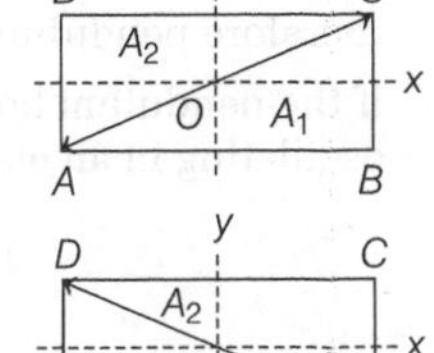

- When $\phi = \pi$, $y = -\frac{A_2}{A_1}x$

- When $\phi = \pi/2$. If $A_1 = A_2 = A$,

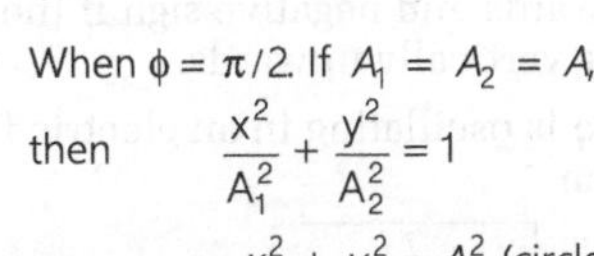

  then $\frac{x^2}{A_1^2} + \frac{y^2}{A_2^2} = 1$

  $x^2 + y^2 = A^2$ (circle)

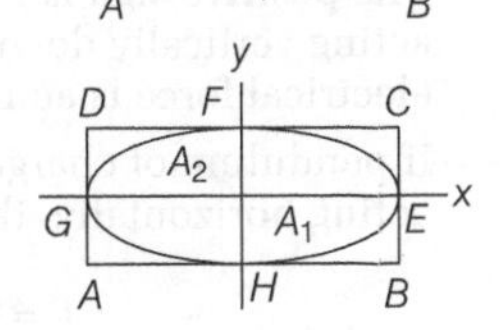

Simple Pendulum

A simple pendulum, in practice, consists of a heavy but small sized metallic bob suspended by a light, inextensible and flexible string. The motion of a simple pendulum is simple harmonic for very small angular displacement (θ) whose time period and frequency are given by

$$T = 2\pi\sqrt{\frac{l}{g}} \quad \text{and} \quad \nu = \frac{1}{2\pi}\sqrt{\frac{g}{l}}$$

where, l is the effective length of the string and g is acceleration due to gravity.

- If a pendulum of length l at temperature $\theta°C$ has a time period T, then on increasing the temperature by $\Delta\theta°C$ its time period changes to ΔT,

  where, $$\frac{\Delta T}{T} = \frac{1}{2}\alpha\,\Delta\theta$$

  where, α is the temperature coefficient of expansion of the string.
- A second's pendulum is a pendulum whose time period is 2s. At a place where $g = 9.8\ \text{ms}^{-2}$, the length of a second's pendulum is 0.9929 m (or 1 m approx).
- If the bob of a pendulum (having density ρ) is made to oscillate in a non-viscous fluid of density σ, then it can be shown that the new period is

$$T = 2\pi\sqrt{\frac{l}{g\left(1 - \frac{\sigma}{\rho}\right)}}$$

- If a pendulum is in a lift or in some other carriage moving vertically with an acceleration a, then the effective value of the acceleration due to gravity becomes $(g \pm a)$ and hence,

$$T = 2\pi\sqrt{\frac{l}{(g \pm a)}}$$

  Here, positive sign is taken for an upward accelerated motion and negative sign for a downward accelerated motion.
- If a pendulum is made to oscillate in a freely falling lift or an orbiting satellite then the effective value of g is zero and hence, the time period of the pendulum will be infinity and therefore pendulum will not oscillate at all.
- If the pendulum bob of mass m has a charge q and is oscillating in an electrical field E, then

$$T = 2\pi\sqrt{\frac{l}{\left(g \pm \frac{qE}{m}\right)}}$$

  The positive sign is to be used if the electrical force is acting vertically downwards and negative sign if the electrical force is acting vertically upwards.
- If pendulum of charge q is oscillating in an electric field E acting horizontally, then

$$T = 2\pi\sqrt{\frac{l}{\sqrt{g^2 + \frac{q^2E^2}{m^2}}}}$$

- If the length of a simple pendulum is increased to such an extent that $l \to \infty$, then its time period is

$$T = 2\pi\sqrt{\frac{R}{g}} = 84.6 \text{ min}$$

  where, R = radius of the earth.
- The graphs l-T and l-T^2 intersect at $T = 1$ s.

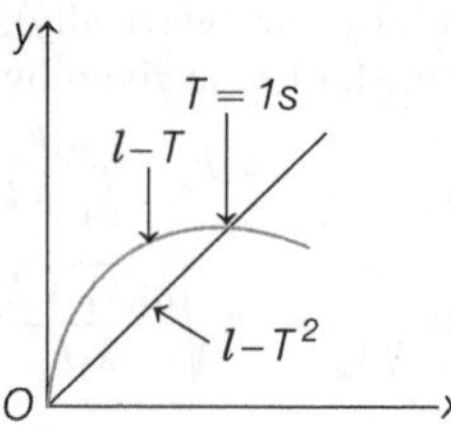

- The graph between T^2 and $1/g$ is a straight line.

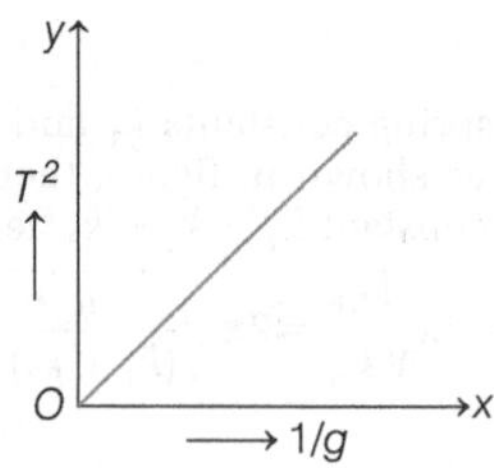

- The graph between T^2 and g is a rectangular hyperbola.

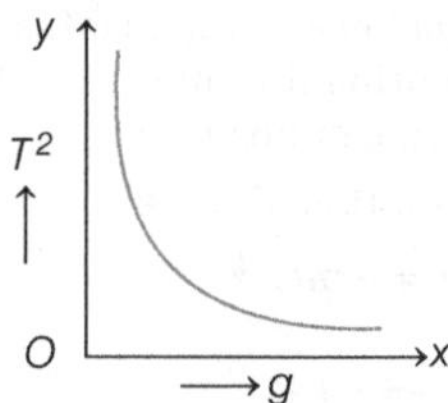

Free, Damped, Forced and Resonant Vibrations

Some of the vibrations are described below.

Free Vibrations

If a body, capable of oscillating, is slightly displaced from its position of equilibrium and then released, it starts oscillating with a frequency of its own.

Such oscillations are called free vibrations.The frequency with which a body oscillates is called the **natural frequency** and is given by

$$\nu_0 = \frac{1}{2\pi}\sqrt{\frac{k}{m}}$$

Here, a body continues to oscillate with a constant amplitude and a fixed frequency.

Damped Vibrations

The oscillations in which the amplitude decreases gradually with the passage of time are called damped vibrations.
Damping force, $F_d = -bv$
where, v is the velocity of the oscillator and b is a damping constant. The displacement of the oscillator is given by

$$x(t) = Ae^{-bt/2m}\sin(\omega' t + \phi)$$

where, ω' = the angular frequency $= \sqrt{\frac{k}{m} - \frac{b^2}{4m^2}}$

The mechanical energy E of the oscillator is given by $E(t) = \frac{1}{2}kA^2e^{-bt/m}$

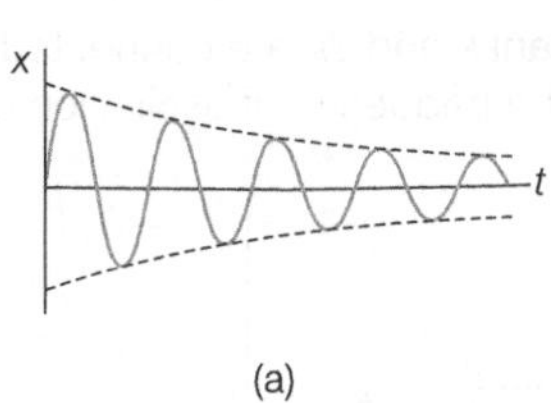

(a)

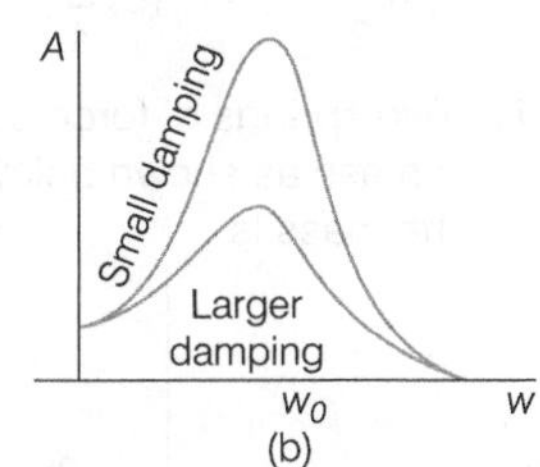

(b)

Forced Vibrations

The vibrations in which a body oscillates under the effect of an external periodic force, whose frequency is different from the natural frequency of the oscillating body, are called forced vibrations.

In forced vibrations the oscillating body vibrates with the frequency of the external force and amplitude of oscillations is generally small.

Resonant Vibrations

It is a special case of forced vibrations in which the frequency of external force is exactly same as the natural frequency of the oscillator.

As a result, the oscillating body begins to vibrate with a large amplitude leading to the phenomenon of resonance to occur. Resonant vibrations play a very important role in music and in tuning of station/channel in a radio/TV etc.

DAY PRACTICE SESSION 1

FOUNDATION QUESTIONS EXERCISE

1 The displacement of a particle is represented by the equation $y = 3\cos\left(\frac{\pi}{4} - 2\omega t\right)$. The motion of the particle is
(a) simple harmonic with period $2\pi/\omega$
(b) simple harmonic with period π/ω
(c) periodic but not simple harmonic
(d) non-periodic

2 The displacement of a particle is represented by the equation $y = \sin^3\omega t$. The motion is
(a) non-periodic
(b) periodic but not simple harmonic
(c) simple harmonic with period $2\pi/\omega$
(d) simple harmonic with period π/ω

3 Motion of an oscillating liquid column in a U-tube is
(a) periodic but not simple harmonic
(b) non-periodic
(c) simple harmonic and time period is independent of the density of the liquid
(d) simple harmonic and time period is directly proportional to the density of the liquid

4 The relation between acceleration and displacement of four particles are given below. Which one of the particle is exempting simple harmonic motion?
(a) $a_x = +2x$ (b) $a_x = +2x^2$
(c) $a_x = -2x^2$ (d) $a_x = -2x$

5 A wave travelling along the x-axis is described by the equation $y(x,t) = 0.005\cos(\alpha x - \beta t)$. If the wavelength and the time period of the wave are 0.08 m and 2.0 s, respectively, then α and β in appropriate units are
(a) $\alpha = 25.00\pi, \beta = \pi$ (b) $\alpha = \frac{0.08}{\pi}, \beta = \frac{2.0}{\pi}$
(c) $\alpha = \frac{0.04}{\pi}, \beta = \frac{1.0}{\pi}$ (d) $\alpha = 12.50\pi, \beta = \frac{\pi}{2.0}$

6 The maximum velocity of a particle executing simple harmonic motion with an amplitude 7 mm, is 4.4 ms^{-1}. The period of oscillation is
(a) 0.01 s (b) 10 s (c) 0.1 s (d) 100 s

7 A point mass oscillates along the x-axis according to the law $x = x_0\cos(\omega t - \pi/4)$. If the acceleration of the particle is written as $a = A\cos(\omega t + \delta)$, then
(a) $A = x_0, \delta = -\frac{\pi}{4}$ (b) $A = x_0\omega^2, \delta = \frac{\pi}{4}$
(c) $A = x_0\omega^2, \delta = -\frac{\pi}{4}$ (d) $A = x_0\omega^2, \delta = \frac{3\pi}{4}$

8 A body is executing SHM when its displacement from the mean position are 4 cm and 5 cm and it has velocity 10 cms^{-1} and 8 cms^{-1}, respectively. Its periodic time t

(a) $\frac{2\pi}{2}$ s (b) π s (c) $\frac{3\pi}{2}$ s (d) 2π s

9 A block rests on a horizontal table, which is executing SHM in the horizontal direction with an amplitude a. If the coefficient of friction is μ, then the block just starts to slip when the frequency of oscillation is

(a) $\frac{1}{2\pi}\sqrt{\frac{\mu g}{a}}$ (b) $2\pi\sqrt{\frac{a}{\mu g}}$ (c) $\frac{1}{2\pi}\sqrt{\frac{a}{\mu g}}$ (d) $\sqrt{\frac{a}{\mu g}}$

10 A coin is placed on a horizontal platform, which undergoes horizontal SHM about a mean position O. The coin placed on the platform does not slip, when angular frequency of the SHM is ω. The coefficient of friction between the coin and platform is μ. The amplitude of oscillation is gradually increased. The coin will be begin to slip on the platform for the first time

(a) at the mean position
(b) at the extreme position of the oscillation
(c) for an amplitude of $\mu g/\omega^2$
(d) for an amplitude of $g/\mu\omega^2$

11 Two particles A and B are oscillating about a point O along a common line such that equation of A is given as $x_1 = a \cos \omega t$ and equation of B is given as $x_2 = b \sin\left(\omega t + \frac{\pi}{2}\right)$

Then, the motion of A w.r.t. B is

(a) a simple harmonic motion with amplitude $(a - b)$
(b) a simple harmonic motion with amplitude $(a + b)$
(c) a simple harmonic motion with amplitude $\sqrt{a^2 + b^2}$
(d) not a simple harmonic motion but oscillatory motion

12 Two particles execute simple harmonic motion on same straight line with same mean position, same time period 6 s and same amplitude 5 cm. Both the particles start SHM from their mean position (in same direction) with a time gap of 1 s. Find the maximum separation between the two particles during their motion.

(a) 2 cm (b) 3 cm (c) 4 cm (d) 5 cm

13 A particle is acted simultaneously by mutually perpendicular simple harmonic motion $x = a \cos \omega t$ and $y = a \sin \omega t$. The trajectory of motion of the particle will be

(a) an ellipse (b) a parabola
(c) a circle (d) a straight line

14 A silver atom in a solid oscillates in simple harmonic motion in some direction with a frequency of 10^{12} per second. What is the force constant of the bonds connecting one atom with the other? (Take, molecular weight of silver = 108 and Avogadro number = 6.02×10^{23} g mol^{-1}) → JEE Main 2018

(a) 6.4 N/m (b) 7.1 N/m (c) 2.2 N/m (d) 5.5 N/m

15 If a spring of stiffness k is cut into two parts A and B of length $l_A : l_B = 2:3$, then the stiffness of spring A is given by → AIEEE 2011

(a) $\frac{5}{2}k$ (b) $\frac{3k}{5}$ (c) $\frac{2k}{5}$ (d) k

16 Two springs of force constants k_1 and k_2, are connected to a mass m as shown. The frequency of oscillation of the mass is ν. If both k_1 and k_2 are made four times their original values, the frequency of oscillation becomes

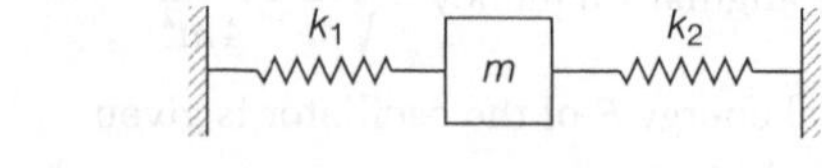

(a) $\frac{\nu}{2}$ (b) $\frac{\nu}{4}$ (c) 4ν (d) 2ν

17 Two springs of force constant k and $2k$ are connected to a mass as shown below. The frequency of oscillation of the mass is

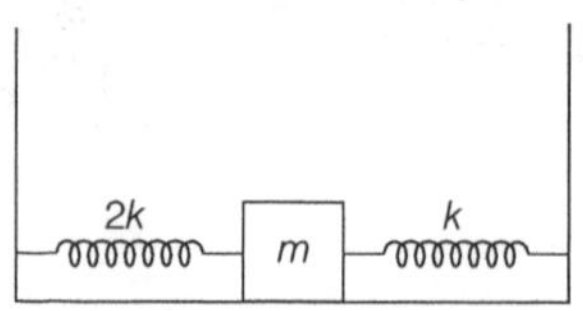

(a) $\frac{1}{2\pi}\sqrt{\frac{k}{m}}$ (b) $\frac{1}{2\pi}\sqrt{\frac{2k}{m}}$ (c) $\frac{1}{2\pi}\sqrt{\frac{3k}{m}}$ (d) $\frac{1}{2\pi}\sqrt{\frac{m}{k}}$

18 A block P of mass m is placed on a horizontal frictionless plane. A second block Q of the same mass m is placed on it and is connected to a spring of spring constant k, the two blocks are pulled by a distance A. Block Q oscillates without slipping. What is the maximum value of frictional force between the two blocks?

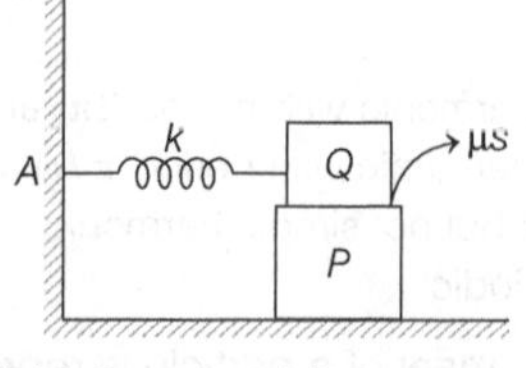

(a) $kA/2$ (b) kA (c) $\mu - mg$ (d) Zero

19 A particle of mass M is attached to three springs A, B and C having equal force constant k. If the particle is pushed a little towards any one of the springs and then left on its own, find the time period of its oscillation.

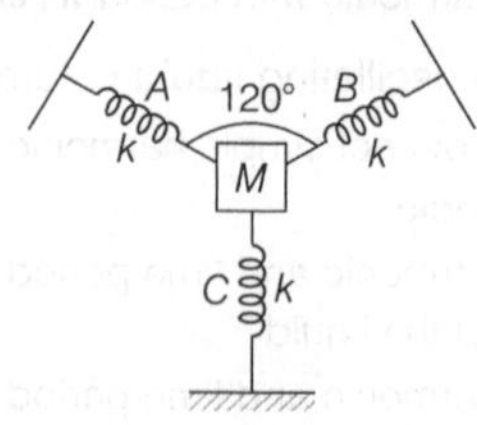

(a) $2\pi\sqrt{(M/k)}$ (b) $2\pi\sqrt{(2M/k)}$
(c) $2\pi\sqrt{(M/2k)}$ (d) $2\pi\sqrt{(M/3k)}$

20 A body performs SHM. Its kinetic energy K varies with time T as indicated in the graph

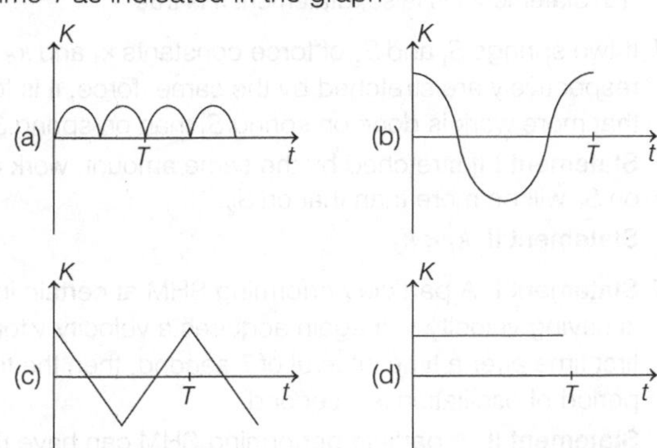

21 A particle is executing simple harmonic motion with a time period T. At time $t = 0$, it is at its position of equilibrium. The kinetic energy-time graph of the particle will look, like → JEE Main 2017 (Offline)

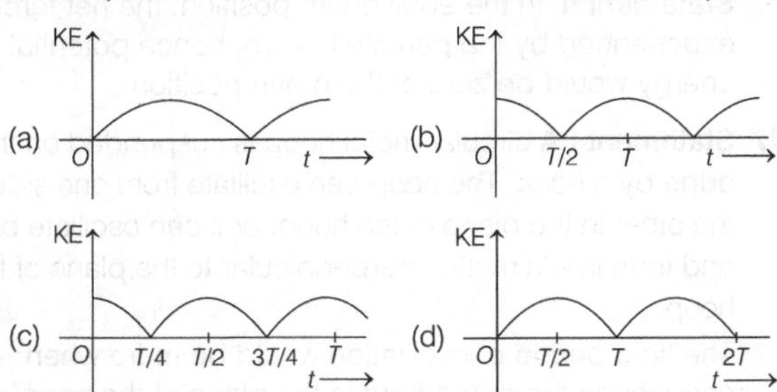

22 For a simple pendulum, a graph is plotted between its Kinetic Energy (KE) and Potential Energy (PE) against its displacement d. Which one of the following represents these correctly? (graphs are schematic and not drawn to scale) → JEE Main 2015

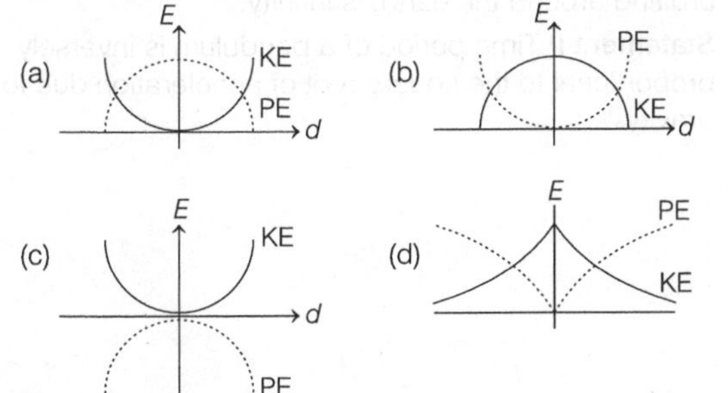

23 The total energy of a particle, executing simple harmonic motion is

(a) $\propto x$
(b) $\propto x^2$
(c) independent of x
(d) $\propto x^{1/2}$
where, x is the displacement from the mean position.

24 For a particle executing SHM, the displacement x is given by $x = A\cos\omega t$. Identify the graph which represents the variation of potential energy (PE) as a function of time t and displacement x.

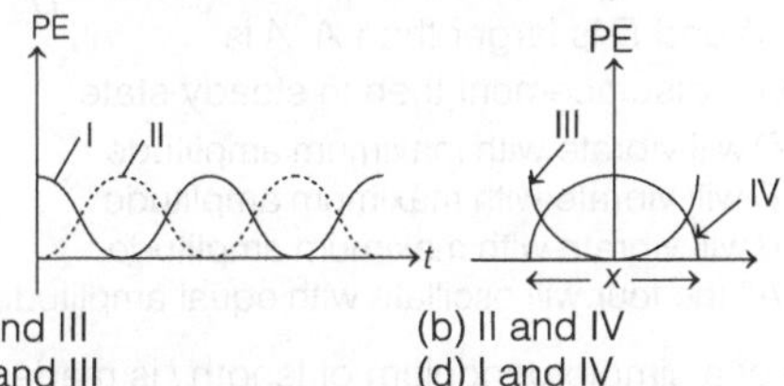

(a) I and III (b) II and IV
(c) II and III (d) I and IV

25 A simple pendulum performs simple harmonic motion about $x = 0$ with an amplitude a, and time period T. The speed of the pendulum at $x = a/2$ will be

(a) $\frac{\pi a\sqrt{3}}{T}$ (b) $\frac{\pi a\sqrt{3}}{2T}$ (c) $\frac{3\pi^2 a}{T}$ (d) $\frac{\pi a}{T}$

26 The value of g decrease by 0.1% on a mountain as compared to sea level. If a simple pendulum is used to record the time, then the length must be

(a) increased by 0.1% (b) decreased by 0.1%
(c) increased by 0.2% (d) decreased by 0.2%

27 Two pendulums have time periods T and $\frac{5T}{4}$. They start SHM at the same time from the mean position. What will be the phase difference between them after the bigger pendulum completes one oscillation?

(a) 45° (b) 90° (c) 60° (d) 30°

28 A simple pendulum of length l is suspended from the roof of a train which is moving in a horizontal direction with an acceleration a. Then, the time period T is given by

(a) $2\pi\sqrt{l/g}$ (b) $2\pi\sqrt{l/(a^2+g^2)^{1/2}}$
(c) $2\pi\sqrt{l/(a+g)}$ (d) $2\pi\sqrt{l/(g-a)}$

29 Two simple pendulums of length 1 m and 4 m respectively are both given small displacement in the same direction. The shorter pendulum has completed number of oscillations equal to → JEE Main (Online) 2013

(a) 2 (b) 7
(c) 5 (d) 3

30 A pendulum of length 2m lift at P. When it reaches Q, it losses 10% of its total energy due to air resistance. The velocity of Q is

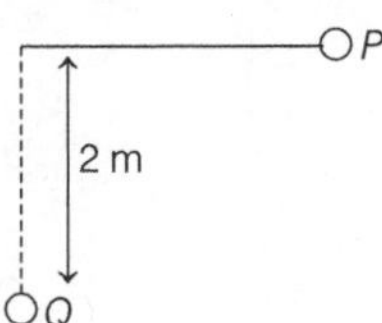

(a) 2 m/s (b) 1 m/s
(c) 6 m/s (d) 8 m/s

31 Four pendulums A, B, C and D are hung from the same elastic support as shown alongside. A and C are of the same length while B is smaller than A and D is larger than A. A is given a displacement then in steady state

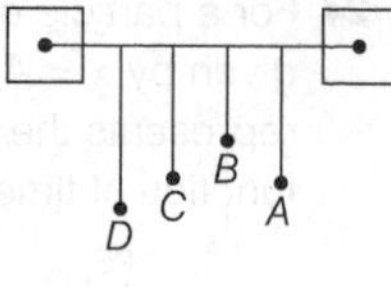

(a) D will vibrate with maximum amplitude
(b) C will vibrate with maximum amplitude
(c) B will vibrate with maximum amplitude
(d) All the four will oscillate with equal amplitude

32 Bob of a simple pendulum of length l is made of iron. The pendulum is oscillating over a horizontal coil carrying direct current. If the time period of the pendulum is T, then → JEE Main (Online) 2013

(a) $T < 2\pi\sqrt{\frac{l}{g}}$ and damping is smaller than in air alone

(b) $T = 2\pi\sqrt{\frac{l}{g}}$ and damping is larger than in air alone

(c) $T > 2\sqrt{\frac{l}{g}}$ and damping is smaller than in air alone

(d) $T < 2\pi\sqrt{\frac{l}{g}}$ and damping is larger than in air alone

33 The amplitude of a damped oscillator decreases to 0.9 times its original magnitude is 5s. In another 10 s it will decreases to α times its original magnitude, where α equals → JEE Main 2013

(a) 0.7 (b) 0.81
(c) 0.729 (d) 0.6

Direction (Q. Nos. 34-38) *Each of these questions contains two statements : Statement I and Statement II. Each of these questions also has four alternative choices, only one of which is the correct answer. You have to select one of the codes (a), (b), (c) and (d) given below.*

(a) Statement I is true, Statement II is true; Statement II is the correct explanation for Statement I
(b) Statement I is true, Statement II is true; Statement II is not the correct explanation for Statement I
(c) Statement I is true; Statement II is false
(d) Statement I is false; Statement II is true

34 If two springs S_1 and S_2 of force constants k_1 and k_2, respectively are stretched by the same force, it is found that more work is done on spring S_1 than on spring S_2.

Statement I If stretched by the same amount, work done on S_1, will be more than that on S_2.

Statement II $k_1 < k_2$

35 **Statement I** A particle performing SHM at certain instant is having velocity v. It again acquires a velocity v for the first time after a time interval of T second, then the time period of oscillation is T second.

Statement II A particle performing SHM can have the same velocity at two instants in one cycle.

36 **Statement I** A particle performing SHM while crossing the mean position is having a minimum potential energy, this minimum potential energy could be non-zero.

Statement II In the equilibrium position, the net force experienced by the particle is zero, hence potential energy would be zero at the mean position.

37 **Statement I** A circular metal hoop is suspended on the edge by a hook. The hoop can oscillate from one side to the other in the plane of the hoop, or it can oscillate back and forth in a direction perpendicular to the plane of the hoop.

The time period of oscillation would be more when oscillations are carried out in the plane of the hoop.

Statement II Time period of physical pendulum is more if the moment of inertia of the rigid body about the corresponding axis, passing through the pivoted point is more.

38 **Statement I** The time period of a pendulum, in a satellite orbiting around the earth, is infinity.

Statement II Time period of a pendulum is inversely proportional to the square root of acceleration due to gravity.

DAY PRACTICE SESSION 2

PROGRESSIVE QUESTIONS EXERCISE

1 A 15 g ball is shot from a spring gun whose spring has a force constant of 600 Nm^{-1}. The spring is compressed by 5 cm. The greatest possible horizontal range of the ball for this compression is ($g = 10 ms^{-2}$).

(a) 10.0 m (b) 6.0 m (c) 12.0 m (d) 8.0 m

2 Two simple harmonic motions are represented by the equations $y_1 = 0.1 \sin\left(100\pi t + \frac{\pi}{3}\right)$ and $y_2 = 0.1 \cos \pi t$. The phase difference of the velocity of particle 1, with respect to the velocity of particle 2 is (at $t = 0$)

(a) $\frac{-\pi}{6}$ (b) $\frac{\pi}{3}$ (c) $\frac{-\pi}{3}$ (d) $\frac{\pi}{6}$

3 A piece of wood has dimension $a \times b \times c$. It is floating in a liquid of density ρ such that side a is vertical. It is now pushed down gently and released. The time period is

(a) $2\pi\sqrt{\rho a/g}$ (b) $2\pi\sqrt{abc/g}$
(c) $2\pi\sqrt{g/\rho a}$ (d) $2\pi\sqrt{bc/\rho g}$

4 The length of a spring is α when a force of 4N is applied on it. The length of a spring is β when a force of 5N is applied on it. Then find the length of the spring when a force of 9N is applied on the spring.

(a) $5\beta - 4\alpha$ (b) $\beta - \alpha$
(c) $5\alpha - 4\beta$ (d) $9(\beta - \alpha)$

5 A simple pendulum of length l has a bob of mass m with a charge q on it. A vertical sheet of charge having surface charge density σ passes through the point of suspension. At equilibrium, the string makes an angle θ with the vertical. If the tension in the string is T then,

(a) $\tan\theta = \frac{\sigma q}{2\varepsilon_0 mg}$ (b) $\tan\theta = \frac{\sigma q}{\varepsilon_0 mg}$
(c) $T > 2\pi\sqrt{\frac{l}{g}}$ (d) $T = 2\pi\sqrt{\frac{l}{g}}$

6 A mass m is suspended from a massless pulley which itself is suspended with the help of a massless extensible spring as shown alongside.

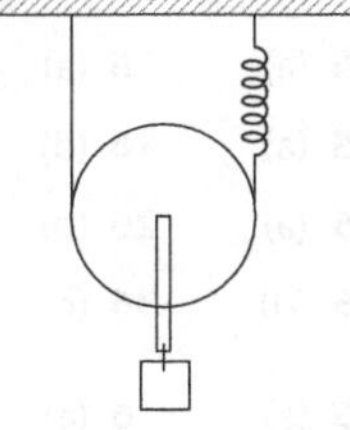

What will be the time period of oscillation of the mass? The force constant of the spring is k.

(a) $\pi\sqrt{m/k}$
(b) $2\pi\sqrt{m/k}$
(c) $4\pi\sqrt{m/k}$
(d) $2\pi\sqrt{m/2k}$

7 If x, v and a denote the displacement, the velocity and the acceleration of a particle executing simple harmonic motion of time period T, then which of the following does not change with time?

(a) $a^2T^2 + 4\pi^2v^2$ (b) $\frac{aT}{x}$
(c) $aT + 2\pi v$ (d) $\frac{aT}{v}$

8 A simple pendulum has time period T_1. The point of suspension is now moved upward according to the relation $y = kt^2$, ($k = 1 ms^{-2}$), where y is the vertical displacement. The time period now becomes T_2. The ratio of $\frac{T_1^2}{T_2^2}$ is (take, $g = 10 ms^{-2}$)

(a) $\frac{6}{5}$ (b) $\frac{5}{6}$ (c) 1 (d) $\frac{4}{5}$

9 A pendulum made of a uniform wire of cross-sectional area A has time period T. When an additional mass M is added to its bob, the time period changes T_M. If the Young's modulus of the material of the wire is Y, then $1/Y$ is equal to (g = gravitational acceleration) **→ JEE Main 2015**

(a) $\left[\left(\frac{T_M}{T}\right)^2 - 1\right]\frac{A}{Mg}$ (b) $\left[\left(\frac{T_M}{T}\right)^2 - 1\right]\frac{Mg}{A}$
(c) $\left[1 - \left(\frac{T_M}{T}\right)^2\right]\frac{A}{Mg}$ (d) $\left[1 - \left(\frac{T}{T_M}\right)^2\right]\frac{A}{Mg}$

10 The bob of a simple pendulum is a spherical hollow ball filled with water. A plugged hole near the bottom of the oscillating bob gets suddenly unplugged. During observation, till the water is coming out, the time period of oscillation would

(a) first increase and then decrease to the original value
(b) first decrease and then increase to the original value
(c) remain unchanged
(d) increase towards a saturation value

11 A pendulum of length $l = 1$ m is released from $\theta_0 = 60°$. The rate of change of speed of the bob at $\theta = 30°$ is (take, $g = 10 m/s^2$).

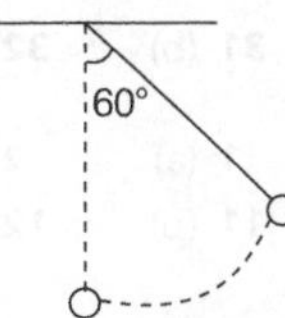

(a) $5\sqrt{3}$ m/s² (b) 5 m/s²
(c) 10 m/s² (d) 2.5 m/s²

12 A particle at the end of a spring executes simple harmonic motion with a period t_1, while the corresponding period for another spring is t_2. If the period of oscillation with the two springs in series is T, Then,

(a) $T = t_1 + t_2$ (b) $T^2 = t_1^2 + t_2^2$
(c) $T^{-1} = t_1^{-1} + t_2^{-1}$ (d) $T^{-2} = t_1^{-2} + t_2^{-2}$

13 A particle performs simple harmonic motion with amplitude A. Its speed is tripled at the instant that it is at a distance $\frac{2}{3}A$ from equilibrium position. The new amplitude of the motion is → JEE Main 2016 (Offline)

(a) $\frac{A}{3}\sqrt{41}$ (b) $3A$
(c) $A\sqrt{3}$ (d) $\frac{7}{3}A$

14 A wooden cube (density of wood d) of side l floats in a liquid of density ρ with its upper and lower surfaces horizontal. If the cube is pushed slightly down and released, it performs simple harmonic motion of period T. Then, T is equal to

(a) $2\pi\sqrt{\frac{l\rho}{(\rho-d)g}}$ (b) $2\pi\sqrt{\frac{ld}{\rho g}}$
(c) $2\pi\sqrt{\frac{l\rho}{dg}}$ (d) $2\pi\sqrt{\frac{ld}{(\rho-d)g}}$

15 Two particles are executing simple harmonic motion of the same amplitude A and frequency ω along the x-axis. Their mean position is separated by distance $X_0 (X_0 > A)$. If the maximum separation between them is $(X_0 + A)$, the phase difference between their motion is

(a) $\frac{\pi}{3}$ (b) $\frac{\pi}{4}$
(c) $\frac{\pi}{6}$ (d) $\frac{\pi}{2}$

16 A particle moves with simple harmonic motion in a straight line. In first τ sec, after starting from rest it travels a distance a and in next τ sec,it travels $2a$, in same direction, then → JEE Main 2014

(a) amplitude of motion is $3a$
(b) time period of oscillations is 8τ
(c) amplitude of motion is $4a$
(d) time period of oscillations is 6τ

17 An ideal gas enclosed in a vertical cylindrical container supports a freely moving piston of mass M. The piston and the cylinder have equal cross sectional area A. When the piston is in equilibrium, the volume of the gas is V_0 and its pressure is P. The piston is slightly displaced from the equilibrium position and released. Assuming that the system is completely, isolated from its surrounding, the piston executes a simple harmonic motion with frequency → JEE Main 2013

(a) $\frac{1}{2\pi}\frac{A\gamma P_0}{V_0 M}$ (b) $\frac{1}{2\pi}\frac{V_0 M P_0}{A^2\gamma}$
(c) $\frac{1}{2\pi}\sqrt{\frac{A^2\gamma P_0}{MV_0}}$ (d) $\frac{1}{2\pi}\sqrt{\frac{MV_0}{A\gamma P_0}}$

18 If a simple pendulum has significant amplitude (up to a factor of $1/e$ of original) only in the period between $t = 0$ s to $t = \tau$ s, then τ may be called the average life of the pendulum. When the spherical bob of the pendulum suffers a retardation (due to viscous drag) proportional to its velocity with b as the constant of proportionality, the average life time of the pendulum is (assuming damping is small) in seconds

(a) $\frac{0.693}{b}$ (b) b
(c) $\frac{1}{b}$ (d) $\frac{2}{b}$

ANSWERS

SESSION 1

1 (b)	**2** (b)	**3** (c)	**4** (d)	**5** (a)	**6** (a)	**7** (d)	**8** (b)	**9** (a)	**10** (c)
11 (a)	**12** (d)	**13** (c)	**14** (b)	**15** (a)	**16** (d)	**17** (c)	**18** (a)	**19** (c)	**20** (a)
21 (c)	**22** (b)	**23** (c)	**24** (a)	**25** (a)	**26** (b)	**27** (b)	**28** (b)	**29** (a)	**30** (c)
31 (b)	**32** (d)	**33** (c)	**34** (d)	**35** (d)	**36** (c)	**37** (a)	**38** (a)		

SESSION 2

1 (a)	**2** (a)	**3** (a)	**4** (a)	**5** (a)	**6** (a)	**7** (b)	**8** (a)	**9** (a)	**10** (a)
11 (b)	**12** (b)	**13** (d)	**14** (b)	**15** (a)	**16** (d)	**17** (c)	**18** (d)		

Hints and Explanations

SESSION 1

1 Given, $y = 3\cos\left(\frac{\pi}{4} - 2\omega t\right)$ ...(i)

Velocity, $v = \frac{dy}{dt}$

$$= 3 \times 2\,\omega \sin\left(\frac{\pi}{4} - 2\omega t\right)$$

Acceleration, $a = \frac{dv}{dt}$

$$= -4\omega^2 \times 3\cos\left(\frac{\pi}{4} - 2\,\omega t\right)$$

$$= -4\,\omega^2 y$$

As $a \propto y$ and negative sign shows that, it is directed towards equilibrium (or mean position), hence particle will execute SHM.

Comparing Eq. (i) with equation

$$y = r\cos(\phi - \omega' t)$$

We have, $\omega' = 2\omega$

or $\frac{2\pi}{T'} = 2\omega$ or $T' = \frac{\pi}{\omega}$

2 Given equation of motion is

$$y = \sin^3 \omega t$$

$$= (3\sin\omega t - \sin 3\omega t)/4$$

$$[\because \sin 3\theta = 3\sin\theta - 4\sin^3\theta]$$

$$\Rightarrow \frac{dy}{dt} = \left[\frac{d}{dt}(3\sin\omega t) - \frac{d}{dt}(\sin 3\omega t)\right]/4$$

$$\Rightarrow 4\frac{dy}{dt} = 3\omega\cos\omega t - [3\omega\cos 3\omega t]$$

$$\Rightarrow 4 \times \frac{d^2y}{dt^2} = -3\omega^2\sin\omega t + 9\omega^2\sin 3\omega t$$

$$\Rightarrow \frac{d^2y}{dt^2} = -\left(\frac{3\omega^2\sin\omega t - 9\omega^2\sin 3\omega t}{4}\right)$$

$\Rightarrow \frac{d^2y}{dt^2}$ is not proportional to y.

Hence, motion is not SHM.

As the expression is involving sine function, hence it will be periodic.

3 The motion of an oscillating liquid column in a U-tube is simple harmonic and the time period is independent of the density of the liquid.

$$T = 2\,\pi\sqrt{\frac{h}{g}}$$

where, h = height of liquid in each column.

4 For motion to be SHM acceleration of the particle must be proportional to negative of displacement.

i.e. $a \propto -(y \text{ or } x)$

We should be clear that y has to be linear.

5 Given, $y = 0.005\cos(\alpha x - \beta t)$

Comparing the equation with the standard form,

$$y = A\cos\left(\frac{x}{\lambda} - \frac{t}{T}\right)2\,\pi$$

we have, $2\pi/\lambda = \alpha$

and $2\pi/T = \beta$

$\Rightarrow$ $\alpha = 2\,\pi/0.08 = 25.00\,\pi$

and $\beta = \pi$

6 Maximum velocity $v = A\omega$,

(where, A is the amplitude and ω is the angular frequency of oscillation).

$\therefore$ $4.4 = (7 \times 10^{-3}) \times 2\,\pi/T$

or $T = \frac{7 \times 10^{-3}}{4.4} \times \frac{2 \times 22}{7} = 0.01$ s

7 Given, $x = x_0\cos\left(\omega\, t - \frac{\pi}{4}\right)$

Acceleration, $a = \frac{d^2x}{dt^2}$

$$= -\omega^2 x_0\cos\left(\omega t - \frac{\pi}{4}\right)$$

$$= \omega^2 x_0\cos\left(\omega t + \frac{3\pi}{4}\right)$$

So, $A = \omega^2 x_0$ and $\delta = \frac{3\pi}{4}$

8 Using $v^2 = \omega^2(a^2 - y^2)$, we have

$$10^2 = \omega^2(a^2 - 4^2)$$

and $8^2 = \omega^2(a^2 - 5^2)$

So, $10^2 - 8^2 = \omega^2(5^2 - 4^2) = (3\omega)^2$

$\Rightarrow$ $6 = 3\omega$ or $\omega = 2$

$\therefore$ $T = 2\,\pi/\omega$

$$= 2\pi/2 = \pi\ \text{s}$$

9 Force of friction $= \mu\, mg = m\omega^2 a$

$$= m\,(2\,\pi\nu)^2 a$$

$\Rightarrow$ $\nu = \frac{1}{2\pi}\sqrt{\frac{\mu\, g}{a}}$

10 Let O be the mean position and x be the distance of the coin from O. The coin will slip, if centrifugal force on the coin just becomes equal to the force of friction i.e.

$$m\,x\omega^2 = \mu\, mg$$

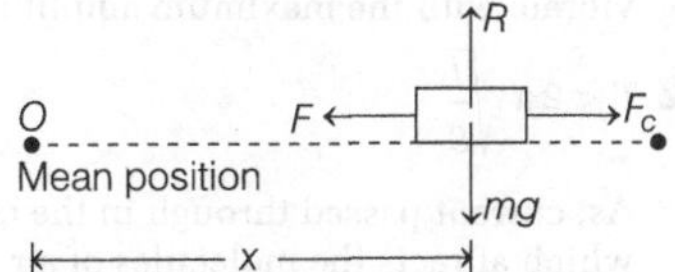

From the diagram,

$m\,A\omega^2 = \mu\, mg$ or $A = \mu g/\omega^2$

11 The displacement of A relative to B is

$x = x_1 - x_2$

$$x = a\cos\omega t - b\sin\left(\omega t + \frac{\pi}{2}\right)$$

$$= a\cos\omega t - b\cos\omega t$$

$$= (a - b)\cos\omega t$$

Which is a simple harmonic motion with amplitude $(a - b)$.

12 Phase difference,

$$\phi = \omega t = \frac{2\pi}{6} \times 1 = \frac{\pi}{3}\ \text{rad}$$

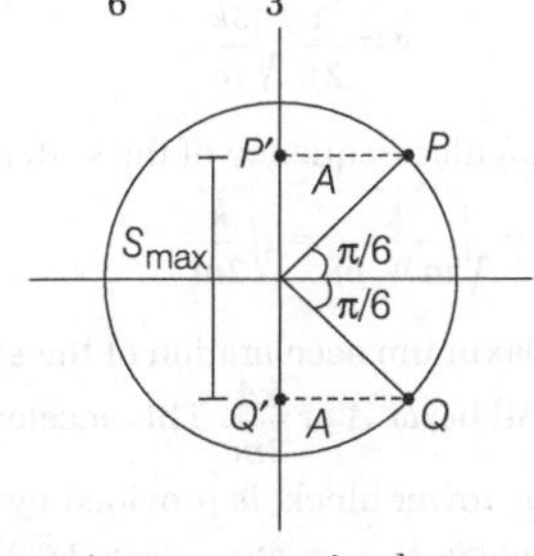

The maximum separation between the two particles is

$$S_{max} = 2A\sin\frac{\pi}{6}$$

or $S_{max} = 2 \times 5 \times \frac{1}{2} = 5$ cm

13 Given, $x = a\cos\omega t$ ...(i)

$Y = a\sin\omega t$ ...(ii)

Squaring and adding Eqs.(i) and (ii), we get

$$x^2 + y^2 = a^2(\cos^2\omega t + \sin^2\omega t)$$

$$= a^2$$

$[\because \cos^2\omega t + \sin^2\omega t = 1]$

This is the equation of a circle.

Clearly, the locus is a circle of constant radius a.

14 For a harmonic oscillator,

$$T = 2\pi\sqrt{\frac{m}{k}}$$

where, k = force constant and $T = \frac{1}{\nu}$

$\therefore$ $k = 4\pi^2\nu^2 m$

$$= 4 \times \left(\frac{22}{7}\right)^2 \times (10^{12})^2 \times \frac{108 \times 10^{-3}}{6.02 \times 10^{23}}$$

$\Rightarrow k = 7.1$ N/m

15 For spring, $k \propto \frac{1}{l}$

$$\therefore \frac{k_A}{k_B} = \frac{l_B}{l_A} \Rightarrow k_A = \frac{l_A + l_B}{l_A}k = \frac{5}{2}k$$

16 We know that, $\nu = \frac{1}{2\pi}\sqrt{\frac{k_1 + k_2}{m}}$

When k_1 and k_2 are made four times their original value.

Then, and

$$\nu' = \frac{1}{2\pi}\cdot 2\sqrt{\frac{k_1 + k_2}{m}} = 2\nu$$

17 The effective spring constant is $K = k + 2k = 3k$.
The time period of oscillation is given by

$$T = 2\pi\sqrt{\frac{m}{3k}} \text{ and } \nu = \frac{1}{T}$$

so, we get

$$\nu = \frac{1}{2\pi}\sqrt{\frac{3k}{m}}$$

18 Angular frequency of the system,

$$\omega = \sqrt{\frac{k}{m+m}} = \sqrt{\frac{k}{2m}}$$

Maximum acceleration of the system will be, $\omega^2 A$ or $\frac{kA}{2m}$. This acceleration of the lower block, is provided by friction.
Hence, $f_{max} = ma_{max} = m\,\omega^2 A$

$$= m\left(\frac{kA}{2m}\right) = \frac{kA}{2}$$

19 When the mass m is pushed in a downward direction through a distance x, the effective restoring force, in magnitude is
$F = kx + kx\cos 60° + kx\cos 60°$
$= 2kx$
$\therefore$ Spring factor, $k' = 2k$
and Inertia factor $= M$
So time period, $T = 2\pi\sqrt{\frac{M}{2k}}$

20 The frequency of kinetic energy is twice that of a particle executive SHM.

21 KE is maximum at mean position and minimum at extreme position $\left(\text{at } t = \frac{T}{4}\right)$.

22 During oscillation, motion of a simple pendulum KE is maximum at the mean position where PE is minimum. At extreme position, KE is minimum and PE is maximum. Thus, correct graph is depicted in option (b).

23 In a simple harmonic motion, when a particle is displaced to a position from its mean position, its kinetic energy is converted into potential energy. Hence, total energy of a particle remains constant or the total energy in simple harmonic motion does not depend on the displacement x.

24 Potential energy is minimum (in this case zero) at the mean position ($x = 0$) and maximum at the extreme positions ($x = \pm A$).
At time $t = 0, x = A$, the potential energy should be maximum. Therefore, graph I is correct. Further in graph III, potential energy is minimum at $x = 0$. Hence, this is also correct.

25 Since, $v = \omega\sqrt{a^2 - y^2}$,
At, x or $y = a/2$

$$\Rightarrow \quad v = \omega\sqrt{a^2 - \frac{a^2}{4}} = \omega\sqrt{\frac{3a^2}{4}}$$

$$= \frac{2\pi}{T} \times \frac{\sqrt{3}\,a}{2} = \frac{\pi\sqrt{3}\,a}{T}$$

26 As, $T = 2\pi\sqrt{l/g}$

Taking log and differentiating the expression, keeping T constant we have

$$\frac{dl}{l} = \frac{dg}{g} = -\frac{0.1}{100}$$

$$\therefore \quad (dl/l)\times 100 = -0.1/100 \times 100$$
$$= -0.1\%$$

27 When bigger pendulum of time period ($5T/4$) completes one oscillation, the smaller pendulum will complete (5/4) oscillation. It means, the smaller pendulum will be leading the bigger pendulum by a phase of $T/4 = \pi/2$ rad $= 90°$.

28 Effective acceleration $= \sqrt{a^2 + g^2}$

$$\therefore \text{Time period, } T = 2\pi\sqrt{\frac{l}{(a^2 + g^2)^{1/2}}}$$

29 Let T_1 and T_2 be the time period of shorter length and larger length pendulums respectively. According to question,

$$nT_1 = (n-1)T_2$$

$$\text{So, } n\,2\pi\sqrt{\frac{1}{8}} = (n-1)2\pi\sqrt{\frac{4}{8}}$$

or $n = (n-1)2 = 2n - 2 \Rightarrow n = 2$

30 By applying conservation of energy between P and Q
$\frac{1}{2}mv^2 = 0.9(mgh)$
$\Rightarrow v^2 = 2 \times 0.9 \times 10 \times 2 = 36 \Rightarrow v = 6$ m/s

31 As A and C are of same length, so they will be in resonance, hence C will vibrate with the maximum amplitude.

32 $T < 2\pi\sqrt{\frac{l}{g}}$

As, current passed through in the coil which attracts the molecules of air closer to it, thus density of air increases which produces larger damping than that in air alone.

33 Amplitude of damped oscillator,
$A = A_0 e^{-\frac{bt}{2m}}$

After 5 s, $\quad 0.9A_0 = A_0 e^{-\frac{b(5)}{2m}}$

$$\Rightarrow \quad 0.9 = e^{-\frac{b(5)}{2m}} \quad \ldots(\text{i})$$

After 10 more second,

$$A = A_0 e^{-\frac{b(15)}{2m}} = A_0\left(e^{-\frac{5b}{2m}}\right)^3 \quad \ldots(\text{ii})$$

From Eqs. (i) and (ii), we get
$A = 0.729A_0$
Hence, $\quad \alpha = 0.729$

34 As no relation between k_1 and k_2 is given in the question, that is why, nothing can be predicted about Statement I. But as in Statement II, $k_1 < k_2$
Then, for same force

$$W = F\cdot x = F\cdot\frac{F}{K} = \frac{F^2}{K}$$

$$\Rightarrow \quad W \propto \frac{1}{k}$$

i.e. $\quad W_1 > W_2$
But for same displacement,
$W = F\cdot x = \frac{1}{2}kx\cdot x = \frac{1}{2}kx^2$
$\Rightarrow W \propto k$, i.e. $W_1 < W_2$
Thus, in the light of Statement II, Statement I is false.

35 Consider the situation as shown in the adjoint figure. Let us say at any instant t_1, the particle crosses A as shown, the particle again acquires the same velocity, when it crosses B let us say at instant t_2. According to statement I, $(t_2 - t_1)$ is the time period of SHM which is wrong.

$|\leftarrow x_1 \rightarrow|\leftarrow x_1 \rightarrow|$
B Equilibrium position A

36 At the mean position,

$$F = 0 = -\frac{dU}{dx} = 0$$

$\Rightarrow U =$ constant which can be zero or non-zero.

37 When the hoop oscillates in its plane, moment of inertia is

$$I_1 = mR^2 + mR^2 \text{ i.e. } I_1 = 2mR^2$$

While when the hoop oscillates in a direction perpendicular to the plane of the hoop, moment of inertia is

$$I_2 = \frac{mR^2}{2} + mR^2 = \frac{3mR^2}{2}$$

Time period of physical pendulum is,
$T = 2\pi\sqrt{\frac{I}{mgd}}$; d is same in both the cases.

38 From the relation of the time period,

$$T = 2\pi\sqrt{\frac{l}{g}} \Rightarrow T \propto \frac{1}{\sqrt{g}}$$

When the satellite is orbiting around the earth, the value of g inside it is zero. Hence, the time period of pendulum in a satellite will be infinity and it is also clear that time period of pendulum is inversely proportional to square root of acceleration due to gravity g.

SESSION 2

1 For getting horizontal range, there must be some inclination of spring with ground to project ball.

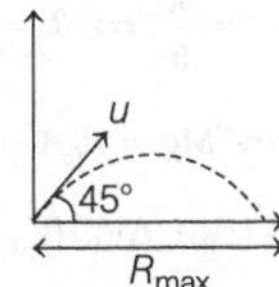

$$R_{max} = \frac{u^2}{g}$$

But KE acquired by ball = PE of spring gun

$$\Rightarrow \frac{1}{2}mu^2 = \frac{1}{2}kx^2 \Rightarrow u^2 = \frac{kx^2}{m}$$

$$\Rightarrow R_{max} = \frac{kx^2}{mg} = \frac{600 \times (5 \times 10^{-2})^2}{15 \times 10^{-3} \times 10} = 10\text{ m}$$

2 Given, $y_1 = 0.1 \sin\left(100\pi t + \frac{\pi}{3}\right)$

$$\Rightarrow \frac{dy_1}{dt} = v_1 = 0.1 \times 100\pi \cos\left(100\pi t + \frac{\pi}{3}\right)$$

or $v_1 = 10\pi \sin\left(100\pi t + \frac{\pi}{3} + \frac{\pi}{2}\right)$

or $v_1 = 10\pi \sin\left(100\pi t + \frac{5\pi}{6}\right)$

and $y_2 = 0.1 \cos \pi t$

$$\Rightarrow \frac{dy_2}{dt} = v_2 = -0.1 \sin \pi t$$

or $v_2 = 0.1 \sin(\pi t + \pi)$

Hence, the phase difference

$$\Delta\phi = \phi_1 - \phi_2 = \left(100\pi t + \frac{5\pi}{6}\right) - (\pi t + \pi)$$

$$= \frac{5\pi}{6} - \pi = -\frac{\pi}{6} \quad (\text{at } t = 0)$$

3 Force of buoyancy $= b \times c \times \rho_w \times g$

$= bcg \quad (\because \rho_w = 1)$

and mass of piece of wood $= abc\rho$

So, acceleration $= -bcg/abc\rho = -(g/a\rho)$

Hence, time period, $T = 2\pi\sqrt{\frac{\rho a}{g}}$

4

$$4 = k(\alpha - l)$$
$$5 = k(\beta - l)$$
$$9 = k(\gamma - l) \Rightarrow \frac{4}{5} = \frac{\alpha - l}{\beta - l}$$

or $4\beta - 4l = 5\alpha - 5l$

$l = 5\alpha - 4\beta$

Now, $9\alpha - 9l = 4\gamma - 4l$

$$4\gamma = 9\alpha - 5l = 9\alpha - 5(5\alpha - 4\beta) = 9\alpha - 25\alpha + 20\beta = 20\beta - 16\alpha$$

$$\gamma = 5\beta - 4\alpha$$

5 In the figure, we represent the electric intensity at B due to the sheet of charge,

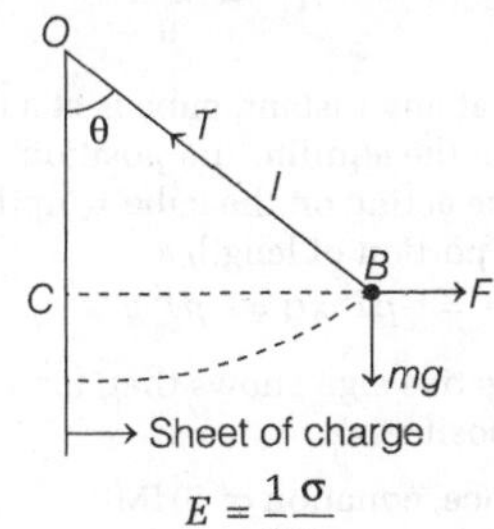

$$E = \frac{1}{2}\frac{\sigma}{\varepsilon_0}$$

Force on bob due to the sheet of charge,

$$F = qE = \frac{1}{2}\frac{\sigma q}{\varepsilon_0}$$

As the bob is in equilibrium,

so $\frac{mg}{OC} = \frac{F}{CB} = \frac{T}{BO}$

or $\frac{CB}{OC} = \frac{F}{mg} = \frac{\sigma q}{2\varepsilon_0 mg} = \tan\theta$

6 If mass m moves down a distance y, then the spring is pulled by $2y$ and the force with which the spring is pulled will be $F = R = mg/2$.

Hence, $mg/2 = k(2y)$

$\Rightarrow y/g = m/4k$

$\Rightarrow T = 2\pi\sqrt{y/g} = 2\pi\sqrt{m/4k} = \pi\sqrt{m/k}$

7 As, $\frac{aT}{x} = \frac{\omega^2 xT}{x} = \frac{4\pi^2}{T^2} \times T = \frac{4\pi^2}{T}$ = constant.

8 Given, $y = kt^2 \Rightarrow a = \frac{d^2y}{dt^2} = 2k$

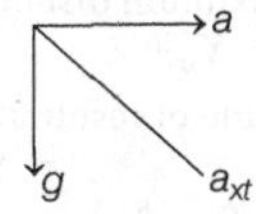

or $a_y = 2\text{ m/s}^2 \quad (\text{as, } k = 1\text{ m/s}^2)$

$\therefore T_1 = 2\pi\sqrt{\frac{l}{g}}$

and $T_2 = 2\pi\sqrt{\frac{l}{g + a_y}}$

$$\therefore \frac{T_1^2}{T_2^2} = \frac{g + a_y}{g} = \frac{10 + 2}{10} = \frac{6}{5}$$

9 We know that time period,

$$T = 2\pi\sqrt{\frac{L}{g}}$$

When additional mass M is added to its bob

$$T_M = 2\pi\sqrt{\frac{L + \Delta L}{g}},$$

where, ΔL is increase in length.

We know that Young modulus of the material

$$Y = \frac{Mg/A}{\Delta L/L} = \frac{MgL}{A\Delta L}$$

$$\Rightarrow \Delta L = \frac{MgL}{AY}$$

$$\therefore T_M = 2\pi\sqrt{\frac{L + \frac{MgL}{AY}}{g}}$$

$$\Rightarrow \left(\frac{T_M}{T}\right)^2 = 1 + \frac{Mg}{AY}$$

or $\frac{Mg}{AY} = \left(\frac{T_M}{T}\right)^2 - 1$

or $\frac{1}{Y} = \frac{A}{Mg}\left[\left(\frac{T_M}{T}\right)^2 - 1\right]$

10

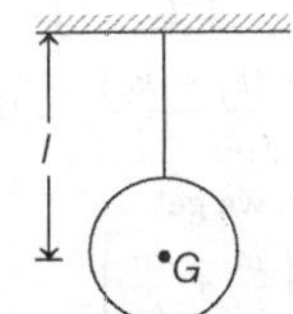

Spherical hollow ball filled with water

Spherical hollow ball half filled with water

$$T = 2\pi\sqrt{\frac{l}{g}} \qquad T_1 = 2\pi\sqrt{\frac{l + \Delta l}{g}}$$

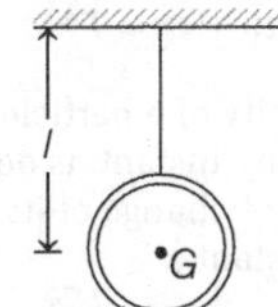

Spherical hollow ball

$$T_2 = 2\pi\sqrt{\frac{l}{g}} \text{ and } T_1 > T_2$$

Hence, time period first increases and then decreases to the original value.

11 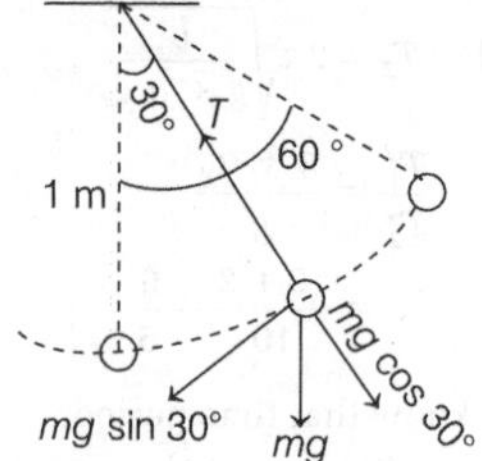

Rate of change of speed $\frac{dv}{dt}$

= tangential acceleration

$= \frac{\text{tangential force}}{\text{mass}} = \frac{mg \sin 30°}{m}$

$= g \sin 30° = 10\left(\frac{1}{2}\right) \text{m/s}^2 = 5 \text{m/s}^2$

12 Time period of the spring,

$$T = 2\pi\sqrt{\left(\frac{m}{k}\right)}$$

Here, k be the force constant of spring. For the first spring,

$$t_1 = 2\pi\sqrt{\left(\frac{m}{k_1}\right)} \quad \text{...(i)}$$

For the second spring,

$$t_2 = 2\pi\sqrt{\left(\frac{m}{k_2}\right)} \quad \text{...(ii)}$$

The effective force constant in the series combination is

$$k = \frac{k_1 k_2}{k_1 + k_2}$$

Time period of combination

$$T = 2\pi\sqrt{\left[\frac{m(k_1 + k_2)}{k_1 k_2}\right]}$$

$$\Rightarrow T^2 = \frac{4\pi^2 m (k_1 + k_2)}{k_1 k_2} \quad \text{...(iii)}$$

From Eqs. (i) and (ii), we get

$$t_1^2 + t_2^2 = 4\pi^2\left(\frac{m}{k_1} + \frac{m}{k_2}\right)$$

or $$t_1^2 + t_2^2 = 4\pi^2 m\left(\frac{1}{k_1} + \frac{1}{k_2}\right)$$

or $$t_1^2 + t_2^2 = \frac{4\pi^2 m (k_1 + k_2)}{k_1 k_2}$$

$$\Rightarrow t_1^2 + t_2^2 = T^2 \quad \text{[from Eq. (iii)]}$$

13 The velocity of a particle executing SHM at any instant, is defined as the time rate of change of its displacement at that instant.

$$v = \omega\sqrt{A^2 - x^2}$$

where, ω is angular frequency, A is amplitude and x is displacement of a particle.

Suppose that the new amplitude of the motion be A'.

Initial velocity of a particle performs SHM,

$$v^2 = \omega^2\left[A^2 - \left(\frac{2A}{3}\right)^2\right] \quad \text{... (i)}$$

where, A is initial amplitude and ω is angular frequency.

Final velocity,

$$(3v)^2 = \omega^2\left[A'^2 - \left(\frac{2A}{3}\right)^2\right] \quad \text{...(ii)}$$

From Eqs. (i) and (ii), we get

$$\frac{1}{9} = \frac{A^2 - \frac{4A^2}{9}}{A'^2 - \frac{4A^2}{9}} \Rightarrow A' = \frac{7A}{3}$$

14 Let at any instant, cube is at a depth x from the equilibrium position, then net force acting on the cube = upthrust on the portion of length x

$\therefore \; F = -\rho l^2 x g = -\rho l^2 g \, x$ ...(i)

Negative sign shows that, force is opposite to x.

Hence, equation of SHM

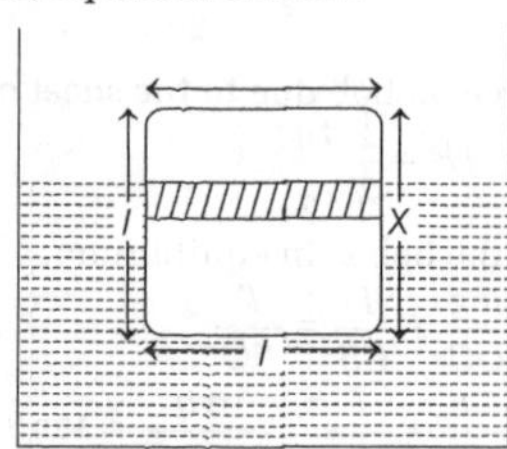

$$F = -kx \quad \text{...(ii)}$$

Comparing Eqs.(i) and (ii), we get

$$k = \rho l^2 g$$

$$\therefore T = 2\pi\sqrt{\frac{m}{k}} = 2\pi\sqrt{\frac{l^3 d}{\rho l^2 g}} = 2\pi\sqrt{\frac{ld}{\rho g}}$$

15 Let $x_1 = A\sin(\omega t + \phi_1)$

and $x_2 = A\sin(\omega t + \phi_2)$

$x_2 - x_1 = A$

$[\sin(\omega t + \phi_2) - \sin(\omega t + \phi_1)]$

$$= 2A\cos\left(\frac{2\omega t + \phi_1 + \phi_2}{2}\right)\sin\left(\frac{\phi_2 - \phi_1}{2}\right)$$

The resultant motion can be treated as a simple harmonic motion with amplitude $2A\sin\left(\frac{\phi_2 - \phi_1}{2}\right)$

Given, maximum distance between the particles $= X_0 + A$

$\therefore$ Amplitude of resultant SHM

$$= X_0 + A - X_0 = A$$

$$\therefore 2A\sin\left(\frac{\phi_2 - \phi_1}{2}\right) = A \Rightarrow$$

$\phi_2 - \phi_1 = \pi/3$

16 In SHM, a particle starts from rest, we have

i.e. $x = A\cos\omega t$, at $t = 0, x = A$

When $t = \tau$, then $x = A - a$ ...(i)

When $t = 2\tau$, then $x = A - 3a$ ...(ii)

On comparing Eqs. (i) and (ii), we get

$$A - a = A\cos\omega\tau$$
$$A - 3a = A\cos 2\omega\tau$$

As $\cos 2\omega\tau = 2\cos^2\omega\tau - 1$

$$\frac{A - 3a}{A} = \frac{2A^2 + 2a^2 - 4Aa - A^2}{A^2}$$

$$A^2 - 3aA = A^2 + 2a^2 - 4Aa$$

$$a^2 = 2aA, \; A = 2a$$

Now, $A - a = A\cos\omega\tau \Rightarrow \cos\omega\tau = 1/2$

$$\Rightarrow \frac{2\pi}{T}\tau = \frac{\pi}{3} \Rightarrow T = 6\tau$$

17 $\frac{Mg}{A} = P_0 \Rightarrow Mg = P_0 A$

$$P_0 V_0^\gamma = (P_0 + \Delta V_0)(V_0 - \Delta V_0)^\gamma$$

$$\Rightarrow P_0 = (P_0 + \Delta P_0)\left(1 - \frac{\Delta V_0}{V_0}\right)^\gamma$$

$$= (P_0 + \Delta P_0)\left(1 - r\frac{\Delta V_0}{V_0}\right)$$

$$= \left(P_0 - VP_0\frac{\Delta V_0}{V_0} + \Delta P_0\right)$$

or $\Delta P_0 = VP_0\frac{\Delta V_0}{V_0}$

But $\Delta V = Ax$,

where, A = area at cross section of piston

$\therefore \Delta P_0 = \frac{\gamma P_0 A}{V_0} x$

Restoring force $F = -\Delta P_0 \times A = -\frac{\gamma P_0 A^2}{V_0} x$

Comparing it with, $F_{res} = -kx$

$$k = \frac{\gamma P_0 A^2}{V_0}$$

$$\therefore f = \frac{1}{2\pi}\sqrt{\frac{k}{M}} = \frac{1}{2\pi}\sqrt{\frac{\gamma P_0 A^2}{MV_0}}$$

18 For damped harmonic motion,

$$ma = -kx - mbv$$

or $$ma + mbv + kx = 0$$

Solution to above equation is

$$x = A_0 e^{-\frac{bt}{2}}\sin\omega t; \text{ with } \omega^2 = \frac{k}{m} - \frac{b^2}{4m}$$

where, amplitude drops exponentially with time.

$$A_\tau = A_0 e^{-\frac{b\tau}{2}}$$

Average time τ is that duration when amplitude drops by 63%, i.e. becomes A_0/e

Thus, $A_\tau = \frac{A_0}{e} A_0 e^{-\frac{b\tau}{2}}$

or $\frac{b\tau}{2} = 1$ or $\tau = \frac{2}{b}$

DAY TWELVE

Waves

Learning & Revision for the Day

- Wave Motion
- Speed of Waves
- Sound Waves
- Displacement Relation for a Progressive or Harmonic Wave
- Principle of Superposition of Waves
- Reflection and Transmission of Waves
- Standing or Stationary Waves
- Beats
- Doppler's Effect

Wave Motion

Wave motion involves transfer of disturbance (energy) from one point to the other with particles of medium oscillating about their mean positions i.e. the particles of the medium do not travel themselves along with the wave. Instead, they oscillate back and forth about the same equilibrium position as the wave passes by. Only the disturbance is propagated.

1. **Longitudinal Waves** When particles of the medium vibrate parallel to the direction of propagation of wave, then wave is called **longitudinal wave**. These waves propagate in the form of compressions and rarefactions. They involve changes in pressure and volume. The medium of propagation must possess elasticity of volume. They are set up in solids, liquids and gases.
2. **Transverse Waves** When the particles of the medium vibrate in a direction perpendicular to the direction of propagation of wave, then wave is called transverse waves. These wave propagtes in the form of crests and troughs. These waves can be set up in solids, on surface of liquids but never in gases.

Terms Used in Wave Motion

- **Angular Wave Number** Number of wavelength in the distance 2π is called the wave number or propagation constant.

$$K = \frac{2\pi}{\lambda} \text{ rad/m}$$

- **Particle velocity** It is the velocity of the particle executing simple harmonic motion.

i.e. $$v = \frac{dy}{dt}$$

where, y denotes displacement at any instant.

- **Wave Velocity** The velocity of transverse wave motion is given by

$$v = \frac{\text{Distance travelled by wave}}{\text{Time taken}}$$

i.e. $$v = \frac{\lambda}{T} = \left(\frac{1}{T}\right)\lambda = \frac{\omega}{x} \quad \text{or} \quad v = \nu\lambda$$

- No. of Questions in Exercises (x)—
- No. of Questions Attempted (y)—
- No. of Correct Questions (z)—
  (Without referring Explanations)
- Accuracy Level ($z/y \times 100$)—
- Prep Level ($z/x \times 100$)—

In order to expect good rank in JEE, your Accuracy Level should be above 85 & Prep Level should be above 75.

- **Differential Equation of Wave Motion**

$$\frac{d^2 y}{dx^2} = \frac{1}{v^2}\frac{d^2 y}{dt^2}$$

Speed of Waves

Speed of waves is divided in two types as per the nature of wave, these are given below

1. Speed of Transverse Wave

The expression for speed of transverse waves in a solid and in case of a stretched string can be obtained theoretically

- In solids, $v = \sqrt{\frac{\eta}{d}}$

  where, η is the modulus of rigidity and d is the density of the medium.

- In a stretched string, $v = \sqrt{\frac{T}{m}} = \sqrt{\frac{Mg}{\pi r^2 d}}$

  where, T = the tension in the string,
  m = the mass per unit length of the string,
  M = mass suspended from the string,
  r = radius of the string and
  d = density of the material of the string.

2. Speed of Longitudinal Wave (or Sound Wave)

Following are the expressions for the speed of longitudinal waves in the different types of media

- If the medium is solid,

$$v = \sqrt{\frac{B + \frac{4}{3}\eta}{\rho}}$$

  where B, η and ρ are values of bulk modulus, modulus of rigidity and density of the solid respectively.
  If the solid is in the form of a long rod, then

$$v = \sqrt{\frac{Y}{\rho}}$$

  where, Y is the Young's modulus of the solid material.

- In a liquid,

$$v = \sqrt{\frac{B}{\rho}}$$

  where B is the bulk modulus of the liquid.

- According to Newton's formula, speed of sound in a gas is obtained by replacing B with initial pressure p of the gas i.e. $B = p$.

$$v = \sqrt{\frac{p}{\rho}}$$

Factors Affecting Speed of Sound

- **Effect of Temperature on Velocity** With rise in temperature, the velocity of sound increases as

$$v = \sqrt{\frac{\gamma RT}{M}}; \quad \text{i.e. } v \propto \sqrt{T}; \quad \frac{v_2}{v_1} = \sqrt{\frac{T_2}{T_1}}$$

  Speed of sound in air increases by 0.61 m/s for every 1°C rise in temperature.

- **Effect of Pressure for Gase Medium** Pressure has no effect on the velocity of sound, provided temperature remains constant.

- **Effect of Humidity** When humidity in air increases, its density decreases and so velocity of sound increases.

  For solids, $v = \sqrt{\frac{Y}{D}}$. For liquids, $v = \sqrt{\frac{B}{D}}$

  where, Y = Young's modulus of elasticity
  B = bulk modulus of elasticity.

Sound Waves

The longitudinal waves which can be heard are called sound waves.

They are classified into following categories

- **Infrasonics** The longitudinal waves having frequencies below 20 Hz are called infrasonics. These waves cannot be heard. These waves can be heard by snakes.
- **Audible waves** The longitudinal waves having the frequency between 20 Hz and 20000 Hz are called audible waves. Human can hear these waves.
- **Ultrasonics** The longitudinal waves having the frequencies above 20000 Hz are called ultrasonics. These waves are also called supersonic waves or supersonics.

Displacement Relation for a Progressive or Harmonic Wave

The equation of a plane progressive or simple harmonic wave travelling along positive direction of x-axis is

$$y = a\sin(\omega t - kx) \Rightarrow y = a\sin\frac{2\pi}{T}\left(t - x\frac{T}{\lambda}\right).$$

$$\Rightarrow \quad y = a\sin\frac{2\pi}{\lambda}(vt - x) \Rightarrow y = a\sin\omega\left(t - \frac{x}{v}\right)$$

$$\Rightarrow \quad y = a\sin 2\pi\left(\frac{t}{T} - \frac{x}{\lambda}\right).$$

- If maximum value of $y = a$, i.e. a is amplitude, then dy/dt = velocity of particle

$$v = \frac{dy}{dt} = a\omega\cos\cdot\frac{2\pi}{\lambda}(vt - x)$$

$$\left(\frac{dy}{dt}\right)_{max} = \frac{2\pi va}{\lambda} = 2\pi na = \omega a \qquad \text{[where, } n = \text{frequency]}$$

- Acceleration of particle
$$\frac{d^2y}{dt^2} = -\omega^2 a \sin\frac{2\pi}{\lambda}(vt - x)$$
- For a wave, velocity of sound wave
$$v = \text{frequency } (n) \times \text{wavelength } (\lambda)$$
$$\Rightarrow \quad v = n\lambda$$
- Angular speed, $\omega = 2\pi n = \frac{2\pi}{T} \Rightarrow \omega = \frac{2\pi v}{\lambda}$

Relation between Phase Difference, Path Difference and Time Difference

- Phase difference $(\phi) = \frac{2\pi}{\lambda} \times$ path difference (x)
$$\Rightarrow \quad \phi = \frac{2\pi x}{\lambda} \Rightarrow x = \frac{\phi\lambda}{2\pi}$$
- Phase difference $(\phi) = \frac{2\pi}{T} \times$ time difference (t)
$$\Rightarrow \quad \phi = \frac{2\pi t}{T} \Rightarrow t = \frac{T\phi}{2\pi}$$
- Time difference $(t) = \frac{T}{\lambda} \times$ path difference (x)
$$\Rightarrow \quad t = \frac{Tx}{\lambda} \Rightarrow x = \frac{\lambda t}{T}$$

Principle of Superposition of Waves

Two or more waves can traverse the same space independently of one another. The resultant displacement of each particle of the medium at any instant is equal to the vector sum of displacements produced by the two waves separately. This principle is called principle of superposition of waves.

$$\boxed{y = y_1 + y_2 + y_3 + \ldots}$$

Interference of Waves

When two waves of same frequency (or same wavelength) travelling along same path superimpose each other, there occurs redistribution of energy in the medium. At a given position (x being constant) displacement due to two waves be

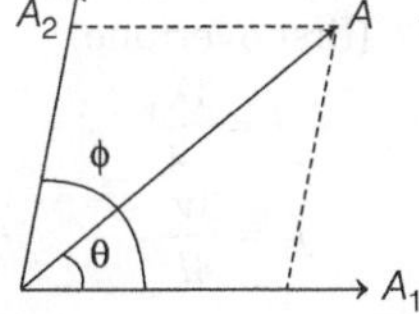

$$y_1 = A_1 \sin \omega t \quad \text{and} \quad y_2 = A_2 \sin(\omega t + \phi)$$

Then, resultant displacement
$$y = y_1 + y_2 = A \sin(\omega t + \phi)$$
where, $$A = \sqrt{A_1^2 + A_2^2 + 2A_1A_2 \cos\phi}$$
and $$\tan\theta = \frac{A_2 \sin\phi}{A_1 + A_2 \cos\phi}$$

Constructive and Destructive Interference

- When the wave meet a point with some phase, **constructuve interference** is obtained at that point.
  (i) Phase difference between the waves at the point of observation $\phi = 0°$ or $2\pi n$.
  (ii) Resultant amplitude at the point of observation will be maximum, $A_{max} = A_1 + A_2$.
- When the waves meet a point with opposite phase, **destructive interference** is obtained at that point.
  (i) Phase difference between the waves at the point of observation $\phi = 180°$ or $(2n-1)\pi$.
  (ii) Resultant amplitude at the point of observation will be minimum, $A_{min} = A_1 - A_2$.

Intensity

The intensity of waves is the average amount of energy transported by the wave per unit area per second normally across a surface at the given point.

Intensity $(I_1) \propto$ (Amplitude $A)^2$
$$\therefore \quad \frac{I_1}{I_2} = \left(\frac{A_1}{A_2}\right)^2$$

If I_1 and I_2 are intensities of the interfering waves and ϕ is the phase difference, then resultant intensity is given by
$$I = I_1 + I_2 + 2\sqrt{I_1 I_2} \cos\phi$$
$$I_{max} = I_1 + I_2 + 2\sqrt{I_1 I_2} = (\sqrt{I_1} + \sqrt{I_2})^2, \text{ for } \phi = 2\pi n$$
and $$I_{min} = I_1 + I_2 - 2\sqrt{I_1 I_2}$$
$$I_{min} = (\sqrt{I_1} - \sqrt{I_2})^2, \text{ for } \phi = (2n+1)\pi$$

Power

If P is power of a sound source, then intensity (I) follows inverse square law of distance (d).
$$I = \frac{P}{4\pi d^2}$$

Reflection and Transmission of Waves

When sound waves are incident on a boundary separating two media, a part of it is reflected back into the initial medium while the remaining is partly absorbed and partly transmitted into the second medium.

Standing or Stationary Waves

Standing or stationary wave is formed due to superposition of two progressive waves of same nature, same frequency (or same wavelength), same amplitude travelling with same speed in a bounded medium in mutually opposite directions.

If the incident wave be represented as $y_1 = A \sin(\omega t - kx)$ and the reflected wave as $y_2 = A \sin(\omega t + kx)$,

then $y = y_1 + y_2 = A \sin(\omega t - kx) + A \sin(\omega t + kx)$

$\Rightarrow \quad y = 2A \cos kx \sin \omega t$

The resultant wave does not represent a progressive wave.

Standing Waves in String

Consider a string of length L stretched under tension T between two fixed points (i.e. clamped at its ends). Transverse wave is set up on the string whose speed is given by $v = \sqrt{T/\mu}$, where μ is the mass per unit length of the string.

Different modes of vibration of stretched string are discussed below

- Let only one anti-node A is formed at the centre and string vibrates in one segment only, it is called **fundamental mode**, then

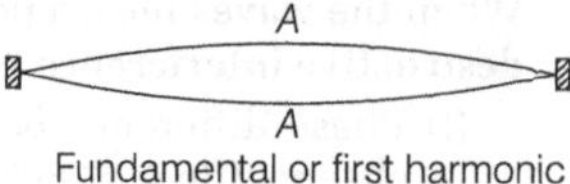

Fundamental or first harmonic

$$L = \frac{\lambda_1}{2} \text{ or } \lambda_1 = 2L$$

Frequency of vibration in fundamental mode

$$\nu_1 = \frac{v}{\lambda_1} = \frac{1}{2L}\sqrt{\frac{T}{\mu}}$$

It is known as the **fundamental frequency** or **first harmonic.**

- If string vibrates in two segments, then

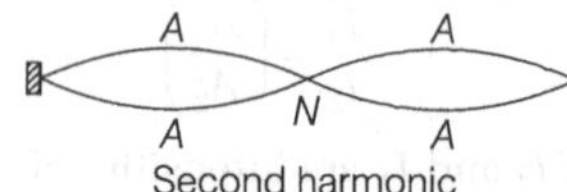

Second harmonic

$$L = \lambda_2$$

and $\nu_2 = \frac{v}{\lambda_2} = \frac{1}{L}\sqrt{\frac{T}{\mu}} = 2\nu_1$

It is known as **first overtone** or **second harmonic.**

- If the string vibrates in three segments,

then $L = \frac{3\lambda_3}{2}$

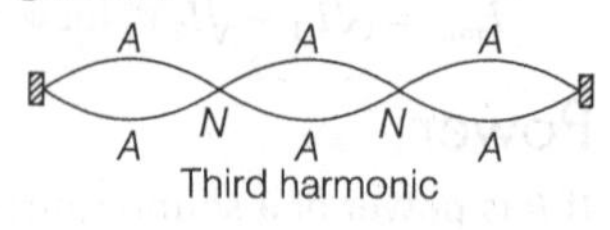

Third harmonic

and $\nu_3 = \frac{v}{\lambda_3} = 3\nu_1$

It is called **second overtone** or **third harmonic.**

- In general, if a string vibrates in p segments [i.e. have $(p+1)$ nodes and p antinodes],

then $\nu_{pth} = \frac{p}{2L}\sqrt{\frac{T}{\mu}} = p\nu_1$

and it is known as pth harmonic or $(p-1)$th overtone.

Standing Waves in Organ Pipes (Air Columns)

Organ pipes are those cylindrical pipes which are used for producing musical (longitudinal) sounds. The standing waves in both organ pipes (i.e. open organ pipe and closed organ pipe) are described below.

1. Open Organ Pipe

In an open organ pipe, always anti-node is formed at both open ends. Various modes of vibration of air column in an open organ pipe are shown below

- First harmonic $l = \frac{\lambda_1}{2} \Rightarrow f_1 = \frac{v}{2l}$

A_1 N A_2; $l = \frac{\lambda_1}{2}$

(a)

- Second harmonic or first overtone $l = \lambda_2$; $f = \frac{2v}{2l}$

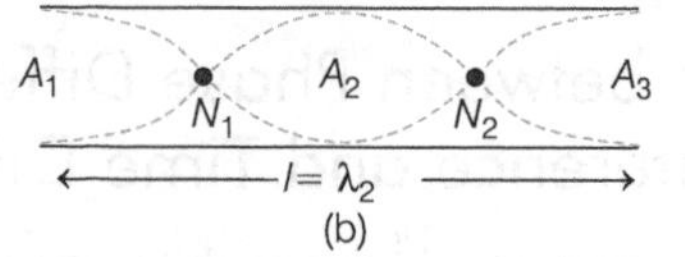

(b)

- Third harmonic or second overtone $l = \frac{3\lambda_3}{2}$; $f = \frac{3v}{2l}$

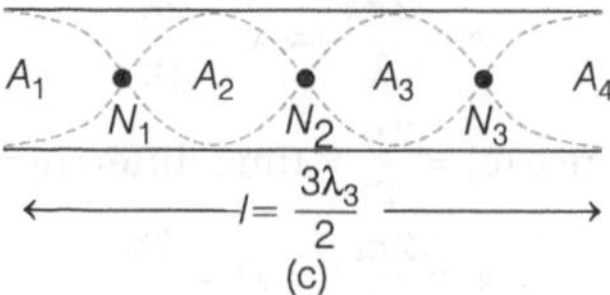

(c)

- All harmonics are present in open pipe with their frequencies in the ratio 1 : 2 : 3 : 4.... and ratio of overtones = 2 : 3 : 4 : 5 ...

Position of nodes from one end $x = \frac{\lambda}{4}, \frac{3\lambda}{4}, \frac{5\lambda}{4} \dots$

Position of anti-nodes from one end

$$x = 0, \frac{\lambda}{2}, \lambda, \frac{3\lambda}{2} \dots$$

2. Closed Organ Pipe

In a chosed organ pipe, always node is formed at the closed end. Various mode of vibration of air column in a closed organ pipe are shown below

- First harmonic

$l = \frac{\lambda_1}{4}$

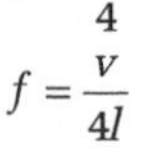

$f = \frac{v}{4l}$

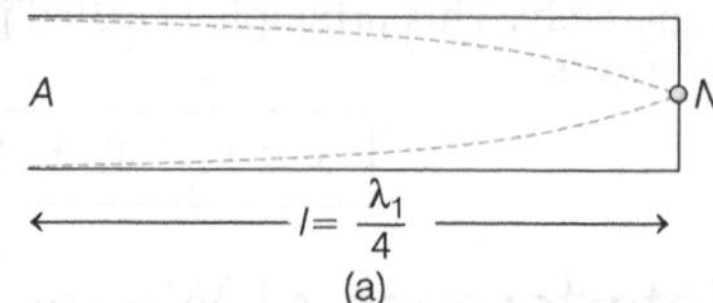

(a)

- Third harmonic (first overtone)

$l = \frac{3\lambda_3}{4}$

$f_3 = \frac{3v}{4l}$

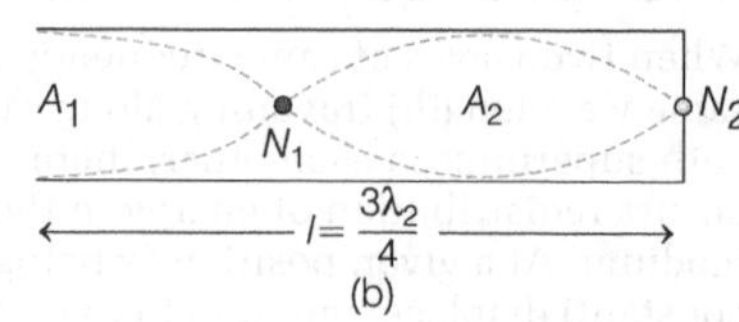

(b)

- Fifth harmonic (second overtone)

$l = \frac{5\lambda_5}{4}$;

$f_5 = \frac{5v}{4l}$

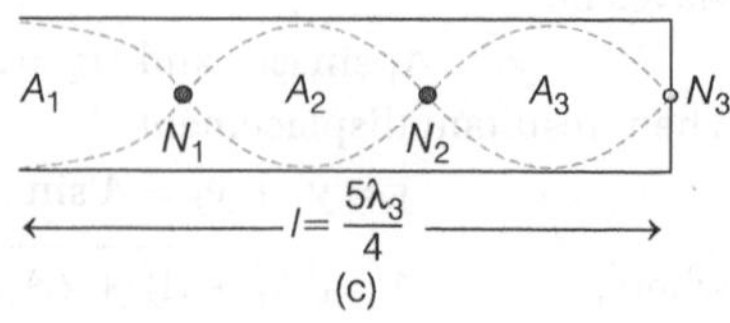

(c)

- In closed organ pipe only odd harmonics are present. Ratio of harmonic is $n_1 : n_3 : n_5 = 1 : 3 : 5$.

- Ratio of overtones = 3 : 5 : 7
- Position of nodes from closed end $x = 0, \frac{\lambda}{2}, \lambda, \frac{3\lambda}{2}, \ldots$
- Position of antinodes from closed end $x = \frac{\lambda}{4}, \frac{3\lambda}{4}, \frac{5\lambda}{4}, \ldots$

Beats

When two sound waves of nearly equal (but never equal) or slightly different frequencies and equal or nearly equal amplitudes travelling along the same direction superimpose at a given point, the resultant sound intensity alternately rises and falls. This alternate rise and fall of sound at a given position is called **beats.**

- Number of beats formed per second is called the frequency of beats. If two sound waves of frequencies ν_1 and ν_2 superimpose, then frequency of beats $= (\nu_1 \sim \nu_2)$, i.e. either $(\nu_1 - \nu_2)$ or $(\nu_2 - \nu_1)$.
- For formation of distinct beats, the difference between the frequencies of two superimposing notes should be less than 10 Hz.
- Our perception of loudness is better co-related with the second level measured in decibel (dB) and defined as follows

$\beta = 10 \log_{10}\left(\frac{I}{I_0}\right)$, where $I_0 = 10^{-12}\ \text{Wm}^2$ at 1kHz.

Tuning Fork

The tuning fork is a metallic device that produces sound of a single frequency.

Suppose, a tuning fork of known frequency n_A is sounded together with another tuning fork of unknown frequency (n_B) and x beats heard per second.

There are two possibilities to know frequency of unknown tuning fork

$$n_A - n_B = x \quad \ldots\text{(i)}$$

$$n_B - n_A = x \quad \ldots\text{(ii)}$$

We can find true frequency of tuning fork B from a pair of tuning forks A and B, in which frequency of A is known and where x is the beats per second.

	When B is loaded (its frequency decreases)		When B is filled (its frequency increases)
(i)	If x increases, then $n_B = n_A - x$	(i)	If x increases, then $n_B = n_A + x$
(ii)	If x decreases, then $n_B = n_A + x$	(ii)	If x decreases, then $n_B = n_A - x$
(iii)	If x remains same, then $n_B = n_A + x$	(iii)	If x remains same, then $n_B = n_A - x$
(iv)	If x becomes zero, then $n_B = n_A + x$	(iv)	If x becomes zero, then $n_B = n_A - x$

Doppler's Effect

The phenomena of apparent change in frequency of source due to a relative motion between the source and observer is called Doppler's effect.

- **When Source is Moving and Observer is at Rest** When source is moving with velocity v_s, towards an observer at rest, then apparent frequency

$$n' = n\left(\frac{v}{v - v_s}\right)$$

S → v_s, → v, O

If source is moving away from observer, then

$$n' = n\left(\frac{v}{v + v_s}\right)$$

- **When Source is at Rest and Observer is Moving** When observer is moving with velocity v_o, towards a source at rest, then apparent frequency.

$$n' = n\left(\frac{v + v_o}{v}\right)$$

S, v_o ←, → v, O

When observer is moving away from source, then

$$n' = n\left(\frac{v - v_o}{v}\right)$$

S, → v, O → v_o

- **When Source and Observer Both are Moving**

(a) When both are moving in same direction along the direction of propagation of sound, then

$$n' = n\left(\frac{v - v_o}{v - v_s}\right)$$

S → v O; → v_s, → v_o

(b) When both are moving in same direction opposite to the direction of propagation of sound, then

$$n' = n\left(\frac{v + v_o}{v + v_s}\right)$$

v_s ← S, v_o ← O, → v

(c) When both are moving towards each other, then

$$n' = n\left(\frac{v + v_o}{v - v_s}\right)$$

S → v O; → v_s, → v_o

(d) When both are moving in opposite direction, away from each other, then

$$n' = n\left(\frac{v - v_o}{v + v_s}\right)$$

v_o ← O, v ←, S → v_s

Transverse Doppler's Effect

- The Doppler's effect in sound does not take place in the transverse direction.
- As shown in figure, the position of a source is S and of observer is O. The component of velocity of source towards the observer is $v \cos \theta$. For this situation, the approach frequency is

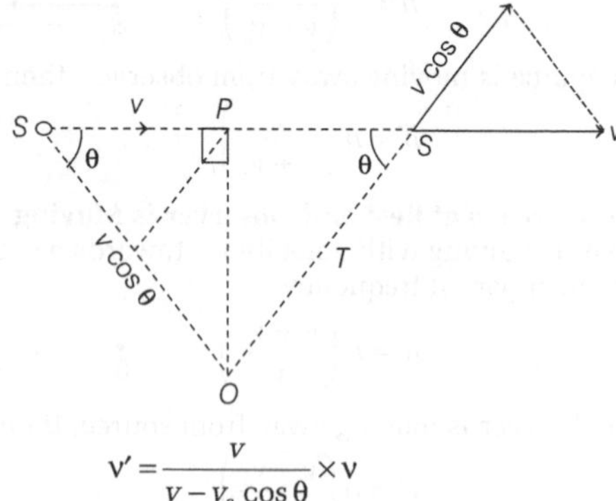

$$\nu' = \frac{v}{v - v_s \cos \theta} \times \nu$$

ν' will now be a function of θ. So, it will no more be constant.

Similarly, if the source is moving away from the observer as shown above, with velocity component $v_s \cos \theta$, then

$$\nu' = \frac{v}{v + v_s \cos \theta} \times \nu$$

- If $\theta = 90°$, the $v_s \cos \theta = 0$ and there is no shift in the frequency. Thus, at point P, Doppler's effect does not occur.

Effect of Wind

If wind is also blowing with a velocity w in the direction of sound, then its velocity is added to the velocity of sound. Hence, in this condition the apparent frequency is given by

$$\nu' = \nu \left(\frac{v + w - v_o}{v + w - v_s} \right)$$

Applications of Doppler's Effect

The measurement of Doppler shift has been used

- by police to check over speeding of vehicles.
- at airports to guide the aircraft.
- to study heart beats and blood flow in different parts of the body.
- by astrophysicist to measure the velocities of plants and stars.

DAY PRACTICE SESSION 1

FOUNDATION QUESTIONS EXERCISE

1 Which of the following statements are true for wave motion?

(a) Mechanical transverse waves can propagate through all mediums
(b) Longitudinal waves can propagate through solids only
(c) Mechanical transverse waves can propagate through solids only
(d) Longitudinal waves can propagate through vacuum

2 A sound wave is passing through air column in the form of compression and rarefaction. In consecutive compression and rarefactions,

(a) density remains constant
(b) Boyle's law is obeyed
(c) bulk modulus of air oscillates
(d) there is no transfer of heat

3 A sound wave of wavelength λ is travelling in a medium with a speed of v m/s enters into another medium where its speed is $2v$ m/s. Wavelength of sound waves in the second medium is

(a) λ (b) $\frac{\lambda}{2}$
(c) 2λ (d) 4λ

4 When tension of a string is increased by 2.5 N, the initial frequency is altered in the ratio of 3 : 2. The initial tension in the string is

(a) 6 N (b) 5 N (c) 4 N (d) 2 N

5 The speed of sound in oxygen (O_2) at a certain temperature is 460 ms^{-1}. The speed of sound in helium (He) at the same temperature will be (assume both gases to be ideal)

(a) 460 ms^{-1} (b) 500 ms^{-1} (c) 650 ms^{-1} (d) 1420 ms^{-1}

6 It takes 2.0 seconds for a sound wave to travel between two fixed points, when the day temperature is 10°C. If the temperature rise to 30°C, the sound wave travel between the same fixed points in

(a) 1.9 s (b) 2.0 s
(c) 2.1 s (d) 2.2 s

7 A wave equation is given by $y = 4 \sin \left[\pi \left(\frac{t}{5} - \frac{x}{9} + \frac{1}{6} \right) \right]$ where, x is in cm and t is in sec. Which of the following is true?

(a) $\lambda = 18$ cm (b) $v = 4\ ms^{-1}$
(c) $a = 0.4$ m (d) $f = 50$ Hz

8 A wave equation which gives the displacement along y-direction is given by $y = 0.001 \sin[100t + x]$, where, x and y are in metre and t is time in second. This represents a wave

(a) of frequency $\frac{100}{\pi}$ Hz
(b) of wavelength 1 m
(c) travelling with a velocity of $\frac{50}{\pi}$ ms^{-1} in the positive x-direction
(d) travelling with a velocity of 100 ms^{-1} in the negative x-direction

9 Which of the following is not true for progressive wave

$$y = 4 \sin 2\pi \left[\frac{t}{0.02} - \frac{x}{100}\right]$$

where, y and x are in cm and t in second.

(a) Its amplitude is 4 cm
(b) Its wavelength is 100 cm
(c) Its frequency is 50 Hz
(d) Its propagation speed is 50×10^{-2} cms^{-1}

10 The equation of a wave on a string of linear mass density 0.04 kg m^{-1} is given by

$$y = 0.02 \text{ (m)} \sin\left[2\pi\left(\frac{t}{0.04 \text{ (s)}} - \frac{x}{0.50 \text{ (m)}}\right)\right].$$

The tension in the string is

(a) 4.0 N (b) 12.5 N (c) 0.5 N (d) 6.25 N

11 When two sound waves travel in the same direction in a medium the displacement of a particle located at X at time t is given by → JEE Main (Online) 2013

$$y_1 = 0.05 \cos(0.50\pi x - 100\pi t)$$
$$y_2 = 0.05 \cos(0.46\pi x - 92\pi t)$$

where y_1, y_2 and x are in metres and t in seconds. The speed of sound in the medium is

(a) 92 m/s (b) 200 m/s (c) 100 m/s (d) 332 m/s

12 In order to double the frequency of the fundamental note emitted by a stretched string, the length is reduced to $\frac{3}{4}$ th of the original length and the tension is changed. The factor, by which the tension is to be changed, is

(a) $\frac{3}{8}$ (b) $\frac{2}{3}$ (c) $\frac{8}{9}$ (d) $\frac{9}{4}$

13 A granite rod of 60 cm length is clamped at its middle point and is set into longitudinal vibrations. The density of granite is 2.7×10^3 kg/m^3 and its Young's modulus is 9.27×10^{10} Pa. What will be the fundamental frequency of the longitudinal vibrations? → JEE Main 2018

(a) 5 kHz (b) 2.5 kHz (c) 10 kHz (d) 7.5 kHz

14 A pipe open at both ends has a fundamental frequency f in air. The pipe is dipped vertically in water, so that half of it is in water. The fundamental frequency of the air column is now → JEE Main 2016 (Offline)

(a) $\frac{f}{2}$ (b) $\frac{3f}{4}$ (c) $2f$ (d) f

15 Third overtone of a closed organ pipe is in unison with fourth harmonic of an open organ pipe. The ratio of the lengths of the pipes are

(a) $\frac{7}{8}$ (b) $\frac{3}{4}$ (c) $\frac{5}{7}$ (d) $\frac{8}{7}$

16 Motion of two particles is given by

$$y_1 = 0.25 \sin(310t), \quad y_2 = 0.25 \sin(316t)$$

Find beat frequency.

(a) 3 (b) $3/\pi$ c) $6/\pi$ (d) 6

17 Three sound waves of equal amplitudes have frequencies $(\nu - 1), \nu, (\nu + 1)$. They superimpose to give beat. The number of beats produced per second will be

(a) 4 (b) 3 (c) 2 (d) 1

18 Two tuning forks P and Q when set vibrating, give 4 beat/s. If a prong of the fork P is filled, the beats are reduced to 2 s^{-1}. What is the frequency of P, if Q is 250 Hz?

(a) 246 Hz (b) 250 Hz (c) 254 Hz (d) 252 Hz

19 16 tuning forks are arranged in the order of increasing frequencies. Any two successive forks give 8 beat/s, when sounded together. If the frequency of the last fork is twice the first, then the frequency of the first fork is

(a) 120 Hz (b) 160 Hz
(c) 180 Hz (d) 220 Hz

20 A vehicle with a horn of frequency n is moving with a velocity of 30 ms^{-1} in a direction perpendicular to the straight line joining the observer and the vehicle. The observer perceives the sound to have a frequency $(n + n_1)$. If velocity of sound in air be 330 ms^{-1}, then

(a) $n_1 = 10n$ (b) $n_1 = 0$ (c) $n_1 = \frac{n}{11}$ (d) $n_1 = -\frac{n}{11}$

21 An observer is moving with half the speed of light towards a stationary microwave source emitting waves at frequency 10 GHz. What is the frequency of the microwave measured by the observer? (speed of light $= 3 \times 10^8$ ms^{-1}) → JEE Main (Offline) 2017

(a) 12.1 GHz (b) 17.3 GHz
(c) 15.3 GHz (d) 10.1 GHz

22 Two trains are moving towards each other at speeds of 20 ms^{-1} and 15 ms^{-1} relative to the ground. The first train sounds a whistle of frequency 600 Hz, the frequency of the whistle heard by a passenger in the second train before the train meet is
(Speed of sound in air = 340 ms^{-1})

(a) 600 Hz (b) 585 Hz (c) 645 Hz (d) 666 Hz

23 A fixed source of sound emitting a certain frequency appears as ν_a when the observer is approaching the source with speed v_o and ν_r when the observer recedes from the source with the same speed. The frequency of the source is

(a) $\frac{\nu_r + \nu_a}{2}$ (b) $\frac{\nu_r - \nu_a}{2}$ (c) $\sqrt{\nu_a \cdot \nu_b}$ (d) $\frac{2\nu_r \cdot \nu_a}{\nu_r + \nu_a}$

Direction (Q. Nos. 24-31) *Each of these questions contains two statements : Statement I and Statement II. Each of these questions also has four alternative choices, only one of which is the correct answer. You have to select one of the codes (a), (b), (c), (d) given below:*

(a) Statement I is true, Statement II is true; Statement II is the correct explanation for Statement I

(b) Statement I is true, Statement II is true; Statement II is not the correct explanation for Statement I

(c) Statement I is true; Statement II is false

(d) Statement I is false; Statement II is true

24 **Statement I** A tuning fork is in resonance with a closed pipe. But the same tuning fork cannot be in resonance with an open pipe of the same length.

Statement II The same tuning fork will not be in resonance with open pipe of same length due to end correction of pipe.

25 **Statement I** In a sound wave, a displacement node is a pressure antinode and *vice-versa.*

Statement II Displacement node is a point of minimum displacement.

26 **Statement I** Velocity of particles while crossing mean position (in stationary waves) varies from maximum at antinodes to zero at nodes.

Statement II Amplitude of vibration at antinodes is maximum and at nodes, the amplitude is zero and all particles between two successive nodes cross the mean position together.

27 **Statement I** We can recognise our friends by listening their voices.

Statement II The quality of sound produced by different persons are different.

28 **Statement I** The basic of Laplace correction was that, exchange of heat between the region of compression and rarefaction in air is not possible.

Statement II Air is a bad conduction of heat and velocity of sound in air is large.

29 **Statement I** If two waves of same amplitude, produce a resultant wave of same amplitude, then the phase difference between them will be 120°.

Statement II The resultant amplitude of two waves is equal to sum of amplitude of two waves.

30 **Statement I** Two longitudinal waves given by equation $y_1(x, t) = 2a \sin(\omega t - kx)$ and $y_2(x, t) = a \sin(2\omega t - 2kx)$ will have equal intensity.

Statement II Intensity of waves of given frequency in same medium is proportional to square of amplitude only.

31. **Statement I** If we see the oscillations of a stretched wire at higher overtone mode, frequency of oscillations increases, but wavelength decreases.

Statement II From $v = \nu \cdot \lambda$, $\lambda \propto \frac{1}{\nu}$ as $v =$ constant.

DAY PRACTICE SESSION 2

PROGRESSIVE QUESTIONS EXERCISE

1 Sound of frequency f passes through a Quinck's tube, adjusted for intensity I_m. What should be the length to which the tube should be moved to reduce intensity to 50% (speed of sound is v)?

(a) $\frac{v}{2f}$ (b) $\frac{v}{4f}$ (c) $\frac{v}{8f}$ (d) $\frac{v}{16f}$

2 A racing car moving towards a cliff sounds its horn. The driver observes that the sound reflected from the cliff has a pitch one octave higher than the actual sound of the horn. If v is the velocity of sound, the velocity of the car is

(a) $\frac{v}{\sqrt{2}}$ (b) $\frac{v}{2}$

(c) $\frac{v}{3}$ (d) $\frac{v}{4}$

3 A train of sound waves is propagated along an organ pipe and gets reflected from an open end. If the displacement amplitude of the waves (incident and reflected) are 0.002 cm, the frequency is 1000 Hz and wavelength is 40 cm. Then, the displacement amplitude of vibration at a point at distance 10 cm from the open end, inside the pipe is

(a) 0.002 cm (b) 0.003 cm (c) 0.001 cm (d) 0.000 cm

4 An engine approaches a hill with a constant speed. When it is at a distance of 0.9 km, it blows a whistle whose echo is heard by the driver after 5 s. If the speed of sound in air is 330 m/s, then the speed of the engine is

→ JEE Main (Online) 2013

(a) 32 m/s (b) 27.5 m/s (c) 60 m/s (d) 30 m/s

5 A travelling wave represented by $y = A\sin(\omega t - kx)$ is superimposed on another wave represented by $y = A\sin(\omega t + kx)$. The resultant is

(a) a standing wave having nodes at $x = \left(n + \frac{1}{2}\right)\frac{\lambda}{2}, n = 0, 1, 2$
(b) a wave travelling along $+x$ direction
(c) a wave travelling along $-x$ direction
(d) a standing wave having nodes at $x = \frac{n\lambda}{2}, n = 0, 1, 2.$

6 A piston fitted in cylindrical pipe is pulled as shown in the figure. A tuning fork is sounded at open end and loudest sound is heard at open length 13 cm, 41 cm and 69 cm. The frequency of tuning fork if velocity of sound is 350 ms^{-1}, is

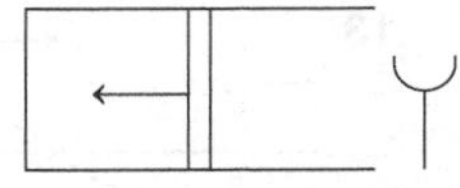

(a) 1250 Hz (b) 625 Hz
(c) 417 Hz (d) 715 Hz

7 A and B are two sources generating sound waves. A listener is situated at C. The frequency of the source at A is 500 Hz. A now, moves towards C with a speed 4 m/s. The number of beats heard at C is 6. When A moves away from C with speed 4 m/s, the number of beats heard at C is 18. The speed of sound is 340 m/s. The frequency of the source at B is → JEE Main (Online) 2013

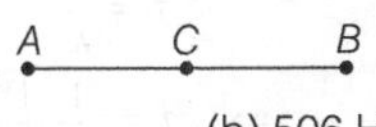

(a) 500 Hz (b) 506 Hz
(c) 512 Hz (d) 494 Hz

8 A pipe of length 85 cm is closed from one end.Find the number of possible natural oscillations of air column in the pipe whose frequencies lie below 1250 Hz. The velocity of sound in air is 340 m/s. → JEE Main 2014

(a) 12 (b) 8 (c) 6 (d) 4

9 A sonometer wire of length 114 cm is fixed at both the ends. Where should the two bridges be placed so as to divide the wire into three segments whose fundamental frequencies are in the ratio 1:3:4? → JEE Main (Online) 2013

(a) At 36 cm and 84 cm from one end
(b) At 24 cm and 72 cm from one end
(c) At 48 cm and 96 cm from one end
(d) At 72 cm and 96 cm from one end

10 The transverse displacement $y(x, t)$ of a wave on a string is given by $y(x, t) = e^{-(ax^2 + bt^2 + 2\sqrt{ab}\,xt)}$. This represents a

(a) wave moving in $-x$ direction with speed $\sqrt{\frac{b}{a}}$
(b) standing wave of frequency $\sqrt{b}$
(c) standing wave of frequency $\frac{1}{\sqrt{b}}$
(d) wave moving in $+x$ direction with speed $\sqrt{a/b}$

11 A train is moving on a straight track with speed 20 ms^{-1}. It is blowing its whistle at the frequency of 1000 Hz. The percentage change in the frequency heard by a person standing near the track as the train passes him is close to (speed of sound = 320 ms^{-1}) → JEE Main 2015

(a) 6% (b) 12% (c) 18% (d) 24%

12 A uniform string of length 20 m is suspended from a rigid support. A short wave pulse is introduced at its lowest end. It starts moving up the string. The time taken to reach the support is (Take, $g = 10\,\text{ms}^{-2}$) → JEE Main (Offline) 2016

(a) $2\pi\sqrt{2}$ s (b) 2 s (c) $2\sqrt{2}$ s (d) $\sqrt{2}$ s

13 A motor cycle starts from rest and accelerates along a straight path at 2 ms^{-2}. At the starting point of the motor cycle, there is a stationary electric siren. How far has the motor cycle gone when the driver hears the frequency of the siren at 94% of its value when the motor cycle was at rest? (Speed of sound = 330 ms^{-1})

(a) 49 m (b) 98 m (c) 147 m (d) 196 m

ANSWERS

SESSION 1

1 (c)	2 (d)	3 (c)	4 (d)	5 (d)	6 (a)	7 (a)	8 (d)	9 (d)	10 (d)
11 (b)	12 (d)	13 (a)	14 (d)	15 (a)	16 (b)	17 (c)	18 (a)	19 (a)	20 (b)
21 (b)	22 (d)	23 (a)	24 (c)	25 (b)	26 (a)	27 (a)	28 (a)	29 (c)	30 (c)
31 (a)									

SESSION 2

1 (c)	2 (c)	3 (d)	4 (d)	5 (a)	6 (b)	7 (c)	8 (c)	9 (b)	10 (a)
11 (b)	12 (c)	13 (b)							

Hints and Explanations

SESSION 1

1 Mechanical transverse wave can be set up in solids but never in liquids and gases.

2 There is no transfer of heat from compression to rarefaction as air is a bad conductor of heat and time of compression and rarefaction is too small.

3 In the first medium, frequency

$$f = \frac{c}{\lambda} = \frac{v}{\lambda}$$

It remains the same in second medium, i.e. $f' = f$

$$\frac{v'}{\lambda'} = \frac{2v}{\lambda'} = \frac{v}{\lambda} \Rightarrow \lambda' = 2\lambda$$

4 Since, frequency, $\nu \propto \sqrt{T}$

$$\therefore \quad \frac{\nu_1}{\nu_2} = \sqrt{\frac{T_1}{T_2}} \quad \text{or} \quad \frac{3}{2} = \sqrt{\frac{T + 2.5}{T}}$$

$$\text{or} \quad \frac{9}{4} = 1 + \frac{2.5}{T} \quad \text{or} \quad \frac{2.5}{T} = \frac{5}{4} \quad \text{or} \quad T = 2\,\text{N}$$

5 Speed of sound is given by

$$v = \sqrt{\frac{\gamma RT}{M}}$$

$$v_{O_2} = \sqrt{\frac{\frac{7}{5}RT}{32}} \quad \text{and} \quad v_{He} = \sqrt{\frac{\frac{5}{3}RT}{4}}$$

$$\therefore \quad \frac{v_{O_2}}{v_{He}} = \sqrt{\frac{7 \times 3 \times 4}{5 \times 32 \times 5}}$$

$$\text{or} \quad v_{He} = 460 \times \sqrt{\frac{5 \times 32 \times 5}{7 \times 3 \times 4}} \approx 1420\,\text{ms}^{-1}$$

6 Let distance between two fixed points be d, then

$$t = \frac{d}{v} \text{ also } v \propto \sqrt{T}$$

$$\Rightarrow \quad \frac{t_1}{t_2} = \frac{v_1}{v_2} = \sqrt{\frac{T_2}{T_1}}$$

$$\Rightarrow \quad \frac{2}{t_2} = \sqrt{\frac{303}{283}} \Rightarrow t_2 = 1.9\,\text{s}$$

7 The given equation be written as

$$y = 4\sin\left[\pi\left(\frac{t}{5} - \frac{x}{9} + \frac{1}{6}\right)\right] \quad \text{...(i)}$$

The standard wave equation can be written as

$$y = a\sin(\omega t - kx + \phi)$$

$$y = a\sin\left(\frac{2\pi}{T}t - \frac{2\pi}{\lambda}x + \phi\right) \quad \text{...(ii)}$$

Equating Eqs. (i) and (ii), we get

Amplitude, $a = 4\,\text{cm}$

Frequency, $f = \frac{1}{T} = \frac{1}{10}\,\text{Hz} = 0.1\,\text{Hz}$

Wavelength, $\lambda = 2 \times 9 = 18\,\text{cm}$

Velocity, $v = f\lambda$

$$= 0.1 \times 18 = 1.8\,\text{cms}^{-1}$$

8 Comparing given equation with standard form of wave equation, we get

$$A = 0.001\,\text{m}, \omega = 100\,\text{s}^{-1}$$

and $k = 1\,\text{m}^{-1}$

$$\therefore \quad \nu = \frac{100}{2\pi} = \frac{50}{\pi}\,\text{Hz}$$

$$\lambda = \frac{2\pi}{k} = 2\pi\,\text{m}$$

$$\text{and } v = \frac{\omega}{k} = \frac{100\,\text{s}^{-1}}{1\,\text{m}^{-1}} = 100\,\text{ms}^{-1}$$

Moreover, as the wave equations of the form $y = A\sin(\omega t + kx)$, the wave is travelling along negative x-direction.

9 From the given wave equation, we find that

$$A = 4\,\text{cm}, \omega = \frac{2\pi}{0.02}\,\text{s}^{-1} = 100\,\pi\text{s}^{-1}$$

$$\text{and } k = \frac{2\pi}{100}\,\text{cm}^{-1}$$

$$\therefore \quad \lambda = \frac{2\pi}{k} = 100\,\text{cm}$$

$$\nu = \frac{\omega}{2\pi} = \frac{100\pi}{2\pi} = 50\,\text{Hz}$$

$$\text{and} \quad v = \frac{\omega}{k} = \frac{100\pi}{\frac{2\pi}{100}} = 50 \times 10^2\,\text{cms}^{-1}$$

Hence, answer of wave speed v is wrong, i.e. option (d) is correct.

10 Tension in the string is given by

$$T = \mu v^2 = \mu\frac{\omega^2}{k^2}$$

$$= \frac{0.04\,(2\pi/0.04)^2}{(2\pi/0.50)^2} = 6.25\,\text{N}$$

11 $y_1 = 0.05\cos(0.50\,\pi\,x - 100\pi\,t)$

and $y_2 = 0.05\cos(0.46\,\pi\,x - 92\,\pi t)$

Comparing these two equations with $y = A\sin(k\,x - \omega t)$

We have, $\omega_1 = 100\,\pi$ and $\omega_2 = 92\,\pi$

$$\text{Now, speeds, } v_1 = \frac{A}{100\pi} = \frac{0.05}{100\,\pi}$$

$$\text{and} \quad v_2 = \frac{A}{92\,\pi} = \frac{0.05}{92\,\pi}$$

Now, the resultant speed,

$$v = \sqrt{\left(\frac{0.05}{92\,\pi}\right)^2 + \left(\frac{0.05}{100\,\pi}\right)^2} \quad [\pi = 3.14]$$

$$= 200\,\text{m/s}$$

12 From law of string, other factors remaining unchanged,

$$\frac{\nu_1}{\nu_2} = \frac{l_2}{l_1}\sqrt{\frac{T_1}{T_2}}$$

Now, $\nu_2 = 2\nu_1, l_2 = \frac{3}{4}l_1$,

Hence, we have

$$\frac{1}{2} = \frac{3}{4} \times \sqrt{\frac{T_1}{T_2}} \Rightarrow \frac{2}{3} = \sqrt{\frac{T_1}{T_2}}$$

$$\Rightarrow \quad \frac{T_2}{T_1} = \frac{9}{4} \quad \text{or} \quad T_2 = \frac{9}{4}T_1$$

13

$\frac{\lambda}{4}$

L

From vibration mode,

$$\frac{\lambda}{2} = L$$

$$\Rightarrow \quad \lambda = 2L$$

$$\therefore \text{ Wave speed, } v = \sqrt{\frac{Y}{\rho}}$$

So, frequency

$$f = \frac{v}{\lambda}$$

$$\Rightarrow \quad f = \frac{1}{2L}\sqrt{\frac{Y}{\rho}}$$

$$= \frac{1}{2 \times 60 \times 10^{-2}}\sqrt{\frac{9.27 \times 10^{10}}{2.7 \times 10^3}}$$

$$\approx 5000\,\text{Hz}$$

$$f = 5\,\text{kHz}$$

14 For open ends, fundamental frequency f in air, we have

$$\frac{\lambda}{2} = l$$

$$\Rightarrow \quad \lambda = 2l$$

$$v = f\lambda$$

$$\Rightarrow \quad f = \frac{v}{\lambda} = \frac{v}{2l} \quad \text{...(i)}$$

l

When a pipe is dipped vertically in water, so that half of it is in water, we have

$l/2$

l

$$\frac{\lambda}{4} = \frac{l}{2}$$

$$\Rightarrow \quad \lambda = 2l \Rightarrow v = f'\lambda$$

$$\Rightarrow \quad f' = \frac{v}{\lambda} = \frac{v}{2l} = f \quad \text{...(ii)}$$

Thus, the fundamental frequency of the air column is now,

$$f = f'$$

15 We know that, third overtone of closed organ pipe means seventh harmonic

$\therefore \quad (\nu_7)_{closed} = (\nu_4)_{open}$

or $\quad 7\left(\frac{v}{4l_c}\right) = 4\left(\frac{v}{2l_o}\right)$

or $\quad \frac{l_o}{l_c} = \frac{8}{7}$

or $\quad \frac{l_c}{l_o} = \frac{7}{8}$

16 $y_1 = 0.25 \sin(310t)$ ...(i)

and $y_2 = 0.25 \sin(316t)$ ...(ii)

We have, $\omega_1 = 310$

$\Rightarrow \quad f_1 = \frac{310}{2\pi}$ unit

and $\quad \omega_2 = 316$

$\Rightarrow \quad f_2 = \frac{316}{2\pi}$ unit

Hence, beat frequency

$= f_2 - f_1 = \frac{316}{2\pi} - \frac{310}{2\pi}$

$= \frac{3}{\pi}$ unit

17 Maximum number of beats $= \nu + 1 - (\nu - 1) = 2$

18 There are four beats between P and Q, therefore the possible frequencies of P are 246 or 254 (i.e. 250 ± 4) Hz.

When the prong of P is filled, its frequency becomes greater than the original frequency.

If we assume that the original frequency of P is 254, then on filling its frequency will be greater than 254. The beats between P and Q will be more than 4. But it is given that the beats are reduced to 2, therefore, 254 is not possible. Therefore, the required frequency must be 246 Hz.

19 As forks have been arranged in ascending order of frequencies, hence if frequency of Ist fork be n, then

$n_2 = n + 8$

$n_3 = n + 2 \times 8 = n + 16$

and $\quad n_{16} = n + 15 \times 8$

$= n + 120 = 2n$

$\Rightarrow \quad n = 120$ Hz

20 Since, vehicle having siren is moving in a perpendicular direction, hence there will be no Doppler shift in frequency and $n_1 = 0$.

21 As the observer is moving towards the source, so frequency of waves emitted by the source will be given by the formula,

$$f_{observed} = f_{actual} \cdot \left(\frac{1 + v/c}{1 - v/c}\right)^{1/2}$$

Here, frequency, $\frac{v}{c} = \frac{1}{2}$

So, $f_{observed} = f_{actual}\left(\frac{3/2}{1/2}\right)^{1/2}$

$\therefore \quad f_{observed} = 10 \times \sqrt{3} = 17.3$ GHz

22 Here, $v = 340$ ms^{-1},

$v_s =$ velocity of 1st train $= 20$ ms^{-1},

$v_o =$ velocity of 2nd train $= 15$ ms^{-1} and

$\nu_0 = 600$ Hz.

As S and O are approaching each other, hence

$$\nu = \left[\frac{v + v_o}{v - v_s}\right] \nu_0$$

$$= \left(\frac{340 + 15}{340 - 20}\right) \times 600 = 666 \text{ Hz}$$

23 Here, $\nu_a = \nu\left(\frac{v + v_o}{v}\right)$

or $\quad \frac{\nu_a}{\nu} = 1 + \frac{v_o}{v}$

or $\quad \frac{v_o}{v} = \frac{\nu_a}{\nu} - 1$ ...(i)

Again, $\nu_r = \nu\left(\frac{v - v_o}{v}\right)$

$\therefore \quad \frac{v_o}{v} = 1 - \frac{\nu_r}{\nu}$ ...(ii)

From Eqs. (i) and (ii), we get

$\frac{\nu_a}{\nu} - 1 = 1 - \frac{\nu_r}{\nu}$

or $\quad \frac{\nu_a}{\nu} + \frac{\nu_r}{\nu} = 2$

or $\quad \frac{\nu_a + \nu_r}{\nu} = 2$

or $\quad \nu = \frac{\nu_a + \nu_r}{2}$

24 If a closed pipe of length L is in resonance with a tuning fork of frequency ν, then $\nu = \frac{v}{4L}$

An open pipe of same length L produces vibrations of frequency $\frac{v}{2L}$. Obviously, it cannot be in resonance with the given tuning fork of frequency $\nu\left(= \frac{v}{4L}\right)$.

25 At the point where a compression and a rarefaction meet, the displacement is minimum and it is called displacement node. At this point, the pressure difference is maximum i.e. at the same time, it is a pressure antinode.

On the other hand, at the mid-point of a compression or a rarefaction, the displacement variation is maximum i.e. such a point is displacement antinode. However such a point is pressure node, as pressure variation is minimum at such a point.

26 Stationary wave is represented as shown in figure.

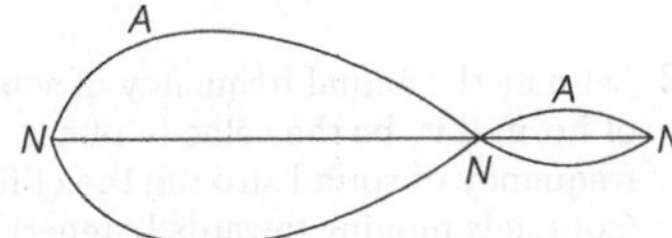

It is quite clear from figure that at nodes the amplitude is zero and velocity of particle is also zero and at antinodes the amplitude is maximum. So that the velocity of particle is also maximum and all particles cross mean position between two successive nodes.

27 Sounds coming from the different sources can be recognised by virtue of their quality which is characteristics of sound. That is why we recognise the voices of our friends.

28 According to Laplace, the changes in pressure and volume of a gas, when sound waves propagated through it, are not isothermal but adiabatic. A gas is a bad conductor of heat. It does not allow the free exchange of heat between compressed layer, rarified layer and its surrounding.

29 The resultant amplitude of two waves is given by

$$A = \sqrt{a_1^2 + a_2^2 + 2a_1 a_2 \cos\theta}$$

Here, $a_1 = a_2 = A = a$

$1/2 = 1 + \cos\theta$

or $\cos\theta = -1/2$ or $\theta = 120°$

30 $I = \frac{1}{2}\rho\omega^2 A^2 v$

Here, $\rho =$ density of medium,

$A =$ amplitude,

$\omega =$ angular frequency and

$v =$ velocity of wave

$\therefore$ Intensity depend upon amplitude, frequency as well as velocity of wave

Also, $\quad I_1 = I_2$

31 $\nu = \eta\left(\frac{v}{2l}\right)$, where $n = 1, 2, 3, \ldots$

As, η increases, frequency increases. Hence, wavelength decreases.

SESSION 2

1 $I_m = 4ka^2, I = 2ka$

$I' = k(a^2 + a^2 + 2a^2 \cos^2 \theta)$

$\Rightarrow \quad \phi = \frac{\pi}{2}$

Path difference $= \frac{\lambda}{4} = 2x$

$\Rightarrow \quad \frac{v}{4f} = 2x$

$\Rightarrow \quad x = \frac{v}{8f}$

2 Let n be the actual frequency of sound of horn. If v_s be the velocity of car, then frequency of sound striking the cliff (source is moving towards listener)

$$n' = \frac{v \times n}{v - v_s} \quad \text{...(i)}$$

The frequency of sound heard on reflection

$$n'' = \frac{(v + v_s)n'}{v} = \frac{(v + v_s)}{v} \times \frac{v \times n}{(v - v_s)}$$

or $\quad \frac{n''}{n} = \frac{v + v_s}{v - v_s} = 2$

$v + v_s = 2v - 2v_s$

or $\quad v_s = \frac{v}{3}$

3 The equation of stationary wave for open organ pipe can be written as,

$y = 2A \cos\left(\frac{2\pi x}{\lambda}\right) \sin\left(\frac{2\pi ft}{v}\right)$,

where x is the open end from where wave gets reflected.

Amplitude of stationary wave is,

$A_s = 2A \cos\left(\frac{2\pi x}{\lambda}\right)$

For $x = 0.1$ m,

$A_s = 2 \times 0.002 \cos\left[\frac{2\pi \times 0.1}{0.4}\right] = 0$

4 We have, $900 + 900 - x$

$= 330 \times 5 = 1650$

(900 – x) m

x 900 m

$\therefore \quad x = 150$ m

$\therefore$ Speed $= \frac{150}{5} = 30$ m/s

5 $y = y_1 + y_2$

$= A \sin(\omega t - kx) + A \sin(\omega t + kx)$

$y = 2A \sin\omega t \cos kx$

Clearly, it is equation of standing wave for position of nodes $y = 0$.

i.e. $\quad x = (2n + 1)\frac{\lambda}{4}$

$= \left(n + \frac{1}{2}\right)\frac{\lambda}{2}$

6 In a closed organ pipe in which length of air-column can be increased or decreased, the first resonance occurs at $\lambda/4$ and second resonance occurs at $3\lambda/4$.

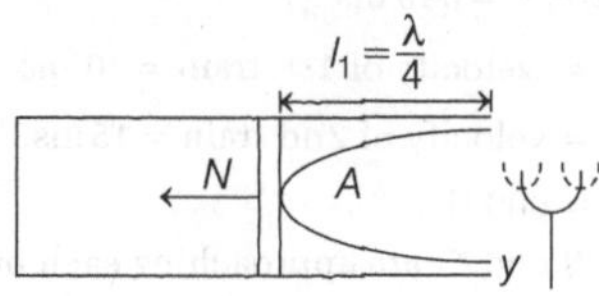

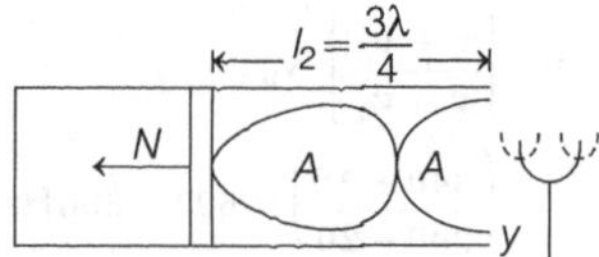

Thus, at first resonance,

$$\frac{\lambda}{4} = 13 \quad \text{...(i)}$$

and at second resonance,

$$\frac{3\lambda}{4} = 41 \quad \text{...(ii)}$$

Subtracting Eq. (i) from Eq. (ii), we have

$\frac{3\lambda}{4} - \frac{\lambda}{4} = 41 - 13$

$\therefore \quad \lambda = 56$ cm

Hence, frequency of tuning fork,

$n = \frac{v}{\lambda} = \frac{350}{56 \times 10^{-2}} = 625$Hz

7 Here, frequency of source = 500 Hz

Speed of source $A = 4$ m/s $= u$

Then, source is moving towards stationary observer,

$v' = \frac{v}{v - u} v_0$

(where, v = speed of sound)

$= \frac{340}{340 - 4} \times 500$

$\Rightarrow \quad v' = \frac{340}{336} \times 500$ Hz

$= 506$ Hz

Now, when source is reciding from the observer

$v' = \frac{v}{v + u} v_0$

$= \frac{340}{344} \times 500$ Hz

$\therefore \quad v' = 494$ Hz

According to question,

Let freqency of source B is Z Hz

$\therefore \quad Z = 506 \pm 6 \Rightarrow Z = 500$ or 512

and $Z = 494 \pm 18 \Rightarrow Z = 512$ or 476

Thus, required frequency = 512 Hz

8 For closed organ pipe $= \frac{(2n + 1)v}{4l}$

[where, n=0,1,2.....]

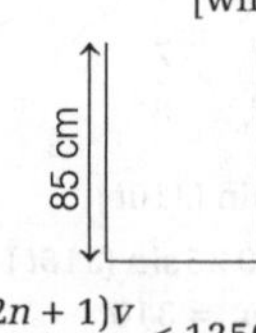

$\frac{(2n + 1)v}{4l} < 1250$

$(2n + 1) < 1250 \times \frac{4 \times 0.85}{340}$

$(2n + 1) < 12.52$

$n < 5.25$

So, $n = 0,1,2,3,...,5$

So, we have 6 possibilities.

9 $v_1 = \frac{1}{2L_1}\sqrt{\frac{f}{\mu}}$

$L_1 \quad L_2 \quad L_3$

Let length of three segments be L_1, L_2 and L_3,

$v_2 = \frac{1}{2L_2}\sqrt{f/\mu}$

$v_3 = \frac{1}{2L_3}\sqrt{f/\mu}$

So, that $\quad v_1 L_1 = v_2 L_2 = v_3 L_3$

As, $\quad v_1 : v_2 : v_3 = 1 : 3 : 4$

$v_1 = \frac{v_2 L_2}{L_1}$

and $\quad v_2 = 3v_1, v_3 = 4v_1$

$L_2 = \frac{v_1}{v_2} L_1 = \frac{1}{3} L_1 = \frac{L_1}{3}$

and $L_3 = \frac{v_1}{v_3} L_1 = \frac{v_1}{4v_1} L_1 = \frac{L_1}{4}$

$L_1 + L_2 + L_3 = 114$

Now, $L_1\left(1 + \frac{1}{3} + \frac{1}{4}\right) = 114$

$\Rightarrow L_1\left(\frac{12 + 4 + 3}{12}\right) = 114$

$L_1\left(\frac{19}{12}\right) = 114$

$\Rightarrow L_1 = \frac{(114 \times 12)}{19} = 72$ cm

$L_2 = \frac{L_1}{3} = \frac{72}{3} = 24$ cm

10 $y(x,t) = e^{-(ax^2 + bt^2 + 2\sqrt{ab}xt)}$

$= e^{-(\sqrt{a}x + \sqrt{b}\,t)^2}$

It is a function of type

$y = f(\omega t + kx)$

$\therefore$ $y(x,t)$ represents wave travelling along $-x$ direction.

Speed of wave $= \dfrac{\omega}{k} = \dfrac{\sqrt{b}}{\sqrt{a}} = \sqrt{\dfrac{b}{a}}$

11 Apparent frequency heard by the person before crossing the train.

$$f_1 = \left(\frac{c}{c - v_s}\right) f_o = \left(\frac{320}{320 - 20}\right) 1000$$

Similarly, apparent frequency heard,after crossing the trains

$$f_2 = \left(\frac{c}{c + v_s}\right) f_o = \left(\frac{320}{320 + 20}\right) 1000$$

[c = speed of sound in air]

$$\Delta f = f_1 - f_2$$

$$= \left(\frac{2cv_s}{c^2 - v_s^2}\right) f_0$$

Percentage change in frequency heard by the person standing near the track as the train passes him is

or $\dfrac{\Delta f}{f_0} \times 100 = \left(\dfrac{2cv_s}{c^2 - v_s^2}\right) \times 100$

$$= \frac{2 \times 320 \times 20}{300 \times 340} \times 100$$

$$= \frac{2 \times 32 \times 20}{3 \times 34}$$

$$= 12.55\% = 12\%$$

12 A uniform string of length 20 m is suspended from a rigid support. Such that the time taken to reach the support,

$$T = \frac{mgx}{l}$$

(l, m) P dx x

So, velocity at point $P = \sqrt{\dfrac{\dfrac{mgx}{l}}{m/l}}$

i.e. $v = \sqrt{gx}$

$$\frac{dx}{dt} = \sqrt{gx}$$

$\Rightarrow$ $\displaystyle\int_0^{20} \frac{dx}{\sqrt{x}} = \int_0^t \sqrt{g}\,dt$

$$[2\sqrt{x}]_0^{20} = \sqrt{10}t$$

$\Rightarrow$ $2\sqrt{20} = \sqrt{10}t$

$$t = 2\sqrt{2}\text{ s}$$

13 For motor cycle, $u = 0, a = 2\text{ ms}^{-2}$

Observer is in motion and source is at rest, then apparent frequency,

$$v' = v\,\frac{v - v_o}{v + v_s}$$

$\Rightarrow$ $\dfrac{94}{100}v = v\,\dfrac{330 - v_o}{330}$

$\Rightarrow$ $330 - v_o = \dfrac{330 \times 94}{100}$

$\Rightarrow$ $v_o = 330 - \dfrac{94 \times 33}{10}$

$$= \frac{33 \times 6}{10}\text{ ms}^{-1}$$

$$s = \frac{v^2 - u^2}{2a}$$

$$= \frac{9 \times 33 \times 33}{100}$$

$$= \frac{9 \times 1089}{100} \approx 98\text{ m}$$

DAY THIRTEEN

Unit Test 2

(Waves and Oscillations)

1 A mass m attached to a spring of spring constant k is stretched a distance x_0 from its equilibrium position and released with no initial velocity. The maximum speed attained by the mass in its subsequent motion and the time at which this speed would be attained are respectively,

(a) $\sqrt{\frac{k}{m}}x_0,\ \pi\sqrt{\frac{m}{k}}$ (b) $\sqrt{\frac{k}{m}}\frac{x_0}{2},\ \frac{\pi}{2}\sqrt{\frac{m}{k}}$

(c) $\sqrt{\frac{k}{m}}x_0,\ \frac{\pi}{2}\sqrt{\frac{m}{k}}$ (d) $\sqrt{\frac{k}{m}}\frac{x_0}{2},\ \pi\sqrt{\frac{m}{k}}$

2 The acceleration-displacement graph of a particle executing SHM is shown in the figure. The time period of SHM is

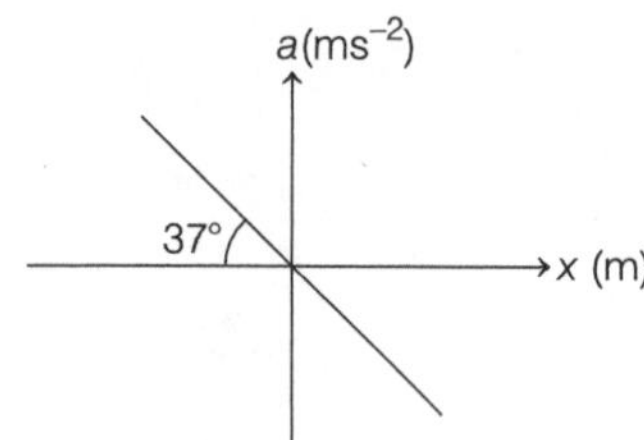

(a) $\frac{4\pi}{\sqrt{3}}$ s

(b) $\frac{2\pi}{\sqrt{3}}$ s

(c) The given graph doesn't represent SHM

(d) Information is insufficient

3 A spring balance has a scale that can read from 0 to 50 kg. The length of the scale is 20 cm. A body suspended from this balance, when displaced and released, oscillates harmonically with a time period of 0.6 s. The mass of the body is (Take, $g = 10\,\text{ms}^{-2}$)

(a) 10 kg (b) 25 kg
(c) 18 kg (d) 22.8 kg

4 A block is kept on a table which performs simple harmonic motion with frequency 5Hz in horizontal plane. The maximum amplitude of the table at which block does not slip on the surface of table is, (if coefficient of friction between the block and surface of table is 0.6.) (Given, $g = 10\text{ m/s}^2$).

(a) 0.06 m (b) 0.006 m
(c) 0.02 m (d) 0.002 m

5 For a particle executing SHM, determine the ratio of average acceleration of the particle between extreme position and the equilibrium position w.r.t. the maximum acceleration

(a) $\frac{4}{\pi}$ (b) $\frac{2}{\pi}$

(c) $\frac{1}{\pi}$ (d) $\frac{1}{2\pi}$

6 Two springs are made to oscillate simple harmonically when the same mass is suspended, individually. The time periods obtained are T_1 and T_2. If both the springs are connected in series and then made to oscillate when suspended by the same mass, the resulting time will be

(a) $T_1 + T_2$ (b) $\frac{T_1T_2}{T_1 + T_2}$

(c) $\sqrt{T_1^2 + T_2^2}$ (d) $\frac{T_1 + T_2}{2}$

7 Find the time period of oscillations of a torsional pendulum, if the torsional constant of wire is $10\pi^2$ in SI units. The moment of inertia of the rigid body is 10 kg - m^2 about the axis of rotation

(a) 1 s (b) 2 s

(c) 4 s (d) $\frac{1}{2}$ s

8 Acceleration-displacement graph for four particles are shown, identify the one which represents SHM for all the values of displacements

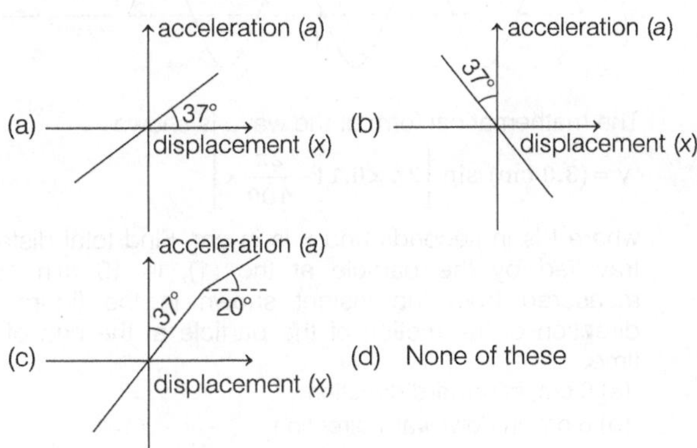

9 In case of simple harmonic motion, if the velocity is plotted along the X-axis and the displacement from the equilibrium position is plotted along the Y-axis, the resultant curve happens to be an ellipse with the ratio $\frac{\text{major axis (along } X)}{\text{minor axis (along } Y)} = 20\pi$.

What is the frequency of the simple harmonic motion?

(a) 100 Hz (b) 20 Hz
(c) 10 Hz (d) $\frac{1}{10}$ Hz

10 A simple pendulum is suspended from the ceiling of a stationary tram car. Now, the car starts accelerating, the time period of a simple pendulum is the least when [Take magnitude of acceleration to be same in all the cases]

(a) car is accelerating up
(b) car is accelerating down
(c) car is accelerating horizontally
(d) car is stationary

11 A particle executes SHM about O with an amplitude A and time period T. The magnitude of its acceleration, at $\frac{T}{8}$ s after the particle reaches the extreme position, would be

(a) $\frac{4\pi^2 A}{\sqrt{2}T^2}$ (b) $\frac{4\pi^2 A}{T^2}$
(c) $\frac{2\pi^2 A}{\sqrt{2}T^2}$ (d) None of these

12 A string of length 1.5 m with its two ends clamped, is vibrating in the fundamental mode. The amplitude at the centre of the string is 4 mm. The minimum distance between two points having amplitude 2 mm, is

(a) 1 m (b) 75 cm
(c) 60 cm (d) 50 cm

13 A spring of negligible mass having a force constant k extends by an amount y when a mass m is hung from it. The mass is pulled down a little and then released. The system begins to execute SHM of amplitude A and angular frequency ω. The total energy of the mass-spring system will be

(a) $\frac{m\omega^2 A^2}{2}$ (b) $\frac{m\omega^2 A^2}{2} - \frac{ky^2}{2}$
(c) $\frac{ky^2}{2}$ (d) $\frac{m\omega^2 A^2}{2} + \frac{ky^2}{2}$

14 Total number of independent harmonic waves in the resultant displacement equation given by

$$y = 3\sin^2 \frac{t}{2} \sin 800t$$

(a) 2 (b) 1 (c) 4 (d) 3

15 A load of mass m falls from a height h on the scale pan hung from a spring as shown. If the spring constant is k, mass of the scale pan is zero and the mass m does not bounce relative to the pan, then the amplitude of vibration is

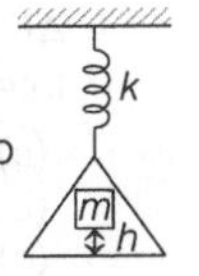

(a) $\frac{mg}{k}$ (b) $\frac{mg}{k}\sqrt{1+\frac{2hk}{mg}}$
(c) $\frac{mg}{k} + \frac{mg}{k}\sqrt{\frac{1+2hk}{mg}}$ (d) None of these

16 A sound absorber attenuates the sound level by 30dB. The intensity of sound is decreased by a factor of

(a) 100 (b) 1000
(c) 10000 (d) 10

17 The velocity of a particle executing a simple harmonic motion is 13 ms^{-1}, when its distance from the equilibrium position is 3 m and its velocity is 12 ms^{-1}, when it is 5 m away from equilibrium position. The frequency of the SHM is

(a) $\frac{5\pi}{8}$ (b) $\frac{5}{8\pi}$
(c) $\frac{8\pi}{5}$ (d) $\frac{8}{5\pi}$

18 Two identical strings A and B, have nearly the same tension. When they both vibrate in their fundamental resonant modes, there is a beat frequency of 3 Hz. When string B is tightened slightly, to increase the tension, the beat frequency becomes 6 Hz. This means

(a) that before tightening A had a higher frequency than B, but after tightening, B has a higher frequency than A
(b) that before tightening B has higher frequency than A, but after tightening A has higher frequency than B
(c) that before and after tightening A has higher frequency than B
(d) that before and after tightening B has higher frequency than A

19 The ratio of densities of oxygen and nitrogen is 16 : 14. At what temperature, the speed of sound in oxygen will be equal to its speed in nitrogen at 14°C?
(a) 50°C (b) 52°C
(c) 48°C (d) 55°C

20 A train is passing by a platform at a constant speed of 40 ms^{-1}. The horn of the train has a frequency of 320 Hz. Find the overall change in frequency detected by a person standing on the platform, i.e. when the train approaching and then precedes from him. (Take, velocity of sound in air as 320 ms^{-1})
(a) 216.4 Hz (b) 81.3 Hz
(c) 365.7 Hz (d) 284.4 Hz

21 A string of length 0.4 m and mass 10^{-2} kg is clamped at one end. The tension in the string is 1.6 N. Identical wave pulses are generated at the free end, after a time interval Δt. The minimum value of Δt, so that a constructive interference takes place between successive pulses is
(a) 0.1s
(b) 0.05 s
(c) 0.2 s
(d) Constructive interference cannot take place

22 A string vibrates according to the equation
$Y = 5 \sin \left(\frac{2\pi x}{3}\right) \times \cos 20\pi t$, where x and y are in cm and t in second. The distance between two adjacent nodes is
(a) 3 cm (b) 4.5 cm (c) 6 cm (d) 1.5 cm

23 A point source of sound is placed in a non-absorbing medium. Two points A and B are at the distance of 1 m and 2 m, respectively from the source. The ratio of amplitudes of waves at A to B is
(a) 1 : 1 (b) 1: 4 (c) 1: 2 (d) 2 : 1

24 The mathematical form of three travelling waves are given by

$$y_1 = (2 \text{ cm}) \sin (3x - 6t),$$
$$y_2 = (3 \text{ cm}) \sin (4x - 12t),$$
and $$y_3 = (4 \text{ cm}) \sin (5x - 11t)$$

of these waves,
(a) wave 1 has greatest wave speed and wave has maximum transverse string speed
(b) wave 2 has greatest wave speed and wave 1 has greatest maximum transverse string speed
(c) wave 3 has greatest wave speed and wave 1 has maximum transverse string speed
(d) wave 2 has greatest wave speed and wave 3 has maximum transverse string speed

25 If the maximum speed of a particle carrying a travelling wave is v_0, then find the speed of a particle when the displacement is half that of the maximum value
(a) $\frac{v_0}{2}$ (b) $\frac{\sqrt{3}v_0}{4}$ (c) $\frac{\sqrt{3}v_0}{2}$ (d) v_0

26 A transverse wave on a string travelling along positive x-axis has been shown in the figure below

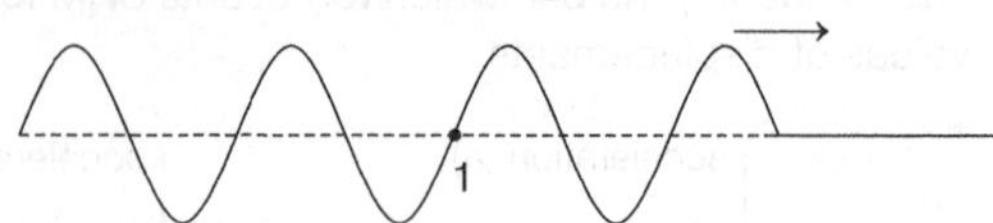

The mathematical form of the wave is shown
$y = (3.0 \text{ cm}) \sin \left[2\pi \times 0.1 t - \frac{2\pi}{100} x\right]$
where t is in seconds and x is in cm. Find total distance travelled by the particle at the (1), in 10 min 15 s, measured from the instant shown in the figure and direction of the motion of the particle at the end of this time.
(a) 6 cm, in upward direction
(b) 6 cm, in downward direction
(c) 738 cm, in upward direction
(d) 732 cm, in upward direction

27 A wire having a linear mass density of 5×10^{-3} kg m^{-1} is stretched between two rigid supports with a tension of 450 N. The wire resonates at a frequency of 420 Hz. The next higher frequency at which the wire resonates is 490Hz. The length of the wire is
(a) 2.5 m (b) 2.14 m
(c) 2.25 m (d) 2.0 m

28 A man stands on a weighing machine placed on a horizontal plateform. The machine reads 50 kg. By means of a suitable mechanism, the plateform is made to execute harmonic vibration up and down with a frequency of 2 vibrations per second. What will be the effect on the reading of the weighing machine? The amplitude of vibration of plateform is 5 cm. (Take, $g = 10 \text{ ms}^{-2}$)
(a) 11 kgf to 93 kgf
(b) 10.5 kgf to 89.5 kgf
(c) 10 kgf to 15.5 kgf
(d) 25.6 kgf to 100.5 kgf

29 A small trolley of mass 2.0 kg resting on a horizontal turn table is connected by a light spring to the centre of the table. When the turn table is set into rotation at speed of 360 rpm, the length of the stretched spring is 43 cm. If the original length of the spring is 36 cm, the force constant is
(a) 17025 Nm^{-1} (b) 16225 Nm^{-1}
(c) 17475 Nm^{-1} (d) 17555 Nm^{-1}

30 At 16°C, two open end organ pipes, when sounded together give 34 beats in 2 s. How many beats per second will be produced, if the temperature is raised to 51°C?
(Neglect increase in length of the pipes)
(a) 18 s^{-1} (b) 15 s^{-1}
(c) 20 s^{-1} (d) 10 s^{-1}

31 During earthquake, both longitudinal and transverse waves are produced having speeds 4.0 km/s and 8.0 km/s, respectively. If the first transverse wave reaches the seismograph 8 minutes after the arrival of first longitudinal wave, then the distance of the position, where the earthquake occurred is

(a) 3440 km (b) 3880 km
(c) 3840 km (d) 3500 km

32 A pendulum has time period T in air. When it is made to oscillate in water, it acquired a time period $T' = \sqrt{2}T$. The density of the pendulum bob is equal to (Take, density of water = 1)

(a) $\sqrt{2}$ (b) 2
(c) $2\sqrt{2}$ (d) None of these

33 On a planet a freely falling body takes 2 s when it is dropped from a height of 8 m. The time period of simple pendulum of length 1 m on that planet is

(a) 3.14 s (b) 16.28 s
(c) 1.57 s (d) None of these

Direction (Q. Nos. 34-40) *Each of these questions contains two statements Statement I and Statement II. Each of these questions also has four alternative choices, only one of which is the correct answer. You have to select one of the codes (a), (b), (c), (d) given below*

(a) Statement I is true, Statement II is true; Statement II is the correct explanation for Statement I
(b) Statement I is true, Statement II is true; Statement II is not the correct explanation for Statement I
(c) Statement I is true, Statement II is false
(d) Statement I is false, Statement II is true

34 **Statement I** Waves on a string can be longitudinal in nature.

Statement II The string cannot be compressed or rarified.

35 **Statement I** A wave of frequency 500 Hz is propagating with a velocity of 350 m/s. Distance between two particles with 60° phase difference is 12 cm.

Statement II $x = \frac{\lambda}{2\pi}\phi$

36 **Statement I** When a wave goes from one medium to other, average power transmitted by the wave may change.

Statement II Due to a change in the medium, amplitude, speed, wavelength and frequency of the wave may change.

37 **Statement I** A particle performs a simple harmonic motion with amplitude A and angular frequency ω. To change the angular frequency of simple harmonic motion to 3ω, and amplitude to $A/2$, we have to supply an extra energy of $\frac{5}{4}m\omega^2A^2$, where m is the mass of the particle executing simple harmonic motion.

Statement II Angular frequency of the simple harmonic motion is independent of the amplitude of oscillation.

38 **Statement I** Time period of spring pendulum is the same whether in an accelerated or in an inertial frame of reference.

Statement II Mass of the bob of the spring pendulum and the spring constant of spring are independent of the acceleration of the frame of reference.

39 **Statement I** The total energy of a particle executing simple harmonic motion, can be negative.

Statement II Potential energy of a system can be negative.

40 **Statement I** A circular metal hoop is suspended on the edge, by a hook. The hoop can oscillate from one side to the other in the plane of the hoop, or it can oscillate back and forth in a direction perpendicular to the plane of the hoop. The time period of oscillation would be more when oscillations are carried out in the plane of hoop.

Statement II Time period of physical pendulum is more, if the moment of inertia of the rigid body about the corresponding axis passing through the pivoted point, is more.

ANSWERS

1. (c)	**2.** (a)	**3.** (d)	**4.** (b)	**5.** (b)	**6.** (c)	**7.** (b)	**8.** (b)	**9.** (c)	**10.** (c)
11. (a)	**12.** (a)	**13.** (d)	**14.** (d)	**15.** (b)	**16.** (b)	**17.** (b)	**18.** (d)	**19.** (d)	**20.** (b)
21. (c)	**22.** (a)	**23.** (d)	**24.** (d)	**25.** (c)	**26.** (c)	**27.** (b)	**28.** (b)	**29.** (c)	**30.** (a)
31. (c)	**32.** (b)	**33.** (a)	**34.** (d)	**35.** (a)	**36.** (c)	**37.** (d)	**38.** (a)	**39.** (a)	**40.** (a)

Hints and Explanations

1 At the mean position, the speed will be maximum.

$$\frac{kx_0^2}{2} = \frac{mv^2}{2}$$

$$\Rightarrow \quad v = v_{max} = \sqrt{\frac{k}{m}}\, x_0$$

and this is attained at $t = \frac{T}{4}$.

Time period of motion is,

$$T = 2\pi\sqrt{\frac{m}{k}}$$

So, the required time is,

$$t = \frac{T}{4} = \frac{\pi}{2}\sqrt{\frac{m}{k}}$$

2 $\frac{da}{dx} = -\tan 37° = -\frac{3}{4} \Rightarrow a = -\frac{3}{4}x$

On comparing with $a = -\omega^2 x$, we get

$$\omega^2 = \frac{3}{4} \Rightarrow \frac{2\pi}{T} = \frac{\sqrt{3}}{2} \Rightarrow T = \frac{4\pi}{\sqrt{3}}\text{ s}$$

3 The scale can read a maximum of 50 kg, for a length of 20 cm. Let spring constant be k then,

$$kx_0 = mg$$

[for $m = 50$ kg, $x_0 = 20$ cm]

$$\Rightarrow k \times 0.2 = 50 \times 10 \Rightarrow k = 2500\text{ Nm}$$

Let mass of the body be m_0, then from

$$T = 2\pi\sqrt{\frac{m_0}{k}} \Rightarrow 0.6 = 2\pi\sqrt{\frac{m_0}{2500}}$$

$$\Rightarrow m_0 = 22.8\text{ kg}$$

4 Maximum force on the block on the surface of table due to simple harmonic motion, $F = m\omega^2 A$, where $A \to$ amplitude.

Friction force on the block, $F_s = \mu mg$

It will not slip on the surface of the table, if

$$F = F_s$$

$$m\omega^2 A = \mu mg$$

$$A = \frac{\mu g}{\omega^2}$$

$$= \frac{0.6 \times 10}{(2 \times 3.14 \times 5)^2}$$

$$= 0.006\text{ m}$$

5 Let the equation of SHM be,

$x = A \sin \omega t$

Average acceleration between extreme position and the equilibrium position, i.e. from time $t = 0$ to $t = \frac{T}{4}$

$$I_m = \frac{\int_0^{T/4} \omega^2 A \sin \omega t\, dt}{T/4}$$

Maximum acceleration $(a_{max}) = \omega^2 A$.

Then, the required ratio is,

$$= \frac{\int_0^{T/4} \omega^2 A \sin \omega t\, dt}{T/4 \times \omega^2 A} = \frac{2}{\pi}$$

6 Let the spring constants of the two springs be k_1 and k_2 respectively, then,

$$T = 2\pi\sqrt{\frac{m}{K}} \text{ and } k = \frac{4\pi^2 m}{T^2}$$

When the two springs are connected in series, then

$$T = 2\pi\sqrt{\frac{m}{k_{eq}}}$$

$$\text{where,} \quad k_{eq} = \frac{k_1 k_2}{k_1 + k_2}$$

$$\Rightarrow \quad T = \sqrt{T_1^2 + T_2^2}$$

7 For torisional pendulum, $\tau = -k\theta$

$$\alpha = -\frac{k}{I}\theta$$

$$\Rightarrow \quad \omega^2 = \frac{k}{I}$$

$$T = 2\pi\sqrt{\frac{I}{k}} = 2\pi\sqrt{\frac{10}{10\pi^2}} = 2\text{ s}$$

8 For a particle to execute SHM,

$a = -\omega^2 x$

So, $\frac{da}{dx} = -\omega^2$, where ω^2 is positive quantity. This means for a particle to execute SHM, the acceleration-displacement curve should be a straight line having a negative slope, which is shown in option (b).

9 In representation of SHM in ellipse as velocity along x-axis and displacement along y-axis.

$$\because \frac{\text{Major axis (along } X\text{-axis)}}{\text{Minor axis (along } Y\text{-axis)}} = 20\,\pi$$

$$\Rightarrow \frac{\omega A}{A} = 20\pi \Rightarrow \omega = 20\,\pi$$

$$\Rightarrow 2\pi f = 20\,\pi$$

The frequency of SHM, $f = 10$ Hz

10 When car is accelerating horizontally

$$g_{eff} = \sqrt{g^2 + a^2}$$

11 Let at $t = 0$, the particle be at the extreme position, then the equation of SHM can be written as

$$x = A\cos(\omega t) = A\cos\left(\frac{2\pi}{T} \times t\right)$$

$$\text{At } t = \frac{T}{8}, \; x = A\cos\frac{\pi}{4} = \frac{A}{\sqrt{2}}$$

$$\text{Acceleration} = -\omega^2 x = -\left(\frac{2\pi}{T}\right)^2 \times \frac{A}{\sqrt{2}}$$

$$\text{Magnitude of acceleration} = \frac{4\pi^2 A}{\sqrt{2}\, T^2}$$

12 $A_s = 2A \sin kx$

$$2\text{ mm} = 4\text{ mm} \sin kx$$

$$\Rightarrow \quad kx = \frac{\pi}{6}, \frac{5\pi}{6}$$

$$\Rightarrow \quad x_2 - x_1 = \left[\frac{5\pi}{6} - \frac{\pi}{6}\right] \times \frac{1}{k} = \frac{\lambda}{3}$$

As the string is vibrating in fundamental mode

$$L = \frac{\lambda}{2} \Rightarrow \lambda = 2L = 3\text{ m}$$

So, required separation between two points,

$$x_2 - x_1 = 1\text{ m}$$

13 From initial equilibrium position,

$$ky = mg$$

When block is at distance x below mean position

Kinetic energy of the block,

$$K = \frac{m\omega^2 A^2}{2}\cos^2(\omega t - \phi)$$

[From SHM theory]

Elastic potential energy of spring-block-earth system,

$$U_e = \frac{k(y + x)^2}{2}$$

where, $x = A\sin(\omega t + \phi)$

Gravitational potential energy of spring-block-earth system is, $U_g = -mgx$

Taking, mean position as reference position for gravitation potential energy.

Total energy,

$$E = K + U_e + U_g = \frac{m\omega^2 A^2}{2} + \frac{ky^2}{2}$$

14 $y = 3\sin^2\frac{t}{2}\sin 800t$

$$= 3\left(\frac{1 - \cos t}{2}\right)\sin 800t$$

$$= \frac{3}{2}\sin 800t - \frac{3}{2}\sin 800t\cos t$$

$$= \frac{3}{2}\sin 800t - \frac{3}{4}[2\sin 800t\cos t]$$

$$= \frac{3}{2}\sin 800t - \frac{3}{4}[\sin 801t + \sin 799t]$$

$$= \frac{3}{2}\sin 800t - \frac{3}{4}\sin 801t - \frac{3}{4}\sin 799t$$

$\therefore$ Number of independent harmonic wave $= 3$.

15 From the conservation principle,

$$mgh = \frac{1}{2}kx_0^2 - mgx_0$$

where, x_0 is maximum elongation in spring.

$$\Rightarrow \frac{1}{2}kx_0^2 - mgx_0 - mgh = 0$$

$$\Rightarrow x_0^2 - \frac{2mg}{k}x_0 - \frac{2mg}{k}h = 0$$

$$\Rightarrow x_0 = \frac{2\frac{mg}{2} \pm \sqrt{\left(\frac{2mg}{k}\right)^2 + 4 \times \frac{2mg}{k}h}}{2}$$

Amplitude = Elongation for lowest extreme position – elongation for equilibrium position.

$$= x_0 - x_1 = \frac{mg}{k}\sqrt{1 + \frac{2hk}{mg}}\left[\because x_1 = \frac{mg}{k}\right].$$

16 If I_1 be initial intensity of sound and I_2 be the final intensity of sound,

then $S_1 = 10\log\frac{I_1}{I_0}$ and $S_2 = 10\log\frac{I_2}{I_0}$

$$S_1 - S_2 = 30 \text{ dB}$$

$$10\log\frac{I_1}{I_0} - 10\log\frac{I_2}{I_0} = 30$$

$$10\log\left(\frac{I_1}{I_0} / \frac{I_2}{I_0}\right) = 30$$

$$10\log\frac{I_1}{I_2} = 30$$

$$\log\frac{I_1}{I_2} = 3$$

$$\frac{I_1}{I_2} = 10^3$$

$$I_2 = \frac{1}{1000}I_1$$

17 Speed in SHM, $v = \omega\sqrt{a^2 - x^2}$

$$\Rightarrow v^2 = \omega^2(a^2 - x^2)$$

According to question,

$$v_1^2 = \omega^2(a^2 - x_1^2)$$

$$v_2^2 = \omega^2(a^2 - x_2^2)$$

$$\Rightarrow v_1^2 - v_2^2 = \omega^2(x_2^2 - x_1^2)$$

$$\Rightarrow \omega = \sqrt{\frac{v_1^2 - v_2^2}{x_2^2 - x_1^2}} = \sqrt{\frac{(13)^2 - (12)^2}{(5)^2 - (3)^2}}$$

$$= \sqrt{\frac{169 - 144}{25 - 9}} = \sqrt{\frac{25}{16}} = \frac{5}{4}$$

Also, $\omega = 2\pi f = \frac{5}{4} \Rightarrow f = \frac{5}{8\pi}$.

18 Let the fundamental frequencies of A and B, before tightening of B are f_1 and f_2, respectively.

Then either $f_1 - f_2 = 3$ or $f_2 - f_1 = 3$.

As tension in B increases (due to tightening), its frequency increases and the beat frequency also increases.

If $f_1 - f_2 = 3$, then according to given condition, when f_2 increases $f_1 - f_2$ the decreases so the frequencies of strings are related by $f_2 - f_1 = 3$.

i.e. before tightening, $f_2 > f_1$

After tightening,

$$f_2' - f_1 = 6,$$

i.e. $$f_2' > f_1$$

19 Speed of sound, $v = \sqrt{\frac{\gamma P}{\rho}} = \sqrt{\frac{\gamma RT}{M}}$

Speed of sound in oxygen at t°C,

$$v_{oxy} = \sqrt{\frac{\gamma R(t + 273)}{M_{oxy}}}$$

Speed of nitrogen at 14°C,

$$v_N = \sqrt{\frac{\gamma R(14 + 273)}{M_N}}$$

As $$v_{oxy} = v_N$$

$$\Rightarrow \frac{M_N}{M_{oxy}} = \frac{14 + 273}{t + 273}$$

$$\Rightarrow \frac{\rho_N}{\rho_{oxy}} = \frac{14 + 273}{t + 273}$$

$$\Rightarrow \frac{14}{16} = \frac{14 + 273}{t + 273}$$

$$\Rightarrow t = \frac{2296 - 1911}{7} = 55° \text{C}.$$

20 For situation 1,

$$f_{ap1} = \frac{v - 0}{v - v_s} \times f$$

$$= \frac{320}{320 - 40} \times 320$$

$$= 365.7 \text{ Hz}$$

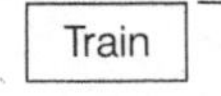

For situation 2,

$$f_{ap2} = \frac{v}{v - (-v_s)} \times f$$

$$= 284.4 \text{ Hz}$$

$$\Delta f = f_{ap1} - f_{ap2}$$

$$= 81.3 \text{ Hz}$$

21 Velocity of wave on the string

$$= \sqrt{\frac{T}{\mu}} = 8 \text{ ms}^{-1}$$

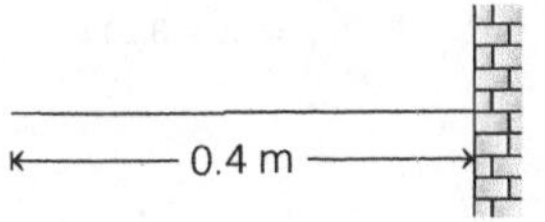

The pulse gets inverted after reflection from the fixed end, so for constructive interference to take place between successive pulses, the first pulse has to undergo two reflections from the fixed end.

So, $\Delta t = \frac{2 \times 0.4 + 2 \times 0.4}{8} = 0.2$ s

22 The node and antinodes are formed in a standing wave pattern as a result of the interference of two waves.

Distance between two nodes is half of wavelength (λ).

So, standard equation of standing wave is

$$y = 2a\sin\frac{2\pi x}{\lambda}\cos\frac{2\pi vt}{\lambda} \quad \text{...(i)}$$

where, a is amplitude, λ is wavelength, v is velocity and t is time.

Given equation

$$y = 5\sin\frac{2\pi x}{3}\cos 20\pi t \quad \text{...(ii)}$$

Comparing Eqs. (i) and (ii), we get

$$\frac{2\pi x}{\lambda} = \frac{2\pi x}{3}$$

$$\Rightarrow \lambda = 3 \text{ cm}$$

23 Let the power of source be P and let it be placed at Q.

Then, intensity at A and at B would be given by

$$I_A = \frac{P}{4\pi \times 1^2}$$

and $$I_B = \frac{P}{4\pi \times 2^2}$$

$$\Rightarrow \frac{(\text{Amp.})_A}{(\text{Amp.})_B} = \sqrt{\frac{I_A}{I_B}} = \sqrt{\frac{2^2}{1^2}} = 2:1$$

24 For the wave, $y = A\sin(kx - \omega t)$, the wave speed is $\frac{\omega}{k}$ and the maximum transverse string speed is $A\omega$.

Hence option (d) is correct.

25 For the wave $y = A\sin(\omega t - kx)$,

$$v_0 = A\omega$$

where A is the maximum displacement.

For the given condition,

$$\frac{A}{2} = A\sin(\omega t - kx)$$

$$\Rightarrow \sin(\omega t - kx) = \frac{1}{2}$$

and $$\frac{\partial y}{\partial t} = A\omega\cos(\omega t - kx)$$

$$= A\omega\frac{\sqrt{3}}{2} = \frac{\sqrt{3}v_0}{2}$$

26 We have, $T = \frac{1}{0.1}$ s = 10 s

In one complete cycle, particle travels a distance, 4 times the amplitude.

So, in a time interval of 10 min 15 s i.e., 615 s i.e., 61 full + 1 half- cycles, the distance travelled

$= (4 \times 3) \times 61 + (2 \times 3) \times 1$
$= 732 + 6 = 738$ cm

At this instant, the particle is moving in an upward direction.

27 The frequency of vibration is

$$\nu = \frac{p}{2l}\sqrt{\frac{T}{m}}$$

As $420 = \frac{p}{2l}\sqrt{\frac{T}{m}}$

and $490 = \frac{p+1}{2l}\sqrt{\frac{T}{m}}$

$\Rightarrow \frac{420}{490} = \frac{p}{p+1} = \frac{6}{7}$

$\Rightarrow 7p = 6p + 6 \Rightarrow p = 6$

$\therefore 420 = \frac{6}{2l}\sqrt{\frac{450}{5\times 10^{-3}}}$

$\Rightarrow l = \frac{900}{420} = 2.14$ m

28 Maximum acceleration,

$A_{max} = \omega^2 a = (2\pi\nu)^2 a = 4\pi^2\nu^2 a$

$= 4 \times \left(\frac{22}{7}\right)^2 \times (2)^2 + 0\cdot 05$

$= 7.9$ ms^{-2}

Maximum force on the man

$= m(g + A_{max})$
$= 50(10 + 7.9) = 895$N
$= 89.5$ kgf

Minimum force on the man

$= m(g - A_{max})$
$= 50(10 - 7.9) = 105$N $= 10.5$ kgf

Hence, the reading of weighing machine varies between 10.5 kgf and 89.5 kgf.

29 Here $m = 2$ kg; $\nu = \frac{360}{60} = 6$ rps

Extension produced in spring,

$y = 43 - 36 = 7$ cm
$= 7 \times 10^{-2}$ m

On rotation the required centripetal force is provided by tension in spring i.e. $ky = mr(2\pi\nu)^2 = 4\pi^2\nu^2 mr$

$\Rightarrow k = \frac{4\pi^2\nu^2 mr}{y}$

where, r is length of stretched spring.

$\Rightarrow k = \frac{4\times(22/7)^2 \times 6^2 \times 2\times(43\times 10^{-2})}{7\times 10^{-2}}$

$= 17475$ Nm^{-1}.

30 Let l_1 and l_2 be the lengths of the two pipes, then

$m = n_1 - n_2 = \frac{\nu_{16}}{2l_1} - \frac{\nu_{16}}{2l_2}$

$\Rightarrow \nu_{16}\left(\frac{1}{2l_1} - \frac{1}{2l_2}\right) = 17$ ... (i)

and $\nu_{51}\left(\frac{1}{2l_1} - \frac{1}{2l_2}\right) = m$ ... (ii)

Divide Eq. (ii) by Eq. (i), we get

$\frac{\nu_{51}}{\nu_{16}} = \frac{m}{17} = \sqrt{\frac{273+51}{273+16}} = \sqrt{\frac{324}{289}}$

$\Rightarrow m = \frac{18}{17} \times 17 = 18$ s^{-1}

31 Let d be the distance, then time taken by longitudinal wave,

$t_1 = \frac{d}{v_1} = \frac{d}{4}$ s

Time taken by transverse wave,

$t_2 = \frac{d}{v_2} = \frac{d}{8}$s

As, $t_1 - t_2 = 8$ min $= 8 \times 60$ s

$\therefore \frac{d}{4} - \frac{d}{8} = 480 \Rightarrow \frac{d}{8} = 480$

$\Rightarrow d = 480 \times 8 = 3840$ km.

32 The effective acceleration of bob in water

$$g' = g\left(1 - \frac{\sigma}{\rho}\right)$$

where, σ = density of water and ρ = density of bob

The time period of bob in air, $T = 2\pi\sqrt{\frac{l}{g}}$

The time period in water, $T' = 2\pi\sqrt{\frac{l}{g'}}$

$\therefore \frac{T}{T'} = \sqrt{\frac{g'}{g}} = \sqrt{\frac{g\left(1-\frac{\sigma}{\rho}\right)}{g}}$

$= \sqrt{1 - \frac{\sigma}{\rho}} = \sqrt{1 - \frac{1}{\rho}}$

Given that, $\frac{T}{T'} = \frac{1}{\sqrt{2}} \Rightarrow \frac{1}{2} = 1 - \frac{1}{\rho}$

or $\rho = 2$

33 The body is dropped from a height h and takes time t, then

$h = ut + \frac{1}{2}g_p t^2$

$h = \frac{1}{2}g_p t^2$ $(\because u = 0)$

$g_p = \frac{2h}{t^2} = \frac{2\times 8}{4} = 4$ ms^{-2}

Time period, $T = 2\pi\sqrt{\frac{l}{g}} = 2\pi\sqrt{\frac{1}{4}} = \frac{2\pi}{2}$

$= \pi = 3.14$ s

34 For longitudinal wave, the medium has to compress and rarify while the string cannot be compressed or rarified.

35 $\lambda = \frac{v}{n} = \frac{350}{500} = 0.7$ m

$\phi = 60° = 60 \times \frac{\pi}{180°} = \frac{\pi}{3}$ radian

As, $x = \frac{\lambda}{2\pi}\phi$

$\Rightarrow x = \frac{0.7}{2\pi} \times \left[\frac{\pi}{3}\right] = 0.12$ m $= 12$ cm

36 $P_{av} = \frac{\rho v \omega^2 A^2}{2}$

37 Total initial energy of particle in SHM,

$E_1 = \frac{1}{2}m\omega^2 A^2$

Energy when amplitude is $\frac{A}{2}$ and angular frequency is 3ω.

$E_2 = \frac{1}{2}m(3\omega)^2\left(\frac{A}{2}\right)^2 = \frac{9}{8}m\omega^2 A^2$

$\therefore$ Extra energy $= E_2 - E_1$

$= \frac{9}{8}m\omega^2 A^2 - \frac{1}{2}m\omega^2 A^2$

$= \frac{5}{8}m\omega^2 A^2 = \frac{5}{4}E_1$(gain)

38 The time period of a spring pendulum is given by, $T = 2\pi\sqrt{\frac{m}{k}}$ and hence is not affected by the acceleration of the frame of reference.

39 Total energy of the particle performing simple harmonic motion is,

$E = K + U = K_{max} + U_{min}$. K is always positive, while U could be positive, negative or zero. If U_{min} is negative and its value is greater than K_{max}, then E would be negative.

40 When the hoop oscillates in its plane, moment of inertia is

$I_1 = mR^2 + mR^2$ i.e., $I_1 = 2mR^2$.

While when the hoop oscillates in a direction perpendicular to the plane of the hoop, moment of inertia is

$I_2 = \frac{mR^2}{2} + mR^2 = \frac{3mR^2}{2}$

The time period of physical pendulum,

$T = 2\pi\sqrt{\frac{I}{mgd}}$.

Here, d is a same in both the cases.

DAY FOURTEEN

Properties of Matter

Learning & Revision for the Day

- Elastic Behaviour
- Stress
- Strain
- Hooke's Law
- Work Done (or Potential Energy Stored) in a Stretched Line
- Thermal Stresses and Strains
- Fluid Statics
- Pascal's Law
- Laws of Floatation
- Viscosity
- Streamline and Turbulent Flow
- Equation of Continuity
- Bernoulli's Principle
- Surface Tension
- Surface Energy
- Angle of Contact
- Excess Pressure Over a Liquid Film
- Capillary Rise or Capillarity

Elastic Behaviour

Elasticity is the property of body by virtue of which a body regains or tends to regain its original configuration (shape as well as size), when the external deforming forces acting on it, is removed.

Stress

The internal restoring force per unit area of cross-section of the deformed body is called stress.

Thus, $\text{Stress, } \sigma = \dfrac{\text{Restoring force}}{\text{Area}} = \dfrac{F}{A}$

SI unit of stress is Nm^{-2} or pascal (Pa).

Different types of stress are given below

1. Normal or Longitudinal Stress

If area of cross-section of a rod is A and a deforming force F is applied along the length of the rod and perpendicular to its cross-section, then stress produced in the rod is called normal or longitudinal stress.

$$\text{Longitudinal stress} = \frac{F_n}{A}$$

- No. of Questions in Exercises (x)—
- No. of Questions Attempted (y)—
- No. of Correct Questions (z)—
  (Without referring Explanations)

- Accuracy Level ($z/y \times 100$)—
- Prep Level ($z/x \times 100$)—

In order to expect good rank in JEE, your Accuracy Level should be above 85 & Prep Level should be above 75.

Longitudinal stress is of two types

(i) **Tensile stress** When length of the rod is increased on application of deforming force over it, then stress produced in rod is called tensile stress.

(ii) **Compressive stress** When length of the rod is decreased on application of deforming force, then the stress produced is called compressive stress.

2. Volumetric Stress

When a force is applied on a body such that it produces a change in volume and density, shape remaining same

(i) at any point, the force is perpendicular to its surface.

(ii) at any small area, the magnitude of force is directly proportional to its area.

Then, force per unit area is called volumetric stress.

$\therefore$ Volumetric stress $= \dfrac{F_v}{A}$

3. Shearing or Tangential Stress

When the force is applied tangentially to a surface, then it is called tangential or shearing stress.

Tangential stress $= \dfrac{F_t}{A}$

It produces a change in shape, volume remaining same.

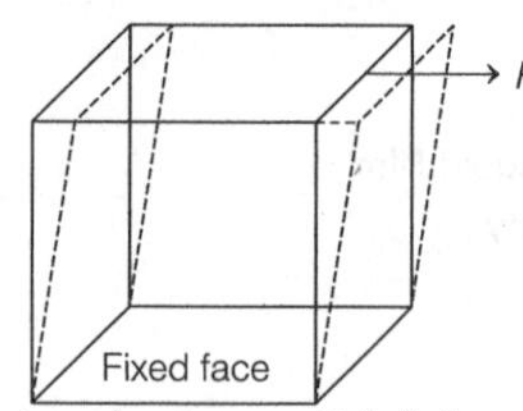

An object under tangential deforming forces

Strain

Strain is the ratio of change in configuration to the original configuration of the body.

Being the ratio of two similar quantities, strain is a unitless and dimensionless quantity.

(i) When the deforming force causes a change in length, it is called **longitudinal strain.** For a wire or rod, longitudinal strain is defined as the ratio of change in length to the original length.

$$\therefore \text{Longitudinal strain} = \frac{\text{Change in length } (\Delta L)}{\text{Original length } (L)}$$

(ii) When the deforming force causes a change in volume, the strain is called **volumetric strain.**

$$\text{Volumetric strain} = \frac{\text{Change in volume } (\Delta V)}{\text{Original volume } (V)}$$

(iii) When the deforming force, applied tangentially to a surface, produces a change in shape of the body, the strain developed is called **shearing strain** or **shear.**

$$\text{Shearing strain, } \phi = \frac{x}{L}$$

Stress-Strain Relationship

For a solid, the graph between stress (either tensile or compressive) and normal strain is shown in figure.

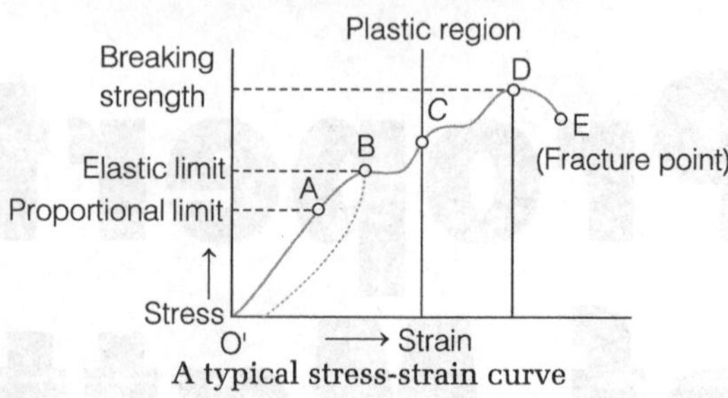

A typical stress-strain curve

- In the above graph, point A is called **proportional limit.** Till this point, stress and strain are proportional to each other.
- From point A to B, stress and strain are not proportional, B is called **elastic limit** and OB is **elastic region.**
- Beyond point B, strain increases without increase in stress, it is called **plastic behaviour**. Region between point C and D is called **plastic region.**
- Finally, at point D, wire may break, maximum stress corresponding to point D is called **breaking stress.**

The materials of the wire, which break as soon as stress is increased beyond the elastic limit are called **brittle.** Graphically, for such materials the portion of graph between B and E is almost zero. While the materials of the wire, which have a good plastic range (portion between B and E) are called **ductile.**

Hooke's Law

According to the Hooke's law, for any body, within the elastic limit, stress developed is directly proportional to the strain produced.

$$\text{Stress} \propto \text{Strain}$$

$$\text{Stress} = E \times \text{Strain}$$

The ratio of stress to strain, within the elastic limit, is called the **coefficient (or modulus) of elasticity** for the given material.

Depending on the type of stress applied and resulting strain, we have the following three of elasticity given as,

$$E = \frac{\text{Stress}}{\text{Strain}}$$

There are three modulus of elasticity.

1. Young's Modulus

Young's modulus of elasticity (Y) is defined as the ratio of normal stress (either tensile or compressive stress) to the longitudinal strain within a elastic limit.

Young's modules,

$$Y = \frac{\text{Normal stress}}{\text{Longitudinal strain}} = \frac{F/A}{\Delta L/L} = \frac{FL}{A\Delta L}$$

2. Bulk Modulus

It is defined as the ratio of the normal stress to the volumetric strain. Coefficient of volume elasticity,

$$B = \frac{F/A}{\Delta V/V} = -\frac{pV}{\Delta V}$$

where, $p = \frac{F}{A}$ = the pressure or stress negative sign signifies that for an increase in pressure, the volume will decrease.

Reciprocal of bulk modulus is called **compressibility.**

3. Modulus of Rigidity (Shear modulus)

It is defined as the ratio of tangential stress to shearing stress.

$$\eta = \frac{\text{Tangential stress}}{\text{Shearing strain}} = \frac{F/A}{\phi} = \frac{F}{A\phi} = \frac{FL}{Ax}$$

- **Breaking force** depends upon the area of cross-section of the wire.

  $\therefore$ Breaking force $\propto A$

  Breaking force $= P \times A$

  Here, P is a constant of proportionality and known as breaking stress.

Poisson's Ratio

For a long bar, the Poisson's ratio is defined as the ratio of lateral strain to longitudinal strain.

$\therefore$ Poisson's ratio, $\sigma = \frac{\text{Lateral strain}}{\text{Longitudinal strain}} = \frac{\Delta D/D}{\Delta L/L} = \frac{\Delta r/r}{\Delta L/L}$

Poisson's ratio is a unitless and dimensionless term. Its value depends on the nature of the material. Theoretically, value of σ must lie between -1 and $+0.5$ but for most metallic solids $0 < \sigma < 0.5$.

Inter-Relations Between Elastic Constants

Y = Young's modulus, η = Rigidity modulus,
B = Bulk modulus, σ = Poisson's ratio

The inter relation between elastic constants are

$$Y = 2\eta(1+\sigma), \quad Y = 3B(1-2\sigma)$$

$$\frac{9}{Y} = \frac{3}{\eta} + \frac{1}{B} \quad \text{or} \quad Y = \frac{9B\eta}{\eta + 3B} \quad \text{or} \quad \sigma = \frac{3B - 2\eta}{6B + 2\eta}$$

Work Done (or Potential Energy Stored) in a Stretched Wire

Work is done against the internal restoring forces, while stretching a wire. This work is stored as elastic potential energy. The work done is given by

- Work done $W = \frac{1}{2} \times$ stretching force $\times$ elongation

$$= \frac{1}{2} F\Delta L = \frac{1}{2}\frac{YA}{L}(\Delta L)^2$$

= Energy stored in the wire (U)

- Energy stored per unit volume (or energy density)

$$= \frac{U}{V} = \frac{1}{2}\frac{F\Delta L}{AL} = \frac{1}{2}\text{ stress} \times \text{strain} = \frac{Y}{2}(\text{strain})^2$$

Thermal Stresses and Strains

When a body is allowed to expand or contract with increase or decrease in temperature, no stresses are induced in the body. But if the deformation of the body is prevented, some stresses are induced in the body. Such stresses are called thermal stresses or temperature stresses. The corresponding strains are called thermal strains or **temperature strains.**

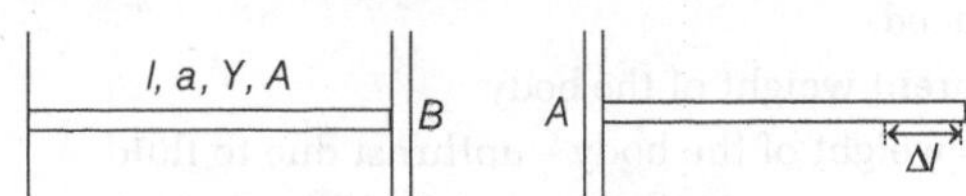

A rod under thermal stress and strain

By definition, coefficient of linear expansion $\alpha = \frac{\Delta l}{l\theta}$

thermal strain $\frac{\Delta l}{l} = \alpha\Delta\theta$

So thermal stress $= Y\alpha\Delta\theta$

Tensile or compressive force produced in the body

$$F = YA\alpha\Delta\theta$$

Fluid Statics

The substances which flow are called fluids, that includes both liquid and gas. The science of fluids at rest is called fluid statics where fluid mass is stationary w.r.t. container, containing the fluid.

Thrust and Pressure

Normal force exerted by fluid at surface in contact is called **thrust** of fluid.

The thrust exerted by a fluid at rest per unit surface area of contact surface is called the **fluid pressure.**

$\therefore$ Pressure $p = \frac{\text{Normal force (thrust) } F}{\text{Surface area } A}$

Pressure is a scalar and its SI unit is Nm^{-2} or pascal (Pa), where, 1 Pa = 1 N m^{-2}.

Pressure due to Fluid Column

Pressure p exerted by the fluids at the bottom of container having height h

$$p = h\rho g$$

where, ρ = density of fluid.

Gauge Pressure

The pressure difference between the real hydrostatic pressure and the atmospheric pressure is known as the gauge pressure.

$\therefore$ Gauge pressure = Real pressure (p) – Atmospheric pressure (p_0)

Pascal's Law

According to Pascal's law of transmission of pressure, the increase in pressure at any one point of the enclosed liquid in equilibrium or at rest, is transmitted equally to all other points of the liquid and also to the walls of the container. Hydraulic lift, hydraulic press, hydraulic brakes etc. are based on the Pascal's law.

Archimedes' Principle and Buoyancy

When a body is immersed in a fluid, it experiences an upthrust due to the fluid and as a result the apparent weight of the body is reduced.

∴ Apparent weight of the body

= weight of the body – upthrust due to fluid

= weight of the body – weight of the fluid displaced

e.g. For a floating body, the volume of a body $(V - V_s)$ remaining outside the liquid will be given by

$$V_0 = V - V_s = V - V\frac{\rho}{\sigma} = V\left(1 - \frac{\rho}{\sigma}\right)$$

where, ρ = density of liquid

and σ = density of body immersed in liquid

Buoyant Force or Buoyancy

Buoyant force, $F = h\rho g A = mg$

where, h = height of body immersed in liquid,

m = mass of body and A = area

- It is an upward force acting on the body immersed in a liquid.
- It is equal to the weight of liquid displaced by the immersed part of the body.
- The buoyant force acts at the centre of buoyancy which is the centre of gravity of the liquid displaced by the body when immersed in the liquid.
- The line joining the centre of gravity and centre of buoyancy is called central line.
- Metacentre, is a point where the vertical line passing through the centre of buoyancy intersects the central line.

Laws of Floatation

When a body of density ρ_B and volume V is immersed in a liquid of density σ, the forces acting on the body are

- The weight of body $W = mg = V\rho_B g$ acting vertically downwards through the centre of gravity of the body.
- The upthrust $F = V\sigma g$ acting vertically upwards through the centre of gravity of the displaced liquid *i.e.*, centre of buoyancy.

So, the following three cases are possible.

Case I The density of body is greater than that of liquid (i.e. $\rho_B > \sigma$). In this case, as weight will be more than upthrust, the body will sink. $(W > F)$

Case II The density of body is equal to the density of liquid (i.e. $\rho_B = \sigma$). In this case, $W = F$. so, the body will float fully submerged in neutral equilibrium anywhere in the liquid.

Case III The density of body is less than that of liquid (i.e. $\rho_B < \sigma$). In this case, $W < F$, so the body will move upwards and **in equilibrium** will float partially immersed in the liquid such that

$$W = V_{in}\sigma g$$

[V_{in} is the volume of body in the liquid]

or $V\rho_B g = V_{in}\rho g$ [as, $W = mg = \rho_B V g$]

or $V\rho_B = V_{in}\sigma$ ...(i)

Viscosity

Viscosity is the property of a fluid due to which it opposes the relative motion between its different layers.

Force between the layers opposing the relative motion is called **viscous force.**

If there are two fluid layers having surface area A and velocity gradient dv/dr, then the viscous force is given by $\mathbf{F} = -\eta A\frac{d\mathbf{v}}{dr}$. where, constant η is called the coefficient of viscosity of the given fluid.

SI unit of coefficient of viscosity is N m^{-2}s or Pa-s or poiseuille (Pl).

Terminal Velocity

If a small spherical body is dropped in a fluid, then initially it is accelerated under the action of gravity. However, with an increase in speed, the viscous force increases and soon it balances the weight of the body.

Now, the body moves with a constant velocity, called the **terminal velocity.**

Terminal velocity v_t is given by

$$v_t = \frac{2}{9}\frac{r^2(\rho - \sigma)g}{\eta}$$

where, r = radius of the falling body,

ρ = density of the falling body

and σ = density of the fluid.

Stokes' Law

Stokes proved that for a small spherical body of radius r moving with a constant speed v called terminal velocity through a fluid having coefficient of viscosity η, the viscous force F is given by

$$F = 6\pi\eta r v$$

It is known as Stokes' law.

Streamline and Turbulent Flow

Flow of a fluid is said to be **streamlined** if each element of the fluid passing through a particular point travels along the same path, with exactly the same velocity as that of the preceding element. A special case of streamline flow is **laminar flow**.

A **turbulent flow** is the one in which the motion of the fluid particles is disordered or irregular.

Critical Velocity

For a fluid, the critical velocity is that limiting velocity of the fluid flow upto which the flow is streamlined and beyond which the flow becomes turbulent. Value of critical velocity for the flow of liquid of density ρ and coefficient of viscosity η, flowing through a horizontal tube of radius r is given by

$$v_c \propto \frac{\eta}{\rho r}$$

where, r = radius of tube.

Reynold's Number (N_R)

Reynold's number as the ratio of the inertial force per unit area to the viscous force per unit area for a fluid.

$$N_R = \frac{v^2\rho}{\eta v / r} = \frac{\rho v r}{\eta}$$

A smaller value of Reynold's number (generally $N_R \leq 1000$) indicates a streamline flow but a higher value ($N_R \geq 1500$) indicates that the flow is turbulent and between 1000 to 1500, the flow is unstable.

Equation of Continuity

Let us consider the streamline flow of an ideal, non-viscous fluid through a tube of variable cross-section.

Let at the two sections, the cross-sectional areas be A_1 and A_2, respectively and the fluid flow velocities are v_1 and v_2, then according to the equation of continuity

$$A_1 v_1 \rho_1 = A_2 v_2 \rho_2$$

where, ρ_1 and ρ_2 are the respective densities of the fluid. Equation of continuity is based on the conservation of mass.

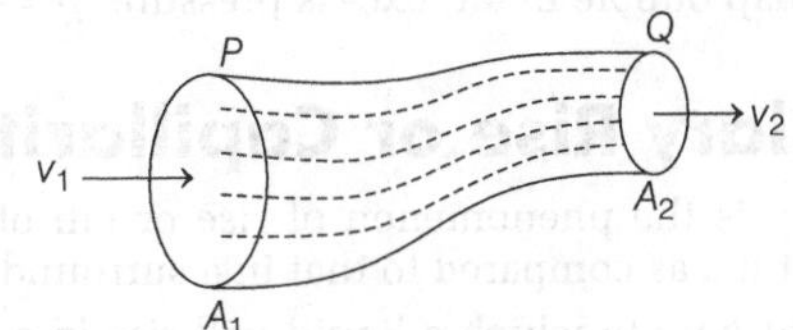

A liquid is flowing through a tube of non-uniform cross-section

If the fluid which is flowing, is incompressible, then

$$\rho_1 = \rho_2$$

So, equation of continuity is simplified as

$$A_1 v_1 = A_2 v_2 \quad \text{or} \quad Av = \text{constant}$$

Energy of a Flowing Liquid

There are three types of energies in a flowing liquid.

- **Pressure Energy** If p is the pressure on the area A of a fluid and the liquid moves through a distance l due to this pressure, then

  Pressure energy of liquid = work done
  = force × displacement = pAl

  The volume of the liquid is Al. Hence, pressure energy per unit volume of liquid $= \frac{pAl}{Al} = p$
- **Kinetic Energy** If a liquid of mass m and volume V is flowing with velocity v, then the kinetic energy $= \frac{1}{2} mv^2$

  ∴ Kinetic energy per unit volume of liquid

  $$= \frac{1}{2}\left(\frac{m}{V}\right)v^2 = \frac{1}{2}\rho v^2$$

  Here, ρ is the density of liquid and V = volume.
- **Potential Energy** If a liquid of mass m is at a height h from the reference line ($h = 0$), then its potential energy is mgh.

  ∴ Potential energy per unit volume of the liquid

  $$= \left(\frac{m}{V}\right)gh = \rho gh$$

Bernoulli's Principle

According to the Bernoulli's principle for steady flow of an incompressible, non-viscous fluid through a tube/pipe, the total energy (i.e. the sum of kinetic energy, potential energy and pressure energy) per unit volume (or per unit mass too) remains constant at all points of flow provided that there is no source or sink of the fluid along the flow.

Mathematically, we have

$$p + \rho gh + \frac{1}{2}\rho v^2 = \text{constant} \quad \text{or} \quad \frac{p}{\rho g} + h + \frac{v^2}{2g} = \text{constant}$$

In this expression, $\frac{v^2}{2g}$ is **velocity head** and $\frac{p}{\rho g}$ is **pressure head**.

Velocity of Efflux

- If a liquid is filled in a vessel up to a height H and a small orifice O is made at a height h, then from Bernoulli's theorem it can be shown that velocity of efflux v of the liquid from the vessel is

  $$v = \sqrt{2g(H-h)}$$
- The flowing fluid describes a parabolic path and hits the base level at a horizontal distance (called the **range**) $R = 2\sqrt{h(H-h)}$.

  The range is maximum, when $h = \frac{H}{2}$ and in that case $R_{\max} = H$.

Applications Based on the Bernoulli's Principle

- The action of carburetor, paintgun, scent sprayer, atomiser and insect sprayer is based on the Bernoulli's principle.
- The action of the Bunsen's burner, gas burner, oil stove and exhaust pump is also based on the Bernoulli's principle.
- Motion of a spinning ball (Magnus effect) is based on Bernoulli's theorem.
- Blowing of roofs by wind storms, attraction between two parallel moving boats moving close to each other, fluttering of a flag etc., are also based on Bernoulli's theorem.

Surface Tension

Surface tension is the property of a liquid due to which its free surface behaves like a stretched elastic membrane and tends to have the least possible surface area.

Surface tension $S = \dfrac{\text{Force}}{\text{Length}} = \dfrac{F}{l}$

Here, F is force acting on the unit length of an imaginary line drawn of the surface of the liquid.

SI unit of surface tension is Nm^{-1} or Jm^{-2}. It is a scalar and its dimensional formula is $[MT^{-2}]$.

Surface Energy

Surface energy of a liquid is the potential energy of the molecules of a surface film of the liquid by virtue of its position.

When the surface area of a liquid is increased, work is done against the cohesive force of molecules and this work is stored in the form of additional surface energy.

Increase in surface potential energy

$$\Delta U = \text{Work done } (\Delta W) = S\Delta A$$

where, ΔA is the increase in surface area of the liquid.

- **Work done in Blowing a Liquid Drop** If a liquid drop is blown up from a radius r_1 to r_2, then work done in the process,
$$W = S(A_2 - A_1) = S \times 4\pi\,(r_2^2 - r_1^2)$$
- **Work done in Blowing a Soap Bubble** As a soap bubble has two free surfaces, hence, work done in blowing a soap bubble, so as to increase its radius from r_1 to r_2, is given by
$$W = S \times 8\pi\,(r_2^2 - r_1^2)$$
- **Work done in Splitting a Bigger Drop into n Smaller Droplets** If a liquid drop of radius R is split up into n smaller droplets, all of the same size, then radius of each droplet
$$r = R(n)^{-1/3}$$
and work done
$$W = S \times 4\pi\,(nr^2 - R^2) = S \times 4\pi R^2\,(n^{1/3} - 1)$$
- **Coalescence of Drops** If n small liquid drops of radius r each, combine together so as to form a single bigger drop of radius $R = n^{1/3} r$ then in the process, energy is released. Release of energy is given by
$$\Delta U = S \times 4\pi\,(nr^2 - R^2) = S \times 4\pi r^2 n(1 - n^{-1/3})$$

Angle of Contact

Angle of contact for a given liquid-solid combination is defined as the angle subtended between the tangents to the liquid surface and the solid surface, inside the liquid, the tangents are drawn at the point of contact.

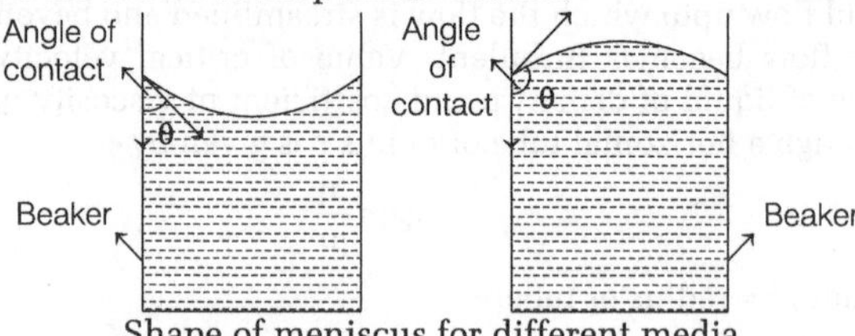

Shape of meniscus for different media

- Value of the angle of contact depends on the nature of liquid and solid both.
- For a liquid having concave meniscus when adhesive force > cohesive force angle of contact θ is acute $(\theta < 90°)$ but for a convex meniscus (when cohesive force > adhesive force) the angle of contact is obtuse $(\theta > 90°)$.
- Value of angle of contact θ decreases with an increase in temperature.

Excess Pressure Over a Liquid Film

If a free liquid surface film is plane, then pressure on the liquid and the vapour sides of the film are the same, otherwise there is always some pressure difference. Following cases arise.

- For a spherical liquid drop of radius r, the excess pressure inside the drop $p = \dfrac{2S}{r}$

  where, S = surface tension of the liquid.
- For an air bubble in a liquid, excess pressure $p = \dfrac{2S}{r}$
- For a soap bubble in air, excess pressure $p = \dfrac{4S}{r}$

Capillary Rise or Capillarity

Capillarity is the phenomenon of rise or fall of a liquid in a capillary tube as compared to that in a surrounding liquid.

The height h up to which a liquid will rise in a capillary tube is given by

$$h = \frac{2S\cos\theta}{r\rho g} = \frac{2S}{R\rho g}$$

where, r = radius of the capillary tube

and $R = \dfrac{r}{\cos\theta}$ = radius of liquid meniscus.

The rise in capillary tube $h \propto \dfrac{1}{r}$ (Jurin's law).

- If a capillary tube, dipped in a liquid is tilted at an angle α from the vertical, the vertical height h of the liquid column remains the same. However, the length of the liquid column (l) in the capillary tube increases to

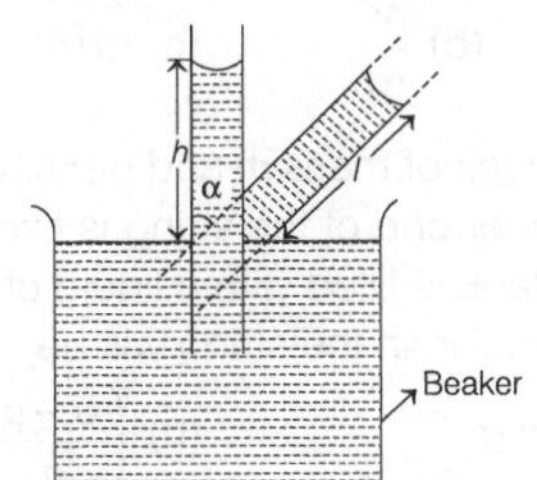

Effect of tilting capillary tube in a liquid

$$l = \frac{h}{\cos \alpha}.$$

- If the capillary tube is of insufficient length, the liquid rises up to the upper end of the tube and then the radius of its meniscus changes from R to R' such that $hR = h'R'$, where h' = insufficient length of the tube.
- After connection due to the weight of liquid contained in the meniscus, the formula for the height is given by

$$h = \frac{2s}{\rho r g} - \frac{r}{3}$$

This is known as **ascent formula.**

DAY PRACTICE SESSION 1

FOUNDATION QUESTIONS EXERCISE

1 A man grows into a giant such that his linear dimensions increase by a factor of 9. Assuming that his density remains same, the stress in the leg will change by a factor of → JEE Main 2017 (Offline)

(a) $\frac{1}{9}$ (b) 81
(c) $\frac{1}{81}$ (d) 9

2 The Young's modulus of brass and steel are respectively 1.0×10^{11} Nm^{-2} and 2.0×10^{11} Nm^{-2}. A brass wire and a steel wire of the same length extend by 1 mm, each under the same force. If radii of brass and steel wires are R_B and R_S respectively, then

(a) $R_S = \sqrt{2}\, R_B$ (b) $R_S = \frac{R_B}{\sqrt{2}}$
(c) $R_S = 4R_B$ (d) $R_S = \frac{R_B}{2}$

3 One end of a horizontal thick copper wire of length $2L$ and radius $2R$ is welded to an end of another horizontal thin copper wire of length L and radius R. When the arrangement is stretched by applying forces at two ends, the ratio of the elongation in the thin wire to that in the thick wire is

(a) 0.25 (b) 0.50
(c) 2.00 (d) 4.00

4 The following four wires are made of the same material. Which of these will have the largest extension when the same tension is applied ?

(a) Length = 50 cm, diameter = 0.5 mm
(b) Length = 100 cm, diameter = 1 mm
(c) Length = 200 cm, diameter = 2 mm
(d) Length = 300 cm, diameter = 3 mm

5 The length of a metal wire is l_1 when the tension in it is T_1 and is l_2 when the tension is T_2. The original length of the wire is

(a) $\frac{l_1 + l_2}{2}$ (b) $\frac{l_1T_2 + l_2T_1}{T_1 + T_2}$
(c) $\frac{l_1T_2 - l_2T_1}{T_2 - T_1}$ (d) $\sqrt{T_1T_2l_1l_2}$

6 Two wires are made of the same material and have the same volume. However, wire 1 has cross-sectional area A and wire 2 has cross-sectional area $3A$. If the length of wire 1 increases by Δx on applying force F, how much force is needed to stretch wire 2 by the same amount? → AIEEE 2009

(a) F (b) $4F$ (c) $6F$ (d) $9F$

7 If the ratio of lengths, radii and Young's modulii of steel and brass wires in the figure are a, b and c respectively, then the corresponding ratio of increase in their lengths is → JEE Main (Online) 2013

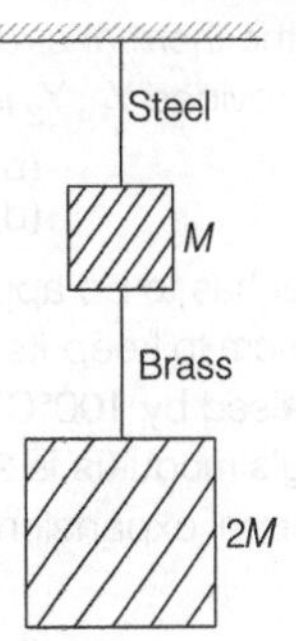

(a) $\frac{3c}{2ab^2}$ (b) $\frac{2a^2c}{b}$ (c) $\frac{3a}{2b^2c}$ (d) $\frac{2ac}{b^2}$

8 The pressure of a medium is changed from 1.01×10^5 Pa to 1.165×10^5 Pa and change in volume is 10% keeping temperature constant. The bulk modulus of the medium is

(a) 204.8×10^5 Pa (b) 102.4×10^5 Pa
(c) 51.2×10^5 Pa (d) 1.55×10^5 Pa

9 A solid sphere of radius r made of a soft material of bulk modulus K is surrounded by a liquid in a cylindrical container. A massless piston of area a floats on the surface of the liquid, covering entire cross-section of cylindrical container. When a mass m is placed on the surface of the piston to compress the liquid, the fractional decrement in the radius of the sphere, $\left(\frac{dr}{r}\right)$ is → JEE Main 2018

(a) $\frac{Ka}{mg}$ (b) $\frac{Ka}{3mg}$
(c) $\frac{mg}{3Ka}$ (d) $\frac{mg}{Ka}$

10 The graph shows the behaviour of a length of wire in the region for which the substance obeys Hooke's law. P and Q represent

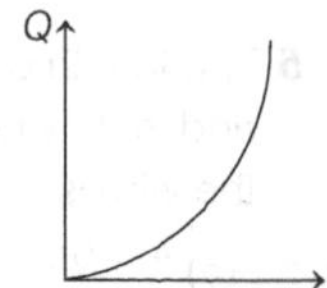

(a) P = applied force, Q = extension
(b) P = extension, Q = applied force
(c) P = extension, Q = stored elastic energy
(d) P = stored elastic energy, Q = extension

11 If work done in stretching a wire by 1mm is 2 J. The work necessary for stretching another wire of same material but with double the radius and half the length, by 1 mm distance, is

(a) 16 J (b) 4 J
(c) 1/4 J (d) 8 J

12 Two rods of different materials having coefficients of thermal expansion α_1, α_2 and Young's modulii Y_1, Y_2 respectively are fixed between two rigid massive walls. The rods are heated such that they undergo the same increase in temperature. There is no bending of the rods. If $\alpha_1 : \alpha_2 = 2 : 3$, the thermal stresses developed in the two rods are equal provided $Y_1 : Y_2$ is equal to

(a) 2 : 3 (b) 1 : 1
(c) 3 : 2 (d) 4 : 9

13 The pressure that has to be applied to the ends of a steel wire of length 10 cm to keep its length constant when its temperature is raised by 100°C is
(For steel, Young's modulus is $2 \times 10^{11} \text{Nm}^{-2}$ and coefficient of thermal expansion is $1.1 \times 10^{-5} \text{ K}^{-1}$) → JEE Main 2014

(a) 2.2×10^8 Pa (b) 2.2×10^9 Pa
(c) 2.2×10^7 Pa (d) 2.2×10^6 Pa

14 A metal rod of Young's modulus Y and coefficient of thermal expansion α is held at its two ends such that its length remains invariant. If its temperature is raised by t°C, the linear stress developed in it is → AIEEE 2011

(a) $\frac{\alpha t}{Y}$ (b) $\frac{Y}{\alpha t}$ (c) $Y\alpha t$ (d) $\frac{1}{Y\alpha t}$

15 A wooden block of mass m and density ρ is tied to a string, the other end of the string is fixed to the bottom of a tank. The tank is filled with a liquid of density σ with $\sigma > \rho$. The tension in the string will be

(a) $\left(\frac{\sigma-\rho}{\sigma}\right)mg$ (b) $\left(\frac{\sigma-\rho}{\rho}\right)mg$
(c) $\frac{\rho mg}{\sigma}$ (d) $\frac{\sigma mg}{\rho}$

16 A ball is made of a material of density ρ where $\rho_{oil} < \rho < \rho_{water}$ with ρ_{oil} and ρ_{water} representing the densities of oil and water, respectively. The oil and water are immiscible. If the above ball is in equilibrium in a mixture of this oil and water, which of the following pictures represents its equilibrium position? → AIEEE 2010

(a)
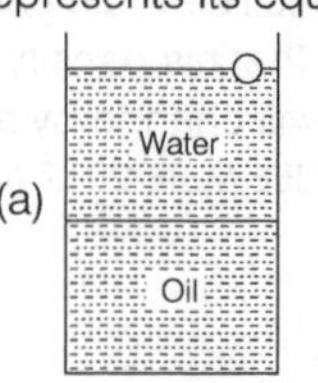

(b)
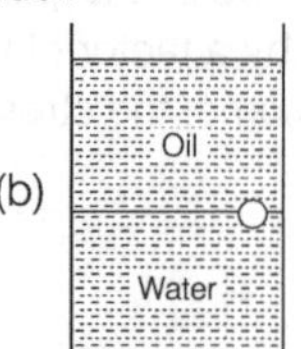

(c)
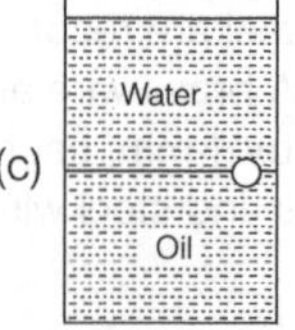

(d)
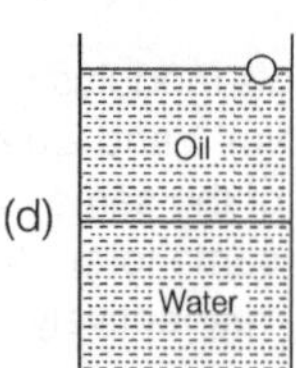

17 Two mercury drops (each of radius r) merge to form a bigger drop. The surface energy of the bigger drop, if T is the surface tension, is → AIEEE 2011

(a) $2^{5/3} \pi r^2 T$ (b) $4\pi r^2 T$
(c) $2\pi r^2 T$ (d) $2^{8/3} \pi r^2 T$

18 A raindrop of radius 0.2 cm is falling through air with a terminal velocity of 8.7 m/s. The viscosity of air in SI units is (take, $\rho_{water} = 1000 \text{ kg/m}^3$ and $\rho_{air} = 1 \text{ kg/m}^3$).

(a) 10^{-4} poise (b) 1×10^{-3} poise
(c) 8.6×10^{-3} poise (d) 1.02×10^{-3} poise

19 If a ball of steel (density $\rho = 7.8 \text{ g cm}^{-3}$) attains a terminal velocity of 10 cms^{-1} when falling in a tank of water (coefficient of viscosity $\eta_{water} = 8.5 \times 10^{-4}$ Pa-s) then its terminal velocity in glycerine ($\rho = 1.2 \text{ gcm}^{-3}$, $\eta = 13.2$ Pa-s) would be nearly → AIEEE 2011

(a) $1.6 \times 10^{-5} \text{ cms}^{-1}$ (b) $6.25 \times 10^{-4} \text{ cms}^{-1}$
(c) $6.45 \times 10^{-4} \text{ cms}^{-1}$ (d) $1.5 \times 10^{-5} \text{ cms}^{-1}$

20 Water is flowing continuously from a tap having an internal diameter 8×10^{-3} m. The water velocity as it leaves the tap is 0.4 ms^{-1}. The diameter of the water stream at a distance 2×10^{-1} m below the tap is close to → AIEEE 2011

(a) 7.5×10^{-3} m (b) 9.6×10^{-3} m
(c) 3.6×10^{-3} m (d) 5.0×10^{-3} m

21 At what speed, the velocity head of water is equal to pressure head of 40 cm of Hg?

(a) 10.3 ms^{-1} (b) 2.8 ms^{-1}
(c) 5.6 ms^{-1} (d) 8.4 ms^{-1}

22 A liquid X of density 3.36 g/cm^3 is poured in a U-tube, which contains Hg. Another liquid Y is poured in the left arm with height 8 cm and upper levels of X and Y are same. What is the density of Y?

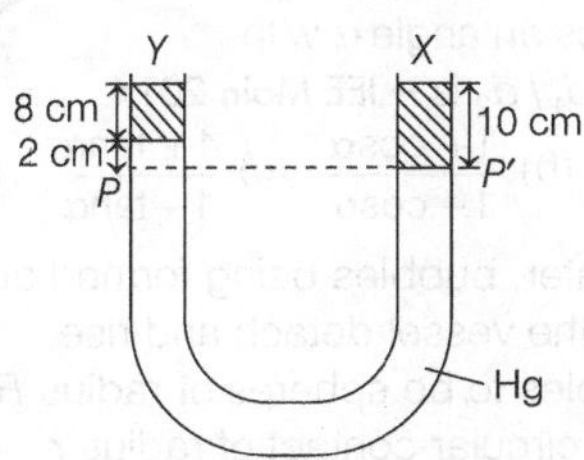

(a) 0.8 g /cc (b) 1.2 g /cc (c) 1.4 g /cc (d) 1.6 g /cc

23 A thin liquid film formed between a U-shaped wire and a light slider supports a weight of 1.5×10^{-2}N (see figure). The length of the slider is 30 cm and its weight negligible. The surface tension of the liquid film is

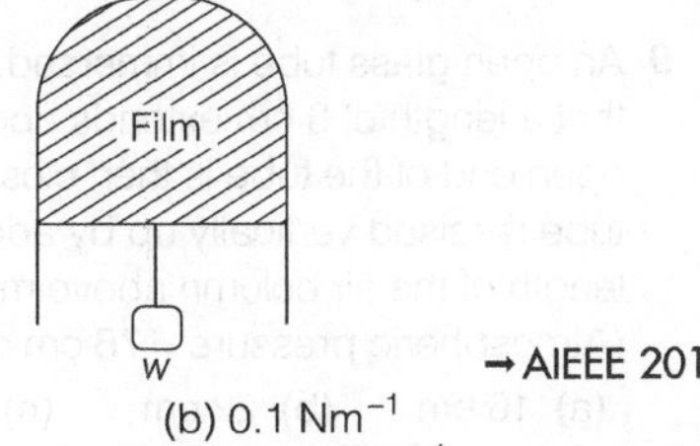

→ AIEEE 2012

(a) 0.0125 Nm^{-1} (b) 0.1 Nm^{-1}
(c) 0.05 Nm^{-1} (d) 0.025 Nm^{-1}

24 Work done in increasing the size of a soap bubble from radius of 3 cm to 5 cm is nearly (surface tension of soap solution = 0.03 Nm^{-1}) → AIEEE 2011

(a) 0.2 π mJ (b) 2 π mJ (c) 0.4 π mJ (d) 4 π mJ

25 While measuring surface tension of water using capillary rise method, height of the lower meniscus from free surface of water is 3 cm while inner radius of capillary tube is found to be 0.5 cm. Then, compute tension of water using this data (take, contact angle between glass and water as 0 and $g = 9.81\,m/s^2$).

(a) 0.72 N/m
(b) 0.77 N/m
(c) 1.67 N/m
(d) None of the above

26 Match the following columns.

	Column I		Column II
A.	Magnus energy	1.	Pascal's law
B.	Loss of energy	2.	Bernoulli's principle
C.	Pressure is same at the same level in a liquid	3.	Viscous force
D.	Gas burner	4.	Spinning ball

Codes

	A	B	C	D		A	B	C	D
(a)	1	4	2	3	(b)	1	2	3	4
(c)	2	2	4	3	(d)	4	3	1	2

Direction (Q. Nos. 27-30) *Each of these questions contains two statements : Statement I and Statement II. Each of these questions also has four alternative choices, only one of which is the correct answer. You have to select one of the codes (a), (b), (c), (d) given below*

(a) Statement I is true, Statement II is true; Statement II is the correct explanation for Statement I
(b) Statement I is true, Statement II is true; Statement II is not the correct explanation for Statement I
(c) Statement I is true; Statement II is false
(d) Statement I is false; Statement II is true

27 **Statement I** Aeroplanes are made to run on the runway before take off, so that they acquire the necessary lift.

Statement II This is as per the Bernoulli's theorem.

28 **Statement I** Finer the capillary, greater is the height to which the liquid rises in the tube.

Statement II This is in accordance with the ascent formula.

29 **Statement I** The bridges are declared unsafe after a long use.

Statement II Elastic strength of bridges decreases with time.

30 **Statement I** A small drop of mercury is spherical but bigger drops are oval in shape.

Statement II Surface tension of liquid decreases with an increase in temperature.

DAY PRACTICE SESSION 2

PROGRESSIVE QUESTIONS EXERCISE

1 A light rod of length L is suspended from a support horizontally by means of two vertical wires A and B of equal lengths as shown in the figure. Cross-section area of A is half that of B and Young's modulus of A is double than that of B. A weight w is hung on the rod as shown. The value of x, so that the stress in A is same as that in B, is

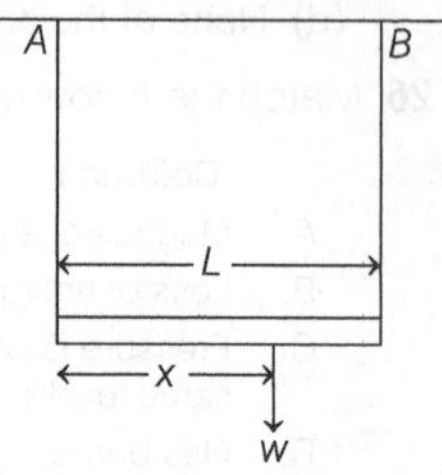

(a) $\frac{L}{3}$ (b) $\frac{L}{2}$ (c) $\frac{2L}{3}$ (d) $\frac{3L}{4}$

2 A pendulum made of a uniform wire of cross-sectional area A has time period T. When an additional mass M is added to its bob, the time period changes to T_M. If the Young's modulus of the material of the wire is Y, then $\frac{1}{Y}$ is equal to (take, g = gravitational acceleration) → JEE Main 2015

(a) $\left[\left(\frac{T_M}{T}\right)^2 - 1\right]\frac{A}{Mg}$ (b) $\left[\left(\frac{T_M}{T}\right)^2 - 1\right]\frac{Mg}{A}$

(c) $\left[1 - \left(\frac{T_M}{T}\right)^2\right]\frac{A}{Mg}$ (d) $\left[1 - \left(\frac{T}{T_M}\right)^2\right]\frac{A}{Mg}$

3 The length of an elastic string is 1m, when the longitudinal tension is 4 N and the length is b metres, when the tension is 5 N. The length of the string (in metre) when the longitudinal tension is 9 N is

(a) $2b - \frac{a}{2}$ (b) $5b - 4a$ (c) $4a - 3b$ (d) $a - b$

4 A rectangular frame is to be suspended symmetrically by two strings of equal length on two supports (figure). It can be done in one of the following three ways:

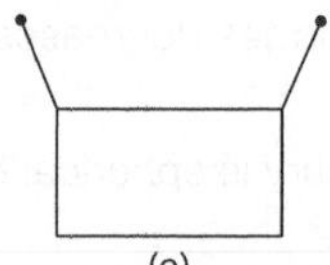

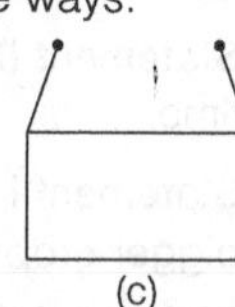

(c)

The tension in the strings will be

(a) the same in all cases (b) least in (a)
(c) least in (b) (d) least in (c)

5 To what depth must a rubber ball be taken in deep sea, so that its volume is decreased by 0.1%. (The bulk modulus of rubber is 9.8×10^8 N/m^2; and the density of sea water is 10^3 kg/m^3.)

(a) 100 m (b) 60 m (c) 75 m (d) 65 m

6 The free surface of oil in a tanker, at rest, is horizontal. If the tanker starts accelerating the free surface will be tilted by an angle θ. If the acceleration is a ms^{-2}, then the slope of the free surface is

(a) g/a (b) a/g (c) ag (d) a^2/g

7 There is a circular tube in a vertical plane. Two liquids which do not mix and of densities d_1 and d_2 are filled in the tube. Each liquid subtends 90° angle at centre. Radius joining their interface makes an angle α with vertical. Ratio d_1/d_2 is → JEE Main 2014

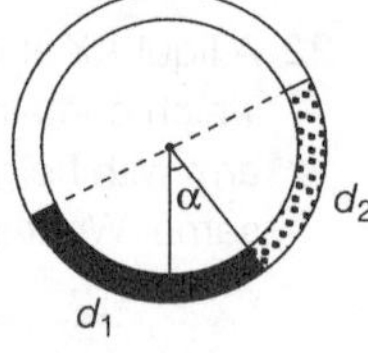

(a) $\frac{1+\sin\alpha}{1-\sin\alpha}$ (b) $\frac{1+\cos\alpha}{1-\cos\alpha}$ (c) $\frac{1+\tan\alpha}{1-\tan\alpha}$ (d) $\frac{1+\sin\alpha}{1-\cos\alpha}$

8 On heating water, bubbles being formed at the bottom of the vessel detach and rise. Take the bubbles to be spheres of radius R and making a circular contact of radius r with the bottom of the vessel. If $r \ll R$ and the surface tension of water is T, value of r just before bubbles detach is (density of water is ρ) → JEE Main 2014

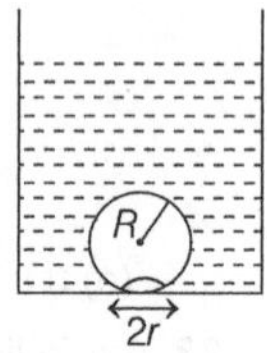

(a) $R^2\sqrt{\frac{2\rho_w g}{3T}}$ (b) $R^2\sqrt{\frac{\rho_w g}{6T}}$ (c) $R^2\sqrt{\frac{\rho_w g}{T}}$ (d) $R^2\sqrt{\frac{3\rho_w g}{T}}$

9 An open glass tube is immersed in mercury in such a way that a length of 8 cm extends above the mercury level. The open end of the tube is then closed and sealed and the tube is raised vertically up by additional 46 cm. What will be length of the air column above mercury in the tube now? (Atmospheric pressure = 76 cm of Hg] → JEE Main 2014

(a) 16 cm (b) 22 cm (c) 38 cm (d) 6 cm

10 Two narrow bores of diameter 3.0 mm and 6.0 mm are joined together to form a U-tube open at both ends. If the U-tube contains water, what is the difference in its levels in the two limbs of the tube? Surface tension of water at the temperature of the experiment is 7.3×10^{-2} N/m. Take the angle of contact to be zero and density of water to be 1.0×10^3 kg/m^3.(take, g = 9.8 m/s^2).

(a) 6 mm (b) 2 mm (c) 5 mm (d) 3 mm

11 A long metal rod of length l and relative density σ, is held vertically with its lower end just touching the surface of water. The speed of the rod when it just sinks in water, is given by

(a) $\sqrt{2gl\sigma}$ (b) $\sqrt{2gl(2\sigma - 1)}$

(c) $\sqrt{2gl\left(1 - \frac{1}{2\sigma}\right)}$ (d) $\sqrt{2gl}$

12 A wooden wheel of radius R is made of two semi-circular parts (see figure). The two parts are held together by a ring made of a metal strip of cross-sectional area S and length L. L is slightly less than $2\pi R$. To fit the ring on the wheel, it is heated so that its temperature rises by ΔT and it just steps over the wheel. As it cools down to surrounding temperature, it presses the semi-circular parts together. If the coefficient of linear expansion of the metal is α and its Young's modulus is Y, the force that one part of the wheel applies on the other part is → AIEEE 2012

(a) $2\pi SY\alpha\Delta T$ (b) $SY\alpha\Delta T$ (c) $\pi SY\alpha\Delta T$ (d) $2SY\alpha\Delta T$

13 A sonometer wire of length 1.5 m is made of steel. The tension in it produces an elastic strain of 1%. What is the fundamental frequency of steel, if density and elasticity of steel are 7.7×10^3 kg/m^3 and 2.2×10^{11} N/m^2, respectively? → JEE Main 2013

(a) 188.5 Hz (b) 178.2 Hz (c) 200.5 Hz (d) 770 Hz

14 A wire of length $2L$ and radius r is stretched and clamped between A and B. If the Young's modulus of the material of the wire be Y, tension in the wire, when stretched in the position ADB will be

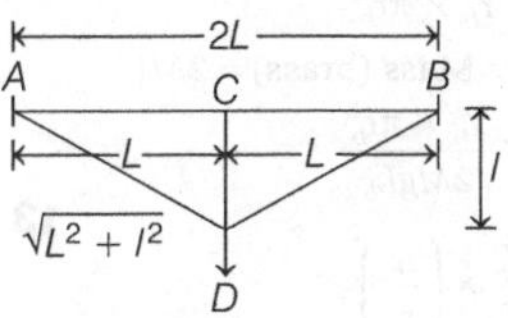

(a) $\pi r^2 YLt$ (b) $\pi r^2 Yl^2/2L^2$
(c) $2\pi r^2 YL^2/l^2$ (d) None of these

15 A uniform cylinder of length L and mass M having cross-sectional area A is suspended, with its length veritcal from a fixed point by a massless spring such that it is half submerged in a liquid of density σ at equilibrium position. The extenstion x_0 of the spring when it is in equilibrium is → JEE Main 2013

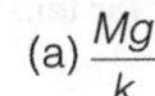

(a) $\frac{Mg}{k}$ (b) $\frac{Mg}{k}\left(\frac{1-LA\sigma}{M}\right)$

(c) $\frac{Mg}{k}\left(\frac{1-LA\sigma}{2M}\right)$ (d) $\frac{Mg}{k}\left(\frac{1+LA\sigma}{M}\right)$

16 A uniform wire (Young's modulus 2×10^{11} Nm^{-2}) is subjected to longitudinal tensile stress of 5×10^7 Nm^{-2}. If the overall volume change in the wire is 0.02%, the fractional decrease in the radius of the wire is close to → JEE Main (Online) 2013

(a) 1.0×10^{-4} (b) 1.5×10^{-4} (c) 0.17×10^{-4} (d) 5×10^{-4}

17 A copper wire of length 1.0 m and a steel wire of length 0.5 m having equal cross-sectional areas are joined end to end. The composite wire is stretched by a certain load which stretches the copper wire by 1 mm.

If the Young's modulii of copper and steel are respectively 1.0×10^{11} Nm^{-2} and 2.0×10^{11} Nm^{-2}, the total extension of the composite wire is → JEE Main (Online) 2013

(a) 1.75 mm (b) 2.0 mm (c) 1.50 mm (d) 1.25 mm

18 A wire of mass m, and length l is suspended from a ceiling. Due to its own weight it elongates, consider cross-section area of wire as A and Young's modulus of material of wire as Y. The elongation in the wire is

(a) $\frac{2mg}{3YA}$ (b) $\frac{mgl}{YA}$ (c) $\frac{mgl}{2YA}$ (d) $\frac{3mg}{YAl}$

Direction (Q. Nos. 19-20) *Each of these questions contains two statements : Statement* I *and Statement* II. *Each of these questions also has four alternative choices, only one of which is the correct answer. You have to select one of the codes (a), (b), (c), (d) given below*

(a) Statement I is true, Statement II is true; Statement II is the correct explanation for Statement I
(b) Statement I is true, Statement II is true; Statement II is not the correct explanation for Statement I
(c) Statement I is true; Statement II is false
(d) Statement I is false; Statement II is true

19 **Statement I** In taking into account the fact that any object which floats must have an average density less than that of water, during world war-I, a number of cargo vessels were made of concrete.

Statement II Concrete cargo vessels were filled with air.

20 **Statement I** The stream of water flowing at high speed from a garden hosepipe tends to spread like a fountain when held vertically up, but tends to narrow down when held vertically down.

Statement II In any steady flow of an incompressible fluid, the volume flow rate of the fluid remains constant.

ANSWERS

SESSION 1	1 (d)	2 (b)	3 (c)	4 (a)	5 (c)	6 (d)	7 (c)	8 (d)	9 (c)	10 (c)
	11 (a)	12 (c)	13 (a)	14 (c)	15 (b)	16 (b)	17 (d)	18 (b)	19 (b)	20 (c)
	21 (b)	22 (a)	23 (d)	24 (c)	25 (b)	26 (d)	27 (a)	28 (a)	29 (a)	30 (b)
SESSION 2	1 (c)	2 (a)	3 (b)	4 (c)	5 (a)	6 (b)	7 (c)	8 (a)	9 (a)	10 (c)
	11 (c)	12 (d)	13 (b)	14 (b)	15 (c)	16 (c)	17 (d)	18 (c)	19 (b)	20 (a)

Hints and Explanations

SESSION 1

1 $\because$ Stress $= \frac{\text{Weight}}{\text{Area}}$

Volume will become (9^3) times.

So weight = volume × density × g will also become $(9)^3$ times.

Area of cross-section will become $(9)^2$ times

$$= \frac{9^3 \times W_0}{9^2 \times A_0} = 9\left(\frac{W_0}{A_0}\right)$$

Hence, the stress increases by a factor of 9.

2 $\Delta L = \frac{FL}{YA} = \frac{FL}{Y\pi R^2}$

As F, L and ΔL are the same, hence

$$YR^2 = \text{constant}$$

$\therefore\ 2.0 \times 10^{11} R_S^2 = 1.0 \times 10^{11} R_B^2$

$\Rightarrow\ R_S = \frac{R_B}{\sqrt{2}}$

3 $\Delta l = \frac{FL}{AY} = \frac{FL}{(\pi r^2)Y}$

$\therefore\ \Delta l \propto \frac{L}{r^2}$

$\therefore\ \frac{\Delta l_1}{\Delta l_2} = \frac{L/R^2}{2L/(2R)^2} = 2$

4 $\Delta l = \frac{Fl}{AY} = \frac{Fl}{\left(\frac{\pi d^2}{4}\right)Y}$ or $(\Delta l) \propto \frac{l}{d^2}$

Now, $\frac{l}{d^2}$ is maximum in option (a).

5 Let l = original length of the wire,

$$\Delta l_1 = l_1 - l$$

Similarly, the change in length of the second wire is

$$\Delta l_2 = l_2 - l$$

Now, $Y = \frac{T_1}{A} \times \frac{l}{\Delta l_1} = \frac{T_2}{A} \times \frac{l}{\Delta l_2}$

$\Rightarrow\ \frac{T_1}{\Delta l_1} = \frac{T_2}{\Delta l_2}$

$\Rightarrow\ \frac{T_1}{l_1 - l} = \frac{T_2}{l_2 - l}$

$\Rightarrow\ T_1 l_2 - T_1 l = T_2 l_1 - T_2 l$

$\Rightarrow\ l = \frac{T_2 l_1 - T_1 l_2}{T_2 - T_1}$

6 As volume is same.

$\therefore\ A_1 l_1 = A_2 l_2$

$\Rightarrow\ l_2 = \frac{A_1 l_1}{A_2} = \frac{A \times l_1}{3A} = \frac{l_1}{3}$

$\Rightarrow\ \frac{l_1}{l_2} = 3$

As, $\Delta x_1 = \frac{F_1}{A\gamma} \times l_1$ ...(i)

and $\Delta x_2 = \frac{F_2}{3A\gamma} l_2$ ...(ii)

Here, $\Delta x_1 = \Delta x_2$

$\therefore\ \frac{F_2}{3A\gamma} l_2 = \frac{F_1}{A\gamma} l_1$

$\Rightarrow\ F_2 = 3F_1 \times \frac{l_1}{l_2} = 3F_1 \times 3 = 9F$

7 For steel wire,

$$\Delta l_s = \frac{F \times L_s}{Y_s \times A_s} = \frac{3MgL_s}{Y_s \times \pi r_s^2}$$

[$\because$ Mass (steel) $= 2M + M = 3M$]

For brass wire,

$$\Delta l_b = \frac{F \times l_b}{Y_b \times A_b} = \frac{2MgL_b}{Y_b \times \pi r_b^2}$$

[$\because$ Mass (brass) $= 2M$]

$\Rightarrow\ \frac{\Delta l_s}{\Delta l_b} = \frac{3Mg\,L_s}{Y_s \times \pi r_s^2} \times \frac{Y_b \times \pi r_b^2}{2MgL_b}$

$= \frac{3}{2} \times \frac{L_s}{L_b} \times \frac{Y_b}{Y_s} \times \left(\frac{r_b}{r_s}\right)^2$

$= \frac{3a}{2b^2c}$

8 From the definition of bulk modulus,

$$B = \frac{-dp}{(dV/V)}$$

Substituting the values, we have

$$B = \frac{(1.165 - 1.01) \times 10^5}{(10\ /\ 100)} = 1.55 \times 10^5 \text{Pa}$$

9 $\because$ Bulk modulus,

$$K = \frac{\text{Volumetric stress}}{\text{Volumetric strain}} = \frac{\Delta p}{\frac{\Delta V}{V}}$$

$\Rightarrow\ K = \frac{mg}{a\left(\frac{3\Delta r}{r}\right)}$

$\left[\because V = \frac{4}{3}\pi r^3,\ \text{so}\ \frac{\Delta V}{V} = \frac{3\Delta r}{r}\right]$

$\Rightarrow\ \frac{\Delta r}{r} = \frac{mg}{3aK}$

10 If Hooke's law is being obeyed, then force extension graph is a straight line. Stored elastic energy extension $\left(U = \frac{1}{2}Fx = \frac{1}{2}\frac{YA}{L}x^2\right)$ should be a parabolic curve symmetric about the U axis. Hence, in the graph P represents extension and Q the stored elastic energy.

11 As work done

$$= \frac{1}{2} Y \times (\text{strain})^2 \times \text{volume}$$

$\Rightarrow\ 2 = \frac{1}{2} \times Y \times \left(\frac{\Delta L}{L}\right)^2 \times AL$

$= \frac{YA(\Delta L)^2}{2L}$

Now, work done,

$$W' = \frac{Y(4A)(\Delta L)^2}{2(L/2)}$$

$$= 8\left(\frac{YA(\Delta L)^2}{2L}\right)$$

$$= 8 \times 2 = 16 \text{ J}$$

12 Thermal stress $\sigma = Y\alpha\Delta\theta$

Given, $\sigma_1 = \sigma_2$

$\therefore\ Y_1\alpha_1\Delta\theta = Y_2\alpha_2\Delta\theta$

or $\frac{Y_1}{Y_2} = \frac{\alpha_2}{\alpha_1} = \frac{3}{2}$

13 If the deformation is small, then the stress in a body is directly proportional to the corresponding strain.

According to Hooke's law, i.e. Young's modulus

$$(Y) = \frac{\text{Tensile stress}}{\text{Tensile strain}}$$

So, $Y = \frac{F/A}{\Delta L/L}$

$= \frac{FL}{A\Delta L}$

If the rod is compressed, then compressive stress and strain appear. Their ratio Y is same as that for tensile case.

Given, length of a steel wire (L) = 10 cm

Temperature $(\theta) = 100°\text{C}$

As length is constant.

$\therefore$ Strain $= \frac{\Delta L}{L} = \alpha\Delta\theta$

Now, pressure = stress = $Y \times$ strain

[Given, $Y = 2 \times 10^{11}\ \text{N/m}^2$ and $\alpha = 1.1 \times 10^{-5}\ \text{K}^{-1}$]

$= 2 \times 10^{11} \times 1.1 \times 10^{-5} \times 100$

$= 2.2 \times 10^8$ Pa

14 Change in length $\Delta L = \alpha L\,\Delta T = \frac{FL}{AY}$

$\Rightarrow$ Stress $= \frac{F}{A} = Y\alpha\Delta T = Y\alpha t$

(as, $\Delta T = t$)

15 From free body diagram of the wooden block,

$$V\sigma g = mg + T$$

[V is the volume of block]

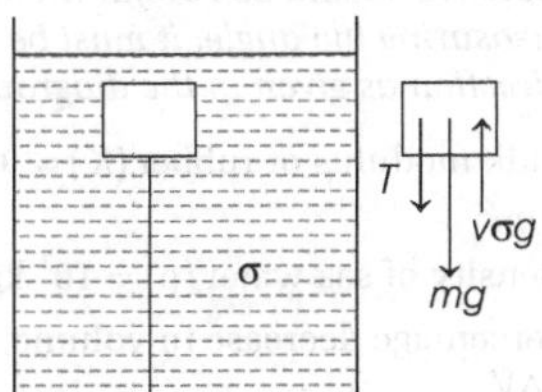

$$T = V\sigma g - mg$$

$$T = \frac{m}{\rho}\sigma g - mg = mg\left(\frac{\sigma - \rho}{\rho}\right)$$

16 Because $\rho_{oil} < \rho < \rho_{water}$

So, it will sink through oil but will not sink in water. So, it will stay at the oil-water interface.

17 Let R be the radius of the bigger drop, then volume of bigger drop

= 2 × volume of small drop

$$\frac{4}{3}\pi R^3 = 2 \times \frac{4}{3}\pi r^3 \Rightarrow R = 2^{1/3} r$$

Surface energy of bigger drop,

$$E = 4\pi R^2 T = 4 \times 2^{2/3}\pi r^2 T$$

$$= 2^{8/3}\pi r^2 T$$

18 We have,

$$6\pi\eta r v = \frac{4}{3}\pi r^3 g\rho - \frac{4}{3}\pi r^3 \sigma g$$

where, $\rho \to \rho_{water}$ and $\sigma \to \rho_{air}$

$$\Rightarrow \eta = \frac{2gr^2(\rho - \sigma)}{9v}$$

$$= \frac{2 \times 9.81 \times (0.2 \times 10^{-2})^2 \times 999}{9 \times 8.7}$$

$$= 1 \times 10^{-3} \text{ poise}$$

19 $v \propto \frac{\rho - \rho_0}{\eta}$

$$\therefore \frac{v_2}{v_1} = \frac{\rho - \rho_{02}}{\rho - \rho_{01}} \times \frac{\eta_1}{\eta_2}$$

$$v_2 = \frac{7.8 - 1.2}{7.8 - 1} \times \frac{8.5 \times 10^{-4} \times 10}{13.2}$$

$$= 6.25 \times 10^{-4} \text{ cms}^{-1}$$

20 From Bernoulli's theorem,

$$\rho gh = \frac{1}{2}\rho(v_2^2 - v_1^2)$$

$$\Rightarrow gh = \frac{1}{2}v_1^2\left[\left(\frac{v_2}{v_1}\right)^2 - 1\right]$$

$$= \frac{1}{2}v_1^2\left[\left(\frac{A_1}{A_2}\right)^2 - 1\right]$$

$(\because A_1 v_1 = A_2 v_2)$

$$\Rightarrow \left(\frac{A_1}{A_2}\right)^2 = 1 + \frac{2hg}{v_1^2}$$

$$\Rightarrow \left(\frac{D_1}{D_2}\right)^4 = 1 + \frac{2gh}{v_1^2}$$

$$\Rightarrow D_2 = \frac{D_1}{\left(1 + \frac{2gh}{v_1^2}\right)^{1/4}}$$

$$= \frac{8 \times 10^{-3}}{\left(1 + \frac{2 \times 10 \times 0.2}{(0.4)^2}\right)^{1/4}}$$

$$= 3.6 \times 10^{-3} \text{ m}$$

21 Velocity head $= \frac{v^2}{2g}$

and pressure head $= \frac{p}{\rho g}$

As velocity of water is equal to the pressure head of 40 cm of Hg column, hence

$$\frac{v^2}{2g} = \frac{h\rho g}{\rho g}$$

$$\Rightarrow v^2 = 2hg$$

$$\Rightarrow v = \sqrt{2gh}$$

$$= \sqrt{2 \times 9.8 \times 0.4}$$

$$= 2.8 \text{ ms}^{-1}$$

22 As shown in the figure, in the two arms of the tube the pressure remains the same on the surface PP'.

Hence,

$$8 \times \rho_Y \times g + 2 \times \rho_{Hg} \times g = 10 \times \rho_X \times g$$

$$\therefore \quad 8\rho_Y + 2 \times 13.6 = 10 \times 3.36$$

Y X
8 cm 2 cm P 10 cm P′
Hg

$$\text{or} \quad \rho_Y = \frac{33.6 - 27.2}{8} = 0.8 \text{ g/cc}$$

23 At equilibrium, weight of the given block is balanced by force due to surface tension, i.e. $2L\ S = w$

$$\text{or} \quad S = \frac{w}{2\text{ L}}$$

$$= \frac{1.5 \times 10^{-2} \text{ N}}{2 \times 0.3 \text{ m}}$$

$$= 0.025 \text{ Nm}^{-1}$$

24 Work done = Change in surface energy

$$\Rightarrow \quad W = 2T \times 4\pi(R_2^2 - R_1^2)$$

$$= 2 \times 0.03 \times 4\pi[(5)^2 - (3)^2] \times 10^{-4}$$

$$W = 0.4\ \pi \text{ mJ}$$

25 As, $T = \frac{r\left(h + \frac{r}{3}\right)\rho g}{2\cos\theta}$

$$= \frac{0.5 \times 10^{-2}\left[3 + \frac{0.5}{3}\right] \times 10^{-2} \times 10^3 \times 9.81}{2}$$

$$= 0.77 \text{ N/m}$$

26 (A) When a spinning ball is thrown it deviates from its ususal path in flight. This is due to magnus effect.

(B) Viscous forces tends to reduce the speed of flowing fluid by virtue of internal frictional force. Hence, the energy of liquid flow reduced.

(C) Pascal's law states that if gravity effect is neglected, the pressure of every point of liquid in equilibrium of rest is the same.

(D) Gas burners are based on Bernoulli's principle.

27 The shape of the aeroplane wings is peculier i.e. its upper surface is more curved than its lower surface. Due to this, when the aeroplane runs on runway the speed of air above the wings is larger than the speed below the wings. Thus, according to Bernoulli's theorem, the pressure above wings becomes lesser than the pressure below the wings. Due to this difference of pressure a vertical lift acts on aeroplane.

28 The height of column is given by ascent formula,

$$h = \frac{2S\cos\theta}{r\rho g} - \frac{r}{3}$$

If the tube is very narrow, $r/3$ can be neglected in comparison with h.

Hence, $h = \frac{2S\cos\theta}{r\rho g}$.

Thus, as the value of r (radius of tube) decreases, the height increases.

$$\left(\because h \propto \frac{1}{r}\right)$$

29 A bridge during its use undergoes alternating stress and strain for a large number of times each day, depending on movement on it. With time it loses its elastic strength and the amount of strain in the bridge for a given stress becomes large and ultimately it may collapse.

30 In case of small drop of mercury the force of gravity is small and the surface tension plays a vital role. Therefore, surface tends to have minimum surface area and sphere has minimum area.

In case of large mercury drop, the gravitational pull becomes more effective than the surface tension and exerts downwards pull on the drop. Hence, the large drop of mercury becomes elliptical or oval.

Also, surface tension of a liquid decreases with rise of temperature, $S_t = S_o(1-\alpha t)$.

SESSION 2

1 Let tension in wire A and B be T_A and T_B, respectively,

$$T_A + T_B = W$$

and $T_A \times x = T_B(L - x)$

Solving the above equations,

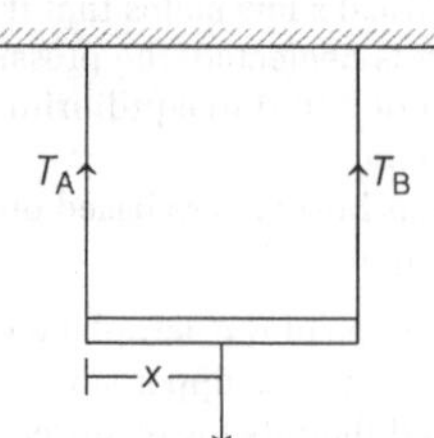

$$T_A = \frac{W(L-x)}{L},$$

$$T_B = \frac{Wx}{L}$$

Stress in $A = \frac{T_A}{A_A}$, where A_A is cross-section area of wire A.

Stress in $B = \frac{T_B}{A_B}$, where A_B is cross-section area of wire B.

It is given,

$$A_A = \frac{A_B}{2},$$

$$\frac{T_A}{A_A} = \frac{T_B}{A_B}$$

which gives $x = \frac{2L}{3}$

2 We know that time period,

$$T = 2\pi\sqrt{\frac{L}{g}}$$

When additional mass M is added to its bob

$$T_M = 2\pi\sqrt{\frac{L+\Delta L}{g}},$$

where, ΔL is increase in length.

We know that Young modulus of the material,

$$Y = \frac{Mg/A}{\Delta L/L} = \frac{MgL}{A\Delta L}$$

$$\Rightarrow \quad \Delta L = \frac{MgL}{AY}$$

$$\therefore \quad T_M = 2\pi\sqrt{\frac{L + \frac{MgL}{AY}}{g}}$$

$$\Rightarrow \quad \left(\frac{T_M}{T}\right)^2 = 1 + \frac{Mg}{AY}$$

or $$\frac{Mg}{AY} = \left(\frac{T_M}{T}\right)^2 - 1$$

or $$\frac{1}{Y} = \frac{A}{Mg}\left[\left(\frac{T_M}{T}\right)^2 - 1\right]$$

3 If L is the initial length, then the increase in length by a tension F is given by $l = \frac{FL}{\pi r^2 Y}$

Hence, $a = L + l = L + \frac{4L}{\pi r^2 Y} = L + 4C$

and $b = L + \frac{5L}{\pi r^2 Y} = L + 5C$

where, $C = \frac{L}{\pi r^2 Y}$

Thus, on solving for L and C, we get

$$L = 5a - 4b \text{ and } C = b - a$$

Hence, for $F = 9$ N, we get

$$x = L + \frac{9L}{\pi r^2 Y}$$
$$= L + 9C$$
$$= (5a - 4b) + 9(b - a)$$
$$= 5b - 4a$$

4 Consider the FBD diagram of the rectangular frame

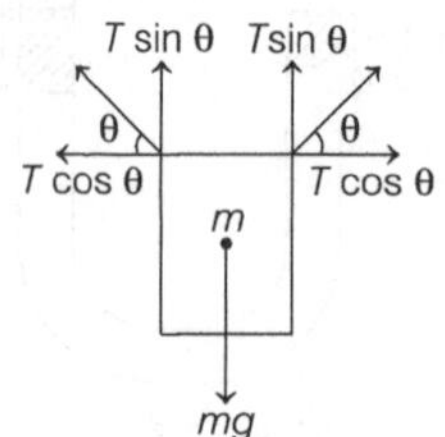

Balancing vertical forces

$2T\sin\theta - mg = 0$

[T is tension in the string]

$$\Rightarrow \quad 2T\sin\theta = mg \quad \text{...(i)}$$

Total horizontal force

$$= T\cos\theta - T\cos\theta = 0$$

Now from Eq. (i), we get

$$T = \frac{mg}{2\sin\theta}$$

As, mg is constant

$$\Rightarrow T \propto \frac{1}{\sin\theta} \Rightarrow T_{max} = \frac{mg}{2\sin\theta_{min}}$$

$$\sin\theta_{min} = 0 \Rightarrow \theta_{min} = 0$$

No option matches with $\theta = 0°$

$$T_{min} = \frac{mg}{2\sin\theta_{max}}$$

(since, $\sin\theta_{max} = 1$)

$$\sin\theta_{max} = 1 \Rightarrow \theta = 90°$$

Matches with option (b).

Hence, tension is least for the case (b).

Note *We should be careful when measuring the angle, it must be in the direction as given in the diagram.*

5 Bulk modulus of rubber $(K) = 9.8 \times 10^8$ N/m^2

Density of sea water $(\rho) = 10^3$ kg/m^3

Percentage decrease in volume,

$$\left(\frac{\Delta V}{V} \times 100\right) = 0.1$$

or $$\frac{\Delta V}{V} = \frac{0.1}{100}$$

or $$\frac{\Delta V}{V} = \frac{1}{1000}$$

Let the rubber ball be taken up to depth h.

$\therefore$ Change in pressure $(p) = h\rho g$

$\therefore$ Bulk modulus

$$(K) = \frac{p}{(\Delta V/V)} = \frac{h\rho g}{(\Delta V/V)}$$

or $$h = \frac{K \times (\Delta V/V)}{\rho g}$$

$$= \frac{9.8 \times 10^8 \times \frac{1}{1000}}{10^3 \times 9.8} = 100 \text{ m}$$

6 As the tanker starts accelerating free surface of the tanker will not be horizontal because pseudo force acts.

Consider the diagram where a tanker is accelerating with acceleration a.

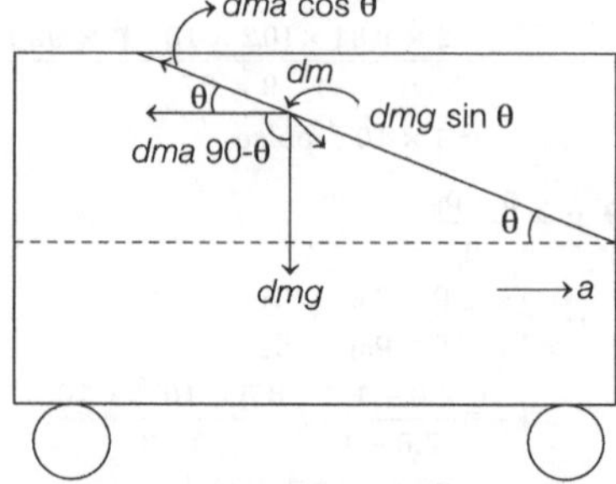

Consider an elementary particle of the fluid of mass dm

The acting forces on the particle with respect to the tanker are shown above.

Now, balancing forces (as the particles is in equilibrium) along the inclined direction, component of weight = component of pseudo force, i.e.

$dmg\sin\theta = dma\cos\theta$ (we have assumed that the surface is inclined at an angle θ)

where, dma is pseudo force

$$\Rightarrow \quad g\sin\theta = a\cos\theta \Rightarrow a = g\tan\theta$$

$$\Rightarrow \quad \tan\theta = \frac{a}{g} = \text{slope}$$

7 Equating pressure at A, we get

$$R\sin\alpha\, d_2 + R\cos\alpha\, d_2 + R(1-\cos\alpha)\,d_1 = R(1-\sin\alpha)\,d_1$$

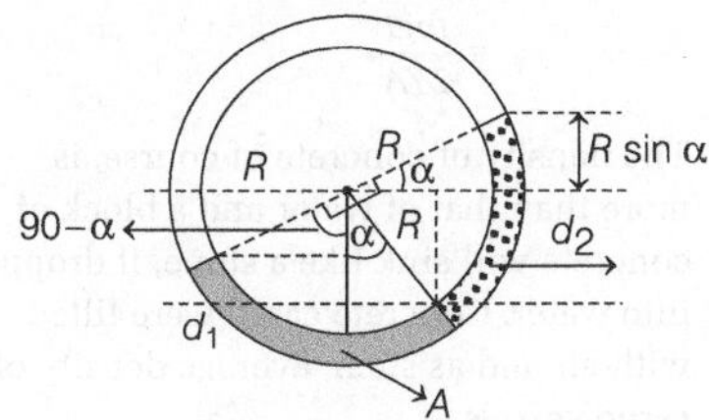

$$(\sin\alpha + \cos\alpha)\,d_2 = d_1(\cos\alpha - \sin\alpha)$$

$$\Rightarrow \quad \frac{d_1}{d_2} = \frac{1+\tan\alpha}{1-\tan\alpha}$$

8 The bubble will detach if, Buoyant force $\geq$ Surface tension force

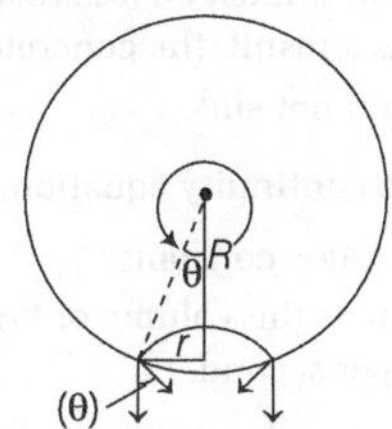

$$\int \sin\theta\, T \times dl = T(2\pi r)\sin\theta$$

$$\frac{4}{3}\pi R^3 \rho_w g \geq \int T \times dl \sin\theta$$

$$(\rho_w)\left(\frac{4}{3}\pi R^3\right)g \geq (T)(2\pi r)\sin\theta$$

As, $\sin\theta = \dfrac{r}{R}$

Solving, $r = \sqrt{\dfrac{2\rho_w R^4 g}{3T}} = R^2\sqrt{\dfrac{2\rho_w g}{3T}}$

9 In this question, the system is accelerating horizontally i.e. no component of acceleration in vertical direction. Hence, the pressure in the vertical direction will remain unaffected.

i.e. $\quad p_1 = p_0 + \rho g h$

Again, we have to use the concept that the pressure in the same level will be same.

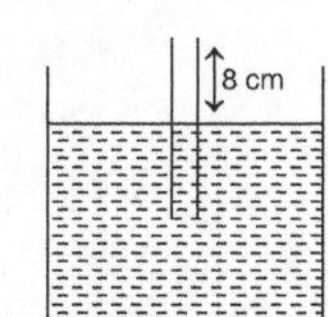

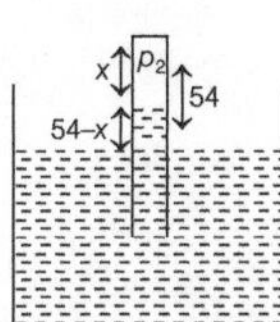

For air trapped in tube,

$$p_1 V_1 = p_2 V_2$$

$$p_1 = p_{atm} = \rho g 76$$

$$V_1 = A \cdot 8$$

[where, A = area of cross-section]

$$p_2 = p_{atm} - \rho g(54-x) = \rho g(22+x)$$

$$V_2 = A \cdot x$$

$$\rho g 76 \times 8A = \rho g(22+x)A\,x$$

$$x^2 + 22x - 76\times 8 = 0 \Rightarrow x = 16\text{ cm}$$

10 Given, surface tension of water

$(S) = 7.3 \times 10^{-2}$ N/m

Density of water $(\rho) = 1.0 \times 10^3$ kg/m^3

Acceleration due to gravity,

$$g = 9.8 \text{ m/s}^2$$

Angle of contact $\theta = 0°$

Diameter of one side, $2r_1 = 3.0$ mm

$\therefore$ $r_1 = 1.5\text{ mm} = 1.5 \times 10^{-3}$ m

Diameter of other side

$$2r_2 = 6.0 \text{ mm}$$

$$r_2 = 3.0\text{ mm} = 3.0 \times 10^{-3}\text{ m}$$

Height of water column rises in first and second tubes

$$h_1 = \frac{2S\cos\theta}{r_1 \rho g}$$

$$h_2 = \frac{2S\cos\theta}{r_2 \rho g}$$

$\therefore$ Difference in levels of water rises in both tubes,

$$\Delta h = h_1 - h_2 = \frac{2S\cos\theta}{\rho g}\left(\frac{1}{r_1} - \frac{1}{r_2}\right)$$

$$= \frac{2 \times 7.3 \times 10^{-2} \times \cos 0°}{1.0 \times 10^3 \times 9.8}\left[\frac{1}{1.5\times 10^{-3}} - \frac{1}{3.0 \times 10^{-3}}\right]$$

$$= \frac{14.6}{9.8} \times 10^{-2}\left[\frac{2-1}{3}\right]$$

$$= 4.9\text{ mm} \approx 5\text{ mm}$$

11 Let the densities of metal and water be ρ and ρ_0 respectively and let x be the length of the rod immersed in water at an instant of time t. Then, acceleration at that instant = apparent weight divided by the mass of the rod, i.e.

$$\frac{dv}{dt} = \frac{\pi r^2 l\rho g - \pi r^2 x\rho_0 g}{\pi r^2 l\rho}$$

$$= g - \frac{gx\rho_0}{l\rho} = g\left(1 - \frac{x}{\sigma l}\right)$$

or $\quad \dfrac{dv}{dx}\dfrac{dx}{dt} = g\left(1 - \dfrac{x}{\sigma l}\right)$

or $\quad v\dfrac{dv}{dx} = g\left(1 - \dfrac{x}{\sigma l}\right)$

On integrating, we get

$$\frac{v^2}{2} = g\left|x - \frac{x^2}{2\sigma l}\right|_0^l = g\left(l - \frac{l}{2\sigma}\right)$$

$$\Rightarrow \quad v = \sqrt{2gl\left(1 - \frac{1}{2\sigma}\right)}$$

12 Elongation due to change in temperature,

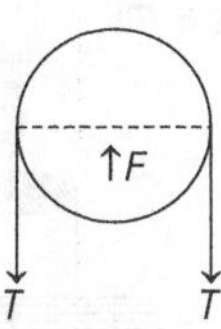

$$\Delta l = L\alpha\Delta T$$

which is compensated by elastic strain, when temperature becomes normal, *i.e.*,

$$\Delta l = \frac{TL}{YS}$$

Thus, $\quad \dfrac{TL}{YS} = L\alpha\Delta T$

$$\Rightarrow \quad T = YS\alpha\Delta T$$

At equilibrium, force exerted by one half on other,

$$F = 2T = 2YS\alpha\Delta T$$

13 Frequency, $f = \dfrac{v}{2l} = \dfrac{1}{2l}\sqrt{\dfrac{T}{\mu}} = \dfrac{1}{2l}\sqrt{\dfrac{T}{Ad}}$

where, T is tension in the wire and μ is the mass per unit length of wire.

Also, Young's modulus,

$$Y = \frac{Tl}{A\Delta l}$$

$$\Rightarrow \quad \frac{T}{A} = \frac{Y\Delta l}{l}$$

Putting this value in expression of frequency, we have

$$f = \frac{1}{2l}\sqrt{\frac{y\Delta l}{ld}}$$

Given, $l = 1.5$ m, $\dfrac{\Delta l}{l} = 0.01$

$$d = 7.7 \times 10^3 \text{ kg/m}^3,$$

$$Y = 2.2 \times 10^{11}\text{ N/m}^2$$

Putting these values we, have

$$f = \frac{1}{2l}\sqrt{\frac{2.2 \times 10^{11} \times 0.01}{7.7 \times 10^3}}$$

$$f = \sqrt{\frac{2}{7}} \times \frac{10^3}{3}$$

$$f \approx 178.2 \text{ Hz}$$

Hence, option (b) is true.

14 As, $Y = \dfrac{\text{Stress}}{\text{Strain}} = \dfrac{F}{\pi r^2 \times \text{strain}}$

$$\Rightarrow F = T = Y\pi r^2 \times \text{strain}$$

Now, strain

$$= \frac{\sqrt{L^2 + l^2}}{L} - L = \left(1 + \frac{l^2}{L^2}\right)^{1/2} - 1$$

$$\approx 1 + \frac{1}{2}\frac{l^2}{L^2} - 1 = \frac{l^2}{2L^2}$$

(by binomial expansion)

$$\therefore T = \frac{Y\pi r^2 \cdot l^2}{2L^2}$$

15 In equilibrium,
upward force = downward force
$kx_0 + F_B = mg$

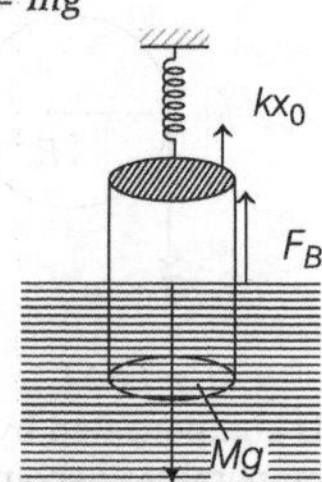

Here, kx_0 is restoring force of spring and F_B is buoyancy force.

$$kx_0 + \sigma \frac{L}{2} Ag = Mg$$

$$x_0 = \frac{Mg - \frac{\sigma LAg}{2}}{k} = \frac{Mg}{k}\left(1 - \frac{\sigma LA}{2M}\right)$$

16 Given, $Y = 2 \times 10^{11}$ N/m^2

Stress $= 5 \times 10^7$ N / m^2

As, $\frac{\text{stress}}{\text{strain}} = Y$

$$\Rightarrow \text{Strain} = \frac{5 \times 10^7}{2 \times 10^{11}} = 2.5 \times 10^{-4}$$

It is symmetric strain.

Now, strain of 2.5×10^{-4} is equivalent.

As, $\frac{\Delta V}{V} = 3\left(\frac{\Delta r}{r}\right)$

$$\therefore \frac{2.5 \times 10^{-4}}{3} = \frac{\Delta r}{r} = 0.83 \times 10^{-4}$$

$\therefore$ Fraction decrease in radius

$= (1.00 - 0.83)\, 10^{-4}$

$= 0.17 \times 10^{-4}$

17 Here, $Y_c = 1 \times 10^{11}$ N/m^2

$Y_s = 2 \times 10^{11}$ N/m^2

$l_c = 1.0$ m, $l_2 = 0.5$ m

and $\Delta l_c = 1 \times 10^{-3}$ m

As, $(\text{strain})_c = \frac{\text{stress}}{Y_c}$

$\Rightarrow 1 \times 10^{-3} = \frac{\text{stress}}{1 \times 10^{11}}$

$\Rightarrow$ stress $= 10^8$ N/m^2

Now, $Y_s = \frac{\text{stress}}{\text{strain}}$

$\Rightarrow$ strain $= \frac{10^8}{2 \times 10^{11}} = 0.5 \times 10^{-3}$

or $\frac{\Delta l_s}{1/2} = 0.5 \times 10^{-3}$

$\Rightarrow \Delta l_s = 0.25 \times 10^{-3}$

$\therefore \Delta l = \Delta l_c + \Delta l_s = 1 + 0.25$

$= 1.25$ mm

18 Consider an element of wire of width dx at a distance x from bottom end of wire. The force experienced by this element is due to the gravitational force of portion of wire lower to it.

So, $Y = \frac{T / A}{\frac{\Delta(dx)}{dx}}$, where $\Delta(dx)$ is the elongation in this element.

Now, $\Delta(dx) = \frac{T}{YA}dx = \frac{mg}{YAl} \times x dx$

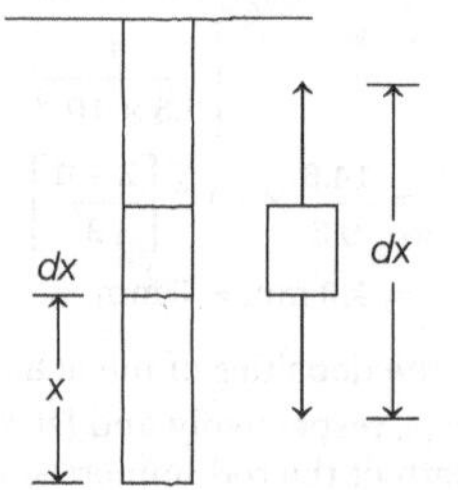

Total elongation,

$$\Delta l = \int \Delta(dx)$$

$$= \int_0^l \frac{mg \times x dx}{YAl}$$

$$= \frac{mgl}{2YA}.$$

19 The density of concrete of course, is more than that of water and a block of concrete will sink like a stone, if dropped into water. Concrete cargo were filled with air and as such, average density of cargo vessels

$$= \frac{[\text{Mass of concrete} + \text{Mass of air}]}{[\text{Volume of concrete} + \text{Volume of air}]}$$

It follows that the average density of cargo vessels must be less than that of water. As a result, the concrete cargo vessels did not sink.

20 From the continuity equation,

Av = constant

where, Av is the volume of liquid flowing per second.

or $A \propto \frac{1}{v}$

As the stream falls, its speed v increases and hence its area of cross-section a will decrease. That is why the stream will become narrow.

When the stream will go up, its speed decreases, hence its area of cross-section will increase and it will become broader and spreads out like a fountain. Hence, option (a) is true.

DAY FIFTEEN

Heat and Thermodynamics

Learning & Revision for the Day

- Heat
- Thermometry
- Thermal Expansion
- Specific Heat Capacity
- Calorimetry
- Change of State
- Heating Curve
- Zeroth Law of Thermodynamics
- First Law of Thermodynamics
- Thermodynamic Processes
- Second Law of Thermodynamics
- Reversible and Irreversible Processes
- Carnot Engine and its Efficiency
- Refrigerator
- Equation of State of Perfect Gas
- Kinetic Theory of Gases
- Degree of Freedom (f)
- Specific Heat Capacities of Gases
- Mean Free Path
- Avogadro's Number

Heat

Heat is a form of energy which characterises the thermal state of matter. It is transferred from one body to the other due to temperature difference between them.

Heat is a scalar quantity with dimensions $[ML^2T^{-2}]$ and its SI unit is joule (J) while practical unit is calorie (cal); 1 cal = 4.18 J

If mechanical energy (work) is converted into heat then, the ratio of work done (W) to heat produced (Q) always remains the same and constant.

i.e. $$\frac{W}{Q} = \text{constant} = J \text{ or } W = JQ$$

The constant J is called **mechanical equivalent of heat.**

Temperature

The factor that determines the flow of heat from one body to another when they are in contact with each other, is called temperature. Its SI unit is kelvin.

Thermometer

An instrument used to measure the temperature of a body is called a thermometer. For construction of thermometer, two fixed reference point **ice point** and **steam point** are taken. Some common types of thermometers are as follows:

1. **Liquid (mercury) thermometer** Range of temperature: –50°C to 350°C
2. **Gas thermometer (Nitrogen gas)** Range of temperature: –200°C to 1600°C
3. **Pyrometers** Range of temperature: –800°C to 6000°C

- No. of Questions in Exercises (x)—
- No. of Questions Attempted (y)—
- No. of Correct Questions (z)—
  (Without referring Explanations)
- Accuracy Level ($z/y \times 100$)—
- Prep Level ($z/x \times 100$)—

In order to expect good rank in JEE, your Accuracy Level should be above 85 & Prep Level should be above 75.

Scales of Temperature

Three most common scales are Celsius scale or Centigrade scale, Fahrenheit scale and Kelvin scale (Absolute scale).

Scale	Ice point/lower reference point	Steam point / Upper reference point	Unit
Celsius	0	100	°C
Fahrenheit	32	212	°F
Kelvin	273.15	373.15	K

Relation between C, F and K scales is

$$\frac{C}{5} = \frac{F-32}{9} = \frac{K-273.15}{5}$$

In general, $$\frac{\text{Temperature of } X - \text{Ice point of } X}{\text{Steam point of } X - \text{Ice point of } X} = \frac{\text{Temperature of } Y - \text{Ice point of } Y}{\text{Steam point of } Y - \text{Ice point of } Y}$$

Thermometry

The branch dealing with measurement of temperature is called thermometry.

Let thermometric properties at temperatures 0°C (ice point), 100°C (steam point) and t°C (unknown temperature) are X_0, X_{100}, and X_t, respectively. Then,

$$\frac{X_t - X_0}{t} = \frac{X_{100} - X_0}{100} \quad \text{or} \quad \frac{X_t - X_0}{X_{100} - X_0} = \frac{t}{100}$$

Thus, $$t = \left(\frac{X_t - X_0}{X_{100} - X_0}\right) \times 100°\text{C}$$

Thermal Expansion

Almost all substances (solid, liquid and gas) expand on heating and contract on cooling. The expansion of a substance on heating is called thermal expansion of substance.

Thermal Expansion of Solids

Thermal expansion in solids is of three types:

1. **Linear Expansion** Thermal expansion along a single dimension of a solid body is defined as the linear expansion.

   If a rod is having length l_0 at temperature T, then elongation in length of rod due to rise in temperature by ΔT is,

   $$\Delta l = l_0 \alpha \Delta T \quad \text{or} \quad \alpha = \frac{\Delta l}{l_0 \times \Delta T}$$

   where, α is the coefficient of linear expansion whose value depends on the nature of the material.

   Final length, $l_f = l_0 + l_0\alpha \Delta T$
   $= l_0(1 + \alpha \Delta T)$

2. **Superficial Expansion or Areal Expansion** Expansion of solids along two dimensions of solid objects is defined as superficial expansion.

   Coefficient of superficial expansion,

   $$\beta = \frac{\Delta A}{A_0 \times \Delta T}$$

   Final area, $A_f = A_0(1 + \beta\Delta T)$
   where, A_0 is the area of the body at temperature T.

3. **Volume or Cubical Expansion** Expansion of solids along three dimensions of solids objects is defined as cubical expansion.

   Coefficient of volume or cubical expansion,

   $$\gamma = \frac{\Delta V}{V_0 \times \Delta T}$$

   Final volume, $V = V_0 (1 + \gamma\Delta T)$.
   where, V_0 is the volume of the body at temperature T.

NOTE • The coefficients α, β and γ for a solid are related to each other.

$$\alpha = \frac{\beta}{2} = \frac{\gamma}{3}$$

• As temperature increases, density decreases according to relation,

$$\rho = \frac{\rho_0}{1 + \gamma\Delta T}$$

or $\rho = \rho_0(1 - \gamma\Delta T)$ [valid for small ΔT]

Thermal Expansion of Liquid

Liquids do not have linear and superficial expansion but these only have volume expansion.

Liquids have two coefficients of volume expansion

1. Coefficient of apparent expansion,

   $$\gamma_a = \frac{\text{Apparent expansion in volume}}{\text{Initial volume} \times \Delta T} = \frac{(\Delta V)_a}{V \times \Delta T}$$

2. Coefficient of real expansion,

   $$\gamma_r = \frac{\text{Real expansion in volume}}{\text{Initial volume} \times \Delta T} = \frac{(\Delta V)_r}{V \times \Delta T}$$

Anomalous/Exceptional Behaviour of Water

Generally, density of liquids decreases with increase in temperature but for water as the temperature increases from 0 to 4°C, its density increases and as temperature increases beyond 4°C, the density decreases.

The variation in the density of water with temperature is shown in the figure given below.

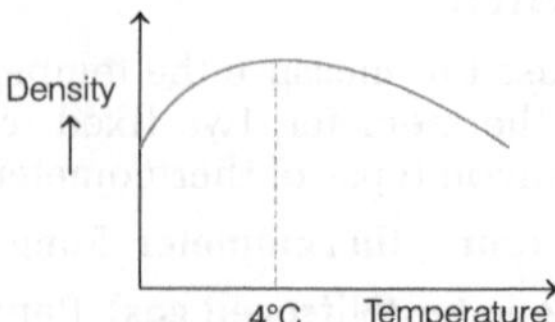

Anomalous behaviour of water

Thermal Expansion of Gases

Gases have no definite shape, therefore, gases have only volume expansion.

1. The coefficient of volume expansion at constant pressure,
$$\alpha = \frac{\Delta V}{V_0} \times \frac{1}{\Delta T}$$
Final volume, $V' = V(1 + \alpha\,\Delta T)$
2. The coefficient of pressure expansion at constant volume,
$$\beta = \frac{\Delta p}{p} \times \frac{1}{\Delta T}$$
Final pressure, $p' = p(1 + \beta\,\Delta T)$.

Specific Heat Capacity

The quantity of heat required to raise the temperature of unit mass of a substance by 1°C is called specific heat.

Specific heat, $s = \dfrac{Q}{m \times \Delta T}$

The SI unit of specific heat is J kg^{-1} k^{-1}.

- Specific heat capacity can have any value from 0 to ∞. For some substances under particular situations, it can have negative values also.
- The product of mass of the body and specific heat is termed as **heat capacity**, $C = m \times s$.

Molar Heat Capacity

The amount of heat required to change the temperature of a unit mole of substance by 1°C is termed as its molar heat capacity,

$$C_m = \frac{Q}{\mu \Delta T}$$

Generally, for gases, two molar heat capacities are very common—molar heat capacity at constant pressure (C_p) and molar heat capacity at constant volume (C_V).

Water Equivalent of a Substance

Water equivalent of certain amount of substance is defined as the amount of water, which when replaced by the substance requires the same amount of heat for the same rise in temperature.

$$m_w = \frac{mS}{S_w},$$

where, m_w = water equivalent of substance whose mass is m,
S = specific heat capacity of substance
and S_w = specific heat capacity of water

Calorimetry

Calorimetry means measurement of heat. When a body at higher temperature is brought in contact with another body at lower temperature, the heat lost by the hot body is equal to the heat gained by the colder body and provided no heat is allowed to escape to the surrounding.

A device in which heat measurement can be made is called a calorimeter.

If temperature changes, heat exchanged is given by

$$Q = ms\Delta T$$

As temperature of the body increases, it means heat is taken by the body, otherwise given out by the body.

Change of State

When we supply heat (energy) to a body and its temperature does not change, then the energy consumed by the body is used up in changing its phase and the process is called change of state.

Latent Heat

In case of phase change, heat is consumed during melting and boiling while released during freezing and condensation.

The heat required to change the phase of a system is proportional to the mass of the system i.e.

$$Q \propto m$$
$$Q = mL$$

where, L is the latent heat, which is defined as the amount of heat required to change the phase of the unit mass of a substance at given temperature.

- In case of ice, the latent heat of fusion of ice is 80 cal/gm.
- In case of water, the latent heat of vapourisation is 536 cal/gm.

Sublimation

- A substance can sometimes change directly from solid to gaseous phase, this process is termed as **sublimation**. Corresponding latent heat is termed as **latent heat of sublimation L_s**. The reverse process can also occur.
- Very pure water can be cooled several degrees below the freezing temperature without freezing, the resulting unstable state is described as **supercooled**. When this supercooled water is disturbed (either by dropping dust particles etc.), it crystallises within a second or less.
- A liquid can sometimes be **superheated** above its normal boiling temperature. Any small disturbance such as agitation causes local boiling with bubble formation.

Heating Curve

If we supply energy to a body in solid state (temperature<melting point) at a constant rate, then the curve drawn between temperature and time is termed as the heating curve.

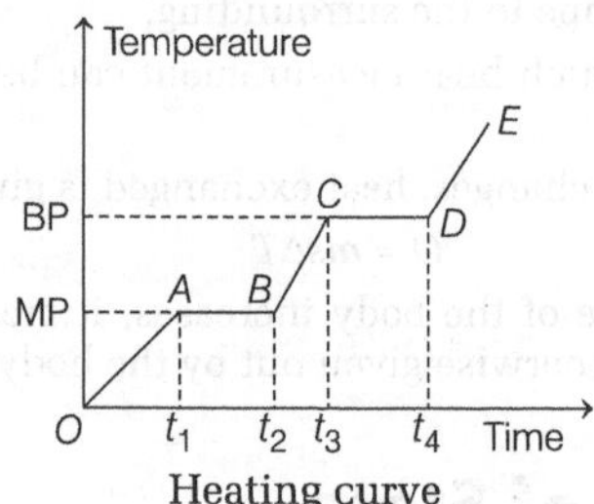

Heating curve

OA represents heating of the solid,

$$S_{\text{solid}} \propto \frac{1}{\text{Slope of } OA}$$

AB represents melting of the solid,

length of $AB \propto L_f$

BC represents heating of the liquid,

$$S_{\text{liquid}} \propto \frac{1}{\text{Slope of } BC}$$

CD represents boiling (vaporisation) of the liquid, length of $CD \propto L_V$, DE represents heating of the gaseous phase,

$$S_{\text{gas}} \propto \frac{1}{\text{Slope of } DE}$$

Zeroth Law of Thermodynamics

When there is no exchange of heat between two objects placed in contact, then both are called in **thermal equilibrium**.

According to this law, if two systems A and B, separated by an adiabatic wall, are separately and independently in thermal equilibrium with a third system C, then the systems A and B are also in a state of thermal equilibrium with each other.

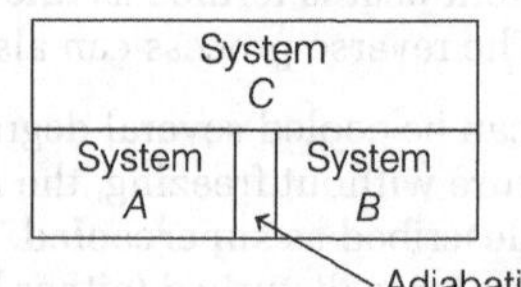

Three system of thermal equilibrium

Basic Terms Used Thermodynamics

Internal Energy

Internal energy of a system is defined as the sum of the total kinetic energy of all its constituent particles and sum of all the potential energies of interaction among these particles.

The internal energy of an ideal gas is totally kinetic and it is given by

$$U = \frac{3}{2}\mu RT$$

and change in internal energy

$$\Delta U = \frac{3}{2}\mu R\Delta T.$$

For non-ideal gases, internal energy depends not only on the temperature but also on the pressure.

Work

Consider a system in a cylinder with movable piston, whose volume can be changed (a gas, liquid or solid). Suppose, the cylinder has a cross-sectional area A and pressure exerted by system on the piston face is p. The work done by the system on the surroundings for small displacement dx is $dW = pAdx$.

$$W = \int dW = \int_{V_i}^{V_f} pdV$$

i.e. work done in a finite change of volume from V_i to V_f.

- Work done by the system depends on the initial and final states.
- If volume of the system increases, then work is done by the system and it is taken as positive work done.
- If volume of the system decreases, then work is done on the system and it is taken as negative work done.

First Law of Thermodynamics

According to this law, the heat given to a system (ΔQ) is equal to the sum of increase in its internal energy (ΔU) and the work done (ΔW) by the system against the surroundings.

Mathematically, $\Delta Q = \Delta U + \Delta W$

Sign Convention

$\Delta Q = +$ ve when heat supplied

$= -$ ve when heat is ejected

$\Delta U = +$ ve when temperature increases

$= -$ ve when temperature decreases

$\Delta W = +$ ve when work is done by the system (expansion)

$= -$ ve when work is done on system (compression)

First law of thermodynamics is based on the energy conservation.

Thermodynamic Processes

A thermodynamic process is the process of change of state of a system involving change of thermodynamic variables, e.g. p, V, T etc. When a system undergo a thermodynamic change, then work done either by system on surrounding or by surroundings on system is called external work.

$$W_{\text{ext}} = \int_{V_1}^{V_2} p\, dV = \text{area under } p\text{-}V \text{ curve.}$$

1. Isothermal Process

It is that process in which temperature remains constant. Here, exchange of heat with the surroundings is allowed.

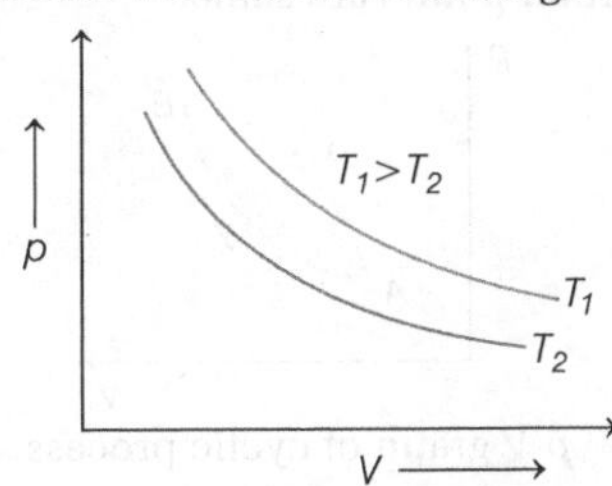

p-V graph of an isothermal expansion process

As temperature T remains constant in an isothermal process, hence as per Boyle's law

$$p \propto \frac{1}{V} \text{ or } pV = \text{constant}$$

Molar specific heat of a gas under isothermal condition

$$C = \frac{\Delta Q}{m\Delta T} = \frac{\Delta Q}{m\,(0)} = \infty$$

Slope of p-V curve at any point is $\frac{dp}{dV} = -\frac{p}{V}$.

Work done in an isothermal process

$$\Delta W = \int_{V_i}^{V_f} pdV = nRT \ln\left(\frac{V_f}{V_i}\right)$$

where, n = number of moles, R = gas constant and T = temperature.

V_f and V_i are final and initial volume of the gas respectively.

As per first law of thermodynamics, since, $\Delta T = 0$, hence, $\Delta U = 0$ for an ideal gas and we have $\Delta Q = \Delta W$.

Thus, heat supplied to the system in an isothermal process is entirely used to do work against external surroundings.

2. Adiabatic Process

It is that process in which there is no exchange of heat of the system with its surroundings. Thus, in an adiabatic process p, V and T change but $\Delta Q = 0$ or entropy remains constant $\left(\Delta S = \frac{\Delta Q}{T} = 0\right)$.

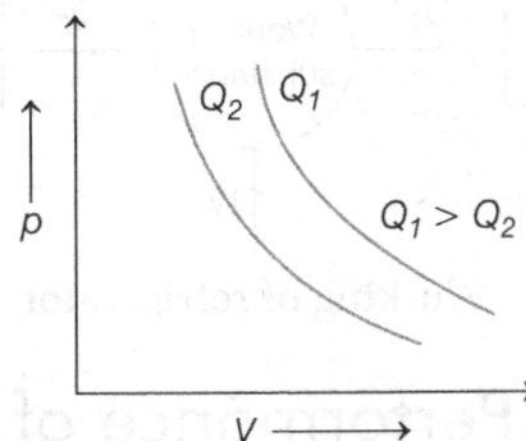

p-V graph for adiabatic process

The equation of state for an adiabatic process is

$$pV^{\gamma} = \text{constant}$$

or $$TV^{\gamma-1} = \text{constant}$$

or $$T^{\gamma}p^{1-\gamma} = \text{constant}$$

- **Molar specific heat** of a gas under adiabatic condition

$$C = \frac{\Delta Q}{m \cdot \Delta T} = \frac{0}{m \cdot \Delta T} = 0$$

- **Slope of an adiabatic curve** at a point is $\frac{dp}{dV} = -\gamma\frac{p}{V}$.
- **Work done in an adiabatic process**

$$\Delta W = \int_{V_i}^{V_f} p\,dV = \frac{\mu R}{(\gamma-1)}(T_i - T_f)$$

During an adiabatic expansion $\Delta W = +$ ve, hence, temperature of gas falls, i.e. an adiabatic expansion is always accompanied with cooling.

As per first law of thermodynamics, since, $\Delta Q = 0$ in an adiabatic process hence,

$$\Delta U = -\Delta W$$

- **Free expansion** is an adiabatic process in which $\Delta W = 0$. Hence, in accordance with first law of thermodynamics $\Delta U = 0$ i.e. the final and initial values of the internal energy are equal in free expansion.

3. Isochoric Process

It is that thermodynamic process in which volume remains constant.

In an isochoric process for a given mass of gas

$$p \propto T \text{ or } \frac{p}{T} = \text{constant}$$

- **Indicator diagram** for an isochoric process is a straight line parallel to p-axis.

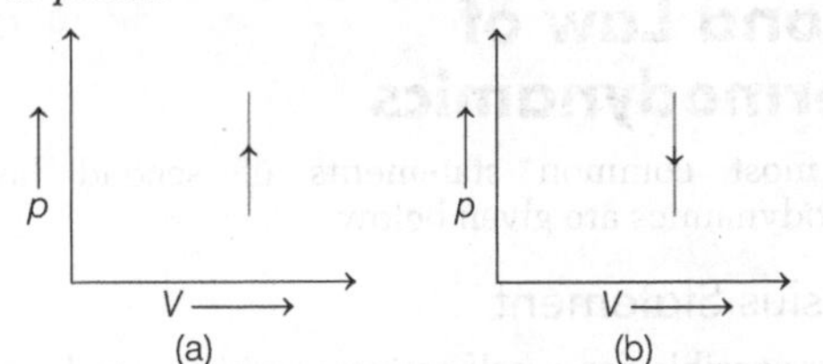

Graph (a) shows isometric heating graph in which pressure increases, temperature increases, ΔQ is positive and ΔU is positive.

Similarly, Graph (b) shows isometric cooling graph in which pressure decreases, temperature decreases, ΔQ is negative and ΔU is negative.

- **Molar specific heat** of a gas under isochoric condition $C_V = \frac{f}{2}R$, where f is the number of degrees of freedom per molecule.
- **Work done in an isochoric process**

$$\Delta W = \int p\,dV = 0$$

As $\Delta W = 0$ hence, according to first law of thermodynamics, we have

$$(\Delta Q)_V = \Delta U = \mu C_V\,\Delta T = \frac{\mu R}{(\gamma-1)}\Delta T$$

4. Isobaric Process

It is that process in which pressure remains constant.
As in an isobaric process for a given mass of gas

$$V \propto T$$

or
$$\frac{V}{T} = \text{constant}$$

- **Indicator diagram** for an isobaric process is a straight line parallel to X-axis.

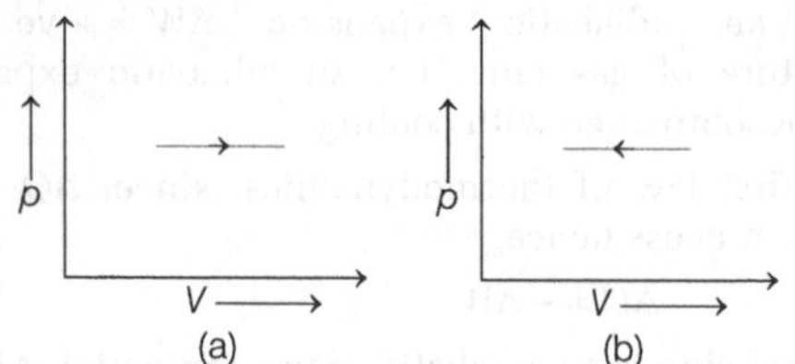

Graph (a) represent isobaric expansion, graph (b) represent isobaric compression.

- **Work done in an isobaric process**

$$\Delta W = \int p\,dV = p\int_{V_i}^{V_f} dV$$
$$= p(V_f - V_i) = p\Delta V$$

- **Molar specific heat** of a gas under isobaric condition

$$C_p = \left(\frac{f}{2} + 1\right) R = C_V + R$$

Second Law of Thermodynamics

Two most common statements of second law of thermodynamics are given below

Clausius Statement

It is impossible for a self-acting machine, working in a cyclic process to transfer heat from a colder body to a hotter body without the aid of an external agency.

Kelvin-Planck's Statement

It is impossible to design an engine which extracts heat from a reservoir and fully converts it into work without producing any other effect.

Reversible and Irreversible Processes

A reversible process is one which can be reversed in such a way that all changes taking place in the direct process are exactly repeated in inverse order and in opposite sense, and no changes are left in any of the bodies taking part in the process or in the surroundings. Any process which is not reversible exactly is an irreversible process.

Cyclic Process

In cyclic process, if the process takes the path AxB, it returns *via* ByA, the initial and final points are same.

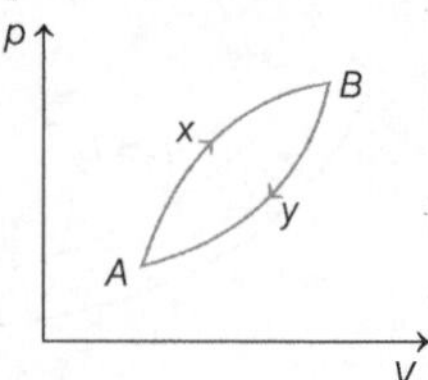

p-V graph of cyclic process

Carnot Engine and its Efficiency

Carnot engine is a theoretical, ideal heat engine working in a reversible cyclic process operating between two temperatures T_1 (heat source) and T_2(heat sink). The Carnot's cycle consists of two isothermal processes connected by two adiabatic processes as shown in the figure.

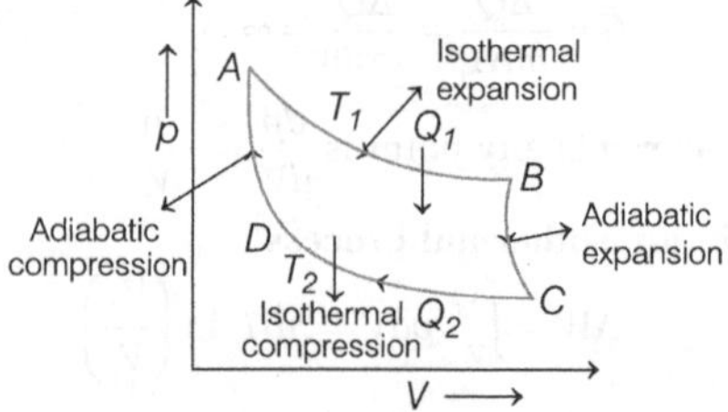

Various process in Carnot cycle

The efficiency of a Carnot's cycle is given by

$$\eta = \frac{W}{Q_1} = 1 - \frac{Q_2}{Q_1} = 1 - \frac{T_2}{T_1}$$

The efficiency does not depend on the nature or quantity of the working substance.

Refrigerator

A refrigerator or heat pump is basically a heat engine running in reverse direction. It takes heat from colder body (sink) and after doing some work gives the rest heat to the hotter body (source). An ideal refrigerator can be regarded as Carnot's ideal heat engine working in the reverse direction.

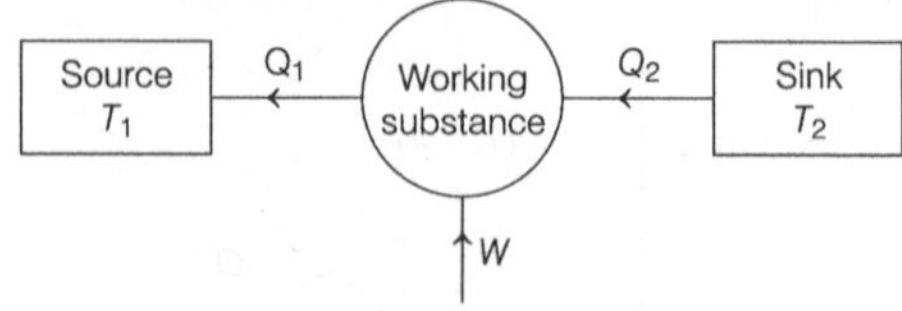

Working of refrigerator

Coefficient of Performance of a Refrigerator

It is defined as the ratio of quantity of heat removed per cycle (Q_2) to the work done (W) on the working substance per cycle to remove this heat.

$$\beta = \frac{Q_2}{W} = \frac{Q_2}{Q_1 - Q_2} \text{ or } \beta = \frac{T_2}{T_1 - T_2} = \frac{1-\eta}{\eta}$$

Equation of State of a Perfect Gas

The equation which relates the pressure (P), volume (V) and temperature (T) of the given state of an ideal gas is known as ideal or perfect gas equation.

For 1 mole of gas, $\frac{pV}{T} = R$ (constant)

$$pV = RT$$

where, R is **universal gas constant**. The SI unit of gas constant is J/mol-K. Its value is 8.314 J/mol-K or 8.314×10^7 erg/mol-K or 2 cal/mol-K. The dimensions of R are $[ML^2T^{-1}\theta^{-1}]$.

Moreover, gas constant $R = \frac{\text{Work}}{\text{Moles} \times \text{Temperature}}$

- The perfect gas equation for 1 molecule of gas is
$$pV = kT$$
- Boltzmann's constant is represented by per mole gas constant
i.e. $k = \frac{R}{N} = \frac{8.31}{6.023 \times 10^{23}} = 1.38 \times 10^{-23}\ \text{JK}^{-1}$
Its dimensions are $[ML^2T^{-2}\theta^{-1}]$.

Kinetic Theory of Gases

Kinetic theory of gases relates the macroscopic properties of gases (such as pressure, temperature etc.) to the microscopic properties of gas molecules (such as speed, momentum, kinetic energy of molecules etc).

Assumptions of Kinetic Theory of Gases

- Every gas is composed of tiny particles known as molecules. The size of molecules is much smaller than the intermolecular spacing.
- The molecules of a gas are identical, spherical, rigid and perfectly elastic point masses.
- Molecules are in a state of random rapid motion. They collide with each other. There is no loss of energy during collision. Only the direction of motion is changed.
- The time spent in collision between two molecules is negligible in comparison to time between two successive collisions.
- The number of collisions per unit volume in a gas remains constant. No attractive or repulsive force acts between gas molecules.
- Gravitational attraction among the molecules is ineffective due to extremely small masses and very high speed of molecules.
- Molecules constantly collide with the walls of container due to which their momentum changes. The change in momentum is transferred to the walls of the container. Consequently, on the walls of container pressure is exerted by gas molecules. The density of gas is constant at all points of the container.

Work Done on Compressing a Gas

Work done $W = p \cdot \Delta V$, where p = pressure of the gas and ΔV = change in volume of the gas.

When two ideal gases having molar masses M_1 and M_2 are mixed, then thermodynamic variables/parameters for mixture would be given by

- M (molar mass) $= \frac{n_1 m_1 + n_2 m_2}{n_1 + n_2}$
- C_V (of the mixture) $= \frac{n_1 C_{V_1} + n_2 C_{V_2}}{n_1 + n_2}$
- C_p (of the mixture) $= \frac{n_1 C_{p_1} + n_2 C_{p_2}}{n_1 + n_2}$
- γ (of the mixture) $= \frac{n_1 C_{p_1} + n_2 C_{p_2}}{n_1 C_{V_1} + n_2 C_{V_2}}$ or is given by

$$\frac{n_1 + n_2}{\gamma - 1} = \frac{n_1}{\gamma_1 - 1} + \frac{n_2}{\gamma_2 - 1}$$

where, symbols have their usual meanings.

Kinetic Energy and Temperature

In ideal gases, the point particles can have only translational motion and thus only translational energy.

- Translational KE of a molecule $= \frac{1}{2} m\overline{c^2} = \frac{3}{2} kT$
- Mean KE per molecule $= \frac{3}{2} kT$
- Mean kinetic energy per gram mole is given by
$$\text{KE}_{\text{mole}} = \left(\frac{1}{2} m\overline{c^2}\right) N = \frac{3}{2} kTN = \frac{3}{2} RT$$
- Average kinetic energy of gas $= \frac{3}{2} pV$
- KE per molecule $= \frac{3pV}{2N} = \frac{3RT}{2N}$
- KE per mole $= \frac{3}{2} kT$
- KE per volume $= \frac{3}{2} p$

Concept of Pressure

Pressure p exerted by a perfect gas on the walls of container is given by

$$p = \frac{1}{3} \frac{mN\overline{c^2}}{V} = \frac{1}{3} \frac{M}{V} \overline{c^2}$$

$$pV = \frac{1}{3}(\rho\overline{c^2}) = \frac{1}{3}\left(\frac{2}{2}\rho\overline{c^2}\right) = \frac{2}{3}\left(\frac{1}{2}\rho\overline{c^2}\right) = \frac{2}{3} E$$

Here, m = mass of each molecule, $\bar{c}$ = root mean square velocity of molecules, ρ = density of gas, M = mass of gas enclosed in volume V of container, and E = Total KE of the ideal gas.

Various Speeds of Gas Molecules

- **Root mean square speed** It is defined as the square root of mean of squares of the speed of different molecules

i.e. $v_{rms} = \sqrt{\frac{v_1^2 + v_2^2 + v_3^2 + \dots}{N}} = \sqrt{\overline{v^2}}$

From the expression of pressure, $p = \frac{1}{3}\rho v_{rms}^2$

$$v_{rms} = \sqrt{\frac{3p}{\rho}} = \sqrt{\frac{3pV}{\text{Mass of gas}}} = \sqrt{\frac{3RT}{M}} = \sqrt{\frac{3kT}{m}}$$

- **Most probable speed** It is defined as the speed which is possessed by maximum fraction of total number of molecules of the gas.

$$v_{mp} = \sqrt{\frac{2p}{\rho}} = \sqrt{\frac{2RT}{M}} = \sqrt{\frac{2kT}{m}}$$

- **Average speed** It is the arithmetic mean of the speeds of molecules in a gas at given temperature.

$$v_{av} = \frac{v_1 + v_2 + v_3 + v_4 + \dots}{N}$$

Average speed, $v_{av} = \sqrt{\frac{8p}{\pi\rho}} = \sqrt{\frac{8}{\pi}\frac{RT}{M}} = \sqrt{\frac{8kT}{\pi m}}$

Degree of Freedom (f)

The term degree of freedom of a system refers to the possible independent motions a system can have

- for monoatomic gas, $(f) = 3$
- for diatomic gas, $(f) = 5$
- for triatomic gas, $(f) = 6$(non-linear)
- for triatomic (linear) gas, $(f) = 7$
- for N-atomic molecule $(f) = 6N - 3$
- for N-atomic linear molecule $(f) = 6N - 5$

Law of Equipartition of Energy

According to law of equipartition of energy for any system in thermal equilibrium, the total energy is equally distributed among its various degree of freedom and each degree of freedom is associated with energy $\frac{1}{2}kT$

(where, $k = 1.38 \times 10^{-23}$ J/K and T = absolute temperature of the system).

Specific Heat Capacities of Gases

The specific heat of gas can have many values, but out of them following two values are important

Specific Heat at Constant Volume

The specific heat of a gas at constant volume is defined as the quantity of heat required to raise the temperature of unit mass of gas through 1°C or 1 K when its volume is kept constant i.e.

$$c_V = \frac{(\Delta Q)_V}{m\Delta T}$$

For one mole of gas,

$$C_V = Mc_V = \frac{M(\Delta Q)_V}{m\Delta T} = \frac{1}{n}\frac{(\Delta Q)_V}{\Delta T} \quad \left[\because n = \frac{m}{M}\right]$$

Specific Heat at Constant Pressure

The specific heat of a gas at constant pressure is defined as the quantity of heat required to raise the temperature of unit mass of gas through 1K, when its pressure is kept constant i.e.

$$c_p = \frac{(\Delta Q)_p}{m\Delta T}.$$

For one mole of gas,

$$\therefore \quad C_p = Mc_p = \frac{M(\Delta Q)_p}{m\Delta T} \quad \dots(i)$$

$$= \frac{1}{n}\frac{(\Delta Q)_p}{\Delta T} \quad [\because n = m/M]$$

Specific heat of a gas at constant pressure is greater than the specific heat at constant volume i.e. $C_p > C_V$.

C_p and C_V are related to each other according to relation,

$$C_p - C_V = \frac{R}{J} \quad \dots(ii)$$

Eq. (ii) is called Mayer's relation. If C_p and C_V are measured in the units of work and R is also in the units of work (or energy), then Eq. (ii) becomes $C_p - C_V = R$.

Specific Heat in Terms of Degree of Freedom

For a gas at temperature T, the internal energy $U = \frac{f}{2}nRT$.

Change in energy, $\Delta U = \frac{f}{2}nR\Delta T \Rightarrow C_V = \frac{1}{2}fR$

Specific heat at constant pressure, $C_p = \left(\frac{f}{2} + 1\right)R$.

Ratio of C_p and C_V, $\gamma = \frac{C_p}{C_V} = \frac{\left(\frac{f}{2} + 1\right)R}{\frac{f}{2}R} = 1 + \frac{2}{f}$.

Mean Free Path

The distance travelled by a gas molecule between two successive collision is known as free path.

$$\text{Mean free path} = \frac{\text{Total distance covered}}{\text{Number of collisions}}$$

The mean free path of a gas molecule is the average distance between two successive collisions. It is represented by λ.

$$\lambda = \frac{1}{\sqrt{2}\pi\sigma^2 n}$$

Here, σ = diameter of the molecule and n = number of molecules per unit volume.

Avogadro's Number

According to Avogadro's hypothesis, gram atomic masses of all elements contain the same number of atoms and this number is called Avogadro's number (N_A) and its value is 6.02×10^{23}.

DAY PRACTICE SESSION 1

FOUNDATION QUESTIONS EXERCISE

1 An aluminium sphere of 20 cm diameter is heated from 0°C to 100°C. Its volume changes by (given that coefficient of linear expansion for aluminium $\alpha_{Al} = 23 \times 10^{-6}$ /°C) → AIEEE 2011

(a) 28.9 cc (b) 2.89 cc
(c) 9.28 cc (d) 49.8 cc

2 C_p and C_V are specific heats at constant pressure and constant volume, respectively. It is observed that $C_p - C_V = a$ for hydrogen gas $C_p - C_V = b$ for nitrogen gas. The correct relation between a and b is → JEE Main 2017

(a) $a = b$ (b) $a = 14\,b$
(c) $a = 28\,b$ (d) $a = \frac{1}{14}\,b$

3 A copper ball of mass 100 g is at a temperature T. It is dropped in a copper calorimeter of mass 100 g, filled with 170 g of water at room temperature. Subsequently, the temperature of the system is found to be 75°C. T is (Take, room temperature = 30°C, specific heat of copper = 0.1 cal/g°C)

(a) 885°C (b) 1250°C
(c) 825°C (d) 800°C

4 100 g of water is heated from 30°C to 50°C. Ignoring the slight expansion of the water, the change in its internal energy is (Take, specific heat of water is 4184 J/kg/K) → AIEEE 2011

(a) 8.4 kJ (b) 84 kJ
(c) 2.1 kJ (d) 4.2 kJ

5 Consider p-V diagram for an ideal gas shown in figure.

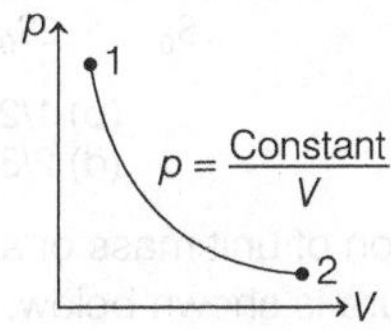

Out of the following diagrams which represents the T-p diagram?

(a) 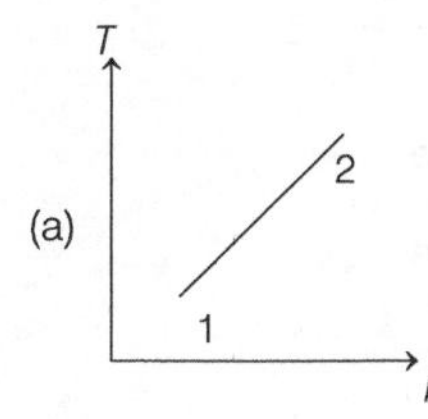

(b) 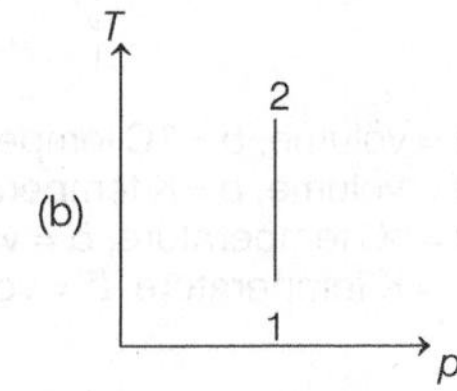

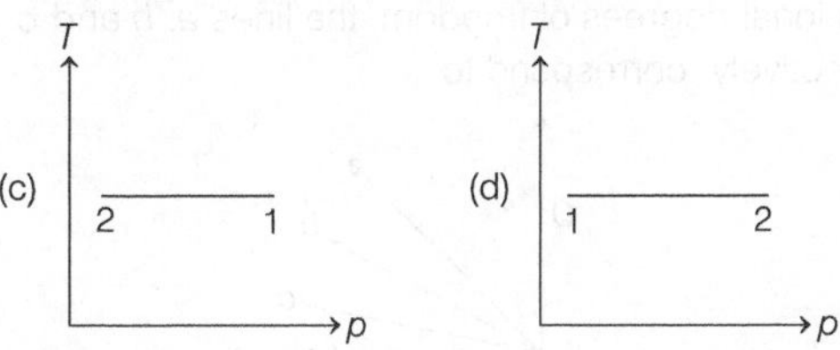

6 An ideal gas undergoes four different processes from the same initial state. Four processes are adiabatic, isothermal, isobaric and isochoric. Out of 1, 2, 3 and 4 which one is adiabatic.

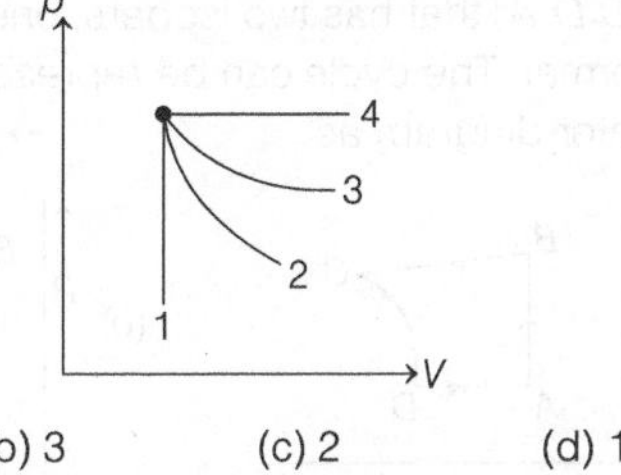

(a) 4 (b) 3 (c) 2 (d) 1

7 Equal masses of two liquids A and B contained in vessels of negligible heat capacity are supplied heat at the same rate. The temperature-time graphs for the two liquids are shown in the figure. If S represents specific heat and L represents latent heat of liquid, then

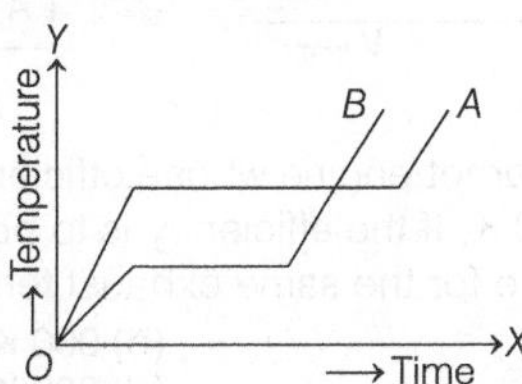

(a) $S_A > S_B; L_A < L_B$
(b) $S_A > S_B; L_A > L_B$
(c) $S_A < S_B; L_A < L_B$
(d) $S_A < S_B; L_A > L_B$

8 p-V plots for two gases during adiabatic processes are shown in the figure. Plots 1 and 2 should correspond respectively to

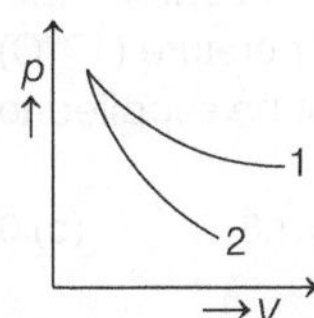

(a) He and O_2 (b) O_2 and He (c) He and Ar (d) O_2 and N_2

9 Figure shows the variation in temperature (ΔT) with the amount of heat supplied (Q) in an isobaric process corresponding to a monoatomic (M), diatomic (D) and a polyatomic (P) gas. The initial state of all the gases are the same and the scale for the axes coincide, ignoring vibrational degrees of freedom, the lines a, b and c respectively, correspond to

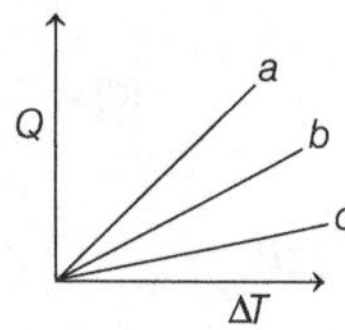

→ JEE Main (Online) 2013

(a) P, M and D (b) M, D and P
(c) P, D and M (d) D, M and P

10 A certain amount of gas is taken through a cyclic process ($A\,B\,C\,D\,A$) that has two isobars, one isochoric and one isothermal. The cycle can be represented on a p-V indicator diagram as → JEE Main (Online) 2013

(a) 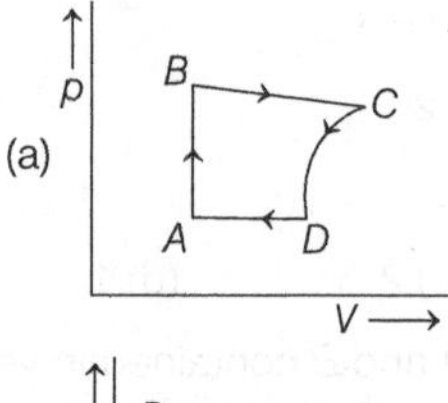

(b) 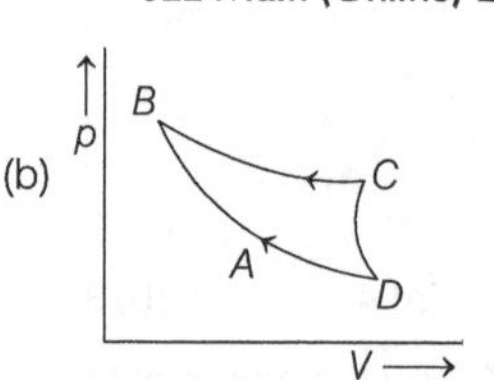

(c) 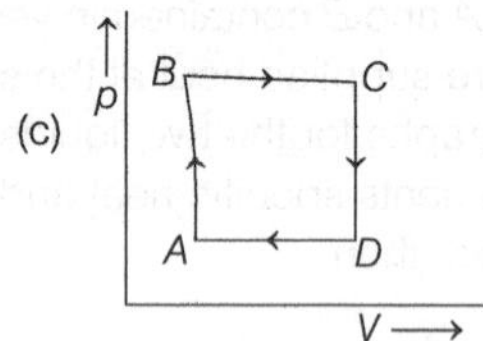

(d) 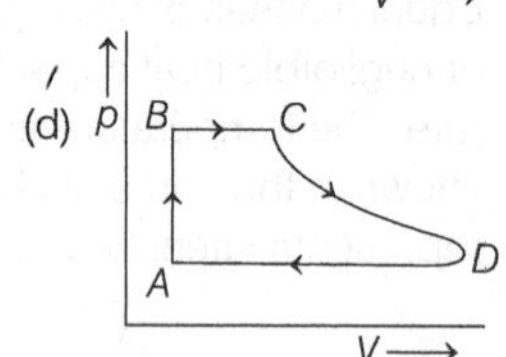

11 An ideal Carnot engine whose efficiency is 40%, receives heat at 500 K. If the efficiency is to be 50%, the intake temperature for the same exhaust temperature is

(a) 600 K (b) 900 K
(c) 700 K (d) 800 K

12 The pressure inside a tyre is 4 atm at 27°C. If the tyre bursts suddenly, its final temperature will be $\left(\text{Given, } r = \frac{7}{5}\right)$

(a) $300(4)^{7/2}$ (b) $300(4)^{2/7}$
(c) $300(2)^{7/2}$ (d) $300(4)^{-2/7}$

13 A refrigerator works between the temperature of melting ice and room temperature (17°C). The amount of energy (in kWh) that must be supplied to freeze 1kg of water at 0°C is

(a) 1.4 (b) 1.8 (c) 0.058 (d) 2.5

14 A gas expands with temperature according to the relation $V = kT^{2/3}$. Calculate the work done when the temperature changes by 60 K ?

(a) 10 R (b) 30 R
(c) 40 R (d) 20 R

15 A Carnot engine takes 3×10^6cal of heat from a reservoir at 627°C and gives it to a sink at 27°C. The work done by the engine is

(a) 4.2×10^6 J (b) 8.4×10^6 J (c) 16.8×10^6 J (d) 3×10^6 J

16 A Carnot engine operating between temperatures T_1 and T_2 has efficiency $\frac{1}{6}$. When T_2 is lowered by 62 K, its efficiency increases to $\frac{1}{3}$. Then, T_1 and T_2 are respectively → AIEEE 2011

(a) 372 K and 330 K (b) 330 K and 268 K
(c) 310 K and 248 K (d) 372 K and 310 K

17 A Carnot engine, whose efficiency is 40%, takes in heat from a source maintained at a temperature of 500 K. It is desired to have an engine of efficiency 60%. Then, the intake temperature for the same exhaust (sink) temperature must be → AIEEE 2012

(a) efficiency of Carnot engine cannot be made larger than 50%
(b) 1200 K
(c) 750 K
(d) 600 K

18 The temperature-entropy diagram of a reversible engine is given in the figure. Its efficiency is

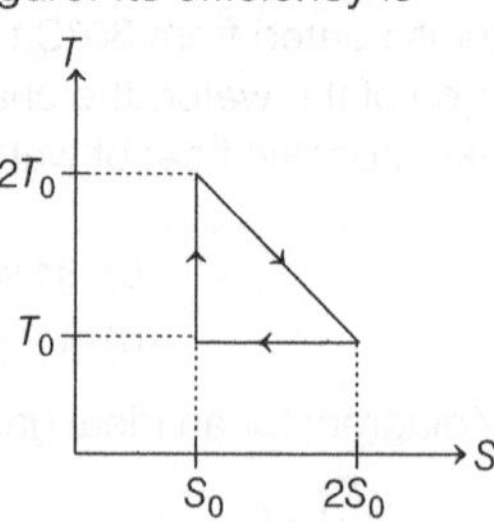

(a) 1/4 (b) 1/2
(c) 1/3 (d) 2/3

19 The expansion on of unit mass of a perfect gas at constant pressure is shown below.

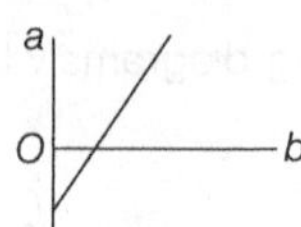

(a) a = volume, b = °C temperature
(b) a = volume, b =K temperature
(c) a = °C temperature, b = volume
(d) a = K temperature, b = volume

20 The temperature of an open room of volume $30\,m^3$ increases from 17°C to 27°C due to the sunshine. The atmospheric pressure in the room remains 1×10^5 Pa. If n_i and n_f are the number of molecules in the room before and after heating, then $n_f - n_i$ will be → JEE Main 2017

(a) 1.38×10^{23} (b) 2.5×10^{25}
(c) -2.5×10^{25} (d) -1.61×10^{23}

21 Consider a spherical shell of radius R at temperature T. The black body radiation inside it can be considered as an ideal gas of photons with internal energy per unit volume $u = \frac{U}{V} \propto T^4$ and pressure $p = \frac{1}{3}\left(\frac{U}{V}\right)$. If the shell now undergoes an adiabatic expansion, the relation between T and R is → JEE Main 2015

(a) $T \propto e^{-R}$
(b) $T \propto e^{-3R}$
(c) $T \propto \frac{1}{R}$
(d) $T \propto \frac{1}{R^3}$

22 A thermally insulated vessel contains an ideal gas of molecular mass M and ratio of specific heats γ. It is moving with speed v and suddenly brought to rest. Assuming no heat is lost to the surroundings, its temperature increases by → AIEEE 2011

(a) $\frac{(\gamma-1)}{2\gamma R} Mv^2 K$ (b) $\frac{\gamma M v^2}{2R} K$
(c) $\frac{(\gamma-1)}{2R} Mv^2 K$ (d) $\frac{(\gamma-1)}{2(\gamma+1)R} Mv^2 K$

23 Three perfect gases at absolute temperatures T_1, T_2 and T_3 are mixed. The masses of molecules are m_1, m_2 and m_3 and the number of molecules are n_1, n_2 and n_3, respectively. Assuming no loss of energy, the final temperature of the mixture is → AIEEE 2011

(a) $\frac{n_1T_1 + n_2T_2 + n_3T_3}{n_1 + n_2 + n_3}$ (b) $\frac{n_1T_1^2 + n_2T_2^2 + n_3T_3^2}{n_1T_1 + n_2T_2 + n_3T_3}$
(c) $\frac{n_1^2T_1^2 + n_2^2T_2^2 + n_3^2T_3^2}{n_1T_1 + n_2T_2 + n_3T_3}$ (d) $\frac{(T_1 + T_2 + T_3)}{3}$

24 The value of molar specific heat at constant volume for 1 mole of polyatomic gas having n number of degrees of freedom at temperature T K is
(here, R = universal gas constant)

(a) $\frac{nR}{2T}$ (b) $\frac{nR}{2}$
(c) $\frac{nRT}{2}$ (d) $2nRT$

25 p-V diagram of a diatomic gas is a straight line passing through origin. The molar heat capacity of the gas in the process will be

(a) $4R$ (b) $3R$ (c) $4R/3$ (d) $2.5R$

Direction (Q. Nos. 26-30) *Each of these questions contains two statements : Statement* I *and Statement* II. *Each of these questions also has four alternative choices, only one of which is the correct answer. You have to select one of the codes (a), (b), (c), (d) given below:*

(a) Statement I is true, Statement II is true; Statement II is the correct explanation for Statement I
(b) Statement I is true, Statement II is true; Statement II is not the correct explanation for Statement I
(c) Statement I is true; Statement II is false
(d) Statement I is false; Statement II is true

26 **Statement I** Work done by a gas in isothermal expansion is more than the work done by the gas in the same expansion adiabatically.
Statement II Temperature remains constant in isothermal expansion, but not in adiabatic expansion.

27 **Statement I** When 1 g of water at 100°C is converted to steam at 100°C, the internal energy of the system does not change.
Statement II From $dU = nC_V\,dT$, if temperature of the system remains constant, then $dU = 0$, i.e. internal energy remains constant.

28 **Statement I** In an isothermal process (quasistatic), the heat exchange between the system and surroundings takes place even though the gas has the same temperature as that of the surrounding.
Statement II There is an infinitesimal difference in temperature between the system and the surroundings.

29 **Statement I** A special type of thermometer (used to measure very high temperatures and calibrated for an ideal black body) measures a value lower than the actual value of the temperature of a red hot iron piece kept in open.
Statement II As the iron piece is kept in open, it loses its heat.

30 **Statement I** The internal energy of a perfect gas is entirely kinetic and depends only on absolute temperature of the gas and not on its pressure or volume.
Statement II A perfect gas is heated keeping pressure constant and later at constant volume. For the same amount of heat the temperature of the gas at constant pressure is lower than that at constant volume.
→ JEE Main (Online) 2013

DAY PRACTICE SESSION 2

PROGRESSIVE QUESTIONS EXERCISE

1 A solid body of constant heat capacity 1 J/°C is being heated by keeping it in contact with reservoirs in two ways.

(i) Sequentially keeping in contact with 2 reservoirs such that, each reservoir supplies same amount of heat.

(ii) Sequentially keeping in contact with 8 reservoirs such that each reservoir supplies same amount of heat.

In both the cases, body is brought from initial temperature 100°C to final temperature 200°C. Entropy change of the body in the two cases respectively, is → JEE Main 2015

(a) ln2, 4ln2
(b) ln2, ln2
(c) ln2, 2ln2
(d) 2ln2, 8ln2

2 Consider a collection of a large number of particles each with speed v. The direction of velocity is randomly distributed in the collection. What is the magnitude of the relative velocity between a pairs in the collection?

(a) $\frac{2v}{\pi}$ (b) $\frac{v}{\pi}$
(c) $\frac{8v}{\pi}$ (d) $\frac{4v}{\pi}$

3 An ideal gas (molar specific heat $C_V = 5R/2$) is taken along paths acb, adb and ab, $p_2 = 2p_1, V_2 = 2V_1$. Along ab, $p = kV$, where k is a constant. The various parameters are shown in the figure. Match the Column I with the corresponding options of Column II and mark the correct option from the codes given below.

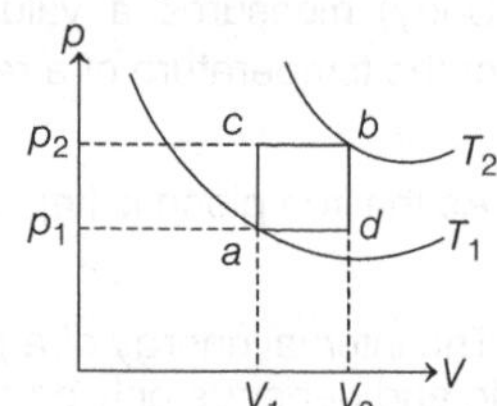

	Column I		Column II
A.	W_{acb}	1.	$15RT_1/2$
B.	W_{adb}	2.	$-15RT_1/2$
C.	ΔU_{ab}	3.	RT_1
D.	ΔU_{bca}	4.	$2RT_1$

Codes

	A	B	C	D		A	B	C	D
(a)	3	4	1	2	(b)	2	3	4	1
(c)	1	2	3	4	(d)	4	3	1	2

4 Diatomic molecules like hydrogen have energies due to both translational as well as rotational motion. From the equation in kinetic theory $pV = \frac{2}{3}E$, E is

(a) the total energy per unit volume
(b) only the translational part of energy, because rotational energy is very small compared to the translational energy
(c) only the translational part of the energy, because during collisions with the wall, pressure related to change in linear momentum
(d) the translational part of the energy, because rotational energies of molecules can be of either sign and its average over all the molecules is zero

5 An ideal monoatomic gas is confined in a cylinder by a spring-loaded piston of cross-section $8\times10^{-3}\text{m}^2$. Initially, the gas is at 300 K and occupies a volume of $2.4\times10^{-3}\text{m}^3$ and the spring is in a relaxed state. The gas is heated by a small heater coil H. The force constant of the spring is 8000 Nm^{-1} and the atmospheric pressure is 1×10^5 Pa. The cylinder and piston are thermally insulated. The piston and the spring are massless and there is no friction between the piston and cylinder. There is no heat loss through heater coil wire and thermal capacity of the heater coil is negligible. With all the above assumptions, if the gas is heated by the heater until the piston moves out slowly by 0.1 m, then the final temperature is

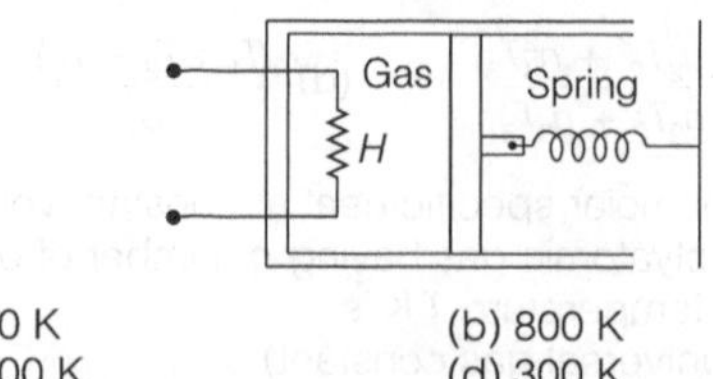

(a) 400 K (b) 800 K
(c) 1200 K (d) 300 K

6 A diatomic ideal gas is used in a car engine as the working substance. If during the adiabatic expansion part of the cycle, volume of the gas increases from V to $32V$, the efficiency of the engine is → AIEEE 2010

(a) 0.5 (b) 0.75
(c) 0.99 (d) 0.25

7 The specific heat capacity of a metal at low temperature (T) is given as $C_p(\text{kJK}^{-1}\text{ kg}^{-1}) = 32\left(\frac{T}{400}\right)^3$. A 100 g vessel of this metal is to be cooled from 20 K to 4 K by a special refrigerator operating at room temperature (27°C). The amount of work required to cool the vessel is → AIEEE 2011

(a) equal to 0.002 kJ
(b) greater than 0.148 kJ
(c) between 0.148 kJ and 0.028 kJ
(d) less than 0.028 kJ

8 A horizontal cylinder with adiabatic walls is closed at both ends and is divided into two parts by a frictionless piston that is also insulating. Initially, the value of pressure and temperature of the ideal gas on each side of the cylinder are V_0, p_0 and T_0, respectively. A heating coil in the right-hand part is used to slowly heat the gas on that side until the pressure in both parts reaches $64p_0/27$. The heat capacity C_V of the gas is independent of temperature and $C_p/C_V = \gamma = 1.5$.
Take, $V_0 = 16\text{m}^3$, $T_0 = 324$ K, $p_0 = 3\times10^5$ Pa
Column I represents the physical parameters of the gas, Column II gives their corresponding values, match the Column I with Column II and mark the correct option from the codes given below.

	Column I		Column II
A.	Final left-hand volume (in m^3)	1.	432
B.	Final left-hand temperature (in K)	2.	9
C.	Final right-hand temperature (in K)	3.	1104
D.	Work done (in kJ) on the left-hand gas	4.	3200

Codes

	A	B	C	D		A	B	C	D
(a)	2	1	3	4	(b)	1	2	3	4
(c)	4	1	2	3	(d)	3	4	1	2

9 The mass of a hydrogen molecule is 3.32×10^{-27} kg. If 10^{23} hydrogen molecules strike per second, a fixed wall of area 2 cm^2 at an angle of 45° to the normal and rebound classically with a speed of 10^3 m/s, then the pressure on the wall is nearly → JEE Main 2018

(a) 2.35×10^3 N/m^2 (b) 4.70×10^3 N/m^2
(c) 2.35×10^2 N/m^2 (d) 4.70×10^2 N/m^2

10 Two moles of an ideal monoatomic gas occupies a volume V at 27°C. The gas expands adiabatically to a volume $2V$. Calculate (i) the final temperature of the gas and (ii) change in its internal energy. → JEE Main 2018

(a) (i) 189 K (ii) 2.7 kJ (b) (i) 195 K (ii) –2.7 kJ
(c) (i) 189 K (ii) –2.7 kJ (d) (i) 195 K (ii) 2.7 kJ

11 An external pressure p is applied on a cube at 0°C, so that it is equally compressed from all sides. K is the bulk modulus of the material of the cube and α is its coefficient of linear expansion. Suppose we want to bring the cube to its original size by heating. The temperature should be raised by → JEE Main 2017

(a) $\frac{p}{\alpha K}$ (b) $\frac{3\alpha}{pK}$ (c) $3pK\alpha$ (d) $\frac{p}{3\alpha K}$

12 Consider an ideal gas confined in an isolated closed chamber. As the gas undergoes an adiabatic expansion, the average time of collision between molecules increases as V^q, where V is the volume of the gas. The value of q is $\left(\gamma = \frac{C_p}{C_V}\right)$ → JEE Main 2015

(a) $\frac{3\gamma+5}{6}$ (b) $\frac{3\gamma-5}{6}$ (c) $\frac{\gamma+1}{2}$ (d) $\frac{\gamma-1}{2}$

13 A pendulum clock loses 12 s a day, if the temperature is 40°C and gains 4 s in a day, if the temperature is 20°C. The temperature at which the clock will show correct time and the coefficient of linear expansion (α) of the metal of the pendulum shaft are respectively → JEE Main 2016

(a) 25°C, $\alpha = 1.85\times10^{-5}$/°C (b) 60°C, $\alpha = 1.85\times10^{-4}$/°C
(c) 30°C, $\alpha = 1.85\times10^{-3}$/°C (d) 55°C, $\alpha = 1.85\times10^{-2}$/°C

14 An ideal gas undergoes a quasistatic, reversible process in which its molar heat capacity C remains constant. If during this process, the relation of pressure p and volume V is given by pV^n = constant, then n is given by (Here, C_p and C_V are molar specific heat at constant pressure and constant volume, respectively) → JEE Main 2016

(a) $n = \frac{C_p}{C_V}$ (b) $n = \frac{C - C_p}{C - C_V}$ (c) $n = \frac{C_p - C}{C - C_V}$ (d) $n = \frac{C - C_V}{C - C_p}$

15 Helium gas goes through a cycle $ABCDA$ (consisting of two isochoric and isobaric lines) as shown in figure. Efficiency of this cycle is nearly (assume the gas to be close to ideal gas) → AIEEE 2012

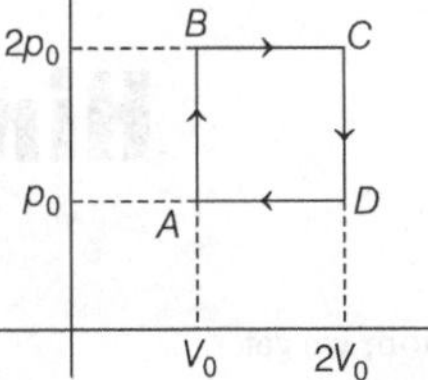

(a) 15.4% (b) 9.1% (c) 10.5%2 (d) 12.5%

16 An ideal gas is taken from the state A (pressure p, volume V) to the state B (pressure $p/2$, volume $2V$), a long straight line path in the p-V diagram. Select the correct statement from the following.

(a) The work done by the gas in the process A to B, exceeds the work that would be done by it, if system were taken along the isothermal
(b) In the T-V diagram, the path AB becomes a part of a hyperbola
(c) In the p-T diagram, the path AB becomes a part of a hyperbola
(d) In going from A to B, the temperature T of the gas first decreases to a minimum value and then increases

17 One mole of diatomic ideal gas undergoes a cyclic process *ABC* as shown in the figure. The process *BC* is adiabatic. The temperatures at *A*, *B* and *C* are 400 K, 800 K and 600 K, respectively. Choose the correct statement. → JEE Main 2014

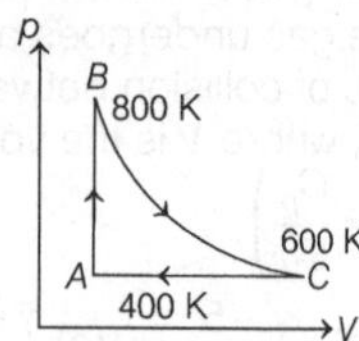

(a) The change in internal energy in whole cyclic process is 250 *R*

(b) The change in internal energy in the process *CA* is 700 *R*

(c) The change in internal energy in the process *AB* is −350 *R*

(d) The change in internal energy in the process *BC* is −500 *R*

18 Two moles of helium are mixed with *n* moles of hydrogen. The root mean square speed of the gas molecules in the mixture is $\sqrt{2}$ times the speed of sound in the mixture. Then, value of *n* is

(a) 1 (b) 3/2 (c) 2 (d) 3

19 *n* moles of an ideal gas undergoes a process *A* and *B* as shown in the figure. The maximum temperature of the gas during the process will be → JEE Main 2016

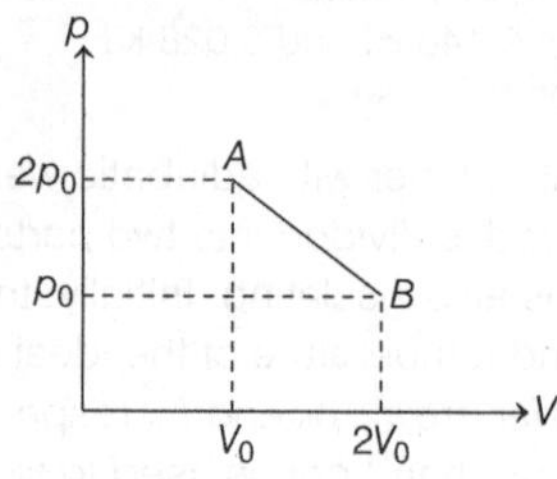

(a) $\frac{9}{4}\frac{p_0V_0}{nR}$ (b) $\frac{3}{2}\frac{p_0V_0}{nR}$ (c) $\frac{9}{2}\frac{p_0V_0}{nR}$ (d) $\frac{9p_0V_0}{nR}$

ANSWERS

SESSION 1	1 (a)	2 (b)	3 (a)	4 (a)	5 (c)	6 (c)	7 (d)	8 (b)	9 (c)	10 (c)
	11 (a)	12 (d)	13 (c)	14 (c)	15 (b)	16 (d)	17 (c)	18 (c)	19 (c)	20 (c)
	21 (c)	22 (c)	23 (a)	24 (b)	25 (b)	26 (a)	27 (a)	28 (a)	29 (c)	30 (b)
SESSION 2	1 (b)	2 (b)	3 (d)	4 (c)	5 (b)	6 (b)	7 (c)	8 (a)	9 (a)	10 (c)
	11 (d)	12 (c)	13 (a)	14 (b)	15 (a)	16 (a)	17 (d)	18 (c)	19 (a)	

Hints and Explanations

SESSION 1

1 Cubical expansion, we get

$$\Delta V = \gamma V\Delta T = 3\alpha V\Delta T$$

$$= 3 \times 23 \times 10^{-6} \times \left(\frac{4}{3}\pi (10)^3\right) \times 100 \quad \left(\because r = \frac{d}{2} = 10\text{cm}\right)$$

$$= 28.9 \text{ cc}$$

2 By Mayer's relation, for 1 g mole of a gas,

$$C_p - C_V = R$$

So, when *n* gram moles are given,

$$C_p - C_V = \frac{R}{n}$$

As per given question,

$$a = C_p - C_V = \frac{R}{2}; \text{for } H_2 \quad \dots\text{(i)}$$

$$b = C_p - C_V = \frac{R}{28}; \text{for } N_2 \quad \dots\text{(ii)}$$

From Eqs. (i) and (ii), we get

$$a = 14b$$

3 Heat gained (water + calorimeter) = Heat lost by copper ball

$$\Rightarrow m_w s_w \Delta T + m_c s_c \Delta T = m_B s_B \Delta T$$

$$\Rightarrow 170 \times 1 \times (75 - 30) + 100 \times 0.1 \times (75 - 30) = 100 \times 0.1 \times (T - 75)$$

$$\therefore \quad T = 885°\text{C}$$

4 As, work done = 0

$$\therefore \Delta U = mC\Delta T$$

$$= 100 \times 10^{-3} \times 4184 \times (50 - 30)$$

$$= 8.4 \text{ kJ}$$

5 In the diagram, *T* is constant and $p_1 > p_2$. This situation is represented by curve (c). In the solution figure, in which $p_1 > p_2$ and straight line parallel to pressure axis represents constant temperature.

6

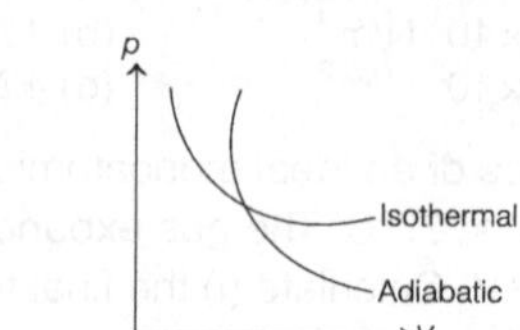

Slope of *p*-*V* curve

Isothermal process, $\frac{dp}{dV} = -\frac{p}{V}$

Adiabatic process, $\frac{dp}{dV} = -\gamma\frac{p}{V}$

Thus, (c) is correct.

7 As temperature of A rises faster than the temperature of B, therefore specific heat of A is less than that of B, i.e. $S_A < S_B$. Horizontal portions of graphs represent conversion of liquid into vapours. The horizontal portion is larger for liquid A, therefore $L_A > L_B$.

8 As it is clear from the figure,
Slope of curve 2 > Slope of curve 1

$$(\gamma p)_2 > (\gamma p)_1$$
$$\Rightarrow \gamma_2 > \gamma_1$$
$$\Rightarrow \gamma_{He} > \gamma_{O_2}$$

Adiabatic curve 2 corresponds to helium and adiabatic curve 1 corresponds to oxygen.

9 We know that,

$$Q = C_p \Delta T$$

$\because$ Degree of freedom $\propto C_p$
So, slope is higher for higher degree of freedom.

10 From given figure, in processes BC and DA, pressure of gas is constant, hence these represent isobaric process.
In process CD, volume is constant, therefore it represents isochoric process.
In process AB, temperature is constant, so it represent isothermal process.

11 As, $\eta = 1 - \dfrac{T_2}{T_1}$

$$\Rightarrow \frac{T_2}{T_1} = 1 - \eta$$
$$\Rightarrow \frac{T_2}{500} = 1 - \frac{40}{100} = \frac{3}{5}$$
$$\Rightarrow T_2 = 300 \text{ K}$$

Now, $\dfrac{T_2}{T_1'} = 1 - \eta'$

$$= 1 - \frac{50}{100} = \frac{1}{2}$$
$$\Rightarrow T_1' = 2T_2 = 2 \times 300$$
$$= 600 \text{ K}$$

12 In an adiabatic process,
$p_2^{(1-\gamma)} T_2^{\gamma} = p_1^{(1-\gamma)} T_1^{\gamma}$

$$T_2 = T_1 \left(\frac{p_1}{p_2}\right)^{(1-\gamma)/\gamma}$$
$$= 300 \left(\frac{4}{1}\right)^{\frac{(1-7/5)}{7/5}} = 300\,(4)^{-2/7}$$

13 $T_2 = 0°\text{C} = 273$ K, $T_1 = 17°\text{C} = 17 + 273 = 290$ K

Coefficient of performance $= \dfrac{Q_2}{W}$

$$= \frac{T_2}{T_1 - T_2}$$
$$\frac{80 \times 1000 \times 4.2}{W} = \frac{273}{290 - 273} = \frac{273}{17}$$
$$\therefore W = \frac{80 \times 1000 \times 4.2 \times 17}{273} \text{ J}$$
$$\text{or } W = \frac{33.6 \times 17 \times 10^4}{273 \times 3.6 \times 10^5} \text{ kWh}$$
$$= 0.058 \text{ kWh}$$

14 $dW = pdV = \dfrac{RT}{V} dV \quad \text{...(i)}$

As, $V = kT^{2/3}$,

$$dV = k\frac{2}{3}T^{-1/3}\, dT$$
$$\therefore \frac{dV}{V} = \frac{k\frac{2}{3}T^{-1/3}\, dT}{kT^{2/3}} = \frac{2}{3}\frac{dT}{T}$$

From Eq. (i), we get

$$W = \int_{T_1}^{T_2} RT \frac{dV}{V} = \int_{T_1}^{T_2} RT \frac{2}{3}\frac{dT}{T}$$
$$W = \frac{2}{3}R(T_2 - T_1) = \frac{2}{3}R \times 60 = 40R$$

15 Here, $T_1 = 627°\text{C}$

$$= 627 + 273 = 900 \text{ K}$$
$$T_2 = 27°\text{C} = 27 + 273 = 300 \text{ K},$$
$$Q_1 = 3 \times 10^6 \text{ cal}$$
$$\because \frac{Q_2}{Q_1} = \frac{T_2}{T_1} = \frac{300}{900} = \frac{1}{3} \Rightarrow Q_2 = \frac{Q_1}{3}$$
$$\therefore W = Q_1 - Q_2 = Q_1 - \frac{Q_1}{3}$$
$$= \frac{2}{3}Q_1 = \frac{2}{3} \times 3 \times 10^6 \text{ cal}$$
$$W = 2 \times 10^6 \text{ cal}$$
$$= 8.4 \times 10^6 \text{ J}$$

16 $\because \eta_1 = 1 - \dfrac{T_2}{T_1}$

$$\Rightarrow \frac{1}{6} = 1 - \frac{T_2}{T_1}$$
$$\Rightarrow \frac{T_2}{T_1} = \frac{5}{6} \quad \text{...(i)}$$
$$\therefore \eta_2 = 1 - \frac{T_2 - 62}{T_1}$$
$$\Rightarrow \frac{1}{3} = 1 - \frac{T_2 - 62}{T_1} \quad \text{...(ii)}$$

On solving Eqs. (i) and (ii), we get
$T_1 = 372$ K and $T_2 = 310$ K

17 Efficiency, $\eta = 1 - \dfrac{T_{sink}}{T_{source}}$

Now, $0.4 = 1 - \dfrac{T_{sink}}{500 \text{ K}}$

$$\Rightarrow T_{sink} = 0.6 \times 500 \text{ K} = 300 \text{ K}$$

Thus, $0.6 = 1 - \dfrac{300 \text{ K}}{T'_{source}}$

$$\Rightarrow T'_{source} = \frac{300 \text{ K}}{0.4} = 750 \text{ K}$$

18 We have,

$$Q_1 = T_0 S_0 + \frac{1}{2}T_0 S_0 = \frac{3}{2}T_0 S_0$$

$$Q_2 = T_0 S_0,\ Q_3 = 0$$
$$\Rightarrow \eta = \frac{W}{Q_1} = \frac{Q_1 - Q_2}{Q_1}$$
$$= 1 - \frac{Q_2}{Q_1} = 1 - \frac{2}{3} = \frac{1}{3}$$

19 In the given graph, line has positive slope with the X-axis and negative intercept on the Y- axis.
So, we can write the equation of lines as

$$y = mx - C \quad \text{...(i)}$$

According to Charles' law,

$$V_t = \frac{V_0}{273}t + V_0$$

By rewriting this equation, we get

$$t = \left(\frac{273}{V_0}\right)V_t - 273 \quad \text{...(ii)}$$

By comparing Eqs. (i) and (ii), we can say that temperature is represented on the Y-axis and volume on the X-axis.

20 From $pV = nRT = \dfrac{N}{N_A}RT$

We have,

$$n_f - n_i = \frac{pVN_A}{RT_f} - \frac{pVN_A}{RT_i}$$
$$\Rightarrow n_f - n_i = \frac{10^5 \times 30}{8.3} \times 6.02 \times 10^{23} \cdot \left(\frac{1}{300} - \frac{1}{290}\right)$$
$$= -2.5 \times 10^{25}$$
$$\therefore \Delta n = -2.5 \times 10^{25}$$

21 According to question,

$$p = \frac{1}{3}\left(\frac{U}{V}\right)$$
$$\Rightarrow \frac{nRT}{V} = \frac{1}{3}\left(\frac{U}{V}\right) \quad [\because pV = nRT]$$
$$\text{or } \frac{nRT}{V} \propto \frac{1}{3}T^4$$
$$\text{or } VT^3 = \text{constant}$$
$$\text{or } \frac{4}{3}\pi R^3 T^3 = \text{constant}$$
$$\text{or } TR = \text{constant} \Rightarrow T \propto \frac{1}{R}$$

22 As no heat is lost.

Loss of kinetic energy

= Gain of internal energy of gas

$$\frac{1}{2}mv^2 = n\,C_V\Delta T$$

$$\Rightarrow \quad \frac{1}{2}mv^2 = \frac{m}{M}\cdot\frac{R}{\gamma-1}\Delta T$$

$$\Rightarrow \quad \Delta T = \frac{Mv^2(\gamma-1)}{2R}K$$

23 $\frac{F}{2}n_1kT_1 + \frac{F}{2}n_2kT_2 + \frac{F}{2}n_3kT_3$

$$= \frac{F}{2}(n_1+n_2+n_3)kT$$

$$\Rightarrow \quad T = \frac{n_1T_1 + n_2T_2 + n_3T_3}{n_1+n_2+n_3}$$

24 According to law of equipartition of energy, average KE per molecule per degree of freedom at temperature T is $\frac{1}{2}kT$. The average KE per molecule of polyatomic gas

molecule $= \frac{n}{2}kT$

The average KE per molecule of polyatomic gas

$$(E) = \frac{n}{2}kT \times N = \frac{n}{2}RT$$

$$C_V = \frac{d}{dt}\left(\frac{n}{2}RT\right) = \frac{n}{2}R$$

25 As p-V diagram is a straight line passing through origin, therefore $p \propto V$ or pV^{-1} = constant.

In the process, pV^x = constant, molar heat capacity is given by

$$C = \frac{R}{\gamma-1} + \frac{R}{1-x}$$

where, $x = -1$ and $\gamma = 1.4$ for diatomic gas

$$C = \frac{R}{1.4-1} + \frac{R}{1-(-1)}$$

$$= \frac{5}{2}R + \frac{R}{2}$$

$\therefore \ C = 3R$

26 

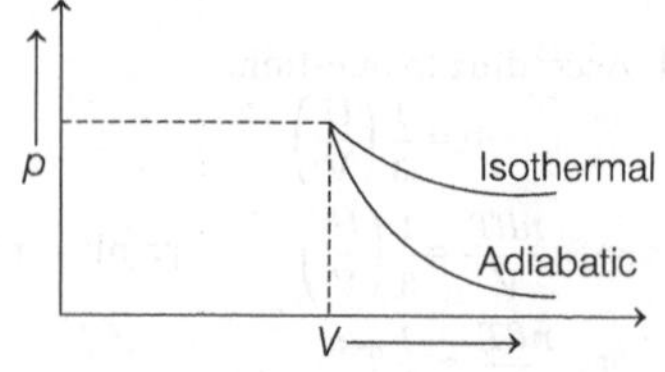

The slope of adiabatic curve is several times the slope of an isothermal curve and slope of both is negative. Thus, area under adiabatic curve is smaller than that under isothermal curve.

27 Since, 1 g of water is converted into steam at constant temperature of 100°C, i.e. $dT = 0$.

$\therefore$ Change in internal energy,

$$dU = nC_VdT = 0 \text{ i.e. } U = \text{constant}$$

28 In isothermal process, the heat exchange between system and surrounding at constant temperature, i.e. there is an infinitesimal difference in temperature between the system and the surrounding.

29 Since, the thermometer is caliberated with an ideal black body, the body that emits or absorbs all the radiations falling on it, shows a lower value of temperature. This is because iron is not a black body, i.e. does not absorb/emit all radiation falling on it.

30 The external energy depends upon absolute temperature of gas. Also, Statement II is correct, but both the statements are independently true.

SESSION 2

1 Since, entropy is a state function, therefore change in entropy in both the processes must be same.

2 Consider any two particles having angle θ between directions of their velocities.

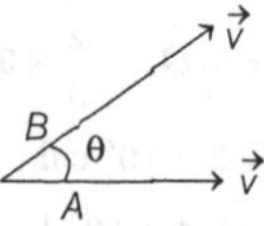

Then, $\vec{v}_{rel} = \vec{v}_B - \vec{v}_A$

i.e. $v_{rel} = \sqrt{v^2 + v^2 - 2v^2\cos\theta}$

$$= \sqrt{2v^2(1-\cos\theta)} = 2v\sin\frac{\theta}{2}$$

So, average v_{rel} over all pairs

$$v_{rel} = \frac{\int_0^{2\pi} v_{rel}\,d\theta}{\int_0^{2\pi} d\theta} = \frac{\int_0^{2\pi} 2v\sin\frac{\theta}{2}d\theta}{\int_0^{2\pi} d\theta}$$

$$= \frac{2v \times 2[-\cos(\theta/2)]_0^{2\pi}}{2\pi} = \frac{4v}{\pi} > v$$

3 A → 4.; B → 3.; C → 1.; D → 2.

$$W_{acb} = W_{ac} + W_{cb} = 0 + p_2(V_2 - V_1)$$

$$= p_2V_1 = 2p_1V_1 = 2RT_1$$

$$W_{adb} = W_{ad} + W_{db}$$

$$= p_1(V_2 - V_1) + 0 = p_1V_1 = RT_1$$

$$\Delta U_{ab} = U_{ac} + U_{cb}$$

$$= (Q_{ac} - W_{ac}) + (Q_{cb} - W_{cb})$$

$$C_V(T_c - T_1) + C_V(T_2 - T_c) = \frac{5R}{2}(T_2 - T_1)$$

$$C_V = \frac{5R}{2} \text{ (given)}$$

For an ideal gas,

$$\frac{p_1V_1}{T_1} = \frac{p_2V_2}{T_2} \Rightarrow T_2 = 4T_1$$

$$\Delta U_{ab} = \frac{5R}{2}(4T_1 - T_1) = \frac{15RT_1}{2}$$

For the process,

$$\Delta U_{bca} = -\Delta U_{ab} = \frac{-15RT_1}{2}$$

4 In the relation $pV = \frac{2}{3}E$, E is only the translational part of energy of molecules. This is because during collision of molecules with the walls, pressure exerted relates to change in linear momentum of gas molecules.

5 $V_1 = 2.4 \times 10^{-3}\,\text{m}^3$, $p_1 = p_0 = 10^5\,\text{Nm}^{-2}$

and $T_1 = 300\,\text{K}$ (given)

If area of cross-section of piston is A and it moves through distance x, then increment in volume of the gas $= Ax$.

If force constant of a spring is k, then force $F = kx$ and

pressure $= F/A = kx/A$.

$$V_2 = V_1 + Ax$$

$$= 2.4 \times 10^{-3} + 8 \times 10^{-3} \times 0.1$$

$$= 3.2 \times 10^{-3}$$

and $p_2 = p_0 + \frac{kx}{A}$

$$= 10^5 + \frac{8000 \times 0.1}{8 \times 10^{-3}} = 2 \times 10^5$$

From ideal gas equation,

$$\frac{p_1V_1}{T_1} = \frac{p_2V_2}{T_2}$$

$$\Rightarrow \quad \frac{10^5 \times 2.4 \times 10^{-3}}{300} = \frac{2 \times 10^5 \times 3.2 \times 10^{-3}}{T_2}$$

$$\Rightarrow \quad T_2 = 800\,\text{K}$$

6 The efficiency of cycle,

$$\eta = 1 - \frac{T_2}{T_1}$$

For adiabatic process,

$$TV^{\gamma-1} = \text{constant}$$

For diatomic gas, $\gamma = \frac{7}{5}$

$$T_1V_1^{\gamma-1} = T_2V_2^{\gamma-1}$$

$$T_1 = T_2\left(\frac{V_2}{V_1}\right)^{\gamma-1}$$

$$T_1 = T_2(32)^{\frac{7}{5}-1}$$

$$= T_2(2^5)^{2/5} = T_2 \times 4$$

$$T_1 = 4T_2$$

$$\therefore \quad \eta = \left(1 - \frac{1}{4}\right)$$

$$= \frac{3}{4} = 0.75$$

7 Heat required to change the temperature of vessel by a small amount dT

$-dQ = mC_p dT$

Total heat required,

$$-Q = m\int_{20}^{4} 32\left(\frac{T}{400}\right)^3 dT$$

$$= \frac{100 \times 10^{-3} \times 32}{(400)^3}\left[\frac{T^4}{4}\right]_{20}^{4}$$

$\Rightarrow Q = 0.001996$ kJ

Work done required to maintain the temperature of sink to T_2,

$$W = Q_1 - Q_2 = \frac{Q_1 - Q_2}{Q_2} Q_2$$

$$= \left(\frac{T_1}{T_2} - 1\right) Q_2$$

$$\Rightarrow W = \left(\frac{T_1 - T_2}{T_2}\right) Q_2$$

For $T_2 = 20$ K,

$$W_1 = \frac{300 - 20}{20} \times 0.001996$$

$= 0.028$ kJ

For $T_2 = 4$ K,

$$W_2 = \frac{300 - 4}{4} \times 0.001996$$

$= 0.148$ kJ

As temperature is changing from 20 K to 4 K, work done required will be more than W_1, but less than W_2.

8 A → 2; B → 1; C → 3; D → 4.

The compression in the left-hand side is adiabatic

$$p_0 V_0^{\gamma} = p_1 V_1^{\gamma}$$

$$V_1 = V_0\left(\frac{p_0}{p_1}\right)^{2/3} = V_0\left(\frac{27}{64}\right)^{2/3}$$

$$= \frac{9V_0}{16} = 9 \text{ m}^3$$

Also, $\frac{p_0V_0}{T_0} = \frac{p_1V_1}{T_1}$

$$\Rightarrow T_1 = \frac{p_1V_1}{p_0V_0} = \frac{4T_0}{3} = 432 \text{ K}$$

$$\therefore p_2 = \frac{64p_0}{27}$$

and $\frac{p_0V_0}{T_0} = \frac{p_2V_2}{T_2}$

$$\Rightarrow V_2 = 2V_0 - V_1$$

$= 23$ m^3

$$\Rightarrow T_2 = \frac{92T_0}{27} = 1104 \text{ K}$$

Work done on the left-hand side gas is

$$W = \frac{p_1V_1 - p_0V_0}{\gamma - 1}$$

$$= \frac{\left(\frac{64}{27} \times \frac{9}{16} - 1\right)p_0V_0}{\frac{3}{2} - 1} = \frac{2}{3}p_0V_0$$

$= 3200$ kJ

9

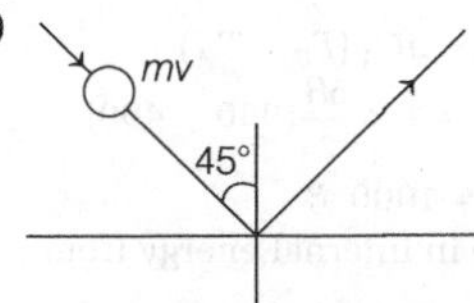

Momentum imparted due to first collision $= 2mv \sin 45° = \sqrt{2}mv$

$$\left[\because \sin 45° = \frac{1}{\sqrt{2}}\right]$$

$$\therefore \text{Pressure on surface} = \frac{n\sqrt{2}mv}{\text{Area}}$$

$$= \frac{10^{23} \times \sqrt{2} \times 3.32 \times 10^{-27} \times 10^3}{(2 \times 10^{-2})^2}$$

$p = 2.35 \times 10^3$ N/m^2

10 For adiabatic process, relation of temperature and volume is,

$$T_2V_2^{\gamma-1} = T_1V_1^{\gamma-1}$$

$$\Rightarrow T_2(2V)^{2/3} = 300(V)^{2/3}$$

$$\left[\gamma = \frac{5}{3} \text{ for monoatomic gases}\right]$$

$$\Rightarrow T_2 = \frac{300}{2^{2/3}} \approx 189 \text{ K}$$

Also, in adiabatic process,

$$\Delta Q = 0, \Delta U = -\Delta W$$

or $$\Delta U = \frac{-nR(\Delta T)}{\gamma - 1}$$

$$= -2 \times \frac{3}{2} \times \frac{25}{3}(300 - 189)$$

≈ -2.7 kJ

11 $K = \frac{p}{(-\Delta V/V)}$

$$\Rightarrow -\frac{\Delta V}{V} = \frac{p}{K}$$

$$\Rightarrow -\Delta V = \frac{pV}{K}$$

Change in volume, $\Delta V = \gamma V \Delta T$

where, γ = coefficient of volume expansion.

Again, $\gamma = 3\alpha$

where, α is coefficient of linear expansion.

$$\therefore \Delta V = V(3\alpha)\Delta T$$

$$\therefore \frac{pV}{K} = V(3\alpha)\Delta T$$

$$\therefore \Delta T = \frac{p}{3\alpha K}$$

12 For an adiabatic process,

$TV^{\gamma-1}$ = constant.

We know that, average time of collision between molecules,

$$\tau = \frac{1}{n\pi\sqrt{2}\, v_{rms}\, d^2}$$

where, n = number of molecules per unit volume

and v_{rms} = rms velocity of molecules.

As, $n \propto \frac{1}{V}$

and $v_{rms} \propto \sqrt{T}$

$$\Rightarrow \tau \propto \frac{V}{\sqrt{T}}$$

Thus, we can write

$$n = K_1V^{-1}$$

and $v_{rms} = K_2 T^{1/2}$

where, K_1 and K_2 are constants.

For adiabatic process,

$TV^{\gamma-1}$ = constant.

Thus, we can write

$$\tau \propto VT^{-1/2} \propto V(V^{1-\gamma})^{-1/2}$$

or $\tau \propto V^{\frac{\gamma+1}{2}}$

13 Time period of a pendulum,

$$T = 2\pi\sqrt{\frac{l}{g}}$$

where, l is length of pendulum and g is acceleration due to gravity.

Such as change in time period of a pendulum,

$$\frac{\Delta T}{T} = \frac{1}{2}\frac{\Delta l}{l}$$

When clock losses 12 s, we get

$$\frac{12}{T} = \frac{1}{2}\alpha(40 - \theta) \quad \text{...(i)}$$

When clock gains 4 s, we get

$$\frac{4}{T} = \frac{1}{2}\alpha(\theta - 20) \quad \text{...(ii)}$$

Comparing Eqs. (i) and (ii), we get

$$3 = \frac{40 - \theta}{\theta - 20}$$

$$\Rightarrow \theta = 25°\text{C}$$

Substituting the value of θ in Eq. (i), we have

$$\frac{12}{T} = \frac{1}{2}\alpha(40 - 25)$$

$$\Rightarrow \frac{12}{24 \times 3600} = \frac{1}{2}\alpha(15)$$

$$\alpha = \frac{24}{24 \times 3600 \times 15}$$

$$\alpha = 1.85 \times 10^{-5}/°\text{C}$$

14 For polytropic process, specific heat for an ideal gas,

$$C = \frac{R}{1-n} + C_V$$

$$\therefore \quad \frac{R}{1-n} + C_V = C$$

$$\Rightarrow \quad \frac{R}{1-n} = C - C_V$$

$$\Rightarrow \quad \frac{R}{C - C_V} = 1 - n$$

(where, $R = C_p - C_V$)

$$\Rightarrow \quad \frac{C_p - C_V}{C - C_V} = 1 - n$$

$$\Rightarrow \quad n = 1 - \frac{C_p - C_V}{C - C_V}$$

$$\Rightarrow \quad n = \frac{C - C_p}{C - C_V}$$

Thus, number of moles n is given by

$$n = \frac{C - C_p}{C - C_V}$$

15 Efficiency of a process is defined as the ratio of work done to energy supplied.

Here,

$$\eta = \frac{\Delta W}{\Delta Q} = \frac{\text{Area under } p\text{-}V \text{ diagram}}{\Delta Q_{AB} + \Delta Q_{BC}}$$

$$\therefore \ \eta = \frac{p_0V_0}{nC_V\Delta T_1 + nC_p\Delta T_2}$$

$$= \frac{p_0V_0}{\frac{3}{2}nR(T_B - T_A) + \frac{5}{2}nR(T_C - T_B)}$$

$$= \frac{p_0V_0}{\left[\frac{3}{2}(2p_0V_0 - p_0V_0) + \frac{5}{4}(4p_0V_0 - 2p_0V_0)\right]}$$

$$= \frac{p_0V_0}{\frac{3}{2}p_0V_0 + \frac{5}{2}\cdot 2p_0V_0}$$

$$= \frac{1}{6.5} = 15.4\%$$

16 Work done $= \frac{1}{2}\left(p + \frac{p}{2}\right)V$

$$= \frac{3}{4}pV = 0.75pV$$

Work done during isothermal process

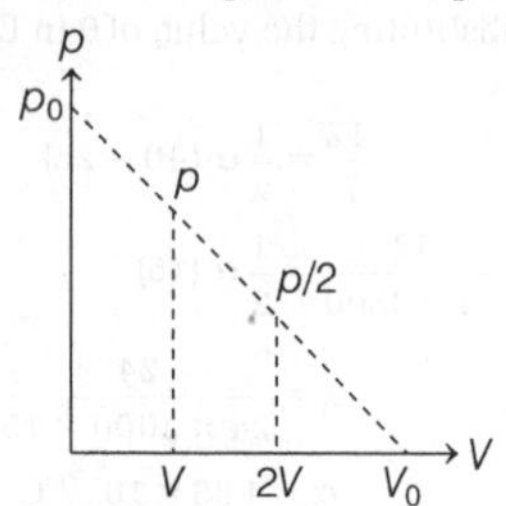

$$= RT \times 2.3026\log_{10}\left(\frac{2V}{V}\right)$$

$$= 0.693\,pV$$

Thus, statement (a) is correct.

17 According to first law of thermodynamics, we get

(i) Change in internal energy from A to B, i.e.

$$\Delta U_{AB} = nC_V(T_B - T_A)$$

$$= 1 \times \frac{5R}{2}(800 - 400)$$

$$= 1000\ R$$

(ii) Change in internal energy from B to C,

$$\Delta U_{BC} = nC_V(T_C - T_B)$$

$$= 1 \times \frac{5R}{2}(600 - 800)$$

$$= -500\,R$$

(iii) $\Delta U_{\text{isothermal}} = 0$

(iv) Change in internal energy from C to A, i.e. ΔU_{CA}

$$\Delta U_{CA} = nC_V(T_A - T_C)$$

$$= 1 \times \frac{5R}{2}(400 - 600)$$

$$= -500\ R$$

18 $v_{\text{rms}} = \sqrt{\frac{3RT}{M}}$,

$v_{\text{sound}} = \sqrt{\frac{\gamma RT}{M}}$,

$v_{\text{rms}} = \sqrt{2}\, v_{\text{sound}}$

Solving it, we get

$$\sqrt{3} = \sqrt{2\gamma}$$

$$\therefore \quad \gamma = \frac{3}{2} \text{ for the mixture.}$$

As, $\gamma = \frac{C_p}{C_V} = \frac{n_1C_{p_1} + n_2C_{p_2}}{n_1 + n_2} \times \frac{n_1 + n_2}{n_1C_{V_1} + n_2C_{V_2}}$

$$\gamma = \frac{n_1C_{p_1} + n_2C_{p_2}}{n_1C_{V_1} + n_2C_{V_2}}$$

For helium,

$$C_{p_1} = \frac{5}{2}R, C_{V_1} = \frac{3}{2}R$$

For hydrogen,

$$C_{p_2} = \frac{7}{2}R, C_{V_2} = \frac{5}{2}R$$

$$\therefore \quad \frac{3}{2} = \frac{2\left(\frac{5}{2}R\right) + n\left(\frac{7}{2}R\right)}{2\left(\frac{3}{2}R\right) + n\left(\frac{5}{2}R\right)}$$

$$= \frac{10 + 7n}{6 + 5n}$$

$$\Rightarrow \quad n = 2$$

19 As, T will be maximum temperature, where product of pV is maximum.

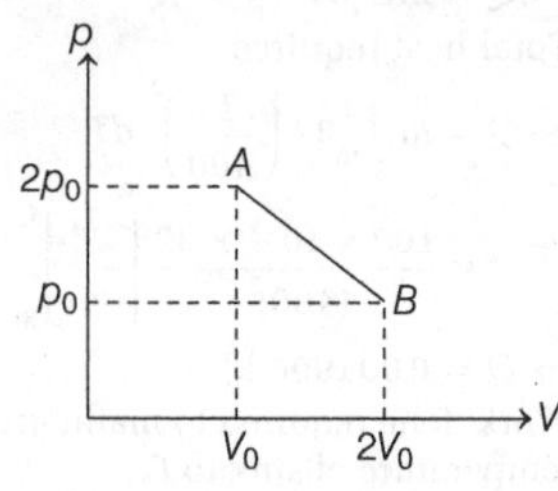

Equation of line AB, we have

$$y - y_1 = \frac{y_2 - y_1}{x_2 - x_1}(x - x_1)$$

$$\Rightarrow \quad p - p_0 = \frac{2p_0 - p_0}{V_0 - 2V_0}(V - 2V_0)$$

$$\Rightarrow \quad p - p_0 = \frac{-p_0}{V_0}(V - 2V_0)$$

$$nRT = \frac{-p_0}{V_0}V^2 + 3p_0V$$

$$T = \frac{1}{nR}\left(\frac{-p_0}{V_0}V^2 + 3p_0V\right)$$

For maximum temperature,

$$\frac{\partial T}{\partial V} = 0$$

$$\frac{-p_0}{V_0}(2V) + 3p_0 = 0$$

$$\frac{-p_0}{V_0}(2V) = -3p_0$$

$$\Rightarrow \quad V = \frac{3}{2}V_0$$

(condition for maximum temperature)

Thus, the maximum temperature of the gas during the process will be

$$T_{\max} = \frac{1}{nR}\left(\frac{-p_0}{V_0} \times \frac{9}{4}V_0^2 + 3p_0 \times \frac{3}{2}V_0\right)$$

$$= \frac{1}{nR}\left(-\frac{9}{4}p_0V_0 + \frac{9}{2}p_0V_0\right)$$

$$= \frac{9}{4}\frac{p_0V_0}{nR}$$

DAY SIXTEEN

Transfer of Heat

Learning & Revision for the Day

- Modes of Heat Transfer
- Some Common Terms and Points
- Perfectly Black Body
- Kirchhoff's Law of Radiation
- Stefan's Law
- Newtons Law of Cooling
- Wien's Displacement Law

Heat is a form of energy which characterises the thermal state of matter. It is transferred from one body to the other due to temperature difference between them.

Heat is a scalar quantity with dimensions $[ML^2T^{-2}]$ and its SI unit is joule (J) while practical unit is calorie (cal); 1 cal = 4.18 J.

The heat can be transferred from one body to the another body, through the following modes

(i) Conduction (ii) Convection (iii) Radiation

Conduction

The process of heat-transmission in which the particles of the body do not leave their position is called conduction.

Thermal Conductivity

The amount of heat transmitted through a conductor is given by $Q = \dfrac{KA\Delta T\, t}{l}$

where, A = area of cross-section,

ΔT = temperature difference $= T_2 - T_1$,

t = time elapsed,

K = thermal conductivity

and l = length of conductor

The rate of transmission of heat by conduction is given by

$$H = \frac{\Delta Q}{\Delta t} = \frac{KA\Delta T}{l}$$

The unit of thermal conductivity is $Wm^{-1}K^{-1}$.

- No. of Questions in Exercises (x)—
- No. of Questions Attempted (y)—
- No. of Correct Questions (z)—
  (Without referring Explanations)
- Accuracy Level ($z/y \times 100$)—
- Prep Level ($z/x \times 100$)—

In order to expect good rank in JEE, your Accuracy Level should be above 85 & Prep Level should be above 75.

Thermal Resistance

$$|H| = \left|\frac{\Delta Q}{\Delta t}\right| = \frac{KA}{l} \cdot \Delta T = \frac{\Delta T}{l / KA}$$

The term $\frac{l}{KA}$ is generally called the **thermal resistance** (R).

- Equation for rate of heat conduction can be written as

$$H = \frac{Q}{t} = \frac{\Delta T}{R_{\text{thermal}}}$$

It is equivalent/analysis to ohm's law which states that

$$I = \frac{V}{R_{(\text{electrical})}}$$

where, $H = \frac{Q}{t}$ is equivalent of electric current and called as heat, ΔT is equivalent of voltage (PD) and R_{thermal} is equivalent of $R_{\text{electrical}}$.

Combination of Metallic Rods

1. **Series Combination** In a series combination of two metal rods, equivalent thermal conductivity is given by

$$K_s = \frac{l_1 + l_2}{\frac{l_1}{K_1} + \frac{l_2}{K_2}}$$

or $K_s = \frac{2K_1K_2}{K_1 + K_2}$ [if $l_1 = l_2$]

K_1 K_2
l_1 l_2

Series combination of rods

If temperature of the interface of the series combination be T, then

$$T = \frac{K_1T_1 + K_2T_2}{K_1 + K_2}$$

2. **Parallel Combinations** In a parallel combination of two metal rods, thermal conductivity is given by

$$K_p = \frac{K_1A_1 + K_2A_2}{A_1 + A_2}$$

or $K_p = \frac{K_1 + K_2}{2}$ [if $A_1 = A_2$]

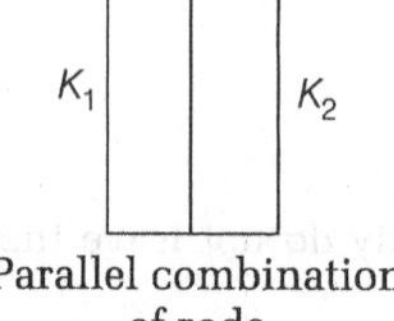

Parallel combination of rods

Formation and Growth of Ice on a Lake

Time required for the thickness of the layer of ice to increase from d_1 to d_2 will be

$$t = \frac{\rho L_f}{2KT}(d_2^2 - d_1^2)$$

where, ρ = density of ice,

L_f = latent heat of fusion of ice

and K = thermal conductivity of ice

Widemann-Franz Law

According to the Widemann-Franz law, the ratio of thermal and electrical conductivities is same for the metals at a particular temperature and is proportional to the absolute temperature of the metal.

i.e. $$\frac{K}{\sigma} \propto T$$

or $$\frac{K}{\sigma T} = \text{constant}$$

Convection

The process of heat-transmission in which the particles of the fluid move is called convenction.

Natural Convection

In natural convection gravity plays an important role. When a fluid is heated, the hot part expands and becomes less dense. Consequently it rises and the upper colder part is replaced. This again gets hot, rises up and is replaced by the colder part of the fluid.

Forced Convection

In a forced convection the material is forced to move up by a pump or by some other physical means. Common examples of forced convection are human circulatory system, cooling system of an automobile engine and forced air heating system in offices, etc.

Radiation

The process of the transfer of heat from one place to another place without heating the intervening medium is called radiation.

Interaction of Radiation with Matter

When radiant energy Q is incident on a body, a part of it Q_a is absorbed, another part Q_r is reflected back and yet another part Q_t is transmitted such that

$$Q = Q_a + Q_r + Q_t$$

or $$\frac{Q_a}{Q} + \frac{Q_r}{Q} + \frac{Q_t}{Q} = 1$$

or $$a + r + t = 1$$

where, $a = \frac{Q_a}{Q}$ = absorbing power or absorptance,

$r = \frac{Q_r}{Q}$ = reflecting power or reflectance

and $t = \frac{Q_t}{Q}$ = transmitting power or transmittance

Some Common Terms and Points

- **Absorptive power** (α) It is defined as the ratio of the radiant energy absorbed by it in a given time to the total radiant energy incident on it in the same interval of time.

$$\alpha = \frac{\text{Energy absorbed}}{\text{Energy incident}}$$

As a perfectly black body absorbs all radiations incident on it, the absorptive power of a perfectly black body is maximum and unity.

- **Spectral absorptive power** (a_λ) It is the ratio of radiant energy absorbed by a surface to the radiant energy incident on it for a particular wavelength λ. The spectral absorptive power a_λ is related to absorptive power a through the relation

$$a = \int_0^\infty a_\lambda d\lambda$$

- **Emissive power** (e) It is the total amount of energy radiated by a body per second per unit area of surface

$$e = \frac{1}{A}\frac{\Delta Q}{\Delta t}$$

- **Spectral emissive power** (e_λ) It is emissive power for a particular wavelength λ. Thus,

$$e = \int_0^\infty e_\lambda d\lambda$$

- **Emissivity** (ε) Emissivity of a body at a given temperature is defined as the ratio of the total emissive power of the body (e) to the total emissive power of a perfect black body (E) at that temperature,

i.e. $$\varepsilon = \frac{e}{E}$$

Perfectly Black Body

A perfectly black body is the one which completely absorbs the radiations of all the wavelengths that are incident on it. Thus, absorbing power of a perfectly black body is 1 (i.e $a = 1$).

When perfectly black body is heated to a suitable high temperature, it emits radiation of all possible wavelengths.

e.g. temperature of the sun is very high (6000 k approx.) it emits all possible radiations. So, it is an example of black body.

- For perfectly black body, $a = 1$, $r = t = 0$
- For a perfect reflector, $a = t = 0, r = 1$
- For a perfect transmitter, $a = r = 0, t = 1$.

Kirchhoff's Law of Radiation

Kirchhoff's law of radiation states that the ratio of emissive power to absorptive power of a body, is same for all surfaces at the same temperature and is equal to the emissive power of a perfectly black body at that temperature.

Mathematically, $\frac{e_1}{a_1} = \frac{e_2}{a_2} = \ldots = E$ (Black body)

- Kirchhoff's law implies that 'a good absorber is a good emitter (or radiator) too'.
- Fraunhoffer's lines (dark lines observed in solar spectrum) can be easily explained on the basis of Kirchhoff's laws.

Stefan's Law

According to the Stefan's law, the emissive power of a perfectly black body (energy emitted by black body per unit surface area per unit time) is directly proportional to the fourth power of its absolute temperature.

Mathematically, $E \propto T^4$

or $$E = \sigma T^4$$

or $$E = \sigma (T^4 - T_0^4)$$

where, σ is a constant known as the **Stefan's constant** and its value is $5.67 \times 10^{-8}\ \text{Wm}^{-2}\ \text{K}^{-4}$ and T_0 is the temperature of surrounding of black body.

- For a body, whose emissivity is ε, Stefan's law is modified as, $e = \varepsilon\sigma T^4$
- The total radiant energy Q emitted by a body of surface area A in time t, is given by

$$Q = Ate = At\varepsilon\sigma T^4$$

- The radiant power (P), i.e. energy radiated by a body per unit time is given by

$$P = \frac{Q}{t} = A\varepsilon\sigma T^4$$

- If a body at temperature T is surrounded by another body at temperature T_0 (where, $T_0 < T$), then according to Stefan's law of power

$$P = \varepsilon\sigma A(T^4 - T_o^4)$$

- If a body at temperature T is surrounded by another body at temperature T_0 (where, $T_0 < T$), then Stefan's law is modified as,

$$E = \sigma (T^4 - T_0^4) \quad \text{[black body]}$$

and $$e = \varepsilon\sigma (T^4 - T_0^4) \quad \text{[any body]}$$

Newton's Law of Cooling

According to the Newton's law of cooling, rate of cooling of a body is directly proportional to the temperature difference between the body and the surroundings, provided the temperature difference is small.

Mathematically, $\frac{dT}{dt} \propto (T - T_0)$ or $-\frac{dT}{dt} = k(T - T_0)$

where, k is a constant.

If a body cools by radiation through a small temperature difference from T_1 to T_2 in a short time t when the surrounding temperature is T_0, then

$$\frac{dT}{dt} \simeq \frac{T_1 - T_2}{t} \quad \text{and } T = \frac{T_1 + T_2}{2}$$

The Newton's law of cooling becomes

$$\left[\frac{T_1 - T_2}{t}\right] = k\left[\frac{T_1 + T_2}{2} - T_0\right]$$

Black Body Spectrum

The black body spectrum is a continuous spectrum as shown in the figure. At a given temperature, initially the intensity of thermal radiation increases with an increase in wavelength and reaches a maximum value at a particular wavelength λ_m. On increasing the wavelength beyond λ_m, the intensity of radiation E_λ starts decreasing.

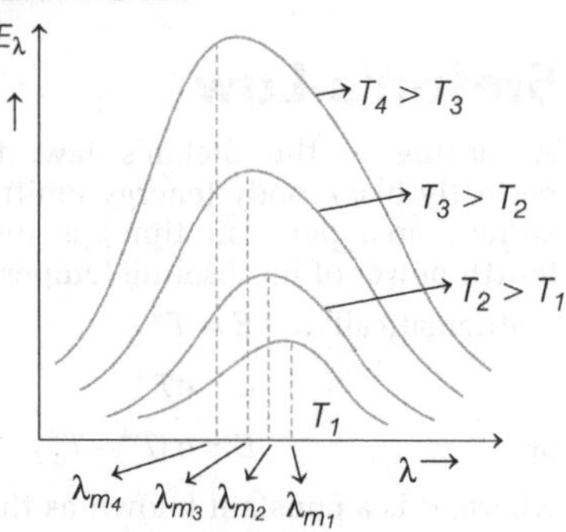

Graph between intensity E_λ and λ

Variation of intensity of thermal radiation with wavelength is shown in fig. The total area under E_λ-λ curve gives the total intensity of radiation at that temperature. The area, in accordance with the Stefan's law of radiation, is directly proportional to the fourth power of the temperature.

Wien's Displacement Law

According to Wien's law, the product of wavelength corresponding to maximum intensity of radiation and temperature of body is constant i.e. $\lambda_m T$ = constant = b, where b is known as the Wien's constant and its value is 2.89×10^{-3} mK.

Solar Constant

The amount of heat received from the sun by one square centimeter area of a surface placed normally to the sun rays at mean distance of the earth from the sun is known as solar constant. It is denoted by S.

$$S = \left(\frac{r}{R}\right)^2 \sigma T^4$$

where, r is the radius of sun and R is the mean earth's distance from sun value of solar constant $S = 1.937$ cal/cm^2/min.

DAY PRACTICE SESSION 1

FOUNDATION QUESTIONS EXERCISE

1 A cylindrical rod is having temperatures T_1 and T_2 at its ends. The rate of flow of heat is Q_1. If all the linear dimensions are doubled keeping the temperature constant, then rate of flow of heat Q_2 will be

(a) $4Q_1$ (b) $2Q_1$ (c) $\frac{Q_1}{4}$ (d) $\frac{Q_1}{2}$

2 A uniform metallic rod rotates about its perpendicular bisector with constant angular speed. If it is heated uniformly to raise its temperature slightly

(a) Its speed of rotation increases
(b) Its speed of rotation decreases
(c) Its speed of rotation remains same
(d) Its speed increases because its moment of inertia increases

3 Two slabs A and B of different materials but with the same thickness are joined as shown in the figure. The thermal conductivities of A and B are K_1 and K_2, respectively. The thermal conductivity of the composite slab will be

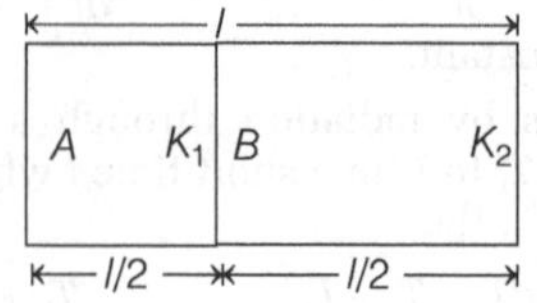

(a) $\frac{1}{2}(K_1 + K_2)$ (b) $\sqrt{K_1K_2}$ (c) $(K_1 + K_2)$ (d) $\frac{2K_1K_2}{(K_1 + K_2)}$

4 The coefficient of thermal conductivity of copper is 9 times that of steel. In the composite cylindrical bar shown in the figure, what will be the temperature at the junction of copper and steel?

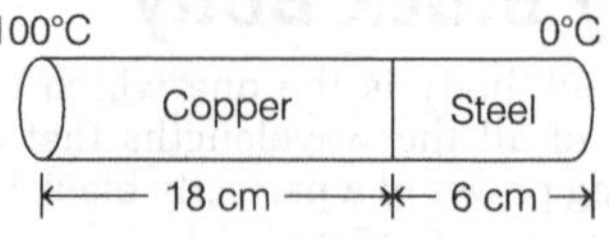

(a) 75°C (b) 67°C (c) 25°C (d) 33°C

5 Three objects coloured black, grey and white can withstand hostile conditions at 2800°C. These objects are thrown into furnace where each of them attains a temperature of 2000°C. Which object will have the brightest glow?

(a) The white object
(b) The black object
(c) All glow with equal brightness
(d) Grey object

6 A black body maintained at a certain temperature radiates heat energy at the rate Q Watt. If its surface is smoothened, so as to lower its emissivity by 10%, what will be the increase in its rate of radiation at double the initial temperature?

(a) $(0.9 \times 2^4 - 1)\,Q$ W (b) $0.9 \times 2^4\,Q$ W
(c) $(0.9 \times 2)^4 Q$ W (d) $(0.9)^4 \times 2Q$ W

7 We consider the radiation emitted by the human body. Which of the following statement is true?
(a) The radiation is emitted during the summers and absorbed during the winters.
(b) The radiation emitted lies in the ultraviolet region and hence is not visible.
(c) The radiation emitted is in the infrared region.
(d) The radiation is emitted only during the day.

8 Parallel rays of light of intensity $I = 912\ \text{Wm}^{-2}$ are incident on a spherical black body kept in surroundings of temperature 300 K. Take, Stefan constant $\sigma = 5.7 \times 10^{-8}\ \text{Wm}^{-2}\ \text{K}^{-4}$ and assume that the energy exchange with the surroundings is only through radiation. The final steady state temperature of the black body is close to → 2014 Main
(a) 330 K (b) 660 K (c) 990 K (d) 1550

9 The spectral energy distribution of a star is maximum at twice temperature as that of the sun. The total energy radiated by the star is
(a) twice as that of the sun
(b) same as that of the sun
(c) sixteen times as that of the sun
(d) one-sixteenth of the sun

10 Newton's law of cooling holds good only, if the temperature difference between the body and the surroundings is
(a) less than 10°C (b) more than 10°C
(c) less than 100°C (d) more than 100°C

11 A liquid in a beaker has temperature $\theta(t)$ at time t and θ_0 is temperature of surroundings, then according to Newton's law of cooling, the correct graph between $\log_e(\theta - \theta_0)$ and t is

(a) 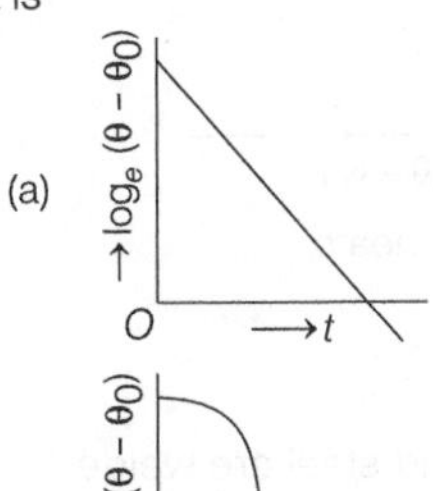
(b) 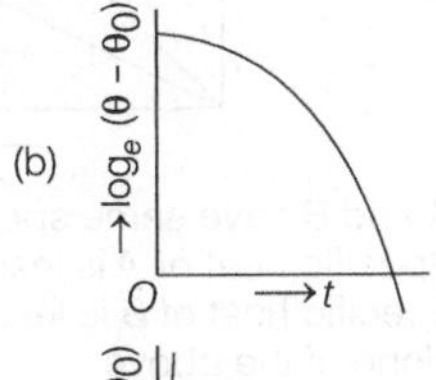
(c) 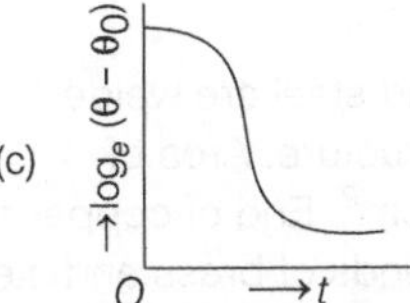
(d)

12 If a piece of metal is heated to temperature θ and then allowed to cool in a room which is at temperature θ_0, the graph between the temperature T of the metal and time will be closed to → JEE Main 2013

(a) 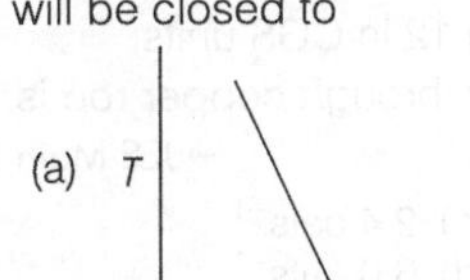
(b) 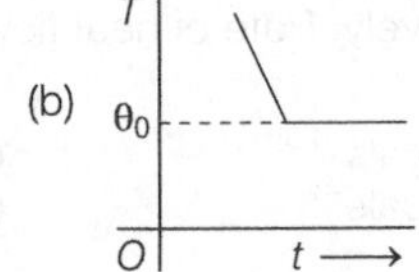
(c) 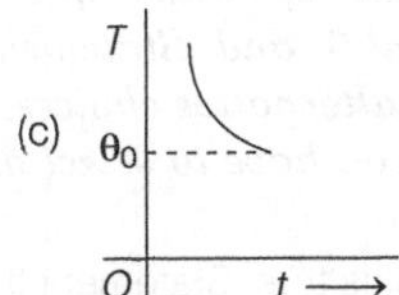
(d) 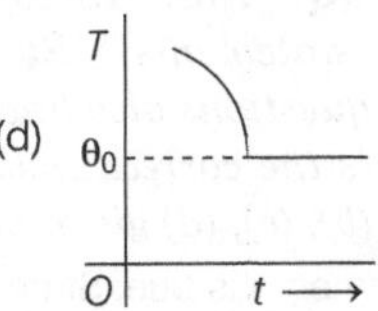

13 A sphere, a cube and a thin circular plate, all of same material and same mass are initially heated to same high temperature. Then
(a) plate will cool fastest and cube the slowest
(b) sphere will cool fastest and cube the slowest
(c) plate will cool fastest and sphere the slowest
(d) cube will cool fastest and plate the slowest

14 Temperatures of two stars are in the ratio 3 : 2. If wavelength for the maximum intensity of the first body is 4000 Å, what is the corresponding wavelength of the second body?
(a) 9000 Å (b) 6000 Å
(c) 2000 Å (d) 8000 Å

15 The energy spectrum of a black body exhibits a maximum around a wavelength λ_0. The temperature of the black body is now changed such that the energy is maximum around a wavelength $\frac{3\lambda_0}{4}$. The power radiated by the two black bodies will now increase by a factor of
(a) 64/27 (b) 256/81 (c) 4/3 (d) 16/9

16 Three discs, A, B and C having radii 2 m, 4 m and 6 m respectively, are coated with carbon black on their outer surfaces. The wavelengths corresponding to maximum intensity are 300 nm, 400 nm and 500 nm, respectively. The power radiated by them are Q_A, Q_B and Q_C, respectively
(a) Q_A is maximum
(b) Q_B is maximum
(c) Q_C is maximum
(d) $Q_A = Q_B = Q_C$

17 Variation of radiant energy emitted by sun, filament of tungsten lamp and welding arc as a function of its wavelength is shown in figure. Which of the following options is the correct match?

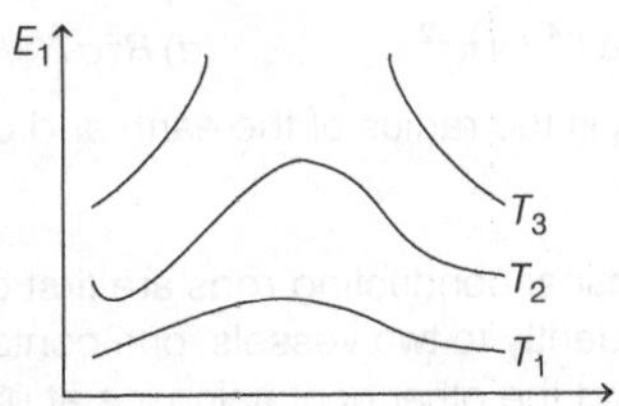

(a) Sun-T_1, tungsten filament-T_2, welding arc-T_3
(b) Sun-T_2, tungsten filament-T_1, welding arc-T_3
(c) Sun-T_3, tungsten filament-T_2, welding arc-T_1
(d) Sun-T_1, tungsten filament-T_3, welding arc-T_2

Direction (Q. Nos. 18-20) *Each of these questions contains two statements : Statement I and Statement II. Each of these questions also has four alternative choices, only one of which is the correct answer. You have to select one of the codes (a), (b), (c), (d) given below.*

(a) Statement I is true, Statement II is true; Statement II is the correct explanation for Statement I
(b) Statement I is true, Statement II is true; Statement II is not the correct explanation for Statement I
(c) Statement I is true; Statement II is false
(d) Statement I is false; Statement II is true

18 **Statement I** A solid sphere of copper of radius R and a hollow sphere of the same material of inner radius r and outer radius R are heated to the same temperature and allowed to cool down in the same environment. The hollow sphere cools faster.

Statement II Rate of cooling follows the Stefan's law which is $E \propto T^4$.

19 **Statement I** A body that is a good radiator is also a good absorber of radiation at a given wavelength.

Statement II According to Kirchhoff's law, the absorptivity of a body is equal to its emissivity at a given wavelength.

20 **Statement I** For higher temperatures, the peak emission wavelength of a black body shifts towards the lower wavelength side.

Statement II Peak emission wavelength of a black body is proportional to the fourth-power of the temperature.

DAY PRACTICE SESSION 2

PROGRESSIVE QUESTIONS EXERCISE

1 A metallic sphere cools from 50°C to 40°C in 300 s. If the room temperature is 20°C, then its temperature in the next 5 min will be

(a) 38°C (b) 33.3°C (c) 30°C (d) 36°C

2 A pan filled with hot food cools from 94° C to 86 °C in 2 min, when the room temperature is at 20 °C, how long will it take to cool from 71 °C to 69 °C?

(a) 14 s (b) 3 s (c) 42 s (d) 13 s

3 Two slabs A and B of equal surface area are placed one over the other such that their surfaces are completely in contact. The thickness of slab A is twice that of B. The coefficient of thermal conductivity of slab A is twice that of B. The first surface of slab A is maintained at 100°C, while the second surface of slab B is maintained at 25°C. The temperature at the contact of their surfaces is

(a) 62.5°C (b) 45°C (c) 55°C (d) 85°C

4 Assuming the sun to be a spherical body of radius R at a temperature of T K, evaluate the total radiant power, incident on the earth, at a distance r from the sun.

(a) $4\pi r_0^2 R^2 \sigma T^4 / r^2$ (b) $\pi r_0^2 R^2 \sigma T^4 / r^2$
(c) $r_0^2 R^2 \sigma T^4 / 4\pi r^2$ (d) $R^2 \sigma T^4 / r^2$

where, r_0 is the radius of the earth and σ is the Stefan's constant.

5 Two identical conducting rods are first connected independently to two vessels, one containing water at 100°C and the other containing ice at 0°C. In the second case, the rods are joined end to end and connected to the same vessels. Let q_1 and q_2 g/s be the rate of melting of ice in two cases respectively. The ratio of q_1 / q_2 is

(a) $\frac{1}{2}$ (b) $\frac{2}{1}$ (c) $\frac{4}{1}$ (d) $\frac{1}{4}$

6 Two circular discs A and B with equal radii are blackened. They are heated to same temperature and are cooled under identical conditions. What inference do you draw from their cooling curves?

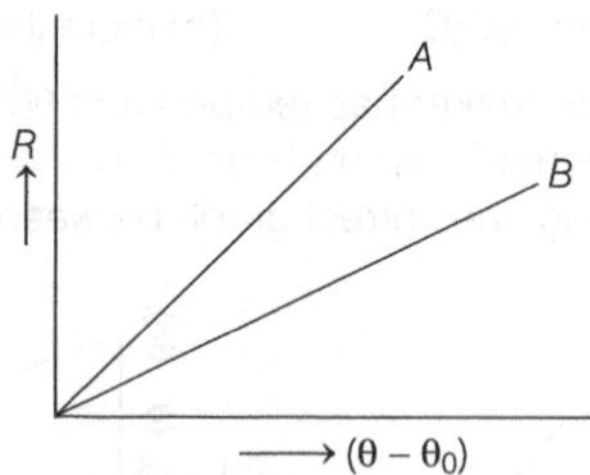

(a) A and B have same specific heats
(b) Specific heat of A is less
(c) Specific heat of B is less
(d) None of the above

7 Three rods of copper, brass and steel are welded together to form a Y-shaped structure. Area of cross-section of each rod is 4 cm^2. End of copper rod is maintained at 100°C whereas ends of brass and steel are kept at 0°C. Lengths of the copper, brass and steel rods are 46, 13 and 12 cm respectively.
The rods are thermally insulated from surroundings except at ends. Thermal conductivities of copper, brass and steel are 0.92, 0.26 and 0.12 in CGS units, respectively. Rate of heat flow through copper rod is

→ JEE Main 2014

(a) 1.2 cals^{-1} (b) 2.4 cals^{-1}
(c) 4.8 cals^{-1} (d) 6.0 cals^{-1}

8 A mass of 50 g of water in a closed vessel, with surroundings at a constant temperature takes 2 min to cool from 30°C to 25°C. A mass of 100g of another liquid in an identical vessel with identical surroundings takes the same time to cool from 30°C to 25°C. The specific heat of the liquid is (The water equivalent of the vessel is 30 g.) → JEE Main (Online) 2013

(a) 2.0 kcal/kg (b) 7 kcal/kg
(c) 3 kcal/kg (d) 0.5 kcal/kg

9 The figure shows a system of two concentric spheres of radii r_1 and r_2 and kept at temperatures T_1 and T_2 respectively. The radial rate of flow of heat in a substance between the two concentric spheres, is proportional to

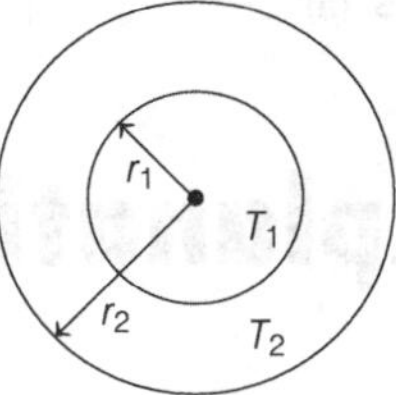

(a) $\frac{(r_2 - r_1)}{(r_1 r_2)}$ (b) $\ln\left(\frac{r_2}{r_1}\right)$ (c) $\frac{r_1 r_2}{(r_2 - r_1)}$ (d) $(r_2 - r_1)$

10 A slab of stone of area 3600 cm^2 and thickness 10 cm is exposed on the lower surface to steam at 100°C. A block of ice at 0°C rests on the upper surface of the slab. If in 1 h 4.8 kg of ice melted the thermal conductivity of the stone is

(a) 1.24 W/m/k (b) 2.24 W/m/k
(c) 0.24 W/m/k (d) 1.54 W/m/k

11 Two spherical stars A and B emit black body radiation. The radius of A is 400 times that of B and A emits 10^4 times the power emitted from B. The ratio $\left(\frac{\lambda_A}{\lambda_B}\right)$ of their wavelengths λ_A and λ_B at which the peaks occur in their respective radiation curves is

(a) 1 (b) 2
(c) 3 (d) 5

12 A metal is heated in a furnace where a sensor is kept above the metal surface to read the power radiated (P) by the metal. The sensor has a scale that displays $\log_2(P/P_0)$, where P_0 is a constant. When the metal surface is at a temperature of 487°C, the sensor shows a value 1. Assume that the emissivity of the metallic surface remains constant. What is the value displayed by the sensor when the temperature of the metal surface is raised to 2767°C?

(a) 1 (b) 4 (c) 7 (d) 9

13 Two spherical bodies A (radius 6 cm) and B (radius 18 cm) are at temperatures T_1 and T_2, respectively. The maximum intensity in the emission spectrum of A is at 500 nm and in that of B is at 1500 nm. Considering them to be black bodies, what will be the ratio of the rate of total energy radiated by A to that of B?

(a) 9 (b) 7 (c) 5 (d) 1

14

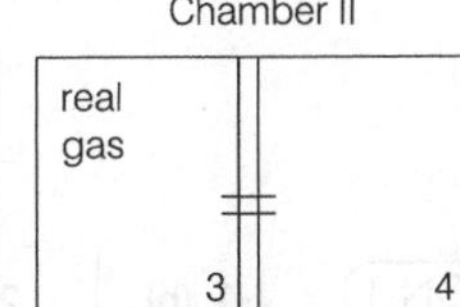

There are two identical chambers, completely thermally insulated from surrounding. Both chambers have a partition wall dividing the chambers in two compartments. Compartment 1 is filled with an ideal gas and compartment 3 is filled with a real gas. Compartments 2 and 4 are vacuum. A small hole (orifice) is made in the partition walls and the gases are allowed to expand in vacuum.

Statement I No change in the temperature of the gas takes place when ideal gas expands in vacuum. However, the temperature of real gas goes down (cooling) when it expands in vacuum.

Statement II The internal energy of an ideal gas is only kinetic. The internal energy of a real gas is kinetic as well as potential. → JEE Main (Online) 2013

(a) Statement I is false and Statement II is true
(b) Statement I and Statement II both are true. Statement II is the correct explanation of Statement I
(c) Statement I is true and Statement II is false
(d) Statement I and Statement II both are true, but Statement II is not the correct explanation of Statement I

15 A rod AB of uniform cross-section consists of four section AC, CD, DE and EB of different metals with thermal conductivities K, (0.8) K, (1.2) K and (1.50) K, respectively. Their lengths are respectively L, (1.2) L, (1.5) L and (0.6) L. They are joined rigidly in succession at C, D and E to form the rod AB. The end A is maintained at 100 °C and the end B is maintained at 0°C. The steady state temperatures of the joints C, D and E are respectively T_C, T_D and T_E. Column I lists the temperature differences $(T_A - T_C)$, $(T_C - T_D)$, $(T_D - T_E)$ and $(T_E - T_B)$ in the four sections and column II their values jumbled up. Match each item in column I with its correct value in column II.

A C D E B

	Column I		Column II
A.	$(T_A - T_C)$	1.	9.6
B.	$(T_C - T_D)$	2.	30.1
C.	$(T_D - T_E)$	3.	24.1
D.	$(T_E - T_B)$	4.	36.2

	A	B	C	D		A	B	C	D
(a)	3	4	2	1	(b)	1	2	4	3
(c)	3	4	1	2	(d)	3	2	1	4

ANSWERS

SESSION 1									
1 (b)	2 (b)	3 (d)	4 (a)	5 (b)	6 (a)	7 (c)	8 (a)	9 (c)	10 (a)
11 (a)	12 (c)	13 (c)	14 (b)	15 (b)	16 (b)	17 (c)	18 (c)	19 (a)	20 (c)

SESSION 2									
1 (b)	2 (c)	3 (a)	4 (b)	5 (c)	6 (b)	7 (c)	8 (d)	9 (c)	10 (a)
11 (b)	12 (d)	13 (a)	14 (c)	15 (a)					

Hints and Explanations

SESSION 1

1 Initially, $Q_1 = \dfrac{KA_1(T_1 - T_2)}{l_1}$ but on doubling all dimensions $l_2 = 2l_1$ and $A_2 = 4A_1$.

Hence, $Q_2 = \dfrac{KA_2(T_1 - T_2)}{l_2}$

$= \dfrac{K4A_1(T_1 - T_2)}{2l_1} = 2\dfrac{KA_1(T_1 - T_2)}{l_1}$

$= 2Q_1$

2 When a metallic rod is heated it expands. Its moment of inertia (I) about a perpendicular bisector increases. According to law of conservation of angular momentum, its angular speed (ω) decreases, since $\omega \propto 1/I$. (According to law of conservation of angular momentum).

3 The thermal resistance of a slab of length l_1, area of cross-section A and thermal conductivity K is given by

$$R = \frac{l}{KA}$$

Since, the slabs are joined in series, the thermal resistance of the composite slab is

$\therefore \quad R_C = R_1 + R_2$

or $\quad \dfrac{l}{K_C A} = \dfrac{l/2}{K_1 A} + \dfrac{l/2}{K_2 A}$

or $\quad K_C = \dfrac{2K_1 K_2}{(K_1 + K_2)}$

4 From temperature of interface,

$$\theta = \frac{K_1\theta_1 l_2 + K_2\theta_2 l_1}{K_1 l_2 + K_2 l_1}$$

It is given that $K_{Cu} = 9K_s$.

So, if $K_s = K_1 = K$, then

$K_{Cu} = K_2 = 9K$

$\Rightarrow \theta = \dfrac{9K \times 100 \times 6 + K \times 0 \times 18}{9K \times 6 + K \times 18}$

$= \dfrac{5400\,K}{72\,K} = 75°\,C$

5 An ideal black body absorbs all the radiations incident upon it and has an emissivity equal to 1. If a black body and an identical body are kept at the same temperature, then the black body will radiate the maximum power.

Hence, the black object at a temperature of 2000°C will have the brightest glow.

6 For black body,

Rate of radiation $Q = \sigma T^4$

After smoothing and doubling the temperature = Rate Q

$= 0.9\,\sigma\,(2T)^4$

$= 0.9 \times 2^4\,Q$

Change $= (0.9 \times 2^4 - 1)\,Q$ W

7 The heat radiation emitted by the human body have wavelength of the order of 7.9×10^{-7} m to 10^{-3} m, which is ofcourse the range of infrared region.

Hence, human body emits radiation in infrared region.

8 In steady state

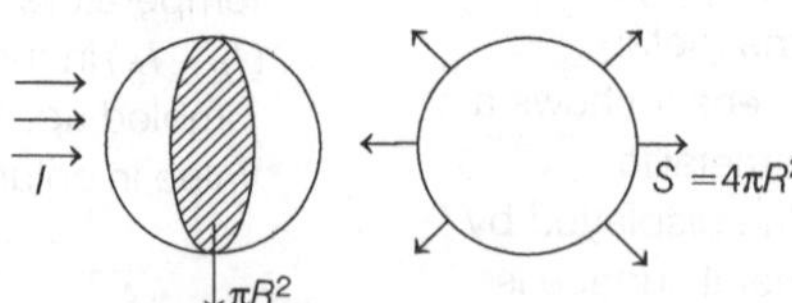

Energy incident per second = Energy radiated per second

$\therefore \; I\pi R^2 = \sigma(T^4 - T_0^4)\,4\pi R^2$

$\Rightarrow \quad I = \sigma(T^4 - T_0^4)\,4$

$\Rightarrow T^4 - T_0^4 = 40 \times 10^8$

$\Rightarrow T^4 - 81 \times 10^8 = 40 \times 10^8$

$\Rightarrow \quad T^4 = 121 \times 10^8$

$\Rightarrow \quad T \approx 330$ K

9 From Stefan's law of radiation, $E = \sigma T^4$

where, σ is Stefan's constant.

Given, $T = 2T_S$

$\therefore \quad E' = \sigma(2T_S)^4 = 16\,\sigma T_S^4 = 16E_S$

Hence, total energy radiated by star is sixteen times as that of the sun.

10 Newton's law of cooling states, that," the rate of cooling of a body is directly proportional to temperature difference between the body and the surroundings, provided the temperature difference is small, (less than 10°C)" and Newton's law of cooling is given by

$$\frac{dT}{dt} \propto (\theta - \theta_0)$$

11 According to Newton's law of cooling, rate of fall in temperature is proportional to the difference in temperature of the body with surrounding, i.e.

$$-\frac{d\theta}{dt} = k(\theta - \theta_0)$$

$\Rightarrow \quad \displaystyle\int \frac{d\theta}{\theta - \theta_0} = \int -k\,dt$

$\Rightarrow \quad \ln(\theta - \theta_0) = kt + C$

which is a straight line with negative slope.

12 According to Newton's cooling law,

$T_2 = T_1 + Ce^{-Kt}$

where, $\quad C = T_i - T_1$

(difference in temperature of body and surrounding)

$\Rightarrow \quad T_2 \propto e^{-Kt}$

Thus, the graph decays exponentially. This is shown in fig. (c).

13 We know that, the rate of loss of heat from a body is directly proportional to the surface area of the body. For a given mass of a material, the surface area of a circular plate is maximum and of sphere is least. Hence, plate will cool fastest and sphere the slowest.

14 According to Wien's displacement law,

$$\lambda_m T = \text{constant}$$

$$\therefore \quad \frac{(\lambda_m)_1}{(\lambda_m)_2} = \frac{T_2}{T_1}$$

Here, $\frac{T_1}{T_2} = \frac{3}{2}$

$$(\lambda_m)_1 = 4000 \text{ Å}$$

$$\therefore \quad (\lambda_m)_2 = \frac{4000 \times 3}{2} = 6000 \text{ Å}$$

15 We know that, $\lambda_m T$ = constant and the power radiated by a black body is proportional to T^4 i.e. $P \propto T^4$, Hence, $P \propto (\lambda_m)^{-4}$

$$\Rightarrow \frac{P_2}{P_1} = \left(\frac{\lambda_{m_1}}{\lambda_{m_2}}\right)^4 = \left(\frac{\lambda_0}{3\lambda_0 4}\right)^4 = \left(\frac{4}{3}\right)^4 = \frac{256}{81}$$

16 $Q \propto AT^4$ and $\lambda_m T$ = constant.

Hence, $Q \propto \frac{A}{(\lambda_m)^4}$ or $Q \propto \frac{r^2}{(\lambda_m)^4}$

$$Q_A : Q_B : Q_C = \frac{(2)^2}{(3)^4} : \frac{(4)^2}{(4)^4} : \frac{(6)^2}{(5)^4} = \frac{4}{81} : \frac{1}{16} : \frac{36}{625} = 0.05 : 0.0625 : 0.0576$$

i.e. Q_B is maximum.

17 $\lambda_m T$ = constant

From the graph $T_3 > T_2 > T_1$

Temperature of sun will be maximum. Therefore, (c) is the correct option.

18 As external radii of both the spheres are equal, the surface areas of the two are also equal. Therefore, when the two spheres are heated to the same temperature, both radiate heat at the same rate.

Now, rate of loss of heat from a sphere $= Mc\frac{d\theta}{dt}$

Therefore, rate of cooling

$$\frac{d\theta}{dt} = \frac{\text{rate of loss of heat}}{Mc}$$

or $\frac{d\theta}{dt} \propto \frac{1}{M}$

Since, mass of a hollow sphere is less, its rate of cooling will be fast.

19 According to Kirchhoff's law of radiation, the ratio of emissive power to absorptive power of a body, is same for all surfaces at the same temperature and at a particular wavelength.

Thus, Kirchhoff's law implies that a good absorber is a good emitter (or radiator) too or *vice-versa*.

20 As the temperature of the black body increases, two distinct behaviours are observed. The first effect is that the peak of the distribution shifts towards the shorter wavelength side. This shift is found to obey the following relationship called the Wien's displacement law, which is given by $\lambda_m T$ = constant.

The second effect is that the total amount of energy, the black body emits per unit area per unit time increases with fourth power of the absolute temperature *T*.

SESSION 2

1 According to the Newton's law of cooling,

$$\frac{50-40}{300} = K\left[\frac{50+40}{2} - 20\right]$$

$$\Rightarrow \frac{10}{300} = K\left[\frac{90}{2} - 20\right] = K \times 25$$

$$\Rightarrow K = \frac{10}{300 \times 25} = \frac{1}{30 \times 25}$$

Similarly,

$$\frac{40-\theta}{300} = K\left[\frac{40+\theta}{2} - 20\right] = K\left[20 + \frac{\theta}{2} - 20\right] = \frac{K\theta}{2} = \frac{\theta}{2 \times 30 \times 25} = \frac{\theta}{1500}$$

$$\Rightarrow 300\theta = 1500(40-\theta) = 60000 - 1500\,\theta$$

$$\Rightarrow 1800\,\theta = 60000$$

$$\Rightarrow \theta = \frac{60000}{1800} = 33.3°\text{C}$$

2 The average temperature of 94 °C and 86 °C is 90 °C, which is 70 °C above the room temperature, under these conditions the pan cools 8°C in 2 min, we have

$$\frac{\text{Change in temperature}}{\text{Time}} = k\Delta T$$

$$\frac{8°\text{C}}{2\text{ min}} = k(70°\text{C}) \quad \ldots\text{(i)}$$

The average of 69 °C and 71°C is 70 °C, which is 50 °C above room temperature. *K* is same for this situation as for the original

$$\frac{2°\text{C}}{\text{Time}} = k(50°\text{C}) \quad \ldots\text{(ii)}$$

On dividing Eq. (i) by Eq. (ii), we get

$$\frac{8°\text{C}/2\text{ min}}{2°\text{C}/\text{ time}} = \frac{k(70°\text{C})}{k(50°\text{C})},$$

$$T = 0.7\text{ min} = 42\text{s}$$

3 According to question, the temperature at the contact of the surface is given by

$$= \frac{K_1 d_2 \theta_1 + K_2 d_1 \theta_2}{K_1 d_2 + K_2 d_1} = \frac{2K_2 d_2 \times 100 + 2d_2 \times K_2 \times 25}{2K_2 d_2 + K_2 2d_2} = \frac{200 + 50}{4} = 62.5°\text{C}$$

4 From Stefan's law, the rate at which energy is radiated by sun from its surface is

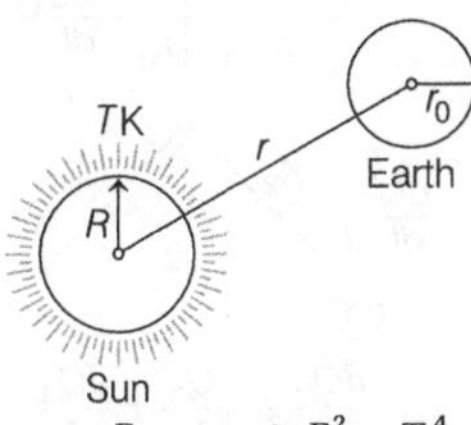

$$P = \sigma \times 4\pi R^2 \times T^4$$

[Sun is a perfect black body as it emits radiations of all wavelengths and so for it, $e = 1$]

The intensity of this power at the surface of the earth

[under the assumption $r >> r_0$] is

$$I = \frac{P}{4\pi r^2} = \frac{\sigma \times 4\pi R^2 T^4}{4\pi r^2} = \frac{\sigma R^2 T^4}{r^2}$$

The area of the earth which receives this energy is only one-half of the total surface area of earth, whose projection would be πr_0^2.

$\therefore$ Total radiant power as received by the earth $= \pi r_0^2 \times I = \frac{\pi r_0^2 R^2 \sigma T^4}{r^2}$

5 When the rods are placed in vessels

$$\frac{\theta}{t} = \frac{(T_1 - T_2)}{R}$$

$$= \left(\frac{\theta}{t}\right)_1 = \frac{mL}{t} = q_1 L = \frac{(100-0)}{R/2} \quad \ldots\text{(i)}$$

When the rods are joined end to end

$$\left(\frac{\theta}{t}\right)_2 = \frac{mL}{t} = q_2 L = \frac{(100-0)}{2R} \quad \ldots\text{(ii)}$$

From Eqs. (i) and (ii), we get

$$\frac{q_1}{q_2} = \frac{4}{1}.$$

6 According to Newton's law of cooling, rate of cooling is given by,

$$\left(\frac{-dT}{dt}\right) = \frac{eA\sigma}{mc}(T^4 - T_0^4)$$

where, *c* is specific heat of material.

or $\left(\frac{-dT}{dt}\right) \propto \frac{1}{c}$

i.e. rate of cooling varies inversely as specific heat. From the graph, for *A*,rate of cooling is larger. Therefore, specific heat of *A* is smaller.

7 In thermal conduction, it is found that in steady state the heat current is directly proportional to the area of cross-section A, which is proportional to the change in temperature $(T_1 - T_2)$.

Then, $\frac{\Delta Q}{\Delta t} = \frac{KA(T_1 - T_2)}{x}$

According to thermal conductivity, we get

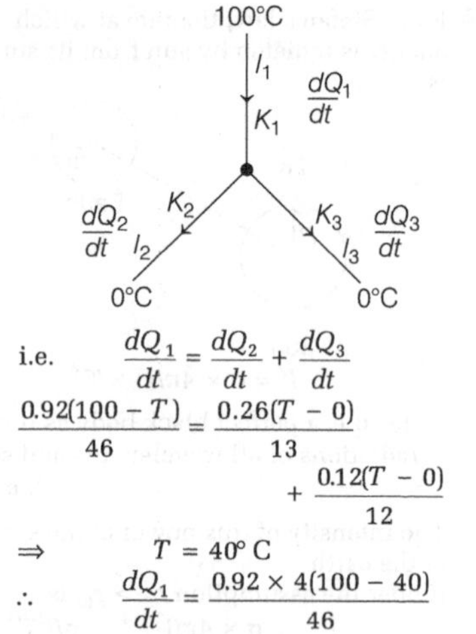

i.e. $\frac{dQ_1}{dt} = \frac{dQ_2}{dt} + \frac{dQ_3}{dt}$

$$\frac{0.92(100 - T)}{46} = \frac{0.26(T - 0)}{13} + \frac{0.12(T - 0)}{12}$$

$\Rightarrow \quad T = 40°\,C$

$\therefore \quad \frac{dQ_1}{dt} = \frac{0.92 \times 4(100 - 40)}{46}$

$= 4.8\ \text{cals}^{-1}$

8 As, $\Delta Q = ms\Delta\theta$ (for water)

$= 50 \times s \times 5$

$\Rightarrow \quad \left(\frac{dQ}{dt}\right)_s$ = rate of cooling

$= \frac{250}{2 \times 60} = \frac{25}{2 \times 6}$

$(\because S_w = 1\ \text{cal/g})$

Now, other liquid $\left(\frac{dQ}{dt}\right)_l$

= rate of cooling

$= \frac{100 \times s \times 5}{2 \times 60} = \frac{50}{2 \times 6} s$

Now, $\left(\frac{dQ}{dt}\right)_l = \left(\frac{dQ}{dt}\right)_s$

$\Rightarrow \quad s = 0.5\ \text{cal/g} = 0.5\ \text{kcal/kg}$

9 To measure the radial rate of heat flow, we have to go for integration technique as here the area of the surface through which heat will flow is not constant.

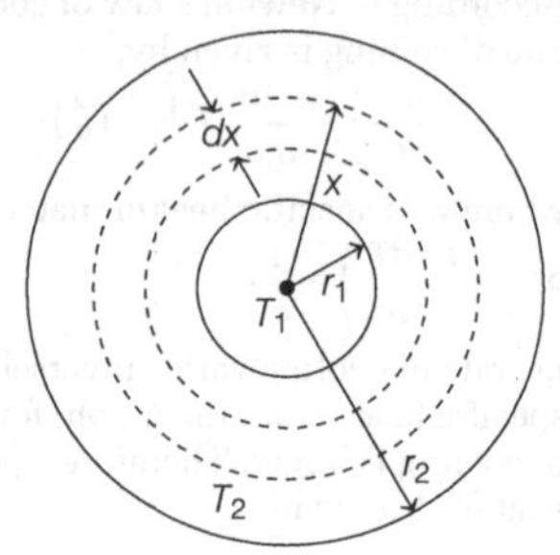

Let us consider an element (spherical shell) of thickness dx and radius x as shown in figure. Let us first find the equivalent thermal resistance between inner and outer sphere.

Resistance of shell $= dR = \frac{dx}{K \times 4\pi x^2}$

$\left(\begin{array}{l}\text{From } R = \frac{1}{KA},\\ \text{where, } K = \text{thermal conductivity}\end{array}\right)$

$\Rightarrow \int dR = R = \int_{r_1}^{r_2} \frac{dx}{4\pi K x^2}$

$= \frac{1}{4\pi K}\left[\frac{1}{r_1} - \frac{1}{r_2}\right] = \frac{r_2 - r_1}{4\pi K(r_1 r_2)}$

Rate of heat flow $= H = \frac{T_1 - T_2}{R}$

$= \frac{T_1 - T_2}{r_2 - r_1} \times 4\pi K(r_1 r_2)$

$\propto \frac{r_1 r_2}{r_2 - r_1}$

10 Assuming that heat loss from the sides of the slab is negligible, the amount of heat flowing through the slab is

$$Q = \frac{kA(T_1 - T_2)t}{d} \quad \text{...(i)}$$

If m is the mass of ice and L the latent heat of fusion, then

$$Q = mL \quad \text{...(ii)}$$

From Eqs. (i) and (ii), we have

$$mL = \frac{kA(T_1 - T_2)t}{d}$$

or $\quad k = \frac{mLd}{A(T_1 - T_2)t} \quad \text{...(iii)}$

Given, $m = 4.8$ kg, $d = 10$ cm $= 0.1$ m,

$A = 3600\ \text{cm}^2 = 0.36\ \text{m}^2$,

$T_1 = 100°\text{C},\ T_2 = 0°\text{C}$

and $\quad t = 1\ \text{h} = (60 \times 60)\text{s}$

We know that,

$L = 80\ \text{cal/g} = 80000\ \text{cal/kg}$

$= 80000 \times 4.2\ \text{J/kg} = 3.36 \times 10^5\ \text{J/kg}$

Substituting these values in Eq. (iii) and solving, we get

$k = 1.24\ \text{J/s/m/°C}$ or 1.24 W/m/k

11 Power, $P = (\sigma T^4 A) = \sigma T^4 (4\pi R^2)$

or, $\quad P \propto T^4 R^2 \quad \text{...(i)}$

According to Wien's law,

$\lambda \propto \frac{1}{T}$

(λ is the wavelength at which peak occurs)

$\therefore$ Eq. (i) will become,

$$P \propto \frac{R^2}{\lambda^4} \text{ or } \lambda \propto \left[\frac{R^2}{P}\right]^{1/4}$$

$\Rightarrow \quad \frac{\lambda_A}{\lambda_B} = \left[\frac{R_A}{R_B}\right]^{1/2}\left[\frac{P_B}{P_A}\right]^{1/4}$

$= [400]^{1/2}\left[\frac{1}{10^4}\right]^{1/4} = 2$

12 Given, $\log_2 \frac{P_1}{P_0} = 1$

Therefore, $\frac{P_1}{P_0} = 2$

According to Stefan's law,

$P \propto T^2$

$\Rightarrow \quad \frac{P_2}{P_1} = \left(\frac{T_2}{T_1}\right)^4 = \left(\frac{2767 + 273}{487 + 273}\right)^4 = 4^4$

$\frac{P_2}{P_1} = \frac{P_2}{2P_0} = 4^4$

$\frac{P_2}{P_0} = 2 \times 4^4$

$\log_2 \frac{P_2}{P_0} = \log_2 [2 \times 4^4] = \log_2 2 + \log_2 4^4$

$= 1 + \log_2 2^8 = 9$

13 We know that, $\lambda_m \propto \frac{1}{T}$

$\therefore \quad \frac{\lambda_A}{\lambda_B} = \frac{T_B}{T_A} = \frac{500}{1500} = \frac{1}{3}$

$E \propto T^4 A$

(where, A = surface area $= 4\pi R^2$)

$\therefore \quad E \propto T^4 R^2$

$$\frac{E_A}{E_B} = \left(\frac{T_A}{T_B}\right)^4 \left(\frac{R_A}{R_B}\right)^2 = (3)^4 \left(\frac{6}{18}\right)^2 = 9$$

14 Intermolecular distance in ideal gases is assume to be large as compared to real one. Hence, the internal energy of an ideal gas and a real gas is kinetic as well as potential.

According to Newton's cooling law, option (c) is correct answer.

15 $A \to 3$; $B \to 4$; $C \to 2$; $D \to 1$

We have four sections, AB, BC, CD and DE with (dQ/dt) as the steady state thermal energy transmitted per second (A being the areas of cross-section)

$$\frac{dQ}{dt} = \frac{KA(100 - T_c)}{L}$$

$$= \frac{A(0.8)K(T_C - T_D)}{(1.2)L}$$

$$= \frac{(1.2)KA(T_D - T_E)}{(1.5)L} = \frac{(1.5)KAT_E}{(0.6)L}$$

These give

$(100 - T_C) = \left(\frac{0.8}{1.2}\right)(T_C - T_D)$

$= \left(\frac{1.2}{1.5}\right)(T_D - T_E) = \left(\frac{1.5}{0.6}\right)T_E$

$6(100 - T_C) = 4(T_C - T_D)$

$= (4.8)(T_D - T_E) = 15T_E$

Solving for the differences $(100 - T_C)$, $(T_C - T_D)$, $(T_D - T_E)$ and T_E remaining that the sum of these differences is 100, we obtain

$(T_A - T_C) = 24.1, (T_C - T_D) = 36.2$

$(T_D - T_E) = 30.1$

and $\quad (T_E - T_B) = 9.6$

DAY SEVENTEEN

Unit Test 3

(General Properties of Matter)

1 A jar is filled with two non-mixing liquids 1 and 2 having densities d_1 and d_2 respectively. A solid ball, made of a material of density d_3, is dropped in the jar. It comes to equilibrium in the position as shown in the figure. Which of the following is true for d_1, d_2 and d_3?

Liquid 1 d_1

d_3

Liquid 2 d_2

(a) $d_1 > d_3 > d_2$ (b) $d_1 < d_2 < d_3$
(c) $d_1 < d_3 < d_2$ (d) $d_3 < d_1 < d_2$

2 A spherical body of volume V and density σ is suspended from a string, the other end of the string is connected to the roof of a sealed container filled with an ideal fluid of density ρ.

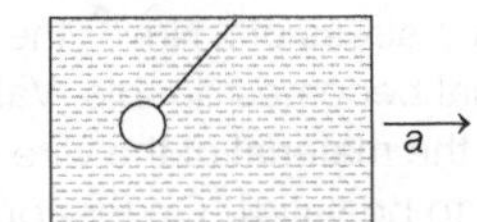

If the container accelerates towards right with a constant acceleration a, then the force exerted by the liquid on the body when it is in equilibrium w.r.t. fluid, is

(a) $V\rho\sqrt{a^2 + g^2} + V\sigma a$ (b) $V\sigma a$
(c) $\sqrt{[V\rho(g + a)]^2 + [V\sigma a]^2}$ (d) $V\rho\sqrt{g^2 + a^2}$

3 A hot metallic sphere of radius r radiates heat. Its rate of cooling is

(a) independent of r
(b) proportional to r
(c) proportional to r^2
(d) proportional to $1/r$

4 A certain ideal gas undergoes a polytropic process pV^n = constant such that the molar specific heat during the process is negative. If the ratio of the specific heat of the gas be γ, then the range of values of n will be

(a) $0 < n < \gamma$
(b) $1 < n < \gamma$
(c) $n = \gamma$
(d) $n > \gamma$

5 Pressure p, volume V and temperature T for a certain material are related by

$$p = \frac{AT - BT^2}{V}$$

where, A and B are constants. Find an expression for the work done by the material if the temperature changes from T_1 to T_2 reduce while the pressure remains constant.

(a) $W = A(T_2 - T_1) - B(T_2^3 - T_1^3)$
(b) $W = A(T_2^2 - T_1^2) - B(T_2 - T_1)$
(c) $W = A(T_2 - T_1) - B\left(T_2 - \frac{1}{2}T_1\right)$
(d) $W = A(T_2 - T_1) - B(T_2^2 - T_1^2)$

6 A soap bubble is very slowly blown on the end of a glass tube by a mechanical pump which supplies a fixed volume of air every minute whatever be the pressure against which it is pumping. The excess pressure Δp inside the bubble varies with time is shown by which of the graph?

(a) 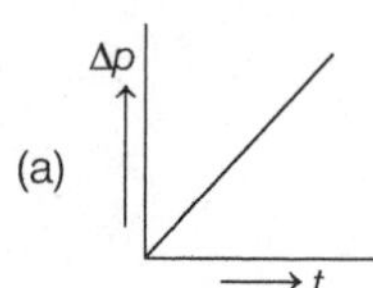

(b) 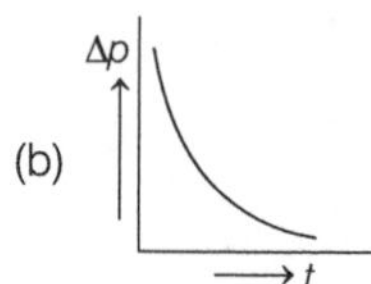

(c) 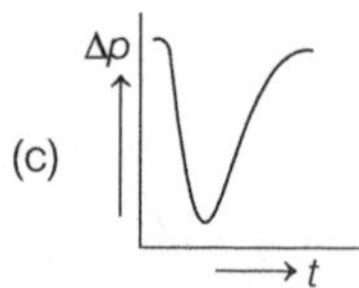

(d) 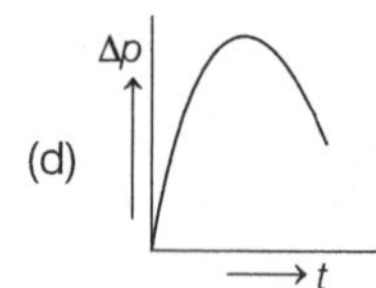

7 A small electric immersion heater is used to heat 100 g of water for a cup of instant coffee. The heater is labelled "200 W," which means that it converts electrical energy to thermal energy at this rate. Calculate the time required to bring all this water from 23°C to 100°C, ignoring any heat losses. [$c = 4190$ J kg^{-1} K^{-1}]

(a) 100 s (b) 200 s (c) 190 s (d) 161 s

8 The average depth of Indian ocean is about 3000 m. Calculate the fractional compression, $\Delta V/V$, of water at the bottom of the ocean. Given that the bulk modulus of water is 2.2×10^9 Nm^{-2}.

(Take, $g = 10$ ms^{-2})

(a) 1.36×10^{-2} (b) 3×10^{-3}
(c) 1.5×10^{-2} (d) 1.36×10^{-6}

9 The temperature of the two outer surfaces of a composite slab, consisting of two materials having coefficients of thermal conductivity K, $2K$ and thickness x, $4x$, respectively are T_2 and T_1 ($T_2 > T_1$). The rate of heat transfer through the slab in a steady state is

$\left[\frac{A(T_2 - T_1)K}{x}\right] f$ with f which is equal to

(a) 1 (b) 1/2
(c) 2/3 (d) 1/3

10 An aeroplane has a mass of 1.60×10^4 kg and each wing has an area of 40 m^2. During level flight, the pressure on the wings's lower surface is 7×10^4 Pa. The pressure on the upper surface of the wing is

(Take, $p_0 = 10^5$ Pa and assume the pressure difference is only on wings and not on body)

(a) 10^5 Pa
(b) 6.8×10^4 Pa
(c) 7×10^4 Pa
(d) 6.6×10^4 Pa

11 The coefficients of thermal conductivity of copper, mercury and glass are K_c, K_m and K_g, respectively, such that $K_c > k_m > K_g$. If the same quantity of heat is to flow per second per unit area of each and corresponding temperature gradients are, X_c, X_m and X_g, respectively, then

(a) $X_c = X_m = X_g$ (b) $X_c > X_m > X_g$
(c) $X_c < X_m < X_g$ (d) $X_m < X_c < X_g$

12 The temperature of the source of a Carnot's heat engine is 1000°C. Its efficiency could be 100% only if the temperature of the sink is

(a) 1000°C (b) 0°C
(c) equal to triple of water (d) – 273.16°C

13 A steel rod is 3.00 cm in diameter at 25°C. A brass ring has an interior diameter of 2.992 cm at 25°C. At what common temperature will the ring just slide onto the rod? (take, $\alpha_s = 11 \times 10^{-6}\,°C^{-1}, \alpha_b = 19 \times 10^{-6}\,°C^{-1}$)

(a) 460°C (b) 260°C (c) 500°C (d) 360°C

14 A diver is hunting for a fish with a water gun. He accidentally fires the gun, so that bullet punctures the side of the ship. The hole is located at a depth of 10 m below the water surface. The speed with which water enter in the ship is

(a) 18 ms^{-1} (b) 14 ms^{-1}
(c) 25 ms^{-1} (d) Cannot be determined

15 A material has a Poisson's ratio 0.3. If a uniform rod of it suffers longitudinal strain $4 \cdot 5 \times 10^{-3}$, then calculate the percentage change in its volume.

(a) 0.15% (b) 0.25% (c) 0.18% (d) 0.5%

16 Compute the number of moles and in 1.00 cm^3 of an ideal gas at a pressure of 100 Pa and at a temperature of 220 K.

(a) 3.35×10^{-8}mol (b) 4.57×10^{-7}mol
(c) 5.47×10^{-8}mol (d) 2.75×10^{-8}mol

17 A slab consists of two parallel layers of copper and brass of the same thickness same area of cross-section and having thermal conductivities in the ratio 1 : 4. If the free face of brass is at 100°C and that of copper is at 0°C, the temperature of the interface is

(a) 80°C (b) 20°C (c) 60°C (d) 40°C

18 On applying a stress of x Nm^{-2}, the length of wire of some material becomes double. Value of the Young's modulus for the material of the wire in Nm^{-2}, is [Assume Hooke's law to be valid] "Go for approx results"

(a) x (b) $2x$
(c) $x/2$ (d) Insufficient information

19 743 J of heat energy is added to raise the temperature of 5 mole of an ideal gas by 2 K at constant pressure. How much heat energy is required to raise the temperature of the same mass of the gas by 2K at constant volume? (Take, $R = 8.3$ J/K-mol)

(a) 826 J (b) 743 J (c) 660 J (d) 620 J

20 Six identical conducting rods are joined as shown in figure given below. Points A and D are maintained at temperatures 200 °C and 20°C, respectively. The temperature of junction B will be

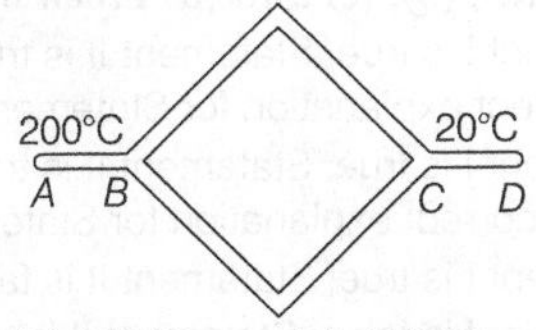

(a) 120°C (b) 100°C (c) 140 °C (d) 80°C

21 One mole of an ideal gas is taken along the process in which pV^x = constant. The graph shown represent the variation of molar heat capacity of such a gas with respect to x. The values of c' and x', respectively, are given by

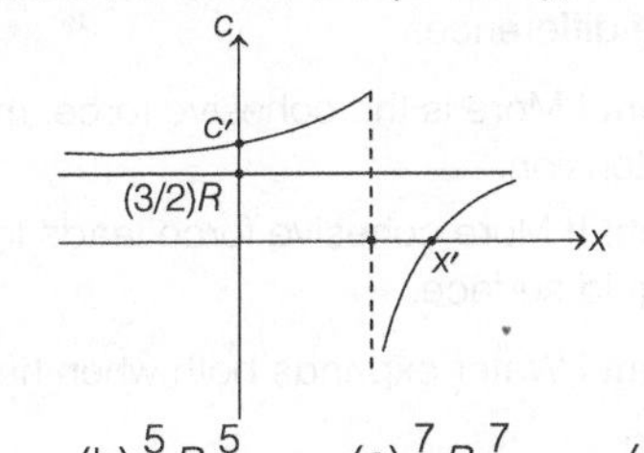

(a) $\frac{5}{2}R, \frac{5}{2}$ (b) $\frac{5}{2}R, \frac{5}{3}$ (c) $\frac{7}{2}R, \frac{7}{2}$ (d) $\frac{5}{2}R, \frac{7}{5}$

22 A wire of length L and radius r is fixed at one end. When a stretching force F is applied at the free end, the elongation in the wire is l. When another wire of the same material but of length $2L$ and radius $2r$, also fixed at one end is stretched by a force $2F$ applied at the free end, then elongation in the second wire will be

(a) $l/2$ (b) l (c) $2l$ (d) $l/4$

23 A Carnot engine has an efficiency of 22.0%. It operates between constant-temperature reservoirs differing in temperature by 75.0°C. What are the temperatures of the two reservoirs?

(a) 58°C, 10°C (b) 78°C, –5°C
(c) 68°C, –7°C (d) 50°C, 0°C

24 A material has a Poisson's ratio 0.50. If a uniform rod of it suffers a longitudinal strain of 2×10^{-3}, then the percentage change in volume is

(a) 0.6 (b) 0.4 (c) 0.2 (d) zero

25 An open vessel full of water is falling freely under gravity. There is a small hole in one face of the vessel, as shown in the figure.

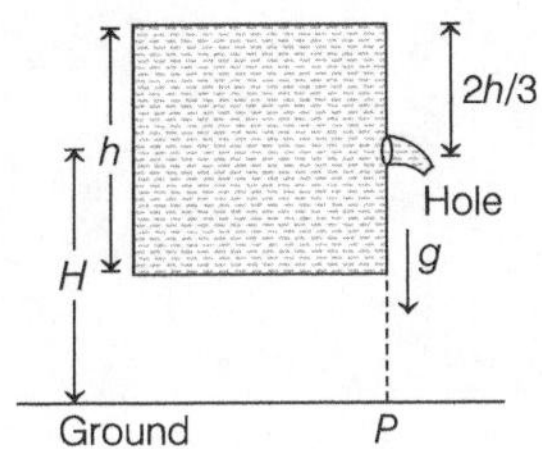

The water which comes out from the hole at the instant when the hole is at height H above the ground, strikes the ground at a distance x from P. Which of the following is correct for the situation described?

(a) The value of x is $2\sqrt{\frac{2hH}{3}}$ (b) The value of x is $\sqrt{\frac{4hH}{3}}$

(c) The value of x cannot be computed from the information provided

(d) The question is irrelevant as no water comes out from the hole

26 Water flows through a horizontal pipe of varying cross-section at the rate of 20 litres per minute. Then the velocity of water at a point where diameter is 4 cm, is

(a) 0.25 ms^{-1} (b) 0.26 ms^{-1} (c) 0.22 ms^{-1} (d) 0.4 ms^{-1}

27 Oxygen gas having a volume of 1000 cm^3 at 40.0°C and 1.01×10^5 Pa expands until its volume is 1500 cm^3 and its pressure is 1.06×10^5 Pa. Find the final temperature of the sample.

(a) 197°C (b) 220 K (c) 300°C (d) 300 K

28 A mercury drop of radius 1.0 cm is sprayed into 10^6 droplets of equal sizes. The energy spent in this process is [Surface tension of mercury is equal to 32×10^{-2} Nm^{-1}]

(a) 3.98×10^{-4} J (b) 8.46×10^{-4} J
(c) 3.98×10^{-2} J (d) 8.46×10^{-2} J

29 A chef, on finding his stove out of order, decides to boil the water for his wife's coffee by shaking it in a thermos flask. Suppose that he uses tap water at 15°C and that the water falls 30 cm in each shake, the chef making 30 shakes each minute. Neglecting any loss of thermal energy by the flask, how long must he shake the flask until the water reaches 100°C ?

(a) 2.25×10^3min (b) 3.97×10^3min
(c) 4.03×10^3min (d) 5.25×10^3min

30 The maximum amount of heat which may be lost per second by radiation by a sphere 14 cm in diameter at a temperature of 227°C, when placed in an enclosure at 27°C. Given, Stefan's constant = 5.7×10^{-8} Wm^{-2}K^{-4}

(a) 45.48 cal/s (b) 40 cal/s
(c) 42.5 cal/s (d) 40.5 cal/s

31 Four moles of an ideal gas undergo a reversible isothermal expansion from volume V_1 to volume $V_2 = 2V_1$ at temperature $T = 400$K. Find the entropy change of the gas.

(a) 8.22×10^3 JK^{-1} (b) 8.22×10^2 JK^{-1}
(c) 23.1 J K^{-1} (d) 10.00×10^3 JK^{-1}

32 A swimmer of mass m rests on top of a styrofoam slab, which has a thickness h and density ρ_s. The area of the slab if it floats in water with its upper surface just awash is (take, density of water to be ρ_w)

(a) $\frac{m}{h(\rho_s + \rho_w)}$ (b) $\frac{m}{h\rho_w}$ (c) $\frac{m}{h(\rho_s - \rho_w)}$ (d) $\frac{m}{h(\rho_w - \rho_s)}$

33 An ice-berg of density 900 kg-m^{-3} is floating in water of density 1000 kg-m^{-3}. The percentage of volume of ice-berg outside the water is

(a) 20% (b) 35% (c) 10% (d) 11%

34 A uniform capillary tube of length l and inner radius r with its upper end sealed is submerged vertically into water. The outside pressure is p_0 and surface tension of water is γ. When a length x of the capillary is submerged into water, it is found that the water level inside and outside the capillary coincide. The value of x is

(a) $\dfrac{l}{\left(1+\dfrac{p_0 r}{4\gamma}\right)}$ (b) $l\left(1-\dfrac{p_0 r}{4\gamma}\right)$ (c) $l\left(1-\dfrac{p_0 r}{2\gamma}\right)$ (d) $\dfrac{l}{\left(1+\dfrac{p_0 r}{2\gamma}\right)}$

35 2 moles of an ideal monoatomic gas is carried from a state (p_0, V_0) to state $(2p_0, 2V_0)$ along a straight line path in a p-V diagram. The amount of heat absorbed by the gas in the process is given by

(a) $3p_0V_0$ (b) $\frac{9}{2}p_0V_0$
(c) $6p_0V_0$ (d) $\frac{3}{2}p_0V_0$

36 The stress along the length of a rod (with rectangular cross-section) is 1% of the Young's modulus of its materials. What is the approximate percentage of change of its volume? (Poisson's ratio of the material of the rod is 0.3)

(a) 3% (b) 1%
(c) 0.7% (d) 0.4%

Direction (Q. Nos. 37-40) *Each of these questions contains two statements : Statement I and Statement II. Each of these questions also has four alternative choices, only one of which is the correct answer. You have to select one of the codes (a), (b), (c) and (d) given below*

(a) Statement I is true, Statement II is true; Statement II is the correct explanation for Statement I
(b) Statement I is true, Statement II is true; Statement II is not the correct explanation for Statement I
(c) Statement I is true; Statement II is false
(d) Statement I is false; Statement II is true

37 **Statement I** A ship floats higher in water on a high pressure day than on a low pressure day.

Statement II Floating of ship in the water is possible because of the buoyant force which is present due to the pressure difference.

38 **Statement I** More is the cohesive force, more is the surface tension.

Statement II More cohesive force leads to more shrinking of the liquid surface.

39 **Statement I** Water expands both when heated or cooled from 4°C.

Statement II Density of water is minimum at 4°C.

40 **Statement I** If the temperature of a star is doubled, then the rate of loss of heat from it becomes 16 times.

Statement II Specific heat varies with temperature.

ANSWERS

1. (c)	**2.** (d)	**3.** (d)	**4.** (b)	**5.** (d)	**6.** (b)	**7.** (d)	**8.** (a)	**9.** (d)	**10.** (b)
11. (c)	**12.** (d)	**13.** (d)	**14.** (b)	**15.** (c)	**16.** (c)	**17.** (a)	**18.** (a)	**19.** (c)	**20.** (c)
21. (b)	**22.** (b)	**23.** (c)	**24.** (b)	**25.** (d)	**26.** (b)	**27.** (a)	**28.** (c)	**29.** (c)	**30.** (a)
31. (c)	**32.** (d)	**33.** (c)	**34.** (d)	**35.** (c)	**36.** (d)	**37.** (d)	**38.** (b)	**39.** (c)	**40.** (b)

Hints and Explanations

1 d_3 floats in d_2 and sinks in $d_1 \Rightarrow d_1 < d_3 < d_2$.

2 The forces exerted by liquid on body is shown in figure, when body is in equilibrium w.r.t. fluid.

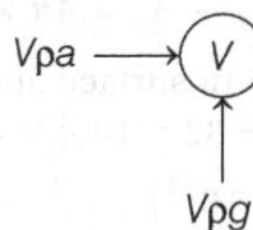

Forces exerted by liquid

So, required force, $F = \rho V\sqrt{a^2 + g^2}$

3 Rate of cooling

$$R_c = \frac{d\theta}{dt} = \frac{A\varepsilon\sigma(T^4 - T_0^4)}{mc}$$

$$\Rightarrow \frac{d\theta}{dt} \propto \frac{A}{V} \propto \frac{r^2}{r^3} \Rightarrow \frac{d\theta}{dt} \propto \frac{1}{r}$$

4 Since, pV^n = constant and also $pV = RT$, taking 1 mol of the gas for simplicity $dU = C_V dt$
where, $C_V \rightarrow$ molar specific heat at constant volume
Now, the molar specific heat in a polytropic process
pV^n = constant is given by

$$C_V = \left(\frac{R}{\gamma - 1}\right) - \left(\frac{R}{n - 1}\right) = \frac{(n - \gamma)R}{(n - 1)(\gamma - 1)} \quad \ldots \text{(i)}$$

From this equation, we see that C_V will be negative when $n < \gamma$ and $n > 1$, simultaneously, i.e. $1 < n < \gamma$. Since, γ for all ideal gases is greater than 1, if $n > \gamma$ or $n < 1$, then C_V will be positive.

5 Work $W = p(V_2 - V_1)$ at constant p

Initial volume, $V_1 = (AT_1 - BT_1^2)/p$

Final volume is $V_2 = (AT_2 - BT_2^2)/p$

$\Rightarrow W = A(T_2 - T_1) - B(T_2^2 - T_1^2)$

6 $\Delta p = \dfrac{2T}{R}$

As, the number of moles of air increases. Radius increases and Δp decreases.

7 Heat required to raise the temperature of water must be equal to the power output of the heater P, multiplied by time t.

$$t = \frac{Q}{P} = \frac{cm(T_f - T_i)}{P}$$

$$= \frac{(4190)\,(0.100)\,(100° - 23°)}{200} = 161 \text{ s}$$

8 Here, $h = 3000$ m,

$B = 2.2 \times 10^9 \text{ Nm}^{-2}$

$p = h\rho g = 3000 \times 10^3 \times 10$

$= 3 \times 10^7 \text{ Nm}^{-2}$

$\therefore$ Compressional strain $= \dfrac{\Delta V}{V} = \dfrac{p}{B}$

$$= \frac{3 \times 10^7}{2.2 \times 10^9} = 1.36 \times 10^{-2}$$

9 Equation of thermal conductivity of the given combination $K_{eq} = \dfrac{l_1 + l_2}{\dfrac{l_1}{K_1} + \dfrac{l_2}{K_2}}$

$$= \frac{x + 4x}{\dfrac{x}{K} + \dfrac{4x}{2K}} = \frac{5}{3}K$$

Hence, rate of flow of heat through the given combination is

$$\frac{\theta}{t} = \frac{K_{eq}A(T_2 - T_1)}{(x + 4x)} = \frac{5/3KA(T_2 - T_1)}{5x}$$

$$= \frac{1/3KA(T_2 - T_1)}{x}$$

On comparing it with given equations, we get $f = \dfrac{1}{3}$.

10 Let p_1 be the pressure on the upper wing surface, then for the vertical equilibrium of the plane

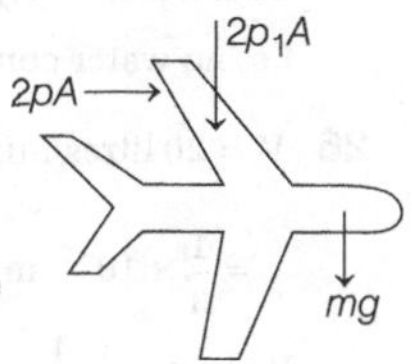

$2p_1A + mg = 2pA$

$$\Rightarrow \quad p_1 = p - \frac{mg}{2A}$$

$$= 7 \times 10^4 - \frac{1.6 \times 10^4 \times 10}{2 \times 40}$$

$= 6.8 \times 10^4$ Pa

11 $\dfrac{dQ/dt}{A} = K\left(\dfrac{\Delta\theta}{\Delta x}\right)$

Rate of flow of heat per unit area
= Thermal conductivity × Temperature gradient

Temperature gradient $(X) \propto \dfrac{1}{\text{Thermal conductivity (K)}}$

$\left(\text{As, } \dfrac{dQ/dt}{A} = \text{constant}\right)$

As, $K_c > K_m > K_g$, therefore $X_c < X_m < X_g$

12 Efficiency of Carnot's heat engine = 100%

$$\frac{T_1 - T_2}{T_1} \times 100 = 100$$

$\Rightarrow \quad T_1 - T_2 = T_1$

$\Rightarrow \quad T_2 = 0\text{ K} = -273.16° \text{ C}$

13 If $D_s = D_b$,

then $D_{s0} + \alpha_s D_{s0}\Delta T = D_{b0} + \alpha_b D_{b0}\Delta T$

So, $\quad \Delta T = \dfrac{D_{s0} - D_{b0}}{\alpha_b D_{b0} - \alpha_s D_{s0}}$

$$= \frac{3.000 - 2.992}{(19 \times 10^{-6})(2.992) - (11 \times 10^{-6})(3.00)}$$

$= 335°\text{C}$

The temperature is
$T = 25°\text{C} + 335°\text{C} = 360°\text{C}$

14 Applying the Bernoulli's theorem for any two convenient points, let us say we are applying just at the water surface and just inside the hole.

$$p_0 + 0 + 0 = p_0 + \frac{\rho v^2}{2} + \rho g(-h)$$

where, v is required speed and water surface is taken as the reference level.

$\Rightarrow \quad v = \sqrt{2gh} = \sqrt{2 \times 9.8 \times 10}$

$= 14 \text{ ms}^{-1}$

15 Here, $\sigma = 0.3$, $\dfrac{\Delta l}{l} = 4.5 \times 10^{-3}$

$$\sigma = -\frac{\Delta R/R}{\Delta l/l} \Rightarrow \frac{\Delta R}{R} = -\sigma\frac{\Delta l}{l}$$

$\Rightarrow \dfrac{\Delta R}{R} = -0.3 \times (4.5 \times 10^{-3})$

$= -1.35 \times 10^{-3}$

Volume, $V = \pi R^2 l$

$$\frac{\Delta V}{V} = \frac{2\Delta R}{R} + \frac{\Delta l}{l}$$

Percentage change in volume $= \dfrac{\Delta V}{V} \times 100$

$= \left(\dfrac{2\Delta R}{R} + \dfrac{\Delta l}{l}\right) \times 100$

$= [2 \times (-1.35) + 4.5] \times 10^{-3} \times 100$

$= 1.8 \times 10^{-3} \times 100 = 0.18\%$

16 From the ideal gas law, $pV = nRT$

$$n = \frac{pV}{RT} = \frac{(100)(1.0 \times 10^{-6})}{(8.31)(220)}$$

$= 5.47 \times 10^{-8}$ mol

17 Temperature of the interface,

$$\theta = \frac{K_1\theta_1 + K_2\theta_2}{K_1 + K_2}$$

$\left(\because \dfrac{K_1}{K_2} = \dfrac{1}{4} \Rightarrow \text{If } K_1 = K, \text{ then } K_2 = 4K\right)$

$$\Rightarrow \quad \theta = \frac{K \times 0 + 4K \times 100}{5K} = 80°\text{C}$$

18 $Y = \dfrac{\text{Stress}}{\text{Strain}} = \dfrac{x}{\dfrac{2l - l}{l}} = \dfrac{x}{1} = x$

In actual, the above expression is not exact for this much elongation.

19 For constant pressure process,
$Q_1 = nC_p\Delta T = 743$ J
For constant volume process,
$$Q_2 = nC_V dT = n(C_p - R)dT$$
$$= nC_p dT - nRdT$$
$$Q_2 = 743 - 5 \times 8.3 \times 2 = 660 \text{ J}$$

20 Let the thermal resistance of each rod be R.
The two resistances connected along two paths from B to C are equivalent to 2 R each and their parallel combination is R.
Effective thermal resistance between B and $D = 2R$

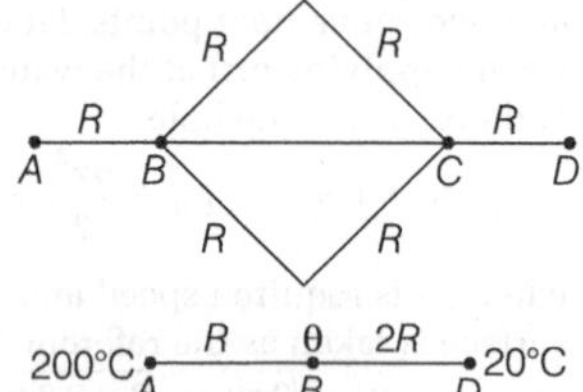

Temperature of interface,
$$\theta = \frac{R_1\theta_2 + R_2\theta_1}{R_1 + R_2}$$
$$\theta = \frac{R \times 20 + 2R \times 200}{R + 2R} = \frac{420}{3} = 140°\text{C}$$

21 At $x = \infty, C = \dfrac{3}{2}R$, from pV^x = constant
$\Rightarrow p^{1/x}V$ = another constant, so at $x = \infty$, V = constant
hence $C = C_v = \dfrac{3}{2}R$ and then
$$C_p = C_V + R = \frac{5}{2}R$$
at $x = 0$, P = constant and $C = C'$
hence, $C' = C_p = \dfrac{5}{2}R$
at $x = x', C = 0$, so the process is adiabatic,hence
$$x' = \frac{C_p}{C_V} = \frac{5}{3}$$

22 $Y = \dfrac{F/A}{\Delta l/l}$

For the first wire, $Y = \dfrac{F \times L}{\pi r^2 \times l}$

For the second wire, $Y = \dfrac{2F \times 2L}{\pi(2r)^2 \times l'}$

From the above two equations, $l' = l$

23 For an ideal engine, the efficiency is related to the reservoir temperatures by $\varepsilon = (T_H - T_C)/T_H$. Thus,
$$T_H = (T_H - T_C)/\varepsilon$$
$$= (75 \text{ K})/(0.22) = 341 \text{ K}$$
$$= 68°\text{C}.$$
The temperature of the cold reservoir is
$$T_C = T_H - 75 = 341\text{K} - 75\text{K}$$
$$= 266\text{K} = -7°\text{C}.$$

24 $\dfrac{dV}{V} = (1 + 2\sigma)\dfrac{dl}{l}$
$$= 2 \times 2 \times 10^{-3} = 4 \times 10^{-3}$$
$$\left[\because \sigma = 0.5 = \frac{1}{2}\right]$$
$\therefore$ Percentage change in volume
$$= 4 \times 10^{-1} = 0.4\%$$

25 As vessel is falling freely under gravity, the pressure at all points within the liquid remains the same as the atmospheric pressure. If we apply Bernoulli's theorem just inside and outside the hole, then
$$p_{\text{inside}} + \frac{\rho v_{\text{inside}}^2}{2} + \rho g H$$
$$= p_{\text{outside}} + \frac{\rho v_{\text{outside}}^2}{2} + \rho g H$$
$v_{\text{inside}} = 0, p_{\text{inside}} = p_{\text{outside}} = p$
[atmospheric pressure]
Therefore, $v_{\text{outside}} = 0$
i.e. no water comes out from the hole.

26 $V = 20$ litres/min $= \dfrac{20 \times 1000}{60 \times (100)^3}$ m^{3}s^{-1}
$$= \frac{1}{3} \times 10^{-3} \text{ m}^3\text{s}^{-1}$$
Radius, $r = \dfrac{4}{2} = 2$ cm $= 0.02$ m
Area of cross-section,
$$a = \pi r^2 = \frac{22}{7} \times (0.02)^2 \text{ m}^2$$
Let v be the velocity of the flow of water at the given point, then
$$V = av$$
$$\Rightarrow \frac{1}{3} \times 10^{-3} = \frac{22}{7} \times (0.02)^2 \times v$$
$$\Rightarrow v = \frac{7 \times 10^{-3}}{3 \times 22 \times (0.02)^2}$$
$$= 0.2639 \simeq 0.26 \text{ ms}^{-1}$$

27 $n = \dfrac{PV}{RT} = \dfrac{1.06 \times 10^5 \times 1000 \times 10^{-6}}{8.31 \times 313}$
$$= 4.07 \times 10^{-2}$$
Using $pV = nRT$
$$T = \frac{pV}{nR} = \frac{(1.06 \times 10^5)(1500 \times 10^{-6})}{(4.07 \times 10^{-2})(8.31)}$$
$$= 470 \text{ K}$$
$$= 197°\text{C}$$

28 Let r be the radius of one droplet.
Now, $\dfrac{4}{3}\pi R^3 = 10^6 \times \dfrac{4}{3}\pi r^3, r = \dfrac{R}{100}$
$$= \frac{1}{100}\text{cm} = 10^{-4}\text{m}$$
$$A_i = 4\pi R^2$$
$$A_f = 10^6 \times 4\pi r^2$$
Change in area,
$\Delta A = A_f - A_i = 4\pi \times 99 \times 10^{-4}$ m^2
Increase in surface energy
$= S\Delta A = 32 \times 10^{-2} \times 4\pi \times 99 \times 10^{-4}$ J
$= 3.98 \times 10^{-2}$ J
The increase in surface energy is at the expense of internal energy, so energy spent $= 3.98 \times 10^{-2}$ J

29 Time taken,
$$t = \frac{Q}{Rmgh} = \frac{cm(T_f - T_i)}{Rmgh} = \frac{c(T_f - T_i)}{Rgh}$$
$$= \frac{(4190)(100 - 15)}{(30)(9.8)(0.30)}$$
$$= 4.03 \times 10^3 \text{ min}$$

30 Temperature of sphere
$$T = 227°\text{C} = 500 \text{ K}$$
Temperature of surroundings
$T_0 = 27°\text{C} = 300$ K
Radius, $r = 7$ cm $= 0.07$ m
Area of sphere,
$$A = 4\pi r^2 = 4 \times \frac{22}{7} \times (0.07)^2$$
$$= 6.16 \times 10^{-2} \text{ m}^2$$
The energy lost from A,
$$AE = A\sigma(T^4 - T_0^4)$$
$$6.16 \times 10^{-2} \times 5.7 \times 10^{-8} \times [(500)^4 - (300)^4]$$
$$= 6.16 \times 5.7 \times 10^{-10} \times 100^4 (5^4 - 3^4)$$
$$= 6.16 \times 5.7 \times 10^{-2} \times 544 \text{ Js}^{-1}$$
The heat lost per sec.
$$H = \frac{AE}{J} = \frac{6.16 \times 5.7 \times 10^{-2} \times 544}{4.2}$$
$$= 45{\cdot}48 \text{ cal/s}$$

31 We have $p = nRT/V$. The work done by the gas during the isothermal expansion is
$$W = \int_{V_1}^{V_2} pdV = nRT\int_{V_1}^{V_2}\frac{dV}{V} = nRT \ln\frac{V_2}{V_1}$$
Substituting $V_2 = 2V_1$ to obtain $W = nRT$
$= (4.00)(8.314)(400) \ln 2$
$= 9.22 \times 10^3$ J
Since, the expansion is isothermal, $\Delta E_{\text{int}} = 0$ and $Q = W$
Thus, $\Delta S = \dfrac{W}{T}$
$$= \frac{9.22 \times 10^3 \text{ J}}{400 \text{ K}} = 23.1 \text{ J K}^{-1}$$

32 From equilibrium,
$mg + Ah\rho_s \times g = Ah\rho_w \times g$
where, A is the required cross-sectional area

$$\Rightarrow \quad A = \frac{m}{h(\rho_w - \rho_s)}$$

33 Let volume of ice-berg is V and its density is ρ. If V_{in} is volume inside the water, then

$$V_{in}\sigma g = V\rho g$$

where, σ = density of water

$$\Rightarrow \quad V_{in} = \left(\frac{\rho}{\sigma}\right)V$$

$$\Rightarrow V_{out} = V - V_{in} = \left(\frac{\sigma - \rho}{\sigma}\right)V = \left(\frac{1000 - 900}{1000}\right)V = \frac{V}{10}$$

$$\Rightarrow \quad \frac{V_{out}}{V} = 0.1 = 10\%.$$

34 The pressure inside tube changes when it is submerged in water. Thus

$$p_1V_1 = p_2V_2$$
$$p_0(lA) = p'(l - x)A$$
$$\therefore \quad p' = \frac{p_0 l}{l - x}$$

As level of water is same inside and outside of capillary tube

$$\therefore \quad p' - p_0 = \frac{2\gamma}{r}$$

or $$\frac{p_0 l}{l - x} - p_0 = \frac{2\gamma}{r}$$

$$\Rightarrow \quad x = \frac{l}{1 + \frac{p_0 r}{2\gamma}}$$

35 The internal energy, $\Delta U = nC_V\Delta T$

C_V = specific heat of gas at constant volume

$$\Rightarrow \quad \Delta U = n\cdot\frac{3R}{2}\left(\frac{4p_0V_0}{nR} - \frac{p_0V_0}{nR}\right) = n\cdot\frac{3R}{2}\cdot\frac{3p_0V_0}{nR} = \frac{9}{2}p_0V_0$$

Work done by the gas,

$$W = (2p_0 + p_0)\frac{V_0}{2} = \frac{3p_0V_0}{2}$$

From first law of thermodynamics,

$$\Delta Q = dW + dU = \frac{3p_0V_0}{2} + \frac{9}{2}p_0V_0 = \frac{12p_0V_0}{2} = 6p_0V_0$$

36 Stress $= \frac{F}{\Delta A} = 1\%$ of $Y = \frac{Y}{100}$

Also, $$Y = \frac{\text{Stress}}{\text{Strain}} = \frac{F/\Delta A}{\Delta l/l} = \frac{\frac{Y}{100}}{\Delta l/l}$$

$$\Rightarrow \quad \frac{\Delta l}{l} = \frac{1}{100}$$

Poisson's ratio, $\sigma = \frac{-\Delta r/r}{\Delta l/l}$

$$\Rightarrow \frac{\Delta r}{r} = -\sigma\frac{\Delta l}{l} = -\frac{0.3}{100}$$

$$\therefore \frac{\Delta V}{V} \times 100 = \left(\frac{2\Delta r}{r} + \frac{\Delta l}{l}\right) \times 100 = \left(2 \times \frac{-0.3}{100} + \frac{1}{100}\right) \times 100 = 0.4\%$$

37 A body of weight $w = mg = V\rho g$ float in a liquid as a upthrust $F = V\sigma g$ acts vertically upwards through the centre of gravity of displaced liquid also called the centre of buoyancy. It is independence of atmospheric pressure.

38 Surface tension, $S = \frac{\text{Force}}{\text{Length}} = \frac{F}{l}$

$\therefore S \propto F$

Thus, more the force, the surface tension is more. Also, this force tends to have the least possible surface area.

39 At 4°C, the volume of water is minimum. When it is cooled below 4°C or heated above 4°C, then it expands or its volume increases. As volume at 4°C is minimum, thus its density $\left(= \frac{\text{mass}}{\text{volume}}\right)$ will be maximum.

40 From Stefan's law, $E = \sigma T^4$ or $E \propto T^4$

$$\therefore \quad \frac{E_1}{E_2} = \left(\frac{T_1}{T_2}\right)^4 = \left(\frac{T}{2T}\right)^4$$

or $$\frac{E_1}{E_2} = \frac{1}{16}$$

or $$E_2 = 16E_1 = 16 \text{ times}$$

Specific heat too varies with temperature. As a matter of fact, specific heat is zero at 0K for all the materials.

DAY EIGHTEEN

Electrostatics

Learning & Revision for the Day

- Electric Charge
- Coulomb's Law of Forces between Two Point Charges
- Superposition Principle
- Electric Field
- Motion of A Charged Particle in An Electric Field
- Electric Field due to a Point Charge
- Continuous Charge Distribution
- Electric Dipole
- Electric Flux (ϕ_E)
- Gauss Law
- Electric Potential
- Electric Potential Energy
- Equipotential Surface
- Conductors and Insulators
- Electrical Capacitance
- Capacitor

If the charge in a body does not move, then the fricitional electricity is known as static electricity. The branch of physics which deals with static electricity is called electrostatics.

Electric Charge

Electric charge is the property associated with matter due to which it produces and experiences electric and magnetic effects.

Conservation of Charge

We can neither create nor destroy electric charge. The charge can simply be transferred from one body to another. There are three modes of charge transfer:

(a) By friction (b) By conduction (c) By induction

Quantisation of Charge

Electric charge is quantised. The minimum amount of charge, which may reside independently is the electronic charge e having a value of 1.6×10^{-19} C, i.e. $Q = \pm ne$, where, n is any integer.

Important properties of charges are listed below

- Like charges repel while opposite charges attract each other.
- Charge is invariant i.e. charge does not change with change in velocity.
- According to theory of relativity, the mass, time and length change with a change in velocity but charge does not change.
- A charged body attracts a lighter neutral body.
- Electronic charge is additive, i.e. the total charge on a body is the algebraic sum of all the charges present in different parts of the body. For example, if a body has different charges as $+2q$, $+4q$, $-3q$, $-q$, then the total charge on the body is $+2q$.

- No. of Questions in Exercises (x)—
- No. of Questions Attempted (y)—
- No. of Correct Questions (z)—
  (Without referring Explanations)
- Accuracy Level ($z/y \times 100$)—
- Prep Level ($z/x \times 100$)—

In order to expect good rank in JEE, your Accuracy Level should be above 85 & Prep Level should be above 75.

Coulomb's Law of Forces between Two Point Charges

- If q_1 and q_2 be two stationary point charges in free space separated by a distance r, then the force of attraction / repulsion between them is

$$F = \frac{K|q_1||q_2|}{r^2} = \frac{1}{4\pi\varepsilon_0}\cdot\frac{|q_1||q_2|}{r^2} \qquad \left[K = \frac{1}{4\pi\varepsilon_0}\right]$$

$$= \frac{9\times10^9 \times |q_1||q_2|}{r^2} \qquad [K = 9\times10^9 \text{ N-m}^2/\text{c}^2]$$

- If some dielectric medium is completely filled between the given charges, then the Coulomb's force between them becomes

$$F_m = \frac{1}{4\pi\varepsilon}\frac{q_1q_2}{r^2} = \frac{1}{4\pi\varepsilon_0\varepsilon_r}\frac{q_1q_2}{r^2} \qquad \left[\because \frac{\varepsilon}{\varepsilon_0} = \varepsilon_r \text{ or } k\right]$$

$$= \frac{1}{4\pi k\varepsilon_0}\cdot\frac{q_1q_2}{r^2}$$

Forces between Multiple Charges

When a number of point charges are present in a region then force acting between any two point charges remains unaffected by the presence of other charges and remains same as according to Coulomb's law. If four identical charges of magnitude q each are placed at the four corners of square of side a, then the force on any one charge due to the rest of the three charges is

$$F = \frac{1}{4\pi\varepsilon_0}\frac{q^2}{a^2}(2\sqrt{2}+1)$$

Superposition Principle

It states that, the net force on any one charge is equal to the vector sum of the forces exerted on it by all other charges. If there are four charges q_1, q_2, q_3 and q_4, then the force on q_1 (say) due to q_2, q_3 and q_4 is given by $\mathbf{F}_1 = \mathbf{F}_{12} + \mathbf{F}_{13} + \mathbf{F}_{14}$, where $\mathbf{F}_{12}$ is the force on q_1 due to q_2, $\mathbf{F}_{13}$ that due to q_3 and $\mathbf{F}_{14}$ that due to q_4.

Electric Field

The space surrounding an electric charge q in which another charge q_0 experiences a force of attraction or repulsion, is called the electric field of charge q. The charge q is called the **source charge** and the charge q_0 is called the **test charge**. The test charge must be **negligibly small** so that it does not modify the electric field of the source charge.

Intensity (or Strength) of Electric Field (E)

The intensity of electric field at a point in an electric field is the ratio of the forces acting on the test charge placed at that point to the magnitude of the test charge.

$$\mathbf{E} = \frac{\mathbf{F}}{q_0}, \qquad \text{where, } \mathbf{F} \text{ is the force acting on } q_0.$$

Electric field intensity (**E**) is a vector quantity.

The direction of electric field is same as that of force acting on the positive test charge. Unit of E is NC^{-1} or Vm^{-1}.

Electric Field Lines

An electric field line in an electric field is a smooth curve, tangent to which, at any point, gives the direction of the electric field at that point.

Properties of electric field lines are given below

- Electric field lines come out of a positive charge and go into the negative charge.
- No two electric field lines intersect each other.
- Electric field lines are continuous but they never form a closed loop.
- Electric field lines cannot exist inside a conductor. **Electric shielding** is based on this property.

Motion of a Charged Particle in an Electric Field

Let a charged particle of mass m and charge q, enters the electric field along X-axis with speed u. The electric field E is along Y-axis is given by

$$F_y = qE$$

and force along X-axis remains zero, i.e.

$$F_x = 0$$

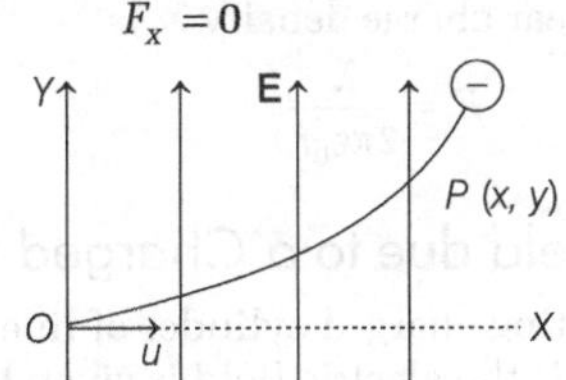

$\therefore$ Acceleration of the particle along Y-axis is given by

$$a_y = \frac{F_y}{m} = \frac{qE}{m}$$

The initial velocity is zero along Y-axis ($u_y = 0$).

$\therefore$ The deflection of charged particle along Y-axis after time t is given by $y = u_y t + \frac{1}{2}a_y t^2$

$$= \frac{qE}{2m}t^2$$

Along X-axis there is no acceleration, so the distance covered by particle in time t along X-axis is given by $x = ut$

Eliminating t, we have

$$y = \left(\frac{qE}{2mu^2}x^2\right)$$

$$y \propto x^2$$

This shows that the path of charged particle in perpendicular field is a **parabola**.

Electric Field due to a Point Charge

1. Electric Field due to a Point Charge at a Distance

Electric field at a distance r from a point charge q is

$$E = \frac{1}{4\pi\varepsilon_0} \cdot \frac{q}{r^2}$$

- If q_1 and q_2 are two like point charges, separated by a distance r, a neutral point between them is obtained at a point distant r_1 from q_1, such that

$$r_1 = \frac{r}{\left[1 + \sqrt{\frac{q_2}{q_1}}\right]}$$

- If q_1 and q_2 are two charges of opposite nature separated by a distance r, a neutral point is obtained in the extended line joining them, at a distance r_1 from q_1, such that,

$$r_1 = \frac{r}{\left[\sqrt{\frac{q_2}{q_1}} - 1\right]}$$

2. Electric Field due to Infinitely Long Uniformly Charged Straight Wire

Electric field at a point situated at a normal distance r, from an infinitely long uniformly charged straight wire having a linear charge density λ, is

$$E = \frac{\lambda}{2\pi\varepsilon_0 r}$$

3. Electric Field due to a Charged Cylinder

- For a conducting charged cylinder of linear charge density λ and radius R, the electric field is given by

$$E = \frac{\lambda}{2\pi\varepsilon_0 r}, \text{ for } r > R,$$

$$E = \frac{\lambda}{2\pi\varepsilon_0 R}, \text{ for } r = R$$

and $E = 0$, for $r < R$

- For a non-conducting charged cylinder, for $r \le R$,

$$E = \frac{\lambda r}{2\pi\varepsilon_0 R^2} \text{ and } E = \frac{\lambda}{2\pi\varepsilon_0 r}, \text{ for } r > R$$

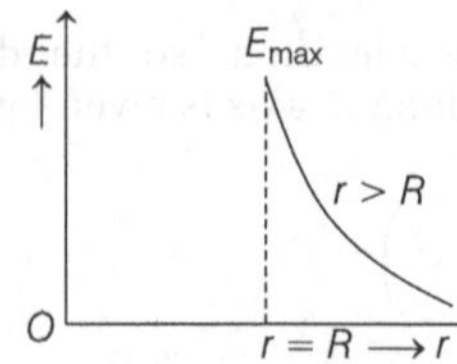

(a) Variation of electric field with distance for conducting cylinder

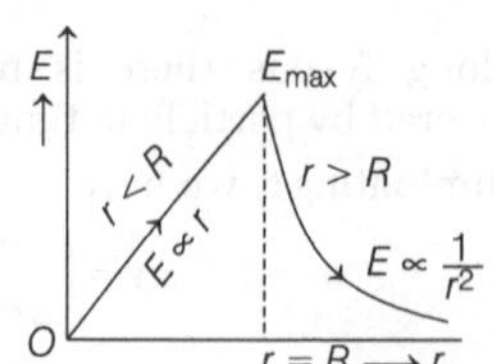

(b) Variation of electric field with distance for non-conducting cylinder

4. Electric Field due to a Uniformly Charged Infinite Plane Sheet

Electric field near a uniformly charged infinite plane sheet having surface charge density σ is given by

$$E = \frac{\sigma}{2\,\varepsilon_0}$$

5. Electric Field due to a Uniformly Charged Thin Spherical Shell

For a charged conducting sphere/shell of radius R and total charge Q, the electric field is given by

***Case* I** $E = 0$, for $r < R$

***Case* II** $E = \dfrac{Q}{4\pi\varepsilon_0 R^2}$, for $r = R$

***Case* III** $E = \dfrac{Q}{4\pi\varepsilon_0 r^2}$, for $r > R$

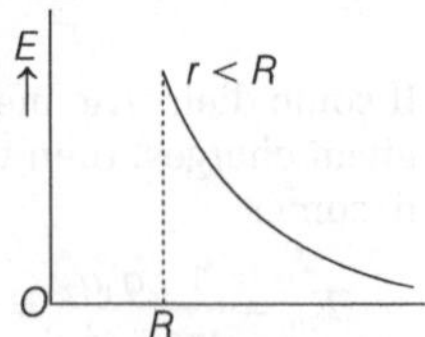

Variation of electric field with distance for uniformly charge spherical shell

Continuous Charge Distribution

The continuous charge distribution may be one dimensional, two dimensional and three-dimensional.

1. **Linear charge density** (λ) If charge is distributed along a line, *i.e.*, straight or curve is called linear charge distribution. The uniform charge distribution q over a length L of the straight rod.

   Then, the linear charge density, $\lambda = \dfrac{q}{L}$

   Its unit is coulomb metre^{-1} (Cm^{-1}).

2. **Surface charge density** (σ) If charge is distributed over a surface is called surface charge density, *i.e.*,

$$\sigma = q/A$$

   Its unit is coulomb m^{-2} (Cm^{-2})

3. **Volume charge density** (ρ) If charge is distributed over the volume of an object, is called volume charge density, *i.e.*, $p = \dfrac{q}{V}$. Its unit is coulomb metre^{-3} (Cm^{-3}).

Electric Dipole

An electric dipole consists of two equal and opposite charges separated by a small distance.

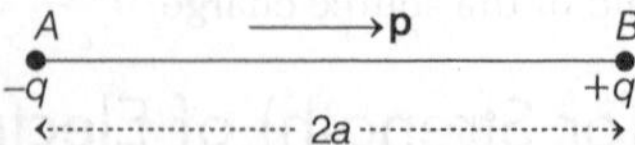

The dipole moment of a dipole is defined as the product of the magnitude of either charges and the distance between them. Therefore, dipole moment

$$\mathbf{p} = q(2\mathbf{a})$$

Electric Field due to a Dipole

- At a point distant r from the centre of a dipole, along its axial line $\mathbf{E} = \frac{1}{4\pi\varepsilon_0} \cdot \frac{2pr}{(r^2 - a^2)^2}$

  [direction of $\mathbf{E}$ is the same as that of $\mathbf{p}$]

  For a short dipole,

  $$\mathbf{E} = \frac{1}{4\pi\varepsilon_0} \frac{2\mathbf{p}}{r^3} \quad [r >> a]$$

- At a point distant r from the centre of a dipole, along its equatorial line

  $$\mathbf{E} = \frac{1}{4\pi\varepsilon_0} \cdot \frac{\mathbf{p}}{(r^2 + a^2)^{3/2}}$$

  [direction of $\mathbf{E}$ is opposite to that of $\mathbf{p}$]

  For a short dipole $\mathbf{E} = \frac{1}{4\pi\varepsilon_0} \frac{\mathbf{p}}{r^3}$ $[r >> a]$

- At a point distant r from the centre of a short dipole, along a line inclined at an angle θ with the dipole axis

  $$E = \frac{1}{4\pi\varepsilon_0} \cdot \frac{p}{r^3} \sqrt{3\cos^2\theta + 1}$$

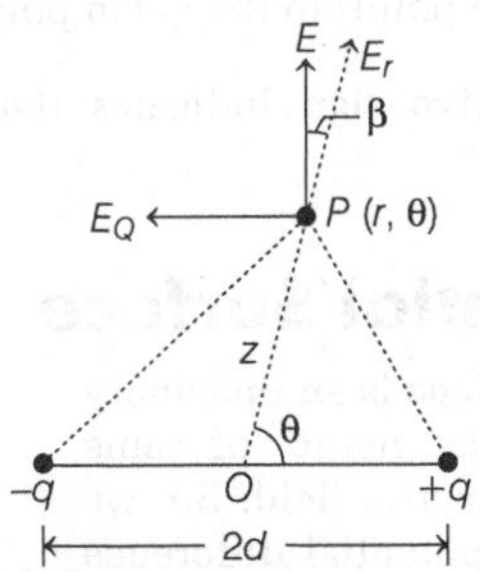

- $\mathbf{E}$ subtends an angle β from $\mathbf{r}$ such that

  $$\tan\beta = \frac{1}{2}\tan\theta$$

Torque on a Dipole in a Uniform Electric Field

When a dipole is placed in an external electric field, making an angle θ with the direction of the uniform electric field E, it experiences a torque given by

$$\tau = qE \times AC$$

$$\tau = \mathbf{p} \times \mathbf{E}$$

$$\tau = pE \sin\theta$$

or $qE \times 2d\sin\theta = (q \times 2d)\, E \sin\theta$

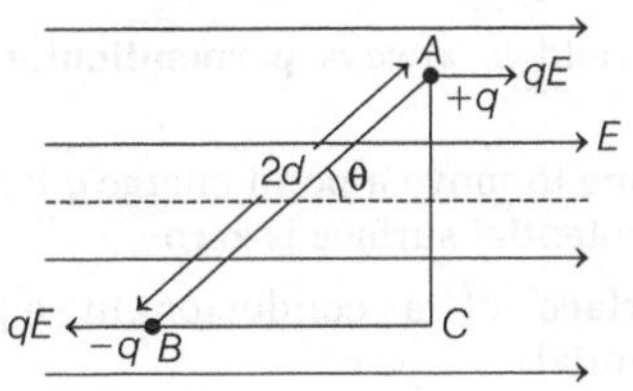

Work Done in Rotating a Dipole

If an electric dipole initially kept in an uniform electric field $\mathbf{E}$, making an angle θ_1, is rotated so as to finally subtend an angle θ_2, then the work done for rotating the dipole is,

$$W = pE(\cos\theta_1 - \cos\theta_2)$$

Potential Energy of a Dipole

It is the amount of work done in rotating an electric dipole from a direction perpendicular to electric field to a particular direction.

Hence, $U = -pE\cos\theta$ or $U = -\mathbf{p}\cdot\mathbf{E}$

Electric Flux (ϕ_E)

It is a measure of the flow of electric field through a surface. It can be defined as the total number of lines of electric field passes through a surface placed perpendicular to direction of field.

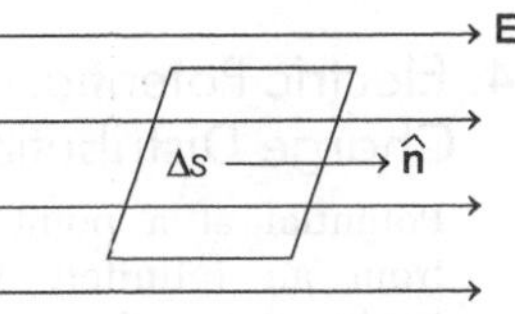

i.e. $\phi_E = \int EdS\cos\theta = \int \mathbf{E}\cdot d\mathbf{S} = \int \mathbf{E}\cdot\hat{\mathbf{n}}\, dS$

Gauss's Law

The total electric flux linked with a closed surface is equal to $\frac{1}{\varepsilon_0}$ times, the net charge enclosed by that surface. Thus,

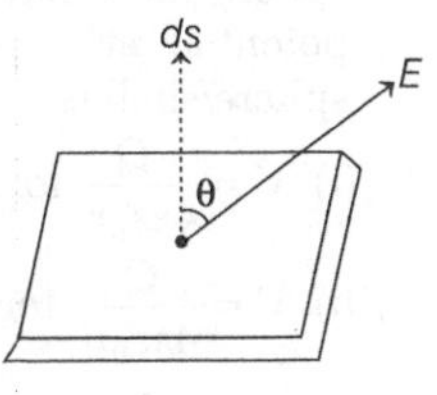

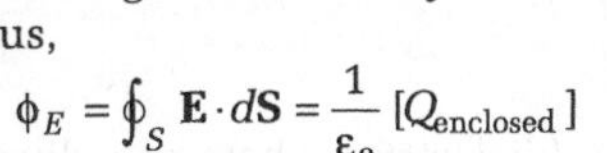

$$\phi_E = \oint_S \mathbf{E}\cdot d\mathbf{S} = \frac{1}{\varepsilon_0}[Q_{\text{enclosed}}]$$

where, $Q_{\text{enclosed}} \sum_{i=1}^{i=n} q_i$ is the algebraic sum of all the charges inside the closed surface.

Electric Potential

The amount of work done in bringing a unit positive charge, without any acceleration, from infinity to that point, along any arbitrary path.

$$V = \frac{W}{q_0}$$

Electric potential is a state function and does not depend on the path followed.

1. Electric Potential Due to a Point Charge

Potential due to a point charge Q, at a distance r is given by $V = \frac{1}{4\pi\varepsilon_0} \cdot \frac{Q}{r}$

2. Electric Potential Due to a System of Charges

If a number of charges $q_1, q_2, q_3, \ldots$ are present in space, then the electric potential at any point will be

$$V = V_1 + V_2 + V_3 + \ldots$$

$$= \frac{1}{4\pi\varepsilon_0}\left[\frac{q_1}{r_1} + \frac{q_2}{r_2} + \frac{q_3}{r_3} + \ldots\right] = \frac{1}{4\pi\varepsilon_0}\sum_{i=1}^{n}\left(\frac{q_i}{r_i}\right)$$

3. Electric Potential Due to an Electric Dipole

At any general point, $V = \frac{1}{4\pi\varepsilon_0}\frac{p\cos\theta}{r^2}$

On the dipole axis, $\theta = 0°$ and $V = \frac{1}{4\pi\varepsilon_0}\cdot\frac{p}{r^2}$

On the equitorial axis, $\theta = 90°$ and $V = 0$

4. Electric Potential due to Some Common Charge Distributions

Potential at a point distant r from an infinitely long wire having linear charge density λ, is

$$V = \frac{\lambda}{2\pi\varepsilon_0}\cdot\ln r$$

For a charged conducting sphere/shell having total charge Q and radius R, the potential at a point distant r from the centre of the sphere/shell is

(i) $V = \frac{Q}{4\pi\varepsilon_0 r}$, for $r > R$ (ii) $V = \frac{1}{4\pi\varepsilon_0}\frac{Q}{R}$, for $r = R$

(iii) $V = \frac{Q}{4\pi\varepsilon_0 R}$, for $r \le R$

For a charged non-conducting (dielectric) sphere of radius R, the charge Q is uniformly distributed over the entire volume.

Hence, (i) $V = \frac{Q}{4\pi\varepsilon_0 r}$, for $r > R$

(ii) $V = \frac{Q}{4\pi\varepsilon_0 R}$, for $r = R$

and (iii) $V = \frac{Q}{4\pi\varepsilon_0}\left[\frac{3R^2 - r^2}{2R^3}\right]$, for $r < R$

At the centre of the sphere $(r = 0)$

$$V = \frac{3Q}{8\pi\varepsilon_0 R} = \frac{3}{2}V_s \qquad \left[V_s = \frac{Q}{4\pi\varepsilon_0 R}\right]$$

Electric Potential Energy

The electric energy of a system of charges is the work that has been done in bringing those charges from infinity to near each other to form the system. For two point charges q_1 and q_2 separated by distance r_{12}, the potential energy is given by $U = \frac{1}{4\pi\varepsilon_0}\frac{q_1 q_2}{r_{12}}$.

In general, for a system of n charges, the electric potential energy is given by

$$U = \frac{1}{2}\Sigma\frac{q_i q_j}{4\pi\varepsilon_0 r_{ij}}, i \ne j$$

$\left[\frac{1}{2}\right.$ is used as each term in summation will appear twice$\left.\right]$

Relation between E and V

Because E is force per unit charge and V is work per unit charge. E and V are related in the same way as work and force.

Work done against the field to take a unit positive charge from infinity (reference point) to the given point $V_P = -\int_{\infty}^{P}\mathbf{E}\cdot d\mathbf{r}$ volt

where, the negative sign indicates that the work is done against the field.

Equipotential Surface

Equipotential surface is an imaginary surface joining the points of same potential in an electric field. So, we can say that the potential difference between any two points on an equipotential surface is **zero**.

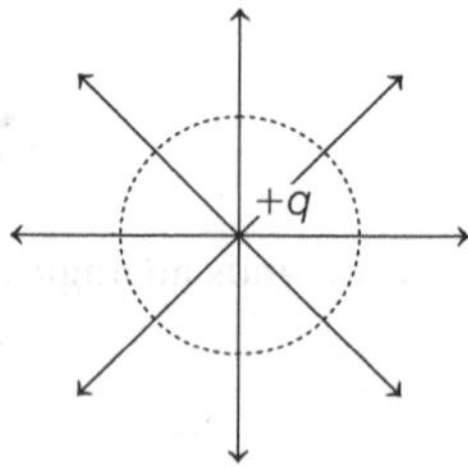

The electric lines of force at each point of an equipotential surface are **normal** to the surface. Figure shows the electric lines of force due to point charge $+q$. The spherical surface will be the equipotential surface and the electrical lines of force emanating from the point charge will be radial and normal to the spherical surface.

Regarding equipotential surface, following points are worth noting

(i) Equipotential surface may be **planar**, solid etc. But equipotential surface can never be **point size.**

(ii) Equipotential surface is single valued. So, equipotential surfaces never **cross** each other.

(iii) Electric field is always **perpendicular** to equipotential surface.

(iv) Work done to move a point charge q between two points on equipotential surface is **zero.**

(v) The **surface** of a conductor in equilibrium is an equipotential.

Conductors and Insulators

Conductors are those materials through which electricity can pass through easily. e.g. metals like copper, silver, iron etc. **Insulators** are those materials through which electricity cannot pass through, e.g. rubber, ebonite, mica etc.

Dielectrics and Polarisation

Dielectrics are insulating materials which transmit electric effect without actually conducting electricity.

e.g. mica, glass, water etc.

When a dielectric is placed in an external electric field, so the molecules of dielectric gain a permanent electric dipole moment. This process is called polarisation.

Electrical Capacitance

Capacitance of a conductor is the amount of charge needed in order to raise the potential of the conductor by unity.

Mathematically, Capacitance $C = \dfrac{Q}{V}$

Sharing of Charges

- Let us have two charged conductors having charges Q_1 and Q_2 (or potentials V_1, V_2 and capacitances C_1, C_2 respectively). If these are joined together. In such a cases

  Common potential, $V = \dfrac{Q_1 + Q_2}{C_1 + C_2} = \dfrac{C_1V_1 + C_2V_2}{C_1 + C_2}$
- During sharing of charges, there is some loss of electrostatic energy, which in turn reappears as heat or light. The loss of electrostatic energy

  $$\Delta U = U_i - U_f = \frac{C_1C_2}{2(C_1 + C_2)}(V_1 - V_2)^2$$
- When charges are shared between any two bodies, their potential become equal. The charges acquired are in the ratio of their capacitances.

Capacitor

A capacitor is a device which stores electrostatic energy. It consists of conductors of any shape and size carrying charges of equal magnitudes and opposite signs and separated by an insulating medium.

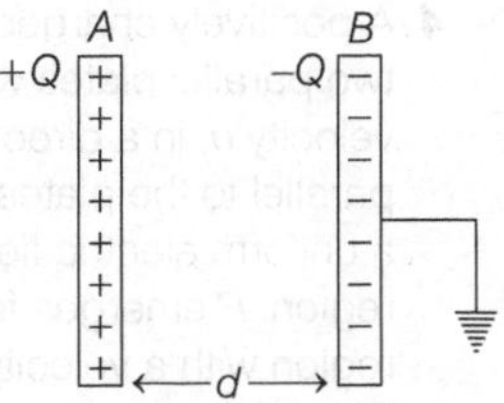

There are two types of combination of capacitors:

1. Series Grouping

In a series arrangement, $V = V_1 + V_2 + V_3 + \dots$

and $V_1 : V_2 : V_3 \dots = \dfrac{1}{C_1} : \dfrac{1}{C_2} : \dfrac{1}{C_3} : \dots$

The equivalent capacitance C_s is given by

$$\frac{1}{C_s} = \frac{1}{C_1} + \frac{1}{C_2} + \frac{1}{C_3} + \dots = \sum_{i=1}^{i=n} \frac{1}{C_i}$$

2. Parallel Grouping

In a parallel arrangement, $Q = Q_1 + Q_2 + Q_3 + \dots$

and $Q_1 : Q_2 : Q_3 \dots = C_1 : C_2 : C_3 \dots$

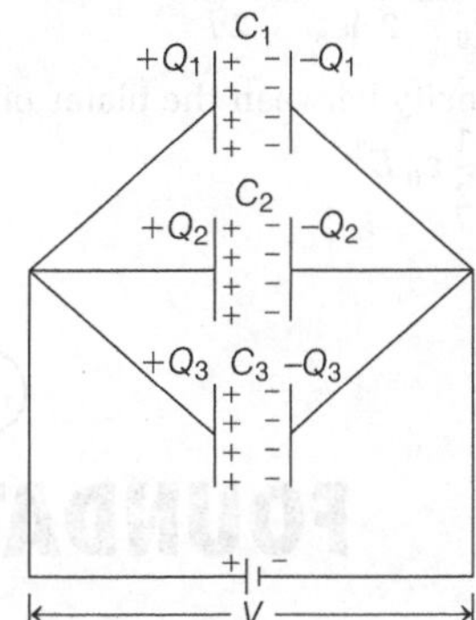

The equivalent capacitance is given by

$$C_p = C_1 + C_2 + C_3 + \dots = \sum_{i=1}^{i=n} C_i$$

Capacitance of a Parallel Plate Capacitor

1. Capacitor without Dielectric Medium between the Plates

If the magnitude of charge on each plate of a parallel plate capacitor be Q and the overlapping area of plates be A, then

- Electric field between the plates,

  $$E = \frac{\sigma}{\varepsilon_0} = \frac{Q}{\varepsilon_0 A}$$
- Potential difference between the plates

  $V = E \cdot d = \dfrac{\sigma d}{\varepsilon_0} = \dfrac{Qd}{\varepsilon_0 A}$, where d = separation between the two plates.
- Capacitance, $C = \dfrac{Q}{V} = \dfrac{\varepsilon_0 A}{d}$

2. Capacitor with Dielectric Medium between the Plates

- If a dielectric medium of dielectric constant K is completely filled between the plates of a capacitor, then its capacitance becomes,

$$C' = \frac{K\varepsilon_0 A'}{d} = KC_0 \qquad \left[\text{where, } C_0 = \frac{\varepsilon_0 A'}{d}\right]$$

- If a dielectric slab/sheet of thickness t (where, $t < d$) is introduced between the plates of the capacitor, then

$$C' = \frac{\varepsilon_0 A}{\left(d - t + \dfrac{t}{K}\right)}$$

- Magnitude of the attractive force between the plates of a parallel plate capacitor is given by

$$F = \frac{\sigma^2 A}{2\varepsilon_0} = \frac{Q^2}{2A\varepsilon_0} = \frac{CV^2}{2d}$$

- The energy density between the plates of a capacitor

$$u = \frac{U}{\text{Volume}} = \frac{1}{2}\varepsilon_0 E^2$$

Energy Stored in a Capacitor

If a capacitor of capacity C is charged to a potential V, the electrostatic energy stored in it is,

$$U = \frac{1}{2}CV^2 = \frac{1}{2}QV = \frac{1}{2}\frac{Q^2}{C}$$

Energy Loss During Parallel Combination

When two capacitor of C_1 capacitance charge to potential V_1, whereas another of C_2 charge to potential of V_2, then after parallel combination.

$$\text{Loss in energy} = \frac{1}{2}\frac{C_1 C_2}{C_1 + C_2}(V_1 - V_2)^2$$

DAY PRACTICE SESSION 1

FOUNDATION QUESTIONS EXERCISE

1 Two balls of same mass and carrying equal charge are hung from a fixed support of length l. At electrostatic equilibrium, assuming that angles made by each thread is small, the separation, X between the balls is proportional → JEE Main (Online) 2013

(a) l (b) l^2 (c) $l^{2/3}$ (d) $l^{1/3}$

2 In the basic CsCl crystal structure, $^+$Cs and $^-$Cl ions are arranged in a bcc configuration as shown below. The net electrostatic force exerted by the 8 Cs^+ on the the Cl^- ion is

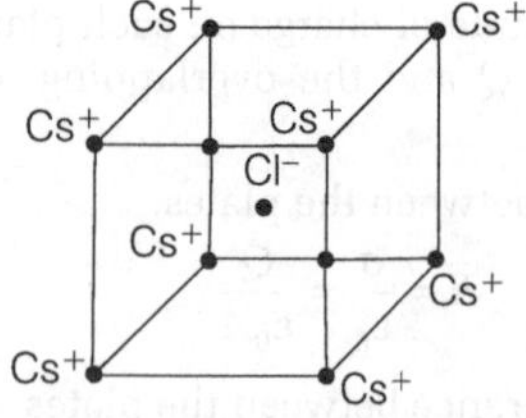

(a) $\dfrac{1}{4\pi\varepsilon_0}\dfrac{4e^2}{3a^2}$ (b) $\dfrac{1}{4\pi\varepsilon_0}\dfrac{16e^2}{3a^2}$ (c) $\dfrac{1}{4\pi\varepsilon_0}\dfrac{32a^2}{3a^2}$ (d) zero

3 A long cylindrical shell carries positive surface charge σ in the upper half and negative surface charge $-\sigma$ in the lower half. The electric field lines around the cylinder will look like figure given in (figures are schematic and not drawn to scale) → JEE Main 2015

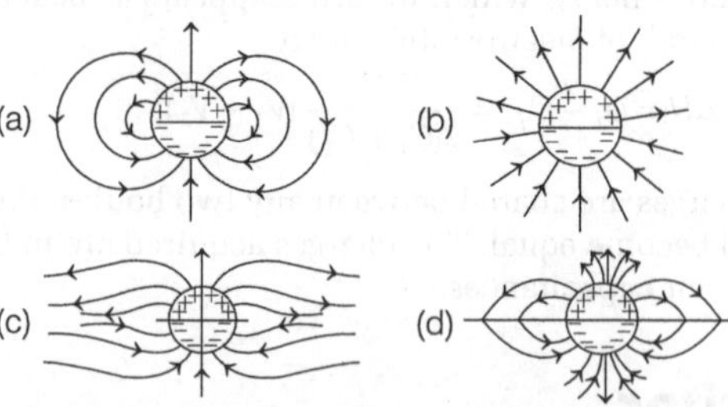

4 A positively charged particle P enters the region between two parallel plates with a velocity u, in a direction parallel to the plates. There is a uniform electric field in this region. P emerges from this region with a velocity v. Taking C as a constant, v will depend on u as

(a) $v = Cu$ (b) $v = \sqrt{u^2 + Cu}$

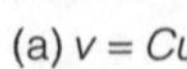

(c) $v = \sqrt{u^2 + \dfrac{C}{u}}$ (d) $v = \sqrt{u^2 + \dfrac{C}{u^2}}$

5 An infinite line charge produces a field of 9×10^4 N/C at a distance of 2 cm. Calculate the linear charge density.
(a) 10^{-3} C/m (b) 10^{-4} C/m (c) 10^{-5}C/m (d) 10^{-7}C/m

6 In a uniformly charged sphere of total charge Q and radius R, the electric field E is plotted as function of distance from the centre. The graph which would correspond to the above will be → AIEEE 2012

(a) 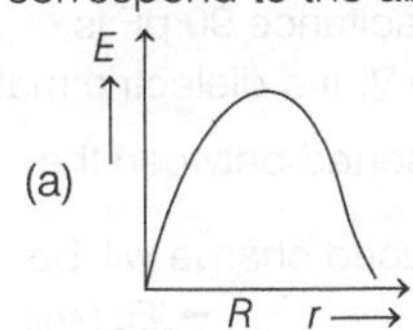

(b) 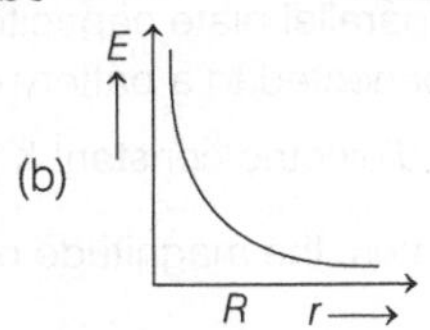

(c) 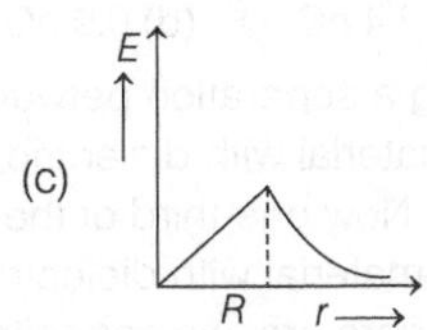

(d) 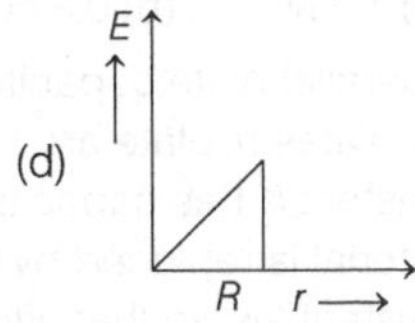

7 Let $\rho(r) = \frac{Q}{\pi R^4} r$ be the charge density distribution for a solid sphere of radius R and total charge Q. For a point P inside the sphere at distance r_1 from the centre of the sphere, the magnitude of electric field is → AIEEE 2009

(a) zero (b) $\frac{Q}{4\pi\varepsilon_0 r_1^2}$ (c) $\frac{Q r_1^2}{4\pi\varepsilon_0 R^4}$ (d) $\frac{Q r_1^2}{3\pi\varepsilon_0 R^4}$

8 Two points dipoles of dipole moment $\mathbf{P}_1$ and $\mathbf{P}_2$ are at a distance x from each other an $\mathbf{P}_1 \parallel \mathbf{P}_2$. The force between the dipoles is → JEE Main (Online) 2013

(a) $\frac{1}{4\pi\varepsilon_0}\frac{4P_3P_2}{x^3}$ (b) $\frac{1}{4\pi\varepsilon_0}\frac{3P_3P_2}{x^3}$
(c) $\frac{1}{4\pi\varepsilon_0}\frac{6P_1P_2}{x^3}$ (d) $\frac{1}{4\pi\varepsilon_0}\frac{3P_1P_2}{x^3}$

9 If the electric flux entering and leaving an enclosed surface respectively are ϕ_1 and ϕ_2, the electric charge inside the surface will be

(a) $\frac{\phi_2 - \phi_1}{\varepsilon_0}$ (b) $\frac{\phi_1 + \phi_2}{\varepsilon_0}$ (c) $\frac{\phi_1 - \phi_2}{\varepsilon_0}$ (d) $\varepsilon_0(\phi_1 + \phi_2)$

10 A cylinder of radius R and length L is placed in a uniform electric field E parallel to the axis of the cylinder, the total electric flux for the surface of the cylinder is

(a) $2\pi R^2 E$ (b) $\frac{\pi R^2}{E}$ (c) $\frac{\pi R^2 + \pi R^2}{E}$ (d) zero

11 A large insulated sphere of radius r, charged with Q units of electricity, is placed in contact with a small insulated uncharged sphere of radius r' and is then separated. The charge on the smaller sphere will now be

(a) $Q(r + r')$ (b) $Q(r - r')$
(c) $\frac{Q}{r' + r}$ (d) $\frac{Qr'}{r' + r}$

12 Consider a finite insulated, uncharged conductor placed near a finite positively charged conductor. The uncharged body must have a potential → JEE Main (Online) 2013

(a) less than the charged conductor and more than at infinity
(b) more than the charged conductor and less than at infinity
(c) more than the charged conductor and more than at infinity
(d) less than the charged conductor and less than at infinity

13 The electrostatic potential inside a charged spherical ball is given by $\phi = ar^2 + b$, where r is the distance from the centre where a, b are constants. Then the charge density inside the ball is → AIEEE 2011

(a) $-6a\,\varepsilon_0 r$ (b) $-24\,\pi a\varepsilon_0$
(c) $-6a\varepsilon_0$ (d) $-24\,\pi a\varepsilon_0 r$

14 Two points P and Q are maintained at the potentials of 10 V and −4 V, respectively. The work done in moving 100 electrons from P to Q is → AIEEE 2009

(a) -19×10^{-17} J (b) 9.60×10^{-17} J
(c) -2.24×10^{-16} J (d) 2.24×10^{-16} J

15 The variation of electric potential with distance from a fixed point is shown in the figure. What is the value of electric field at $x = 2$ m?

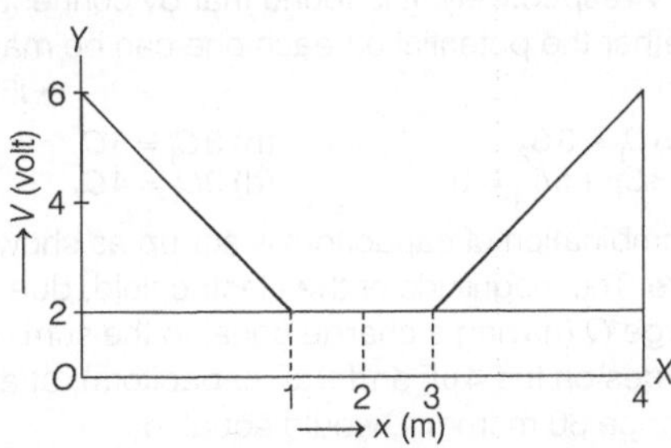

(a) Zero (b) 6/2
(c) 6/1 (d) 6/3

16 The electric potential V at any point (x, y, z) in space is given by $V = 4x^2$ volt. The electric field at (1, 0, 2) m in Vm^{-1} is

(a) 8, along the positive x-axis
(b) 8, along the negative x-axis
(c) 16, along the x-axis
(d) 16, along the z-axis

17 Assume that an electric field $\mathbf{E} = 30x^2\,\hat{\mathbf{i}}$ exists in space. Then, the potential difference $V_A - V_O$, where V_O is the potential at the origin and V_A the potential at $x = 2$ m/s → JEE Main 2014

(a) 120 J/C (b) −120 J/C
(c) −80 J/C (d) 80 J/C

18 Figure shows some equipotential lines distributed in space. A charged object is move from point A to point B.

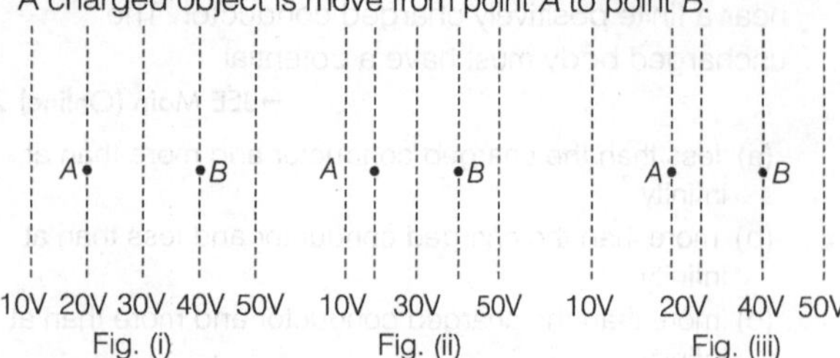

(a) The work done in Fig. (i) is the greatest
(b) The work done in Fig. (ii) is least
(c) The work done is the same in Fig.(i), Fig. (ii) and Fig. (iii)
(d) The work done in Fig. (iii) is greater that Fig. (ii) but equal to that in

19 Two condensers C_1 and C_2 in a circuit are joined as shown in the figure. The potential of point A is V_1 and that of B is V_2. The potential of point D will be

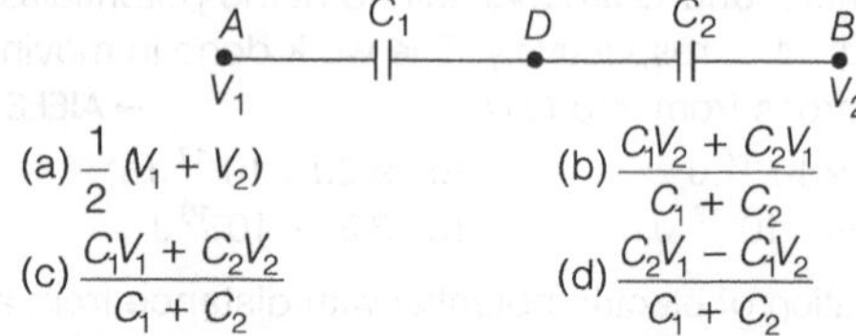

(a) $\frac{1}{2}(V_1 + V_2)$ (b) $\frac{C_1V_2 + C_2V_1}{C_1 + C_2}$
(c) $\frac{C_1V_1 + C_2V_2}{C_1 + C_2}$ (d) $\frac{C_2V_1 - C_1V_2}{C_1 + C_2}$

20 Two capacitors C_1 and C_2 are charged to 120 V and 200 V respectively. It is found that by connecting them together the potential on each one can be made zero. Then → JEE Main 2013

(a) $5C_1 = 3C_2$ (b) $3C_1 = 5C_2$
(c) $3C_1 + 5C_2 = 0$ (d) $9C_1 = 4C_2$

21 A combination of capacitors is set-up as shown in the figure. The magnitude of the electric field, due to a point charge Q (having a charge equal to the sum of the charges on the 4 μF and 9 μF capacitors), at a point distance 30 m from it, would equal to

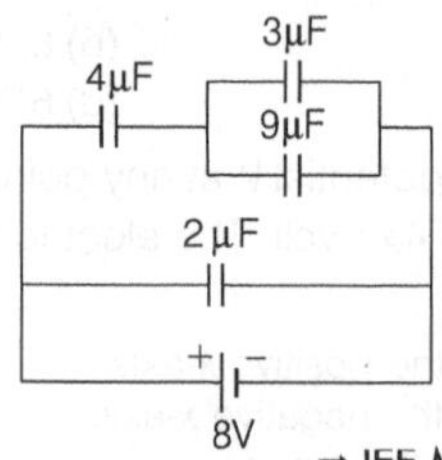

→ JEE Main 2016 (Offline)

(a) 240 N/C (b) 360 N/C (c) 420 N/C (d) 480 N/C

22 An uncharged parallel plate capacitor having a dielectric of constant K is connected to a similar air cored parallel capacitor charged to a potential V. The two share the charge and the common potential is V'. The dielectric constant K is

(a) $\frac{V' - V}{V' + V}$ (b) $\frac{V' - V}{V'}$ (c) $\frac{V' - V}{V}$ (d) $\frac{V - V'}{V'}$

23 A parallel plate capacitor is made of two circular plates separated by a distance of 5 mm and with a dielectric of dielectric constant 2.2 between them. When the electric field in the dielectric is 3×10^4 V/m, the charge density of the positive plate will be close to → JEE Main 2014

(a) 6×10^{-7} C/m^2 (b) 3×10^{-7} C/m^2
(c) 3×10^4 C/m^2 (d) 6×10^4 C/m^2

24 A parallel plate capacitor of capacitance 90 pF is connected to a battery of emf 20 V. If a dielectric material of dielectric constant $K = \frac{5}{3}$ is inserted between the plates, the magnitude of the induced charge will be → JEE Main 2018

(a) 1.2 nC (b) 0.3 nC (c) 2.4 nC (d) 0.9 nC

25 A parallel plate capacitor having a separation between the plates d, plate area A and material with dielectric constant K has capacitance C_0. Now one-third of the material is replaced by another material with dielectric constant $2K$, so that effectively there are two capacitors one with area $\frac{1}{3}A$, dielectric constant $2K$ and another with area $\frac{2}{3}A$ and dielectric constant K. If the capacitance of this new capacitor is C then C/C_0 is → JEE Main (Online) 2013

(a) 1 (b) $\frac{4}{3}$ (c) $\frac{2}{3}$ (d) $\frac{1}{3}$

26 A parallel plate capacitor of area 60 cm^2 and separation 3 mm is charged initially to 90 μC. If the medium between the plate gets slightly conducting and the plate loses the charge initially at the rate of 2.5×10^{-8} C/s, then what is the magnetic field between the plates? → JEE Main (Online) 2013

(a) 2.5×10^{-8} T (b) 2.0×10^{-7} T
(c) 1.63×10^{-11} T (d) Zero

27 **Case I** Identical point charges of magnitude Q are kept at the corners of a regular pentagon of side a.
Case II One charge is now removed. Match the following for above two cases.

Column I	Column II
A. Electric field as the centre of pentagon in case I	1. $\frac{1}{4\pi\varepsilon_0}\frac{Q \times 5}{a}$
B. Electric potential at the centre of pentagon in case I	2. Zero
C. Electric field at the centre of pentagon in case II	3. $\frac{1}{4\pi\varepsilon_0}\frac{Q}{a^2}$
D. Electric potential at the centre of pentagon in case II	4. $\frac{1}{4\pi\varepsilon_0}\frac{Q}{a} \times 4$

Codes

	A	B	C	D		A	B	C	D
(a)	2	1	3	4	(b)	1	2	3	4
(c)	2	3	4	2	(d)	4	1	2	4

Direction (Q. Nos. 28-29) *Each of these questions contains two statements : Statement I and Statement II. Each of these questions also has four alternative choices, only one of which is the correct answer. You have to select one of the codes (a), (b), (c), (d) given below:*

(a) Statement I is true, Statement II is false
(b) Statement I is true, Statement II is true; Statement II is not the correct explanation of Statement I
(c) Statement I is true, Statement II is true; Statement II is the correct explanation of Statement I
(d) Statement I is false, Statement II is true

28 **Statement I** For a charged particle moving from point P to point Q, the net work done by an electrostatic field on the particle is independent of the path connecting point P to point Q.

Statement II The net work done by a conservative force on an object moving along a closed loop is zero.

29 **Statement I** No work is required to be done to move a test charge between any two points on an equipotential surface.

Statement II Electric lines of force at the equipotential surfaces are mutually perpendicular to each other.
→ JEE Main (Online) 2013

DAY PRACTICE SESSION 2

PROGRESSIVE QUESTIONS EXERCISE

1 A capacitance of 2μF is required in an electrical circuit across a potential difference of 1kV. A large number of 1μF capacitors are available which can withstand a potential difference of not more than 300 V. The minimum number of capacitors required to achieve this is
→ JEE Main 2017 (Offline)

(a) 16 (b) 24 (c) 32 (d) 2

2 Three concentric metal shells A, B and C of respective radii a, b and c $(a < b < c)$ have surface charge densities $+\sigma$, $-\sigma$ and $+\sigma$, respectively. The potential of shell B is
→ JEE Main 2018

(a) $\frac{\sigma}{\varepsilon_0}\left[\frac{a^2-b^2}{a}+c\right]$ (b) $\frac{\sigma}{\varepsilon_0}\left[\frac{a^2-b^2}{b}+c\right]$

(c) $\frac{\sigma}{\varepsilon_0}\left[\frac{b^2-c^2}{b}+a\right]$ (d) $\frac{\sigma}{\varepsilon_0}\left[\frac{b^2-c^2}{c}+a\right]$

3 An electric dipole has a fixed dipole moment $\mathbf{p}$, which makes angle θ with respect to X-axis. When subjected to an electric field $\mathbf{E}_1 = E\,\hat{\mathbf{i}}$, it experiences a torque $\mathbf{E}_1 = \tau\,\hat{\mathbf{k}}$. When subjected to another electric field $\mathbf{E}_2 = \sqrt{3}E_1\,\hat{\mathbf{j}}$, it experiences a torque $\mathbf{T}_2 = -\mathbf{T}_1$. The angle θ is
→ JEE Main 2017 (Offline)

(a) 45° (b) 60° (c) 90° (d) 30°

4 The region between two concentric spheres of radii a and b, respectively (see the figure), has volume charge density $\rho = \frac{A}{r}$, where A is a constant and r is the distance from the centre. At the centre of the spheres is a point charge Q. The value of A such that the electric field in the region between the spheres will be constant is
→ JEE Main 2016 (Offline)

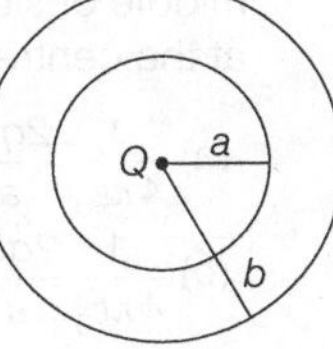

(a) $\frac{Q}{2\pi a^2}$ (b) $\frac{Q}{2\pi(b^2-a^2)}$

(c) $\frac{2Q}{\pi(a^2-b^2)}$ (d) $\frac{2Q}{\pi a^2}$

5 A charge Q is placed at each of the opposite corners of a square. A charge q is placed at each of the other two corners. If the net electrical force on Q is zero, then $\frac{Q}{q}$ equals
→ AIEEE 2009

(a) $-2\sqrt{2}$ (b) -1 (c) 1 (d) $-\frac{1}{\sqrt{2}}$

6 In the given circuit, charge Q_2 on the 2μF capacitor changes as C is varied from 1μF to 3μF. Q_2 as a function of C is given properly by (figures are drawn schematically and are not to scale)
→ JEE Main 2015

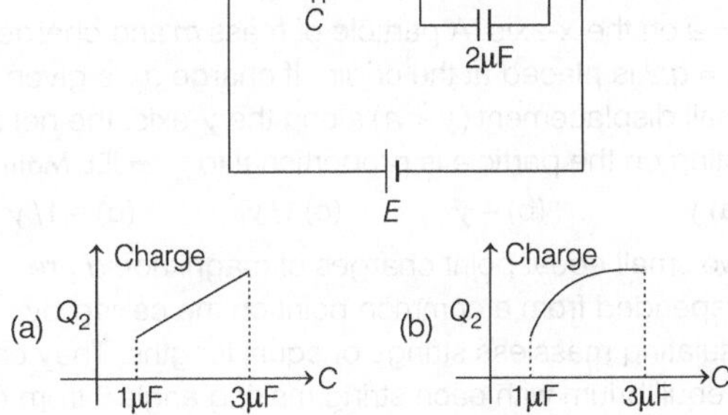

(c) 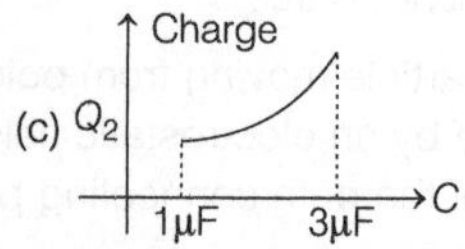

(d) 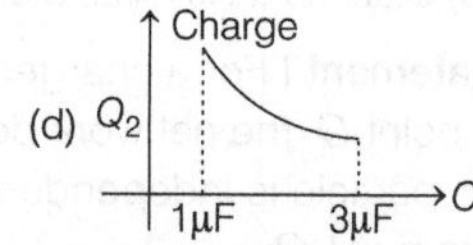

7 A uniform electric field **E** exists between the plates of a charged condenser. A charged particle enters the space between the plates and perpendicular to **E**. The path of the particle between the plates is a → JEE Main (Online) 2013

(a) straight line (b) hyperbola
(c) parabola (d) circle

8 The surface charge density of a thin charged disc of radius R is σ. The value of the electric field at the centre of the disc is $\frac{\sigma}{2\varepsilon_0}$. With respect to the field at the centre, the electric field along the axis at a distance R from the centre of the disc → JEE Main (Online) 2013

(a) reduces by 71% (b) reduces by 29.3%
(c) reduces by 9.7% (d) reduces by 14.6%

9 A uniformly charged solid sphere of radius R has potential V_0 (measured with respect to ∞) on its surface. For this sphere, the equipotential surfaces with potentials $\frac{3V_0}{2}, \frac{5V_0}{4}, \frac{3V_0}{4}$ and $\frac{V_0}{4}$ have radius R_1, R_2, R_3, and R_4 respectively. Then, → JEE Main 2015

(a) $R_1 = 0$ and $R_2 > (R_4 - R_3)$
(b) $R_1 \neq 0$ and $(R_2 - R_1) > (R_4 - R_3)$
(c) $R_1 = 0$ and $R_2 < (R_4 - R_3)$
(d) $2R < R_4$

10 A charge Q is uniformly distributed over a long rod AB of length L as shown in the figure. The electric potential at the point O lying at distance → JEE Main 2013

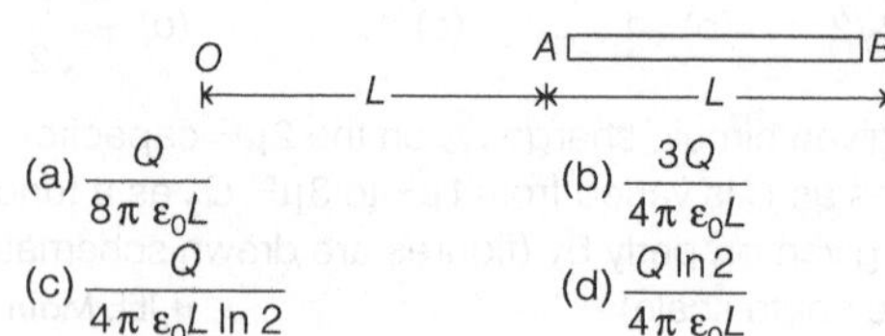

(a) $\frac{Q}{8\pi\varepsilon_0 L}$ (b) $\frac{3Q}{4\pi\varepsilon_0 L}$
(c) $\frac{Q}{4\pi\varepsilon_0 L \ln 2}$ (d) $\frac{Q \ln 2}{4\pi\varepsilon_0 L}$

11 Two charges, each equal to q, are kept at $x = -a$ and $x = a$ on the x-axis. A particle of mass m and charge $q_0 = q/2$ is placed at the origin. If charge q_0 is given a small displacement $(y < a)$ along the y-axis, the net force acting on the particle is proportional to → JEE Main 2013

(a) y (b) $-y$ (c) $1/y$ (d) $-1/y$

12 Two small equal point charges of magnitude q are suspended from a common point on the ceiling by insulating massless strings of equal lengths. They come to equilibrium with each string making angle θ from the vertical. If the mass of each charge is m, then the electrostatic potential at the centre of line joining them will be $\left(\frac{1}{4\pi\varepsilon_0} = k\right)$. → JEE Main (Online) 2013

(a) $2\sqrt{k\,mg\tan\theta}$ (b) $\sqrt{k\,mg\tan\theta}$
(c) $4\sqrt{k\,mg/\tan\theta}$ (d) $4\sqrt{k\,mg\tan\theta}$

13 A point charge of magnitude $+1\,\mu C$ is fixed at (0, 0, 0). An isolated uncharged spherical conductor, is fixed with its centre at (4, 0, 0). The potential and the induced electric field at the centre of the sphere is → JEE Main (Online) 2013

(a) 1.8×10^5 V and -5.625×10^6 V/m
(b) 0 V and 0 V / m
(c) 2.25×10^3 V and 5.625×10^2 V/m
(d) 2.25×10^5 V and 0 V/m

14 Two identical charged spheres are suspended by strings of equal lengths. The strings make an angle of 30° with each other. When suspended in a liquid of density 0.8 g cm^{-3}, the angle remains the same. If density of the material of the sphere is 16 gcm^{-3}, the dielectric constant of the liquid is → AIEEE 2010

(a) 4 (b) 3 (c) 2 (d) 1

15 A circuit is shown in figure for which $C_1 = (3 \pm 0.011)\,\mu F$, $C_2 = (5 \pm 0.01)\,\mu F$ and $C_3 = (1 \pm 0.01)\,\mu F$. If C is the equivalent capacitance across AB, then C is given by

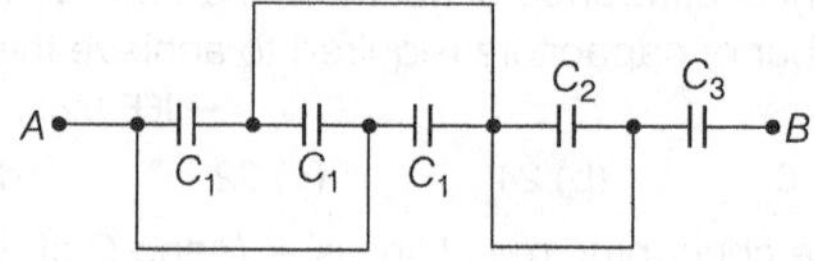

(a) $(0.9 \pm 0.114)\,\mu F$ (b) $(0.9 \pm 0.01)\,\mu F$
(c) $(0.9 \pm 0.023)\,\mu F$ (d) $(0.9 \pm 0.09)\,\mu F$

16 A thin semi-circular ring of radius r has a positive charge q distributed uniformly over it. The net field **E** at the centre O is → AIEEE 2010

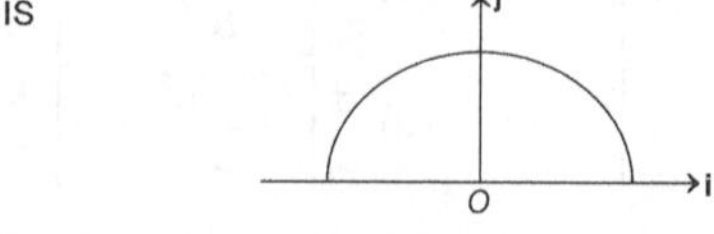

(a) $\frac{q}{4\pi^2\varepsilon_0 r^2}\,\mathbf{j}$ (b) $-\frac{q}{4\pi^2\varepsilon_0 r^2}\,\mathbf{j}$
(c) $-\frac{q}{2\pi^2\varepsilon_0 r^2}\,\mathbf{j}$ (d) $\frac{q}{2\pi^2\varepsilon_0 r^2}\,\mathbf{j}$

17 Two positive charges of magnitude q are placed at the ends of a side 1 of a square of side $2a$. Two negative charges of the same magnitude are kept at the other corners. Starting from rest, if a charge Q moves from the middle of side 1 to the centre of square, its kinetic energy at the centre of square is → AIEEE 2011

(a) $\frac{1}{4\pi\varepsilon_0}\frac{2qQ}{a}\left(1 - \frac{1}{\sqrt{5}}\right)$ (b) zero
(c) $\frac{1}{4\pi\varepsilon_0}\frac{2qQ}{a}\left(1 + \frac{1}{\sqrt{5}}\right)$ (d) $\frac{1}{4\pi\varepsilon_0}\frac{2qQ}{a}\left(1 - \frac{2}{\sqrt{5}}\right)$

18 Two identical charged spheres suspended from a common point by two massless strings of length l are initially a distance d ($d << l$) apart because of their mutual repulsion. The charge begins to leak from both the spheres at a constant rate. As a result charges approach each other with a velocity v. Then as a function of distance x between them, → AIEEE 2011

(a) $v \propto x^{-1}$ (b) $v \propto x^{1/2}$ (c) $v \propto x$ (d) $v \propto x^{-1/2}$

19 Combination of two identical capacitors, a resistor R and a DC voltage source of voltage 6 V is used in an experiment on C-R circuit. It is found that for a parallel combination of the capacitor, the time in which the voltage of the fully charged combination reduces to half its original voltage is 10 s. For series combination, the time needed for reducing the voltage of the fully charged series combination by half is → AIEEE 2011

(a) 20 s (b) 10 s (c) 5 s (d) 2.5 s

20 A resistor R and 2 μF capacitor in series is connected through a switch to 200 V direct supply. Across the capacitor is a neon bulb that lights up at 120 V. Calculate the value of R to make the bulb light up 5 s after the switch has been closed (take, $\log_{10} 2.5 = 0.4$) → AIEEE 2011

(a) $1.7 \times 10^5 \,\Omega$ (b) $2.7 \times 10^6 \,\Omega$
(c) $3.3 \times 10^7 \,\Omega$ (d) $1.3 \times 10^4 \,\Omega$

21 Let there be a spherically symmetric charge distribution with charge density varying as $\rho(r) = \rho_0 \left(\frac{5}{4} - \frac{r}{R}\right)$ upto $r = R$, and $\rho(r) = 0$ for $r > R$, where, r is the distance from the origin. The electric field at a distance $r (r < R)$ from the origin is given by → AIEEE 2010

(a) $\frac{4\pi\rho_0 r}{3\varepsilon_0}\left(\frac{5}{3} - \frac{r}{R}\right)$ (b) $\frac{\rho_0 r}{4\varepsilon_0}\left(\frac{5}{3} - \frac{r}{R}\right)$

(c) $\frac{4\rho_0 r}{3\varepsilon_0}\left(\frac{5}{4} - \frac{r}{R}\right)$ (d) $\frac{\rho_0 r}{3\varepsilon_0}\left(\frac{5}{4} - \frac{r}{R}\right)$

22 Let C be the capacitance of a capacitor discharging through a resistor R. Suppose t_1 is the time taken for the energy stored in the capacitor to reduce to half its initial value and t_2 is the time taken for the charge to reduce to one-fourth its initial value. Then, the ratio t_1 / t_2 will be → AIEEE 2010

(a) 1 (b) 1/2
(c) 1/4 (d) 2

23 The question has statement I and statement II. Of the four choices given after the statements, choose the one that best describes the two statements.

An insulating solid sphere of radius R has a uniform positive charge density ρ. As a result of this uniform charge distribution, there is a finite value of electric potential, at the surface of the sphere, at the surface of the sphere and also at a point outside the sphere. The electric potential at infinite is zero.

Statement I When a charge q is taken from the centre of the surface of the sphere its potential energy changes by $\frac{q\rho}{3\varepsilon_0}$.

Statement II The electric field at a distance $r (r < R)$ from the centre of the sphere is $\frac{\rho r}{3\varepsilon_0}$. → AIEEE 2012

(a) Statement I is false, Statement II is true
(b) Statement I is true, Statement II is false
(c) Statement I is true, Statement II is true, Statement II is the correct explanation for Statement II
(d) Statement I is true, Statement II is true, Statement II is not the correct explanation of Statement I

ANSWERS

SESSION 1	**1** (d)	**2** (d)	**3** (a)	**4** (d)	**5** (d)	**6** (c)	**7** (c)	**8** (c)	**9** (d)	**10** (d)
	11 (d)	**12** (a)	**13** (c)	**14** (d)	**15** (a)	**16** (b)	**17** (c)	**18** (c)	**19** (b)	**20** (b)
	21 (c)	**22** (d)	**23** (a)	**24** (a)	**25** (b)	**26** (d)	**27** (a)	**28** (b)	**29** (b)	
SESSION 2	**1** (c)	**2** (b)	**3** (b)	**4** (a)	**5** (a)	**6** (d)	**7** (c)	**8** (a)	**9** (c,d)	**10** (d)
	11 (a)	**12** (c)	**13** (c)	**14** (c)	**15** (c)	**16** (c)	**17** (a)	**18** (d)	**19** (d)	**20** (b)
	21 (b)	**22** (c)	**23** (a)							

Hints and Explanations

SESSION 1

1 $F = \frac{1}{4\pi\varepsilon_0}\frac{q^2}{x^2} = T\sin\theta$

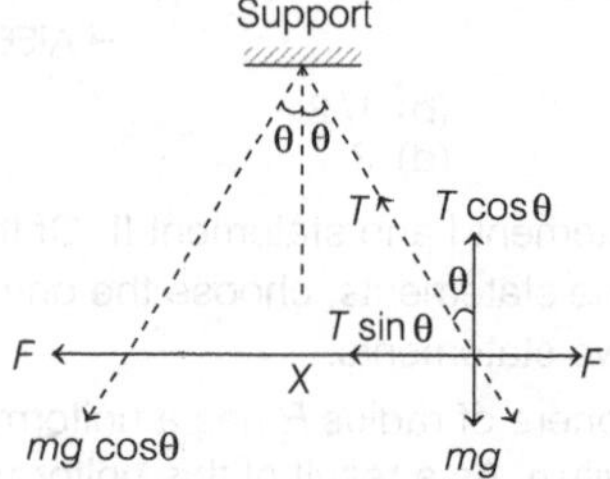

The resultant of components $mg\cos\theta$ and force of repulsion balances, the tension in the string for the equilibrium massive change.

$$T\cos\theta = mg$$

$$\tan\theta = \frac{1}{4\pi\varepsilon_0}\cdot\frac{q^2}{x^2 mg}$$

$$\frac{x}{2l} = \frac{q^2}{4\pi\varepsilon_0 x^2 mg}$$

$$\Rightarrow \quad \frac{x}{2l} \propto \frac{ql}{x^2} \text{ or } x^3 \propto l \text{ or } x \propto l^{1/3}$$

Thus, we find the separation between the balls is proportional to $l^{1/3}$, where, l is length of string.

2. By symmetry resultant force applied by eight charges on corners is zero.

3 Field lines should originate from positive charge and terminate to negative charge. Thus, (b) and (c) are not possible. Electric field lines cannot form corners as shown in (d). Thus, correct option is (a).

4 $v_x = u, t = \frac{l}{u}, \; a_y = \frac{Eq}{m} = a$

$$v_y = at = \frac{al}{u}$$

As, $$v^2 = v_x^2 + v_y^2$$

$$v^2 = u^2 + \frac{a^2 l^2}{u^2}$$

$$= u^2 + \frac{C}{u^2},$$

$$v = \sqrt{u^2 + \frac{C}{u^2}}$$

5 Let λ be the linear charge density.

Given, distance $r = 2\text{ cm} = 2\times10^{-2}$ m

Electric field $E = 9\times10^4$ N/C

Using the formula of electric field due to an infinite line charge.

$$E = \frac{\lambda}{2\pi\varepsilon_0 r}$$

Dividing and multiplying by 2 to get $\frac{1}{4\pi\varepsilon_0}$ because, we have the value of $\frac{1}{4\pi\varepsilon_0}$

$$E = \frac{2}{2}\times\frac{\lambda}{2\pi\varepsilon_0 r} = \frac{2\lambda}{4\pi\varepsilon_0 r}$$

Putting the values, we get

$$9\times10^4 = \frac{2\times9\times10^9\times\lambda}{2\times10^{-2}}$$

$$\lambda = \frac{9\times10^4\times2\times10^{-2}}{2\times9\times10^9} = 10^{-7}\text{ C/m}$$

Thus, the linear charge density is 10^{-7} C/m.

6 Electric field inside the uniformly charged sphere varies linearly, as $E = \frac{kQ}{R^3}\cdot r, (r \le R)$, while outside the sphere, it varies as inverse square of distance, $E = \frac{kQ}{r^2}; (r \ge R)$ which is correctly represented in option (c).

7 According to Gauss' theorem,

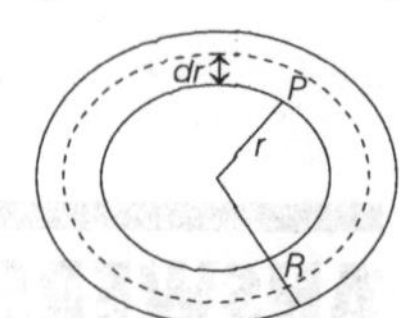

$$E4\pi r_1^2 = \frac{\int_0^{r_1}\frac{Q}{\pi R^4}r4\pi r^2\,dr}{\varepsilon_0}$$

$$\Rightarrow \quad E = \frac{Qr_1^2}{4\pi\varepsilon_0 R^4}$$

8 We know that, $F = \frac{1}{4\pi\varepsilon_0}\frac{2P_1P_2}{r^3}$

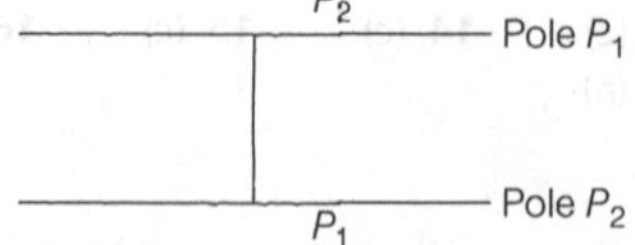

With the help of this relation, we find the force between dipole is $\frac{1}{4\pi\varepsilon_0}\frac{6P_1P_2}{x^3}$.

9 Let $-q_1$ be the charge, due to which flux ϕ_1 is entering the surface.

$$\phi_1 = \frac{-q_1}{\varepsilon_0} \text{ or } q_1 = -\phi_1\varepsilon_0$$

Let $+q_2$ be the charge, due to which flux ϕ_2, is leaving the surface.

$$\phi_2 = \frac{q_2}{\varepsilon_0} \text{ or } q_2 = \varepsilon_0\phi_2$$

Electric charge inside the surface

$$= q_2 - q_1 = \varepsilon_0\phi_2 + \varepsilon_0\phi_1 = \varepsilon_0(\phi_2 + \phi_1)$$

10 As uniform electric field is parallel to the cylindrical axis

$$\int \mathbf{E}\cdot d\mathbf{S} = \int E\,dS\cos 90^\circ = 0$$

Further flux entering the cylinder at one end = flux leaving the cylinder at other end. Therefore, total electric flux is zero.

11 Common potential,

$$V = \frac{Q+0}{4\pi\varepsilon_0(r+r')}$$

Charge on smaller sphere,

$$4\pi\varepsilon_0 r'\times V = \frac{Qr'}{r+r'}$$

12 The uncharged body must have a potential less than the charged conductor and more than at infinity.

i.e. $V_\infty < V$ or $V > V_\infty$

13 Electric field, $E = \frac{d\phi}{dr} = -2ar$

By Gauss' theorem,

$$E(4\pi r^2) = \frac{q}{\varepsilon_0}$$

$$\Rightarrow \quad q = -8\pi\varepsilon_0 ar^3$$

$$\rho = \frac{dq}{dV} = \frac{dq}{dr}\times\frac{dr}{dV}$$

$$= (-24\pi\varepsilon_0 ar^2)\left(\frac{1}{4\pi r^2}\right)$$

$$= -6\varepsilon_0 a$$

14 $W = QdV = Q(V_q - V_p)$

$$= -100\times(1.6\times10^{-19})\times(-4-10)$$

$$= +100\times1.6\times10^{-19}\times14$$

$$= +2.24\times10^{-16}\text{ J}$$

15 As, $E = \frac{dV}{dr}$ and around $x = 2$ m,

$$V = \text{constant}$$

$$\therefore \; dV = 0 \text{ and } E = 0$$

16 As E and V are related as,

$$E = \frac{-dv}{dx} = \frac{-d}{dx}(4x^2) = -8x$$

At (1, 0, 2), $E = -8(1) = -8$

17 As we know that, potential difference $V_A - V_O$ is

$$dV = -Edx, \int_{V_0}^{V_A} dV = -\int_0^2 30x^2 dx$$

$$V_A - V_O = -30 \times \left[\frac{x^3}{3}\right]_0^2$$
$$= -10 \times [2^3 - (0)^3]$$
$$= -10 \times 8 = -80 \text{ J/C}$$

18 The work done by a electrostatics force is given by $W_{12} = q(V_2 - V_1)$. Here, initial and final potentials are same in all three cases and same charge is moved, so work done is same in all three cases.

19 Let the potential of point D be V. If q is charge on each condenser, then

$$V_1 - V = qC_1$$
$$\Rightarrow \quad V - V_2 = qC_2$$

Divide $\quad \dfrac{V_1 - V}{V - V_2} = \dfrac{C_1}{C_2}$

$$VC_1 - V_2C_1 = V_1C_2 - VC_2$$
$$V(C_1 + C_2) = C_1V_2 + C_2V_1;$$
$$V = \frac{C_1V_2 + C_2V_1}{C_1 + C_2}$$

20

C_1, 120 V

C_2, 200 V

For potential to be made zero after connection, the charge on both the capacitors are equal.

i.e. $\quad q_1 = q_2$

$\therefore \quad C_1V_1 = C_2V_2, 120C_1 = 200C_2$

$\Rightarrow \quad 3\,C_1 = 5\,C_2$

21 Resultant circuit,

3μF, 4μF, 9μF ≡ 4μF 12μF ≡ 3μF

As, charge on 3μF = 3μF × 8V = 24 μC

$\therefore$ Charge on 4μ F = Charge on 12 μF = 24μC

Charge on 3μF = 3μF × 2V = 6μC

Charge on 9μF = 9μF × 2V = 18μC

Charge on 4μF + Charge on 9μF = (24 + 18)μC = 42μC

$\therefore$ Electric field at a point distant 30 m

$$= \frac{9 \times 10^3 \times 42 \times 10^{-6}}{30 \times 30} = 420 \text{N/C}$$

22 Initial charge = CV and

Final charge = $CV' + KCV'$

Since, initial charge = final charge

$$K = \frac{V - V'}{V'} K = \frac{V - V'}{V'}$$

23 When free space between parallel plates of capacitor, the

eletric field, $E = \dfrac{\sigma}{\varepsilon_0}$

When dielectric is introduced between parallel plates of capacitor,

$$E' = \frac{\sigma}{K\varepsilon_0}$$

Electric field inside dielectric

$$\frac{\sigma}{K\varepsilon_0} = 3 \times 10^4$$

where, K = dielectric constant of medium = 2.2

ε_0 = permittivity of free space = 8.85×10^{-12}

$$\Rightarrow \sigma = 2.2 \times 8.85 \times 10^{-12} \times 3 \times 10^4$$
$$= 6.6 \times 8.85 \times 10^{-8}$$
$$= 5.841 \times 10^{-7} = 6 \times 10^{-7} \text{ C/m}^2$$

24 Magnitude of induced charge is given by

$$Q' = (K - 1)CV_0$$
$$= \left(\frac{5}{3} - 1\right) 90 \times 10^{-12} \times 20$$
$$= 1.2 \times 10^{-9}\text{C}$$
$$\Rightarrow Q' = 1.2 \text{ nC}$$

25 As capacitance,

$$C_0 = k\frac{\varepsilon_0 A}{d}$$

Now equivalent capacitance,

$$C = \frac{2K}{3}\frac{\varepsilon_0 A}{d} + \frac{2k}{3}\frac{\varepsilon_0 A}{d} = \frac{4k}{3}\frac{\varepsilon_0 A}{d}$$

$$\therefore \text{ Ratio} = \frac{C}{\varepsilon_0} = \frac{\frac{4k}{3}\frac{\varepsilon_0 A}{d}}{k\frac{\varepsilon_0 A}{d}} = \frac{4}{3}$$

26

60 cm, Q=90 μC, 3 mm

Given, $\quad \dfrac{dQ}{dt} = 2.5 \times 10^{-8}$ C/s

But in case of capacitor, there is no magnetic field inside the capacitor i.e. zero.

27 Point charges are uniformly distributed around the centre O. Hence, electric field E is zero.

$\therefore \quad A \to 2$

Electric potential $V = 5 \times \left(\dfrac{Q}{4\pi\varepsilon_0 a}\right)$

$\therefore \quad B \to 1$

In case II, the electric field of three identically placed charges is zero. Net field at the centre O is due to a single point charge and is given as

$$E = \frac{Q}{4\pi\varepsilon_0 a^2}$$

$\therefore \quad C \to 3$

In case II, the electric potential, being scalar, becomes

$$V = 4 \times \frac{Q}{4\pi\varepsilon_0 a}$$

$\therefore \quad D \to 4$

28 Work done by conservative force does not depend on the path. Electrostatic force is a conservative force.

29 As, $W = q(V_A - V_B)$. At equipotential surface $V_A = V_B$ so, $W = 0$

Now, we know that field lines makes an angle of 90° with the equipotential surfaces but these are parallel to one-another.

SESSION 2

1 As each capacitors cannot withstand more than 300 V, so there should be four capacitors in each row became in this condition 1 kV i.e. 1000 V will be divided by 4 (i.e. 250 not more than 300 V).

Now, equivalent capacitance of one row

$$= \frac{1}{4} \times 1 \,\mu\text{F} = 0.25\mu\text{F}$$

[$\because$ in series combination, $C_{eq} = C/n$]

Now, we need equivalent of 2μF, so let we need n such rows

$\therefore \quad n \times 0.25 = 2\mu$F

[$\because$ in parallel combination $C_{eq} = nC$]

$$n = \frac{2}{0.25} = 8$$

$\therefore$ Total number of capacitors = number of rows × number of capacitors in each row = 8 × 4 = 32

2 Potential of B = Potential due to charge on A + Potential due to charge on B + Potential due to charge on C.

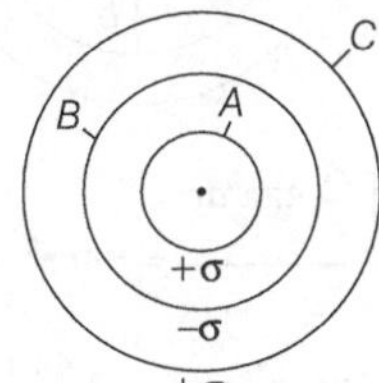

$\therefore V_B = \frac{k(Q_A + Q_B)}{b} + \frac{kQ_C}{c}$

$= \frac{1}{4\pi\varepsilon_0}\left[\frac{\sigma 4\pi a^2}{b} - \frac{\sigma 4\pi b^2}{b} + \frac{\sigma 4\pi c^2}{c}\right]$

$= \frac{\sigma\varepsilon}{\varepsilon_0}\left(\frac{a^2 - b^2}{b} + \frac{c^2}{c}\right)$

$= \frac{\sigma}{\varepsilon_0}\left(\frac{a^2 - b^2}{b} + c\right)$

$$V_B = \frac{\sigma}{\varepsilon_0}\left(\frac{a^2 - b^2}{b} + c\right)$$

3

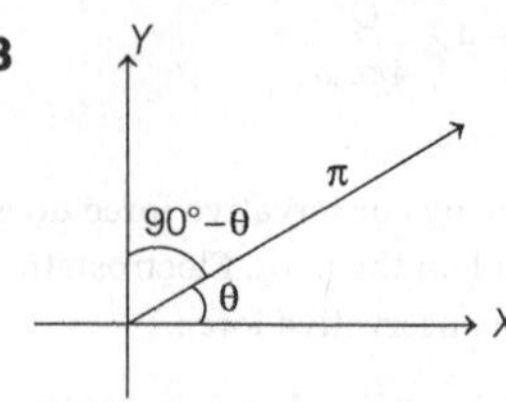

Torque applied on a dipole $\tau = pE \sin\theta$ where, θ = angle between axis of dipole and electric field.

For electric field $E_1 = E\hat{i}$

it means field is directed along positive X direction, so angle between dipole and field will remain θ, therefore torque in this direction

$$E_1 = pE_1 \sin\theta$$

In electric field $E_2 = \sqrt{3}\,E\hat{j}$,

it means field is directed along positive Y-axis, so angle between dipole and field will be $90° - \theta$.

Torque in this direction

$$\tau_2 = pE \sin(90° - \theta)$$
$$= p\sqrt{3}\,E_1 \cos\theta$$

According to question,

$\tau_2 = -\tau_1 \Rightarrow |\tau_2| = |\tau_1|$

$\therefore pE_1 \sin\theta = p\sqrt{3}\,E_1 \cos\theta$

$\tan\theta = \sqrt{3}$

$\Rightarrow \tan\theta = \tan 60°$

$\therefore \theta = 60°$

4 As, Gaussian surface at distance r from centre,

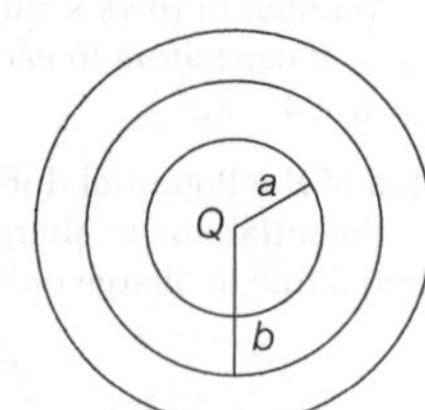

$$\frac{Q + \int_a^r \frac{A}{r} 4\pi r^2 dr}{\varepsilon_0} = E 4\pi r^2$$

$$E\, 4\pi\varepsilon_0 r^2 = Q + A\frac{4\pi}{r^2}\left(\frac{r^2 - a^2}{2}\right)$$

$$E = \frac{1}{4\pi\varepsilon_0}\left[\frac{Q}{r^2} + A\,2\pi\left(\frac{r^2 - a^2}{r^2}\right)\right]$$

$$E = \frac{1}{4\pi\varepsilon_0}\left(\frac{Q}{r^2} + A2\pi - \frac{A2\pi a^2}{r^2}\right)$$

$$E = \frac{1}{4\pi\varepsilon_0} \times A \times 2\pi$$

At the centre of the spheres is a point charge Q. The value of A such that the electric field in the region between the spheres will be constant is

As, $Q = 2\pi Aa^2$ i.e. $A = \frac{Q}{2\pi a^2}$

5 Three forces F_{41}, F_{42} and F_{43} acting on Q as shown.

Resultant of $F_{41} + F_{43} = \sqrt{2}F_{\text{each}}$

$$= \sqrt{2}\,\frac{1}{4\pi\varepsilon_0}\frac{Qq}{d^2}$$

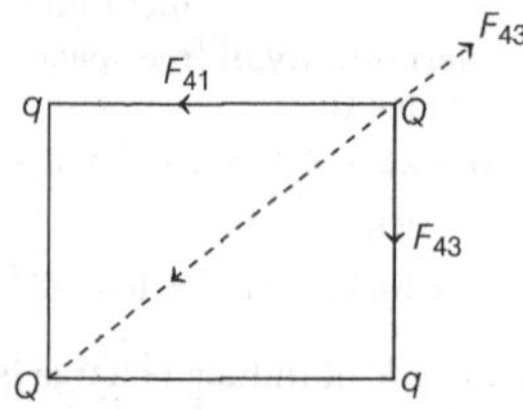

Resultant on Q becomes zero only when q charges are of negative nature.

$\therefore F = \frac{1}{4\pi\varepsilon_0}\frac{Q \times Q}{(\sqrt{2}d)^2}$

$\Rightarrow \sqrt{2}\frac{dQ}{d^2} = \frac{Q \times Q}{2d^2}$

$\Rightarrow \sqrt{2} \times q = \frac{Q \times Q}{2}$

$\therefore q = \frac{Q}{2\sqrt{2}}$ or $\frac{Q}{q} = -2\sqrt{2}$

6 Assume negative terminal of the battery as grounded (0 V). Suppose, potential of point x is V.

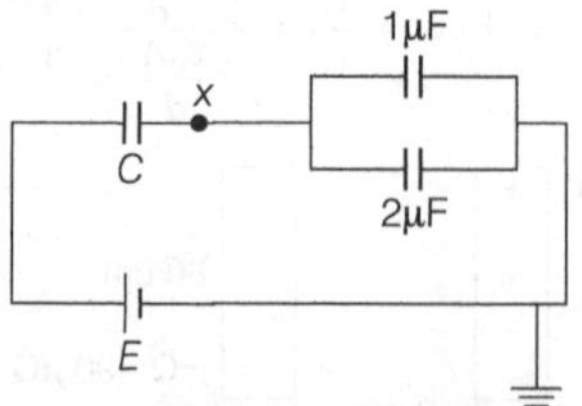

From the circuit diagram, we can write

$Q_C = Q_1 + Q_2$

or $C(E - V) = 1 \times V + 2 \times V$

or $V[C + 3] = CE$ or $V = \frac{CE}{3 + C}$

$\therefore Q_2 = C_2(V) = \frac{2CE}{3 + C} = \frac{2E}{1 + 3/C}$

As, C varied from 1μF to 3 μF, charge increases with decreasing slope.

Note As $C \to \infty, Q_2 \to 2E$ = constant

7

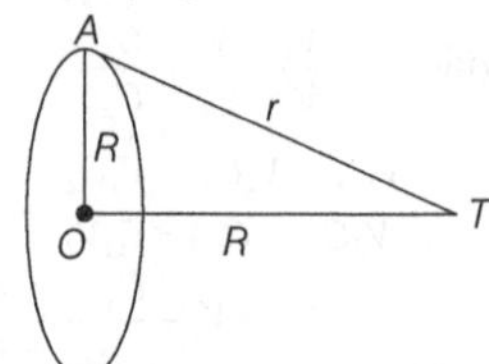

As the speed of particle is far to the intensity vector, and there is no acceleration in the direction for to **E**, but there is a electric force exerting on the particle (charge) whenever, it motion is in electric field.

Hence, continuously a force exerting on the particle for to its velocity or speed. Hence, path of particle must be parabola. As in the projectile motion.

8 Given, $E_0 = \frac{\sigma}{2\varepsilon_0}$

Also, we know that

$$E_p = \frac{1}{4\pi\varepsilon_0}\frac{qR}{(R^2 + R^2)^{3/2}}$$

but $q = \sigma A = \sigma\pi R^2$

$\Rightarrow E_p = \frac{1}{4\pi\varepsilon_0}\frac{\sigma\pi R^3}{2\sqrt{2}R^3} = \frac{\sigma}{8\sqrt{2}\varepsilon_0}$

$\therefore$ Reduction in electric field

$$= \frac{\sigma}{2\varepsilon_0}\left[1 - \frac{1}{4\sqrt{2}}\right]$$

Clearly, reduction is approx 71%.

9 Potential at the surface of the charged sphere

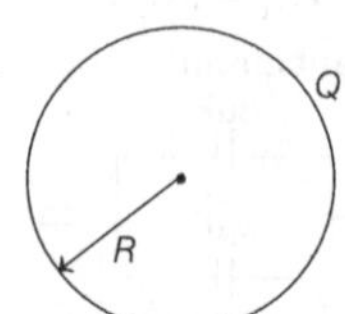

Charged sphere

$V_0 = \frac{KQ}{R}$, $V = \frac{KQ}{r}, r \geq R$

$= \frac{KQ}{2R^3}(3R^2 - r^2); r \leq R$

$V_{\text{centre}} = V_c = \frac{KQ}{2R^3} \times 3R^2$

$= \frac{3KQ}{2R} = \frac{3V_0}{2} \Rightarrow R_1 = 0$

As potential decreases for outside points. Thus, according to the question, we can write

$$V_{R_2} = \frac{5V_0}{4} = \frac{KQ}{2R^3}(3R^2 - R_2^2)$$

$$\Rightarrow \quad \frac{5V_0}{4} = \frac{V_0}{2R^2}(3R^2 - R_2^2)$$

or $\quad \frac{5}{2} = 3 - \left(\frac{R_2}{R}\right)^2$

or $\left(\frac{R_2}{R}\right)^2 = 3 - \frac{5}{2} = \frac{1}{2}$ or $R_2 = \frac{R}{\sqrt{2}}$

Similarly,

$$V_{R_3} = \frac{3V_0}{4} \Rightarrow \frac{KQ}{R_3} = \frac{3}{4} \times \frac{KQ}{R}$$

or $\quad R_3 = \frac{4}{3}R$

$$V_{R4} = \frac{KQ}{R_4} = \frac{V_0}{4}$$

$$\Rightarrow \quad \frac{KQ}{R_4} = \frac{1}{4} \times \frac{KQ}{R} \text{ or } R_4 = 4R$$

Thus, both options (d) and (c) are correct.

10

$$V = \int_L^{2L} \frac{kd\,Q}{x} = \int_L^{2L} \frac{\left(\frac{Q}{L}\right)dx}{x}$$

$$= \frac{Q}{4\pi\varepsilon_0 L}\int_L^{2L}\left(\frac{1}{x}\right)dx$$

$$= \frac{Q}{4\pi\varepsilon_0 L}[\log_e x]_L^{2L}$$

$$= \frac{Q}{4\pi\varepsilon_0 L}[\log_e 2L - \log_e L]$$

$$= \frac{Q}{4\pi\varepsilon_0 L}\left[\log_e \frac{2L}{L}\right]$$

$$= \frac{Q}{4\pi\varepsilon_0 L}\ln(2)$$

11 $F_{\text{net}} = 2F\cos\theta$

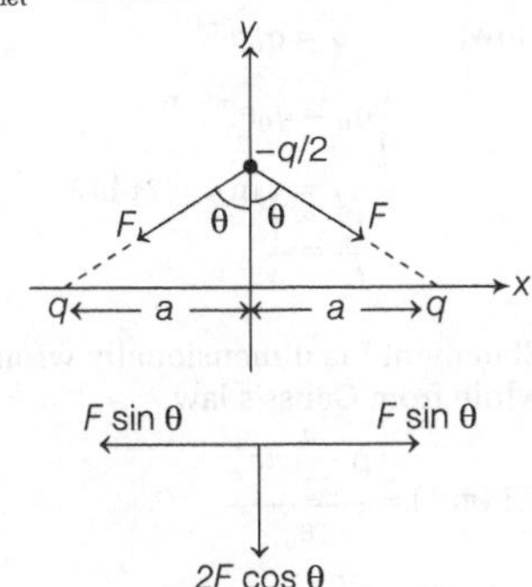

$$F_{\text{net}} = \frac{2kq\left(\frac{q}{2}\right)}{(\sqrt{y^2 + a^2})^2} \cdot \frac{y}{\sqrt{y^2 + a^2}}$$

$$F_{\text{net}} = \frac{2kq\left(\frac{q}{2}\right)y}{(y^2 + a^2)^{3/2}} \Rightarrow \frac{kq^2y}{a^3} \propto y$$

12

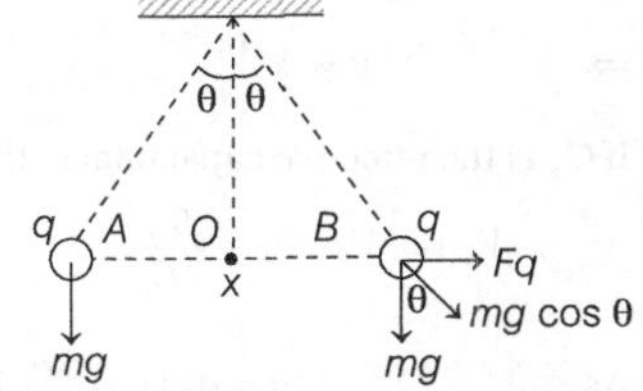

Force of attraction $= \frac{1}{4\pi\varepsilon_0}\frac{q^2}{x^2}$

Potential at the mid-point O

$$= \frac{1}{4\pi\varepsilon_0}\frac{q}{x} + \frac{1}{4\pi\varepsilon_0}\frac{q}{x}$$

$$= \frac{1}{2\pi\varepsilon_0}\frac{q}{x} \quad \text{...(i)}$$

From the figure,

$$\tan\theta = \frac{\frac{1}{4\pi\varepsilon_0}\frac{q^2}{x^2}}{mg}$$

$$\therefore mg\tan\theta = k\frac{q^2}{x^2},\ x = \sqrt{\frac{kq^2}{mg\tan\theta}}$$

$$\Rightarrow \quad x = q\sqrt{\frac{k}{mg\tan\theta}} \quad \text{...(ii)}$$

From Eqs. (i) and (ii), we get

Potential $= 4\sqrt{K\,mg/\tan\theta}$

13

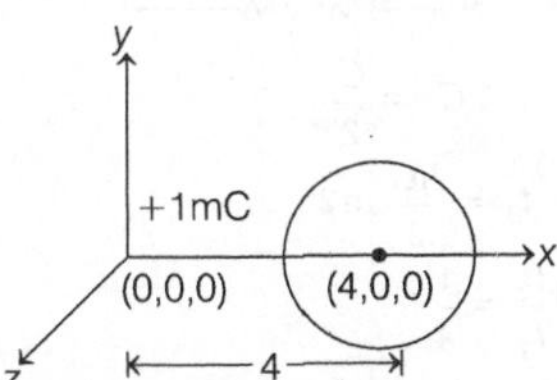

Potential at the centre of the sphere.

$$V = \frac{1}{4\pi\varepsilon_0}\frac{Q}{r} = \frac{1}{4\pi\varepsilon_0}\frac{1 \times 10^{-6}}{4}$$

$$= \frac{9 \times 10^9 \times 10^{-6}}{4} = 2.25 \times 10^3 \text{ V}$$

Now, electric field at the centre of sphere

$$E = \frac{1}{4\pi\varepsilon_0}\frac{Q}{r^2} = 9 \times 10^9 \frac{1 \times 10^{-6}}{4^2}$$

$$= 0.5625 \times 10^3$$

$$= 5.625 \times 10^2 \text{ V/m}$$

14 Each ball in equilibrium have following three forms

(i) tension (ii) repulsive force

(iii) weight

By applying Lami's theorem,

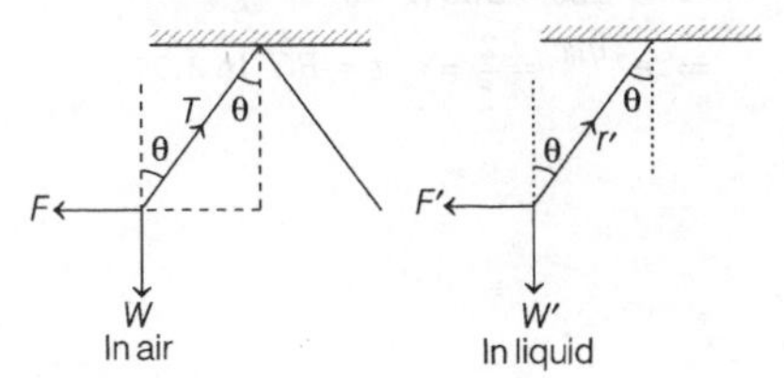

In liquid $F' = \frac{F}{K}$

where, K = dielectric constant of liquid

$$W' = W - \text{upthrust} = V\rho g - V\sigma g$$

where, ρ = density of material,

σ = density of liquid

In air, using Lami's theorem,

$$\frac{W}{\sin(90° - \theta)} = \frac{F}{\sin(180° - \theta)}$$

or $\quad \frac{W}{\cos\theta} = \frac{F}{\sin\theta} \quad \text{...(i)}$

In liquid,

$$\frac{W'}{\cos\theta} = \frac{F'}{\sin\theta}$$

(As angles are same)...(ii)

On dividing Eq. (i) by Eq. (ii), we get

$$\frac{W}{W'} = \frac{F}{F'}$$

or $\quad K = \frac{W}{W - \text{upthrust}}$

$$\Rightarrow K = \frac{V\rho g}{V\rho g - V\sigma g} = \frac{\rho}{\rho - \sigma} = \frac{1.6}{1.6 - 0.8} = 2$$

15 The capacitor C_2 is shorted, so it is not playing any role in circuit and can be removed. The 3 capacitors each of C_1 are connected in parallel and this is connected to C_3 in series.

$$C_{\text{eq}} = \frac{3C_1C_3}{3C_1 + C_3} = C = \frac{3 \times 3 \times 1}{3 \times 3 + 1}$$

$$= 0.9\ \mu\text{F}$$

So, $\frac{\Delta C}{C} = \frac{3\Delta C_1}{C_1} + \frac{\Delta C_3}{C_3} + \frac{3\Delta C_1 + \Delta C_3}{3C_1 + C_3}$

[For computation of errors worst has to be taken]

$$\Rightarrow \quad \frac{\Delta C}{0.9} = \frac{3 \times 0.011}{3} + \frac{0.01}{1} + \frac{(0.033 + 0.01)}{10}$$

$$\Rightarrow \quad \Delta C = \pm\ 0.023\ \mu\text{F}$$

16 Linear charge density $\lambda = \left(\frac{q}{\pi r}\right)$

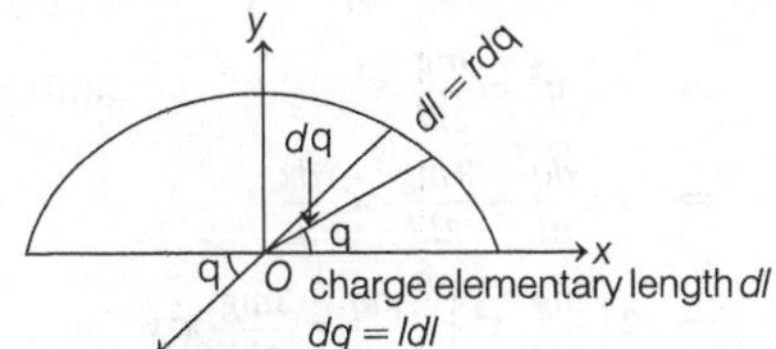

Net field at O,

$$E = \int dE\sin\theta\,(-\mathbf{j})$$

$$= \int \frac{K \cdot dq}{r^2}\sin\theta\,(-\mathbf{j})$$

$$E = \frac{K}{r^2}\int \frac{qr}{\pi r}d\theta\sin\theta\,(-\mathbf{j}) \quad (\because dl = rd\theta)$$

$$= \frac{K}{r^2}\frac{q}{\pi}\int_0^{\pi}\sin\theta\,(-\mathbf{j}) = \frac{q}{2\pi^2\varepsilon_0 r^2}(-\mathbf{j})$$

17 $U_i = \frac{2kqQ}{a} + \frac{2k(-q)Q}{\sqrt{5}a}$

$= \frac{1}{4\pi\varepsilon_0} \cdot \frac{2qQ}{a}\left[1 - \frac{1}{\sqrt{5}}\right]$

$U_f = 0$

By conservation of energy,

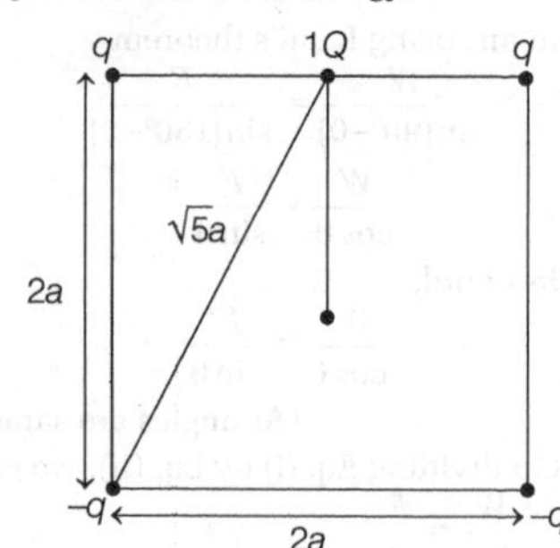

Gain in KE = loss in PE,

$$K = \frac{1}{4\pi\varepsilon_0} \cdot \frac{2qQ}{a}\left[1 - \frac{1}{\sqrt{5}}\right]$$

18 At any instant,

$T\cos\theta = mg$ ...(i)

$T\sin\theta = F_e = \frac{kq^2}{x^2}$ ...(ii)

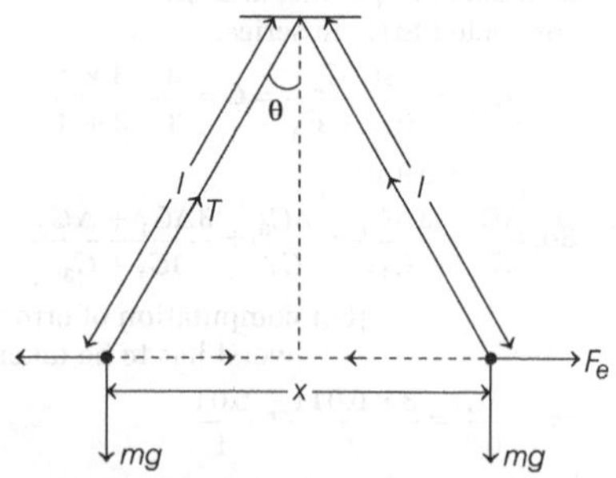

From Eqs. (i) and (ii), we get

$\frac{kq^2}{x^2} = mg\tan\theta$

(q = magnitude of charge)

$\Rightarrow q^2 = \frac{mg}{k} \cdot \frac{x}{2l} \cdot x^2 \left(\because \tan\theta = \frac{x}{2l}\right)$

$\Rightarrow q^2 = \frac{mg}{2kl}x^3$ ...(iii)

$\Rightarrow 2q\frac{dq}{dt} = \frac{3mg}{2kl}x^2\frac{dx}{dt}$

$\Rightarrow 2\left(\frac{mg}{2kl}x^3\right)^{1/2}\frac{dq}{dt} = \frac{3mg}{2kl}x^2 v$

$\left[\because q = \left(\frac{mg}{2kl}x^3\right)^{1/2}\right]$

$\Rightarrow vx^{1/2} = \text{constant}$

$\Rightarrow v \propto x^{-1/2}$

19 If C_e is the effective capacitance, then

$V_C = \frac{1}{2}V_0, \frac{q}{C_e} = \frac{q_0}{2C_e}$

As $q = q_0(1 - e^{\frac{-t}{RC_e}})$

$\Rightarrow q_0(1 - e^{-t/RC_e}) = \frac{q_0}{2}$

$\Rightarrow t = RC_e \ln 2$

For parallel grouping,

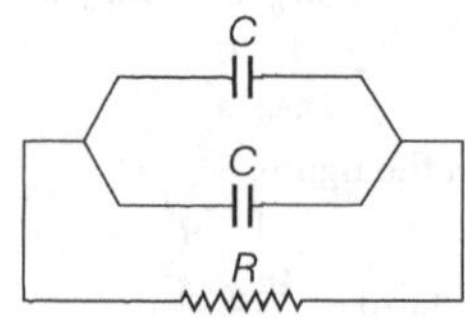

$C_e = 2C$

$\therefore t_2 = 2RC\ln 2$

For series grouping,

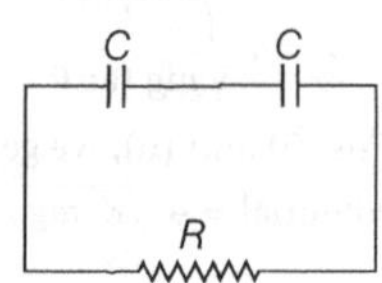

$C_e = \frac{C}{2}$

$\therefore t_1 = \frac{RC}{2}\ln 2$

$\therefore \frac{t_2}{t_1} = \frac{1}{4}$

$\Rightarrow t_2 = \frac{t_1}{4} = \frac{10}{4}$

$\Rightarrow t_2 = 2.5\text{ s}$

20 Neon bulb is filled with gas, so its resistance is infinite, hence no current flows through it.

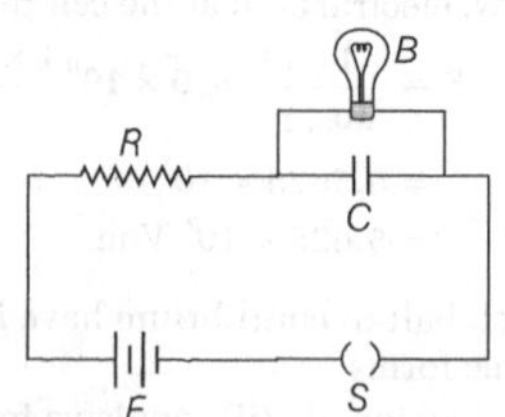

Now, $V_c = E(1 - e^{-t/RC})$

$\Rightarrow 120 = 200(1 - e^{-t/RC})$

$\Rightarrow e^{-t/RC} = \frac{2}{5} \Rightarrow t = RC\ln 2.5$

$\Rightarrow R = \frac{t}{C\ln 2.5}$

$= \frac{t}{2.303C\log 2.5}$

$= 2.7 \times 10^6\Omega$

21 Apply shell theorem, the total charge upto distance r can be calculated as followed

$dq = 4\pi r^2 \cdot dr \cdot \rho$

$= 4\pi r^2 \cdot dr \cdot \rho_0\left[\frac{5}{4} - \frac{r}{R}\right]$

$= 4\pi\rho_0\left[\frac{5}{4}r^2 dr - \frac{r^3}{R}dr\right]$

$\int dq = q = 4\pi\rho_0\int_0^r\left(\frac{5}{4}r^2 dr - \frac{r^3}{R}dr\right)$

$= 4\pi\rho_0\left[\frac{5}{4}\frac{r^3}{3} - \frac{1}{R}\frac{r^4}{4}\right]$

$E = \frac{kq}{r^2}$

$= \frac{1}{4\pi\varepsilon_0}\frac{1}{r^2} \cdot 4\pi\rho_0\left[\frac{5}{4}\left(\frac{r^3}{3}\right) - \frac{r^4}{4R}\right]$

$E = \frac{\rho_0 r}{4\varepsilon_0}\left[\frac{5}{3} - \frac{r}{R}\right]$

22 Energy stored in capacitor,

$U = \frac{1}{2}\frac{q^2}{C} = \frac{1}{2C}(q_0 e^{-t/\tau})^2$

$= \frac{q_0^2}{2C}e^{-2t/\tau}$ [where, $\tau = CR$]

$U = U_i e^{-2t/\tau}$

$\frac{1}{2}U_i = U_i e^{-2t_1/\tau}$

$\frac{1}{2} = e^{-2t_1/\tau}$

$\Rightarrow t_1 = \frac{\tau}{2}\ln 2$

Now, $q = q_0 e^{-t/\tau}$

$\frac{1}{4}q_0 = q_0 e^{-t_2/\tau}$

$t_2 = \tau\ln 4 = 2\tau\ln 2$

$\therefore \frac{t_1}{t_2} = \frac{1}{4}$

23 Statement I is dimensionally wrong while from Gauss's law,

$E(4\pi r^2) = \frac{\rho \cdot \frac{4}{3}\pi r^3}{\varepsilon_0}$

$\Rightarrow E = \frac{\rho r}{3\varepsilon_0}$

Given Statement II is correct.

DAY NINETEEN

Current Electricity

Learning & Revision for the Day

- Electric Current
- Ohm's Law
- Electrical Resistivity
- Electrical Resistance
- Resistance of Different Materials
- Series and Parallel Combinations of Resistors
- Electric Energy and Power
- Electric Cell
- Potential Difference and emf of a Cell
- Kirchhoff's Laws and their Applications
- Wheatstone's Bridge
- Meter Bridge (Special Case of Wheatstone Bridge)

Electric Current

Electric current is defined as the amount of charge flowing across any section of wire per unit time. If charge Δq passes through the area in time interval Δt at uniform rate, then current i is defined as

$$i = \frac{\Delta q}{\Delta t} \qquad \ldots(i)$$

SI unit of electric current is ampere (A).

- Conventional direction of flow of current is taken to be the direction of flow of positive charge or opposite to the direction of flow of negative charge.
- Electric current is a scalar as it does not follow the law vector of addition.

Current Density

Current per unit area is termed as current density.

$$J = \frac{I}{A}\ (\text{Am}^{-2})$$

It is a vector quantity.

Drift Velocity

- Drift velocity is the average uniform velocity acquired by conduction electrons inside a metallic conductor on application of an external electric field.

  The drift velocity is given by the relation

$$\mathbf{v}_d = -\frac{e\,\mathbf{E}}{m}\tau$$

  where, τ known as relaxation time.

PREP MIRROR
Your Personal Preparation Indicator

- No. of Questions in Exercises (x)—
- No. of Questions Attempted (y)—
- No. of Correct Questions (z)—
  (*Without referring Explanations*)

- Accuracy Level ($z/y \times 100$)—
- Prep Level ($z/x \times 100$)—

In order to expect good rank in JEE, your Accuracy Level should be above 85 & Prep Level should be above 75.

- Drift velocity per unit electric field is called the **mobility** of the electrons. Thus, mobility,

$$\mu = \left|\frac{\mathbf{v}_d}{\mathbf{E}}\right| = \frac{e}{m}\tau$$

- In terms of drift speed, electric current flowing through a conductor is expressed as $I = nAev_d$

where, A = cross-section area of conductor,

n = number of conduction electrons per unit volume,

v_d = drift velocity of electrons

and e = charge of one electron.

Ohm's Law

Ohm's law states that the physical conditions such as temperature, mechanical strain, etc., are kept constant, then current (i) flowing through a conductor is directly proportional to the potential difference across its two ends.

i.e. $i \propto V$ or $V \propto i$ or $V = Ri$ or $\frac{V}{i} = R =$ a constant,

where R depends on the nature of material and it given dimension.

Electrical Resistivity

The resistance of a resistor (an element in a circuit with some resistance R) depends on its geometrical factors (length, cross-sectional area) as also on the nature of the substance of which the resistor is made. Electrical resistance of a rectangular slab depends on its length (l) and its cross-sectional area (A).

i.e., $$R \propto l \quad \text{...(i)}$$

and $$R \propto \frac{1}{A} \quad \text{...(ii)}$$

Combining the two relations, we get

$$R \propto \frac{l}{A} \quad \text{...(iii)}$$

or $$R = \frac{\rho l}{A} \quad \text{...(iv)}$$

where, ρ is a constant of proportionality called resistivity.

$$\rho = \frac{m}{ne^2\tau}$$

Resistance of Different Materials

Resistance offered by the conductors is minimum while resistance offered by an insulator is maximum. Semiconductors have resistance which is intermediate to conductor and insulator.

Electrical Resistance

Electrical resistance is defined as the ratio in the potential difference (v) across the ends of the conductor to the current (i) flowing through it, i.e. $R = \frac{V}{i}$

The SI unit of electrical restristance is Ω (ohm) and its dimension is $[ML^2T^{-3}A^{-2}]$.

Resistance of Different Materials

A perfect conductor would have zero resistivity and a perfect insulator would have infinite resistivity. Though these are ideal limits, the electrical resistivity of substances has a very wide range. Metals have low resistivity of $10^{-8}\ \Omega$m to $10^{-6}\ \Omega$m, while insulators like glass or rubber have resistivity, some 10^{18} times (or even more) greater, Generally, good electrical conductors like metals are also good conductors of heat, while insulators like ceramic or plastic materials are also poor thermal conductors.

V-I Characteristics of Ohmic and Non-ohmic Conductors

Substances obeying Ohm's law are called **Ohmic resistors**, e.g. metals and their alloys. Substances which do not obey Ohm's law are called **non-ohmic resistors**, e.g. electrolytes, gases, thermionic tubes, transistors, rectifiers, etc.

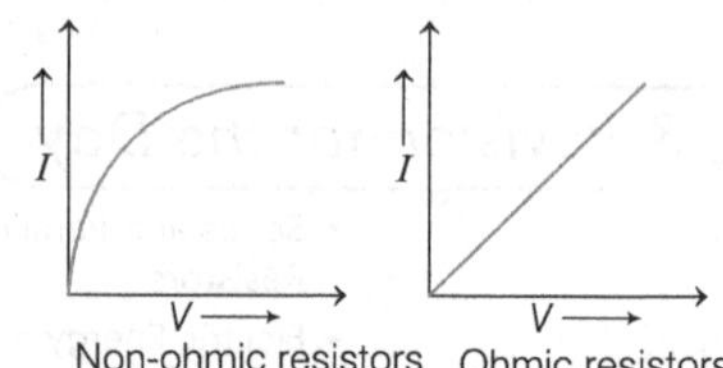

Non-ohmic resistors Ohmic resistors

Colour Code for Resistors

The value of resistance used in electrical and electronic circuits vary over a wide range. These resistances are usually carbon resistances and a colour code is used to indicate the value of resistance.

Their value ranges from kilo-ohm to mega-ohm. Their percentage accuracy is indicated by a colour code printed on them. Carbon resistors are compact, inexpensive and are used in electronic circuits.

Colour Code for Carbon Resistors

Colour	Letter of remember	Number	Multiplie r	Tolerance
Black	B	0	10^0	—
Brown	B	1	10^1	—
Red	R	2	10^2	—
Orange	O	3	10^3	—
Yellow	Y	4	10^4	—
Green	G	5	10^5	—
Blue	B	6	10^6	—
Violet	V	7	10^7	—
Grey	G	8	10^8	—
White	W	9	10^9	—
Gold	—	—	10^{-1}	5%
Silver	—	—	10^{-2}	10%
No colour	—	—	—	20%

Now to find the colour coding of carbon resistor, we must remember the bold capital letters of the following sentences:

Black **B**rown **ROY** **G**reat **B**ritain **V**ery **G**ood **W**ife **W**earing **G**old **S**ilver **N**ecklace

Or

Black **B**rown **R**ods of **Y**our **G**ate **B**ecame **V**ery **G**ood **W**hen **G**iven **S**ilver **C**olour

The colours of first two bands A and B correspond to figures and the colour of the third band C represents multipliers, respectively. The fourth band represents the tolerance.

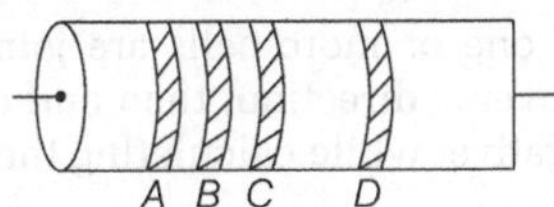

e.g. Consider a carbon resistor of bands A and B of black and red colour having figures 0 and 2. The third band C of green colour having multiplier 10^5.

$\therefore$ Resistance of the value is given by

$$R = 02 \times 10^5 \Omega$$

But the fourth band D having gold colour, which represents a tolerance of $\pm 5\%$.

Hence, the value of carbon resistance is

$$R = 02 \times 10^5 \Omega \pm 5\%$$

Series and Parallel Combinations of Resistors

Series Combination

A series circuit is a circuit in which resistors are arranged in a chain, so the current has only one path to take. The current is the same through each resistor. The total resistance of the circuit is found by simply adding up the resistance values of the individual resistors. Equivalent resistance of resistors in series

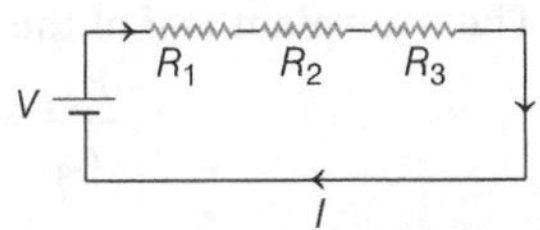

$$R = R_1 + R_2 + R_3 + \ldots$$

Parallel Combination

A parallel circuit is a circuit in which the resistors are arranged with their heads connected together and their tails connected together. The current in a parallel circuit breaks up, with some flowing along each parallel branch and recombining, when the branches meet again. The voltage across each resistor is parallel is the same.

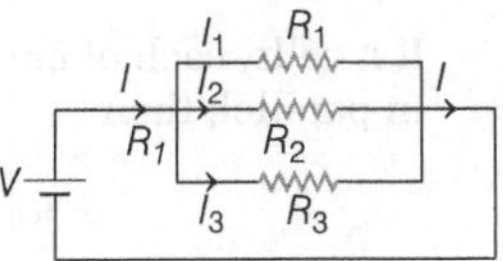

The total resistance of a set of resistors in parallel is found by adding up the reciprocals of the resistance values, and then, taking the reciprocal of the total.

The equivalent resistance of resistors in parallel,

$$\frac{1}{R} = \frac{1}{R_1} + \frac{1}{R_2} + \frac{1}{R_3} + \ldots$$

Temperature Dependence of Resistance

Resistance and resistivity of metallic conductors increases with increase in temperature. The relation is written as

$$R_\theta = R_0 (1 + \alpha\theta + \beta\theta^2) \text{ and } \rho_\theta = \rho_0 (1 + \alpha\theta + \beta\theta^2)$$

where, R_0 and ρ_0 are values of resistance and resistivity at 0°C and R_θ and ρ_θ at θ°C. α and β are two constants whose value vary from metal to metal.

Electric Energy and Power

Whenever the electric current is passed through a conductor, it becomes hot after short time. This effect is known as **heating effect** of current or **Joule heating effect.**

$$H = W = I^2 Rt \text{ joule} = \frac{I^2 Rt}{418} \text{ cal}$$

The rate at which work is done by the source of emf in maintaining the effect of current in a circuit is called electric power of the circuit,

$$P = VI \text{ watt}$$

Other expressions for power,

$$P = I^2 R \text{ watt} \Rightarrow P = \frac{V^2}{R}$$

Electric Cell

An electric cell is a device which maintains a continuous flow of charge (or electric current) in a circuit by a chemical reaction. In an electric cell, there are two rods of different metals called electrodes.

Internal Resistance of a Cell

Thus, when a current is drawn through a source, the potential difference between the terminal of the source is

$$V = E - ir$$

This can also be shown as below

A —E—r—I— B

$$V_A - E + Ir = V_B$$

or $$V_A - V_B = E - Ir$$

Following three special cases are possible

(i) If the current flows in opposite direction (as in case of charging of battery), then $V = E + Ir$

(ii) $V = E$, if the current through the cell is zero.

(iii) $V = 0$, if the cell is short circuited.

This is because current in the circuit,

$$I = \frac{E}{r} \quad \text{or} \quad E = Ir$$

$$\therefore \qquad E - Ir = 0 \quad \text{or} \quad V = 0$$

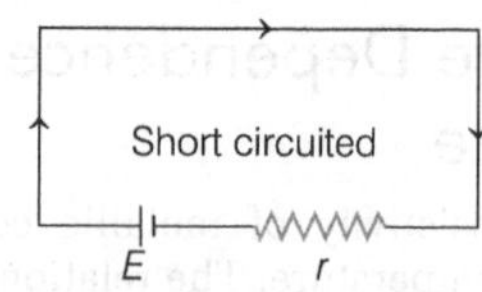

Thus, we can summarise, it was follows

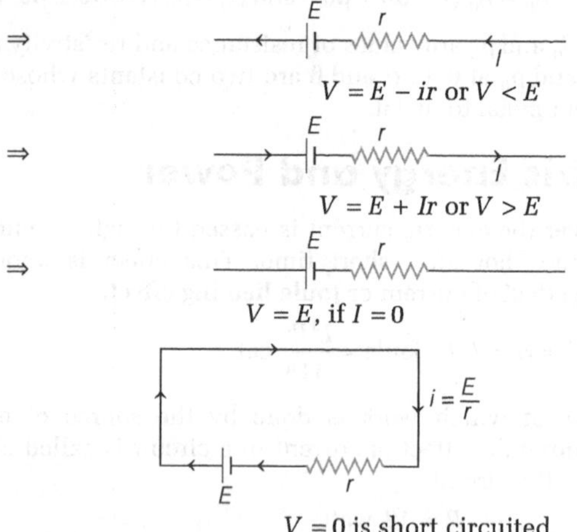

$\Rightarrow$ $V = E - ir$ or $V < E$

$\Rightarrow$ $V = E + Ir$ or $V > E$

$\Rightarrow$ $V = E$, if $I = 0$

$V = 0$ is short circuited

Potential Difference and emf of a Cell

Electromotive force (emf) of a cell is the terminal potential difference of cell when it is in an **open circuit**, i.e. it is not supplying any current to the external circuit. However, when it is supplying a current to an external resistance, the voltage across the terminals of cell is called the **terminal voltage** or **terminal potential difference.**

If E be the emf and r the internal resistance of a cell and a resistance R is joined with it, then current in the circuit, $I = \frac{E}{R + r}$ and terminal potential difference,

$$V = IR = \frac{ER}{(R + r)}$$

or $\qquad V = E - Ir$

Internal resistance of cell,

$$r = \left(\frac{E - V}{V}\right)R = R\left(\frac{E}{V} - 1\right)$$

Terminal voltage is more than emf of cell when cell is charged and it is given by $V = E + Ir$.

Combination of Cells in Series and in Parallel

A group of cells is called a battery. Two common grouping of cells are

1. Series Grouping

In series grouping, if all the cells are joined so as to supply current in the same direction, then resultant emf,

$$E_{eq} = E_1 + E_2 + E_3 + \ldots$$

However, if one or more cells are joined so as to supply current in reverse direction, then emf of that/those cells is taken as negative, while calculating the equivalent emf.

+ve E_1 E_2 E_3 –ve
r_1 r_2 r_3

The equivalent internal resistance of the cell,

$$r_{eq} = r_1 + r_2 + r_3 + \ldots$$

If n cells, each of emf E and internal resistance r, are joined in series, then

$$E_{eq} = nE \quad \text{and} \quad r_{eq} = nr$$

2. Parallel Grouping

In parallel grouping, if positive terminals of all cells have been joined at one point and all negative terminals at another point, then $\frac{1}{r_{eq}} = \frac{1}{r_1} + \frac{1}{r_2} + \frac{1}{r_3} + \ldots$

The equivalent emf of the parallel grouping is given by

$$\frac{E_{eq}}{r_{eq}} = \frac{E_1}{r_1} + \frac{E_2}{r_2} + \frac{E_3}{r_3} + \ldots$$

E_1 r_1 E_2 r_2 E_3 r_3 +ve –ve

If n cells, each of emf E and internal resistance r, all joined in parallel, then

$$r_{eq} = \frac{r}{n}$$

But $\qquad E_{eq} = E$

Kirchhoff's Laws and their Applications

Many electric circuits cannot be reduced to simple series parallel combinations. For example, two circuits that cannot be so broken down are shown in figure

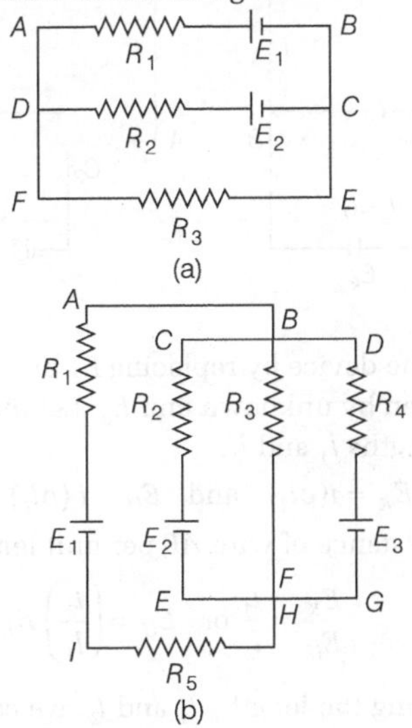

(a)
(b)

However, it is always possible to analyze such circuits by applying two rules, derived by Kirchhoff.

Junction Rule

The algebraic sum of the currents at any junction is zero.

i.e. $\sum_{\text{junction}} i = 0$

This law can also be written as, "the sum of all the currents directed towards a point in circuit is equal to the sum of all the currents directed away from that point."

Thus, in figure, $i_1 + i_2 = i_3 + i_4$

The junction rule is bases on **conservation of electric charge.**

Loop Rule

The algebraic sum of the potential difference in any loop including those associated emf's and those of resistive elements, must be equal to zero.

That is, $\sum_{\text{closed loop}} \Delta V = 0$

This law represent **conservation of energy.**

Applying Kirchhoff's law for the following circuit, we have Resulting equation is

$$V_{r_1} + V_{r_2} + V_{r_3} - 10 = 0.$$

Wheatstone's Bridge

For measuring accurately any resistance Wheatstone bridge is widely used. It is an arrangement of four resistances used to measure one of them in terms of the other three.

Consider four resistances P, Q, R and S are connected in the four arms of a quadrilateral according to figure, the bridge is said to be balanced when galvanometer gives zero deflection, i.e. potential at point B and D is same $(V_B = V_D)$.

In this condition,

$$\frac{P}{Q} = \frac{R}{S}$$

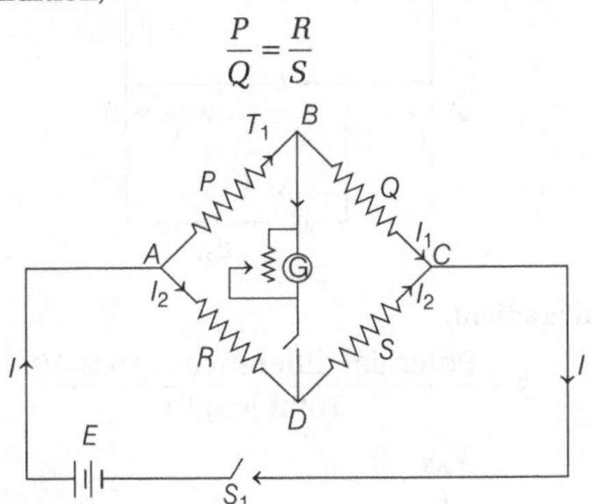

Meter Bridge (Special Case of Wheatstone Bridge)

This is the simplest form of Wheastone bridge and is specially useful for comparing resistances more accurately. The construction of the meter bridge is shown in the figure. It consists of one metre resistance wire clamped between two metallic strips bent at right angles and it has two points for connection.

There are two gaps; in one of whose value is to be determined is connected. The galvanometer is connected with the help of jockey across BD and the cell is connected across AC. After making connections, the jockey is moved along the wire and the null point is found. Wheatstone bridge, wire used is of uniform material and cross-section. the resistance can be found with the help of the following relation

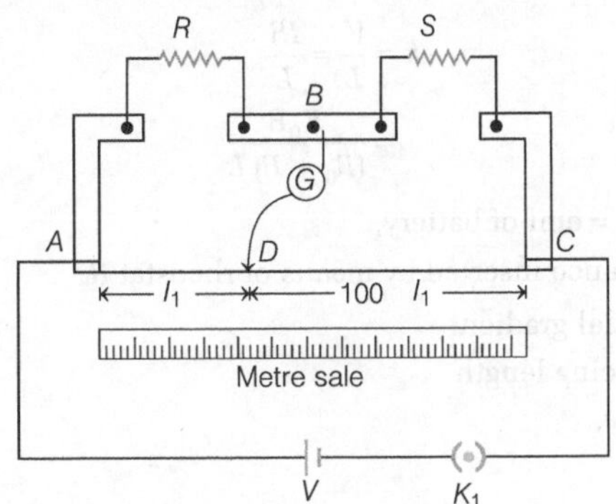

$$\frac{R}{S} = \frac{l_1}{(100 - l_1)} \quad \text{or} \quad R = S\,\frac{l_1}{100 - l_1}$$

Potentiometer

Principle Potentiometer is an ideal device to measure the potential difference between two points. It consists of a long resistance wire AB of uniform cross-section in which a steady direct current is set up by means of a battery.

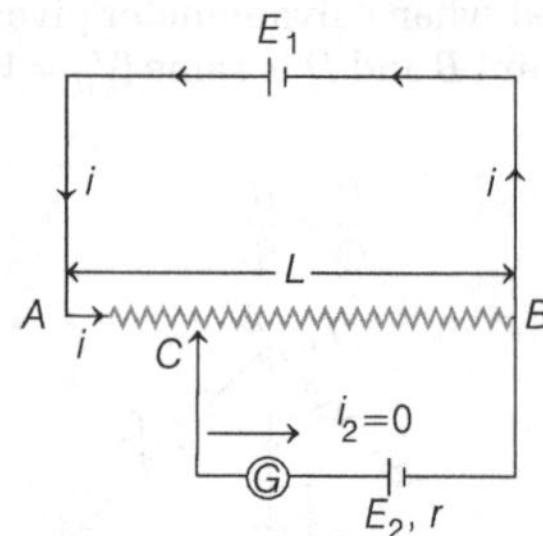

Potential gradient,

$$k = \frac{\text{Potential difference across } AB}{\text{Total length}}$$

$$= \frac{V_{AB}}{L}$$

$$= \frac{iR_{AB}}{L} = i\lambda$$

where, $\lambda = \frac{R_{AB}}{L}$ = resistance per unit length of potentiometer wire.

The emf of source balanced between points B and C

$$E_2 = kl$$

$$= i\,\frac{R_{AB}}{l} \times l = iR_{CB}$$

Here, AB is a long uniform resistance wire (length AB may be ranging from 1 m to 10 m). E_0 is a battery whose emf is known supplying a constant current I for flow through the potentiometer wire. If R be the total resistance of potentiometer wire and L its total length, then potential gradient, i.e. fall in potential per unit length along the potentiometer will be

$$k = \frac{V}{L} = \frac{IR}{L}$$

$$= \frac{E_0 R}{(R_0 + R)\,L}$$

where, E_0 = emf of battery,

R_0 = resistance inserted by means of rheostat R_h

k = potential gradient.

$L \rightarrow$ balancing length

$J \rightarrow$ jockey.

Applications of Potentiometer

(i) To find emf of an unknown battery

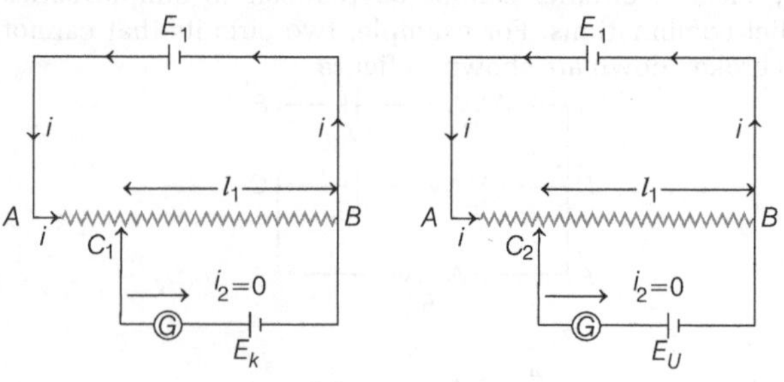

We calibrate the device by replacing E_2 by a source of known emf E_k and then by unknown emf E_u. Let the null points are obtained at lengths l_1 and l_2.

Then, $\quad E_K = i(\rho l_1)$ and $E_U = i\,(\rho l_2)$

Here, ρ = resistance of wire AB per unit length

$$\therefore \quad \frac{E_K}{E_U} = \frac{l_1}{l_2} \text{ or } E_U = \left(\frac{l_2}{l_1}\right) E_K$$

So, by measuring the lengths l_1 and l_2, we can find the emf of an unknown battery.

(ii) To find the internal resistance of a cell

Firstly, the emf E of the cell is balanced against a length $AD = l_1$. For this, the switch S' is left opened and S is closed. A known resistance R is then connected to the cell as shown. The terminal voltage V is now balanced against a smaller length $AD' = l_2$. Here, now switch is opened and S' is closed.

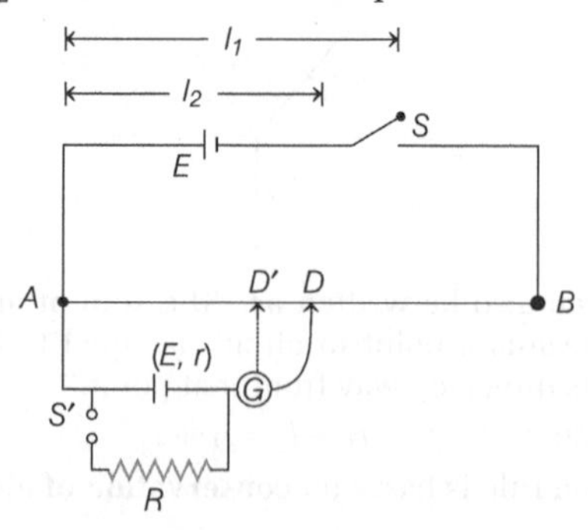

Then, $$\frac{E}{V} = \frac{l_1}{l_2}$$

Since, $$\frac{E}{V} = \frac{R + r}{R} \qquad \{\because E = i(R + r) \text{ and } V = iR\}$$

or $$\frac{R + r}{R} = \frac{l_1}{l_2} \Rightarrow r = \left(\frac{l_1}{l_2} - 1\right) R$$

DAY PRACTICE SESSION 1

FOUNDATION QUESTIONS EXERCISE

1 A wire of length L and 3 identical cells of negligible internal resistance are connected in series.Due to the current, the temperature of the wire is raised by ΔT in a time t. A number N of similar cells is now connected in series with a wire of the same material and cross-section but of length $2L$. The temperature of the wire is raised by the same amount ΔT in the same time t. The value of N is

(a) 4 (b) 6 (c) 8 (d) 9

2 In a large building, there are 15 bulbs of 40W, 5 bulbs of 100W, 5 fans of 80 W and 1 heater of 1kW. The voltage of the electric mains is 220 V. The minimum capacity of the main fuse is → JEE Main 2014

(a) 8A (b) 10A (c) 12A (d) 14A

3 Two electric bulbs marked 25 W-220 V and 100 W-220 V are connected in series to a 440 V supply. Which of the bulbs will fuse? → AIEEE 2012

(a) Both (b) 100 W (c) 25 W (d) Neither

4 A given resistor has the following colour scheme of the various strips on it brown, black, green and silver. Its value in ohm is

(a) $1.0 \times 10^4 \pm 10\%$
(b) $1.0 \times 10^5 \pm 10\%$
(c) $1.0 \times 10^6 \pm 10\%$
(d) $1.0 \times 10^7 \pm 10\%$

5 Correct set up to verify Ohm's law is → JEE Main (Online) 2013

(a) 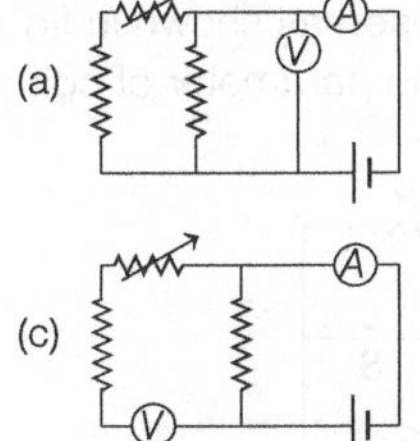

(b) 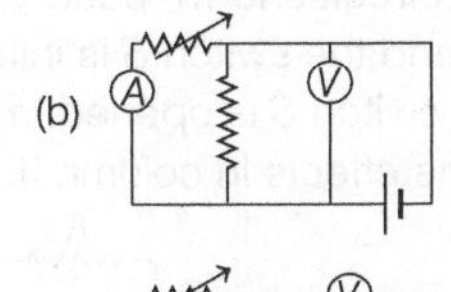

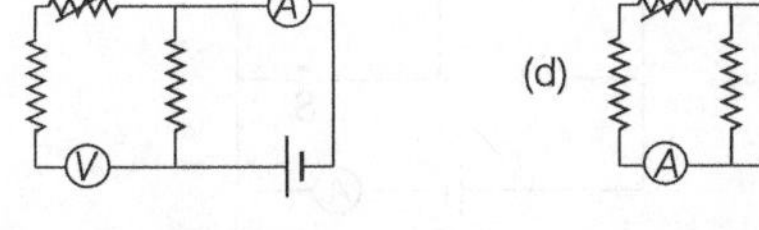

6 Six equal resistances are connected between points P, Q and R as shown in figure. Then net resistance will be maximum between → JEE Main (Online) 2013

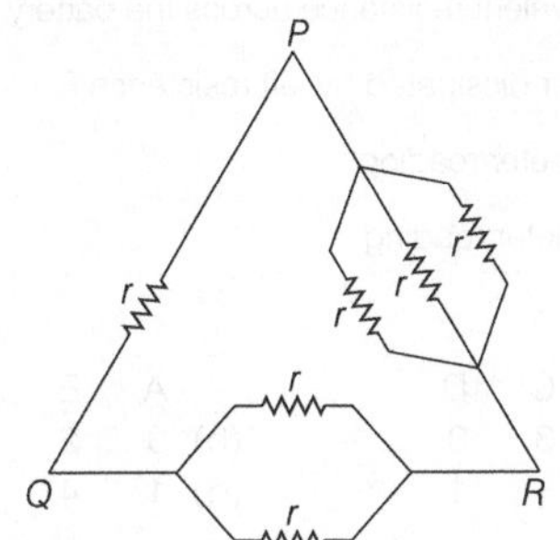

(a) P and R
(b) P and Q
(c) Q and R
(d) Any two points

7 An infinite ladder network of resistances is constructed with 1 Ω and 2 Ω resistance as shown in figure. The 6 V battery between A and B has negligible internal resistance, then effective resistance between A and B is

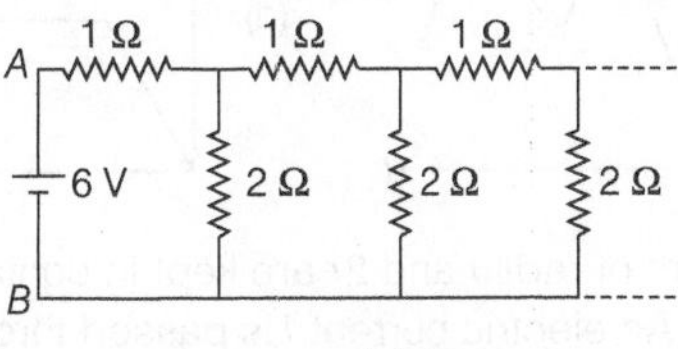

(a) 3 Ω
(b) 2 Ω
(c) 6/5 Ω
(d) 5/6 Ω

8 In the circuit in the figure below, the potential difference across P and Q will be nearest to

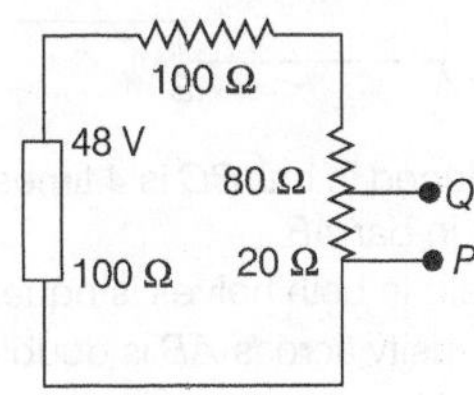

(a) 9.6 V
(b) 6.6 V
(c) 4 V
(d) 3.2 V

9 The charge on the 4 μF capacitor in the steady state, is

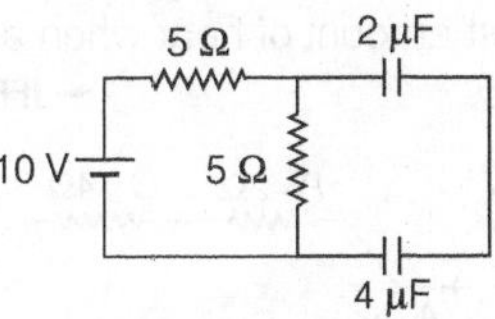

(a) $\frac{20}{3}$ μC
(b) $\frac{10}{3}$ μC
(c) $\frac{5}{6}$ μC
(d) 10μC

10 The temperature dependence of resistances of Cu and undoped Si in the temperature range 300-400 K, is best described by → JEE Main 2016 (Offline)

(a) linear increase for Cu, linear increase for Si
(b) linear increase for Cu, exponential increase for Si
(c) linear increase for Cu, exponential decrease for Si
(d) linear decrease for Cu, linear decrease for Si

11 When an electric heater is switched ON, the current flowing through it is plotted against time (t). Taking into account the variation of resistance with temperature, which of the following best represents the variations

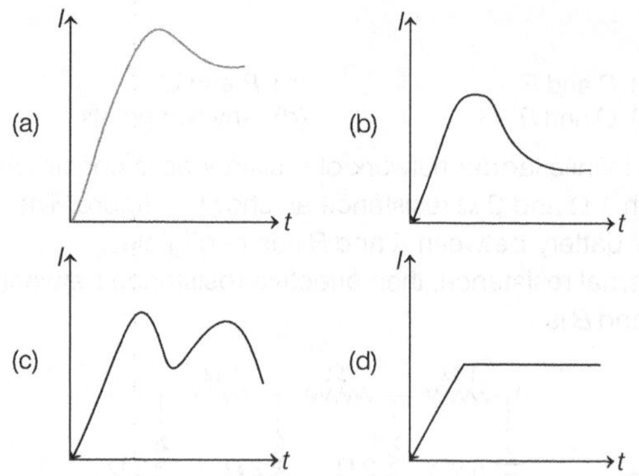

12 Two bars of radii r and $2r$ are kept in contact as shown. An electric current i is passed through the bars. Which one of the following is correct?

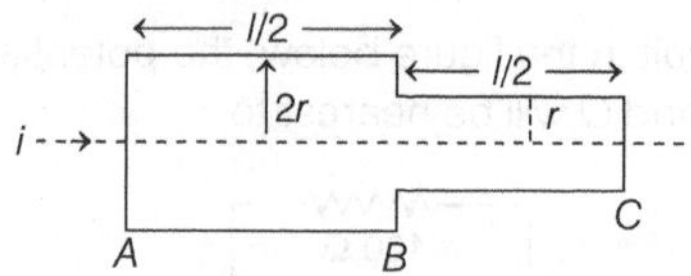

(a) Heat produced in bar BC is 4 times the heat produced in bar AB

(b) Electric field in both halves is equal

(c) Current density across AB is double that of across BC

(d) Potential difference across AB is 4 times that of across BC

13 Which of the four resistances P, Q, R and S generate the greatest amount of heat when a current flows from A to B? → JEE Main (Online) 2013

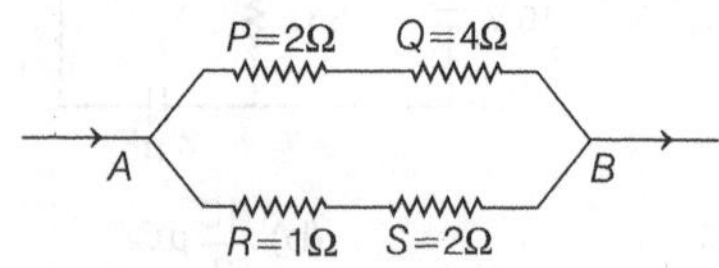

(a) Q (b) S (c) P (d) R

14 A DC source of emf $E_1 = 100$ V and internal resistance $r = 0.5\ \Omega$, a storage battery of emf $E_2 = 90$ V and an external resistance R are connected as shown in figure. For what value of R no current will pass through the battery? → JEE Main (Online) 2013

E_1 $r=0.5\ \Omega$ E_2 R

(a) $5.5\ \Omega$ (b) $3.5\ \Omega$
(c) $4.5\ \Omega$ (d) $2.5\ \Omega$

15 Two batteries of emf E_1 and $E_2 (E_2 > E_1)$ and internal resistances r_1 and r_2, respectively are connected in parallel as shown in figure.

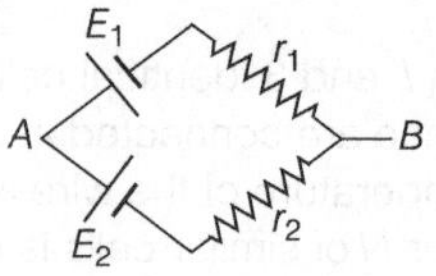

(a) Two equivalent emf E_{eq} of the two cells is between E_1 and E_2, i.e. $E_1 < E_{eq} < E_2$

(b) The equivalent emf E_{eq} is smaller than E_1

(c) The E_{eq} is given by $E_{eq} = E_1 + E_2$ always

(d) E_{eq} is independent of internal resistance r_1 and r_2

16 Match the following columns.

	Column I		Column II
A.	Smaller the resistance greater the current	1.	If the same voltage is applied
B.	Greater the resistance smaller the current	2.	If the same current is passed
C.	Greater the resistance smaller the power	3.	When resistances are connected in series
D.	Greater the resistance greater the power	4.	When resistances are connected in parallel

Codes

	A	B	C	D		A	B	C	D
(a)	1	2	3	4	(b)	4	1	2	3
(c)	4	3	1	2	(d)	3	4	2	1

17 In the circuit shown, battery, ammeter and voltmeter are ideal and the switch S is initially closed as shown in figure. When switch S is opened, match the parameter of column I with the effects in column II.

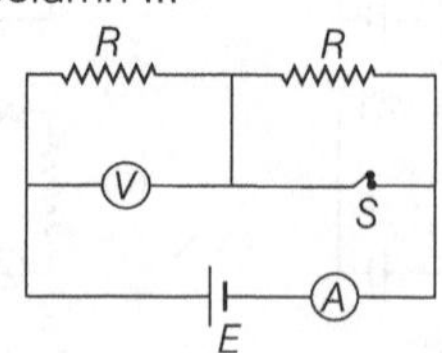

	Column I		Column II
A.	Equivalent resistance across the battery	1.	Remains same
B.	Power dissipated by left resistance R	2.	Increases
C.	Voltmeter reading	3.	Decreases
D.	Ammeter reading	4.	Becomes zero

Codes

	A	B	C	D		A	B	C	D
(a)	2	3	3	3	(b)	3	2	2	2
(c)	4	1	1	1	(d)	1	4	4	4

18 In the circuit shown in figure. The point F is grounded. Which of the following is wrong statement?

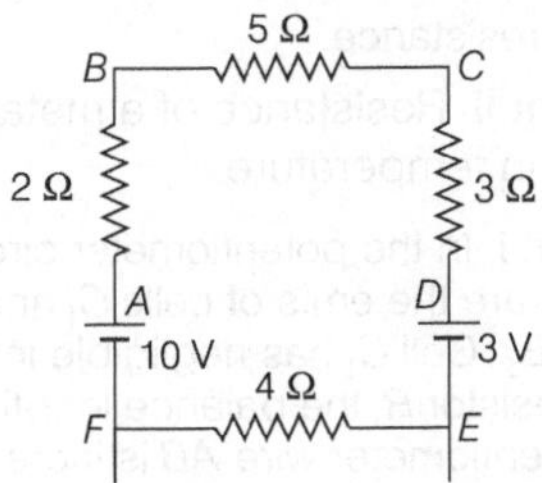

(a) D is at 5 V
(b) E is at zero potential
(c) The current in the circuit will be 0.5 A
(d) None of the above

19 The reading of ammeter as shown in figure is,

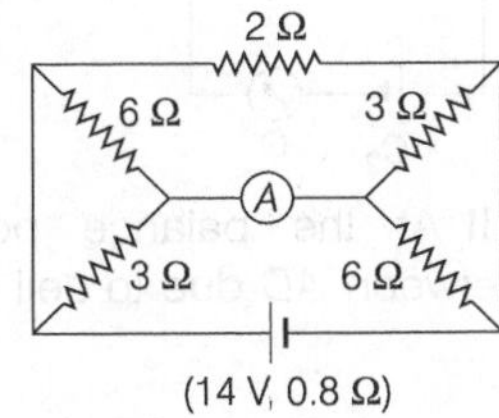

(a) 6.56 A (b) 3.28 A
(c) 2.18 A (d) 1.09 A

20 What is the potential difference between points A and D of circuit as shown in figure?

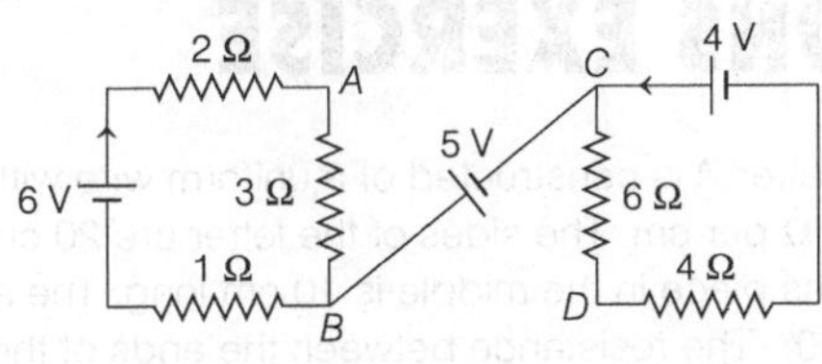

(a) 5 V (b) 9 V
(c) 10.4 V (d) 11.4 V

21 In the circuit shown below, the current in the 1 Ω resistor is → JEE Main 2014

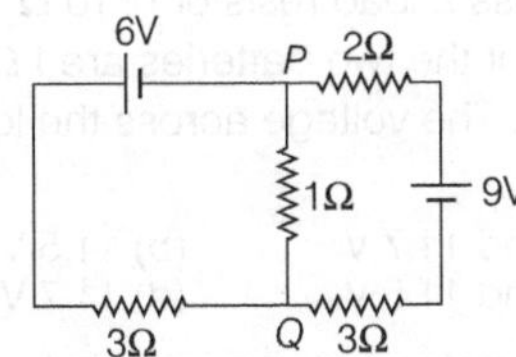

(a) 1.3A, from P to Q
(b) 0A
(c) 0.13 A, from Q to P
(d) 0.13 A, from P to Q

22 In the below circuit, the current in each resistance is → JEE Main 2017 (Offline)

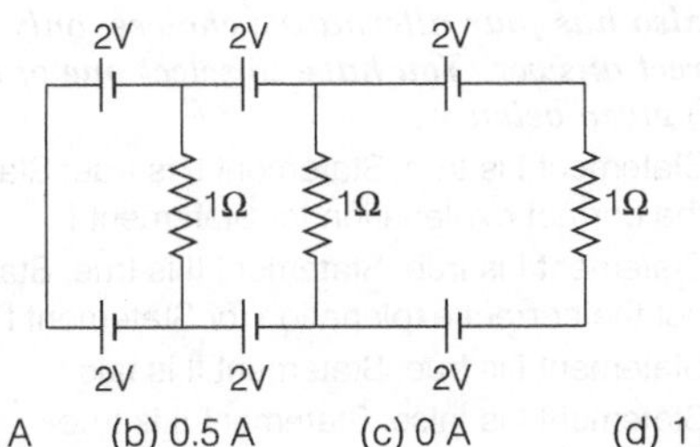

(a) 0.25 A (b) 0.5 A (c) 0 A (d) 1 A

23 Which of the following statements is false? → JEE Main 2017 (Offline)

(a) In a balanced Wheatstone bridge, if the cell and the galvanometer are exchanged, the null point is disturbed
(b) A rheostat can be used as a potential divider
(c) Kirchhoff's second law represents energy conservation
(d) Wheatstone bridge is the most sensitive when all the four resistances are of the same order of magnitude

24 In a metre bridge experiment, null point is obtained at 40 cm from one end of the wire when resistance X is balanced against another resistance Y. If $X < Y$, then the new position of the null points from the same end, if one decides to balance a resistance of $3X$ against Y, will be close to → JEE Main (Online) 2013

(a) 80 cm (b) 75 cm (c) 67 cm (d) 50 cm

25 On interchanging the resistances, the balance point of a meter bridge shifts to the left by 10 cm. The resistance of their series combination is 1 kΩ. How much was the resistance on the left slot before interchanging the resistances? → JEE Main 2018

(a) 990 Ω (b) 505 Ω (c) 550 Ω (d) 910 Ω

26 In an experiment to measure the internal resistance of a cell by potentiometer, it is found that the balance point is at a length of 2 m when the cell is shunted by a 4 Ω resistance; and is at a length of 3 m when the cell is shunted by a 8 Ω resistance. The internal resistance of the cell is, then

(a) 12 Ω (b) 8 Ω (c) 16 Ω (d) 1 Ω

27 A 6 V battery is connected to the terminals of a 3 m long wire of uniform thickness and resistance of 100 Ω. The difference of potential between two points on the wire separated by a distance of 50 cm will be

(a) 2 V (b) 3 V (c) 1 V (d) 15 V

28 The current in the primary circuit of potentiometer is 0.2A. The specific resistance and cross-section of the potentiometer wire are 4×10^{-7} Ω m and 8×10^{-7} m^2 respectively. The potential gradient will be equal to → AIEEE 2011

(a) 0.2 V/m (b) 1 V/m (c) 0.3 V/m (d) 0.1 V/m

Direction (Q. Nos. 29-32) *Each of these questions contains two statements : Statement I and Statement II. Each of these question also has four alternative choices, only one of which is the correct answer. You have to select one of the codes (a), (b), (c), (d) given below*

(a) Statement I is true, Statement II is true; Statement II is the correct explanation for Statement I
(b) Statement I is true, Statement II is true; Statement II is not the correct explanation for Statement I
(c) Statement I is true; Statement II is false
(d) Statement I is false; Statement II is true

29 **Statement I** As temperature decreases, the relaxation time of a conducting material increases.

Statement II Number of collisions per unit time of electrons with lattice ions increases as the temperature increases.

30 **Statement I** Potential difference across the terminals of a battery can be greater than its emf.

Statement II When current is taken from battery, $V = E - Ir$ (Symbols have their usual meaning).

31 **Statement I** In a meter bridge experiment, null point for an unknown resistance is measured. Now, the unknown resistance is put inside an enclosure maintained at a higher temperature. The null point can be obtained at the same point as before by decreasing the value of the standard resistance.

Statement II Resistance of a metal increases with increase in temperature.

32 **Statement I** In the potentiometer circuit shown in figure, E_1 and E_2 are the emfs of cells C_1 and C_2 respectively with $E_1 > E_2$. Cell C_1 has negligible internal resistance. For a given resistor R, the balance length is x. If the diameter of the potentiometer wire AB is increased, the balance length x will decrease.

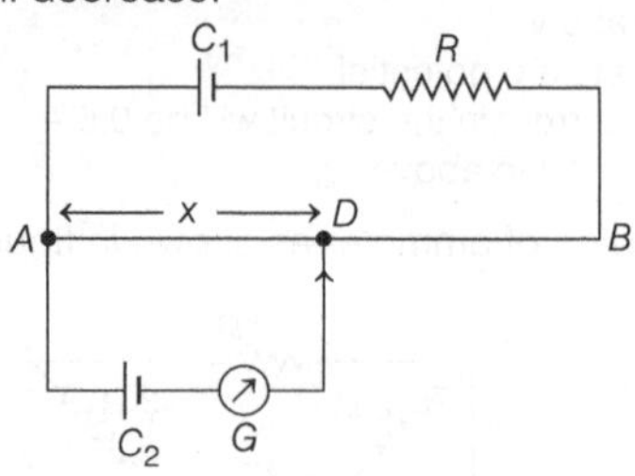

Statement II At the balance point, the potential difference between AD due to cell $C_1 = E_2$, the emf of cell C_2.

DAY PRACTICE SESSION 2

PROGRESSIVE QUESTIONS EXERCISE

1 If a wire is stretched to make it 0.1% longer, its resistance will → AIEEE 2012

(a) increase by 0.2% (b) decrease by 0.2%
(c) decrease by 0.05 % (d) increase by 0.05%

2 In the given circuit diagram, when the current reaches steady state in the circuit, the charge on the capacitor of capacitance C will be

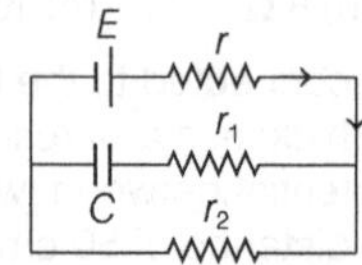

→ JEE Main 2017 (Offline)]

(a) $CE\dfrac{r_1}{(r_2 + r)}$ (b) $CE\dfrac{r_2}{(r + r_2)}$
(c) $CE\dfrac{r_1}{(r_1 + r)}$ (d) CE

3 When 5V potential difference is applied across a wire of length 0.1 m, the drift speed of electrons is 2.5×10^{-4} ms^{-1}. If the electron density in the wire is 8×10^{28} m^{-3} the resistivity of the material is close to → JEE Main 2015

(a) $1.6 \times 10^{-8}\,\Omega$m (b) $1.6 \times 10^{-7}\,\Omega$m
(c) $1.6 \times 10^{-6}\,\Omega$m (d) $1.6 \times 10^{-5}\,\Omega$m

4 A letter A is constructed of a uniform wire with resistance 1.0 Ω per cm. The sides of the letter are 20 cm and the cross piece in the middle is 10 cm long. The apex angle is 60°. The resistance between the ends of the legs is close to → JEE Main (Online) 2013

(a) 50.0 Ω (b) 10 Ω
(c) 36.7 Ω (d) 26.7 Ω

5 Two batteries with emf 12 V and 13 V are connected in parallel across a load resistor of 10 Ω. The internal resistances of the two batteries are 1 Ω and 2 Ω, respectively. The voltage across the load lies between → JEE Main 2018

(a) 11.6 V and 11.7 V (b) 11.5 V and 11.6 V
(c) 11.4 V and 11.5 V (d) 11.7 V and 11.8 V

6 In a potentiometer experiment, it is found that no current passes through the galvanometer when the terminals of the cell are connected across 52 cm of the potentiometer wire. If the cell is shunted by a resistance of 5 Ω, a balance is found when the cell is connected across 40 cm of the wire. Find the internal resistance of the cell. → JEE Main 2018

(a) 1 Ω (b) 1.5 Ω (c) 2 Ω (d) 2.5 Ω

7 Two conductors have the same resistance at 0°C but their temperature coefficients of resistance are α_1 and α_2. The respective temperature coefficients of their series and parallel combinations are nearly → AIEEE 2010

(a) $\frac{\alpha_1+\alpha_2}{2}, \alpha_1+\alpha_2$ (b) $\alpha_1+\alpha_2, \frac{\alpha_1+\alpha_2}{2}$

(c) $\alpha_1+\alpha_2, \frac{\alpha_1\alpha_2}{\alpha_1+\alpha_2}$ (d) $\frac{\alpha_1+\alpha_2}{2}, \frac{\alpha_1+\alpha_2}{2}$

8 There are two concentric spheres of radius a and b respectively. If the space between them is filled with medium of resistivity ρ, then the resistance of the intergap between the two spheres will be

(a) $\frac{\rho}{4\pi(b+a)}$ (b) $\frac{\rho}{4\pi}\left(\frac{1}{b}-\frac{1}{a}\right)$

(c) $\frac{\rho}{4\pi}\left(\frac{1}{a^2}-\frac{1}{b^2}\right)$ (d) $\frac{\rho}{4\pi}\left(\frac{1}{a}-\frac{1}{b}\right)$

9 To find the resistance of a galvanometer by the half deflection method the following circuit is used with resistances $R_1 = 9970\ \Omega$, $R_2 = 30\ \Omega$ and $R_3 = 0$. The deflection in the galvanometer is d. With $R_3 = 107\ \Omega$ the deflection changed to $\frac{d}{2}$. The galvanometer resistance is approximately → JEE Main (Online) 2013

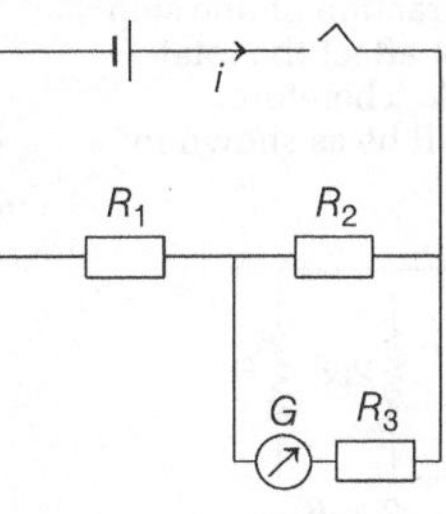

(a) 107 Ω (b) 137 Ω (c) 107/2 Ω (d) 77 Ω

10 A source of emf $E = 10$ V and having negligible internal resistance is connected to a variable resistance. The resistance varies as shown in figure. The total charge that has passed through the resistor R during the time interval from t_1 to t_2 is

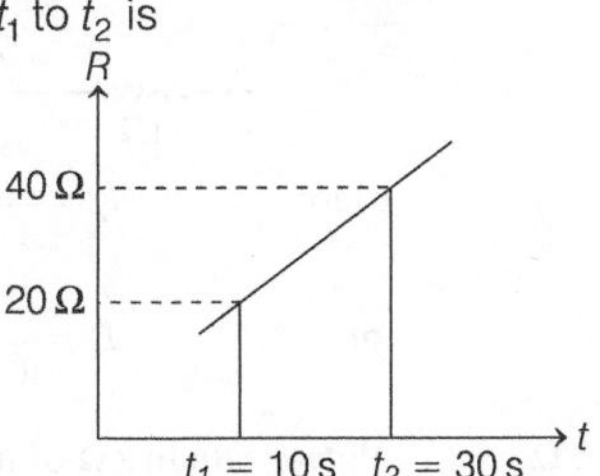

(a) $40 \log_e 4$ (b) $30 \log_e 3$

(c) $20 \log_e 2$ (d) $10 \log_e 2$

11 The potential difference across 8Ω resistance is 48V as shown in figure. The value of potential difference across points A and B will be

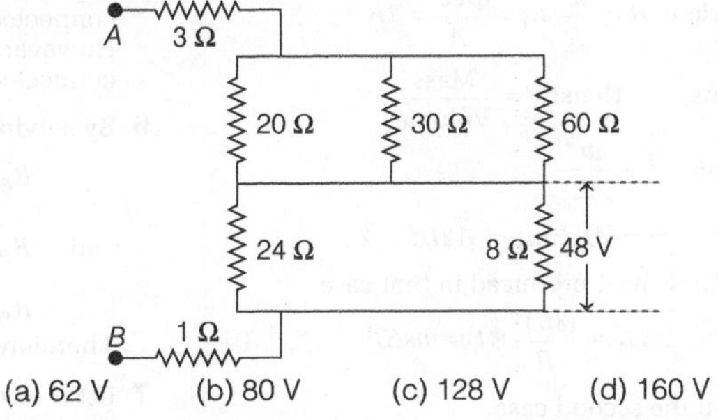

(a) 62 V (b) 80 V (c) 128 V (d) 160 V

12 The V-I graph for a conductor at temperatures T_1 and T_2 are as shown in the figure, The term $T_2 - T_1$ is proportional to

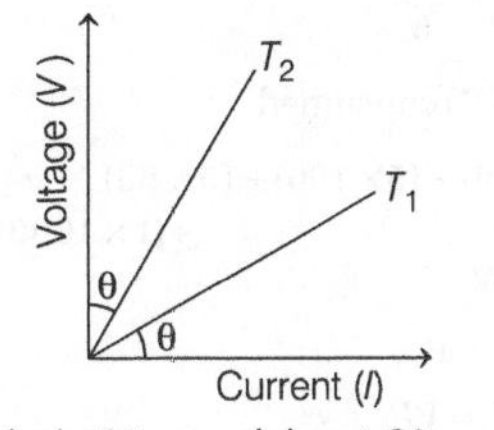

(a) cos 2θ (b) sin 2θ (c) cot 2θ (d) tan 2θ

13 The charge supplied by source varies with time t as $Q = at - bt^2$. The total heat produced in resistor $2R$ is

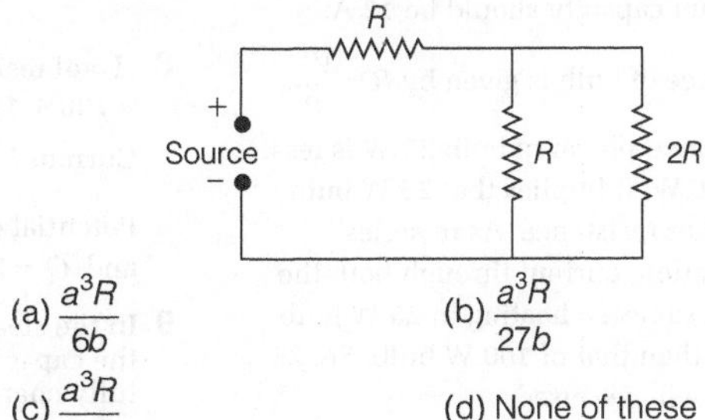

(a) $\frac{a^3R}{6b}$ (b) $\frac{a^3R}{27b}$

(c) $\frac{a^3R}{3b}$ (d) None of these

ANSWERS

SESSION 1	1 (b)	2 (c)	3 (c)	4 (c)	5 (a)	6 (b)	7 (b)	8 (d)	9 (a)	10 (c)
	11 (b)	12 (a)	13 (b)	14 (c)	15 (a)	16 (a)	17 (a)	18 (b)	19 (c)	20 (c)
	21 (c)	22 (c)	23 (a)	24 (c)	25 (c)	26 (b)	27 (c)	28 (d)	29 (a)	30 (b)
	31 (d)	32 (d)								
SESSION 2	1 (a)	2 (b)	3 (d)	4 (d)	5 (b)	6 (b)	7 (d)	8 (d)	9 (d)	10 (d)
	11 (d)	12 (c)	13 (b)							

Hints and Explanations

SESSION 1

1 Here, $R = \frac{\rho L}{A}; R_1 = \frac{\rho 2L}{A} = 2R$

As, $\text{Density} = \frac{\text{Mass}}{\text{Volume}}$

i.e. $d = \frac{m}{AL}$

$\therefore \quad m = ALd; m_1 = A2Ld = 2m$

Now, heat produced in first case

$$H_1 = \frac{(3E)^2}{R} \times t = ms\Delta T \quad \text{...(i)}$$

In the second case,

$$H_2 = \frac{(NE)^2}{2R} \times t = 2ms\Delta T \quad \text{...(ii)}$$

On solving Eqs. (i) and (ii), we get

$$N = 6$$

2 Total power (P) consumed

$= (15 \times 40) + (5 \times 100) + (5 \times 80) + (1 \times 1000)$

$= 2500\,\text{W}$

As we know that,

Power, i.e. $P = VI$

$\Rightarrow \quad I = \frac{2500}{220}\text{A} = \frac{125}{11}$

$= 11.3\,\text{A}$

Minimum capacity should be 12 A.

3 Resistance of bulb is given by $R = \frac{V^2}{P}$.

As the rated power of bulb 25 W is less than 100 W, it implies that 25 W bulb has higher resistance. As in series combination, current through both the bulbs is same, so heating in 25 W bulb is more than that of 100 W bulb. So, 25 W bulb will get fused.

4 Numbers attached for brown, black, green and silver are 1, 0, 5, ± 10%. Therefore, the resistance of given resistor

$= 10 \times 10^5\ \Omega \pm 10\%$

$= 1.0 \times 10^6\ \Omega \pm 10\%$

5 Ohm's law states that the current (I) flowing through a conductor is directly proportional to the potential difference (V) across its ends provided its physical conditions such as temperature, mechanical strain, etc. kept constant,

i.e. $\quad I \propto V$ or $V \propto I$

or $\quad V = RI$ (where, R is constant)

Thus, in order to study Ohm's law experimentally, voltmeter (V) should be connected parallel to the resistor. However, ammeter (A) should be connected in series with the resistor.

6 By solving this, we get net resistance as,

$$R_{PQ} = \frac{5}{11}r,\ R_{QR} = \frac{4}{11}r$$

and $\quad R_{PR} = \frac{3}{11}r$

$$R_{PQ} > R_{QR} > R_{PR}$$

Therefore, R_{PQ} is maximum.

7 Let R be the resistance of infinite ladder. The addition or subtraction of one step in the ladder will not affect the total resistance of network. Therefore, equivalent circuit will be as shown in figure.

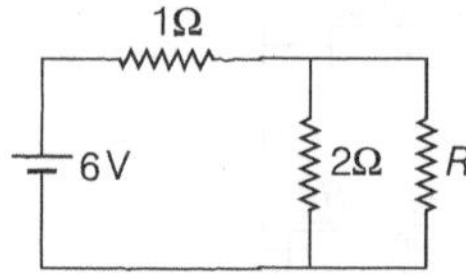

Total resistance $= 1 + \frac{2 \times R}{R + 2} = R$

or $\quad R + 2 + 2R = R^2 + 2R$

$\Rightarrow \quad R^2 - R - 2 = 0$

On solving, we get

$$R = 2\,\Omega$$

8 Total resistance of circuit

$= 100 + 100 + 80 + 20 = 300\Omega$

Current $I = \frac{48}{300} = 0.16\,\text{A}$

Potential difference across P and $Q = 20 \times 0.16 = 3.2$ V

9 In the steady state, the capacitors are fully charged and acts as open circuit, so the equivalent circuit in steady state would be as shown alongside figure.

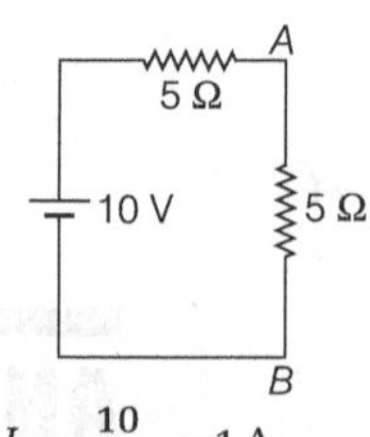

Steady state current $I = \frac{10}{5 + 5} = 1\,\text{A}$

So, potential drop across AB is

$$V = 5\text{V}$$

Sum of potential difference across 2μF and 4 μF capacitors is 5V. As capacitors are in series, charges on them would be same, let us say it is q.

From KVL, $\frac{q}{2} + \frac{q}{4} = 5$

$\Rightarrow \quad q = \frac{20}{3}\,\mu\text{C}$

10 As, we know copper is a conductor, so increase in temperature, increases the resistance. Then, silicon (Si) is semiconductor, so with increase in temperature, resistance will decrease.

11 The filament of the heater reaches its steady resistance when the heater reaches the steady temperature, which is much higher than room temperature. The resistance at room temperature is then much lower than the resistance of its steady state.

When the heater is switched ON, it draws a larger current than its steady state current as the filament heats up, its resistance increases and current falls to steady state value.

12 Current flowing through both the bars is equal. Now, the heat produced is given by

$$H = I^2Rt$$

or $\quad H \propto R$

or $\quad \frac{H_{AB}}{H_{BC}} = \frac{R_{AB}}{R_{BC}} = \frac{(1/2r)^2}{(1/r)^2} \quad \left(\because R \propto \frac{1}{A} \propto \frac{1}{r^2}\right)$

$= \frac{1}{4}$

or $\quad H_{BC} = 4H_{AB}$

13 We know that, $I \propto \frac{1}{R}$,

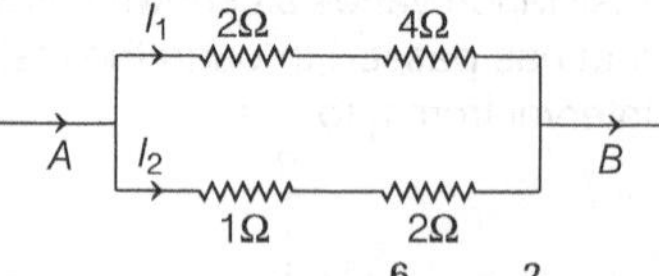

Here, $\quad I_2 = \frac{6}{3 + 6}I = \frac{2}{3}I$

or $\quad I_1 = \frac{3}{6 + 3}I = \frac{1}{3}I$

Power rate in 2Ω of upper series

$= 2 \times \left(\frac{1}{3}I\right)^2 = \frac{2}{9}I^2$

Power rate in 4Ω of upper series

$= 4 \times \left(\frac{1}{3}I\right)^2 = \frac{4}{9}I^2$

Power rate of 1 Ω in lower series

$= 1 \times \left(\frac{2}{3}I\right)^2 = \frac{4}{9}I^2$

Power rate of 2Ω in lower series

$= 2 \times \left(\frac{2}{3}I\right)^2 = \frac{8}{9}I^2$

$\therefore$ Greatest amount of heat is generated by S.

14 Given, $E_1 = 100\,\text{V}$,

$$r = 0.5\,\Omega,$$
$$E_2 = 90\,\text{V}$$

External resistance $= R$

For no current pass through the battery

$$\frac{100}{R + r} = \frac{90}{R}$$

$$\Rightarrow \quad \frac{10}{R + \frac{1}{2}\,\Omega} = \frac{9}{R}$$

$$\Rightarrow \quad 10R = 9R + 4.5\,\Omega$$

$$\therefore \quad R = 4.5\,\Omega$$

15 The equivalent emf of this combination is given by

$$E_{eq} = \frac{E_2 r_1 + E_1 r_2}{r_1 + r_2}$$

This suggest that the equivalent emf E_{eq} of the two cells is given by $E_1 < E_{eq} < E_2$.

16 Ohm's law, $I = \frac{V}{R}$

and power $= I^2 R = \frac{V^2}{R}$

When the resistors are connected in series, the effective resistance is more than that as when they are connected in parallel.

17 When switch S is opened, then right side resistance R which was short circuited earlier contributes to equivalent resistance. Hence, equivalent resistance across the battery increases, power dissipated by left resistance R decreases, voltmeter reading decreases and ammeter reading decreases.

18 Effective emf of circuit $= 10 - 3 = 7\,\text{V}$

Total resistance of circuit

$$= 2 + 5 + 3 + 4$$
$$= 14\,\Omega$$

Current, $I = 7/14 = 0.5\,\text{A}$

Potential difference between A and $D = 0.5 \times 10 = 5\,\text{V}$

Potential at $D = 10 - 5 = 5\,\text{V}$

Potential at $E = 5 - 3 = 2\,\text{V}$

Hence, E cannot be at zero potential, as there is potential drop at E.

19 The equivalent circuit of the given circuit will be reduced to as shown in figure.

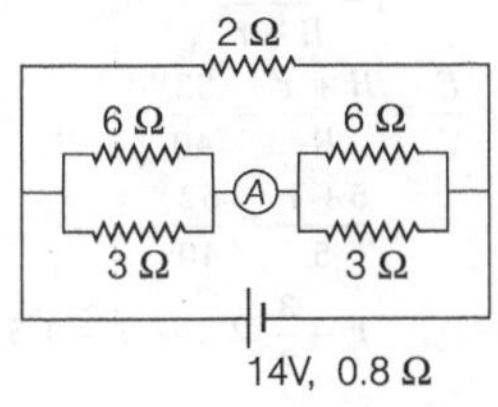

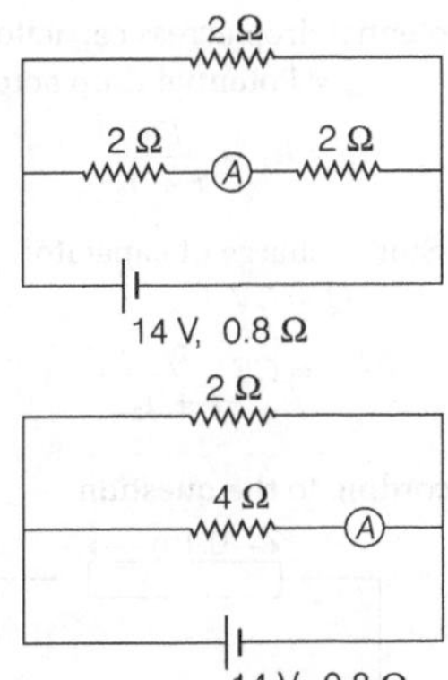

Total resistance of the circuit

$$= \frac{2 \times 4}{2 + 4} + 0.8 = \frac{8}{6} + 0.8 = \frac{12.8}{6}\,\Omega$$

Main current in the circuit

$$= \frac{14}{(12.8\,/\,6)} = \frac{84}{12.8}\,\text{A}$$

Reading of ammeter

$$= \frac{84}{12.8} \times \frac{2}{6} = 2.18\,\text{A}$$

20 Let I_1 and I_2 be the currents drawn from cells of emf 6 V and 4 V in the circuits, respectively. Then,

$$I_1 = \frac{6}{2 + 3 + 1} = 1\,\text{A}$$

and $$I_2 = \frac{4}{6 + 4} = 0.4\,\text{A}$$

$$V_A - V_B = 1 \times 3 = 3\,\text{V};$$
$$V_B - V_C = 5\,\text{V}$$

and $V_C - V_D = 0.4 \times 6 = 2.4\,\text{V}$

$$\therefore \quad V_A - V_D = (V_A - V_B) + (V_B - V_C) + (V_C - V_D)$$
$$= 3 + 5 + 2.4 = 10.4\,\text{V}$$

21 Connect point Q to ground and apply KCL. Consider the grounded circuit as shown below.

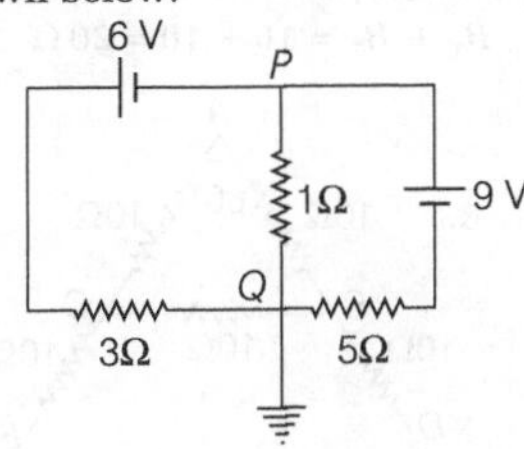

Applying KCL at point Q, we can write

Incoming current at Q = outgoing current from Q

$$\Rightarrow \quad \frac{V + 6}{3} + \frac{V}{1} = \frac{9 - V}{5}$$

or $$V\left[\frac{1}{3} + \frac{1}{5} + 1\right] = \frac{9}{5} - 2$$

or $$V\left[\frac{5 + 3 + 15}{15}\right] = \frac{9 - 10}{5}$$

or $$V\left[\frac{23}{15}\right] = \frac{-1}{5}$$

or $$V = \frac{-3}{23} = -0.13\,\text{V}$$

$\therefore$ Current $= -1 \times 0.13$

$$= -0.13\,\text{A}$$

22 Each resistance is connected with two cells combined in opposite direction, so potential drop across each resistor is zero. Hence, the current through each of resistor is zero.

23 In a balanced Wheatstone bridge, there is no effect on position of null point, if we exchange the battery and galvanometer. So, option (a) is incorrect.

24 As, $\frac{x}{y} = \frac{40}{100 - 40} = \frac{40}{60}$

$$\frac{3x}{y} = 3\left(\frac{40}{60}\right) = \frac{120}{60} = \frac{2}{1}$$

Now, the total length $= 100$

$\therefore$ Required length $= \frac{100}{3} \times 2 = 67\,\text{cm}$

25 We have, $X + Y = 1000\,\Omega$

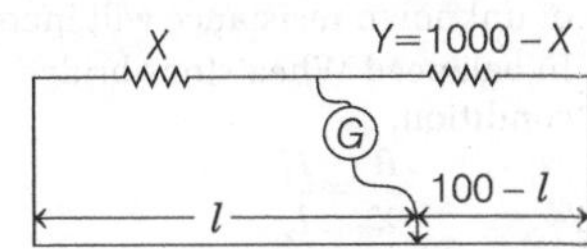

Initially, $$\frac{X}{l} = \frac{1000 - X}{100 - l} \quad \text{...(i)}$$

When X and Y are interchanged, then

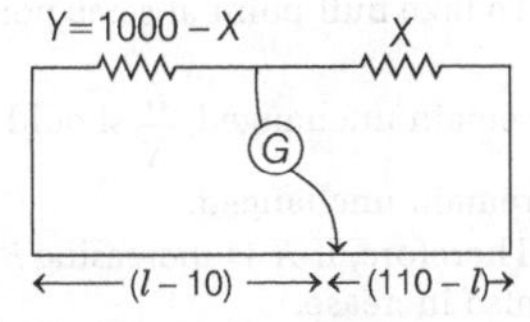

$$\frac{1000 - X}{l - 10} = \frac{X}{100 - (l - 10)}$$

or $$\frac{1000 - X}{l - 10} = \frac{X}{110 - l} \quad \text{...(ii)}$$

From Eqs. (i) and (ii), we get

$$\frac{100 - l}{l} = \frac{l - 10}{110 - l}$$

$$(100 - l)(110 - l) = (l - 10)l$$

$$11000 - 100l - 110l + l^2 = l^2 - 10l$$

$$\Rightarrow \quad 11000 = 200l$$

$$\therefore \quad l = 55\,\text{cm}$$

Substituting the value of l in Eq. (i), we get

$$\frac{X}{55} = \frac{1000 - X}{100 - 55}$$

$$\Rightarrow \quad 20X = 11000$$

$$\therefore \quad X = 550\,\Omega$$

26 As, $\left(\frac{l-2}{2}\right)4 = \left(\frac{l-3}{3}\right)8$

$\Rightarrow l = 6$

Therefore, $r = \left(\frac{l-2}{2}\right)4 = 8\,\Omega$

27 Potential gradient along the wire,

$K = \frac{6}{300}$ V/cm

Potential difference across 50 cm length is

$$V = k \times 50 = \frac{6}{300} \times 50 = 1\text{ V}$$

28 Potential gradient of a potentiometer,

$$K = \frac{I\rho}{A} = \frac{0.2 \times 4 \times 10^{-7}}{8 \times 10^{-7}} = 0.1\text{ V/m}$$

30 When the battery is undergoing charging processes, then

$$V = E + Ir > E$$

So, Statement I is correct.

Statement II is also correct but not explaining Statement I.

31 With increase in temperature, the value of unknown resistance will increase.

In balanced Wheatstone bridge condition,

$$\frac{R}{X} = \frac{l_1}{l_2}$$

Here,

R = value of standard resistance,

X = value of unknown resistance.

To take null point at same point or $\frac{l_1}{l_2}$ to remain unchanged, $\frac{R}{X}$ should also remain unchanged.

Therefore, if X is increasing R, should also increase.

32 If the diameter of wire AB is increased, its resistance will decrease. Hence, the potential difference between A and B due to cell C_1 will decrease. Therefore, the null point will be obtained at a higher value of x.

SESSION 2

1 $R = \frac{\rho l}{A} = \frac{\rho l^2}{V}$ (V = volume)

$\therefore \frac{\Delta R}{R} = 2\frac{\Delta l}{l} = +\ 0.2\%$

2 In steady state, no current flows through the capacitor. So, resistance r_1 becomes ineffective.

So, the current in circuit,

$$I = \frac{E}{r + r_2 \text{ (Total Resistance)}}$$

$\because$ Potential drop across capacitor

= Potential drop across r_2

$$= Ir_2 = \frac{Er_2}{r + r_2}$$

$\therefore$ Stored charge of capacitor,

$$Q = CV$$

$$= CE\frac{r_2}{r + r_2}$$

3 According to the question,

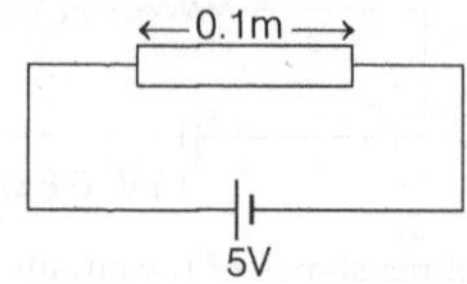

$v_d = 2.5 \times 10^{-4}$ m/s

$\Rightarrow n = 8 \times 10^{28}$ / m^3

We know that,

$J = nev_d$

or $I = n_e v_d A$

where, symbols have their usual meaning.

$\Rightarrow \frac{V}{R} = nev_d A$

or $\frac{VA}{\rho L} = nev_d A$

or $\frac{V}{\rho L} = nev_d$

or $\rho = \frac{V}{nev_d L}$

$$= \frac{5}{8 \times 10^{28} \times 1.6 \times 10^{-19} \times 2.5 \times 10^{-4} \times 0.1}$$

$$\rho = 1.6 \times 10^{-5}\ \Omega\text{ m}$$

4 We have, equivalent resistance in series

$R_1 + R_2 = 10 + 10 = 20\,\Omega$

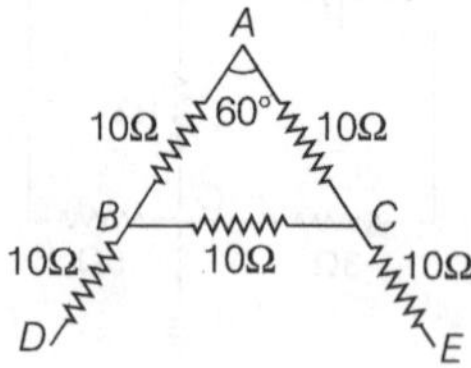

and in parallel $= \frac{1}{R_P} = \frac{1}{20} + \frac{1}{10} = \frac{3}{20}$

$\Rightarrow R_P = \frac{20}{3}$

Therefore, $R_{eq} = \frac{20}{3} + 10 + 10$

$$= \frac{20 + 30 + 30}{3}$$

$$= \frac{80}{3} = 26.66 = 26.7\,\Omega$$

5

For parallel combination of cells,

$$E_{eq} = \frac{\frac{E_1}{r_1} + \frac{E_2}{r_2}}{\frac{1}{r_1} + \frac{1}{r_2}}$$

$$\therefore E_{eq} = \frac{\frac{12}{1} + \frac{13}{2}}{\frac{1}{1} + \frac{1}{2}} = \frac{37}{3}\text{ V}$$

Potential drop across 10 Ω resistance,

$$V = \left(\frac{E}{R_{\text{total}}}\right) \times 10 = \frac{37/3}{\left(10 + \frac{2}{3}\right)} \times 10$$

$= 11.56$ V

$\therefore V = 11.56$ V

6 With only the cell,

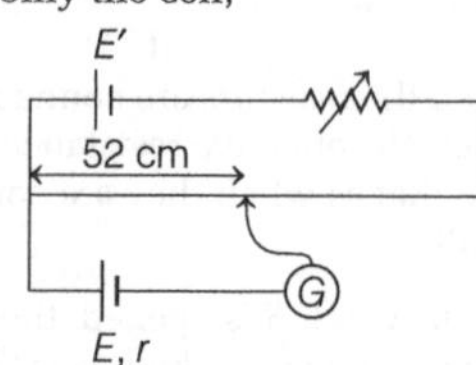

On balancing,

$E = 52 \times x$ ...(i)

where, x is the potential gradient of the wire.

When the cell is shunted,

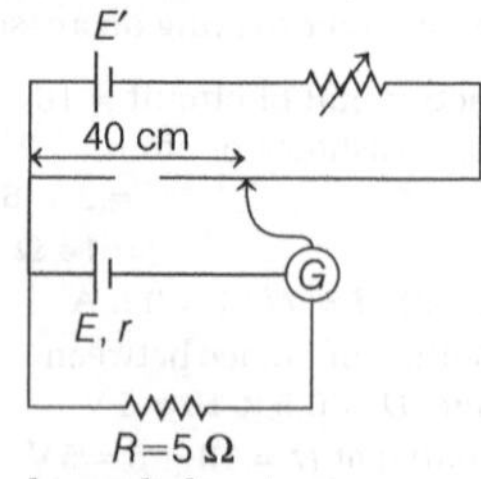

Similarly, on balancing,

$V = E - \frac{Er}{(R+r)} = 40 \times x$ ...(ii)

Solving Eqs. (i) and (ii), we get

$$\frac{E}{V} = \frac{1}{1 - \frac{r}{R+r}} = \frac{52}{40}$$

$\Rightarrow \frac{E}{V} = \frac{R+r}{R} = \frac{52}{40}$

$\Rightarrow \frac{5+r}{5} = \frac{52}{40}$

$\Rightarrow r = \frac{3}{2}\,\Omega \Rightarrow r = 1.5\,\Omega$

7 Let R_0 be the initial resistance of both conductors

$\therefore$ At temperature θ their resistances will be,

$$R_1 = R_0(1+\alpha_1\theta)$$

and $$R_2 = R_0(1+\alpha_2\theta)$$

For series combination,

$$R_s = R_1 + R_2$$

$$R_{s0}(1+\alpha_s\theta) = R_0(1+\alpha_1\theta) + R_0(1+\alpha_2\theta)$$

where, $R_{s0} = R_0 + R_0 = 2R_0$

$\therefore 2R_0(1+\alpha_s\theta) = 2R_0 + R_0\theta(\alpha_1+\alpha_2)$

or $$\alpha_s = \frac{\alpha_1+\alpha_2}{2}$$

For parallel combination,

$$R_{p0}(1+\alpha_p\theta) = \frac{R_0(1+\alpha_1\theta)R_0(1+\alpha_2\theta)}{R_0(1+\alpha_1\theta)+R_0(1+\alpha_2\theta)}$$

where, $R_{p0} = \frac{R_0R_0}{R_0+R_0} = \frac{R_0}{2}$

$$\therefore \frac{R_0}{2}(1+\alpha_p\theta) = \frac{R_0^2(1+\alpha_1\theta+\alpha_2\theta+\alpha_1\alpha_2\theta^2)}{R_0(2+\alpha_1\theta+\alpha_2\theta)}$$

as α_1 and α_2 are small quantities.

$\therefore \alpha_1, \alpha_2$ is negligible.

So, neglect $\alpha_1, \alpha_2, \theta^2$

or $$\alpha_p = \frac{\alpha_1+\alpha_2}{2+(\alpha_1+\alpha_2)\theta}$$

$$= \frac{\alpha_1+\alpha_2}{2}\left[1-\left(\frac{\alpha_1+\alpha_2}{2}\right)\theta\right]$$

[Binomial expansion]

as $(\alpha_1+\alpha_2)^2$ is negligible

$$\therefore \alpha_p = \frac{\alpha_1+\alpha_2}{2}$$

8 Consider a concentric spherical shell of radius x and thickness dx as shown in figure. Its resistance, dR is

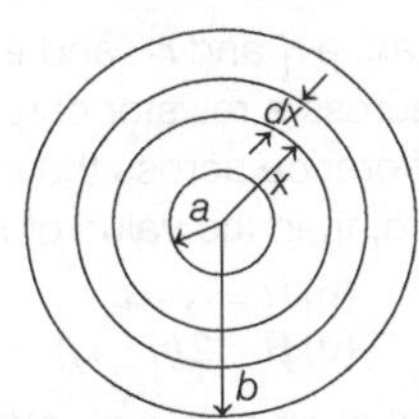

$$dR = \frac{\rho\,dx}{4\pi x^2} \quad \left(\because R = \frac{\rho l}{A}\right)$$

$\therefore$ Total resistance,

$$R = \frac{\rho}{4\pi}\int_a^b \frac{dx}{x^2}$$

$$= \frac{\rho}{4\pi}\left[\frac{1}{a} - \frac{1}{b}\right]$$

9 As at initial condition the deflection is d while $R_3 = 0$,

then equivalent resistance of R_2 and R_3 $= R_2 + R_3 = R_2 = 30\,\Omega$

Now, when $R_3 = 107\,\Omega$

and $R_2 = 30\,\Omega$

Then, if the deflection is $\frac{d}{2}$, so equivalent resistance should be

$$\frac{30}{2} = 15\,\Omega$$

It is only when equivalent resistance and R_3 and R_g will be parallel to R_2 giving resistance 15 Ω.

Let $R_3 - R_g$ = equivalent

$= 30\,\Omega = R$

$$\therefore \frac{1}{R_2} + \frac{1}{R} = \frac{1}{30} + \frac{1}{30} = \frac{1}{15}$$

$$\therefore R_{eq} = 15\,\Omega$$

Thus, R_g must will be 77 Ω in order to maintain

$$R_3 - R_g = 30$$

$$\Rightarrow 107 - R_g = 30$$

$$= R_g = 77\,\Omega$$

10 Let $R = at + b$

At $t = 10$ s, $R = 20\,\Omega$

$\therefore 20 = 10a + b$ ...(i)

At $t = 30a + b$ ...(ii)

Solving Eqs. (i) and (ii), we get

$$a = 1.0\,\Omega/\text{s}$$

and $$b = 10\,\Omega$$

$$\therefore R = (t+10)$$

$$I = \frac{E}{R} = \frac{10}{t+10}$$

$$\Delta q = \int_{10}^{30} I\,dt$$

$$= \int_{10}^{30}\left(\frac{10}{t+10}\right)dt$$

$$= 10\log_e(2)$$

11 Effective value of resistance of parallel combination of 20 Ω, 30 Ω, 60 Ω is R_1, where

$$\frac{1}{R_1} = \frac{1}{20} + \frac{1}{30} + \frac{1}{60}$$

$$= \frac{3+2+1}{60} = \frac{6}{60} = \frac{1}{10}$$

$$R_1 = 10\,\Omega$$

Similarly, effective value of parallel combination of 24 Ω and 8 Ω resistance is given by

$$R_2 = \frac{24\times 8}{24+8} = 6\,\Omega$$

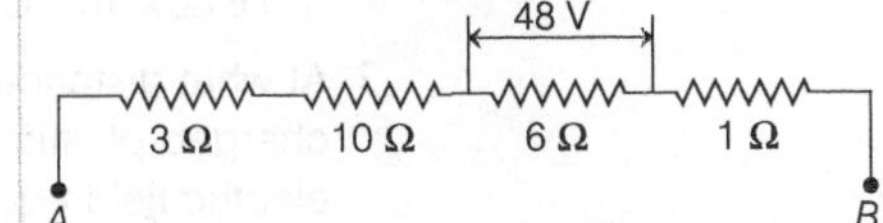

Hence, the circuit may be redrawn as shown in the adjacent figure, where total resistance across A and B,

$$R = 3 + 10 + 6 + 1 = 20\,\Omega.$$

As potential difference across R_2 (= 6 Ω) is 48 V, hence

$$V_{AB} = 48\times\frac{R}{R_2}$$

$$= \frac{48\times 20}{6} = 160\text{ V}$$

12 $R_1 = \tan\theta = R_0(1+\alpha T_1)$

and $R_2 = \cot\theta = R_0(1+\alpha T_2)$

$\cot\theta - \tan\theta$

$$= R_0(1+\alpha T_2) - R_0(1+\alpha T_1)$$

$$= R_0\alpha(T_2 - T_1)$$

or $$T_2 - T_1 = \frac{1}{\alpha R_0}(\cot\theta - \tan\theta)$$

$$= \frac{1}{\alpha R_0}\left(\frac{\cos\theta}{\sin\theta} - \frac{\sin\theta}{\cos\theta}\right)$$

$$= \frac{2\cos 2\theta}{\alpha R_0 \sin 2\theta}$$

$$= \frac{2}{\alpha R_0}\cot 2\theta$$

$$T_2 - T_1 \propto \cot 2\theta$$

13 $Q = at - bt^2$

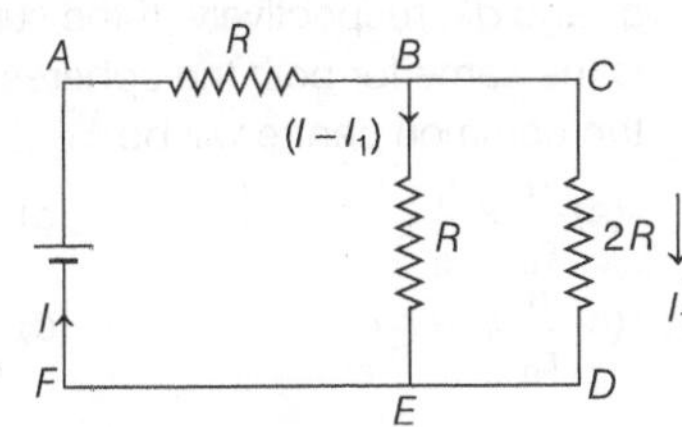

$$\therefore I = \frac{dQ}{dt} = a - 2bt$$

when, $t = t_0, I = 0,$

i.e., $a - 2bt = 0$ ...(i)

In loop $BCDEB$,

$I_1(2R) - (I - I_1)R = 0$ or $3I_1 = I$

$$I_1 = \frac{i}{3} = \frac{a-2bt}{3}$$

$$H = \int_0^{t_0}(I_1^2(2h))$$

$$= \frac{2R}{9}\int_0^{t_0}(a-2bt)^2\,dt$$

$$= \frac{2R}{9}\left[\int_0^{t_0}(a^2 - 4b^2t^2 - 4dt)dt\right]$$

$$= \frac{2R}{9}\left[\left\{a^2t + \frac{4b^2t^2}{3} - \frac{4bat^2}{2}\right\}_0^{t_0}\right]$$

$$= \frac{2R}{9}\left[a^2t_0 + \frac{4b^2t_0^3}{3} - 2bat_0^2\right]$$

$$t_0 = \frac{a}{2b}$$ [from Eq. (i)]

$$H = \frac{2R}{9}\left[a^2\times\frac{a}{2b} + \frac{4b^2}{3}\cdot\frac{a^3}{8b^3} - 2ab\frac{a^2}{4b^2}\right]$$

$$= \frac{2R}{9}\left[\frac{a^3}{2b} + \frac{a^3}{6b} - \frac{a^3}{2b}\right]$$

$$= \frac{a^3R}{27b}$$

DAY TWENTY

Unit Test 4

(Electrostatics and Current Electricity)

1 Two concentric spheres of radii r_1 and r_2 carry charges q_1 and q_2, respectively. If the surface charge density (σ) is the same for both the spheres, the electric potential at the common centre will be

(a) $\frac{\sigma}{\varepsilon_0} \times \frac{r_1}{r_2}$ (b) $\frac{\sigma}{\varepsilon_0} \times \frac{r_2}{r_1}$

(c) $\frac{\sigma}{\varepsilon_0}(r_1 - r_2)$ (d) $\frac{\sigma}{\varepsilon_0}(r_1 + r_2)$

2 Point charge q moves from point P to point S along the path $PQRS$ (figure shown) in a uniform electric field E pointing coparallel to the positive direction of the X-axis. The coordinates of the points P, Q, R and S $(a,b,0)$ $(2a,0,0)$ $(a,-b,0)$ and $(0,0,0)$ respectively. The work done by the field in the above process is given by the expression

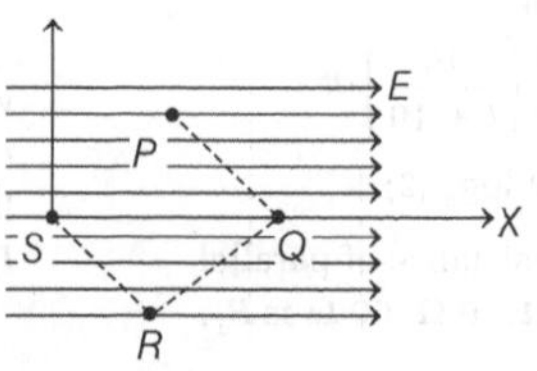

(a) qEa (b) $-qEa$

(c) $qEa\sqrt{2}$ (d) $qE\sqrt{[(2a)^2 + b^2]}$

3 The variation of potential with distance R from a fixed point is as shown below. The electric field at $R = 5\text{m}$ is

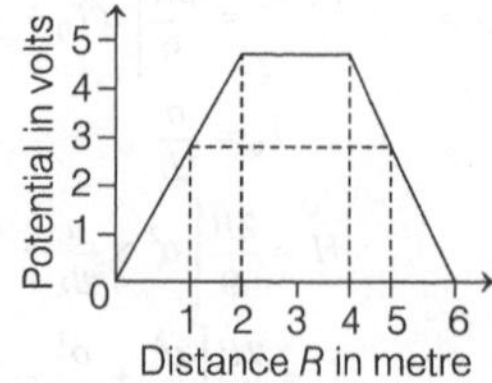

(a) 2.5 V/m (b) −2.5 V/m

(c) 2 / 5 V/m (d) −2 / 5 V/m

4 In the given circuit, it is observed that the current I is independent of the value of resistance R_6. Then, the resistance value must satisfy

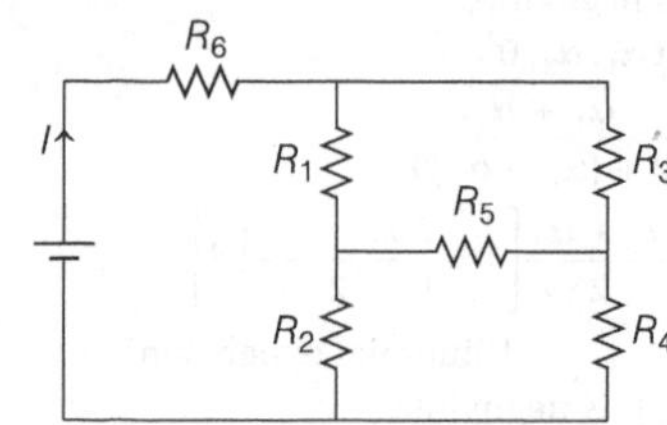

(a) $\frac{1}{R_5} + \frac{1}{R_6} = \frac{1}{R_1 + R_2} \times \frac{1}{R_3 + R_4}$

(b) $R_1R_4 = R_2R_3$

(c) $R_1R_2R_5 = R_3R_4R_6$

(d) $R_1R_3 = R_2R_4 = R_5R_6$

5 Two cells of internal resistance r_1 and r_2; and at same emf are connected in series, across a resistor of resistance R. If the terminal potential difference across the cells of internal resistance r_1 is zero, then the value of R is

(a) $R = 2(r_1 + r_2)$ (b) $R = r_2 - r_1$

(c) $R = r_1 - r_2$ (d) $R = 2(r_1 - r_2)$

6 The electric dipole moment of an electron and proton 4.30 nm apart is

(a) 6.88×10^{-28} C-m (b) 5.88×10^{-28} C-m

(c) 6.88×10^{28} C-m (d) 5.88×10^{28} C-m

7 At what distance along the central axis of a uniformly charged plastic disk of radius R is the magnitude of the electric field equal to one-half the magnitude of the field at the centre of the surface of the disk?

(a) $\frac{R}{\sqrt{2}}$ (b) $\frac{R}{\sqrt{3}}$ (c) $\sqrt{2}R$ (d) $\sqrt{3}R$

8 Work done in placing a charge of 8×10^{-18} C on a condenser of capacity 100 μF is

(a) 16×10^{-32} J (b) 3.1×10^{-28} J
(c) 64×10^{-32} J (d) 32×10^{-32} J

9 A drop, having a mass of 4.8×10^{-10}g and a charge of 2.4×10^{-18}C is suspended between two charged horizontal plates at a distance 1.0 cm apart. Find the potential difference between the plates. If polarity of the plates be changed, then calculate the instantaneous acceleration of the drop.

(a) 1.96×10^{6}V, 18.6 ms^{-2} (b) 1.86×10^{4}V, 18.6 ms^{-2}
(c) 1.96×10^{4}V, 19.6 ms^{-2} (d) 2.96×10^{4}V, 17.6 ms^{-2}

10 What is the direction of the electric field at the centre of the square of figure, if $q = 1.0 \times 10^{-8}$C and $a = 5.0$ cm?

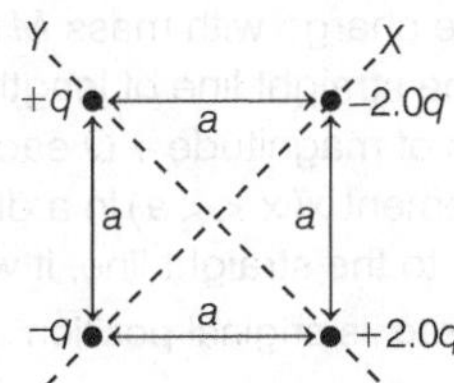

(a) 30° with X-axis (b) 45° with X-axis
(c) 60° with X-axis (d) 90° with X-axis

11 In a potentiometer experiment, the balancing with a cell is at length 240 cm. On shunting the cell with a resistance of 2 Ω, the balancing length becomes 120 cm. The internal resistance of the cell is

(a) 4 Ω (b) 2 Ω (c) 1 Ω (d) 0.5 Ω

12 The resistance of a wire at 20°C is 20 Ω and at 500°C is 60 Ω. At which temperature its resistance will be 25Ω?

(a) 50°C (b) 60°C
(c) 70°C (d) 80°C

13 A charged cloud system produces an electric field in the air near earth's surface. A particle of charge -2×10^{-9}C is acted on by a downward electrostatic force of 3×10^{-6}N, when placed in this field. The ratio of the magnitude of the electrostatic force to the magnitude of the gravitational force in the case of proton is

(a) 1.6×10^{-19} (b) 1.5×10^{-10} (c) 1.6×10^{19} (d) 1.4×10^{10}

14 An infinite non-conducting sheet has a surface charge density $\sigma = 0.10\ \mu$C m^{-2} on one side. How far apart are equipotential surfaces whose potentials differ by 50 V?

(a) 5.8×10^{-3}m (b) 6.8×10^{-3}m
(c) 7.8×10^{-3}m (d) 8.8×10^{-3}m

15 An electron is released from rest in a uniform electric field of magnitude 2.00×10^{4} NC^{-1}. Acceleration of the electron is (ignore gravitation)

(a) 2.51×10^{15}ms^{-2} (b) 2.51×10^{-15}ms^{-2}
(c) 3.51×10^{15}ms^{-2} (d) 3.51×10^{-15}ms^{-2}

16 In figure, the net potential at point P due to the four point charges, if $V = 0$ at infinity is

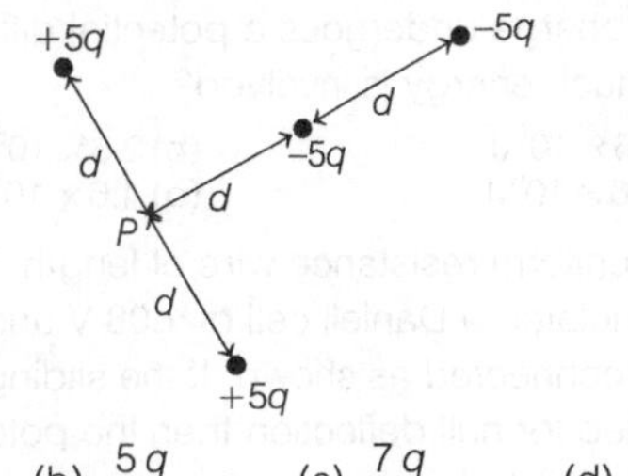

(a) $\frac{3q}{8\pi\varepsilon_0}$ (b) $\frac{5q}{8\pi\varepsilon_0}$ (c) $\frac{7q}{8\pi\varepsilon_0}$ (d) $\frac{9q}{8\pi\varepsilon_0}$

17 V-I graphs for parallel and series combination of two identical resistors are as shown in figure. Which graph represent parallel combination?

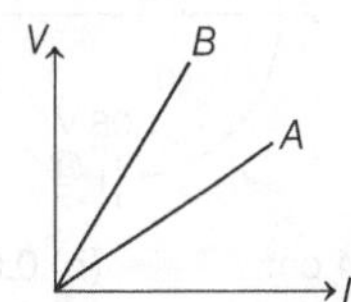

(a) Graph-A (b) Graph-B
(c) Both graph A and B (d) None of these

18 The equivalent resistance between points A and B in the following diagram is

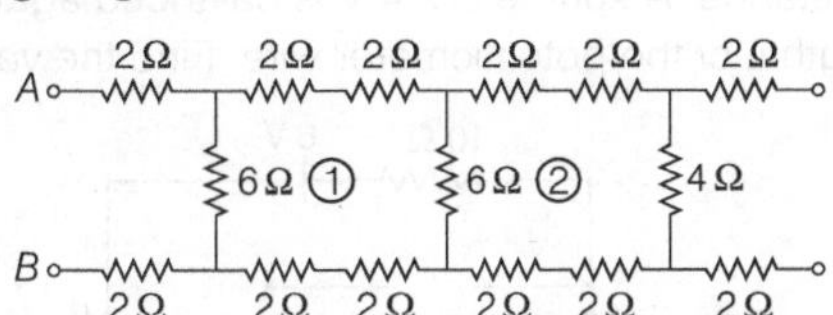

(a) 2 Ω (b) 8 Ω (c) 9 Ω (d) 10 Ω

19 Two cells connected in series have electromotive force of 1.5 V each. Their internal resistance are 0.5 Ω and 0.25 Ω respectively. This combination is connected to a resistance of 2.25 Ω . Potential difference across the terminals of each cell

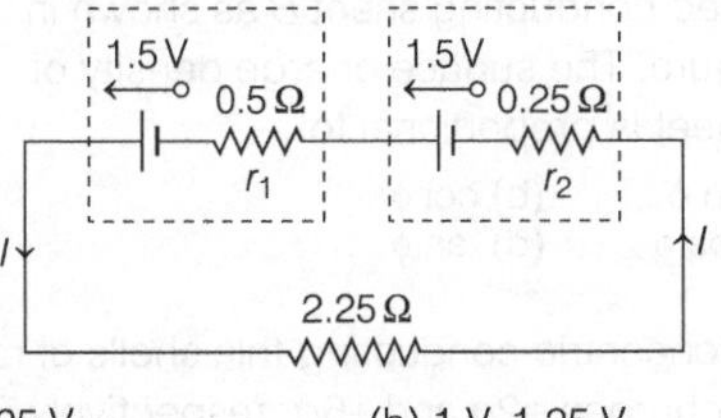

(a) 1 V, 0.25 V (b) 1 V, 1.25 V
(c) 1.5 V, 2.25 V (d) 1.5 V, 2.56 V

20 A charge of 0.8 C is divided into two charges Q_1 and Q_2. These are kept at a separation of 30 cm. The force on Q is maximum, when

(a) $Q_1 = Q_2 = 0.4$ C (b) $Q \approx 0.8$ C, Q_2 is negligible
(c) Q_1 is negligible, $Q_2 \approx 0.8$ C (d) $Q_1 = 0.2$ C, $Q_2 = 0.6$ C

21 A particular 12 V car battery can send a total charge of 84 A-h through a circuit, from one terminal to other. If this entire charge undergoes a potential difference of 12 V, how much energy is involved?

(a) 1.6×10^6 J (b) 2.6×10^6 J
(c) 3.6×10^6 J (d) 4.6×10^6 J

22 *AB* is uniform resistance wire of length 1 m. A 2 V accumulator, a Daniell cell of 1.08 V and a galvanometer *G* are connected as shown. If the sliding contact is adjusted for null deflection then the potential gradient in *AB* and the balancing length, measured from end *A* are respectively.

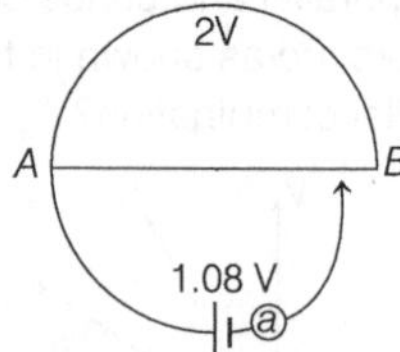

(a) 0.02 V/cm, 54 cm (b) 0.0308 V/cm, 46 cm
(c) 0.0092 V/cm, 49.6 cm (d) 0.02 V/cm, 50.4 cm

23 A potentiometer wire of length 200 cm has a resistance of 20 Ω. It is connected in series with a resistance of 10 Ω and an accumulator of emf 6 V having negligible internal resistance. A source of 2.4 V is balanced against a length *L* of the potentiometer wire. Find the value of *L*.

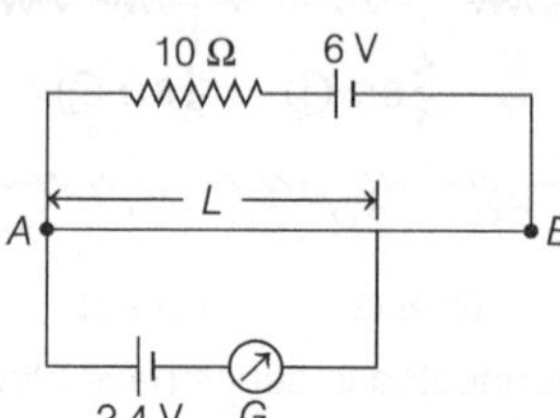

(a) 100 cm (b) 120 cm (c) 110 cm (d) 140 cm

24 A charged ball *A* hangs from a silk thread, which makes an angle ϕ with a large charged conducting sheet *B* as shown in the figure. The surface charge density of the sheet is proportional to

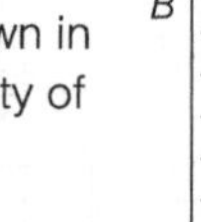

(a) $\sin\phi$ (b) $\cot\phi$
(c) $\cos\phi$ (d) $\tan\phi$

25 Two concentric conducting thin shells of radius *r* and 2*r* carry charges +2*q* and +6*q*, respectively. The magnitude of electric field at a distance *x* outside and inside from the surface of outer sphere is same, then the volume of *x* is

(a) $\frac{r}{2}$ (b) $\frac{2r}{3}$ (c) $\frac{r}{3}$ (d) $\frac{r}{6}$

26 There are two electric bulbs rated 60 W, 120 V and 90 W, 120 V. They are connected in parallel with 240 V supply, then

(a) both bulbs work properly
(b) both bulbs will fuse
(c) Only 60 W bulb will fuse
(d) Only 90 W bulb will fuse

27 Masses of the three wires of same material are in the ratio of 1 : 2 : 3 and their lengths in the ratio of 3 : 2 : 1. Electrical resistance of these wires will be in the ratio of

(a) 1 : 1 : 1 (b) 1 : 2 : 3 (c) 9 : 4 : 1 (d) 27 : 6 : 1

28 Six wires, each of resistance r_1 are connected so as to form a tetrahedron. The equivalent resistance of the combination when current enters through one corner and leaves through other corner is

(a) r (b) $2r$ (c) $\frac{r}{3}$ (d) $\frac{r}{2}$

29 A unit negative charge with mass *M* resides at the mid-point of the straight line of length 2*a* adjoining two fixed charges of magnitude + *Q* each. If it is given a very small displacement $x(x << a)$ in a direction perpendicular to the straight line, it will

(a) came back to its original position and stay there
(b) execute oscillations with frequency $\frac{1}{2\pi}\sqrt{\frac{Q}{4\pi\varepsilon_0 Ma^3}}$
(c) execute oscillations with frequency $\frac{1}{2\pi}\sqrt{\frac{Q}{4\pi\varepsilon_0 Ma^2}}$
(d) execute oscillations with frequency $\frac{1}{2\pi}\sqrt{\frac{Q}{2\pi\varepsilon_0 Ma^3}}$

30 A 28 μF capacitor is charged to 100 V and another 2μ F capacitor to 200 V, they are connected in parallel. Then, the total final energy is

(a) 0.1537 J (b) 0.0155 J (c) 0.1865 J (d) 0.123 J

31 A hollow copper tube of 1m length has got external diameter equal to 10cm and its walls are 5mm thick. Then the resistance of tube, if its specific resistance is 1.7×10^{-8} Ω m, is

(a) 1.139×10^{-5} Ω (b) 1.327×10^{-6} Ω
(c) 1.150×10^{-5} Ω (d) 1.125×10^{-4} Ω

32 Manjeet's room heater is marked as 1000 W-200V. If the voltage drops to 160 V, the percentage change in the power of the heater is

(a) 40% (b) 42% (c) 36% (d) 50%

Direction (Q. Nos. 33-40) *Each of these questions contains two statements : Statement I and Statement II. Each of these questions also has four alternative choices, only one of which is the correct answer. You have to select one of the codes (a), (b), (c), (d) given below*

(a) Statement I is true, Statement II is true; Statement II is the correct explanation for Statement I
(b) Statement I is true, Statement II is true; Statement II is not the correct explanation for Statement I
(c) Statement I is true; Statement II is false
(d) Statement I is false; Statement II is true

33 Statement I A and B are two conducting spheres of same radius. A being solid and B hollow. Both are charged to the same potential. Then,

Charge on A = charge on B.

Statement II Potentials on both are same.

34 Statement I A thin metallic wire is bent into semicircular shape, then its resistivity decreases.

Statement II On bending, the drift of electron in the wire remains same.

35 Statement I The circuits containing capacitor be handed cautiously, even when there is no current.

Statement II A dielectric differs from an insulator.

36 Statement I In the following circuit, the net resistance between points A and B is R

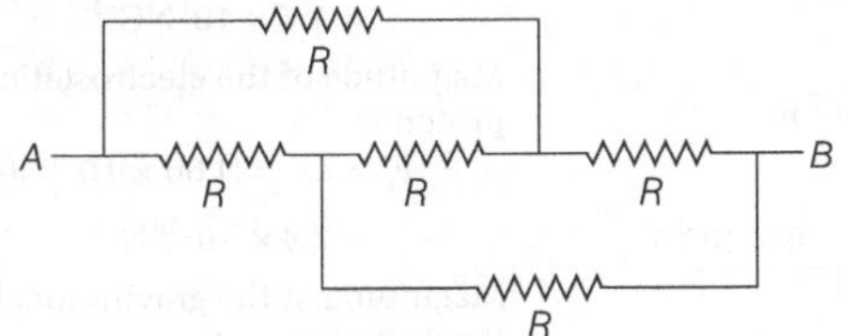

Statement II All the resistances are in parallel to each other.

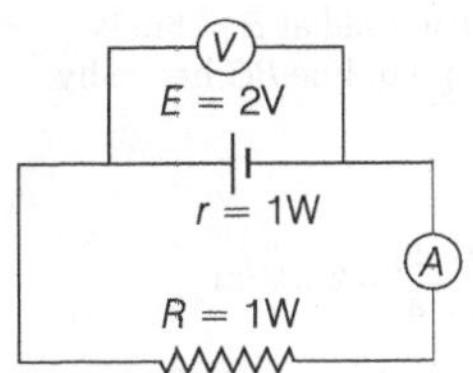

37 Statement I In the following circuit, emf is 2V and internal resistance of the cell is $1\,\Omega$ and $R = 1\,\Omega$, then reading of the voltmeter is 1V.

Statement II $V = E - Ir$, where $E = 2\,V$, $I = \frac{2}{2} = 1$ A and $R = 1\,\Omega$.

38 Statement I The power delivered to a light bulb is more just after it is switched ON and the glow of the filament is increasing, as compared to when the bulb is glowing steadily, i.e. after sometime of switching ON.

Statement II As temperature increases, resistance of conductor increases.

39 Statement I When a wire is stretched, so that its diameter is halved, its resistance becomes 16 times.

Statement II Resistance of wire decreases with increase in length.

40 Statement I A potentiometer is preferred over that of a voltmeter for measurement of emf of a cell.

Statement II Potentiometer is preferred as it does not draw any current from the cell.

ANSWERS

1. (d)	**2.** (b)	**3.** (a)	**4.** (b)	**5.** (c)	**6.** (a)	**7.** (b)	**8.** (d)	**9.** (c)	**10.** (b)
11. (b)	**12.** (d)	**13.** (d)	**14.** (d)	**15.** (c)	**16.** (b)	**17.** (a)	**18.** (b)	**19.** (b)	**20.** (a)
21. (c)	**22.** (a)	**23.** (b)	**24.** (d)	**25.** (b)	**26.** (b)	**27.** (d)	**28.** (d)	**29.** (d)	**30.** (c)
31. (a)	**32.** (c)	**33.** (a)	**34.** (d)	**35.** (b)	**36.** (c)	**37.** (a)	**38.** (a)	**39.** (c)	**40.** (a)

Hints and Explanations

1 Electric potential of the common centre, is

$$V = \frac{q_1}{4\pi\varepsilon_0 r_1} + \frac{q_2}{4\pi\varepsilon_0 r_2}$$

$$V = \frac{\sigma}{\varepsilon_0} \times r_1 + \frac{\sigma}{\varepsilon_0} \times r_2$$

$$= \frac{\sigma}{\varepsilon_0}(r_1 + r_2)$$

$$\left[\because \begin{array}{l} q_1 = 4\pi r_1^2 \times \sigma \\ q_2 = 4\pi r_2^2 \times \sigma \end{array}\right]$$

2 As electric field is a conservative field

Hence, the work done does not depend on path

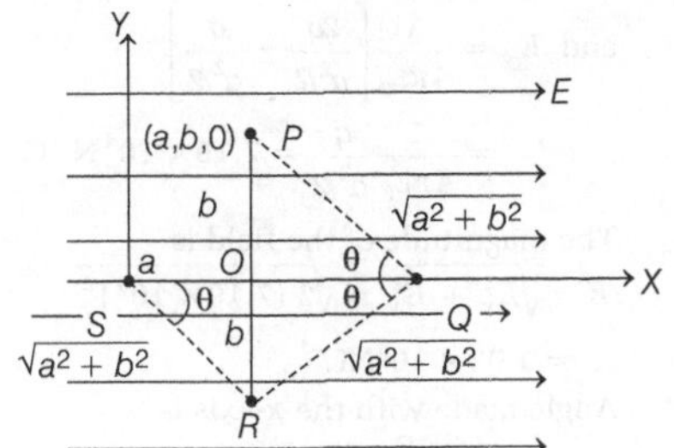

$$\therefore \quad W_{PQRS} = W_{PQS} = W_{PO} + W_{OS}$$

$$= Fb\cos 90° + Fa\cos 180°$$

$$= 0 + q\,Ea(-1) = -q\,Ea$$

3 Intensity at 5m is same as at any point between B and C because the slope of BC is same throughout (i.e. electric field between B and C is uniform).

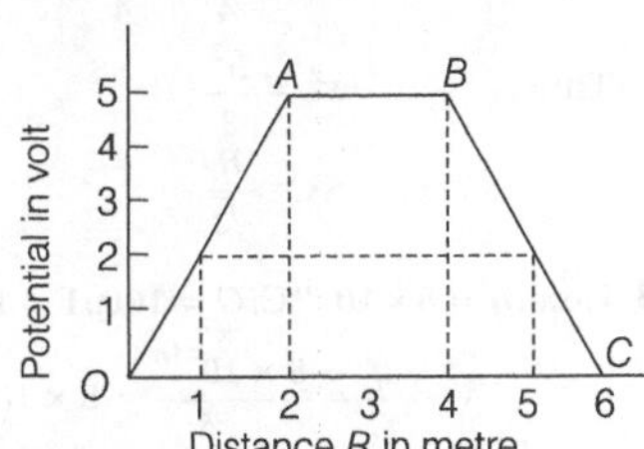

Therefore electric field at $R = 5$m is equal to the slope of line BC hence by

$$E = \frac{-dV}{dr};$$

$$E = -\frac{(0-5)}{6-4} = 2.5\,\text{V/m}$$

4 From figure, it can be seen that I is independent of resistance R_5, so no current flow through it. This require that the R_1 and R_2 junction is at same potential of the junction at R_3 and R_4. So, according to Wheatstone bridge condition,

$$\frac{R_1}{R_2} = \frac{R_3}{R_4} \Rightarrow R_1R_4 = R_2R_3$$

5 Given, $V_1 = 0$

$V_1 + V_2 = IR$

$\Rightarrow \quad V_2 = IR$

$-E - Ir_2 = IR$

But $V_1 = E - Ir_1 \Rightarrow E = Ir_1 \qquad [\because V_1 = 0]$

$\Rightarrow Ir_1 - Ir_2 = IR$

or $\quad R = r_1 - r_2$

6 Magnitude of a dipole moment is

$$p = qd$$
$$= (1.60 \times 10^{-19})(4.30 \times 10^{-9})$$
$$= 6.88 \times 10^{-28}\ \text{C-m}$$

7 At a point on the axis of a uniformaly charged disk at a distance x above the centre of the disk, the magnitude of the electric field is

$$E = \frac{\sigma}{2\varepsilon_0}\left[1 - \frac{x}{\sqrt{x^2 + R^2}}\right]$$

But $E_c = \dfrac{\sigma}{2\varepsilon_0}$ such that $\dfrac{E}{E_c} = \dfrac{1}{2}$

Then, $1 - \dfrac{x}{\sqrt{x^2 + R^2}} = \dfrac{1}{2}$

or $\quad \dfrac{x}{\sqrt{x^2 + R^2}} = \dfrac{1}{2}$

Squaring both side, we get

$$\frac{x^2}{x^2 + R^2} = \frac{1}{4}$$

or $\quad x^2 = \dfrac{x^2}{4} + \dfrac{R^2}{4}$

Thus, $\quad x^2 = \dfrac{R^2}{3}$

$$x = \frac{R}{\sqrt{3}}$$

8 Here, $q = 8\times10^{-18}$C, $C = 100\mu\text{F} = 10^{-4}$ F

$$V = \frac{q}{C} = \frac{8 \times 10^{-18}}{10^{-4}} = 8 \times 10^{-14}\ \text{V}$$

Energy stored $= \dfrac{1}{2}qV$

$$= \frac{1}{2} \times 8 \times 10^{-18} \times 8 \times 10^{-14}$$
$$= 32 \times 10^{-32}$$

9 Let m and q be the mass and the charge of the drop and E the intensity of electric field between the plates. Since, the drop is in equilibrium, the electric force qE acting on it balances its weight mg, i.e.

$$qE = mg$$

If the potential difference between the plates is V and the distance between them is d, then $E = V/d$.

$\therefore \quad q(V/d) = mg$ or $V = mgd/q$

Here, $m = 4.8 \times 10^{-10}$ g $= 4.8 \times 10^{-13}$kg,

$g = 9.8\,\text{N kg}^{-1}$,

$d = 1.0$ cm$= 1.0\times 10^{-2}$ m

and $\quad q = 2.4 \times 10^{-18}$C

$$\therefore \quad V = \frac{(4.8 \times 10^{-13}) \times 9.8 \times (1.0 \times 10^{-2})}{2.4 \times 10^{-18}}$$
$$= 1.96 \times 10^4\ \text{V}$$

On changing the polarity of the plates, the electric force qE will also be directed downwards. Then, the acceleration of the drop is

$$a = \frac{qE + mg}{m}$$

But $\quad qE = mg$

$\therefore \quad a = 2g = 19.6\ \text{ms}^{-2}$

10 Since, each charge distance from centre

$$d = \frac{\sqrt{2}\,a}{2} = \frac{a}{\sqrt{2}}$$

Net field due to these two charges is

$$E_x = \frac{1}{4\pi\varepsilon_0}\left[\frac{2q}{d^2/2} - \frac{q}{d^2/2}\right] = \frac{1}{4\pi\varepsilon_0}\frac{q}{d^2/2}$$

$$= \frac{(9\times 10^9)(1.0\times 10^{-8})}{\frac{(0.050)^2}{2}}$$

$$= 7.19\times 10^4\,\text{NC}^{-1}$$

and $E_y = \dfrac{1}{4\pi\varepsilon_0}\left[\dfrac{2q}{d^2/2} - \dfrac{q}{d^2/2}\right]$

$$= \frac{1}{4\pi\varepsilon_0}\frac{q}{d^2/2} = 7.19\times 10^4\,\text{N/C}$$

The magnitude of the field is

$$E = \sqrt{E_x^2 + E_y^2} = \sqrt{2\,(7.19\times 10^4)^2}$$
$$= 1.02\times 10^5\text{NC}^{-1}$$

Angle made with the x-axis is

$$\theta = \tan^{-1}\frac{E_y}{E_x}$$
$$= \tan^{-1}(1) = 45^\circ$$

It is upward in the diagram, from the centre of the square towards the centre of the upper side.

11 Here, $\quad r = \dfrac{l_1 - l_2}{l_2} \times 2\Omega$

where, $l_1 = 240$ cm, $l_2 = 120$ cm

$$= \frac{240 - 120}{120} \times 2 = \frac{120}{120} \times 2 = 2\Omega$$

12 Use $\quad R_t = R_0(1 + \alpha t)$

Here, $\quad 20 = R_0(1 + 20\,\alpha)$

$60 = R_0(1 + 500\,\alpha)$

Here, $\quad R_t = 25\,\Omega$

Solving, we find $t = 80°$C

13 We have, $F = qE$

Thus, $E = \dfrac{F}{q} = \dfrac{3 \times 10^{-6}}{2 \times 10^{-9}}$

$$= 1.5\times 10^3\,\text{NC}^{-1}$$

Magnitude of the electrostatic force on a proton is

$$F_e = eE = (1.60 \times 10^{-19})(1.5\times 10^3)$$
$$= 2.4\times 10^{-16}\text{N}$$

Magnitude of the gravitational force on the proton is

$$F_g = mg = (1.67 \times 10^{-27})(9.8)$$
$$= 1.63\times 10^{-26}\text{N}$$

The ratio of the force is

$$\frac{F_e}{F_g} = \frac{2.4 \times 10^{-16}}{1.63 \times 10^{-26}}$$
$$= 1.4 \times 10^{10}$$

14 Electric field, $E = \dfrac{\sigma}{2\varepsilon_0}$

and electric potential, $V = V_s - \int_0^x E\,dx$

$$= V_s - Ex$$

Here, two surfaces are separated by Δx, then their potentials difference in magnitude by

$$\Delta V = E\Delta x = \left(\frac{\sigma}{2\varepsilon_0}\right)\Delta x$$

Thus, $\quad \Delta x = \dfrac{2\varepsilon_0\Delta V}{\sigma}$

$$= \frac{2(8.85 \times 10^{-12})(50)}{0.10 \times 10^{-6}}$$
$$= 8.8 \times 10^{-3}\ \text{m}$$

15 We know that, $F = eE$

By Newton's second law,

$$a = \frac{F}{m} = \frac{eE}{m}$$

$$= \frac{(1.60 \times 10^{-19})(2.00 \times 10^4)}{9.11 \times 10^{-31}}$$
$$= 3.51\times 10^{15}\text{ms}^{-2}$$

16 Net potential at point P,

$$V = \frac{q}{4\pi\varepsilon_0}\left[-\frac{5}{2d} - \frac{5}{d} + \frac{5}{d} + \frac{5}{d}\right] = \frac{5q}{8\pi\varepsilon_0}$$

17 $R_{Parallel} < R_{Series}$. From graph A it is clear that slope of the line A is lower than the slope of the line B. Also, slope = resistance, so line A represents the graph for parallel combination.

18 The resistances 2Ω and 2Ω at the last terminals are outside the circuit and so they may be ignored. Now, in loop 2, the resistances $(2\Omega + 2\Omega)$, 4Ω and $(2\Omega + 2\Omega)$ are in series. Their equivalent resistance is 12Ω, which is in parallel with 6Ω. The equivalent resistance is

$$R' = \frac{6 \times 12}{6 + 12} = 4\Omega$$

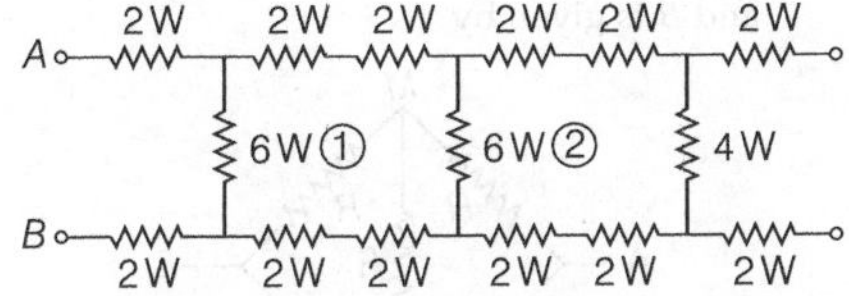

Similarly, in loop 1, the resistances $(2\Omega + 2\Omega), R'(= 4\Omega)$ and $(2\Omega + 2\Omega)$ are in series and these are in parallel with 6Ω. Hence, their equivalent resistance is

$$R'' = 4\Omega$$

Lastly, between the points A and B, the resistances 2Ω, $R''= (4\Omega)$ and 2Ω are in series. Hence, their equivalent resistance is

$$2\Omega + 4\Omega + 2\Omega = 8\Omega$$

19 The arrangement is shown in the figure. The effective emf in the circuit is

$E = 1.5 + 1.5 = 3.0$ V

and the total resistance is

$$R = 0.5 + 0.25 + 2.25 = 3.0\ \Omega$$

Hence, the current in the circuit is

$$I = \frac{E}{R} = \frac{3.0}{3.0} = 1.0\text{ A}$$

Potential difference across the terminals of the first cell is

$$V_1 = E - Ir_1 = 1.5 - (1.0) \times (0.5) = 1.0\text{ V}$$

Potential difference across the terminals of the second cell is

$$V_2 = E - Ir_2 = 1.5 - (1.0) \times (0.25) = 1.25\text{ V}$$

20 Given,

$$Q_1 \xleftrightarrow{30\text{ cm}} Q_2$$

(a) $Q_1 = Q_2 = 0.4$ C

The force of Q_1 due to Q_2,

$$F = k_P \frac{Q_1 Q_2}{30 \times 10^{-2}} = K\frac{Q_1 Q_2 \times 100}{30}$$

$$= K \times \frac{0.4 \times 0.4 \times 100}{30} = \frac{8}{15}K$$

(b) When $Q_1 = 0.8$ C, $Q_2 \approx 0$

$$F = k \times \frac{0.8 \times 0}{30 \times 10^{-2}} = 0$$

(c) When $Q_1 \approx 0$, $Q_2 = 0.8$ C

$$F = k \times \frac{0 \times 0.8}{30 \times 10^{-2}} = 0$$

(d) When $Q_1 = 0.2$C, $Q_2 = 0.6$ C

$$F = k \times \frac{0.2 \times 0.6\ 100}{30} = \frac{2}{5}k$$

Hence, for $Q_1 = Q_2 = 0.4$C, the force will be maximum.

21 An ampere is coulomb per second, so

84 A - h $= 84 \times 3600 = 3.0 \times 10^5$C

The change in potential energy is

$$\Delta U = q\Delta V = 3.0 \times 10^5 \times 12 = 3.6 \times 10^6\text{ J}$$

22 Potential difference per cm

$$= \frac{2\text{ V}}{100\text{ cm}} = 0.02\text{ V/cm}$$

$$\text{Balancing length} = \frac{100}{2} \times 1.08 = 54\text{ cm}$$

23 The current in the potentiometer wire AB is

$$I = \frac{6}{20 + 10} = 0.2\text{ A}$$

The potential difference across the potentiometer wire is

V = current × resistance

$= 0.2 \times 20 = 4$ V

The length of the wire is $l = 200$ cm. So, the potential gradient along the wire is

$$k = \frac{V}{l} = \frac{4}{200}$$

$= 0.02\text{ Vcm}^{-1}$

The emf 2.4 V is balanced against a length L of the potentiometer wire.

i.e. $2.4 = kL$ or $L = \frac{2.4}{k}$

$$= \frac{2.4}{0.02} = 120\text{ cm}$$

24 Electric field at point A

$$E = \frac{\sigma}{\varepsilon_0} \quad \text{...(i)}$$

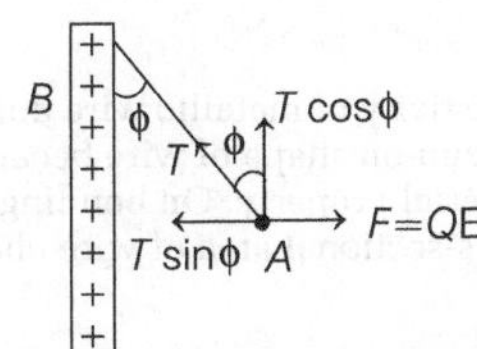

For equilibrium of forces at A,

$$T\sin\phi = QE$$

$$T\sin\phi = Q \cdot \frac{\sigma}{\varepsilon_0}, \quad \text{[from Eq. (i)]}$$

$$T\sin\phi = \frac{Q\sigma}{\varepsilon_0} \quad \text{...(ii)}$$

and $\quad T\cos\phi = mg \quad$ ...(iii)

On dividing Eq. (ii) by Eq. (iii), we get

$$\tan\phi = \frac{Q\sigma}{\varepsilon_0 mg}$$

or $\quad \sigma = \frac{\varepsilon_0 mg}{Q}\tan\phi$

$\Rightarrow \quad \sigma \propto \tan\phi$

25 Electric field at a distance x outside from surface of outer shell

= Electric field at a distance x inside from surface of outer shell

i.e. $\frac{1}{4\pi\varepsilon_0} \cdot \frac{2q + 6q}{(2r + x)^2} = \frac{1}{4\pi\varepsilon_0} \cdot \frac{2q}{(2r - x)^2}$

$\Rightarrow \quad \frac{4}{(2r + x)^2} = \frac{1}{(2r - x)^2}$

$\Rightarrow \quad \frac{2}{2r + x} = \frac{1}{2r - x}$

$\Rightarrow \quad 4r - 2x = 2r + x$

$\Rightarrow \quad 2r = 3x$

$\Rightarrow \quad x = \frac{2r}{3}$

26 $R_1 = \frac{V_1^2}{P_1} = \frac{120^2}{60} = 240\Omega$

$R_2 = \frac{V_2^2}{P_2} = \frac{120^2}{90} = 160\Omega$

$\therefore I_1 = \frac{P_1}{V_1} = \frac{60}{120} = 0.5$ A

$I_2 = \frac{P_2}{V_2} = \frac{90}{120} = 0.75$ A

When both bulbs are connected in parallel, then equivalent resistance

$$R_P = \frac{R_1 R_2}{R_1 + R_2}$$

$$= \frac{240 \times 160}{240 + 160} = 96\Omega$$

$\therefore$ When they are connected with 240 V supply, then

$$I = \frac{240}{96} = 2.5\text{ A}$$

Now, current in 60 W bulb,

$$I_1' = I \cdot \frac{160}{400}$$

$$= 2.5 \times \frac{160}{400} = 1\text{ A}$$

Current in 90 W bulb,

$$I_2' = 2.5 \times \frac{240}{400} = 1.5\text{ A}$$

Since, $I_1' > I$ and $I_2' > I_2$

Hence, both bulbs will fuse.

27 Mass, M = Volume × Density

$= Al \times d$ or $A = \dfrac{M}{ld}$

Resistance, $R = \dfrac{\rho l}{A} = \dfrac{\rho l}{\left(\dfrac{M}{ld}\right)} = \dfrac{\rho l^2 d}{M}$

So, $R \propto \dfrac{l^2}{M}$

Thus, $R_1 : R_2 : R_3 = \dfrac{l_1^2}{M_1} : \dfrac{l_2^2}{M_2} : \dfrac{l_3^2}{M_3}$

$= \dfrac{3^2}{1} : \dfrac{2^2}{2} : \dfrac{1^2}{3} = 27 : 6 : 1$

28 Six wires each of resistance r from a tetrahedron as shown in the following figure.

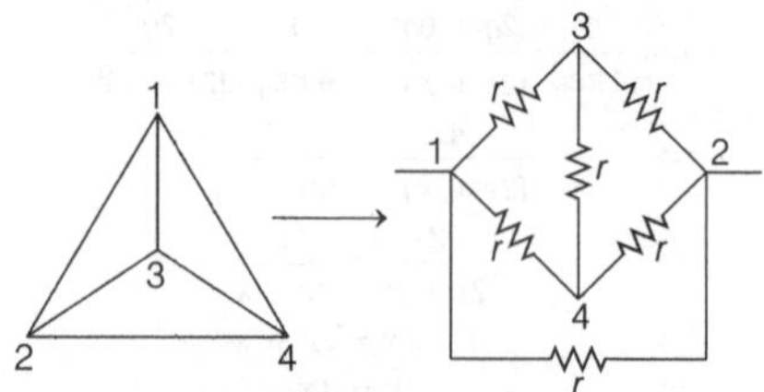

In Wheatstone circuit, the equivalent resistance of upper circuit

$\dfrac{I}{R} = \dfrac{I}{2r} + \dfrac{1}{2r} = \dfrac{2}{2r} = \dfrac{1}{r}$

$\Rightarrow$ $R = r$

It will be in parallel with outer resistance,

$\dfrac{1}{R_{eq}} = \dfrac{1}{r} + \dfrac{1}{r} = \dfrac{2}{R}$

$\Rightarrow$ $R_{eq} = \dfrac{r}{2}$.

29 From figure the net force,

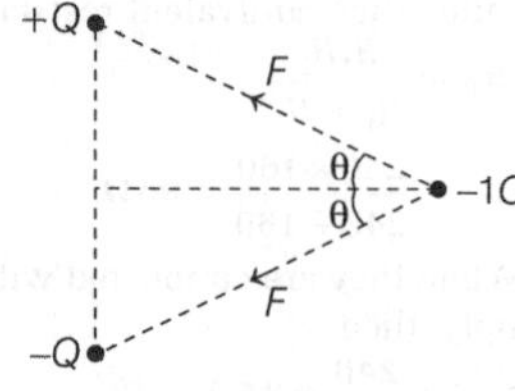

$F_{net} = -F\cos\theta + (-F\cos\theta)$

$= -2F\cos\theta$

$= -2\dfrac{kQ(-1)}{x^2 + a^2} \times \dfrac{x}{\sqrt{x^2 + a^2}}$

$= \dfrac{2kQ}{(x^2 + a^2)^{3/2}} \cdot x$

Or $F_{net} = +\left(\dfrac{2kQ}{a^3}\right) \cdot x$ $[\because x << a]$

Frequency of oscillation,

$= \dfrac{1}{2\pi}\sqrt{\dfrac{2KQ}{Ma^3}} = \dfrac{1}{2\pi}\sqrt{\dfrac{2 \times 1 \times Q}{4\pi\varepsilon_0 Ma^3}}$

$= \dfrac{1}{2\pi}\sqrt{\dfrac{Q}{2\pi\varepsilon_0 Ma^3}}$

30 Here $C_1 = 28\,\mu\text{F} = 28 \times 10^{-6}$F

$C_2 = 3\,\mu\text{F} = 3 \times 10^{-6}$F

$V_1 = 100$V and $V_2 = 200$V

$\Rightarrow$ Charge of C_1,

$q_1 = C_1V_1 = 28 \times 10^{-6} \times 100$

$= 28 \times 10^{-4}$C

Charge of C_2

$q_2 = C_2V_2 = 3 \times 10^{-6} \times 200$

$= 6 \times 10^{-4}$C

Potential,

$V = \dfrac{q_1 + q_2}{C_1 + C_2} = \dfrac{28\times10^{-4} + 6\times10^{-4}}{28\times10^{-6} + 3\times10^{-6}}$

$= 109.68$ V

Total final energy, $U = \dfrac{1}{2}(C_1 + C_2)V^2$

$= \dfrac{1}{2}(28\times10^{-6} + 3\times10^{-6})(109.68)^2$

$= 0.1865$J

31 External radius,

$r_2 = \dfrac{10}{2} = 5\text{cm} = 0.05\text{m}$

Internal radius $r_1 = r_2$ – thickness of tube

$= 0.05 - 0.005$

$= 0.045$ m

Area of cross-section $= \pi(r_2^2 - r_1^2)$

$= \pi[(0.05)^2 - (0.045)^2]$

$= 1.492 \times 10^{-3}\,\text{m}^2$

Resistance of copper tube,

$R = \rho\dfrac{l}{a} = \dfrac{1.7\times10^{-8}\times1}{1.492\times10^{-3}}$

$= 1.139 \times 10^{-5}\,\Omega$

32 Resistance of heater,

$R = \dfrac{V^2}{P} = \dfrac{(200)^2}{1000} = 40\Omega$

Power of heater at $V' = 160$V

$P' = \dfrac{V'^2}{R} = \dfrac{(160)^2}{40} = 640$ W

Percentage fall in the power of the heater,

$\dfrac{P - P'}{P} \times 100 = \dfrac{1000 - 640}{1000} \times 100 = 36\%$

34 Resistivity of metallic wire does not depend on shape of wire because it is a material property. On bending, the cross-sectional area of wire changes but drift velocity of electron does not depend on area of cross-section, so it remains same.

35 A capacitor does not discharge itself. In case the capacitor is connected in a circuit containing a source of high voltage, the capacitor charges itself to a very high potential. So, if a person handles it without discharging, he may get a severe shock.Dielectrics and insulators cannot conduct electricity but in case of a dielectric, when an external field is applied, induced charges appear on the faces of the dielectric. In other words, the dielectric have the property of transmitting electric effects without conducting.

36 The equivalent circuit is represented as, This is balanced Wheatstone bridge hence, resistance in branch MN is not taken into consideration. Hence, the equivalent resistance between points A and B is given by

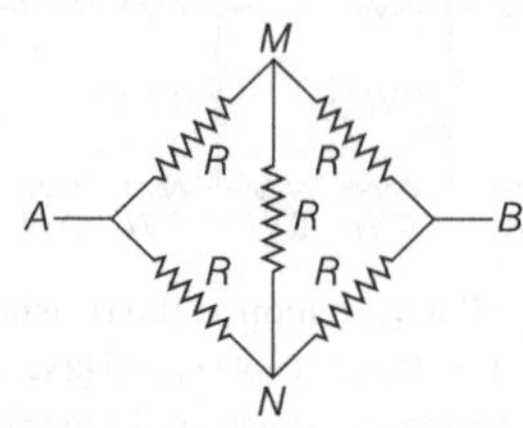

$\dfrac{1}{R_{AB}} = \dfrac{1}{(R + R)} + \dfrac{1}{(R + R)}$

or $\dfrac{1}{R_{AB}} = \dfrac{2}{2R} = \dfrac{1}{R}$

$\therefore$ $R_{AB} = R$

39 The resistance of a wire is

$R = \rho\dfrac{l}{A}$, ρ being specific resistance

or $R \propto \dfrac{Al}{A^2}$

or $R \propto \dfrac{1}{r^4}$ $(\because A = \pi r^2)$

Hence, when diameter is halved the resistance of the wire is

$R \propto \dfrac{1}{\left(\dfrac{r}{2}\right)^4} = 16R$ ...(i)

Hence, its resistance will become 16 times.

Again from Eq. (i), we get

$R \propto \dfrac{l}{A}$ or $R \propto \dfrac{l^2}{Al}$ or $R \propto l^2$

Therefore, on increasing the length resistance increases.

DAY TWENTY ONE

Magnetic Effect of Current

Learning & Revision for the Day

- Concept of Magnetic Field
- Biot-Savart's Law and its Applications
- Ampere's Circuital Law
- Force on a Moving Charge in Uniform Magnetic Field
- Cyclotron
- Magnetic Force on a Current Carrying Conductor
- Moving Coil Galvanometer

Concept of Magnetic Field

If a magnet is placed in a magnetic field, then it experiences a force on it, Also, when a magnet is placed near a current carrying conductor, then it experiences the similar force, it means that current carrying conductor produces a magnetic field around it. This effect of current is called **magnetic effect of current**.

Biot-Savart's Law and its Applications

The magnetic field $d\mathbf{B}$ at a point P, due to a current element $I\,d\mathbf{l}$ is given by

$$d\mathbf{B} = \frac{\mu_0}{4\pi}\frac{I(d\mathbf{l} \times \mathbf{r})}{r^3}$$

where, θ is the angle between $d\mathbf{l}$ and $\mathbf{r}$.

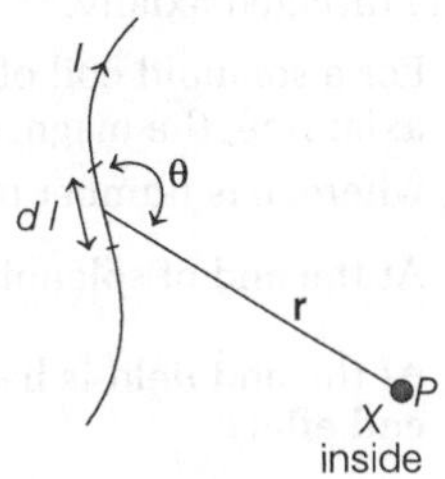

Direction of magnetic field produced due to a current carrying straight wire can be obtained by the right hand thumb rule.

- No. of Questions in Exercises (x)—
- No. of Questions Attempted (y)—
- No. of Correct Questions (z)—
  (*Without referring Explanations*)
- Accuracy Level ($z/y \times 100$)—
- Prep Level ($z/x \times 100$)—

In order to expect good rank in JEE, your Accuracy Level should be above 85 & Prep Level should be above 75.

Magnetic Field due to Circular Current Loop

- If there is a circular coil of radius R and N number of turns, carrying a current I through the turns, then magnetic field at the centre of coil is given by

$$B = \frac{\mu_0 NI}{2R}$$

- If there is a circular arc of wire subtending an angle θ at the centre of arc, then the magnetic field at the centre point

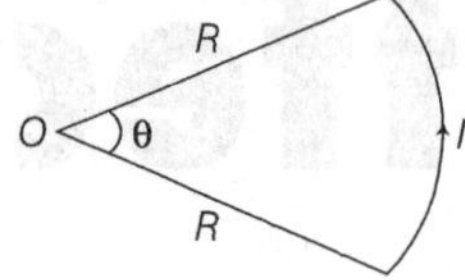

$$B = \frac{\mu_0 I}{2R}\left(\frac{\theta}{2\pi}\right)$$

- At a point P situated at a distance r from centre of a current carrying circular coil along its axial line.

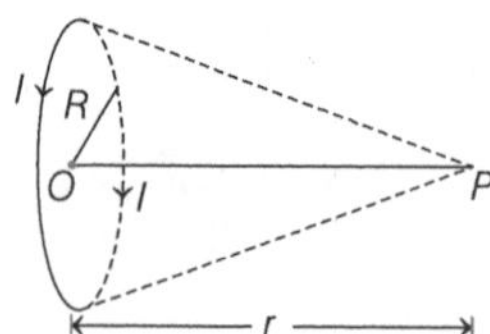

The magnetic field is

$$B = \frac{\mu_0 NIR^2}{2(R^2 + r^2)^{3/2}}$$

If $r >> R$, then at a point along the axial line, $B = \frac{\mu_0 NIR^2}{2r^3}$

Ampere's Circuital Law

The line integral of the magnetic field **B** around any closed path is equal to μ_0 times the net current I threading through the area enclosed by the closed path.

Mathematically, $\oint \mathbf{B}\cdot d\mathbf{l} = \mu_0 \Sigma I$

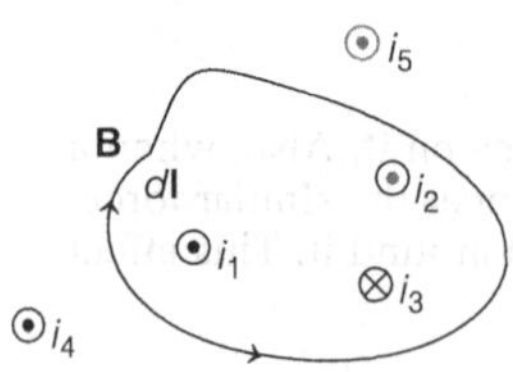

Now, consider the diagram above.

Here, $\Sigma I = i_1 + i_2 - i_3$

Hence, $\oint \mathbf{B}\cdot d\mathbf{l} = \mu_0 \cdot (i_1 + i_2 - i_3)$

Applications of Ampere's law

1. Magnetic field due to Straight Current Carrying Wire

The magnetic field due to a current carrying wire of finite length at a point P situated at a normal distance r is

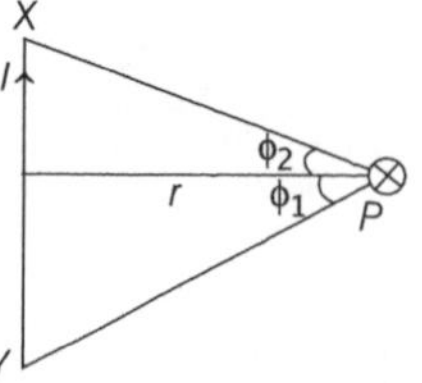

$$B = \frac{\mu_0 I}{4\pi r}(\sin\phi_1 + \sin\phi_2)$$

- If point P lies symmetrically on the perpendicular bisector of wire XY, then $\phi_1 = \phi_2 = \phi$ (say) and hence

$$B = \frac{\mu_0 I}{4\pi r}\cdot 2\sin\phi = \frac{\mu_0 I \sin\phi}{2\pi r}$$

- For a wire of infinite length $\phi_1 = \phi_2 = 90°$ and hence

$$B = \frac{\mu_0 I}{2\pi r}$$

- When the wire XY is of infinite length, but the point P lies near the end X or Y, then $\phi_1 = 0°$ and $\phi_2 = 90°$ and hence,

$$B = \frac{\mu_0 I}{4\pi r}$$

- When point P lies on axial position of current carrying conductor, then magnetic field at P,

$$B = 0.$$

- When wire is of infinite length, then magnetic field near the end will be half, that of at the perpendicular bisector.

2. Magnetic Field due to a Thick (Cylindrical) Wire

Magnetic field at a point outside the wire

$B = \frac{\mu_0 I}{2\pi r}$, where r is the distance of given point from centre of wire and $r > R$.

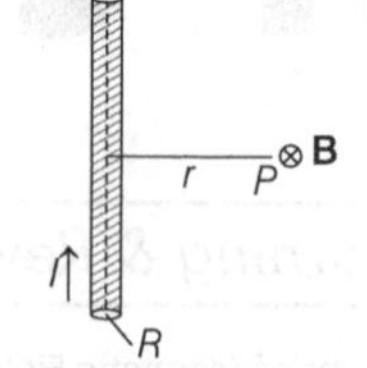

Thick cylindrical wire

- Magnetic field at a point inside the wire at a distance r from centre of wire ($r < R$) is

$$B = \frac{\mu_0 I}{2\pi}\cdot\frac{r}{R^2}$$

- Magnetic field inside a hollow current carrying conductor is zero.

3. Magnetic Field due to a Solenoid

A current carrying solenoid behaves as a bar magnet. The face, where current is flowing clockwise behaves as South pole and the face, where current is seen flowing anti-clockwise, behaves as North pole.

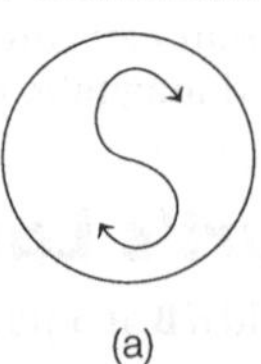

(a)

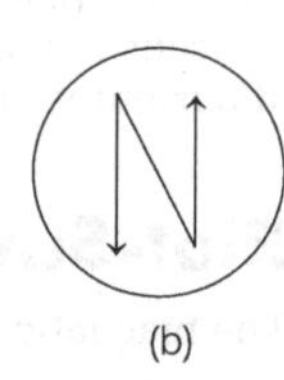

(b)

For such a solenoid, the magnetic field inside it is uniform and directed axially.

- For a solenoid coil of infinite length at a point on its axial line, the magnetic field, $B = \mu_0 nI$ where, n is number of turns per unit length.
- At the end of solenoid, $B = \frac{1}{2}\mu_0 nI$
- At the end field is half of at the centre, this is called **end effect.**

4. Toroidal Solenoids

For a toroid (i.e. a ring shaped closed solenoid) magnetic field at any point within the core of toroid $B = \mu_0 nI$,

where $n = \frac{N}{2\pi R}$, R = radius of toroid.

Force on a Moving Charge in Uniform Magnetic Field and Electric Field

- If a charge q is moving with velocity $\mathbf{v}$ enters in a region in which electric field $\mathbf{E}$ and magnetic field $\mathbf{B}$ both are present, it experiences force due to both fields simultaneously. The force experienced by the charged particle is given by the expression

$$\mathbf{F} = q(\mathbf{v} \times \mathbf{B}) + q\mathbf{E}$$

Here, magnetic force $\mathbf{F}_m = q(\mathbf{v} \times \mathbf{B}) = Bqv\sin\theta$ and electric force $\mathbf{F}_e = q\mathbf{E}$.

- The direction of magnetic force is same as $\mathbf{v} \times \mathbf{B}$ if charge is positive and opposite to $\mathbf{v} \times \mathbf{B}$, if charge q is negative.

Motion of a Charged Particle in a Uniform Magnetic Field

- (i) If a charge particle enters a uniform magnetic field B with a velocity v in a direction perpendicular to that of B (i.e. $\theta = 90°$), then the charged particle experiences a force $F_m = qvB$. Under its influence, the particle describes a circular path, such that

Radius of circular path, $r = \dfrac{mv}{qB}$

In general,

$$r = \frac{mv}{qB} = \frac{p}{qB} = \frac{\sqrt{2mK}}{qB} = \frac{\sqrt{2mqV}}{qB} = \frac{1}{B}\sqrt{\frac{2mV}{q}}$$

where, $p = mv$ = momentum of charged particle, K = kinetic energy of charged particle and V = accelerating potential difference.

(ii) The **period of revolution** of charged particle $T = \dfrac{2\pi m}{qB}$,

the **frequency of revolution** $\nu = \dfrac{qB}{2\pi m}$

or **angular frequency** $\omega = \dfrac{qB}{m}$

- (i) If a charged particle is moving at an angle θ, to the magnetic field (where θ, is other than 0°, 90° or 180°), it describes a helical path, where **radius of helical path**,

$$r = \frac{mv\sin\theta}{qB}.$$

(ii) **Revolution period**, $T = \dfrac{2\pi m}{qB}$

or **Frequency**, $\nu = \dfrac{qB}{2\pi m}$

(iii) Moreover, pitch (the linear distance travelled during one complete revolution) of helical path is given by

$$p = v\cos\theta \cdot T = \frac{2\pi mv\cos\theta}{qB}$$

- If the direction of a $\mathbf{v}$ is parallel or anti-parallel to $\mathbf{B}$, $\theta = 0$ or $\theta = 180°$ and therefore $F = 0$. Hence, the trajectory of the particle is a straight line.

If the velocity of the charged particle is not perpendicular to the field, we will break the velocity in parallel ($v_{||}$) and perpendicular ($v_\perp$) components.

$$r = \frac{mv_\perp}{qB}$$

$$\text{Pitch, } p = (v_{||})T$$

Cyclotron

It is a device used to accelerate positively charged particles, e.g. proton, deuteron, α-particle and other heavy ions to high energy of 100 MeV or more.

Cyclotron frequency, $\nu = \dfrac{Bq}{2\pi m}$.

Maximum energy gained by the charged particle

$$E_{\max} = \left[\frac{q^2B^2}{2m}\right]r^2$$

where, r = maximum radius of the circular path followed by the positive ion.

Maximum energy obtained by the particle is in the form of kinetic energy.

Magnetic Force on a Current Carrying Conductor

If a current carrying conductor is placed in a magnetic field $\mathbf{B}$, then a small current element $I\,d\mathbf{l}$ experiences a force given by

$$d\mathbf{F}_m = Id\mathbf{l} \times \mathbf{B}$$

and the total force experienced by whole current carrying conductor will be

$$\mathbf{F}_m = \int d\mathbf{F}_m = \int I(d\mathbf{l} \times \mathbf{B})$$

The direction of force can also be determined by applying Fleming's left hand rule or right hand thumb rule.

Force between Two Parallel Current Carrying Conductors

- Two parallel current carrying conductors exert magnetic force on one another.

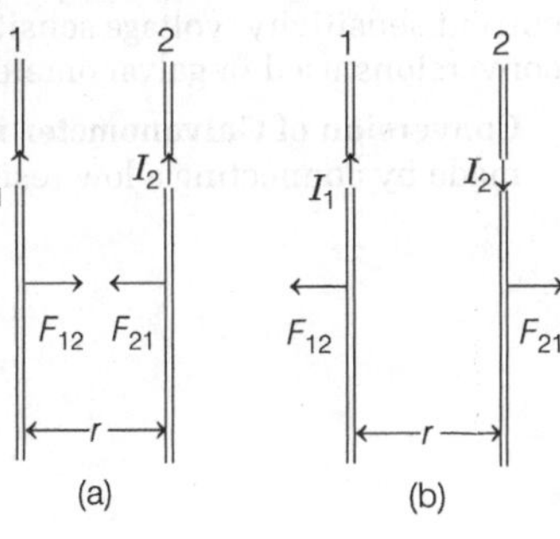

- Magnetic force experienced by length l of any one conductor due to the other current carrying conductor is

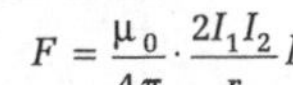

$$F = \frac{\mu_0}{4\pi} \cdot \frac{2I_1I_2}{r} l$$

Force per unit length, $\dfrac{F}{l} = F_0 = \dfrac{\mu_0}{4\pi} \cdot \dfrac{2I_1I_2}{r}$

- If the conductors carries current in same direction, then force between them will be attractive.
- If the conductor carries current in opposite direction, then force will be repulsive.

Torque

When a current carrying loop placed in a uniform magnetic field, it experience torque,

$$\tau = NIAB\sin\theta$$

where, NiA is defined as the magnitude of the dipole moment of the coil

$$(p_m) \cdot \tau = p_m B \sin\theta$$

$$\Rightarrow \quad \tau = \mathbf{p}_m \times \mathbf{B}$$

- A current carrying loop (of any shape) behaves as a magnetic dipole whose magnetic moment is given by
  $$(p_m) = IA$$
- If we have a current carrying coil having N turns, then magnetic moment P_m of dipole will be
  $$(p_m) = NIA$$
- Magnetic moment of a current carrying coil is a vector and its direction is given by right hand thumb rule.

Moving Coil Galvanometer (MCG)

MCG is used to measure the current upto nanoampere. The deflecting torque of MCG,

$$\tau_{def} = NBIA$$

A restoring torque is set up in the suspension fibre. If α is the angle of trust, the restoring torque is

$$\tau_{restoring} = KI$$

where, K is galvanometer constant.

Some Important Concepts Related to Moving Coil Galvanometer

Some of the important concepts related to galvanometer, i.e. current sensitivity, voltage sensitivity and some of conversions used in galvanometer are given below.

- **Conversion of Galvanometer into Ammeter** An ammeter is made by connecting a low resistance S in parallel with a pivoted type moving coil galvanometer G. S is known as shunt.

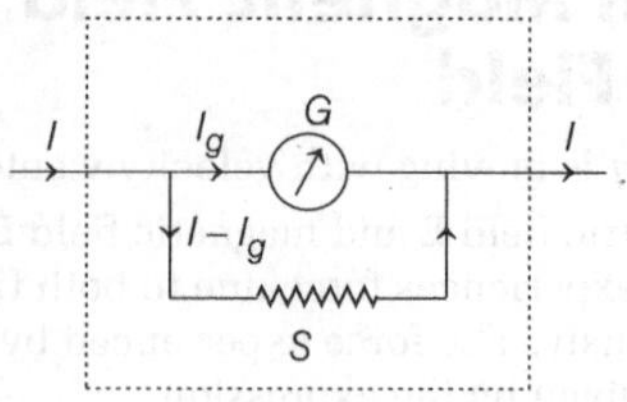

Ammeter

Then, from circuit,

$$I_g \times G = (I - I_g) \times S$$

$$\Rightarrow \quad S = \left(\frac{I_g}{I - I_g}\right) G$$

So, $S << G$, only a small fraction of current goes through the galvanometer.

- **Conversion of Galvanometer into Voltmeter**
  A voltmeter is made by connecting a resistor of high resistance R in series with a pivoted type moving coil galvanometer G.

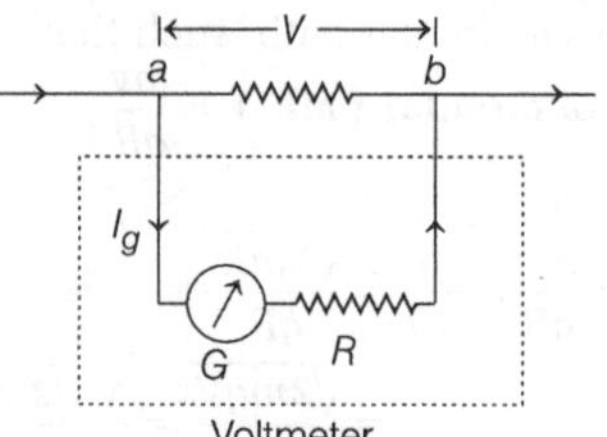

Voltmeter

From the circuit, $\quad I_g = \dfrac{V}{G+R} \Rightarrow R = \dfrac{V}{I_g} - G$

- **Current Sensitivity** The current sensitivity of a galvanometer is defined as the deflection produced in the galvanometer per unit current flowing through it.
  $$S_I = \frac{\alpha}{I} = \frac{NBA}{C}$$
- **Voltage Sensitivity** Voltage sensitivity of a galvanometer is defined as the deflection produced in the galvanometer per unit voltage applied to it.
  $$S_V = \frac{\alpha}{V} = \frac{\alpha}{IR} = \frac{S_I}{R} = \frac{NBA}{RC}$$

DAY PRACTICE SESSION 1

FOUNDATION QUESTIONS EXERCISE

1 A loosely wound helix made of stiff wire is mounted vertically with the lower end just touching a dish of mercury. When a current from a battery is started in the coil through the mercury

(a) the wire oscillates
(b) the wire continues making contact
(c) the wire breaks contact just as current is passed
(d) the mercury will expand by heating due to passes of current

2 A current I flows through a closed loop as shown in figure.

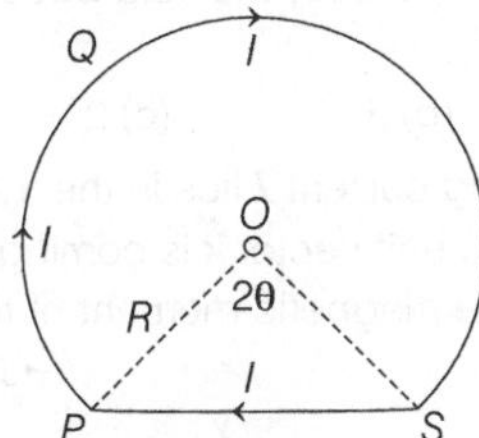

The magnetic field induction at the centre O is

(a) $\frac{\mu_0 I}{4\pi R}\theta$ (b) $\frac{\mu_0 I}{4\pi R}(\theta + \sin\theta)$

(c) $\frac{\mu_0 I}{4\pi R}(\pi - \theta + \sin\theta)$ (d) $\frac{\mu_0 I}{2\pi R}(\pi - \theta + \tan\theta)$

3 The magnetic induction at the centre O in the figure as shown is

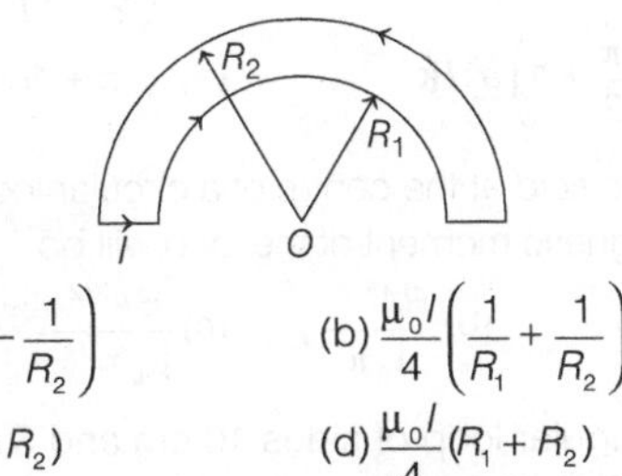

(a) $\frac{\mu_0 I}{4}\left(\frac{1}{R_1} - \frac{1}{R_2}\right)$ (b) $\frac{\mu_0 I}{4}\left(\frac{1}{R_1} + \frac{1}{R_2}\right)$

(c) $\frac{\mu_0 I}{4}(R_1 - R_2)$ (d) $\frac{\mu_0 I}{4}(R_1 + R_2)$

4 An electron moves in a circular orbit with a uniform speed v. It produces a magnetic field B at the centre of the circle. The radius of the circle is proportional to

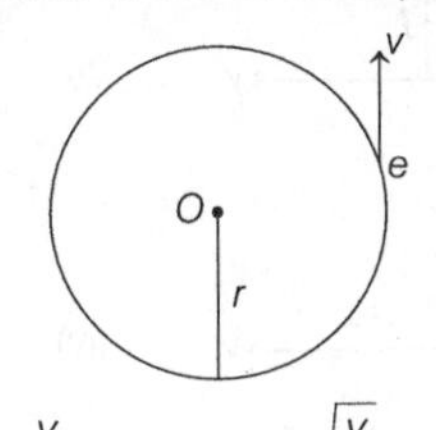

(a) $\frac{B}{v}$ (b) $\frac{v}{R}$ (c) $\sqrt{\frac{v}{B}}$ (d) $\sqrt{\frac{B}{v}}$

5 The magnitude of the magnetic field (B) due to loop $ABCD$ at the origin (O) is **→ AIEEE 2009**

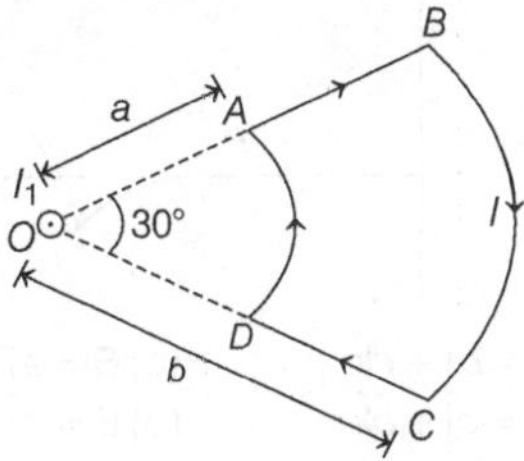

(a) zero (b) $\frac{\mu_0 I(b-a)}{24ab}$

(c) $\frac{\mu_0 I}{4\pi}\left[\frac{b-a}{ab}\right]$ (d) $\frac{\mu_0 I}{4\pi}\left[2(b-a) + \frac{\pi}{3}(a+b)\right]$

6 A current I flows in an infinity long wire with cross-section in the form of a semicircular ring of radius R. The magnitude of the magnetic induction along its axis is **→ AIEEE 2012**

(a) $\frac{\mu_0 I}{2\pi^2 R}$ (b) $\frac{\mu_0 I}{2\pi R}$ (c) $\frac{\mu_0 I}{4\pi R}$ (d) $\frac{\mu_0 I}{\pi^2 R}$

7 Two coaxial solenoids of different radii carry current I in the same direction. Let $\mathbf{F}_1$ be the magnetic force on the inner solenoid due to the outer one and $\mathbf{F}_2$ be the magnetic force on the outer solenoid due to the inner one. Then, **→ JEE Main 2015**

(a) $\mathbf{F}_1 = \mathbf{F}_2 = 0$
(b) $\mathbf{F}_1$ is radially inwards and $\mathbf{F}_2$ is radially outwards
(c) $\mathbf{F}_1$ is radially inwards and $\mathbf{F}_2 = 0$
(d) $\mathbf{F}_1$ is radially outwards and $\mathbf{F}_2 = 0$

8 An infinitely long conductor PQR is bent to form a right angle as shown. A current I flows through PQR. The magnetic field due to this current at the point M is H_1. Now another infinitely long straight conductor QS is connected at Q, so that the current is $I/2$ in QR as well as in QS, the current in PQ remaining unchanged. The magnetic field at M is now H_2. The ratio $H_1 : H_2$ is given by

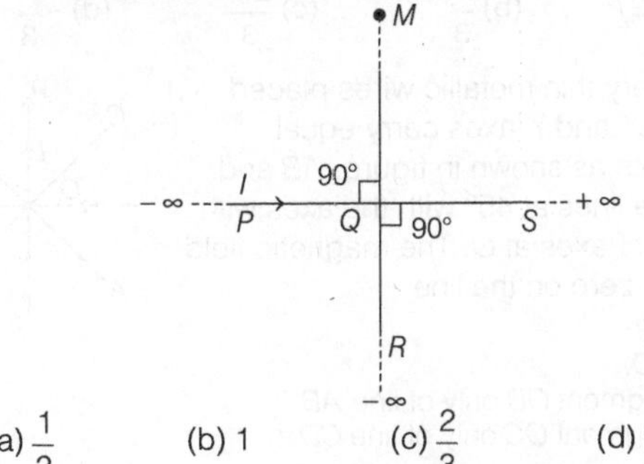

(a) $\frac{1}{2}$ (b) 1 (c) $\frac{2}{3}$ (d) 2

9 For a positively charged particle moving in a *xy*-plane initially along the *x*-axis, there is a sudden change in its path due to the presence of electric and/or magnetic fields beyond *P*. The curved path is shown in the *xy*-plane and is found to be non-circular. Which one of the following combinations is possible?

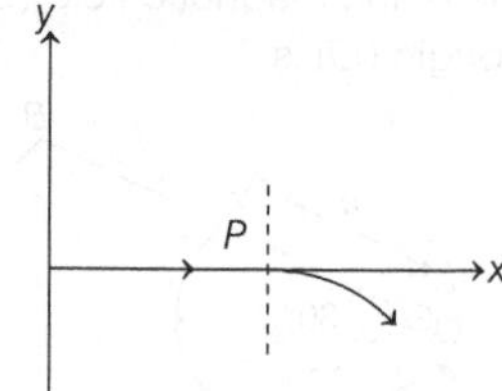

(a) $\mathbf{E} = 0, \mathbf{B} = b\hat{\mathbf{i}} + c\hat{\mathbf{k}}$ (b) $\mathbf{E} = a\hat{\mathbf{i}}; \mathbf{B} = c\hat{\mathbf{k}} + a\hat{\mathbf{i}}$
(c) $\mathbf{E} = 0; \mathbf{B} = c\hat{\mathbf{j}} + b\hat{\mathbf{k}}$ (d) $\mathbf{E} = a\hat{\mathbf{i}}; \mathbf{B} = c\hat{\mathbf{k}} + b\hat{\mathbf{j}}$

10 A magnetic field 4×10^{-3} kT exerts a force $(4\hat{\mathbf{i}} + 3\hat{\mathbf{j}}) \times 10^{-10}$ N on a particle having a charge 10^{-9}C and going on the *xy*-plane. The velocity of the particle is
(a) $-75\,\hat{\mathbf{i}} + 100\,\hat{\mathbf{j}}$ (b) $-100\,\hat{\mathbf{i}} + 75\,\hat{\mathbf{j}}$
(c) $25\,\hat{\mathbf{i}} + 2\,\hat{\mathbf{j}}$ (d) $2\,\hat{\mathbf{i}} + 25\,\hat{\mathbf{i}}$

11 An electron, a proton and an alpha particle having the same kinetic energy are moving in circular orbits of radii r_e, r_p, r_α respectively, in a uniform magnetic field *B*. The relation between r_e, r_p, r_α is → JEE Main 2018
(a) $r_e > r_p = r_\alpha$ (b) $r_e < r_p = r_\alpha$ (c) $r_e < r_p < r_\alpha$ (d) $r_e < r_\alpha < r_p$

12 The cyclotron frequency of an electron gyrating in a magnetic field of 1 T is approximately
(a) 28 MHz (b) 280 MHz (c) 2.8 GHz (d) 28 GHz

13 A proton and an α-particle enters a uniform magnetic field perpendicularly with the same speed. If proton takes 25 μs to make 5 revolutions, then the periodic time for the α-particle would be
(a) 50 μs (b) 25 μs (c) 10 μs (d) 5 μs

14 Two long conductors separated by a distance *d* carry current I_1 and I_2 in the same direction. They exert a force *F* on each other. Now the current in one of them is increased to two times and its direction is reversed. The distance is also increased to 3*d*. The new value of the force between them is
(a) $-2F$ (b) $\frac{F}{3}$ (c) $\frac{2F}{3}$ (d) $-\frac{F}{3}$

15 Two very thin metallic wires placed along *X* and *Y*-axes carry equal currents as shown in figure. *AB* and *CD* are lines at 45° with the axes with origin of axes at *O*. The magnetic field will be zero on the line

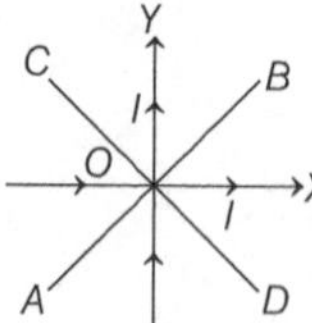

(a) *AB*
(b) *CD*
(c) segment *OB* only of line *AB*
(d) segment *OC* only of line *CD*

16 A thin flexible wire of length *L* is connected to two adjacent fixed points and carries a current *I* in the clockwise direction, as shown in the figure. When the system is put in a uniform magnetic field of strength *B* going into the plane of the paper, the wire takes the shape of a circle. The tension in the wire is → AIEEE 2011

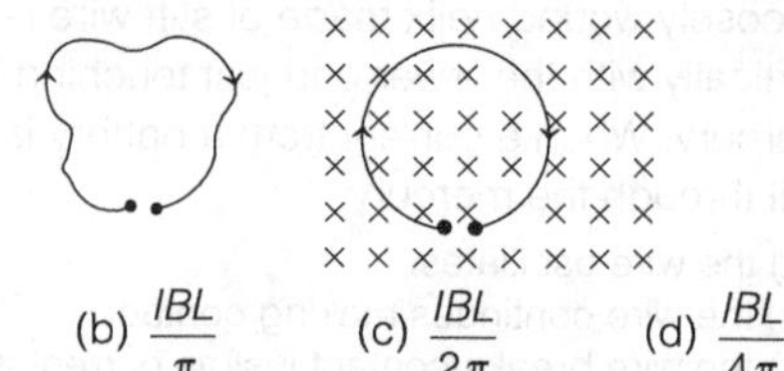

(a) *IBL* (b) $\frac{IBL}{\pi}$ (c) $\frac{IBL}{2\pi}$ (d) $\frac{IBL}{4\pi}$

17 Two parallel long wires *A* and *B* carry currents I_1 and $I_2 (< I_1)$. When I_1 and I_2 are in the same direction, the magnetic field at a point mid-way between the wires is 10 μT. If I_2 is reversed, the field becomes 30 μT. The ratio I_1 / I_2 is
(a) 1 (b) 3 (c) 2 (d) 4

18 A loop carrying current I lies in the *xy*- plane as shown in the figure. The unit vector $\hat{\mathbf{k}}$ is coming out of the plane of the paper. The magnetic moment of the current loop is → JEE Main (Online) 2013

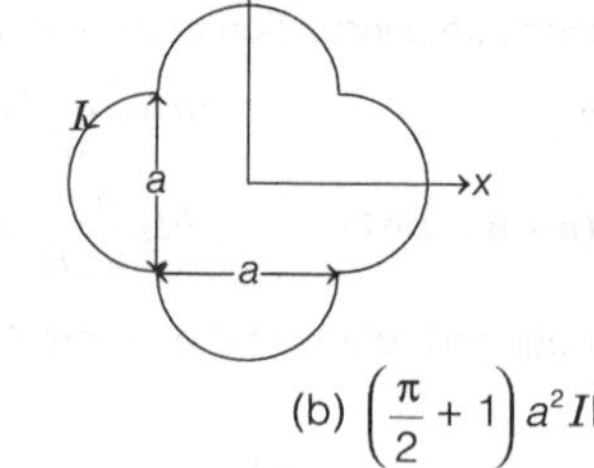

(a) $a^2 I\,\hat{\mathbf{k}}$ (b) $\left(\frac{\pi}{2} + 1\right) a^2 I\hat{\mathbf{k}}$
(c) $-\left(\frac{\pi}{2} + 1\right) a^2 I\hat{\mathbf{k}}$ (d) $(2\pi + 1)\, a^2 I\hat{\mathbf{k}}$

19 Magnetic field at the centre of a circular loop of area *A* is *B*. The magnetic moment of the loop will be
(a) $\frac{BA^2}{\mu_0 \pi}$ (b) $\frac{BA^{3/2}}{\mu_0 \pi}$ (c) $\frac{BA^{3/2}}{\mu_0 \pi^{1/2}}$ (d) $\frac{2BA^{3/2}}{\mu_0 \pi^{1/2}}$

20 A rectangular loop of sides 10 cm and 5 cm carrying a current I of 12 A is placed in different orientations as shown in the figures below. → JEE Main 2015

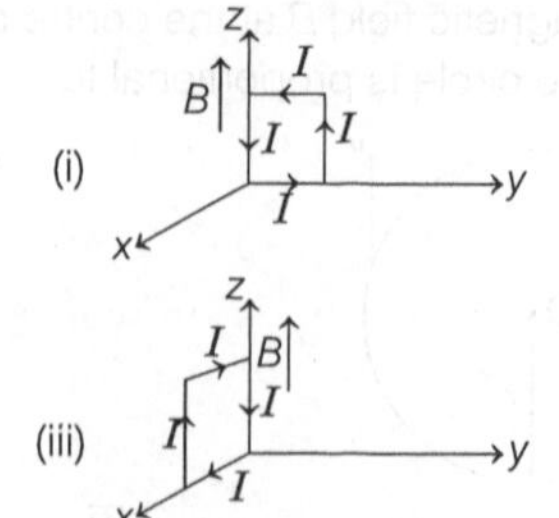

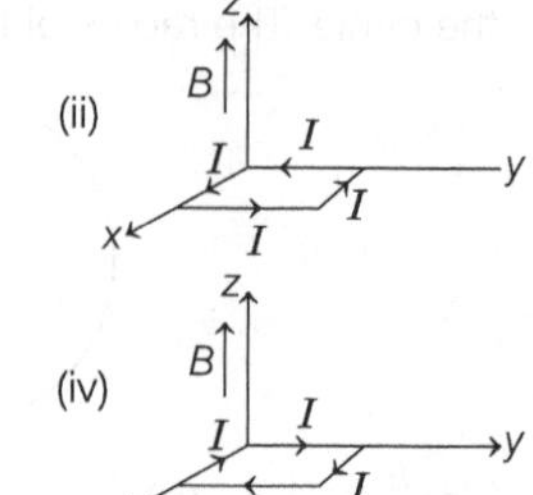

If there is a uniform magnetic field of 0.3T in the positive z-direction, then in which orientations the loop would be in 1. stable equilibrium and 2. unstable equilibrium?

(a) (i) and (ii) respectively (b) (i) and (iii) respectively
(c) (ii) and (iv) respectively (d) (ii) and (iii) respectively

21 When a current of 5 mA is passed through a galvanometer having a coil of resistance 15 Ω, it shows full scale deflection. The value of the resistance to be put in series with the galvanometer to convert it into a voltmeter of range 0-10 V is → **JEE Main 2017 (Offline)**

(a) $2.045 \times 10^3\ \Omega$ (b) $2.535 \times 10^3\ \Omega$
(c) $4.005 \times 10^3\ \Omega$ (d) $1.985 \times 10^3\ \Omega$

22 A galvanometer having a coil resistance of 100 Ω gives a full scale deflection when a current of 1 mA is passed through it. The value of the resistance which can convert this galvanometer into ammeter giving a full scale deflection for a current of 10 A, is → **JEE Main 2016 (Offline)**

(a) 0.01 Ω (b) 2 Ω (c) 0.1 Ω (d) 3 Ω

Direction (Q. Nos. 23-28) *Each of these questions contains two statements : Statement* I *and Statement* II. *Each of these questions also has four alternative choice, only one of which is the correct answer. you have to select one of the codes (a), (b), (c), (d) given below*

(a) Statement I is true, Statement II is true; Statement II is the correct explanation for Statement I
(b) Statement I is true, Statement II is true; Statement II is not the correct explanation
(c) Statement I is true; Statement II is false
(d) Statement I is false; Statement II is true

23 **Statement I** If a charged particle is projected in a region where **B** is perpendicular to velocity of projection, then the net force acting on the particle is independent of its mass.

Statement II The particle is performing uniform circular motion and net force acting on it is $\frac{mv^2}{r}$.

24 **Statement I** A uniformly moving charged particle in a magnetic field, may follow a path along magnetic field lines.

Statement II The direction of magnetic force experienced by a charged particle is perpendicular to its velocity and **B**.

25 **Statement I** The magnetic force experienced by a moving charged particle in a magnetic field is invariant in nature just like any other force.

Statement II Magnetic force experienced by a charged particle is given by $\mathbf{F} = q(\mathbf{v} \times \mathbf{B})$, where **v** is the velocity of charge particle w.r.t. frame of reference in which we are taking **F**.

26 **Statement I** Cyclotron is a device which is used to accelerate the positive ion.

Statement II Cyclotron frequency depends upon the velocity.

27 **Statement I** Magnetic field due to a infinite straight conductor varies inversely as the distance from it.

Statement II The lines of electric force due to a straight current carrying conductor are concentric circles.

28 **Statement I** If a proton and α-particle enter a uniform magnetic field perpendicularly with the same speed, the time period of revolution of α-particle will be double than that of proton.

Statement II Time period of charged particle is given by $T = \frac{2\pi m}{Bq}$.

DAY PRACTICE SESSION 2

PROGRESSIVE QUESTIONS EXERCISE

1 A cell is connected between two points of a uniformly thick circular conductor. I_1 and I_2 are the currents flowing in two parts of the circular conductor of radius a. The magnetic field at the centre of the loop will be

(a) zero (b) $\frac{\mu_0}{4\pi}(I_1 - I_2)$
(c) $\frac{\mu_0}{2a}(I_1 + I_2)$ (d) $\frac{\mu_0}{a}(I_1 + I_2)$

2 A coil having N turns is wound tightly in the form of a spiral with inner and outer radii a and b respectively. When a current I passes through the coil, the magnetic field at the centre is

(a) $\frac{\mu_0 NI}{b}$ (b) $\frac{2\mu_0 NI}{a}$
(c) $\frac{\mu_0 NI}{2(b-a)} \log_e \frac{b}{a}$ (d) $\frac{\mu_0 I^N}{2(b-a)} \log_e \frac{b}{a}$

3 A particle of mass m and charge q moves with a constant velocity v along the positive x-direction. It enters a region containing a uniform magnetic field B directed along the negative z-direction, extending from $x = a$ to $x = b$. The minimum value of v required, so that the particle can just enter the region $x > b$ is

(a) qbB/m (b) $q(b-a)B/m$
(c) qaB/m (d) $q(b+a)B/2m$

4 The magnetic field normal to the plane of a wire of n turns and radius r which carries a current I is measured on the axis of the coil at a small distance is from the centre of the coil. This is smaller than the magnetic field at the centre by the fraction

(a) $(2/3)r^2/h^2$ (b) $(3/2)r^2/h^2$
(c) $(2/3)h^2/r^2$ (d) $(3/2)h^2/r^2$

5 A coil having N turns is wound tightly in the form of a spiral with inner and outer radii a and b respectively. When a current I passes through the coil, the magnetic field at the centre is → AIEEE 2012

(a) $\frac{\mu_0 NI}{b}$ (b) $\frac{2\mu_0 NI}{a}$
(c) $\frac{\mu_0 NI}{2(b-a)}\ln\frac{b}{a}$ (d) $\frac{\mu_0 I}{2(b-a)}\ln\frac{b}{a}$

6 Two long current carrying thin wires, both with current I, are held by insulating threads of length L and are in equilibrium as shown in the figure, with threads making an angle θ with the vertical. If wires have mass λ per unit length, then the value of I is
(g = gravitational acceleration)

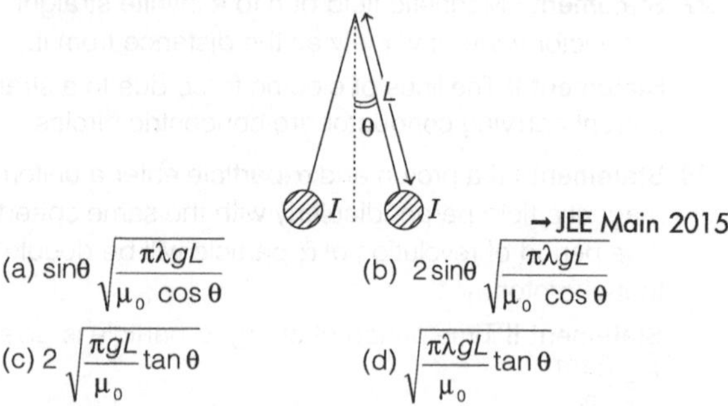

→ JEE Main 2015

(a) $\sin\theta\sqrt{\frac{\pi\lambda gL}{\mu_0\cos\theta}}$ (b) $2\sin\theta\sqrt{\frac{\pi\lambda gL}{\mu_0\cos\theta}}$

(c) $2\sqrt{\frac{\pi gL}{\mu_0}\tan\theta}$ (d) $\sqrt{\frac{\pi\lambda gL}{\mu_0}\tan\theta}$

7 Two identical wires A and B, each of length l, carry the same current I. Wire A is bent into a circle of radius R and wire B is bent to form a square of side a. If B_A and B_B are the values of magnetic field at the centres of the circle and square respectively, then the ratio $\frac{B_A}{B_B}$ is

→ JEE Main 2016 (Offline)

(a) $\frac{\pi^2}{8}$ (b) $\frac{\pi^2}{16\sqrt{2}}$ (c) $\frac{\pi^2}{16}$ (d) $\frac{\pi^2}{8\sqrt{2}}$

8 An infinitely long hollow conducting cylinder with inner radius $R/2$ and outer radius R carries a uniform current density along its length. The magnitude of the magnetic field, $|\mathbf{B}|$ as a function of the radial distance r from the axis is best represented by → JEE Main (Online) 2013

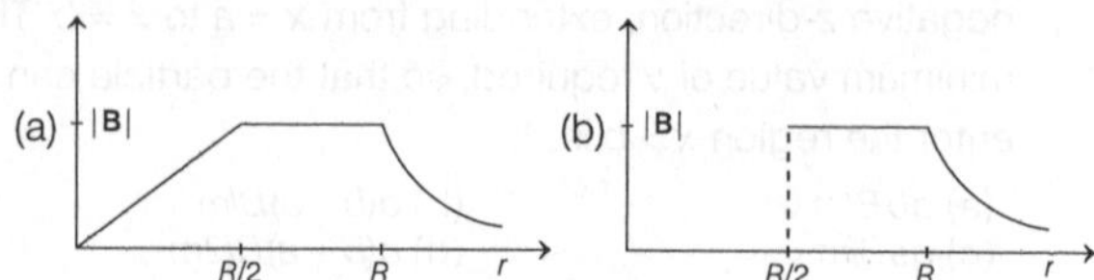

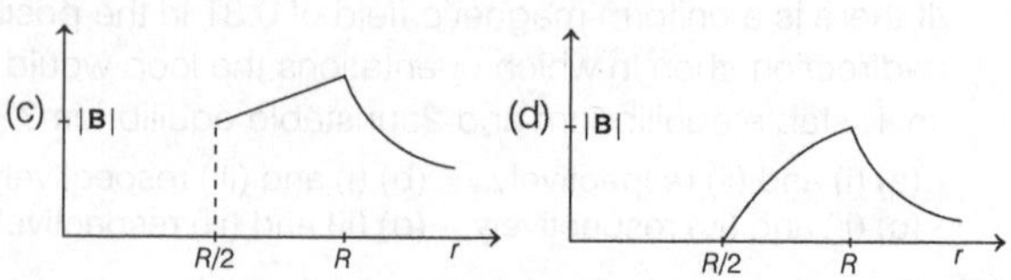

9 Electrons emitted with negligible speed from an electron gun are accelerated through a potential difference V_0 along the x-axis. These electrons emerge from a narrow hole into a uniform magnetic field of strength B directed along x-axis.
Some electrons emerging at slightly divergent angles as shown. These paraxial electrons are refocussed on the x-axis at a distance.

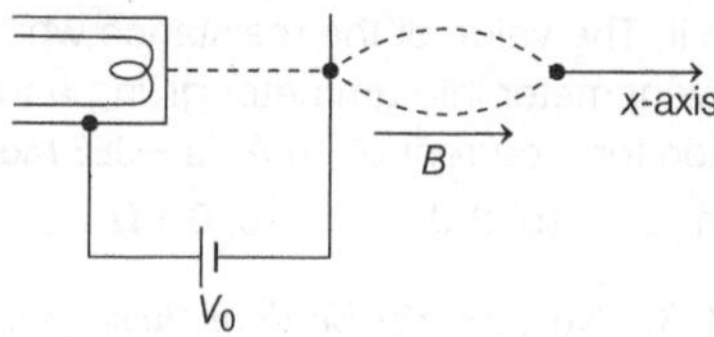

(a) $\frac{8\pi^2 mV_0}{3eB}$ (b) $\sqrt{\frac{8\pi^2 mV_0}{eB^2}}$ (c) $\sqrt{\frac{2\pi^2 mV_0}{eB^2}}$ (d) $\sqrt{\frac{4\pi^2 mV_0}{eB^2}}$

10 A conductor lies along the z-axis at $-1.5 \le z < 1.5$ m and carries a fixed current of 10.0 A in $-a_z$ direction (see figure). For a field $\mathbf{B} = 3.0\times10^{-4}e^{-0.2x}a_y$ T, find the power required to move the conductor at constant speed to $x = 2.0$ m, $y = 0$ in 5×10^{-3}s. Assume parallel motion along the x-axis. → JEE Main 2014

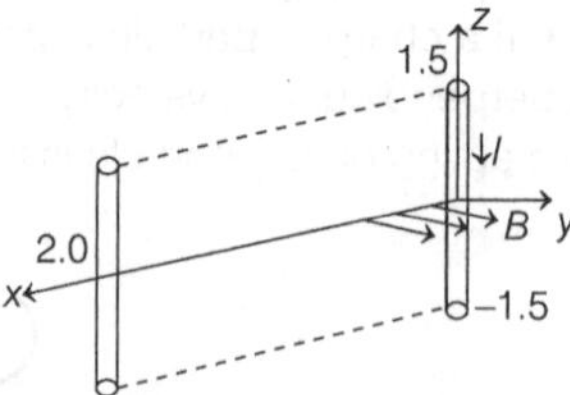

(a) 1.57 W (b) 2.97 W (c) 14.85 W (d) 29.7 W

11 A long straight wire carries a current I. A particle of charge $+q$ and mass m is projected with a speed v from a distance x_0 as shown. The minimum separation between the wire and particle is

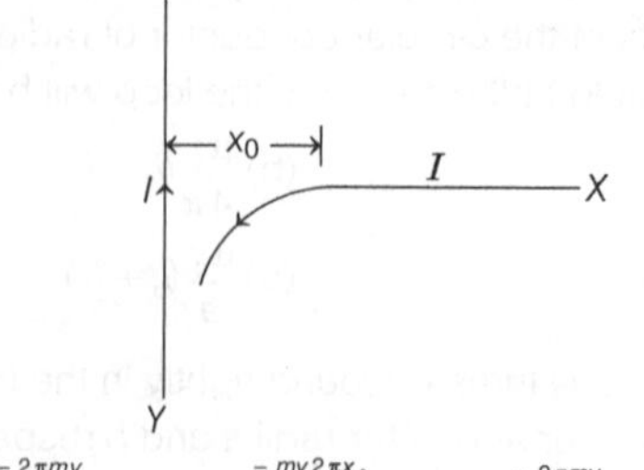

(a) $x_0 e^{\frac{-2\pi mv}{\mu_0 qI}}$ (b) $x_0 e^{\frac{-mv2\pi x_0}{\mu_0 qI}}$ (c) $x_0 e^{\frac{-2\pi mv}{qI}}$ (d) zero

12 A current carrying circular loop of radius R is placed in the xy-plane with centre at the origin. Half of the loop with $x > 0$ is now bent, so that it now lies in the yz-plane.

(a) The magnitude of magnetic moment now diminishes
(b) The magnetic moment does not change
(c) The magnitude of B at $(0, 0, z)$, $z >> R$ increases
(d) The magnitude of B at$(0, 0, z)$, $z >> R$ is unchanged

13 A 100 turn rectangular coil $ABCD$ (in xy-plane) is hung from one arm of a balance (shown in figure). A mass 500 g is added to the other arm to balance the weight of the coil. A current of 4.9 A passes through the coil and a constant magnetic field of 0.2 T acting inward (in xz-plane) is switched on such that only arm CD of length 1cm lies in the field. How much additional mass m must be added to regain the balance?

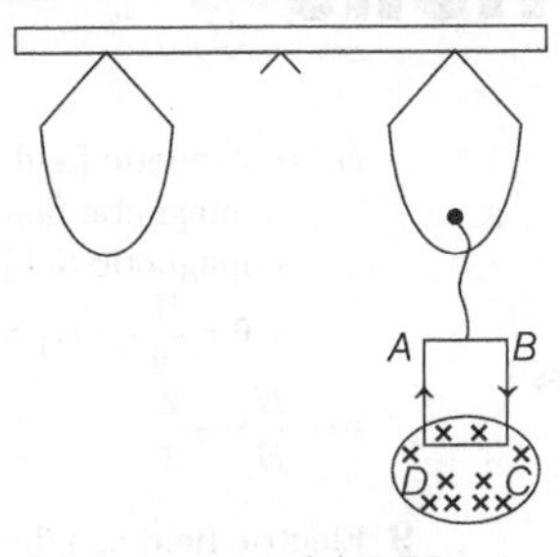

(a) 2 g (b) 1 g (c) 0.75 g (d) 1.5 g

14 A charge Q is uniformly distributed over the surface of non-conducting disc of radius R. The disc rotates about an axis perpendicular to its plane and passing through its centre with an angular velocity ω. As a result of this rotation, a magnetic field of induction B is obtained at the centre of the disc. If we keep both the amount of charge placed on the disc and its angular velocity to be constant and vary the radius of the disc then the variation of the magnetic induction at the centre of the disc will be represented by the figure. → JEE Main (Online) 2013

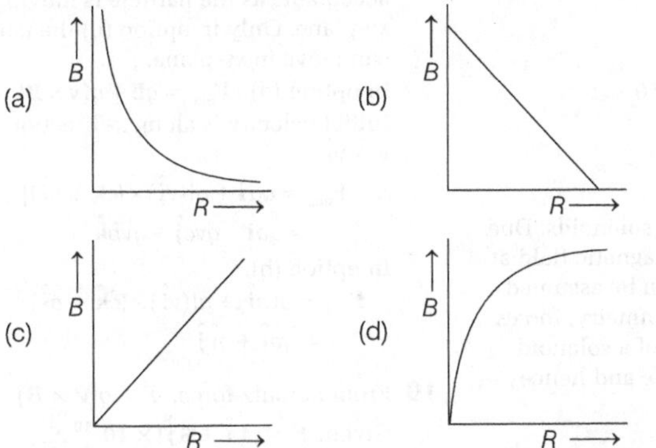

15 A beam consisting of four types of ions a, b, c and d enters a region that contains a uniform magnetic field as shown in figure. The field is perpendicular to the plane of the paper, but its precise direction is not given. All ions in the beam travel with the same speed.

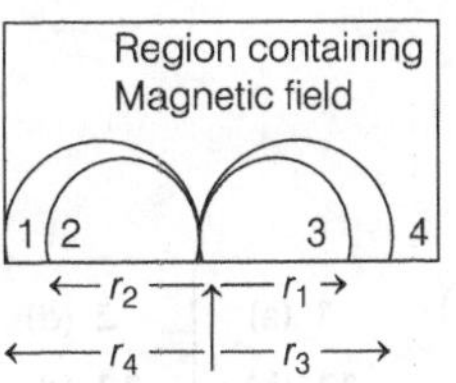

The table below gives the masses and charges of the ions

ION	MASS	CHARGE
A	$2m$	$+e$
B	$4m$	$-e$
C	$2m$	$-e$
D	m	$+e$

The ions fall at different positions 1,2,3 and 4 as shown. Correctly match the ions with respective falling positions

Column I	Column II
A	1
B	2
C	3
D	4

	A	B	C	D
(a)	4	3	2	1
(b)	1	2	3	4
(c)	4	1	2	3
(d)	3	4	1	2

16 Two long parallel wires are at a distance $2d$ apart. They carry steady equal current flowing out of the plane of the paper as shown. The variation of the magnetic field along the line XX' is given by → AIEEE 2010

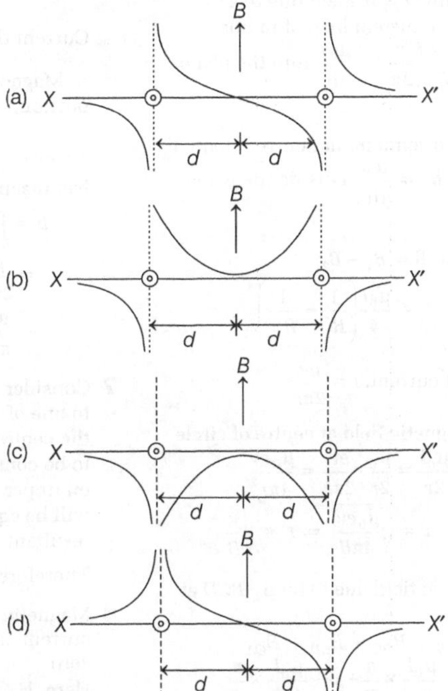

ANSWERS

SESSION 1	**1** (a)	**2** (d)	**3** (a)	**4** (c)	**5** (b)	**6** (d)	**7** (a)	**8** (c)	**9** (b)	**10** (a)
	11 (b)	**12** (d)	**13** (c)	**14** (c)	**15** (a)	**16** (c)	**17** (c)	**18** (b)	**19** (d)	**20** (c)
	21 (d)	**22** (a)	**23** (c)	**24** (d)	**25** (d)	**26** (c)	**27** (b)	**28** (a)		
SESSION 2	**1** (a)	**2** (c)	**3** (b)	**4** (d)	**5** (c)	**6** (b)	**7** (d)	**8** (d)	**9** (b)	**10** (b)
	11 (a)	**12** (a)	**13** (b)	**14** (a)	**15** (c)	**16** (a)				

Hints and Explanations

SESSION 1

1 Loosely wound helix get compressed when possess current. So, connection of mercury lost and demagnetisation takes place. So, result a oscillatory motion.

2 $B_0 = B_{PQS} + B_{SP}$

$$= \frac{\mu_0}{4\pi}\frac{I}{R}(2\pi - 2\theta) + \frac{\mu_0}{4\pi}\frac{1}{R\cos\theta}(\sin\theta + \sin\theta)$$

$$= \frac{\mu_0 I}{2\pi R}(\pi - \theta + \tan\theta)$$

3 Magnetic field due to straight parts of wire at point $O = 0$. Field due to a semicircular current loop of radius R_1, $B_1 = \frac{\mu_0 I}{2R_1}\left(\frac{\pi}{2\pi}\right) = \frac{\mu_0 I}{4R_1}$ into the plane of paper.

Field due to semicircular current loop of radius R_2, $B_2 = \frac{\mu_0 I}{4R_2}$ outside the plane of paper.

$\therefore$ Net field $B = B_1 - B_2$

$$= \frac{\mu_0 I}{4}\left(\frac{1}{R_1} - \frac{1}{R_2}\right)$$

4 Equivalent current, $I = \frac{ev}{2\pi r}$

Hence, magnetic field at centre of circle

$$B = \frac{\mu_0 I}{2r} = \frac{\mu_0}{2r}\cdot\frac{ev}{2\pi r} = \frac{\mu_0 ev}{4\pi r^2}$$

$$\Rightarrow \quad r = \sqrt{\frac{\mu_0 ev}{4\pi B}} \Rightarrow r \propto \sqrt{\frac{v}{B}}$$

5 Net magnetic field due to loop $ABCD$ at O is

$$B = B_{AB} + B_{BC} + B_{CD} + B_{OA}$$

$$= 0 + \frac{\mu_0 I}{4\pi a}\times\frac{\pi}{6} + 0 - \frac{\mu_0 I}{4\pi b}\times\frac{\pi}{6}$$

$$= \frac{\mu_0 I}{24a} - \frac{\mu_0 I}{24b} = \frac{\mu_0 I}{24ab}(b - a)$$

6 Consider the wire to be made up of large number of thin wires of infinite length. Consider such wire of thickness dl subtending an angle $d\theta$ at centre.

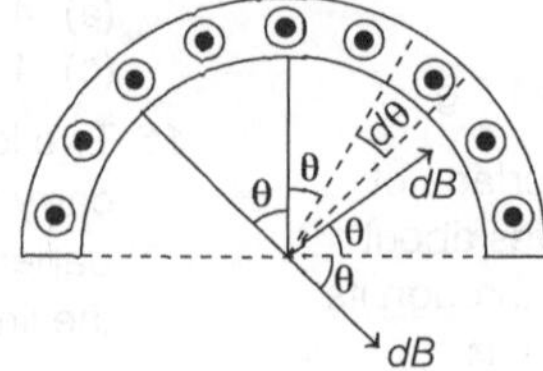

Current through this wire, $dI = \frac{d\theta}{\pi}I$

$\therefore$ Magnetic field at centre due to this portion,

$$dB = \frac{\mu_0}{4\pi}\cdot\frac{2dI}{R} = \frac{\mu_0 I}{2\pi^2 R}d\theta$$

Net magnetic field at the centre

$$B = \int_{-\pi/2}^{\pi/2} dB\cos\theta$$

$$= \frac{\mu_0 I}{2\pi^2 R}\int_{-\pi/2}^{\pi/2}\cos\theta\, d\theta$$

$$= \frac{\mu_0 I}{\pi^2 R}$$

7 Consider the two coaxial solenoids. Due to one of the solenoids magnetic field at the centre of the other can be assumed to be constant. Due to symmetry, forces on upper and lower part of a solenoid will be equal and opposite and hence resultant is zero.

Therefore, $\mathbf{F}_1 = \mathbf{F}_2 = 0$

8 Magnetic field at any point lying on the current carrying straight conductor is zero.

Here, H_1 = Magnetic field at M due to current in PQ

H_2 = Magnetic field at M due to QR + magnetic field at M due to QS + magnetic field at M due to PQ

$$= 0 + \frac{H_1}{2} + H_1 = \frac{3}{2}H_1$$

$$\Rightarrow \quad \frac{H_1}{H_2} = \frac{2}{3}$$

9 Electric field can deviate the path of the particle in the shown direction only when it is along negative y-direction. In the given option **E** is either zero or along x-direction. Hence, it is the magnetic field which is really responsible for its curved path. Options (a) and (c) cannot be accepted as the path will be helix in that case (when the velocity vector makes an angle other than 0°,180 or 90° with the magnetic field, path is a helix). Option (d) is wrong because in that case component of net force on the particle also comes in k direction which is not acceptable as the particle is moving in xy-plane. Only in option (b) the particle can move in xy-plane.

In option (d) : $\mathbf{F}_{net} = q\mathbf{E} + q(\mathbf{v}\times\mathbf{B})$

Initial velocity is along x-direction. So let $\mathbf{v} = v\hat{\mathbf{i}}$

$$\therefore \quad \mathbf{F}_{net} = qa\hat{\mathbf{i}} + q[(v\hat{\mathbf{i}})\times(c\hat{\mathbf{k}} + b\hat{\mathbf{j}})]$$

$$= qa\hat{\mathbf{i}} - qvc\hat{\mathbf{j}} + qvb\hat{\mathbf{k}}$$

In option (b),

$$\mathbf{F}_{net} = q(a\hat{\mathbf{i}}) + q[(v\hat{\mathbf{i}})\times(c\hat{\mathbf{k}} + a\hat{\mathbf{i}})]$$

$$= qa\hat{\mathbf{i}} + qv\hat{\mathbf{j}}$$

10 From Lorentz force, $F = q(v \times B)$

Given, $F = (4\hat{\mathbf{i}} + 3\hat{\mathbf{j}})\times 10^{-10}$ N,

$q = 10^{-9}$C, $B = 4\times 10^{-3}\,\hat{\mathbf{k}}$T

$$\therefore \quad (4\hat{\mathbf{i}} + 3\hat{\mathbf{j}})\times 10^{-10} = 10^{-9}(a\hat{\mathbf{i}} + b\hat{\mathbf{j}})\times(4\times 10^{-3})$$

Solving, we get

$$a = -75, b = 100 \Rightarrow v = -75\hat{\mathbf{i}} + 100\hat{\mathbf{j}}$$

11 From $Bqv = \frac{mv^2}{r}$, we have

$$r = \frac{mv}{Bq} = \frac{\sqrt{2mK}}{Bq}$$

where, K is the kinetic energy.
As, kinetic energies of particles are same;

$$r \propto \frac{\sqrt{m}}{q}$$

$$\Rightarrow \quad r_e : r_p : r_\alpha = \frac{\sqrt{m_e}}{e} : \frac{\sqrt{m_p}}{e} : \frac{\sqrt{4m_p}}{2e}$$

Clearly, $r_p = r_\alpha$ and r_e is least $[\because m_e < m_p]$

So, $r_e < r_p = r_\alpha$

12 Cyclotron frequency,

$$\nu = \frac{Bq}{2\pi m} = \frac{1 \times 1.6 \times 10^{-19}}{2\pi \times 9.1 \times 10^{-31}}$$

$$= 2.8 \times 10^{10}\ \text{Hz} = 28\ \text{GHz}$$

13 Time taken by to make one revolution $= \frac{25}{5} = 5\,\mu\text{s}$

As $T = \frac{2\pi m}{qB}$; so $\frac{T_2}{T_1} = \frac{m_2}{m_1} \times \frac{q_1}{q_2}$

or $T_2 = T_1 \frac{m_2 q_1}{m_1 q_2} = \frac{5 \times 4m_1}{m_1} = \frac{q}{2q}$

$= 10\,\mu\text{s}$

14 Initial force $F = \frac{\mu_0}{4\pi} \cdot \frac{2I_1 I_2}{d}$ (attractive)

and final force $F' = \frac{\mu_0}{4\pi} \cdot \frac{2(2I_1)I_2}{(3d)}$

(repulsive) $\Rightarrow F' = \frac{2}{3}F$

15 Along the line AB, magnetic field due to two wires is equal but in mutually opposite directions (as per right hand thumb rule). Hence, net magnetic field will be zero along the line AB.

16

$$2T \sin\left(\frac{d\theta}{2}\right) = IdlB$$

$$\Rightarrow \quad 2T\frac{d\theta}{2} = IRd\theta B \qquad [\because dl = Rd\theta]$$

$$\Rightarrow \quad T = BIR = \frac{BIL}{2\pi} \qquad [\because L = 2\pi R]$$

17 $\frac{\mu_0}{4\pi}\frac{2I_1}{r} - \frac{\mu_0}{4\pi}\frac{2I_2}{r} = 10\,\mu\text{T}$

$\frac{\mu_0 2I_1}{4\pi r} + \frac{\mu_0}{4\pi}\frac{2I_2}{r} = 30\,\mu\text{T}$

On solving, $I_1 = 20$ A and $I_2 = 10$ A

So, $I_1 / I_2 = 2$

18 As, $M = I \times$ Area of loop

$$= I \times \left[a^2 + \frac{\pi a^2}{4 \times 2} \times 4\right]\mathbf{k}$$

$$= I \times a^2\left[\frac{\pi}{2} + 1\right]\mathbf{k} = \left(\frac{\pi}{2} + 1\right)a^2 I\,\mathbf{k}$$

19 As, $B = \frac{\mu_0}{4\pi}\frac{2\pi I}{r} = \frac{\mu_0 I}{2r}$

$\Rightarrow \quad I = \frac{2Br}{\mu_0}$

Also $A = \pi r^2$ or $r = \left(\frac{A}{\pi}\right)^{1/2}$

Magnetic moment, $M = IA = \frac{2Br}{\mu_0}A$

$$= \frac{2BA}{\mu_0} \times \left(\frac{A}{\pi}\right)^{1/2} = \frac{2BA^{3/2}}{\mu_0 \pi^{1/2}}$$

20 Since, **B** is uniform only torque acts on a current carrying loop.
As, $\tau = \mathbf{M} \times \mathbf{B} \Rightarrow |\tau| = |\mathbf{M}||\mathbf{B}| \sin\theta$
For orientation shown in figure (ii),
$\theta = 0^0, \tau = 0$ (stable equilibrium)
and for figure (iv), $\theta = \pi, \tau = 0$
(unstable equilibrium)

21 Suppose a resistance R_s is connected in series with galvanometer to convert it into voltmeter.

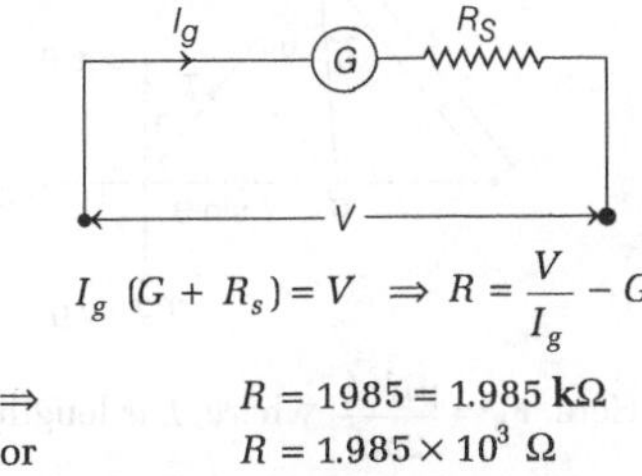

$$I_g\,(G + R_s) = V \ \Rightarrow R = \frac{V}{I_g} - G$$

$\Rightarrow \quad R = 1985 = 1.985\ \text{k}\Omega$

or $\quad R = 1.985 \times 10^3\ \Omega$

22 Maximum voltage that can be applied across the galvanometer coil
$= 100\ \Omega \times 10^{-3}\ \text{A} = 0.1\ \text{V}$

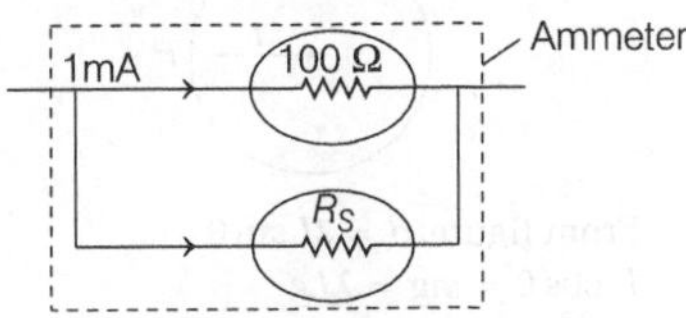

If R_s is the shunt resistance, then

$R_s \times 10\,\text{A} = 0.1\ \text{V}$

$\Rightarrow \quad R_s = 0.01\ \Omega$

23 In this case, the charged particle performs uniform circular motion and magnetic force is providing the necessary centripetal force,

i.e. $\quad \frac{mv^2}{r} = qvB$

$\frac{mv^2}{r}$ is not the force acting on charged particle it is simply equal to net force acting on the particle.

24 Statement I is false, as **B** is perpendicular to **F**, so particle cannot follow magnetic field lines (tangent to which gives the direction of magnetic field).

25 Statement I is false and Statement II is true. Magnetic force may have different value in different frames of reference, that's why it is not invariant in nature.

26 Cyclotron frequency is given by

$$\nu = \frac{1}{T} = \frac{Bq}{2\pi m}$$

It is obvious that cyclotron frequency does not depend upon velocity of charged particle.

27 The magnetic field at a point due to current flowing through an infinitely long conductor is given by

$$B = \frac{\mu_0}{4\pi} \cdot \frac{2I}{a}.$$

where, a is the distance of that point from conductor. Now, according to right hand thumb rule it follows that magnetic field is in the form of concentric circles, whose centres lie on the straight conductor.

28 We know that,

$$T \propto \frac{m}{q} \qquad \left[\because T = \frac{2\pi m}{Bq}\right]$$

For α-particle,

$$T_\alpha \propto \frac{4m}{2q}$$

For proton,

$$T_{p^+} \propto \frac{m}{q}$$

So, $\quad 2T_p = T_\alpha$

SESSION 2

1 The resistance of the portion PRQ will be
$R_1 = l_1 \rho$

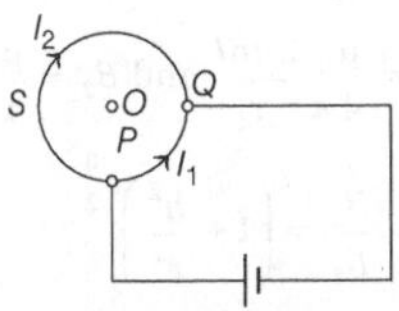

Resistance of the portion PSQ will be
$R_2 = l_2 \rho$
Potential difference across P and Q
$= I_1 R_1 = I_2 R_2$
$I_1 l_1 \rho = I_2 l_2 \rho$ or $I_1 l_1 = I_2 l_2$ ...(i)
Magnetic field induction at the centre O due to currents through circular conductors PRQ and PSQ will be
$= B_1 - B_2$

$$= \frac{\mu_0}{4\pi}\frac{I_1 l_1 \sin 90^\circ}{r^2} - \frac{\mu_0}{4\pi}\frac{I_2 l_2 \sin 90^\circ}{r^2} = 0$$

2 Refer to the figure, number of turns in dx, $n = \frac{N \cdot dx}{b-a}$

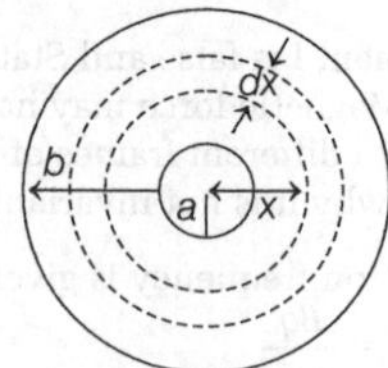

The magnetic field induction at the centre O due to current I through the entire spiral is

$$B = \int_a^b \frac{\mu_0}{4\pi} \frac{2\pi n I dx}{x}$$

$$= \frac{\mu_0}{4\pi} 2\pi \int_a^b \left(\frac{N}{b-a}\right) \frac{dx}{x} I$$

$$= \frac{\mu_0}{4\pi} \frac{NI}{(b-a)} (\log_e x)_a^b = \frac{\mu_0}{2} \frac{NI}{(b-a)} \log_e \frac{b}{a}$$

3 In the figure, the z-axis points out of the paper and the magnetic field is directed into the paper, existing in the region between PQ and RS. The particle moves in a circular path of radius r in the magnetic field. It can just enter the region $x > b$ for $r \geq (b-a)$.

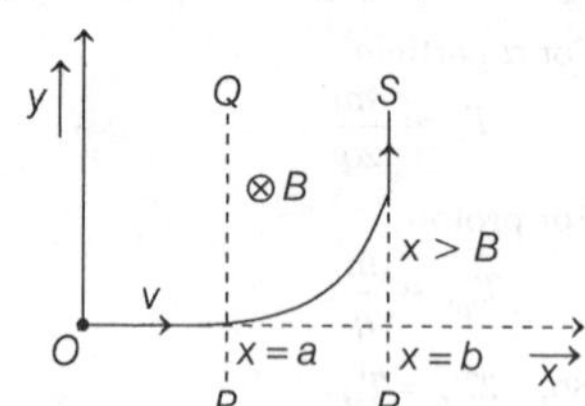

Now, $r = \frac{mv}{qB} \geq (b-a)$

or $v \geq \frac{q(b-a)B}{m} \Rightarrow v_{\min} = \frac{q(b-a)B}{m}$

4 $B_1 = \frac{\mu_0}{4\pi} \frac{2\pi n I}{r}$ and $B_2 = \frac{\mu_0}{4\pi} \frac{2\pi n I r^2}{(r^2 + h^2)^{3/2}}$

So, $\frac{B_2}{B_1} = \left(1 + \frac{h^2}{r^2}\right)^{-\frac{3}{2}}$

Fractional decrease in the magnetic field will be $= \frac{B_1 - B_2}{B_1} = \left(1 - \frac{B_2}{B_1}\right)$

$$= \left[1 - \left(1 + \frac{h^2}{r^2}\right)^{-3/2}\right]$$

$$= 1 - \left(1 - \frac{3}{2} \cdot \frac{h^2}{r^2}\right) = \frac{3}{2} \frac{h^2}{r^2}$$

5 Number of turns per unit width $= \frac{N}{b-a}$

Consider an elemental ring of radius x and thickness dx.

Number of turns in the ring

$$= dN = \frac{Ndx}{b-a}$$

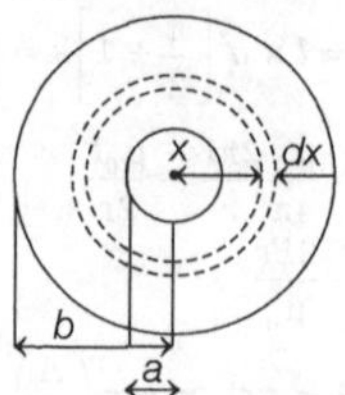

Magnetic field at the centre due to the ring element

$$dB = \frac{\mu_0 (dN) I}{2x} = \frac{\mu_0 I}{2} \cdot \frac{Ndx}{(b-a)} \cdot \frac{1}{x}$$

$\therefore$ Field at the centre

$$= \int dB = \frac{\mu_0 NI}{2(b-a)} \int_a^b \frac{dx}{x}$$

$$= \frac{\mu_0 NI}{2(b-a)} \ln \frac{b}{a}$$

6 Consider free body diagram of the wire. As the wires are in equilibrium, they must carry current in opposite direction.

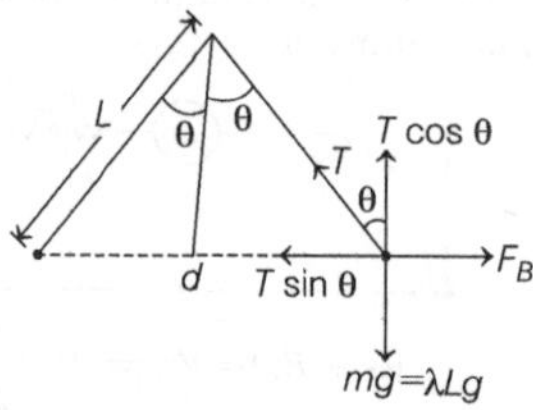

Here, $F_B = \frac{\mu_0 I^2 L}{2\pi d}$, where, L is length of each wire are d is separation between wires.

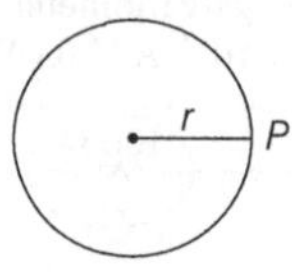

From figure, $d = 2L \sin\theta$

$T \cos\theta = mg = \lambda Lg$

(in vertical direction) ...(i)

$$T \sin\theta = F_B = \frac{\mu_0 I^2 L}{4\pi L \sin\theta}$$

(In horizontal direction) ...(ii)

From Eqs. (i) and (ii), we get

$$\frac{T \sin\theta}{T \cos\theta} = \frac{\mu_0 I^2 L}{4\pi L \sin\theta \times \lambda Lg}$$

$$\therefore I = \sqrt{\frac{4\pi \lambda Lg \sin^2\theta}{\mu_0 \cos\theta}} = 2\sin\theta \sqrt{\frac{\pi \lambda Lg}{\mu_0 \cos\theta}}$$

7 Magnetic field in case of circle of radius R, we have

$$B_A = \frac{\mu_0 I}{2R}$$

As, $2\pi R = l$ (l is length of a wire)

$$R = \frac{l}{2\pi}$$

$$\Rightarrow B_A = \frac{\mu_0 I}{2 \times \frac{l}{2\pi}}$$

$$= \frac{\mu_0 I \pi}{l} \quad \text{...(i)}$$

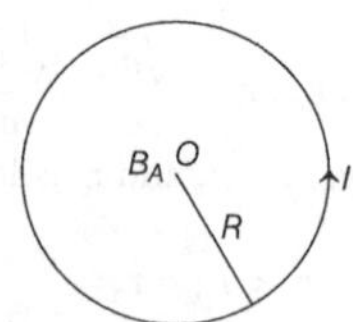

Magnetic field in case of square of side a, we get

$$B_B = 4 \times \frac{\mu_0}{4\pi} \times \frac{I}{\left(\frac{a}{2}\right)} \left(\frac{1}{\sqrt{2}} + \frac{1}{\sqrt{2}}\right)$$

$$\Rightarrow B_B = \frac{4I\mu_0}{\pi a \sqrt{2}} = \frac{\mu_0 2\sqrt{2} I}{a\pi}$$

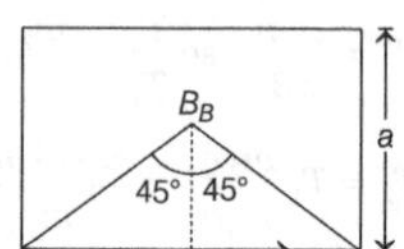

As, $4a = l$, $a = \frac{l}{4} \Rightarrow B_B = \frac{8\sqrt{2}\, \mu_0 I}{\pi l}$ ...(ii)

Dividing Eq. (i) by Eq. (ii), we get

$$\frac{B_A}{B_B} = \frac{\pi^2}{8\sqrt{2}}$$

8 ***Case* I** $x < \frac{R}{2}$,

$|\mathbf{B}| = 0$

***Case* II** $\frac{R}{2} \leq x < R$

$\int \mathbf{B} \cdot d\mathbf{l} = \mu_0 I$

$|\mathbf{B}| 2\pi x = \mu_0$

$$|\mathbf{B}| = \frac{\mu_0 J}{2x}\left(x^2 - \frac{R^2}{4}\right) \quad \left[\pi x^2 - \pi\left(\frac{R}{2}\right)^2\right] J$$

***Case* III** $x \geq R$

$\int \mathbf{B} \cdot d\mathbf{l} = \mu_0 I$

$$|\mathbf{B}| 2\pi x = \mu_0 \left[\pi R^2 - \pi\left(\frac{R}{2}\right)^2\right] J$$

$$|\mathbf{B}| = \frac{\mu_0 J}{2x} \frac{3}{2} R^2,$$

$$|\mathbf{B}| = \frac{3\mu_0 J R^2}{8x}$$

So,

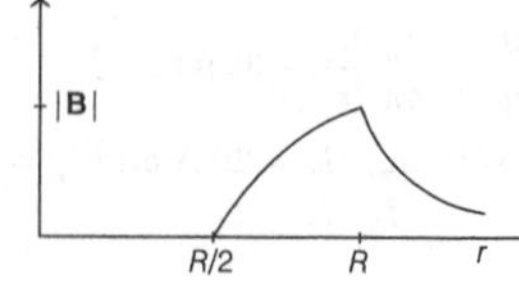

9 The electrons will be refocussed after distance equal to pitch.

$$\text{Pitch} = V_x T = V\frac{2\pi m}{eB}$$

$$= \sqrt{\frac{2eV_0}{m}}\cdot\frac{2\pi m}{eB}$$

$$= \sqrt{\frac{8\pi^2 mV_0}{eB^2}} \qquad \left[\because eV_0 = \frac{1}{2}mv^2\right]$$

10 When force exerted on a current carrying conductor

$$F_{\text{ext}} = BIL$$

$$\text{Average power} = \frac{\text{Work done}}{\text{Time taken}}$$

$$P = \frac{1}{t}\int_0^2 F_{\text{ext.}}\cdot dx = \frac{1}{t}\int_0^2 B(x)IL\,dx$$

$$= \frac{1}{5\times 10^{-3}}\int_0^2 3\times 10^{-4} e^{-0.2x}\times 10\times 3\,dx$$

$$= 9\,[1 - e^{-0.4}]$$

$$= 9\left[1 - \frac{1}{e^{0.4}}\right] = 2.967 \approx 2.97\text{ W}$$

11 $F = q(\hat{i}\,v_x + \hat{j}\,v_y)\times\left[\frac{\mu_0 I}{2\pi x}\hat{k}\right]$

$$= \hat{j}q\,v_x\frac{\mu_0 I}{2\pi x} - \hat{i}q\,v_y\frac{\mu_0 I}{2\pi x}$$

$$\therefore\quad a_x = \frac{F_x}{m} = -\frac{\mu_0 Iqv_y}{2\pi xm}$$

Also, $a_x = \frac{dv_x}{dt} = \frac{dv_x}{dx}\cdot\frac{dx}{dt} = \frac{v_x\,dv_x}{dx}$

Since, $v_x^2 + v_y^2 = v^2$

or $\quad 2v_x\,dv_x + 2v_y dv_y = 0$

$\Rightarrow \quad v_x dv_x = -v_y dv_y$

Hence,

$$\frac{v_x dv_x}{dx} = -\frac{v_y dv_y}{dx} = -\frac{\mu_0 Iq\,v_y}{2\pi\,xm}$$

$$\Rightarrow \quad \frac{dx}{x} = \frac{dv_y 2\pi\,m}{\mu_0 qI}$$

Initially, $x = x_0$ and $v_y = 0$

At minimum separation $v_x = 0,\ v_y = v$

Thus, $\int_{x_0}^{x}\frac{dx}{x} = \frac{2\pi m}{\mu_0 qI}\int_0^{-v} dv_y$

$$\Rightarrow \quad \log\frac{x}{x_0} = -\frac{2\pi\,m\,v}{\mu_0 qI}$$

$$\Rightarrow \quad x = x_0 e^{\frac{-2\pi m\,v}{\mu_0 qI}}$$

12 For a circular loop of radius R, carrying current I in xy-plane, the magnetic moment $M = I\times\pi R^2$. It acts perpendicular to the loop along z-direction. When half of the current loop is bent in yz- plane, then magnetic moment due to half current loop in xy-plane, $M_1 = I\,(\pi R^2/2)$ acting along z-direction. Magnetic moment due to half current loop in yz-plane, $M_2 = I\,(\pi R^2/2)$ along x-direction.

Effective magnetic moment due to entire bent current loop,

$$M' = \sqrt{M_1^2 + M_2^2}$$

$$= \sqrt{(I\pi R^2/2)^2 + (I\pi\,R^2/2)^2}$$

$$= \frac{I\pi R^2}{2}\sqrt{2} < M$$

i.e. magnetic moment diminishes

The magnitude of **B** at a point on the axis of loop, distance z from the centre of current loop in xy-plane is

$$B = \frac{\mu_0}{4\pi}\frac{2\pi/R^2}{(R^2 + z^2)^{3/2}}$$

The magnitude of **B** at a point distance z from the centre of bent current loop, whose half part is in xy-plane and half part is in yx- plane, is

$$B = \sqrt{\left[\frac{\mu_0}{4\pi}\frac{\pi/R^2}{(R^2 + z^2)^{3/2}}\right]^2 + \left[\frac{\mu_0}{4\pi}\frac{\pi/R^2}{(R^2 + z^2)^{3/2}}\right]^2}$$

$$= \frac{\mu_0}{4\pi}\frac{\pi/R^2}{(R^2 + z^2)^{3/2}}\sqrt{2} < B$$

13 Let the mass of coil $= M$

Mass added other arm $= 500$ g

$= 500\times 10^{-3}$ kg

Mass of coil = Mass in other arm (for balancing)

$$M = 500\times 10^{-3}\text{ kg}$$

$$M = 0.5\text{kg}$$

When the current is switched on,

Current, $I = 4.9$ A

Magnetic field $B = 0.2$ T (xz-plane)

Length of arm $CD = 1$ cm,

Mass added to balance $= m$

Let F be the force due to magnetic field.

The direction of magnetic field is inward in xz-plane, the length vector is left, so by using the Fleming's left hand rule, the direction of force is downwards in the plane of paper

$$F = I\,(l\times B) = 4.9\,(0.01\times 0.2\sin 90^\circ)$$

$$F = 4.9\times 0.01\times 0.2 \qquad \text{...(i)}$$

For balancing,

mass of coil $\times\,g$ + force due to magnetic field

$$= 500\times 10^{-3}g + m\times g$$

$$0.5\times 9.8 + 4.9\times 0.01\times 0.2 = 500\times 10^{-3}\times 9.8 + m\times 9.8$$

$$9.8\,(0.5 + 0.001) = 9.8\,(0.5 + m)$$

[From Eq. (i)]

$$m = 0.001\text{ kg} = 1\text{ g}$$

Thus, 1 g of mass must be added to regain the balance.

14 $dB = \frac{\mu_0(dq)}{2r}\left(\frac{\omega}{2\pi}\right)$

$$B = \int dB = \frac{\mu_0\omega}{4\pi}\cdot\frac{Q}{\pi R^2}2\pi\int_0^R\frac{rdr}{r}$$

$$B = \frac{\mu_0\omega Q}{2\pi R^2}\cdot R,\ B = \frac{\mu_0\omega Q}{2\pi R} \Rightarrow B\propto\frac{1}{R}$$

15 A $\rightarrow$ 4, B $\rightarrow$ 1, C $\rightarrow$ 2, D $\rightarrow$ 3

$$R = mv/qB$$

$$R_B > R_A$$

and $R_A = R_C$ (in opposite sense)

and R_D is smallest.

16 The magnetic field in between because of each will be in opposite direction

$$B_{\text{in between}} = \frac{\mu_0 I}{2\pi x}\hat{j} - \frac{\mu_0 I}{2\pi\,(2d - x)}(-\hat{j})$$

$$= \frac{\mu_0 I}{2\pi}\left[\frac{1}{x} - \frac{1}{2d - x}\right](\hat{j})$$

at $x = d,\ B_{\text{in between}} = 0$

For $x < d,\ B_{\text{in between}} = (\hat{j})$ and

For $x > d,\ B_{\text{in between}} = (-\hat{j})$

Towards x, net magnetic field will add up and direction will be $(-\hat{j})$ and towards x', net magnetic field will add up and direction will be $(\hat{j})$.

DAY TWENTY TWO

Magnetism

Learning & Revision for the Day

- Current Loop as a Magnetic Dipole
- Bar Magnet
- Magnetic Field Lines
- The Earth's Magnetism
- Magnetic Behaviour of Materials
- Hysteresis Curve
- Electromagnet

Current Loop as a Magnetic Dipole

A current loop is equivalent to a magnetic dipole. If $A(=\pi a^2)$ be the area of the loop, then the magnitude of its dipole moment is

$$p_m = iA = i\pi a^2$$

where, a is radius of coil, i is current flowing through it

$$i = \frac{p_m}{\pi a^2} \quad \ldots(i)$$

Magnetic field at the centre of a circular current loop is given by

$$B = \frac{\mu_0 i}{2a} \quad \ldots(ii)$$

Putting the value of i from Eq. (i) in Eq. (ii), we get

$$B = \frac{\mu_0}{2\pi}\frac{p_m}{a^3}$$

This is the expression for the magnetic field at the centre of the current loop in terms of its dipole moment. Instead of circular loop, if there is a circular coil having n turns, its dipole moment would be $p_m = niA = ni\pi a^2$.

Bar Magnet

A bar magnet may be viewed as a combination of two magnetic poles, North pole and South pole, separated by some distance. The distance is known as the magnetic length of the given bar magnet.

A bar magnet exhibits two important properties, namely

(i) the attractive property and (ii) the directive property.

- If m is the pole strength and $2l$ the magnetic length of the bar magnet, then its magnetic moment is $\mathbf{M} = m(2\mathbf{l})$.
- Magnetic moment is a vector quantity whose direction is from S pole towards N pole.
- SI unit of magnetic pole strength (m), is **ampere metre** (Am) and of magnetic dipole moment (**M**) is **ampere metre^2** (Am^2).
- If a bar magnet is broken, the fragments are independent magnetic dipoles and not isolated magnetic poles.

- No. of Questions in Exercises (x)—
- No. of Questions Attempted (y)—
- No. of Correct Questions (z)—
(Without referring Explanations)

- Accuracy Level ($z/y \times 100$)—
- Prep Level ($z/x \times 100$)—

In order to expect good rank in JEE, your Accuracy Level should be above 85 & Prep Level should be above 75.

Magnetic Field due to a Bar Magnet

The magnetic field in free space, at a point having distance r from the given bar magnet (or magnetic dipole) is calculated in two conditions, along axial line and along equatorial line.

- **Along axial line** $\mathbf{B} = \frac{\mu_0}{4\pi} \frac{2\mathbf{M}r}{(r^2 - l^2)^2}$

  and the direction of **B** is the same as the direction of **M**. For a short dipole (or for a far away point on the axis) when $r >> l$, the above relation is simplified as

  $$\mathbf{B} = \frac{\mu_0}{4\pi} \frac{2\mathbf{M}}{r^3}$$

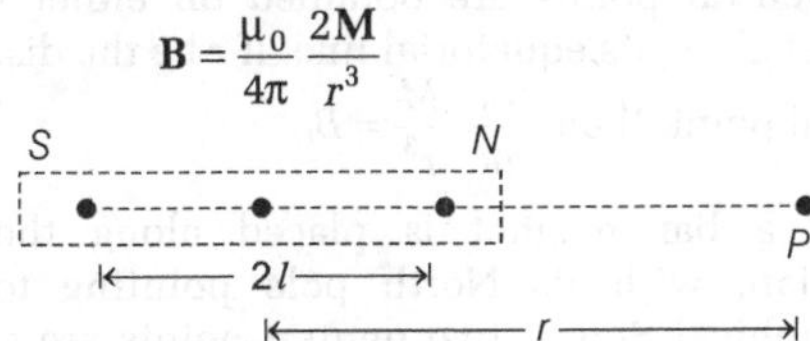

- **Along the equatorial line**

  $$\mathbf{B} = \frac{\mu_0}{4\pi} \frac{\mathbf{M}}{(r^2 + l^2)^{3/2}}$$

  If $r >> l$, the relation is modified as, $\mathbf{B} = \frac{\mu_0}{4\pi} \frac{\mathbf{M}}{r^3}$

  However, along the equatorial line, the direction of **B** is opposite to that of **M**.

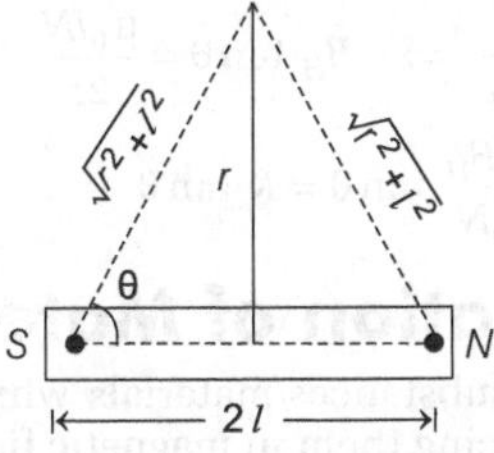

- In general, in a direction making an angle θ from with the magnetic axis, the magnetic field is given by

  $$\mathbf{B} = \frac{\mu_0}{4\pi} \frac{\mathbf{M}}{r^3} \sqrt{(3\cos^2\theta + 1)}$$

  In these relations, μ_0 is a constant having a value of $4\pi \times 10^{-7}$ T mA^{-1} and it is known as the **magnetic permeability of free space.**

  For solenoid $\mathbf{B} = \mu_0 ni$

  where, n is number of turns per unit length of solenoid and i the current through it.

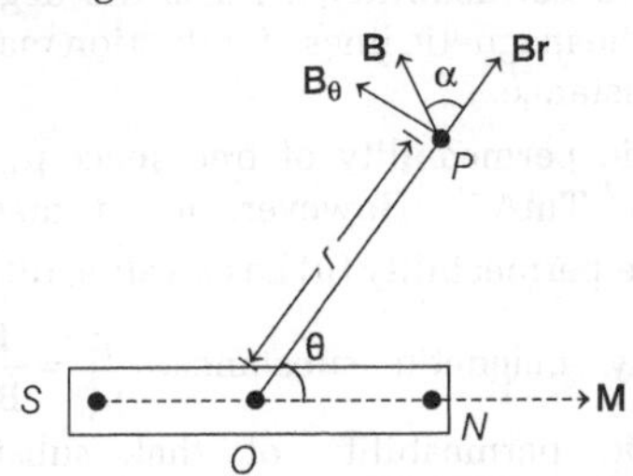

Bar Magnet as an Equivalent Solenoid

The magnetic field (axial) at a point at a distance r and radius a of solenoid is given by $\mathbf{B} = \frac{\mu_0 nIla^2}{r^3}$

and magnetic moment of solenoid is $\mathbf{M} = n(2l) I\pi a^2$

NOTE
- Even a single electron moving in its orbit behaves as a magnetic dipole and has a definite magnetic moment.
- Bohr magnetron is the magnetic moment due to the orbital motion of an electron revolving in the inner most orbit ($n = 1$). Its value is $m_B = \frac{eh}{4\pi m_e} = 9.27 \times 10^{-24}$ A-m^2

Torque on a Magnetic Dipole in a Magnetic Field

A magnetic dipole when placed in an uniform magnetic field, does not experience any net force. However, it experiences a torque given by $\tau = \mathbf{M} \times \mathbf{B}$ or $\tau = MB\sin\theta$

where, θ is the angle from the magnetic field, along which the dipole has been placed.

- **Work done** in rotating a magnetic dipole in a uniform magnetic field **B** from an initial orientation θ_1 to the final orientation θ_2, is given by $W = MB(\cos\theta_1 - \cos\theta_2)$.

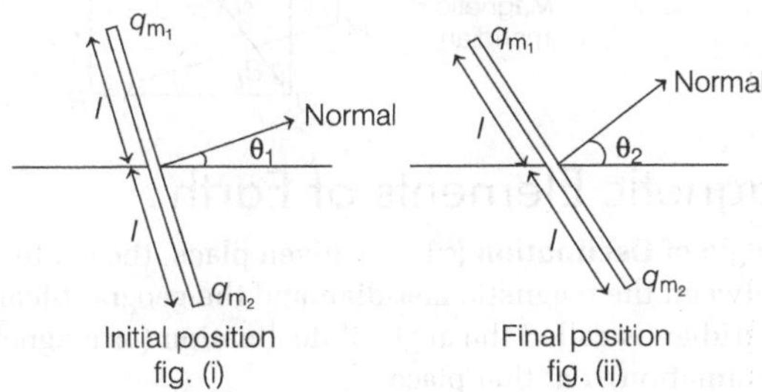

Initial position
fig. (i)

Final position
fig. (ii)

- **Potential energy** of a magnetic dipole placed in a uniform magnetic field, is given by $U_B = -\mathbf{M} \cdot \mathbf{B} = -MB\cos\theta$ where, θ is the angle from the direction of magnetic field and the axis of dipole.
- **The magnetic compass** (needle) of magnetic moment M and moment of inertia I and allowing it to oscillate in the magnetic field. Then, its time-period is $T = 2\pi\sqrt{I/MB}$
- Behaviour of a magnetic dipole in a magnetic field, is similar to the behaviour of an electric dipole in an electric field. However, the constant $\frac{1}{4\pi\varepsilon_0}$ is replaced by $\frac{\mu_0}{4\pi}$.
- If a magnetic dipole is in the form of a wire or a thin rod, when bent, its magnetic dipole moment **M** changes because the separation between its poles has changed.

Magnetic Field Lines

The magnetic field lines is defined as the path along which the compass needles are aligned. They are used to represent magnetic field in a region.

- Magnetic field lines are closed continuous curves.
- Tangent drawn at any point on magnetic field lines gives the direction of magnetic field at that point.
- Two magnetic field lines cannot intersect each other.
- Outside a magnet, they are directed from north to south pole and inside a magnet they are directed from south to north.

The Earth's Magnetism

The earth is a natural source of magnetic field, thus a magnetic field is always present everywhere near the surface of the earth. A freely suspended magnet always points in the north-south direction even in the absence of any other magnet. This suggests that the earth itself behaves as a magnet which causes a freely suspended magnet (or magnetic needle) to point always in a particular direction : north and south. The shape of earth's magnetic field resembles that of a bar magnet of length one-fifth of earth's diameter buried at its centre. Magnetic field of earth is shown in the figure given below.

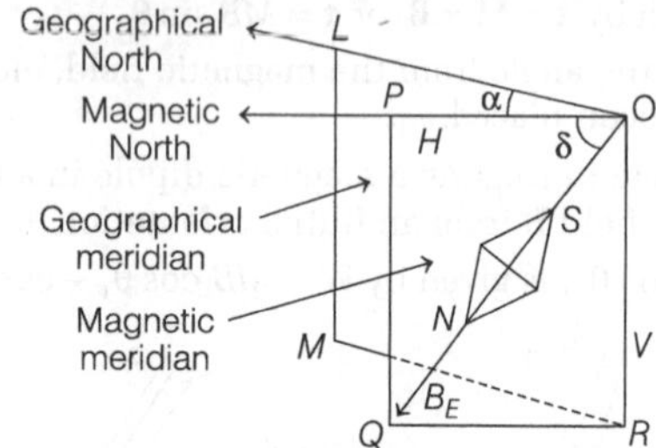

Magnetic Elements of Earth

- **Angle of Declination (α)** At a given place, the acute angle between the magnetic meridian and the geographical meridian is called the angle of declination (or magnetic declination) α at that place.
- **Angle of Inclination or Dip (δ)** The angle of dip δ at a place is the angle which the direction of the earth's total magnetic field B_E subtends with the horizontal direction.
- **Horizontal Component of the Earth's Magnetic Field (B_H)** As earth's magnetic field, in general, is inclined at an angle δ with the horizontal direction, it may be resolved into horizontal component B_H and a vertical component B_V, where $B_H = B_E \cos \delta$ and $B_V = B_E \sin \delta$

$$\Rightarrow \quad B_E = \sqrt{B_H^2 + B_V^2} \text{ and } \tan \delta = \frac{B_V}{B_H}$$

Variation of Magnetic Elements of the Earth

At the magnetic equator, angle of dip is zero. Value of the angle of dip gradually increases, on going from equator to magnetic poles. At the magnetic poles, value of the dip angle is 90°.

At the magnetic equator, $B_H = B_E \cos 0° = B_E$

and at poles, $B_H = B_E \cos 90° = 0$.

Similarly, at the magnetic equator, $B_V = B_E \sin 0° = 0$

and at the poles, $B_V = B_E \sin 90° = B_E$.

Magnetic elements of the earth at a place change with time also.

Neutral Points

A neutral point is a point at which the resultant magnetic field is zero. Following two cases are of special interest.

1. When a bar magnet is placed along the magnetic meridian with its North pole pointing towards geographical North, two neutral points are obtained on either side of the magnet along its equatorial line. If r be the distance of the neutral point, then $\frac{\mu_0}{4\pi}\frac{M}{r^3} = B_H$.
2. When a bar magnet is placed along the magnetic meridian, with its North pole pointing towards the geographical South, two neutral points are obtained on either side of the magnet along its axial line.

   Hence, we have $\frac{\mu_0}{4\pi}\frac{2M}{r^3} = B_H$.

Tangent Galvanometer

It is an instrument to measure electric current. The essential parts are a vertical coil of conducting wire and a small compass needle pivoted at centre of coil. The deflection, θ of needle is given by,

$$\tan \theta = \frac{B}{B_H} \Rightarrow B_H \tan \theta = \frac{\mu_0 IN}{2r}$$

or $$i = \frac{2r\,B_H}{\mu_0 N} \tan \theta = K \tan \theta$$

Magnetisation of Materials

There are some substances/materials which acquire magnetic properties on placing them in magnetic field. This phenomena is called magnetisation of materials.

To describe the magnetic properties of material, we have to understand the following terms:

(i) **Magnetic Induction or Magnetic Flux Density (B)** Whenever a piece of magnetic substance is placed in an external magnetising field, the substance becomes magnetised. If $\mathbf{B}_0$ is the magnetic field in free space, then

$$\mathbf{B} = \mu_r \mathbf{B}_0.$$

$\oint \mathbf{B} \cdot d\mathbf{S}$ is magnetic flux which is equal to $\mu_0 m_{\text{inside}}$, where m_{inside} is the net pole strength inside a close surface.

(ii) **Magnetic Permeability (μ)** It is the degree or extent to which the magnetic lines of induction may pass through a given distance.

Magnetic permeability of free space μ_0 has a value of $4\pi \times 10^{-7}$ TmA^{-1}. However, for a magnetic material, absolute permeability (μ) has a value, different than μ_0.

For any magnetic substance, $\frac{\mu}{\mu_0} = \frac{\mathbf{B}}{\mathbf{B}_0} = \mu_r$ = relative magnetic permeability of that substance. **Relative magnetic permeability** μ_r is a unitless and dimensionless term.

(iii) **Intensity of Magnetisation (I)** Intensity of magnetisation of a substance is defined as the magnetic moment induced in the substance per unit volume, when placed in the magnetising field. Thus, $\mathbf{I} = \frac{\mathbf{M}}{V}$

It is a vector quantity and its SI unit is Am^{-1}.

(iv) **Intensity of Magnetising Field or Magnetic Intensity (H)** It is a measure of the capability of external magnetising field to magnetise the given substance and is mathematically defined as

$$\mathbf{H} = \frac{\mathbf{B}_0}{\mu_0} \text{ or } \mathbf{H} = \frac{\mathbf{B}}{\mu}$$

Magnetic intensity **H** is a vector quantity and its SI unit is Am^{-1}.

(v) **Magnetic Susceptibility (χ_m)** Magnetic susceptibility of a substance is the ratio of the intensity of magnetisation I induced in the substance to the magnetic intensity H. Thus, $\chi_m = \frac{I}{H}$. It is a scalar quantity and it has no units or dimensions.

Relation between μ_r and χ_m we have, $B = \mu_0(I + H)$

or $B = \mu_0 H\left(\frac{I}{H} + 1\right)$ or $B = B_0(\chi_m + 1)$ or $\frac{B}{B_0} = \chi_m + 1$

But $\frac{B}{B_0} = \frac{\mu}{\mu_0} = \mu_r$ = relative permeability

$\therefore \quad \mu_r = \chi_m + 1$

Magnetic Materials

According to behaviour of magnetic substances, they are classified into three cases:

Diamagnetic Materials

These are materials which show a very small decrease in magnetic flux, when placed in a strong magnetising field. Hydrogen, water, copper, zinc, antimony, bismuth, etc., are examples of diamagnetic materials.

- In a diamagnetic material, the net magnetic moment (sum of that due to orbital motion and spin motion of electrons) of an atom is zero. The external magnetic field **B** distorts the electron orbit and thus induces a small magnetic moment in the opposite direction.
- Diamagnetic materials are feebly repelled in an external magnetic field and thus have a tendency to shift from the stronger to weaker regions of the magnetic field.
- The relative permeability of any diamagnetic substance is slightly less than 1 (i.e. $\mu_r < 1$) and susceptibility has a small negative value.
- Diamagnetism is an intrinsic property and does not vary with magnetic field **B** or temperature.

Paramagnetic Materials

These are the materials which show a small increase in the magnetic flux when placed in a magnetising field. Oxygen, air, platinum, aluminium, etc., are examples of paramagnetic materials.

- In a paramagnetic material, the net magnetic moment of every atom is non-zero.
- Paramagnetic materials are feebly attracted in an external magnetic field and thus, have a tendency to shift from the weaker to the stronger regions of magnetic field.
- The relative permeability μ_r of a paramagnetic material is slightly greater than one ($\mu_r > 1$). Magnetic susceptibility χ_m of paramagnetic materials is positive.
- Paramagnetism is temperature dependent. According to the **Curie's law,** the magnetic susceptibility of a paramagnetic substance is inversely proportional to its temperature T.

Mathematically, $\chi_m = \frac{C}{T}$, where C is the **Curie constant.**

Ferromagnetic Materials

These are the materials which are strongly attracted by a magnetic field and can themselves be magnetised even in a weak magnetising field. Iron, steel,nickel and cobalt are ferromagnetic.

- These materials show a large increase in the magnetic flux, when placed in a magnetic field. Thus, for them $\mu_r >> 1$. Accordingly, χ_m is positive and large.
- Ferromagnetic materials exhibit all properties exhibited by paramagnetic substances and by a much larger measure.
- Magnetic susceptibility of ferromagnetic materials decreases steadily with a rise in temperature. Above a certain temperature T_c (known as **Curie temperature**), the substance loses its ferromagnetic character and begins to behave as a paramagnetic substance.
- Above the Curie temperature T_c, the magnetic susceptibility of a ferromagnetic material varies as

$$\chi_m \propto \frac{1}{(T - T_c)} \quad \text{or} \quad \chi_m = \frac{C}{(T - T_c)}$$

where, C is a constant. It is known as the Curie-Weiss law.

Hysteresis Curve

The lag of intensity of magnetisation behind the magnetising field during the process of magnetisation and demagnetisation of a ferromagnetic material is called **hysteresis.** The whole graph $ABCDEFA$ is a closed loop and known as **hysteresis loop.**

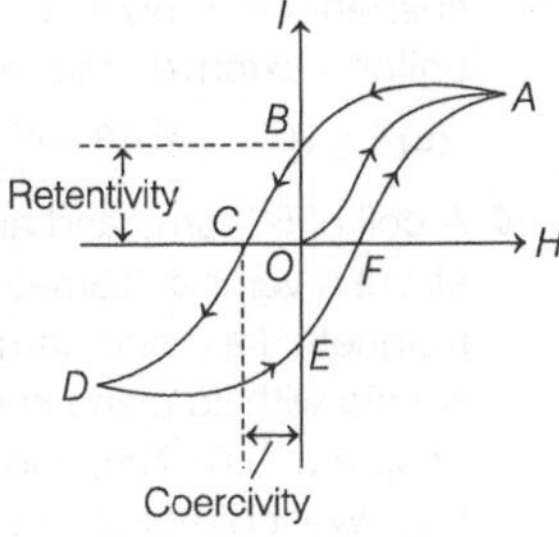

Electromagnet

Electromagnets are usually in the form of iron core solenoids. The ferromagnetic property of the iron core causes the internal magnetic domains of the iron to line up with the smaller driving magnetic field produced by the current in the solenoid.

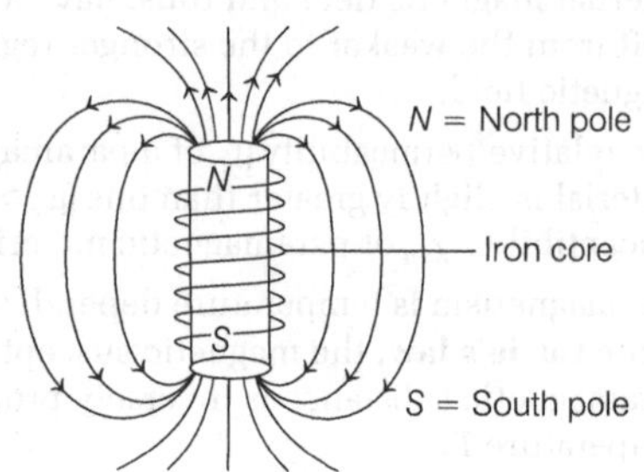

The effect is the multiplication of the magnetic field by factors of ten to eleven thousands. The solenoid field relationship is $B = k\,\mu_0 nI$, where $\mu = k\,\mu_0$ and k is the relative permeability of the iron, the figure shows the magnetic effect of the iron core.

Permanent Magnet

Substances which at room temperature retain their ferromagnetic property for a long period of time are called permanent magnets. Permanent magnets can be made in a variety of ways.

An efficient way to make a permanent magnet is to place a ferromagnetic rod in a solenoid and pass a current. The magnetic field of the solenoid magnetises the rod.

DAY PRACTICE SESSION 1

FOUNDATION QUESTIONS EXERCISE

1 A short bar magnet placed with its axis at 30° with a uniform external magnetic field of 0.16 T, experiences a torque of magnitude 0.032 J. The magnetic moment of the bar magnet will be

(a) 0.23 JT^{-1} (b) 0.40 JT^{-1}
(c) 0.80 JT^{-1} (d) zero

2 A bar magnet of length 10 cm and having pole strength equal to 10^{-3} Wb, is kept in a magnetic field having magnetic induction B equal to $4\pi \times 10^{-3}$ T. It makes an angle of 30° with the direction of magnetic induction. The value of the torque acting on the magnet is

(a) 0.5 N-m (b) $2\pi \times 10^{-5}$ N-m
(c) $\pi \times 10^{-5}$ N-m (d) 0.5×10^{2} N-m

3 M and $M\sqrt{3}$ are the magnetic dipole moments of the two magnets, which are joined to form a cross figure. The inclination of the system with the field, if their combination is suspended freely in a uniform external magnetic field B is

M
θ
B
M√3

(a) $\theta = 30°$ (b) $\theta = 45°$ (c) $\theta = 60°$ (d) $\theta = 15°$

4 A coil of 50 turns and area 1.25×10^{-3} m^2 is pivoted about a vertical diameter in an uniform horizontal magnetic field and carries a current of 2 A. When the coil is held with its plane in the N-S direction, it experiences a couple of 0.04 Nm, and when its plane is along the East-West direction, it experiences a couple of 0.03 Nm. The magnetic induction is

(a) 0.2 T (b) 0.3 T (c) 0.5 T (d) 0.4 T

5 A magnetic needle of magnetic moment 6.7×10^{-2} Am^2 and moment of inertia 7.5×10^{-6} kg m^2 is performing simple harmonic oscillations in a magnetic field of 0.01 T. Time taken for 10 complete oscillations is → JEE Main 2017 (Offline)

(a) 8.89 s (b) 6.98 s
(c) 8.76 s (d) 6.65 s

6 Hysteresis loops for two magnetic materials A and B are as given below

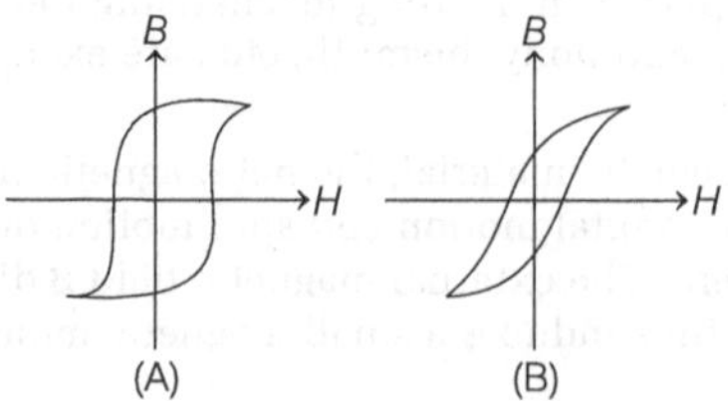

These materials are used to make magnets for electric generators, transformer core and electromagnet core. Then, it is proper to use → JEE Main 2016 (Offline)

(a) A for electric generators and transformers
(b) A for electromagnets and B for electric generators
(c) A for transformers and B for electric generators
(d) B for electromagnets and transformers

7 The coercivity of a small magnet where the ferromagnet gets demagnetised is 3×10^{3} Am^{-1}. The current required to be passed in a solenoid of length 10 cm and number of turns 100, so that the magnet gets demagnetised when inside the solenoid is → JEE Main 2014

(a) 30 mA (b) 60 mA (c) 3 A (d) 6 A

8 The figure illustrates how B, the flux density inside a sample of unmagnetised ferromagnetic material, varies with B_0, the magnetic flux density in which the sample is kept. For the sample to be suitable for making a permanent magnet.

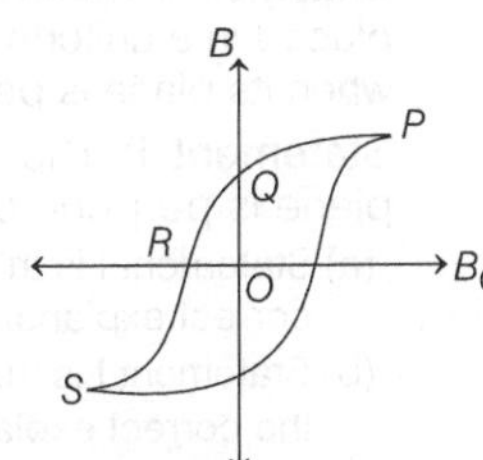

(a) OQ should be large, OR should be small
(b) OQ and OR should both be large
(c) OQ should be small and OR should be large
(d) OQ and OR should both be small

9 Match the following columns.

	Column I		Column II
A.	Magnetic moment	1.	$[ML^0T^{-2}A^{-1}]$
B.	Permeability	2.	Vector
C.	Intensity of magnetisation	3.	Nm^3 / Wb
D.	Magnetic induction	4.	Scalar

Codes

	A	B	C	D		A	B	C	D
(a)	1	2	3	4	(b)	3	4	2	1
(c)	4	3	1	2	(d)	2	1	3	4

10 A current carrying coil is placed with its axis perpendicular to N-S direction. Let horizontal component of earth's magnetic field be H_0 and magnetic field inside the loop be H. If a magnet is suspended inside the loop, it makes angle θ with H. Then, θ equal to

(a) $\tan^{-1}\left(\frac{H_0}{H}\right)$ (b) $\tan^{-1}\left(\frac{H}{H_0}\right)$
(c) $\operatorname{cosec}^{-1}\left(\frac{H}{H_0}\right)$ (d) $\cot^{-1}\left(\frac{H_0}{H}\right)$

11 Two short bar magnets of length 1 cm each have magnetic moments 1.20 Am^2 and 1.00 Am^2, respectively. They are placed on a horizontal table parallel to each other with their N poles pointing towards the South. They have a common magnetic equator and are separated by a distance of 20.0 cm. The value of the resultant horizontal magnetic induction at the mid-point O of the line joining their centres is close to (Horizontal component of the earth's magnetic induction is 3.6×10^{-5} Wb/m^2) → **JEE Main 2013**

(a) 3.6×10^{-5} Wb/m^2 (b) 2.56×10^{-4} Wb/m^2
(c) 3.50×10^{-4} Wb/m^2 (d) 5.80×10^{-4} Wb/m^2

12 The magnetic susceptibility of a paramagnetic substance at –73°C is 0.0060, then its value at –173°C will be

(a) 0.0030 (b) 0.0120 (c) 0.0180 (d) 0.0045

13 Consider the two idealised systems (i) a parallel plate capacitor with large plates and small separation and (ii) a long solenoid of length $L >> R$, radius of cross-section. In (i) **E** is ideally treated as a constant between plates and zero outside. In (ii) magnetic field is constant inside the solenoid and zero outside. These idealised assumptions, however contradict fundamental laws as below

(a) case (i) contradicts Gauss' law for electrostatic fields
(b) case (ii) contradicts Gauss' law for magnetic fields
(c) case (i) agrees with $\oint \mathbf{E}\cdot d\mathbf{l} = 0$
(d) case (ii) contradicts $\oint \mathbf{H}\cdot d\mathbf{l} = I_{en}$

14 A paramagnetic sample shows a net magnetisation of $8Am^{-1}$ when placed in an external magnetic field of 0.6 T at a temperature of 4 K. When the same sample is placed in an external magnetic field of 0.2 T at a temperature of 16 K, the magnetisation will be

(a) $\frac{32}{3}Am^{-1}$ (b) $\frac{2}{3}Am^{-1}$ (c) $6Am^{-1}$ (d) $2.4\,Am^{-1}$

15 If the areas under the I-H hysteresis loop and B-H hysteresis loop are denoted by A_1 and A_2, then

(a) $A_2 = \mu_0 A_1$ (b) $A_2 = A_1$
(c) $A_2 = \frac{A_1}{\mu_0}$ (d) $A_2 = \mu_0^2 A_1$

16 A short magnet oscillates with a time period 0.1 s at a place, where horizontal magnetic field is 24 μT. A downward current of 18 A is established in a vertical wire 20 cm East of the magnet. The new time period of oscillator

(a) 0.1 s (b) 0.089 s (c) 0.076 s (d) 0.057 s

17 Two bar magnets of the same mass, same length and breadth but having magnetic moments M and $2M$, respectively are joined together pole to pole and suspended by a string. The time period of the assembly in a magnetic field of strength H is 3 s. If now the polarity of one of the magnets is reversed and the combination is again made to oscillate in the same field, the time of oscillation is

(a) 3 s (b) $3\sqrt{3}$ s (c) $3/\sqrt{3}$ s (d) 6 s

18 Two magnets A and B are identical and these are arranged as shown in the figure. Their lengths are negligible in comparision to the separation between them. A magnetic needle is placed between the magnets at point P and it gets deflected by an angle θ. The ratio of distances d_1 and d_2, will be

(a) $(2\cot\theta)^{1/3}$ (b) $(2\tan\theta)^{1/3}$
(c) $(2\cot\theta)^{-1/3}$ (d) $(2\tan\theta)^{-1/3}$

19 The plane of a dip circle is set in the geographic meridian and the apparent dip is δ_1. It is then set in a vertical plane perpendicular to the geographic meridian.

The apparent dip angle is δ_2. The declination θ at the place is

(a) $\theta = \tan^{-1}(\tan\delta_1 \tan\delta_2)$ (b) $\theta = \tan^{-1}(\tan\delta_1 + \tan\delta_2)$
(c) $\theta = \tan^{-1}\left(\frac{\tan\delta_1}{\tan\delta_2}\right)$ (d) $\theta = \tan^{-1}(\tan\delta_1 - \tan\delta_2)$

20 This question contains two statements : Statement I *(Assertion) and Statement* II *(Reason). The question also has four alternative choices, only one of which is the correct answer. You have to select one of the codes (a), (b), (c), (d).*

Statement I A current carrying loop is free to rotate. It is placed in a uniform magnetic field. It attains equilibrium when its plane is perpendicular to the magnetic field.

Statement II The torque on the coil is zero when its plane is perpendicular to the magnetic field.

(a) Statement I is true, Statement II is true; Statement II is the correct explanation for Statement I
(b) Statement I is true, Statement II is true; Statement II is not the correct explanation for Statement I
(c) Statement I is true; Statement II is false
(d) Statement I is false; Statement II is true

DAY PRACTICE SESSION 2

PROGRESSIVE QUESTIONS EXERCISE

1 The dipole moment of a circular loop carrying a current I, is m and the magnetic field at the centre of the loop is B_1. When the dipole moment is doubled by keeping the current constant, the magnetic field at the centre of the loop is B_2. The ratio $\frac{B_1}{B_2}$ is → JEE Main 2018

(a) 2 (b) $\sqrt{3}$
(c) $\sqrt{2}$ (d) $\frac{1}{\sqrt{2}}$

2 The current on the winding of a toroid is 2A. It has 400 turns and mean circumferential length is 40 cm. With the help of search coil and charge measuring instrument the magnetic field is found to be 1T. The susceptibility is

(a) 100 (b) 290
(c) 398 (d) 397

3 The length of a magnet is large compared to its width and breadth. The time period of its oscillation in a vibration magnetometer is 2 s. The magnet is cut along its length into three equal parts and three parts are then placed on each other with their like poles together. The time period of this combination will be

(a) 2 s (b) $\frac{2}{3}$ s
(c) $2\sqrt{3}$ s (d) $\frac{2}{\sqrt{3}}$ s

4 A thin circular disc of radius R is uniformly charged with density $\sigma > 0$ per unit area. The disc rotates about its axis with a uniform angular speed ω. The magnetic moment of the disc is → AIEEE 2011

(a) $2\pi R^4\sigma\omega$ (b) $\pi R^4\sigma\omega$
(c) $\frac{\pi R^4}{2}\sigma\omega$ (d) $\frac{\pi R^4}{4}\sigma\omega$

5 Consider the plane S formed by the dipole axis and the axis of the earth. Let P be point on the magnetic equator and in S. Let Q be the point of intersection of the geographical and magnetic equators. The declination and dip angles at P and Q are

(a) 0° and 11.3° (b) 0° and 0°
(c) 11.3° and 6.5° (d) 11.3° and 11.3°

6 A bar magnet has pole strength 3.6 A-m and length 12 cm. Its area of cross-section is 0.9 cm^2. The magnetic field B at the centre of the bar magnet is

(a) 6×10^{-3} T (b) 5×10^{-2} T
(c) 2.5×10^{-2} T (d) 2.5×10^{-8} T

7 A bar magnet suspended by a suspension fibre, is placed in the magnetic meridian with no twist in the suspension fibre. On turning the upper end of the suspension fibre by an angle of 120° from the meridian, the magnet is deflected by an angle of 30° from the meridian. Then, the angle by which the upper end of the suspension fibre has to be twisted, so as to deflect the magnet through 90° from the meridian is

(a) 270° (b) 240°
(c) 330° (d) 180°

8 A bar magnet 8 cm long, is placed in the magnetic meridian with the N pole, pointing towards the geographical North. Two neutral points, separated by a distance of 6 cm are obtained on the equatorial axis of the magnet. If $B_H = 3.2 \times 10^{-5}$ T, then the pole strength of the magnet is

(a) 5 A-cm^2 (b) 10 A-cm^2
(c) 2.5 A-cm^2 (d) 20 A-cm^2

9 There are two current carrying planer coils made each from wire of length L. C_1 is circular coil (radius R) and C_2 is square (side a). These are so constructed that they have same frequency of oscillation when they are placed in the same uniform magnetic field **B** and carry the same current. The value of a in terms of R is

(a) $3R$ (b) $\sqrt{3}R$
(c) $\sqrt{2}R$ (d) $2R$

ANSWERS

SESSION 1	1 (b)	2 (a)	3 (c)	4 (d)	5 (d)	6 (d)	7 (c)	8 (b)	9 (b)	10 (a)
	11 (b)	12 (b)	13 (b)	14 (b)	15 (a)	16 (c)	17 (b)	18 (a)	19 (c)	20 (c)
SESSION 2	1 (c)	2 (d)	3 (b)	4 (d)	5 (a)	6 (b)	7 (a)	8 (a)	9 (a)	

Hints and Explanations

SESSION 1

1 As, $M = \dfrac{\tau}{B \sin \theta}$

$$= \frac{0.032}{0.16 \times \sin 30^\circ} = 0.40 \text{ JT}^{-1}$$

2 As, $\mu_0 m = 10^{-3}$ Wb

$$m = \frac{10^{-3}}{\mu_0}$$

Magnetic moment of the magnet,

$$M = m \times 2l = \frac{10^{-3}}{\mu_0}(0.1) = \frac{10^{-4}}{\mu_0}$$

Now, $\tau = MB \sin \theta$

$$= \left(\frac{10^{-4}}{\mu_0}\right) \times 4\pi \times 10^{-3} \sin 30^\circ$$

$$= 0.5 \text{ N-m}$$

3 Torque (τ) acting on the magnet (1) is

$$\tau_1 = MB \sin\theta$$

$$\tau_2 = \sqrt{3}\, MB \sin\theta$$

For equilibrium, $\tau_1 = \tau_2$

$\therefore \quad MB \sin\theta = \sqrt{3}\, MB \cos\theta$

$\Rightarrow \tan\theta = \sqrt{3} = \tan 60^\circ \Rightarrow \theta = 60^\circ$

4 Here, $N = 50$, $A = 1.25 \times 10^{-3}$ m^2,

$I = 2$A

$$M = NIA$$

$$= 50 \times 2 \times 1.25 \times 10^{-3}$$

$$= 0.125 \text{ A-m}^2$$

If the normal to the face of the coil makes an angle θ with the magnetic induction B, the torque

$$\tau = MB \cos\theta = 0.04 \quad \ldots\text{(i)}$$

Now, when the plane of the coil is turned through 90°, the torque becomes,

$$\tau = MB \sin \theta = 0.03 \quad \ldots\text{(ii)}$$

Squaring and adding Eqs. (i) and (ii), we get

$$\tau = 0.05 \Rightarrow MB = 0.05$$

$$\Rightarrow \quad B = \frac{0.05}{M} = \frac{0.05}{0.125} = 0.4 \text{ T}$$

5 Time period of oscillation is

$$T = 2\pi \sqrt{\frac{I}{MB}}$$

$$\Rightarrow \quad T = 2\pi \sqrt{\frac{7.5 \times 10^{-6}}{6.7 \times 10^{-2} \times 0.01}} = 0.665 \text{ s}$$

Time taken to complete 10 oscillations

$$= 10 \times 0.665 = 6.65 \text{ s}$$

Hence, time for 10 oscillations is $t = 6.65$ s.

6 Area of hysteresis loop is proportional to the net energy absorbed per unit volume by the material, as it is taken over a complete cycle of magnetisation. For electromagnets and transformers, energy loss should be low. i.e. thin hysteresis curves.

Also, $|B| \to 0$ when $H = 0$ and $|H|$ should be small when $B \to 0$.

7 For solenoid, the magnetic field needed to be magnetised the magnet.

$$H = nl$$

$$\left[\text{where, } N = 100,\; l = 10\text{cm} = \frac{10}{100}\text{m} = 0.1 \text{ m}\right]$$

$$H = \frac{N}{l} I \Rightarrow 3 \times 10^3 = \frac{100}{0.1} \times I$$

$$I = 3 \text{ A}$$

8 For making permanent magnet, the material should have high residual magnetism and high coercivity i.e. *OQ* and *OR* should be large.

9 A - SI unit of magnetic moment is Nm^3 / Wb.

B - Permeability (μ) is a scalar.

C - $I = \dfrac{M}{V} = \dfrac{\text{A}}{\text{m}^{-1}}$, it is a vector.

D - Magnetic induction

$$B = \text{ML}^0\text{T}^{-2}\text{A}^{-1}$$

10 In given case H and H_0 are perpendicular to each other.

From figure,

$$\tan\theta = \frac{H_0}{H}$$

H_0

θ H

$$\Rightarrow \quad \theta = \tan^{-1}\left(\frac{H_0}{H}\right)$$

11 $B_{\text{net}} = B_1 + B_2 + B_H$

$$B_{\text{net}} = \frac{\mu_0}{4\pi} \frac{(M_1 + M_2)}{r^3} + B_H$$

$$= \frac{10^{-7}(1.2 + 1)}{(0.1)^3} + 3.6 \times 10^{-5}$$

$$= 2.56 \times 10^{-4} \text{ Wb/m}^2$$

12 As, $\quad \chi_m \propto \dfrac{1}{T}$

$$\Rightarrow \quad \frac{\chi_2}{\chi_1} = \frac{T_1}{T_2}$$

or $\quad \dfrac{\chi_2}{0.0060} = \dfrac{273 - 73}{273 - 173} = \dfrac{200}{100} = 2$

or $\quad \chi_2 = 0.0120$

13 As Gauss' law states,

$\oint_S \mathbf{E} \cdot d\mathbf{S} = \dfrac{q}{\varepsilon_0}$ for electrostatic field. It does not contradict for electrostatic fields as the electric field lines do not form continuous closed path. According to Gauss's law in magnetic field,

$$\oint_S \mathbf{E} \cdot d\mathbf{S} = 0$$

It contradicts for magnetic field, because there is a magnetic field inside the solenoid and no field outside the solenoid carrying current but the magnetic field lines form the closed path.

14 As Curie's law explains, we can deduce a formula for the relation between magnetic field induction, temperature and magnetisation.

i.e. I (magnetisation) $\propto \dfrac{B \text{ (magnetic field induction)}}{t\text{(temperature in kelvin)}}$

$$\Rightarrow \frac{I_2}{I_1} = \frac{B_2}{B_1} \times \frac{t_1}{t_2}$$

Let us suppose, here $I_1 = 8\,\text{Am}^{-1}$

$$B_1 = 0.6\text{T}, t_1 = 4\text{K}$$
$$B_2 = 0.2\text{T}, t_2 = 16\text{K}$$
$$\Rightarrow \frac{0.2}{0.6} \times \frac{4}{16} = \frac{I_2}{8}$$
$$\Rightarrow I_2 = \frac{2}{3}\text{Am}^{-1}$$

15 As, $B = \mu_0(H + I)$

$\Rightarrow dB = \mu_0\, dH + \mu_0\, dI$

or $\oint HdB = \mu_0 \oint HdH + \mu_0 \oint H\, dI$

$\because \oint HdH = 0$

$\oint HdB = \mu_0 \oint H \cdot dI$

Area of the B-H loop $= \mu_0 \times$ area of I-H loop

i.e. $A_2 = \mu_0 A_1$

16 Given, $B_H = 24\mu\text{T}$

$$B_{\text{current}} = \frac{\mu_0 I}{2\pi d} = \frac{2\times 10^{-7} \times 18}{0.2} = 18\,\mu\,\text{T}$$

Now, $T' = T\sqrt{\dfrac{24}{42}} = \dfrac{0.1 \times 2}{\sqrt{7}} = 0.076\text{ s}$

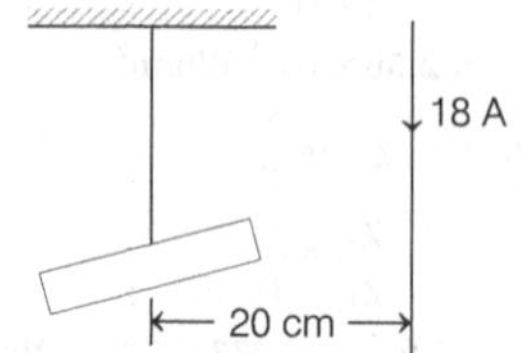

17 At the pole, for the combination

$M_1 = 2M + M = 3M, T_1 = 3\text{ s}$

When the polarity of one is reversed, then

$M_2 = 2M - M = M$

Thus, we have, from

$$\frac{T_2^2}{T_1^2} = \frac{M_1}{M_2} \quad \left(\because T = 2\pi\sqrt{\frac{I}{MB}}\right)$$
$$\Rightarrow \frac{T_2^2}{T_1^2} = \frac{3M}{M} = 3$$
$$\therefore T_2^2 = 3T_1^2 = 3 \times 9 = 27$$
$$\therefore T_2 = \sqrt{27} = 3\sqrt{3}\text{ s}$$

18 $B = B_H \tan\theta$

$$\frac{\mu_0}{4\pi}\left(\frac{2M}{d_1^3}\right) = \left(\frac{\mu_0}{4\pi}\frac{M}{d_2^3}\right)\tan\theta$$
$$\frac{2}{d_1^3} = \frac{\tan\theta}{d_2^3}$$
$$\left(\frac{d_1}{d_2}\right)^3 = \frac{2}{\tan\theta} = 2\cot\theta$$
$$\frac{d_1}{d_2} = (2\cot\theta)^{1/3}$$

19 $\tan\delta_1 = \dfrac{V}{H\cos\theta}$

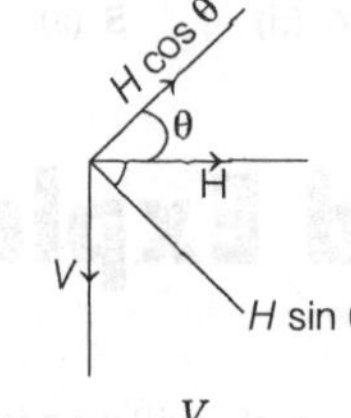

$$\tan\delta_2 = \frac{V}{H\cos(90° - \theta)} = \frac{V}{H\sin\theta}$$
$$\frac{\tan\delta_1}{\tan\delta_2} = \frac{\sin\theta}{\cos\theta} = \tan\theta$$

or $\theta = \tan^{-1}\left(\dfrac{\tan\delta_1}{\tan\delta_2}\right)$

20 The torque experienced by a coil in a magnetic field is,

$$\tau = M \times B = MB\sin\theta$$

when plane is perpendicular,

$$\theta = 90°$$
$$\tau = MB\sin 90° = MB \text{ [maximum]}$$

Hence, Statement I is true, while statement II is false.

SESSION 2

1 As $m = IA$, so to change dipole moment (current is kept constant), we have to change radius of loop.

Initially, $m = I\pi R^2$ and $B_1 = \dfrac{\mu_0 I}{2R_1}$

Finally, $m' = 2m = I\pi R_2^2$

$\Rightarrow 2I\pi R_1^2 = I\pi R_2^2$

or $R_2 = \sqrt{2}R_1$

So, $B_2 = \dfrac{\mu_0 I}{2(R_2)} = \dfrac{\mu_0 I}{2\sqrt{2}R_1}$

Hence, ratio $\dfrac{B_1}{B_2} = \dfrac{\left(\dfrac{\mu_0 I}{2R_1}\right)}{\left(\dfrac{\mu_0 I}{2\sqrt{2}R_1}\right)} = \sqrt{2}$

$\therefore$ Ratio $\dfrac{B_1}{B_2} = \sqrt{2}$

2 As, $n = \dfrac{400}{2\pi R} = \dfrac{400}{40 \times 10^{-2}} = 1000$

$$\mu = ni = 1000 \times 2 = 2000$$
$$B = \mu_0 \mu_r \mu$$
$$\Rightarrow \mu_0 \mu_r = \frac{1}{2000} = 5 \times 10^{-4}$$
$$\Rightarrow \mu_r = \frac{5\times 10^{-4}}{\mu_0} = \frac{5 \times 10^{-4}}{4\pi \times 10^{-7}} = 398$$
$$\Rightarrow \chi = \mu_r - 1 = 397$$

3 The time period of oscillations of magnet,

$$T = 2\pi\sqrt{\left(\frac{I}{MH}\right)} \quad \text{...(i)}$$

where, I = moment of inertia of magnet $= \dfrac{mL^2}{12}$

[m, being the mass of magnet]

M = pole strength $\times L$

and H = horizontal component of the earth's magnetic field.

When the three equal parts of magnet are placed on one another with their like poles together, then

$$I' = \frac{1}{12}\left(\frac{m}{3}\right) \times \left(\frac{L}{3}\right)^2 \times 3 = \frac{1}{12}\frac{mL^2}{9} = \frac{I}{9}$$

and M' = pole strength $\times \dfrac{L}{3} \times 3 = M$

Hence, $T' = 2\pi\sqrt{\left(\dfrac{I/9}{MH}\right)}$

or $T' = \dfrac{1}{3} \times T$ or $T' = \dfrac{2}{3}\text{ s}$

4 Let us consider the disc to be made up of large number of concentric elementary rings. Consider one such ring of radius x and thickness dx. Charge on this elementary ring,

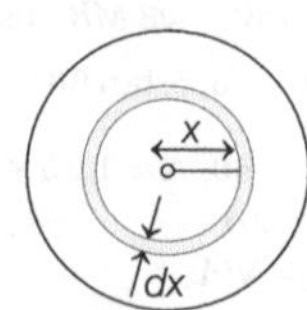

$$dq = \sigma \times 2\pi x dx = 2\pi\sigma x dx$$

Current associated with this elementary ring,

$$dI = \frac{dq}{dt} = dq \times f = \sigma\omega x dx$$

[$\because f$ is frequency and $\omega = 2\pi f$]

Magnetic moment of this elementary ring,

$$dM = dI\pi x^2 = \pi\sigma\omega x^3 dx$$

$\therefore$ Magnetic moment of the entire disc,

$$M = \int_0^R dM = \pi\sigma\omega\int_0^R x^3 dx = \frac{1}{4}\pi R^4 \sigma\omega$$

5 P is in the plane S, needle is in North, so the declination is zero.

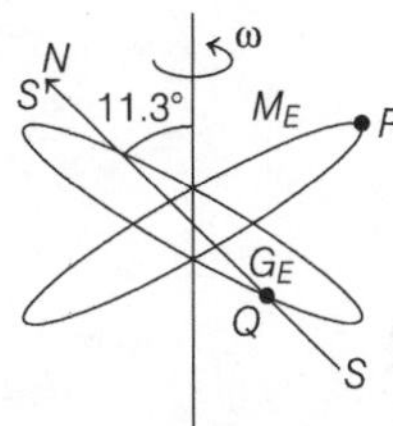

P is also on the magnetic equator, so the angle of dip = 0, because the value of angle of dip at equator is zero Q is also on the magnetic equator, thus the angle of dip is zero. As the earth tilted on its axis by 11.3°, thus the declination at Q is 11.3°.

6 As, $I = \frac{m}{A} = \frac{3.6}{0.9 \times 10^{-4}}$

$= 4 \times 10^4 \text{ Am}^{-1}$

$H_N = \frac{m}{4\pi d^2}$

$= \frac{3.6}{4\pi \times (6 \times 10^{-2})^2}$

$= 79.6 \text{ Am}^{-1}$

$H = H_N + H_S$

$\Rightarrow \quad H = H_N + H_S = 159.2 \text{ A/m}$, towards S pole

$B = \mu_0(H + I) = 4\pi \times 10^{-7}$ $(4 \times 10^4 - 159.2)$

$= 5 \times 10^{-2}$ T, towards N pole

7 As, $\tau = C\phi = MH \sin\theta$

Case I

$\theta = 30°, \phi° = 120° - 30° = 90°$ ...(i)

Case II $C(\phi - 90°) = MH \sin 90°$ ...(ii)

On dividing Eq. (ii) by Eq. (i), we get

$$\frac{\phi - 90°}{90°} = \frac{MH \sin 90°}{MH \sin 30°}$$

$$= \frac{1}{1/2} = 2$$

or $\phi - 90° = 180°$

or $\phi = 180° + 90° = 270°$

8 At the neutral point,

$$B = B_H$$

$$\Rightarrow \frac{\mu_0}{4\pi} \times \frac{M}{(r^2 + l^2)^{3/2}} = B_H$$

In CGS system,

$$\frac{M}{(r^2 + l^2)^{3/2}} = B_H$$

$$\Rightarrow \frac{m \times 2l}{(r^2 + l^2)^{3/2}} = B_H$$

$$\Rightarrow \frac{m \times 2 \times 4}{(3^2 + 4^2)^{3/2}} = 0.32$$

$$\Rightarrow \frac{8m}{(25)^{3/2}} = 0.32$$

$$\Rightarrow m = \frac{125 \times 0.32}{8}$$

$$= 5 \text{ A-cm}^2$$

9

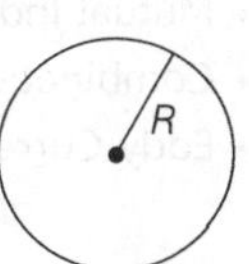

C_1 = Circular coil of radius R, length L, number of turns per unit length $n_1 = \frac{L}{2\pi R}$

Magnetic moment of C_1

$$\Rightarrow m_1 = n_1 I A_1$$

$$m_1 = \frac{L \cdot I \cdot \pi R^2}{2\pi R},$$

$$m_1 = \frac{LIR}{2} \quad \text{...(i)}$$

Moment of inertia of C_1

$$\Rightarrow I_1 = \frac{MR^2}{2} \quad \text{...(ii)}$$

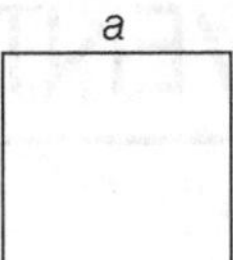

C_2 = Square of side a and perimeter L, number of turns per unit length $n_2 = \frac{L}{4a}$

Magnetic moment of C_2

$$\Rightarrow m_2 = n_2 I A_2$$

$$m_2 = \frac{L}{4a} \cdot I \cdot a^2$$

$$m_2 = \frac{LIa}{4} \quad \text{...(iii)}$$

Moment of inertia of C_2

$$\Rightarrow I_2 = \frac{Ma^2}{12} \quad \text{...(iv)}$$

Frequency of C_1

$$\Rightarrow f_1 = 2\pi\sqrt{\frac{I_1}{m_1 B}}$$

Frequency of C_2

$$\Rightarrow f_2 = 2\pi\sqrt{\frac{I_2}{m_2 B}}$$

According to question,

$$f_1 = f_2$$

$$2\pi\sqrt{\frac{I_1}{m_1 B}} = 2\pi\sqrt{\frac{I_2}{m_2 B}}$$

$$\frac{I_1}{m_1} = \frac{I_2}{m_2}$$

or $\frac{m_2}{m_1} = \frac{I_2}{I_1}$

Putting the values by Eqs. (i), (ii), (iii) and (iv), we get

$$\frac{LIa \cdot 2}{4 \times LIR} = \frac{Ma^2 \cdot 2}{12 \cdot MR^2}$$

$$3R = a$$

Thus, the value of a is $3R$.

DAY TWENTY THREE

Electromagnetic Induction

Learning & Revision for the Day

- Magnetic Flux (ϕ_B)
- Faraday's Law of Electromagnetic Induction
- Motional Emf
- Rotational Emf
- Self-Induction
- Mutual Induction
- Combination of Inductors
- Eddy Currents

Magnetic Flux (ϕ_B)

The flux associated with a magnetic field is defined in a similar manner to that used to define electric flux. Consider an element of area ds on an arbitrary shaped surface as shown in figure. If the magnetic field at this element is $\mathbf{B}$, the magnetic flux through the element is,

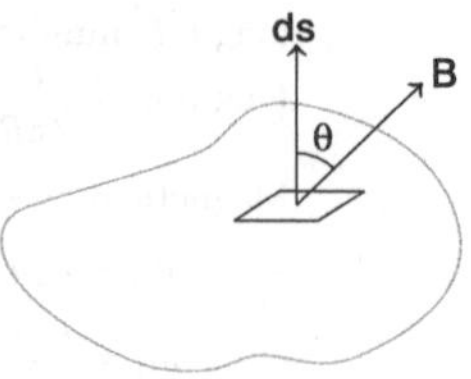

$$d\phi_B = \mathbf{B} \cdot d\mathbf{s} = Bds \cos\theta$$

Here, $d\mathbf{s}$ is a vector that is perpendicular to the surface and has a magnitude equal to the area ds and θ is the angle between $\mathbf{B}$ and $d\mathbf{s}$ at that element.

Magnetic flux is a scalar quantity. Outward magnetic flux is taken as positive (i.e. $\theta < 90°$) and inward flux is taken as negative (i.e. $\theta > 90°$).

SI unit of magnetic flux is 1 weber (1 Wb).

where, $\quad 1\text{ Wb} = 1\text{ T} \times 1\text{ m}^2 = 1\text{ T-m}^2$

Dimensional formula of magnetic flux is $[ML^2 T^{-2} A^{-1}]$.

- No. of Questions in Exercises (x)—
- No. of Questions Attempted (y)—
- No. of Correct Questions (z)— (Without referring Explanations)
- Accuracy Level ($z/y \times 100$)—
- Prep Level ($z/x \times 100$)—

In order to expect good rank in JEE, your Accuracy Level should be above 85 & Prep Level should be above 75.

Faraday's Law of Electromagnetic Induction

This law states that, the induced emf in a closed loop equals the negative of the time rate of change of magnetic flux through the loop.

Induced emf, $\quad |e| = \dfrac{d\phi_B}{dt}$

- For N turns, $\quad |e| = N\dfrac{d\phi_B}{dt}$

  However, if we consider the direction of induced emf, then

$$e = -N\frac{d\phi_B}{dt} = -\frac{Nd(BA\cos\theta)}{dt} = \frac{-NBA(\cos\theta_2 - \cos\theta_1)}{\Delta t}$$

- If the given electric circuit is a closed circuit having a total resistance R, then the induced current,

$$I = \frac{e}{R} = -\frac{N}{R}\frac{d\phi_B}{dt}$$

Induced charge, $dq = Idt = -\frac{N}{R}\, d\phi_B$

and induced power, $P = \frac{e^2}{R} = \frac{N^2}{R}\left(\frac{d\phi_B}{dt}\right)^2$

Lenz's Law

The negative sign in Faraday's equations of electromagnetic induction describes the direction in which the induced emf drives current around a circuit. However, that direction is most easily determined with the help of Lenz's law. This law states that **the direction of any magnetic induction effect is such as to oppose the cause of the effect.**

Later, we will see that Lenz's law is directly related to **energy conservation.**

Motional Emf

Let a conducting rod of length l be moving with a uniform velocity **v** perpendicular to a uniform magnetic field **B**, an induced emf is set up.

The magnitude of the induced emf will be

$$|e| = B\,l\,v$$

- If the rod is moving such that it makes an angle θ with the direction of the magnetic field, then

$$|e| = B\,l\,v \sin\theta$$

Hence, for the motion parallel to **B**, the induced emf is zero.

- When a conducting rod moves horizontally, then an induced emf is set up between its ends due to the vertical component of the earth's magnetic field. However, at the magnetic equator, induced emf will be zero, because $B_V = 0$.
- If during landing or taking off, the wings of an aeroplane are along the East-West direction, an induced emf is set up across the wings (due to the effect of B_H).

Motional Emf in a Loop

If a conducting rod moves on two parallel conducting rails, then an emf is induced whose magnitude is $|e| = B\,l\,v$ and the direction is given by the Fleming's right hand rule.

- Induced current, $|I| = \frac{|e|}{R} = \frac{B\,l\,v}{R}$
- Magnetic force, $F_m = BIl = \frac{B^2 l^2 v}{R}$

Rotational Emf

Let a conducting rod of length l rotate about an axis passing through one of its ends (that end may be fixed), with an angular velocity ω in a plane perpendicular to the magnetic field B, then an induced emf is set up between the ends of the rod, whose magnitude is given by

$$|e| = \frac{1}{2} B l^2 \omega$$

Self-Induction

Self-induction is the phenomenon due to which an induced emf is set up in a coil or a circuit whenever the current passing through it changes. The induced emf opposes the change that causes it and is thus known as **back emf.**

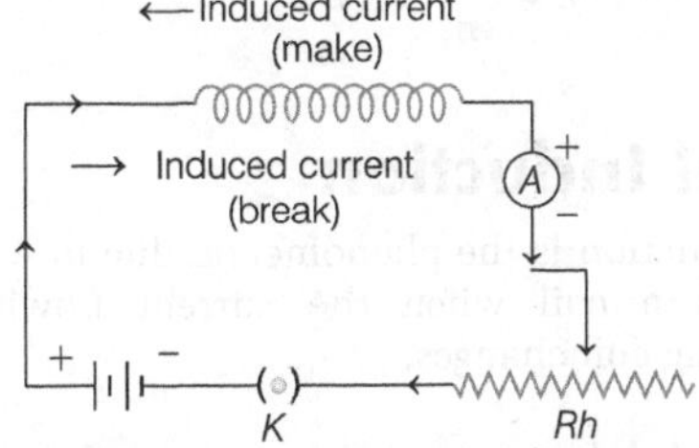

- Inductance is the inherent property of electrical circuits and is known as the **electrical inertia.**
- An inductor is said to be an **ideal inductor** if its resistance is zero.
- An inductor does not oppose current but opposes changes (growth or decay of current) in the circuit.

Self-Inductance

Flux linked with the coil is

$$N\phi_B \propto I \quad \text{or} \quad N\phi_B = LI,$$

where the constant L is known as the **coefficient of self-induction** or **self-inductance** of the given coil.

It may be defined as the magnetic flux linked with the coil, when a constant current of 1 A is passed through it.

Induced emf due to self-induction,

$$e = -N\frac{d\phi}{dt} = -L\frac{dI}{dt}$$

SI unit of inductance is **henry.**

Magnetic Potential Energy of an Inductor

- In building, a steady current in an electric circuit, some work is done by the emf of the source, against the self-inductance of the coil.

  The work done, $W = \frac{1}{2} LI^2$

- The work done is stored as the magnetic potential energy of that inductor.

  Thus, $U = \frac{1}{2} L I^2$

Formulae for Self-Inductance

- For a circular coil of radius R and N turns, the self-inductance,

$$L = \frac{1}{2}\mu_0 \pi N^2 R$$

- For a solenoid coil having length l, total number of turns N and cross-sectional area A,

$$L = \frac{\mu_0 N^2 A}{l} = \mu_0 n^2 Al \qquad \left[\text{where, } n = \frac{N}{l}\right]$$

- For a toroid of radius R and number of turns N,

$$L = \frac{1}{2}\mu_0 N^2 R$$

- For a square coil of side a and number of turns N,

$$L = \frac{2\sqrt{2}}{\pi}\mu_0 N^2 a$$

Mutual Induction

Mutual induction is the phenomenon due to which an emf is induced in a coil when the current flowing through a neighbouring coil changes.

Mutual Inductance

Mutual inductance of a pair of coils is defined as the magnetic flux linked with one coil, when a constant current of unit magnitude, flows through the other coil.

Mathematically, $N\phi_{B_2} = MI_1$

where, M is known as the **mutual inductance** for the given pair of coils.

Induced emf due to mutual inductance,

$$e_2 = -N\frac{d\phi_{B_2}}{dt} = -M\frac{dI_1}{dt}$$

Hence, mutual inductance for a pair of coils is numerically equal to the magnitude of induced emf in one coil when current in the other coil changes at a rate of 1 As^{-1}.

SI unit of mutual inductance M, is **henry.**

Mutual inductance of a pair of coils is maximum, when the two coils are wound on the same frame. However, mutual inductance is negligible when the two coils are oriented mutually perpendicular to each other (see figure). In this context, we define a term **coupling coefficient** k.

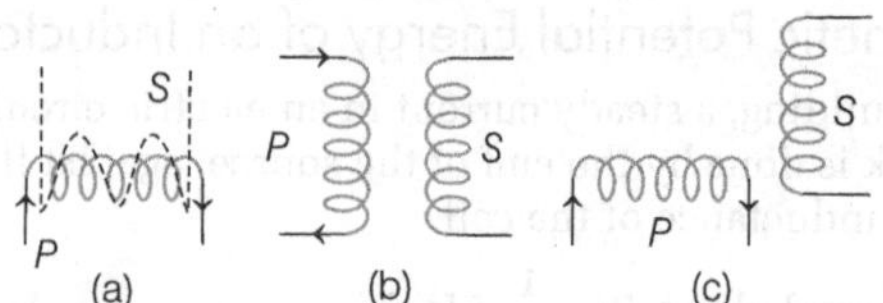

Coupling coefficient is given by

$$k = \frac{\text{Magnetic flux linked with secondary coil}}{\text{Magnetic flux developed in primary coil}}$$

It is observed that $0 \le k \le 1$.

For a pair of two magnetically coupled coils of self-inductances L_1 and L_2 respectively, the mutual inductance,

$$M_{12} = M_{21} = M = k\sqrt{L_1 L_2}$$

where, k is the coupling coefficient.

Formulae for Mutual Inductance

- Assuming the coupling coefficient $k = 1$ and medium to be a free space or air. Mutual inductance of a pair of concentric circular coils is

$$M = \frac{\mu_0 N_1 N_2 \pi r^2}{2R}$$

where, r = radius of the coil (of small radius)
and R = radius of the coil (of larger radius).

- For a pair of two solenoid coils, wound one over the other, $M = \frac{\mu_0 N_1 N_2 A}{l}$

For a pair of concentric coplanar square coils,

$$M = \frac{2\sqrt{2}\,\mu_0 N_1 N_2 a^2}{\pi b}$$

where, a = side of the smaller coil and b = side of the larger coil.

- For a given pair of coils, mutually coupled, then according to theorem of reciprocity,

$$M_{12} = M_{21} = M$$

Combination of Inductors

- If two coils of self-inductances L_1 and L_2 are placed quite far apart and are arranged in series, then their equivalent inductance,

$$L_s = L_1 + L_2$$

- If the coils are placed quite close to each other, so as to mutually affect each other, then their equivalent inductance,

$$L_s = L_1 + L_2 \pm 2M$$

Here, M has been written with $\pm$ sign depending on the fact whether currents in the two coils are flowing in same sense or opposite sense.

- If two coils of self-inductances L_1 and L_2 are connected in parallel, then equivalent inductance L_p is given by

$$\frac{1}{L_p} = \frac{1}{L_1} + \frac{1}{L_2} \Rightarrow L_p = \frac{L_1 L_2}{L_1 + L_2}$$

Eddy Currents

Currents induced in the body of bulk of the conductors due to change in magnetic flux linked to them, are called the eddy currents. The production of eddy currents in a metallic conductor leads to a loss of electric energy in the form of heat energy.

Eddy currents can be minimised by taking the metal (generally soft iron) core in the form of a combination of thin laminated sheets or by slotting process.

DAY PRACTICE SESSION 1

FOUNDATION QUESTIONS EXERCISE

1 The magnetic field in a certain region is given by $B = (40\hat{i} - 18\hat{k})$ gauss. How much flux passes through a loop of area 5 cm^2, in this region, if the loop lies flat on the *xy*-plane?

(a) -900×10^{-9} Wb (b) 900×10^{-9} Wb
(c) Zero (d) 9 Wb

2 The flux linked with a coil at any instant *t* is given by $\phi = 10t^2 - 50t + 250$. The induced emf at $t = 3$ s is

(a) −190 V (b) −10 V (c) 10 V (d) 190 V

3 A coil having *n* turns and resistance $R\ \Omega$ is connected with a galvanometer of resistance $4R\ \Omega$. This combination is moved for time *t* seconds from a magnetic field W_1 weber to W_2 weber. The induced current in the circuit is

(a) $\frac{W_2 - W_1}{5Rnt}$ (b) $-\frac{n(W_2 - W_1)}{5Rt}$
(c) $-\frac{(W_2 - W_1)}{Rnt}$ (d) $-\frac{n(W_2 - W_1)}{Rt}$

4 In a coil of resistance 100 Ω, a current is induced by changing the magnetic flux through it as shown in the figure. The magnitude of change in flux through the coil is **→ JEE Main 2017 (Offline)**

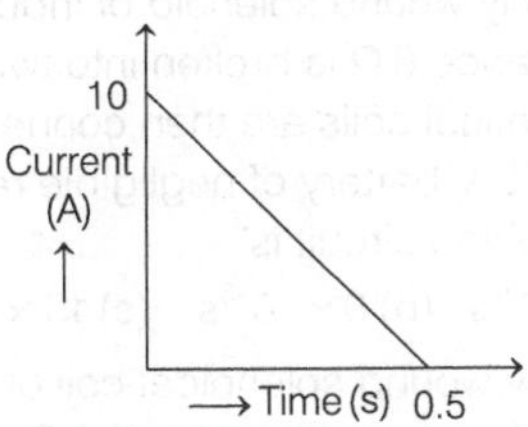

(a) 225 Wb (b) 250 Wb (c) 275 Wb (d) 200 Wb

5 A coil has an area of 0.05 m^2 and has 800 turns. After placing the coil in a magnetic field of strength 4×10^{-5} Wb m^{-2}, perpendicular to the field, the coil is rotated by 90° in 0.1 s. The average emf induced is

(a) zero (b) 0.016 V (c) 0.01 V (d) 0.032 V

6 A cylindrical bar magnet is rotated about its axis. A wire is connected from the axis and is made to touch the cylindrical suface through a contact. Then,

(a) a direct current flows in the ammeter *A*
(b) no current flows through the ammeter *A*
(c) an alternating sinusoidal current flows through the ammeter *A* with a time period $T = \frac{2\pi}{\omega}$
(d) a time varying non- sinusoidal current flows through the ammeter *A*

7 Coil *A* is made to rotate about a vertical axis (figure). No current flows in *B*, if *A* is at rest. The current in coil *A*, when the current in *B* (at $t = 0$) is counter clockwise and the coil *A* is as shown at this instant, ($t = 0$), is

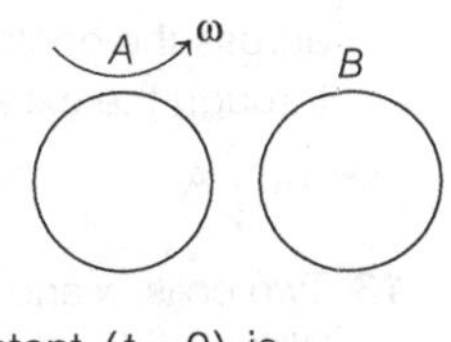

(a) constant current clockwise
(b) varying current clockwise
(c) varying current counter clockwise
(d) constant current counter clockwise

8 One conducting U-tube can slide inside another as shown in figure, maintaining electrical contacts between the tubes. The magnetic field *B* is perpendicular to the plane of the figure. If each tube moves towards the other at a constant speed *v*, then the emf induced in the circuit in terms of *B*, *l* and *v*, where *l* is the width of each tube, will be

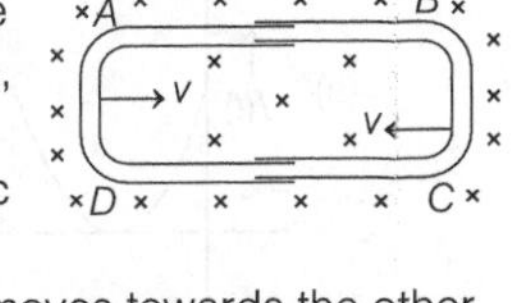

(a) *Blv* (b) −*Blv*
(c) zero (d) 2*Blv*

9 A boat is moving due to East in a region where the earth's magnetic field is 5.0×10^{-5} $NA^{-1}m^{-1}$ due to North and horizontal. The boat carries a vertical aerial 2m long. If the speed of the boat is 1.50 ms^{-1}, the magnitude of the induced emf in the wire of aerial is **→ AIEEE 2011**

(a) 1mV (b) 0.75 mV
(c) 0.50mV (d) 0.15mV

10 A helicopter rises vertically with a speed of 10 ms^{-1}. If helicopter has a length of 10 m and the horizontal component of the earth's magnetic field is 1.5×10^{-3} Wbm^{-2}, the emf induced between the tip of the nose and the tail of the helicopter, is

(a) 0.15 V (b) 125 V
(c) 130 V (d) 5 V

11 The magnetic field as shown in the figure is directed into the plane of paper. *A X C A* is a semi-circular conducting loop of radius *a* with centre *O*. The loop rotates clockwise with velocity ω about an axis fixed at *O* and perpendicular to the plane of the paper. The resistance of the loop is *R*. The induced current is

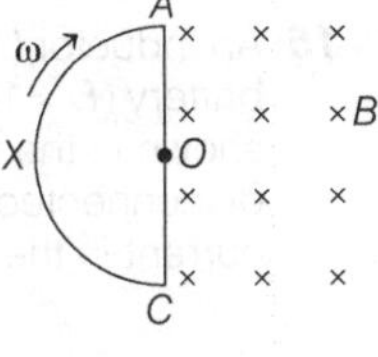

(a) $\frac{\omega r^2}{2R}$ (b) $-\frac{B\omega r^2}{2R}$ (c) $\frac{-2R}{B\omega r}$ (d) $\frac{2R}{\omega r^2}$

12 A metal rod of resistance 20 Ω is fixed along the diameter of a conducting ring of radius 0.1 m and lies on the *xy*-plane. There is a magnetic field $B = |50\mathbf{k}|$. The ring rotates with an angular velocity $\omega = 20 \text{ rads}^{-1}$ about its axis. An external resistance of 10 Ω is connected across the centre of the ring and the rim. The current through the external resistance is

(a) $\frac{1}{2}$ A (b) $\frac{1}{3}$ A (c) $\frac{1}{4}$ A (d) zero

13 Two coils, *x* and *y* are kept in close vicinity of each other. When a varying current, $I(t)$ flows through coil *x*, the induced emf [$V(t)$] in coil *Y*, varies in the manner shown here. The variation if $I(t)$, with time can then be represented by the graph labelled as graph. → JEE Main (Online) 2013

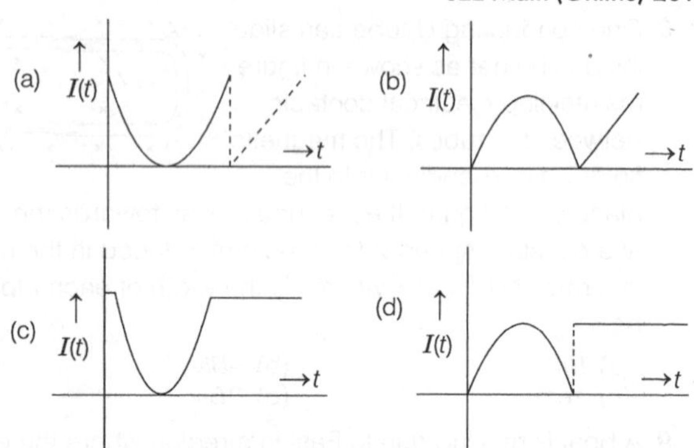

14 In the circuit shown here, the point *C* is kept connected to point *A* till the current flowing through the circuit becomes constant. Afterward, suddenly point *C* is disconnected from point *A* and connected to point *B* at time $t = 0$. Ratio of the voltage across resistance and the inductor at $t = L/R$ will be equal to → JEE Main 2014

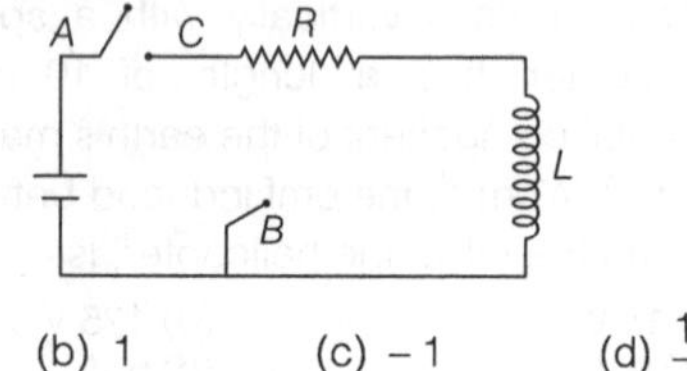

(a) $\frac{e}{1-e}$ (b) 1 (c) -1 (d) $\frac{1-e}{e}$

15 An inductor ($L = 100$ mH), a resistor ($R = 100\ \Omega$) and a battery ($E = 100$ V) are initially connected in series as shown in the figure. After a long time, the battery is disconnected after short circuiting the points *A* and *B*. The current in the circuit 1 millisecond after the short circuit is

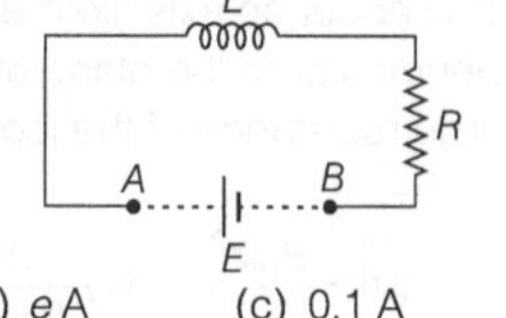

(a) $1/e$ A (b) e A (c) 0.1 A (d) 1 A

16 An inductor ($L = 0.03$ H) and a resistor ($R = 0.15$ kΩ) are connected in series to a battery of 15V emf in a circuit shown below. The key K_1 has been kept closed for a long time. Then at $t = 0$, K_1 is opened and key K_2 is closed simultaneously. At $t = 1$ ms, the current in the circuit will be ($e^5 \cong 150$) → JEE Main 2015

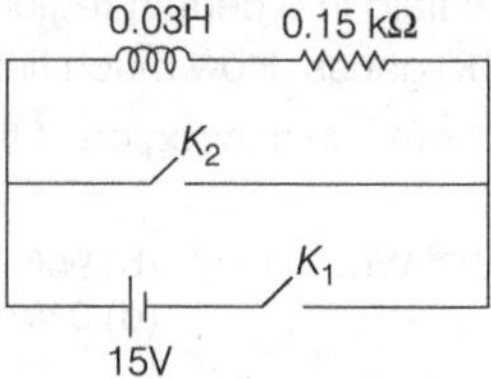

(a) 100 mA (b) 67 mA (c) 6.7 mA (d) 0.67mA

17 An inductor coil stores 32J of magnetic energy and dissipates it as heat at the rate of 320 W when a current of 4 A is passed through it. The time constant of the circuit is

(a) 0.2 s (b) 0.1 s (c) 0.3 s (d) 0.4 s

18 In series *R*-*L* circuit, switch is closed at $t = 0$. The charge which passes through the battery in one time constant is

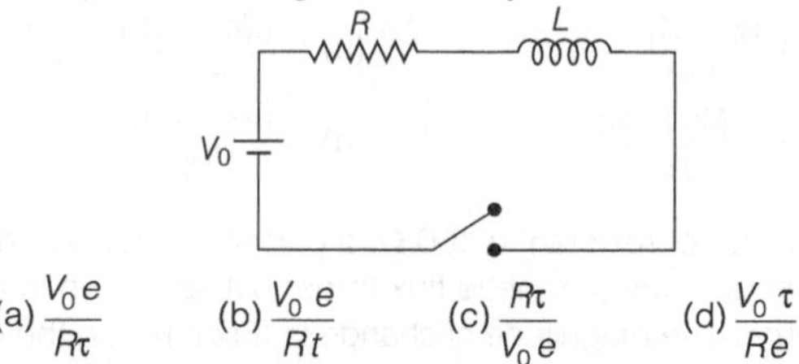

(a) $\frac{V_0 e}{R\tau}$ (b) $\frac{V_0 e}{Rt}$ (c) $\frac{R\tau}{V_0 e}$ (d) $\frac{V_0 \tau}{Re}$

19 An uniformly wound solenoid of inductance 1.8×10^{-4}H and resistance 6 Ω is broken into two identical parts. These identical coils are then connected in parallel across a 15 V battery of negligible resistance. The time constant of the circuit is

(a) 3×10^{-5}s (b) 6×10^{-5}s (c) 1.5×10^{-5}s (d) 1.8×10^{-5}s

20 A uniformly wound solenoidal coil of self-inductance 1.8×10^{-4} H and a resistance of 6 Ω is broken up into two identical coils. These identical coils are then connected in parallel across a 120 V battery of negligible resistance. The time constant of the current in the circuit and the steady state current through the battery is

(a) 3×10^{-5} s, 8 A (b) 1.5×10^{-5} s, 8 A
(c) 0.75×10^{-4}s, 4 A (d) 6×10^{-5} s, 2 A

21 A coil is suspended in a uniform magnetic field with the plane of the coil parallel to the magnetic lines of force. When a current is passed through the coil, it starts oscillating; it is very difficult to stop. But if an aluminium plate is placed near to the coil, it stops. This is due to → AIEEE 2011

(a) development of air current when the plate is placed
(b) induction of electrical charge on the plate
(c) shielding of magnetic lines of force as aluminium is a paramagnetic material
(d) electromagnetic induction in the aluminium plate giving rise to electromagnetic damping

Direction (Q. Nos. 22-26) *Each of these questions contains two statements : Statement I and Statement II. Each of these questions also has four alternative choices, only one of which is the correct answer. You have to select one of the codes (a), (b), (c), (d) given below*

(a) Statement I is true, Statement II is true; Statement II is the correct explanation for Statement I
(b) Statement I is true, Statement II is true; Statement II is not the correct explanation for Statement I
(c) Statement I is true; Statement II is false
(d) Statement I is false; Statement II is true

22 **Statement I** The mutual inductance of two coils is doubled if the self-inductance of the primary and the secondary coil is doubled.

Statement II Mutual inductance is proportional to the square root of self-inductance of primary and secondary coils.

23 **Statement I** The energy stored in the inductor of 2 H, when a current of 10 A flows through it, is 100 J.

Statement II Energy stored in an inductor is directly proportional to its inductance.

24 **Statement I** An artificial satellite with a metal surface, is moving about the earth in a circular orbit. A current is induced when the plane of the orbit is inclined to the plane of the equator.

Statement II The satellite cuts the magnetic field of earth.

25 **Statement I** A coil *A* is connected to a voltmeter *V* and the other coil *B* is connected to an alternating current source. If a large copper sheet *C* is placed between the two coils, the induced emf in the coil *A* is reduced.

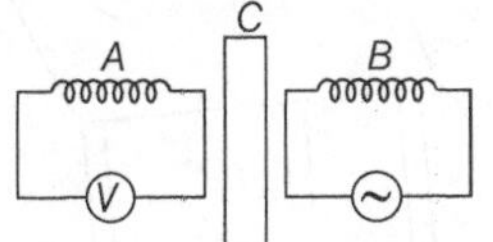

Statement II Copper sheet between the coils, has no effect on the induced emf in coil *A*.

26 **Statement I** When a DC current is made to flow in a soft wire loop of arbitrary shape, it tend to acquire a circular shape.

Statement II Flux linked with a wire loop is maximum when loop is a circle.

DAY PRACTICE SESSION 2

PROGRESSIVE QUESTIONS EXERCISE

1 The inductance per unit length of a double tape line as shown in the figure.

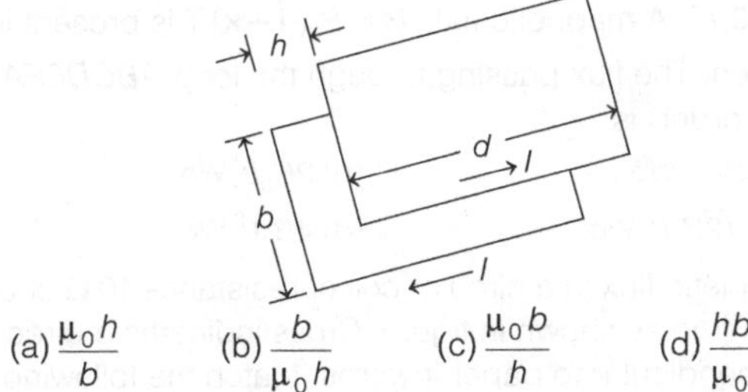

(a) $\frac{\mu_0 h}{b}$ (b) $\frac{b}{\mu_0 h}$ (c) $\frac{\mu_0 b}{h}$ (d) $\frac{hb}{\mu_0}$

2 An air-cored solenoid with length 30 cm, area of cross-section 25 cm^2 and number of turns 500, carries a current of 2.5 A. The current is suddenly switched off in a brief time of 10^{-3} s. How much is the average back emf induced across the ends of the open switch in the circuit? Ignore the variation in magnetic field near the ends of the solenoid.

(a) 6.5 V (b) 7.4 V (c) 8.2 V (d) 9.3 V

3 A short circuited coil is placed in a time varying magnetic field. Electrical power is dissipated due to the current induced in the coil. If the number of turns are quadrupled and the wire radius is halved, the electrical power dissipated in the coil, would be

(a) halved (b) the same (c) doubled (d) quadrupled

4 A circular coil of radius 8.0 cm and 20 turns is rotated about its vertical diameter with an angular speed of 50 rad/s in a uniform horizontal magnetic field of magnitude 3.0×10^{-2} T. Obtain the maximum and average emf induced in the coil. If the coil forms a closed-loop of resistance 10 Ω, calculate the maximum value of current in the coil. Calculate the average power loss due to Joule heating.

(a) 0.012 W (b) 0.1 W
(c) 0.018 W (d) 0.42 W

5 A long straight solenoid with cross-sectional radius *a* and number of turns per unit length *n* has a current varying with time as I As^{-1}. The magnitude of the eddy current as a function of distance *r* from the solenoid axis is

(a) $\frac{-n\mu_0 a^2 I}{2r}$ (b) $\frac{\mu_0 In}{2a}$ (c) $\frac{-na^2 I}{2\mu_0 r}$ (d) $\frac{\mu_0 Ia}{2n}$

6 A circular loop of radius 0.3 cm lies parallel to a much bigger circular loop of radius 20 cm. The centre of the small loop is on the axis of the bigger loop. The distance between their centres is 15 cm. If a current of 2.0 A flows through the smaller loop, then the flux linked with bigger loop is

(a) 9.2×10^{-11}Wb (b) 6×10^{-11}Wb
(c) 3.3×10^{-11}Wb (d) 6.6×10^{-9}Wb

7 A metallic rod of length l is tied to a string of length $2l$ and made to rotate with angular speed ω on a horizontal table with one end of the string fixed. If there is a vertical magnetic field B in the region, the emf induced across the ends of the rod is

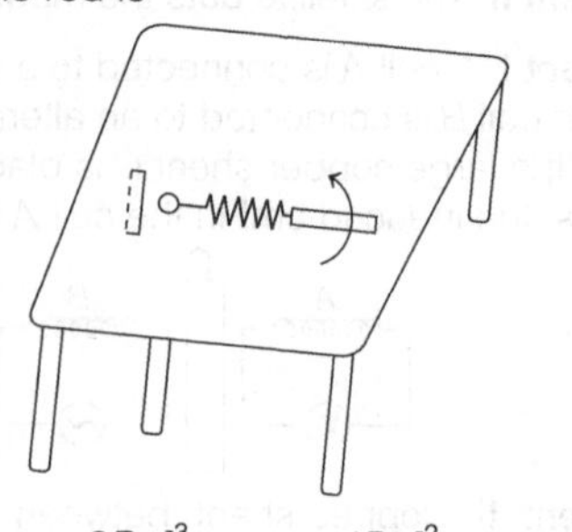

(a) $\frac{2B\omega l^3}{2}$ (b) $\frac{3B\omega l^3}{2}$ (c) $\frac{4B\omega l^2}{2}$ (d) $\frac{5B\omega l^2}{2}$

8 A rectangular loop has a sliding connector PQ of length l and resistance $R\ \Omega$ and it is moving with a speed v as shown. The setup is placed in a uniform magnetic field going into the plane of the paper. The three currents I_1, I_2 and I are

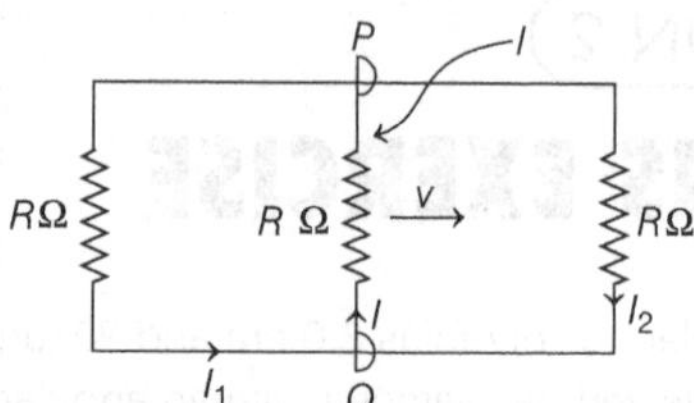

(a) $I_1 = -I_2 = \frac{Blv}{R}, I = \frac{2Blv}{R}$

(b) $I_1 = I_2 = \frac{Blv}{3R}, I = \frac{2Blv}{3R}$

(c) $I_1 = I_2 = I = \frac{Blv}{R}$

(d) $I_1 = I_2 = \frac{Blv}{6R}, I = \frac{Blv}{3R}$

9 An ideal coil of 10 H is connected in series with a resistance of 5 Ω and a battery of 5 V. 2s after the connection is made, the current flowing (in ampere) in the circuit is

(a) $(1-e)$ (b) e (c) e^{-1} (d) $(1-e^{-1})$

10 An inductor of inductance $L = 400$ mH and resistors of resistances $R_1 = 4\ \Omega$ and $R_2 = 2\ \Omega$ are connected to battery of emf 12 V as shown in the figure. The internal resistance of the battery is negligible. The switch S is closed at $t = 0$. The potential drop across L as a function of time is → AIEEE 2009

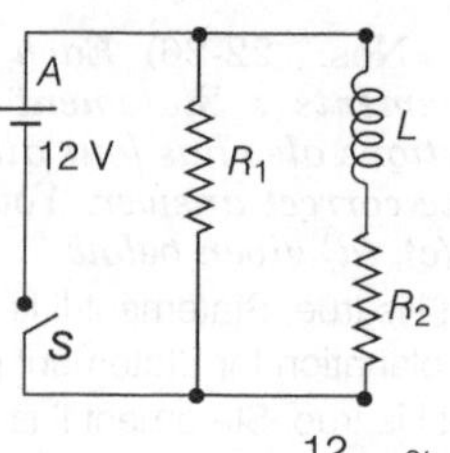

(a) $6e^{-5t}$ V (b) $\frac{12}{t}e^{-3t}$ V

(c) $6(1 - e^{-t/0.2})$ V (d) $12e^{-5t}$ V

11 In a uniform magnetic field of induction B, a wire in the form of semi-circle of radius r rotates about the diameter of the circle with angular frequency ω. If the total resistance of the circuit is R, the mean power generated per period of rotation is

(a) $\frac{B\pi r^2\omega}{2R}$ (b) $\frac{(B\pi r^2\omega)^2}{8R}$

(c) $\frac{(B\pi r\omega)^2}{2R}$ (d) $\frac{(B\pi r\omega^2)^2}{8R}$

12 The inductance between A and D is

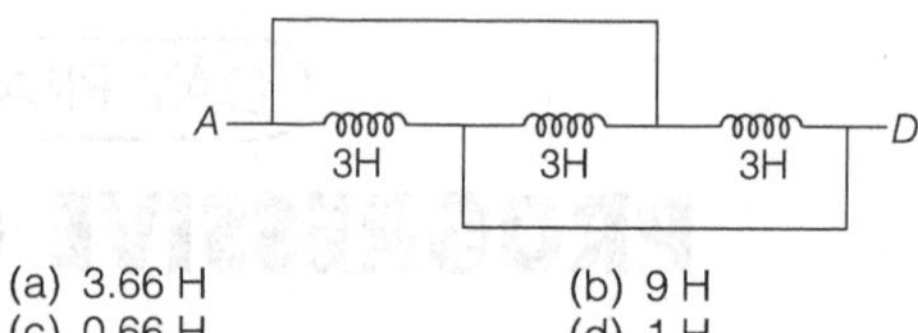

(a) 3.66 H (b) 9 H
(c) 0.66 H (d) 1 H

13 A loop, made of straight edges has six corners at $A(0,0,0), B(L,0,0), C(L,L,0), D(0,L,0), E(0,L,L)$ and $F(0,0,L)$. A magnetic field $\mathbf{B} = B_0(\hat{\mathbf{i}} + \hat{\mathbf{k}})$ T is present in the region. The flux passing through the loop $ABCDEFA$ (in that order) is

(a) B_0L^2 Wb (b) $2B_0L^2$ Wb

(c) $\sqrt{2}B_0L^2$ Wb (d) $4B_0L^2$ Wb

14 Magnetic flux in a circular coil of resistance 10 Ω changes with time as shown in figure. Cross indicates a direction perpendicular to paper inwards. Match the following.

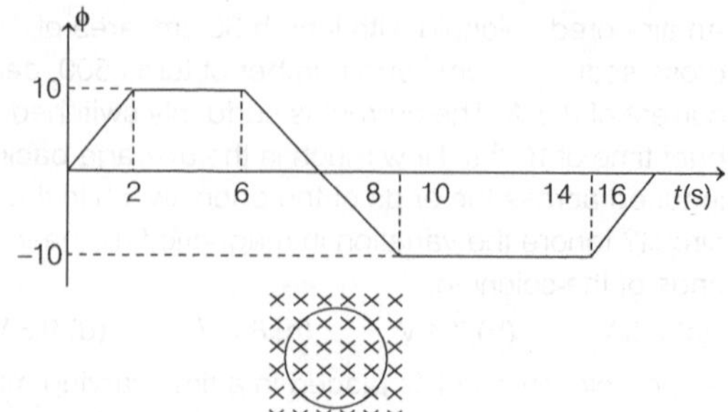

Column I	Column II
A. at $t = 1$ s	1. emf induced is zero
B. at $t = 5$s	2. emf induced in anti-clockwise direction
C. at $t = 9$s	3. emf induced in clockwise direction
D. at $t = 15$s	4. 2 A

Codes

	A	B	C	D		A	B	C	D
(a)	2	1	1	2	(b)	1	2	3	4
(c)	4	3	2	1	(d)	2	4	1	3

15 A square loop is symmetrically placed between two infinitely long current carrying wires in the same direction. Magnitude of currents in both the wires are same.Now match the following two columns.

Column I	Column II
A. Loop is moved towards right	1. Induced current in the loop is clockwise
B. Loop is moved towards left	2. Induced current in the loop is anti-clockwise
C. Wire-1 is moved towards left	3. Induced current in the loop is zero
D. Wire-2 is moved towards right	4. Induced current in the loop is non-zero

Codes

	A	B	C	D		A	B	C	D
(a)	1	2	1	2	(b)	1	3	2	4
(c)	1	2	3	4	(d)	4	2	2	4

ANSWERS

SESSION 1	1 (a)	2 (b)	3 (b)	4 (b)	5 (b)	6 (b)	7 (b)	8 (d)	9 (d)	10 (a)
	11 (b)	12 (b)	13 (c)	14 (c)	15 (a)	16 (d)	17 (a)	18 (d)	19 (a)	20 (a)
	21 (d)	22 (a)	23 (a)	24 (a)	25 (c)	26 (b)				
SESSION 2	1 (a)	2 (a)	3 (b)	4 (c)	5 (a)	6 (a)	7 (d)	8 (b)	9 (d)	10 (d)
	11 (b)	12 (d)	13 (b)	14 (a)	15 (a)					

Hints and Explanations

SESSION 1

1 As loop is in the xy-plane, only the z-component of the magnetic field, is effective.

$B = -18$ gauss $= -18 \times 10^{-4}$ T

$A = 5 \times 10^{-4} \text{m}^2$

$\phi = BA\cos 0° = -18 \times 10^{-4} \times 5 \times 10^{-4}$

$= -90 \times 10^{-8}$ Wb

$= -900 \times 10^{-9}$ Wb

2 $\phi = 10t^2 - 50t + 250$

From Faraday's law of electromagnetic induction,

$e = -d\phi/dt$

$\therefore \quad e = -[10 \times 2t - 50]$

$\therefore e|_{t=3s} = -[10 \times 6 - 50] = -10$ V

3 The rate of change of flux or emf induced in the coil is $e = -n\dfrac{d\phi}{dt}$.

$\therefore$ Induced current,

$$I = \frac{e}{R'} = -\frac{n}{R'}\frac{d\phi}{dt} \quad \text{...(i)}$$

Given, $R' = R + 4R = 5R$,

$d\phi = W_2 - W_1, dt = t$

[here, W_1 and W_2 are flux associated with one turn]

Putting the given values in Eq. (i), we get

$$I = -\frac{n}{5R}\frac{(W_2 - W_1)}{t}$$

4 Induced constant, $I = \dfrac{e}{R}$

Here, e = induced emf $= \dfrac{d\phi}{dt}$

$$I = \frac{e}{R} = \left(\frac{d\phi}{dt}\right)\cdot\frac{1}{R}$$

$d\phi = IRdt$

$\Rightarrow \quad \phi = \int IRdt$

$\therefore$ Here, R is constant

$\therefore \quad \phi = R\int Idt$

$\int I \cdot dt$ = Area under I-t graph

$= \dfrac{1}{2} \times 10 \times 0.5 = 2.5$

$\therefore \quad \phi = R \times 2.5 = 100 \times 2.5$

$= 250$ Wb

5 As from Faraday's rule,

$$e = N\frac{d\phi}{dt} = \frac{NAdB}{dt}$$

$$= 800 \times 0.05 \times \frac{4 \times 10^{-5}}{0.1}$$

$= 0.016$ V

6 When cylindrical bar magnet is rotated about its axis,no change in flux linked with the circuit takes place, consequently no emf induces and hence,no current flows through the ammeter A.

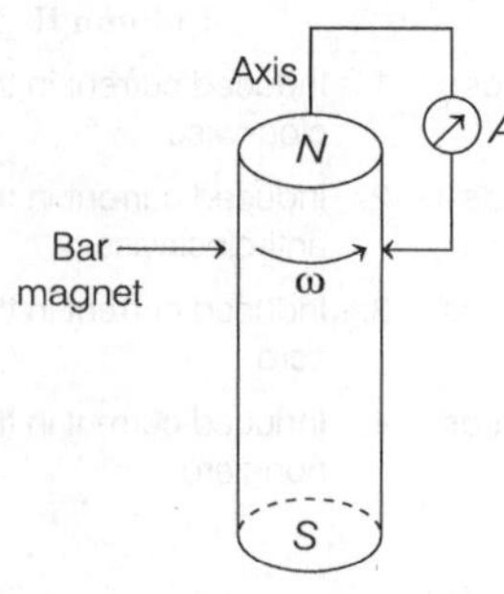

7 When the current in B(at $t = 0$) is counter clockwise and the coil A is considered above to it. The counter clockwise flow of the current in B is equivalent to North pole of magnet and magnetic field lines are emerging upward to coil A. When coil A start rotating at $t = 0$, the current in A is time varying along clockwise direction by Lenz's rule.

8 Relative velocity $= v - (-v) = 2v = \dfrac{dl}{dt}$

Now, $e = \dfrac{d\phi}{dt} \Rightarrow e = \dfrac{Bldl}{dt}$ $[\because \phi = BA]$

Induced emf, $e = 2Blv$ $\left[\because \dfrac{dl}{dt} = 2v\right]$

9 $E_{ind} = B \times v \times l = 5.0 \times 10^{-5} \times 1.50 \times 2$

$= 10.0 \times 10^{-5} \times 1.5$

$= 15 \times 10^{-5} = 0.15\text{mV}$

10 $e = Bl\, v = B_H l v$

$= 1.5 \times 10^{-3} \times 10 \times 10 = 0.15\text{ V}$

11 $A = \dfrac{\theta}{\pi}\dfrac{\pi r^2}{2} = \dfrac{\theta r^2}{2} = \dfrac{\omega t r^2}{2}$

Flux, $\phi = BA = B\dfrac{\omega t r^2}{2}$

Emf, $e = -\dfrac{d\phi}{dt} = -\dfrac{B\omega r^2}{2}$

$I = \dfrac{-B\omega r^2}{2R}$

After half rotation $A(t) = \pi r^2 - \dfrac{\omega t r^2}{2}$

will give same current but in opposite direction.

12 Here, resistance of rod $= 20\Omega$, $r = 0.1$ m, $B = 50$ T, acting along the z-axis and $\omega = 20 \text{ rad s}^{-1}$.

Potential difference between the centre of the ring and the rim is

$V = \dfrac{1}{2}B\omega r^2$

$= \dfrac{1}{2} \times 50 \times 20 \times (0.1)^2 = 5\text{ V}$

The equivalent circuit of the arrangement is shown in figures.

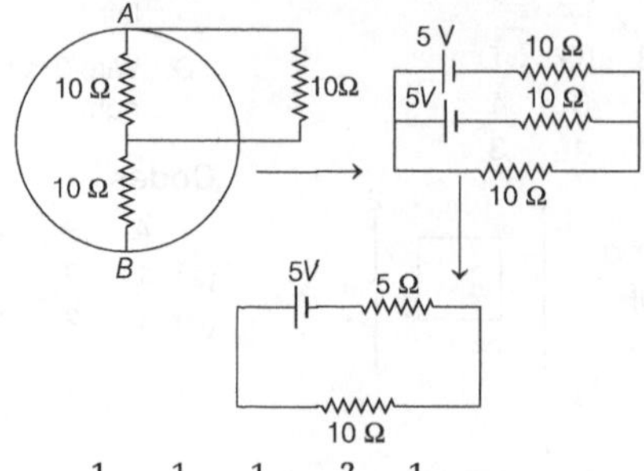

$\dfrac{1}{R_p} = \dfrac{1}{10} + \dfrac{1}{10} \Rightarrow \dfrac{2}{10} = \dfrac{1}{5} \Rightarrow R_p = 5\,\Omega$

Current through the external resistance,

$I = \dfrac{E}{R+r} = \dfrac{5}{10+5} = \dfrac{1}{3}\text{ A}$

13 Firstly, the current decreases due to electrical inertia goes to zero, but due to back emf induced in the easily, the induced current in the coil decreases off a point when back emf is equal to the applied emf induced in the another coil. The value of emf of two current is zero.Then, current is regularly increased, after that time became it is continuously by supplied by the source (variable).

14 After connecting C to B hanging the switch, the circuit will act like L-R discharging circuit.

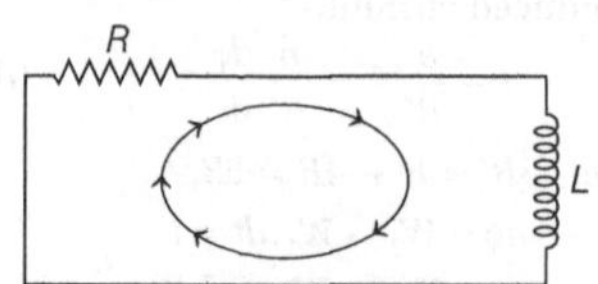

Applying Kirchhoff's loop equation,

$V_R + V_L = 0 \Rightarrow V_R = -V_L$

$\therefore \dfrac{V_R}{V_L} = -1$

15 This is a combined example of growth and decay of current in an L-R circuit.

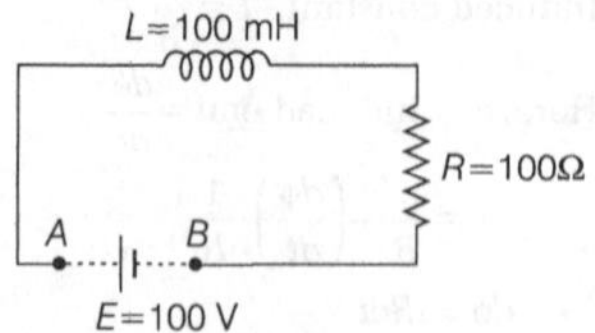

The current through circuit just before shorting the battery,

$I_0 = \dfrac{E}{R} = \dfrac{100}{100} = 1\text{ A}$

[as inductor would be shorted in steady state]

After this decay of current starts in the circuit, according to the equation,

$I = I_0 e^{-t/\tau}$

where, $\tau = L/R$

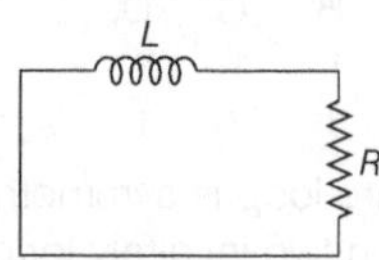

$I = 1 \times e^{-(1\times10^{-3})/(100\times10^{-3}/100)} = \left(\dfrac{1}{e}\right)\text{ A}$

[$\because t = 1$ millisecond $= 1 \times 10^{-3}$ s and $L = 100 \times 10^{-3}$ H]

16 After long time inductor behaves as short-circuit.

At $t = 0$, the inductor behaves as short-circuited. The current

$I_0 = \dfrac{E_0}{R} = \dfrac{15\text{V}}{0.15\text{ k}\Omega} = 100\text{ mA}$

As K_2 is closed, current through the inductor starts decay, which is given at any time t as

$I = I_0 e^{\frac{-tR}{L}} = (100\text{ mA})e^{\frac{-t\times15000}{3}}$

At $t = 1$ ms

$I = (100\text{ mA})e^{-\frac{1\times10^{-3}\times15\times10^{3}}{3}}$

$I = (100\text{ mA})e^{-5} = 0.6737\text{ mA}$

or $I = 0.67$ mA

17 $U = \dfrac{LI^2}{2}$ or $32 = \dfrac{L(4)^2}{2} = L = 4\text{H}$

$P = I^2R,\ R = \dfrac{320}{(4)^2} = 20\,\Omega$

$\therefore \tau = \dfrac{L}{R} = \dfrac{4}{20} = 0.2\text{ s}$

18 $I = \dfrac{V_0}{R}(1 - e^{-t/\tau})$

$Q = \int I dt = \int_0^{\tau}\dfrac{V_0}{R}(1 - e^{-t/\tau})\,dt$

$\Rightarrow Q = \dfrac{V_0}{R}\tau + \tau(e^{-1} - 1)\dfrac{V_0}{R} = \dfrac{V_0\tau}{Re}$

19 Inductance of each part,

$L_1 = L_2 = \dfrac{L}{2} = 0.9 \times 10^{-4}\text{ H}$

Resistance of each part,

$R_1 = R_2 = \dfrac{R}{2} = 3\Omega$

Time constant,

$\tau = \dfrac{L_1||L_2}{R_1||R_2} = \dfrac{L_1L_2}{L_1 + L_2} \times \dfrac{R_1 + R_2}{R_1R_2}$

$\tau = \dfrac{1.8 \times 10^{-4} \times 0.9 \times 10^{-4}}{1.8 \times 10^{-4} + 0.9 \times 10^{-4}} \times \dfrac{6+3}{6 \times 3}$

$= \dfrac{1.62 \times 10^{-4}}{5.4} = 0.3 \times 10^{-4} = 3 \times 10^{-5}\text{s}$

20 Since, the self-inductance in parallel is given by

$$\frac{1}{L_p} = \frac{1}{L} + \frac{1}{L} = \frac{2}{L} \Rightarrow L_p = \frac{L}{2}$$

and $L = \frac{1.8 \times 10^{-14}}{2} = 0.9 \times 10^{-4}$ H

$\therefore \quad L_p = 0.45 \times 10^{-4}$ H

Resistance of each part, $r = 6/2 = 3\Omega$

Now, $\frac{1}{r_p} = \frac{1}{3} + \frac{1}{3} = \frac{2}{3}$

$\therefore \quad r_p = \frac{3}{2}\,\Omega$

So, the time constant of the circuit is given by

$$\tau = \frac{L_p}{r_p} = \frac{0.45 \times 10^{-4}}{3/2} = 3 \times 10^{-5}\ \text{s}$$

and the steady current is,

$$I = \frac{V}{r_p} = \frac{12}{3/2} = 8\ \text{A}$$

21 According to Lenz's law, electromagnetic induction takes place in the aluminium plate for which eddy current is developed. This causes loss in energy which results in damping of oscillatory motion of the coil.

22 If two coils of inductances L_1 and L_2 are joined together, then their mutual inductance is given by

$$M = k\sqrt{L_1 L_2}$$

It is clear from the relation, if the self-inductance of the primary and the secondary coil is doubled, the mutual inductance of the coils, will also be doubled.

23 The energy stored in the inductor is given by

$$U = \frac{1}{2}LI_0^2 = \frac{1}{2} \times 2 \times (10)^2 = 100\ \text{J}$$

It is obvious that energy stored in the inductor, is directly proportional to its inductance.

24 It is concept of eddy current losses.

25 In the absence of the copper sheet, induced emf will be produced in the coil A due to the mutual induction between the coils A and B. As a result, voltmeter will show deflection depending on the magnitude of the induced emf.

When the copper sheet is placed between the two coils, eddy currents will be setup in the coil. Since, the eddy currents have an opposing effect, the magnetic flux linked with A due to eddy current will always be opposite to that due to the alternating current through B. Thus, induced current will be reduced.

26 Each section of wire repels diametrically opposite section as current flows in opposite direction.

SESSION 2

1 Neglecting end effects of magnetic field, we have

$$B = \frac{\mu_0 I}{b}$$

Flux ϕ per unit length of the plates is

$$\frac{\mu_0 I}{b} \times h \times 1 = \frac{\mu_0 hI}{b}$$

Also, $\phi = LI \Rightarrow L = \frac{\mu_0 h}{b}$

2 Given, length of solenoid $l = 30$ cm $= 30 \times 10^{-2}$ m

Area of cross-section $A = 25\,\text{cm}^2 = 25 \times 10^{-4}\ \text{m}^2$

Number of turns $N = 500$

Current $I_1 = 2.5$ A, $I_2 = 0$

Brief time $dt = 10^{-3}$ s

Induced emf in the solenoid

$$e = \frac{d\phi}{dt} = \frac{d}{dt}(BA) \qquad (\because \phi = BA)$$

Magnetic field induction B at a point well inside the long solenoid carrying current I is

$B = \mu_0 nI$

$\left(\text{where, } n = \text{Number of turns per unit length} = \frac{N}{l}\right)$

$$e = NA\frac{dB}{dt} = A\frac{d}{dt}\left(\mu_0 \frac{N}{l} I\right)$$

$$= A\frac{\mu_0 N}{l} \cdot \frac{dI}{dt}$$

$$e = 500 \times 25 \times 10^{-4} \times 4 \times 3.14 \times 10^{-7} \times \frac{500}{30 \times 10^{-2}} \times \frac{2.5}{10^{-3}}$$

$= 6.5$ V

3 The magnitude of the induced voltage is proportional to the rate of change of magnetic flux which, in turns depends on the number of turns in the coil, i.e. $V \propto n$.

So, resistance of a wire is given by

$$R = \frac{\rho l}{\pi r^2} \qquad [A = \pi r^2]$$

i.e. $R \propto \frac{l}{r^2}$

[ρ is a resistivity of a wire]

$$\therefore \quad P = \frac{V^2}{R} \propto \frac{n^2}{l/r^2} \Rightarrow P = \frac{(nr)^2}{l}$$

$$\therefore \quad \frac{P_2}{P_1} = \left(\frac{n_2}{n_1}\right)^2 \times \left(\frac{r_2}{r_1}\right)^2 \times \left(\frac{l_1}{l_2}\right)$$

or $\frac{l_1}{l_2} = \left(\frac{r_2}{r_1}\right)^2 \Rightarrow \frac{P_2}{P_1} = \left(\frac{n_2}{n_1}\right)^2 \times \left(\frac{r_2}{r_1}\right)^4$

$\left[\text{Given, } \frac{n_2}{n_1} = 4 \text{ and } \frac{r_2}{r_1} = \frac{1}{2}\right]$

So, $\frac{P_2}{P_1} = (4)^2 \times \left(\frac{1}{2}\right)^4 = 16 \times \frac{1}{16} = 1$

4 Average induced emf

$$e_{av} = \frac{1}{T}\int_0^{2\pi} e\,dt = \frac{1}{T}\int_0^{\pi} NBA\,\omega \sin\omega t\,dt$$

$$e_{av} = \frac{1}{T} \cdot NAB\,\omega\left[\frac{\cos \omega t}{\omega}\right]_0^{2\pi}$$

$$= \frac{NBA}{T}[\cos 2\pi - \cos 0°]$$

$$e_{av} = \frac{NBA}{T}[1 - 1] = 0$$

For full cycle average emf, $e_{av} = 0$

Average power loss due to heating

$$= \frac{E_0 I_0}{2} = \frac{0.603 \times 0.0603}{2} = 0.018\ \text{W}$$

5 $B = n\mu_0 I$ and $\oint E \cdot dl = -\frac{d\phi}{dt}$

For $r < a$, $E\,(2\pi r) = -\pi r^2\, n\mu_0 I$

or $E = -\frac{n\mu_0 Ir}{2}$ (for $r < a$)

where, $I = \frac{dI}{dt}$. (for $r > a$)

$$E(2\pi r) = -\pi a^2 n\mu_0 I \Rightarrow E = \frac{-n\mu_0 a^2 I}{2r}$$

6 Mutual inductance of two coils,

$$M = \frac{\mu_0 R_1^2 \pi R_2^2}{2(R_1^2 \times x^2)^{3/2}}$$

Flux through the bigger coil,

$$M = \frac{\mu_0}{4\pi} \cdot \frac{\pi^2 R_1^2 R_2^2}{(R_1^2 + x^2)}$$

Substituting the values

$$M = \frac{\mu_0 (2)(20 \times 10^{-2})^2}{2[(0.2)^2 + (0.15)^2]} \times \pi(0.3 \times 10^{-2})^2$$

On solving,

$M = 9.216 \times 10^{-11}$

$= 9.216 \times 10^{-11} \approx 9.2 \times 10^{-11}$ Wb

7 Consider a small element of length dx as shown below emf induced due to whole rod

$$e = \int_{2l}^{3l}(\omega x)B\,dx = B\omega\frac{[(3l)^2 - (2l)^2]}{2}$$

$$= \frac{5Bl^2\omega}{2}$$

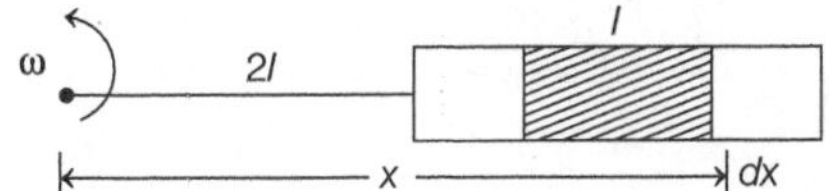

8 A moving conductor is equivalent to a battery of emf = vBl (motion emf)
Equivalent circuit, $I = I_1 + I_2$

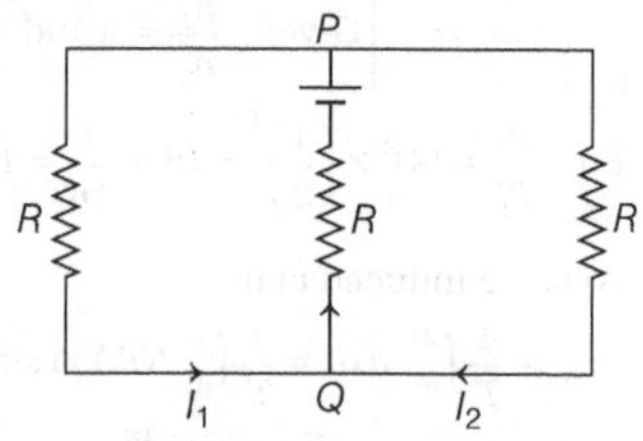

Applying Kirchhoff's law,

$$I_1 R + IR - vBl = 0 \quad ...(i)$$

$$I_2 R + IR - vBl = 0 \quad ...(ii)$$

Adding Eqs. (i) and (ii), we get

$$2IR + IR = 2vBl$$

$$I = \frac{2vBl}{3R}$$

$$I_1 = I_2 = \left|\frac{Blv}{3R}\right|$$

9 Rise of current in L-R circuit is given by

$$I = I_0(1 - e^{-t/\tau})$$

where, $I_0 = \frac{E}{R} = \frac{5}{5} = 1\,\text{A}$

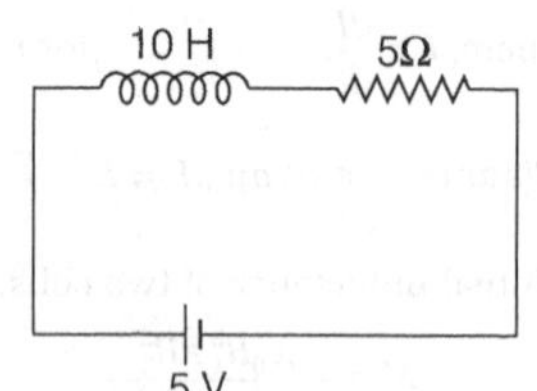

Now, $\tau = \frac{L}{R} = \frac{10}{5} = 2\,\text{s}$

After 2s, i.e. at $t = 2$s

Rise of current $I = (1 - e^{-1})$ A

10 $I_1 = \frac{E}{R_1} = \frac{12}{4} = 3\,\text{A}$

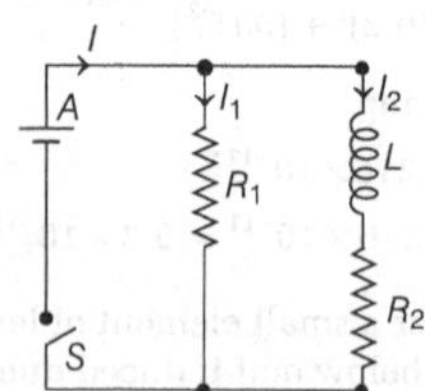

$\because$ Potential drop $= E - I_2 R_2$

$$I_2 = I_0(1 - e^{-T/t_c})$$

[current as a function of time]

$$\Rightarrow \quad I_0 = \frac{E}{R_2} = \frac{12}{2} = 6\,\text{A}$$

and $$t_c = \frac{L}{R_2} = \frac{400 \times 10^{-3}}{2} = 0.2$$

$$I_2 = 6(1 - e^{-t/0.2})$$

$$\Rightarrow \quad I_2 = 6(1 - e^{-5t})$$

Potential drop across

$$L = E - R_2 I_2$$

$$= 12 - 2 \times 6(1 - e^{-5t})$$

$$= 12e^{-5t}\,\text{V}$$

11 The flux associated with coil of area A and magnetic induction B is

$$\phi = BA\cos\theta = \frac{1}{2}B\pi r^2 \cos\omega t$$

$$\left[\because A = \frac{1}{2}\pi r^2\right]$$

$$\therefore \quad e_{\text{induced}} = -\frac{d\phi}{dt}$$

$$= -\frac{d}{dt}\left(\frac{1}{2}B\pi r^2 \cos\omega t\right)$$

$$= \frac{1}{2}B\pi r^2 \omega \sin\omega t$$

$$\therefore \text{ Power, } P = \frac{e^2_{\text{induced}}}{R} \quad [\because P = V^2/R]$$

$$= \frac{B^2\pi^2 r^4 \omega^2 \sin^2\omega t}{4R}$$

Hence, $P_{\text{mean}} = \langle P \rangle$

$$= \frac{B^2\pi^2 r^4 \omega^2}{4R} \cdot \frac{1}{2}$$

$$\left[\because \langle \sin^2\omega t \rangle = \frac{1}{2}\right]$$

$$= \frac{(B\pi r^2 \omega)^2}{8R}$$

12

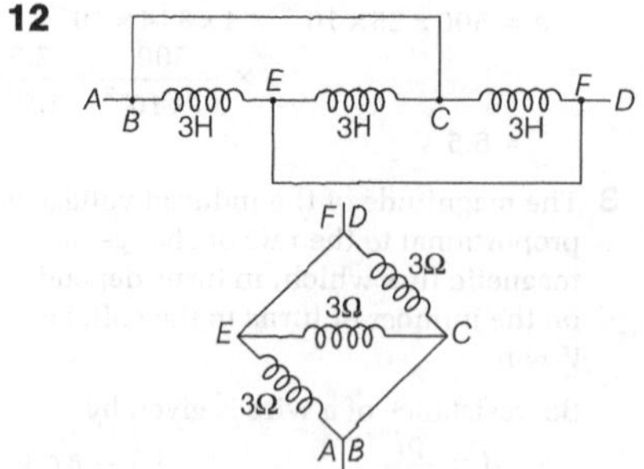

Here, inductors are in parallel.

$$\therefore \quad \frac{1}{L} = \frac{1}{3} + \frac{1}{3} + \frac{1}{3}$$

or $\quad L = 1\,\text{H}$

13 Also, the magnetic flux linked with uniform surface of area A in uniform magnetic field is given by

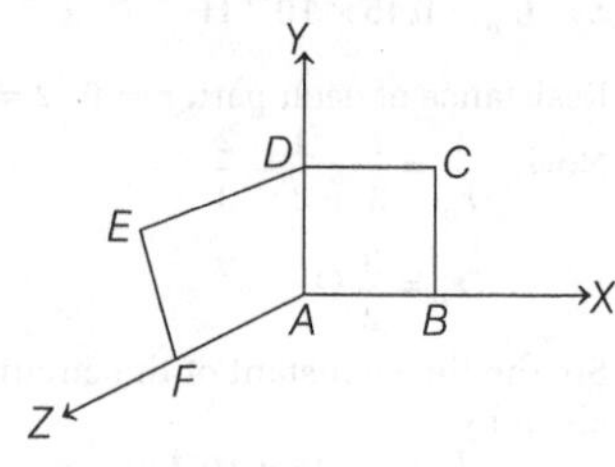

$$\phi = \mathbf{B} \cdot \mathbf{A}$$

$$\mathbf{A} = \mathbf{A}_1 + \mathbf{A}_2$$

$$= (L^2\,\hat{\mathbf{k}} + L^2\,\hat{\mathbf{i}})$$

and $\mathbf{B} = B_0(\hat{\mathbf{i}} + \hat{\mathbf{k}})$T

Now,

$$\phi = \mathbf{B} \cdot \mathbf{A}$$

$$= B_0(\hat{\mathbf{i}} + \hat{\mathbf{k}}).(L^2\,\hat{\mathbf{k}} + L^2\,\hat{\mathbf{i}})$$

$$= 2B_0 L^2\,\text{Wb}$$

14 A→ 2, B→ 1, C→ 1, D→ 2

A. At $t = 1$ s, flux is increasing in the inward direction, hence induced emf will be in anti-clockwise direction.

B. At $t = 5$ s, there is no change in flux, so induced emf is zero.

C. At $t = 9$ s, flux is constant, hence induced emf will be zero.

D. At $t = 15$ s, flux is decreasing in upward direction, so induced emf will be in anti-clockwise direction.

15 A - 1, B - 2, C - 1, D - 2

When loop is moved towards right, upward flux increases, so current in loop is clockwise. When loop is moved towards left, downward flux increases, so current in loop is anti-clockwise. When wire 1 is moved left, upward flux through loop increases, so current is clockwise. When wire 2 is moved right, downward flux increases, so current is anti-clockwise.

DAY TWENTY FOUR

Alternating Current

Learning & Revision for the Day

- Peak and RMS Values of Alternating Current/Voltage
- Different Types of AC Circuits
- Series AC Circuits
- Power in an AC Circuit
- AC Generator
- Transformer

An **alternating current** is the current (or voltage) whose magnitude keeps on changing continuously with time, between zero and a maximum value and its direction also reverses periodically.

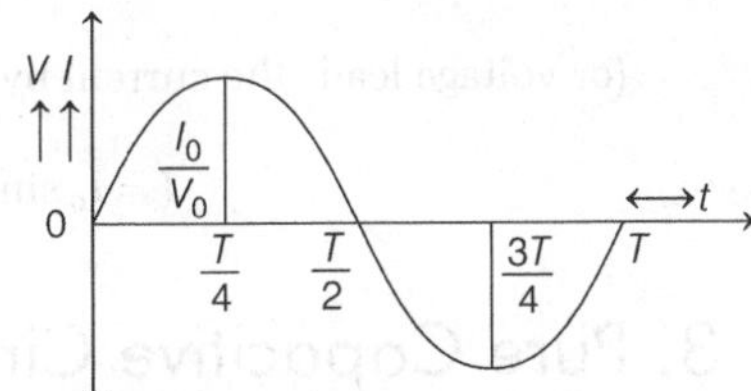

Peak and RMS Values of Alternating Current/ Voltage

RMS value of alternating voltage is equal to $\frac{1}{\sqrt{2}}$ times of peak value. i.e. $V_{rms} = \frac{V_0}{\sqrt{2}}$

Similarly, $I_{rms} = \frac{I_0}{\sqrt{2}}$

Mean Value or Average Value

The steady current, which when passes through a circuit for half the time period of alternating current, sends the same amount of charge as done by the alternating current in the same time through the same circuit, is called mean or average value of alternating current. It is denoted by i_m or i_{av}

$$i_m \text{ or } i_{av} = \frac{2i_0}{\pi} = 0.636 i_0$$

Mean or average value of alternating current during a half cycle is 0.636 times (or 63.6% of) its peak value (i_0).

- No. of Questions in Exercises (x)—
- No. of Questions Attempted (y)—
- No. of Correct Questions (z)—
(Without referring Explanations)

- Accuracy Level ($z/y \times 100$)—
- Prep Level ($z/x \times 100$)—

In order to expect good rank in JEE, your Accuracy Level should be above 85 & Prep Level should be above 75.

Similarly, mean or average value of alternating emf

$$V_m \text{ or } V_{av} = \frac{2V_0}{\pi} = 0.636$$

Peak Value

The maximum value (amplitude) of alternating current and voltage is called peak value.

RMS Value

The steady current, which when passes through a resistance for a given time will produce the same amount of heat as the alternating current does in the same resistance and in the same time, is called rms value of alternating current. It is denoted by i_{rms} or $i_v = \frac{i_0}{\sqrt{2}} = 0.707\, i_0$

where, i_0 = peak value of alternating current

Similarly, rms value of alternating emf

$$V_{rms} = \frac{V_0}{\sqrt{2}} = 0.707\, V_0$$

Reactance and Impedance

- The opposition offered by a pure inductor or capacitor or both to the flow of AC, through it, is called **reactance** (X). Its unit is ohm (Ω) and dimensional formula is $[ML^2T^{-3}A^{-2}]$.
- **Reactance is of two types**
  (i) Inductive reactance, $X_L = L\omega$ and
  (ii) Capacitive reactance, $X_C = \frac{1}{C\omega}$
- Reciprocal of reactance is known as **susceptance**.
  Thus, $S = \frac{1}{X}$
- Total opposition offered by an AC circuit to the flow of current through it, is called its **impedance** (Z). Its unit is ohm and dimensional formula is $[ML^2T^{-3}A^{-2}]$.
  For an AC circuit, $Z = \sqrt{X^2 + R^2} = \sqrt{(X_L - X_C)^2 + R^2}$
- Reciprocal of impedance is known as **admittance**.
  Thus, $Y = \frac{1}{Z}$. Its unit is Siemens (S).

Different Types of AC Circuits

The circuit consists of only resistor or only capacitor or only inductor are called pure resistive, pure inductive and pure capacitive circuit.

1. Pure Resistive Circuit

Let an alternating voltage $V = V_0 \sin \omega t$ be applied across a pure resistance R. Then,

Current, $I = \frac{V}{R}$ or $I_{rms} = \frac{V_{rms}}{R}$

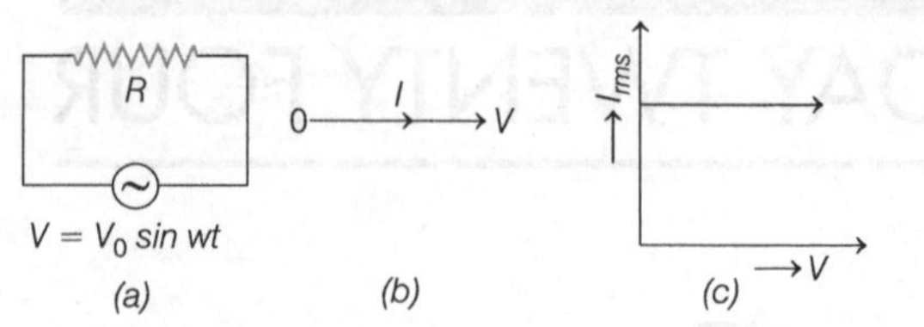

Current and voltage are in the same phase, i.e. current is given by $I = I_0 \sin \omega t$.

2. Pure Inductive Circuit

Let an alternating voltage $V = V_0 \sin \omega t$ be applied across a pure inductance L.

Then, the average power $= V_{rms} I_{rms} \cos \frac{\pi}{2} = 0$

The inductance offers some opposition to the flow of AC, known as **inductive reactance** $X_L = 2\pi\nu L = L\omega$.

Thus, a pure inductance does not oppose the flow of DC ($\omega = 0$) but opposes the flow of AC.

Current flowing, $I = \frac{V}{X_L}$

(a) L, $V = V_0 \sin \omega t$ (b) V, $\frac{\pi}{2}$, 0, I (c) X_L, 0, V

In pure inductive circuit, current decreases with an increase in frequency, it lags behind the voltage by $\frac{\pi}{2}$ (or voltage leads the current by $\frac{\pi}{2}$) and is thus given by

$$I = I_0 \sin\left(\omega t - \frac{\pi}{2}\right)$$

3. Pure Capacitive Circuit

Let an alternating voltage $V = V_0 \sin \omega t$ be applied across a pure capacitance C. Then, the capacitance offers some opposition to the flow of current, but allows AC to pass through it. The opposition offered is known as the **capacitive reactance**.

$$X_C = \frac{1}{C\omega}\,\Omega$$

$$= \frac{1}{C \times 2\pi\nu}\,\Omega$$

Current flowing, $I = \frac{V}{X_C}$

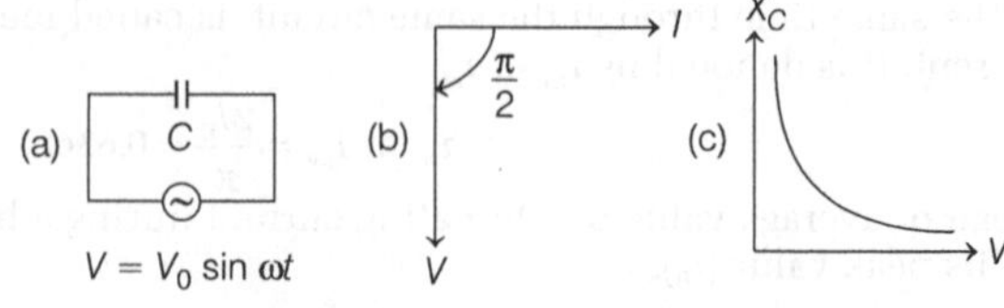

In pure capacitive circuit, current increases with an increase in frequency and leads the voltage by $\frac{\pi}{2}$ (or voltage lags behind the current by $\frac{\pi}{2}$) and is thus, given by

$$I = I_0 \sin\left(\omega t + \frac{\pi}{2}\right)$$

Series AC Circuits

Some of the series AC circuits are given below

1. Series *L-R* Circuit

The potential difference across the resistance in an AC circuit is in phase with current and it leads in phase by 90° with current across the inductor.

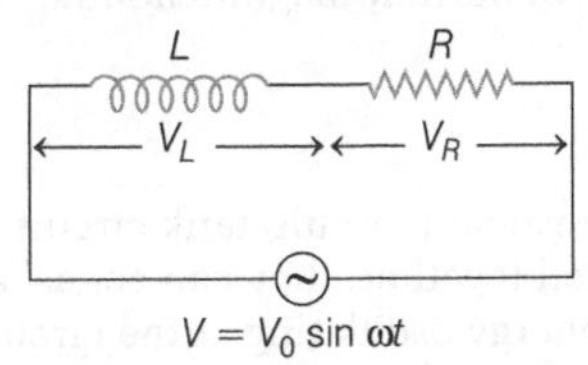

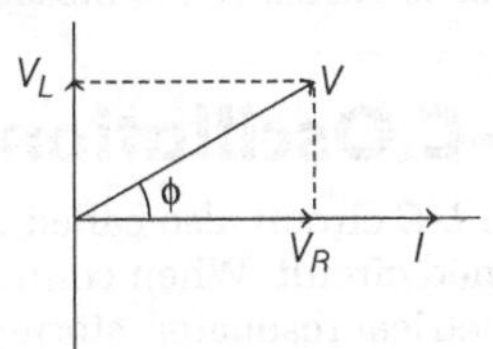

$$V = V_0 \sin \omega t \text{ and } I = \frac{V_0}{Z} \sin(\omega t - \phi)$$

where, $Z = \sqrt{R^2 + (\omega L)^2}$

Current lags behind the voltage by ϕ.

and $\tan \phi = \frac{\omega L}{R}$

$\therefore \quad V = \sqrt{V_R^2 + V_L^2}$

where, V_R = voltage across resistor R

and V_L = voltage across inductor.

2. Series *R-C* Circuit

The potential difference across a resistance in AC circuit is in phase with current and it lags in phase by 90° with the current in the capacitor.

$$V = V_0 \sin \omega t \text{ and } I = \frac{V_0}{Z} \sin (\omega t + \phi)$$

where, $Z = \sqrt{R^2 + \left(\frac{1}{\omega C}\right)^2}$

Current leads the voltage by ϕ.

and $\tan \phi = \frac{-1/\omega C}{R}$

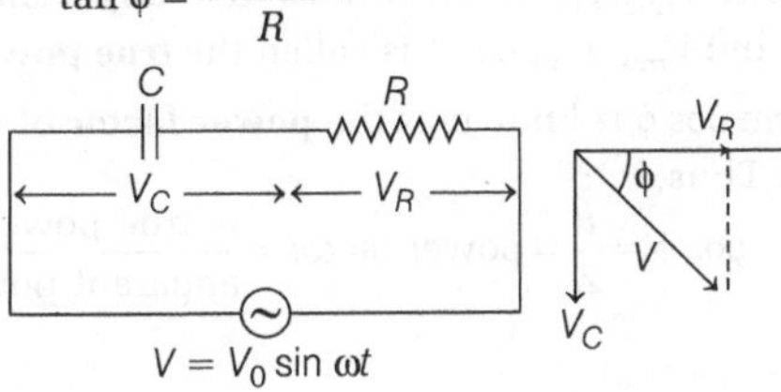

$\therefore \quad V = \sqrt{V_R^2 + V_C^2}$

where, V_R = voltage across resistor R

and V_C = voltage across capacitor.

3. Series *L-C* Circuit

The potential difference across a capacitor in AC lags in phase by 90° and leads in phase by 90° across inductor with the current in the circuit.

$$V = V_0 \sin \omega t,\ I = \frac{V_0}{Z} \sin(\omega t - \phi)$$

where, $Z = X_L - X_C$ and $\tan \phi = \frac{X_L - X_C}{0} = \infty$

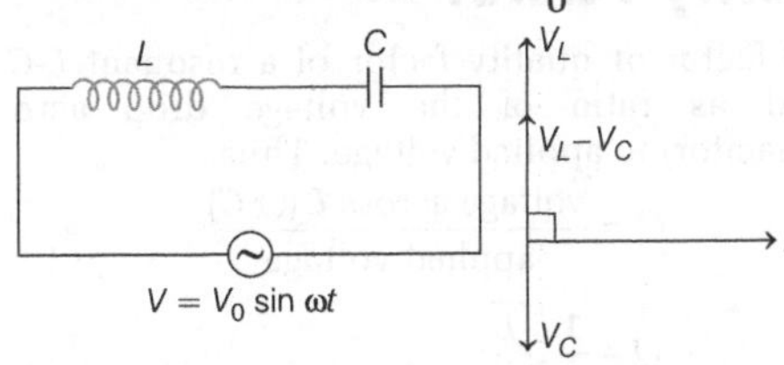

For $X_L > X_C$, $\phi = \frac{\pi}{2}$ and for $X_L < X_C$, $\phi = -\frac{\pi}{2}$

If $X_L = X_C$ i.e. at $\omega = \frac{1}{\sqrt{LC}}$, $Z = 0$ and I_0 becomes infinity.

This condition is termed as the **resonant condition** and this frequency is termed as natural frequency of the circuit.

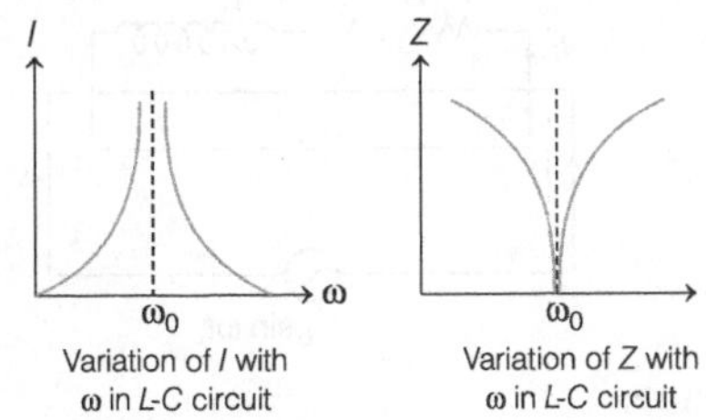

Variation of I with ω in *L-C* circuit
Variation of Z with ω in *L-C* circuit

4. Series *L-C-R* Circuit

For *L-C-R* circuit $V = V_0 \sin \omega t$, $I = \frac{V_0}{Z} \sin (\omega t - \phi)$

where, $Z = \sqrt{R^2 + \left(\omega L - \frac{1}{\omega C}\right)^2}$

and $\tan \phi = \frac{X_L - X_C}{Z}$

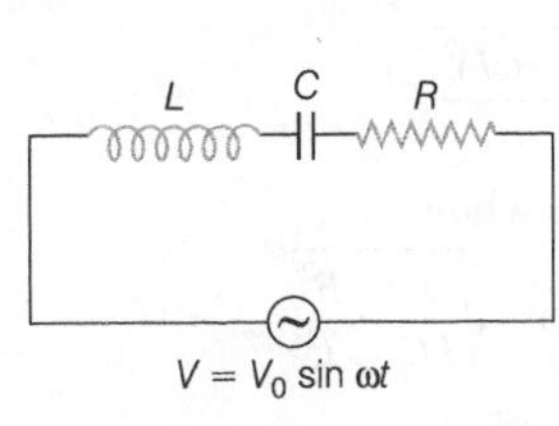

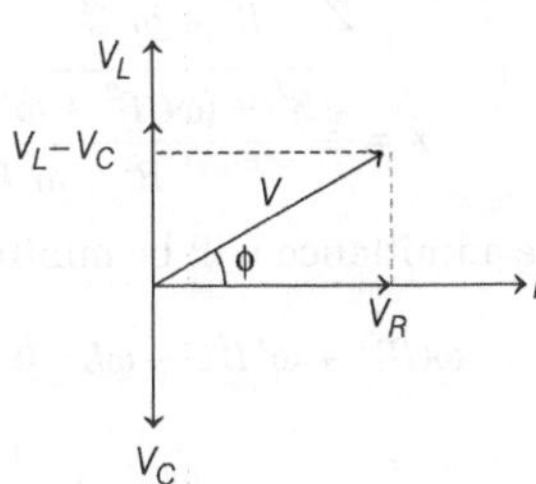

Phasor diagram for *L-C-R* series circuit for $X_L > X_C$

For $X_L > X_C$, current lags voltage.

$X_L < X_C$, current leads voltage.

$X_L = X_C$, current and voltage are in phase.

If $X_L = X_C \Rightarrow \omega_0 = \frac{1}{\sqrt{LC}}$, i.e. the natural frequency of the circuit is equal to the applied frequency, then the circuit is said to be in **resonance.**

At resonance, the current in the circuit is maximum and the impedance is minimum and equal to R.

Resonance frequency, $\nu = \frac{1}{2\pi\sqrt{LC}}$

Quality Factor

The Q-factor or quality factor of a resonant L-C-R circuit is defined as ratio of the voltage drop across inductor (or capacitor) to applied voltage. Thus,

$$Q = \frac{\text{voltage across } L \text{ (or } C)}{\text{applied voltage}}$$

$$Q = \frac{1}{R}\sqrt{\frac{L}{C}}$$

Parallel Resonant Circuit

In this combination, a capacitor is connected in parallel with a series combination of inductor and resistor.

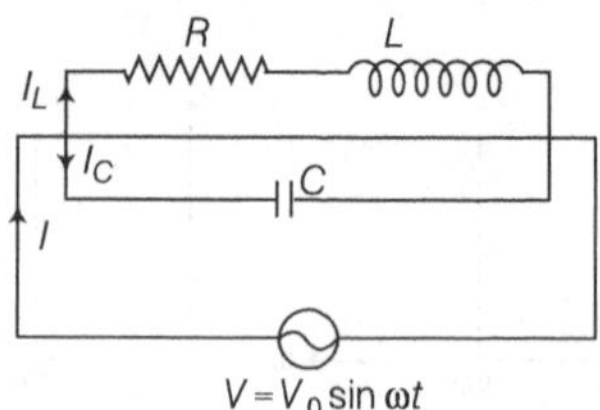

From the figure,

$$I = I_L + I_C$$

or $$\frac{V}{Z} = \frac{V}{R + j\omega L} + \frac{V}{-j/\omega C}$$

$\therefore$ $$\frac{1}{Z} = \frac{1}{R + j\omega L} + j\omega C$$

$\frac{1}{Z}$ is known as admittance (Y).

Thus, $$Y = \frac{1}{Z} = \frac{R - j\omega L}{R^2 + \omega^2 L^2} + j\omega C$$

$\therefore$ $$Y = \frac{\sqrt{R^2 + (\omega CR^2 + \omega^3 L^2 C - \omega L)^2}}{R^2 + \omega^2 L^2}$$

The admittance will be minimum, when

$$\omega CR^2 + \omega^3 L^2 C - \omega L = 0 \text{ or } \omega = \sqrt{\frac{1}{LC} - \frac{R^2}{L^2}}$$

$\therefore$ $$f = \frac{\omega}{2\pi} = \frac{1}{2\pi}\sqrt{\frac{1}{LC} - \frac{R^2}{L^2}}$$

It is known as **resonance frequency.**

At resonance frequency admittance is minimum of the impedance is maximum. Thus, the parallel circuit does not allow this frequency from the source to pass in the circuit. Due to this reason, the circuit with such a frequency is known as **rejector circuit.**

- Dynamic resistance, $Z_{\max} = \frac{1}{Y_{\max}} = \frac{L}{CR}$
- Peak current through the supply $= \frac{V_0}{L/CR} = \frac{V_0 CR}{L}$
- The peak current through capacitor $= \frac{V_0}{1/\omega C} = V_0 \omega C$
- Q-factor $= \frac{V_0 \omega C}{V_0 CR/L} = \frac{\omega L}{R}$

This is basically the measure of current magnification.

L-C Oscillations

An L-C circuit also called a resonant circuit, tank circuit or tuned circuit. When connected together, they can act as an electrical resonator, storing energy oscillating at the circuits resonant frequency.

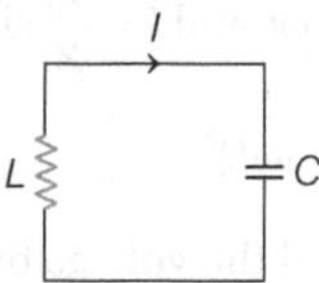

The energy oscillates back and forth between the capacitor and inductor until internal resistance makes the oscillations die out. The oscillation frequency is determined by the capacitance and inductance values,

$$f = \frac{\omega_0}{2\pi} = \frac{1}{2\pi\sqrt{LC}}$$

Power in an AC Circuit

Let a voltage $V = V_0 \sin \omega t$ be applied across an AC and consequently a current $I = I_0 \sin(\omega t - \phi)$ flows through the circuit. Then,

- **Instantaneous power** $= VI = V_0 I_0 \sin \omega t \sin(\omega t - \phi)$ and its value varies with time. Here, ϕ is known as **phase difference** between V and I.
- **Average power** over a full cycle of AC is
  $$P_{av} = V_{rms} I_{rms} \cos \phi = \frac{1}{2} V_0 I_0 \cos \phi$$
  The term $V_{rms} I_{rms}$ is known as the **apparent** or **virtual power**, but $V_{rms}\ I_{rms} \cos \phi$ is called the **true power.**
- The term $\cos \phi$ is known as the **power factor** of the given circuit. Thus,
  $$\cos \phi = \frac{R}{Z} = \text{power factor} = \frac{\text{true power}}{\text{apparent power}}$$

- For a pure resistive circuit, V and I are in phase ($\phi = 0°$), hence $\cos\phi = 1$ and average power $= V_{rms} I_{rms}$

  For a pure inductive or a pure capacitive circuit, current and voltage differ in phase by $\frac{\pi}{2}\left(\text{i.e. } \phi = \frac{\pi}{2}\right)$.

  $\therefore \quad P_{avg} = 0$

- Power loss $= I^2 R = \frac{V^2}{R}$

Wattless Current

Average power is given by $P_{av} = E_{rms} I_{rms} \cos\phi$

The phase difference between E_{rms} and I_{rms} is ϕ. We can resolve I_{rms} into two components

$$I_{rms} \cos\phi \text{ and } I_{rms} \sin\phi$$

Here, the component $I_{rms} \cos\phi$ contributes towards power dissipation and the component $I_{rms} \sin\phi$ does not contribute towards power dissipation. Therefore, it is called **wattless current.**

Choke Coil

A low resistance inductor coil used to suppress or limit the flow of alternating current without affecting the flow of direct current is called **choke coil.**

Let us consider a choke coil (used in tube lights) of large inductance L and low resistance R. The power factor for such a coil is given by

$$\cos\phi = \frac{R}{\sqrt{R^2 + \omega^2 L^2}} \approx \frac{R}{\omega L} \quad [\text{as, } R << \omega L]$$

As $R << \omega L$, $\cos\phi$ is very small. Thus, the power absorbed by the coil $V_{rms} I_{rms} \cos\phi$ is very small. On account of its large impedance $Z = \sqrt{R^2 + \omega^2 L^2}$, the current passing through the coil is very small. Such a coil is used in AC circuits for the purpose of adjusting current to any required value without wastage of energy.

The only loss of energy is due to hysteresis in the iron core, which is much less than the loss of energy in the resistance that can also reduce the current, if placed instead of the choke coil.

AC Generator

An electric generator or dynamo is a device used to produce electrical energy at the expense of mechanical/thermal energy. It works on the principle of electromagnetic induction, when a coil is rotated in a uniform magnetic field, an induced emf is set up between its ends. The induced emf is given by

$$e = e_0 \sin\omega t = NBA\omega \sin\omega t.$$

The direction of the induced emf is alternating in nature.

Transformer

It is a device which works in AC circuits only and is based on the principle of mutual induction.

Transformer is used to suitably increase or decrease the voltage in an AC circuit. Transformer which transforms strong AC at low voltage into a weaker current at high alternating voltage is called a step-up transformer. A step-down transformer transforms weak current at a higher alternating voltage into a strong current at a lower alternating voltage.

For an **ideal transformer** $\frac{e_s}{e_p} = \frac{V_s}{V_p} = \frac{N_s}{N_p} = \frac{I_p}{I_s} = k$

where, k is known as the transformation ratio.

For a step-up transformer, $k > 1$ but for a step-down transformer, $k < 1$.

In a transformer, the input emf and the output emf differ in phase by π radians.

The **efficiency of a transformer** is given by

$$\eta = \frac{\text{output power}}{\text{input power}} = \frac{V_s I_s}{V_p I_p}$$

For an ideal transformer, $\eta = 100\%$ or 1. However, for practical transformers, $\eta \approx 85 - 90\%$.

Possible causes of energy loss in transformer are

- Heating due to winding resistance
- Eddy current losses
- Magnetic flux leakage and
- Hysteresis loss. To minimise these losses, the transformer core is made up of a laminated soft iron strips.

DAY PRACTICE SESSION 1

FOUNDATION QUESTIONS EXERCISE

1 The alternating current in a circuit is described by the graph as shown in figure. The rms current obtained from the graph would be

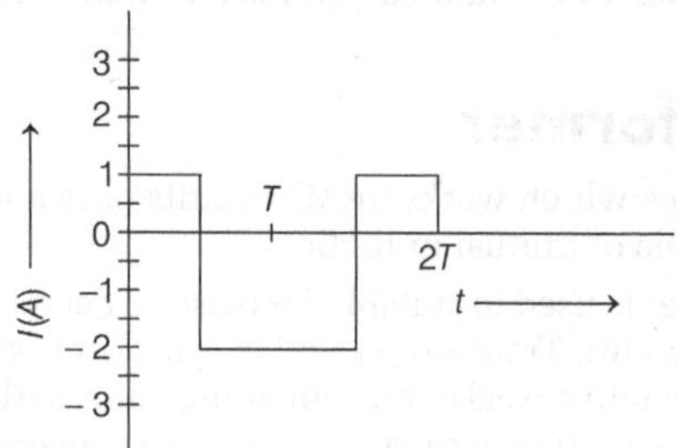

(a) 1.414 A (b) 2.2 A
(c) 1.9 A (d) 2.6 A

2 An alternating voltage (in volts) given by

$$V = 200\sqrt{2}\,\sin(100\,t)$$

is connected to a 1μF capacitor through an AC ammeter. The reading of the ammeter will be

(a) 10 mA (b) 20 mA
(c) 40 mA (d) 80 mA

3 The figure shows an experimental plot discharging of a capacitor in an *R*-*C* circuit. The time constant τ of this circuit lies between → AIEEE 2012

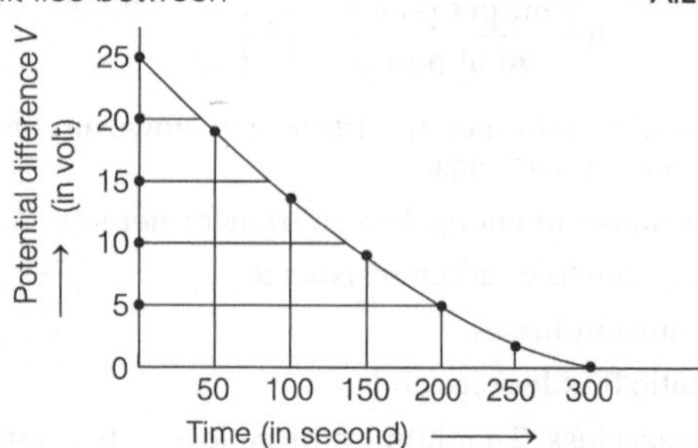

(a) 150 s and 200 s (b) 0 and 50 s
(c) 50 s and 100 s (d) 100 s and 150 s

4 In the circuit shown below, the key *K* is closed at $t = 0$. The current through the battery is → AIEEE 2010

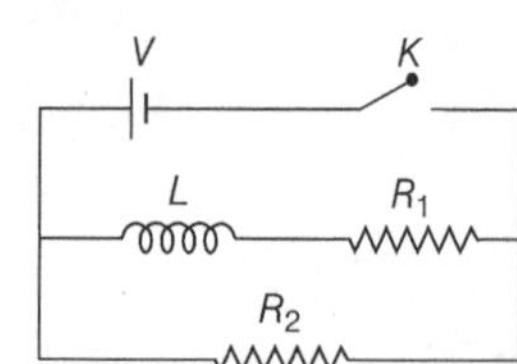

(a) $\dfrac{VR_1R_2}{\sqrt{R_1^2 + R_2^2}}$ at $t = 0$ and $\dfrac{V}{R_2}$ at $t = \infty$

(b) $\dfrac{V}{R_2}$ at $t = 0$ and $\dfrac{V(R_1 + R_2)}{R_1R_2}$ at $t = \infty$

(c) $\dfrac{V}{R_2}$ at $t = 0$ and $\dfrac{VR_1R_2}{\sqrt{R_1^2 + R_2^2}}$ at $t = \infty$

(d) $\dfrac{V(R_1 + R_2)}{R_1R_2}$ at $t = 0$ and $\dfrac{V}{R_2}$ at $t = \infty$

5 An *L*-*C* circuit is in a state of resonance. If $C = 0.1\mu F$ and $L = 0.25$ H, then neglecting the ohmic resistance of the circuit, find the frequency of oscillations.

(a) 1007 Hz (b) 100 Hz (c) 109 Hz (d) 500 Hz

6 An *L*-*C*-*R* circuit as shown in the figure is connected to a voltage source V_{AC} whose frequency can be varied. The frequency, at which the voltage across the resistor is maximum, is → JEE Main (Online) 2013

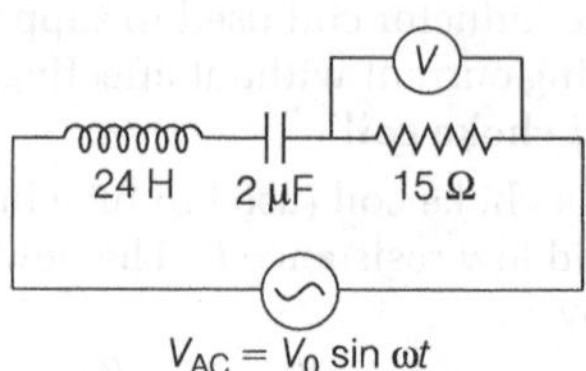

$V_{AC} = V_0 \sin \omega t$

(a) 902 Hz (b) 143 Hz (c) 23 Hz (d) 345 Hz

7 In the circuit shown here, the voltage across *L* and *C* are respectively, 300 V and 400 V. The voltage *E* of the AC source is → JEE Main Online 2013

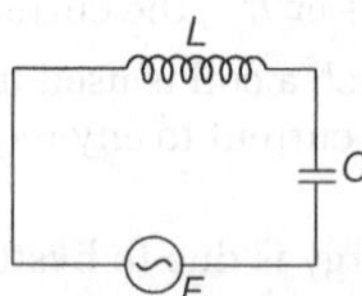

(a) 400 V (b) 500 V (c) 100 V (d) 700 V

8 A fully charged capacitor *C* with initial charge q_0 is connected to a coil of self-inductance *L* at $t = 0$. The time at which the energy is stored equally between the electric and the magnetic fields is → AIEEE 2011

(a) $\dfrac{\pi}{4}\sqrt{LC}$ (b) $2\pi\sqrt{LC}$ (c) $\sqrt{LC}$ (d) $\pi\sqrt{LC}$

9 In an *L*-*C*-*R* circuit, if *V* is the effective value of the applied voltage, V_R is the voltage across *R*, V_L is the effective voltage across *L*, V_C is the effective voltage across *C*, then

(a) $V = V_R + V_L + V_C$ (b) $V^2 = V_R^2 + V_L^2 + V_C^2$
(c) $V^2 = V_R^2 + (V_L - V_C)^2$ (d) $V^2 = V_L^2 + (V_R - V_C)^2$

10 An AC circuit of variable frequency f is connected to an L-C-R series circuit. Which one of the graphs in the figure, represents the variation of current I in the circuit with frequency f?

(a) 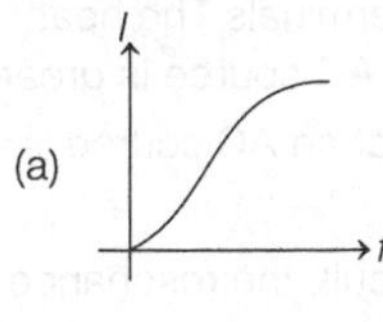
(b) 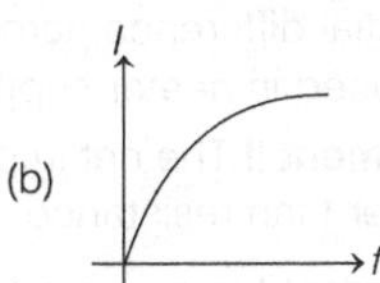
(c) 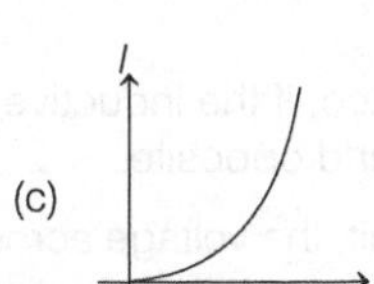
(d) 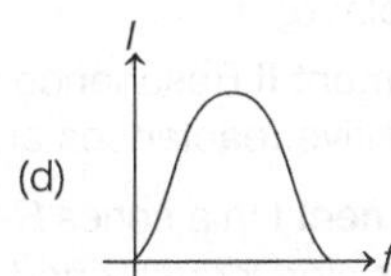

11 In a series L-C-R circuit, $C = 10^{-11}$ F, $L = 10^{-5}$ H and $R = 100\ \Omega$, when a constant DC voltage E is applied to the circuit, the capacitor acquires a charge 10^{-9} C. The DC source is replaced by a sinusoidal voltage source in which the peak voltage E_0 is equal to the constant DC voltage E. At resonance, the peak value of the charge acquired by the capacitor will be → **JEE Main (Online) 2013**

(a) 10^{-15} C (b) 10^{-6} C
(c) 10^{-10} C (d) 10^{-8} C

12 In a L-C-R circuit as shown below, both switches are open initially. Now, switch S_1 and S_2, kept open (q is charge on the capacitor and $\tau = RC$ is capacitance time constant). Which of the following statement is correct? → **JEE Main 2013**

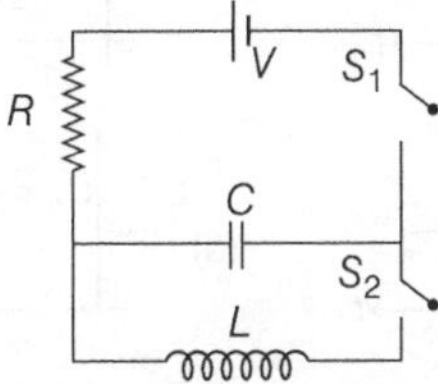

(a) Work done by the battery is half of the energy dissipated in the resistor
(b) At $t = \tau$, $q = \dfrac{CV}{2}$
(c) At $t = 2\tau$, $q = CV(1 - e^{-2})$
(d) At $t = \dfrac{\tau}{2}$, $q = CV(1 - e^{-1})$

13 In a series resonant L-C-R circuit, the voltage across R is 100 V and $R = 1\ \text{k}\Omega$ with $C = 2\ \mu$F. The resonant frequency ω is 200 rad/s. At resonance, the voltage across L is

(a) 2.5×10^{-2} V (b) 40 V
(c) 250 V (d) 4×10^{-3} V

14 The plot given below is of the average power delivered to an L-R-C circuit *versus* frequency. The quality factor of the circuit is → **JEE Main (Online) 2013**

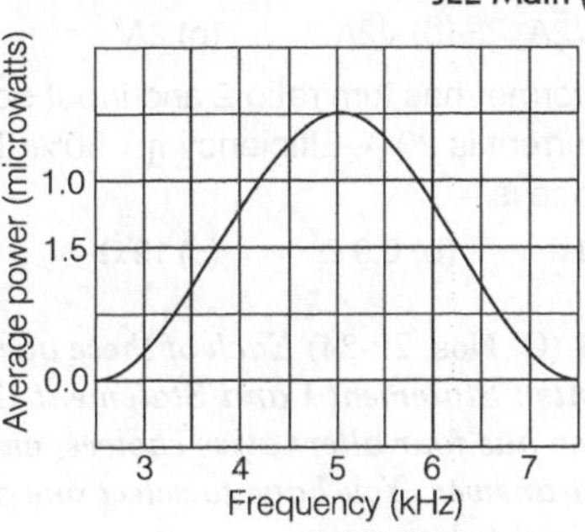

(a) 5.0 (b) 2.0
(c) 2.5 (d) 0.4

15 For an R-L-C circuit driven with voltage of amplitude v_m and frequency $\omega_0 = \dfrac{1}{\sqrt{LC}}$, the current exhibits resonance. The quality factor Q is given by → **JEE Main 2018**

(a) $\dfrac{\omega_0 L}{R}$ (b) $\dfrac{\omega_0 R}{L}$
(c) $\dfrac{R}{\omega_0 C}$ (d) $\dfrac{CR}{\omega_0}$

16 In a series L-C-R circuit, $R = 200\ \Omega$ and the voltage; and the frequency of the main supply is 220 V and 50 Hz, respectively. On taking out the capacitance from the circuit, the current lags behind the voltage by 30°. On taking out the inductor from the circuit, the current leads the voltage by 30°. The power dissipated in the L-C-R circuit is → **AIEEE 2010**

(a) 305 W (b) 210 W (c) zero (d) 242 W

17 In an AC circuit, the voltage applied is $E = E_0 \sin \omega t$. The resulting current in the circuit is $I = I_0 \sin\left(\omega t - \dfrac{\pi}{2}\right)$. The power consumption in the circuit is given by

(a) $P = \dfrac{E_0 I_0}{\sqrt{2}}$ (b) $P =$ zero
(c) $P = \dfrac{E_0 I_0}{2}$ (d) $P = \sqrt{2} E_0 I_0$

18 Which of the following components of an L-C-R circuit, with an AC supply, dissipates energy?

(a) L (b) R (c) C (d) All of these

19 An AC circuit consists of a 220 Ω resistance and a 0.7 H choke. The power absorbed from 220 V and 50 Hz source connected in this circuit, if the resistance and choke are joined in series is

(a) 110 W (b) 50 W (c) 220 W (d) 440 W

20 The output of a step-down transformer is measured to be 24 V when connected to a 12 W light bulb. The value of the peak current is

(a) $1/\sqrt{2}$A (b) $\sqrt{2}$A (c) 2A (d) $2\sqrt{2}$A

21 A transformer has turn ratio 2 and input power 3600 W. Load current is 20 A. Efficiency $\eta = 90\%$. The internal resistance is

(a) 1 Ω (b) 0.9 Ω (c) 19 Ω (d) 3 Ω

Direction (Q. Nos. 22-24) *Each of these questions contains two statements : Statement* I *and Statement* II. *Each of these questions also has four alternative choices, only one of which is the correct answer. You have to select one of the codes (a), (b), (c), (d) given below.*

(a) Statement I is true, Statement II is true; Statement II is the correct explanation for Statement I
(b) Statement I is true, Statement II is true; Statement II is not the correct explanation for Statement I
(c) Statement I is true; Statement II is false
(d) Statement I is false; Statement II is true

22 Statement I Two identical heaters are connected to two different sources one DC and other AC having same potential difference across their terminals.The heat produced in heater supplied with AC source is greater.

Statement II The net impedance of an AC source is greater than resistance.

23 Statement I In a series *L*- *C*-*R* circuit, the resonance can take place.

Statement II Resonance takes place, if the inductive and capacitive reactances are equal and opposite.

24 Statement I In a series *R*-*L*-*C* circuit, the voltage across the resistor, inductor and capacitor are 8V, 16V and 10V, respectively. The resultant emf in the circuit is 10V.

Statement II Resultant emf of the circuit is given by the relation $E = \sqrt{V_R^2 + (V_L - V_C)^2}$.

DAY PRACTICE SESSION 2

PROGRESSIVE QUESTIONS EXERCISE

1 In an AC circuit, the instantaneous emf and current are given by

$$e = 100 \sin 30\,t,\ i = 20 \sin\left(30\,t - \frac{\pi}{4}\right)$$

In one cycle of AC, the average power consumed by the circuit and the wattless current are respectively, → JEE Main 2018

(a) 50 W, 10 A (b) $\frac{1000}{\sqrt{2}}$ W, 10 A
(c) $\frac{50}{\sqrt{2}}$ W, 0 A (d) 50 W, 0 A

2 An arc lamp requires a direct current of 10 A at 80 V to function. If it is connected to a 220 V (rms), 50 Hz AC supply, the series inductor needed for it to work is close to → JEE Main 2016

(a) 80 H (b) 0.08 H (c) 0.044 H (d) 0.065 H

3 An *L*-*C*-*R* circuit is equivalent to a damped pendulum. In an *L*-*C*-*R* circuit, the capacitor is charged to Q_0 and then connected to the *L* and *R* as shown below

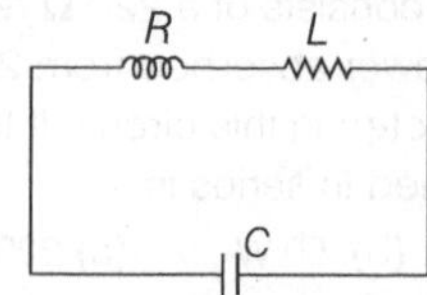

If a student plots graphs of the square of maximum charge (Q_{max}^2) on the capacitor with time (t) for two different values L_1 and L_2 ($L_1 > L_2$) of *L*, then which of the following represents this graph correctly? (Plots are schematic and not drawn to scale) → JEE Main 2015

(a) 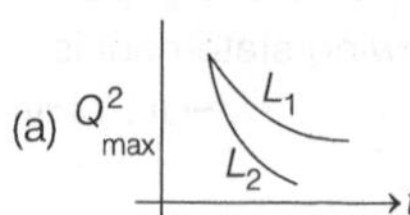

(c) 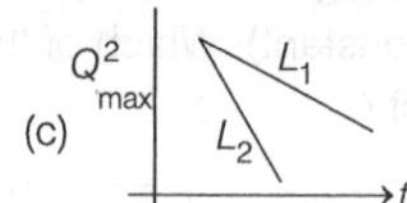

(b) Q^2_{max}, L_2, L_1, t

(d) 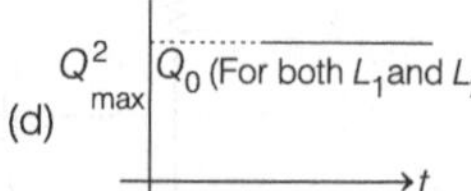

4 A resistor *R* and 2 μF capacitor in series is connected through a switch to 200 V direct supply. Across the capacitor is a neon bulb that lights up at 120 V. Calculate the value of *R* to make the bulb light up 5 s after the switch has been closed (Take, $\log_{10} 2.5 = 0.4$) → AIEEE 2011

(a) $1.7 \times 10^5\ \Omega$ (b) $2.7 \times 10^6\ \Omega$
(c) $3.3 \times 10^7\ \Omega$ (d) $1.3 \times 10^4\ \Omega$

5 Let *C* be the capacitance of a capacitor discharging through a resistor *R*. Suppose t_1 is the time taken for the energy stored in the capacitor to reduce to half its initial value and t_2 is the time taken for the charge to reduce to one-fourth its initial value. Then, the ratio $\frac{t_1}{t_2}$ will be → AIEEE 2010

(a) 1 (b) $\frac{1}{2}$ (c) $\frac{1}{4}$ (d) 2

6 The average current in terms of I_0 for the waveform as shown is

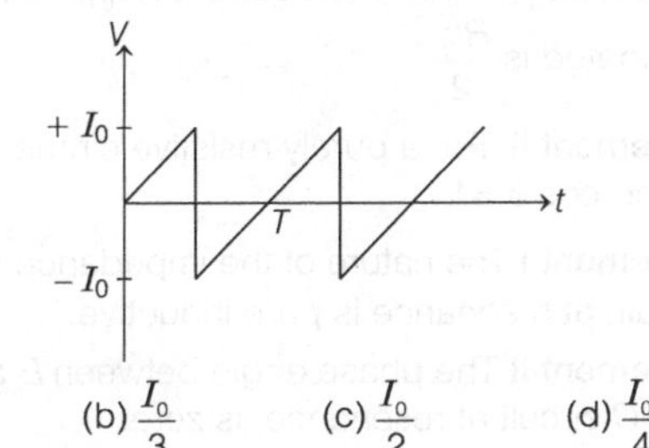

(a) I_0 (b) $\frac{I_0}{3}$ (c) $\frac{I_0}{2}$ (d) $\frac{I_0}{4}$

7 A bulb is rated 55 W/110 V. It is to be connected to a 220 V/50 Hz with inductor in series. The value of inductance, so that bulb gets correct voltage is

(a) 200 Ω (b) 110 Ω (c) 50 Ω (d) 220 Ω

8 A coil of 0.01 H inductance and 1 Ω resistance is connected to 200 V, 50 Hz AC supply. The impedance of the circuit and time lag between maximum alternating voltage and current would be

(a) 3.3 Ω and $\frac{1}{250}$ s (b) 3.9 Ω and $\frac{1}{160}$ s
(c) 4.2 Ω and $\frac{1}{100}$ s (d) 2.8 Ω and $\frac{1}{120}$ s

9 The bandwidth in a series *L-C-R* circuit is

(a) $\frac{LC}{2\sqrt{R^2C^2 + 4LC}}$ (b) $\frac{2LC}{\sqrt{R^2C^2 + 4LC}}$
(c) $\frac{\sqrt{R^2C^2 + 4LC}}{LC}$ (d) zero

10 An *L-C-R* circuit, consists of an inductor, a capacitor and a resistor driven by a battery and connected by two switches S_1 and S_2, as shown in the figure.

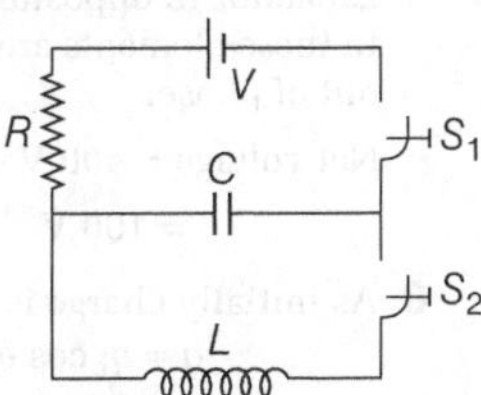

At time $t = 0$, when the charge on the capacitor plates is q, switch S_1 is opened and S_2 is closed. The maximum charge the capacitor can hold, is q_0. Choose the correct equation

(a) $q = q_0 \cos\left(\frac{t}{\sqrt{LC}} + \frac{\pi}{2}\right)$ (b) $q = q_0 \cos\left(\frac{t}{\sqrt{LC}} - \frac{\pi}{2}\right)$
(c) $q = -LC\frac{d^2q}{dt^2}$ (d) $q = -\frac{1}{\sqrt{LC}}\frac{d^2q}{dt^2}$

11 An alternating emf of angular frequency ω is applied across an inductance. The instantaneous power developed in the circuit has an angular frequecy

(a) ω/4 (b) ω/2 (c) ω (d) 2ω

12 An inductor of reactance 1 Ω and a resistor of 2Ω are connected in series to the terminal of a 6V(rms) AC source.The power dissipated in the circuit is

(a) 8 W (b)12 W (c)14.4 W (d)18 W

13 You are given many resistances, capacitors and inductors. These are connected to a variable DC voltage source (the first two circuits) or an AC voltage source of 50Hz frequency (the next three circuits)in different ways as shown in Column II. When a current *I* (steady state for DC or rms for AC) flows through the circuit, the corresponding voltage V_1 and V_2 (indicated in circuits) are related as shown in Column I. Match the Column I with Column II and mark the correct option from the codes given below.

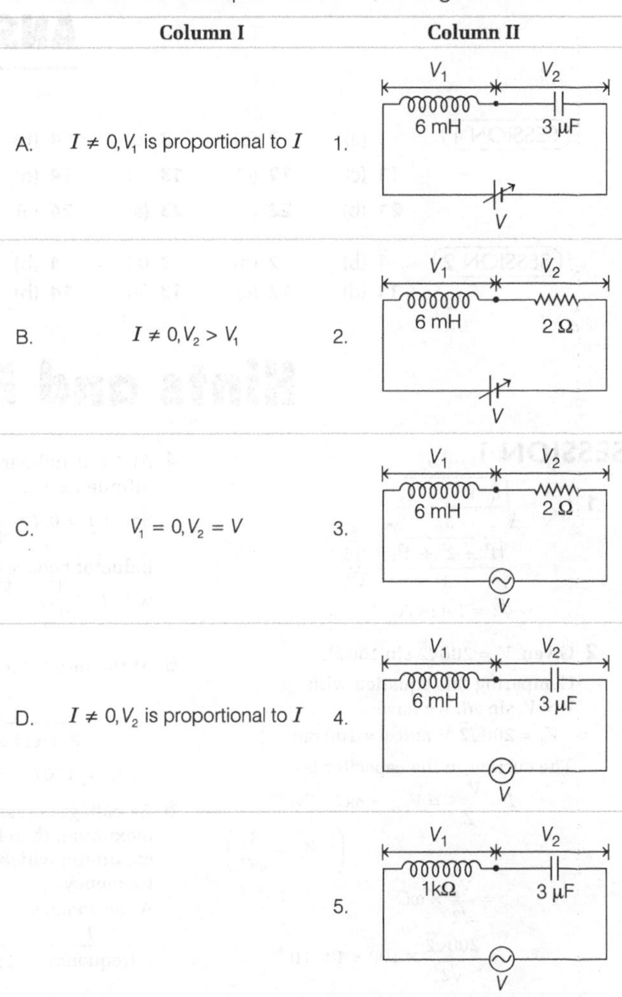

	Column I		Column II
A.	$I \neq 0, V_1$ is proportional to I	1.	V_1, V_2; 6 mH, 3 μF; V
B.	$I \neq 0, V_2 > V_1$	2.	V_1, V_2; 6 mH, 2 Ω; V
C.	$V_1 = 0, V_2 = V$	3.	V_1, V_2; 6 mH, 2 Ω; V
D.	$I \neq 0, V_2$ is proportional to I	4.	V_1, V_2; 6 mH, 3 μF; V
		5.	V_1, V_2; 1kΩ, 3 μF; V

Codes

	A	B	C	D
(a)	(2,3,4,5)	(1,2)	(2,3,4,5)	(3,4,5)
(b)	(3,4,5)	(2,3,4,5)	(1,2)	(2,3,4,5)
(c)	(1,2)	(3,4,5)	(2,3,4,5)	(2,3,4,5)
(d)	(3,4,5)	(1,2)	(2,3,4,5)	(2,3,4,5)

Direction (Q. Nos. 14 *and* 15) *Each of these questions contains two statements : Statement* I *and Statement* II. *Each of these questions also has four alternative choices, only one of which is the correct answer. You have to select one of the codes (a), (b), (c), (d) given below.*

(a) Statement I is true, Statement II is true; Statement II is the correct explanation for Statement I
(b) Statement I is true, Statement II is true; Statement II is not the correct explanation for Statement I
(c) Statement I is true; Statement II is false
(d) Statement I is false; Statement II is true

14 **Statement I** A sinusoidal AC current flows through a resistance R. If the peak current is I_0, then the power dissipated is $\frac{RI_0^2}{2}$.

Statement II For a purely resistive circuit, the power factor, $\cos\phi = 1$.

15 **Statement I** The nature of the impedance of L-C-R circuit, at resonance is pure inductive.

Statement II The phase angle between E and I in a R-L-C circuit at resonance, is zero.

ANSWERS

SESSION 1	1 (a)	2 (b)	3 (d)	4 (b)	5 (a)	6 (c)	7 (c)	8 (a)	9 (c)	10 (d)
	11 (c)	12 (c)	13 (c)	14 (d)	15 (a)	16 (d)	17 (b)	18 (b)	19 (a)	20 (a)
	21 (b)	22 (a)	23 (a)	24 (a)						

SESSION 2	1 (b)	2 (d)	3 (a)	4 (b)	5 (c)	6 (c)	7 (d)	8 (a)	9 (c)	10 (c)
	11 (d)	12 (c)	13 (a)	14 (b)	15 (d)					

Hints and Explanations

SESSION 1

1 $I_{rms} = \sqrt{\frac{I_1^2 + I_2^2 + I_3^2}{3}}$

$= \sqrt{\frac{1^2 + 2^2 + 1^2}{3}} = \sqrt{\frac{6}{3}}$

$= \sqrt{2} = 1.414$ A

2 Given, $V = 200\sqrt{2}\sin(100t)$.

Comparing this equation with $V = V_0 \sin\omega t$, we have

$V_0 = 200\sqrt{2}$ V and $\omega = 100$ rad s^{-1}

The current in the capacitor is

$I = \frac{V_{rms}}{Z_C} = V_{rms} \times \omega C$ $\left(\because Z_C = \frac{1}{\omega C}\right)$

$= \frac{V_0}{\sqrt{2}} \times \omega C$

$= \frac{200\sqrt{2}}{\sqrt{2}} \times 100 \times 1 \times 10^{-6}$

$= 20 \times 10^{-3}$ A $= 20$ mA

3 Time constant τ is the duration when the value of potential drops by 63% of its initial maximum value (i.e. V_0/e). Here, 37% of 25 V = 9.25 V which lies between 100 s to 150 s in the graph.

4 At $t = 0$, inductor behaves like an infinite resistance.

So, at $t = 0$, $I = \frac{V}{R_2}$ and at $t = \infty$, inductor behaves like a conducting wire $I = \frac{V}{R_{eq}} = \frac{V(R_1 + R_2)}{R_1 R_2}$

5 At the resonance, $\nu = \frac{1}{2\pi\sqrt{LC}}$

$= \frac{1}{2 \times 3.14 \times \sqrt{0.25 \times 0.1 \times 10^{-6}}}$

$= 1007$ Hz

6 As voltage across resistance is maximum, therefore a power is maximum which is at the resonance frequency.

At resonance,

$\frac{1}{\text{frequency}} = \frac{1}{2\pi\sqrt{LC}}$

$= \frac{1}{2\pi} \frac{1}{\sqrt{24 \times 2 \times 10^{-6}}}$

$= \frac{1}{2\pi} \frac{1000}{\sqrt{48}}$

$= \frac{1}{2\pi} \frac{1000}{6.9V}$

$= \frac{1000}{2 \times 3.14 \times 6.92}$

$= \frac{1000}{43.45} = 23$ Hz

7 Since, reactances produced by inductor and capacitor in opposite direction. So, voltage in these elements are distributed at 180°, i.e. out of phase.

Net voltage = 400 V − 300 V

= 100 V

8 As initially charge is maximum,

$q = q_0 \cos\omega t$

$\Rightarrow I = \frac{dq}{dt} = -\omega q_0 \sin\omega t$

Given, $\frac{1}{2} L I^2 = \frac{q^2}{2C}$

$\Rightarrow \frac{1}{2} L (\omega q_0 \sin\omega t)^2 = \frac{(q_0 \cos\omega t)^2}{2C}$

But $\omega = \frac{1}{\sqrt{LC}}$

$\Rightarrow \tan\omega t = 1$

$\omega t = \frac{\pi}{4}$

$\Rightarrow t = \frac{\pi}{4\omega} = \frac{\pi}{4}\sqrt{LC}$

9 Phaser diagram of *L-C-R* series circuit is shown in figure.

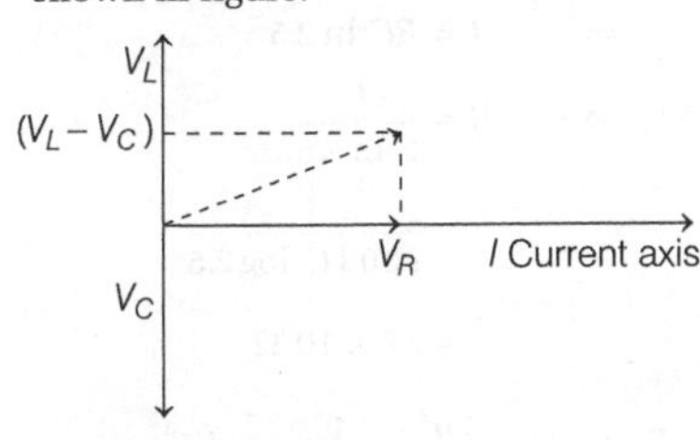

$$V^2 = V_R^2 + (V_L - V_C)^2$$

10 The current in an *L-C-R* circuit is given by

$$I = \frac{V}{\left[R^2 + \left(\omega L - \frac{1}{\omega C}\right)^2\right]^{1/2}}$$

where, $\omega = 2\pi f$

Thus, I increases with an increase in ω upto a value given by $\omega = \omega_c$, i.e. at $\omega = \omega_c$, we have

$$\omega L = \frac{1}{\omega C}$$

$$\Rightarrow \quad \omega_c = \frac{1}{\sqrt{LC}}$$

where, I is maximum.

At $\omega > \omega_c$, I again starts decreasing with an increase in ω.

11 As energy stored in capacitor $= \frac{1}{2}\frac{q^2}{C}$

Now, when AC is connected to the circuit energy speed $= \frac{1}{2}LI^2$

By equating the energies, we get

$$\frac{1}{2}\frac{q^2}{C} = \frac{1}{2}LI^2$$

$$\frac{(10^{-9})^2}{10^{-11}} = \frac{1}{2} \times 10^{-5} I^2$$

$$\Rightarrow \quad I = \frac{1}{10}\text{ A}$$

Now, $V = IR$

$$= \frac{1}{10} \times 100 = 10\text{ V}$$

Therefore,

$$Q = CV$$
$$= 10^{-11} \times 10 = 10^{-10}\text{ C}$$

12 For charging of capacitor,

$q = CV(1 - e^{-t/\tau})$ at $t = 2\tau$,

$q = CV(1 - e^{-2})$

13 At resonance, $\omega L = 1/\omega C$

Current flowing through the circuit,

$$I = \frac{V_R}{R} = \frac{100}{1000} = 0.1\text{ A}$$

So, voltage across L is given by

$$V_L = I X_L = I\omega L$$

But $\omega L = 1/\omega C$

$$\therefore \quad V_L = \frac{I}{\omega C} = V_C$$

$$= \frac{0.1}{200 \times 2 \times 10^{-6}} = 250\text{ V}$$

14 As quality factor, $Q = \frac{\omega_0}{B}$

where, ω_0 = resonant frequency

and B = bandwidth.

From the graph, $B = 2.5$ kHz,

$Q = 0.4$

(by observing the curve)

15 Sharpness of resonance of a resonant *L-C-R* circuit is determined by the ratio of resonant frequency with the selectivity of circuit. This ratio is also called "quality factor" or *Q*-factor.

$$Q\text{-factor} = \frac{\omega_0}{2\Delta\omega} = \frac{\omega_0 L}{R}$$

16 The given circuit is under resonance as $X_L = X_C$

Hence, power dissipated in the circuit is

$$P = \frac{V^2}{R} = 242\text{ W}$$

17 For given circuit, current is lagging the voltage by $\frac{\pi}{2}$, so circuit is purely inductive and there is no power consumption in the circuit. The work done by battery is stored as magnetic energy in the inductor.

18 In an AC circuit, only resistor R dissipates energy. L and C do not dissipate energy, because for both of them current is wattless ($\phi = 90°$).

19 In series impedance of circuit is

$$Z = \sqrt{R^2 + \omega^2 L^2} = \sqrt{R^2 + (2\pi f L)^2}$$
$$= \sqrt{(220)^2 + (2 \times 3.14 \times 50 \times 0.7)^2}$$
$$= 311\,\Omega$$

$$\therefore \quad I_{\text{rms}} = \frac{V_{\text{rms}}}{Z}$$

$$= \frac{220}{311} = 0.707\text{ A} = 0.707\text{ A}$$

and $\cos\phi = \frac{R}{Z} = \frac{220}{311} = 0.707$

Now, power absorbed in the circuit is

$$P = V_{\text{rms}} I_{\text{rms}} \cos\phi$$
$$= (220)(0.707)(0.707)$$
$$= 109.96 \approx 110\text{ W}$$

20 Secondary voltage, $V_S = 24$ V

Power associated with secondary

$P_S = 12$ W

$$\therefore \quad I_S = \frac{P_S}{V_S} = \frac{12}{24} = \frac{1}{2}\text{ A} = 0.5\text{ A}$$

Peak value of the current in the secondary,

$$I_0 = I_S\sqrt{2}$$
$$= (0.5)(1.414)$$
$$= 0.707 = \frac{1}{\sqrt{2}}\text{ A}$$

21 As, $V_2 = \frac{3600}{20} = 180$ V

$\therefore \quad V_0 = V_2 \times \eta = 180 \times 0.9 = 162$ V

Now, $V_0 = V_2 - I_2 r$

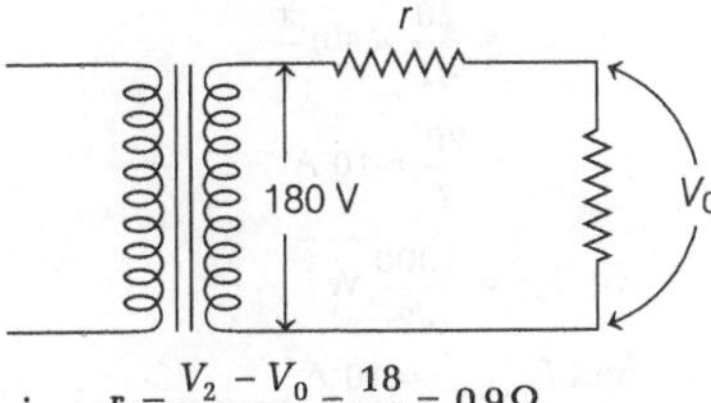

$$\therefore \quad r = \frac{V_2 - V_0}{I} = \frac{18}{20} = 0.9\,\Omega$$

22 For the case of DC, the frequency is zero and the net impedance is equal to the resistance.For the case of AC, the impedance of the AC circuit is given by

$Z = \sqrt{R^2 + \omega^2 L^2}$

where, R = resistance,

ω = angular frequency

and L = inductance.

23 At a particular value of angular frequency, the inductive reactance and capacitive reactance will becomes just equal to each other and opposite in value. So that, the impedance of circuit is minimum, i.e. equal to R.

$$X_L = X_C$$

$$\Rightarrow \quad \omega_0 L = \frac{1}{\omega_0 C}$$

$$\Rightarrow \quad \omega = \frac{1}{\sqrt{LC}}$$

24 In a series *R-L-C* circuit, the resultant emf is

$$E = \sqrt{V_R^2 + (V_L - V_C)^2}$$
$$= \sqrt{8^2 + (16 - 10)^2}$$
$$= \sqrt{64 + 36} = \sqrt{100}$$
$$= 10 \text{ V}$$

SESSION 2

1 Given, $e = 100\sin 30t$

and $i = 20 \sin\left(30t - \frac{\pi}{4}\right)$

$\therefore$ Average power,

$$P_{av} = e_{rms} I_{rms} \cos\phi$$
$$= \frac{100}{\sqrt{2}} \times \frac{20}{\sqrt{2}} \times \cos\frac{\pi}{4}$$
$$= \frac{1000}{\sqrt{2}} \text{ W}$$

Wattless current is

$$I = I_{rms} \sin\phi$$
$$= \frac{20}{\sqrt{2}} \times \sin\frac{\pi}{4}$$
$$= \frac{20}{2} = 10 \text{ A}$$

$\therefore$ $P_{av} = \frac{1000}{\sqrt{2}}$ W

and $I_{wattless} = 10$ A

2 Given, $I = 10$ A, $V = 80$ V

$\therefore$ $R = \frac{V}{I} = \frac{80}{10} = 8\,\Omega$

and $\omega = 50$ Hz

For AC circuit, we have

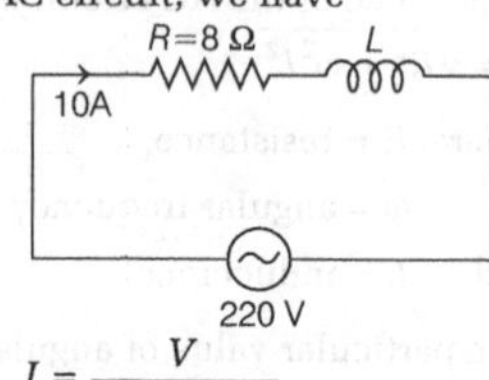

$$I = \frac{V}{\sqrt{8^2 + X_L^2}}$$

$\Rightarrow$ $10 = \frac{220}{\sqrt{64 + X_L^2}}$

$\Rightarrow$ $\sqrt{64 + X_L^2} = 22$

Squaring on both sides, we get

$$64 + X_L^2 = 484$$

$\Rightarrow$ $X_L^2 = 484 - 64 = 420$

$$X_L = \sqrt{420}$$

$\Rightarrow$ $2\pi \times \omega L = \sqrt{420}$

Series inductor on an arc lamp,

$$L = \frac{\sqrt{420}}{(2\pi \times 50)} = 0.065\text{H}$$

3 Consider the *L-C-R* circuit at any time *t*

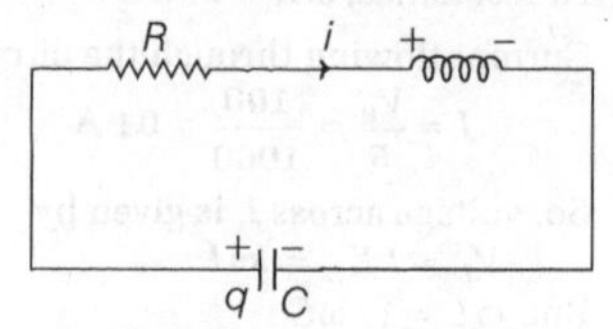

Now, applying KVL, we have

$$\frac{q}{C} - iR - \frac{Ldi}{dt} = 0$$

As current is decreasing with time we can write $i = -\frac{dq}{dt}$

$\Rightarrow$ $\frac{q}{C} + \frac{dq}{dt} R + \frac{Ld^2q}{dt^2} = 0$

or $\frac{d^2q}{dt^2} + \frac{R}{L}\frac{dq}{dt} + \frac{q}{LC} = 0$

This equation is equivalent to that of a damped oscillator.

Thus, we can write the solution as

$$Q_{max}(t) = Q_0 \cdot e^{-Rt/2L}$$

or $Q_{max}^2 = Q_0^2 e^{-\frac{Rt}{L}}$

As $L_1 > L_2$ damping is faster for L_2.

Aliter Inductance is inertia of circuit. It means inductance opposes the flow of charge, more inductance means decay of charge is slow.

In option (a), in a given time to, $Q_1^2 > Q_2^2$.

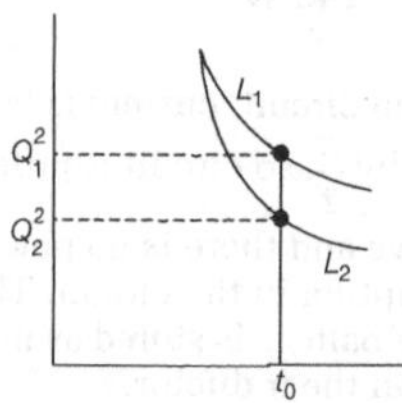

So, $L_1 > L_2$.

Hence, option (a) is correct.

4 Neon bulb is filled with gas, so its resistance is infinite, hence no current flows through it.

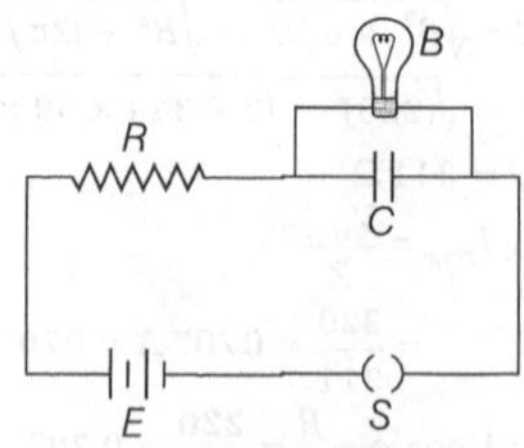

Now, $V_C = E(1 - e^{-t/RC})$

$\Rightarrow$ $120 = 200(1 - e^{-t/RC})$

$\Rightarrow$ $e^{-t/RC} = \frac{2}{5}$

$\Rightarrow$ $t = RC \ln 2.5$

$\Rightarrow$ $R = \frac{t}{C \ln 2.5}$

$$= \frac{t}{2.303\,C \log 2.5}$$
$$= 2.7 \times 10^6 \Omega$$

5 As, $U = \frac{1}{2}\frac{q^2}{C} = \frac{1}{2C}(q_0 e^{-t/\tau})^2$

$$= \frac{q_0^2}{2C} e^{-2t/\tau} \quad \text{(where, } \tau = CR)$$

$$U = U_I e^{-2t/\tau}$$

$$\frac{1}{2}U_I = U_I e^{-2t_1/\tau}, \frac{1}{2} = e^{-2t_1/\tau}$$

$\Rightarrow$ $t_1 = \frac{\tau}{2} \ln 2$

Now, $q = q_0 e^{-t/\tau}$, $\frac{1}{4}q_0 = q_0 e^{-t_2/\tau}$,

$$t_2 = \tau \ln 4 = 2\tau \ln 2$$

$\therefore$ $\frac{t_1}{t_2} = \frac{1}{4}$

6 As, $I = 2I_0 \frac{t}{T_0}$

where, $0 < t < \frac{T_0}{2}$

and $I = 2I_0\left(\frac{t}{T_0} - 1\right)$

where, $\frac{T_0}{2} < t < T_0$

$\therefore$ $I_{av} = \frac{2}{T}\int_0^{T_0} I\,dt$

$$= \frac{2}{T_0}\left[\int_0^{\frac{T}{2}} \frac{2I_0 t}{T_0} dt\right]$$

$$= \frac{2}{T_0^2}\left[\frac{2I_0 T_0^2}{2 \times 4}\right] = \frac{I_0}{2}$$

7 We have, $110 = \frac{V_{app} R}{\sqrt{R^2 + L^2\omega^2}}$

$\Rightarrow$ $110 = \frac{220R}{\sqrt{R^2 + L^2\omega^2}}$

$\Rightarrow$ $4R^2 = R^2 + L^2\omega^2$

$\Rightarrow$ $L\omega = \sqrt{3}R$

$\Rightarrow$ $L(100\pi) = 220\sqrt{3}$

$\Rightarrow$ $L = \frac{2.2\sqrt{3}}{3.14}$

$$= \frac{2.2 \times (1.732)}{3.14} = 1.2 \text{ H}$$

$\Rightarrow$ $R = \frac{110 \times 110}{55} = 220\Omega$

8 Given, inductance, $L = 0.01$ H, resistance, $R = 1\Omega$, voltage, $V = 200$ V and frequency, $f = 50$ Hz

Impedance of the circuit,

$$Z = \sqrt{R^2 + X_L^2} = \sqrt{R^2 + (2\pi f L)^2}$$

$$= \sqrt{1^2 + (2\times 3.14\times 50\times 0.01)^2}$$

or $Z = \sqrt{10.86} = 3.3\Omega$

$$\therefore \tan\phi = \frac{\omega L}{R} = \frac{2\pi f L}{R} = \frac{2\times 3.14\times 50\times 0.01}{1} = 3.14$$

$$\phi = \tan^{-1}(3.14) = 72^\circ$$

Phase difference, $\phi = \frac{72\times\pi}{180}$ rad

Time lag between alternating voltage and current,

$$\Delta t = \frac{\phi}{\omega} = \frac{72\pi}{180\times 2\pi\times 50} = \frac{1}{250}\text{ s}$$

9 At cut-off frequency, $Z = \sqrt{2}\,R$

$$R^2 + \left(L\omega - \frac{1}{C\omega}\right)^2 = 2R^2$$

$$\Rightarrow \quad L\omega - \frac{1}{C\omega} = R$$

$$\Rightarrow \quad LC\omega^2 - RC\omega - L = 0$$

$$\Rightarrow \quad \omega = \frac{RC \pm \sqrt{R^2C^2 + 4LC}}{2LC}$$

$$\Delta\omega = \omega_{01} - \omega_{02} = \frac{2\sqrt{R^2C^2 + 4LC}}{2LC} = \frac{\sqrt{R^2C^2 + 4LC}}{LC}$$

10 When S_2 is closed and S_1 is open, the charge oscillates in the L-C circuit at an angular frequency is given by

$$\omega = \frac{1}{\sqrt{LC}} \quad \text{...(i)}$$

Now, $q \neq 0$ at $t = 0$. Hence, options (a) and (b) are wrong. The charge q varies with time t as

$$q = q_0\cos(\omega t + \phi) \quad \text{...(ii)}$$

where, ϕ is not equal to $\pi/2$.

Differentiating Eq. (ii) twice with respect to t, we get

$$\frac{d^2q}{dt^2} = -\omega^2 q_0\cos(\omega t + \phi) = -\omega^2 q$$

$$q = -\frac{1}{\omega^2}\frac{d^2q}{dt^2} = -LC\frac{d^2q}{dt^2} \quad \text{[using Eq. (i)]}$$

11 The instantaneous value of emf and current in inductive circuit are given by $E = E_0\sin\omega t$ and $I = I_0\sin\left(\omega t - \frac{\pi}{2}\right)$ respectively.

So, $P_{inst} = EI$

$$= E_0\sin\omega t\times I_0\sin\left(\omega t - \frac{\pi}{2}\right)$$

$$= E_0I_0\sin\omega t\left(\sin\omega t\cos\frac{\pi}{2} - \cos\omega t\sin\frac{\pi}{2}\right)$$

$$= E_0I_0\sin\omega t\cos\omega t$$

$$= \frac{1}{2}E_0I_0\sin 2\omega t$$

$$(\because \sin 2\omega t = 2\sin\omega t\cos\omega t)$$

Hence, angular frequency of instantaneous power is 2ω.

12 Given, $X_L = 1\Omega, R = 2\Omega$

$$E_{rms} = 6\text{ V}, P_{av} = ?$$

Average power dissipated in the circuit,

$$P_{av} = E_{rms}\cos\phi$$

$$I_{rms} = \frac{I_0}{\sqrt{2}} = \frac{E_{rms}}{Z}$$

$$Z = \sqrt{R^2 + X_L^2} = \sqrt{4+1} = \sqrt{5}$$

$$\therefore \quad I_{rms} = \frac{6}{\sqrt{5}}\text{ A}$$

$$\cos\phi = \frac{R}{Z} = \frac{2}{\sqrt{5}}$$

$$P_{av} = 6\times\frac{6}{\sqrt{5}}\times\frac{2}{\sqrt{5}} \quad \text{[from Eq.(i)]}$$

$$= \frac{72}{\sqrt{5}\sqrt{5}} = \frac{72}{5} = 14.4\text{ W}$$

13 In circuit 1,

In steady state, $I = 0$

So, no option matches for circuit 1.

In circuit 2,

$V_1 = 0$ and $V_2 = 2I = V$

$\therefore$ B, C, D $\rightarrow$ 2

In circuit 3,

$$V_1 = X_LI = 2\pi fLI = 2\pi\times 50\times 6\times 10^{-3} = 1.88I$$

and $V_2 = 2I$

$\therefore$ A, B, D $\rightarrow$ 3

In circuit 4,

$V_1 = X_LI = 1.88I$

$V_2 = X_CI = 1061I$

$\therefore$ A, B, D $\rightarrow$ 4

In circuit 5,

$V_1 = IR = 1000I$

$V_2 = X_CI = 1016I$

$\therefore$ A, B, D $\rightarrow$ 5

14 Power dissipated is given by $P = E_{rms}I_{rms}\cos\phi$

We know that for a purely resistive circuit, the power factor,

$$\cos\phi = \frac{R}{Z} = \frac{R}{R} = 1$$

Hence, $P = E_{rms}\times I_{rms}$

$$= (RI_{rms})\times I_{rms} = R(I_{rms})^2$$

$$= R\left(\frac{I_0}{\sqrt{2}}\right)^2 = \frac{RI_0^2}{2}$$

15 Since, at resonance,

$$\omega L = \frac{1}{\omega C}$$

and impedance,

$$Z = \sqrt{R^2 + \left(\omega L - \frac{1}{\omega C}\right)^2}$$

or $Z = \sqrt{R^2} = R$

Hence, nature of impedance at resonance is resistive.

Also, in a L-C-R circuit, phase angle between the emf and the alternating current I, is zero at resonance.

DAY TWENTY FIVE

Electromagnetic Waves

Learning & Revision for the Day

- Electromagnetic Waves and their Characteristics
- Maxwell's Equations
- Transverse Nature of Electromagnetic Waves
- Spectrum of Electromagnetic Radiation

Electromagnetic Waves and their Characteristics

Electromagnetic waves are those waves, in which electric and magnetic fields vary sinusoidally in space with time. The electric and magnetic fields are mutually perpendicular to each other and each field is perpendicular to the direction of propagation of the wave.

- Maxwell's theory predicted that electromagnetic waves of all frequencies (and hence all wavelengths) propagate in vacuum, with a speed given by

$$c = \frac{1}{\sqrt{\mu_0 \varepsilon_0}}.$$

where, μ_0 is the magnetic permeability and ε_0 is the electric permittivity of vacuum. Now, for the vacuum, $\mu_0 = 4\pi \times 10^{-7}\ \text{TmA}^{-1}$ and $\varepsilon_0 = 8.85 \times 10^{-12}\,\text{C}^2\text{N}^{-1}\text{m}^{-2}$. Substituting these values in the above relation, we have

$$c = \frac{1}{[(4\pi \times 10^{-7})(8.85 \times 10^{-12})]^{1/2}} \simeq 3.0 \times 10^8\,\text{ms}^{-1}$$

- All the electromagnetic waves are of the transverse nature whose speed depends upon the medium, but their frequency does not depend on the medium.
- Transverse waves can be polarised.
- Energy is being transported with the electromagnetic waves.

Conduction Current

It is a current in the electric circuit, which arises due to the flow of electrons in the connecting wires of the circuit, in a definite closed path.

- No. of Questions in Exercises (x)—
- No. of Questions Attempted (y)—
- No. of Correct Questions (z)—
(*Without referring Explanations*)

- Accuracy Level ($z/y \times 100$)—
- Prep Level ($z/x \times 100$)—

In order to expect good rank in JEE, your Accuracy Level should be above 85 & Prep Level should be above 75.

Maxwell's Displacement Current

It is that current which comes into play in the region, whenever the electric field and hence the electric flux is changing with time.

$$i_d = \varepsilon_0 \frac{d\phi_E}{dt}$$

The generalised form of the Ampere's law is

$$\oint \mathbf{B} \cdot d\mathbf{l} = \mu_0(i_c + i_d) = \mu_0\left(i_c + \varepsilon_0 \frac{d\phi_E}{dt}\right)$$

where, i_c is conduction current.

Maxwell's Equations

Maxwell in 1862, gave the basic laws of electricity and magnetism in the form of four fundamental equations, which are known as Maxwell's equations.

- **Gauss's law for electrostatics** This law states that the total electric flux through any closed surface is always equal to $\frac{1}{\varepsilon_0}$ times the net charged enclosed by that surface.

  Mathematically, $\oint_S \mathbf{E} \cdot d\mathbf{S} = \frac{q}{\varepsilon_0}$

- **Gauss's law for magnetism** This law also predicts that the isolated magnetic monopole does not exist.

  i.e. net magnetic flux through any closed surface is always zero.

  Mathematically, $\oint_S \mathbf{B} \cdot d\mathbf{S} = 0$

- **Faraday's law of electromagnetic induction** It states that the induced e.m.f. produced in a circuit is numerically equal to time rate of change of magnetic flux through it.

  Mathematically, $\oint \mathbf{E} \cdot d\mathbf{l} = -\frac{d\phi_B}{dt}$

- **Ampere-Maxwell's law** At an instant in a circuit, the conduction current is equal to displacement current.

  Mathematically, $\oint \mathbf{E} \cdot d\mathbf{l} = \mu_0\left(I_c + \varepsilon_0 \frac{d\phi_E}{dt}\right)$

  These equations are collectively called Maxwell's equations.

Properties of Electromagnetic Waves

- If the electromagnetic wave is travelling along the positive direction of the X-axis, the electric field is oscillating parallel to the Y-axis and the magnetic field is oscillating parallel to the Z-axis.

$$E = E_0 \sin(\omega t - kx) \Rightarrow B = B_0 \sin(\omega t - kx)$$

In this, E_0 and B_0 are the amplitudes of the fields.

Further, $c = \frac{E_0}{B_0} = \frac{1}{\sqrt{\varepsilon_0 \mu_0}}$ = speed of light in vacuum

- The rate of flow of energy in an electromagnetic wave, is described by the vector **S** called the **Poynting vector**, which is defined by the expression,

$$\mathbf{S} = \frac{1}{\mu_0} \mathbf{E} \times \mathbf{B}$$

- The time average of S over one cycle is known as the wave intensity. When the average is taken, we obtain an expression involving the time average of $\cos^2(kx - \omega t)$ which equals $\frac{1}{2}$.

Thus, $$I = S_{av} = \frac{E_0 B_0}{2\mu_0} = \frac{E_0^2}{2\mu_0 c} = \frac{c\,B_0^2}{2\mu_0} \text{ Wm}^{-2}$$

- The total average energy per unit volume is,

$$u = u_E + u_B = \frac{\varepsilon_0 E_0^2}{2} = \frac{B^2}{2\mu_0}$$

- The radiation pressure p exerted on a perfectly absorbing surface, $p = \frac{S}{c}$
- If the surface is a perfect reflector and incidence is normal, then the momentum transported to the surface in a time t is given by, $p = \frac{2u}{c}$ and the radiation pressure will be, $p = \frac{2S}{c}$.
- Energy density of electromagnetic wave,

$$u_e = \frac{1}{2} \varepsilon_0 E_{u_B}^2 = \frac{1}{2} \frac{B^2}{\mu_0}$$

- Momentum delivered, $p = \frac{u}{c}$ (absorbing surface)

  $p = \frac{2u}{c}$ (reflecting surface)

- Energy of wave $= \frac{hc}{\lambda} = h\nu$

Transverse Nature of Electromagnetic Waves

According to Maxwell, electromagnetic waves consist of time varying electric and magnetic fields, which are perpendicular to each other, as well as direction of wave propagation.

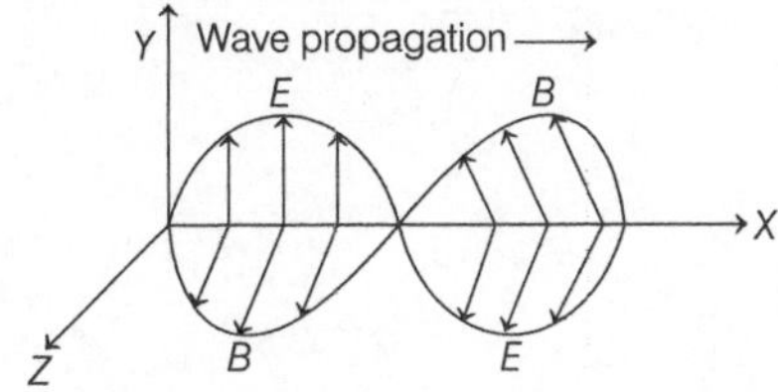

Spectrum of Electromagnetic Radiation

The array obtained on arranging all the electromagnetic waves in an order on the basis of their wavelength is called the electromagnetic spectrum.

The Electromagnetic Spectrum

Name	Frequency Range (Hz)	Wavelength Range (m)	Source
Radio waves	10^4 to 10^8	0.1 to 600	Oscillating electric circuits
Microwaves	10^9 to 10^{12}	10^{-3} to 0.3	Oscillating current in special vacuum tubes
Infrared	10^{11} to 5×10^{14}	10^{-6} to 5×10^{-3}	Outer electrons in atoms and molecules
Visible light	4×10^{14} to 7×10^{14}	4×10^{-7} to 8×10^{-7}	Outer electrons in atoms
Ultraviolet	10^{15} to 10^{17}	1.5×10^{-7} to 3.5×10^{-7}	Outer electrons in atoms
X-rays	10^{18} to 10^{20}	10^{-11} to 10^{-8}	Inner electrons in atoms and sudden deacceleration of high energy free electrons
Gamma rays	10^{19} to 10^{24}	10^{-16} to 10^{-13}	Nuclei of atoms and sudden deacceleration of high energy free electrons

Various Electromagnetic Radiations

- **Gamma rays** The main sources of gamma rays are the natural and artificial radioactive substances. These rays affect the photographic plate and mainly used in the treatment of cancer disease.
- **X-rays** X-rays are produced, when highly energetic cathode rays are stopped by a metal target of high melting point. They affect the photographic plate and can penetrate through the transparent materials. They are mainly used in detecting the fracture of bones, hidden bullet, needle, costly material etc. inside the body, and also used in the study of crystal structure.
- **Ultraviolet Rays** The major part of the radiations received from sun consists of the ultraviolet radiation. Its other sources are the electric discharge tube, carbon arc, etc.

  These radiations are mainly used in excitation of photoelectric effect and to kill the bacteria of many diseases.
- **Visible Light** Visible light is obtained from the glowing bodies, while they are white hot. The light obtained from the electric bulbs, sodium lamp, fluorescent tube is the visible light.
- **Thermal or Infrared Waves** A body on being heated, emits out the infrared waves. These radiations have the maximum heating effect. The glass absorbs these radiations, therefore for the study of these radiations, rock salt prism is used instead of a glass prism. These waves are mainly used for therapeutic purpose by the doctors because of their heating effect.
- **Microwaves** These waves are produced by the spark discharge or magnetron valve. They are detected by the crystal or semiconductor detector. These waves are used mainly in radar and long distance communication.
- **Radio waves** They can be obtained by the flow of high frequency alternating current in an electric conductor.

  These waves are detected by the tank circuit in a radio receiver or transmitter.

Applications of Electromagnetic Spectrum

- Radio waves are used in radar and radio broadcasting.
- Microwaves are used in long distance wireless communications *via* satellites.
- Infrared, visible and ultraviolet radiations are used to know the structure of molecules.
- Diffraction of X-rays by crystals, gives the details of the structure of crystals.

DAY PRACTICE SESSION 1

FOUNDATION QUESTIONS EXERCISE

1 During the porpagation of electromagnetic waves in a medium, → JEE Main 2014

(a) electric energy density is double of the magnetic energy density
(b) electric energy density is half of the magnetic energy density
(c) electric energy density is equal to the magnetic energy density
(d) Both electric and magnetic energy densities are zero

2 A perfectly reflecting mirror has an area of 1 cm^2. Light energy is allowed to fall on it for 1 h at the rate of 10 W cm^{-2}. The force acting on the mirror is

(a) 6.7×10^{-8} N (b) 2.3×10^{-4} N
(c) 10^{-3} N (d) zero

3 The magnetic field between the plates of radius 12 cm, separated by a distance of 4 mm of a parallel plate capacitor of capacitance 100 pF along the axis of plates having conduction current of 0.15 A, is

(a) zero (b) 1.5 T (c) 15 T (d) 0.15 T

4 Instantaneous displacement current of 1.0 A in the space between the parallel plates of a 1μF capacitor, can be established by changing potential difference of

(a) 10^{-6} Vs^{-1} (b) 10^{6} Vs^{-1}
(c) 10^{-8} Vs^{-1} (d) 10^{8} Vs^{-1}

5 A large parallel plate capacitor, whose plates have an area of 1 m^2 and are separated from each other by 1 mm, is being charged at a rate of 25 Vs^{-1}. If the dielectric between the plates has the dielectric constant 10, then the displacement current at this instant is

(a) 25 μA (b) 11μA (c) 2.2 μA (d) 1.1 μA

6 A parallel plate capacitor with plate area A and separation between the plates d, is charged by a constant current I. Consider a plane surface of area $A/2$ parallel to the plates and drawn simultaneously between the plates. The displacement current through this area is

(a) I (b) $\frac{I}{2}$ (c) $\frac{I}{4}$ (d) $\frac{I}{8}$

7 Select the correct statement from the following → JEE Main (Online) 2013

(a) Electromagnetic waves cannot travel in vacuum
(b) Electromagnetic waves are longitudinal waves
(c) Electromagnetic waves are produced by charges moving with uniform velocity
(d) Electromagnetic waves carry both energy and momentum as they propagate through space

8 In an apparatus, the electric field was found to oscillate with an amplitude of 18 Vm^{-1}. The magnitude of the oscillating magnetic field will be

(a) 4×10^{-6} T (b) 6×10^{-8} T
(c) 9×10^{-9} T (d) 11×10^{-11} T

9 The magnetic field in a travelling electromagnetic wave has a peak value of 20 nT. The peak value of electric field strength is → JEE Main 2013

(a) 3 V/m (b) 6 V/m (c) 9 V/m (d) 12 V/m

10 The rms value of the electric field of the light coming from the sun is 720 NC^{-1}. The average total energy density of the electromagnetic wave is

(a) 4.58×10^{-6} Jm^{-3} (b) 6.37×10^{-9} Jm^{-3}
(c) 81.35×10^{-12} Jm^{-3} (d) 3.3×10^{-3} Jm^{-3}

11 A radiation of energy E falls normally on a perfectly reflecting surface. The momentum transferred to the surface is

(a) $\frac{E}{c}$ (b) $\frac{2E}{c}$ (c) Ec (d) $\frac{E}{c^2}$

12 An electromagnetic wave in vacuum has the electric and magnetic fields **E** and **B**, which are always perpendicular to each other. The direction of polarisation is given by **X** and that of wave propagation by $\hat{\mathbf{k}}$. Then, → AIEEE 2012

(a) $\mathbf{X} \parallel \mathbf{B}$ and $\hat{\mathbf{k}} \parallel \mathbf{B}\times\mathbf{E}$ (b) $\mathbf{X} \parallel \mathbf{E}$ and $\hat{\mathbf{k}} \parallel \mathbf{E}\times\mathbf{B}$
(c) $\mathbf{X} \parallel \mathbf{B}$ and $\hat{\mathbf{k}} \parallel \mathbf{E}\times\mathbf{B}$ (d) $\mathbf{X} \parallel \mathbf{B}$ and $\hat{\mathbf{k}} \parallel \mathbf{B}\times\mathbf{E}$

13 An electromagnetic wave travels in vacuum along z-direction $\mathbf{E} = (E_1\,\hat{\mathbf{i}} + E_2\,\hat{\mathbf{j}})\cos(kz - \omega t)$. Choose the correct option from the following

(a) The associated magnetic field is given as
$B = \frac{1}{C}(E_1\,\hat{\mathbf{i}} - E\,\hat{\mathbf{j}})\cos(kz - \omega t)$
(b) The associated magnetic field is given as
$B = \frac{1}{C}(E_1\,\hat{\mathbf{i}} - E\,\hat{\mathbf{j}})\cos(kz - \omega t)$
(c) The given electromagnetic field is circularly polarised
(d) The given electromagnetic wave is plane polarised

14 Match List I (Electromagnetic wave type) with List II (Its association/application) and select the correct option from the choices given below the lists. → JEE Main 2014

	List I		List II
A.	Infrared waves	1.	To treat muscular strain
B.	Radio waves	2.	For broadcasting
C.	X-rays	3.	To detect fracture of bones
D.	Ultraviolet waves	4.	Absorbed by the ozone layer of the atmosphere

Codes

	A	B	C	D		A	B	C	D
(a)	4	3	2	1	(b)	1	2	4	3
(c)	3	2	1	4	(d)	1	2	3	4

15 Arrange the following electromagnetic radiations per quantum in the order of increasing energy.

A. Blue light B. Yellow light
C. X-ray D. Radio wave

→ JEE Main 2016 (Offline)

(a) D, B, A, C (b) A, B, D, C
(c) C, A, B, D (d) B, A, D, C

Direction (Q. Nos. 16-20) *Each of these questions contains two statements : Statement I (Assertion) and Statement II (Reason). Each of these questions also has four alternative choices, only one of which is the correct answer. You have to select one of the codes (a), (b), (c), (d) given below*

(a) Statement I is true, Statement II is true; Statement II is the correct explanation for Statement I
(b) Statement I is true, Statement II is true; Statement II is not the correct explanation for Statement I
(c) Statement I is true; Statement II is false
(d) Statement I is false; Statement II is true

16 **Statement I** Ultraviolet radiation being higher frequency waves are dangerous to human being.

Statement II Ultraviolet radiations are absorbed by the atmosphere.

17 **Statement I** If the earth did not have atmosphere, its average surface temperature would be lower than what is now.

Statement II Greenhouse effect of the atmosphere would be absent, if the earth did not have atmosphere.

18 **Statement I** Electromagnetic waves exert radiation pressure.

Statement II Electromagnetic waves carry energy.

19 **Statement I** Light is a transverse wave, but not an electromagnetic wave.

Statement II Maxwell showed that speed of electromagnetic waves is related to the permeability and the permittivity of the medium through which it travels.

20 **Statement I** Out of radio waves and microwaves, the radio waves undergo more diffraction.

Statement II Radio waves have greater frequency compared to microwaves. → JEE Main (Online) 2013

DAY PRACTICE SESSION 2

PROGRESSIVE QUESTIONS EXERCISE

1 You are given a 2 μF parallel plate capacitor. How would you establish an instantaneous displacement current of 1 mA in the space between its plates?

(a) By applying a varying potential difference of 500 V/s
(b) By applying a varying potential difference of 400 V/s
(c) By applying a varying potential difference of 100 V/s
(d) By applying a varying potential difference of 300 V/s

2 A uniform but time varying magnetic field $B(t)$ exists in a circular region of radius a and is directed into the plane of the paper as shown in the figure. The magnitude of the induced electric field at a point P, a distance r from the centre of the circular region

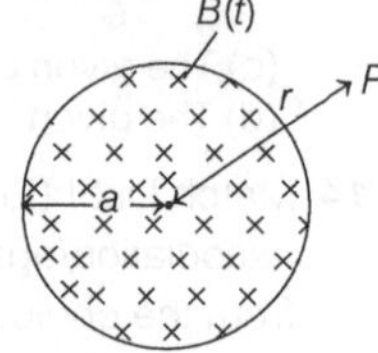

(a) increases with r
(b) decreases with r
(c) decreases as $\frac{1}{r^2}$
(d) zero

3 The magnetic field of a beam emerging from a filter facing a flood light is given by $B = 12 \times 10^{-8} \sin(1.20 \times 10^7 z - 3.60 \times 10^{15} t)$ T. What is the average intensity of the beam?

(a) 1.7 W/m^2 (b) 2.3 W/m^2
(c) 2.7 W/m^2 (d) 3.2 W/m^2

4 The ratio of contributions made by the electric field and magnetic field components to the intensity of an electromagnetic wave is

(a) c:1 (b) c^2:1 (c) 1:1 (d) $\sqrt{c}$:1

5 An FM radio station, antenna radiates a power of 10 kW at a wavelength of 3 m. Assume the radiated power is confined to and is uniform over a hemisphere with antenna at its centre. E_{max} at a distance of 10 km from antenna is

(a) 0.62 NC^{-1} (b) 0.41 NC^{-1}
(c) 0.31 NC^{-1} (d) 0.10 NC^{-1}

6 Assume that all the energy from a 1000 W lamp is radiated uniformly, then the amplitude of electric field of radiation at a distance of 2 m from the lamp is

(a) 245.01 V/m (b) 17 V/m
(c) 0 (d) 2.96 V/m

7 A red LED emits light at 0.1W uniformly around it. The amplitude of the electric field of the light at a distance of 1m from the diode is → JEE Main 2015

(a) 1.73 V/m
(b) 2.45 V/m
(c) 5.48 V/m
(d) 7.75 V/m

8 In a transverse wave, the distance between a crest and neighbouring through at the same instant is 4.0 cm and the distance between a crest and trough at the same place is 1.0 cm. The next crest appears at the same place after a time interval of 0.4 s. The maximum speed of the vibrating particles in the medium is → JEE Main (Online) 2013

(a) $\frac{3\pi}{2}$ cm/s (b) 5π cm/s
(c) $\frac{\pi}{2}$ cm/s (d) 2π cm/s

9 An EM wave from air enters a medium. The electric fields are $\mathbf{E}_1 = E_{01}\hat{\mathbf{x}}\cos\left[2\pi\nu\left(\frac{z}{c}-t\right)\right]$ in air and $\mathbf{E}_2 = E_{02}\hat{\mathbf{x}}\cos[k(2z-ct)]$ in medium, where the wave number k and frequency ν refer to their values in air. The medium is non-magnetic.

If ε_{r_1} and ε_{r_2} refer to relative permittivities of air and medium respectively, which of the following options is correct? → JEE Main 2018

(a) $\frac{\varepsilon_{r_1}}{\varepsilon_{r_2}} = 4$ (b) $\frac{\varepsilon_{r_1}}{\varepsilon_{r_2}} = 2$ (c) $\frac{\varepsilon_{r_1}}{\varepsilon_{r_2}} = \frac{1}{4}$ (d) $\frac{\varepsilon_{r_1}}{\varepsilon_{r_2}} = \frac{1}{2}$

10 An electromagnetic wave of frequency $\nu = 3.0$ MHz passes from vacuum into a dielectric medium with permittivity $\varepsilon = 4.0$. Then,

(a) wavelength is doubled and the frequency remains unchanged
(b) wavelength is doubled and frequency becomes half
(c) wavelength is halved and frequency remains unchanged
(d) wavelength and frequency both remain unchanged

11 The magnetic field at a point between the plates of a capacitor at a perpendicular distance R from the axis of the capacitor plate radius R, having the displacement current I_D is given by

(a) $\frac{\mu I_D r}{2\pi R^2}$ (b) $\frac{\mu_0 I_D}{2\pi R}$
(c) $\frac{\mu_0 I_D}{\pi r^2}$ (d) zero

ANSWERS

SESSION 1	1 (c)	2 (a)	3 (a)	4 (b)	5 (c)	6 (b)	7 (d)	8 (b)	9 (b)	10 (a)
	11 (b)	12 (b)	13 (d)	14 (d)	15 (a)	16 (b)	17 (a)	18 (b)	19 (d)	20 (c)
SESSION 2	1 (a)	2 (b)	3 (a)	4 (c)	5 (d)	6 (a)	7 (b)	8 (b)	9 (c)	10 (c)
	11 (b)									

Hints and Explanations

SESSION 1

1 Both the energy densities are equal, i.e. energy is equally divided between electric and magnetic field.

2 Let E = energy falling on the surface per second = 10 J

Momentum of photons,

$$p = \frac{h}{\lambda} = \frac{h}{c/f}$$

$$= \frac{hf}{c} = \frac{E}{c}$$

On reflection, change in momentum per second = force

$$= 2p = \frac{2E}{c} = \frac{2\times 10}{3\times 10^8}$$

$$= 6.7\times 10^{-8}\text{ N}$$

3 As $B \propto r$, since the point is on the axis, where $r = 0$, so $B = 0$.

4 As, $I_d = C\left(\frac{V}{t}\right)$ $\left(\because I = \frac{dQ}{dt}\right)$

or $\frac{V}{t} = \frac{I_d}{C}$

$$= \frac{1.0}{10^{-6}} = 10^6\text{ Vs}^{-1}$$

5 As, $C = \frac{\varepsilon_0 KA}{d}$

$$= \frac{(8.85\times 10^{-12})\times 10\times 1}{10^{-3}}$$

$$= 8.85\times 10^{-8}\text{ F}$$

$$\therefore \quad I = \frac{d}{dt}(CV) = C\frac{dV}{dt}$$

$$= 8.85\times 10^{-8}\times 25$$

$$= 2.2\times 10^{-6} = 2.2\,\mu\text{A}$$

6 Charge on the capacitor plates, at time t is,

$$q = It$$

Electric field between the plates at this instant,

$$E = \frac{q}{A\varepsilon_0} = \frac{It}{A\varepsilon_0}$$

Electric flux through the given area,

$$\phi_E = \left(\frac{A}{2}\right)E = \frac{It}{2\varepsilon_0}$$

Therefore, displacement current,

$$I_d = \varepsilon_0\frac{d\phi_E}{dt}$$

$$= \varepsilon_0\frac{d}{dt}\left(\frac{It}{2\varepsilon_0}\right) = \frac{I}{2}$$

7 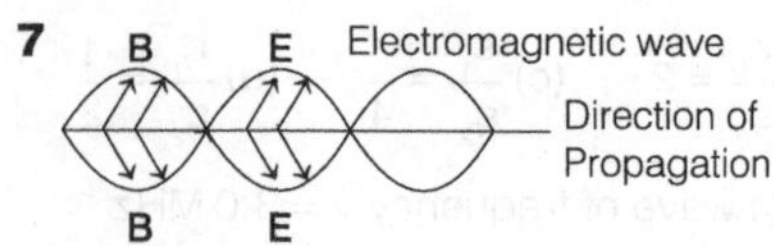

As electromagnetic waves contains both electric field and magnetic field. It carry both energy and momentum according to de-Broglie wave particle duality of radiations.

8 Here, $E_0 = 18 \text{ Vm}^{-1}$

$$\therefore \quad B_0 = \frac{E_0}{c} = \frac{18}{3 \times 10^8} = 6 \times 10^{-8} \text{ T}$$

9 $\mathbf{E} = \mathbf{B} \times c$ keep value of electric field

$$|\mathbf{E}| = |\mathbf{B}||c| = 20 \times 10^{-9} \times 3 \times 10^8 = 6 \text{ V/m}$$

10 Total average energy $= \varepsilon_0 E_{rms}^2$

$$= 8.85 \times 10^{-12} \times (720)^2 = 4.58 \times 10^{-6} \text{ Jm}^{-3}$$

11 Initial momentum of surface, $p_i = E/c$ where, c = velocity of light (constant). Since, the surface is perfectly reflecting, so the same momentum will be reflected completely. Final momentum,

$$p_f = E/c \quad \text{[negative value]}$$

$\therefore$ Change in momentum,

$$\Delta p = p_f - p_i = -\frac{E}{c} - \frac{E}{c} = -\frac{2E}{c}$$

Thus, momentum transferred to the surface is

$$\Delta p' = |\Delta p| = \frac{2E}{c}$$

12 In electromagnetic wave, the direction of propagation of wave, electric field and magnetic field are mutually perpendicular, i.e. wave propagates perpendicular to **E** and **B** or along **E**×**B**. While polarisation of wave takes place parallel to electric field vector.

13 Here, in electromagnetic wave, the electric field vector is given as

$$\mathbf{E} = (E_1\hat{\mathbf{i}} + E_2\hat{\mathbf{j}})\cos(kz - \omega t)$$

In electromagnetic wave, the associated magnetic field vector,

$$\mathbf{B} = \frac{E}{c} = \frac{E_1\hat{\mathbf{i}} + E_2\hat{\mathbf{j}}}{c}\cos(kz - \omega t)$$

Also, **E** and **B** are perpendicular to each other and the propagation of electromagnetic wave is perpendicular to **E** as well as **B**, so the given electromagnetic wave is plane polarised.

14 (a) Infrared waves are used to treat muscular strain.
(b) Radio waves are used for broadcasting purposes.
(c) X-rays are used to detect fracture of bones.
(d) Ultraviolet waves are absorbed by ozone.

15 As, we know energy liberated,

$$E = \frac{hc}{\lambda}$$

i.e. $$E \propto \frac{1}{\lambda}$$

So, lesser the wavelength, than greater will be energy liberated by electromagnetic radiations per quantum.
As, order of wavelength is given by
X-ray, VIBGYOR, **Radio waves**
(C) (A) (B) (D)
$\therefore$ Order of increasing energy of electromagnetic radiations per quantum.

$$\Rightarrow \quad D < B < A < C$$

16 Ultraviolet radiations are electromagnetic waves. The wavelength of these waves ranges between 4000 Å to 100 Å, i.e. of smaller wavelength and higher frequency. They are absorbed by ozone layer of stratosphere in atmosphere. They cause skin diseases and they are harmful to eye and may cause permanent blindness.

17 Earth is heated by sun's infrared radiation.The earth also emits radiation most in infrared region. These radiations are reflected back by heavy gases like CO_2 in atmosphere. These back radiation keep the earth's surface warm at night. This phenomenon is called greenhouse effect. When the atmosphere were absent, then temperature of the earth falls.

18 Electromagnetic waves have linear momentum as well as energy. From this we conclude that, we can exert radiation pressure by making a beam of electromagnetic radiation fall on an object. Let us assume that object is free to move and that the radiation is entirely absorbed in the object during time interval Δt. The object gains an energy ΔU from the radiation. Maxwell showed that the object also gains linear momentum, the magnitude Δp of the change in momentum of the object is related to the energy change ΔU as

$$\Delta p = \frac{\Delta U}{c} \quad \text{(total absorption)}$$

19 In free space or vacuum, the speed of electromagnetic waves is

$$c = \frac{1}{\sqrt{\mu_0 \varepsilon_0}} \quad \ldots\text{(i)}$$

Here, $\mu_0 = 4\pi \times 10^{-7} \text{ Ns}^2 \text{ C}^{-2}$ is permeability (constant) of free space. $\varepsilon_0 = 8.85418 \times 10^{-12} \text{ C}^2\text{N}^{-1}\text{m}^{-2}$ is the permittivity of free space. On substituting the values in Eq. (i), we have

$$c = \frac{1}{\sqrt{4\pi \times 10^{-7} \times 8.85418 \times 10^{-12}}} = 2.99792 \times 10^8 \text{ms}^{-1}$$

This is same as the speed of light in vacuum. From this we conclude that light is an electromagnetic wave.

20 The frequency of radio waves less than the frequency of microwaves.
$\because$ Frequency of radio waves $= 3 \times 10^8$ Hz and frequency of microwaves $= 10^{10}$ Hz
$\therefore \quad \nu_{\text{radio waves}} < \nu_{\text{microwaves}}$

SESSION 2

1 Given, capacitance of capacitor, $C = 2\mu\text{F}$
Displacement current, $I_d = 1$ mA
Charge, $q = CV$

$$I_d dt = C dV \quad [\because q = It]$$

or $$I_d = C\frac{dV}{dt}$$

$$1 \times 10^{-3} = 2 \times 10^{-6} \times \frac{dV}{dt}$$

or $$\frac{dV}{dt} = \frac{1}{2} \times 10^{+3} = 500 \text{ V/s}$$

Clearly, by applying a varying potential difference of 500 V/s, we would produce a displacement current of desired value.

2 A time varying magnetic field produces an electric field. The magnitude of the electric field at a distance r from the centre of a circular region of radius a, where a time varying field B exists, is given by

$$E = \frac{a^2}{2r}\frac{dB}{dt}$$

At $r = a$,

$$E = \left(\frac{a}{2}\right)\frac{dB}{dt}$$

This is the value of E at the edge of the circular region. For $r > a$, E decreases with r.

3 Magnetic field, $\mathbf{B} = B_0 \sin \omega t$.
Given equation,

$$B = 12 \times 10^{-8} \sin(1.20 \times 10^7 z - 3.60 \times 10^{15} t)\text{T}$$

On comparing this equation with standard equation, we get

$B_0 = 12 \times 10^{-8}$

The average intensity of the beam,

$$I_{av} = \frac{1}{2}\frac{B_0^2}{\mu_0}\cdot c$$

$$= \frac{1}{2} \times \frac{(12 \times 10^{-8})^2 \times 3 \times 10^8}{4\pi \times 10^{-7}}$$

$$= 1.7 \text{ W/m}^2$$

4 Intensity in terms of electric field,

$$U_{av} = \frac{1}{2}\varepsilon_0 E_0^2$$

Intensity in terms of magnetic field,

$$U_{av} = \frac{1}{2}\frac{B_0^2}{\mu_0}$$

Now, taking the intensity in terms of electric field,

$$(U_{av})\text{ electric field} = \frac{1}{2}\varepsilon_0 E_0^2$$

$$= \frac{1}{2}\varepsilon_0 (cB_0)^2 \quad (\because E_0 = cB_0)$$

But, $c = \frac{1}{\sqrt{\mu_0 \varepsilon_0}}$

$$\therefore (U_{av})_{\text{electric field}} = \frac{1}{2}\varepsilon_0 \times \frac{1}{\mu_0 \varepsilon_0} B_0^2$$

$$= \frac{1}{2} \cdot \frac{B_0^2}{\mu_0}$$

Thus, the energy in electromagnetic wave is divided equally between electric field vector and magnetic field vector. Therefore, the ratio of contributions by the electric field and magnetic field components to the intensity of an electromagnetic wave is 1:1.

5 As, $I = \frac{\rho}{2\pi r^2} = \frac{10^4}{2\pi (10 \times 10^3)^2} = \frac{10^{-4}}{2\pi}$

and $I = \frac{1}{2}\varepsilon_0 E_0^2 c$

$$\Rightarrow E_0 = \sqrt{\frac{2I}{\varepsilon_0 c}}$$

$$= \sqrt{\frac{10^{-4}}{\pi \times 8.85 \times 10^{-12} \times 3 \times 10^8}}$$

or $E_0 = 0.10 \text{ NC}^{-1}$

6 Poynting vector, $S = E \times H$

$$= EH \sin 90^\circ = EH$$

$$\text{Energy of lamp} = \frac{1000 \text{ W}}{\pi r^2}$$

$$= \frac{1000}{\pi \times 2^2} \text{ Jm}^{-2} \text{ s}^{-1}$$

S represents energy flow per unit area per second, we have

$$EH = \frac{1000}{\pi \times 2^2} = 79.61,$$

$$\frac{E}{H} = 377$$

$$EH \times \frac{E}{H} = 79.61 \times 377 = 30015.9$$

$$\Rightarrow E = \sqrt{30015.9}$$

$$= 173.25 \text{ V/m}$$

Amplitude of electric field of radiation is

$$E_0 = E\sqrt{2} = 245.01 \text{ V/m}$$

7 Consider the LED as a point source of light. Let power of the LED is P. Intensity at r from the source,

$$I = \frac{P}{4\pi r^2} \quad \text{...(i)}$$

As we know that ,

$$I = \frac{1}{2}\varepsilon_0 E_0^2 c \quad \text{...(ii)}$$

From Eqs.(i) and (ii), we can write

$$\frac{P}{4\pi r^2} = \frac{1}{2}\varepsilon_0 E_0^2 c$$

or $E_0^2 = \frac{2P}{4\pi\varepsilon_0 r^2 c}$

$$= \frac{2 \times 0.1 \times 9 \times 10^9}{1 \times 3 \times 10^8}$$

or $E_0^2 = 6$

$$\Rightarrow E_0 = \sqrt{6} = 2.45 \text{ V/m}$$

8 Given, $\frac{\lambda}{4} = 4$ cm

$\therefore \lambda = 16$ cm and $T = 0.4$ s

As, $f\lambda \times T = 2\pi$

$$\Rightarrow f = \frac{2\pi}{16 \times 0.4} = \frac{5\pi}{16}\text{s}^{-1}$$

Now, $v = f\lambda = \frac{5\pi}{16} \times 16 = 5\pi$ cm/s

9 Speed of progressive wave is given by,

$$v = \frac{\omega}{k}$$

As electric field in air is,

$$E_1 = E_{01}\hat{\mathbf{x}}\cos\left(\frac{2\pi vz}{c} - 2\pi vt\right)$$

$$\therefore \text{Speed in air} = \frac{2\pi v}{\left(\frac{2\pi v}{c}\right)} = c$$

Also, $c = \frac{1}{\sqrt{\mu_0 \varepsilon_{r_1} \varepsilon_0}}$ ...(i)

In medium,

$$E_2 = E_{02}\hat{\mathbf{x}}\cos(2kz - kct)$$

$$\therefore \text{Speed in medium} = \frac{kc}{2k} = \frac{c}{2}$$

Also, $\frac{c}{2} = \frac{1}{\sqrt{\mu_0 \varepsilon_{r_2} \varepsilon_0}}$ ...(ii)

As medium is non-magnetic medium,

$\mu_{\text{medium}} = \mu_{\text{air}}$

On dividing Eq. (i) by Eq. (ii), we have

$$2 = \sqrt{\frac{\varepsilon_{r_2}}{\varepsilon_{r_1}}} \Rightarrow \frac{\varepsilon_{r_1}}{\varepsilon_{r_2}} = \frac{1}{4}$$

10 In vacuum, $\varepsilon_0 = 1$

In medium, $\varepsilon = 4$

$\therefore$ Refractive index,

$$\mu = \sqrt{\frac{\varepsilon}{\varepsilon_0}} = \sqrt{\frac{4}{1}} = 2$$

Wavelength,

$$\lambda' = \frac{\lambda}{\mu} = \frac{\lambda}{2}$$

and wave velocity,

$$v = \frac{c}{\mu} = \frac{c}{2} \quad \left[\because \mu = \frac{c}{v}\right]$$

Hence, it is clear that wavelength and velocity will become half, but frequency remains unchanged when the wave is passing through any medium.

11 According to the Ampere-Maxwell's law, for a closed surface,

$$\oint \mathbf{B} \cdot d\mathbf{l} = \mu_0 I_D$$

As, $B(2\pi R) = \mu_0 I_D$

$$\Rightarrow B = \frac{\mu_0 I_D}{2\pi R}$$

DAY TWENTY SIX

Unit Test 5

(Magnetostatic, EMI & AC, EM Wave)

1 A solenoid of some length and radius 2 cm has a layer of winding. A 2 cm long wire of mass 5 g lies inside the solenoid along the axis of solenoid. The wire is connected to some external circuit, so that a current of 5 A flows through the wire. The value of current to be in the winding, so that magnetic force supports the wire weight is (take, $g = 10 \text{ms}^{-2}$)

(a) zero (b) $\simeq$ 400A
(c) $\simeq$ 32000A (d) Not possible

2 A circular coil of 20 turns and radius 10 cm is placed in a uniform magnetic field of 0.1 T normal to the plane of the coil. The coil carries a current of 5 A. The coil is made up of copper wire of cross-sectional area 10^{-5} m^2 and the number of free electrons per unit volume of copper is 10^{29}. The average force experienced by an electron in the coil due to magnetic field is

(a) 5×10^{-25} N (b) zero
(c) 8×10^{-24} N (d) None of these

3 A wire *AB* of length 5 m carrying a current of 2A is placed in a region of uniform magnetic field $B = 0.5$ T as shown in figure. The magnetic force experienced by wire is

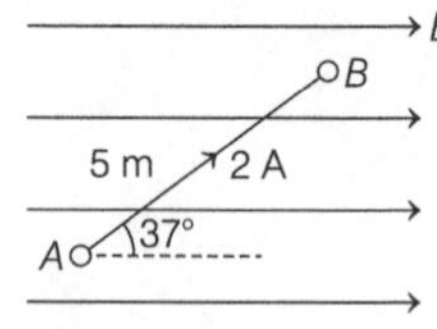

(a) 5 N (b) 4 N (c) 3 N (d) 8 N

4 A coil of metal wire is kept stationary in a non-uniform magnetic field, then

(a) an emf and current both induced in the coil
(b) a current but no emf is induced in the coil
(c) an emf but no current is induced in the coil
(d) Neither emf nor current is induced in the coil

5 A thin circular ring of area *A* is held perpendicular to a uniform magnetic field of induction *B*. A small cut is made in the ring and a galvanometer is connected across the ends such that the total resistance of the circuit is *R*. When the ring is suddenly squeezed to zero area, the charge flowing through the galvanometer is

(a) $\frac{BR}{A}$ (b) $\frac{AB}{R}$ (c) ABR (d) $\frac{B^2A}{R^2}$

6 Two infinite long current carrying wires *A* and *B* are placed as shown in figure. Each wire carries same current *I*. The resultant magnetic field intensity at point *P* is

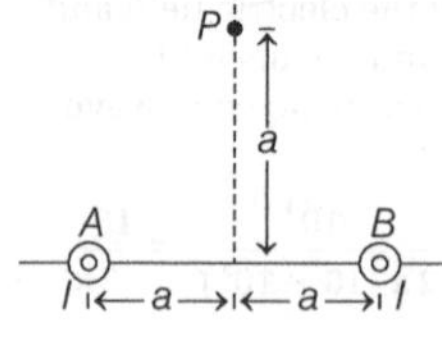

(a) $\frac{\mu_0 I}{2\pi a}$ (b) $\frac{\sqrt{2}\mu_0 I}{2\pi a}$
(c) $\frac{\mu_0 I}{2\sqrt{2}\pi a}$ (d) $\frac{\mu_0 I}{4\sqrt{2}\pi a}$

7 An equilateral triangular loop is kept near to a current carrying long wire as shown in figure. Under the action of magnetic force alone, the loop

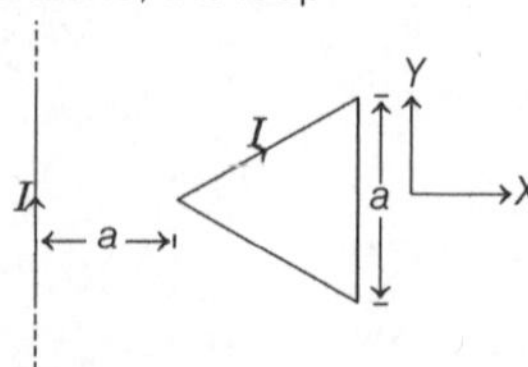

(a) must move away from wire along *X*-axis
(b) must move towards wire along *X*-axis
(c) must move along *Z*-axis
(d) must move along *Y*-axis

8 A current carrying loop is placed in the non-uniform magnetic field whose variation in space is shown in figure. Direction of magnetic field is into the plane of paper. The magnetic force experienced by loop is

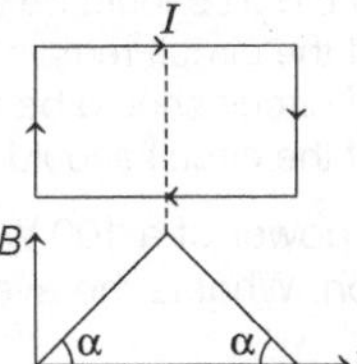

(a) non-zero (b) zero
(c) can't say anything (d) None of these

9 An electron is launched with velocity **v** in a uniform magnetic field **B**. The angle θ between **v** and **B** lies between 0 and $\frac{\pi}{2}$. Its velocity vector **v** returns to its initial value in a time interval of

(a) $\frac{2\pi m}{eB}$
(b) $\frac{2\times 2\pi m}{eB}$
(c) $\frac{\pi m}{eB}$
(d) depends upon angle between **v** and **B**

10 A straight conductor carrying a direct current of $3I$ is split into $2I$ and I as shown in the figure.

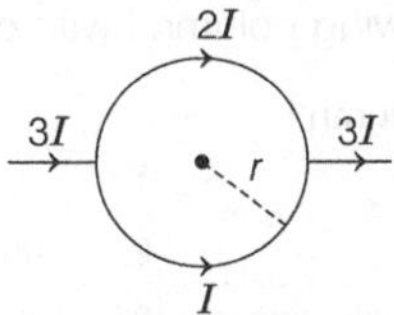

Magnetic induction at centre of loop of radius r is

(a) 0 (b) $\frac{\mu_0 I}{2\pi r}$
(c) $\frac{\pi_0 I}{2r}$ (d) $\frac{\mu_0 I}{4r}$

11 A wire in the form of a circular loop of radius r lies with its plane normal to a magnetic field B. If the wire is pulled to take a square shape in the same plane in time t, the emf induced in the loop is given by

(a) $\frac{\pi Br^2}{t}\left(1-\frac{\pi}{10}\right)$ (b) $\frac{\pi Br^2}{t}\left(1-\frac{\pi}{8}\right)$
(c) $\frac{\pi Br^2}{t}\left(1-\frac{\pi}{6}\right)$ (d) $\frac{\pi Br^2}{t}\left(1-\frac{\pi}{4}\right)$

12 The mutual inductance between two planar concentric rings of radii r_1 and r_2 (with $r_1 > r_2$) placed in air is given by

(a) $\frac{\mu_0 \pi r_2^2}{2r_1}$ (b) $\frac{\mu_0 \pi r_1^2}{2r_2}$
(c) $\frac{\mu_0 \pi (r_1 + r_2)^2}{2r_1}$ (d) $\frac{\mu_0 \pi (r_1 + r_2)^2}{2r_2}$

13 A conducting loop is placed in a magnetic field (uniform) as shown in the figure. For this situation mark the correct statement.

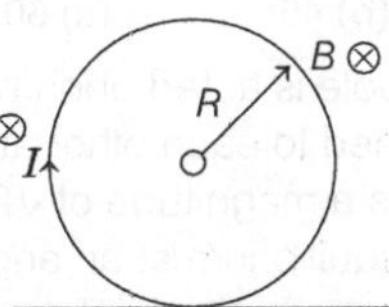

(a) The force of compression experienced by loop is IRB
(b) The force of compression experienced by loop is $2\pi IRB$
(c) The force of expansion experienced by loop is IRB
(d) The force of expansion experienced by loop is $2\pi IRB$

14 Mark the correct statement.

(a) Ideal inductor does not dissipate power in an AC circuit
(b) Ideal inductor dissipates maximum power in an AC circuit
(c) In an inductor, current lags behind the voltage by π
(d) In inductor, current leads voltage by π

15 Two very long straight parallel wires carry steady currents I and $-I$, respectively. The distance between the wires is d. At a certain instant of time, a point charge q is at a point equidistant from the wires in the plane of the wires. Its instantaneous velocity v is perpendicular to this plane. The magnitude of the force due to the magnetic field acting on the charge at this instant is

(a) $\frac{\mu_0 Iqv}{2\pi d}$ (b) $\frac{\mu_0 Iqv}{\pi d}$ (c) $\frac{2\mu_0 Iqv}{\pi d}$ (d) zero

16 An aeroplane is moving North horizontally, with a speed of $200\ \text{ms}^{-1}$, at a place where the vertical component of the earth's magnetic field is 0.5×10^{-4} T. What is the induced emf set up between the tips of the wings if they are 10 m apart?

(a) 0.01 V (b) 0.1 V (c) 1 V (d) 10 V

17 A square metal wire loop of side 10 cm and resistance 1 Ω is moved with a constant velocity v in a uniform magnetic field $B = 2$ T as shown in the figure. The magnetic field is perpendicular to the plane of the loop and directed into the paper. The loop is connected to a network of resistors, each equal to 3 Ω. What should be the speed of the loop, so as to have a steady current of 1 mA in the loop?

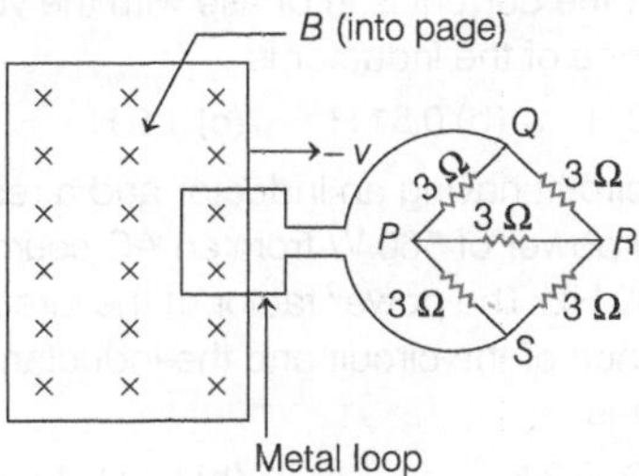

(a) $1\ \text{cm s}^{-1}$ (b) $2\ \text{cm s}^{-1}$ (c) $3\ \text{cm s}^{-1}$ (d) $4\ \text{cm s}^{-1}$

18 At a certain place, the horizontal component of earth's magnetic field is $\sqrt{3}$ times the vertical component. The angle of dip at that place is
(a) 30° (b) 45° (c) 60° (d) 90°

19 A magnetic dipole is acted upon by two magnetic fields which are inclined to each other at an angle of 75°. One of the fields has a magnitude of $\sqrt{2} \times 10^{-2}$T. The dipole attains stable equilibrium at an angle of 30° with this field. What is the magnitude of the other fields?
(a) 0.01 T (b) 0.02 T (c) 0.03 T (d) 0.04 T

20 A charge of 4μC is placed on a small conducting sphere that is located at the end of thin insulating rod of length 0.5 m. The rod rotates in horizontal plane with a constant angular velocity of 100 rads^{-1} about a vertical axis that passes through its other end. The magnetic moment of the rotating charge is
(a) zero
(b) 0.5×10^{-4} Am^2
(c) 1.25×10^{-4} Am^2
(d) magnetic moment is not defined for this case

21 A parallel plate capacitor is moving with a velocity of 25 ms^{-1} through a uniform magnetic field of 1.5 T as shown in figure. If the electric field within the capacitor plates is 175 NC^{-1} and plate area is 25×10^{-7} m^2, then the magnetic force experienced by positive charge plate is

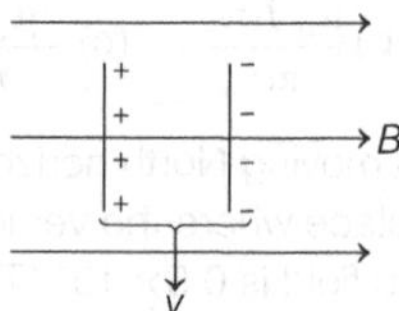

(a) 1.45×10^{-13} N (b) zero
(c) 8.67×10^{-15} N (d) 3.87×10^{-15} N

22 In an AC circuit, the potential difference V and current I are given respectively by $V = 100 \sin(100t)$ V and $I = 100 \sin\left(100t + \frac{\pi}{3}\right)$ mA. The power dissipated in the circuit will be
(a) 10^4 W (b) 10 W (c) 2.5 W (d) 5 W

23 An inductor L, a capacitor of 20 μF and a resistor of 10 Ω are connected in series with an AC source of frequency 50 Hz. If the current is in phase with the voltage, then the inductance of the inductor is
(a) 2.00 H (b) 0.51 H (c) 1.5 H (d) 0.99 H

24 An AC circuit having an inductor and a resistor in series draws a power of 560 W from an AC source marked 210 V-60 Hz. The power factor of the circuit is 0.8, the impedance of the circuit and the inductance of the inductor is
(a) 65 Ω, 0.2 H (b) 64 Ω, 1.0 H
(c) 63 Ω, 0.1 H (d) 50 Ω, 1.5 H

25 The resonant frequency and Q-factor of a series L-C-R circuit with $L = 3.0$ H, $C = 27\ \mu$F and $R = 7.4\ \Omega$. How will you improve the sharpness of resonance of the circuit by reducing its full width at half maximum by a factor of 2?
(a) Resistance of circuit should be increased
(b) Resistance of the circuit remain same
(c) Resistance of circuit should be increased by 3.7 Ω
(d) Resistance of the circuit should be reduced to 3.7 Ω

26 About 5% of the power of a 100 W light bulb is converted to visible radiation. What is the average intensity of visible radiation
(i) at a distance of 1 m from the bulb?
(ii) at a distance of 10 m?
Assume that the radiation is emitted isotropically and neglect reflection.
(a) 0.2 Wm^{-2}, 0.002 Wm^{-2} (b) 0.6 Wm^{-2}, 0.006 Wm^{-2}
(c) 0.3 Wm^{-2}, 0.003 Wm^{-2} (d) 0.4 Wm^{-2}, 0.004 Wm^{-2}

27 A charged particle oscillates about its mean equilibrium position with a frequency of 10^9Hz. Frequency of the electromagnetic waves produced by the oscillator is
(a) 10 Hz (b) 10^5 Hz (c) 10^9 Hz (d) 10^{10} Hz

28 The amplitude of the magnetic field part of a electromagnetic wave in vacuum is $B_0 = 510$ nT. What is the amplitude of the electric field part of the wave?
(a) 140 NC^{-1} (b) 153 NC^{-1}
(c) 163 NC^{-1} (d) 133 NC^{-1}

29 Match the following column I with column II.

	Column I		Column I
A.	Capacitor	1.	increases AC
B.	Inductor	2.	reduces AC
C.	Energy dissipation is due to	3.	is conductor for DC
D.	A transformer	4.	resistance only

Code

	A	B	C	D
(a)	2	2, 3	4	1, 2
(b)	4	3,4	2	2,3
(c)	1	2,3	4	2
(d)	3	2	4	1

30 Match the following of column I with column II.

	Column I		Column II
A.	Lorentz force	1.	$\oint \mathbf{E}.d\mathbf{A} = \frac{q}{\varepsilon_0}$
B.	Gauss's law	2.	$d\mathbf{B} = \frac{\mu_0}{4\pi} \frac{Id\mathbf{l} \times \mathbf{r}}{r^3}$
C.	Biot-Savart law	3.	$\mathbf{F} = q[\mathbf{E} + (\mathbf{v} \times \mathbf{B})]$
D.	Coulomb's law	4.	$F = \frac{1}{4\pi\varepsilon_0} \frac{q_1 q_2}{r^2}$

Code

	A	B	C	D		A	B	C	D
(a)	3	1	2	4	(b)	1	2	3	4
(c)	4	3	2	1	(d)	2	1	4	3

31 Match the following of column I with column II.

Column I		Column II	
A.	Ultraviolet	1.	Radar system
B.	Infrared	2.	Roengten
C.	X-rays	3.	Heat radiation
D.	Microwaves	4.	Water purification

Code

	A	B	C	D		A	B	C	D
(a)	4	3	2	1	(b)	1	2	3	4
(c)	2	1	4	3	(d)	3	2	4	1

32 **Assertion** (A) Two identical heaters are connected to two different sources one DC and other AC having same potential difference across their terminals. The heat produced in heater supplied with DC source is greater.

Reason (R) The net impedance of an AC source is greater than resistance.

(a) Both Assertion and Reason are true and Reason is the correct explanation of Assertion
(b) Both Assertion and Reason are false
(c) Assertion is true but Reason is false
(d) Both Assertion and Reason are true but Reason is not true explanation of Assertion

33 **Assertion** (A) Induction coils are made of copper.

Reason (B) Induced current is more in wire having less resistance

(a) Both Assertion and Reason are true and Reason is the correct explanation of Assertion
(b) Both Assertion and Reason are false
(c) Assertion is true but Reason is false
(d) Both Assertion and Reason are true but Reason is not true explanation of Assertion

Direction (Q. Nos. 34-40) *Each of these questions contains two statements : Statement* I *and Statement* II. *Each of these questions also has four alternative choices, only one of which is the correct answer. You have to select one of the codes (a), (b), (c), (d) given below*

(a) Statement I is true, Statement II is true; Statement II is the correct explanation for Statement I
(b) Statement I is true, Statement II is true; Statement II is not the correct explanation for Statement I
(c) Statement I is true; Statement II is false
(d) Statement I is false; Statement II is true

34 **Statement I** A flexible wire loop of irregular shape carrying current when placed in a uniform external magnetic field acquires circular shape.

Statement II Any current carrying loop when placed in external magnetic field tries to acquires minimum energy and hence maximum magnetic flux and for a given perimeter circular shape is having greatest area.

35 **Statement I** A hanging spring is attached to a battery and switch. On closing the switch a current suddenly flows in the spring, as a result spring compresses.

Statement II When two current carrying coils are placed close to each other in same plane, then they attract each other if sense of current is same in both.

36 **Statement I** No current is induced in a metal loop if it is rotated in an electric field.

Statement II The electric flux through the loop does not change with time.

37 **Statement I** The power factor of an inductor is zero.

Statement II In the inductor, the emf and current differ in phase by $\frac{\pi}{2}$.

38 **Statement I** In a series *R-L-C* circuit the voltage across-resistor, inductor and capacitor are 8 V, 16 V and 10 V respectively. The resultant emf in the circuit is 10 V.

Statement II Resultant emf of the circuit is given by the relation $E = \sqrt{V_R^2 + (V_L - V_C)^2}$.

39 **Statement I** X-rays travel faster than light waves in vacuum.

Statement II The energy of X-rays photon is greater than the light photon.

40 **Statement I** An electron moving in the positive *x*-direction enters a region where uniform electric and magnetic fields exist perpendicular to each other. The electric field is in the negative *y*-direction. If the electron travels undeflected in this region, the direction of the magnetic field is along the negative *z*-axis.

Statement II If a charged particle moves in a direction perpendicular to a magnetic field, the direction of the force acting on it is given by Fleming's left hand rule.

ANSWERS

1. (d)	**2.** (a)	**3.** (c)	**4.** (d)	**5.** (b)	**6.** (a)	**7.** (a)	**8.** (b)	**9.** (a)	**10.** (d)
11. (d)	**12.** (a)	**13.** (d)	**14.** (a)	**15.** (d)	**16.** (b)	**17.** (b)	**18.** (a)	**19.** (a)	**20.** (b)
21. (a)	**22.** (c)	**23.** (b)	**24.** (c)	**25.** (d)	**26.** (d)	**27.** (c)	**28.** (b)	**29.** (a)	**30.** (a)
31. (a)	**32.** (d)	**33.** (a)	**34.** (a)	**35.** (a)	**36.** (c)	**37.** (a)	**38.** (a)	**39.** (d)	**40.** (b)

Hints and Explanations

1 Whatever be the current through solenoid winding, the direction of magnetic field is along the axis of solenoid and hence the magnetic force experienced by wire is zero and hence its weight cannot be supported by magnetic force.

2 Drift speed of electron,

$$v_d = \frac{I}{neA} = \frac{5}{10^{29} \times 1.6 \times 10^{-19} \times 10^{-5}}$$
$$= 3.125 \times 10^{-5}\ \text{ms}^{-1}$$

Average magnetic force experienced by each electron is,

$$F = qvB$$
$$= 1.6 \times 10^{-19} \times 3.125 \times 10^{-5} \times 0.1$$
$$= 0.5 \times 10^{-24}\ \text{N} = 5 \times 10^{-25}\ \text{N}$$

3 Length of wire AB, $l = 5$ m

Current, $i = 2$ A, $\theta = 37°$ and $B = 0.5$ T

Magnetic force on wire, $F = iBl \sin\theta$

$$= 2 \times 0.5 \times 5 \times \sin 37° = 3\ \text{N}$$

5 $\therefore$ Induced charge, $q = \dfrac{\text{Change of flux}}{\text{Resistance}}$

$$= \frac{\phi_f - \phi_i}{R}$$

But final area = 0, therefore, $\phi_f = 0$. Numerically, $\phi_i = BA$. Therefore, $q = BA/R$.

6 $B_A = B_B = \dfrac{\mu_0 I}{2\pi \times \sqrt{2}a}$

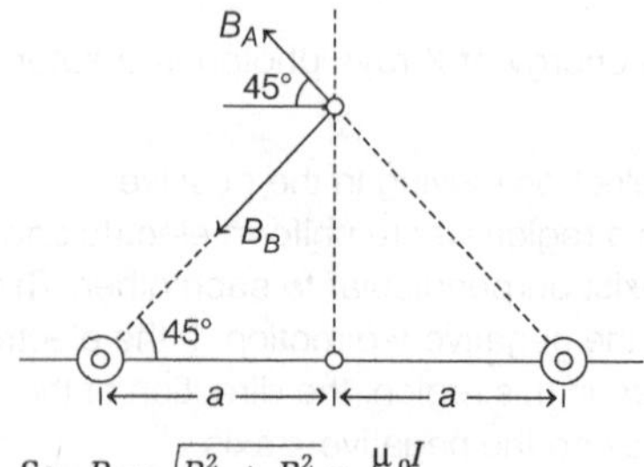

So, $B = \sqrt{B_A^2 + B_B^2} = \dfrac{\mu_0 I}{2\pi a}$

7 The net force acting on loop would be along X-axis (to determine whether it is along positive or negative, X-axis calculation has to be carried out) as shown in figure.

8 Each and every pair of loop elements located symmetrically w.r.t. central line experiences zero net force. So, total magnetic force experienced by loop is zero.

9 Time interval in which its velocity v returns to its initial value is same as time period of the particle, to execute the circle.

Since, it does not depend upon θ. So,

$$\text{time required} = \frac{2\pi m}{eB}.$$

10 $\therefore B_{\text{net}} = \dfrac{\mu_0 2I}{4r} - \dfrac{\mu_0 I}{4r} = \dfrac{\mu_0 I}{4r}$

11 Induced emf

$$(e) = \frac{\text{Magnetic field} \times \text{change in area}}{\text{Time}}$$
$$= \frac{B\Delta A}{t}$$

Since, the circumference of the circular loop $= 2\pi r$,

The side of the square loop $= \dfrac{2\pi r}{4} = \dfrac{\pi r}{2}$

Therefore,

$$\Delta A = \pi r^2 - \left(\frac{\pi r}{2}\right)^2 = \pi r^2\left(1 - \frac{\pi}{4}\right)$$
$$\therefore \quad e = \frac{B(\pi r^2)}{t}\left(1 - \frac{\pi}{4}\right)$$

12 Magnetic field due to the larger coil at its centre is

$$B = \frac{\mu_0 I}{2r_1}$$

where, I is the current in the larger coil. Flux through the inner coil is

$$\phi = B \times \pi r_2^2 = \frac{\mu_0 I}{2r_1} \times \pi r_2^2$$

But $\quad \phi = MI$

Therefore, $\quad M = \dfrac{\mu_0 \pi r_2^2}{2r_1}$

13 By right hand thumb rule force of expansion would act on it whose magnitude is given by

$$dT = I(dl \times B)$$
$$\Rightarrow T = IB\int dl$$
$$= 2\pi IRB$$

14 The instantaneous voltage and current in an AC circuit containing an ideal inductance only are

$$E = E_0 \sin\omega t,\ I = I_0 \sin\left(\omega t - \frac{\pi}{2}\right)$$
$$P_{\text{ins}} = EI = E_0 I_0 \sin\omega t \sin\left(\omega t - \frac{\pi}{2}\right)$$
$$= -E_0 I_0 \sin\omega t \cos\omega t = -\frac{1}{2} E_0 I_0 \sin 2\omega t$$

Average power for one complete cycle is

$$P = \frac{1}{T}\left(-\frac{1}{2}E_0 i_0\right)\int_0^T \sin 2\omega t\, dt = 0$$

15 Since, **B** and v are anti-parallel to each other, angle between the two is zero or 180°.

$$\mathbf{F} = q\,v \times \mathbf{B} = qv\,B \sin\theta$$

For $\quad \theta = 0, \sin 0 = 0 \Rightarrow F = 0$

16 $\therefore$

$$e = Blv = 0.5 \times 10^{-4} \times 10 \times 200 = 0.1\ \text{V}$$

17 The network $PQRS$ is a balanced Wheatstone's bridge. Hence, the resistance of 3 Ω between P and R is ineffective. The net resistance of the network, therefore, is 3 Ω. Total resistance $R = 3\ \Omega + 1\ \Omega = 4\ \Omega$.

Now, induced emf is

$$e = Blv = 2 \times 0.1 \times v = 0.2v.$$

$\therefore$ Induced current $I = \dfrac{e}{R} = \dfrac{0.2v}{4}$

Given, $\quad I = 1 \times 10^{-3}$ A

Hence, $\quad 1 \times 10^{-3} = \dfrac{0.2v}{4}$

Which gives, $v = 2 \times 10^{-2}\ \text{ms}^{-1} = 2\ \text{cms}^{-1}$

18 $\therefore B_H = B\cos\theta$ and $B_V = B\sin\theta$

Hence, $\quad \dfrac{B_V}{B_H} = \tan\theta$

Given, $\quad \dfrac{B_V}{B_H} = \dfrac{1}{\sqrt{3}}$

Therefore, $\quad \tan\theta = \dfrac{1}{\sqrt{3}}$, i.e. $\theta = 30°$

19 Refer to figure. Let θ_1 (= 30°) be the angle between the magnetic moment vector **m** and the field vector $\mathbf{B}_1$ (= 1.5×10^{-2} T). Then, as shown in figure, the angle between m and the other field $\mathbf{B}_2$ will be $\theta_2 = 75° - 30° = 45°$.

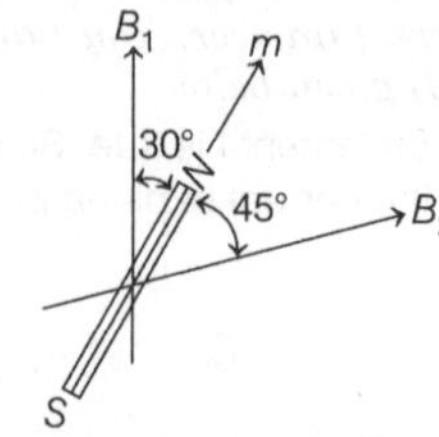

The field $\mathbf{B}_1$ exerts a torque $\tau_1 = \mathbf{m} \times \mathbf{B}_1$ on the dipole and the field $\mathbf{B}_2$ exerts a torque $\tau_2 = \mathbf{m} \times \mathbf{B}_2$, where **m** is the magnetic moment of the dipole.

Since, the dipole is in stable equilibrium, the net torque τ $(= \tau_1 + \tau_2)$ must be zero, i.e. the two torques must be equal and opposite. In terms of magnitudes, we have

$$mB_1 \sin\theta_1 = mB_2 \sin\theta_2$$

or $$B_2 = \frac{B_1 \sin\theta_1}{\sin\theta_2}$$

$$= \frac{\sqrt{2}\times 10^{-2} \times \sin 30^\circ}{\sin 45^\circ} = 0.01\,\text{T}$$

20 A moving charge along a circle is equivalent to a current carrying ring, whose current is given by

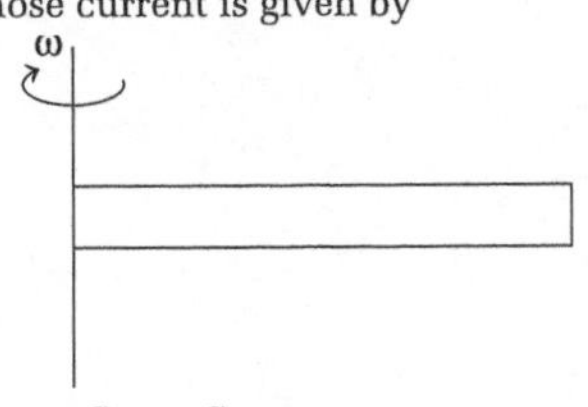

$$I = \frac{q}{T} = \frac{q}{2\pi/\omega}$$

$$\Rightarrow I = \frac{4\times 10^{-6}}{2\pi} \times 100 = 0.64 \times 10^{-4}\,\text{A}$$

Magnetic moment of rotating charge is $M = IA$

$$\Rightarrow M = 0.64 \times 10^{-4} \times \pi \times (0.5)^2$$

$$= 0.5 \times 10^{-4}\,\text{Am}^2$$

21 Electric field in between the capacitor plates is given by

$$E = \frac{q}{\varepsilon_0 A}$$

where, q is the charge on capacitor.

$$q = \varepsilon_0 A \times E$$

$$= 8.85 \times 10^{-12} \times 25 \times 10^{-7} \times 175$$

$$= 3.87 \times 10^{-15}\,\text{C}$$

Magnetic force experienced by positive plate is,

$F_m = qvB = 1.45 \times 10^{-13}$ N in a direction out of plane of paper.

22 Voltage amplitude $V_0 = 100$ V,

current amplitude $I_0 = 100$ mA

$= 100 \times 10^{-3}$ A

and phase difference between I and V is $\phi = \frac{\pi}{3} = 60^\circ$.

Now, power dissipated is given by

$$P = \frac{V_0 I_0}{2} \cos\phi$$

$$= \frac{100 \times 100 \times 10^{-3}}{2} \times \cos 60^\circ = 2.5\,\text{W}$$

23 In an L-C-R circuit, the current and the voltage are in phase $(\phi = 0)$, when

$$\tan\phi = \frac{\omega L - \frac{1}{\omega C}}{R} = 0$$

or $$\omega L = \frac{1}{\omega C} \text{ or } L = \frac{1}{\omega^2 C}$$

Here, $\omega = 2\pi f = 2 \times 3.14 \times 50\,\text{s}^{-1}$

$= 314\,\text{s}^{-1}$ and $C = 20$

$\mu\text{F} = 20 \times 10^{-6}$ F

$$\therefore L = \frac{1}{(314\,\text{s}^{-1})^2 \times (20 \times 10^{-6}\,\text{F})} = 0.51\,\text{H}$$

24 The average power over a complete cycle is

$$P = E_{rms} \times I_{rms} \times \cos\phi$$

where, $\cos\phi$ is the power factor

$$\therefore I_{rms} = \frac{P}{E_{rms} \times \cos\phi} = \frac{560}{210 \times 0.8}$$

$$= \frac{10}{3}\,\text{A}$$

The impedance of the circuit is

$$Z = \frac{E_{rms}}{I_{rms}} = \frac{210\,\text{V}}{(10/3)\,\text{A}} = 63\,\Omega$$

The power is consumed in R only. Therefore,

$$P = (I_{rms})^2 R \text{ or } R = \frac{560}{\left(\frac{10}{3}\right)^2} = 50.4\,\Omega$$

Now, the impedance of an L-R circuit is

$$Z = \sqrt{R^2 + (\omega L)^2}$$

$$\therefore (\omega L)^2 = Z^2 - R^2 = (63)^2 - (50.4)^2$$

$$= 1428.84\,\Omega^2$$

or $$\omega L = \sqrt{1428.84} = 37.8\,\Omega$$

$$\therefore L = \frac{37.8}{2\pi f} = \frac{37.8}{2 \times 3.14 \times 60} = 0.1\,\text{H}$$

25 Given, $L = 3.0$ H,

$C = 27\,\mu\text{F} = 27 \times 10^{-6}$F

and $R = 7.4\,\Omega$

The resonant frequency is given by

$$\omega_0 = \frac{1}{\sqrt{LC}} = \frac{1}{\sqrt{3.0 \times (27 \times 10^{-6})}}$$

$$= 111\,\text{rads}^{-1}$$

The Q-factor of the circuit is given by

$$Q = \frac{\omega_0 L}{R} = \frac{111 \times 3.0}{7.4} = 45$$

The "bandwidth" (the difference of half-power frequencies is given by

$$\omega_2 - \omega_1 = R/L$$

Smaller the value of $(\omega_2 - \omega_1)$, sharper is the resonance. To reduce $(\omega_2 - \omega_1)$ by a factor of 2, the resistance R should be halved that is, the resistance of the circuit should be reduced to $3.7\,\Omega$.

26 Power converted into visible radiation

$$P = \frac{5}{100} \times 100 = 5\,\text{W}$$

$$\text{Intensity} = \frac{\text{Energy}}{\text{Area} \times \text{Time}}$$

$$= \frac{\text{Power}}{\text{Area}} = \frac{P}{4\pi r^2}$$

(i) Intensity $= \frac{5}{4 \times 3.14 \times 1 \times 1}$

$= 0.4\,\text{Wm}^{-2}$

(ii) Intensity $= \frac{5}{4 \times 3.14 \times 10 \times 10}$

$= 0.004\,\text{Wm}^{-2}$

27 Frequency of electromagnetic wave = Frequency of oscillation of charge.

28 $\frac{E_0}{B_0} = c$ or $E_0 = cB_0$

32 Heat (AC) $= V_{rms} I_{rms} = \frac{V_{max} I_{max}}{2}$

Heat (DC) $= V_{max} I_{max}$

33 Since, copper consists of a very small ohmic resistance so, inductance coils are made of copper. A large induced current is produced in such.

34 Each section of irregular wire loop experiences an outward force.

35 Opposite polarities are produced in the forces of spring loops.

36 A current is induced in a loop only if magnetic flux linked with the coil changes.

37 $\therefore \cos\phi = \cos\frac{\pi}{2} = 0$

38 The resultant emf in the L-C-R circuit is given by

$$E = \sqrt{V_R^2 + (V_L - V_C)^2}$$

$$E = \sqrt{(8)^2 + (16-10)^2} = \sqrt{64+36}$$

$$E = 10\,\text{V}$$

39 Since, X-rays are electromagnetic waves, we know that electromagnetic wave travels with same velocity of light in vacuum.

Now, from the formula

$$E = \frac{hc}{\lambda}$$

The wavelength of X-rays are small than light waves and energy is inversely proportional to the wavelength. Hence, the energy of X-rays photon will be greater than light waves.

40 Because electron has a negative charge, an electric field in the negative y-direction will deflect it in the positive y-direction. It will travel undeflected if the magnetic field imparts an equal deflection in the negative y-direction. Since, the magnetic force is perpendicular to the magnetic field and the charge of electron is negative, the direction of the magnetic field (according to Fleming's left hand rule) should be along the negative z-direction.

DAY TWENTY SEVEN

Ray Optics

Learning & Revision for the Day

- Reflection of Light
- Mirror formula
- Refraction of Light
- Lens
- Total Internal Reflection (TIR)
- Deviation by a Prism
- Dispersion by a Prism
- Refraction Through a Prism
- Scattering of Light

Reflection of Light

The phenomena of bouncing back of light on striking a smooth surface is called reflection of light.

- According to the laws of reflection, (i) the incident ray, reflected ray and the normal drawn on the reflecting surface at the point of incidence lie in the same plane and (ii) the angle of incidence $\angle i =$ angle of reflection $\angle r$.
- Laws of reflection are true for reflection from a polished mirror or from an unpolished surface or for diffused reflection.
- Whenever reflection takes place from a denser medium, the reflected rays undergo a phase change of π.

Reflection from a Plane Mirror

- If a ray is incident on a plane mirror at an angle of incidence i, then it suffers a deviation of $(\pi - 2i)$ and for two inclined plane mirrors deviation is $(360° - 2\theta)$.
- While keeping an object fixed, a plane mirror is rotated in its plane by an angle θ, then the reflected ray rotates in the same direction by an angle 2θ.
- Focal length as well as the radius of curvature of a plane mirror is infinity. Power of a plane mirror is zero.
- If two plane mirrors are inclined to each other at an angle θ, the total number of images formed of an object kept between them, is $n = \dfrac{2\pi}{\theta}$ or $\left(\dfrac{2\pi}{\theta} - 1\right)$, when it is odd.
- The minimum size of a plane mirror fixed on a wall of a room, so that a person at the centre of the room may see the full image of the wall behind him, should be 1/3rd the size of the wall.

Reflection from a Spherical Mirror

- A spherical mirror is a part of a hollow sphere whose one surface is polished, so that it becomes reflecting. The other surface of the mirror is made opaque.

PREP MIRROR

Your Personal Preparation Indicator

- No. of Questions in Exercises (x)—
- No. of Questions Attempted (y)—
- No. of Correct Questions (z)—
  (Without referring Explanations)

- Accuracy Level $(z/y \times 100)$—
- Prep Level $(z/x \times 100)$—

In order to expect good rank in JEE, your Accuracy Level should be above 85 & Prep Level should be above 75.

- Image formed by a concave mirroris is virtual and erect, when the object is placed between the pole and the principal focus of concave mirror. In all other cases, the image formed is real and inverted one.
- Image formed by a convex mirror is virtual, erect and diminished in size irrespective of the position of the object. Moreover, image is formed in between the pole and the principal focus of the mirror.
- The focal length of a spherical mirror is half of its radius of curvature, i.e. $f = \frac{R}{2}$.

Mirror Formula

Let an object be placed at a distance u from the pole of a mirror and its image is formed at a distance v from the pole. Then, according to mirror formula, $\frac{1}{v} + \frac{1}{u} = \frac{1}{f}$

- The power of a mirror (in dioptre), is given as

$$P = \frac{1}{f(\text{in metre})}$$

- If a thin object of height h is placed perpendicular to the principal axis of a mirror and the height of its image be h', then the transverse or lateral magnification produced is given by

$$m = \frac{h'}{h} = -\frac{v}{u} = \frac{f}{f-u} = \frac{f-v}{f}$$

Negative sign of magnification means the inverted image and positive sign means an erect image.

- When a small sized object is placed along the principal axis, then its longitudinal (or axial) magnification is given by

$$\text{Axial magnification} = -\frac{dv}{du} = \left(\frac{v}{u}\right)^2 = \left(\frac{f}{f-u}\right)^2 = \left(\frac{f-v}{f}\right)^2$$

Refraction of Light

When light passes from one medium, say air, to another medium, say glass, a part is reflected back into the first medium and the rest passes into the second medium. When it passes into the second medium, it either bends towards the normal or away from the normal.

This phenomenon is known as refraction.

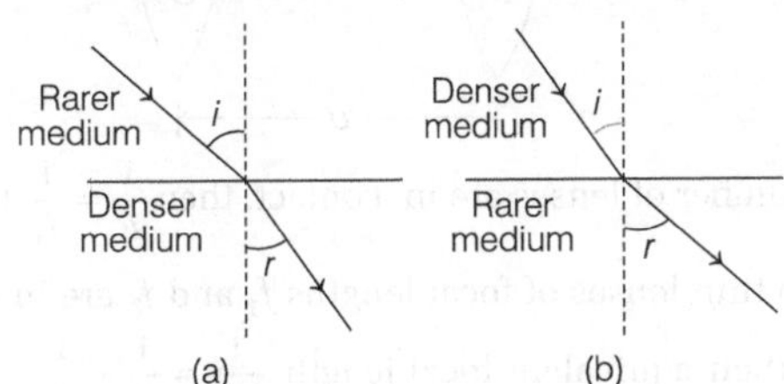

Fig. (a) and (b) shows refraction of light

Refractive Index

For a given pair of media, the ratio of the sine of angle of incidence (i) to the sine of angle of refraction (r) is a constant, which is called the refractive index of second medium, w.r.t. first medium.

Thus, $$\frac{\sin i}{\sin r} = \text{constant} = n_{21} = \frac{n_2}{n_1}$$

This is also called **Snell's law**.

- Refractive index is a unitless, dimensionless and a scalar quantity.
- The refractive index of a medium w.r.t. vacuum (or free space) is known as its absolute refractive index. It is defined as the ratio of the speed of light in vacuum (c) to the speed of light in a given medium (v).

$$\therefore \quad n = \frac{c}{v}$$

Value of absolute refractive index of a medium can be 1 or more than 1, but never less than 1.

- When light travels from one material medium to another, the ratio of the speed of light in the first medium to that in the second medium is known as the relative refractive index of second medium, w.r.t. the first medium. Thus,

$$n_{21} = \frac{v_1}{v_2} = \frac{c/v_2}{c/v_1} = \frac{n_2}{n_1}$$

- When light undergoing refraction through several media finally enters the first medium itself, then

$$n_{21} \times n_{32} \times n_{13} = 1 \text{ or } n_{32} = \frac{n_{31}}{n_{21}}$$

- When the object is in denser medium and the observer is in rarer medium, then real and apparent depth have the relationship, $\frac{\text{Real depth}}{\text{Apparent depth}} = n_{21}$

i.e. real depth > apparent depth shift, $y = h - h' = \left(1 - \frac{1}{n_{21}}\right)h$

where, h and h' are real and apparent depths.

Refraction from a Spherical Surface

Let an object be placed in a medium of refractive index n_1 at a distance u from the pole of a spherical surface of radius of curvature R and after refraction, its image is formed in a medium of refractive index n_2 at a distance v, then

$$\frac{n_2}{v} - \frac{n_1}{u} = \frac{n_2 - n_1}{R}$$

The relation is true for all surfaces, whether the image formed is real or virtual.

Lens

A lens is part of a transparent refracting medium bound by two surfaces, with atleast one of the two surfaces being a curved one. The curved surface may be spherical or cylindrical.

The lens formula is given by $\frac{1}{v} - \frac{1}{u} = \frac{1}{f}$

For a thin object of height h placed perpendicular to the principal axis at a distance u, if the height of image formed is h', then lateral or transverse magnification m is given by

$$m = \frac{h'}{h} = \frac{v}{u} = \frac{f}{f+u} = \frac{f-v}{f}$$

For a small sized object placed linearly along the principal axis, its **axial** or **longitudinal magnification** is given by

$$\text{Axial magnification} = -\frac{dv}{du} = \left(\frac{v}{u}\right)^2 = \left(\frac{f}{f+u}\right)^2 = \left(\frac{f-v}{f}\right)^2$$

Silvering of Lens

When one surface of a lens is silvered, it behaves as a mirror.

The focal length of silvered lens is $\frac{1}{F} = \frac{2}{f_l} + \frac{1}{f_m}$

where, f_l is focal length of plane convex lens and f_m is focal length of corresponding mirror.

In case of plano-convex lens

- **When curved surface is silvered**, then focal length of silvered lens is $F = \frac{R}{2\mu}$

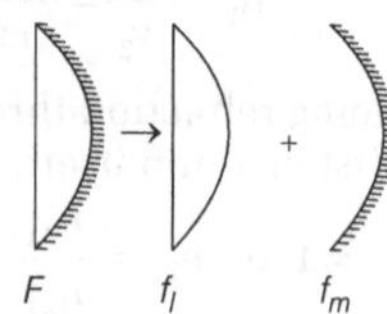

where, $R = 2f_m$ or $R = f_l(\mu - 1)$

- **When plane surface is silvered**, then

$$F = \frac{R}{2(\mu-1)} \text{ and } f_l = \frac{R}{(\mu-1)}, f_m = \infty$$

F f_l f_m

- **When double convex lens is silvered**, then

$$F = \frac{R}{2(2\mu-1)} \text{ and } f_l = \frac{R}{2(\mu-1)} \Rightarrow f_m = \frac{R}{2}.$$

F f_l f_m

Power of a Lens

The power of a lens is mathematically given by the reciprocal of its focal length, i.e. power $P = \frac{1}{f(\text{m})}$

SI unit of power is dioptre (D). Power of a converging lens is positive and that of a diverging lens is negative.

Lens Maker's Formula

For a lens having surfaces with radii of curvature R_1 and R_2 respectively, its focal length is given by

$$P = \frac{1}{f} = (n_{21} - 1)\left(\frac{1}{R_1} - \frac{1}{R_2}\right)$$

where, $n_{21} = \frac{n_2}{n_1}$ = refractive index of the lens material w.r.t. the surroundings.

Cutting and Combination of a Lens

If a symmetrical convex lens of focal length f is cut into two parts along its optic axis, then focal length of each part (a plano-convex lens) is $2f$. However, if the two parts are joined as shown in the figure, the focal length of the combination is again f.

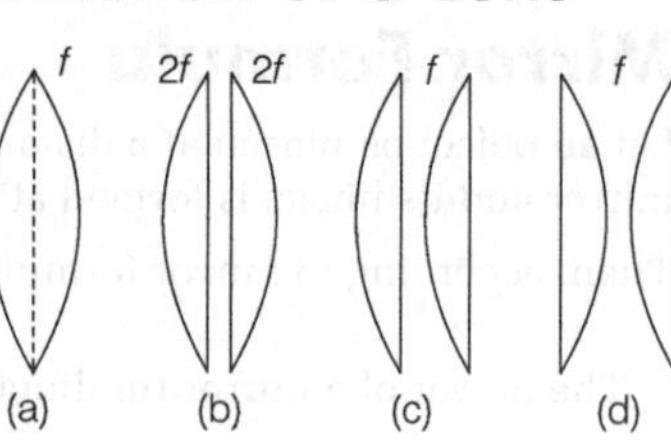

If a symmetrical convex lens of focal length f is cut into two parts along the principal axis, then the focal length of each part remains unchanged, as f [Fig. f]. If these two parts are joined with the curved ends on one side, the focal length of the combination is $\frac{f}{2}$ [Fig. g]. But on joining the two parts in opposite sense, the net focal length becomes ∞ (or net power = 0) (Fig. h).

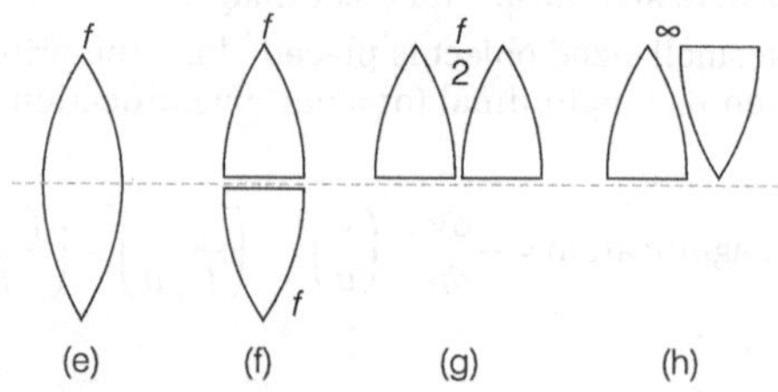

- The equivalent focal length of co-axial combination of two lenses is given by $\frac{1}{F} = \frac{1}{f_1} + \frac{1}{f_2} - \frac{d}{f_1 f_2}$

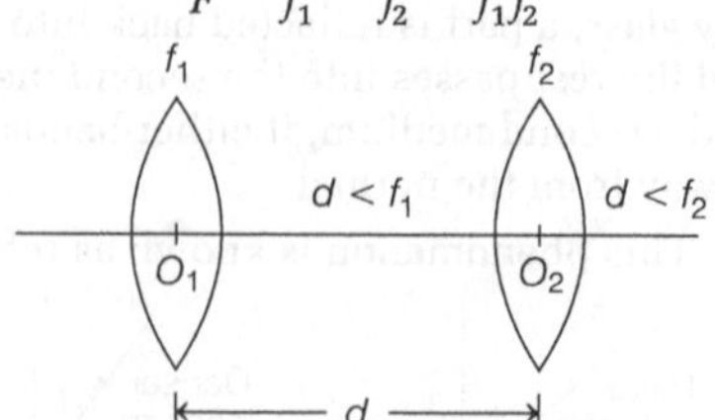

- If a number of lenses are in contact, then $\frac{1}{F} = \frac{1}{f_1} + \frac{1}{f_2} + \ldots$
- If two thin lenses of focal lengths f_1 and f_2 are in contact, then their equivalent focal length $\frac{1}{f_{eq}} = \frac{1}{f_1} + \frac{1}{f_2}$

In terms of power, $P_{eq} = P_1 + P_2$

Important Results

- Effective diameter of light transmitting area is called aperture. Intensity of image $\propto$ (Aperture)^2
- Relation between object and image speed, $v_i = \left(\frac{f}{f+u}\right)^2 v_o$
- Newton's formula states, $f^2 = x_1 x_2$ where, x_1 and x_2 = distance of object; and image from first and second principal foci. This formula is also called thin lens formula.
- If lens immersed in a liquid, then $\frac{f_l}{f_a} = \frac{({}_a\mu_g - 1)}{({}_l\mu_g - 1)} = \frac{({}_a\mu_g - 1)}{\left(\frac{{}_a\mu_g}{{}_a\mu_l} - 1\right)}$

Total Internal Reflection (TIR)

When a ray of light goes from a denser to a rarer medium, it bends away from the normal. For a certain angle of incidence i_C, the angle of refraction in rarer medium becomes 90°. The angle i_C is called the **critical angle.**

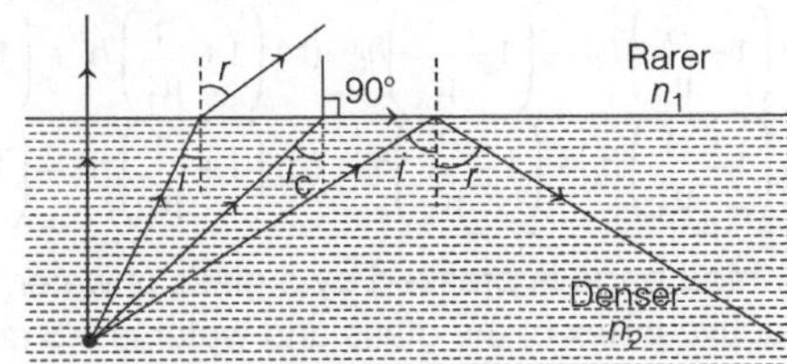

TIR in a medium

$$\sin i_C = \frac{n_1}{n_2} = \frac{1}{n_{21}}$$

For the angle of incidence greater than the critical angle ($i > i_C$) in the denser medium, the light ray is totally internally reflected back into the denser medium itself.

Conditions for Total Internal Reflection

- The light ray should travel from the denser medium towards the rarer medium.
- The angle of incidence should be the greater than the critical angle.

Common Examples of Total Internal Reflection

- **Looming** An optical illusion in cold countries.
- **Mirage** An optical illusion in deserts.
- **Brilliance of diamond** Due to repeated internal reflections diamond sparkles.
- **Optical fibre** Each fibre consists of core and cladding. The refractive index of core material is higher than that of cladding. Light entering at small angle on one end undergoes repeated total internal reflections along the fibre and finally comes out.

Deviation by a Prism

A prism is a homogeneous, transparent medium bounded by two plane surfaces inclined at an angle A with each other. These surfaces are called as refracting surfaces and the angle between them is called angle of prism A.

Deviation produced by a prism is,

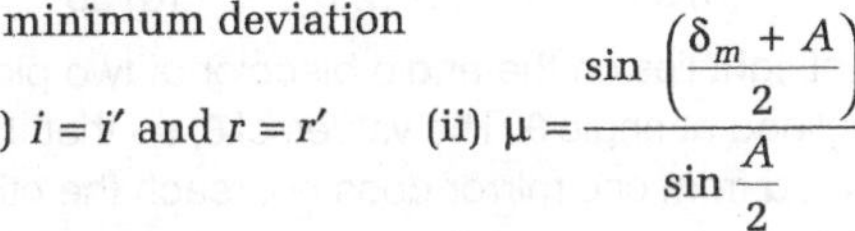

$$\delta = i + i' - A$$

$\Rightarrow r + r' = A$

For grazing incidence $i = 90°$

and grazing emergence $i' = 90°$

For minimum deviation

(i) $i = i'$ and $r = r'$ (ii) $\mu = \frac{\sin\left(\frac{\delta_m + A}{2}\right)}{\sin\frac{A}{2}}$

In case of minimum deviation, ray is passing through prism symmetrically.

For maximum deviation (δ_{max}), $i = 90°$ or $i' = 90°$

For thin prism, $\delta = (\mu - 1)A$

Dispersion by a Prism

Dispersion of light is the phenomenon of splitting of white light into its constituent colours on passing light through a prism. This is because different colours have different wavelength and hence different refractive indices.

Angular dispersion $= \delta_v - \delta_r = (n_v - n_r)A$

where, n_v and n_r represent refractive index for violet and red lights.

Dispersive power, $\omega = \frac{n_v - n_r}{n - 1}$, where $n = \frac{n_v + n_r}{2}$ is the mean refractive index.

By combining two prisms with angle A and A' and refractive index n and n' respectively, we can create conditions of

- Dispersion without deviation when, $A' = -\frac{(n-1)A}{(n'-1)}$
- Deviation without dispersion when,
$$A' = -\left[\frac{n_v - n_r}{n'_v - n'_r}\right]A$$

Refraction Through a Prism

A ray of light suffers two refractions at the two surfaces on passing through a prism. Angle of deviation through a prism $\delta = i + e - A$. where, i is the angle of incidence, e is the angle of emergence and A is the angle of prism.

Scattering of Light

Molecules of a medium after absorbing incoming light radiations, emit them in all directions. This phenomenon is called scattering. According to Rayleigh, intensity of scattered light $\propto \frac{1}{\lambda^4}$.

There are some phenomenon based on scattering

- Sky looks blue due to scattering.
- At the time of sunrise and sunset, sun looks reddish.
- Danger signals are made of red colour.

DAY PRACTICE SESSION 1

FOUNDATION QUESTIONS EXERCISE

1 To get three images of a single object, one should have two plane mirrors at an angle of

(a) 60° (b) 90° (c) 120° (d) 30°

2 A source of light lies on the angle bisector of two plane mirrors inclined at angle θ. The values of θ, so that the light reflected from one mirror does not reach the other mirror will be

(a) $\theta \geq 120°$ (b) $\theta \geq 90°$
(c) $\theta \leq 120°$ (d) $\theta < 30°$

3 A beam of light composed of red and green rays is incident obliquely at a point on the face of a rectangular glass slab. When coming out of the opposite parallel face, the red and green rays emerge from

(a) two points propagating in two different non-parallel directions
(b) two points propagating in two different parallel directions
(c) one point propagating in two different directions
(d) one point propagating in the same direction

4 The optical path of a monochromatic light is same if it goes through 4.0 cm of glass or 4.5 cm of water. If the refractive index of glass is 1.53, the refractive index of the water is

(a) 1.30 (b) 1.36
(c) 1.42 (d) 1.46

5 On a hot summer night, the refractive index of air is smallest near the ground and increases with height from the ground. When a light beam is directed horizontally as it travels, the light beam → JEE Main 2015

(a) becomes narrower
(b) goes horizontally without any deflection
(c) bends downwards
(d) bends upwards

6 A transparent solid cylinder rod has a refractive index of $\frac{2}{\sqrt{3}}$. It is surrounded by air. A light ray is incident at the mid-point of one end of the rod as shown in the figure.

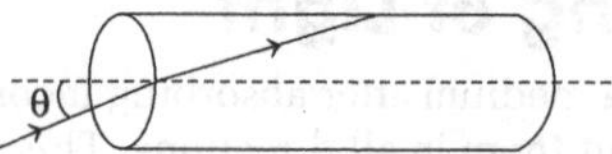

The incident angle θ for which the light ray grazes along the wall of the rod is

(a) $\sin^{-1}\left(\frac{1}{2}\right)$ (b) $\sin^{-1}\left(\frac{\sqrt{3}}{2}\right)$
(c) $\sin^{-1}\left(\frac{2}{\sqrt{3}}\right)$ (d) $\sin^{-1}\left(\frac{1}{\sqrt{3}}\right)$

7 A ray of light is incident on the surface of separation of a medium at an angle 45° and is refracted in the medium at an angle 30°. What will be the speed of light in the medium?

(a) 1.96×10^8 ms^{-1} (b) 2.12×10^8 ms^{-1}
(c) 3.18×10^8 ms^{-1} (d) 3.33×10^8 ms^{-1}

8 A beaker contains water upto a height h_1 and kerosene of height h_2 above water so that the total height of (water + kerosene) is $(h_1 + h_2)$. Refractive index of water is μ_1 and that of kerosene is μ_2. The apparent shift in the position of the bottom of the beaker when viewed from above is → AIEEE 2011

(a) $\left(1-\frac{1}{\mu_1}\right)h_2 + \left(1-\frac{1}{\mu_2}\right)h_1$ (b) $\left(1+\frac{1}{\mu_1}\right)h_1 + \left(1+\frac{1}{\mu_2}\right)h_2$
(c) $\left(1-\frac{1}{\mu_1}\right)h_1 + \left(1-\frac{1}{\mu_2}\right)h_2$ (d) $\left(1+\frac{1}{\mu_1}\right)h_2 - \left(1+\frac{1}{\mu_2}\right)h_1$

9 A printed page is pressed by a glass of water. The refractive index of the glass and water is 1.5 and 1.33, respectively. If the thickness of the bottom of glass is 1 cm and depth of water is 5 cm, how much the page will appear to be shifted if viewed from the top? → JEE Main (Online) 2013

(a) 1.033 cm (b) 3.581 cm
(c) 1.3533 cm (d) 1.90 cm

10 A fish looking up through the water sees the outside world, contained in a circular horizon. If the refractive index of water is $\frac{4}{3}$ and the fish is 12 cm below the water surface, the radius of this circle in cm, is

(a) $36\sqrt{7}$ (b) $\frac{36}{\sqrt{7}}$ (c) $36\sqrt{5}$ (d) $4\sqrt{5}$

11 A green light is incident from the water to the air-water interface at the critical angle (θ). Select correct statement. → JEE Main 2014

(a) The spectrum of visible light whose frequency is more than that of green light will come out to the medium
(b) The entire spectrum of visible light will come out of the water at various angles to the normal
(c) The entire spectrum of visible light come out of the water at an angle of 90° to the normal
(d) The spectrum of visible light whose frequency is less than that of green light will come out to the medium

12 Light is incident from a medium into air at two possible angles of incidence (a) 20° and (b) 40°. In the medium, light travels 3.0 cm in 0.2 ns. The ray will → JEE Main (Online) 2013

(a) suffer total internal reflection in both cases (a) and (b)
(b) suffer total internal reflection in case (b) only
(c) have partial reflection and partial transmission in case (b)
(d) have 100% transmission in case (a)

13 The graph between angle of deviation (δ) and angle of incidence (i) for a triangular prism is represented by

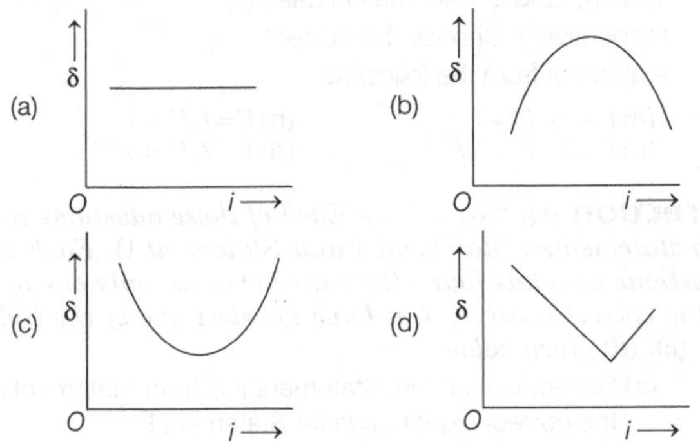

14 The refractive index of glass is 1.520 for red light and 1.525 for blue light. Let D_1 and D_2 be the angles of minimum deviation for the red and blue light respectively in a prism of this glass. Then,
(a) $D_1 < D_2$
(b) $D_1 = D_2$
(c) D_1 can be less than or greater than D_2 depending upon the angle of prism
(d) $D_1 > D_2$

15 Two beams of red and violet colours are made to pass separately through a prism (angle of the prism is 60°). In the position of minimum deviation, the angle of refraction will be
(a) 30° for both the colours
(b) greater for the violet colour
(c) greater for the red colour
(d) equal but not 30° for both the colours

16 The maximum magnification that can be obtained with a convex lens of focal length 2.5 cm is (The least distance of distinct vision is 25 cm)
(a) 10 (b) 0.1 (c) 62.5 (d) 11

17 A wire mesh consisting of very small squares is viewed at a distance of 8 cm through a magnifying lens of focal length 10 cm, kept close to the eye. The magnification produced by the lens is
(a) 5 (b) 8 (c) 10 (d) 20

18 When the distance between the object and the screen is more than 4*F*, we can obtain image of an object on the screen for the two positions of a lens. It is called displacement method. In one case, the image is magnified and in the other case it is diminished. Then, the ratio of the size of image to the diminished image is
(a) $\frac{(D+d)^2}{(D-d)^2}$ (b) $\frac{D}{d}$ (c) $\frac{D^2}{d^2}$ (d) $\frac{D+d}{D-d}$

19 A biconvex lens of focal length *f* forms a circular image of radius *r* of sun in the focal plane. Then, which option is correct?
(a) $\pi r^2 \propto f$
(b) $\pi r^2 \propto f^2$
(c) If lower half part is covered by black sheet, then area of the image is equal to $\pi r^2/2$
(d) If *f* is doubled, intensity will increase

20 In an optics experiments, with the position of the object fixed, a student varies the position of a convex lens and for each position, the screen is adjusted to get a clear image of the object. A graph between the object distance *u* and the image distance *v*, from the lens, is plotted using the same scale for the two axes. A straight line passing through the origin and making an angle of 45° with the *x*-axis meets the experimental curve at *P*. The coordinates of *P* will be → AIEEE 2009
(a) $(2f, 2f)$ (b) $\left(\frac{f}{2}, \frac{f}{2}\right)$
(c) (f, f) (d) $(4f, 4f)$

21 A student measures the focal length of a convex lens by putting an object pin at a distance *u* from the lens and measuring the distance *v* of the image pin. The graph between *u* and *v* plotted by the student should look like

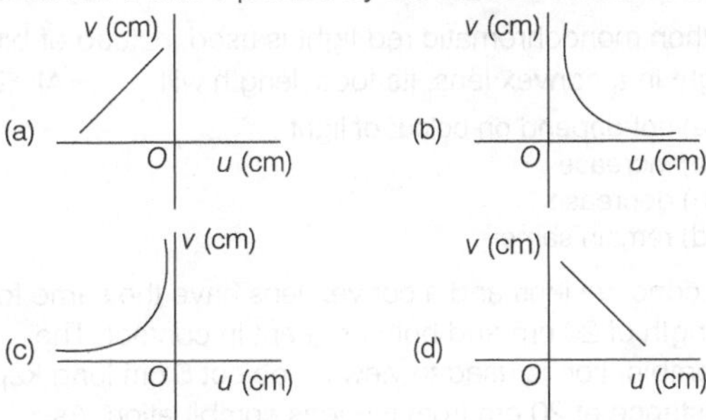

22 An object approaches a convergent lens from the left of the lens with a uniform speed 5 m/s and stops at the focus. The image
(a) moves away from the lens with a uniform speed 5 m/s
(b) moves away from the lens with a uniform acceleration
(c) moves away from the lens with a non-uniform acceleration
(d) moves towards the lens with a non-uniform acceleration

23 In an experiment to determine the focal length (*f*) of a concave mirror by the *u*-*v* method, a student places the object pin *A* on the principal axis at a distance *x* from the pole *P*. The student looks at the pin and inverted image from a distance keeping his/her eye in line with *PA*. When the student shifts his/her eye towards left, the image appears to the right of the object pin. Then,
(a) $x < f$ (b) $f < x < 2f$
(c) $x = 2f$ (d) $x > 2f$

24 The image of an illuminated square is obtained on a screen with the help of a converging lens. The distance of the square from the lens is 40 cm. The area of the image is 9 times that of the square. The focal length of the lens is →JEE Main (Online) 2013

(a) 36 cm (b) 27 cm
(c) 60 cm (d) 30 cm

25 Which one of the following spherical lenses does not exhibit dispersion? The radii of curvature of the surfaces of the lenses are as given in the diagrams.

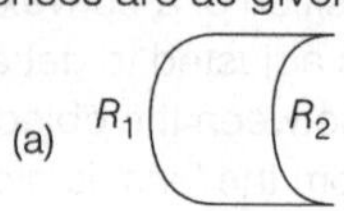

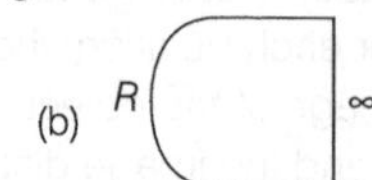

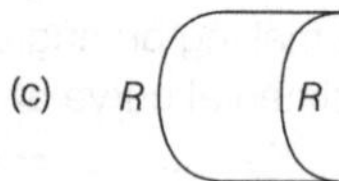

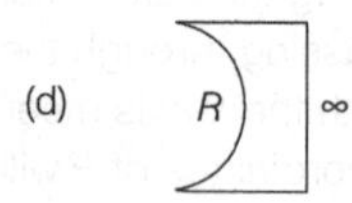

26 A double convex lens made of glass (refractive index $n = 1.5$) has the radii of curvature of both the surfaces as 20 cm. Incident light rays parallel to the axis of the lens will converge at a distance L such that

(a) $L = 20$ cm (b) $L = 10$ cm
(c) $L = 40$ cm (d) $L = \frac{20}{3}$ cm

27 When monochromatic red light is used instead of blue light in a convex lens, its focal length will →AIEEE 2011

(a) not depend on colour of light
(b) increase
(c) decrease
(d) remain same

28 A concave lens and a convex lens have the same focal length of 20 cm and both are kept in contact. The combination is used to view an object 5 cm long kept at a distance of 20 cm from the lens combination. As compared to the object, the image will be

(a) magnified and inverted
(b) diminished and erect
(c) of the same size and erect
(d) of the same size and inverted

29 To make an achromatic combination, a convex lens of focal length 42 cm having dispersive power of 0.14 is placed in contact with a concave lens of dispersive power 0.21. The focal length of the concave lens should be

(a) 63 cm (b) 21 cm
(c) 42 cm (d) 14 cm

30 A convex lens is in contact with a concave lens. The magnitude of the ratio of their focal length is $\frac{3}{2}$. Their equivalent focal length is 30 cm. What are their individual focal lengths?

(a) $-75, 50$ (b) $-10, 15$ (c) $75, 50$ (d) $-15, 10$

31 An equiconvex lens is cut into two halves along (i) XOX' and (ii) along YOY' as shown in figure. Let f, f' and f'' be the focal lengths of the complete lens, of each half in case (i), and of each half in case (ii) respectively. Choose the correct statement from the following.

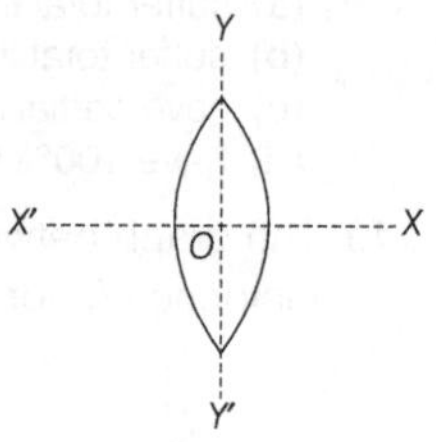

(a) $f' = 2f, f'' = f$ (b) $f' = f, f'' = f$
(c) $f' = 2f, f'' = 2f$ (d) $f' = f, f'' = 2f$

Direction

(Q. Nos. 32-36) *Each of these questions contains two statements : Statement I and Statement II. Each of these questions also has four alternative choices, only one of which is the correct answer. You have to select one of the codes (a), (b), (c), (d) given below*

(a) Statement I is true, Statement II is true; Statement II is the correct explanation for Statement I
(b) Statement I is true, Statement II is true; Statement II is not the correct explanation for Statement I
(c) Statement I is true; Statement II is false
(d) Statement I is false; Statement II is true

32 **Statement I** The formula connecting u, v and f for a spherical mirror is valid only for mirrors whose sizes are very small as compared to their radii of curvature.

Statement II Laws of reflection are strictly valid for plane surfaces, but not for large spherical surfaces.

33 **Statement I** Endoscopy involves use of optical fibres to study internal organs.

Statement II Optical fibres are based on the phenomenon of total internal reflection.

34 **Statement I** The refractive index of diamond is $\sqrt{6}$ and that of liquid is $\sqrt{3}$. If the light travels from diamond to the liquid, it will be totally reflected when the angle of incidence is 30°.

Statement II $n = \frac{1}{\sin C}$, where n is the refractive index of diamond with respect to liquid.

35 **Statement I** A double convex lens ($n = 1.5$) has a focal length 10 cm. When the lens is immersed in water ($n = 4/3$), its focal length becomes 40 cm.

Statement II $\frac{1}{f} = \frac{n_l - n_m}{n_m}\left(\frac{1}{R_1} - \frac{1}{R_2}\right)$.

36 A thin air film is formed by putting the convex surface of a plane-convex lens over a plane glass plate. With monochromatic light, this film gives an interference pattern due to light reflected from the top (convex) surface and the bottom (glass plate) surface of the film.

Statement I When light reflects from the air-glass plate interface, the reflected wave suffers a phase change of π.

Statement II The centre of the interference pattern is dark. →AIEEE 2011

DAY PRACTICE SESSION 2

PROGRESSIVE QUESTIONS EXERCISE

1 Two plane mirrors are inclined at 90°. An object is placed between them whose coordinates are (a, b). The position vectors of all the images formed is

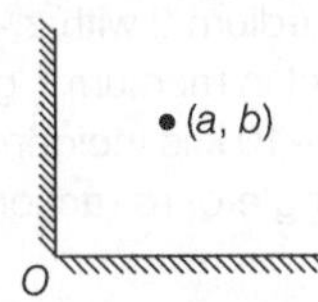

(a) $-a\hat{i}-b\hat{j}, a\hat{i}+b\hat{j}, -a\hat{i}+b\hat{j}$
(b) $-a\hat{i}+b\hat{j}, -a\hat{j}-b\hat{j}, a\hat{i}-b\hat{j}$
(c) $a\hat{i}+b\hat{j}, -a\hat{i}-b\hat{j}, a\hat{i}-b\hat{j}$
(d) None of the above

2 A small coin is resting on the bottom of a beaker filled with a liquid. A ray of light from the coin travels upto the surface of the liquid and moves along its surface (see figure).

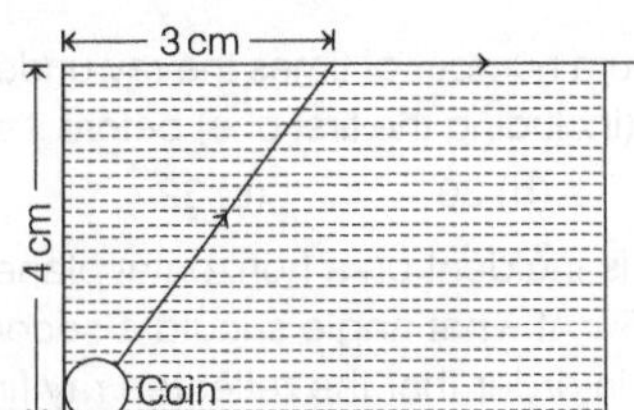

How fast is the light travelling in the liquid?

(a) $1.8\times10^{8}\ \text{ms}^{-1}$ (b) $2.4\times10^{8}\ \text{ms}^{-1}$
(c) $3.0\times10^{8}\ \text{ms}^{-1}$ (d) $1.2\times10^{8}\ \text{ms}^{-1}$

3 The reflective surface is given by $y = 2\sin x$ and it is facing positive axis. What is the least value of coordinate of the point where a ray parallel to positive x-axis becomes parallel to positive y-axis after reflection?

(a) $\left(\frac{\pi}{3}, \sqrt{3}\right)$ (b) $\left(\frac{\pi}{2}, \sqrt{2}\right)$ (c) $\left(\frac{\pi}{3}, \sqrt{2}\right)$ (d) $\left(\frac{\pi}{4}, \sqrt{3}\right)$

4 A light ray is incident perpendicular to one face of a 90° prism and is totally internally reflected at the glass-air interface. If the angle of reflection is 45°, we conclude that for the refractive index n as

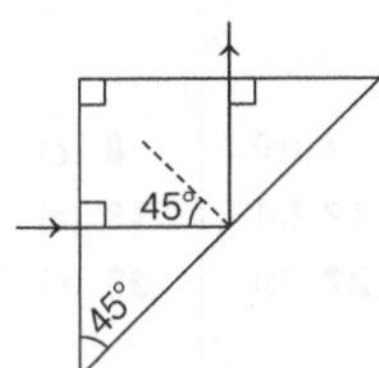

(a) $n<\frac{1}{\sqrt{2}}$ (b) $n>\sqrt{2}$ (c) $n>\frac{1}{\sqrt{2}}$ (d) $n<\sqrt{2}$

5 The speed at which the image of the luminous point object is moving, if the luminous point object is moving at speed v_0 towards a spherical mirror, along its axis is (Given, R = radius of curvature, u = object distance)

(a) $v_i = -v_0$ (b) $v_i = -v_0\left[\frac{R}{2u-R}\right]^2$
(c) $v_i = -v_0\left(\frac{2u-R}{R}\right)$ (d) $v_i = -v_0\left(\frac{R}{2u-R}\right)$

6 Diameter of a plano-convex lens is 6 cm and thickness at the centre is 3 mm. If speed of light in material of lens is 2×10^{8} m/s, the focal length of lens

(a) 15 cm (b) 20 cm (c) 30 cm (d) 10 cm

7 A plano-convex lens of refractive index 1.5 and radius of curvature 30 cm is silvered at the curved surface. Now, this lens has been used to form the image of an object. At what distance from this lens, an object be placed in order to have a real image of the size of the object?

(a) 20 cm (b) 30 cm (c) 60 cm (d) 80 cm

8 Monochromatic light is incident on a glass prism of angle A. If the refractive index of the material of the prism is μ, a ray incident at an angle θ, on the face AB would get transmitted through the face AC of the prism provided

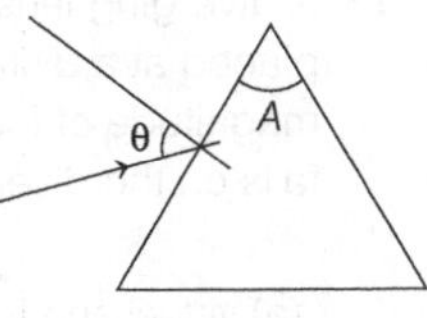

→ **JEE Main 2015**

(a) $\theta > \sin^{-1}\left[\mu\sin\left(A-\sin^{-1}\frac{1}{\mu}\right)\right]$
(b) $\theta < \sin^{-1}\left[\mu\sin\left(A+\sin^{-1}\frac{1}{\mu}\right)\right]$
(c) $\theta > \cos^{-1}\left[\mu\sin\left(A+\sin^{-1}\frac{1}{\mu}\right)\right]$
(d) $\theta < \cos^{-1}\left[\mu\sin\left(A-\sin^{-1}\frac{1}{\mu}\right)\right]$

9 A thin convex lens made from crown glass $\left(\mu=\frac{3}{2}\right)$ has focal length f. When it is measured in two different liquid having refractive indices $\frac{4}{3}$ and $\frac{5}{3}$, it has focal lengths f_1 and f_2, respectively. The correct relation between focal length is → **JEE Main 2014**

(a) $f_2 > f$ and f_1 becomes one
(b) f_1 and f_2 both becomes one
(c) $f_1 = f_2 < f$
(d) $f_1 > f$ and f_2 becomes one

10 A light ray falls on a square glass slab as showin in the diagram. The index of refraction of the glass, if total internal reflection is to occur at the vertical face, is equal to → JEE Main 2013

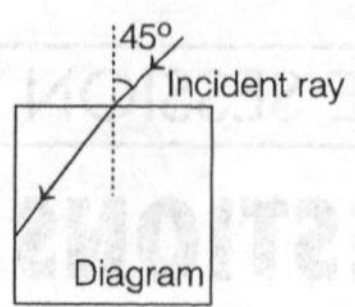

(a) $\frac{(\sqrt{2}+1)}{2}$ (b) $\sqrt{\frac{5}{2}}$ (c) $\frac{3}{2}$ (d) $\sqrt{\frac{3}{2}}$

11 A spectrometer gives the following reading when used to measure the angle of a prism.

Main scale reading : 58.5 degree
Vernier scale reading : 09 divisions

Given that 1 division on main scale corresponds to 0.5 degree. Total divisions on the vernier scale is 30 and match with 29 divisions of the main scale. The angle of the prism from the above data is → AIEEE 2012

(a) 58.59° (b) 58.77°
(c) 58.65° (d) 59°

12 A car is fitted with a convex side-view mirror of focal length 20 cm. A second car 2.8 m behind the first car is overtaking the first car at a relative speed of 15 m/s. The speed of the image of the second car as seen in the mirror of the first one is → AIEEE 2011

(a) $\frac{1}{15}$ m/s (b) 10 m/s (c) 15 m/s (d) $\frac{1}{10}$ m/s

13 A diverging lens with magnitude of focal length 25 cm is placed at a distance of 15 cm from a converging lens of magnitude of focal length 20 cm. A beam of parallel light falls on the diverging lens. The final image formed is → JEE Main 2017 (Offline)

(a) virtual and at a distance of 40 cm from convergent lens
(b) real and at a distance of 40 cm from the divergent lens
(c) real and at a distance of 6 cm from the convergent lens
(d) real and at a distance of 40 cm from convergent lens

14 In an experiment for determination of refractive index of glass of a prism by i - δ, plot, it was found that a ray incident at an angle 35° suffers a deviation of 40° and that it emerges at an angle 79°. In that case, which of the following is closest to the maximum possible value of the refractive index? → JEE Main 2016 (Offline)

(a) 1.5 (b) 1.6 (c) 1.7 (d) 1.8

15 An object 2.4 m infront of a lens forms a sharp image on a film 12 cm behind the lens. A glass plate 1 cm thick, of refractive index 1.50 is interposed between lens and film with its plane faces parallel to film. At what distance (from lens) should object shifted to be in sharp focus on film? → AIEEE 2012

(a) 7.2 m (b) 2.4 m (c) 3.2 m (d) 5.6 m

16 Let the zx-plane be the boundary between two transparent media. Medium 1 in $z \geq 0$ has a refractive index of $\sqrt{2}$ and medium 2 with $z < 0$ has a refractive index of $\sqrt{3}$. A ray of light in medium 1 given by the vector $\mathbf{A} = 6\sqrt{3}\,\mathbf{i} + 8\sqrt{3}\,\mathbf{j} - 10\,\mathbf{k}$ is incident on the plane of separation. The angle of refraction in medium 2 is → AIEEE 2011

(a) 45° (b) 60° (c) 75° (d) 30°

17 Two plane mirrors A and B are aligned parallel to each other, as shown in the figure. A light ray is incident at an angle 30° at a point just inside one end of A. The plane of incidence coincides with the plane of the figure.

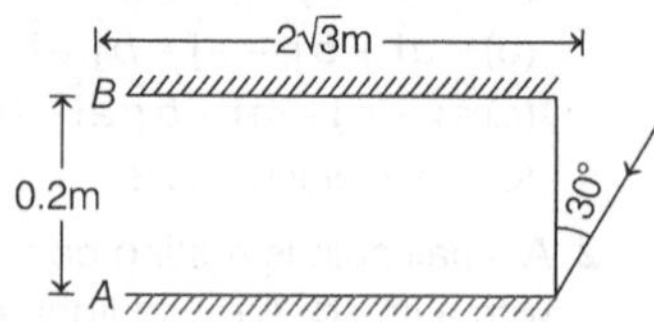

The maximum number of times the ray undergoes reflections (including the first one) before it emerges out is

(a) 28 (b) 30 (c) 32 (d) 34

18 A light ray is incident on a horizontal plane mirror at an angle of 45°. At what angle should a second plane mirror be placed in order that the reflected ray finally be reflected horizontally from the second mirror as shown in figure, is

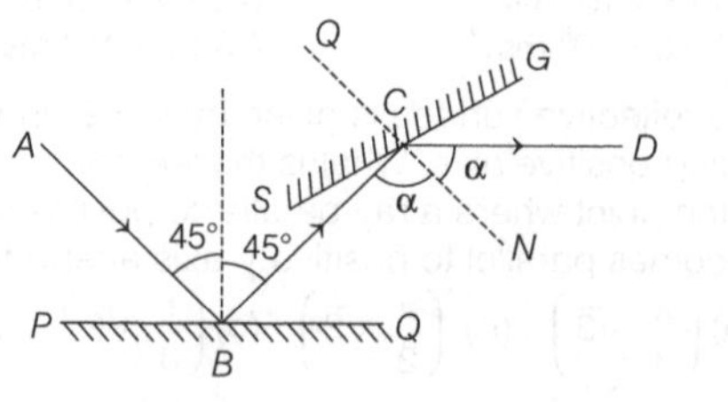

(a) θ = 45° (b) θ = 60° (c) θ = 22.5° (d) θ = 15.3°

ANSWERS

SESSION 1									
1 (b)	2 (a)	3 (b)	4 (b)	5 (d)	6 (d)	7 (b)	8 (c)	9 (d)	10 (b)
11 (d)	12 (b)	13 (c)	14 (a)	15 (a)	16 (d)	17 (a)	18 (a)	19 (b)	20 (a)
21 (c)	22 (c)	23 (b)	24 (a)	25 (c)	26 (a)	27 (b)	28 (c)	29 (a)	30 (d)
31 (d)	32 (c)	33 (a)	34 (d)	35 (a)	36 (b)				

SESSION 2									
1 (b)	2 (a)	3 (a)	4 (b)	5 (b)	6 (c)	7 (a)	8 (a)	9 (d)	10 (d)
11 (c)	12 (a)	13 (d)	14 (a)	15 (d)	16 (a)	17 (b)	18 (c)		

Hints and Explanations

SESSION 1

1 Number of images

$$n = \frac{360^\circ}{\theta} - 1$$

where, θ is angle between mirrors.

$$\therefore \quad 3 = \frac{360^\circ}{\theta} - 1 \quad \text{or} \quad \theta = 90^\circ$$

2 $n = \frac{360}{\theta}$

Number of images, $N = n$, which is odd $= n - 1$

For the given condition, no successive reflection takes place. So, the number of images will be $N \le 2$

$$n - 1 \le 2$$

$$n \le 3$$

$$\frac{360}{\theta} \le 3$$

$$120 \le \theta \Rightarrow \theta \ge 120^\circ$$

3 In any medium other than air or vacuum, the velocities of different colours are different. Therefore, both red and green colours are refracted at different angles of refractions.

Hence, after emerging from glass slab through opposite parallel face, they appear at two different points and move in two different parallel directions.

4 As optical paths are equal, hence

$$n_g \cdot x_g = n_w \cdot x_w$$

$$\Rightarrow \quad n_w = n_g \cdot \frac{x_g}{x_w}$$

$$= 1.53 \times \frac{4.0}{4.5} = 1.36$$

5 A horizontal beam is travelling from a denser to a rarer medium, so it bends upwards (away from normal).

6 $\sin C = \frac{\sqrt{3}}{2}$ ...(i)

$$\sin r = \sin(90^\circ - C) = \cos C = \frac{1}{2}$$

$$\frac{\sin \theta}{\sin r} = \frac{\mu_2}{\mu_1}$$

$$\sin \theta = \frac{2}{\sqrt{3}} \times \frac{1}{2}$$

$$\theta = \sin^{-1}\left(\frac{1}{\sqrt{3}}\right)$$

7 $n = \frac{\sin i}{\sin r} = \frac{c}{v}$

Hence, $v = \frac{c \sin r}{\sin i}$

$$= \frac{3 \times 10^8 \times \sin 30^\circ}{\sin 45^\circ}$$

$$= \frac{3 \times 10^8}{\sqrt{2}}$$

$$= 2.12 \times 10^8 \text{ ms}^{-1}$$

8 We know, apparent shift,

$$\Delta h = \left(1 - \frac{1}{\mu}\right)h$$

$\therefore$ Apparent shift produced by water,

$$\Delta h_1 = \left(1 - \frac{1}{\mu_1}\right)h_1$$

and apparent shift produced by kerosene,

$$\Delta h_2 = \left(1 - \frac{1}{\mu_2}\right)h_2$$

$$\therefore \quad \Delta h = \Delta h_1 + \Delta h_2 = \left(1 - \frac{1}{\mu_1}\right)h_1 + \left(1 - \frac{1}{\mu_2}\right)h_2$$

9 Given, $\mu = 1.5, t_1 = 5\,\text{cm}$,

$\mu_2 = 1.33$ and $t_2 = 1$ cm

Change in path $= \Delta t_1 + \Delta t_2$

$$= \left(1 - \frac{1}{\mu_1}\right) \times t_1 + \left(1 - \frac{1}{\mu_2}\right) \times t_2$$

$$= \left(1 - \frac{1}{1.5}\right) \times 5 + \left(1 - \frac{1}{1.33}\right) \times 1$$

$$\approx 1.90 \text{ cm}$$

10 The situation is shown in figure.

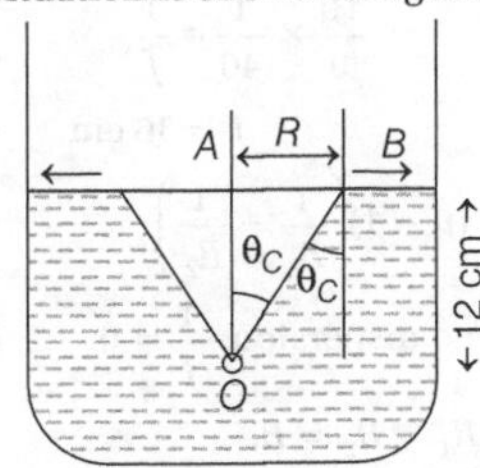

$$\sin \theta_C = \frac{1}{\mu}$$

$$\tan \theta_C = \frac{AB}{OA}$$

$$\therefore \quad AB = OA \tan \theta_C$$

$$\text{or } AB = \frac{OA}{\sqrt{n^2 - 1}} = \frac{12}{\sqrt{\left(\frac{4}{3}\right)^2 - 1}} = \frac{36}{\sqrt{7}}$$

11 $\mu_r < \mu_g < \mu_v$

$\Rightarrow \sin(\theta_r) > \sin(\theta_g) > \sin \theta_v$

$\theta_r > \theta_g > \theta_v$ [$\theta_r, \theta_g, \theta_v$ are critical angles of corresponding colours]. Thus, all colours from red to green will emerge out of water.

12 Speed of light in medium

$$= \frac{3 \times 10^{-2}}{0.2 \times 10^{-9}}$$

$$= \frac{3}{2} \times 10^8 \text{ ms}^{-1} = 1.5 \times 10^8 \text{ ms}^{-1}$$

As, $\frac{\mu_2}{\mu_1} = \frac{v_1}{v_2}$

$$\frac{\mu}{1} = \frac{3 \times 10^8}{1.5 \times 10^8} \Rightarrow \mu = 2$$

We have, $\sin C = \frac{1}{2}$

$$\Rightarrow \quad C = \sin^{-1}\left(\frac{1}{2}\right) = 30^\circ$$

When the value of incidence angle is greater than critical angle than total internal reflection take place.

In the second case we get total interval reflection.

13 Angle of deviation depends upon the angle of incidence. If we determine experimentally the angles of deviation corresponding to different angles of incidence, then the plot between i and δ that we will get is shown below

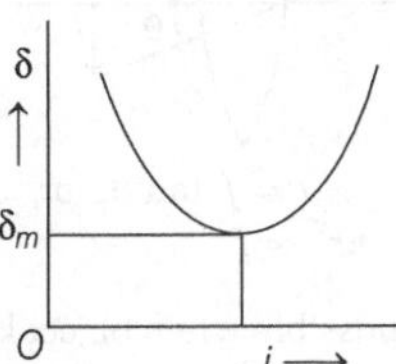

14 $D = (n - 1)A$

For blue light n is greater than that for red light, so $D_2 > D_1$.

15 At minimum deviation ($\delta = \delta_m$)

$$r_1 = r_2 = \frac{A}{2} = \frac{60^\circ}{2} = 30^\circ$$

(For both colours)

16 Maximum magnification is obtained when image is formed at near point of eye. In that case,

$$m = 1 + \frac{D}{f} = 1 + \frac{25}{2.5} = 11$$

17 $\frac{1}{f} = \frac{1}{v} - \frac{1}{u}$ ...(i)

Given, $f = 10\,\text{cm}$ (as lens is converging)

$u = -8\,\text{cm}$ (as object is placed on left side of the lens)

On substituting these values in Eq. (i), we get

$$\frac{1}{10} = \frac{1}{v} - \frac{1}{-8}$$

$$\Rightarrow \frac{1}{v} = \frac{1}{10} - \frac{1}{8}$$

$$\Rightarrow \frac{1}{v} = \frac{8-10}{80}$$

$$\therefore v = \frac{80}{-2} = -40\,\text{cm}$$

Hence, magnification produced by the lens

$$m = \frac{v}{u} = \frac{-40}{-8} = 5$$

18 In displacement method, the ratio of the diminished image to the object is

$$m_2 = \frac{I_2}{O} = \frac{D-d}{D+d}$$

and the ratio of image to the object is $\frac{D+d}{D-d}$

Hence, $\frac{m_1}{m_2} = \frac{I_1}{O} \times \frac{O}{I_2} = \frac{(D+d)^2}{(D-d)^2}$

19

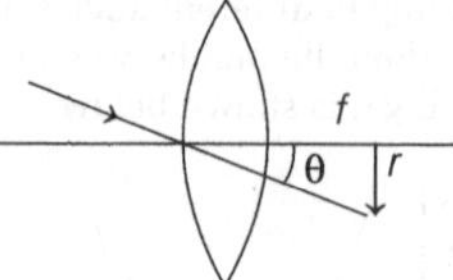

$$r = f \tan\theta \quad \text{or} \quad r \propto f$$

$$\therefore \pi r^2 \propto f^2$$

20 It is possible when object kept at centre of curvature because then only position of object and image would be same. i.e. $u = v$ [which is the point of intersection between curve and straight line]

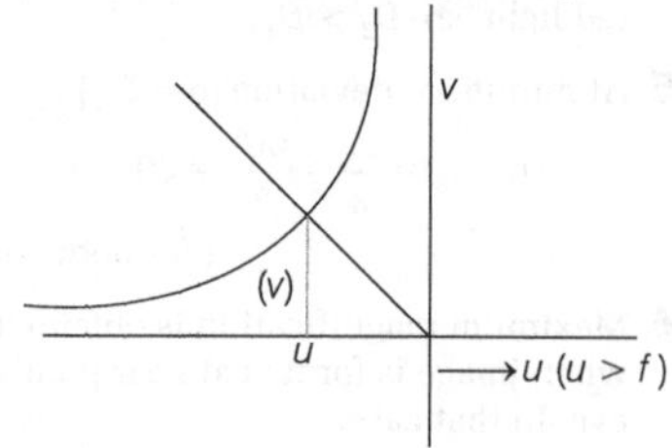

$u = v$

$u = 2f, v = 2f$

21 $\therefore \frac{1}{v} - \frac{1}{u} = \frac{1}{f}$ = constant,

which is a rectangular hyperbola.

22 When object would approach the lens then image would move away from the lens with non-uniform acceleration. This can be seen from the fact that when object is far away, image would be formed at focus P. When object approaches the lens then image would move away such that when object approaches to focus, then image would approach to infinity.

23 Since, object and image move in opposite directions, the positioning should be as shown in the figure. Object lies between focus and centre of curvature $f < x < 2f$.

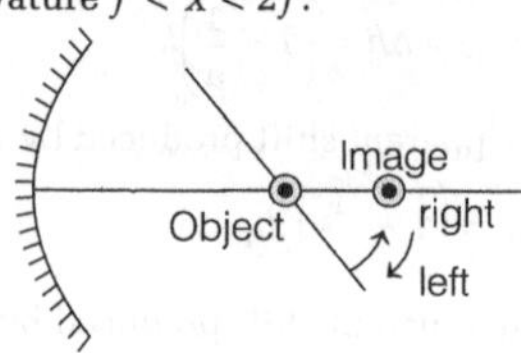

24 As magnification $= 9 = \frac{v}{u}$

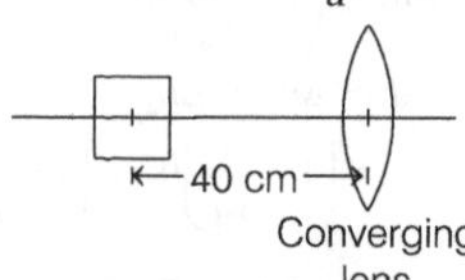

$\therefore v = 9u = 9 \times 40 = 360$

Now, $\frac{1}{v} - \frac{1}{u} = \frac{1}{f}$

$$\frac{1}{v} - \frac{1}{-40} = \frac{1}{f}$$

$$\frac{1}{360} + \frac{1}{40} = \frac{1}{f}$$

$$\Rightarrow \left[\frac{1}{9} + 1\right]\frac{1}{40} = \frac{1}{f}$$

$$\Rightarrow \frac{10}{9} \times \frac{1}{40} = \frac{1}{f}$$

$$\Rightarrow f = 36\,\text{cm}$$

25 $\frac{1}{f} = (\mu - 1)\left(\frac{1}{R_1} - \frac{1}{R_2}\right)$

For no dispersion, $\frac{1}{f} = 0$

or $R_1 = R_2 = R$.

So, (c) is correct option.

26 Here, $n = 1.5$, as per sign convention followed

$R_1 = +20\,\text{cm}$ and $R_2 = -20\,\text{cm}$

$$\therefore \frac{1}{f} = (n-1)\left(\frac{1}{R_1} - \frac{1}{R_2}\right)$$

$$= (1.5-1)\left[\frac{1}{(+20)} - \frac{1}{(-20)}\right]$$

$$= 0.5 \times \frac{2}{20} = \frac{1}{20}$$

$\Rightarrow f = 20\,\text{cm}$

Incident rays travelling parallel to the axis of lens will converge at its second principal focus.

Hence, $L = +20\,\text{cm}$

27 $\frac{1}{f} = (\mu - 1)\left(\frac{1}{R_1} - \frac{1}{R_2}\right)$

Also, by Cauchy's formula,

$$\mu = A + \frac{B}{\lambda^2} + \frac{C}{\lambda^4} + \ldots$$

$$\Rightarrow \mu \propto \frac{1}{\lambda}$$

As $\lambda_{\text{blue}} < \lambda_{\text{red}} \Rightarrow \mu_{\text{blue}} > \mu_{\text{red}}$

Hence, $f_{\text{red}} > f_{\text{blue}}$

28 When a concave lens is joined in contact with a convex lens of same focal length, the combination behaves as a glass plate of infinite focal length or zero power. Hence, the image of an object will be of the same size and erect.

29 For an achromatic combination, the condition is

$$\frac{\omega_1}{f_1} + \frac{\omega_2}{f_2} = 0$$

Here, $\omega_1 = 0.14$, $f_1 = 42\,\text{cm}$

thus, we get $f_2 = -63\,\text{cm}$

30 Let focal length of convex lens is $+f$, then focal length of concave lens would be $-\frac{3}{2}f$.

From the given condition,

$$\frac{1}{30} = \frac{1}{f} - \frac{2}{3f} = \frac{1}{3f}$$

$$\therefore f = 10\,\text{cm}$$

Therefore, focal length of convex lens $= +10$ cm and that of concave lens $= -15$ cm.

31 (i) Cutting along YOY'

Focal length get doubled.

$$f'' = 2f$$

(ii) Cutting along XOX'

Focal length unchanged.

32 Laws of reflection are valid for plane surfaces, irregular surface and curved surface.

33 Optical fibre consists of a very long and thin fibre of quartz glass. When a light ray is incident at one end of the fibre making a small angle of incidence, it suffers refraction from air to quartz and strikes the fibre-layer interface at an angle of incidence greater than the critical angle.

It therefore, suffers total internal reflection and strikes its opposite interface. At this interface also, the angle of incidence is greater than the critical angle. So, it again suffers total internal reflection. Thus, optical fibre is based on total internal reflection. Endoscopy is a process for viewing internal organs of human body. This process use a device endoscope which is based on total internal reflection.

34 Refraction index of diamond w.r.t. liquid

$$_{l}n_{d} = \frac{1}{\sin C}$$

$$\therefore \quad \frac{\sqrt{6}}{\sqrt{3}} = \frac{1}{\sin C}$$

or $\sin C = \frac{1}{\sqrt{2}} = \sin 45°$

$$\therefore \quad C = 45°$$

For total internal reflection angle should be greater than critical angle.

But here angle of incidence is lower than critical angle, so total internal reflection does not occur in light.

35 In water,

$$\frac{1}{f_w} = \left(\frac{n_l - n_m}{n_m}\right)\left(\frac{1}{R_1} - \frac{1}{R_2}\right) \quad \text{...(i)}$$

$$\frac{1}{f_a} = (n_l - 1)\left(\frac{1}{R_1} - \frac{1}{R_2}\right) \quad \text{...(ii)}$$

$$\frac{f_w}{f_a} = \frac{(n_l - 1)}{\left(\frac{n_l - n_m}{n_m}\right)} = \frac{(1.5 - 1)(4/3)}{(1.5 - 4/3)}$$

$$= \frac{0.5 \times 4/3}{\frac{3}{2} - \frac{4}{3}} = \frac{(2/3)6}{1}$$

$$\therefore \quad f_w = 4f_a = 4 \times 10 = 40\text{cm}$$

36 Both statements I and II are correct but statement II does not explain statement I.

SESSION 2

1 The images formed by combination of two plane mirrors are lying on a circle whose radius is equal to OS.

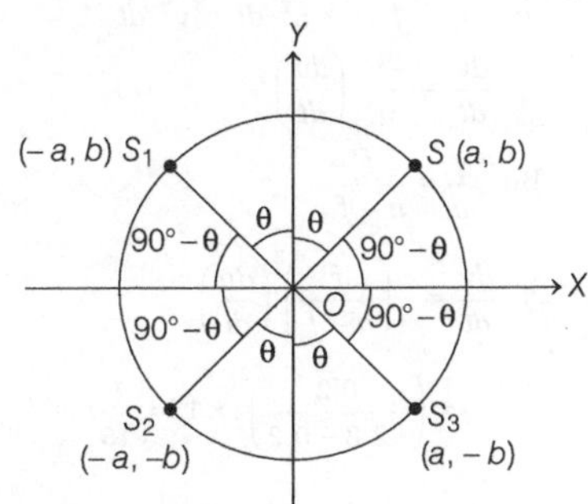

The centre of the circle is lying on meeting point of mirrors (i.e. O). The position of images from diagram is for S_1, $r_1 = -a\hat{\mathbf{i}} + b\hat{\mathbf{j}}$, $r_2 = -a\hat{\mathbf{i}} - b\hat{\mathbf{j}}$, $r_3 = a\hat{\mathbf{i}} - b\hat{\mathbf{j}}$.

Hence, option (b) is true.

2 As shown in figure, a light ray from the coin will not emerge out of liquid, if $i > C$.

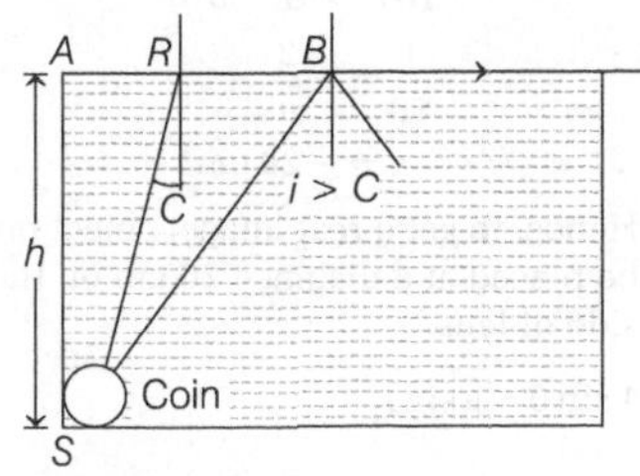

Therefore, minimum radius R corresponds to $i = C$.

In ΔSAB,

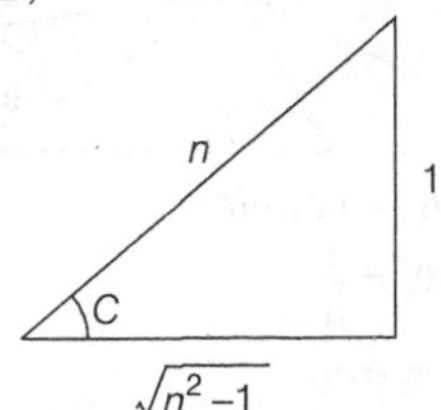

$$\frac{R}{h} = \tan C$$

or $R = h \tan C$

or $R = \frac{h}{\sqrt{n^2 - 1}}$

$$\left[\because \sin C = \frac{1}{n} \text{ and } \tan C = \frac{\sin C}{\sqrt{1 - \sin^2 C}}\right]$$

Given, $R = 3\text{cm}$, $h = 4\text{cm}$

Hence, $\frac{3}{4} = \frac{1}{\sqrt{n^2 - 1}}$

or $n^2 = \frac{25}{9}$ or $n = \frac{5}{3}$

But $n = \frac{c}{v}$

or $v = \frac{c}{n} = \frac{3 \times 10^8}{5/3} = 1.8 \times 10^8 \text{ ms}^{-1}$

3 Let the incidence point is $P(x, y)$

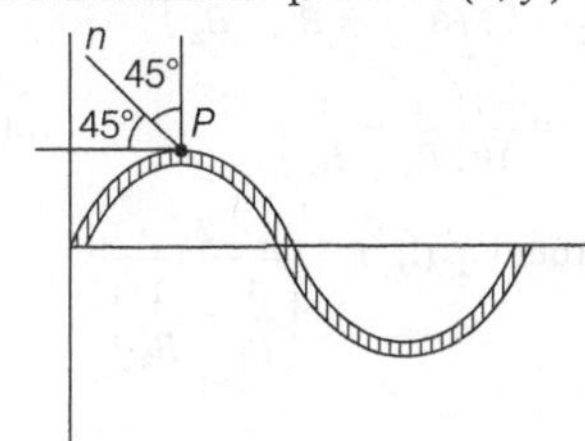

$$m = \tan 45° = 1$$

(from law of reflection)

$$y = 2 \sin x$$

$$\therefore \quad m = \frac{dy}{dx} = 2\cos x = 1$$

$$\cos x = \frac{1}{2}$$

$$x = \frac{\pi}{3}$$

The corresponding value of y is $2 \sin \frac{\pi}{3} = \sqrt{3}$

4 For total internal reflection from glass-air interface, critical angle C must be less than angle of incidence.

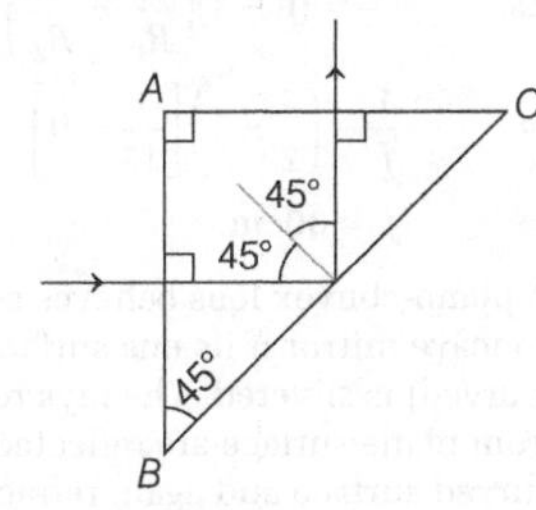

i.e. $C < i$

or $C < 45°$ $\quad (\because \angle i = 45°)$

But $n = \frac{1}{\sin C} \Rightarrow C = \sin^{-1}\left(\frac{1}{n}\right)$

or $\sin^{-1}\left(\frac{1}{n}\right) < 45°$

or $\frac{1}{n} < \sin 45°$

or $n > \frac{1}{\sin 45°}$

or $n > \frac{1}{(1/\sqrt{2})}$

or $n > \sqrt{2}$

5 $\frac{1}{v} + \frac{1}{u} = \frac{1}{f}$

Differentiating both sides

$$-\frac{1}{v^2}\frac{dv}{dt} = \frac{1}{u^2}\frac{du}{dt}$$

$$\frac{dv}{dt} = v_i = -\left(\frac{v}{u}\right)^2 \frac{du}{dt}$$

$$= -\left(\frac{v}{u}\right)^2 v_0$$

Again, $\frac{1}{v} = \frac{1}{f} - \frac{1}{u} = \frac{2}{R} - \frac{1}{u} = \frac{2u - R}{Ru}$

$$v = \frac{uR}{2u - R}$$

$$v_i = -\left(\frac{v}{u}\right)^2 v_0 = -v_0\left(\frac{R}{2u - R}\right)^2$$

6

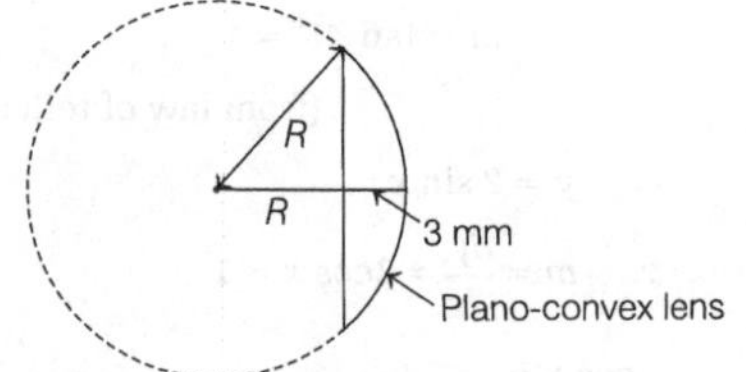

By Pythagoras theorem,

$$R^2 = 3^2 + (R - 3\,\text{mm})^2$$

$$\Rightarrow R^2 = 3^2 + R^2 - 2R(3\,\text{mm}) + (3\,\text{mm})^2$$

$$\Rightarrow \quad R \simeq 15\,\text{cm}$$

Also, $\mu = \frac{c}{v} \Rightarrow \mu = \frac{3\times10^8}{2\times10^8} = \frac{3}{2}$

As, $\frac{1}{f} = (\mu - 1)\left(\frac{1}{R_1} - \frac{1}{R_2}\right)$

$$\Rightarrow \frac{1}{f} = \left(\frac{3}{2} - 1\right)\left[\frac{1}{15} - 0\right]$$

$$\Rightarrow \quad f = 30\text{cm}$$

7 A plano-convex lens behaves as a concave mirror if its one surface (curved) is silvered. The rays refracted from plane surface are reflected from curved surface and again refract from plane surface. Therefore, in this lens two refractions and one reflection occur.

Let the focal length of silvered lens is F.

$$\frac{1}{F} = \frac{1}{f} + \frac{1}{f} + \frac{1}{f_m} = \frac{2}{f} + \frac{1}{f_m}$$

where, f = focal length of lens before silvering,

f_m = focal length of spherical mirror.

$$\frac{1}{F} = \frac{2}{f} + \frac{2}{R} \quad \text{...(i)}$$

$(\because R = 2f_m)$

Now,

$$\frac{1}{f} = (n-1)\left(\frac{1}{R_1} - \frac{1}{R_2}\right) \quad \text{...(ii)}$$

Here, $R_1 = \infty$, $R_2 = 30$ cm

$$\therefore \quad \frac{1}{f} = (1.5 - 1)\left(\frac{1}{\infty} - \frac{1}{30}\right)$$

or $\frac{1}{f} = \frac{0.5}{30} = -\frac{1}{60}$

or $f = -60$ cm

Hence, from Eq. (i), we get

$$\frac{1}{F} = \frac{2}{60} + \frac{2}{30} = \frac{6}{60}$$

$$F = 10\text{ cm}$$

Again given that,

size of object = size of image

i.e. $O = I$

$$\therefore \quad m = -\frac{v}{u} = \frac{I}{O}$$

$$\Rightarrow \quad \frac{v}{u} = -1$$

or $v = -u$

Thus, from lens formula,

$$\frac{1}{F} = \frac{1}{v} - \frac{1}{u}$$

$$\frac{1}{10} = \frac{1}{-u} - \frac{1}{u}$$

$$\frac{1}{10} = -\frac{2}{u}$$

$$\therefore \quad u = -20\text{ cm}$$

Hence, to get a real image, object must be placed at a distance 20 cm on the left side of lens.

8 $1\sin\theta = \mu\sin r_1$

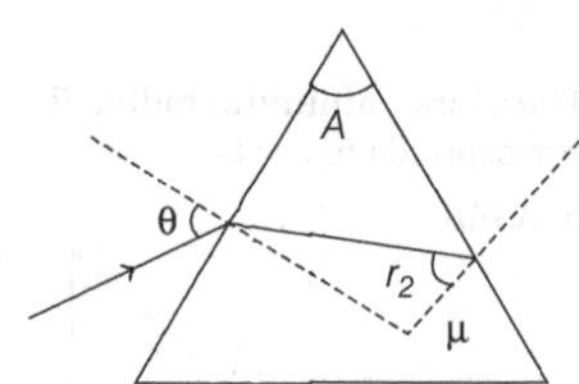

$$\mu\sin\theta_c = 1\sin 90^\circ$$

$$\sin\theta_c = \frac{1}{\mu}$$

$$r_2 < \theta_c$$

$$\sin r_2 < \sin\theta_c$$

$$r_1 + r_2 = A,\ r_1 = A - r_2,\ r_1 = A - \theta_c$$

$$\sin r_1 > \sin(A - \theta_c)$$

$$\Rightarrow \quad \mu\sin r_1 > \mu\sin(A - \theta_c)$$

$$\Rightarrow \quad \sin\theta > \mu\sin(A - \theta_c)$$

$$\theta > \sin^{-1}\left[\mu\sin\left(A - \sin^{-1}\frac{1}{\mu}\right)\right]$$

9 $\frac{1}{f} = \left(\frac{3}{2} - 1\right)\left(\frac{1}{R_1} - \frac{1}{R_2}\right)$

$$= \frac{1}{2}\left(\frac{1}{R_1} - \frac{1}{R_2}\right) \quad \text{...(i)}$$

$$\frac{1}{f_1} = \left(\frac{3/2}{4/3} - 1\right)\left(\frac{1}{R_1} - \frac{1}{R_2}\right)$$

$$= \frac{1}{8}\left(\frac{1}{R_1} - \frac{1}{R_2}\right) \quad \text{...(ii)}$$

$$\frac{1}{f_2} = \left(\frac{3/2}{5/3} - 1\right)\left(\frac{1}{R_1} - \frac{1}{R_2}\right)$$

$$= \frac{-1}{10}\left(\frac{1}{R_1} - \frac{1}{R_2}\right) \quad \text{...(iii)}$$

From Eq. (i), $f = \frac{2}{\left(\frac{1}{R_1} - \frac{1}{R_2}\right)}$

From Eq. (ii), $f_1 = \frac{8}{\left(\frac{1}{R_1} - \frac{1}{R_2}\right)}$

From Eq. (iii), $f_2 = \frac{-10}{\left(\frac{1}{R_1} - \frac{1}{R_2}\right)}$

10 From figure, we get $r = \sin^{-1}\left(\frac{1}{\sqrt{3}}\right)$

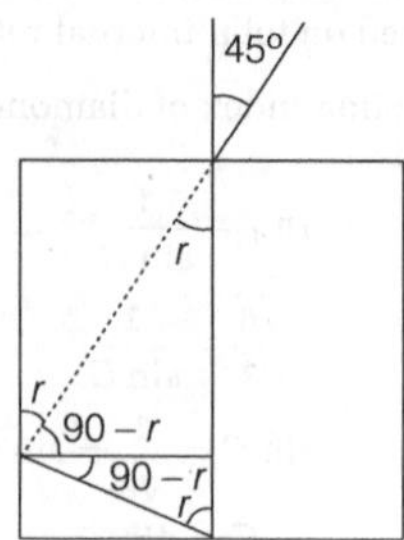

For critical angle, $\sin C = \frac{1}{\mu}$

Now, by Snell's law, we have

$$\frac{\mu}{1} = \frac{\sin i}{\sin r} = \frac{\sin 45^\circ}{\sin\left(\sin^{-1}\left(\frac{1}{\sqrt{3}}\right)\right)}$$

$$= \frac{\frac{1}{\sqrt{2}}}{\frac{1}{\sqrt{3}}},\quad \mu = \sqrt{\frac{3}{2}}$$

11 1 Vernier scale division = $\frac{29}{30}$ main scale division

$$1\text{ VSD} = \frac{29}{30} \times 0.5^\circ = \left(\frac{29}{60}\right)^\circ$$

Thus, least count = 1 MSD − 1VSD

$$= \left(\frac{1}{2}\right)^\circ - \left(\frac{29}{60}\right)^\circ = \left(\frac{1}{60}\right)^\circ$$

So, reading = main scale reading + vernier scale reading

$$= \text{MSR} + n \times \text{LC}$$

$$= 58.5^\circ + 9 \times \left(\frac{1}{60}\right)^\circ = 58.65^\circ$$

12 $\frac{1}{u} + \frac{1}{v} = \frac{1}{f} \Rightarrow \frac{-1}{u^2}\frac{du}{dt} - \frac{1}{v^2}\frac{dv}{dt} = 0$

$$\frac{dv}{dt} = \frac{-v^2}{u^2}\left(\frac{du}{dt}\right)$$

But $\frac{v}{u} = \frac{f}{u - f}$

$$\therefore \quad \frac{dv}{dt} = -\left(\frac{f}{u - f}\right)^2\left(\frac{du}{dt}\right)$$

$$= \left(\frac{0.2}{-2.8 - 0.2}\right)^2 \times 15 = \frac{1}{15}\text{ms}^{-1}$$

13 Focal length of diverging lens is 25 cm.

As the rays are coming parallel, so the image (I_1) will be formed at the focus of diverging lens i.e. at 25 cm towards left of diverging lens.

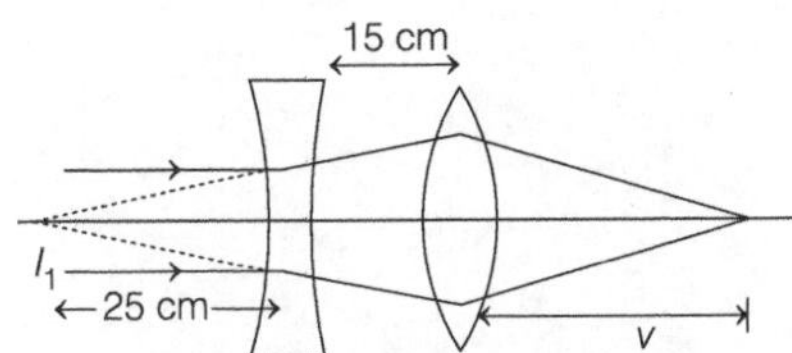

Now, the image (I_1) will work as object for converging lens.

For converging lens, distance of object u (i.e. distance of I_1)

$$= -(25+15) = -40 \text{ cm}$$

$$f = 20 \text{ cm}$$

$\therefore$ From len's formula $\frac{1}{f} = \frac{1}{v} - \frac{1}{u}$

$$\frac{1}{20} = \frac{1}{v} - \frac{1}{-40} \Rightarrow \frac{1}{v} = \frac{1}{20} - \frac{1}{40}$$

$$\frac{1}{v} = \frac{1}{90}$$

$$\Rightarrow \quad v = 40 \text{ cm}$$

v is positive so image will be real and will form at right side of converging lens at 40 cm.

14 If μ is refractive index of material of prism, then from Snell's law

$$\mu = \frac{\sin i}{\sin r} = \frac{\sin(A+\delta_m)/2}{\sin A/2} \quad \ldots\text{(i)}$$

where, A is angle of prism and δ_m is minimum deviation through prism.

Given, $i = 35^\circ, \delta = 40^\circ, e = 79^\circ$.

So, angle of deviation by a glass prism,

$$\delta = i + e - A$$

$$\Rightarrow 40^\circ = 35^\circ + 79^\circ - A$$

i.e. Angle of prism $\Rightarrow A = 74^\circ$.

Such that, $r_1 + r_2 = A = 74^\circ$.

Let us put $\mu = 1.5$ in Eq. (i), we get

$$1.5 = \frac{\sin\left(\frac{A+\delta_{\min}}{2}\right)}{\sin A/2}$$

$$\Rightarrow \quad 1.5 = \frac{\sin\left(\frac{74^\circ+\delta_{\min}}{2}\right)}{\sin 37^\circ}$$

$$\Rightarrow \quad 0.9 = \sin\left(37^\circ + \frac{\delta_{\min}}{2}\right)$$

$$(\because \sin 37^\circ \approx 0.6)$$

$$\sin 64^\circ = \sin\left(37^\circ + \frac{\delta_{\min}}{2}\right)$$

$$(\because \sin 64^\circ = 0.9)$$

$$37^\circ + \frac{\delta_{\min}}{2} = 64^\circ \Rightarrow \delta_{\min} \approx 54^\circ$$

This angle is greater than the 40° deviation angle already given. For greater μ, deviation will be even higher. Hence, μ of the given prism should be lesser than 1.5. Hence, the closest option will be 1.5.

15 Shift in image position due to glass plate,

$$S = \left(1 - \frac{1}{\mu}\right)t = \left(1 - \frac{1}{1.5}\right) \times 1 \text{ cm}$$

$$= \frac{1}{3} \text{ cm}$$

For focal length of the lens,

$$\frac{1}{f} = \frac{1}{v} - \frac{1}{u} = \frac{1}{12} - \frac{1}{-240}$$

or $$\frac{1}{f} = \frac{20+1}{240} \Rightarrow f = \frac{240}{21} \text{ cm}$$

Now, to get back image on the film, lens has to form image at $\left(12 - \frac{1}{3}\right)$ cm $= \frac{35}{3}$ cm such that the glass plate will shift the image on the film.

As, $\frac{1}{f} = \frac{1}{v} - \frac{1}{u}$

$$\Rightarrow \quad \frac{1}{u} = \frac{1}{v} - \frac{1}{f} = \frac{3}{35} - \frac{21}{240}$$

$$= \frac{48 \times 3 - 7 \times 21}{1680} = -\frac{1}{560}$$

$$\Rightarrow \quad u = -5.6 \text{ m}$$

16 As refractive index for $z > 0$ and $z \le 0$ is different, xy-plane should be boundary between two media.

Angle of incidence,

$$\cos i = \left|\frac{A_z}{\sqrt{A_x^2 + A_y^2 + A_z^2}}\right| = \frac{1}{2}$$

$$\therefore \quad i = 60^\circ$$

From Snell's law,

$$\frac{\sin i}{\sin r} = \frac{\mu_2}{\mu_1} = \frac{\sqrt{3}}{\sqrt{2}}$$

$$\sin r = \frac{\sqrt{2}}{\sqrt{3}} \times \sin 60^\circ$$

$$= \frac{\sqrt{2}}{\sqrt{3}} \times \frac{\sqrt{3}}{2}$$

$$= \frac{1}{\sqrt{2}} = 45^\circ$$

$$\Rightarrow \quad r = 45^\circ$$

17 $d = 0.2 \tan 30^\circ = \frac{0.2}{\sqrt{3}}$

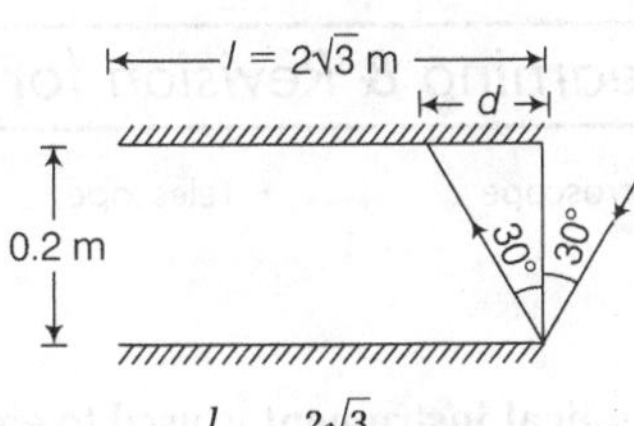

$$\frac{l}{d} = \frac{2\sqrt{3}}{0.2/\sqrt{3}} = 30$$

Therefore, maximum number of reflections are 30.

18 From the figure CD = emergent ray, and CD is parallel to PQ and BC is a line intersecting these parallel lines.

So, $\quad \angle DCB + \angle CBQ = 180^\circ$

$$\angle DCN + \angle NCB + \angle CBQ = 180^\circ$$

$$\alpha + \alpha + 45^\circ = 180^\circ$$

$$\alpha = 67.5^\circ$$

But $\quad \angle NCS = 90^\circ$

So, second mirror is at angle of 22.5° with horizontal.

DAY TWENTY EIGHT

Optical Instruments

Learning & Revision for the Day

- Microscope
- Telescope
- Resolving Power of an Optical Instrument

An **optical instrument** is used to enhance and analyses the light waves. The light waves are in the form of photons, hence optical instruments also determine the characteristics properties of light waves.

Microscope

It is an optical instrument which forms a magnified image of a small nearby object and thus, increases the visual angle subtended by the image at the eye, so that the object is seen to be bigger and distinct.

1. Simple Microscope (Magnifying Glass)

It consists of a single convex lens of small focal length and forms a magnified image of an object placed between the optical centre and the principal focus of the lens.

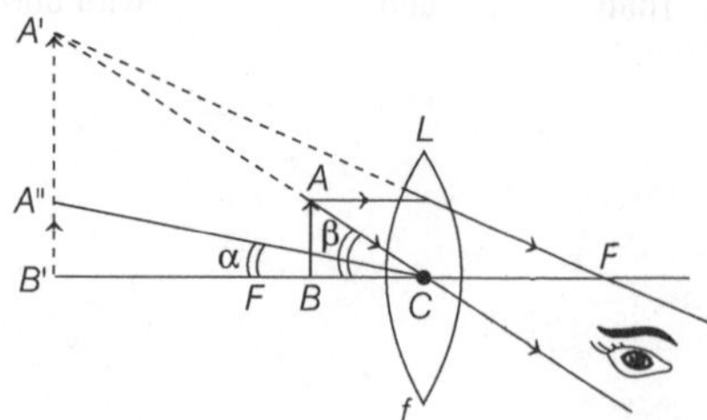

If the image is formed at the near point of eye, then $m = \left(1 + \frac{D}{f}\right)$

But, if the image is formed at infinity, then $m = \frac{D}{f}$

where, D = normal viewing distance (25 cm),

f = focal length of magnifying lens.

PREP MIRROR
Your Personal Preparation Indicator

- No. of Questions in Exercises (x)—
- No. of Questions Attempted (y)—
- No. of Correct Questions (z)—
  (*Without referring Explanations*)

- Accuracy Level ($z/y \times 100$)—
- Prep Level ($z/x \times 100$)—

In order to expect good rank in JEE, your Accuracy Level should be above 85 & Prep Level should be above 75.

2. Compound Microscope

It consists of two lenses of small focal length and small apertures. Also, the focal length and aperture of objective lens are smaller than that of eyepiece. The image formed by the objective lens is real, inverted and magnified. This image acts as the object for the eyepiece and the final image is highly magnified, virtual and inverted w.r.t. the original object.

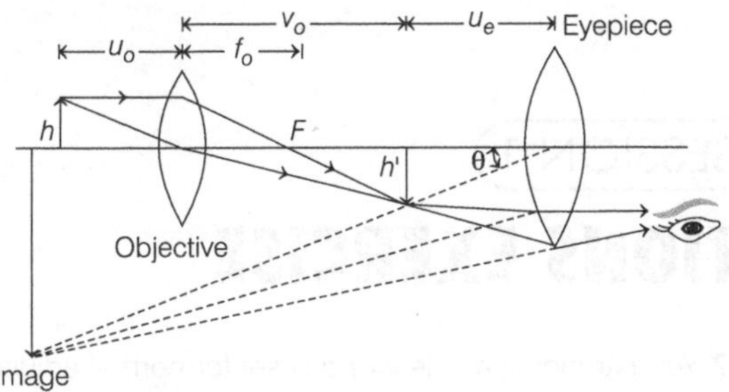

If m_o and m_e be the magnifications produced by the objective and the eyepiece respectively, then total magnification of microscope $m = m_o \times m_e$.

If final image is formed at the near point *(D)* of the eye, then

$$m = -\frac{v_o}{u_o}\left(1 + \frac{D}{f_e}\right)$$

or

$$m = -\frac{L}{f_o}\left(1 + \frac{D}{f_e}\right) \text{(approx)}$$

If final image is formed at infinity, then

$$m = -\frac{v_o}{u_o} \cdot \frac{D}{f_e} = -\frac{L}{f_o} \cdot \frac{D}{f_e} \text{ (approx)}$$

Length of tube of microscope, $L = v_o + u_e$.

NOTE
- Huygens' eyepiece is free from chromatic and spherical aberration, but it cannot be used for measurement purposes.
- Ramsden's eyepiece can be used for precise measurement as cross wires can be fixed in this eyepiece. It slightly suffers from spherical and chromatic aberrations.

Telescope

Telescope is an optical instrument which increases, the visual angle at the eye by forming the image of a distant object at the least distance of distinct vision, so that the object is seen distinct and bigger.

Refracting Telescope

It consists of an objective lens of large focal length f_o and large aperture. The eyepiece consists of a convex lens of small aperture and small focal length f_e.

Distance between the two lenses is set as,

$$L = f_o + f_e$$

In normal adjustment, the final image is formed at infinity and magnifying power of the telescope is

$$m = -\frac{f_o}{f_e}$$

In practical adjustment, the final image is formed at the near point of the observer's eye. In this arrangement,

$$m = -\frac{f_o}{f_e}\left(1 + \frac{f_e}{D}\right)$$

Reflecting Telescope

It consists of an objective which is a large paraboloid concave mirror of maximum possible focal length f_o and the eyepiece is a convex lens of small focal length and small aperture, then

Magnifying power, $\quad m = -\dfrac{f_o}{f_e}$

Reflecting type telescope is considered superior as it is free from spherical and chromatic aberrations, is easy to install and maintain, and can produce image of greater intensity.

NOTE
- The large aperture of telescope objective, helps in forming a brighter image.
- If diameter of pupil of human eye is *d* and that of telescope be *D*, then image formed by telescope will be $\left(\frac{D}{d}\right)^2$ times brighter than the image of the same object, seen directly by the unaided eye.

Resolving Power of an Optical Instrument

Resolving power of an optical instrument is its ability to produce distinct images of two points of an object (or two nearby objects) very close together. Resolving power of an optical instrument is inverse of its limit of resolution. Smaller the limit of resolution of a device, higher is its resolving power. Limit of resolution of a normal human eye is 1′.

The minimum distance (or angular distance) between two points of an object whose images can be formed distinctly by the lens of an optical instrument, is called its **limit of resolution.**

Resolving Power of a Telescope

If the aperture (diameter) of the telescope objective be the *D*, then the minimum angular separation ($d\theta$) between two distant objects, whose images are just resolved by the telescope, is

$$d\theta = \frac{1.22\lambda}{D}$$

and resolving power of the telescope,

$$\text{RP} = \frac{1}{d\theta} = \frac{D}{1.22\lambda}$$

Resolving Power of a Microscope

The least distance (d) between two points, whose images are just seen distinctly by a microscope, is given by

$$d = \frac{1.22\lambda}{2n_m \sin\theta}$$

where, λ = wavelength of light used to illuminate the object,

n_m = refractive index of the medium between the object and the objective lens, and

θ = semi angle of the cone of light from the point object.

The term $n_m \sin\theta$ is generally called the numerical aperture of the microscope.

$\therefore$ Resolving power of the microscope,

$$\text{RP} = \frac{1}{d} = \frac{2\,n_m \sin\theta}{1.22\lambda} = \frac{n_m \sin\theta}{0.61\lambda}$$

DAY PRACTICE SESSION 1

FOUNDATION QUESTIONS EXERCISE

1 To obtain a magnified image at the distance of distinct vision with simple microscope, where should the object be placed?
(a) Away from the focus
(b) At focus
(c) Between the focus and the optical centre
(d) None of the above

2 To obtain the maximum magnification with a simple microscope, where should the eye be placed?
(a) Close to the lens
(b) Half way between the focus and the optical centre
(c) Close to the focus
(d) None of the above

3 A man can see clearly upto 3 m. To see upto 12 m, he must use a lens of
(a) $-\frac{3}{4}$ D (b) 3 D (c) -3 D (d) $-\frac{1}{4}$ D

4 A magnifying glass is used as the object to be viewed can be brought closer to the eye than the normal near point. This results in
(a) the formation of virtual erect image and larger angle to be subtended by the object at the eye and hence viewed in greater detail
(b) increase in field of view
(c) infinite magnification at near point
(d) a diminished but clear image

5 In a compound microscope, the objective produces a magnification of 10, while the eyepiece produces a magnification of 5, then the over all magnification achieved by a compound microscope is
(a) 2 (b) 50 (c) 0.5 (d) 25.00

6 A telescope consists of two thin lenses of focal lengths 0.3 m and 3 cm, respectively. It is focussed on moon which subtends on angle of 0.5° at the objective. Then, the angle subtended at the eye by the final image will be
(a) 5° (b) 0.25° (c) 0.5° (d) 0.35°

7 An astronomical telescope is set for normal adjustment and the distance between its objective and eyepiece is 1.05 cm. The magnifying power of the telescope is 20. What is the focal length of the objective?
(a) 2m (b) 1 m (c) 0.05 m (d) 0.25 m

8 The magnifying power of a telescope is 9. When it is adjusted for parallel rays, the distance between the objective and the eyepiece is found to be 20cm. The focal length of lenses are
(a) 18 cm, 2 cm (b) 11cm, 9cm
(c) 10cm, 10cm (d) 15 cm, 5 cm

9 A simple telescope, consisting of an objective of focal length 60 cm and a single eye lens of focal length 5 cm is focussed on a distant object in such a way that parallel rays emerge from the eye lens. If the object subtends an angle of 2° at the objective, the angular width of the image is
(a) 10° (b) 24° (c) 50° (d) (1/6)°

10 In a laboratory four convex lenses L_1, L_2, L_3 and L_4 of focal lengths 2, 4, 6 and 8cm respectively are available. Two of these lenses form a telescope of length 10 cm and magnifying power 4. The objective and eye lenses are respectively,
(a) L_2, L_3 (b) L_1, L_4 (c) L_1, L_2 (d) L_4, L_1

11 The magnifying power of an astronomical telescope is 8 and the distance between the two lenses is 54cm. The focal length of eye lens and objective lens will be respectively
(a) 6 cm and 48 cm (b) 48 cm and 6 cm
(c) 8 cm and 64 cm (d) 6 cm and 60 cm

12 In a refracting astronomical telescope, the final image is
(a) real, inverted and magnified
(b) real, erect and magnified
(c) virtual, erect and magnified
(d) virtual, inverted and magnified

13 The magnifying power of a telescope is high, if
(a) both the objective and eyepiece have short focal lengths
(b) both the objective and the eyepiece have long focal lengths
(c) the objective has a short focal length and the eyepiece has a long focal length
(d) the objective has a long focal length and the eyepiece has a short focal length

14 An astronomical telescope has a large aperture to
(a) reduce spherical aberration
(b) have high resolution
(c) increase span of observation
(d) have low dispersion

15 Two point white dots are 1 mm apart on a black paper. They are viewed by eye of pupil diameter 3 mm. Approximately, what is the maximum distance at which these dots can be resolved by the eye?
(take, wavelength of light = 500 nm)
(a) 5 m (b) 1 m
(c) 6 m (d) 3 m

16 Wavelength of light used in an optical instrument are $\lambda_1 = 4000$ Å and $\lambda_2 = 5000$ Å, then ratio of their respective resolving powers (corresponding to λ_1 and λ_2) is
(a) 16 : 25
(b) 9 : 1
(c) 4 : 5
(d) 5 : 4

Direction (Q. Nos. 17-20) *Each of these questions contains two statements Statement I and Statement II. Each of these questions also has four alternative choices, only one of which is the correct answer. You have to select one of the codes (a), (b), (c), (d) given below*
(a) Statement I is true, Statement II is true; Statement II is the correct explanation for Statement I
(b) Statement I is true, Statement II is true; Statement II is not the correct explanation for Statement I
(c) Statement I is true; Statement II is false
(d) Statement I is false; Statement II is true

17 **Statement I** Very large size telescopes are reflecting telescopes instead of refracting telescopes.
Statement II It is easier to provide mechanical support to large size mirrors than large size lenses.
→ JEE Main (Online) 2013

18 **Statement I** The resolving power of a telescope is more if the diameter of the objective lens is more.
Statement II Objective lens of large diameter collects more light.

19 **Statement I** The focal length of the objective of telescope is larger than that of eyepiece.
Statement II The resolving power of telescope increases when the aperture of objective is small.

20 **Statement I** Resolving power of an optical instrument is reciprocal to its limit of resolution.
Statement II Smaller the distance between two point objects the instrument can resolve, higher is its resolving power.

DAY PRACTICE SESSION 2

PROGRESSIVE QUESTIONS EXERCISE

1 An observer looks at a distance tree of height 10 m with a telescope of magnifying power of 20. To the observer the tree appears → JEE Main 2016 (Offline)
(a) 10 times taller (b) 10 times nearer
(c) 20 times taller (d) 20 times nearer

2 Assuming human pupil to have a radius of 0.25 cm and a comfortable viewing distance of 25 cm, the minimum separation between two objects that human eye can resolve at 500 nm wavelength is → JEE Main 2015
(a) 30 μm (b) 100 μm (c) 300 μm (d) 1 μm

3 For compound microscope $f_o = 1$ cm, $f_e = 2.5$ cm. An object is placed at distance 1.2 cm from objective lens. What should be the length of microscope for normal adjustment?
(a) 8.5 cm (b) 8.3 cm (c) 6.5 cm (d) 6.3 cm

4 A telescope consists of two lenses of focal length 10 cm and 1 cm. The length of the telescope when an object is kept at a distance of 60 cm from the objective and the final image is formed at least distant of distinct vision is
(a) 15.05 cm
(b) 12.96 cm
(c) 13.63 cm
(d) 14.44 cm

5 A light source is placed at a distance b from a screen. The power of the lens required to obtain k-fold magnified image is
(a) $\frac{k+1}{kb}$ (b) $\frac{(k+1)^2}{kb}$
(c) $\frac{kb}{k+1}$ (d) $\frac{kb}{(k-1)^2}$

6 Two white dots are 1 mm apart on a black paper. They are viewed by naked eye of pupil size 3 mm diameter. Upto what distance, the dots are clearly and separately visible?

(a) 3 m (b) 6 m
(c) 1 m (d) 5 m

7 Resolution of human eye is about 1 min. From what distance a normal human eye can just resolve two objects which are 3m apart.

(a) 10 km (b) 15 km (c) 20 km (d) 30 km

8 A student make a compound microscope by using two lenses of focal lengths 1.5 cm and 6.25 cm. She kept an object at 2 cm from objective and forms its final image at 25 cm from eye lens. Distance between two lenses is

(a) 6 cm (b) 7 cm
(c) 9 cm (d) 11 cm

9 Orbital radius of moon is 3.8×10^5 km and its diameter is 3.5×10^2 km. It is seen from a telescope with lenses of 4 m and 10 m. Angle subtended by the moon's image on eye of observer will be

(a) 15° (b) 20°
(c) 30° (d) 35°

10 Match the following column I with column II.

Column I	Column II
A. Magnification of simple microscope	1. $\frac{-v_0}{u_0}\left(1+\frac{D}{f_e}\right)$
B. Magnification of compound microscope when image formed at least distance of vision	2. $\left(\frac{D}{1.22\lambda}\right)$
C. Tube length of telescope	3. $\left(1+\frac{D}{f}\right)$
D. Resolving power of a telescope	4. $f_0 + f_e$

Codes

	A	B	C	D		A	B	C	D
(a)	1	2	3	4	(b)	2	1	4	3
(c)	3	1	4	2	(d)	3	2	4	1

ANSWERS

SESSION 1	1 (c)	2 (a)	3 (d)	4 (a)	5 (b)	6 (a)	7 (b)	8 (a)	9 (b)	10 (d)
	11 (a)	12 (d)	13 (d)	14 (b)	15 (a)	16 (d)	17 (a)	18 (a)	19 (c)	20 (c)
SESSION 2	1 (c)	2 (a)	3 (b)	4 (b)	5 (b)	6 (d)	7 (a)	8 (d)	9 (b)	10 (c)

Hints and Explanations

SESSION 1

1 Magnification is obtained in a simple microscope when the object is close to the lens between the focus and the optical centre, as lens is converging.

2 The magnification of simple microscope is given by

$$m = 1 + \frac{D-a}{f}$$

where, D = least distance of vision,
f = focal length and
a = distance of lens from eye

So, lesser the distance between eye and lens, greater is the magnification.

3 Using, $f = \frac{xy}{x-y}$

$\Rightarrow f = \frac{3 \times 12}{3-12} = -4$ m

Hence, power of lens required,

$$P = \frac{1}{f} = \frac{-1}{4} \text{D}$$

4 This results in the formation of virtual, erect image and the object subtends a larger angle at the eye and the image is viewed in greater detail.

5 $m = m_o \times m_e = 10 \times 5 = 50$

6 For a telescope, $\frac{\beta}{\alpha} = \frac{f_o}{f_e}$

$\therefore \quad \frac{\beta}{0.5°} = \frac{0.3}{0.03} \Rightarrow \beta = 5°$

7. Here, $m = f_o / f_e$

and $L = f_o + f_e = 1.05$

or $f_e = 1.05 - f_o$

and $20 = f_o/(1.05 - f_o)$

This gives $f_o = 1$ m and $f_e = 0.05$ m.

8 $f_o + f_e = 20$ cm,

$m = f_o / f_e = 9.$

This gives

$f_e = 2$ cm and $f_o = 18$ cm.

9 It is a case of normal adjustment.

Hence, $m = f_o / f_e$

Also, $m = \beta/\alpha$

Therefore, $\frac{\beta}{\alpha} = \frac{f_o}{f_e}$

Here, $f_o = 60$ cm, $f_e = 5$ cm, $\alpha = 2°$

Hence, $\beta = 24°$.

10 Length of tube = 10 cm

$f_o + f_e = 10$ cm ...(i)

Magnification, $m = \frac{f_o}{f_e} = 4$

$f_o = 4 f_e$

On putting in Eq. (i), we get

$$5f_e = 10 \text{ cm}$$

or $f_e = 2 \text{ cm}$

and $f_o = 8 \text{ cm}$

Hence, L_4 and L_1 will be used.

Note *In telescope, objective always have larger focal length than eyepiece.*

11 As $L = f_o + f_e = 54 \text{ cm}$

and $|m| = \frac{f_o}{f_e} = 8$

On simplification, we get

$f_e = + 6 \text{ cm}$

and $f_o = + 48 \text{ cm}$

12 The image formed by the objective lens is real, inverted and larger object and the eyepiece forms a second image but virtual, inverted and larger than the first.

13 The magnifying power of a telescope (if the object is at infinity) is given by

$$M = \frac{f_o}{f_e} \cdot \frac{D + f_e}{D}$$

where, D = least distance of distinct vision, where the final image is formed.

14 An astronomical telescopic as a large aperture to have high resolution.

15 We know that,

$$\frac{y}{D} \geq 1.22\frac{\lambda}{d}$$

$\Rightarrow$ $D \leq \frac{yd}{1.22\,\lambda}$

$\therefore$ $D \leq \frac{10^{-3} \times 3 \times 10^{-3}}{1.22 \times 5 \times 10^{-7}} = 5 \text{ m}$

$\Rightarrow$ $D_{max} = 5\text{m}$

16 Resolving power of an optical instrument is inversely proportional to λ, i.e.

$$\text{RP} \propto \frac{1}{\lambda}$$

$\therefore$ $\frac{\text{Resolving power at } \lambda_1}{\text{Resolving power at } \lambda_2} = \frac{\lambda_2}{\lambda_1}$

$= \frac{5000}{4000}$

$= 5:4$

17 As very large size telescope needs mechanical support to large size mirror than size of lens. So, in order to fulfil this mechanical support telescope is reflecting instead of refracting telescope.

18 Resolving power of telescope is

$$= \frac{a}{1.22\lambda}$$

where, a is the diameter of objective lens and λ is the wavelength of light used. It is obvious that on increasing a, more light is collected by objective lens and so, the image formed is more bright. Thus, resolving power of telescope increases.

19 The magnifying power of telescope in relaxed state is

$$m = \frac{f_o}{f_e}$$

So, for high magnification, the focal length of objective length should be larger than that of eyepiece.

Resolving power of a telescope

$$= \frac{d}{1.22\lambda}$$

For high resolving power, diameter (d) of objective should be higher.

20 Resolving power of an optical instrument is its ability to produce distinct image of two points of an object very close together. Resolving power of an optical instrument is inverse of its limit of resolution. Smaller the limit of resolution of a device higher is its resolving power.

SESSION 2

1 Height of image depends upon the magnifying power to see a 20 times taller object, as the angular magnification should be 20 and we observe angular magnification. Option (c) would not be very correct as the telescope can be adjusted to form the image anywhere between infinity and least distance for distinct vision.

Suppose that the image is formed at infinity. Then, the observer will have to focus the eyes at infinity to observe the image. Hence, it is incorrect to say that the image will be appear nearer to the observer.

2 We can write resolving angle of naked eye as

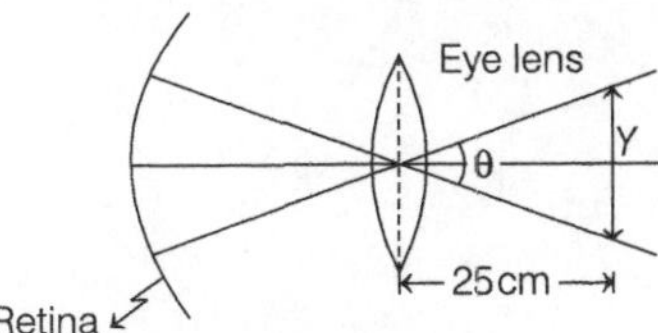

Assuming human pupil to have a radius $r = 0.25 \text{ cm}$ or diameter $d = 2r = 2 \times 0.25 = 0.5 \text{ cm}$, the wavelength of light $\lambda = 500 \text{ nm}$

$= 5 \times 10^{-7} \text{ m}$

We have the formula, $\sin\theta = \frac{1.22\,\lambda}{d}$

$\therefore$ $\sin\theta = \frac{1.22 \times 5 \times 10^{-7}}{0.5 \times 10^{-2}}$

$= \frac{1.22 \times 5 \times 10^{-7}}{5 \times 10^{-3}}$

$= 1.22 \times 10^{-4}$

The distance of comfortable viewing is $D = 25 \text{ cm} = 0.25 \text{ m}$

Let Y be the minimum separation between two objects that human eye can resolve i.e. $\sin\theta = \frac{Y}{D}$

$\therefore$ $y = D \sin\theta = 0.25 \times 1.22 \times 10^{-4}$

$= 3 \times 10^{-5} \text{ m} = 30\,\mu\text{m}$

3 When final image is formed at normal adjustment, then length of compound microscope,

$$L = v_o + u_e = \frac{u_o f_o}{(u_o + f_o)} + \frac{f_e D}{f_e + D}$$

$= \frac{-1.2 \times 1}{-1.2 + 1} + \frac{2.5 \times 25}{2.5 + 25}$

$= 6 + 2.27$

$= 8.27 \approx 8.3 \text{ cm}$

4 Two lenses used are eyepiece and objective.

For eyepiece, $f_e = 1 \text{ cm}$,

$D = v_e = 25 \text{cm}$

$$\frac{1}{v_e} - \frac{1}{u_e} = \frac{1}{f_e}$$

$\Rightarrow$ $-\frac{1}{25} - \frac{1}{u_e} = 1$

$\Rightarrow$ $u_e = -\frac{25}{26} \text{ cm}$

For objective $u_o = -60 \text{ cm}$,

$f_o = 10 \text{ cm}$

$$\frac{1}{v_o} - \frac{1}{u_o} = \frac{1}{f_o}$$

$$\frac{1}{v_o} + \frac{1}{60} = \frac{1}{10}$$

$\Rightarrow$ $\frac{1}{v_o} = \frac{1}{10} - \frac{1}{60}$

$\Rightarrow$ $\frac{1}{v_o} = \frac{5}{60}$

$\Rightarrow$ $v_o = \frac{60}{5}$

$= 12 \text{ cm}$

Length of telescope,

$L = v_o + u_e$

$= 12 + \frac{25}{26}$

$= 12.96 \text{ cm}$

5 Distance of light source from lens is x_0 and distance of screen from lens is $(b - x_0)$.

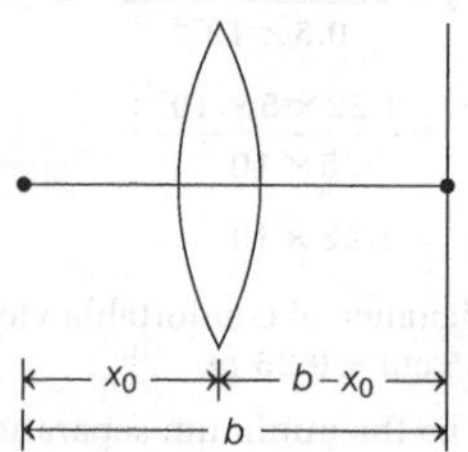

$\therefore$ Image is formed on screen, hence $(b - x_0)$ is also the image distance.

Image is formed on the screen, so m will be negative.

$$m = -k = \frac{v}{u}$$

$$\therefore \quad v = -ku$$

Here, $u = -x_0$, $v = b - x_0$

$$\therefore \quad b - x_0 = kx_0, \quad x_0 = \frac{b}{1+k}$$

From lens formula,

$$\frac{1}{v} - \frac{1}{u} = \frac{1}{f}$$

$$\frac{1}{b - x_0} - \frac{1}{-x_0} = p$$

$$\Rightarrow \quad p = \frac{b}{x_0(b - x_0)}$$

Putting the value of x_0, we have

$$p = \frac{(k+1)^2}{kb}$$

6 Using, $\frac{1.22\lambda}{a} = \frac{x}{d}$

$$\Rightarrow \quad d = \frac{xa}{1.22\lambda}$$

$$= \frac{1 \times 10^{-3} \times 3 \times 10^{-3}}{1.22 \times 500 \times 10^{-9}} \approx 5\text{ m}$$

Note *In case, wavelength of light is not given, we take mean value of visible region ≈ 500 nm.*

7 Using, $\theta = \frac{d}{r}$

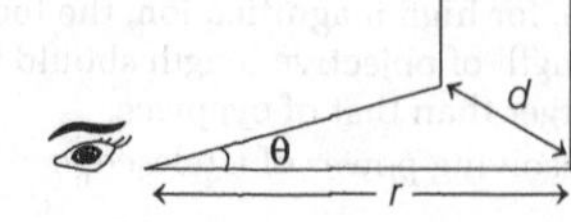

We have, $r = \frac{d}{\theta}$

$$\left[\text{where, } \theta \text{ is in radians} \quad = 1' = \left(\frac{1}{60}\right)^\circ = \frac{1}{60} \times \frac{\pi}{180}\right]$$

$$= \frac{3}{\left(\frac{1}{60}\right) \times \frac{\pi}{180}} = \frac{3 \times 60 \times 180}{\pi}$$

$$\approx 10.3\text{ km} \approx 10\text{ km}$$

8 As, $f_o < f_e$, for a compound microscope.

So, $f_o = 1.5\text{ cm}$ and $f_e = 6.25\text{ cm}$

Now, length of tube

= distance between lenses

$$= \frac{u_o f_o}{u_o - f_o} + \frac{f_e D}{f_e + D}$$

$$= \frac{2 \times 1.5}{2 - 1.5} + \frac{6.25 \times 25}{6.25 + 25} = 11\text{ cm}$$

9 Angle subtended by moon on the objective of telescope is,

$$\alpha = \frac{3.5 \times 10^3}{3.8 \times 10^5}$$

$$= \frac{3.5}{3.8} \times 10^{-2}\text{ rad}$$

As, $m = \frac{f_o}{f_e} = \frac{\beta}{\alpha}$

$$\therefore \quad \beta = \frac{400}{10} \times \alpha$$

$$= 40\alpha = 40 \times \frac{3.5}{3.8} \times 10^{-2} \times \frac{180}{\pi} \approx 20^\circ$$

10 A - Magnification of simple microscope is given by,

$$m = \left(1 + \frac{D}{F}\right)$$

B - Magnification of compound microscope, when image is formed at least distance of vision is,

$$m = \frac{-v_0}{u_0}\left(1 + \frac{D}{f_e}\right)$$

C - Tube length of telescope,

$$L = f_0 + f_e$$

D - Resolving power of telescope

$$= \left(\frac{D}{1.22\lambda}\right)$$

DAY TWENTY NINE

Wave Optics

Learning & Revision for the Day

- Wavefront
- Interference of Light
- Young's Double Slit Experiment
- Coherent Sources
- Interference in Thin Films
- Diffraction
- Polarisation of Light
- Brewster's Law
- Law of Malus
- Polaroids

According to Huygens', light is a form of energy, which travels in the form of waves through a hypothetical medium 'ether'. The medium was supposed to be all pervading, transparent, extremely light, perfectly elastic and an ideal fluid.

Light waves transmit energy as well as momentum and travel in the free space with a constant speed of $3 \times 10^8 \text{ ms}^{-1}$. However, in a material medium, their speed varies from medium to medium depending on the refractive index of the medium.

Wavefront

A wavefront is the locus of all those points (either particles) which are vibrating in the same phase. The shape of the wavefront depends on the nature and dimension of the source of light.

- In an isotropic medium, for a point source of light, the wavefront is spherical in nature.
- For a line (slit) source of light, the wavefront is cylindrical in shape.
- For a parallel beam of light, the wavefront is a plane wavefront.

Huygens' Principle

Every point on a given wavefront, acts as secondary source of light and emits secondary wavelets which travel in all directions with the speed of light in the medium. A surface touching all these secondary wavelets tangentially in the forward direction, gives the new wavefront at that instant of time.

Laws of reflection and refraction can be determined by using Huygens' principle.

Interference of Light

Interference of light is the phenomenon of redistribution of light energy in space when two light waves of same frequency (or same wavelength) emitted by two coherent sources, travelling in a given direction, superimpose on each other. If a_1 and a_2 be the amplitudes of two light waves of same frequency and ϕ be the phase difference between them, then the amplitude of resultant wave is given by

$$A_R = \sqrt{a_1^2 + a_2^2 + 2a_1a_2 \cos \phi}$$

- No. of Questions in Exercises (x)—
- No. of Questions Attempted (y)—
- No. of Correct Questions (z)—
  (Without referring Explanations)

- Accuracy Level ($z/y \times 100$)—
- Prep Level ($z/x \times 100$)—

In order to expect good rank in JEE, your Accuracy Level should be above 85 & Prep Level should be above 75.

and in terms of intensity of light,

$$I_R = I_1 + I_2 + 2\sqrt{I_1 I_2}\cos\phi.$$

Condition for Constructive Interference

If at some point in space, the phase difference between two waves, $\phi = 0°$ or $2n\pi$ or path difference between two waves, $\Delta = 0$ or $n\lambda$, where n is an integer, then $A_R = a_1 + a_2$ or $I_R = I_1 + I_2 + 2\sqrt{I_1 I_2}$ is maximum. Such an interference is called **constructive interference**.

Condition for Destructive Interference

If at some point in space, the phase difference between two waves, $\phi = (2n-1)\pi$ or path difference, $\delta\Delta = (2n-1)\frac{\lambda}{2}$, then at such points $A_R = (a_1 - a_2)$ and $I_R = I_1 + I_2 - 2\sqrt{I_1 I_2}$ is minimum leading to a **destructive interference**.

Amplitude Ratio

$$\frac{I_{max}}{I_{min}} = \frac{I_1 + I_2 + 2\sqrt{I_1 I_2}}{I_1 + I_2 - 2\sqrt{I_1 I_2}} = \left[\frac{\sqrt{I_1} + \sqrt{I_2}}{\sqrt{I_1} - \sqrt{I_2}}\right]^2$$

$$= \left[\frac{a_1 + a_2}{a_1 - a_2}\right]^2 = \left[\frac{r+1}{r-1}\right]^2$$

where, $r = \frac{a_1}{a_2}$ = amplitude ratio.

NOTE
- For identical sources, $I_1 = I_2 = I_0$
- For constructive interference, $I_{max} = 4I_0$ and $I = 4I_0 \cos^2\frac{\phi}{2}$
- For destructive interference, $I_{min} = 0$

Young's Double Slit Experiment

The arrangement is shown in figure monochromatic light of one wavelength is used.

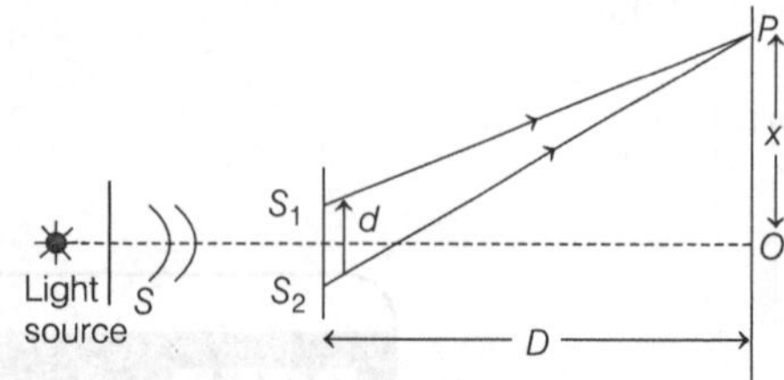

Young's experimental arrangement to produce interference pattern

Bright and dark fringes are formed on the screen with central point O behaving as the central bright fringe, because for O, the path difference $\Delta = 0$.

For light waves reaching a point P, situated at a distance x from central point Δ, the path difference,

$$\Delta = S_2P - S_1P = \frac{xd}{D}$$

Case I If $\frac{xd}{D} = n\lambda$, then we get nth bright fringe. Hence, position of bright fringes on the screen are given by the relation, $x = \frac{nD\lambda}{d}$.

Case II If $\frac{xd}{D} = (2n-1)\frac{\lambda}{2}$, then we get nth dark fringe.

Hence, for nth dark fringe,

$$x = \frac{(2n-1)D\lambda}{2d}$$

where, $n = 1, 2, 3, \ldots\ldots$.

Fringe Width

The separation between any two consecutive bright or dark fringes is called fringe width β.

Thus, $$\beta = \frac{D\lambda}{d}$$

and for a given arrangement, it is constant, i.e. all fringes are uniformly spaced.

Moreover, fringe width β is

(i) $\beta \propto D$, (ii) $\beta \propto \lambda$ and (iii) $\beta \propto \frac{1}{d}$.

Angular fringe width of interference pattern,

$$\alpha = \frac{\beta}{D} = \frac{\lambda}{d}$$

If in a given field of view n_1, fringes of light of wavelength λ_1 are visible and n_2 fringes of wavelength λ_2 are visible, then $n_1\lambda_1 = n_2\lambda_2$

NOTE
- If whole apparatus of Young's double slit experiment is immersed in a transparent medium of refractive index n_m, then fringe width in the medium, $\beta_m = \frac{D\lambda}{n_m d}$.

Displacement of Fringe Pattern

When a thin transparent plate is introduced in the path of one of the interfering waves trains, it is found that the entire fringe pattern is shifted through a constant distance. This shift takes place towards the wave train, in the path of which the plate is introduced.

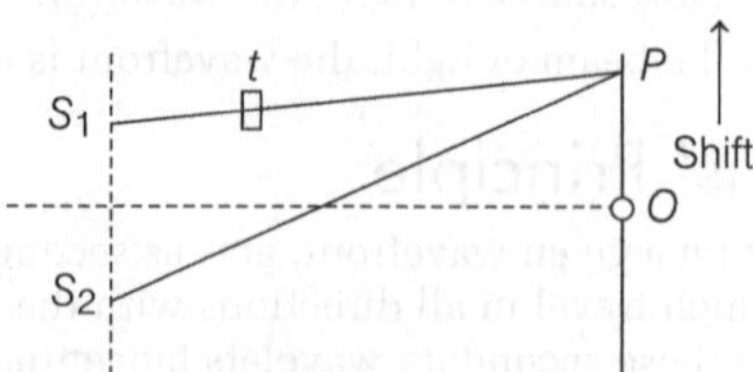

Young's double slit experiment

The fringe width of the patterns remains same. Effective optical path that is equivalently covered in air is $S_1P + t(\mu - 1)$.

Thus, the path difference $= S_2P - S_1P - t(\mu - 1)$

$$= \frac{xd}{D} - t(\mu - 1)$$

[Optical path = refractive index× width of the material]

$\therefore$ Extra path $= \mu t - t = (\mu - 1)t$

Coherent Sources

Two light sources are said to be coherent, if they emit light of exactly same frequency (or wavelength), such that the originating phase difference between the waves emitted by them is either zero or remains constant. For sustained interference pattern, the interfering light sources must be coherent one.

There are two possible techniques for obtaining coherent light sources.

- In division of wavefront technique, we divide the wavefront emitted by a narrow source in two parts by reflection, refraction or diffraction.
- In division of amplitude technique, a single extended light beam of large amplitude is splitted into two or more waves by making use of partial reflection or refraction.

NOTE • Two independent sources of light can never be coherent. Two light sources can be coherent only, if these have been derived from a single parental light source.

Interference in Thin Films

In white light thin films, whose thickness is comparable to wavelength of light, show various colours due to interference of light waves reflected from the two surfaces of thin film.

For interference in reflected light condition of constructive interference (maximum intensity),

$$\Delta = 2n_m\, t \cos r = (2n-1)\frac{\lambda}{2}$$

Condition of destructive interference (minimum intensity),

$$\Delta = 2n_m\, t \cos r = (2n)\frac{\lambda}{2}$$

For interference in refracted light condition of constructive interference (maximum intensity),

$$\Delta = 2n_m t \cos r = (2n)\frac{\lambda}{2}$$

Condition of destructive interference (minimum intensity),

$$\Delta = 2n_m t \cos r = (2n-1)\frac{\lambda}{2}, \text{ where } n = 1, 2, 3, \dots$$

Shift in Interference Pattern

If a transparent thin sheet of thickness t and refractive index n_m is placed in the path of one of the superimposing waves (say in front of slit S_2 of Young's double slit experiment), then it causes an additional path difference due to which interference pattern shifts.

- Additional path difference due to sheet $= (n_m - 1)\,t$
- Fringe shift $= \frac{D}{d}(n_m - 1)t = \frac{\beta}{\lambda}(n_m - 1)t$
- If due to presence of thin film, the fringe pattern shifts by n fringes, then

$$n = \frac{(n_m - 1)\,t}{\lambda} \text{ or } t = \frac{n\lambda}{(n_m - 1)}$$

Shift is independent of the order of fringe and wavelength.

NOTE • **Fresnel's biprism** is a device to produce coherent sources by division of wavefront,

$$d = 2a(n-1)\,\alpha$$

The distance between the coherent sources and screen,

$$D = a + b$$

The fringe width is given by $\beta = \frac{D\lambda}{d} = \frac{\lambda(a+b)}{2a(n-1)\,\alpha}$

Diffraction

Diffraction of light is the phenomenon of bending of light around the edges of an aperture or obstacle and entry of light even in the region of geometrical shadow, when size of aperture or obstacle is comparable to wavelength of light used. Diffraction is characteristic of all types of waves. Greater the wavelength, more pronounced is the diffraction effect. It is due to this reason that diffraction effect is very commonly observed in sound.

Diffraction due to a Single Slit

Fraunhofer's arrangement for studying diffraction at a single narrow slit (width = a) is shown in adjoining figure. Lenses L_1 and L_2 are used to render incident light beam parallel and to focus parallel light beam.

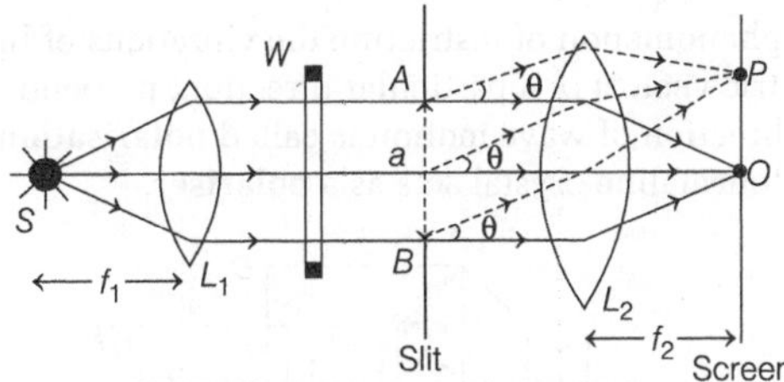

Fresnel diffraction through a slit

As a result of diffraction, we obtain a broad, bright maxima at symmetrical centre point O and on either side of it, we get secondary diffraction maxima of successively falling intensity and poor contrast, as shown in figure.

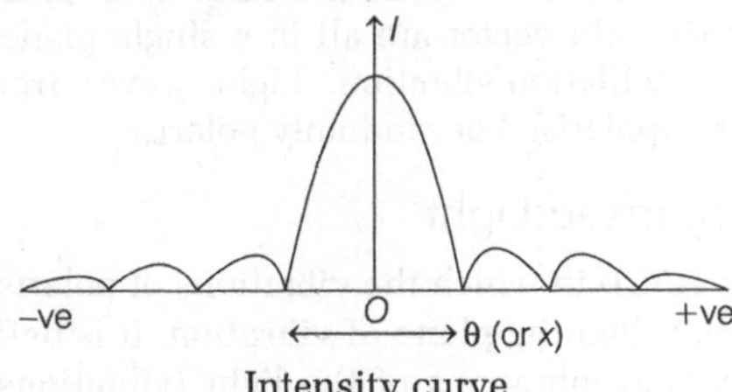

Intensity curve

- Condition of diffraction minima is given by

$$a \sin\theta = n\lambda$$

where, $n = 1, 2, 3, 4, \dots$

But the condition of secondary diffraction maxima is

$$a \sin\theta = (2n+1)\frac{\lambda}{2}$$

where, $n = 1, 2, 3, 4, \dots$

- Angular position of nth secondary minima is given by

$$\sin\theta = \theta = n\frac{\lambda}{a}$$

and linear distance, $x_n = D\theta = \frac{nD\lambda}{a} = \frac{nf_2\lambda}{a}$

where, f_2 is focal length of lens L_2 and $D = f_2$.

- Similarly, for nth maxima, we have

$$\sin\theta = \theta = \frac{(2n+1)\lambda}{2a} \text{ and } x_n = \frac{(2n+1)D\lambda}{2a} = \frac{(2n+1)f_2\lambda}{2a}$$

The angular width of central maxima is $2\theta = \frac{2\lambda}{a}$

or linear width of central maxima $= \frac{2D\lambda}{a} = \frac{2f_2\lambda}{2}$

NOTE • The angular width of central maxima is double as compared to angular width of secondary diffraction maxima.

Polarisation of Light

- Light is an electromagnetic wave in which electric and magnetic field vectors very sinusoidally, perpendicular to each other as well as perpendicular to the direction of propagation of wave of light.
- The phenomenon of restricting the vibrations of light (electric vector) in a particular direction, perpendicular to the direction of wave motion is called **polarisation of light**. The tourmaline crystal acts as a **polariser**.

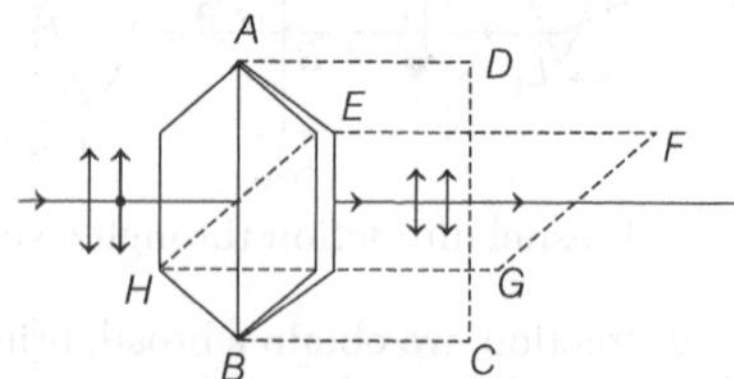

Polarisation of Light

Thus, electromagnetic waves are said to be polarised when their electric field vector are all in a single plane, called the plane of oscillation/vibration. Light waves from common sources are upolarised or randomly polarised.

Plane Polarised Light

The plane $ABCD$ in which the vibrations of polarised light are confined is called the **plane of vibration**. It is defined as The light, in which vibrations of the light (vibrations of electric vector) when restricted to a particular plane the light itself is called plane polarised light. The vibrations in a plane polarised light are perpendicular to the plane of polarisation.

NOTE • Only transverse waves can be polarised. Thus, it proved that light waves are transverse waves.

Brewster's Law

According to this law, when unpolarised light is incident at an angle called polarising angle, i_p on an interface separating air from a medium of refractive index μ, then the reflected light is fully polarised (perpendicular to the plane of incidence), provided

$$\mu = \tan i_p$$

This relation represents Brewster's law. Note that the parallel components of incident light do not disappear, but refract into the medium, with the perpendicular components.

Law of Malus

When a beam of completely plane polarised light is incident on an analyser, the resultant intensity of light (I) transmitted from the analyser varies directly as the square of cosine of angle (θ) between plane of transmission of analyser and polariser.

i.e. $I \propto \cos^2\theta$

If intensity of plane polarised light incidenting on analyser is I_0, then intensity of emerging light from analyser is $I_0\cos^2\theta$.

NOTE • We can prove that when unpolarised light of intensity I_0 gets polarised on passing through a polaroid, its intensity becomes half, i.e. $I = \frac{1}{2}I_0$.

Polaroids

Polaroids are thin and large sheets of crystalline polarising material (made artifically) capable of producing plane polarised beams of large cross-section.

The important uses are

- These reduce excess glare and hence sun glasses are fitted with polaroid sheets.
- These are also used to reduce headlight glare of cars.
- They are used to improve colour contrast in old oil paintings.
- In wind shields of automobiles.
- In window panes.
- In three dimensional motion pictures.

DAY PRACTICE SESSION 1

FOUNDATION QUESTIONS EXERCISE

1 The wavefront of a distant source of light is of shape

(a) spherical (b) cylindrical (c) elliptical (d) plane

2 Which of the following cannot be explained on the basis of wave nature of light?

(i) Polarisation (ii) Optical activity
(iii) Photoelectric effect (iv) Compton effect

(a) III and IV (b) II and III
(c) I and III (d) II and IV

3 A Young's double slit experiment uses a monochromatic source. The shape of the interference fringes formed on a screen is → AIEEE 2005

(a) hyperbola (b) circle
(c) straight line (d) parabola

4 Two light rays having the same wavelength λ in vacuum are in phase initially. Then, the first ray travels a path L_1 through a medium of refractive index μ_1, while the second ray travels a path of length L_2 through a medium of refractive index μ_2. The two waves are then combined to observe interference.
The phase difference between the two waves is

(a) $\frac{2\pi}{\lambda}(\mu_1 L_1 - \mu_2 L_2)$ (b) $\frac{2\pi}{\lambda}(L_2 - L_1)$
(c) $\frac{2\pi}{\lambda}\left(\frac{L_1}{\mu_1} - \frac{L_2}{\mu_2}\right)$ (d) $\frac{2\pi}{\lambda}(\mu_2 L_1 - \mu_1 L_2)$

5 The Young's double slit experiment is performed with blue and with green light of wavelengths 4360 Å and 5460 Å respectively. If x is the distance of the 4th maxima from the central one, then

(a) $x_{(\text{blue})} = x_{(\text{green})}$
(b) $x_{(\text{blue})} > x_{(\text{green})}$
(c) $x_{(\text{blue})} < x_{(\text{green})}$
(d) $x_{(\text{blue})} / x_{(\text{green})} = 5460/4360$

6 In Young's double slit experiment, the length of band is 1mm. The fringe width is 0.021 mm. The number of fringes is

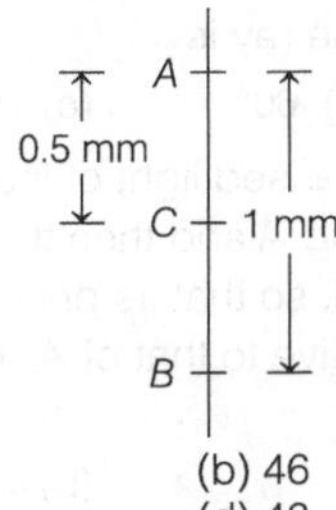

(a) 45 (b) 46
(c) 49 (d) 48

7 In a Young's double slit experiment, two coherent sources are placed 0.90 mm apart and the fringes are observed one metre away. If it produces the second dark fringe at a distance of 1 mm from the central fringe, the wavelength of monochromatic light used will be

(a) 60×10^{-4} cm (b) 10×10^{-4} cm
(c) 10×10^{-5} cm (d) 6×10^{-5} cm

8 In Young's double slit experiment, the two slits act as coherent sources of waves of equal amplitude A and wavelength λ. In another experiment with the same arrangement the two slits are made to act as incoherent sources of waves of same amplitude and wavelength. If the intensity at the middle point of the screen in the first case is I_1 and in the second case I_2, then the ratio $\frac{I_1}{I_2}$ is → AIEEE 2012

(a) 4 (b) 2 (c) 1 (d) 0.5

9 A mixture of light consisting of wavelength 590 nm and an unknown wavelength, illuminates the Young's double slit and gives rise to two overlapping interference patterns on the screen. The central maximum of both lights coincide. Further, it is observed that the third bright fringe of known light coincides with the 4th bright fringe of the unknown light. From this data, the wavelength of the unknown light is → AIEEE 2003

(a) 885.0 nm (b) 442.5 nm
(c) 776.8 nm (d) 393.4 nm

10 Two coherent point sources S_1 and S_2 are separated by a small distance d as shown. The fringes obtained on the screen will be → JEE Main 2013

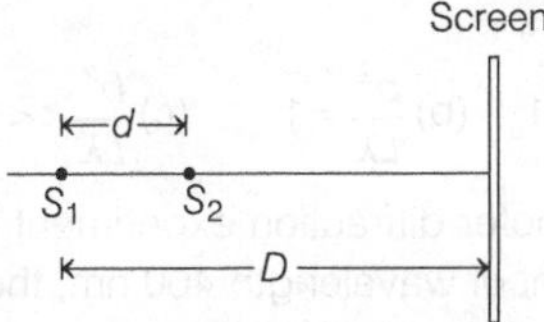

(a) points (b) straight lines
(c) semi-circle (d) concentric circles

11 The source that illuminates the double-slit in 'double-slit interference experiment' emits two distinct monochromatic waves of wavelength 500 nm and 600 nm, each of them producing its own pattern on the screen. At the central point of the pattern when path difference is zero, maxima of both the patterns coincide and the resulting interference pattern is most distinct at the region of zero path difference.

But as one moves out of this central region, the two fringe systems are gradually out of step such that maximum due to one wave length coincides with the minimum due to the other and the combined fringe system becomes completely indistinct. This may happen when path difference in nm is

→JEE Main (Online) 2013

(a) 2000 (b) 3000 (c) 1000 (d) 1500

12 In Young's double slit experiment, the intensity at a point where the path difference is $\frac{\lambda}{6}$ (λ being the wavelength of the light used) is I. If I_0 denotes the maximum intensity, I/I_0 is equal to →AIEEE 2007

(a) $\frac{1}{\sqrt{2}}$ (b) $\frac{\sqrt{3}}{2}$ (c) $\frac{1}{2}$ (d) $\frac{3}{4}$

13 In Young's double slit experiment, the intensity at a point is 1/4 of the maximum intensity. Angular position of this point is →AIEEE 2005

(a) $\sin^{-1}\left(\frac{\lambda}{d}\right)$ (b) $\sin^{-1}\left(\frac{\lambda}{2d}\right)$ (c) $\sin^{-1}\left(\frac{\lambda}{3d}\right)$ (d) $\sin^{-1}\left(\frac{\lambda}{4d}\right)$

14 The first diffraction minimum due to single slit diffraction is θ, for a light of wavelength 5000 Å. If the width of slit is 1×10^{-4} cm. Then, the value of θ is

(a) 30° (b) 45° (c) 60° (d) 15°

15 A parallel monochromatic beam of light is incident normally on a narrow slit. A diffraction pattern is formed on a screen placed perpendicular to the direction of incident beam. At the first maximum of the diffraction pattern, the phase difference between the rays coming from the edges of the slit is

(a) 0 (b) $\pi/2$ (c) π (d) 2π

16 In Fraunhofer diffraction experiment, L is the distance between screen and the obstacle, b is the size of obstacle and λ is wavelength of incident light. The general condition for the applicability of Fraunhofer diffraction is

(a) $\frac{b^2}{L\lambda}>>1$ (b) $\frac{b^2}{L\lambda}=1$ (c) $\frac{b^2}{L\lambda}<<1$ (d) $\frac{b^2}{L\lambda}\neq1$

17 In a Fraunhofer diffraction experiment at a single slit using a light of wavelength 400 nm, the first minimum is formed at an angle of 30°. The direction θ of the first secondary maximum is given by

(a) $\sin^{-1}\left(\frac{2}{3}\right)$ (b) $\sin^{-1}\left(\frac{3}{4}\right)$

(c) $\sin^{-1}\left(\frac{1}{4}\right)$ (d) $\tan^{-1}\left(\frac{2}{3}\right)$

18 An unpolarised beam of light is incident on a group of four polarising sheets which are arranged in such a way that the characteristic direction of each polarising sheet makes an angle of 30° with that of the preceding sheet. The percentage of incident light transmitted by the first polariser will be

(a) 100% (b) 50% (c) 25% (d) 125%

19 A beam of ordinary unpolarised light passes through a tourmaline crystal C_1 and then it passes through another tourmaline crystal C_2 which is oriented such that its principal plane is parallel to that of C_2. The intensity of emergent light is I_0. Now, C_2 is rotated by 60° about the ray. The emergent ray will have an intensity

(a) $2I_0$ (b) $I_0/2$ (c) $I_0/4$ (d) $I_0/\sqrt{2}$

20 A beam of light AO is incident on a glass slab ($\mu = 1.54$) in a direction as shown in the figure. The reflected ray OB is passed through a nicol prism. On viewing through a nicol prism, we find on rotating the prism that

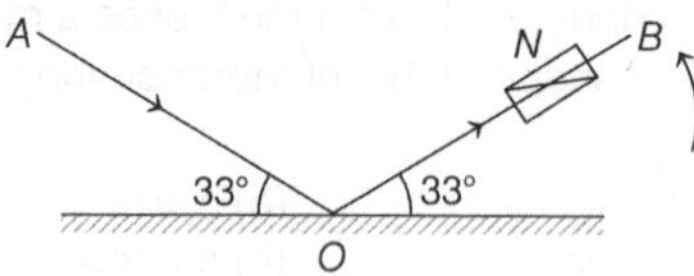

(a) the intensity is reduced down to zero and remains zero
(b) the intensity reduces down somewhat and rises again
(c) there is no change in intensity
(d) the intensity gradually reduces to zero and then again increases

21 When an unpolarised light of intensity I_0 is incident on a polarising sheet, the intensity of the light which does not get transmitted is →AIEEE 2005

(a) $\frac{1}{2}I_0$ (b) $\frac{1}{4}I_0$ (c) zero (d) I_0

22 Two beams, A and B of plane polarised light with mutually perpendicular planes of polarisation are seen through a polaroid. From the position when the beam A has maximum intensity (and beam B has zero intensity), a rotation of polaroid through 30° makes the two beams appear equally bright.If the initial intensities of the two beams are I_A and I_B respectively, then I_A/I_B equals →JEE Main 2014

(a) 3 (b) $\frac{3}{2}$ (c) 1 (d) $\frac{1}{3}$

23 A ray of light is incident on the surface of a glass plate of refractive index 1.732 at the polarising angle. The angle of refraction of the ray is

(a) 45° (b) 60° (c) 15° (d) 30°

24 A beam of unpolarised light of intensity I_0 is passed through a polaroid A and then through another polaroid B which is oriented, so that its principal plane makes an angle of 45° relative to that of A. The intensity of the emergent light is →JEE Main 2013

(a) I_0 (b) $I_0/2$
(c) $I_0/4$ (d) $I_0/8$

Direction (Q. Nos. 25-28) *Each of these questions contains two statements : Statement I and Statement II. Each of these questions also has four alternative choices, only one of which is the correct answer. You have to select one of the codes (a), (b), (c) and (d) given below:*

(a) Statement I is true, Statement II is true; Statement II is the correct explanation for Statement I
(b) Statement I is true, Statement II is true; Statement II is not the correct explanation for Statement I
(c) Statement I is true; Statement II is false
(d) Statement I is false; Statement II is true

25 **Statement I** The thick film shows no interference pattern. Take thickness of the order of a few cms.

Statement II For interference pattern to be observed path difference between two waves is of the order of few wavelengths.

26 **Statement I** To observe diffraction of light, the size of obstacle/aperture should be of the order of 10^{-7} m.

Statement II 10^{-7}m is the order of wavelength of the visible light.

27 **Statement I** For a given medium, the polarising angle is 60°. The critical angle for this medium is 35°.

Statement II $\mu = \tan i_p$.

28 **Statement I** In Young's double slit experiment, the number of fringes observed in the field of view is small with longer wavelength of light and is large with shorter wavelength of light.

Statement II In the double slit experiment the fringe width depends directly on the wavelength of light.

→ JEE Main (Online) 2013

DAY PRACTICE SESSION 2

PROGRESSIVE QUESTIONS EXERCISE

1 A parallel beam of light of intensity I_0 is incident on a glass plate, 25% of light is reflected by upper surface and 50% of light in reflected from lower surface. The ratio of maximum to minimum intensity in interference region of reflected ray is

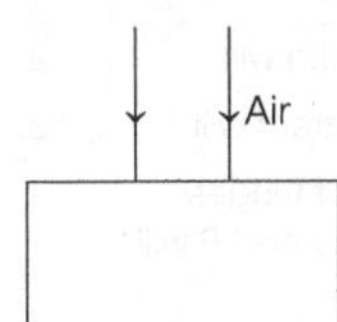

(a) $\left(\dfrac{\frac{1}{2}+\sqrt{\frac{3}{8}}}{\frac{1}{2}-\sqrt{\frac{3}{8}}}\right)^2$ (b) $\left(\dfrac{\frac{1}{4}+\sqrt{\frac{3}{8}}}{\frac{1}{2}-\sqrt{\frac{3}{8}}}\right)^2$ (c) $\dfrac{5}{8}$ (d) $\dfrac{8}{5}$

2 White light is used to illuminate the two slits in a Young's double slit experiment. The separation between the slits is b and the screen is at a distance d ($>> b$) from the slits. At a point on the screen directly in front of one of the slits, certain wavelengths are missing. Some of these missing wavelengths are

(a) $\lambda = \dfrac{2b^2}{3d}$ (b) $\lambda = \dfrac{b^2}{3d}$ (c) $\lambda = \dfrac{2b^2}{d}$ (d) $\lambda = \dfrac{3b^2}{d}$

3 A small transparent slab containing material of $\mu = 1.5$ is placed along AS_2 (figure). What will be the distance from O of the principal maxima and of the first minima on either side of the principal maxima obtained in the absence of the glass slab.

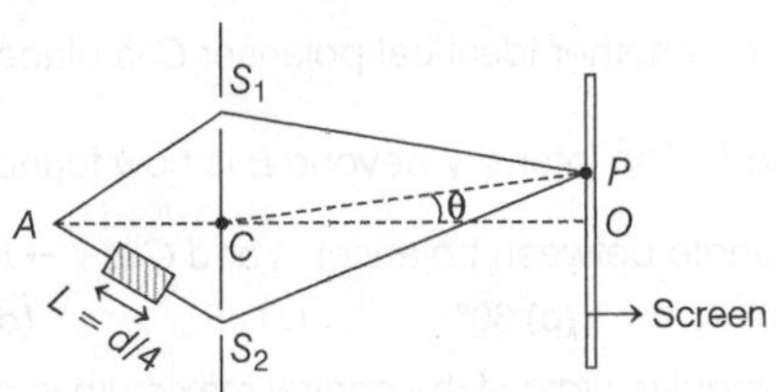

$AC = CO = D, S_1C = S_2C = d << D$

(a) $\dfrac{5}{\sqrt{235}}$ below point O (b) $\dfrac{5}{\sqrt{231}}$ below point O
(c) $\dfrac{5}{\sqrt{220}}$ below point O (d) $\dfrac{5}{\sqrt{110}}$ below point O

4 n identical waves each of intensity I_0 interfere with each other. The ratio of maximum intensities if the interference is (i) coherent and (ii) incoherent is

→ JEE Main (Online) 2013

(a) n^2 (b) $\dfrac{1}{n}$ (c) $\dfrac{1}{n^2}$ (d) n

5 In a Young's double slit experiment, one of the slit is wider than other, so that amplitude of the light from one slit is double of that from other slit. If I_m be the maximum intensity, the resultant intensity I when they interfere at phase difference ϕ is given by → AIEEE 2012

(a) $\dfrac{I_m}{9}(4 + 5\cos\phi)$ (b) $\dfrac{I_m}{3}\left(1 + 2\cos^2\dfrac{\phi}{2}\right)$
(c) $\dfrac{I_m}{5}\left(1 + 4\cos^2\dfrac{\phi}{2}\right)$ (d) $\dfrac{I_m}{9}\left(1 + 8\cos^2\dfrac{\phi}{2}\right)$

6 An object 2.4 m in front of a lens forms a sharp image on a film 12 cm behind the lens. A glass plate 1 cm thick, of refractive index 1.50 is interposed between lens and film with its plane faces parallel to film. At what distance (from lens) should object shifted to be in sharp focus on film? → AIEEE 2012

(a) 7.2 m (b) 2.4 m (c) 3.2 m (d) 5.6 m

7 In a YDSE, bichromatic light of wavelength 400 nm and 560 nm are used. The distance between the slits is 0.1 mm and the distance between the plane of the slits and the screen is 1 m. The minimum distance between two successive regions of complete darkness is → AIEEE 2004

(a) 4 mm (b) 5.6 mm (c) 14 mm (d) 28 mm

8 The maximum number of possible interference maxima for slit-separation equal to twice the wavelength in Young's double slit experiment, is → AIEEE 2004

(a) infinite (b) five (c) three (d) zero

9 Unpolarised light of intensity I passes through an ideal polariser A. Another identical polariser B is placed behind A. The intensity of light beyond B is found to be $\frac{I}{2}$. Now, another identical polariser C is placed between A and B. The intensity beyond B is now found to be $\frac{1}{8}$. The angle between polariser A and C is → JEE Main 2018

(a) 0° (b) 30° (c) 45° (d) 60°

10 The angular width of the central maximum in a single slit diffraction pattern is 60°. The width of the slit is 1 μm. The slit is illuminated by monochromatic plane waves. If another slit of same width is made near it, Young's fringes can be observed on a screen placed at a distance 50 cm from the slits. If the observed fringe width is 1 cm, what is slit separation distance? (i.e. distance between the centres of each slit.) → JEE Main 2018

(a) 25 μm (b) 50 μm (c) 75 μm (d) 100 μm

11 The box of a pin hole camera of length L, has a hole of radius a. It is assumed that when the hole is illuminated by a parallel beam of light of wavelength λ the spread of the spot (obtained on the opposite wall of the camera) is the sum of its geometrical spread and the spread due to diffraction. The spot would then have its minimum size (say b_{min}) when → JEE Main 2016 (Offline)

(a) $a = \frac{\lambda^2}{L}$ and $b_{min} = \left(\frac{2\lambda^2}{L}\right)$ (b) $a = \sqrt{\lambda L}$ and $b_{min} = \left(\frac{2\lambda^2}{L}\right)$

(c) $a = \sqrt{\lambda L}$ and $b_{min} = \sqrt{4\lambda L}$ (d) $a = \frac{\lambda^2}{L}$ and $b_{min} = \sqrt{4\lambda L}$

12 In the YDSE apparatus shown in figure Δx is the path difference between S_1P and S_2P.

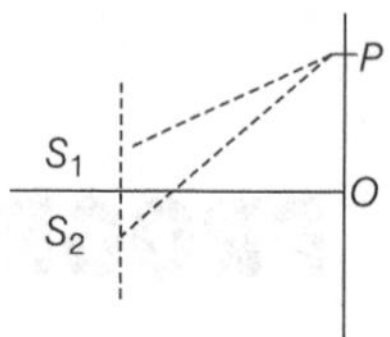

Now a glass slab is introduced in front of S_2, then match the following columns.

	Column I		Column II
A	Δx at P will	1.	increase
B	Fringe width will	2.	decrease
C	Fringe pattern will	3.	remain same
D	Number of fringes between O and P will	4.	shift upward
		5.	shift downward

	A	B	C	D
(a)	1	3	5	3
(b)	2	3	5	3
(c)	3	4	1	3
(d)	5	5	1	2

ANSWERS

SESSION 1									
1 (d)	**2** (a)	**3** (a)	**4** (a)	**5** (c)	**6** (c)	**7** (d)	**8** (b)	**9** (b)	**10** (d)
11 (d)	**12** (d)	**13** (c)	**14** (a)	**15** (d)	**16** (c)	**17** (b)	**18** (b)	**19** (c)	**20** (d)
21 (a)	**22** (d)	**23** (d)	**24** (c)	**25** (a)	**26** (a)	**27** (a)	**28** (a)		

SESSION 2									
1 (a)	**2** (b)	**3** (b)	**4** (d)	**5** (d)	**6** (d)	**7** (d)	**8** (b)	**9** (c)	**10** (a)
11 (c)	**12** (a)								

Hints and Explanations

SESSION 1

1 When the point source or linear source of light is at very large distance, wavefronts are plane wavefronts.

2 Photoelectric effect and Compton effect cannot be explained on the basis of wave nature of light while polarisation and optical activity can be explained.

3 Shape of interference fringes formed on a screen is hyperbolic in nature.

4 Optical path for 1st ray $= \mu_1 L_1$

Optical path for 2nd ray $= \mu_2 L_2$

$\therefore$ Path difference $= (\mu_1 L_1 - \mu_2 L_2)$

Now, phase difference

$$= \frac{2\pi}{\lambda} \times (\text{path difference})$$

$$= \frac{2\pi}{\lambda}(\mu_1 L_1 - \mu_2 L_2)$$

5 Distance of nth maxima,

$$x = n\lambda \frac{D}{d} \propto \lambda$$

As, $\lambda_b < \lambda_g$

$\therefore$ $x_{\text{blue}} < x_{\text{green}}$

6 The number of fringes on either side of C of screen is

$$n_1 = \left[\frac{AC}{\beta}\right] = \left[\frac{0.5}{0.021}\right] = [23.8] \approx 24$$

Total number of fringes

$$= 2n_1 + \text{fringe at centre} = 2n_1 + 1$$
$$= 2 \times 24 + 1 = 48 + 1 = 49$$

7 $x = (2n-1)\dfrac{\lambda}{2}\dfrac{D}{d}$

or $\lambda = \dfrac{2xd}{(2n-1)D}$

$$= \frac{2 \times 10^{-3} \times 0.9 \times 10^{-3}}{(2 \times 2 - 1) \times 1}$$

$$= 6 \times 10^{-7}\ \text{m} = 6 \times 10^{-5}\ \text{cm}$$

8 For coherent sources

$I_1 = 4I_0 \cos^2 \phi/2 = 4I_0$

For incoherent sources

$$I_2 = I_0 + I_0 = 2I_0$$

$\therefore$ $\dfrac{I_1}{I_2} = 2$

9 We have, $\dfrac{3D\lambda_k}{d} = \dfrac{4D\lambda_\mu}{d}$

where, λ_k is the known wavelength and λ_μ is the unknown wavelength. Thus, we get

$$\lambda_\mu = \frac{3\lambda_k}{4} = \frac{3}{4} \times 590 = 442.5\ \text{nm}$$

10 It will be concentric circles.

11 $\because n\lambda_1 = \left(n + \dfrac{1}{2}\right)\lambda_1$

$\Rightarrow$ $n \times 500 \times 10^{-9}$

$$= \left(n + \frac{1}{2}\right) \times 600 \times 10^{-9}$$

$\Rightarrow$ $n = \dfrac{3}{4}$

Now from the formula, $\Delta x = n\lambda$

or $\left(n + \dfrac{1}{2}\right)\lambda$

we get, $\Delta x = 1500\ \text{nm}$

12 Phase difference

$$= \frac{2\pi}{\lambda} \times \text{path difference}$$

i.e. $\phi = \dfrac{2\pi}{\lambda} \times \dfrac{\lambda}{6} = \dfrac{\pi}{3}$

As, $I = I_{\max} \cos^2\left(\dfrac{\phi}{2}\right)$

or $\dfrac{I}{I_{\max}} = \cos^2\left(\dfrac{\phi}{2}\right)$

or $\dfrac{I}{I_0} = \cos^2\left(\dfrac{\pi}{6}\right) = \dfrac{3}{4}$

13 $I = I_{\max} \cos^2\left(\dfrac{\phi}{2}\right)$

$\therefore$ $\dfrac{I_{\max}}{4} = I_{\max} \cos^2\left(\dfrac{\phi}{2}\right)$

$\cos\dfrac{\phi}{2} = \dfrac{1}{2}$ or $\dfrac{\phi}{2} = \dfrac{\pi}{3}$

$\therefore$ $\phi = \dfrac{2\pi}{3} = \left(\dfrac{2\pi}{\lambda}\right)\Delta x$ ...(i)

where, $\Delta x = d \sin\theta$

Substituting in Eq. (i), we get

$$\sin\theta = \frac{\lambda}{3d}$$

or $\theta = \sin^{-1}\left(\dfrac{\lambda}{3d}\right)$

14 The distance of first diffraction minimum from the central principal maximum is

$$x = \frac{D\lambda}{d}$$

$$\frac{x}{D} = \frac{\lambda}{d}$$

$\Rightarrow$ $d = \dfrac{\lambda}{\sin\theta}$

$\Rightarrow$ $\sin\theta = \dfrac{\lambda}{d} = \dfrac{5000 \times 10^{-8}}{1 \times 10^{-4}}$

$$= 0.5 = \sin 30^\circ$$

$\Rightarrow$ $\theta = 30^\circ$

15 The phase difference (ϕ) between the wavelets from the top edge and the bottom edge of the slit is $\phi = \dfrac{2\pi}{\lambda}(d \sin\theta)$

where, d is the slit width. The first minima of the diffraction pattern occurs at $\sin\theta = \dfrac{\lambda}{d}$,

So, $\phi = \dfrac{2\pi}{\lambda}\left(d \times \dfrac{\lambda}{d}\right) = 2\pi.$

16 The general condition for Frounhofer diffraction is $\dfrac{b^2}{L\lambda} << 1.$

17 For first diffraction minimum,

$$a \sin\theta = \lambda$$

$\Rightarrow$ $a = \dfrac{\lambda}{\sin\theta}$

For first secondary maximum,

$$a \sin\theta' = \frac{3\lambda}{2}$$

or $\sin\theta' = \dfrac{3\lambda}{2} \times \dfrac{1}{a} = \dfrac{3\lambda}{2} \times \dfrac{\sin\theta}{\lambda}$

$$= \frac{3}{2} \times \sin 30^\circ = \frac{3}{4}$$

or $\theta' = \sin^{-1}\left(\dfrac{3}{4}\right)$

18 First polariser just polarises the unpolarised light. Therefore, intensity of polarised light transmitted from first polariser is

$$\frac{1}{2}I_0 = 50\%\ I_0$$

19 Intensity of light from $C_2 = I_0$

On rotating through 60°,

$$I = I_0 \cos^2 60^\circ$$

$$= I_0\left(\frac{1}{2}\right)^2 = I_0/4$$

20 As, $i_p = \tan^{-1}(1.54) = 57^\circ$

and in the figure given in question

$$i = 90^\circ - 33^\circ = 57^\circ = i_p$$

$\therefore$ Reflected light along OB is plane polarised. On rotating the nicol prism, intensity gradually reduces to zero and then increases again.

21 $I = I_0 \cos^2\theta$

Intensity of polarised light $= \dfrac{I_0}{2}$

$\therefore$ Intensity of untransmitted light

$$= I_0 - \frac{I_0}{2} = \frac{I_0}{2}$$

22 By law of Malus i.e. $I = I_0 \cos^2 \theta$

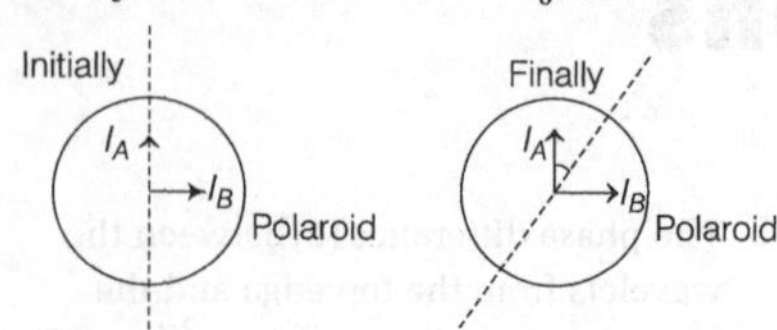

Now, $I_{A'} = I_A \cos^2 30°$

$I_{B'} = I_B \cos^2 60°$

As, $I'_A = I'_B$

$I_A \cos^2 30° = I_B \cos^2 60°$

$\Rightarrow \quad I_A \frac{3}{4} = I_B \frac{1}{4} \Rightarrow \frac{I_A}{I_B} = \frac{1}{3}$

23 From Brewster's law,

$\mu = \tan \theta_p$

$1.732 = \tan \theta_p$

$\Rightarrow \quad \theta_p = \tan^{-1}(1.732) = 60°$

Since, the angle between i_p and r is 90° when the ray is incident at polarising angle, then

$\theta_p + r = 90°$

$r = 90° - \theta_p = 90° - 60° = 30°$

24 Relation between intensities is

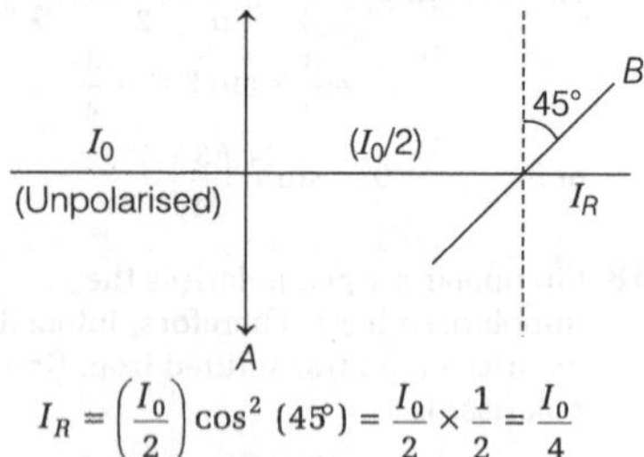

$I_R = \left(\frac{I_0}{2}\right) \cos^2 (45°) = \frac{I_0}{2} \times \frac{1}{2} = \frac{I_0}{4}$

25 For interference to occur, the path difference between two waves is of the order of few wavelength.

26 For diffraction to occur, the size of an obstacle/aperture is comparable to the wavelength of light wave. The order of wavelength of light wave is 10^{-7} m, so diffraction occurs.

27 From the relation,

$\mu = \tan i_p = \tan 60° = \sqrt{3}$ ...(i)

and $\sin C = \frac{1}{\mu} = \frac{1}{\sqrt{3}}$ [from Eq. (i)]

$\therefore \; C = \sin^{-1}\left(\frac{1}{\sqrt{3}}\right) = \sin^{-1}(0.577) = 35°$

28 The number of fringe is smaller in case of larger wavelength is used while in case of smaller wavelength is used the number of fringe is larger.
Also, fringe width is given by

$\beta = \frac{\lambda D}{d} \Rightarrow \beta \propto \lambda$

SESSION 2

1 The intensity of light reflected from upper surface is

$I_1 = I_0 \times 25\% = I_0 \times \frac{25}{100} = \frac{I_0}{4}$

Intensity of transmitted light from upper surface is

$I = I_0 - \frac{I_0}{4} = \frac{3I_0}{4}$

$\therefore$ Intensity of reflected light from lower surface is

$I_2 = \frac{3I_0}{4} \times \frac{50}{100} = \frac{3I_0}{8}$

$\frac{I_{max}}{I_{min}} = \frac{(\sqrt{I_1} + \sqrt{I_2})^2}{(\sqrt{I_1} - \sqrt{I_2})^2}$

$\frac{I_{max}}{I_{min}} = \frac{\left(\sqrt{\frac{I_0}{4}} + \sqrt{\frac{3I_0}{8}}\right)^2}{\left(\sqrt{\frac{I_0}{4}} - \sqrt{\frac{3I_0}{8}}\right)^2}$

$= \frac{\left(\frac{1}{2} + \sqrt{\frac{3}{8}}\right)^2}{\left(\frac{1}{2} - \sqrt{\frac{3}{8}}\right)^2}$

2 Path difference $= S_2P - S_1P$.
From the figure,

$(S_2P)^2 - (S_1P)^2 = b^2$

or $(S_2P - S_1P)(S_2P + S_1P) = b^2$

or $(S_2P - S_1P) = \frac{b^2}{2d}$

For dark fringes,

$\frac{b^2}{2d} = (2n+1)\frac{\lambda}{2}$

For $n = 0$, $\frac{b^2}{2d} = \frac{\lambda}{2}$ or $\lambda = \frac{b^2}{d}$

For $n = 1$, $\frac{b^2}{2d} = \frac{3\lambda}{2}$ or $\lambda = \frac{b^2}{3d}$

3 In case of transparent glass slab of refractive index μ, the path difference $= 2d \sin \theta + (\mu - 1)L$.
For the principal maxima, (path difference is zero)
i.e. $2d \sin \theta_0 + (\mu - 1)L = 0$

or $\sin \theta_0 = -\frac{L(\mu - 1)}{2d} = \frac{-L(0.5)}{2d}$

$[\therefore \; L = d/4]$

or $\sin \theta_0 = \frac{-1}{16}$

or $OP = D \tan \theta_0 = D \sin \theta_0 = \frac{-D}{16}$

For the first minima, the path difference is $\pm \frac{\lambda}{2}$

$\therefore \quad 2d \sin \theta_1 + 0.5L = \pm \frac{\lambda}{2}$

or $\sin \theta_1 = \frac{\pm \lambda/2 - 0.5L}{2d}$

$= \frac{\pm \lambda/2 - d/8}{2d}$

$= \frac{\pm \lambda/2 - \lambda/8}{2\lambda} = \pm \frac{1}{4} - \frac{1}{16}$

[$\because$ The diffraction occurs if the wavelength of waves is nearly equal to the slit width (d)]

On the positive side

$\sin \theta'_1 = +\frac{1}{4} - \frac{1}{16} = \frac{3}{16}$

On the negative side

$\sin \theta''_1 = -\frac{1}{4} - \frac{1}{16} = -\frac{5}{16}$

The first principal maxima on the positive side is at distance

$D \tan \theta'_1 = D \frac{\sin \theta'_1}{\sqrt{1 - \sin^2 \theta'_1}}$

$= D \frac{3}{\sqrt{16^2 - 3^2}} = \frac{3D}{\sqrt{247}}$ above point O.

The first principal minima on the negative side is at distance

$D \tan \theta''_1 = \frac{5}{\sqrt{16^2 - 5^2}} = \frac{5}{\sqrt{231}}$ below point O.

4 When interference is coherent
When two waves of intensities I_1 and I_2 having a phase difference ϕ interfere, the resultant intensity is given as

$I = I_1 + I_2 + 2\sqrt{I_1 I_2} \cos \phi$ ...(i)

The intensity will maximum, then $\phi = 0$ or $\cos \phi = 1$
maximum intensity.

$\therefore \quad I_{max} = I_1 + I_2 + 2\sqrt{I_1 I_2}$

$= (\sqrt{I_1} + \sqrt{I_1})^2$

In case, n identical waves each of intensities I_0 interfere,

$I_{max} = (\sqrt{I_0} + \sqrt{I_0} + \sqrt{I_0} + \dots + n \text{ times})^2$

$= (n\sqrt{I_0})^2$ ...(i)

$\therefore \; I_{max} = n^2 I_0$ ...(ii)

When interference is incoherent
Since, the average value of $\cos \phi$, over a complete cycle is zero
The Eq. (i), becomes,

$I = I_1 + I_2 + 2\sqrt{I_1 I_2} \times 0$

$= I_1 + I_2$ ...(iii)

In case, n identical waves, each of intensities I_0 interfere,
Minimum intensity,

$I_{min} = I_0 + I_0 + I_0 + \dots n$ times

$I_{min} = nI_0$ ...(iv)

$\therefore \quad$ Ratio $\frac{I_{max}}{I_{min}} = \frac{n^2 I_0}{nI_0} = n$

5 Given, $a_1 = 2a_2$

$$\Rightarrow \quad I_1 = 4I_2 = 4I_0$$

$$\therefore \quad I_m = (\sqrt{I_1} + \sqrt{I_2})^2 = (3\sqrt{I_2})^2$$

$$= 9I_2 = 9I_0 = I_0 = \frac{I_m}{9}$$

Now, resultant intensity,

$$I = I_1 + I_2 + 2\sqrt{I_1 I_2} \cos \phi$$

$$= 4I_0 + I_0 + 2\sqrt{4I_0 I_0} \cos \phi$$

$$= 5I_0 + 4I_0 \cos \phi$$

$$= \frac{I_m}{9}(5 + 4\cos \phi)$$

$$= \frac{I_m}{9}[1 + 4(1 + \cos \phi)]$$

$$= \frac{I_m}{9}(1 + 8\cos^2 \phi/2)$$

$$\left[(1 + \cos \theta) = 2\cos^2 \frac{\phi}{2}\right]$$

6 Shift in image position due to glass plate,

$$S = \left(1 - \frac{1}{\mu}\right)t = \left(1 - \frac{1}{1.5}\right) \times 1 \text{ cm} = \frac{1}{3} \text{ cm}$$

For focal length of the lens,

$$\frac{1}{f} = \frac{1}{v} - \frac{1}{u} = \frac{1}{12} - \frac{1}{-240} = \frac{20 + 1}{240}$$

$$\Rightarrow \quad f = \frac{240}{21} \text{ cm}$$

Now, to get back image on the film, lens has to form image at $\left(12 - \frac{1}{3}\right) \text{cm} = \frac{35}{3}$ cm such that the glass plate will shift the image on the film.

As, $\frac{1}{f} = \frac{1}{v} - \frac{1}{u}$

$$\Rightarrow \quad \frac{1}{u} = \frac{1}{v} - \frac{1}{f}$$

$$= \frac{3}{35} - \frac{21}{240}$$

$$\frac{1}{u} = \frac{48 \times 3 - 7 \times 21}{1680} = -\frac{1}{560}$$

$$\Rightarrow \quad u = -5.6 \text{ m}$$

7 Let nth minima of 400 nm coincides with mth minima of 560 nm, then

$$(2n - 1)\left(\frac{400}{2}\right) = (2m - 1)\left(\frac{560}{2}\right)$$

$$\text{or} \quad \frac{2n - 1}{2m - 1} = \frac{7}{5} = \frac{14}{10} = \ldots$$

i.e. 4th minima of 400 nm coincides with 3rd minima of 560 nm.
Location of this minima is,

$$Y_1 = \frac{(2 \times 4 - 1)(1000)(400 \times 10^{-6})}{2 \times 0.1}$$

$$= 14 \text{ mm}$$

Next 11th minima of 400 nm will coincide with 8th minima of 560 nm.
Location of this minima is,

$$Y_2 = \frac{(2 \times 11 - 1)(1000)(400 \times 10^{-6})}{2 \times 0.1}$$

$$= 42 \text{ mm}$$

$\therefore$ Required distance $= Y_2 - Y_1 = 28$ mm

8 For possible interference maxima of the screen, the condition is

$$d \sin \theta = n\lambda \quad \ldots\text{(i)}$$

Given, $\quad d$ = slit width $= 2\lambda$

$$\therefore \quad 2\lambda \sin \theta = n\lambda$$

$$\text{or} \quad 2 \sin \theta = n$$

The maximum value of $\sin \theta$ is 1, hence,

$$n = 2 \times 1 = 2$$

Thus, Eq. (i) must be satisfied by 5 integer values, i.e. $-2, -1, 0, 1, 2$. Hence, the maximum number of possible interference maxima is 5.

9

A C B
I → $I/2$ → → $I/8$
α β

Using Malus's law, intensity available after $C = \frac{I}{2} \times \cos^2 \alpha$

and intensity available after

$$B = \frac{I}{2}\cos^2 \alpha \times \cos^2 \beta$$

$$= \frac{I}{8} \text{ (given)}$$

$$\text{So,} \quad \frac{I}{2} \times \cos^2 \alpha \cdot \cos^2 \beta = \frac{I}{8}$$

$$\Rightarrow \quad \cos^2 \alpha \cdot \cos^2 \beta = \frac{1}{4}$$

This is satisfied with $\alpha = 45°$ and $\beta = 45°$

So, angle between A and C is $45°$.

10 Angular width of diffraction pattern $= 60°$

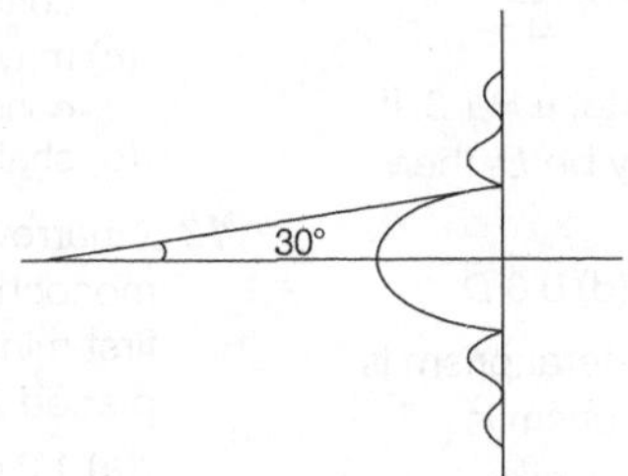

For first minima,

$$\frac{a}{2}\sin \theta = \frac{\lambda}{2},$$

[here, $a = 10^{-6}$ m, $\theta = 30°$]

$$\Rightarrow \quad \lambda = 10^{-6} \times \sin 30°$$

$$\Rightarrow \quad \lambda = \frac{10^{-6}}{2} \text{ m}$$

Now, in case of interference caused by bringing second slit,

$\therefore$ Fringe width,

$$\beta = \frac{\lambda D}{d}$$

[here, $\lambda = \frac{10^{-6}}{2}$ m, $\beta = 1$ cm $= \frac{1}{100}$ m, d = ? and $D = 50$ cm $= \frac{50}{100}$ m]

$$\text{So,} \quad d = \frac{\lambda D}{\beta} = \frac{10^{-6} \times 50}{2 \times \frac{1}{100} \times 100}$$

$$= 25 \times 10^{-6} \text{ m}$$

$$\text{or} \quad d = 25 \mu\text{m}$$

11 In diffraction, first minima, we have

$$\sin \theta = \frac{\lambda}{a}$$

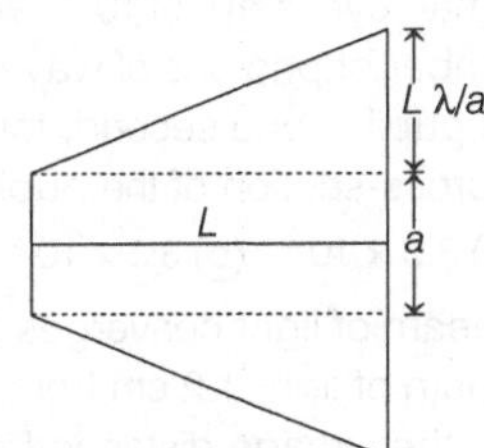

So, size of a spot,

$$b = 2a + \frac{2L\lambda}{a} \quad \ldots\text{(i)}$$

Then, minimum size of a spot, we get

$$\frac{\partial b}{\partial a} = 0 \Rightarrow 1 - \frac{L\lambda}{a^2} = 0$$

$$\Rightarrow \quad a = \sqrt{\lambda L} \quad \ldots\text{(ii)}$$

$$\text{So,} \quad b_{\min} = 2\sqrt{\lambda L} + 2\sqrt{\lambda L}$$

[by substituting the value of a from Eq. (ii) in Eq. (i)]

$$= 4\sqrt{\lambda L}$$

So, the radius of the spot,

$$b_{\min} = \frac{4}{2}\sqrt{\lambda L} = \sqrt{4\lambda L}$$

12 From YDSE,

(A) Path difference $\Delta x = \frac{xd}{D}$

So, as x increases, Δx also increases.

(B) Fringe width $(\beta) = \frac{Dx}{d}$ independent of Δx.

(C) Fringe pattern will shift downward.

(D) β is constant, so number of fringes unaffected.

DAY THIRTY

Unit Test 6 (Optics)

1 Minimum light intensity that can be perceived by normal human eye is about 10^{-10} W m^{-2}. What is the minimum number of photons of wavelength 660 nm that must enter the pupil in one second, for one to see the object? Area of cross-section of the pupil is 10^{-4} m^2.

(a) 3.3×10^2 (b) 3.3×10^3 (c) 3.3×10^4 (d) 3.3×10^5

2 A beam of light converges to a point P. A lens is placed in path of light 1.2 cm from P. If focal length of lens is +20 cm, then image distance from lens is

(a) 4.8 cm (b) 20 cm (c) 7.5 cm (d) 5.2 cm

3 A ray of light incident at an angle θ on a refracting face of a prism emerges from the other face normally. If the angle of the prism is 5° and the prism is made of a material of refractive index 1.5, the angle of incidence is

(a) $\sin^{-1}(0.13)$ (b) $\sin^{-1}(0.52)$ (c) $\sin^{-1}(0.17)$ (d) $\sin^{-1}(0.86)$

4 Two coherent point sources S_1 and S_2 vibrating in phase emit light of wavelength λ. The separation between the sources is 2λ. The smallest distance from S_2 on a line passing through S_2 and perpendicular to S_1S_2, where a minimum of intensity occurs is

(a) $\frac{7\lambda}{12}$ (b) $\frac{15\lambda}{4}$ (c) $\frac{\lambda}{2}$ (d) $\frac{3\lambda}{4}$

5 A thin glass prism $\mu = 1.5$ is immersed in water $\mu = 1.3$. If the angle of deviation in air for particular ray be D, then that in water will be

(a) $0.2\,D$ (b) $0.3\,D$ (c) $0.5\,D$ (d) $0.6\,D$

6 The refractive index of the material of equilateral prism is $\sqrt{3}$. The angle of minimum deviation for the prism is

(a) 30° (b) 41° (c) 49° (d) 60°

7 A thin convergent glass lens $\mu = 1.5$ has a power of + 5.0 D. When this lens is immersed in a liquid of refractive index μ_l, it acts as a diverging lens of focal length 100 cm. The value of μ_l should be

(a) 3/2 (b) 4/3 (c) 5/3 (d) 2

8 A thin convex lens of crown glass having refractive index 1.5 has power 1 D. What will be the power of similar convex lens of refractive index 1.6 ?

(a) 0.6 D (b) 0.8 D (c) 1.2 D (d) 1.6 D

9 A short pulse of white light is incident from air to a glass slab at normal incidence. After travelling through the slab, the first colour to emerge is

(a) blue (b) green (c) violet (d) red

10 An object approaches a converging lens from the left of the lens with a uniform speed 5 m/s and stops at the focus. The image

(a) moves away from the lens with uniform speed 5 m/s
(b) moves away from the lens with uniform acceleration
(c) moves away from the lens with a non-uniform acceleration
(d) moves towards the lens with a non-uniform acceleration

11 A passenger in an aeroplane shall

(a) never see a rainbow
(b) may see a primary and a secondary rainbow as concentric circles
(c) may see a primary and a secondary rainbow as concentric arcs
(d) shall never see a secondary rainbow

12 A narrow slit of width 1 mm is illuminated by monochromatic light $\lambda = 600$ nm. The distance between first minima on either side of center line of a screen placed 2 m away is

(a) 1.2 cm (b) 1.2 mm (c) 2.4 mm (d) 2.4 cm

13 A myopic person having far point 80 cm uses spectacles of power −1.0 D. How far can he see clearly?

(a) 1 m (b) 2 m
(c) 4 m (d) More than 4 m

14 The radius of curvature of the curved surface of a plano-convex lens is 20 cm. If the refractive index of the material of the lens be 1.5, it will

(a) act as a convex lens only for the objects that lie on its curved side
(b) act as a concave lens for the objects that lie on its curved side
(c) act as a convex lens irrespective of the side on which the object lies
(d) act as a concave lens irrespective of the side on which the object lies

15 Monochromatic light of wavelength 800 nm is used in double slit experiment. One of the slit is covered with a transparent slab of thickness 2.4×10^{-5} m. The refractive index of the material of slab is 1.4. What is the number of fringes that will shift due to introduction of the sheet?

(a) 14 (b) 12 (c) 16 (d) 10

16 Two coherent sources are 4.4 mm apart and 4.4 m from the screen as shown in the figure. If the sources emit light of wavelength 440 nm which produce an interference pattern on the screen. The pattern of the interference at point O is

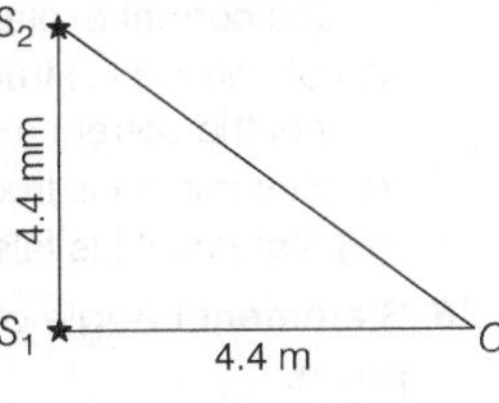

(a) constructive only
(b) destructive only
(c) cannot be predicted
(d) may be constructive or destructive

17 A lens behaves as a converging lens in air and a diverging lens in water. The refractive index of the material of the lens is

(a) equal to unity (b) equal to 1.33
(c) between unity and 1.33 (d) greater than 1.33

18 Angular width of central maximum in the Fraunhoffer's diffraction pattern is measured. Slit is illuminated by the light of wavelength 6000 Å. If slit is illuminated by light of another wavelength, angular width decreases by 30%. Wavelength of light used is

(a) 3500 Å (b) 4200 Å (c) 4700 Å (d) 6000 Å

19 A convex mirror and a concave mirror of radius 10 m each are placed 15 m apart facing each other. An object is placed mid-way between them. If the reflection first take place in the concave mirror and then in another mirror, the position of the final image is

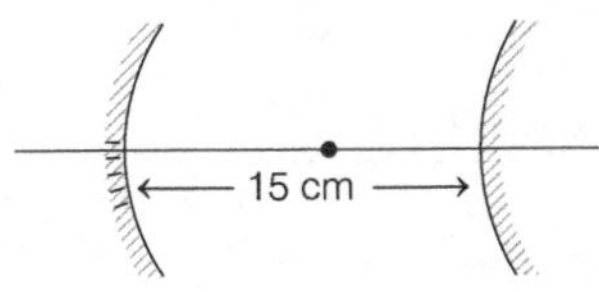

(a) at infinity
(b) at the mid-point of distance between two mirrors
(c) on the pole of concave mirror
(d) on the pole of convex mirror

20 A cube of side 3 m is placed in front of a concave mirror of focal length 2 m with its face P at a distance 4 m and face Q at a distance 7 m from the mirror. What is distance between the images of face P and Q ?

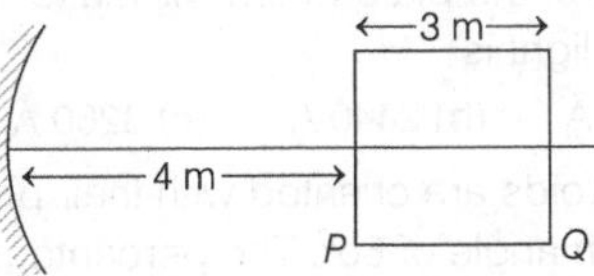

(a) 1.2 m (b) 2.4 m (c) 2.1 m (d) 2.2 m

21 A diminished image of an object is to be obtained on a screen 1 m from it. This can be achieved by approximately placing.

(a) a concave mirror of suitable focal length
(b) a convex mirror of suitable focal length
(c) a concave lens of suitable focal length
(d) a convex lens of focal length less than 0.25 m

22 A convex lens and a concave lens are placed in contact. The ratio of the magnitude of the power of the convex lens to that of the concave lens is 4 : 3. If the focal length of the convex lens is 12 cm, the focal length of the combination will be

(a) 16 cm (b) 24 cm (c) 32 cm (d) 48 cm

23 The radius of curvature of a thin plano-convex lens is 10 cm and the refractive index of its glass is 1.5. If the plane surface is silvered, then it will behave like a

(a) concave mirror of focal length 10 cm
(b) concave mirror of focal length 20 cm
(c) convex mirror of focal length 10 cm
(d) convex mirror of focal length 20 cm

24 When the plane surface of a plano-convex lens of refractive index 1.5 is silvered, it behaves like a concave mirror of focal length 30 cm. When its convex surface is silvered, it will behave like a concave mirror of focal length

(a) 10 cm (b) 20 cm (c) 30 cm (d) 45 cm

25 Two stars are situated at a distance of 8 light years from the earth. Their images are just resolved by a telescope of diameter 0. 25 m. If the wavelength of light from stars is 5000 Å, then the distance between the stars is around

(a) 3×10^{10} m (b) 3.35×10^{11} m
(c) 1.95×10^{11} m (d) 4.32×10^{10} m

26 The refractive index of air is 1.0003. The thickness of air column which will have one more wavelength of yellow light ($\lambda = 6000$ Å), then in the same thickness in vacuum is

(a) 2 mm (b) 2 cm (c) 2 m (d) 2 km

27 A thin symmetric convex lens of refractive index 1.5 and radius of curvature 0.3 m is immersed in water of refractive index 4/3. Its focal length in water is

(a) 0.15 m (b) 0.30 m (c) 0.60 m (d) 1.20 m

28 A parallel beam of sodium light of wavelength 6000 Å is incident on a thin glass plate of $\mu = 1.5$, such that the angle of incidence in the plate is 60°. The smallest thickness of the plate which will make it appear dark by reflected light is

(a) 1260 Å (b) 2440 Å (c) 3260 Å (d) 4000 Å

29 Two polaroids are oriented with their principal planes making an angle of 60°. The percentage of incident unpolarised light which passes through the system is

(a) 50% (b) 100% (c) 12.5% (d) 37.5%

30 In the visible region of the spectrum the rotation of the plane of polarisation is given by $\theta = a + \frac{b}{\lambda^2}$.

The optical rotation produced by a particular material is found to be 30° per mm at $\lambda = 5000$ Å and 50° per mm at $\lambda = 4000$ Å. The value of constant a will be

(a) $+\frac{50^\circ}{9}$ per mm (b) $-\frac{50^\circ}{9}$ per mm

(c) $+\frac{9^\circ}{50}$ per mm (d) $-\frac{9^\circ}{50}$ per mm

31 The refracting angle of a prism is A and the refractive index of the prism is $\cot(A/2)$. The angle of minimum deviation is

(a) $180°-3A$ (b) $180° + 2A$ (c) $90°-A$ (d) $180°-2A$

32 Cross-section of a glass prism is an isosceles triangle. One of refracting faces is silvered. A ray of light falls normally on the other refracting face. After being reflected twice, it emerges through the base of the prism perpendicular to it. The angles of prism are

(a) 54°, 54°, 72° (b) 72°, 72°, 36°

(c) 45°, 45°, 90° (d) 57°, 57°, 76°

33 A spherical surface of radius of curvature R, separates air and glass ($n_{air} = 1.0$, $n_{glass} = 1.5$). The centre of curvature is in glass. A point object P placed in air is found to have a real image in the glass. The line PQ cuts the surface at a point O such that $PO = OQ$. Distance PQ is

(a) $5R$ (b) $3R$ (c) $2R$ (d) R

34 Polarising angle for water is 53°4'. If light is incident at this angle on the surface of water and reflected, the angle of refraction is

(a) 53° 4′ (b) 126°56′

(c) 36°56′ (d) 30° 4′

35 The distance between the first and the sixth minima in the diffraction pattern of a single slit is 0.5 mm. The screen is 0.5 m away from the slit. If the wavelength of light used is 5000 Å, then the slit width will be

(a) 5 mm (b) 2.5 mm

(c) 1.25 mm (d) 1.0 mm

Direction (Q. Nos. 36-40) *Each of these questions contains two statements : Statement I and Statement II. Each of these questions also has four alternative choices, only one of which is the correct answer. You have to select one of the codes (a), (b), (c) and (d) given below*

(a) Statement I is true, Statement II is true; Statement II is the correct explanation for Statement I

(b) Statement I is true, Statement II is true; Statement II is not the correct explanation for Statement I

(c) Statement I is true; Statement II is false

(d) Statement I is false; Statement II is true

36 **Statement I** Angle of deviation depends on the angle of prism.

Statement II For thin prism $\delta = (\mu - 1)A$.

where, δ = angle of deviation, μ = refractive index, A = angle of prism.

37 **Statement I** Glass is transparent but its powder seems opaque. When water is poured over it, it becomes transparent.

Statement II Light gets refracted through water.

38 **Statement I** If convex lens is kept in water its convergent power decreases.

Statement II Focal length of convex lens in water increases.

39 **Statement I** Danger signals are made of red colours.

Statement II Velocity of red light is maximum and thus, more visibility in dark.

40 **Statement I** The clouds in sky generally appear to be whitish.

Statement II Diffraction due to clouds is efficient in equal measure for all wavelengths.

ANSWERS

1. (c)	**2.** (c)	**3.** (a)	**4.** (a)	**5.** (b)	**6.** (d)	**7.** (c)	**8.** (c)	**9.** (d)	**10.** (c)
11. (b)	**12.** (c)	**13.** (c)	**14.** (c)	**15.** (b)	**16.** (a)	**17.** (c)	**18.** (b)	**19.** (d)	**20.** (a)
21. (d)	**22.** (d)	**23.** (a)	**24.** (a)	**25.** (c)	**26.** (a)	**27.** (d)	**28.** (b)	**29.** (c)	**30.** (b)
31. (d)	**32.** (b)	**33.** (a)	**34.** (c)	**35.** (b)	**36.** (a)	**37.** (a)	**38.** (a)	**39.** (c)	**40.** (c)

Hints and Explanations

1 $I = 10^{-10}\ \text{Wm}^{-2} = 10^{-10}\text{Js}^{-1}\text{m}^2$. Let the number of photons required per second be n.

Then, $\dfrac{nh\nu}{10^{-4}} = 10^{-10}$

Hence, $n = 10^{-10} \times 10^{-4}/h\nu$

$$= 10^{-14}\frac{\lambda}{hc}$$

$$= \frac{10^{-14} \times 660 \times 10^{-9}}{6.6 \times 10^{-34} \times 3 \times 10^{8}}$$

$$= 3.3 \times 10^{4}$$

2 $\dfrac{1}{v} = \dfrac{1}{f} + \dfrac{1}{u}$

Here, $f = +20$ and $u = +12$

$\therefore \quad v = 7.5\,\text{cm}$

3 Here, $A = 5°$, $i_1 = ?$

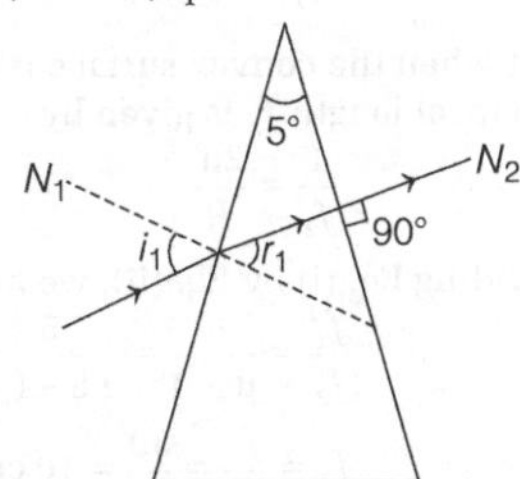

As the ray emerges from the other face of prism normally,

$i_2 = 0°$

$\therefore \quad r_2 = 0°$,

As $\quad r_1 + r_2 = A$

$r_1 = A - r_2 = 5 - 0 = 5°$

From $\mu = \dfrac{\sin i_1}{\sin r_1}$,

$\sin i_1 = \mu \sin r_1$

$\sin i_1 = 1.5 \times \sin 5°$

$= 1.5 \times 0.087$

$\sin i_1 = 0.1305$

$i_1 = \sin^{-1}(0.1305)$

4 Path difference at S_2 is 2λ. Therefore, for minimum intensity at P.

$$S_1P - S_2P = \frac{3\lambda}{2} \neq \frac{\lambda}{2} \qquad \ldots\text{(i)}$$

or $\sqrt{4\lambda^2 + x^2} - x = \dfrac{3\lambda}{2}$

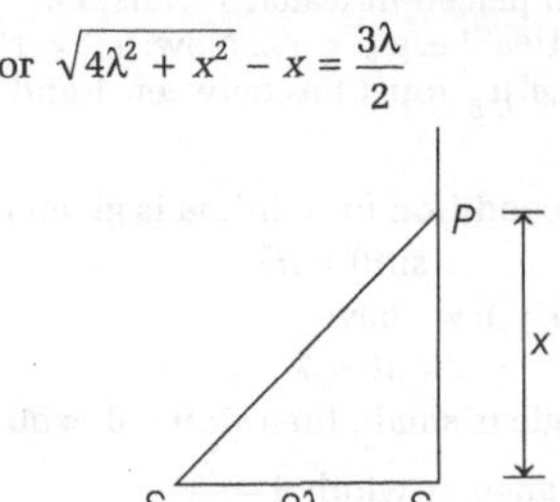

Solving this equation, we get

$$x = \frac{7\lambda}{12}$$

5 $\delta \simeq (\mu - 1)A$

For air, $\quad D = (1.5 - 1)A$

For water, $\delta = ({}_g\mu_w - 1)A = \left(\dfrac{1.5}{1.3} - 1\right)A$

Hence, $\quad \delta = \dfrac{0.2}{1.3} \times \dfrac{D}{0.5} \cong 0.3D$

6
$$\mu = \frac{\sin\left(\dfrac{A + \delta_m}{2}\right)}{\sin\left(\dfrac{A}{2}\right)}$$

$$\Rightarrow \sqrt{3} = \frac{\sin\left(\dfrac{A + \delta_m}{2}\right)}{\sin 30°},$$

because $\quad A = 60°$

or $\sin\dfrac{A + \delta_m}{2} = \dfrac{\sqrt{3}}{2} = \sin 60°$

or $\quad A + \delta_m = 2 \times 60 = 120$

This gives $\delta_m = 60°$.

where, δ_m is minimum deviation.

7 When the lens is in air, we have

$$P_a = \frac{1}{f_a} = \frac{\mu_g - \mu_a}{\mu_a}\left[\frac{1}{R_1} - \frac{1}{R_2}\right]$$

When the lens is in liquid, we have

$$P_l = \frac{1}{f_l} = \frac{\mu_g - \mu_l}{\mu_l}\left[\frac{1}{R_1} - \frac{1}{R_2}\right]$$

Here, $\quad P_a = 5, P_l = -1, \mu_a = 1, \mu_g = 1.5$

On solving, we get

$$\mu_l = \frac{5}{3}$$

8 $\quad P = (\mu - 1)\left[\dfrac{1}{R_1} - \dfrac{1}{R_2}\right]$

Hence, $\dfrac{P_2}{P_1} = \left(\dfrac{\mu_2 - 1}{\mu_1 - 1}\right)$ i.e. $\dfrac{P_2}{1} = \dfrac{1.6 - 1}{1.5 - 1}$

Hence, $\quad P_2 = 1.2\ \text{D}$

9 In air, all the colours of light travel with the same velocity, but in glass, velocities of different colours are different. Velocity of red colour is largest and velocity of violet colour is smallest. Therefore, after travelling through the glass slab, red colour will emerge first.

10 When an object approaches a convergent lens from the left of the lens with a uniform speed of 5 m/s, the image moves away from the lens with a non-uniform acceleration. For example, $f = 20$ m and $u = -50$ m; -45 m, -40 m and -35 m; we get $v = 33.3$ m; 36 m; 40 m and 46.7 m. Clearly, image moves away from the lens with a non-uniform acceleration. Option (c) is correct.

11 In an aeroplane, a passenger may observe a primary and a secondary rainbow as concentric circles.

12 For first dark fringe on either side,

$$d\sin\theta = \lambda$$

and $\quad \sin\theta = \dfrac{y}{D}$

So, $\quad \dfrac{dy}{D} = \lambda$ or $y = \dfrac{\lambda D}{d}$

Distance between two minima $= 2y$

$$= \frac{2 \times 600 \times 10^{-6} \times 2 \times 10^{3}}{1.0}\ \text{mm}$$

$$= 2.4\ \text{mm}$$

13 Use $\dfrac{1}{v} - \dfrac{1}{u} = \dfrac{1}{f}$

Here, $\quad v = -80$ cm, $f = -100$ cm

Hence, $\quad \dfrac{1}{-80} - \dfrac{1}{u} = -\dfrac{1}{100}$

or $\quad -\dfrac{1}{u} = -\dfrac{1}{100} + \dfrac{1}{80} = \dfrac{-80 + 100}{80 \times 100}$

This gives $u = -400$ cm $= -4$ m

14 Here, $\mu = 1.5$

If object lies on plane side;

$R_1 = \infty, R_2 = -20$ cm

$$\frac{1}{f} = (\mu - 1)\left(\frac{1}{R_1} - \frac{1}{R_2}\right)$$

$$= (1.5 - 1)\left(\frac{1}{\infty} + \frac{1}{20}\right) = \frac{1}{40}$$

$f = +40$ cm.

The lens behaves as convex.

If object lies on its curved side,

$R_1 = 20$ cm, $R_2 = \infty$,

$$\frac{1}{f'} = (\mu - 1)\left(\frac{1}{R_1} - \frac{1}{R_2}\right)$$

$$= (1.5 - 1)\left(\frac{1}{20} - \frac{1}{\infty}\right) = \frac{1}{40}$$

$f' = 40$ cm

The lens behaves as convex.

15 The total fringe shift is $H = \frac{\beta}{\lambda}(\mu - 1)t$

The number of fringes that will shift

$$= \frac{\text{total fringe shift}}{\text{fringe width}}$$

or $n = \frac{\frac{\beta}{\lambda}(\mu - 1)t}{\beta} = \frac{(\mu - 1)t}{\lambda}$

or $n = \frac{(1.4 - 1) \times 2.4 \times 10^{-5}}{800 \times 10^{-9}}$

or $n = \frac{0.4 \times 2.4 \times 10^{-5}}{8 \times 10^{-7}}$ or $n = 12$

16 The path difference at point of observation is given by

$$\Delta = S_2O - S_1O$$

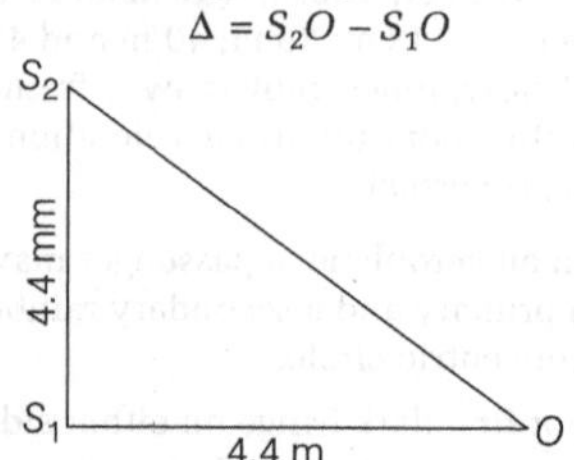

From the figure,

$S_2O = [(4.4)^2 + (4.4 \times 10^{-3})^2]^{1/2}$

or $S_2O = 4.4[1 + (10^{-3})^2]^{1/2}$

or $S_2O = 4.4\left[1 + \frac{1}{2}(10^{-3})^2\right]$

Therefore, $\Delta = 4.4\left[1 + \frac{1}{2}(10^{-3})^2 - 1\right]$

$$= \frac{4.4 \times 10^{-6}}{2} = 2.2 \times 10^{-6} \text{ m}$$

Interference will be constructive, if path difference is an integral multiple of wavelength.

$$n = \frac{\Delta}{\lambda} = \frac{2.2 \times 10^{-6}}{440 \times 10^{-9}} = 5$$

Hence, pattern of interference at point O is constructive.

17 The focal length f of the lens in air is given by

$$\frac{1}{f} = (\mu_g - 1)\left[\frac{1}{R_1} - \frac{1}{R_2}\right]$$

where, μ_g is the refractive index of the lens. If μ_w is the refractive index of water, the focal length in water (f') is given by

$$\frac{1}{f'} = \left(\frac{\mu_g - \mu_w}{\mu_w}\right)\left(\frac{1}{R_1} - \frac{1}{R_2}\right)$$

Since, the lens placed in air is convergent, f is positive. Therefore, $\mu_g > 1$. For the lens to be divergent when placed in water, f' must be negative, i.e. $\mu_g < \mu_w$. Now, $\mu_w = 1.33$. Hence, μ_g must lies between 1 and 1.33.

18 The condition for minima is given by

$$d \sin\theta = n\lambda$$

For $n = 1$, we have

$$d \sin\theta = \lambda$$

If angle is small, then $\sin\theta = \theta \Rightarrow d\theta = \lambda$

Half angular width $\theta = \frac{\lambda}{d}$

Full angular width, $2\theta = 2\frac{\lambda}{d}$

Also, $\omega' = \frac{2\lambda'}{d}$

$\therefore \quad \frac{\lambda'}{\lambda} = \frac{\omega'}{\omega}$

or $\lambda' = \lambda\frac{\omega'}{\omega} = 6000 \times 0.7 = 4200$ Å

19 For reflection from concave mirror

Applying, $\frac{1}{f} = \frac{1}{v} + \frac{1}{u}$, we get

$$\frac{1}{-5} = \frac{1}{v} - \frac{1}{7.5}$$

or $\frac{1}{v} = -\frac{1}{5} + \frac{1}{7.5}$ or $v = -15$ m

The image is formed on the pole of the convex mirror, which will be the position of the object for convex mirror. Therefore,

$$u = 0 \text{ and } f = 5 \text{ m}$$

Hence, $\frac{1}{v} = \frac{1}{5} - \frac{1}{0} = \frac{1}{5} - \infty = \infty$

or $v = 0$

Therefore, final image is formed on the pole of convex mirror.

20 For surface P, we have

$$\frac{1}{v_1} = \frac{1}{f} - \frac{1}{u}$$

$= \frac{1}{2} - \frac{1}{4} = \frac{1}{4}$ or $v_1 = 4$ m

For surface Q, we have

$$v_2 = \frac{1}{2} - \frac{1}{7} = \frac{5}{14}$$

or $v_2 = \frac{14}{5} = 2.8$ m

Therefore, $v_1 - v_2 = (4 - 2.8) = 1.2$ m

21 Convex mirror and concave lens do not form real image. For concave mirror $v > u$, so image will be enlarged, hence only convex lens can be used for the purpose.

22 Given, $f_1 = +12$ cm and $\frac{|P_1|}{|P_2|} = \frac{4}{3}$

Since f_2 is negative, $\frac{f_2}{f_1} = -\frac{4}{3}$

Hence, $f_2 = -\frac{4}{3}f_1 = -\frac{4}{3} \times 12$

$= -16$ cm

The focal length F of the combination is given by

$$\frac{1}{F} = \frac{1}{f_1} + \frac{1}{f_2} = \frac{1}{12} + \frac{1}{-16} = \frac{1}{48}$$

$F = 48$ cm

23 When the plane surface of a plano-convex lens is silvered, it behaves like a concave mirror of focal length f given by

$$\frac{1}{f} = \frac{2(\mu - 1)}{R}$$

or $f = \frac{R}{2(\mu - 1)} = \frac{10}{2(1.5 - 1)} = 10$ cm

24 When the plane surface is silvered the focal length f_1 is given by

$$\frac{1}{f_1} = \frac{2(\mu - 1)}{R} \quad \text{... (i)}$$

But when the convex surface is silvered, the focal length f_2 is given by

$$\frac{1}{f_2} = \frac{2\mu}{R} \quad \text{... (ii)}$$

Dividing Eq. (i) by Eq. (ii), we have

$$\frac{f_1}{f_2} = \frac{\mu}{\mu - 1} = \frac{1.5}{1.5 - 1} = 3$$

or $f_2 = \frac{f_1}{3} = \frac{30}{3} = 10$ cm

25 Limit of resolution of the telescope

$$\alpha = \frac{1.22\lambda}{a} = \frac{d}{x}$$

or $d = \frac{1.22\lambda x}{a}$

$$= \frac{1.22 \times 5 \times 10^{-7} \times 8 \times 10^{16}}{0.25}$$

$= 1.95 \times 10^{11}$ m

26 Let d in cm be the thickness of air column = thickness of vacuum column (given). Number of waves of wavelength $\lambda = 6000$ Å $= 6 \times 10^{-5}$ cm in a thickness d cm in vacuum is

$$n_v = \frac{d}{\lambda}$$

Since, the refractive index of air $\mu = 1.0003$, the wavelength in air will be

$\lambda_a = \dfrac{\lambda}{\mu}$

Therefore, number of waves of wavelength λ_a in d cm of air is

$$n_a = \frac{d}{\lambda_a} = \frac{d\mu}{\lambda}$$

Given that, $n_a - 1 = n_v$

Hence, $\dfrac{d\mu}{\lambda} - 1 = \dfrac{d}{\lambda}$

$$d = \frac{\lambda}{\mu - 1} = \frac{6\times10^{-5}\text{cm}}{1.0003 - 1}$$

$$= 0.2 \text{ cm} = 2 \text{ mm}$$

27 $\dfrac{1}{f} = \dfrac{\mu_g - \mu_w}{\mu_w}\left[\dfrac{1}{R_1} - \dfrac{1}{R_2}\right]$

For a symmetric lens

$$R_2 = -R_1 = -0.3 \text{ m}$$

Therefore, $\dfrac{1}{f} = \dfrac{(1.5 - 4/3)}{4/3}\left[\dfrac{1}{0.3} + \dfrac{1}{0.3}\right]$

which gives $f = 1.20$ m

28

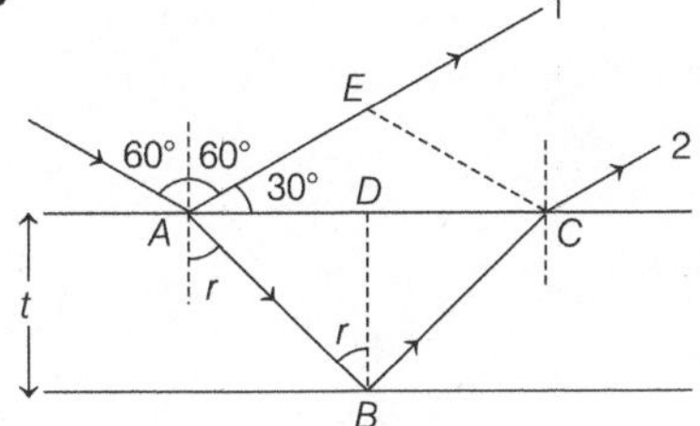

$$\sin r = \frac{\sin 60°}{\mu} = \frac{\sqrt{3}/2}{1.5}$$

$\therefore \quad r = 30°$

$$AB = t\sec r = 1.15t$$

$$AC = 2(AD) = 2(t\tan r) = 1.15t$$

$$AE = AC\cos 30° = 0.99t$$

Now, net path difference between 1 and 2,

$$\Delta X = \mu(2AB) - AE$$

$$= (1.5)(2.3t) - 0.99t = 2.45t$$

For minimum intensity,

$$\Delta X = \lambda$$

or $(2.45t) = 6000$

or $\quad t = 2448 \text{ Å} \approx 2440 \text{ Å}$

29 Intensity of polarised light from first polariser $= \dfrac{100}{2} = 50$

From law of Malus intensity from second nicol

$$I = 50\cos^2 60°$$

$$= \frac{50}{4} = 12.5\%$$

30 $\theta = a + \dfrac{b}{\lambda^2}$

$$30 = a + \frac{b}{(5000)^2}$$

and $\quad 50 = a + \dfrac{b}{(4000)^2}$

Solving for a, we get

$$a = -\frac{50°}{9} \text{ per mm}$$

31 Given, $\mu = \dfrac{\sin\left(\dfrac{A+\delta_m}{2}\right)}{\sin\left(\dfrac{A}{2}\right)} = \cot\left(\dfrac{A}{2}\right)$

$$= \frac{\cos(A/2)}{\sin(A/2)}$$

or $\sin\left(\dfrac{A+\delta_m}{2}\right) = \cos\left(\dfrac{A}{2}\right)$

$$= \sin\left(90° - \frac{A}{2}\right)$$

$\therefore \quad \dfrac{A+\delta_m}{2} = 90° - \dfrac{A}{2}$

or $\quad \delta_m = 180° - 2A$

32

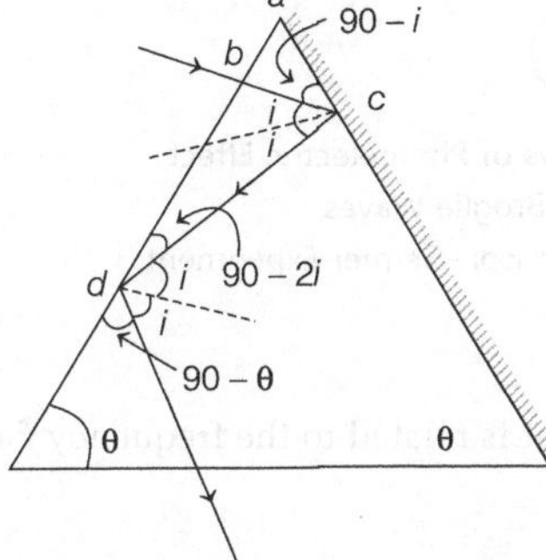

In Δabc,

$$A + 90° + 90° - i = 180°$$

$\Rightarrow \quad i = A$

Also, angle $\theta = 2i = 2A$

This condition is satisfied by option (b) only.

33 For given condition,

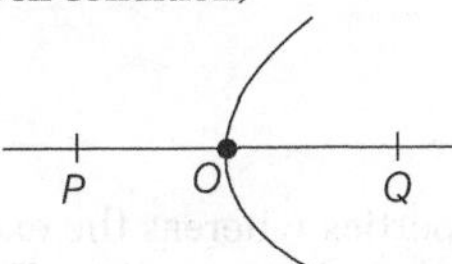

$$\frac{\mu_1}{-u} + \frac{\mu_2}{v} = \frac{\mu_2 - \mu_1}{R}$$

$$\frac{1}{-(-x)} + \frac{1.5}{x} = \frac{1.5 - 1}{R}$$

$\Rightarrow \quad x = 5R$

34 $i_p + r = 90°$

or $\quad r = 90° - i_p = 90° - 53°4' = 36°56'$

35 For the first minimum,

$$a\sin\theta_1 = \lambda \approx a\theta_1 = \frac{ad_1}{D}$$

For the sixth minimum,

$$a\sin\theta_6 = 6\lambda = a\theta_6 = \frac{ad_6}{D}$$

$\therefore \quad (6\lambda - \lambda) = \dfrac{a}{D}(d_6 - d_1)$

or $\quad a = \dfrac{5D\lambda}{(d_6 - d_1)}$

$$= \frac{5\times0.5\times5\times10^{-7}}{0.5\times10^{-3}}$$

$$= 2.5\times10^{-3} \text{ m}$$

$$= 2.5 \text{ mm}$$

36 The relation between angle of deviation δ for a thin prism an angle of prism and refractive index of material of prism is given by $\delta = (\mu - 1)A$

37 We know very well that, glass is transparent. But when the glass is powdered, the irregular reflections occur from the surface of powdered glass and finally the light returns back into the same medium. Because of it the powdered glass looks opaque. When we pour water over the powdered glass, refraction of light takes place and it becomes transparent.

38 The focal length f_w of convex lens in water of refractive index μ_w is given by

$$\frac{1}{f_w} = \left(\frac{\mu_g - \mu_w}{\mu_w}\right)\left(\frac{1}{R_1} - \frac{1}{R_2}\right) \quad \text{...(i)}$$

Here, μ_g is the refractive index of lens (glass).

The focal length of lens in air is given by

$$\frac{1}{f} = (\mu_g - 1)\left(\frac{1}{R_1} - \frac{1}{R_2}\right) \quad \text{...(ii)}$$

From Eqs. (i) and (ii), we conclude the focal length of lens in water (f_w) is greater than focal length of lens in air (f_a). Therefore, the focal length of lens in water get increased consequently power decreases.

39 Red colour consists of longest wavelength and it scatteres least. Therefore, signals of red colour are being seen from the long distances. Hence, the signals are made of red colour light.

40 The clouds consist of dust particles and water droplets. The scattering of sun light by the big dust particles and water drops is not in accordance with the Rayleigh law. But they scatter the light of all the colours by the same amount. Hence, the clouds are seen generally white.

DAY THIRTY ONE

Dual Nature of Matter

Learning & Revision for the Day

- Photon
- Particle Nature of Light
- Photoelectric Effect
- Laws of Photoelectric Effect
- de-Broglie Waves
- Davisson-Germer Experiment

Photon

A particle of light called a **photon** has energy E that is related to the frequency f and wavelength λ of light wave.

By the Einstein equation, $E = hf = \dfrac{hc}{\lambda}$ ...(i)

where, c is the speed of light (in vacuum) and h is **Planck's constant**.

$$h = 6.626 \times 10^{-34} \text{ J-s} = 4.136 \times 10^{-15} \text{ eV-s}$$

Since, energies are often given in electron volt $(1\text{eV} = 1.6 \times 10^{-19} \text{ J})$ and wavelengths are in Å, it is convenient to the combination hc in eV-Å. We have,

$$hc = 12375 \text{ eV-Å}$$

Hence, Eq. (i), in simpler form can be written as,

$$E \text{ (in eV)} = \frac{12375}{\lambda \text{ (in Å)}} \quad \text{...(ii)}$$

The propagation of light is governed by its wave porperties whereas the exchange of energy between, light with matter is governed by its particle properties. The wave-particle duality is a general property in nature. For example, electrons (and other so called particles) also propagate as waves and exchange energy as particles.

Particle Nature of Light

Photoelectric effect gave evidence to the strange fact that, light in interaction with matter behaved as if it was made of quanta or packets of energy, each of energy $h\nu$. Einstein stated that the light quantum can also be associated with momentum $\left(\dfrac{h\nu}{c}\right)$.

This particle like behaviour of light was further confirmed, in 1924, by the Compton experiment of scattering of X-rays from electrons.

- No. of Questions in Exercises (x)—
- No. of Questions Attempted (y)—
- No. of Correct Questions (z)—
  (Without referring Explanations)

- Accuracy Level ($z/y \times 100$)—
- Prep Level ($z/x \times 100$)—

In order to expect good rank in JEE, your Accuracy Level should be above 85 & Prep Level should be above 75.

Photoelectric Effect

- Photoelectric effect is the phenomenon of emission of electrons (known as photoelectrons) from the surface of metals when light radiation of suitable frequency are incident on them.
- The minimum energy of incident radiation needed to eject the electrons from metal surface is known as **work function** (ϕ_0) of that surface.
- The frequency or wavelength corresponding to the work function is called **threshold frequency** or **threshold wavelength**. Work function is related to threshold frequency as,

$$\phi_0 = h\nu_0 = \frac{hc}{\lambda_0}$$

where, λ_0 = threshold wavelength.

- In electron volt units, $\phi\,(\text{eV}) = \frac{hc}{e\lambda_0} = \frac{12400}{\lambda\,(\text{Å})}$
- For photoemission to take place energy of incident light (E) is related as, $E \geq p_0$
- According to Einstein's photoelectric equation,

$$h\nu = \phi_0 + K_{\max}$$

where, $K_{\max} = \frac{1}{2} mv^2_{\max}$ = maximum kinetic energy of ejected photoelectron.

Effect of Intensity on Photoelectric Emission

For a light of given frequency $\nu > \nu_0$ (or given wavelength $\lambda < \lambda_0$), if the intensity of light incident on photosensitive metal surface is increased, the number of photoelectrons and consequently the photoelectric current I increases. However, the stopping potential V_0 remain constant.

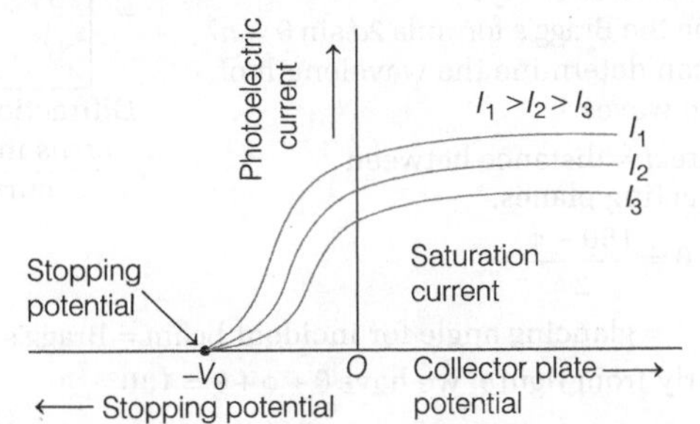

Photoelectric current *versus* stopping potential curve

Effect of Frequency on Photoelectric Emission

If keeping the intensity of incident light constant, the frequency of incident light is increased, then the stopping potential V_0 (and hence, $K_{\max}$) increases, but the photoelectric current I remains unchanged.

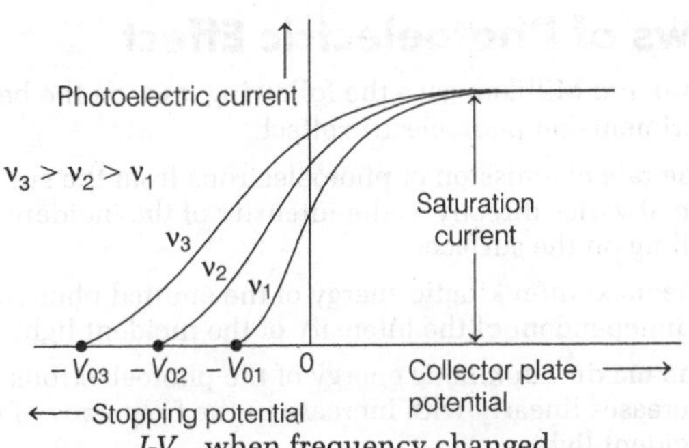

I-V_0, when frequency changed

A photon may collide with a material particle. The total energy and the total momentum remain conserved in such a collision. Photoelectric emission is an instantaneous phenomenon.

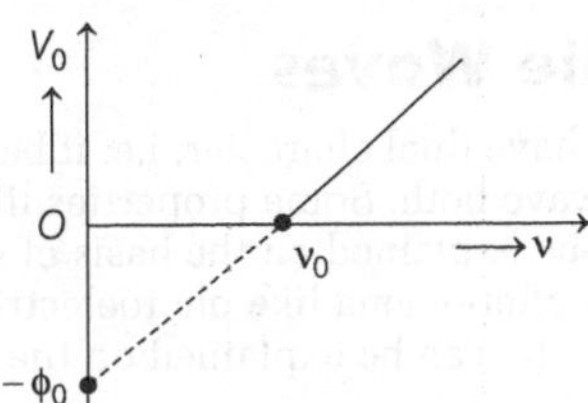

Cut-off voltage *versus* frequency of incident light

Variation of stopping potential V_0 with frequency ν of incident radiation is as shown in above figure.

As, $eV_0 = h(\nu - \nu_0) = h\nu - \phi_0 \Rightarrow V_0 = \frac{h}{e}\nu - \frac{\phi_0}{e}$

Thus, V_0- ν graph is a straight line whose slope is $\frac{h}{e}$ and intercept is $-\phi_0$ eV. The graph meets the ν-axis at ν_0.

Photocurrent $\propto \frac{1}{\nu} \propto \lambda$

Energy and Momentum of Photon

- From Einstein's mass-energy relation $E = h\nu = mc^2$

  Kinetic mass of photon is $m = \frac{h\nu}{c^2}$

  But $\nu = \frac{c}{\lambda}$, where λ is wavelength of the photon.

  $\therefore$ Kinetic mass of photon, $m = \frac{h}{c^2}\left(\frac{c}{\lambda}\right) = \frac{h}{c\lambda}$

  Kinetic mass of photon, $m = \frac{h\nu}{c^2} = \frac{h}{c\lambda}$

- Momentum of photon,

$$p = \text{kinetic mass of photon} \times \text{velocity of photon}$$
$$= \frac{h\nu}{c^2} \times c = \frac{h\nu}{c}$$

  Also, $\nu = \frac{c}{\lambda}$

  $\therefore$ Momentum of photon, $p = \frac{h}{c}\left(\frac{c}{\lambda}\right) = \frac{h}{\lambda}$

Laws of Photoelectric Effect

Lenard and Millikan gave the following laws on the basis of experiments on photoelectric effect.

- The rate of emission of photoelectrons from the surface of a metal varies directly as the intensity of the incident light falling on the surface.
- The maximum kinetic energy of the emitted photoelectrons is independent of the intensity of the incident light.
- The maximum kinetic energy of the photoelectrons increases linearly with increase in the frequency of the incident light.
- As soon as, the light is incident on the surface of the metal, the photoelectrons are emitted instantly, i.e. there is no time lag between incidence of light and emission of electrons ($\approx 10^{-9}$ s).

de-Broglie Waves

Light is said to have dual character, i.e. it behaves like matter (particle) and wave both. Some properties like interference, diffraction can be explained on the basis of wave nature of light, while the phenomena like photoelectric effect, black body radiation, etc. can be explained on the basis of particle nature of light.

In 1942, Louis de-Broglie explained that like light, matter also show dual behaviour, there is a wave associated with moving particle, known as **matter waves or de-Broglie waves.**

de-Broglie Relation

According to quantum theory, energy of photon

$$E = h\nu \quad \text{...(i)}$$

If mass of the photon is taken as m, then as per Einstein's equation

$$E = mc^2 \quad \text{...(ii)}$$

From Eqs. (i) and (ii), we get, $h\nu = mc^2$

$$h\frac{c}{\lambda} = mc^2,$$

where, λ = wavelength of photon

$$\lambda = \frac{h}{mc}$$

de-Broglie asserted that the above equation is completely a general function and applies to photon as well as all other moving particles.

So, $$\lambda = \frac{h}{mv} = \frac{h}{\sqrt{2mE}}$$

where, m is mass of particle and v is its velocity.

- **de-Broglie wavelength associated with charged particle**

$$\lambda = \frac{h}{p} = \frac{h}{\sqrt{2mE}} = \frac{h}{\sqrt{2mqV}}$$

- **de-Broglie wavelength of a gas molecule**

$$\lambda = \frac{h}{\sqrt{3mkT}}$$

where, T = absolute temperature

and k = Boltzmann's constant $= 1.38 \times 10^{-23}$ J/K

- **Ratio of wavelength of photon and electron** The wavelength of photon of energy E is given by $\lambda_p = \frac{hc}{E}$ while the wavelength of an electron of kinetic energy K is given by $\lambda_e = \frac{h}{\sqrt{2mK}}$. Therefore for same energy, the ratio

$$\frac{\lambda_p}{\lambda_e} = \frac{c}{E}\sqrt{2mK} = \sqrt{\frac{2mc^2K}{E^2}}$$

Davisson-Germer Experiment

- The de-Broglie hypothesis was confirmed by Davisson-Germer experiment. It is used to study the scattering of electron from a solid or to verify the wave nature of electron.
- A beam of electrons emitted by electron gun is made to fall on nickel crystal cut along cubical axis at a particular angle. Ni crystal behaves like a three dimensional diffraction grating and it diffracts the electron beam obtained from electron gun.
- The diffracted beam of electrons is received by the detector which can be positioned at any angle by rotating it about the point of incidence.
- The energy of the incident beam of electrons can also be varied by changing the applied voltage to the electron gun.
- According to classical physics, the intensity of scattered beam of electrons at all scattering angle will be same but Davisson and Germer found that the intensity of scattered beam of electrons was not same but different at different angles of scattering.
- It is maximum for diffracting angle 50° at 54 V potential difference.
- If the de-Broglie waves exist for electrons, then these should be diffracted as X-rays.
Using the Bragg's formula $2d\sin\theta = n\lambda$, we can determine the wavelength of these waves,
where d = distance between diffracting planes,

Incident beam 54 V 50°

Diffraction angle *versus* intensity curve

$$\theta = \frac{180 - \phi}{2}$$

= glancing angle for incident beam = Bragg's angle.

Clearly from figure, we have $\theta + \phi + \theta = 180°$

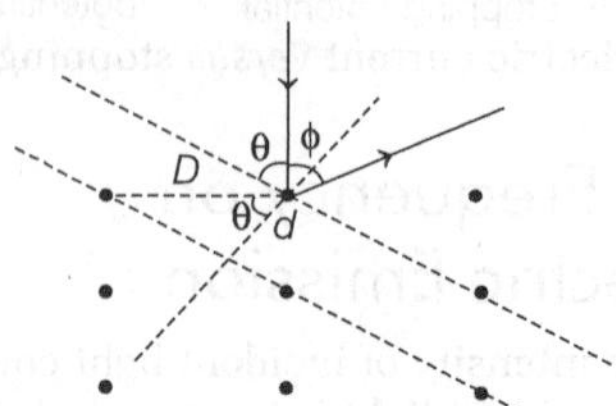

Reflection of electron (beam) by atoms

DAY PRACTICE SESSION 1

FOUNDATION QUESTIONS EXERCISE

1 If a source of power 4kW produces 10^{20} photons/second, the radiation belong to a part of the spectrum called
(a) X-rays (b) ultraviolet rays
(c) microwaves (d) γ-rays

2 What will be the number of photons emitted per second by a 10 W sodium vapour lamp assuming that 90% of the consumed energy is converted into light? (Wavelength of sodium light is 590 nm and $h = 6.63 \times 10^{-34}$ J-s)
(a) 0.267×10^{18} (b) 0.267×10^{19}
(c) 0.267×10^{20} (d) 0.267×10^{17}

3 Two monochromatic beams A and B of equal intensity I, hit a screen. The number of photons hitting the screen by beam A is twice that by beam B. Then, what inference can you make about their frequencies?
(a) $\nu_B = 2\nu_A$ (b) $\nu_B = \nu_A$
(c) $\nu_A = 2\nu_B$ (d) $\nu_B > \nu_A$

4 The eye can detect 5×10^4 photons $m^{-2}s^{-1}$ of light of wavelength 500 nm. The ear can hear intensity upto 10^{-13} Wm^{-2}. As a power detector, which is more sensitive?
(a) Sensitivity of eye is one-fifth of the ear
(b) Sensitivity of eye is five times that of the ear
(c) Both are equally sensitive
(d) Eye cannot be used as a power detector

5 The threshold wavelength for photoelectric emission from a material is 5200Å. Photoelectrons will be emitted when this material is illuminated with monochromatic radiation from a
(a) 50 W infrared lamp (b) 1 W infrared lamp
(c) 50 W ultraviolet lamp (d) 1 W ultraviolet lamp

6 The wavelength of the photoelectric threshold for silver is λ_0. The energy of the electron ejected from the surface of silver by an incident light of wavelength λ ($\lambda < \lambda_0$) will be
(a) $hc(\lambda_0 - \lambda)$ (b) $\frac{hc}{\lambda_0 - \lambda}$
(c) $\frac{h}{c} + \left(\frac{1}{\lambda} - \frac{1}{\lambda_0}\right)$ (d) $hc\left(\frac{\lambda_0 - \lambda}{\lambda_0 \lambda}\right)$

7 In photoelectric effect match the following column I with column II.

	Column I		Column II
A.	If frequency of incident light is increased	1.	Stopping potential may increase
B.	If intensity of incident light is increased	2.	Stopping potential must increase
C.	If work function of metal is increased	3.	Photo effect may stop

Codes

	A	B	C		A	B	C
(a)	1	2	3	(b)	3	2	1
(c)	2	1	3	(d)	2	3	1

8 When a point source of monochromatic light is at a distance of 0.2 m form a photoelectric cell the cut-off voltage and saturation current are 0.6 V and 18 mA respectively. If the same source is placed 0.6 m away from the photoelectric cell, then
(a) the stopping potential will be 0.2 V
(b) the stopping potential will be 0.6 V
(c) the saturation current will be 6 mA
(d) the saturation current will be 18 mA

9 The figure shows the variation of photocurrent (I) with anode potential (V_a) of a photosensitive surface for three different radiations. Let I_a, I_b and I_c be the intensities and f_a, f_b and f_c the frequencies for the waves a, b and c respectively

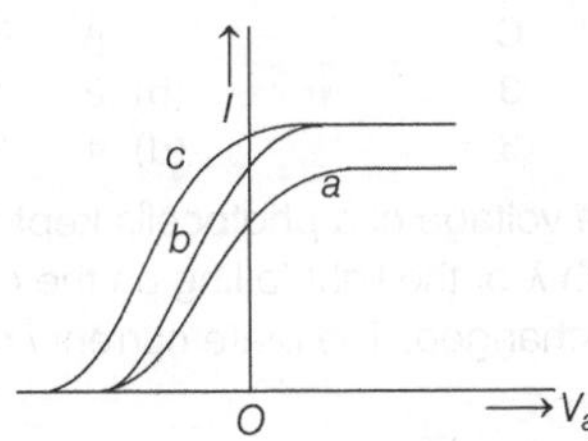

(a) $f_a = f_b$ and $I_a \neq I_b$ (b) $f_a = f_c$ and $I_a = I_c$
(c) $f_a = f_b$ and $I_a = I_c$ (d) $f_b = f_c$ and $I_b = I_c$

10 In a photoelectric effect measurement, the stopping potential for a given metal is found to be V_0 volt when radiation of wavelength λ_0 is used. If radiation of wavelength $2\lambda_0$ is used with the same metal, then the stopping potential (in volt) will be
(a) $\frac{V_0}{2}$ (b) $2V_0$
(c) $V_0 + \frac{hc}{2e\lambda_0}$ (d) $V_0 - \frac{hc}{2e\lambda_0}$

11 In an experiment on photoelectric effect, a student plots stopping potential V_0 against reciprocal of the wavelength λ of the incident light for two different metals A and B. These are as shown in the figure.

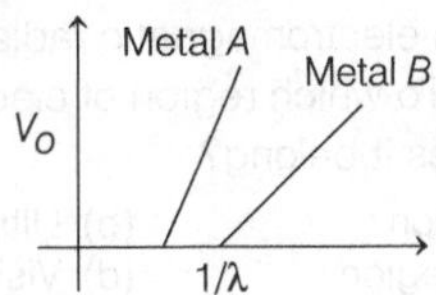

Looking at the graphs, you can most appropriately say that

(a) Work function of metal *B* is greater than that of metal *A*
(b) Work function of metal *A* is greater than that of metal *B*
(c) Students data is not correct
(d) None of the above

12 A copper ball of radius 1cm and work function 4.47 eV is irradiated with ultraviolet radiation of wavelength 2500 Å. The effect of irradiation results in the emission of electrons from the ball. Further the ball will acquire charge and due to this there will be a finite value of the potential on the ball. The charge acquired by the ball is

(a) 5.5×10^{-13} C (b) 7.5×10^{-13}C
(c) 4.5×10^{-12} C (d) 2.5×10^{-11} C

13 Match List I (fundamental experiment) with List II (its conclusion) and select the correct option from the choices given below the list.

List I	List II
A. Franck-Hertz experiment	1. Particle nature of light
B. Photo-electric experiment	2. Discrete energy levels of atom
C. Davisson-Germer experiment	3. Wave nature of electron
	4. Structure of atom

	A	B	C		A	B	C
(a)	1	4	3	(b)	2	4	3
(c)	2	1	3	(d)	4	3	2

14 The anode voltage of a photocells kept fixed. The wavelength λ of the light falling on the cathode is gradually changed. The plate current *I* of photocell varies as follows

(a)
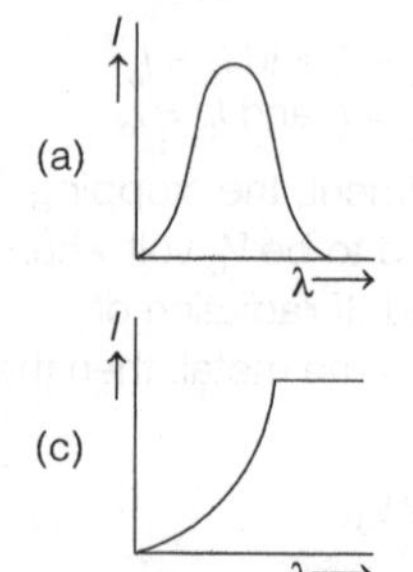

(b)
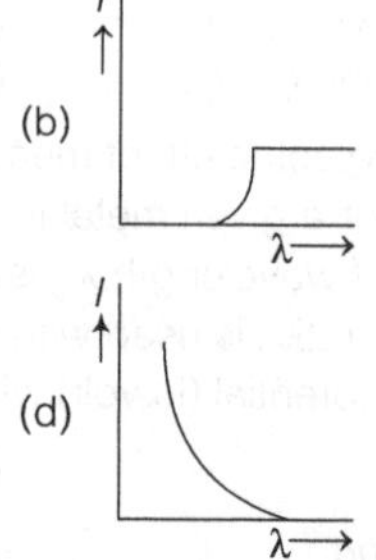

(c) I λ→

(d) I λ→

15 de-Broglie wavelength of an electron accelerated by a voltage of 50V is close to ($e = 1.6\times 10^{-19}$ C, $m_e = 9.1\times 10^{-31}$kg, $h = 6.6\times 10^{-34}$ J-s).

(a) 0.5 Å (b) 1.7Å
(c) 2.4 Å (d) 1.2 Å

16 Photons of an electromagnetic radiation has an energy 11keV each. To which region of electromagnetic spectrum does it belong?

(a) X-ray region (b) Ultraviolet region
(c) Infrared region (d) Visible region

17 The voltage applied to an electron microscope to produce electrons of wavelength 0.50 Å is

(a) 602 V (b) 50 V (c) 138 V (d) 812 V

18 The energy of photon is equal to the kinetic energy of a proton. The energy of photon is *E*. Let λ_1 be the de-Broglie wavelength, of the proton and λ_2 be the wavelength of photon. The ratio λ_1/λ_2 is proportional to

(a) E^0 (b) $E^{\frac{1}{2}}$
(c) E^{-1} (d) E^{-2}

19 An electron is moving with an initial velocity $v = v_0\hat{i}$ and is in a magnetic field $B = B_0\hat{j}$. Then its de-Broglie wavelength

(a) remains constant
(b) increase with time
(c) decrease with time
(d) increases and decreases periodically

20 Orbits of a particle moving in a circle are such that the perimeter of the orbit equals an integer number of de-Broglie wavelengths of the particle. For a charged particle moving in a plane perpendicular to a magnetic field, the radius of the *n*th orbital will therefore be proportional to

(a) n^2 (b) n
(c) $n^{1/2}$ (d) $n^{1/4}$

Direction (Q. Nos. 21-27) *Each of these questions contains two statements : Statement* I *and Statement* II. *Each of these questions also has four alternative choices, only one of which is the correct answer. You have to select one of the codes (a), (b), (c), (d) given below.*

(a) Statement I is true, Statement II is true; Statement II is the correct explanation for Statement I
(b) Statement I is true, Statement II is true; Statement II is not the correct explanation for Statement I
(c) Statement I is true; Statement II is false
(d) Statement I is false; Statement II is true

21 **Statement I** As intensity of incident light (in photoelectric effect) increases, the number of photoelectrons emitted per unit time increases.

Statement II More intensity of light means more energy per unit area per unit time.

22 **Statement I** The relative velocity of two photons travelling in opposite directions is the velocity of light.

Statement II The rest mass of photon is zero.

23 **Statement I** Work function of copper is greater than the work function of sodium but both have same value of threshold frequency and threshold wavelength.

Statement II The frequency is inversely proportional to wavelength.

24 **Statement I** The de-Broglie wavelength of a molecules varies inversely as the square root of temperature.

Statement II The root mean square velocity of molecule depends on the temperature.

25 **Statement I** Davisson-Germer experiment established the wave nature of electrons.

Statement II If electrons have wave nature, they can interfere and show diffraction.

26 **Statement I** A metallic surface is irradiated by a monochromatic light of frequency $\nu > \nu_0$ (the threshold frequency). The maximum kinetic energy and the stopping potential are K_{max} and V_0, respectively. If the frequency incident on the surface is doubled, both the K_{max} and V_0 are also doubled.

Statement II The maximum kinetic energy and the stopping potential of photoelectrons emitted from a surface are linearly dependent on the frequency of incident light.

27 **Statement I** When ultraviolet light is incident on a photocell, its stopping potential is V_0 and the maximum kinetic energy of the photoelectrons is K_{max}. When the ultraviolet light is replaced by X-rays, both V_0 and K_{max} increase.

Statement II Photoelectrons are emitted with speeds ranging from zero to a maximum value because of the range of frequencies present in the incident light.

DAY PRACTICE SESSION 2

PROGRESSIVE QUESTIONS EXERCISE

1 An electron of mass m and charge e are initially at rest. It gets accelerated by a constant electric field E. The rate of change of de-Broglie wavelength of this electron at time t is

(a) $\frac{-h}{eEt^2}$ (b) $\frac{-nh}{eEt^2}$ (c) $\frac{-h}{eE}$ (d) $\frac{-eht}{E}$

2 When a surface 1cm thick is illuminated with light of wavelength λ, the stopping potential is V_0, but when the same surface is illuminated by light of wavelength 3λ, the stopping potential is $V_0/6$, the threshold wavelength for metallic surface is

(a) 4λ (b) 5λ (c) 3λ (d) 2λ

3 A photocell with a constant potential difference of V volt across it is illuminated by a point source from a distance of 25 cm. When the source is moved to a distance of 1m, the electrons emitted by the photocell

(a) carry 1/4th their previous energy
(b) are 1/16th as numerous as before
(c) are 1/4th as numerous as before
(d) carry 1/4th their previous momentum

4 Consider a metal exposed to light of wavelength 600 nm. The maximum energy of the electron doubles when light of wavelength 400 nm is used. Find the work function in eV.

(a) 2.83 eV (b) 2 eV (c) 1.02 eV (d) 3.42 eV

5 A metallic surface is illuminated with monochromatic light of wavelength λ, the stopping potential for photoelectric current is $3V_0$ and when the same surface is illuminated with light of wavelength 2λ, the stopping potential is V_0. The threshold wavelength of this surface for photoelectric effect is

(a) $4\lambda/3$ (b) 6λ (c) 8λ (d) 4λ

6 The graph between $1/\lambda$ and stopping potential (V) of three metals having work functions ϕ_1, ϕ_2 and ϕ_3 in an experiment of photoelectric effect is plotted as shown in the figure. Which of the following statement (s) is/are correct? (Here, λ is the wavelength of the incident ray)

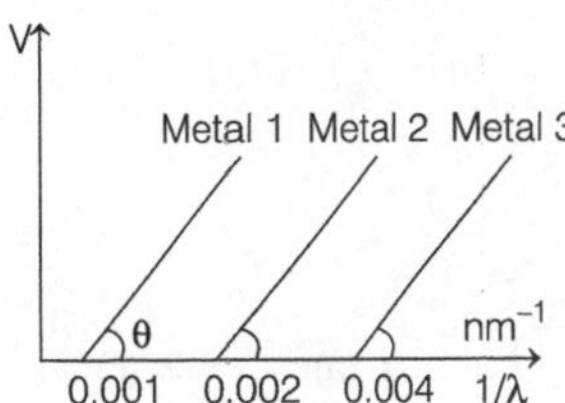

(a) Ratio of work function $\phi_1:\phi_2:\phi_3 = 1:2:4$
(b) Ratio of work function $\phi_1:\phi_2:\phi_3 = 4:2:1$
(c) $\tan\theta$ is directly proportional to hc/e where, h is Planck's constant and c is speed of light
(d) The violet colour light can eject photoelectrons from metal 2 and 3

7 Electrons are accelerated through a potential difference V_0 and protons are accelerated through a potential difference 4V. The de-Broglie wavelength are λ_e and λ_p for electrons and protons respectively.

The ratio of $\frac{\lambda_e}{\lambda_p}$ is given by (Given, m_e is mass of electrons and m_p is mass of proton).

(a) $\frac{\lambda_e}{\lambda_p} = \sqrt{\frac{m_p}{m_e}}$ (b) $\frac{\lambda_e}{\lambda_p} = \sqrt{\frac{m_e}{m_p}}$

(c) $\frac{\lambda_e}{\lambda_p} = \frac{1}{2}\sqrt{\frac{m_e}{m_p}}$ (d) $\frac{\lambda_e}{\lambda_p} = 2\sqrt{\frac{m_p}{m_e}}$

8 An electron from various excited states of hydrogen atom emit radiation to come to the ground state. Let λ_n, λ_g be the de-Broglie wavelength of the electron in the nth state and the ground state, respectively. Let Λ_n be the wavelength of the emitted photon in the transition from the nth state to the ground state. For large n, (A, B are constants)

(a) $\Lambda_n \approx A + \dfrac{B}{\lambda_n^2}$ (b) $\Lambda_n \approx A + B\lambda_n^2$

(c) $\Lambda_n^2 \approx A + B\lambda_n^2$ (d) $\Lambda_n^2 \approx \lambda$

9 An electron beam is accelerated by a potential difference V to hit a metallic target to produce X-rays. It produces continuous as well as characteristic X-rays. If λ_{min} is the smallest possible wavelength of X-rays in the spectrum, the variation of $\log \lambda_{min}$ with $\log V$ is correctly represented in

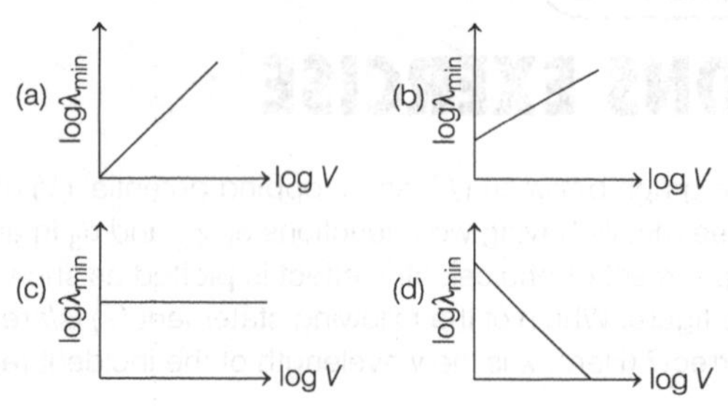

10 A particle A of mass m and initial velocity v collides with a particle B of mass $\dfrac{m}{2}$ which is at rest. The collision is head on, and elastic. The ratio of the de-Broglie wavelengths λ_A to λ_B after the collision is

(a) $\dfrac{\lambda_A}{\lambda_B} = 2$ (b) $\dfrac{\lambda_A}{\lambda_B} = \dfrac{2}{3}$ (c) $\dfrac{\lambda_A}{\lambda_B} = \dfrac{1}{2}$ (d) $\dfrac{\lambda_A}{\lambda_B} = \dfrac{1}{3}$

11 Radiation of wavelength λ is incident on a photocell. The fastest emitted electron has speed v. If the wavelength is changed to $3\lambda/4$, the speed of the fastest emitted electron will be

(a) $> v\left(\dfrac{4}{3}\right)^{1/2}$ (b) $< v\left(\dfrac{4}{3}\right)^{1/2}$ (c) $= v\left(\dfrac{4}{3}\right)^{1/2}$ (d) $= v\left(\dfrac{3}{4}\right)^{1/2}$

12 The radiation corresponding to $3 \rightarrow 2$ transition of hydrogen atom falls on a metal surface to produce photoelectrons. These electrons are made to enter a magnetic field of 3×10^{-4} T. If the radius of the largest circular path followed by these electrons is 10.0 mm, the work function of the metal is close to

(a) 1.8 eV (b) 1.1 eV (c) 0.8 eV (d) 1.6 eV

13 The surface of a metal is illuminated with the light of 400 nm. The kinetic energy of the ejected photoelectrons was found to be 1.68 eV. The work function of the metal is (hc = 1240 eV-nm)

(a) 3.09 eV (b) 1.42 eV (c) 151 eV (d) 1.68 eV

ANSWERS

SESSION 1	1 (a)	2 (c)	3 (a)	4 (b)	5 (c,d)	6 (d)	7 (c)	8 (b)	9 (a)	10 (d)
	11 (c)	12 (a)	13 (c)	14 (d)	15 (b)	16 (b)	17 (a)	18 (b)	19 (a)	20 (c)
	21 (a)	22 (b)	23 (d)	24 (b)	25 (a)	26 (d)	27 (d)			
SESSION 2	1 (a)	2 (b)	3 (b)	4 (c)	5 (d)	6 (a,c)	7 (d)	8 (a)	9 (d)	10 (a)
	11 (a)	12 (b)	13 (b)							

Hints and Explanations

SESSION 1

1 $4 \times 10^3 = 10^{20} \times hf$

$$f = \frac{4 \times 10^3}{10^{20} \times 6.023 \times 10^{-34}}$$

$$f = 6.64 \times 10^{16} \text{ Hz}$$

The obtained frequency lies in the band of X-rays.

2 Energy of photon,

$$E_1 = \frac{hc}{\lambda} = \frac{6.63 \times 10^{-34} \times 3 \times 10^8}{590 \times 10^{-9}}$$

$$= \frac{6.63 \times 3}{59} \times 10^{-18}$$

Light energy produced per second

$$E = \frac{90}{100} \times 10$$

$$= 9 \text{ W}$$

Number of photons emitted per sec $= \dfrac{E}{E_1}$

$$= \frac{9 \times 59}{6.63 \times 3 \times 10^{-18}}$$

$$= 2.67 \times 10^{19} = 0.267 \times 10^{20}$$

3 Intensity A = Intensity B

The number of photons of beam $A = n_A$

The number of photons of beam $B = n_B$

According to question, $n_A = 2n_B$

Let ν_A be the frequency of beam A and ν_B be the frequency of beam B.

$\therefore$ Intensity $\propto$ Energy of photons

$I \propto (h\nu) \times$ Number of photons

$$\therefore \quad \frac{I_A}{I_B} = \frac{n_A \nu_A}{n_B \nu_B}$$

According to question, $I_A = I_B$

$\therefore \quad n_A v_A = n_B v_B \Rightarrow \frac{v_A}{v_B} = \frac{n_B}{n_A} = \frac{1}{2}$

So, $v_B = 2v_A$

4 Sensitivity of eye = energy detected per square meter

$$= \frac{5 \times 10^4 \times 6.6 \times 10^{-34} \times 3 \times 10^8}{500 \times 10^{-9}}$$

$$= 0.2 \times 10^{-13} \text{ Wm}^{-2}$$

The sensitivity of ear = 1×10^{-13} Wm^{-2}

Thus, the sensitivity of eye is five times more than that of the ear.

5 For photo emission to take place, wavelength of incident light should be less than the threshold wavelength of ultraviolet light < 5200 Å while that of infrared radiations > 5200Å.

6 $E_k = \frac{hc}{\lambda} - \frac{hc}{\lambda_0} = hc\left[\frac{\lambda_0 - \lambda}{\lambda_0 \lambda}\right]$

7 As we know,

$eV_0 = h(v - v_0)$ and $hv = \text{KE} + W_0$

So, A → 2, B → 1, C → 3

8 By changing distance the intensity changes but frequency remains same, so stopping potential remains same.

9 Threshold voltage for a and b is same and it depends on frequency. So, $f_a = f_b$.

Photo current of a and b are unequal and photo current depends on intensity.

So, $I_a \neq I_b$

10 $eV_0 = \frac{hc}{\lambda_0} - W_0$ and $eV' = \frac{hc}{2\lambda_0} - W_0$

Subtracting them, we have

$$e(V_0 - V') = \frac{hc}{\lambda_0}\left[1 - \frac{1}{2}\right] = \frac{hc}{2\lambda_0}$$

or $\quad V' = V_0 - \frac{hc}{2e\lambda_0}$

11 We have, $eV_0 = \frac{hc}{\lambda} - \phi \Rightarrow V_0 = \frac{hc}{e\lambda} - \frac{\phi}{e}$

$$V_0 = mx + c$$

∵ Data is not sufficient.

12 As, $\frac{1}{4\pi\varepsilon_0} \frac{Q}{1 \times 10^{-2}} = \frac{\frac{hc}{\lambda} - \phi}{Q}$

$\Rightarrow \quad Q = 5.5 \times 10^{-13}$ C

13 (A) Franck-Hertz experiments is associated with discrete energy levels of atom.

(B) Photo-electric experiment is associated with particle nature of light.

(C) Davisson-Germer experiment is associated with wave nature of electron.

14 As λ is increased, there will be a value of λ above which photoelectron will be cease to come out, so photocurrent will becomes zero.

15 de-Broglie wavelength is,

$$\lambda = \frac{h}{mv} = \frac{h}{\sqrt{2mqV}} = 1.7 \text{ Å}$$

16 As, $\quad E = \frac{hc}{\lambda}$

and $\quad \lambda = \frac{hc}{E} = \frac{6.6 \times 10^{-34} \times 3 \times 10^8}{11 \times 1.6 \times 10^{-19}}$

$$= 1.125 \times 10^{-7} \text{ m}$$

Hence, UV region.

17 de-Broglie wavelength is

$$\lambda = \frac{h}{mv} = \frac{h}{\sqrt{2mE}}$$

But $\quad E = eV$

$$\lambda = \frac{h}{\sqrt{2meV}} \Rightarrow V = \frac{h^2}{2me\lambda^2}$$

$$V = \frac{(6.62 \times 10^{-34})^2}{(0.5 \times 10^{-10})^2 \times 2 \times 9.1 \times 10^{-31} \times 1.6 \times 10^{-19}}$$

$\Rightarrow \quad V = 601.98 \text{ V} \approx 602 \text{ V}$

18 As, $\lambda_1 = \frac{h}{\sqrt{2m_p E}}$, $\lambda_2 = \frac{hc}{E}$

$\therefore \quad \frac{\lambda_1}{\lambda_2} = \frac{h/\sqrt{2m_p E}}{hc/E} = \frac{\sqrt{E}}{\sqrt{2m_p c^2}}$

$\Rightarrow \quad \frac{\lambda_1}{\lambda_2} \propto \sqrt{E} \Rightarrow \frac{\lambda_1}{\lambda_2} \propto E^{1/2}$

19 Here, $v = v_o \mathbf{i}$ $B = B_o \mathbf{j}$

Force on moving electron due to magnetic field is

$$\text{F} = -e(v \times \text{B}) = -e[v_o \mathbf{i} \times B_o \mathbf{j}] = -e\, v_o B_o \mathbf{k}$$

As this force is perpendicular to v and **B**, so the magnitude of v will not change i.e. momentum (mv) will remain constant is magnitude. Hence, de-Broglie wavelength $\lambda = h/mv$ remains constant.

20 As, $2\pi r = n\lambda \Rightarrow r = \frac{n\lambda}{2\pi}$

Now, de-Broglie equation $\lambda = \frac{h}{p}$

$\Rightarrow \quad mv_n = \frac{h}{\lambda} = \frac{h}{\frac{2\pi r_n}{n}} = \frac{nh}{2\pi r_n}$

Also, for charged particle moving in a magnetic field

$$r_n = \frac{mv_n}{qB} = \frac{nh}{(2\pi r_n)qB}$$

$\Rightarrow \quad r_n^2 = \frac{nh}{2\pi qB}$

$\therefore \quad r_n \propto n^{1/2}$

21 From quantum theory of light, as intensity of light increases means number of photons/area/time increases and hence more photons take part in ejecting the photoelectron, thus increasing the number of photoelectrons.

22 Velocity of first photon = $u = c$

Velocity of second photon = $v = -c$

Now relative velocity of first photon with respect to second photon

$$= \frac{u - v}{1 - \frac{uv}{c^2}} = \frac{c - (-c)}{1 - \frac{c \times (-c)}{c^2}}$$

$$= \frac{2c}{1 + \frac{c^2}{c^2}}$$

$$= \frac{2c}{1 + 1} = \frac{2c}{2} = c$$

Also, the rest of mass of photon is zero.

23 When work function of copper is greater than the work function of sodium,then

$$\phi_{\text{Cu}} > \phi_{\text{Na}}$$

$$(hv_o)_{\text{Cu}} > (hv_o)_{\text{Na}} \qquad \text{...(i)}$$

But we know that, $v_o = \frac{c}{\lambda_c}$

Hence, Eq. (i), becomes

$$\left(\frac{hc}{\lambda_o}\right)_{\text{Cu}} > \left(\frac{hc}{\lambda_o}\right)_{\text{Na}} (\lambda_o)_{\text{Na}} > (\lambda_o)_{\text{Cu}}$$

24 de-Broglie wavelength associated with gas molecules varies

$$\lambda \propto \frac{1}{\sqrt{T}}$$

Also, root mean square velocity of gas molecules is $v_{\text{rms}} = \sqrt{\frac{3RT}{M}}$.

25 Davisson and Germer experimentally established wave nature of electron by observing diffraction pattern while bombarding electrons on Ni crystal.

26 Maximum kinetic energy $(\text{KE})_{\text{max}}$ is given by $(\text{KE})_{\text{max}} = hv - hv_0$.

When frequency is increased $(\text{KE})_{\text{max}}$ will increase stopping potential is that negative voltage given to the anode at which photocurrent stops, hence doubling frequency will not effect it, also

If $\quad v' = 2v$

$\therefore \quad K'_{\text{max}} = ev'_0 = h(2v_1 - v_0)$

$\quad K'_{\text{max}} = 2K_{\text{max}} + hv_0$

$\therefore \quad K'_{\text{max}} > 2K_{\text{max}} \Rightarrow v'_0 > 2v_0$

Hence, $(\text{KE})_{\text{max}}$ and stopping potential are linearly dependent on the frequency of incident light.

27 Since, the frequency of ultraviolet light is less than the frequency of X-rays, the energy of each incident photon will be more for X-rays

$$KE_{photoelectron} = h\nu - \phi$$

Stopping potential is to stop the fastest photoelectron

$$V_0 = \frac{h\nu}{e} - \frac{\phi}{e}$$

So, KE_{max} and V_0 both increases.

But KE ranges from zero to KE_{max} because of loss of energy due to subsequent collisions before getting ejected and not due to range of frequencies in the incident light.

SESSION 2

1 Here, the initial velocity is $u = 0$

Since, $a = \frac{eE}{m}$, $v = ?$ at $t = t$

So we get, using $v = u + at = 0 + \frac{eE}{m}t$

This gives $\lambda = \frac{h}{mv} = \frac{h}{m(eEt/m)} = \frac{h}{eEt}$

The rate of change of de-Broglie wavelength is

$$\frac{d\lambda}{dt} = \frac{h}{eE} \times \left(\frac{-1}{t^2}\right) = \frac{-h}{eEt^2}$$

2 From Einstein's photoelectric equation, we have

$$eV_0 = hc\left[\frac{1}{\lambda} - \frac{1}{\lambda_0}\right] \quad \text{...(i)}$$

$$\frac{eV_0}{6} = hc\left[\frac{1}{3\lambda} - \frac{1}{\lambda_0}\right] \quad \text{...(ii)}$$

On dividing Eq. (i) by Eq. (ii), we get

$$6 = \frac{\left(\frac{1}{\lambda} - \frac{1}{\lambda_0}\right)}{\left(\frac{1}{3\lambda} - \frac{1}{\lambda_0}\right)}$$

$$\frac{6}{3\lambda} - \frac{6}{\lambda_0} = \frac{1}{\lambda} - \frac{1}{\lambda_0} \Rightarrow \frac{1}{\lambda} = \frac{5}{\lambda_0}$$

$$\Rightarrow \frac{\lambda_0}{\lambda} = 5 \Rightarrow \lambda_0 = 5\lambda$$

3 Photoelectric current I is directly proportional to intensity of light and intensity $\propto \frac{1}{(\text{distance})^2}$

$$I \propto \frac{1}{r^2}$$

$$I_{25} \propto \frac{1}{(25)^2} \quad \text{...(i)}$$

$$I_{100} \propto \frac{1}{(100)^2} \quad [1 \text{ m} = 100 \text{ cm}] \text{...(ii)}$$

$$\therefore \frac{I_{25}}{I_{100}} = \frac{(100)^2}{(25)^2} = 16 \Rightarrow I_{100} = \frac{I_{25}}{16}$$

4 Given, wavelength

$\lambda_1 = 600 \text{nm} = 600 \times 10^{-9} \text{m}$

Energy correspond to $\lambda_1 = E_1$

Again, wavelength

$\lambda_2 = 400 \text{nm} = 400 \times 10^{-9} \text{m}$

Energy correspond to $\lambda_2 = E_2$

Let the work function of metal is ϕ.

According to question $2E_1 = E_2$

i.e. $$2\left(\frac{hc}{\lambda_1} - \phi_0\right) = \frac{hc}{\lambda_2} - \phi_0$$

or $$\frac{2hc}{\lambda_1} - \frac{hc}{\lambda_2} = 2\phi_0 - \phi_0$$

or $$hc\left[\frac{2}{\lambda_1} - \frac{1}{\lambda_2}\right] = \phi_0$$

$$\phi_0 = 6.63 \times 10^{-34} \times 3 \times 10^8 \left[\frac{2}{600 \times 10^{-9}} - \frac{1}{400 \times 10^{-9}}\right]$$

or $$= \frac{6.63 \times 10^{-34} \times 3 \times 10^8}{10^{-7}}\left[\frac{1}{3} - \frac{1}{4}\right]$$

or $$= 6.63 \times 3 \times 10^{-19}\left[\frac{4-3}{12}\right] \text{J}$$

$$= \frac{6.63 \times 3 \times 10^{-19}}{1.6 \times 10^{-19} \times 12} = 1.02 \text{ eV}$$

5 According to the Einstein's photoelectric effect

$$h\nu - h\nu_0 = \frac{1}{2}mv^2 = eV$$

$$\Rightarrow \left(\frac{hc}{\lambda} - \frac{hc}{\lambda_0}\right) = eV$$

where, λ_0 = threshold wavelength.

Now for the first case,

$$\frac{hc}{\lambda} - \frac{hc}{\lambda_0} = e(3V_0) \quad \text{...(i)}$$

For the second case,

$$\frac{hc}{2\lambda} - \frac{hc}{\lambda_0} = e(V_0) \quad \text{...(ii)}$$

On dividing Eq. (i) by Eq.(ii), we get

$$\frac{hc\left(\frac{1}{\lambda} - \frac{1}{\lambda_0}\right)}{hc\left(\frac{1}{2\lambda} - \frac{1}{\lambda_0}\right)} = \frac{3eV_0}{eV_0}$$

$$\Rightarrow \frac{\frac{\lambda_0 - \lambda}{\lambda\lambda_0}}{\frac{\lambda_0 - 2\lambda}{2\lambda \cdot \lambda_0}} = \frac{3eV_0}{eV_0}$$

$$\lambda_0 = 4\lambda$$

6 From the relation,

$$eV = \frac{hc}{\lambda} - \phi \text{ or } V = \left(\frac{hc}{e}\right)\frac{1}{\lambda} - \frac{\phi}{e}$$

This is the equation of straight line slope is $\tan\theta = hc/e$

$$\phi_1 : \phi_2 : \phi_3 = \frac{hc}{\lambda_{01}} : \frac{hc}{\lambda_{02}} : \frac{hc}{\lambda_{03}}$$

$$\frac{1}{\lambda_{01}} : \frac{1}{\lambda_{02}} : \frac{1}{\lambda_{03}} = 1:2:4$$

$$\frac{1}{\lambda_0} = 0.001 \text{nm}^{-1}$$

$$\lambda_0 = 10{,}000 \text{Å}$$

$$\frac{1}{\lambda_{02}} = 0.002 \text{nm}^{-1}$$

or $$\lambda_{02} = 5000 \text{Å}$$

$$\frac{1}{\lambda_{03}} = 0.004 \text{nm}^{-1}$$

$$\lambda_{03} = 2500 \text{Å}$$

Violet colour has wavelength 4000Å.

So, violet colour can eject photoelectrons from metal 1 and metal 2.

7 We have, $E = qV$, we know that

$$E = \frac{1}{2}mv^2 \Rightarrow v = \sqrt{\frac{2E}{m}}$$

as, $$\lambda = \frac{h}{mv} = \frac{h}{m\sqrt{\frac{2E}{m}}}$$

$$\Rightarrow \lambda = \frac{h}{\sqrt{2mqV}} \quad \text{...(i)}$$

For electron, $$\lambda_e = \frac{h}{\sqrt{2m_e qV}} \quad \text{...(ii)}$$

For protron $$\lambda_p = \frac{h}{\sqrt{2m_p qV}}$$

$$\therefore \lambda_p = \frac{h}{\sqrt{2m_p q \cdot 4V}}$$

$$(\because V = 4V) \text{...(iii)}$$

On dividing Eq. (ii) by Eq. (iii), we get

$$\frac{\lambda_e}{\lambda_p} = 2\sqrt{\frac{m_p}{m_e}}$$

8 If wavelength of emitted photon in de-excitation is Λ_n;

n

Λ_n

1

Then, $$\frac{hc}{\Lambda_n} = E_n - E_g$$

$$\frac{hc}{\Lambda_n} = \frac{p_n^2}{2m} - \frac{p_g^2}{2m} \quad \left[\because E = \frac{p^2}{2m}\right]$$

As energies are negative, we get

$$\frac{hc}{\Lambda_n} = \frac{p_g^2}{2m} - \frac{p_n^2}{2m} = \frac{p_g^2}{2m}\left(1 - \left(\frac{p_n}{p_g}\right)^2\right)$$

$$= \frac{h^2}{2m\lambda_g^2}\left(1 - \frac{\lambda_g^2}{\lambda_n^2}\right) \left[\because p \propto \lambda^{-1}, p = \frac{h}{\lambda}\right]$$

$$\Rightarrow \Lambda_n = \frac{2m\lambda_g^2 c}{h}\left(1 - \frac{\lambda_g^2}{\lambda_n^2}\right)^{-1}$$

$$\Rightarrow \quad \Lambda_n = \frac{2m\lambda_g^2 c}{h}\left(1+\frac{\lambda_g^2}{\lambda_n^2}\right)$$

$[\because (1-x)^{-n} = 1 + nx]$

$$\Rightarrow \quad \Lambda_n \simeq A + \frac{B}{\lambda_n^2}$$

where, $A = \left[\frac{2mc\lambda_g^2}{h}\right]$ and $B = \left[\frac{2mc\lambda_g^4}{h}\right]$ are constants.

9 $\lambda_{\min} = \frac{hc}{eV}$

$\log(\lambda_{\min}) = \log\left(\frac{hc}{e}\right) - \log V$

$y = c - mx$

So, the required graph is given in option (d).

10 For elastic collision,

$p_{\text{before collision}} = p_{\text{after collision}}$.

$mv = mv_A + \frac{m}{2}v_B$

$2v = 2v_A + v_B \quad \ldots \text{(i)}$

Now, coefficient of restitution,

$$e = \frac{v_B - v_A}{u_A - v_B}$$

Here, $u_B = 0$ (Particle at rest) and for elastic collision $e = 1$

$\therefore 1 = \frac{v_B - v_A}{v} \Rightarrow v = v_B - v_A \quad \ldots \text{(ii)}$

From Eq. (i) and Eq. (ii),

$v_A = \frac{v}{3}$ and $v_B = \frac{4v}{3}$

Hence, $\frac{\lambda_A}{\lambda_B} = \frac{\left(\frac{h}{mV_A}\right)}{\left(\frac{h}{\frac{m}{2}.V_B}\right)}$

$$= \frac{V_B}{2V_A} = \frac{4/3}{2/3} = 2$$

11 According to Einstein's photoelectric emission of light, $E = (\text{KE})_{\max} + \phi$

As, $\frac{hc}{\lambda} = (\text{KE})_{\max} + \phi$

If the wavelength of radiation is changed to $\frac{3\lambda}{4}$, then

$$\Rightarrow \frac{4}{3}\frac{hc}{\lambda} = \left(\frac{4}{3}(\text{KE})_{\max} + \frac{\phi}{3}\right) + \phi$$

For fastest emitted electron,

$(\text{KE})_{\max.} = \frac{1}{2}mv'^2 + \phi$

$$\Rightarrow \frac{1}{2}mv'^2 = \frac{4}{3}\left(\frac{1}{2}mv^2\right) + \frac{\phi}{3}$$

i.e. $v' > v\left(\frac{4}{3}\right)^{1/2}$

12 When an electron moves in a circular path, then

Radius, $r = \frac{mv}{eB} \Rightarrow \frac{r^2e^2B^2}{2} = \frac{m^2v^2}{2}$

$\text{KE}_{\max} = \frac{(mv)^2}{2m} \Rightarrow \frac{r^2e^2B^2}{2m} = (\text{KE})_{\max}$

Work function of the metal (W),

i.e. $W = h\nu - \text{KE}_{\max}$

$1.89 - \phi = \frac{r^2e^2B^2}{2m}\frac{1}{2}\text{eV}$

$= \frac{r^2eB^2}{2m}\text{eV}$

[$h\nu \rightarrow 1.89$ eV, for the transition on from third to second orbit of H-atom]

$$= \frac{100\times10^{-6}\times1.6\times10^{-19}\times9\times10^{-8}}{2\times9.1\times10^{-31}}$$

$\phi = 1.89 - \frac{1.6\times9}{2\times9.1}$

$= 1.89 - 0.79 = 1.1$ eV

13 $\because \text{KE}_{\max} = eV_0$

$\Rightarrow \frac{1}{2}mv^2 = eV_0 = 1.68$ eV

$\Rightarrow h\nu = \frac{hc}{\lambda} = \frac{1240 \text{ eV -nm}}{400 \text{ nm}} = 3.1$ eV

$\Rightarrow \quad 3.1 \text{ eV} = W_0 + 1.68$ eV

[From Einstein equation, $E = W_0 + K_{\max}$]

$W_0 = 1.42$ eV

DAY THIRTY TWO

Atoms

Learning & Revision for the Day

- Scattering of α-particles
- Rutherford's Model of an Atom
- Bohr's Model
- Energy Levels and Hydrogen Spectrum

Atom is the smallest particle of an element which contains all properties of element. Molecule is a single atom or a group of atoms joined by chemical bonds. It is the smallest unit of a chemical compound that can have an independent existence. Nuclei refers to a nucleus of an atom, having a given number of nucleons. It is a general term referring to all known isotopes–both stable and unstable of the chemical elements. Thus, O^{16} and O^{17} are different nuclides.

Scattering of α-particles

In 1911, Rutherford successfully explained the scattering of α-particles on the basis of nuclear model of the atom.

Number of α-particles scattered through angle θ is given by

$$N(\theta) \propto \frac{Z^2}{\sin^4 (\theta/2) K^2}$$

where, K is the kinetic energy of α-particle and Z is the atomic number of the metal.

At distance of closest approach the entire initial kinetic energy is converted into potential energy, so

$$\frac{1}{2} mv^2 = \frac{1}{4\pi\varepsilon_0} \frac{Ze(2e)}{r_0}$$

$$\Rightarrow \qquad r_0 = \frac{Ze^2}{mv^2 \pi\varepsilon_0} = \frac{4KZe^2}{mv^2}$$

Rutherford's Model of an Atom

On the basis of scattering of α-particles, Rutherford postulated the following model of the atom

- Atom is a sphere of diameter about 10^{-10} m. Whole of its positive charge and most of its mass is concentrated in the central part called the nucleus.
- The diameter of the nucleus is of the order of 10^{-5} m.
- The space around the nucleus is virtually empty with electrons revolving around the nucleus in the same way as the planets revolve around the sun.
- The electrostatic attraction of the nucleus provides centripetal force to the orbiting electrons.

- No. of Questions in Exercises (x)—
- No. of Questions Attempted (y)—
- No. of Correct Questions (z)—
  (Without referring Explanations)

- Accuracy Level ($z/y \times 100$)—
- Prep Level ($z/x \times 100$)—

In order to expect good rank in JEE, your Accuracy Level should be above 85 & Prep Level should be above 75.

- Total positive charge in the nucleus is equal to the total negative charge of the orbiting electrons.
- Rutherford's model suffers from the following drawbacks
  (a) stability of the atomic model.
  (b) nature of energy spectrum.

Bohr's Model

Bohr added the following postulates to the Rutherford's model of the atom

- The electrons revolve around the nucleus only in certain permitted orbits, in which the angular momentum of the electron is an integral multiple of $h/2\pi$, where h is the Planck's constant
$$\left(L = mv_n r_n = \frac{nh}{2\pi}\right)$$
- The electrons do not radiate energy while revolving in the permitted orbits. That is, the permitted orbits are stationary, non-radiating orbits.
- The energy is radiated only when the electron jumps from an outer permitted orbit to some inner permitted orbit. (Absorption of energy makes the electron jump from inner orbit to outer orbit).
- If energy of the electron in nth and mth orbits be E_n and E_m respectively, and when the electron jumps from nth to mth orbit the radiation frequency ν is emitted, such that $E_n - E_m = h\nu$. This is called the **Bohr's frequency equation.**

NOTE
- Radius of the orbit of electron in a hydrogen atom in its stable state, corresponding to $n = 1$, is called Bohr's radius. Value of Bohr's radius is $r_0 = 0.529$ Å ≈ 0.53 Å.
- The time period of an electron in orbital motion in the Bohr's orbit is
$$T = \frac{2\pi r}{v} = \frac{2\pi \times 0.53\text{Å}}{\frac{c}{137}} = 1.52 \times 10^{-6}\ \text{s} \quad \left[\because v = \frac{c}{137}\right]$$
and the frequency of revolution is $f = \frac{1}{T} = 6.5757 \times 10^{15}$ cps

Some Characteristics of an Atom

- The **orbital radius** of an electron is
$$r_n = 4\pi\varepsilon_0 \frac{n^2h^2}{4\pi^2 Zme^2} = 0.53\frac{n^2}{Z}\ \text{Å}$$
- The **orbital velocity** of an electron is
$$v_n = \frac{1}{4\pi\varepsilon_0}\frac{2Z\pi e^2}{nh} = \left(\frac{c}{137}\right)\frac{Z}{n} = 2.2\times 10^6\left(\frac{Z}{n}\right)\ \text{m/s}$$
- **Orbital frequency** is given by
$$f = \frac{1}{T} = \frac{v}{2\pi r} = \frac{me^4}{4\varepsilon_0^2 n^3 h^3}$$
- The **total energy** of the orbital electron is
$$E = -\left(\frac{me^4Z^2}{8\varepsilon_0^2h^2n^2}\right) = -\left(\frac{me^4}{8\varepsilon_0^2ch^3}\right)ch\frac{Z^2}{n^2}$$
$$= -Rch\frac{Z^2}{n^2} = -13.6\frac{Z^2}{n^2}\text{eV}$$
$$\text{KE} = \frac{me^4Z^2}{8n^2h^2\varepsilon_0^2},\quad \text{PE} = -\frac{me^4Z^2}{4n^2h^2\varepsilon_0^2}$$
- The kinetic, potential and total energies of the electron with r as the radius of the orbit are as follows
$$\text{KE} = \frac{1}{2}\left[\frac{1}{4\pi\varepsilon_0}\frac{Ze^2}{r}\right]$$
$$\text{PE} = -\frac{1}{4\pi\varepsilon_0}\frac{Ze^2}{r}$$
and $$E = -\frac{1}{2}\left[\frac{1}{4\pi\varepsilon_0}\frac{Ze^2}{r}\right]$$
Therefore, they are related to each other as follows
$$\text{KE} = -E \text{ and PE} = 2E$$
- For a hydrogen atom, $r_n \propto n^2$, $v_n \propto \frac{1}{n}$ and $|E| \propto \frac{1}{n^2}$
- The difference in angular momentum associated with the electron in the two successive orbits of hydrogen atom is
$$\Delta L = (n+1)\frac{h}{2\pi} - \frac{nh}{2\pi} = \frac{h}{2\pi}$$

Energy Levels and Hydrogen Spectrum

Hydrogen spectrum consists of spectral lines classified as five spectral series of hydrogen atom. Out of these five, Lyman series lies in the ultraviolet region of spectrum, Balmer series lies in the visible region and the remaining three series, lie in the infrared region of spectrum.

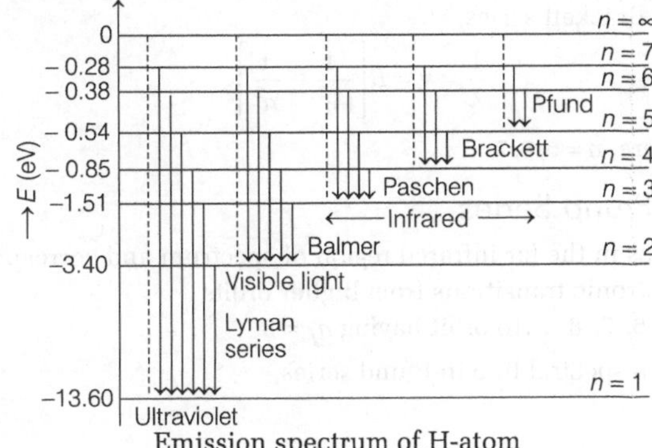

Emission spectrum of H-atom

Total number of emission spectral lines from some excited state n_1 to another energy $n_2 (< n_1)$ is given by

$$\frac{(n_1 - n_2)(n_1 - n_2 + 1)}{2}$$

e.g. total number of lines from $n_1 = n$ to $n_2 = 1$ are $\frac{n(n-1)}{2}$.

The five spectral series of hydrogen atom are given below

1. Lyman Series

Spectral lines of Lyman series correspond to the transition of electron from higher energy levels (orbits) $n_i = 2, 3, 4, \ldots$ to ground energy level (1st orbit) $n_f = 1$.

For Lyman series, $\frac{1}{\lambda} = \bar{v} = R\left[\frac{1}{(1)^2} - \frac{1}{n^2}\right]$

where, $n = 2, 3, 4, \ldots$

It is found that a term $Rch = 13.6 \text{ eV} = 2.17 \times 10^{-18}$ J. The term Rch is known as Rydberg's energy.

2. Balmer Series

Electronic transitions from $n_i = 3, 4, 5, \ldots$ to $n_f = 2$, give rise to spectral lines of Balmer series.

For a Balmer series line,

$$\frac{1}{\lambda} = \bar{v} = R\left[\frac{1}{(2)^2} - \frac{1}{n^2}\right]$$

where, $n = 3, 4, 5, \ldots$

3. Paschen Series

Lines of this series lie in the infrared region and correspond to electronic transition from $n_i = 4, 5, 6, \ldots$ to $n_f = 3$.

$$\frac{1}{\lambda} = \bar{v} = R\left[\frac{1}{(3)^2} - \frac{1}{n^2}\right]$$

where, $n = 4, 5, 6, \ldots$

4. Brackett Series

Spectral lines in the infrared region which corresponds to transition from $n_i = 5, 6, 7, \ldots$ to $n_f = 4$.

For Brackett series,

$$\frac{1}{\lambda} = \bar{v} = R\left[\frac{1}{(4)^2} - \frac{1}{n^2}\right]$$

where, $n = 5, 6, 7, \ldots$

5. Pfund Series

It lies in the far infrared region of spectrum and corresponds to electronic transitions from higher orbits $n_i = 6, 7, 8, \ldots$ to orbit having $n_f = 5$.

For a spectral line in Pfund series,

$$\frac{1}{\lambda} = \bar{v} = R\left[\frac{1}{(5)^2} - \frac{1}{n^2}\right]$$

where, $n = 6, 7, 8, \ldots$

NOTE
- Energy of emitted radiation,

$$\Delta E = E_2 - E_1 = RchZ^2\left(\frac{1}{n_1^2} - \frac{1}{n_2^2}\right) = 13.6\, Z^2\left(\frac{1}{n_1^2} - \frac{1}{n_2^2}\right)$$

- Frequency, $v = \frac{\Delta E}{h} = RcZ^2\left(\frac{1}{n_1^2} - \frac{1}{n_2^2}\right)$

Ionisation Energy and Potential

Ionisation energy of an atom is defined as the energy required to ionise it i.e. to make the electron jump from its present orbit to infinity.

Thus, ionisation energy of hydrogen atom in the ground state $= E_\infty - E_1$

$$= 0 - (-13.6 \text{ eV}) = +\ 13.6 \text{ eV}$$

The potential through which an electron is to be accelerated so that it acquires energy equal to the ionisation energy is called the **ionisation potential**.

Therefore, ionisation potential of hydrogen atom in its ground state is 13.6V.

In general, $E_{\text{ion}} = 13.6 \frac{Z^2}{n^2}$ eV or $V_{\text{ion}} = \frac{E_{\text{ion}}}{e}$

Excitation Energy and Ionisation Potential

Excitation energy is the energy required to excite an electron from a lower energy level to a higher energy level. The potential through which an electron is accelerated so as to gain requisite ionisation energy is called the **ionisation potential**.

Thus, first excitation energy of hydrogen atom

$$= E_2 - E_1$$
$$= -3.4 - (-13.6) \text{ eV}$$
$$= +\ 10.2 \text{ eV}$$

Similarly, second excitation energy of hydrogen atom

$$= E_3 - E_1$$
$$= -1.51 - (-13.6)$$
$$= 12.09 \text{ eV}$$

NOTE
- Total energy of a closed system is always negative and its magnitude is the binding energy of the system.
- Kinetic energy of a particle can't be negative, while the potential energy can be zero, positive or negative.

DAY PRACTICE SESSION 1

FOUNDATION QUESTIONS EXERCISE

1 An α-particle of energy 5 MeV is scattered through 180° by a fixed uranium nucleus. The distance of the closest approach is of the order of

(a) 1Å (b) 10^{-10}cm
(c) 10^{-12}cm (d) 10^{-15} cm

2 To explain theory of hydrogen atom, Bohr considered

(a) quantisation of linear momentum
(b) quantisation of angular momentum
(c) quantisation of angular frequency
(d) quantisation of energy

3 For the Bohr's first orbit of circumference $2\pi r$, the de-Broglie wavelength of revolving electron will be

(a) $2\pi r$ (b) πr (c) $\frac{1}{2\pi r}$ (d) $\frac{1}{4\pi r}$

4 Which of the following transitions in hydrogen atoms emit photons of highest frequency?

(a) $n = 2$ to $n = 6$ (b) $n = 6$ to $n = 2$
(c) $n = 2$ to $n = 1$ (d) $n = 1$ to $n = 2$

5 In the Bohr's model of the hydrogen atom, let r, V and E represents the radius of the orbit, the speed of electron and the total energy of the electron, respectively. Which of the following quantities is proportional to the quantum number n?

(a) E/v (b) r/E (c) vr (d) rE

6 The ratio of the kinetic energy to the energy of an electron in a Bohr's orbit is

(a) −1 (b) 2
(c) 1 : 2 (d) None of these

7 If the atom $_{100}Fm^{257}$ follows the Bohr's model and the radius of last orbit of $_{100}Fm^{257}$ is n times the Bohr's radius, then find the value of n?

(a) 100 (b) 200 (c) 4 (d) 1/4

8 Taking the Bohr's radius as $a_0 = 53$ pm, the radius of Li^{2+} ion in its ground state, on the basis of Bohr's model, will be about

(a) 53 pm (b) 27 pm
(c) 18 pm (d) 13 pm

9 In a hypothetical Bohr's hydrogen atom, the mass of the electron is doubled. The energy E_0 and radius r_0 of the first orbit will be (a_0 is the Bohr radius)

(a) $E_0 = -27.2$ eV ; $r_0 = a_0/2$
(b) $E_0 = -27.2$eV ; $r_0 = a_0$
(c) $E_0 = -13.6$ eV ; $r_0 = a_0/2$
(d) $E_0 = -13.6$ eV ; $r_0 = a_0$

10 As an electron makes a transition from an excited state to the ground state of a hydrogen like atom/ion → JEE Main 2015

(a) its kinetic energy increases but potential energy and total energy decrease
(b) kinetic energy, potential energy and total energy decrease
(c) kinetic energy decreases, potential energy increases but total energy remains same
(d) kinetic energy and total energy decrease but potential energy increases

11 An electron jumps from the 4th orbit to the 2nd orbit of hydrogen atom. Given the Rydberg's constant $R = 10^7 \text{cm}^{-1}$, the frequency (in hertz) of the emitted radiation will be

(a) $\frac{3}{16} \times 10^5$ (b) $\frac{3}{16} \times 10^{15}$ (c) $\frac{9}{16} \times 10^{15}$ (d) $\frac{3}{4} \times 10^{15}$

12 In figure, the energy levels of the hydrogen atom have been shown along with some transitions marked A, B and C. The transitions A, B and C, respectively represents

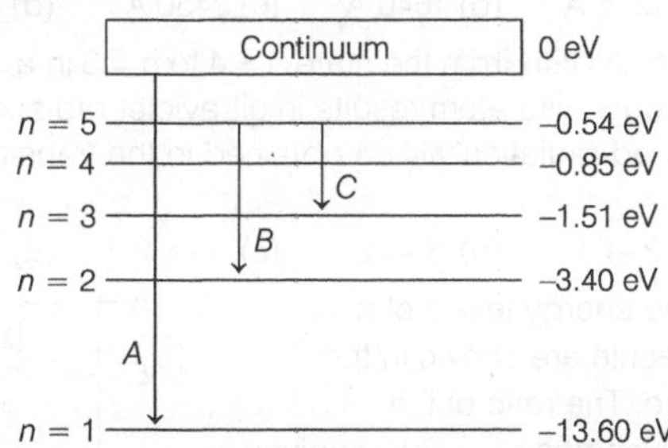

(a) the first member of the Lyman series, third member of Balmer series and second member of Paschen series
(b) the ionisation potential of H, second member of Balmer series and third member of Paschen series
(c) the series limit of Lyman series, second member of Balmer series and second member of Paschen series
(d) the series limit of Lyman series, third member of Balmer series and second member of Paschen series

13 The given diagram indicates the energy levels of a certain atom. When the system moves from $2E$ level to E, a photon of wavelength λ is emitted. The wavelength of photon produced during its transition from $4E/3$ level to E is

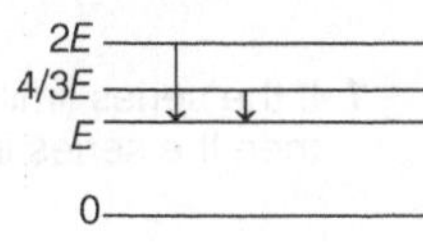

(a) $\lambda/3$ (b) $3\lambda/4$ (c) $4\lambda/3$ (d) 3λ

14 Energy E of a hydrogen atom with principal quantum number n is given by $E = -\frac{13.6}{n^2}$ eV. The energy of a photon ejected when the electron jumps from $n = 3$ state to $n = 2$ state of hydrogen, is approximately

(a) 1.5 eV (b) 0.85 eV (c) 3.4 eV (d) 1.9 eV

15 The ionisation potential of H-atom is 13.6 V. When it is excited from ground state by monochromatic radiations of 970.6 Å, the number of emission lines on deexcitation will be (according to Bohr's theory)

(a) 10 (b) 3 (c) 6 (d) 4

16 In a hydrogen like atom electron makes transition from an energy level with quantum number n to another with quantum number $(n-1)$. If $n >> 1$, the frequency of radiation emitted is proportional to → JEE Main 2013

(a) $\frac{1}{n}$ (b) $\frac{1}{n^2}$ (c) $\frac{1}{n^3/2}$ (d) $\frac{1}{n^3}$

17 Hydrogen atom is excited from ground state to another state with principal quantum number equal to 4. Then, the number of spectral lines in the emission spectra will be → AIEEE 2012

(a) 2 (b) 3 (c) 5 (d) 6

18 The wavelength of the first spectral line in the Balmer series of hydrogen atom is 6561 Å. The wavelength of the second spectral line in Balmer series of single ionised helium atom is → AIEEE 2011

(a) 1215 Å (b) 1640 Å (c) 2450 Å (d) 4687 Å

19 The transition from the state $n = 4$ to $n = 3$ in a hydrogen-like atom results in ultraviolet radiation. Infrared radiation will be obtained in the transition from → AIEEE 2010

(a) $2 \to 1$ (b) $3 \to 2$ (c) $4 \to 2$ (d) $5 \to 3$

20 Some energy levels of a molecule are shown in the figure. The ratio of the wavelengths $r = \lambda_1 / \lambda_2$ is given by → JEE Main 2017 (Offline)

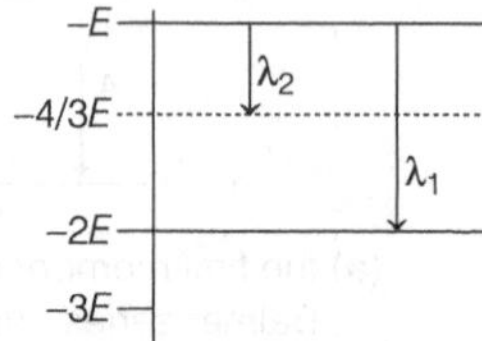

(a) $r = \frac{2}{3}$ (b) $r = \frac{3}{4}$

(c) $r = \frac{1}{3}$ (d) $r = \frac{4}{3}$

21 Energy required for the electron excitation in Li^{2+} from the first to the third Bohr orbit is → AIEEE 2011

(a) 36.3 eV (b) 108.8 eV

(c) 122.4 eV (d) 12.1 eV

22 In hydrogen atom, if the difference in the energy of the electron in $n = 2$ and $n = 3$ orbits is E, the ionisation energy of hydrogen atom is

(a) 13.2 E (b) 7.2 E

(c) 5.6 E (d) 3.2 E

23 Excitation energy of a hydrogen like ion in its first excitation state is 40.8 eV. Energy needed to remove the electron from the ion in ground state is

(a) 54.4 eV (b) 13.6 eV

(c) 40.8 eV (d) 27.2 eV

Direction (Q. Nos. 24-25) *Each of these questions contains two statements : Statement* I *and Statement* II. *Each of these questions also has four alternative choices, only one of which is the correct answer. You have to select one of the codes (a), (b), (c) and (d) given below*

(a) Statement I is true; Statement II is true; Statement II is the correct explanation for Statement I

(b) Statement I is true; Statement II is true; Statement II is not the correct explanation for Statement I

(c) Statement I is true; Statement II is false

(d) Statement I is false; Statement II is true

24 **Statement I** Bohr had to postulate that the electrons in stationary orbits around the nucleus do not radiate.

Statement II According to classical physics all moving electrons radiate electromagnetic radiation.

25 **Statement I** The different lines of emission spectra (like Lyman, Balmer, etc) of atomic hydrogen gas are produced by different atoms.

Statement II The sample of atomic hydrogen gas consists of millions of atoms.

DAY PRACTICE SESSION 2

PROGRESSIVE QUESTIONS EXERCISE

1 If the series limit frequency of the Lyman series is v_L, then the series limit frequency of the Pfund series is → JEE Main 2018

(a) 25 v_L (b) 16 v_L (c) $\frac{v_L}{16}$ (d) $\frac{v_L}{25}$

2 In Rutherford's experiment, the number of alpha particles scattered through an angle of 90° is 28 per minute. Then, the number of particles scattered through an angle of 60° per minute by the same nucleus is

(a) 28 per minute (b) 112 per minute

(c) 12.5 per minute (d) 7 per minute

3 A hydrogen like ion having wavelength difference between first Balmer and Lyman series equal 593 Å has Z equal to

(a) 2 (b) 3 (c) 4 (d) 1

4 The largest wavelength in the ultraviolet region of the hydrogen spectrum is 122 nm. The smallest wavelength in the infrared region of the hydrogen spectrum (to the nearest integer) is

(a) 802 nm (b) 823 nm (c) 1882 nm (d) 1648 nm

5 Energy levels A, B and C of a certain atom corresponding to increasing values of energy i.e. $E_A < E_B < E_C$. If λ_1, λ_2 and λ_3 are the wavelengths of radiations corresponding to the transitions C to B, B to A and C to A respectively, which of the following statements is correct?

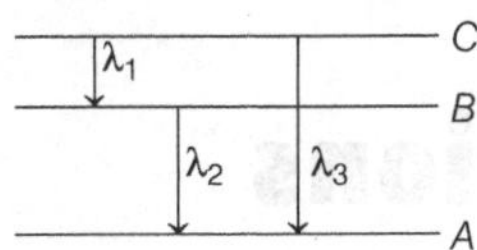

(a) $\lambda_3 = \lambda_1 + \lambda_2$ (b) $\lambda_3 = \dfrac{\lambda_1\lambda_2}{\lambda_1 + \lambda_2}$

(c) $\lambda_1 + \lambda_2 + \lambda_3 = 0$ (d) $\lambda_3^2 = \lambda_1^2 + \lambda_2^2$

6 Hydrogen $({}_1H^1)$, deuterium $({}_1H^2)$, singly ionised helium $({}_2He^4)^+$ and doubly ionised lithium $({}_3Li^6)^{2+}$ all have one electron around the nucleus. Consider an electron transition from $n = 2$ to $n = 1$. If the wavelengths of emitted radiations are $\lambda_1, \lambda_2, \lambda_3$ and λ_4 respectively, then approximately which one of the following is correct? → JEE Main 2014

(a) $4\lambda_1 = 2\lambda_2 = 2\lambda_3 = \lambda_4$ (b) $\lambda_1 = 2\lambda_2 = 2\lambda_3 = \lambda_4$
(c) $\lambda_1 = \lambda_2 = 4\lambda_3 = 9\lambda_4$ (d) $\lambda_1 = 2\lambda_2 = 3\lambda_3 = 4\lambda_4$

7 The potential energy between a proton and an electron is $PE = \dfrac{e^2}{4\pi\varepsilon_0(3R^3)}$, then the radius of the Bohr's orbit is

(a) $\dfrac{4\pi^2e^2m}{4\pi\varepsilon_0 n^2h^2}$ (b) $\dfrac{6\pi^2e^2m}{4\pi\varepsilon_0 n^3h^3}$

(c) $\dfrac{e^2m}{4\pi\varepsilon_0 h^3}$ (d) $\dfrac{2\pi e^2m}{4\pi\varepsilon_0 nh^3}$

8 A small particle of mass m moves such that potential energy $PE = \dfrac{1}{2}mr^2\omega^2$. Assuming Bohr's model of quantisation of angular momentum and circular orbit, radius of nth orbit is proportional to

(a) $\sqrt{n}$ (b) $\sqrt{n^3}$ (c) $\dfrac{1}{\sqrt{n}}$ (d) $\dfrac{1}{\sqrt{n^3}}$

9 The electric potential between a proton and an electron is given by $V = V_0 \ln \dfrac{r}{r_0}$, where r_0 is a constant. Assuming Bohr's model to be applicable, write variation of r_n with n, n being the principal quantum number.

(a) $r_n \propto n$ (b) $r_n \propto \dfrac{1}{n}$

(c) $r_n \propto n^2$ (d) $r_n \propto \dfrac{1}{n^2}$

10 A hydrogen atom moves with a velocity u and makes a head on inelastic collision with another stationary H-atom. Both atoms are in ground state before collision. The minimum value of u if one of them is to be given a minimum excitation energy is

(a) 2.64×10^4 ms^{-1} (b) 6.24×10^4 ms^{-1}
(c) 2.02×10^6 ms^{-1} (d) 6.24×10^8 ms^{-1}

11 In the Bohr's model an electron moves in a circular orbit around the proton. Considering the orbiting electron to be a circular current loop, the magnetic moment of the hydrogen atom, when the electron is in nth excited state, is

(a) $\left(\dfrac{e}{2m}\right)\dfrac{n^2h}{\pi}$ (b) $\left(\dfrac{e}{m}\right)\dfrac{nh}{2\pi}$ (c) $\left(\dfrac{e}{2m}\right)\dfrac{nh}{2\pi}$ (d) $\left(\dfrac{e}{m}\right)\dfrac{n^2h}{2\pi}$

12 A diatomic molecule is made of two masses m_1 and m_2 which are separated by a distance r. If we calculate its rotational energy by applying Bohr's rule of angular momentum quantisation, its energy will be given by (n is an integer) → AIEEE 2012

(a) $\dfrac{(m_1 + m_2)^2n^2h^2}{2m_1^2m_2^2r^2}$ (b) $\dfrac{n^2h^2}{2(m_1 + m_2)r^2}$

(c) $\dfrac{2n^2h^2}{(m_1 + m_2)r^2}$ (d) $\dfrac{(m_1 + m_2)n^2h^2}{2m_1m_2r^2}$

13 In the Bohr's model of hydrogen-like atom the force between the nucleus and the electron is modified as $F = \dfrac{e^2}{4\pi\varepsilon_0}\left(\dfrac{1}{r^2} + \dfrac{\beta}{r^3}\right)$, where β is a constant. For this atom, the radius of the nth orbit in terms of the Bohr's radius $\left(a_0 = \dfrac{\varepsilon_0 h^2}{m\pi e^2}\right)$ is → AIEEE 2010

(a) $r_n = a_0 n - \beta$ (b) $r_n = a_0 n^2 + \beta$
(c) $r_n = a_0 n^2 - \beta$ (d) $r_n = a_0 n + \beta$

Direction (Q. Nos. 14-15) *Each of these questions contains two statements : Statement I and Statement II. Each of these questions also has four alternative choices, only one of which is the correct answer. You have to select one of the codes (a), (b), (c) and (d) given below*

(a) Statement I is true; Statement II is true; Statement II is the correct explanation for Statement I
(b) Statement I is true; Statement II is true; Statement II is not the correct explanation for Statement I
(c) Statement I is true; Statement II is false
(d) Statement I is false; Statement II is true

14 Statement I Balmer series lies in the visible region of electromagnetic spectrum.

Statement II $\dfrac{1}{\lambda} = R\left(\dfrac{1}{2^2} - \dfrac{1}{n^2}\right)$, where $n = 3, 4, 5, \ldots\ldots$

15 Statement I The ionisation potential of hydrogen is found to be 13.6 eV, the ionisation potential of doubly ionised lithium is 122.4 eV.

Statement II Energy in the nth state of hydrogen atom is $E_n = -\dfrac{13.6}{n^2}$.

ANSWERS

SESSION 1	1 (c)	2 (b)	3 (a)	4 (b)	5 (c)	6 (a)	7 (d)	8 (c)	9 (a)	10 (a)
	11 (c)	12 (d)	13 (d)	14 (d)	15 (c)	16 (d)	17 (d)	18 (a)	19 (d)	20 (c)
	21 (b)	22 (b)	23 (a)	24 (b)	25 (b)					
SESSION 2	1 (d)	2 (b)	3 (b)	4 (b)	5 (b)	6 (c)	7 (a)	8 (a)	9 (a)	10 (b)
	11 (c)	12 (d)	13 (c)	14 (a)	15 (b)					

Hints and Explanations

SESSION 1

1 Here, $\frac{1}{2}mv^2 = \frac{1}{4\pi\varepsilon_0}\frac{q_1 q_2}{r}$

$\therefore \quad 5\text{ MeV} = \frac{9\times10^9\times(2e)\times(92e)}{r}$

$\left(\because \frac{1}{2}mv^2 = 5\text{ MeV}\right)$

$\Rightarrow r = \frac{9\times10^9\times2\times92\times(1.6\times10^{-19})^2}{5\times10^6\times1.6\times10^{-19}}$

$r = 53\times10^{-14}$ m

$\approx 10^{-12}$ cm

2 While proposing his theory of hydrogen atom, Bohr considered quantisation of angular momentum as the essential condition for the stationary orbits.

3 According to Bohr's first postulate,

$mvr = \frac{nh}{2\pi}$

$\therefore \quad 2\pi r = n\left(\frac{h}{mv}\right) = n\lambda$

For $n = 1$, $\lambda = 2\pi r$

4 Emission spectrum would rises when electron makes a jump from higher energy level to lower energy level. Frequency of emitted photon is proportional to change in energy of two energy levels, i.e.

$\nu = RcZ^2\left(\frac{1}{n_1^2} - \frac{1}{n_2^2}\right)$

5 According to the Bohr's postulate.

$mvr = \frac{nh}{2\pi} \Rightarrow vr = n\left(\frac{h}{2\pi m}\right)$

$\Rightarrow \quad vr \propto n$

Thus, vr is directly proportional to the principal quantum number.

6 The kinetic energy and total energy of an electron are related as, KE $= -E$

$\therefore \quad \frac{\text{KE}}{E} = -1$

7 $r_m = \left(\frac{m^2}{Z}\right)(0.53\text{ Å}) = (n\times0.53)$ Å

$\therefore \quad \frac{m^2}{Z} = n$

$m = 5$ (for $_{100}\text{fm}^{257}$ the outermost shell)

and $Z = 100$

$\therefore \quad n = \frac{(5)^2}{100} = \frac{1}{4}$

8 On the basis of Bohr's model,

$r = \frac{n^2\lambda^2}{4\pi^2 mKZe^2} = \frac{a_0 n^2}{Z}$

Let Li^{2+} ion, $Z = 3$,

$n = 1$ for ground state.

Given, $a_0 = 53$ pm

$r = \frac{53\times(1)^2}{3}$

$\simeq 18$ pm

9 As, $\quad r \propto \frac{1}{m}$

$\therefore \quad r_0 = \frac{1}{2}a_0$

As, $\quad E \propto m$

$\therefore \quad E_0 = 2(-13.6)$

$= -27.2$ eV

10 As we know that, kinetic energy of an electron is KE$\propto (Z/n)^2$. When the electron makes transition from an excited state to the ground state, then n decreases and KE increases. We know that PE is lowest for ground state.

Also, TE $= -$KE. TE also decreases.

11 $\nu = \frac{c}{\lambda} = c.R\left(\frac{1}{n_1^2} - \frac{1}{n_2^2}\right)$

$= 3\times10^8\times10^7\left(\frac{1}{2^2} - \frac{1}{4^2}\right)$

$= \frac{9}{16}\times10^{15}$ Hz

12 A represents series limit of Lyman series, B represents third member of Balmer series and C represents second member of Paschen series.

13 $\lambda \propto \frac{1}{\Delta E}$

$\therefore \quad \frac{\lambda'}{\lambda} = \frac{(2E-E)}{(4E/3-E)} = \frac{1}{1/3} = 3$

$\therefore \quad \lambda' = 3\lambda$

14 Given, $E_n = -\frac{13.6}{n^2}$ eV

$E_3 = -\frac{13.6}{(3)^2}$ eV $= -\frac{13.6}{9}$ eV

and $E_2 = -\frac{13.6}{(2)^2}$ eV $= -\frac{13.6}{4}$ eV

So, $\Delta E = E_3 - E_2 = -\frac{13.6}{9} - \left(-\frac{13.6}{4}\right)$

$= 1.9$ eV (approximately)

15 Using $E = \frac{hc}{\lambda} = \frac{12400}{970.6}$ eV $= 12.77$ eV

So, electron is excited upto $n = 4$

$\therefore \quad n_2 = 4$

On deexcitation number of emission lines produced $= \frac{n(n-1)}{2} = 6$

16 $\Delta E = h\nu$

$\nu = \frac{\Delta E}{h} = K\left[\frac{1}{(n-1)^2} - \frac{1}{n^2}\right]$

$= \frac{K2n}{n^2(n-1)^2} = \frac{2K}{n^3}$

17 In emission spectrum, number of bright lines is given by

$$\frac{n(n-1)}{2} = \frac{4(4-1)}{2} = 6$$

18 $\frac{1}{6561} = R\left(\frac{1}{4} - \frac{1}{9}\right) = \frac{5R}{36}$

$$\frac{1}{\lambda} = 4R\left(\frac{1}{4} - \frac{1}{16}\right) = \frac{3R \times 4}{16}$$

$$\lambda = 1215 \text{Å}$$

19 Infrared radiation corresponds to least value of $\left(\frac{1}{n_1^2} - \frac{1}{n_2^2}\right)$, i.e. from Paschen, Brackett and Pfund series. Thus, the transition corresponds to $5 \rightarrow 3$.

20 We have, $\lambda = \frac{hc}{\Delta E}$

$\therefore$ So, ratio of wavelengths

$$\frac{\lambda_1}{\lambda_2} = \frac{hc / \Delta E_1}{hc / \Delta E_2} = \frac{\Delta E_2}{\Delta E_1} = \frac{\left(\frac{4}{3}E - E\right)}{2E - E} = \frac{1}{3}$$

21 $\Delta E = 13.6 Z^2 \left(\frac{1}{n_1^2} - \frac{1}{n_2^2}\right)$

$$\left[\because \Delta E = Rhc\left(\frac{1}{n_1^2} - \frac{1}{n_2^2}\right)\right]$$

$$= 13.6(3)^2 \left(\frac{1}{1^2} - \frac{1}{3^2}\right) = 108.8 \text{ eV}$$

22 $E_3 - E_2 = E$

or $\frac{E_1}{9} - \frac{E_1}{4} = E$

or $E_1 = -7.2E$

$\therefore$ Ionisation energy of hydrogen atom is $7.2 E$.

23 Excitation energy,

$$\Delta E = E_2 - E_1 = 13.6 Z^2 \left(\frac{1}{1^2} - \frac{1}{2^2}\right)$$

$$40.8 = 13.6 Z^2 \times \frac{3}{4}$$

$$\therefore \quad Z = 2$$

So, required energy to remove the electron from ground state

$$= + \frac{13.6 Z^2}{(1)^2} = 13.6 (Z)^2 = 54.4 \text{ eV}$$

24 Bohr's postulated that, electron instead of revolving in any orbit around the nucleus, revolves only in some specific orbits. These orbits are called the non-radiating orbits or the stationary orbits. The electrons revolving in these orbits do not radiate any energy. They radiate only when they go from one orbit to the next lower orbit.

25 A single atom can have only one transition at time, we are observing different lines due to large number of transitions taking place simultaneously that occurred in different atoms of the sample.

SESSION 2

1 *Series limit occurs in the transition* $n_2 = \infty$ to $n_1 = 1$ in Lyman series and $n_2 = \infty$ to $n_1 = 5$ in Pfund series. For Lyman series,

$n_2 = \infty$

$h\nu_L = E_g = E_0\left[\frac{1}{1^2} - \frac{1}{\infty}\right] = 13.6 \text{ eV}$

$n_1 = 1$

$$h\nu_L = 13.6 \quad \text{...(i)}$$

In Pfund series

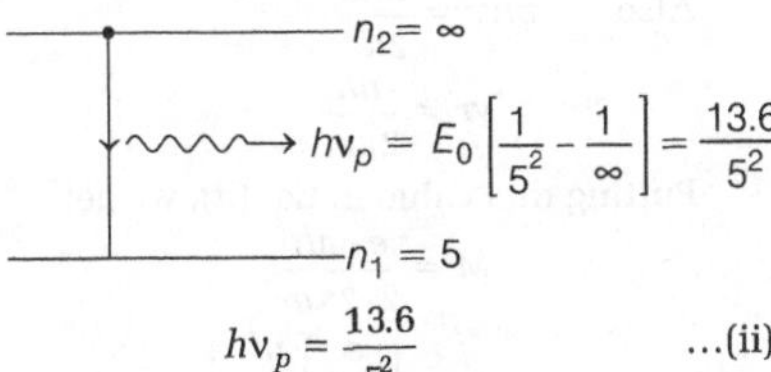

$$h\nu_p = \frac{13.6}{5^2} \quad \text{...(ii)}$$

From Eqs. (i) and (ii), we get

$$25 h\nu_p = h\nu_L$$

$$\therefore \quad \nu_p = \frac{\nu_L}{25}$$

2 According to Rutherford's scattering formula, if the α-particles scattered at an angle θ is directly proportional to $\frac{1}{\sin^4(\theta/2)}$, then $N_\theta = \frac{K}{\sin^4(\theta/2)}$

when $\theta = 90°$,

$$N_\theta = 28 \text{ min}^{-1}$$

$$\Rightarrow \quad 28 = \frac{K}{\sin^4(45°)} = 4K \Rightarrow K = 7$$

Thus $\quad N_\theta = \frac{7}{\sin^4(\theta/2)}$

Hence, the number of α-particles scattered at an angle of 60° per minute is $N'_\theta = \frac{7}{\sin^4 30°} = \frac{7}{(1/2)^4}$

$$= 7 \times 16 = 112 \text{ per minute}$$

3 $\frac{1}{\lambda} = RZ^2 \left(\frac{1}{n_1^2} - \frac{1}{n_2^2}\right)$,

$$\Delta\lambda = \frac{1}{RZ^2\left(\frac{1}{4} - \frac{1}{9}\right)} - \frac{1}{RZ^2\left(\frac{1}{1} - \frac{1}{4}\right)}$$

$$593 = \frac{36}{5RZ^2} - \frac{4}{3RZ^2}$$

$$Z^2 = \frac{88}{15R\,\Delta\lambda}$$

$$= \frac{88}{15(1.097 \times 10^7)(593 \times 10^{-10})}$$

$$= 9$$

$$\Rightarrow Z = 3$$

4 The series in UV-region is Lyman series. Longest wavelength corresponds to minimum energy which occurs in transition from $n = 2$ to $n = 1$.

$$\therefore \quad 122 = \frac{\frac{1}{R}}{\frac{1}{(1)^2} - \frac{1}{(2)^2}} \quad \text{...(i)}$$

The smallest wavelength in the infrared region corresponds to maximum energy of Paschen series.

$$\therefore \quad \lambda = \frac{\frac{1}{R}}{\frac{1}{(3)^2} - \frac{1}{\infty}} \quad \text{...(ii)}$$

Solving Eqs. (i) and (ii), we get

$$\lambda = 823.5 \text{ nm} \approx 823$$

5 Let energy corresponding to state A, B and C be E_A, E_B and E_C.

So, from figure

$$(E_C - E_B) + (E_B - E_A) = (E_C - E_A)$$

or $\quad \frac{hc}{\lambda_1} + \frac{hc}{\lambda_2} = \frac{hc}{\lambda_3}$

$$\Rightarrow \quad \lambda_3 = \frac{\lambda_1 \lambda_2}{\lambda_1 + \lambda_2}$$

C
λ_1
B
λ_2 λ_3
A

6 As we know that,

$$\frac{1}{\lambda} = RZ^2\left(\frac{1}{1^2} - \frac{1}{2^2}\right)$$

$$\lambda = \frac{4}{3RZ^2}$$

$$\lambda_1 = \frac{4}{3R}, \lambda_2 = \frac{4}{3R}$$

$$\lambda_3 = \frac{4}{12R}, \quad \lambda_4 = \frac{4}{27R}$$

$$\Rightarrow \quad \lambda_1 = \lambda_2 = 4\lambda_3 = 9\lambda_4$$

7 $F = \frac{-dPE}{dR} = -\frac{e^2}{4\pi\varepsilon_0 R^4}$

$\Rightarrow \quad \frac{mv^2}{R} = \frac{e^2}{4\pi\varepsilon_0 R^4}.$

Also, $mvR = \frac{nh}{2\pi}$

$\therefore \quad \frac{m}{R}\left(\frac{nh}{2\pi mR}\right)^2 = \frac{e^2}{4\pi\varepsilon_0 R^4}$

$\Rightarrow \quad R = \frac{4\pi^2 e^2 m}{4\pi\varepsilon_0 n^2 h^2}$

8 $F = \frac{-dPE}{dr} = -m\omega^2 r.$

Since, $mvr = \frac{nh}{2\pi}$

or $\quad mr^2\omega = \frac{nh}{2\pi} \quad [\because v = r\omega]$

$\Rightarrow \quad r^2 = \frac{nh}{2\pi m\omega}$

$\Rightarrow \quad r = \sqrt{\frac{nh}{2\pi m\omega}}$

$\Rightarrow \quad r \propto \sqrt{n}$

9 $U = eV = eV_0 \ln\left(\frac{r}{r_0}\right)$

and $|F| = \left|-\frac{dU}{dr}\right| = \frac{eV_0}{r}$

This force will provide the necessary centripetal force. Hence,

$\frac{mv^2}{r} = \frac{eV_0}{r}$

or $\quad v = \sqrt{\frac{eV_0}{m}} \quad \ldots(i)$

Moreover, $mvr = \frac{nh}{2\pi} \quad \ldots(ii)$

On dividing Eq. (ii) by Eq. (i), we have

$mr = \left(\frac{nh}{2\pi}\right)\sqrt{\frac{m}{eV_0}}$

or $\quad r_n \propto n$

10 Momentum, $mu = 2mv$

$\Rightarrow \quad v = \frac{u}{2}$

$\Delta E = \frac{1}{2}mu^2 - \frac{1}{2}(2m)\left(\frac{u}{2}\right)^2$

$= \frac{mu^2}{4}$

$\frac{1}{4}mu^2 = 13.6\left(\frac{1}{1^2} - \frac{1}{2^2}\right)$

$\frac{1}{4}(1.0078)(1.66 \times 10^{-27})u^2$

$= 10.2 \times 1.6 \times 10^{-19}$

$\Rightarrow \quad u = 6.24 \times 10^4 \text{ ms}^{-1}$

11 As, $i = \frac{e}{T}$

and magnetic moment $M = iA$

$(\because A = \pi r^2)$

$\therefore \quad M = \frac{e}{T}\cdot \pi r^2 \quad \ldots(i)$

Now, $\quad T = \frac{2\pi r}{v}$

It becomes,

$M = \frac{e \cdot \pi r^2}{2\pi r / v} = \frac{evr}{2} \quad \ldots(ii)$

Also, $\quad mvr = \frac{nh}{2\pi}$

$vr = \frac{nh}{2\pi m}$

Putting this value in Eq. (ii), we get

$M = \frac{e \cdot nh}{2 \cdot 2\pi m}$

$= \left(\frac{e}{2m}\right)\frac{nh}{2\pi}$

12 Rotational kinetic energy of the two body system rotating about their centre of mass is

$\text{RKE} = \frac{1}{2}\mu\omega^2 r^2,$

$\therefore \quad \text{RKE} = \frac{1}{2}\mu \cdot \left(\frac{nh}{2\pi\mu r^2}\right)^2 r^2$

$\left(\begin{array}{l}\mu = \text{reduced mass} = \frac{m_1 m_2}{m_1 + m_2} \\ L = \frac{nh}{2\pi} = \mu\omega r^2\end{array}\right)$

$= \frac{n^2h^2}{8\pi^2\mu r^2} = \frac{n^2h^2}{2\mu r^2}$

$= \frac{(m_1 + m_2)n^2h^2}{2m_1 m_2 r^2}$

$\left[\text{here, } h^2 = \frac{\lambda}{4\pi}\right]$

13 According to question, the force between nucleus and electron provide necessary centripetal force,

$\frac{e^2}{4\pi\varepsilon_0}\left(\frac{r+\beta}{r^3}\right) = \frac{mv^2}{r} \quad \ldots(i)$

Also, $mvr = \frac{h}{2\pi} \quad \ldots(ii)$

From Eqs. (i) and (ii), we get

$\frac{e^2}{4\pi\varepsilon_0}\left(\frac{r+\beta}{r^3}\right) = \frac{mn^2h^2}{4\pi^2 m^2 r^3}$

$\Rightarrow \quad r + \beta = \frac{\varepsilon_0 n^2 h^2}{\pi m e^2}$

$\Rightarrow \quad r_n = a_0 n^2 - \beta$

$\left(\because a_0 = \frac{\varepsilon_0 h^2}{m\pi e^2}\right)$

14 The wavelength in Balmer series is given by

$\frac{1}{\lambda} = R\left(\frac{1}{2^2} - \frac{1}{n^2}\right), n = 3, 4, 5\ldots$

$\frac{1}{\lambda_{max}} = R\left(\frac{1}{2^2} - \frac{1}{3^2}\right) \Rightarrow \lambda_{max} = \frac{36}{5R}$

$\lambda_{max} = \frac{36 \times 1}{5 \times 1.097 \times 10^7} = 6563\text{Å}$

and $\frac{1}{\lambda_{min}} = R\left(\frac{1}{2^2} - \frac{1}{\infty^2}\right)$

$\lambda_{min} = \frac{4}{R} = \frac{4}{1.097 \times 10^7}$

$= 3646\text{ Å}$

The wavelengths 6563 Å and 3646 Å lie in visible region. Therefore, Balmer series lies in visible region.

15 From Bohr's theory, the energy of hydrogen atom in the nth state is given by $E_n = -\frac{13.6}{n^2}$ eV. For an atom of atomic number Z, with one electron in the outer orbit (singly ionised He or doubly lithium) we use. $E_n = -\frac{13.6Z^2}{n^2}$eV, where Z is the atomic number. Hence, ground state energy of doubly ionised lithium is $\frac{-13.6 \times 9}{(1)^2} = -122.4\text{eV}$

Ionisation potential (potential to be applied to electron to overcome this energy) is 122.4 V.

DAY THIRTY THREE

Nuclei

Learning & Revision for the Day

- Concept of Nucleus
- Radioactivity
- Mass Energy Relation
- Mass Defect and Binding Energy
- Nuclear Fission
- Nuclear Fusion

Concept of Nucleus

In every atom, the positive charge and mass is densely concentrated at the centre of the atom forming its **nucleus**. In nucleus, the number of protons is equal to the atomic number of that element and the remaining particles to fulfil the mass number are the neutrons.

Composition of Nucleus

Nucleus consists of protons and neutrons. Electrons cannot exist inside the nucleus. A proton is a positively charged particle having mass (m_p) of 1.007276 u and charge $(+e) = +1.602 \times 10^{-19}$ C.

For a neutral atom, **Number of proton (Z) = Number of electron**

This number is called the **atomic number**. A neutron is a neutral particle having mass $m_n = 1.008665$ u. The number of neutrons in the nucleus of an atom is called the **neutron number** N. The sum of the number of protons and neutrons is called the **mass number** A. Thus, $A = N + Z$.

Properties of Nucleus

Nuclear size

(a) Size of the nucleus is of the order of fermi (1 fermi $= 10^{-15}$ m).

(b) The radius of the nucleus is given by $R = R_0 A^{1/3}$,

where, $R_0 = 1.3$ fermi and A is the mass number.

Volume

The volume of nucleus is $V = \frac{4}{3}\pi (R_0 A^{1/3})^3$, where, R_0 = radius of the nucleus.

Density

(a) $$\text{Density} = \frac{\text{Mass of nucleus}}{\text{Volume of the nucleus}} = \frac{Am_p}{\frac{4}{3}\pi (R_0 A^{1/3})^3} = \frac{m_p}{\frac{4}{3}\pi R_0^3}$$

where, $m_p = 1.6 \times 10^{-27}$ kg = mass of proton and $R_0 = 1.3$ fermi.

(b) Density of nuclear matter is of the order of 10^{17}kg/m^3.

(c) Density of nuclear matter is independent of the mass number.

- No. of Questions in Exercises (x)—
- No. of Questions Attempted (y)—
- No. of Correct Questions (z)—
  (Without referring Explanations)

- Accuracy Level (z/y × 100)—
- Prep Level (z/x × 100)—

In order to expect good rank in JEE, your Accuracy Level should be above 85 & Prep Level should be above 75.

Isotopes, Isobars and Isotones

Isotopes

Isotopes of an element are nuclides having same atomic number Z, but different mass number A (or different neutron number N) is called isotopes. ${}_1^1H$, ${}_1^2H$, ${}_1^3H$ and ${}_6^{11}C$, ${}_6^{12}C$, ${}_6^{14}C$, etc., are isotopes.

Isobars

Nuclides having same mass number A, but different atomic number Z are called isobars. In isobars number of protons Z as well as number of neutrons N differ but total nucleon (or mass) number $A = N + Z$ is the same. ${}_1^3H$, ${}_2^3He$ and ${}_6^{14}C$, ${}_7^{14}N$ are isobars.

Isotones

Nuclides with different atomic number Z and different mass number A, but same neutron number are called isotones. ${}_1^3H$, ${}_2^4He$ and ${}_{80}^{198}Hg$, ${}_{79}^{197}Au$ are examples of isotones.

Radioactivity

Radioactivity is the phenomenon of spontaneous emission of radiations by heavier nucleus. Some naturally occurring radioactive substances are uranium, thorium, polonium, radium, neptunium, etc. In fact, all elements having atomic number $Z > 82$ are radioactive in nature.

Radiations emitted by radioactive substances are of three types, namely (i) α-particles, (ii) β-particles and (iii) γ- rays.

- **α-particles** are positively charged particles with charge $q_\alpha = +2e$ and mass $m_\alpha = 4m_p$. Thus, α-particles may be considered as helium nuclei (or doubly charged helium ions). Ionising power of α-particles is maximum, but their penetrating power is minimum.
- **β-particles** are negatively charged particles with rest mass as well as charge same as that of electrons. But origin of β-particles is from the nucleus. Their ionising power is lesser than that of α-particles, but speed as well as penetrating power is much greater than that of α-particles. Generally, β-decay means β^--decay.
- **γ-rays** are electromagnetic radiations of extremely short wavelengths. Thus, γ-rays travel with the speed of light. Their ionising power is least, but penetrating power is extremely high. These are not deflected either in an electric or a magnetic field.

Law of Radioactive Decay

According to Rutherford-Soddy's law for radioactive decay, 'The rate of decay of a radioactive material at any instant is proportional to the quantity of that material actually present at that time.'

Mathematically, $\frac{dN}{dt} \propto N$ or $\frac{dN}{dt} = -\lambda N$

Here, λ is a proportionality constant, known as the **decay constant** (or disintegration constant). Unit of λ is s^{-1} or day^{-1} or $year^{-1}$, etc.

It can be shown that number of nuclei present after time t is given by

$$N = N_0 e^{-\lambda t}$$

where, N_0 = number of nuclei present at time $t = 0$.

Again, number of nuclei decayed in time t will be

$$N - N_0 = N_0 [e^{-\lambda t} - 1]$$

= number of **daughter nuclei** produced at time t.

Half-Life Period ($T_{1/2}$)

It is the time in which, activity of the sample falls to one-half of its initial value.

Thus, for $t = \frac{T}{2}$, $N = \frac{N_0}{2}$ and $R = \frac{R_0}{2}$

- The half-life period is related to decay constant λ as
$$T_{1/2} = \frac{0.693}{\lambda}$$
- After n half-lives, the quantity of a radioactive substance left intact (undecayed) is given by
$$N = N_0 \left(\frac{1}{2}\right)^n = N_0 \left(\frac{1}{2}\right)^{\frac{t}{T_{1/2}}}$$

Mean Life Period (τ)

- Mean life of a radioactive sample is the time, at which both N and R have been reduced to $\frac{1}{e}$ or e^{-1} or 36.8% of their initial values. It is found that $\tau = \frac{1}{\lambda}$.
- Half-life $T_{1/2}$ and mean life τ of a radioactive sample are correlated as, $T_{1/2} = 0.693\,\tau$ or $\tau = 1.44 T_{1/2}$.

Activity

The activity of a radioactive substance is defined as the rate of disintegration (or the count rate) of that substance.

Mathematically, activity is defined as

$$R = -\frac{dN}{dt} = \lambda N = \lambda N_0 e^{-\lambda t} = R_0\, e^{-\lambda t}$$

where, $R_0 = \lambda N_0$ = initial value of activity.

Units of activity are

- 1 becquerel = Bq = 1 disintegration per second (SI unit)
- 1 curie = 1 Ci = 3.7×10^{10} Bq
- 1 rutherford = 1 Rd = 10^6 Bq

Mass Energy Relation

In nuclear physics, mass is measured in **unified atomic mass units** (u), 1 u being one-twelfth of the mass of carbon-12 atom and equals 1.66×10^{-27} kg. It can readily be shown using $E = mc^2$ that, 1 u mass has energy 931.5 MeV

$$\text{Thus, } 1\text{ u} \equiv 931.5\text{ MeV} \quad \text{or} \quad c^2 = 931.5\text{ MeV/u}$$

A unit of energy may therefore be considered to be a unit of mass. For example, the electron has a rest mass of about 0.5 MeV.

If the principle of conservation of energy is to hold for nuclear reactions it is clear that mass and energy must be regarded as equivalent. The implication of $E = mc^2$ is that any reaction producing an appreciable mass decrease is a possible source of energy.

- At the rest, mass energy of each of electron and positron, is
$$E_0 = m_0 c^2 = 9.1 \times 10^{-31} \times (3 \times 10^8)^2 \text{ J} = 0.51 \text{ MeV}$$
Therefore, an energy of atleast 1.02 MeV is needed for pair production.

Mass Defect and Binding Energy

- The difference in mass of a nucleus and its constituent nucleons is called the mass defect of that nucleus. Thus,
Mass defect, $\Delta M = Zm_p + (A - Z)m_n - M$
where, M is the mass of a given nucleus.
- **Packing fraction** of an atom is the difference between mass of nucleus and its mass number per nucleon. Thus,
Packing fraction $= \dfrac{M - A}{A}$.
- The energy equivalent of the mass defect of a nucleus is called its **binding energy.**
Thus, binding energy, $\Delta E_b = \Delta M c^2$
$$= [Zm_p + (A - Z)m_n - M]c^2$$
If masses are expressed in atomic mass units, then
$$\Delta E_b = \Delta M \times 931.5 \text{ MeV}$$
$$= [Zm_p + (A - Z)m_n - M] \times 931.5 \text{ MeV}$$
- **Binding energy per nucleon** (ΔE_{bn}) is the average energy needed to separate a nucleus into its individual nucleons.
Thus, $\Delta E_{bn} = \dfrac{\Delta E_b}{A} = \dfrac{\Delta M \times 931}{A} \dfrac{\text{MeV}}{\text{Nucleon}}$

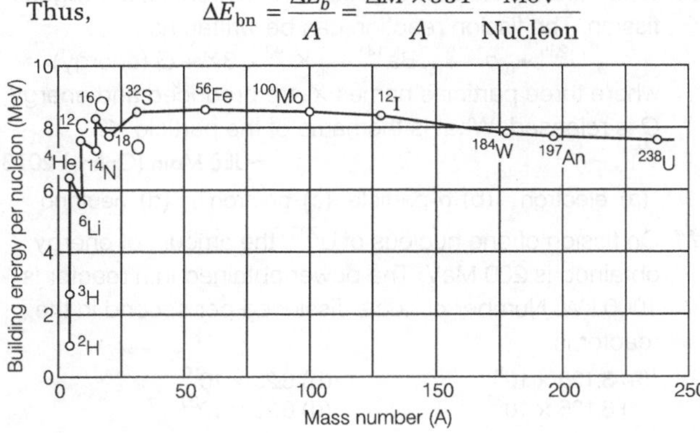

- The shown figure show binding energy per nucleon *versus* mass number. The nuclides showing binding energy per nucleon greater than 7.5 MeV/nucleon are stable.

NOTE
- Nucleons attract each other when they are separated by a distance of 10^{-14} m.
- The density of nucleus is of the order of 10^{17}.

Nuclear Fission

Nuclear fission is the process of splitting of a heavy nucleus ($^{235}_{92}$U or $^{239}_{94}$Pu) into two lighter nuclei of comparable masses along with the release of a large amount of energy ($\simeq$ 200 MeV) after bombardment by slow neutrons.

A characteristic nuclear fission reaction equation for $^{235}_{92}$U is

$$^1_0n(\text{slow}) + {}^{235}_{92}\text{U} \longrightarrow {}^{236}_{92}\text{U} \longrightarrow {}^{144}_{56}\text{Ba} + {}^{89}_{36}\text{Kr} + 3\,{}^1_0n + Q$$

In the fission of uranium, the percentage of mass converted into energy is about 0.1%.

Controlled Chain Reaction and Nuclear Reactor

- In the fission of one nucleus of $^{235}_{92}$U, on an average, $2\frac{1}{2}$ neutrons are released. These released neutrons may further, trigger more fissions causing more neutrons being formed, which in turn may cause more fission. Thus, a self sustained nuclear chain reaction is formed. To maintain the nuclear chain reaction at a steady (sustained) level, the extra neutrons produced, are absorbed by suitable neutron absorbents like cadmium or boron.
- Neutrons formed as a result of fission have an energy of about 2 MeV, whereas for causing further fission, we need slow thermal neutrons having an energy of about 0.3 eV. For this purpose, suitable material called a **moderator** is used, which slow down the neutrons. Water, heavy water and graphite are commonly used as moderators.
- A **nuclear reactor** is an arrangement in which nuclear fission can be carried out through a sustained and a controlled chain reaction and can be employed for producing electrical power, for producing different isotopes and for various other uses.
- Power of a reactor, $P = \dfrac{nE}{t}$, where n = number of atoms undergone fission in time t seconds and E = energy released in each fission.

Reproduction Factor

Reproduction factor (k) of a nuclear chain reaction is defined as

$$k = \frac{\text{Rate of production of neutrons}}{\text{Rate of loss / Absorption of neutrons}}$$

- If $k = 1$, then the chain reaction will be steady and the reactor is said to be **critical**.
- If $k > 1$, then the chain reaction is accelerated and it may cause explosion in the reactor. Such a reactor is called **super-critical**.
- If $k < 1$, then chain reaction gradually slows down and comes to a halt. Such a reactor is called **sub-critical**.

The reactors giving fresh nuclear fuel which often exceeds the nuclear fuel used is known as **breeder reactor**.

Nuclear Fusion

Nuclear fusion is the process, in which two or more light nuclei combine to form a single large nucleus.

The mass of the single nucleus, so formed is less than the sum of the masses of parent nuclei and this difference in mass, results in the release of tremendously large amount of energy. The fusion reaction going on in the central core of sun is a multistep process, but the net reaction is

$$4\,{}_1^1\text{H} + 2\varepsilon^- \longrightarrow {}_2^4\text{He} + 2\nu + 6\gamma + 26.7 \text{ MeV}$$

When two positively charged particles (protons or deuterons) combine to form a larger nucleus, the process is hindered by the Coulombian repulsion between them.

To overcome the Coulombian repulsion, the charged particles are to be given an energy of atleast 400 keV.

For this, proton/deuterons must be heated to a temperature of about 3×10^9 K. Nuclear fusion reaction is therefore, known as **thermo nuclear fusion** reaction.

DAY PRACTICE SESSION 1

FOUNDATION QUESTIONS EXERCISE

1 Two nucleons are at a separation of 1 fm. The net force between them is F_1 if both are neutrons, F_2 if both are protons and F_3 if one is a proton and the other is a neutron.

(a) $F_1 > F_2 > F_3$ (b) $F_2 > F_1 > F_3$
(c) $F_1 = F_3 > F_2$ (d) $F_1 = F_2 > F_3$

2 A radioactive nucleus (initial mass number A and atomic number Z) emits 3α-particles and 2 positrons. The ratio of number of neutrons to that of protons in the final nucleus will be → AIEEE 2010

(a) $\dfrac{A-Z-8}{Z-4}$ (b) $\dfrac{A-Z-4}{Z-8}$
(c) $\dfrac{A-Z-12}{Z-4}$ (d) $\dfrac{A-Z-4}{Z-2}$

3 The sequence of decay of a radioactive nucleus is $N_0 \xrightarrow{\alpha} N_1 \xrightarrow{\beta} N_2 \xrightarrow{\alpha} N_3 \xrightarrow{\alpha} N_4$. If nucleon number and atomic number of N_2 are 176 and 71 respectively, then what are their values for N_4 and N_0?

(a) 168, 67 and 180, 71 (b) 67, 168 and 180, 72
(c) 180, 67 and 72, 180 (d) None of these

4 A radioactive nucleus undergoes a series of decays according to the scheme

$$A \xrightarrow{\alpha} A_1 \xrightarrow{\beta} A_2 \xrightarrow{\alpha} A_3 \xrightarrow{\gamma} A_4$$

If the mass number and atomic number of A are 180 and 72 respectively, these numbers of A_4 are

(a) 172, 69 (b) 177, 69 (c) 171, 69 (d) 172, 68

5 If $N_{t_1} = N_0 e^{-\lambda t_1}$, then the number of atoms decayed during the time interval from t_1 and t_2 ($t_1 > t_2$), will be

(a) $N_{t_1} - N_{t_2} = N_0 [e^{-\lambda t_1} - e^{-\lambda t_2}]$
(b) $N_{t_2} - N_{t_1} = N_0 [e^{-\lambda t_2} - e^{-\lambda t_1}]$
(c) $N_{t_2} - N_{t_1} = N_0 [e^{-\lambda t_2} - e^{\lambda t_1}]$
(d) None of the above

6 A radioactive sample decays by two different processes. Half-life for the first process is t_1 and for the second process is t_2. The effective half-life is

(a) $t_1 + t_2$ (b) $t_1 - t_2$ (c) $(t_1 + t_2)/2$ (d) $\dfrac{t_1 t_2}{t_1 + t_2}$

7 Half-life of a radioactive substance A is 4 days. The probability that a nucleus will decay in two half-lives is

(a) $\dfrac{1}{4}$ (b) $\dfrac{3}{4}$ (c) $\dfrac{1}{2}$ (d) 1

8 The half-life of a radioactive substance is 20 min. The approximate time interval $(t_2 - t_1)$ between the time t_2 when $\dfrac{2}{3}$ of it has decayed and time t_1 when $\dfrac{1}{3}$ of it had decayed is → AIEEE 2011

(a) 14 min (b) 20 min (c) 28 min (d) 7 min

9 A sample of a radioactive element has a mass of 10 g at an instant $t = 0$. The approximate mass of this element in the sample after two mean lives is

(a) 1.35 g (b) 2.50 g (c) 3.70 g (d) 6.30 g

10 When uranium is bombarded with neutrons, it undergoes fission. The fission reaction can be written as

$${}_{92}\text{U}^{235} + {}_0n^1 \rightarrow {}_{56}\text{Ba}^{141} + {}_{36}\text{Kr}^{92} + 3X + Q \text{ (energy)}$$

where three particles names X are produced and energy Q is released. What is the name of the particle X? → JEE Main (Online) 2013

(a) electron (b) α-particle (c) neutron (d) neutrino

11 On fission of one nucleus of U^{235}, the amount of energy obtained is 200 MeV. The power obtained in a reactor is 1000 kW. Number of nuclei fissioned per second in the reactor is

(a) 3.125×10^{16} (b) 6.25×10^{10}
(c) 3.125×10^{32} (d) 6.25×10^{20}

12 If M_o is the mass of an oxygen isotope $_8O^{17}$, M_p and M_n are the masses of a proton and a neutron, respectively the nuclear binding energy of the isotope is
(a) $(M_o - 8M_p)c^2$ (b) $(M_o - 8M_p - 9M_n)c^2$
(c) M_oc^2 (d) $(M_o - 17M_n)c^2$

13 The binding energies per nucleon of Li^7 and He^4 are 5.6 MeV and 7.06 MeV respectively, then the energy of the reaction $Li^7 + p = 2\,[_2He^4]$ will be
(a) 17.28 MeV (b) 39.2 MeV (c) 28.24 MeV (d) 1.46 MeV

14 The below is a plot of binding energy per nucleon E_b, against the nuclear mass M; A, B, C, D, E, F correspond to different nuclei.

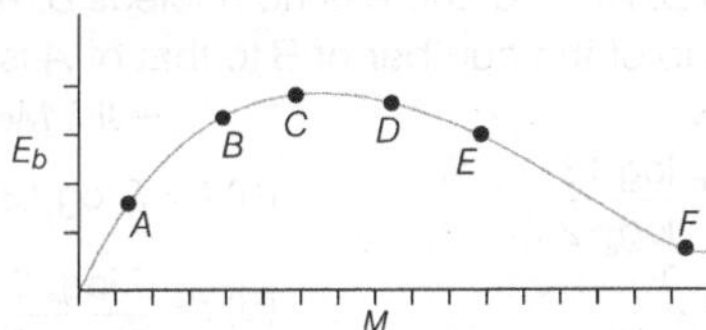

Consider four reactions →[AIEEE 2010]
(i) $A + B \rightarrow C + \varepsilon$ (ii) $C \rightarrow A + B + \varepsilon$
(iii) $D + E \rightarrow F + \varepsilon$ and (iv) $F \rightarrow D + E + \varepsilon$
where ε is the energy released. In which reactions is ε positive?
(a) (i) and (iv) (b) (i) and (iii)
(c) (ii) and (iv) (d) (ii) and (iii)

Direction (Q. Nos. 15-21) *Each of these questions contains two statements : Statement I and Statement II. Each of these questions also has four alternative choices, only one of which is the correct answer. You have to select one of the codes (a), (b), (c) and (d) given below*
(a) Statement I is true, Statement II is true; Statement II is the correct explanation for Statement I
(b) Statement I is true, Statement II is true; Statement II is not the correct explanation for Statement I
(c) Statement I is true; Statement II is false
(d) Statement I is false; Statement II is true

15 Statement I A certain radioactive substance has a half-life period of 30 days. Its disintegration constant is 0.0231 day^{-1}.
Statement II Decay constant varies inversely as half-life.

16 Statement I Half-life of a certain radioactive element is 100 days. After 200 days, fraction left undecayed will be 50%.
Statement II $\frac{N}{N_0} = \left(\frac{1}{2}\right)^n$, where symbols have usual meaning.

17 Statement I In a decay, daughter nucleus shifts two places to the left from the parent nucleus.
Statement II An alpha particle carries four units of mass.

18 Statement I Energy is released in nuclear fission.
Statement II Total binding energy of the fission fragments is larger than the total binding energy of the parent nucleus.

19 Statement I If half-life period and the mean-life of a radioactive element are denoted by T and T_m respectively, then $T < T_m$.
Statement II Mean-life $= \frac{1}{\text{decay constant}}$

20 Statement I Energy is released when heavy nuclei undergo fission or light nuclei undergo fusion.
Statement II For heavy nuclei, binding energy for per nucleon increases with increasing Z while for light nuclei. It decreases with increasing Z.

21 Statement I A nucleus having energy E_1 decays by β^- emission to daughter nucleus having energy E_2, but β^- rays are emitted with a continuous energy spectrum having end point energy $E_1 - E_2$.
Statement II To conserve energy and momentum in β-decay, atleast three particles must take part in the transformation. →AIEEE 2011

DAY PRACTICE SESSION 2

PROGRESSIVE QUESTIONS EXERCISE

1 Consider $x \xrightarrow{-\alpha} y \xrightarrow{-\alpha} z$, where half-lives of x and y are z year and one month. The ratio of atoms of x and y when transient equilibrium $[T_{1/2}(x) > T_{1/2}(y)]$ has been established is
(a) 1 : 22 (b) 1 : 26 (c) 26 : 1 (d) 23 : 1

2 In a nuclear reactor, U^{235} undergoes fission liberating 200 MeV of energy per fission. The reactor has 10% efficiency and produces 1000 MW power. If the reactor is to function for 10 yr, the total mass of uranium required is (Avogadro's number $= 6.02 \times 10^{26}$/K-mol, 1 eV $= 1.6 \times 10^{-19}$ J)
(a) 3.84×10^4 kg (b) 9.28×10^6 kg
(c) 3.84×10^8 kg (d) 9.28×10^4 kg

3 The half-life of a radioactive sample is 10 h. The total number of disintegration in 10th hour measured from a time when the activity was one Ci is
(a) 0.53×10^{-3} (b) 6.91×10^{13}
(c) 2.63×10^{-3} (d) 9.91×10^{13}

4 A piece of wood from the ruins of an ancient building was found to have a ^{14}C activity of 12 disintegrations per minute per gram of its carbon content. The ^{14}C activity of the living wood is 16 disintegrations per minute per gram. How long ago did the tree, from which the wooden sample came, die? Given, half-life of ^{14}C is 5760 yr?
(a) 2391 yr (b) 2300 yr (c) 2250 yr (d) 2261 yr

5 A radioactive sample S_1 having an activity of 5 μCi has twice the number of nuclei as another sample S_2 which has an activity of 10μCi. The half-lives of S_1 and S_2 can be
(a) 20 yr and 5 yr, respectively
(b) 20 yr and 10 yr, respectively
(c) 10 yr each
(d) 5 yr each

6 The half-life period of a radioactive element *X* is same as the mean life time of another radioactive element *Y*. Initially, they have the same number of atoms. Then,
(a) *X* will decay faster than *Y*
(b) *Y* will decay faster than *X*
(c) *Y* and *X* have same decay rate initially
(d) *X* and *Y* decay at same rate always

7 Deuteron is a bound state of a neutron and a proton with a binding energy $B = 2.2$ MeV. A γ-ray of energy *E* is aimed at a deuteron nucleus to try to break it into a (neutron + proton) such that the *n* and *p* move in the direction of the incident γ-ray. Where $E \neq B$. Then, calculate how much bigger that *B* must *E* be for such a process to happen?
(a) $\frac{B^2}{2mc^2}$ (b) $\frac{B}{2mc^2}$ (c) $\frac{B^2}{4mc^2}$ (d) $\frac{3B}{4mc^2}$

8 Assume that a neutron breaks into a proton and an electron. The energy released during this process is (mass of neutron = 1.6725×10^{-27} kg, mass of proton = 1.6725×10^{-27} kg, mass of electron = 9×10^{-31} kg)
→ AIEEE 2012
(a) 0.9 MeV (b) 7.10 MeV (c) 6.30 MeV (d) 5.4 MeV

9 A radioactive nucleus *A* with a half-life *T*, decays into a nucleus *B*. At $t = 0$, there is no nucleus *B*. After sometime *t*, the ratio of the number of *B* to that of *A* is 0.3. Then, *t* is given by
→ JEE Main 2017 (Offline)
(a) $t = T\frac{\log 1.3}{\log_e 2}$ (b) $t = T \log 1.3$
(c) $t = \frac{T}{\log 1.3}$ (d) $t = \frac{T \log_e 2}{2 \log 1.3}$

10 Half-lives of two radioactive elements *A* and *B* are 20 min and 40 min, respectively. Initially, the samples have equal number of nuclei. After 80 min, the ratio of decayed numbers of *A* and *B* nuclei will be
→ JEE Main 2016 (Offline)
(a) 1 : 16 (b) 4 : 1 (c) 1 : 4 (d) 5 : 4

ANSWERS

SESSION 1	1 (c)	2 (b)	3 (a)	4 (a)	5 (a)	6 (d)	7 (b)	8 (b)	9 (a)	10 (c)
	11 (a)	12 (b)	13 (a)	14 (a)	15 (a)	16 (c)	17 (b)	18 (a)	19 (b)	20 (c)
	21 (a)									
SESSION 2	1 (d)	2 (a)	3 (b)	4 (a)	5 (a)	6 (b)	7 (c)	8 (a)	9 (a)	10 (d)

Hints and Explanations

SESSION 1

1 Nuclear force of attraction between any two nucleons (n-n, p-p, p-n) is same. The difference comes up only due to electrostatic force of repulsion between two protons.

$\therefore \quad F_1 = F_3 \neq F_2$

As, $\quad F_2 < F_3$ or F_1

$\therefore \quad F_1 = F_3 > F_2$

2 In positive β-decay a proton is transformed into a neutron and a positron is emitted.

$$p^+ \longrightarrow n^0 + e^+$$

Number of neutrons initially was $A - Z$

Number of neutrons after decay is $(A - Z) - 3 \times 2$ (due to α-particles) $- 2 \times 1$ (due to positive β-decay)

As $[3 \times 2$ (due to α-particles) $+ 2$ (due to positive β-decay)]

Hence, atomic number reduces by 8.

So, the ratio of number of neutrons to that of protons $= \frac{A - Z - 4}{Z - 8}$

3 As mass number of each α-particle is 4 units and its charge is 2 units, therefore for N_4

$$A = 176 - 8 = 168$$

and $\quad Z = 71 - 4 = 67$

Now, the charge of β is -1 and its mass number is zero.

So, $\quad A = 176 + 0 + 4 = 180$

and $\quad Z = 71 - 2 + 2 = 71$

4 As the mass number of each α-particle is 4 units and its charge is 2 unit. Therefore for A_4,
Mass number = 180 − 8 = 172
and $Z = 72 - 4 + 1$ (due to β^-) = 69

5 Since, $N_{t_1} = N_0 e^{-\lambda t_1}$ and $N_{t_2} = N_0 e^{-\lambda t_2}$
Then, the number of atoms decayed during the time interval t_1 to t_2 is
$= N_{t_1} - N_{t_2} = N_0[e^{-\lambda t_1} - e^{-\lambda t_2}]$

6 As, $\lambda = \lambda_1 + \lambda_2$

$$\Rightarrow \frac{1}{t} = \frac{1}{t_1} + \frac{1}{t_2} = \frac{t_2 + t_1}{t_1 t_2}$$

or $$t = \frac{t_1 t_2}{t_1 + t_2}$$

7 After two half-lives 1/4th fraction of nuclei will remain undecayed. Or, 3/4th fraction will decay. Hence, the probability that a nucleus decays in two half-lives is 3/4.

8 $N_1 = N_0 - \frac{1}{3}N_0 = \frac{2}{3}N_0$,

$N_2 = N_0 - \frac{2}{3}N_0 = \frac{1}{3}N_0$

We have, $\frac{N_2}{N_1} = \left(\frac{1}{2}\right)^n$

Here, $n = 1$
$\therefore$ $t_2 - t_1$ = one half-life = 20 min

9 The relation of mean-life and decay constant is,

$$t = 2\tau = \frac{2}{\lambda}, \text{ where } \tau = \frac{1}{\lambda}$$

Then we get from the equation,

$$m = m_0 e^{-\lambda t}$$

$$\Rightarrow \quad m = 10 \times e^{-\lambda \times 2/\lambda} = 10 \times e^{-2}$$

$$= 10 \times 0.135 = 1.35 \text{ g}$$

10 The fission of $_{92}U^{235}$ is represented by

$$_{92}U^{235} + {_0n^1} \rightarrow {_{56}Ba^{141}} + {_{36}Kr^{92}} + 3\,{_0n^1} + Q$$

The name of the particle X is neutron ($_0n^1$).

11 Power received from the reactor is

$$P = 1000 \text{ kW}$$
$$= 1000 \times 1000 = 10^6 \text{ J s}^{-1}$$

Also, 1 MeV = 1.6×10^{-13} J
Number of nuclei fissioned per second

$$= \frac{10^6}{200 \times 1.6 \times 10^{-13}}$$
$$= 3.125 \times 10^{16} \text{ s}^{-1}$$

12 Binding energy
$BE = (M_{\text{nucleus}} - M_{\text{nucleons}})c^2$
$= (M_o - 8M_p - 9M_n)c^2$

13 The reaction is $_3Li^7 + {_1p^1} \longrightarrow 2\,(_2He^4)$

$\therefore$ $E_p = 2E\,(_2He^4) - E\,(\text{Li})$
$= 2\,(4 \times 7.06) - 7 \times 5.6$
$= 56.48 - 39.2 = 17.28$ MeV

14 Both fusion and fission reaction results into tremendous amount of energy release and nucleus/nuclei which has higher binding energy per nucleon than parent nuclei. So, option (a) is correct.

15 Half-life and decay constant for a nuclear reaction are related by a relation, which is

$$T_{1/2} = \frac{0.693}{\lambda}$$

$$\Rightarrow \lambda = \frac{0.693}{T_{1/2}} = \frac{0.693}{30} = 0.0231 \text{ day}^{-1}$$

16 Number of half-lives

$$n = \frac{t}{T} = \frac{200}{100} = 2$$

The fraction left undecayed is given by

$$\therefore \quad \frac{N}{N_0} = \left(\frac{1}{2}\right)^n = \left(\frac{1}{2}\right)^2$$

$$= \frac{1}{4} = 25\%$$

17 On α decay, charge number of parent nucleus decreases by 2 units. As classification or grouping of elements is based on charge number, hence daughter nucleus shifts two places to the left from the parent nucleus.

18 According to concept of binding energy, fission can occur because the total mass energy will decrease; that is ΔE_{bn} (binding energy) will increase. We see that for high mass nuclide ($A = 240$), the binding energy per nucleon is about 7.6 MeV/nucleon. For the middle weight nuclides ($A = 120$), it is about 8.5 MeV/nucleon. Thus, binding energy of fission fragments is larger than the total binding energy of the parent nucleus.

19 We know that half-life period T and decay constant λ are related by the equation.

$$T = \frac{0.6931}{\lambda} \quad \ldots\text{(i)}$$

While mean-life T_m is related with λ by the equation

$$T_m = \frac{1}{\lambda} \quad \ldots\text{(ii)}$$

From Eqs. (i) and (ii), we get

$$T = 0.6931\,T_m$$

or $$T < T_m$$

20 Here, Statement I is correct and Statement II is wrong, which can be directly concluded from binding energy nucleon curve.

21 In practicle situation, atleast three particles take place in transformation, so Energy of β-particle + Energy of third particle = $E_1 - E_2$
Hence, energy of β-particle ≤ $E_1 - E_2$

SESSION 2

1 $$N_2 = \frac{\lambda_1 N_1}{\lambda_2 - \lambda_1}(e^{-\lambda_1 t} - e^{-\lambda_2 t})$$

When $(T_{1/2})_1 > (T_{1/2})_2$ at transient equilibrium, $\lambda_1 < \lambda_2$

$$e^{-\lambda_2 t} << e^{-\lambda_1 t}$$

$$\therefore \quad N_2 = \frac{\lambda_1 N_1 e^{-\lambda_1 t}}{\lambda_2 - \lambda_1}$$

$$= \frac{\lambda_1 N_1}{\lambda_2 - \lambda_1}$$

$$\therefore \quad \frac{N_1}{N_2} = \frac{\lambda_2 - \lambda_1}{\lambda_1}$$

$$= \frac{\frac{0.693}{1} - \frac{0.693}{2 \times 12}}{\frac{0.693}{2 \times 12}} = \frac{23}{1}$$

2 Energy generated by the reactor
1000×10^6 W = 10^9 Js^{-1}
Total energy generated in 10 yr is

$$E = (10^9 \text{ Js}^{-1}) \times 10 \times 365 \times 24 \times 60 \times 60$$
$$= 1.97 \times 10^{30} \text{ MeV}$$

In the reactor 200 MeV energy is liberated in the fission of nucleus of U^{235} atom.
$\therefore$ Total number of U^{235} atoms required is

$$\frac{1.97 \times 10^{30}}{200} = 0.985 \times 10^{28}$$

1 kmol that is 235 kg of U^{235} has 6.02×10^{26} atoms Therefore, total mass of U^{235} having 0.985×10^{28} atoms is

$$\frac{235}{6.02 \times 10^{26}} \times (0.985 \times 10^{28})$$
$$= 3.84 \times 10^3 \text{ kg}$$

Since, efficiency of reactor is 10%, actual mass of U^{235} required is

$$(3.84 \times 10^3) \times \frac{100}{10} = 3.84 \times 10^4 \text{ kg}$$

3 As, $-\frac{dN}{dt} = \lambda N$,

$$N = \frac{3.7 \times 10^{10} \times 3.6 \times 10^4}{0.693}$$

$$\Delta N = -(N_0 e^{-\lambda \times 10 \times 3600} - N_0 e^{-\lambda \times 9 \times 3600})$$

$$(\because \Delta N = N_1 - N_2)$$

$$= \frac{3.7 \times 10^{14} \times 3.6}{0.693}[0.535 - 0.5]$$
$$= 6.91 \times 10^{13}$$

4 Given, $R = 12$ dis/min per g,

$R_0 = 16$ dis/min per g

$T_{1/2} = 5760$ yr

Let t be the time span of the tree.
According to radioactive decay law,

$$R = R_0 e^{-\lambda t} \quad \text{or} \quad e^{\lambda t} = \frac{R_0}{R}$$

Taking log on both the sides

$$\lambda t \log_e e = \log_e \frac{R_0}{R}$$

$$\lambda t = \left(\log_{10}\frac{16}{12}\right) \times 2.303$$

$$t = \frac{2.303\,(\log 4 - \log 3)}{\lambda}$$

$$= 2391.20 \text{ yr} \approx 2391 \text{ yr}$$

5 We know that,
Activity $(A) = \lambda N_0$

For S_1, $A_{s_1} = 5\mu\text{Ci} = \lambda_1 2N_0$ ...(i)

For S_2, $A_{s_2} = 10\mu\text{Ci} = \lambda_2 N_0$ ...(ii)

As we know,

$$T_{s_1\,1/2} = \frac{0.693}{\lambda_1}$$

and $$T_{s_2\,1/2} = \frac{0.693}{\lambda_2}$$

Therefore, by dividing Eqs. (i) and (ii), we get

$$\frac{5}{10} = \frac{T_{s_2\,1/2}\, 2N_0}{T_{s_1\,1/2}\, N_0}$$

$$\Rightarrow \frac{T_{s_2\,1/2}}{T_{s_1\,1/2}} = 4$$

So, only option (a) can be satisfied.

6 According to question,

$$T_{1/2}(X) = \tau(\lambda)$$

$$\Rightarrow \frac{0.693}{\lambda_X} = \frac{1}{\lambda_Y}$$

or $$\lambda_Y = \frac{\lambda_X}{0.693}$$

$$\Rightarrow \lambda_Y > \lambda_X$$

So, Y will decay faster than X.

7 Binding energy $B = 2.2$ MeV
From the energy conservation law,

$$E - B = K_n + K_p = \frac{p_n^2}{2m} + \frac{p_p^2}{2m} \quad \text{...(i)}$$

From conservation of momentum,

$$p_n + p_p = \frac{E}{c} \quad \text{...(ii)}$$

As $E = B$, Eq. (i), $p_n^2 + p_p^2 = 0$

It only happen if $p_n = p_p = 0$

So, the Eq. (ii) cannot satisfy and the process cannot take place.

Let $E = B + X$, where $X << B$ for the process to take place.

Put value of p_n from Eq. (ii) in Eq. (i), we get

$$X = \frac{\left(\frac{E}{c} - p_p\right)^2}{2m} + \frac{p_p^2}{2m}$$

or $$2p_p^2 - \frac{2Ep_p}{c} + \frac{E^2}{c^2} - 2mX = 0$$

Using the formula of quadratic equation, we get

$$p_p = \frac{\frac{2E}{c} + \sqrt{\frac{4E^2}{c^2} - 8\left(\frac{E^2}{c^2} - 2mX\right)}}{4}$$

For the real value p_p, the discriminant is positive

$$\frac{4E^2}{c^2} = 8\left(\frac{E_2}{C_2} - 2mX\right)$$

$$16mX = \frac{4E^2}{c^2}$$

$$X = \frac{E^2}{4mc^2} \simeq \frac{B^2}{4mc^2}$$

8 According to given data, mass of neutron and proton are equal which do not permit the breaking up of neutron and proton. But if we take standard mass of neutron as 1.6750×10^{-27} kg, then

Energy released = mass defect $\times c^2$

$= (m_n - m_p - m_e) \times c^2$

$$= \frac{(1.6750 \times 10^{-27} - 1.6725 \times 10^{-27} - 9 \times 10^{-31})}{1.66 \times 10^{-27}} \times 931.5 \text{ MeV}$$

≈ 0.9 MeV

9 Decay scheme is ,

A ⟶ A, B; N_0 at $t=0$; Let N atoms decays into B in time t; N atoms of B; $N_0 - N$ atoms of A

Given, $$\frac{N_B}{N_A} = 0.3 = \frac{3}{10}$$

$$\Rightarrow \frac{N_B}{N_A} = \frac{30}{100}$$

So, $N_0 = 100 + 30 = 130$ atoms

By using $N = N_0 e^{-\lambda t}$

We have, $100 = 130 e^{-\lambda t}$

$$\Rightarrow \frac{1}{1.3} = e^{-\lambda t}$$

$$\Rightarrow \log 1.3 = \lambda t$$

If T is half-life, then

$$\lambda = \frac{\log_e 2}{T}$$

$$\Rightarrow \log 1.3 = \frac{\log_e 2}{T} \cdot t$$

$$\therefore t = \frac{T \cdot \log(1.3)}{\log_e 2}$$

10 Given, 80 min = 4 half-lives of A = 2 half-lives of B.
Let the initial number of nuclei in each sample be N.

For radioactive element A,

N_A after 80 min $= \frac{N}{2^4}$

⇒ Number of A nuclides decayed

$$= N - \frac{N}{16} = \frac{15}{16}N$$

For radioactive element B,

N_B after 80 min $= \frac{N}{2^2}$

⇒ Number of B nuclides decayed

$$= N - \frac{N}{4} = \frac{3}{4}N$$

∴ Ratio of decayed numbers of A and B nuclei will be

$$\frac{(15/16)N}{(3/4)N} = \frac{5}{4}$$

DAY THIRTY FOUR

Electronic Devices

Learning & Revision for the Day

- Energy Bands in Solids
- Semiconductors
- Semiconductor Diode
- *I-V* Characteristics Semiconductor Diode in Forward and Reverse Bias
- Diode as a Rectifier
- Special Purpose Diodes
- Transistor

Energy Bands in Solids

According to band theory of solids, in a crystalline solid due to mutual interaction among valence electrons of neighbouring atoms, instead of sharp energy levels, energy bands are formed. Energy bands are of the following three types

(i) **Valence band** It is the energy band formed by a series of energy levels of valence electrons actually present. Ordinarily, valence band is completely filled and electrons in this band are unable to gain energy from external electric field. The highest energy level in a valence band at 0 K is called **fermi energy level.**

(ii) **Conduction band** The energy band having just higher energy than the valence band is called conduction band. Electrons in conduction band are commonly called the free electrons.

(iii) **Forbidden band** The energy gap between the valence band and the conduction band of a solid is called the **forbidden energy gap** E_g or forbidden band. Width of forbidden energy gap depends upon the nature of substance.

Semiconductors

- In semiconducting solids, the valence band is completely filled but conduction band is completely empty and the energy gap between them is small enough ($E_g < 3$ eV). At absolute zero temperature, it behaves as an **insulator**.
- A pure semiconductor, in which no impurity of any sort has been mixed, is called **intrinsic semiconductor**. Germanium ($E_g = 0.72$ eV) and silicon ($E_g = 1.1$ eV) are examples of intrinsic semiconductors.
- Electrical conductivity of pure semiconductor is very small. To increase the conductivity of a pure semiconducting material, it is doped with a controlled quantity (1 in 10^5 or 10^6) of suitable impurity. Such a doped semiconductor is called an **extrinsic semiconductor**.

- No. of Questions in Exercises (x)—
- No. of Questions Attempted (y)—
- No. of Correct Questions (z)—
  (Without referring Explanations)
- Accuracy Level ($z/y \times 100$)—
- Prep Level ($z/x \times 100$)—

In order to expect good rank in JEE, your Accuracy Level should be above 85 & Prep Level should be above 75.

- The number of electrons reaching from valence band to conduction band,

$$n = AT^{3/2} e^{-E_g/2kT}$$

where, $k =$ Boltzmann's constant, $T =$ absolute temperature and $A =$ atomic weight.

Superconductors

When few metals are cooled, then below a certain critical temperature, their electrical resistance suddenly becomes zero. In this state, these substances are called **superconductors** and this phenomena is called **superconductivity**. Mercury become superconductor at 4.2 K, lead at 7.25 K and niobium at 9.2 K.

Types of Extrinsic Semiconductor

According to type of doping impurities, extrinsic semiconductor are of two types

1. *n*-type Semiconductor

To prepare an *n*-type semiconductor, a pentavalent impurity, e.g. P, As, Sb is used as a dopant with Si or Ge.

Such an impurity is called **donor impurity**, because each dopant atom provides one **free electron**.

In *n*-type semiconductor $n_e >> n_h$, i.e. electrons are majority charge carriers and the holes are minority charge carriers, such that $n_e \cdot n_h = n_i^2$. An *n*-type semiconductor is electrically neutral and is not negatively charged.

Conductivity, $\sigma \approx n_e \mu_e e$

2. *p*-type Semiconductor

To prepare a *p*-type semiconductor, a trivalent impurity, e.g. B, Al, In, Ga, etc., is used as a dopant with Si or Ge. Such an impurity is called **acceptor impurity** as each impurity atom wants to accept an electron from the crystal lattice. Thus, effectively each dopant atom provides a **hole**.

In *p*-type semiconductor $n_h >> n_e$, i.e. holes are majority charge carriers and electrons minority charge carriers, such that $n_h \cdot n_e = n_i^2$. A *p*-type semiconductor is electrically neutral and is not positively charged.

The number of free electrons in a semiconductor varies with temperature as $T^{3/2}$.

Conductivity, $\sigma \approx n_h \mu_h e$

Semiconductor Diode

A ***p-n* junction** is obtained by joining a small *p*-type crystal with a small *n*-type crystal without employing any other binding material in between them. Whenever a *p-n* junction is formed, electrons from *n*-region diffuse through the junction into *p*-region and the holes from *p*-region diffuse into *n*-region.

As a result of which neutrality of both *n* and *p*-regions is disturbed, and a thin layer of immobile negative charged ions appear near the junction in the *p*-crystal and a layer of positive ions appear near the junction in *n*-crystal.

This layer containing immobile ions is called **depletion layer.** The thickness of depletion layer is approximately of the order of 10^{-6} m.

The potential difference developed across the *p-n* junction due to diffusion of electrons and holes is called the **potential barrier** V_b (or emf of fictitious battery). For germanium diode barrier potential is 0.3 V, but for Si diode, its value is 0.7 V. The barrier electric field developed due to it, is of the order of 10^5 Vm^{-1}.

Mobility of Charge Carriers

The mobility of a charge carrier is defined as the velocity gained by its per unit electric field, i.e. $\mu = V_d/E$.

- Current in semiconductor is, $i = i_e + i_h = eA(n_e V_e + n_h V_h)$

Conductivity, $\sigma = \dfrac{J}{E} = \dfrac{i}{AE} = e(n_e \mu_e + n_h \mu_h)$

(where, $J =$ current density $= nqV$)

I-V Characteristics of Semiconductor Diode in Forward and Reverse Bias

When we join an external potential source, such that *p*-side of *p-n* junction is joined to positve terminal of voltage source and *n*-side to negative terminal of voltage source, the junction is said to be **forward biased** and applied electric field E opposes the barrier electric field E_b.

As a result, width of depletion layer is reduced and on applying a voltage $V > V_b$, a forward current begins to flow. Resistance offered by *p-n* junction in forward bias is small (about 10-50 Ω).

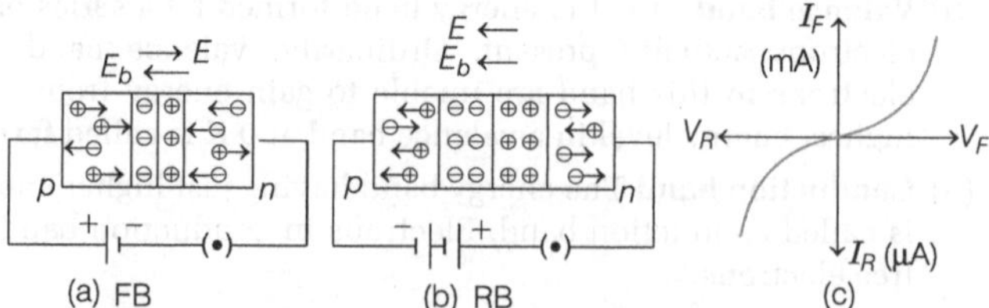

(a) FB (b) RB (c)

If connections of potential source are reversed [Fig. (b)], i.e. *p* -side is connected to negative terminal of battery and *n*-side to positive terminal, the junction is said to be **reverse biased** and in this case E and E_b, being in same direction, are added up. So, the depletion layer broadens and potential barrier is fortified. Consequently, an extremely small leakage current flows across the junction due to minority charge carriers and junction resistance is extremely high ($\simeq 10^5$ Ω). For a sufficiently high reverse bias voltage (25 V or even more), the reverse current suddenly increases. This voltage is called **Zener voltage** or **breakdown voltage** or **avalanche voltage**.

NOTE
- A *p-n* junction behaves as a voltage controlled switch. In forward bias, it acts like ON switch and in reverse bias as OFF switch.
- The *p-n* junction can be presumed as a capacitor, in which the depletion layer acts as dielectric.

Diode as a Rectifier

Junction diode allows current to pass only when it is forward biased. So, if an alternating voltage is applied across a diode, the current flows only in that part of the cycle, when the diode is forward biased.

This property is used to rectify alternating voltages and the circuit used for this purpose is called a rectifier, and the process is known as rectification.

There are two types of rectifier diode as given below

1. **Half Wave Rectifier** A rectifier, which rectifies only one-half of each AC input supply cycle, is called a **half wave rectifier.**

   A half wave rectifier gives discontinuous and pulsating DC output. As no output is obtained corresponding to alternate half cycles of the AC input supply, its efficiency is quite low.

2. **Full Wave Rectifier** A rectifier, which rectifies both halves of each AC input cycle is called a full wave rectifier.

   The output of a full wave rectifier is continuous, but pulsating in nature. However, it can be made smooth by using a filter circuit.

   As output is obtained corresponding to both the half cycles of the AC input supply, its efficiency is more than that of half wave rectifier.

NOTE • The ripple factor is defined as the ratio of rms value of AC component in the output of the rectifier to the DC component in the input.

Special Purpose Diodes

There are few diodes which are designed to serve some special purpose and application.

Zener Diode

It is a highly doped *p-n* junction diode which is not damaged by high reverse current. It is always used in reverse bias in breakdown voltage region and is chiefly used as a voltage regulator.

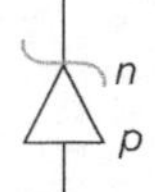

- **Zener Diode as Voltage Regulator** The following circuit is used for stabilising voltage across a load R_L. The circuit consists of a series voltage-dropping resistance R and a Zener diode in parallel with the load R_L.

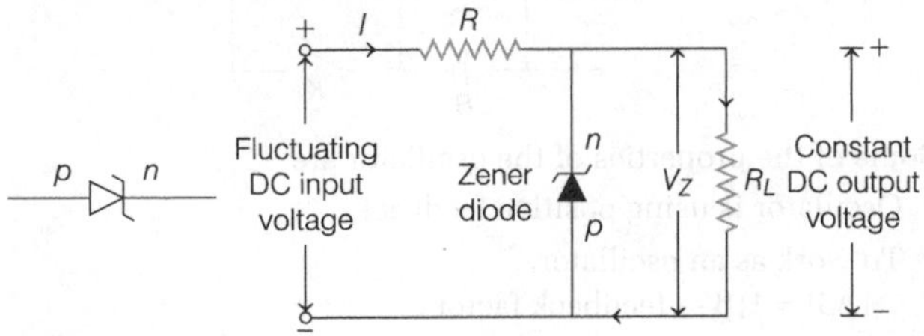

The Zener diode is selected with Zener voltage V_z equal to the voltage desired across the load.

Light Emitting Diode (LED)

It is a specially designed diode made of GaAsP, GaP, etc. When used in forward biased, it emits characteristic, almost monochromatic light. In reverse biased, it works like a normal diode.

- ***I-V* Characteristics** LEDs are current dependent devices with its forward voltage drop (V_F) depending on the forward biased LED current. Characteristics of light emitting diode *I-V* are shown below

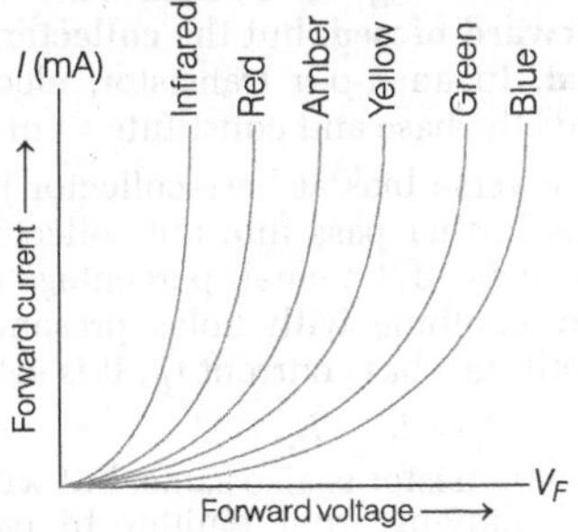

Photodiode

It is a special diode used in reverse bias which conducts only when light of suitable wavelengths is incident on the junction of diode. The energy of incident light photon must be greater than the band gap of semiconductor (i.e. $h\nu > E_g$). Materials used are Cds, Se, Zns, etc.

Solar Cell

It is a special *p-n* junction, in which one of the semiconductors is made extremely thin, so that solar radiation falling on it reaches junction of diode without any absorption. A solar cell directly converts, solar energy into electrical energy. Popularly used solar cells, Ni-cd, PbS cell, etc.

Transistor

A transistor is a combination of two *p-n* junctions joined in series. A junction transistor is known as **Bipolar Junction Transistor** (BJT). It is a three terminal device.

Transistors are of two types

(i) *n-p-n* transistor,

(ii) *p-n-p* transistor

A transistor has three regions

(i) An **emitter** (*E*), which is most heavily doped, and is of moderate size. It supplies large number of charge carriers, which are free electrons in an *n-p-n* transistor and holes in a *p-n-p* transistor.

(ii) A **base** (*B*), which is very lightly doped and is very thin (thickness $\approx 10^{-5}$m).

(iii) A **collector** (*C*), which is moderately doped and is thickest.

A transistor is symbolically represented as shown in figures.

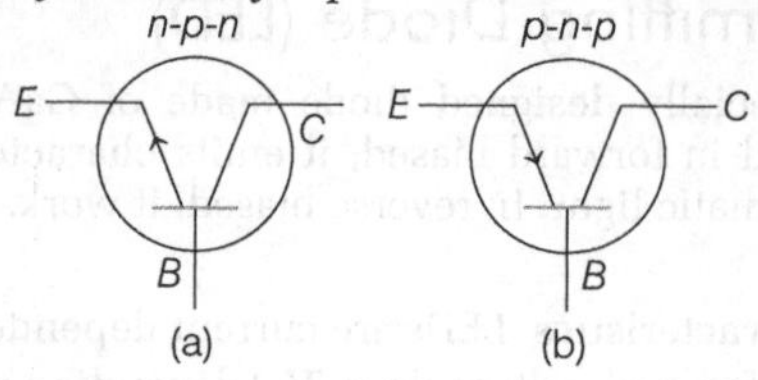

Transistor Action

For proper functioning of a transistor, the **emitter-base junction is forward biased**, but the **collector-base junction is reverse biased.** In an *n-p-n* transistor, electrons flow from emitter towards the base and constitute a current I_E.

Due to larger reverse bias at base-collector junction, most of these electrons further pass into the collector, constituting a collector current I_C. But a small percentage of electrons (less than 5%) may combine with holes present in base. These electrons constitute a base current I_B. It is self evident, that

$$I_E = I_C + I_B.$$

Action of *p-n-p* transistor is also same, but with one difference that holes are moving from emitter to base and then to collector.

A transistor can be connected in either of the following three configurations

(i) Common Emitter (CE) configuration

(ii) Common Base (CB) configuration

(iii) Common Collector (CC) configuration.

Generally, we prefer common emitter configuration, because power gain is maximum in this configuration.

Characteristics of a Transistor

In common emitter configuration, variation of current on the input side with input voltage (I_B *versus* V_{BE}) is known as the **input characteristics**, and the variation in the output current with output voltage (I_C *versus* V_{CE}) is known as **output characteristics**. From these characteristics, we obtain the values of following parameters

- Input resistance, $r_i = \left|\frac{\Delta V_{BE}}{\Delta I_B}\right|_{V_{CE} = \text{constant}}$
- Output resistance, $r_o = \left|\frac{\Delta V_{CE}}{\Delta I_C}\right|_{I_B = \text{constant}}$
- AC current gain, $\beta = \left|\frac{\Delta I_C}{\Delta I_B}\right|_{V_{CE} = \text{constant}}$

  The current gain for common-emitter configuration β ranges from 20 to 200.
- Transconductance, $g_m = \frac{\Delta I_C}{\Delta V_{BE}} = \frac{\beta}{r_i}$
- A transistor can be used as an amplifier. The voltage gain of an amplifier will be given by

  $$A_V = \frac{V_o}{V_i} = \beta \cdot \frac{R_C}{R_B}$$

  where, R_C and R_B are net resistances in collector and base circuits, respectively.
- In common base configuration, AC current gain is defined as $\alpha = \left|\frac{\Delta I_C}{\Delta I_E}\right|_{V_{CE} = \text{constant}}$
- Value of α is slightly less than 1. In fact, $0.95 \le \alpha \le 1$.
- Power gain $= \frac{\Delta P_o}{\Delta P_i} = \beta_{AC}^2 \times$ Resistance gain

NOTE • Current gains α and β are correlated as

$$\beta = \frac{\alpha}{1-\alpha} \text{ or } \alpha = \frac{\beta}{1+\beta}$$

Transistor as an Amplifier

A transistor consisting of two *p-n* junctions, one forward biased and the other reverse biased can be used to amplify a weak signal. The forward biased junction has a low resistance path, whereas the reverse biased junction has a high resistance path. The weak input signal is applied across the forward biased junction, and the output signal is taken across the reverse biased junction.

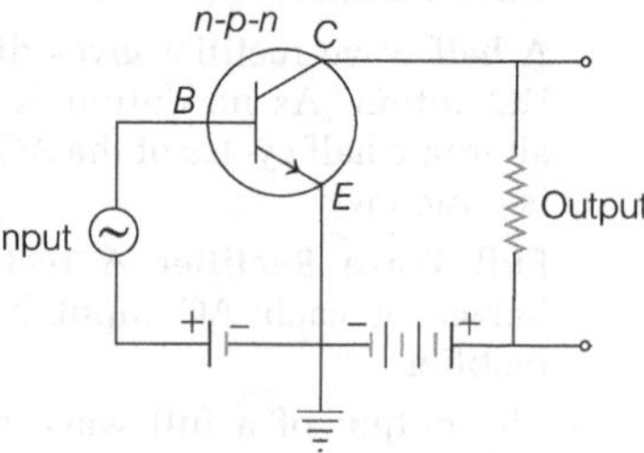

Since, the input and output currents are almost equal, the output signal appears with a much higher voltage. The transistor, thus acts as an amplifier. Common-emitter configuration of transistor amplifier is given alongside.

Transistor as an Oscillator

An electronic oscillator is a device that generates electrical oscillations of constant amplitude and of a desired frequency, without any external input.

The circuit providing such oscillation, is known as a tank oscillator, is using positive feedback.

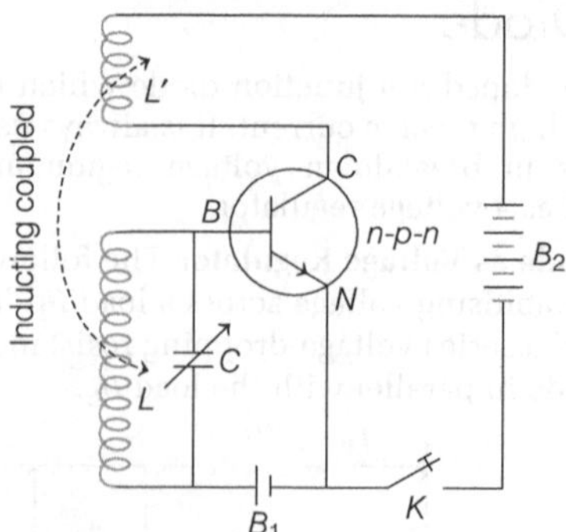

Some of the properties of the oscillator are

- Oscillator is using positive feedback.
- To work as an oscillator,

  $|A\beta| = 1; \beta \rightarrow$ feedback factor
- f = frequency of oscillation $= \frac{1}{2\pi} \times \frac{1}{\sqrt{LC}}$.

DAY PRACTICE SESSION 1

FOUNDATION QUESTIONS EXERCISE

1 The conductivity of a semiconductor increases with increase in temperature because

(a) number density of free current carriers increases
(b) relaxation time increases
(c) Both number density of carriers and relaxation time increases
(d) number density of current carriers increases, relaxation time decreases but effect of decrease in relaxation time is much less than increase in number of density

2 Carbon, silicon and germanium atoms have four valence electrons each. Their valence and conduction bands are separated by energy band gaps represented by $(E_g)_C$, $(E_g)_{Si}$ and $(E_g)_{Ge}$, respectively. Which one of the following relationship is true in their case?

(a) $(E_g)_C > (E_g)_{Si}$ (b) $(E_g)_C = (E_g)_{Si}$
(c) $(E_g)_C < (E_g)_{Ge}$ (d) $(E_g)_C < (E_g)_{Si}$

3 In *n*-type silicon, which of the following statement is true

(a) Electrons are majority carriers and trivalent atoms are the dopants
(b) Electrons are minority carriers and pentavalent atoms are the dopants
(c) Holes are minority carriers and pentavalent atoms are the dopants
(d) Holes are majority carriers and trivalent atoms are the dopants

4 Carbon, silicon and germanium have four valence electrons each. At room temperature, which one of the following statements is most appropriate? → AIEEE 2012

(a) The number of free conduction electrons is significant in C but small in Si and Ge
(b) The number of free conduction electrons is negligibly small in all the three
(c) The number of free electrons for conduction is significant in all the three
(d) The number of free electrons for conduction is significant only in Si and Ge but small in C

5 If the lattice constant of this semiconductor is decreased, then which of the following is correct? → AIEEE 2010

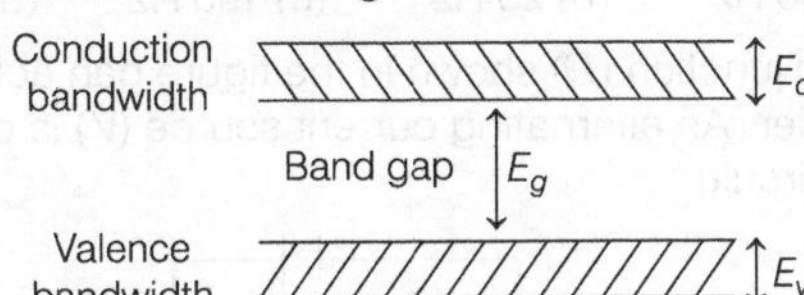

(a) All E_c, E_g and E_v increases
(b) E_c and E_v increases, but E_g decreases
(c) E_c and E_v decreases, but E_g increases
(d) All E_c, E_g and E_v decreases

6 In an unbiased *p-n* junction, holes diffuse from the *p*-region to *n*-region because

(a) free electrons in the *n*-region attract them
(b) they move across the junction by the potential difference
(c) hole concentration in *p*-region is more as compared to *n*-region
(d) All of the above

7 Application of a forward bias to a *p-n* junction

(a) increases the number of donors on the *n*-side
(b) increases the electric field in the depletion zone
(c) increases the potential difference across the depletion zone
(d) widens the depletion zone

8 When forward bias is applied to a *p-n* junction, what happens to the potential barrier V_B and the width of charge depleted region x?

(a) V_B increases, x decreases
(b) V_B decreases, x increases
(c) V_B increases, x increases
(d) V_B decreases, x decreases

9 The temperature (T) dependence of resistivity (ρ) of a semiconductor is represented by

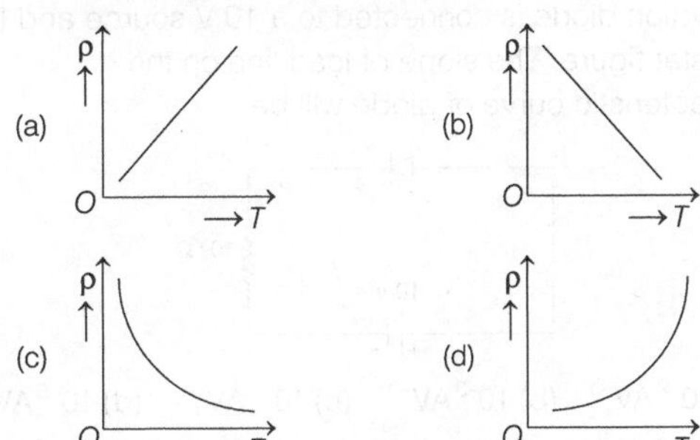

10 In figure, V_0 is the potential barrier across a *p-n* junction, when no battery is connected across the junction.

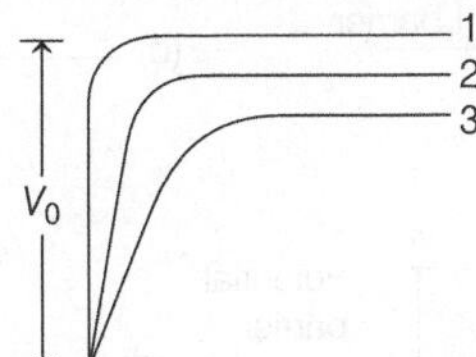

(a) 1 and 3 both corresponds to forward bias of junction
(b) 3 corresponds to forward bias of junction and 1 correspond to reverse bias of junction
(c) 1 corresponds to forward bias and 3 corresponds to reverse bias of junction
(d) 3 and 1 both corresponds to reverse bias of junction

11 For the given circuit of *p-n* junction diode, which of the following statements is correct?

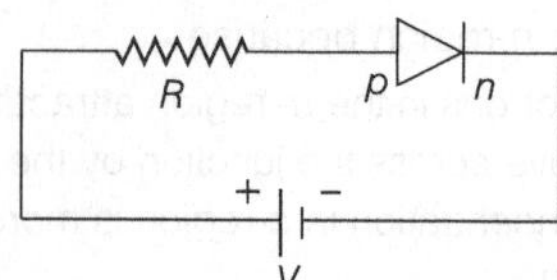

(a) In forward biasing the voltage across *R* is *V*
(b) In forward biasing the voltage across *R* is 2 *V*
(c) In reverse biasing the voltage across *R* is *V*
(d) In reverse biasing the voltage across *R* is 2 *V*

12 In the case of forward biasing of *p-n* junction, which one of the following figures correctly depicts the direction of the flow of charge carriers?

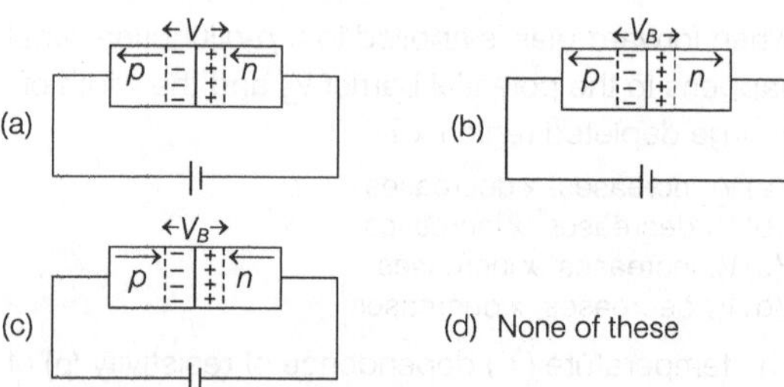

(d) None of these

13 A junction diode is connected to a 10 V source and $10^3 \Omega$ rheostat figure. The slope of load line on the characteristic curve of diode will be

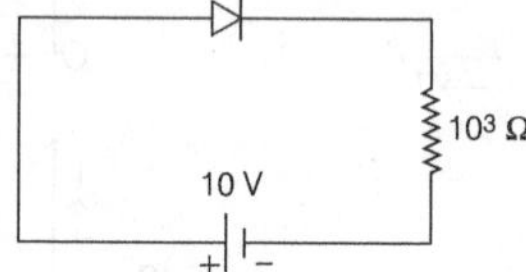

(a) 10^{-2} AV^{-1} (b) 10^{-3} AV^{-1} (c) 10^{-4} AV^{-1} (d) 10^{-5} AV^{-1}

14 In a forward biased *p-n* junction diode, the potential barrier in the depletion region will be of the form

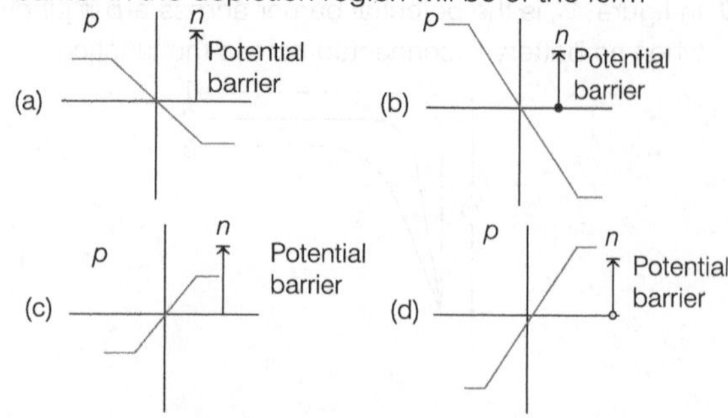

15 The circuit has two oppositely connected ideal diodes in parallel. What is the current flowing in the circuit?

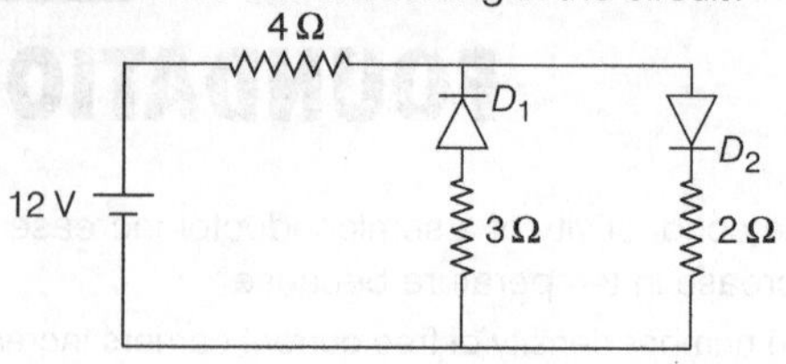

(a) 1.71 A (b) 2.00 A (c) 2.31 A (d) 1.33 A

16 In the following circuits, which one of the diodes is reverse biased?

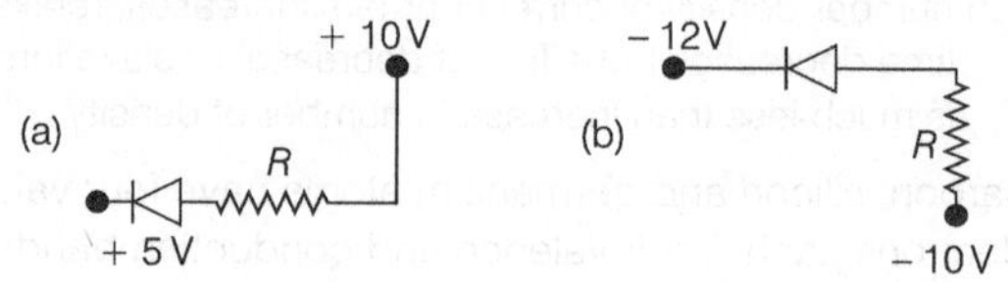

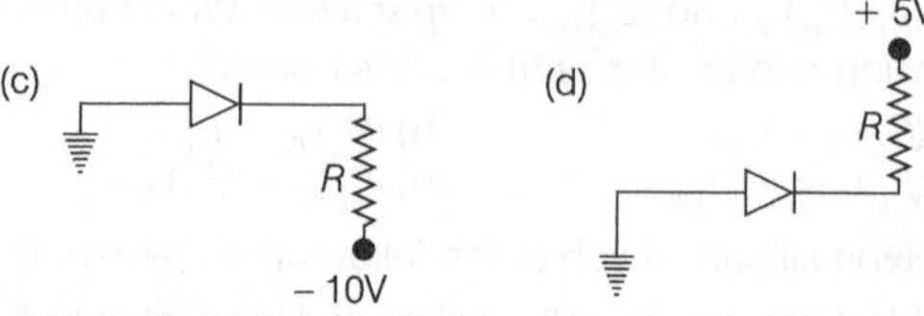

17. The forward biased diode connection is →**JEE Main 2014**

(a) 2V —▷|—www— +4V

(b) −2V —▷|—www— +2V

(c) +2V —▷|—www— −2V

(d) −3V —▷|—www— −3V

18 The reading of the ammeter for a silicon diode in the given circuit is →**JEE Main 2018**

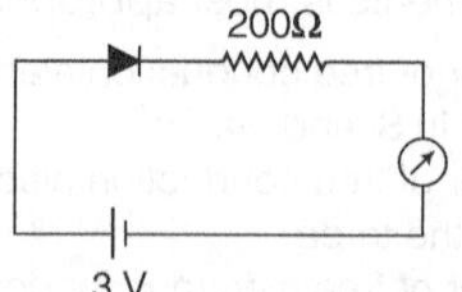

(a) 0 (b) 15 mA (c) 11.5 mA (d) 13.5 mA

19 In a full-wave rectifier circuit operating from 50 Hz mains frequency, the fundamental frequency in the output would be

(a) 50 Hz (b) 25 Hz (c) 100 Hz (d) 70.7 Hz

20 A *p-n* junction (*D*) shown in the figure can act as a rectifier. An alternating current source (*V*) is connected in the circuit. →**AIEEE 2013**

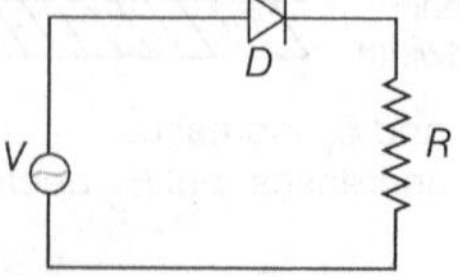

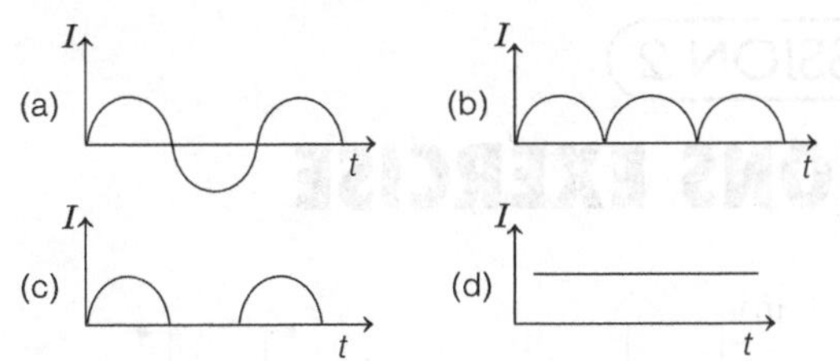

21 Zener breakdown in a semiconductor diode occurs, when
(a) forward current exceeds certain value
(b) reverse bias exceeds certain value
(c) forward bias exceeds certain value
(d) potential barrier is reduced to zero

22 The I-V characteristics of an LED is → AIEEE 2013

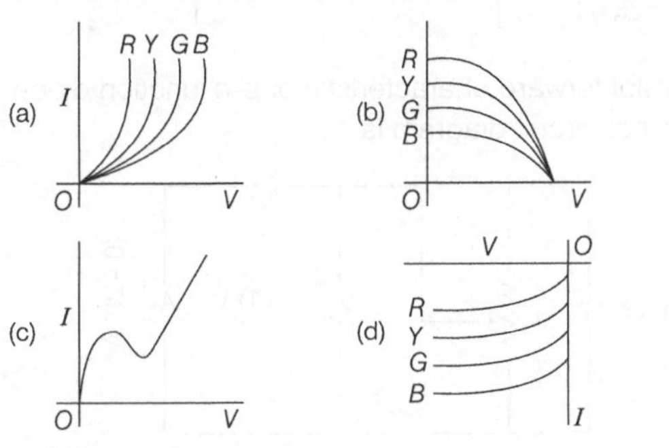

23 Identify the semiconductor devices whose characteristics are as given below, in the order (p),(q),(r),(s). → JEE Main 2016 (Offline)

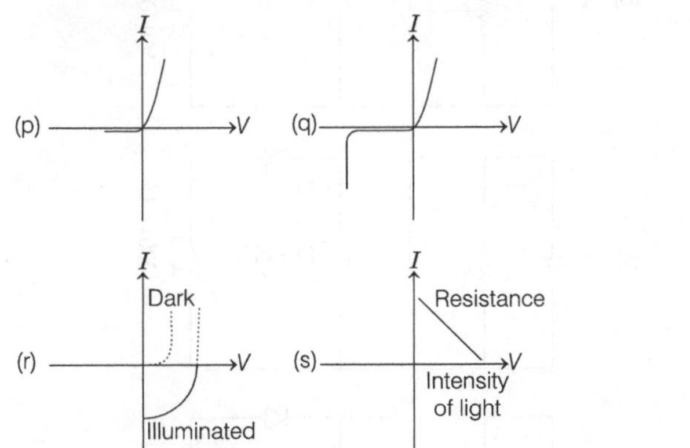

Choose the correct order
(a) Simple diode, Zener diode, Solar cell, Light dependent resistance
(b) Zener diode, Simple diode, Light dependent resistance, Solar cell
(c) Solar cell, Light dependent resistance, Zener diode, Simple diode
(d) Zener diode, Solar cell, Simple diode, Light dependent resistance

24 In a common emitter amplifier circuit using an *n-p-n* transistor, the phase difference between the input and the output voltages will be → JEE Main 2017 (Offline)
(a) 90° (b) 135° (c) 180° (d) 45°

25 When A is the internal stage gain of an amplifier and B is the feedback ratio, then the amplifier becomes as oscillator if
(a) B is negative and magnitude of $B = A/2$
(b) B is negative and magnitude of $B = 1/A$
(c) B is negative and magnitude of $B = A$
(d) B is positive and magnitude of $B = 1/A$

Direction (Q. Nos. 26-30) *Each of these questions contains two statements : Statement I and Statement II. Each of these questions also has four alternative choices, only one of which is the correct answer. You have to select one of the codes (a), (b), (c) and (d) given below.*
(a) Statement I is true, Statement II is true; Statement II is the correct explanation for Statement I
(b) Statement I is true, Statement II is true; Statement II is not the correct explanation for Statement I
(c) Statement I is true; Statement II is false
(d) Statement I is false; Statement II is true

26 **Statement I** If forward current changes by 1.5 mA when forward voltage in semiconductor diode is changed from 0.5 V to 2 V, the forward resistance of diode will be 1 Ω.
Statement II The forward resistance is given by $R_f = \dfrac{\Delta V_f}{\Delta I_f}$

27 **Statement I** A Zener diode is used to get constant voltage at variable current under reverse bias.
Statement II The most popular use of Zener diode is as voltage regulator.

28 **Statement I** Light Emitting Diode (LED) emits spontaneous radiation.
Statement II LED are forward biased *p-n* junctions.

29 **Statement I** When base region has larger width, the collector current increases.
Statement II Electron-hole combination in base results in increases of base current.

30 **Statement I** In a common-emitter transistor amplifier the input current is much less than output current.
Statement II The common-emitter transistor amplifier has very high input impedance.

DAY PRACTICE SESSION 2

PROGRESSIVE QUESTIONS EXERCISE

1 The input resistance of a common-emitter transistor amplifier, if the output resistance is 500 k Ω, the current gain $\alpha = 0.98$ and the power gain is 6.0625×10^6 is

(a) 198 Ω (b) 300 Ω (c) 100 Ω (d) 400 Ω

2 If the resistivity of copper is 1.7×10^{-6} Ω-m, then the mobility of electrons in copper, if each atom of copper contributes one free electron for conduction is [the atomic weight of copper is 63.54 and density is 8.96 g/cc]

(a) 23.36 cm^2/ Vs (b) 503.03 cm^2/ Vs
(c) 43.25 cm^2/ Vs (d) 88 cm^2/ Vs

3 A red LED emits light at 0.1 W uniformly around it. The amplitude of the electric field of the light at a distance of 1 m from the diode is → JEE Main 2015

(a) 1.73 V/m (b) 2.45 V/m
(c) 5.48 V/m (d) 7.75 V/m

4 A working transistor with its three legs marked *P*, *Q* and *R* is tested using a multimeter.

No conduction is found between *P* and *Q*. By conneting the common (negative) terminal of the multimeter to *R* and the other (positive) terminal to *P* or *Q*. Some resistance is seen on the multimeter. Which of the following is the true for the transistor? → AIEEE 2013

(a) It is an *n-p-n* transistor with *R* as base
(b) It is a *p-n-p* transistor with *R* as collector
(c) It is a *p-n-p* transistor with *R* as emitter
(d) It is an *n-p-n* transistor with *R* as collector

5 A piece of copper and another of germanium are cooled from room temperature to 77 K, the resistance of

(a) each of them increases
(b) each of them decreases
(c) copper decreases and germanium increases
(d) copper increases and germanium decreases

6 If in a *p-n* junction diode, a square input signal of 10 V is applied as shown.

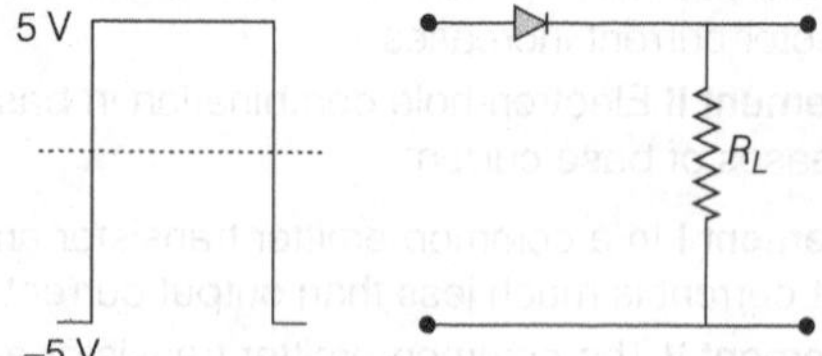

Then, the output signal across R_L will be

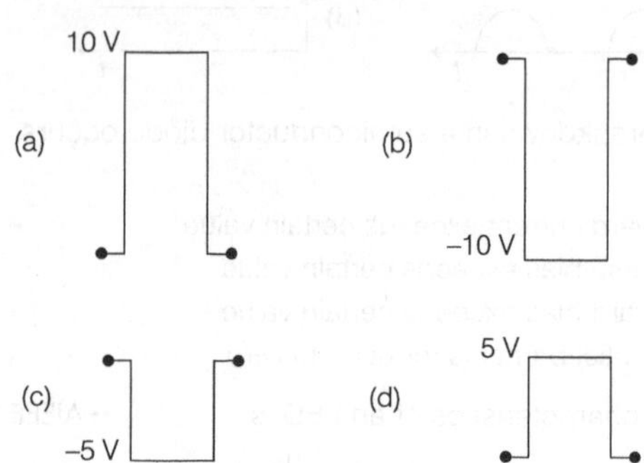

7 To plot forward characteristic of *p-n* junction diode, the correct circuit diagram is

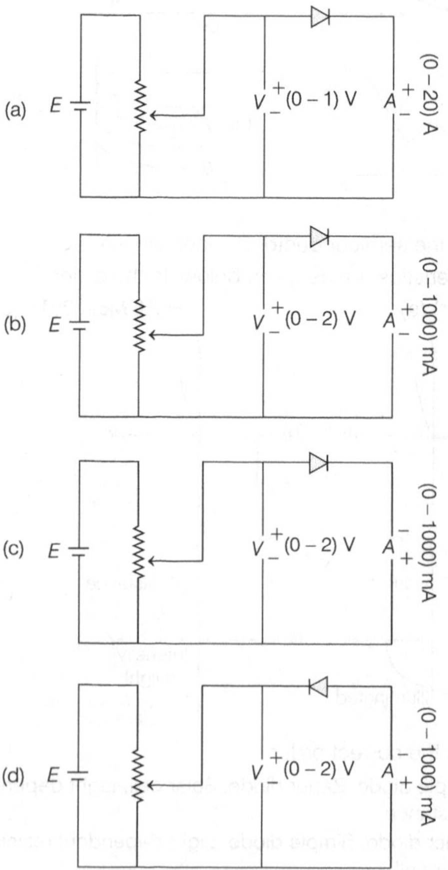

8 A figure is given below

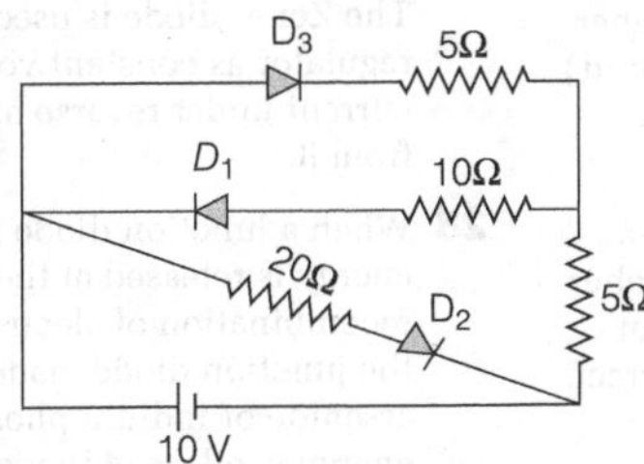

The current through the battery is

(a) 0.5 A (b) 1 A
(c) 1.5 A (d) 2 A

9 In the given circuit, the current through the zener diode is

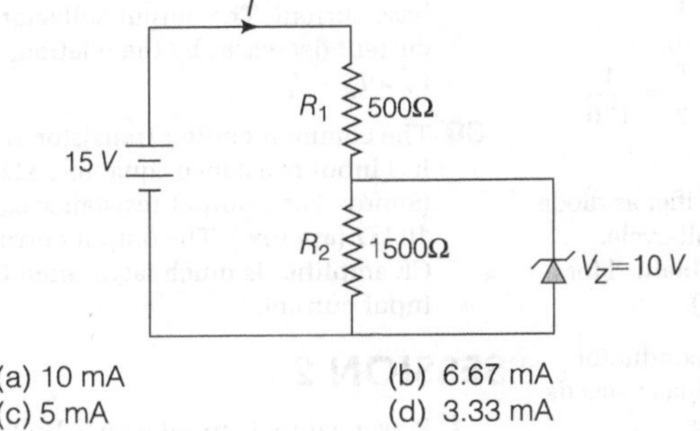

(a) 10 mA (b) 6.67 mA
(c) 5 mA (d) 3.33 mA

10 The length of germanium rod is 0.928 cm and its area of cross-section is 1 mm^2. If for germanium $n_i = 2.5 \times 10^{19} \text{m}^{-3}$, $\mu_h = 0.19 \text{m}^2 \text{V}^{-1} \text{s}^{-1}$, $\mu_e = 0.39 \text{m}^2 \text{V}^{-1} \text{s}^{-1}$, then resistance is

(a) 2.5 kΩ (b) 4.0 kΩ (c) 5.0 kΩ (d) 10.0 kΩ

11 In the figure, potential difference between A and B is

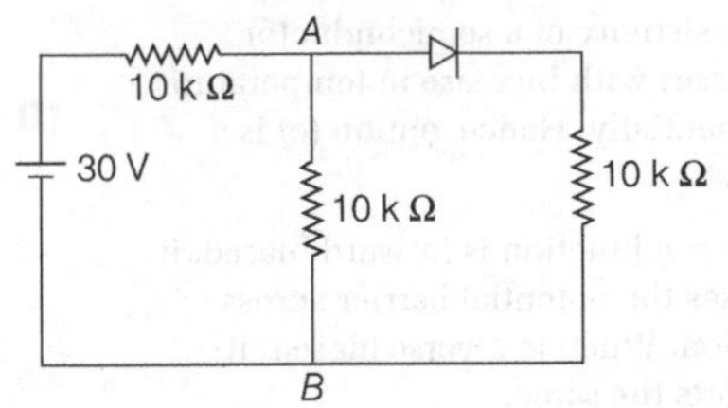

(a) 10 V (b) 20 V (c) 30 V (d) 40 V

12 In a common-base mode of a transistor, the collector current is 5.488 mA for an emitter current of 5.60 mA. The value of the base current amplification factor (β) will be

→ AIEEE 2011

(a) 49 (b) 50 (c) 51 (d) 48

13 For a common emitter configuration, if α and β have their usual meanings, the incorrect relationship between α and β is

→ JEE Main 2016 (Offline)

(a) $\frac{1}{\alpha} = \frac{1}{\beta} + 1$ (b) $\alpha = \frac{\beta}{1-\beta}$ (c) $\alpha = \frac{\beta}{1+\beta}$ (d) $\alpha = \frac{\beta^2}{1+\beta^2}$

ANSWERS

SESSION 1	1 (d)	2 (a)	3 (c)	4 (d)	5 (c)	6 (c)	7 (a)	8 (d)	9 (c)	10 (b)
	11 (a)	12 (c)	13 (b)	14 (d)	15 (b)	16 (d)	17 (c)	18 (c)	19 (c)	20 (c)
	21 (b)	22 (a)	23 (a)	24 (c)	25 (d)	26 (d)	27 (a)	28 (b)	29 (d)	30 (c)
SESSION 2	1 (a)	2 (c)	3 (b)	4 (a)	5 (c)	6 (d)	7 (b)	8 (c)	9 (d)	10 (b)
	11 (a)	12 (a)	13 (a,c)							

Hints and Explanations

SESSION 1

1 Based on the theory discussed, we can conclude that when temperature increases, number density of current carriers increases in the semiconductor, relaxation time decreases but effect of relaxation time will be ignored.

2 Carbon, silicon and germanium are semiconductors.

$(E_g)_C = 5.2 \text{ eV}$,
$(E_g)_{Si} = 1.21 \text{ eV}$
and $(E_g)_{Ge} = 0.75 \text{ eV}$
Thus, $(E_g)_C > (E_g)_{Si}$
and $(E_g)_C > (E_g)_{Ge}$

3 n-type is obtained by doping the Ge or Si with pentavalent atoms. In n-type semiconductor, electrons are majority carriers and holes are minority carriers, hence option (c) is correct.

4 The number of free electrons for conduction is significant only in Si and Ge but small in C, as C is an impurity.

5 If lattice constant of semiconductor is decreased, then E_c and E_v decreases but E_g increases.

6 In an unbiased p-n junction, the diffusion of charge carriers across the junction takes place from higher concentration to lower concentration. Thus, option (c) is correct.

7 In forward biasing more number of electrons enter in n-side from battery thereby increasing the number of donors on the n-side.

8 In a p-n junction in forward bias potential barrier V_B as well as the width of charge depleted region x decreases.

9 The resistivity of a semiconductor decreases with increase in temperature exponentially. Hence, option (c) is correct.

10 When p-n junction is forward biased, it opposes the potential barrier across junction. When is reverse biased, it supports the same.

11 In forward biasing for an ideal diode resistance of diode is zero and whole resistance in the circuit is R. Hence, voltage across R is V.

12 Forward bias is obtained when the negative terminal of the battery is connected to the n-side and the positive terminal to the p-side of the semiconductor. Then, the negative terminal will repel free electrons in the n-section towards the junction and the positive terminal on the p-side will push the holes towards the junction.

13 If V is the voltage across the junction and I is the circuit current, then

$$V + IR = E \quad \text{or} \quad I = \frac{E}{R} - \frac{V}{R} = -\frac{V}{R} + \frac{E}{R}$$

Slope of load line

$$= -\frac{1}{R} = \frac{1}{1000} = 10^{-3}\,\text{AV}^{-1}$$

14 The diode is forward biased, hence the potential barrier decreases and becomes less. Though in both the options (c) and (d). The diode is forward biased but in (d) the barrier width is less.

15 In the given circuit diode D_1 is reverse biased while D_2 is forward biased, so the circuit can be redrawn as

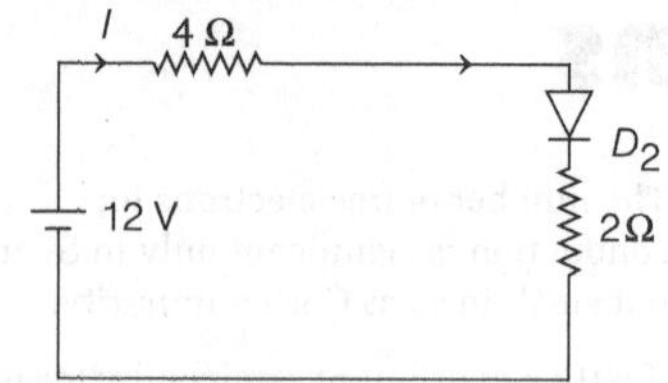

Apply KVL to get current flowing through the circuit

$$-12 + 4I + 2I = 0$$

or $$I = \frac{12}{6} = 2\text{A}$$

16 For reverse biasing of an ideal diode, the potential of n-side should be higher than potential of p-side. Only option (d) is satisfying the criterion for reverse biasing.

17 For forward biased condition of a p-n junction, p-junction should be at higher potential and n-junction should be at lower potential. So, option (c) is correct.

18 Potential drop in a silicon diode in forward bias is around 0.7 V.

In given circuit, potential drop across 200 Ω resistor is

$$I = \frac{\Delta V_{\text{net}}}{R} = \frac{3 - 0.7}{200}$$

$$\Rightarrow I = 0.0115\text{ A} \Rightarrow I = 11.5\text{ mA}$$

19 Given, $f = 50$ Hz and $T = \frac{1}{50}$

For full-wave rectifier, $T_1 = \frac{T}{2} = \frac{1}{100}$

and $f_1 = 100$ Hz

20 Given figure is half wave rectifier as diode conducts only for positive half-cycle. Hence, output waveform is obtained for half cycle only as in figure (c).

21 Zener breakdown in a semiconductor diode occurs when reverse bias exceeds certain value, which is known as breakdown or Zener or Avalanche voltage.

22 For same value of current higher value of voltage is required for higher frequency.

23 Zener diode works in breakdown region.

So, Simple diode → (p)
Zener diode → (q)
Solar cell → (r)
Light dependent resistance → (s)

24 V_i → t (Input) V_o → t (Output)

In a CE n-p-n transistor amplifier output is 180° out of phase with input.

25 The condition for a circuit to oscillate are

(i) feedback should be positive

(ii) output voltage feedback $B = \frac{1}{A}$

26 $$R_f = \frac{\Delta V_f}{\Delta I_f} = \frac{(2 - 0.5)}{1.5 \times 10^{-3}} = 10^3\ \Omega$$
$$= 1\text{ k}\Omega$$

27 Zener diodes are specially designed junction diodes, which can operate in the reverse breakdown voltage region continuously without being damaged. The Zener diode is used as a voltage regulator as constant voltage at variable current under reverse bias is obtained from it.

28 When a junction diode is forward biased energy is released at the junction due to recombination of electrons and holes. In the junction diode made up of gallium arsenide or indium phosphide, the energy is released in visible region. Such a junction diode is called Light Emitting Diode or LED. The radiated energy emitted by LED is equal or less than band gap of semiconductor.

29 When base region has larger width, electron-hole combination increases the base current. The output collector current decreases by the relation, $I_E = I_B + I_C$.

30 The common-emitter transistor amplifier has input resistance equal to 1 kΩ (approx.) and output resistance equal to 10 kΩ (approx.). The output current in CE amplifier is much larger than the input current.

SESSION 2

1 Power gain = Current gain × Voltage gain

$$\text{Current gain} = \beta = \frac{\alpha}{1-\alpha} = \frac{0.98}{1 - 0.98}$$
$$= 49$$

$$A_V = 49\left(\frac{500 \times 10^3}{R_1}\right)\left[\text{As, } A_v = \beta\frac{R_2}{R_1}\right]$$

$$\text{Power gain} = 6.0625 \times 10^6$$
$$= 49 \times \left(\frac{500 \times 10^3}{R_1}\right) \times 49$$

$$R_1 = 198\ \Omega$$

2 Mobility of electron $(\mu) = \frac{\sigma}{ne}$ ...(i)

Resistivity $(\rho) = \frac{1}{\sigma}$ ...(ii)

From Eqs. (i) and (ii), we get

$$\mu = \frac{1}{ne\rho} \quad \text{...(iii)}$$

where, n = number of free electrons per unit volume

$$n = \frac{N_0 \times d}{\text{atomic weight}}$$
$$= \frac{6.023 \times 10^{23} \times 8.96}{63.54}$$
$$= 8.5 \times 10^{22} \quad \text{...(iv)}$$

From Eqs. (iii) and (iv), we get

$$\mu = \frac{1}{8.5 \times 10^{22} \times 1.6 \times 10^{-19} \times 1.7 \times 10^{-6}}$$
$$= 43.25\,\text{cm}^2/\text{Vs}$$

3 Consider the LED as a point source of light.

Let power of the LED is P.

Intensity at r from the source

$$I = \frac{P}{4\pi r^2} \quad ...(i)$$

As we know that, $I = \frac{1}{2}\varepsilon_0 E_0^2 c \quad ...(ii)$

From Eqs. (i) and (ii), we can write

$$\frac{P}{4\pi r^2} = \frac{1}{2}\varepsilon_0 E_0^2 c$$

or $E_0^2 = \frac{2P}{4\pi\varepsilon_0 r^2 c} = \frac{2 \times 0.1 \times 9 \times 10^9}{1 \times 3 \times 10^8}$

or $E_0^2 = 6 \Rightarrow E_0 = \sqrt{6} = 2.45\,\text{V/m}$

4 Since, no conduction is found when multimeter is connected across P and Q, it means either both P and Q are n-type or p-type. So, it means R is base, when R is connected to common terminal and conduction is seen when other terminal is connected to P or Q. So, it means transistor is n-p-n with R as base.

5 We know that resistance of conductor is directly proportional to temperature (i.e. $R \propto \Delta t$), while resistance of semiconductor is inversely proportional to temperature $\left(\text{i.e. } R \propto \frac{1}{\Delta t}\right)$.

Therefore, it is clear that resistance of conductor decreases with decrease in temperature or *vice-versa*, while in case of semiconductor, resistance increases with decrease in temperature or *vice-versa*. Since, copper is pure conductor and germanium is a semiconductor, hence due to decrease in temperature, resistance of conductor decreases while that of semiconductor increases.

6 During – ve cycle, diode will not allow the signal to pass.

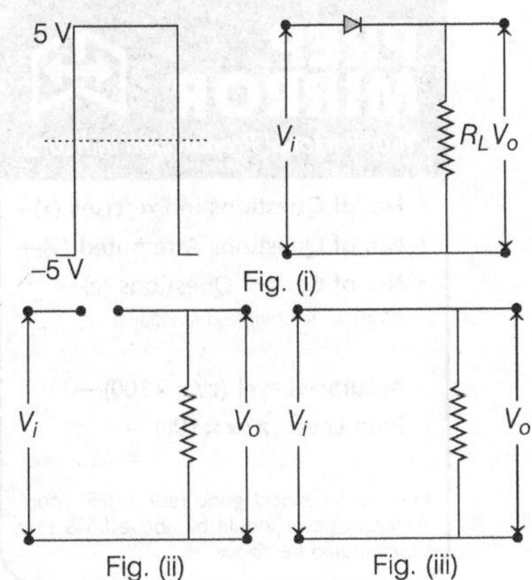

Fig. (i)

Fig. (ii) Fig. (iii)

For $V_i < 0$, the diode is reverse biased and hence offer infinite resistance, so circuit would be like as shown in Fig. (ii) and $V_o = 0$.

For $V_i > 0$, the diode is forward biased and circuit would be as shown in Fig. (iii) and $V_o = V_i$.

Hence, the option (d) is correct.

7 For forward bias mode, the p-side of diode has to be at higher potential than n-side. The meters used are DC, so we have to be careful while connecting them w.r.t. polarity.

Last point is to decide the range of meters, the range of meters has to be in such a way that we can have the readings which leads to plot on realistic scale. If we take 0-20 A ammeter, then reading we read from this is tending to 0 to 5 divisions which is not fruitful.

In options (c) and (d), the polarity of ammeter is not correct. Hence, (b) is correct circuit.

8 In the given circuit, diode D_1 is reverse biased, so it will not conduct. Diode D_2 and D_3 are forward biased, so they conduct. The equivalent circuit is as shown below:

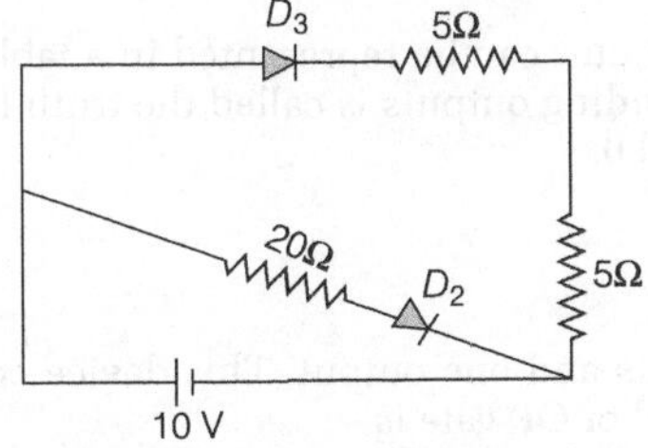

Now, the equivalent resistance of circuit is

$$R_{eq} = \frac{(5+5) \times 20}{(5+5)+20} = \frac{10 \times 20}{10+20} = \frac{200}{30} = \frac{20}{3}\,\Omega$$

Then, current through the battery,

$$I = \frac{V}{R_{eq}} = \frac{10}{20/3} = \frac{3}{2} = 1.5\text{A}$$

9 The voltage drop across R_2 is

$$V_{R_2} = V_2 = 10\,\text{V}$$

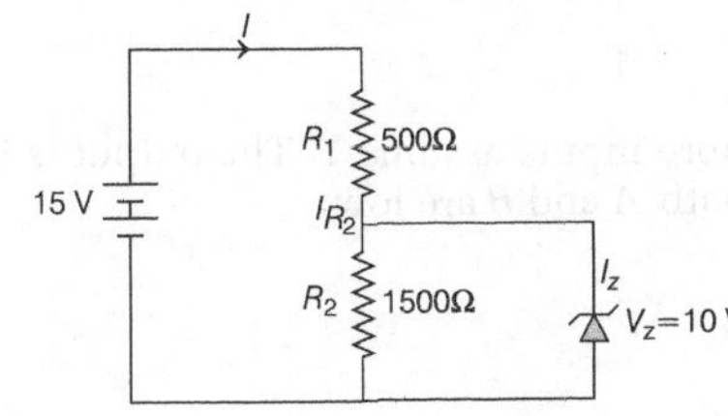

The current through R_2 is

$$I_{R_2} = \frac{V_{R_2}}{R_2} = \frac{10}{1500} = 0.667 \times 10^{-2}\,\text{A} = 6.67\,\text{mA}$$

The voltage drop across R_1 is

$$V_{R_1} = 15\text{V} - V_{R_2} = 15 - 10 = 5\text{V}$$

The current through R_1 is

$$I_{R_1} = \frac{V_{R_1}}{R_1} = \frac{5}{500} = 10^{-2}\,\text{A} = 10\,\text{mA}$$

The current through the zener diode is

$$I_z = I_{R_1} - I_{R_2} = (10 - 6.67)\text{mA} = 3.33\,\text{mA}$$

10 $\therefore R = \frac{\rho l}{A} = \frac{L}{n_i e(\mu_e + \mu_h)A} \quad \left(\because \rho = \frac{1}{\sigma}\right)$

$$= \frac{0.928 \times 10^{-2}}{[2.5 \times 10^{19} \times 1.6 \times 10^{-19} (0.39 + 0.19) \times 10^{-6}]}$$

$$= 4000\,\Omega \text{ or } 4\,\text{k}\Omega$$

11 For forward biased p-n junction diodes its resistance is zero.

So, net resistance of circuit

$$= 10 + \frac{10 \times 10}{10 + 10} = 15\,\text{k}\Omega$$

Net current $I = \frac{V}{R} = \frac{30}{15 \times 10^3} = 2 \times 10^{-3}\,\text{A}$

So, potential difference across

$$AB = 2 \times 10^{-3} \times 5 \times 10^3 = 10\text{V}$$

12 $\therefore \beta = \frac{I_C}{I_B}$ and $I_E = I_C + I_B$

$$\therefore \beta = \frac{I_C}{I_E - I_C} = \frac{5.488}{5.60 - 5.488} = 49$$

13 As, we know, in case of a common-emitter configuration, DC current gain,

$$\alpha = \frac{I_c}{I_e}.$$

where, I_c is collector current and I_e is emitter current

and AC current gain,

$$\beta = \frac{I_c}{I_b}.$$

where, I_b is base current.

Also, $I_e = I_b + I_c$

Dividing whole equation by I_c, we get

$$\Rightarrow \frac{I_e}{I_c} = \frac{I_b}{I_c} + 1$$

$$\Rightarrow \frac{1}{\alpha} = \frac{1}{\beta} + 1$$

$$\Rightarrow \alpha = \frac{\beta}{1 + \beta}$$

DAY THIRTY FIVE

Gate Circuit

Learning & Revision for the Day

- Logic Gates and Truth Table
- The OR Gate
- The AND Gate
- The NOT Gate
- Combination of Logic Gates
- Transistor as a Switch

Logic Gates and Truth Table

Logic Gates The digital circuit that can be analysed with the help of Boolean algebra is called logic gate or logic circuit. A logic gate has two or more inputs but only one input. There are primarily three logic gates namely the OR gate, the AND gate and the NOT gate.

Truth Table The operation of a logic gate or circuit can be represented in a table which contains all possible inputs and their corresponding outputs is called the truth table. To write the truth table, we use binary digits 1 and 0.

The OR Gate

The OR gate is a device has two or more inputs and one output. This device combines two inputs to give one output. The logic symbol of OR gate is

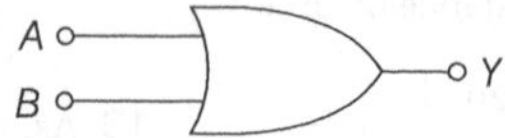

The Boolean expression for OR gate is

$$Y = A + B$$

This indicates Y equals A OR B.

Truth table for OR gate ($Y = A + B$)

A	B	Y
0	0	0
0	1	1
1	0	1
1	1	1

The output of an OR gate assumes 1 if one of more inputs assume 1. The output is high when either of inputs A or B is high, but not if both A and B are low.

PREP MIRROR

Your Personal Preparation Indicator

- No. of Questions in Exercises (x)—
- No. of Questions Attempted (y)—
- No. of Correct Questions (z)—
  (Without referring Explanations)
- Accuracy Level ($z/y \times 100$)—
- Prep Level ($z/x \times 100$)—

In order to expect good rank in JEE, your Accuracy Level should be above 85 & Prep Level should be above 75.

The AND Gate

The AND gate a device has also two or more inputs and one output. The output of AND gate is equal to product of its inputs. The logic symbol of AND gate is

The logic symbol of AND gate is given as under. The Boolean expression for AND gate is $Y = A \cdot B$, this indicates Y equals to A AND B.

Truth table for AND gate ($Y = A \cdot B$)

A	B	Y
0	0	0
0	1	0
1	0	0
1	1	1

The output of an AND gate is 1 only, when all the inputs assume 1.

The NOT Gate

The NOT gate is a device which has only one input and only one output. Its output is complement of input. The logic symbol of NOT gate is as shown in figure.

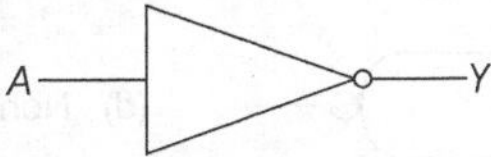

The Boolean expression for NOT gate is

$$Y = \overline{A},$$

which indicates Y equals NOT A.

Truth Table for NOT gate ($Y = \overline{A}$)

A	Y
0	1
1	0

The output of a NOT gate assumes 1, if input is 0 and *vice-versa*. These basic gates (OR, AND and NOT) can be combined in various ways to provide large number of complicated digital circuits.

Combination of Logic Gates

NAND gate and NOR gate are used to make any gate.

1. NAND Gate

In this type of gate, the output of AND gate is fed to input of a NOT gate and final output is obtained at output of NOT gate.

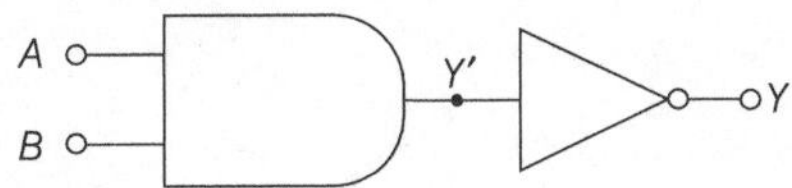

The logic symbol of NAND gate is shown as

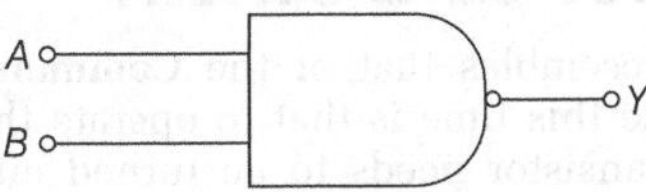

The Boolean expression of NAND gate is $Y = \overline{A \cdot B}$, which indicates 'A AND B are negated'.

Truth table for NAND gate

A	B	Y'	Y
0	0	0	1
1	0	0	1
0	1	0	1
1	1	1	0

2. NOR Gate

In this type of gate, the output of OR gate is fed to input of the NOT gate and final output is obtained at output of the NOT gate.

The logic symbol of NOR gate is shown as

The Boolean expression for NOR gate is $Y = \overline{A + B}$, which indicates that 'A OR B are negated'

Truth table for NOR gate

A	B	Y'	Y
0	0	0	1
1	0	1	0
0	1	1	0
1	1	1	0

NOTE
- NAND and NOR gates are known as universal gate.
- The Boolean expressions obey the commutative law, associative law as well as distributive law.

Commutative law

(i) $A + B = B + A$ (ii) $A \cdot B = B \cdot A$

Associative law

(iii) $A + (B + C) = (A + B) + C$ (iv) $(A \cdot B) \cdot C = A \cdot (B \cdot C)$

Distributive law

(v) $A \cdot (B + C) = A \cdot B + A \cdot C$

(vi) $A + \overline{A} \cdot B = A + B$

(vii) $A + A \cdot B = A$

(viii) $A \cdot (A + B) = A$

(ix) $A \cdot (\overline{A} + B) = A \cdot B$

(x) $\overline{A \cdot B} = \overline{A} + \overline{B}$

(xi) $\overline{A + B} = \overline{A} \cdot \overline{B}$

(xii) $\overline{\overline{A}} = A$

Transistor as a Switch

The circuit resembles that of the **Common-Emitter** circuit. The difference this time is that to operate the transistor as a switch the transistor needs to be turned either fully "OFF" (Cut-off) fully "ON" (Saturated).

An ideal transistor switch would have an infinite resistance when turned 'OFF' resulting in zero current flow and zero resistance, when turned "ON", resulting in maximum current flow.

In practice, when turned "OFF", small leakage currents flow through the transistor and when fully "ON" the device has a low resistance value causing a small saturation voltage (V_{ce}) across it. In both the cut-off and saturation regions, the power dissipated by the transistor is at its minimum.

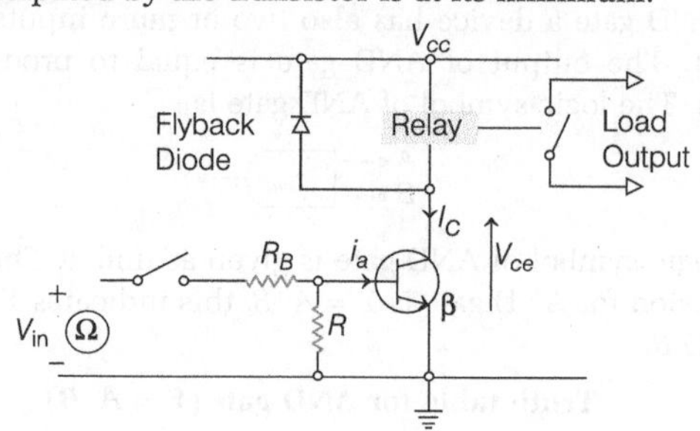

DAY PRACTICE SESSION 1

FOUNDATION QUESTIONS EXERCISE

1 The output of a two input OR gate is fed to a NOT gate, the new gate obtained is

(a) OR gate (b) NOT gate (c) NOR gate (d) NAND gate

2 Which of the following is the truth table for NOT gate?

(a) $\begin{bmatrix} 1 & 1 \\ 0 & 0 \end{bmatrix}$ (b) $\begin{bmatrix} 1 & 0 \\ 0 & 0 \end{bmatrix}$ (c) $\begin{bmatrix} 0 & 1 \\ 1 & 0 \end{bmatrix}$ (d) $\begin{bmatrix} 0 & 1 \\ 1 & 1 \end{bmatrix}$

3 The output of OR gate is high

(a) if either or both the inputs are 1
(b) only if both inputs are 1
(c) if either input is zero
(d) if both inputs are zero

4 To get an output 1 from the circuit as shown in the figure, the input must be

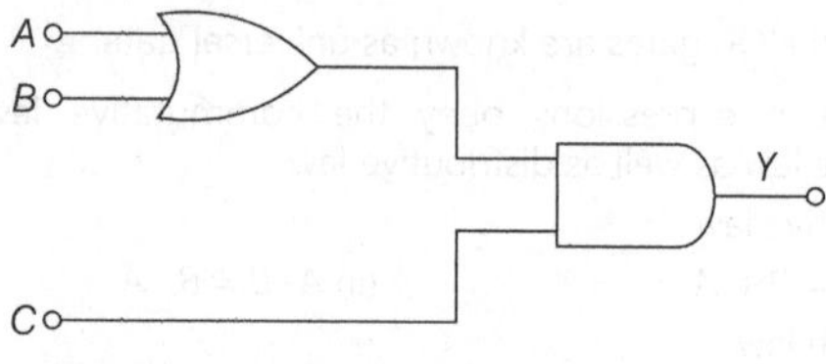

(a) $A = 0, B = 1, C = 0$
(b) $A = 1, B = 0, C = 0$
(c) $A = 1, B = 0, C = 1$
(d) $A = 1, B = 1, C = 0$

5 Digital circuit can be made by the repetitive use of

(a) OR gates (b) AND gates (c) NOT gates (d) NAND gates

6 Which of the following gates will have an output of 1 ?

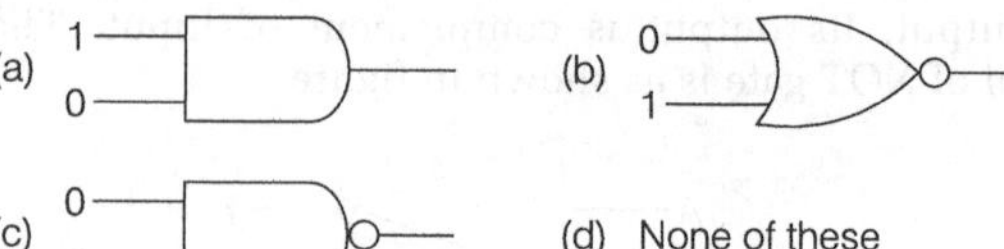

(d) None of these

7 The following truth table is for

A	B	Y
1	1	0
1	0	1
0	1	1
0	0	1

(a) NAND (b) AND (c) XOR (d) NOT

8 Which of the following is the truth table for the circuit below?

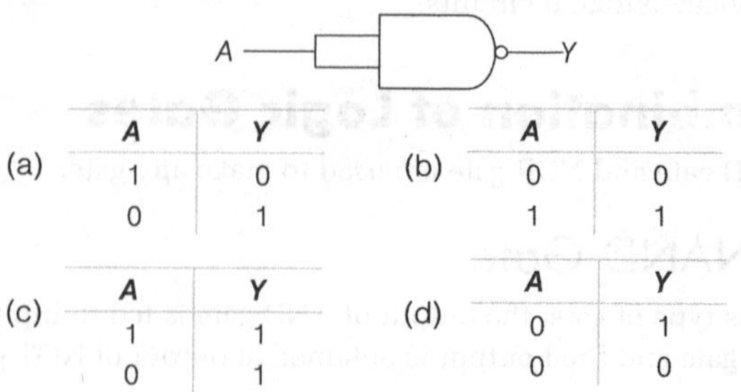

(a)

A	Y
1	0
0	1

(b)

A	Y
0	0
1	1

(c)

A	Y
1	1
0	1

(d)

A	Y
0	1
0	0

9 The circuit as shown in figure below will act as

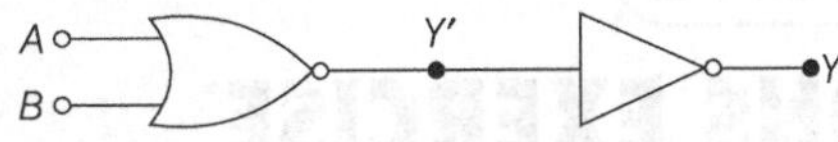

(a) OR gate (b) AND gate
(c) XOR gate (d) None of these

10 The circuit as shown below will acts as

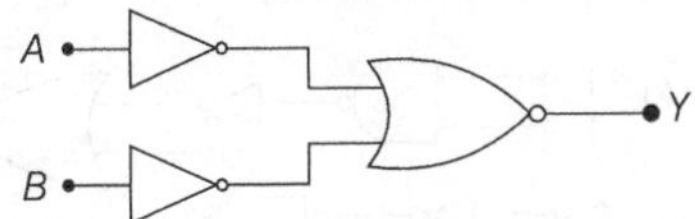

(a) AND gate (b) OR gate
(c) NAND gate (d) NOR gate

11 The circuit is equivalent to

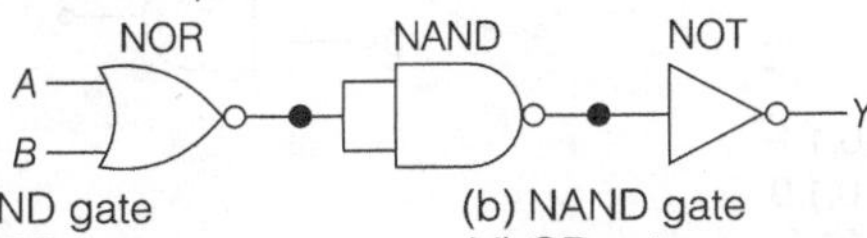

(a) AND gate (b) NAND gate
(c) NOR gate (d) OR gate

12 The output of an OR gate is connected to both the inputs of a NAND gate. The combination will serve as a → AIEEE 2011

(a) OR gate (b) NOT gate
(c) NOR gate (d) AND gate

13 Truth table for system of four NAND gates as shown in figure is → AIEEE 2012

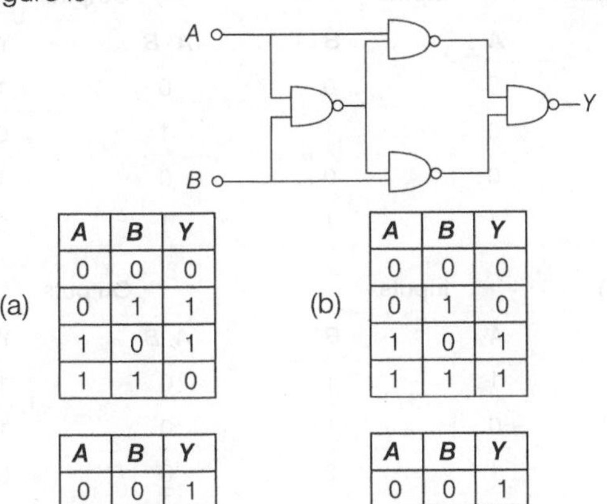

(a)

A	B	Y
0	0	0
0	1	1
1	0	1
1	1	0

(b)

A	B	Y
0	0	0
0	1	0
1	0	1
1	1	1

(c)

A	B	Y
0	0	1
0	1	1
1	0	0
1	1	0

(d)

A	B	Y
0	0	1
0	1	0
1	0	0
1	1	1

14 The combination of gates shown below yields

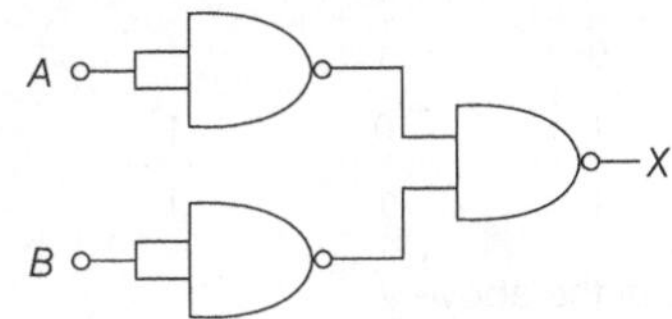

→ AIEEE 2010

(a) OR gate (b) NOT gate
(c) NOR gate (d) NAND gate

15 The diagram of a logic circuit is given below. The output F of the circuit is given by

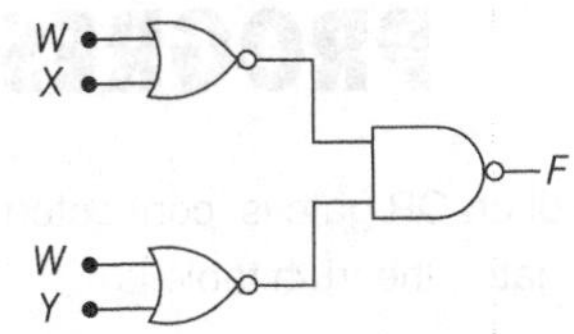

(a) $W\cdot(X+Y)$ (b) $W\cdot(X\cdot Y)$
(c) $W+(X\cdot Y)$ (d) $W+(X+Y)$

16 The output, Y of given logic circuit is

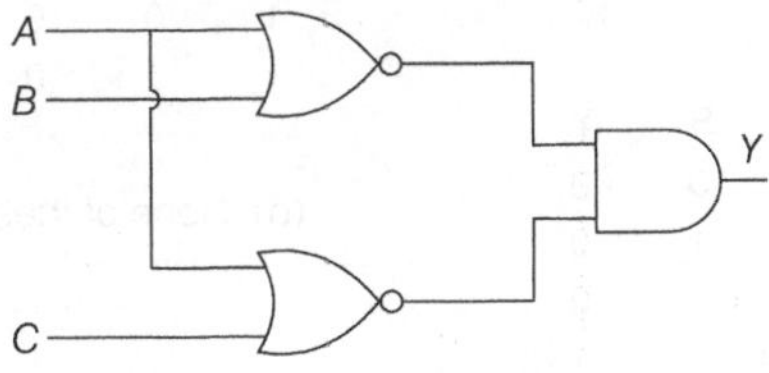

(a) $A\cdot(B+C)$ (b) $A\cdot(B\cdot C)$
(c) $\overline{(A+B)}\cdot\overline{(A+C)}$ (d) $A+B+C$

17 What will be the input of A and B for the Boolean expression $(\overline{A+B})\cdot(\overline{A\cdot B})=1$?

(a) 0, 0 (b) 0, 1
(c) 1, 0 (d) 1, 1

18 Which of the following is not equal to 1 in Boolean algebra?

(a) $A+1$ (b) $A\cdot\overline{A}$
(c) $A+\overline{A}$ (d) $\overline{\overline{A\cdot A}}$

Direction (Q.Nos. 19-21) *Each of these questions contains two statements: Statement I and Statement II. Each of these questions also has four alternative choices, only one of which is the correct answer. You have to select one of the codes (a),(b), (c),(d) given below.*

(a) Statement I is true, Statement II is true; Statement II is the correct explanation for Statement I
(b) Statement I is true, Statement II is true; Statement II is not the correct explanation for Statement I
(c) Statement I is true; Statement II is false
(d) Statement I is false; Statement II is true

19 **Statement I** The logic gate NOT cannot be built using diode.

Statement II The output voltage and the input voltage of the diode have 180° phase difference.

20 **Statement I** NOT gate is also called invertor.

Statement II NOT gate inverts the input signal.

21 **Statement I** NAND or NOR gates are called digital building blocks.

Statement II The repeated use of NAND or NOR gates can produce all the basic or complicated gates.

DAY PRACTICE SESSION 2

PROGRESSIVE QUESTIONS EXERCISE

1 The output of an OR gate is connected to both the inputs of a NAND gate, the truth table is

A	B	Y
1	0	0
1	1	1
0	1	0

A	B	Y
0	0	1
0	1	0
1	0	0
1	1	0

A	B	Y
0	0	0
0	1	0
1	0	0
1	1	1

(d) None of these

2 For the given combination of gates, if the logic states of inputs *A*, *B* and *C* are as follows. $A = B = C = 0$ and $A = B = 1$, $C = 0$, then the logic states of output *D* are

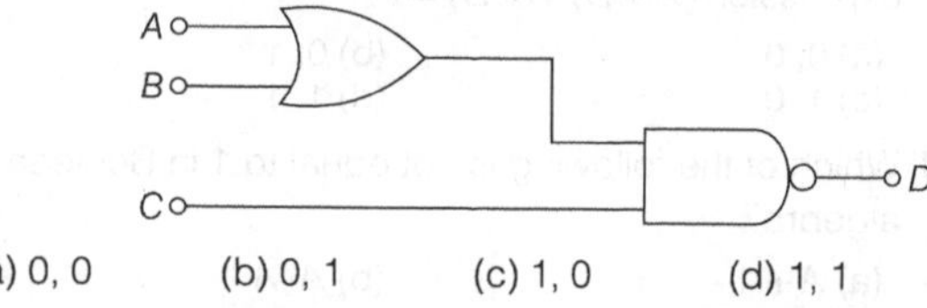

(a) 0, 0 (b) 0, 1 (c) 1, 0 (d) 1, 1

3 The real time variation of input signals *A* and *B* are as shown below. If the inputs are fed into NAND gate, then select the output signal from the following

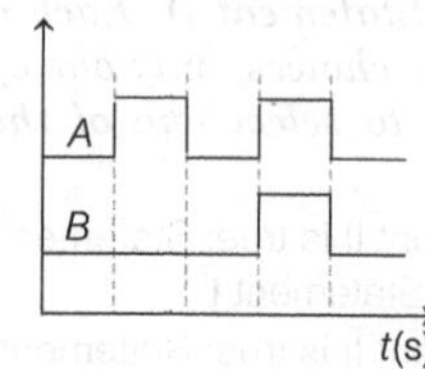

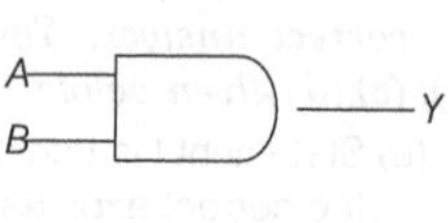

(a)
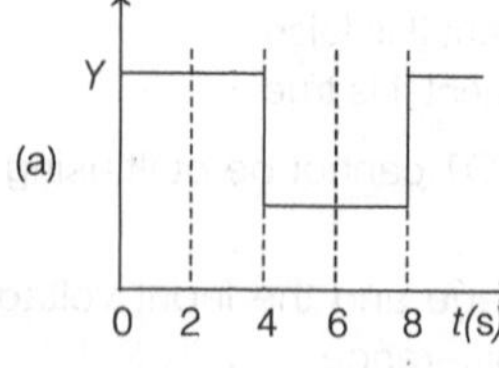

(b)
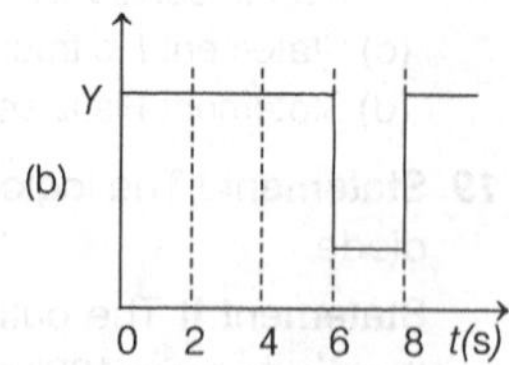

(c)
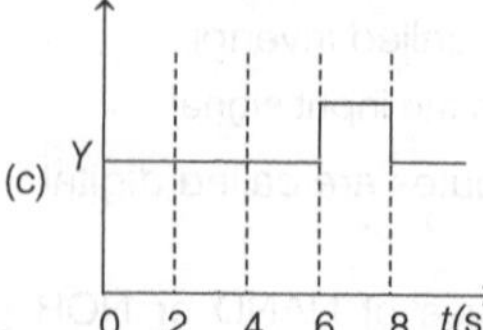

(d)
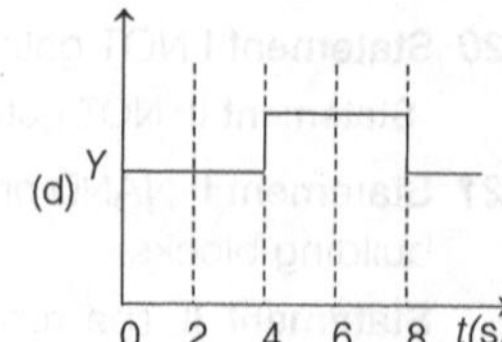

4 In the following combinations of logic gates, the outputs of *A*, *B* and *C* are respectively

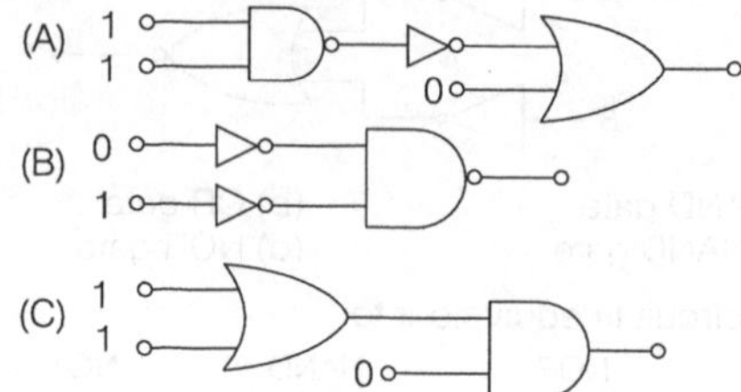

(a) 0,1,1
(b) 0,1,0
(c) 1,1,0
(d) 1,0,1

5 The truth table of the following combination of gates is

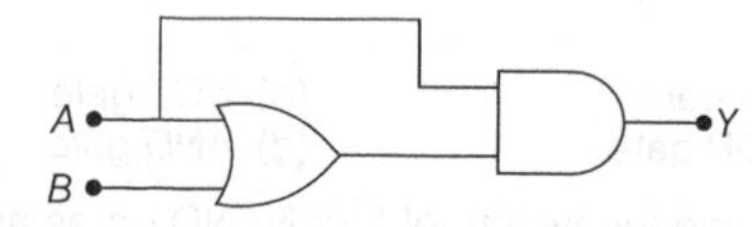

(a)

Inputs		Outputs	
A	B	$A \cdot B$	Y
0	0	0	1
0	1	1	0
0	0	0	1
1	1	1	0

(b)

Inputs		Outputs	
A	B	$A \cdot B$	Y
1	1	0	1
0	1	0	1
1	0	0	0
0	1	1	1

(c)

Inputs		Outputs	
A	B	$A + B$	$Y = A \cdot (A + B)$
0	0	0	0
0	1	1	0
1	0	1	1
1	1	1	1

(d) None of the above

6 The logic circuit shown below has the input waveforms A and B as shown. Pick out the correct output waveform.

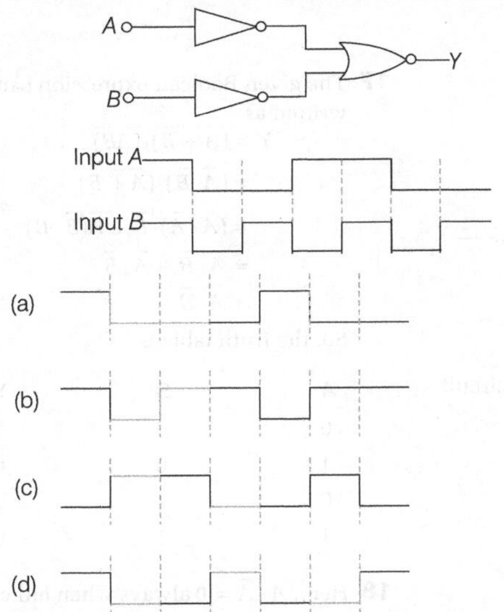

7 In the adjacent circuit, A and B represent two inputs and C represents the output.

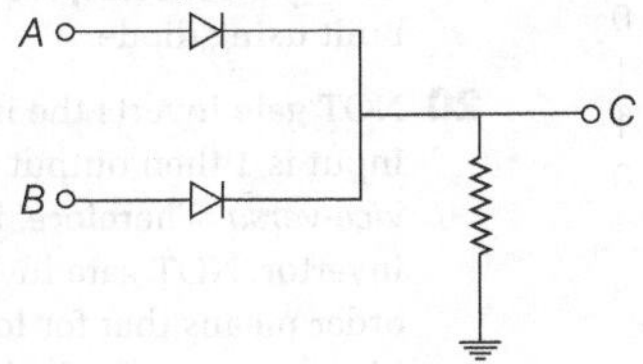

The circuit represents

(a) NOR gate (b) AND gate
(c) NAND gate (d) OR gate

8 Which of the following circuits has given outputs?

A	B	C
0	0	0
0	1	0
1	0	1
1	1	0

→ JEE Main (Online) 2013

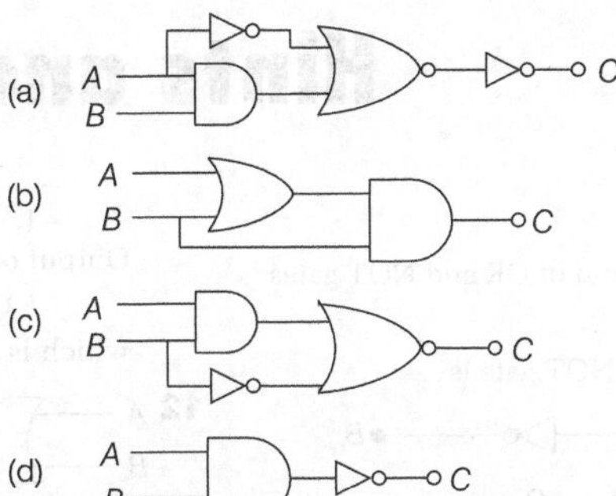

9 Consider two n-p-n transistors as shown in figure. If 0 V corresponds to false and 5 V corresponds to true, then the output at C corresponds to

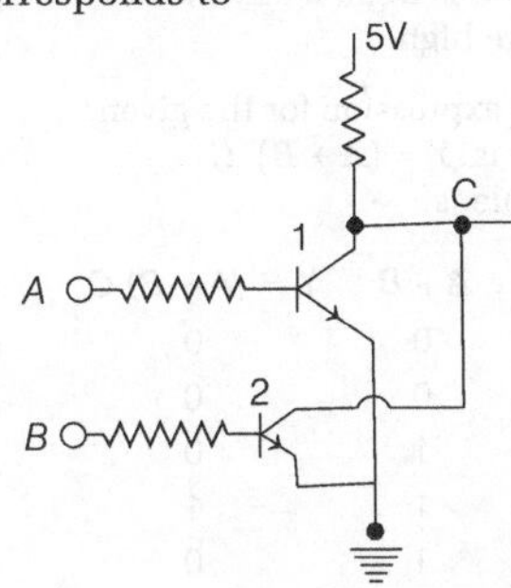

→ JEE Main (Online) 2013

(a) A NAND B (b) A OR B
(c) A AND B (d) A NOR B

10 If a, b, c and d are inputs to a gate and x is its output every time, then as per the following time graph, the gate is → JEE Main 2016 (Offline)

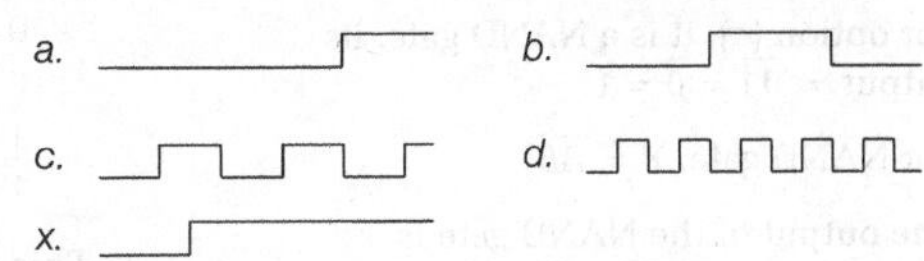

(a) NOT
(b) AND
(c) OR
(d) NAND

ANSWERS

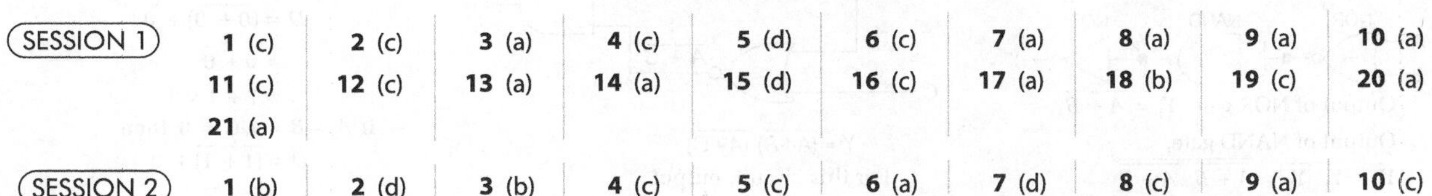

SESSION 1	1 (c)	2 (c)	3 (a)	4 (c)	5 (d)	6 (c)	7 (a)	8 (a)	9 (a)	10 (a)
	11 (c)	12 (c)	13 (a)	14 (a)	15 (d)	16 (c)	17 (a)	18 (b)	19 (c)	20 (a)
	21 (a)									
SESSION 2	1 (b)	2 (d)	3 (b)	4 (c)	5 (c)	6 (a)	7 (d)	8 (c)	9 (a)	10 (c)

Hints and Explanations

SESSION 1

1 The combination of OR and NOT gates is NOR gate.

2 Truth table of NOT gate is

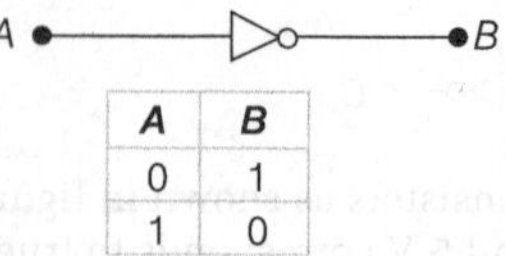

A	B
0	1
1	0

3 OR gate output is high, if anyone or both input are high.

4 The Boolean expression for the given combination is $Y = (A + B) \cdot C$

The truth table is

A	B	C	A + B	Y = (A + B) C
0	0	0	0	0
0	0	1	0	0
0	1	0	1	0
0	1	1	1	1
1	0	0	1	0
1	0	1	1	1
1	1	0	1	0
1	1	1	1	1

Hence, $A = 1$, $B = 0$ and $C = 1$

5 The repetitive use of NAND and NOR gate gives digital circuits.

6 For option (c), it is a NAND gate, its output $= \overline{0.1} = \overline{0} = 1$

7 For NAND gate, $Y = \overline{AB}$

8 The output of the NAND gate is

$$Y = \overline{A \cdot A} = \overline{A} + \overline{A} = \overline{A}$$

9 The output of NOR gate is made input for NOT gate.

$$Y = \overline{\overline{A + B}} = A + B$$

10 The output of two NOT gate is input for NOR gate.

Hence, $Y = \overline{\overline{A} + \overline{B}} = \overline{\overline{A}} \cdot \overline{\overline{B}} = A \cdot B$ (AND gate)

11 The gate circuit can be shown by given two points A and B.

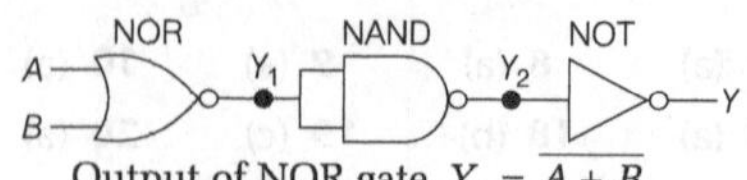

Output of NOR gate, $Y_1 = \overline{A + B}$

Output of NAND gate,

$$Y_2 = \overline{Y_1 \cdot Y_1} = \overline{\overline{A + B} \cdot \overline{A + B}}$$

$$= \overline{\overline{A + B}} + \overline{\overline{A + B}}$$

$$= (A + B) + (A + B) = A + B$$

Output of NOT gate,

$$Y = \overline{Y_2} = \overline{A + B}$$

which is output of NOR gate.

12

$Y' = A + B$ and $Y = \overline{Y'} = \overline{A + B}$

i.e. output of a NOR gate.

13 Boolean expression for the given circuit

$$Y = \overline{\overline{(A \cdot \overline{(A \cdot B)})} \cdot \overline{(B \cdot \overline{(A \cdot B)})}}$$

$$= \overline{(\overline{A} + A \cdot B) \cdot (\overline{B} + A \cdot B)}$$

$$= \overline{(\overline{A} + A \cdot B)} + \overline{(\overline{B} + (A \cdot B))}$$

$$= A \cdot (\overline{A \cdot B}) + B \cdot (\overline{A \cdot B})$$

$$= A \cdot (\overline{A} + \overline{B}) + B \cdot (\overline{A} + \overline{B})$$

$$= A \cdot \overline{B} + B \cdot \overline{A}$$

A	B	$\overline{A}$	$\overline{B}$	$A \cdot \overline{B}$	$B \cdot \overline{A}$	Y
0	0	1	1	0	0	0
0	1	1	0	0	1	1
1	0	0	1	1	0	1
1	1	0	0	0	0	0

So, option (a) is correct.

14 Truth table for given combination is

A	B	X
0	0	0
0	1	1
1	0	1
1	1	1

This comes out to be truth table of OR gate.

15 The output $F = \overline{\overline{(W + X)} \cdot \overline{(W + Y)}}$

$$= \overline{\overline{(W + X)}} + \overline{\overline{(W + Y)}}$$

$$= W + X + W + Y$$

$$= W + X + Y$$

16 The gate circuit is given as

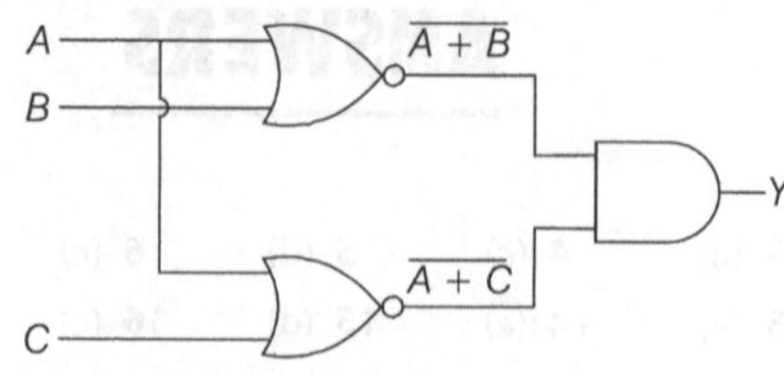

$Y = \overline{(A+B)}\ \overline{(A+C)}$

For this circuit, output

$$Y = (\overline{A + B})(\overline{A + C})$$

17 The given Boolean expression can be written as

$$Y = \overline{(A + B)} \cdot \overline{(AB)}$$

$$= (\overline{A} \cdot \overline{B}) \cdot (\overline{A} + \overline{B})$$

$$= (\overline{A} \cdot \overline{A}) \cdot \overline{B} + \overline{A}(\overline{B} \cdot \overline{B})$$

$$= \overline{A} \cdot \overline{B} + \overline{A} \cdot \overline{B}$$

$$= \overline{A} \cdot \overline{B}$$

So, the truth table is

A	B	Y
0	0	1
1	0	0
0	1	0
1	1	0

18 Here, $A \cdot \overline{A} = 0$ always when either $A = 0$ or 1.

19 NOT gate inverts the signal applied to it. But in diode, the input and output are in same phase. Thus, NOT gate cannot be built using diode.

20 NOT gate inverts the input signal i.e. if input is 1 then output will be zero or *vice-versa*. Therefore, it is called as invertor. NOT gate inverts the input order means that for low input, it gives high output or for high input, it gives low output.

21 NAND or NOR gates are called universal (digital) building blocks because using these two types of gates we can produce all the basic gates namely OR, AND or other complex gates.

SESSION 2

1 When two inputs of a NAND gate are joined together, it works as a NOT gate. The OR gate connected to this NOT gate results is a NOR gate.

2 The output D for the given combination

$$D = \overline{(A + B) \cdot C} = \overline{(A + B)} + \overline{C}$$

If $A = B = C = 0$, then

$$D = \overline{(0 + 0)} + \overline{0}$$

$$= \overline{0} + \overline{0}$$

$$= 1 + 1 = 1$$

If $A = B = 1, C = 0$, then

$$D = \overline{(1 + 1)} + \overline{0}$$

$$= \overline{1} + \overline{0}$$

$$= 0 + 1 = 1$$

3 From real time variation of input signals,we can from truth table for A and B and conclude output from NAND gate.

Inputs		Output
A	B	Y
0	0	1
1	0	1
0	0	1
1	1	0
0	0	1

From output, we can show real time variation of output signal as below.

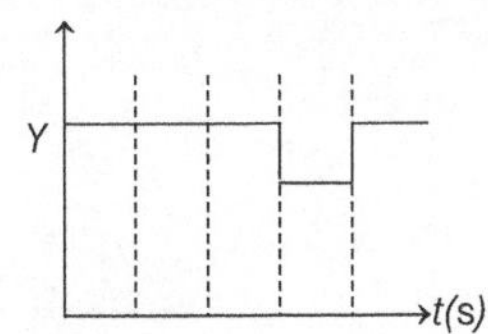

4 A. NAND operation on (1, 1) = 0

NOT operation on (0) = 1

OR operation on (1, 0) = 1

B. NOT operation on (0, 1) = (1, 0)

NAND operation on (1, 0) = 1

C. OR operation on (1, 1) = 1

AND operation on (1, 0) = 0

5 Let us draw the given combination pointing out that the first gate is OR gate second gate is AND gate. The inputs of the OR gate, are A and B, and its output is $A + B$ that is A OR B.

The inputs of the AND gate are A and $A + B$ and its output is $A \cdot (A + B)$ that is A AND (A or B). The truth table for the output is $Y = A.(A + B)$ is as follows

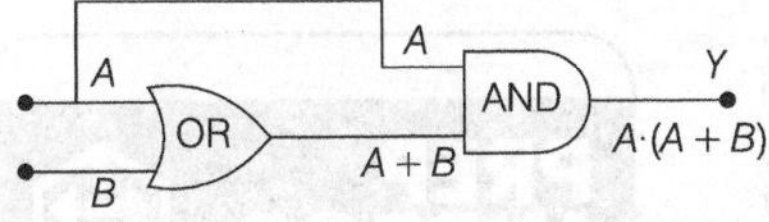

Inputs		Outputs	
A	B	A + B	Y = A·(A + B)
0	0	0	0
0	1	1	0
1	0	1	1
1	1	1	1

6 Truth table

A	B	Y
0	0	0
0	1	0
1	0	0
1	1	1
0	0	0

7 If we give the following inputs to A and B, then corresponding output is shows in table

A	B	C
0	0	0
0	1	1
1	0	1
1	1	1

The above table is similar to OR gate.

8 Observing the given gate we observe that gate would be same as given in option in which.
The values $A = 0$, $B = 0$ gives output 0
The values $A = 0$, $B = 1$ gives output 0
The values $A = 1$, $B = 0$ gives output 1
The value $A = 1$, and $B = 1$ gives output 0

9 From the figure of AND gate

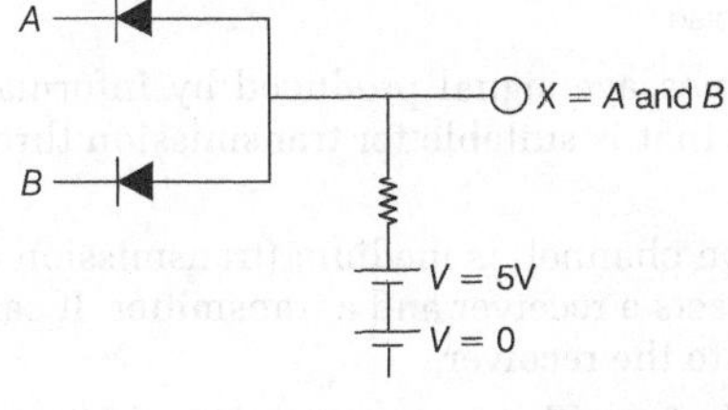

and NOT gate

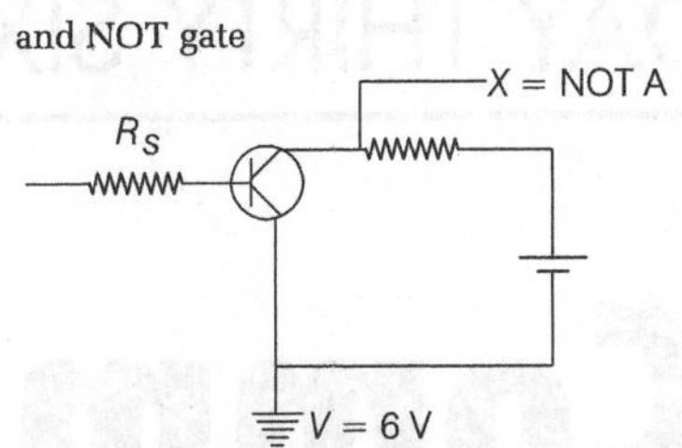

Clearly, the function X = NOT (A AND B) of the logical variables A AND B is called NAND gate.

10 Output of OR gate is 0 when all inputs are 0 and output is 1 when atleast one of the inputs is 1.

Observing output x It is 0 when all inputs are 0 and it is 1 when atleast one of the inputs is 1.

∴ The gate is OR.

Alternative Method

OR Gate

a	b	c	d	x
0	0	0	0	0
0	0	0	1	1
0	0	1	0	1
0	0	1	1	1
0	1	0	0	1
0	1	0	1	1
0	1	1	0	1
0	1	1	1	1
1	0	0	0	1

DAY THIRTY SIX

Communication Systems

Learning & Revision for the Day

- Basic Elements of a Communication System
- Line Communication
- Optical Communication
- Modulation
- Demodulation or Detection
- Propagation of Electromagnetic Waves
- Satellite Communication

Basic Elements of a Communication System

A communication system is a set up used to transmit information from one point to another. The essential parts of a communication system are transmitter, communication channel and receiver as shown in the following block diagram.

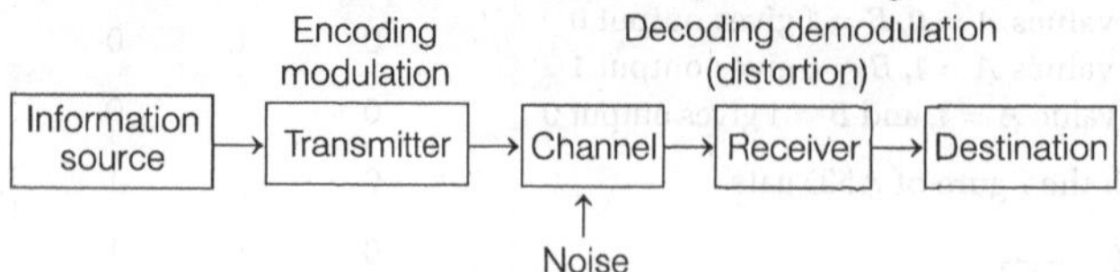

(i) **Transmitter** Transmitter converts the message signal produced by information source into a form (*e.g.*, electrical signal) that is suitable for transmission through the channel to the receiver.

(ii) **Communication Channel** Communication channel is medium (transmission line, an optical fibre or free space) which connects a receiver and a transmitter. It carries the modulated wave from the transmitter to the receiver.

(iii) **Receiver** This set up receives the signals from the communication channel and converts these signals into their original form.

Important Terms Used in Communication System

Signal Signal represents the electrical analog of the information. It can be analog or digital.

(a) **Analog Signal** Signal which varies continuously with respect to time is called analog signal.

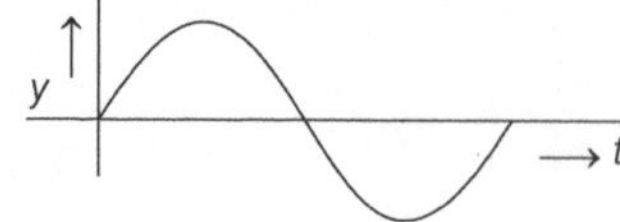

- No. of Questions in Exercises (*x*)—
- No. of Questions Attempted (*y*)—
- No. of Correct Questions (*z*)— (*Without referring Explanations*)

- Accuracy Level ($z/y \times 100$)—
- Prep Level ($z/x \times 100$)—

In order to expect good rank in JEE, your Accuracy Level should be above 85 & Prep Level should be above 75.

(b) **Digital Signal** A digital signal has two levels of current or voltage represented by 0 or 1. It is usually in the form of pulses.

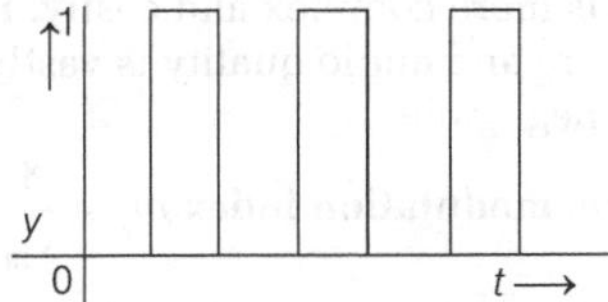

Transducer A device that converts the message signal into electrical signal before feeding it to transmitter. In other words, transducer converts one form of energy into another.

Noise It refers to the unwanted signals that tend to disturb the transmission and hence a distorted version of the transmitted signal reaches at receiver.

Bandwidth Bandwidth refers to the range over, which the frequencies in a signal vary.

Amplification It is the process of increasing the amplitude and thus strength of an electrical signal.

Line Communication

In line communication, there is a physical connection between source and destination. The wired connections between two points are known as **transmission lines.**

The wires that are most popular for wired communication or line communication are

(i) Co-axial (ii) Parallel wire lines (iii) Twisted pair cables

Optical Communication

Optical communication uses light waves in the frequency range 10^{12} to 10^{16} Hz as the guided wave medium for propagation of audio frequency signal.

Main advantage of optical communication system lies in the fact that here very high bandwidths of MHz and even GHz are possible. Consequently, a large number of messages can be transmitted through a single cable without any risk of their intermixing.

Moreover due to very little line loss the quality of reception is vastly superior.

An optical communication system consists of mainly three parts which are

(i) optical source and modulator

(ii) optical fibre cable, and (iii) optical signal detector.

Optical Fibre

Optical fibre make use of the principle of total internal reflection of light. The refractive index n_1 of central core is higher than refractive index n_2 of cladding.

Total internal reflection will take place at core-cladding interface if angle of incidence there is

$$i = \sin^{-1}\left(\frac{n_2}{n_1}\right)$$

For above condition to be fulfilled the light ray must enter the optical fibre at a maximum acceptance angle θ_0 from the axis of fibre such that

$$(\theta_0)_{\max} = \sin^{-1}\left[\frac{\sqrt{n_1^2 - n_2^2}}{n_0}\right]$$

where, n_0 is the refractive index of the outer medium. For air, $n_0 = 1$ and then

$$(\theta_0)_{\max} = \sin^{-1}(\sqrt{n_1^2 - n_2^2})$$

NOTE
- Numerical Aperture (NA) $= n_0 \sin(\theta_0)_{\max} = \sqrt{n_1^2 - n_2^2}$
- All the information in optical fibre is carried out by the principal of total internal reflection and all the information is carried in core of the optical fibre.
- If angle of incidence is greater than $(\theta_0)_{\max}$, then total internal reflection will not take place and some information will be lossed.

Modulation

The phenomenon of superposition of information signal over a high frequency carrier wave is called modulation. In this process, amplitude, frequency or phase angle of a high frequency carrier wave is modified in accordance with the instantaneous value of the low frequency information.

Need for Modulation

Digital and analog signals to be transmitted are usually of low frequency and hence, cannot be transmitted as such. These signals require some carrier to be transported.

(i) **Frequency of Signal** The audio frequency signals (20 Hz to 20 kHz) cannot be transmitted without distortion over long distances due to less energy carried by low frequency audio waves. The energy of a wave is directly proportional to square of its frequency. Even if the audio signal is converted into electrical signal, the later cannot be sent very far without employing large amount of power.

(ii) **Height of antenna** For efficient radiation and reception, the height of transmitting and receiving antennas should be comparable to a quarter wavelength of the frequency used.

$$\text{Wavelength} = \frac{\text{velocity}}{\text{frequency}} = \frac{3 \times 10^8}{\text{frequency (Hz)}} \text{ metre}$$

For 1 MHz it is 75 m and for 15 kHz frequency, the height of antennas has to be about 5 km which size is unthinkable.

(iii) **Number of channels** Audio frequencies are concentrated in the range 20 Hz to 20 kHz. This range is so narrow that there will be overlapping of signals. In order to separate, the various signals it is necessary to convert all of them to different portions of the electromagnetic spectrum.

There are two types of modulation

1. Amplitude Modulation (AM)

- Amplitude Modulation (AM) is the process of changing the amplitude (A_c) of a carrier wave linearly in accordance with the amplitude of message signal (A_m).
- The ratio $\mu = \frac{A_m}{A_c} = \frac{A_{max} - A_{min}}{A_{max} + A_{min}}$ is called the modulation index and in practice we maintain $\mu \le 1$, so as to avoid distortion.
- In AM modulated wave signal we have carrier wave of frequency ω_c and two side bands of frequencies $(\omega_c - \omega_m)$ and $(\omega_c + \omega_m)$, respectively. Thus, total bandwidth of AM signals is $2\omega_m$.
- Upper Side Band Frequency (USB) $= \nu_c + \nu_m$ where, ν_c = carrier frequency, ν_m = signal frequency.
- Lower Side Band Frequency (LSB) $= \nu_c - \nu_m$ where, ν_c = carrier frequency, ν_m = signal frequency
- Bandwidth $= \nu_{USB} - \nu_{LSB} = (\nu_c + \nu_m) - (\nu_c - \nu_m) = 2\,\nu_m$
- Power of carrier, $P_C = \frac{(A_c/\sqrt{2})^2}{R} = \frac{A_c^2}{2R}$
- Power of side band, $P_{sb} = \frac{1}{R}\left(\frac{\mu A_c}{2\sqrt{2}}\right)^2 + \frac{1}{R}\left(\frac{\mu A_c}{2\sqrt{2}}\right)^2 = \frac{\mu^2 A_c^2}{8R}$
- AM technique is simpler and cost effective. However, it suffers from noisy reception, low efficiency, small operating range and poor audio quality.
- **Power dissipated** in AM wave, $P = P_c\left[1 + \frac{\mu^2}{2}\right]$ where, $P_c = \frac{A_c^2}{2R}$ is power dissipated by unmodulated carrier wave and μ = modulation index.

2. Frequency Modulation (FM)

- Frequency modulation is the process of changing the frequency of a carrier wave in accordance with the frequency of message signal.
- In FM modulated wave the amplitude of wave and total transmitted power remains constant.
- Frequency of modulated signal consists of central band of frequency ω_c and side bands of frequencies $(\omega_c \pm \omega_m), (\omega_c \pm 2\omega_m), (\omega_c \pm 3\,\omega_m)$ .....

  The number of side bands depends on the modulation index.
- Total **bandwidth** of modulated signal $= 2n \cdot \omega_m$, where, n = number of significant side band pairs.
- FM technique is more complex and costly. However, efficiency is more and audio quality is vastly improved. Noise level is negligible.

  In the FM wave, **modulation index** $m_f = \frac{\delta}{\nu_m}$

  where, δ = maximum frequency of deviation $= (\nu_c - \nu_{min})$ or $(\nu_{max} - \nu_c)$

  ν_m = modulation frequency.

Demodulation or Detection

The process of recovering the original audio signal from the modulated wave is called demodulation.

Demodulation can be done by using a diode and a capacitor filter as shown under

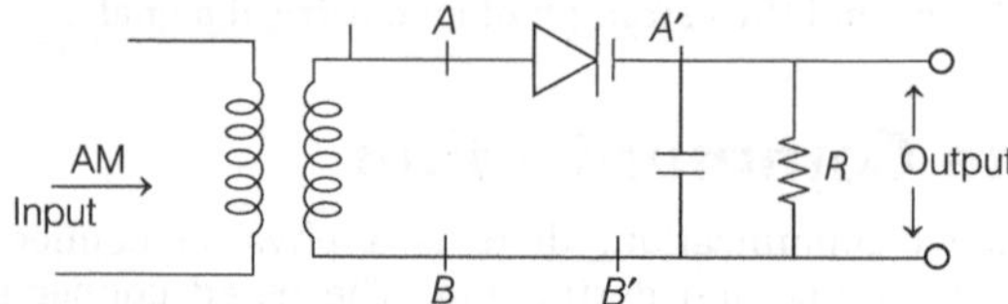

Working of *R-C* Circuits The value of *R-C* is so selected such that $\frac{1}{f_c} << RC$, where f_c = frequency of carrier wave.

For AM modulated wave, a *p-n* junction diode or a vacuum tube diode is used as a demodulator. A diode basically acts as a rectifier and thus, reduces the modulated carrier wave into positive envelope only. This positive envelope is sent through a *R-C* circuit. Carrier wave passes through the capacitor and *AF* signal is regenerated across *R*.

Propagation of Electromagnetic Waves

It is that category of communication in which no line or cable is used as a communication channel and the modulated signal is propagated through free space.

Different types of propagation depending upon properties are

1. Ground Wave Propagation

In this type of communication, transmitting and receiving antenna are close to surface of the earth. This type of propagation can be sustained only at low frequency ($\approx$ 500 kHz to 1500 kHz). Due to such less frequency range, it is also called medium wave propagation.

2. Sky Wave Propagation

Sky wave is the radiowave which is directed towards the sky and reflected back by the ionosphere towards the desired location on the earth. Radiowaves of frequencies 2 MHz to 20 MHz can be reflected by the ionosphere.

- **Critical Frequencies** It is the maximum frequency of the radiowaves which can be reflected from ionosphere and returns to the earth. The radiowave will penetrate the ionosphere above this frequency. It is given by

$$f_c = 9(N_{max})^{1/2}$$

where, N_{max} is the maximum electron density of the ionosphere.

The sky waves being electromagnetic in nature, changes the dielectric constant and refractive index of the ionosphere. The effective refractive index of ionosphere is

$$\mu = \mu_0 \sqrt{1 - \frac{81.45\, N}{\nu^2}}$$

where, N = electron density of ionosphere,

ν = frequency of electromagnetic wave in Hz and

μ_0 = refractive index of free space

- **Bandwidth of a Communication Channel** The difference between the highest and lowest frequencies that a communication channel allows to pass through it is called its bandwidth.

$$\text{Number of channels} = \frac{\text{Total bandwidth of channel}}{\text{Bandwidth per channel}}$$

- **Maximum Usable Frequency (MUF)** In this the sky waves with maximum frequencies are sent at some angles towards the ionosphere. Then these rays will again be reflected by the ionosphere to the earth.

$$\text{MUF} = \frac{\text{Critical Frequency (CF)}}{\cos\theta} = CF \sec\theta$$

This is the angle which is formed along the direction of the incident wave and the normal.

- **Skip distance** When the sky wave is reflected back from the ionosphere having a constant frequency, but less than that of the critical frequency, then the smallest distance from the transmitter to the earth's surface covered by the sky wave is known as skip distance.

3. Space Wave Propagation

The transmitted signal is received by the direct interception of the signal by receiving antenna. The, frequency range is (100 MHz to 220 MHz). The maximum range of this transmission depends upon the height of transmitting antenna and is given by,

$$d = \sqrt{2hR},\ R >> h.$$

where, h = height of the antenna and R = radius of the earth.

Satellite Communication

- It is mainly done with the help of a geostationary satellite orbiting the earth in the equatorial plane from West to East at a height of about 36000 km above the surface of the earth, so that its revolution time is 24 h.
- The transmitted signal from the earth station is uplinked to satellite. The satellite receives it, demodulate it, amplify it and remodulate it and transmit it back. Now it is downlinked to the earth station. A radio transponder does all these jobs in a satellite.
- Uplink frequency and downlink frequency are kept widely different, so as to avoid their interference in free space.
- For world wide coverage three geostationary satellites are required at 120° apart from each other.

DAY PRACTICE SESSION 1

FOUNDATION QUESTIONS EXERCISE

1 The minimum length of antenna required to transmit a radio signal of frequency 20 MHz is

(a) 5 m (b) 7.5 m (c) 2 m (d) 3.75 m

2 Repeaters are required for transmitting microwave terrestrial communication system over a 40-50 km distance because

(a) microwave power decreases rapidly with distance
(b) the curvature of the earth limits the distance over which the line of sight can be established
(c) signal to noise ratio decreases rapidly with distance
(d) signal distortion creeps in rapidly with distance

3 The characteristic impedance of a coaxial cable is of the order of

(a) 50 Ω (b) 200 Ω
(c) 270 Ω (d) None of these

4 If μ_1 and μ_2 are the refractive indices of the materials of core and cladding of an optical fibre, then the loss of light due to its leakage can be minimised by having

(a) $\mu_1 > \mu_2$
(b) $\mu_1 < \mu_2$
(c) $\mu_1 = \mu_2$
(d) None of the above

5 Which of the following four alternatives is not correct? We need modulation → AIEEE 2011

(a) to increase the selectivity
(b) to reduce the time lag between transmission and reception of the information signal
(c) to reduce the size of antenna
(d) to reduce the fractional band width, i.e. the ratio of the signal band width to the centre frequency

6 A message signal of frequency ω_m is superposed on a carrier wave of frequency ω_c to get an Amplitude Modulated wave (AM). The frequency of the AM wave will be

(a) ω_m (b) ω_c (c) $\frac{\omega_c + \omega_m}{2}$ (d) $\frac{\omega_c - \omega_m}{2}$

7 A speech signal of 3 kHz is used to modulate a carrier signal of frequency 1 MHz, using amplitude modulation. The frequencies of the side bands will be

(a) 1·003 MHz and 0·997 MHz
(b) 3001 kHz and 2997 kHz
(c) 1003 kHz and 1000 kHz
(d) 1 MHz and 0.997 MHz

8 In an amplitude modulated wave for audio frequency of 500 cycle/s, the appropriate carrier frequency will be

(a) 50 cycle/s (b) 100 cycle/s
(c) 500 cycle/s (d) 50000 cycle/s

9 For what value of m_a will the total power per cycle be maximum in the modulated wave?

(a) 0 (b) 1 (c) 1/2 (d) greater than 1

10 In amplitude modulation, sinusoidal carrier frequency used is denoted by ω_c and the signal frequency is denoted by ω_m. The bandwidth $(\Delta\omega_m)$ of the signal is such that $\Delta\omega_m << \omega_c$. Which of the following frequencies is not contained in the modulated wave? → **JEE Main 2017 (Offline)**

(a) ω_c (b) $\omega_m + \omega_c$
(c) $\omega_c - \omega_m$ (d) ω_m

11 Choose the correct statement. → **JEE Main 2016 (Offline)**

(a) In amplitude modulation, the amplitude of the high frequency carrier wave is made to vary in proportion to the amplitude of the audio signal.
(b) In amplitude modulation, the frequency of the high frequency carrier wave is made to vary in proportion to the amplitude of the audio signal.
(c) In frequency modulation, the amplitude of the high frequency carrier wave is made to vary in proportion to the amplitude of the audio signal.
(d) In frequency modulation, the amplitude of the high frequency carrier wave is made to vary in proportion to the frequency of the audio signal.

12 A signal of 5 kHz frequency is amplitude modulated on a carrier wave of frequency 2MHz. The frequencies of the resultant signal is/are → **JEE Main 2015**

(a) 2 MHz only
(b) 2005 kHz and 1995 kHz
(c) 2005 kHz, 2000 kHz and 1995 kHz
(d) 2000 kHz and 1995 kHz

13 An EM wave of maximum frequency 300 kHz and critical frequency 100 kHz is to be transmitted to a height equal to 150 km. Calculate the skip distance.

(a) 624 km (b) 849 km
(c) 636 km (d) 942 km

14 The electron density of E, F_1 and F_2 layers of ionosphere is $2\times10^{11}\text{m}^{-3}$, $5\times10^{11}\text{m}^{-3}$ and $8\times10^{11}\text{m}^{-3}$, respectively. What is the ratio of critical frequency for reflection of radiowaves?

(a) 2 : 4 : 3 (b) 4 : 3 : 2 (c) 2 : 3 : 4 (d) 3 : 2 : 4

15 On a particular day, the maximum frequency reflected from the ionosphere is 9 MHz. On another day, it was found to increase by 1MHz. What is the ratio of the maximum electron densities of the ionosphere on the two days?

(a) 1.23 (b) 1.0 (c) 1.43 (d) 0.75

16 Maximum Usable Frequency (MUF) in F-region layers is x, when the critical frequency is 60 MHz and the angle of incidence is 70°, then x is

(a) 150 MHz (b) 170 MHz (c) 175 MHz (d) 190 MHz

17 Frequencies higher than 10 MHz were found not being reflected by the ionosphere on a particular day at a place. The maximum electron density of the ionosphere on the day was near to

(a) $1.5\times10^{10}\text{ m}^{-3}$ (b) $1.24\times10^{12}\text{ m}^{-3}$
(c) $3\times10^{12}\text{ m}^{-3}$ (d) None of these

18 How the sound waves can be sent from one place to another in space communication?

(a) Through wires
(b) Through space
(c) By superimposing it on undamped electromagnetic waves
(d) By superimposing it on damped electromagnetic waves

19 To cover a population of 20 lakh, a transmitter tower should have a height of (Given radius of the earth = 6400 km, population per square km = 1000) is

(a) 25 m (b) 50 m (c) 75 m (d) 100 m

20 The TV transmission tower in Delhi has a height of 240 m. The distance up to which the broadcast can be received. (Taking the radius of the earth to be 6.4×10^6 m) is

(a) 100 km (b) 60 km (c) 55 km (d) 50 km

21 A radar has a power of 1 kW and is operating at a frequency of 10 GHz. It is located on a mountain top of height 500 m. The maximum distance upto which it can detect object located on the surface of the earth (Radius of earth = 6.4×10^6 m) is → **AIEEE 2012**

(a) 80 km (b) 16 km (c) 40 km (d) 64 km

Direction (Q. Nos. 22-25) *Each of these questions contains two statements : Statement* I *and Statement* II. *Each of these questions also has four alternative choices, only one of which is the correct answer. You have to select one of the codes (a), (b), (c) and (d) given below*

(a) Statement I is true, Statement II is true; Statement II is the correct explanation for Statement I
(b) Statement I is true, Statement II is true; Statement II is not the correct explanation for Statement I
(c) Statement I is true; Statement II is false
(d) Statement I is false; Statement II is true

22 **Statement I** Optical fibre communication has immunity to cross-talk.

Statement II Optical interference between fibres is zero.

23 **Statement I** Transducer in communication system converts electrical signal into a physical quantity.

Statement II For information signal is to be transmitted directly to long distances, modulation is necessary.

24 **Statement I** FM broadcast is preferred over AM broadcast.

Statement II Process of combining the message signals with carrier wave is called demodulation.

25 **Statement I** Modem is a demodulator.

Statement II It works only in a transmitting and receiving mode.

DAY PRACTICE SESSION 2

PROGRESSIVE QUESTIONS EXERCISE

1 A telephonic communication service is working at carrier frequency of 10 GHz. Only 10% of it is utilised for transmission. How many telephonic channels can be transmitted simultaneously, if each channel requires a bandwidth of 5 kHz? → JEE Main 2018

(a) 2×10^3 (b) 2×10^4 (c) 2×10^5 (d) 2×10^6

2 A diode AM detector with the output circuit consisting of $R = 1\,k\Omega$ and $C = 1\,\mu F$ would be more suitable for detecting a carrier signal of

(a) 10 kHz (b) 1 kHz (c) 0.75 kHz (d) 0.5 kHz

3 In optical communication system operating at 1200 nm, only 2% of the source frequency is available for TV transmission having a bandwidth of 5 MHz. The number of TV channels that can be transmitted is

(a) 2 million (b) 10 million (c) 0.1 million (d) 1 million

4 If sky wave with a frequency of 50 MHz is incident on D region at an angle of 30°, then angle of refraction is

(a) 15° (b) 30° (c) 60° (d) 45°

5 Three waves A, B and C of frequencies 1600 kHz, 5 MHz and 60 MHz, respectively are to be transmitted from one place to another. Which of the following is the most appropriate mode of communication?

(a) A is transmitted *via* space wave while B and C are transmitted *via* sky wave
(b) A is transmitted *via* ground wave, B *via* sky wave and C *via* space wave
(c) B and C are transmitted *via* ground wave while A is transmitted *via* sky wave
(d) B is transmitted *via* ground wave while A and C are transmitted *via* space wave

6 Consider telecommunication through optical fibres. Which of the following statements is not true?

(a) Optical fibres can be graded refractive index
(b) Optical fibres are subjected to electromagnetic interference from outside
(c) Optical fibres have extremely low transmission loss
(d) Optical fibres may have homogeneous core with a suitable cladding

7 A diode detector is used to detect an amplitude modulated wave of 60% modulation by using a condenser of capacity 250 pF in parallel with a load resistance 100 kΩ. Find the maximum modulated frequency which could be detected by it. → JEE Main 2013

(a) 10.61 MHz (b) 10.61 kHz (c) 5.31 MHz (d) 5.31 kHz

8 What is the modulation index if an audio signal of amplitude one half of the carrier amplitude is used in AM?

(a) 1 (b) 0
(c) 0.5 (d) greater than 1

9 For 100% modulation, the power carried by the side bands (P_{SB}) is given by

(a) $P_{SB} = P$ (b) $P_{SB} = 3P$ (c) $P_{SB} = \frac{1}{3}P$ (d) $P_{SB} = 0$

Direction (Q. Nos. 10-12) *Each of these questions contains two statements : Statement* I *and* Statement II. *Each of these questions also has four alternative choices, only one of which is the correct answer. You have to select one of the codes* (a), (b), (c) *and* (d) *given below*

(a) Statement I is true, Statement II is true; Statement II is the correct explanation for Statement I
(b) Statement I is true, Statement II is true; Statement II is not the correct explanation for Statement I
(c) Statement I is true; Statement II is false
(d) Statement I is false; Statement II is true

10 **Statement I** Sky wave signals are used for long distance radio communication. These signals are in general, less stable than ground wave signals.

Statement II The state of ionosphere varies from hour to hour, day to day and season to season. → AIEEE 2011

11 **Statement I** Higher the modulation index, the reception will be strong and clear.

Statement II The degree, to which the carrier wave is modulated is called modulation index.

12 **Statement I** Television signals are received through sky wave propagation.

Statement II The ionosphere reflects electromagnetic waves frequencies less than a certain critical frequency.

ANSWERS

SESSION 1	1 (d)	2 (b)	3 (a)	4 (a)	5 (b)	6 (b)	7 (a)	8 (d)	9 (b)	10 (d)
	11 (a)	12 (c)	13 (b)	14 (c)	15 (a)	16 (c)	17 (b)	18 (c)	19 (b)	20 (c)
	21 (a)	22 (a)	23 (d)	24 (c)	25 (d)					
SESSION 2	1 (c)	2 (a)	3 (d)	4 (b)	5 (b)	6 (b)	7 (b)	8 (c)	9 (c)	10 (b)
	11 (b)	12 (d)								

Hints and Explanations

SESSION 1

1 $\nu = 20\text{ MHz} = 20 \times 10^6\text{ Hz}$

Wavelength of antenna is,

$$\lambda = \frac{c}{\nu} = \frac{3 \times 10^8}{20 \times 10^6} = 15\text{ m}$$

The minimum length of antenna

$$= \frac{\lambda}{4} = \frac{15}{4} = 3.75\text{ m}$$

2 To increases the range of transmission of microwaves, a number of antennas are erected in between the transmitting and receiving antennas. Such antennas in between the transmitting and receiving antennas are known as repeaters.

3 Coaxial cable have a characteristic impedance from 40 Ω to 150 Ω. So, option (a) is correct.

4 Refractive index of core is always greater than refractive index of cladding, to minimise the loss of light.

5 Modulation does not change time lag between transmission and reception.

6 In amplitude modulation the frequency of modulated wave is equal to the frequency of carrier wave. Thus, option (b) is correct.

7 Here, $\Delta\nu = 3\text{ kHz} = 0.003\text{ MHz}$

Using amplitude modulation, the frequencies of the side band

$$= (\nu + \Delta\nu) \text{ and } (\nu - \Delta\nu)$$
$$= (1 + 0.003) \text{ and } (1 - 0.003)$$
$$= 1.003\text{ MHz and } 0.997\text{ MHz}$$

Thus, option (a) is correct.

8 Carrier frequency is always greater than modulating frequency (i.e. audio frequency), so option (d) is appropriate carrier frequency.

9 Since, $P = P_c\left[1 + \frac{m_a^2}{2}\right]$

Power will be maximum, if $m_a = 1$

Therefore, $P_{\max} = P_c\left[1 + \frac{1}{2}\right]$

$$= \frac{3}{2}P_c = 1.5\,P_c$$

10 Frequency spectrum of modulated wave is

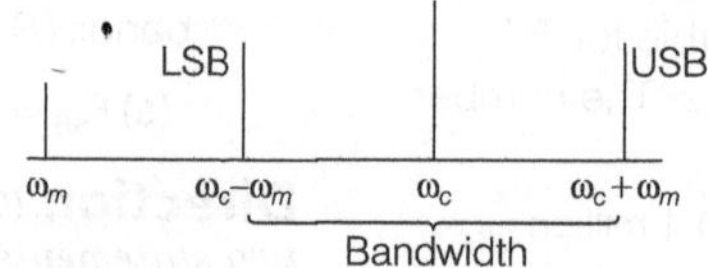

Clearly, ω_m is not included in the spectrum.

11 In amplitude modulation, the amplitude of the high frequency carrier wave is made to vary in proportion to the amplitude of the audio signal.

In frequency modulation, the frequency of the high frequency carrier signal varies with the frequency of audio signal.

12 Frequency associated with AM are $f_c - f_m, f, f_c + f_m$

According to the question

$$f_c = 2\text{ MHz} = 2000\text{ kHz}$$
$$f_m = 5\text{ kHz}$$

Thus, frequency of the resultant signal is/are carrier frequency

$f_c = 2000\text{ kHz}$,

LSB frequency

$$f_c - f_m = 2000\text{ kHz} - 5\text{ kHz} = 1995\text{ kHz}$$

and USB frequency

$$f_c + f_m = 2005\text{ kHz}$$

13 $\therefore D_{\text{skip}} = 2h\sqrt{\left(\frac{\nu}{\nu_c}\right)^2 - 1}$

$$= 2 \times 150\sqrt{\left(\frac{300}{100}\right)^2 - 1}$$
$$= 2 \times 150 \times 2\sqrt{2}$$
$$= 300 \times 2 \times 1.414$$
$$= 2.828 \times 300 = 848.4 \simeq 849\text{ km}$$

14 Critical frequency for reflection of radiowaves is given by

$$\nu_c \propto N^{1/2}$$
$$\nu_{CE} : \nu_{CF_1} : \nu_{CF_2}$$
$$= (2 \times 10^{11})^{1/2} : (5 \times 10^{11})^{1/2} : (8 \times 10^{11})^{1/2}$$
$$= 2 : 3 : 4$$

15 $\therefore \frac{N'_{\max}}{N_{\max}} = \left(\frac{V'_c}{V_c}\right)^2 = \left(\frac{9+1}{9}\right)^2$

$$= \left(\frac{10}{9}\right)^2 = 1.23$$

16 $\therefore$ MUF $= \nu_c / \cos i = 60 \times 10^6 / \cos 70°$

$$= 60 \times 10^6 \times \frac{1}{0.342}$$
$$= 175.43 \times 10^6$$
$$= 175.43\text{ MHz} \approx 175\text{ MHz}$$

17 The critical frequency of a sky wave for reflection from a layer of atmosphere is,

$\nu_c = 9(N_{\max})^{1/2}$

$N_{\max}$ = number density of ionosphere

$$\Rightarrow N_{\max} = \frac{\nu_c^2}{81} = \frac{(10 \times 10^6)^2}{81} = \frac{10^{14}}{81}\text{ m}^{-3}$$
$$= 1.24 \times 10^{12}\text{ m}^{-3}$$

18 In space communication signals are sent directly from transmitting antenna to receiving antenna by superimposing it on undamped electromagnetic waves.

19 Area of region covered = $\pi(2hR)$

In 1 km^2 = 1000 people

$$\frac{1}{1000} \times 20 \times 10^6 = 2 \times 10^3 = A$$

$$2 \times 10^3 = \pi(2 \times h \times 6400)$$

$$\Rightarrow h = \frac{2 \times 10^3}{\pi \times 2 \times 6400} = 0.0497 \text{ km}$$

$$= 49.7 \text{ m} \approx 50 \text{ m}$$

20 $\therefore d = \sqrt{2Rh}$

$$= \sqrt{2 \times 6.4 \times 10^6 \times 240} \text{ m} = 55 \text{km}$$

21 Range of radar on earth surface (optical distance, for space wave, i.e. line of view).

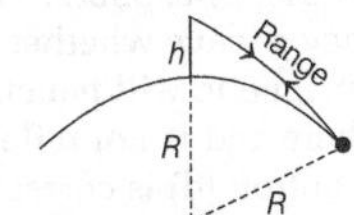

$$\text{Range} = \sqrt{(R + h)^2 - R^2}$$

$$= \sqrt{2Rh + h^2} \approx \sqrt{2Rh}$$

$$= \sqrt{2 \times 6400 \times \frac{1}{2}} \text{ km} = 80 \text{ km}$$

22 Optical communication is a system by which we transfer the informations on any distance from one location to other through optical range of frequency using optical fibre. The optical interference between fibres is zero. Hence, optical fibre communication has immunity to cross-talk.

23 In any communication system information (a physical quantity) is first converted into an electrical signal by a device called transducer. Most of the speech or information signal cannot be directly transmitted to long distances. For this an intermediate step of modulation is necessary in which the information signal is loaded or superimposed on a high frequency wave which acts as a carrier wave.

24 In AM modulation, the amplitude of the carrier signal varies in accordance with the information signal. AM signals are noisy because electrical noise signals significantly affect this. In FM modulation, amplitude of carrier wave is fixed while its frequency is changing. FM gives better quality transmission. It is preferred for transmission of music.

Demodulation is the process in which the original modulating voltage is recovered from the modulated wave.

25 Modem is a modulating and demodulating device. It acts as a modulator in transmitting mode and as demodulator in receiving mode.

SESSION 2

1 Only 10% of 10 GHz is utilised for transmission.

$\therefore$ Band available for transmission

$$= \frac{10}{100} \times 10 \times 10^9 \text{ Hz}$$

$$= 10^9 \text{ Hz}$$

Now, if there are n channels each using 5 kHz, then

$$n \times 5 \times 10^3 = 10^9$$

$$\Rightarrow \quad n = 2 \times 10^5$$

2 Given, $R = 1 \text{ k}\Omega$

$$R = 1 \times 10^3 \,\Omega, C = 1\mu\text{F} = 1 \times 10^{-6} \text{ F}$$

In this condition frequency of carrier signal, $\frac{1}{RC} << f_c$

$$\frac{1}{1 \times 10^3 \times 10^{-6}} << f_c$$

$$\Rightarrow \quad f_c >> 1 \text{ kHz}$$

Because frequency is greater than 1

$$f_c = 10 \text{ kHz}$$

3 The frequency optical communication

$$\nu = \frac{c}{\lambda}$$

$$\Rightarrow \nu = \frac{3 \times 10^8}{1200 \times 10^{-9}}$$

$$= 25 \times 10^{13} \text{ Hz}$$

But only 2% of the source frequency is available for TV transmission

$$\nu' = 2.5 \times 10^{14} \times 2\%$$

$$\nu' = 2.5 \times 10^{14} \times \frac{2}{100}$$

$$\nu' = 5 \times 10^{12} \text{ Hz}$$

$$\text{Number of channels} = \frac{\nu'}{\text{bandwidth}}$$

$$\text{Number of channels} = \frac{5 \times 10^{12}}{5 \times 10^6} = 10^6$$

$$= 1 \text{ million}$$

4 For D-region, $N = 10^9 \text{m}^{-3}$

$$\mu = \sqrt{1 - \frac{81.45 N}{\nu^2}}$$

$$= \sqrt{1 - \frac{81.45 \times 10^9}{(50 \times 10^6)^2}} \approx 1$$

$$\mu = \frac{\sin i}{\sin r} = 1$$

or $\quad \sin r = \sin i$

or $\quad r = i = 30°$

5 For ground wave propagation, the frequency range is 530 kHz to 1710 kHz. For sky wave propagation, the frequency range is 1710 kHz to 40 MHz. For space wave propagation, the frequency range is 54 MHz to 4.2 GHz. Thus, option (b) is correct.

6 Some of the characteristics of an optical fibre are as follows

(i) It works on the principle of total internal reflection.

(ii) It consists of core made up of glass/silica/plastic with refractive index n_1, which is surrounded by a glass or plastic cladding with refractive index n_2 ($n_2 > n_1$). The refractive index of cladding can be either changing abruptly or gradually changing (graded index fibre).

(iii) There is a very little transmission loss through optical fibres.

(iv) There is no interference from stray electric and magnetic fields to the signals through optical fibres.

7

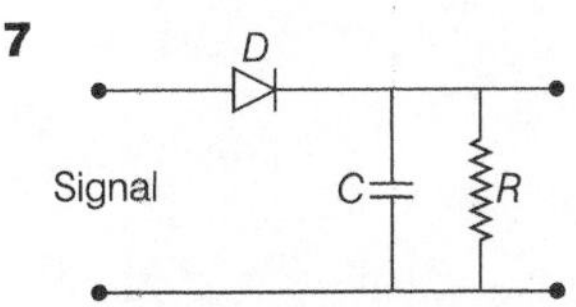

$$\tau = RC = 100 \times 10^3 \times 250 \times 10^{-12} \text{ s}$$

$$= 2.5 \times 10^7 \times 10^{12} \text{ s}$$

$$= 2.5 \times 10^{-5} \text{ s}$$

The highest frequency which can be detected with tolerable distortion is

$$f = \frac{1}{2\pi m_a RC}$$

[where, m_a is modulation]

$$= \frac{1}{2\pi \times 0.6 \times 2.5 \times 10^{-5}} \text{ Hz}$$

$$= \frac{100 \times 10^4}{25 \times 1.2\,\pi} \text{ Hz} = \frac{4}{1.2\,\pi} \times 10^4 \text{ Hz}$$

$$= 10.61 \text{ kHz}$$

This condition is obtained by applying the condition that rate of decay of capacitor voltage must be equal or less than the rate of decay modulated signal voltage for proper detection of modulated signal.

8 Here, $E_m = \frac{1}{2} E_c$

Therefore,

$$E_{\max} = E_c + E_m = E_c + \frac{1}{2} E_c = 1.5\, E_c$$

$$E_{\min} = E_c - E_m = E_c - \frac{1}{2} E_c = 0.5\, E_c$$

Also, $m_a = \frac{E_{\max} - E_{\min}}{E_{\max} + E_{\min}}$

$$= \frac{1.5\, E_c - 0.5\, E_c}{1.5 E_c + 0.5\, E_c}$$

$$m_a = \frac{E_c}{2.0 E_c} = 0.5$$

9 $P_{SB} = \frac{1}{R}\left(\frac{m_a E_c}{2\sqrt{2}}\right)^2 + \frac{1}{R}\left(\frac{m_a E_c}{2\sqrt{2}}\right)^2$

$\left[\because P_c = \frac{E_c^2}{2R}\right]$

$P_{SB} = \frac{m_a^2 E_c^2}{4R} = \frac{m_a^2 P_c}{2} = \frac{P_c}{2}$ $[\because m_a = 1]$

Also, $P = P_c\left[1 + \frac{m_a^2}{2}\right]$

Here, $m_a = 1$

$\Rightarrow$ $P = P_c\left[1 + \frac{1}{2}\right] = \frac{3}{2}P_c$

Hence, $\frac{P_{SB}}{P} = \frac{P_c/2}{\frac{3}{2}P_c} = \frac{1}{3}$

or $P_{SB} = \frac{1}{3}P$

10 In radio communication, sky wave refers to the propagation of radio waves reflected or refracted back towards earth from the ionosphere.

Since, it is not limited by the curvature of the earth, sky wave propagation can be used to communicate beyond horizon. Ionosphere is a region of upper atmosphere and induces the thermosphere and parts of mesosphere and exosphere. It is distinguished because it is ionised by solar radiation. It plays an important part in atmospheric electricity.

11 The modulation index determines the strength and quality of the transmitted signal.

If the modulation index is small the amount of variation in the carrier amplitude will be small consequently the audio signal being transmitted will not be strong.

Hence, for high modulation index or greater degree of modulation, the audio signal reception will be clear and strong.

12 In sky wave propagation, the radiowaves which have frequency between 2 MHz to 30 MHz, are reflected back to the ground by the ionosphere. But radio waves having frequency greater than 30 MHz cannot be reflected by the ionosphere because at this frequency they penetrates the ionosphere. It makes the sky wave propagation less reliable for propagation of TV signal having frequency greater than 30 MHz.

Critical frequency is defined as the highest frequency that is returned to the earth by the ionosphere. Thus, about this frequency a wave whether it is electromagnetic will penetrate the ionosphere and is not reflected by it. Hence, option (d) is correct.

DAY THIRTY SEVEN

Unit Test 7

(Modern Physics)

1 The wavelength of incident light falling on a photosensitive surface is charged from 2000 Å to 2100 Å the corresponding change in stopping potential is

(a) 0.03 V (b) 0.3 V
(c) 3 V (d) 3.3 V

2 Ultraviolet light of wavelength 350 nm and intensity 1.00 Wm^{-2} is incident on a potassium surface. If 0.5% of the photons participate in ejecting the photoelectrons, how many photoelectrons, are emitted per second, if the potassium surface has an area of 1 cm^2 ?

(a) 1.76×10^{18} photoelectrons/s
(b) 1.76×10^{14} photoelectrons/s
(c) 8.8×10^{11} photoelectrons/s
(d) The value of work function is required to complete the value of emitted photoelectrons/s

3 Electric field of an electromagnetic wave in vacuum is;

$$E = \left(3.1 \frac{N}{C}\right) . \cos\left[\left(1.8 \frac{rad}{m}\right) y + \left(5.4 \times 10^{8} \frac{rad}{s}\right) . t\right] \hat{i}$$

Wavelength of this wave as it passes through a medium of refractive index $\frac{3}{2}$ will be

(a) 3.55 m (b) 2.33 m
(c) 1.44 m (d) 3.22 m

4 Taking the Bohr radius as $a_0 = 53$ pm, the radius of Li^{++} ion in its ground state, on the basis of Bohr's model, will be about

(a) 53 pm (b) 27 pm (c) 18 pm (d) 13 pm

5 Consider inputs A and B;

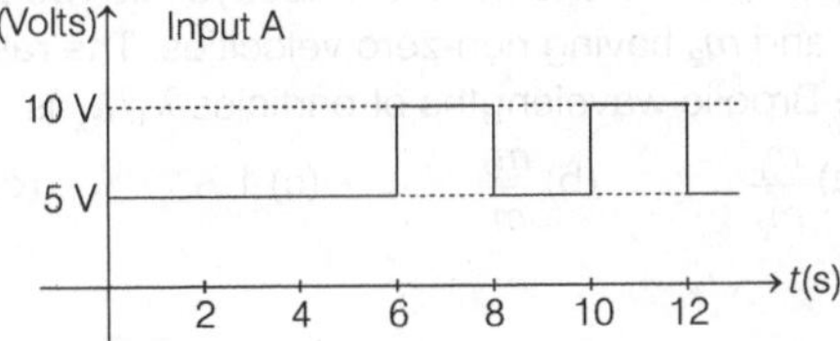

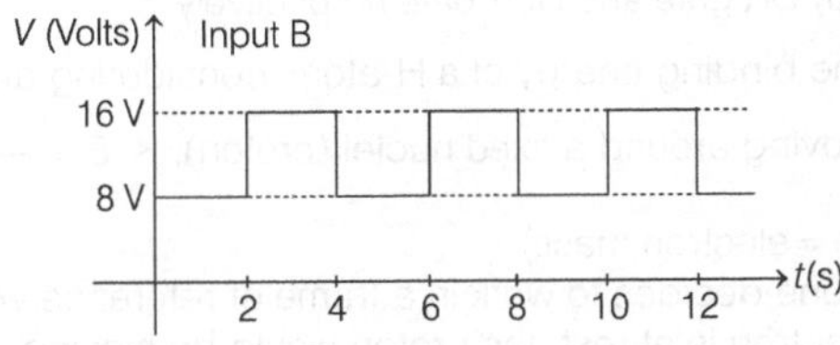

Output of a NAND gate on these inputs will be

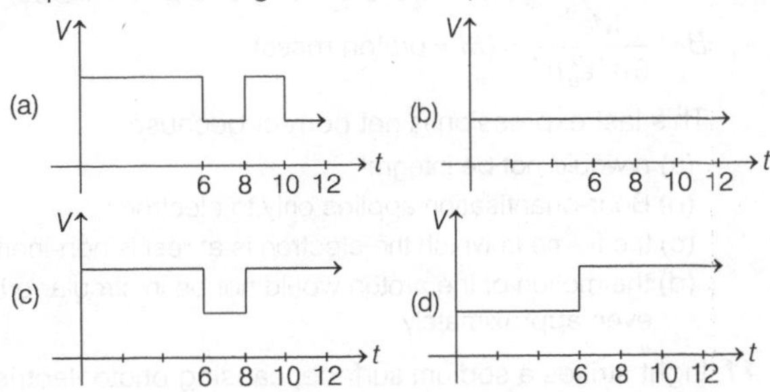

6 A neutron collides with a hydrogen atom in its ground state and excites it to $n = 3$. The energy given to hydrogen atom in this inelastic collision is (Neglect the recoiling of hydrogen atom and assume that energy is not absorbed as KE of H-atom)

(a) 10.2 eV (b) 12.1 eV (c) 12.5 eV (d) None of these

7 An X-ray tube operates at 50 kV. Consider that at each collision, an electron converts 50% of its energy into photons and 10% energy would be dissipated as thermal energy due to the collision then the wavelength of emitted by photons during 2nd collision is
(Take, $hc = 1242$ eV-nm)

(a) 1.242 nm (b) 1.242 Å (c) 4.968 nm (d) 4.968 Å

8 For the nuclear reaction, ${}_{88}Ra^{226} \longrightarrow {}_{86}Rn^{222} + {}_{2}He^{4}$ the radium nucleus is initially at rest and the alpha particle carries the energy 5.3 MeV. The energy released in the reaction is

(a) 5.4 MeV (b) 5.0 MeV (c) 300 MeV (d) 286 MeV

9 The combinations (M) and (N) of the NAND gates are as shown below.

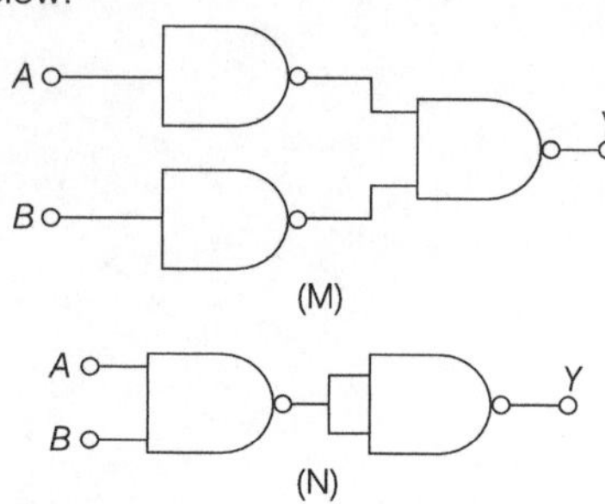

The output (Y) of (M) and (N) are equivalent to the output of

(a) OR gate and AND gate respectively
(b) AND gate and NOT gate respectively
(c) AND gate and OR gate respectively
(d) OR gate and NOT gate respectively

10 The binding energy of a H-atom, considering an electron moving around a fixed nuclei (proton), is $B = -\frac{me^4}{4n^2\varepsilon_0^2h^2}$

(m = electron mass).
If one decides to work in a frame of reference where the electron is at rest, the proton would be moving around it. By similar arguments, the binding energy would be

$B = \frac{Me^4}{8n^2\varepsilon_0^2h^2} = (M = \text{proton mass})$

This last expression is not correct because

(a) n would not be integral
(b) Bohr-quantisation applies only to electron
(c) the frame in which the electron is at rest is non-inertial
(d) the motion of the proton would not be in circular orbits, even approximately

11 Light strikes a sodium surface, causing photoelectric emission. The stopping potential for the ejected electrons is 5.0 V, and the work function of sodium is 2.2 eV. What is the wavelength of the incident light?

(a) 100 nm (b) 170 nm
(c) 150 nm (d) 200 nm

12 Consider a given circuit

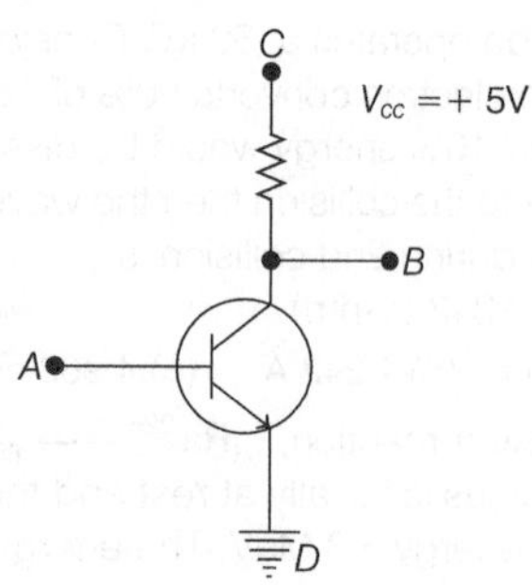

Point C is kept at a constant voltage of 5 V and point D is earthed. If input given at A is

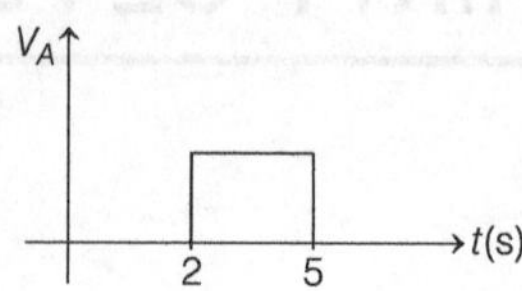

Then output obtained at B is

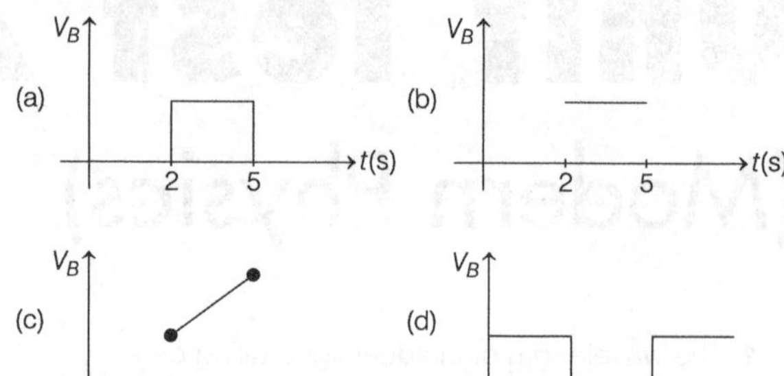

13 The simple Bohr's model cannot be directly applied to calculate the energy levels of an atom with many electrons. This is because

(a) of the electrons not being subject to a central force
(b) of the electrons colliding with each other
(c) of screening effects
(d) the force between the nucleus and an electron will no longer be given by Coulomb's law

14 An electron is trapped in a one dimensional infinite well of width 250 pm and is in its ground state. What is the longest wavelengths of light that can excite the electron from the ground state *via* a single photon absorption?

(a) $\lambda = \frac{4mL^2c}{h(n_i^2 - n_f^2)}$ (b) $\lambda = \frac{2mLc}{h(n_f^2 - n_i^2)}$
(c) $\lambda = \frac{8mL^2c}{h(n_f^2 - n_i^2)}$ (d) $\lambda = \frac{8mL}{h(n_f^2 - n_i^2)}$

15 What is the ratio of the shortest wavelength of the Balmer series to the shortest wavelength of the Lyman series?

(a) 1 : 4 (b) 2 : 4
(c) 4 : 1 (d) 1 : 1

16 In the ground state of the hydrogen atom, the electron has a total energy of −13.6 eV, its kinetic energy is

(a) 12.5 eV (b) 13.6 eV
(c) 14.9 eV (d) −27.2 eV

17 A particle of mass m at rest decays into two particles of m_1 and m_2 having non-zero velocities. The ratio of de-Broglie wavelengths of particles λ_1/λ_2 is

(a) $\frac{m_1}{m_2}$ (b) $\frac{m_2}{m_1}$ (c) 1 (d) $\sqrt{m_2/m_1}$

18 If there are 2 bulbs of same power, one of them gives red colour light, while other gives blue colour light.
If n_r and n_b are the number of photons per unit time emitted by bulbs, then choose correct option;
(r : red; b : blue)
(a) $n_r = n_b$ (b) $n_r < n_b$ (c) $n_r > n_b$ (d) $n_r \cdot n_b = c^2$

19 An α-particle makes an elastic head-on collision with a proton initially at rest.
Ratio of de-Broglie wavelength associated with α-particle and proton after collision will be
(a) 2 : 1 (b) 4 : 3 (c) 1 : 2 (d) 2 : 3

20 A beam of light is allowed to fall over cathode of a photocell after passing through two polaroids. None of the polaroid is rotated keeping other fixed.
Variation of photocurrent is best given by

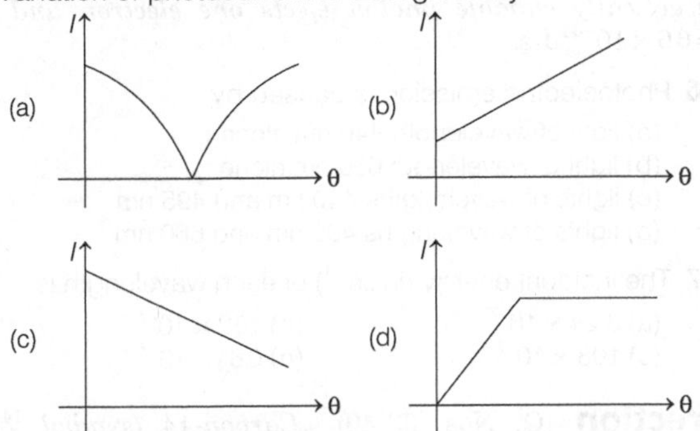

21 Two particles A_1 and A_2 of masses m_1, m_2 $(m_1 < m_2)$ have the same de-Broglie wavelength. Then
(a) then masses are the same
(b) thin energies are the same
(c) energy of A_1 is less than the energy of A_2
(d) energy of A_1 is more than the energy of A_2

22 Two students makes observations of stopping potentials (V_0) and frequencies (f) and plotted their observations graphically as shown below

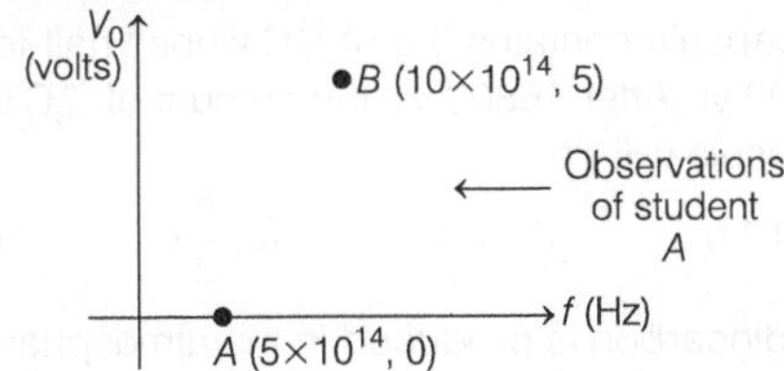

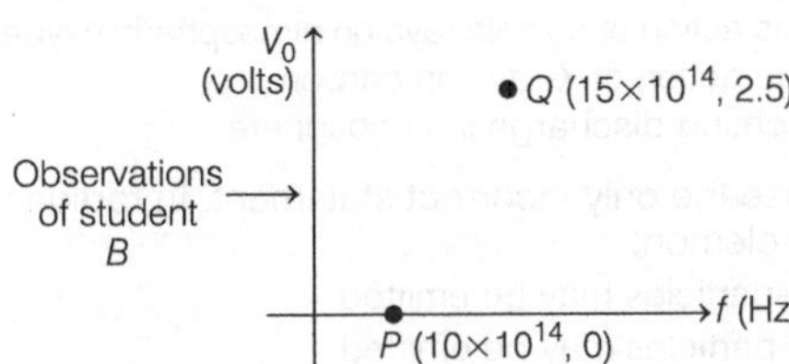

Now, choose the correct option;
(a) Both students gives accurate observation
(b) Readings of A are correct
(c) Readings of B are correct
(d) Reading of Both A and B are incorrect

23 Let potential energy of electron in Bohr's first orbit of hydrogen atom is zero.
Then, total energy of electron in IInd orbit is
(a) 23.80 eV (b) 27.20 eV (c) 13.6 eV (d) 26.25 eV

24 Consider the process;
$${}^{232}_{92}\text{U} \longrightarrow {}^{228}_{90}\text{Th} + {}^{4}_{2}\text{He}$$
Energy released in above process is 5.40 MeV. If this energy remains mainly with 'α' and daughter nucleus then, kinetic energy of ${}^{228}_{90}$Th, nucleus will be
(a) 5.4 MeV (b) 5.3 MeV (c) 0.1 MeV (d) 0.4 MeV

25 The following circuit represents

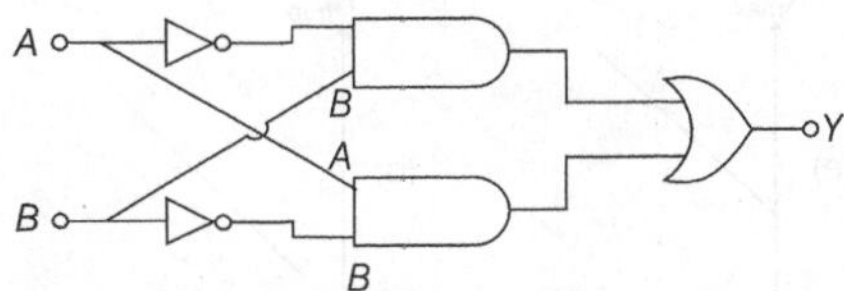

(a) OR gate (b) XOR gate (c) AND gate (d) NAND gate

26 Let a sample of a radioactive substance contains N_0 number of active nuclei at $t = 0$. Then, probability that a randomly choosen nucleus is disintegrated in time t is
(a) $1 - e^{-\lambda t}$ (b) $N_0 e^{-\lambda t}$ (c) $\frac{e^{-\lambda t}}{N_0}$ (d) $e^{-\lambda t}$

27 Match Column I and Column II and mark correct option.

	Column I		Column II
A.	α-decay	p.	Large nucleus.
B.	β^+decay	q.	More neutrons in nucleus.
C.	β^-decay	r.	More protons in nucleus.
D.	γ - decay	s.	More energy in nucleus.
E.	k - capture	t.	Proton number is more than 83 in nucleus.

	A	B	C	D	E
(a)	p	q	r	s	t
(b)	p,t	r,t	q,t	s	r,t
(c)	p	t	q	s	r
(d)	t	s	r	q	p

28 On a particular day, the maximum frequency reflected from the ionosphere is 10 MHz. On another day, it was found to decrease to 8 MHz. What is the ratio of the maximum electron densities of the ionosphere on the two days?
(a) 20 : 10 (b) 30 : 15 (c) 25 : 16 (d) 24 : 11

29 A transmitting antenna at the top of a tower has a height 32 m and that of the receiving antenna is 50 m. What is the maximum distance between them for satisfactory communication in line of sight mode? Given radius of the earth 6.4×10^6m.
(a) 50.0 km (b) 45.5 km (c) 35.5 km (d) 30.2 km

Direction (Q. Nos. 30-31) *According to Einstein, when a photon or light of frequency ν or wavelength λ is incident on photosensitive metal surface of work function ϕ_0, where $\phi_0 < h\nu$ (here h is planck's constant) then the emission of photoelectrons takes place.The maximum kinetic energy of the emitted photoelectrons is given by $K_{max} = h\nu - \phi_0$ If the frequency of the incident light is ν_0 called threshold frequency. The photoelectrons are emitted from metal without any kinetic energy.So $h\nu_0 = \phi_0$*

30 Stopping potential of emitted photoelectron is given by

(a) $\frac{h\nu - \phi_0}{e}$ (b) $h\nu - \phi_0$ (c) $\frac{h\nu}{e}$ (d) $\frac{\phi_0 + h\nu}{e}$

31 The variation of maximum kinetic energy (K_{max}) of the emitted photoelectrons with frequency (ν) of the incident radiations can be represented by

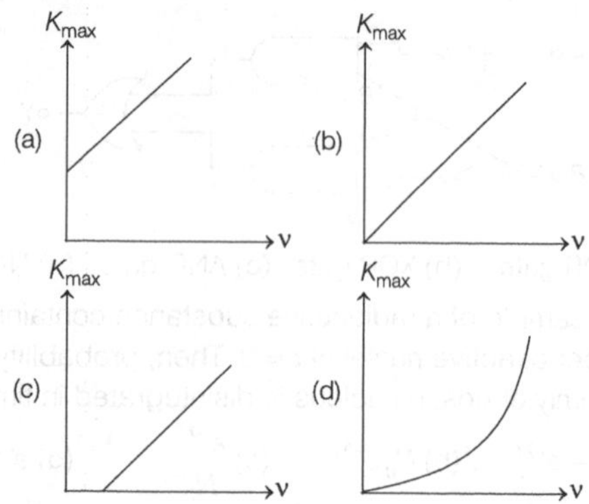

32 Frequencies higher than 10 MHz are found not to be reflected by the ionosphere on a particular day at a place. What is the maximum electron density of the ionosphere?

(a) $\frac{10^{14}}{9}\text{em}^{-3}$ (b) 10^{14}em^{-3}

(c) $\frac{10^{14}}{81}\text{em}^{-3}$ (d) $\frac{10^{14}}{7}\text{em}^{-3}$

Direction (Q. Nos. 33-35) *Each of these questions contains two statements : Statement I and Statement II. Each of these questions also has four alternative choices, only one of which is the correct answer. You have to select one of the codes (a), (b), (c), (d) given below*

(a) Statement I is true, Statement II is true; Statement II is the correct explanation for Statement I
(b) Statement I is true, Statement II is true; Statement II is not the correct explanation for Statement I
(c) Statement I is true; Statement II is false
(d) Statement I is false; Statement II is true

33 Statement I As intensity of incident light (in photoelectric effect) increases, the number of photoelectrons emitted per unit time increases.

Statement II More intensity of light means more energy per unit area per unit time.

34 Statement I Targets in X-ray tubes are made from high melting point metals.

Statement II Most of the energy of striking electrons is lost into collisions which simply appears as thermal energy.

35 Statement I The different lines of emission spectra (like Lyman, Balmer etc) of atomic hydrogen gas are produced by different atoms.

Statement II The sample of atomic hydrogen gas consists of millions of atoms.

Direction (Q. Nos. 36-37) *A beam of light has three wavelengths 440 nm, 495 nm and 660 nm with a total intensity of 3.24×10^{-3} Wm^{-2} equally distributed amongst the three wavelengths. The beam falls normally on an area of 1.0 cm^2 of a clean metallic surface of work function 2.2 eV. Assume that there is no loss of light by reflection and each energetically capable photon ejects one electron and take, $h = 6.6 \times 10^{-34}$ J-s.*

36 Photoelectric emission is caused by

(a) light of wavelength 440 nm alone
(b) light of wavelength 660 nm alone
(c) lights of wavelengths 440 nm and 495 nm
(d) lights of wavelengths 495 nm and 660 nm

37 The incident energy (in Js^{-1}) of each wavelength is

(a) 3.24×10^{-7} (b) 1.62×10^{-7}
(c) 1.08×10^{-7} (d) 0.81×10^{-7}

Direction (Q. Nos. 38-40) *Carbon-14 (symbol ${}^{14}_{6}C$) is produced by the bombardment of atmospheric nitrogen with high energy neutrons according to the equation.*

$${}^{14}_{7}N + {}^{1}_{0}n \longrightarrow {}^{14}_{6}C + {}^{1}_{1}H$$

Radiocarbon is unstable and decays to nitrogen with a half-life of 5600 yr. The carbon-14 is incorporated into atmospheric carbon dioxide molecules which are taken in by plants when they breathe in carbon dioxide. Animals which eat the plants also take in carbon-14. By measuring the ratio of the concentration of ${}^{14}C$ to ${}^{12}C$ in any ancient organism, say a tree, one can determine the date when the organism died.

38 A capsule contains 8 g of ${}^{14}_{6}C$ whose half-life is 5600 yr. After 16800 yr, the amount of ${}^{14}_{6}C$ left in the capsule will be

(a) 4 g (b) 2 g (c) $\frac{8}{3}$ g (d) 1 g

39 Radiocarbon is produced in the atmosphere as a result of

(a) collisions between fast neutrons and nitrogen nuclei
(b) the action of cosmic rays on atmospheric oxygen
(c) the action of X-rays on carbon
(d) lighting discharge in atmosphere

40 Choose the only incorrect statement. In radioactive decay of an element

(a) α-particles may be emitted
(b) β-particles may be emitted
(c) γ-rays may be emitted
(d) the nucleus does not undergo any change

ANSWERS

1. (b)	**2.** (c)	**3.** (b)	**4.** (c)	**5.** (a)	**6.** (b)	**7.** (b)	**8.** (a)	**9.** (a)	**10.** (c)
11. (b)	**12.** (d)	**13.** (a)	**14.** (c)	**15.** (c)	**16.** (d)	**17.** (c)	**18.** (c)	**19.** (d)	**20.** (a)
21. (d)	**22.** (d)	**23.** (a)	**24.** (c)	**25.** (c)	**26.** (a)	**27.** (b)	**28.** (c)	**29.** (b)	**30.** (a)
31. (c)	**32.** (c)	**33.** (b)	**34.** (a)	**35.** (d)	**36.** (c)	**37.** (c)	**38.** (d)	**39.** (a)	**40.** (d)

Hints and Explanations

1 Given, $\lambda_1 = 2000\text{Å} = 2\times10^{-7}\text{ m}$

$\lambda_2 = 2100\text{Å} = 2.1\times10^{-7}\text{ m}$

$$\frac{hc}{\lambda_1} = W + eV_1 \quad \ldots\text{(i)}$$

$$\frac{hc}{\lambda_2} = W + eV_2 \quad \ldots\text{(ii)}$$

Subtracting Eq. (ii) from Eq. (i)

$$hc\left(\frac{1}{\lambda_1} - \frac{1}{\lambda_2}\right) = e(V_1 - V_2)$$

Change in stopping potential,

$$\Delta V = V_1 - V_2 = \frac{hc}{e}\left(\frac{1}{\lambda_1} - \frac{1}{\lambda_2}\right)$$

$$= \frac{6.6\times10^{-34}\times3\times10^8}{1.6\times10^{-19}}\left(\frac{1}{2\times10^{-7}} - \frac{1}{2.1\times10^{-7}}\right)$$

$$= \frac{6.6\times3\times0.1}{1.6\times2\times2.1}\text{ V} = 0.3\text{ V}$$

2 Energy of photon

$$E = \frac{hc}{\lambda} = \frac{1242}{350}\text{ eV} = 3.55\text{ eV}$$

$$= 5.68\times10^{-19}\text{J}$$

Let n photons, per unit area per unit time are reaching the potassium surface, then

$$n = \frac{1.00}{5.68\times10^{-19}} = 1.76\times10^{18}$$

So, number of photons received by potassium surface per unit time is,

$n\times$ Area of potassium surface

$= 1.76\times10^{18}\times1\times10^{-4} = 1.76\times10^{14}$

Required number of photoelectrons emitted per unit time

$$= 1.76\times10^{14}\times\frac{0.5}{100} = 8.8\times10^{11}$$

3 Phase of wave is '$ky + \omega t$',

So, $\quad k = 1.8$ or $\dfrac{2\pi}{\lambda} = 1.8$

$$\Rightarrow \quad \lambda_1 = \frac{2\pi}{1.8}$$

When this wave passes through a medium of refractive index, $\frac{3}{2}$; its wavelength will be

$$\lambda_2 = \frac{\lambda_1}{n} = \frac{2\pi/1.8}{3/2} = \frac{2}{3}\times2\times\frac{22}{7}\times\frac{10}{18}$$

$$= 2.33\text{ m}$$

4 On the basis of Bohr's model,

$$r = \frac{n^2h^2}{4\pi^2mKZe^2} = a_0\frac{n^2}{Z}$$

For Li^{++}ion, $Z = 3$; $n = 1$ for ground state.

Given, $a_0 = 53$ pm

$$\therefore \quad r = \frac{53\times1^2}{3} = 18\text{ pm}$$

5 For a NAND gate, output is

A	B	Y
0	0	1
1	0	1
0	1	1
1	1	0

So, output waveform is like option (a).

6 The energy taken by hydrogen atom corresponds to its transition from $n = 1$ to $n = 3$ state.

ΔE (given to hydrogen atom)

$$= 13.6\left(1 - \frac{1}{9}\right) = 13.6\times\frac{8}{9} = 12.1\text{ eV}$$

7 During first collision, Initial energy of electron = 50 keV

Energy appearing as photon

= 50% of 50 keV = 25 keV

Energy lost in collision

= 10% of 50 keV = 5 keV

Energy left for second collision

= (50 − 25 − 5) keV = 20 keV

For second collision, Initial energy

= 20 keV

Energy of the emitted photon

= 50% of 20 keV = 10000 eV

So, required wavelength, $\lambda = \dfrac{1242}{10000}$ nm

= 1.242 Å

8 As parent nucleus is at rest and emitted particle (α) carries some energy, daughter nucleus (Rn) recoils to conserve the momentum.

The energy released in the reaction appears in the form of kinetic energy of α-particle and the daughter nucleus.

$$Q = K_\alpha + K_D$$

From momentum conservation,

$$p_\alpha = p_D$$

Solving above equation, we have

$$K_\alpha = \frac{M_D}{M_D + M_\alpha}\times Q$$

$$\Rightarrow \quad Q = \frac{M_D + M_\alpha}{M_D}\times K_\alpha$$

$$= \frac{222 + 4}{222}\times5.3$$

$= 5.4$ MeV

9 It follows from the logic symbol (A) that $X = \overline{\overline{A}\,\overline{B}}$

for which the truth table is as follows

A	B	$\overline{A}$	$\overline{B}$	$\overline{A}\cdot\overline{B}$	$X = \overline{\overline{A}\,\overline{B}}$
0	0	1	1	1	0
1	0	0	1	0	1
0	1	1	0	0	1
1	1	0	0	0	1

This truth table satisfies the Boolean expression $X = A + B$, which is the OR gate. Hence, the logic symbol (A) is equivalent to an OR gate. It follows from logic symbol (B) that

$$X = \overline{\overline{A\cdot B}} = A\cdot B$$

which is the Boolean expression for AND gate.

10 If electron is considered at rest, then photons are accelerating around COM of system. Thus, with respect to COM electrons are also accelerating and hence frame is non-inertial.

11 The energy of an incident photon $E = hf = hc/\lambda$, kinetic energy of the most energetic electron emitted $K_m = E - \phi = (hc/\lambda) - \phi, eV_0$ is related to kinetic energy by $eV_0 = K_m$,
so, $eV_0 = (hc/\lambda) - \phi$ and

$$\lambda = \frac{hc}{eV_0 + \phi}$$

$$= \frac{6.6\times10^{-34}\times3\times10^{8}}{5\times1.6\times10^{-19} + 2.2\times1.6\times10^{-19}}$$

$$= 171.8\times10^{-9}\text{ m}$$

$$= 171.8\text{ nm}$$

$$\approx 170\text{ nm}$$

12 Given circuit is a transistorized 'NOT' gate. When A is made positive, transistor is ON and it draws maximum current to collector.
So, $V_B = 0$ for the time A remains positive.

14 Energy levels are $E_n = n^2h^2 8/mL^2$, $f = \Delta E/h = (h/8mL^2)(n_f^2 - n_i^2)$ and the wavelength of the light is

$$\lambda = \frac{c}{f} = \frac{8mL^2c}{h(n_f^2 - n_i^2)}$$

15 The energy E of the photon emitted

$$E = 13.6\left(\frac{1}{n_1^2} - \frac{1}{n_2^2}\right)$$

Frequency f of the electromagnetic wave $f = E/h$ and the wavelength $\lambda = c/f$.

$$\text{Thus, } \frac{1}{\lambda} = \frac{f}{c} = \frac{E}{hc}$$

$$= \frac{13.6}{hc}\left(\frac{1}{n_1^2} - \frac{1}{n_2^2}\right)$$

For which $n_2 = \infty$. For the Balmer series, $n_1 = 2$ and the shortest wavelength is $\lambda_B = 4hc/13.6$. For the Lyman series, $n_1 = 1$ and the shortest wavelength is $\lambda_L = hc/13.6$. The ratio is $\lambda_B/\lambda_L = 4$

17 Initial momentum = 0
$\therefore$ Final momentum = 0

$$\mathbf{p} + \mathbf{p} = 0$$

$$p_1 = p_2 \quad \text{(numerically)}$$

$$\lambda = \frac{h}{p} \propto \frac{1}{p}$$

$$\Rightarrow \quad \frac{\lambda_1}{\lambda_2} = \frac{p_2}{p_1} = 1$$

18 Power, $P = \dfrac{\text{Energy radiated}}{\text{time}}$

So, number of photons,

$$n = \frac{E}{hf\times\Delta t} = \frac{P}{hf}$$

As, $f_{\text{blue}} > f_{\text{red}}$

$\Rightarrow \quad n_{\text{red}} > n_{\text{blue}}$

19 Let initial speed of α is v_0 and final speeds of α and proton are v_α and v_p. Then, momentum conservation gives

$$4m_pv_i = 4m_pv_\alpha + m_pv_p \quad \{\because m_\alpha = 4m_p\}$$

Also, $e = 1$; elastic collision

$\Rightarrow \quad v_i - 0 = v_p - v_\alpha$

Elimination of v_i gives,

$$v_\alpha = \frac{3}{8}v_p$$

$$\text{Now, } \frac{\lambda_\alpha}{\lambda_p} = \frac{h}{m_\alpha v_\alpha}\times\frac{m_pv_p}{h}$$

$$= \frac{m_p}{m_\alpha}\cdot\frac{v_p}{v_\alpha}$$

$$= \frac{1}{4}\times\frac{8}{3} = \frac{2}{3}$$

20 In figure (a) following Malus' law, intensity reduces upto $\theta = \frac{\pi}{2}$ and then increases.
Also, intensity $\propto$ photocurrent.

21 Masses of particles A_1 and A_2 are m_1 and m_2.
where, $m_1 < m_2$

$$\Rightarrow \quad \frac{m_1}{m_2} < 1 \quad \text{...(i)}$$

Since, both particles have same de-Broglie wavelength.

Hence, momentum, $p = \dfrac{h}{\lambda}$ will be same for both particles.

$$\therefore \text{ Energy, } E = \frac{p^2}{2m}$$

$$\frac{E_2}{E_1} = \frac{m_1}{m_2}$$

$$\frac{E_2}{E_1} < 1 \quad \text{[from Eq. (i)]}$$

$$E_2 < E_1$$

22 As slope of V_0 and f graph $= \dfrac{h}{e}$

So, $h = e\times$ slope of graph.
From readings of A,

$$h = \frac{1.6\times10^{-19}\times(5-0)}{(10-5)\times10^{14}}$$

$$= 16\times10^{-34}\text{ J-s}$$

From readings of B,

$$h = \frac{1.6\times10^{-19}\times(2.5-0)}{(15-10)\times10^{14}}$$

$$= 8\times10^{-34}\text{ J-s}$$

Hence, both of the readings are incorrect.

23 PE of electron in Ist orbit = -27.20 eV
Now, in IInd orbit,

$$\text{Total energy of electron} = -\frac{13.6}{n^2}$$

$$= -\frac{13.6}{2^2}$$

$$= -3.4\text{ eV}$$

In IInd orbit, KE = 3.4 eV and PE = $-2\times3.4 = -6.8$ eV
To make PE = 0 in Ist orbit, energy must be increased by 27.20 eV.
So, PE in IInd orbit = 27.20 + (−6.8) = 20.40 eV
Hence total energy = PE + KE = 20.40 + 3.4 = 23.80 eV.

24 Energy is distributed in inverse ratio of masses of products,

$$\text{So, } \frac{k_\alpha}{k_{\text{Th}}} = \frac{228}{4}$$

where $(228 + 4)x = 540$ MeV

$$\therefore \; x = \frac{540}{232}$$

$$\text{So, } k_\alpha = \frac{228}{232}\times5.40$$

$$= 5.30\text{ MeV}$$

So, $k_{\text{Th}} = 0.1$ MeV

25 Output of upper AND gate = $A\overline{B}$
Output of lower AND gate = $B\overline{A}$,
Output $Y = A\overline{B} + B\overline{A}$

26 Probability of surviving after time t

$$= \frac{\text{Number undecayed in time }(t)}{\text{Total number}}$$

$$= \frac{N_0e^{-\lambda t}}{N_0} = e^{-\lambda t}$$

$\therefore$ Probability of decay

$$= 1 - \text{Probability of survival}$$

$$= 1 - e^{-\lambda t}$$

27 For α-decay, electrostatic repulsion must be greater than nuclear force; this happens when nucleus is large.
In β^+decay, a proton is changed to a neutron. This occurs when protons are more.
In β^- decay, a neutron is changed to a proton. This happens when neutrons are more.
In k-capture, if protons are large, a proton and an electrons forms a neutron.

28 $f_c = 10$ MHz, $f'_c = 8$ MHz

$$\frac{N_{\max}}{N'_{\max}} = \left(\frac{f_c}{f'_c}\right)^2 = \left(\frac{10}{8}\right)^2 = 25:16$$

29 $d_m = \sqrt{2\times64\times10^5\times32} + \sqrt{2\times64\times10^5\times50}$ m

$$= 64\times10^2\times\sqrt{10} + 8\times10^3\times\sqrt{10}\text{ m}$$

$$= 144\times10^2\sqrt{10}\text{ m} = 45.5\text{ km}$$

32 Critical frequency,

$$f_c = 10 \text{ MHz} = 10^7 \text{ Hz}$$

$$f_c = 9(N_{max})^{1/2}$$

or $(N_{max})^{1/2} = \frac{f_c}{9}$

or $N_{max} = \left(\frac{f_c}{9}\right)^2$

or $N_{max} = \left(\frac{10^7}{9}\right)^2 \text{ em}^{-3}$

$$= \frac{10^{14}}{81} \text{ em}^{-3}$$

33 Intensity $\propto$ number of photons.

34 When energy lost in a collision is less, it appears in form of heat radiation.

35 In one particular sample, atoms can be excited only upto a particular level.

36 The threshold wavelength is

$$\lambda_0 = \frac{hc}{W_0}$$

$$= \frac{(6.6 \times 10^{-34}) \times (3 \times 10^8)}{2.2 \times 1.6 \times 10^{-19}}$$

$$= 6 \times 10^{-7} \text{ m}$$

$$= 600 \text{ nm}$$

Out of the three given wavelength, two wavelengths λ_1 = 440 nm and λ_2 = 495 nm will cause photoelectric emission as these wavelengths are less than λ_0.

37 Intensity of each wavelength is

$$I = \frac{1}{3} \times 3.24 \times 10^{-3}$$

$$= 1.08 \times 10^{-3} \text{ Wm}^{-2}.$$

Area of metal surface is

$$A = 1 \text{cm}^2$$

$$= 1 \times 10^{-4} \text{ m}^2.$$

Therefore, energy of each wavelength is

$$E = I \times A$$

$$= 1.08 \times 10^{-7} \text{ J-s}^{-1}$$

38 Number of half-lives $n = \frac{16800}{5600} = 3$

Therefore, the amount of C - 14 left

$$= \frac{8 \text{ g}}{(2)^n}$$

$$= \frac{8 \text{ g}}{(2)^3}$$

$$= 1 \text{ g}$$

DAY THIRTY EIGHT

Mock Test 1

(Based on Complete Syllabus)

Instructions

1. The test consists of 30 questions.
2. Candidates will be awarded marks for correct response of each question. 1/4 (one-fourth) marks will be deducted for indicating incorrect response of each question. No deduction from the total score will be made, if no response is indicated for an item in the answer sheet.
3. There is only one correct response for each question. Filling up more than one response in each question will be treated as wrong response.

1 A running man has half the kinetic energy of a boy of half his mass. The man speeds up by 1.0 ms^{-1} and then has the same kinetic energy as the boy. The original speed of the boy was

(a) 2.4 ms^{-1} (b) 9.6 ms^{-1} (c) 4.8 ms^{-1} (d) 7.2 ms^{-1}

2 The length of the string of a simple pendulum is measured with a meter scale, is found to be 92.0 cm, the radius of the bob plus the hook is measured with the help of vernier calliper to be 2.17 cm. Mark out the correct statement.

(a) Least count of meter scale is 0.1 cm
(b) Least count of vernier callipers is 0.01 cm
(c) Effective length of simple pendulum is 94.2 cm
(d) All of the above

3 Two bodies *A* and *B* of equal mass are suspended from two separate massless springs of spring constants k_1 and k_2 respectively. If the bodies oscillate vertically such that their maximum velocities are equal, the ratio of the amplitudes of *A* to that of *B* is

(a) $\frac{k_1}{k_2}$ (b) $\sqrt{\frac{k_1}{k_2}}$

(c) $\frac{k_2}{k_1}$ (d) $\sqrt{\frac{k_2}{k_1}}$

4 The length of a simple pendulum executing simple harmonic motion is increased by 21%. The percentage increase in the time period of the pendulum of increased length is

(a) 11% (b) 21% (c) 42% (d) 10%

5 As the object moves from infinity to focus, then which is true, about the image formed by a single concave mirror?

(a) Always real and speed of image continuously increases
(b) Always real and speed is initially smaller and finally larger than object speed
(c) Initially real and moving with speed smaller than object speed but later on image becomes virtual and moving with speed of object
(d) Always virtual and speed is less than object speed

6 Two bodies of different masses has been released from the top of tower. One is thrown in the horizontal direction while other is dropped, then which will reach the ground first?

(a) The body which has been thrown horizontally
(b) The body which has been dropped
(c) Both will reach the ground simultaneously
(d) Depends on the velocity with which the first body has been projected horizontally

7 A raft of wood of density 600 kgm^{-3} and mass 120 kg floats in water. How much weight can be put on the raft to make it just sink?

(a) 200 kg (b) 120 kg (c) 80 kg (d) 40 kg

8 A 'double star' is a composite system of two stars rotating about their centre of mass under their mutual gravitational attraction. Let us consider such a 'double star' which has two stars of masses m and $2m$ at separation l. If T is the time period of rotation about their centre of mass, then

(a) $T = 2\pi\sqrt{\dfrac{l^3}{mG}}$ (b) $T = 2\pi\sqrt{\dfrac{l^3}{2mG}}$

(c) $T = 2\pi\sqrt{\dfrac{l^3}{3mG}}$ (d) $T = 2\pi\sqrt{\dfrac{l^3}{4mG}}$

9 In a hall, a person receives direct sound waves from a source 120 m away. He also receives waves from the same source which reach him after being reflected from the 25 m high ceiling at a point half-way between them. The two waves interfere constructively for wave lengths (in metre) of

(a) $10, 5, \frac{5}{2}, \ldots$ (b) $20, \frac{20}{3}, \frac{20}{5}, \ldots$

(c) 30, 20, 10, ... (d) 35, 25, 15, ...

10 An AC source producing emf

$$e = e_0\,[\cos(100\,\pi\,s^{-1})\,t + \cos(500\,\pi\,s^{-1})\,t]$$

is connected in series with a capacitor and resistor. The steady state current in the circuit is found to be

$I = I_1\cos[(100\,\pi s^{-1})\,t + \phi_1] + I_2\cos[(500\,\pi\,s^{-1})t + \phi_2]$

(a) $I_1 > I_2$
(b) $I_1 = I_2$
(c) $I_1 < I_2$
(d) The information is insufficient to find the relation between I_1 and I_2

11 A SONAR system fixed in a submarine operates at a frequency 40.0 kHz. An enemy submarine moves towards the SONAR with a speed of 360 km/h. What is the frequency of sound reflected by the submarine? Take the speed of sound in water to be 1450 m/s.

(a) 40.00 kHz (b) 50.53 kHz (c) 45.93 kHz (d) 55.63 kHz

12 If electric potential due to some charge distribution is given by $V = 3/r^2$, where **r** is radial distance, then find electric field at (1, 1, 1)

(a) $\dfrac{2}{\sqrt{3}}$ (b) $\dfrac{2(\hat{i}+\hat{j}+\hat{k})}{\sqrt{3}}$

(c) $\dfrac{2}{8(\hat{i}+\hat{j}+\hat{k})}$ (d) $\dfrac{3}{2(\hat{i}+\hat{j}+\hat{k})}$

13 In the diagram, a plot between δ (deviation) *versus* i (angle of incidence) for a triangular prism is given. From the observed plot, some conclusions can be drawn. Mark out the correct conclusions.

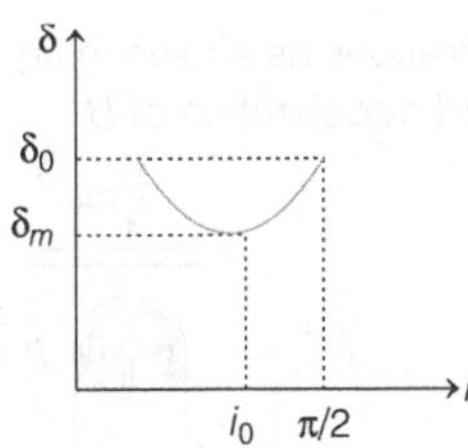

(a) The range of deviation for which two angles of incidence are possible for same deviation is $\delta_0 - \delta_m$
(b) The curve is unsymmetrical about i_0
(c) For a given δ, i is unique
(d) Both (a) and (b) are correct

14 A photon of 10.2 eV energy collides with hydrogen atom in ground state inelastically. After few microseconds one more photon of energy 15 eV collides with same hydrogen atom. Then what can be detected by a suitable detector?

(a) 1 photon of 10.2 eV and an electron of energy 1.4 eV
(b) 2 photons of energy 10.2 eV
(c) 2 photons of energy 3.4 eV
(d) 1 photons of energy 3.4 eV and 1 electron of 1.4 eV

15 A non-conducting plate (infinite plane plate) is given a charge in such a way that Q_1 appears on one side and Q_2 on other side. The face area of plate is A. Find the electric field at points 1 and 2.

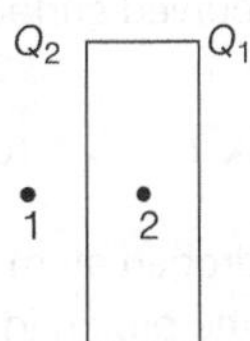

(a) $\dfrac{Q_1+Q_2}{2\varepsilon_0 A}, \dfrac{Q_2-Q_1}{2\varepsilon_0 A}$ (b) $\dfrac{Q_1-Q_2}{2\varepsilon_0 A}, \dfrac{Q_1+Q_2}{2\varepsilon_0 A}$

(c) $\dfrac{Q_1+Q_2}{\varepsilon_0 A}, \dfrac{Q_2-Q_1}{\varepsilon_0 A}$ (d) $\dfrac{Q_1-Q_2}{\varepsilon_0 A}, \dfrac{Q_1+Q_2}{\varepsilon_0 A}$

16 The emf and internal resistance of the battery as shown in figure are 4.3 V and 1 Ω respectively. The external resistance R is 50 Ω. The resistance of the voltmeter and ammeter are 200 Ω and 2 Ω respectively. Find the readings of the two meters.

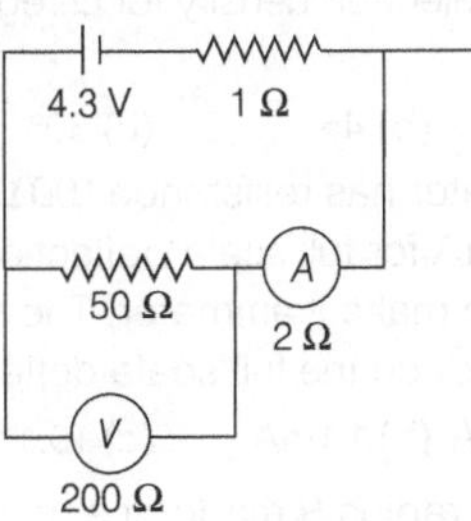

(a) 0.1 A, 2 V (b) 0.1 A, 4 V
(c) 0. 4 A, 1 V (d) 0.4 A, 4 V

17 All the accelerations as shown in figure are with respect to ground, find acceleration of B.

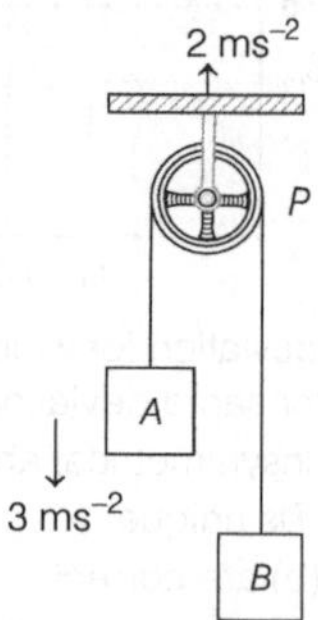

(a) 3 ms^{-2}, upward (b) 5 ms^{-2}, upward
(c) 3 ms^{-2}, downward (d) None of these

18 Three dielectric slabs of thickness $d/4$, $d/7$ and $d/2$ having dielectric constants 2, 8/7 and 4 respectively are inserted between the plates of a parallel plate capacitor having plate separation d and plate area A. Find the capacitance of the system.

(a) $\frac{118\varepsilon_0 A}{75\,d}$ (b) $\frac{88\varepsilon_0 A}{63\,d}$ (c) $\frac{226\varepsilon_0 A}{135\,d}$ (d) $\frac{284\,\varepsilon_0 A}{75\,d}$

19 Light is incident at an angle α on one planer end of a transparent cylindrical rod of refractive index n. The least value of n for which the light entering the rod will not emerge from the curved surface of rod, irrespective of value of α is

(a) $\frac{1}{\sqrt{2}}$ (b) $\sqrt{2}$ (c) $\frac{1}{\sqrt{3}}$ (d) $\sqrt{3}$

20 An electron of hydrogen atom is considered to be revolving around the proton in the circular orbit of radius $\frac{h^2}{4\pi^2 me^2}$ with velocity $\frac{2\pi e^2}{h}$. The equivalent current due to circulating charge is

(a) $\frac{4\pi^2 me^4}{h^3}$ (b) $\frac{4\pi^2 me^5}{h^3}$
(c) $\frac{4\pi^2 m^2 e^4}{h^3}$ (d) None of these

21 A sky wave with a frequency 55 MHz is incident on D-region of earth's atmosphere at 45°. The angle of refraction is (electron density for D-regions is 400 electron/cm^3)

(a) 60° (b) 45° (c) 30° (d) 15°

22 A galvanometer has resistance 100 Ω and it requires current 100 μA for full scale deflection. A resistor 0.1 Ω is connected to make it ammeter. The smallest current in circuit to produce the full scale deflection is

(a) 1000.1 mA (b) 1.1 mA (c) 10.1 mA (d) 100.1 mA

23 A cylinder of radius R made of a material of thermal conductivity K_1 is surrounded by a cylindrical shell of inner radius R and outer radius $2R$ made of a material of thermal conductivity K_2. The two ends of the combined system are maintained at two different temperatures. There is no loss of heat across the cylindrical surface and the system is in steady state. Find the effective thermal conductivity of the system.

(a) $K_1 + K_2$ (b) $\frac{K_1 K_2}{(K_1 + K_2)}$
(c) $\frac{(3K_1 + K_2)}{4}$ (d) $\frac{(K_1 + 3K_2)}{4}$

24 A 20 g bullet pierces through plate of mass $m_1 = 1$ kg and then comes to rest inside a second plate of mass $m_2 = 2.98$ kg. It is found that the two plates, initially at rest, now move with equal velocities. The percentage loss in the initial velocity of bullet when it is between m_1 and m_2 (neglect any loss of material of the bodies, due to action of bullet) will be

(a) 20% (b) 25%
(c) 30% (d) 45%

25 A hole is bored along the diameter of the earth and a particle is dropped into it. If R is the radius of the earth and g is the acceleration due to gravity at the surface of the earth, then the time period of oscillation of the particle is

(a) $2\pi\sqrt{\frac{R}{g}}$ (b) $2\pi\sqrt{\frac{R}{2g}}$
(c) $2\pi\sqrt{\frac{2R}{g}}$ (d) $2\pi\sqrt{\frac{R}{3g}}$

26 Five rods of same dimensions are arranged as shown in the figure. They have thermal conductivities K_1, K_2, K_3, K_4 and K_5. When the points A and B are maintained at different temperatures, no heat flows through the central rod if

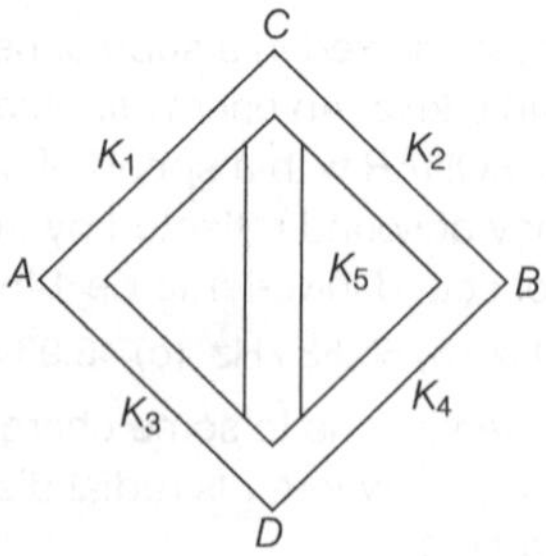

(a) $K_1 K_4 = K_2 K_3$ (b) $K_1 = K_4$ and $K_2 = K_3$
(c) $\frac{K_1}{K_4} = \frac{K_2}{K_3}$ (d) $K_1 K_2 = K_3 K_4$

27 A block of wood has a mass of 25 g. When a 5 g metal piece with a volume of 2 cm^3 is attached to the bottom of the block, the wood barely floats in water. What is the volume V of the wood?

(a) 28 cm^3 (b) 35 cm^3
(c) 48 cm^3 (d) 12 cm^3

28 A body dropped from a height H reaches the ground with a speed of $1.2\sqrt{gH}$. Calculate the work done by air-friction.

(a) $2.8\ mgH$ (b) $-1.3\ mgH$
(c) $1.3\ mgH$ (d) $-0.28\ mgH$

Direction (Q. Nos. 29-30) *Each of these questions contains two statements : Statement I (Assertion) and Statement II (Reason). Each of these questions also has four alternative choices, only one of which is the correct answer. You have to select one of the codes (a), (b), (c), (d) given below*

(a) Statement I is true, Statement II is true; Statement II is the correct explanation for Statement I
(b) Statement I is true, Statement II is true; Statement II is not the correct explanation for Statement I
(c) Statement I is true; Statement II is false
(d) Statement I is false; Statement II is true

29 Statement I The rocket works on the principle of conservation of linear momentum.

Statement II Whenever there is the change in momentum of one body, the same change occurs in the momentum of the second body of the same system (having two bodies only) but in opposite direction.

30 Statement I If the half-life of a radioactive substance is 40 days, then 75% substance decays in 20 days.

Statement II $N = N_0\left(\frac{1}{2}\right)^n$, where $n = \frac{\text{Time elapsed}}{\text{Half-life period}}$

ANSWERS

1. (c)	**2.** (c)	**3.** (d)	**4.** (d)	**5.** (b)	**6.** (c)	**7.** (c)	**8.** (c)	**9.** (b)	**10.** (c)
11. (c)	**12.** (b)	**13.** (d)	**14.** (a)	**15.** (a)	**16.** (b)	**17.** (d)	**18.** (d)	**19.** (b)	**20.** (b)
21. (b)	**22.** (d)	**23.** (d)	**24.** (a)	**25.** (a)	**26.** (a)	**27.** (a)	**28.** (d)	**29.** (a)	**30.** (d)

Hints and Explanations

1 $(KE)_{man} = \frac{1}{2}(KE)_{boy}$

$\Rightarrow \frac{1}{2}mv_m^2 = \frac{1}{2}\left(\frac{1}{2}\times\frac{m}{2}v_b^2\right)$

$\Rightarrow \quad v_m = \frac{v_b}{2}$ ...(i)

Further,

$\frac{1}{2}m(v_m+1)^2 = \frac{1}{2}\left(\frac{m}{2}\right)v_b^2$

$\Rightarrow \quad v_m + 1 = \frac{v_b}{\sqrt{2}}$ ...(ii)

From Eqs. (i) and (ii), we get

$v_b = 2(\sqrt{2}+1) = 4.82\ \text{ms}^{-1}$

$v_m = \sqrt{2}+1 = 2.41\ \text{ms}^{-1}$

2 Effective length of the simple pendulum is (92.0 + 2.17) cm

= 94.2 cm after rounding off to 3 significant digits.

3 Maximum velocity $= a\omega = a\sqrt{\frac{k}{m}}$

Given that, $a_1\sqrt{\frac{k_1}{m}} = a_2\sqrt{\frac{k_2}{m}}$

$\Rightarrow \quad \frac{a_1}{a_2} = \sqrt{\frac{k_2}{k_1}}$

4 $\because T = 2\pi\sqrt{\frac{l}{g}}, T^2 = 4\pi^2\frac{l}{g}, l = \left(\frac{g}{4\pi^2}\right)T^2$

$\therefore$ % change $= x + y + \frac{xy}{100}$

Vaild only for two variables in terms of percentage.

$x \to$ % change in first variable

$x \to$ % change in second variable

% increase in length $= x + x + \frac{x^2}{100}$

$21 = 2x + \frac{x^2}{100}$

On solving, $x = 10\%$

[by cross-check method]

5 As the object moves from infinity to centre of curvature, the image formed by a concave mirror would be real and is moving from focus to centre of curvature, but as the object crosses centre of curvature and moves towards focus the image is still real but moves from centre of curvature towards infinity and when the object is at focus the real image would be formed at infinity.

So, image speed is smaller in beginning when the object is moving from infinity to centre of curvature and increases thereafter.

6 Since, vertical displacement is same, as well as initial velocity in vertical downward direction is zero for both the bodies.

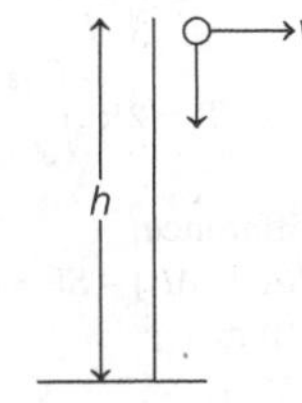

So, $h = \frac{gt_1^2}{2}$ (for horizontal throwing)

$h = \frac{gt_2^2}{2}$ (for dropping)

$\therefore \quad t_1 = t_2$

7 Volume of raft $= \frac{120}{600} = \frac{1}{5}\ \text{m}^3$

Fraction of volume inside water is

$\frac{\rho_{wood}}{\rho_{water}}$ = Relative density $= \frac{600}{1000} = \frac{3}{5}$

So, fraction of volume outside water is

$= \left(1 - \frac{3}{5}\right) = \frac{2}{5}$

$\Rightarrow$ Volume outside water is,

$V_{out} = \frac{2}{5}\times\frac{1}{5} = \frac{2}{25}\ \text{m}^3$

When the raft just sinks, the additional upthrust is

$U = \frac{2}{25}\times 10^3 \times g$

The weight m put on the raft is

$$mg = \frac{2}{25} \times 10^3 \times g$$

$$\therefore \quad m = 80 \text{ kg}$$

8 The system will revolve/rotate about an axis passing through the centre of mass of the combined system. Considering origin at the particle of mass $2m$, we have the centre of mass at a distance $l/3$ from $2l$ and $\frac{2l}{3}$ from m

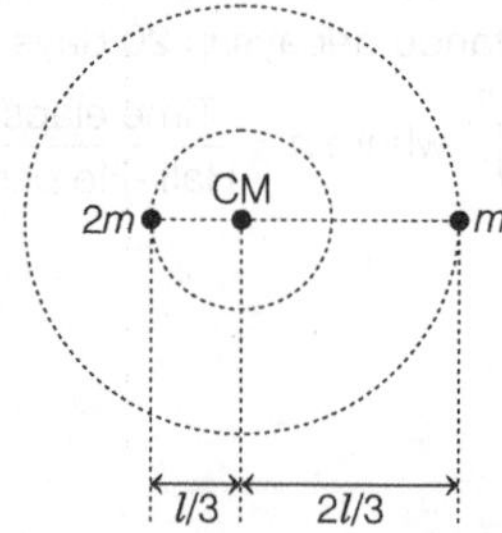

The gravitational force of attraction between $2m$ and m provides the necessary centripetal force to the mass to revolve in a circle of radius $\frac{2l}{3}$ for m or $\frac{l}{3}$ for $2m$.

$$\Rightarrow m\left(\frac{2l}{3}\right)\omega^2 = \frac{Gm\,(2m)}{l^2}$$

$$\Rightarrow \quad \omega = \sqrt{\frac{3Gm}{l^3}}$$

$$\Rightarrow \quad T = 2\pi\sqrt{\frac{l^3}{3Gm}}$$

9 $\therefore$ Path difference,

$\Delta x = (SA + AP) - SP = (65 + 65) - 120$

$\Rightarrow \Delta x = 10$ m

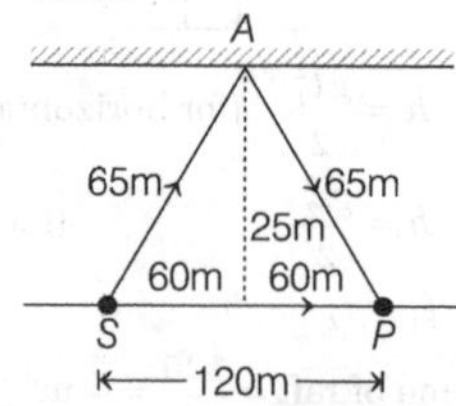

But at A, the wave suffers reflection at the surface of rigid/fixed end or denser medium. Hence, the wave must suffer an additional path change of $\frac{\lambda}{2}$ or a phase change of π.

$\Rightarrow$ Net path difference $= \left(10 - \frac{\lambda}{2}\right)$

For maxima (constructive interference),

Net path difference $= (2n)\frac{\lambda}{2}$;

where, $n = 0, 1, 2, \ldots$

$$\Rightarrow \quad 10 - \frac{\lambda}{2} = 2n\left(\frac{\lambda}{2}\right);$$

where, $n = 0, 1, 2, \ldots$

$\Rightarrow 10 = (2n + 1)\frac{\lambda}{2}$; where, $n = 0, 1, 2, \ldots$

$\Rightarrow \lambda = 20\,(2n + 1)$; where, $n = 0, 1, 2, \ldots$

or $\lambda = 20, \frac{20}{3}, \frac{20}{5}, \ldots$

10 $$I_1 = \frac{e_0}{\sqrt{R^2 + \left(\frac{1}{\omega_1 C}\right)^2}} = \frac{e_0}{Z_1},$$

where, $\omega_1 = 100\,\pi$

$$I_2 = \frac{e_0}{\sqrt{R^2 + \left(\frac{1}{\omega_2 C}\right)^2}} = \frac{e_0}{Z_2},$$

where, $\omega_2 = 500\,\pi$

So, $Z_1 > Z_2$, therefore $I_1 < I_2$.

11 SONAR frequency,

$v_s = 40 \text{ kHz} = 40 \times 10^3 \text{Hz}$

Speed of enemy submarine

$$v_e = 360 \text{ km/h} = 360 \times \frac{5}{18} \text{ m/s}$$

$$= 100 \text{ m/s} \quad \left(\because 1 \text{ km/h} = \frac{5}{18} \text{m/s}\right)$$

Speed of sound in water = 1450 m/s

Apparent frequency received by the submarine,

$$v' = \left(\frac{v + v_e}{v}\right)v_s$$

$$= \left(\frac{1450 + 100}{1450}\right) \times 40 \times 10^3$$

$$= 42.76 \times 10^3 \text{ Hz}$$

Now, the reflected waves have a different frequency,

$$v'' = \left(\frac{v}{v - v_s}\right)v'$$

Here, $v_s = 100$ m/s is velocity of enemy submarine,

$$v'' = \left(\frac{1450}{1450 - 100}\right) \times 42.76 \times 10^3$$

$$= 45.93 \times 10^3 \text{ Hz} = 45.93 \text{ kHz}$$

12 As, $V = \frac{3}{r^2}$

$$\therefore \ \mathbf{E} = -\left(\frac{dV}{dr}\right)\mathbf{r} = -\frac{\partial}{\partial r}\left[\frac{3}{r^2}\right]\mathbf{r} = \frac{6}{r^3}\,\mathbf{r}$$

$$\Rightarrow \ E = 6\frac{(\hat{\mathbf{i}} + \hat{\mathbf{j}} + \hat{\mathbf{k}})}{(\sqrt{3})^3} = \frac{2}{\sqrt{3}}(\hat{\mathbf{i}} + \hat{\mathbf{j}} + \hat{\mathbf{k}})$$

14 When photon strickes the hydrogen atom, the photon is absorbed and H atom reaches in (n=2 state) or first excited state, emitting a photon of energy 10.2 eV. Ionisation energy of H-atom =13.6 eV, so the second photon of energy 15 eV will ionise the H atom and extra energy (15 -13.6)eV = 1.4 eV will be retained by the electron. Thus finally we have one photon of energy 10.2 eV and one electron of energy 1.4eV.

15 At 1; $E_1 = \frac{\sigma_2}{2\varepsilon_0} + \frac{\sigma_1}{2\varepsilon_0} = \frac{Q_1 + Q_2}{2A\varepsilon_0}$

towards left

At 2; $E_2 = \frac{\sigma_2}{2\varepsilon_0} - \frac{\sigma_1}{2\varepsilon_0} = \frac{Q_2 - Q_1}{2A\varepsilon_0}$

towards right

where, $\sigma_2 = \frac{Q_2}{A}$ and $\sigma_1 = \frac{Q_1}{A}$

16 First of all draw the equivalent circuit diagram, current flowing through circuit

$$= \frac{4.3}{(50 \parallel 200 + 2 + 1)} = 0.1 \text{ A}$$

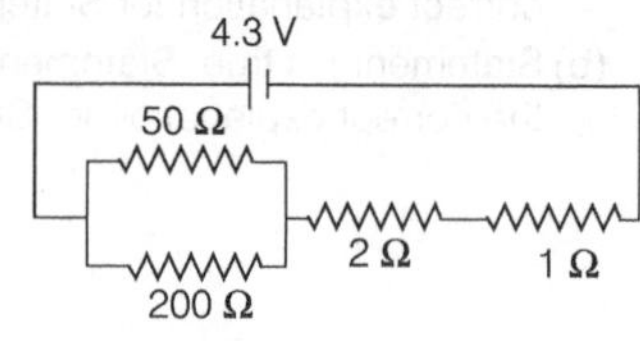

Voltmeter reading $= 4.3 - 0.1 \times 3 = 4$ V

17 Consider downward direction as positive

$$\mathbf{a}_{AP} = -\,\mathbf{a}_{BP}$$

$$\mathbf{a}_{AG} = \mathbf{a}_{AP} + \mathbf{a}_{PG}$$

$$3 = \mathbf{a}_{AP} - 2$$

$$\Rightarrow \quad \mathbf{a}_{AP} = 5 \text{ ms}^{-2}$$

$$\therefore \quad \mathbf{a}_{BG} = \mathbf{a}_{BP} + \mathbf{a}_{PG} = -5 - 2 = -7 \text{ ms}^{-2}$$

18 Here, three slabs are in series

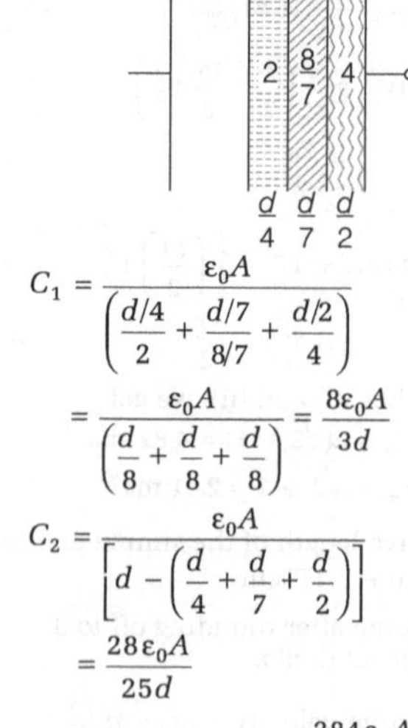

$$C_1 = \frac{\varepsilon_0 A}{\left(\frac{d/4}{2} + \frac{d/7}{8/7} + \frac{d/2}{4}\right)}$$

$$= \frac{\varepsilon_0 A}{\left(\frac{d}{8} + \frac{d}{8} + \frac{d}{8}\right)} = \frac{8\varepsilon_0 A}{3d}$$

$$C_2 = \frac{\varepsilon_0 A}{\left[d - \left(\frac{d}{4} + \frac{d}{7} + \frac{d}{2}\right)\right]} = \frac{28\varepsilon_0 A}{25d}$$

Now, $C_{eq} = C_1 + C_2 = \frac{284\varepsilon_0 A}{75d}$

19 $\because r + i = 90°$

$\Rightarrow \quad i = 90° - r$

For ray not to emerge from curved surface,

$$i > C$$

$$\Rightarrow \quad \sin i > \sin C$$

$$\Rightarrow \quad \sin (90^\circ - r) > \sin C$$

$$\Rightarrow \quad \cos r > \sin C$$

$$\Rightarrow \quad \sqrt{1 - \sin^2 r} > \frac{1}{n} \qquad \left[\because \sin C = \frac{1}{n}\right]$$

$$\Rightarrow \quad \frac{1 - \sin^2 i}{n^2} > \frac{1}{n^2}$$

$$\Rightarrow \quad 1 > \frac{1}{n^2}(1 + \sin^2 i)$$

$$\Rightarrow \quad n^2 > 1 + \sin^2 i \Rightarrow n > \sqrt{2}$$

$$[\because \sin i = 1 \text{ for } i = 90^\circ]$$

$\therefore$ Least value is $\sqrt{2}$.

20 As, $T = \frac{2\pi r}{v} = \frac{2\pi \times h^2}{4\pi^2 me^2} \times \frac{h}{2\pi e^2}$

$$= \frac{h^3}{4\pi^2 me^4}$$

$\therefore$ Current, $I = \frac{e}{T} = \frac{4\pi^2 me^5}{h^3}$

21 $n_{\text{eff}} = n_0 \sqrt{1 - \left(\frac{80.5\text{ N}}{v^2}\right)}$

$$= 1\sqrt{1 - \frac{80.5 \times (400 \times 10^6)}{(55 \times 10^6)^2}} = 1$$

Also, $n_{\text{eff}} = \frac{\sin i}{\sin r} \Rightarrow \sin i = \sin r$

$$\Rightarrow \quad r = i = 45^\circ$$

22 $\frac{i_g}{i} = \frac{S}{S + G}$

$$i = \frac{S + G}{S} \cdot i_g$$

$$= \frac{0.1 + 100}{0.1} \times 100 \times 10^{-6}\text{ A}$$

$$= 100.1 \times 10^{-3}\text{ A} = 100.1\text{ mA}$$

23 This can be considered as a parallel combination of two, one the inner cylinder and the other surrounding cylinder.

$$A_1 = \pi R^2,$$

$$A_2 = \pi (4R^2 - R^2) = 3\pi R^2,$$

$$L_1 = L_2 = L$$

Heat is flowing only along the length of tube

$$H_{\text{eq}} = H_1 + H_2$$

$$= \frac{K_1 A_1 (T_1 - T_2)}{L} + \frac{K_2 A_2 (T_1 - T_2)}{L}$$

$$\Rightarrow K_{\text{eq}} \times 4 = K_1 + 3K_2$$

$$\Rightarrow \quad K_{\text{eq}} = \frac{(K_1 + 3K_2)}{4}$$

24 The situation is as shown in figure. Firstly take first sheet and bullet as the system,

$$mu = mu_1 + m_1 v$$

Now, take second sheet and bullet as the system,

$$mu_1 = (m_1 + m_2)v$$

Solving this equation, we get

Percentage loss in

$$u = \frac{(u - u_1)}{u} \times 100\% = 20\%$$

25 $F = -\frac{GM'm}{x^2}$

$$\Rightarrow F = -\frac{G\left(\frac{4}{3}\pi x^3 \rho\right) m}{x^2}$$

$$= -\left(\frac{4}{3}\pi G\rho m\right)x$$

$$\Rightarrow x' = -\left(\frac{4}{3}\pi G\rho\right)x$$

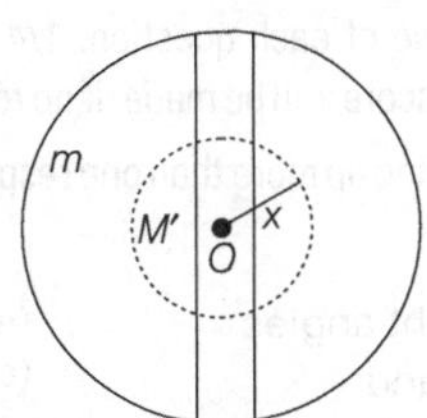

Time period of oscillation,

$$T = 2\pi\sqrt{\frac{x}{|x'|}}$$

$$= 2\pi\sqrt{\frac{3}{4\pi G\rho}}$$

$$\Rightarrow \quad T = 2\pi\sqrt{\frac{R}{g}}$$

$$\left[\because g = \frac{Gm}{R^2} = \frac{4}{3}\pi R\rho G\right]$$

26 The arrangement of rods is analogous to the arrangement of resistances in a Wheatstone bridge balanced condition. Thus, no heat flows through the rod conductivity K_5, then

$$\frac{K_1}{K_3} = \frac{K_2}{K_4}$$

$$\Rightarrow \quad K_1 K_4 = K_2 K_3$$

27 Let volume of wood is V cm^3, then total volume of displaced water is $(V + 2)$ cm^3, then for translational equilibrium, $(V + 2)\rho g = (25g + 5g)$

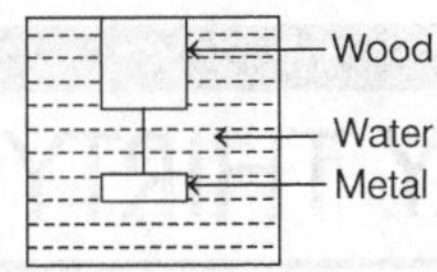

where all the quantities are in CGS unit and ρ is the density of water.

$$\Rightarrow \quad (V + 2) \times 1 = 30$$

$$\therefore \quad V = 28\text{ cm}^3$$

28 The forces acting on the body are force of gravity and air-friction

According to work-energy theorem, total work done on the body = Gain in Kinetic energy

$$W = \frac{1}{2}mv^2$$

$$= \frac{1}{2}m(1.2\sqrt{gH})^2$$

$$= 0.72\ mgH$$

As work done by gravity,

$$W_1 = mgH$$

$\therefore$ Work done by friction,

$$W_2 = W - W_1$$

$$= 0.72\,mgH - mgH$$

$$= -0.28\,mgH$$

29 Since, in the rocket fuel is undergoing combustion, the gases produced in this process leave the body of the rocket with large velocity and produce upthrust to the rocket. Let us assume that the fuel is undergoing combustion at the constant rate, then rate of change of momentum of the rocket will be constant. Since, more and more fuel will be burnt the mass of rocket will go on decreasing, so it will lead to increase the velocity of the rocket more and more rapidly.

30 Here, $\quad N = N_0\left(\frac{1}{2}\right)^{t/T}$

or $\quad \frac{N}{N_0} = \left(\frac{1}{2}\right)^{t/T} \quad$ ...(i)

where, T is the half-life period and $\frac{N}{N_0}$ is fraction of atoms left after time t. Here,

$$T = 40\text{ days and } \frac{N}{N_0} = \frac{25}{100}$$

$$= \frac{1}{4} = 0.25$$

Putting the values of T and $\frac{N}{N_0}$ in Eq. (i), we get

$$\frac{1}{4} = \left(\frac{1}{2}\right)^{t/40} \text{ or } \left(\frac{1}{2}\right)^2 = \left(\frac{1}{2}\right)^{t/40}$$

or $\quad \frac{t}{40} = 2$ or $t = 80$ days

DAY THIRTY NINE

Mock Test 2

(Based on Complete Syllabus)

Instructions

1. The test consists of 30 questions.
2. Candidates will be awarded marks for correct response of each question. 1/4 (one-fourth) marks will be deducted for indicating incorrect response of each question. No deduction from the total score will be made, if no response is indicated for an item in the answer sheet.
3. There is only one correct response for each question. Filling up more than one response in each question will be treated as wrong response.

1 A particle projected with velocity v_0 strikes at right angles a plane passing through the point of projection and having inclination β with the horizontal. Find the height (from horizontal plane) of the point, where the particle strikes the plane

(a) $y = \frac{2v_0^2}{g(4+\cot^2\beta)}$ (b) $y = \frac{v_0^2}{g(4+\cot^2\beta)}$

(c) $y = \frac{v_0^2}{(4+\cot^2\beta)}$ (d) $y = \frac{v_0^2}{2g(4+\cot^2\beta)}$

2 If the charge of $10\mu C$ and $-2\mu C$ are given to two plates of a capacitor, which are connected across a battery of 12 V, find the capacitance of the capacitor.

(a) 0.33 μF (b) 0.5 μF (c) 0.41 μF (d) 0.66 μF

3 A man can swim with a speed of 4 Kmh^{-1} in still water. How long does he take to cross a river 1 km wide, if the river flows steadily 3 Kmh^{-1} and he makes his strokes normal to the river current. How far down the river does he go when he reaches the other bank?

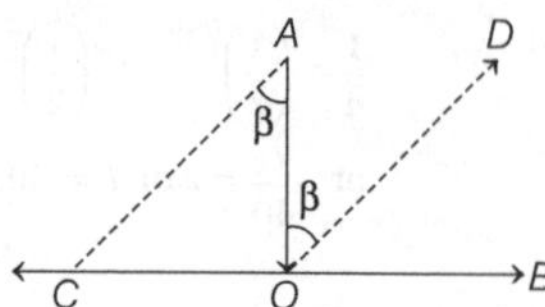

(a) 600 km (b) 750 km
(c) 800 km (d) 850 km

4 A body when projected vertically up, covers a total distance s, during its time of flight. If we neglect gravity then how much distance the particle will travel during the same time. Will it fall back?

(a) s, Yes
(b) s, No
(c) $2s$, Yes
(d) $2s$, No

5 Number of neutrons in a gold nucleus with $A = 197$ and $Z = 79$ is (gold, $_Z Au^A$)

(a) 79
(b) 197
(c) 118
(d) None of these

6 The density of hydrogen nucleus with $Z = 1$ is 2.29×10^{17} kgm^{-3}. The density of gold nucleus $Z = 79$ would be

(a) $\frac{2.29}{79} \times 10^{17} kgm^{-3}$

(b) $2.29 \times 79 \times 10^{17} kgm^{-3}$

(c) $2.29 \times 10^{17} kgm^{-3}$

(d) $\frac{2.29}{\sqrt{79}} \times 10^{17} kgm^{-3}$

7 If mass of a proton is 1.007825 amu and mass of a neutron is 1.008665 amu, then mass of ${}_3Li^7$ nucleus approximately be

(a) 7.058075 amu (b) 7.000000 amu
(c) 7.023475 amu (d) 7.034600 amu

8 Find the equivalent thermal resistance of the combination of rods as shown in the figure. Every rod has the same length l and cross-sectional area A. Thermal conductivities are mentioned in figure. Assume that there is no heat loss due to radiation or convection.

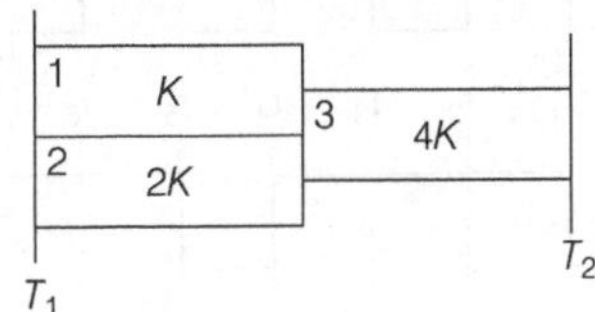

(a) $\frac{l}{4KA}$ (b) $\frac{7l}{4KA}$ (c) $\frac{7l}{12KA}$ (d) $\frac{l}{12KA}$

9 Two steel balls A and B are placed inside a right circular cylinder, of diameter 54 cm making contacts at points P, Q and R as shown in the figure. The radius $r_A = 12$ cm and $r_B = 18$ cm. The masses are $m_A = 15$ kg and $m_B = 60$ kg. The force exerted by the floor at the point Q and the wall at R are respectively (taking, $g = 10\ ms^{-2}$)

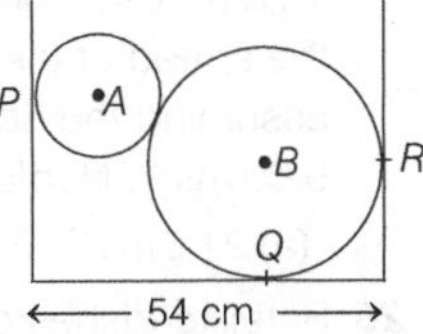

(a) 600 N, 150 N (b) 750 N, 150 N
(c) 600 N, 200 N (d) 750 N, 200 N

10 In J J Thomson's experiment, a potential difference of 320 V is accelerating the electron. The electron beam is entering a region having uniform magnetic field 6×10^{-5} T, acting perpendicular to it. Find the value of electric field in this region, so that the electron does not experience any deflection. (Take, $m_e = 9.1 \times 10^{-31}$ kg)

(a) 640 Vm^{-1} (b) 642 Vm^{-1}
(c) 637 Vm^{-1} (d) 644 Vm^{-1}

11 A wire of length 100 cm is connected to a cell of emf 2V and negligible internal resistance. The resistance of the wire is 3 Ω. The additional resistance required to produce a potential difference of 1 mV/cm is

(a) 47 Ω (b) 57 Ω
(c) 60 Ω (d) 55 Ω

12 A particular piano string is supposed to vibrate at a frequency of 440 Hz. In order to check its frequency, a tuning fork known to vibrate at a frequency of 440 Hz is sounded at the same time the piano key is struck, and a beat frequency of 4 beats/s is heard. Find the possible frequencies at which the string could be vibrating.

(a) 444 Hz, 436 Hz (b) 440 Hz, 436 Hz
(c) 444 Hz, 440 Hz (d) 449 Hz, 440 Hz

13 A uniform rectangular marble slab is 3.4 m long and 2.0 m wide. It has a mass of 180 kg. If it is originally lying on the flat ground, how much work is needed to stand it on one end?

(a) 2.0 kJ (b) 3.0 J (c) 3.0 kJ (d) 3000 kJ

Direction (Q. Nos. 14-15) *Each of these questions contains two statements : Statement I (Assertion) and Statement II (Reason). Each of these questions also has four alternative choices, only one of which is the correct answer. You have to select one of the codes (a), (b), (c) and (d) given below.*

(a) Statement I is true, Statement II is true; Statement II is the correct explanation for Statement I
(b) Statement I is true, Statement II is true; Statement II is not the correct explanation for Statement I
(c) Statement I is true; Statement II is false
(d) Statement I is false; Statement II is true

14 Statement I When the range of projectile is maximum, the time of flight is the largest.

Statement II Range is maximum, when angle of projection is 45°.

15 Statement I 1 amu is equivalent to 931 MeV.

Statement II Energy equivalent (E) or mass (m) is $E = mc^2$

16 The two blocks as shown in the figure, have equal masses and $\mu_s = \mu_k = 0.3$ for both blocks. Wedge W is fixed and block A is given initial speed of 1 ms^{-1}, down the plane. How far will it move before coming to rest, if inclines and strings are quite long? (take, $g = 10\ ms^{-2}$)

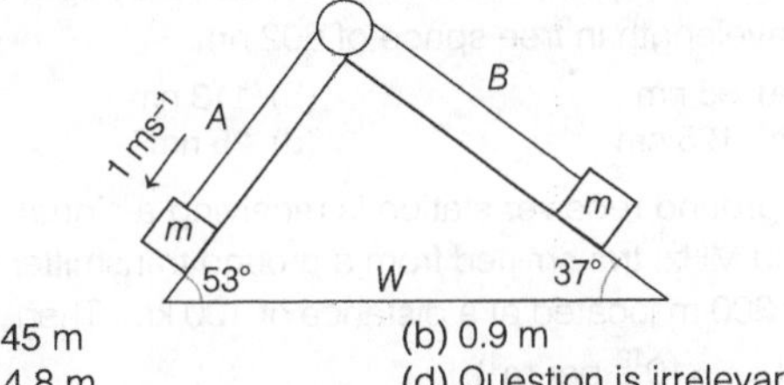

(a) 0.45 m (b) 0.9 m
(c) + 4.8 m (d) Question is irrelevant

17 As shown in figure, a uniform solid sphere rolls on a horizontal surface at 20 ms^{-1}. It then rolls up the incline shown. What will be the value of h, where the ball stops?

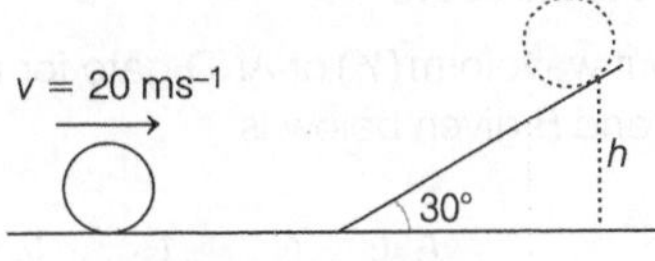

(a) 28.6 m (b) 8.6 m (c) 6 m (d) 18.6 m

18 A 1.6 kg block on a horizontal surface is attached to a spring with a spring constant of $1.0 \times 10^3\ Nm^{-1}$. The spring is compressed to a distance of 2.0 cm, and the block is released from rest. Calculate the speed of the block as it passes through the equilibrium position, $x = 0$, if the surface is frictionless.

(a) 0.75 ms^{-1} (b) 0.50 ms^{-1} (c) 0.25 ms^{-1} (d) 2.25 ms^{-1}

19 A small conducting circular loop is placed inside a long solenoid carrying a current. The plane of the loop contains the axis of the solenoid. If the current in the solenoid is varied, the current induced in the loop is

(a) clockwise
(b) anti-clockwise
(c) zero
(d) clockwise or anti-clockwise depending on whether the current is increased or decreased

20 An inductor coil joined to a 6 V battery draws a steady current of 12 A. This coil is connected to a capacitor and an AC source of rms voltage 6V in series. If the current in the circuit is in phase with the emf, the rms current will be

(a) 16.9 A (b) 12 A
(c) 8 A (d) 9.87 A

21 Two wires, each having a weight per unit length of 1.0×10^{-4} Nm^{-1}, are strung parallel to one another above earth's surface, one directly above the other. The wires are aligned in a North-South direction, so that earth's magnetic field will not affect them. When their distance of separation is 0.10 m, what must be the current in each in order for the lower wire to levitate the upper wire? Assume that the wires carry the same currents, travelling in opposite directions.

(a) 2.7 A (b) 0.1 A
(c) 3.5 A (d) 7.1 A

22 Calculate the minimum thickness of a soap-bubble film ($n = 1.33$) that will result in constructive interference in the reflected light, if the film is illuminated by light with a wavelength in free space of 602 nm.

(a) 98 nm (b) 113 nm
(c) 125 nm (d) 25 nm

23 A ground receiver station is receiving a signal at 100 MHz, transmitted from a ground transmitter at a height of 300 m located at a distance of 100 km. Then, ($N_{max} = 10^{12}$ per m^3)

(a) signal is coming *via* space wave
(b) signal is coming *via* sky wave
(c) signal is coming *via* satellite transponder
(d) None of the above

24 The output waveform (Y) of AND gate for the following inputs A and B given below is

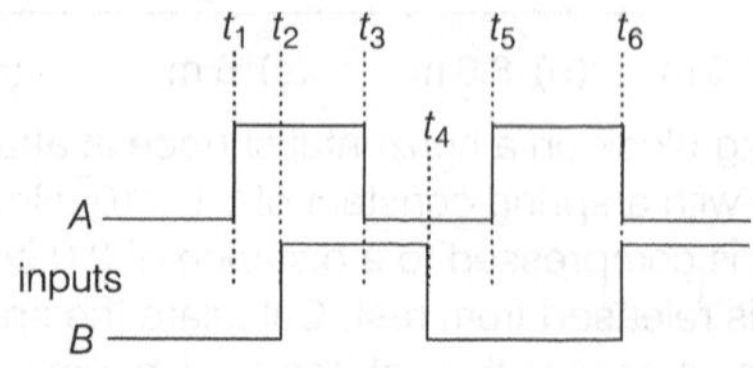

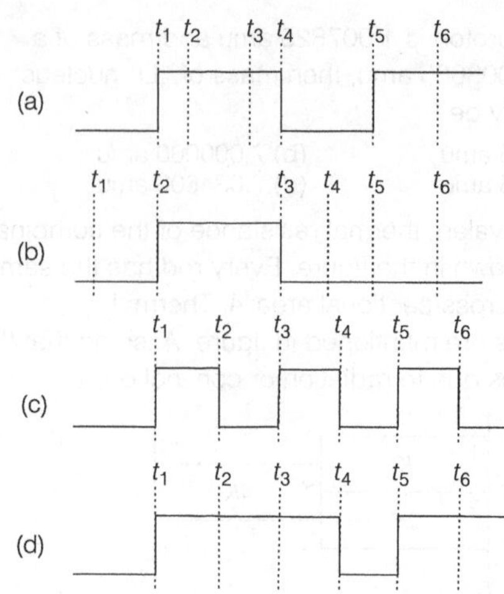

25 A sledge and its rider together weight 800 N. They move down on a frictionless hill through a vertical distance of 10.0 m. Use conservation of mechanical energy to find the speed of the sledge at the bottom of the hill, assuming the rider pushes off with an initial speed of 5.00 ms^{-1}. Neglect air resistance,

(a) 21.5 ms^{-1} (b) 14.9 ms^{-1} (c) 4.9 ms^{-1} (d) 20.3 ms^{-1}

26 Satellite dishes do not have to change directions in order to stay focussed on a signal from a satellite. This means that the satellite always has to be found at the same location with respect to the surface of the earth. For this to occur, the satellite must be at a height such that its revolution period is the same as that of earth, 24 h. At what height must the satellite be so to achieve this?

(a) $\left(\frac{T^2}{4\pi^2} GM_e\right)^{1/3}$ (b) $\left(\frac{T^2}{4\pi^2} GM_e\right)$
(c) $\left(\frac{T^2}{4\pi^2} GM_e\right)^{1/2}$ (d) $\left(\frac{T}{4\pi^2} GM_e\right)^{1/3}$

27 Water with a mass of 2.0 kg is held at constant volume in a container while 10.0 kJ of energy is slowly added by a flame. The container is not well insulated, and as a result 2.0 kJ of energy leaks out to the surroundings. What is the temperature increase of water?

(a) 0.28° C (b) 27° C (c) 0.96° C (d) 1.27° C

28 A 20 g bullet is fired horizontally with a speed of 600 ms^{-1} into a 7 kg block on a table top. The bullet b lodges in the block B. If the coefficient of kinetic friction between the block and the table top is 0.4, what is the distance the block will slide?

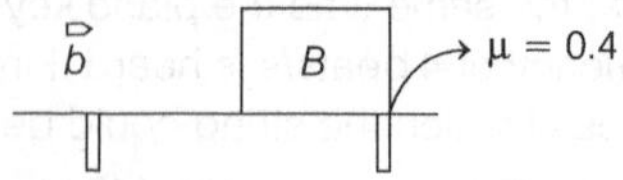

(a) 0.5 m (b) 1.2 m (c) 0.37 m (d) 0.85 m

29 A uniform rope, of mass m per unit length, hangs vertically from a support, so that the lower end just touches the table top. If it is released, then at the time a length y of the rope has fallen, the force on the table is equivalent to the weight of the length ky of the rope. Find the value of k.

(a) 1 (b) 2 (c) 3 (d) 3.5

30 Two converging lenses are placed 20.0 cm apart, as shown in the figure. If the first lens has a focal length of 10.0 cm and the second has a focal length of 20.0 cm, locate the final image formed of an object 30.0 cm in front of the first lens.

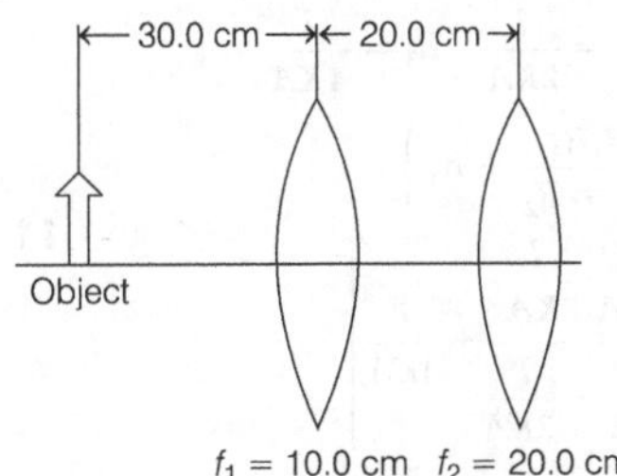

(a) 6.67 cm left (b) 6.67 cm right
(c) 15.0 cm left (d) 15.0 cm right

ANSWERS

1. (a)	**2.** (b)	**3.** (b)	**4.** (d)	**5.** (c)	**6.** (c)	**7.** (a)	**8.** (c)	**9.** (d)	**10.** (c)
11. (b)	**12.** (a)	**13.** (c)	**14.** (d)	**15.** (a)	**16.** (a)	**17.** (a)	**18.** (b)	**19.** (c)	**20.** (b)
21. (d)	**22.** (b)	**23.** (c)	**24.** (b)	**25.** (b)	**26.** (a)	**27.** (c)	**28.** (c)	**29.** (c)	**30.** (a)

Hints and Explanations

1 Let α be the angle between the velocity of projection and the inclined plane

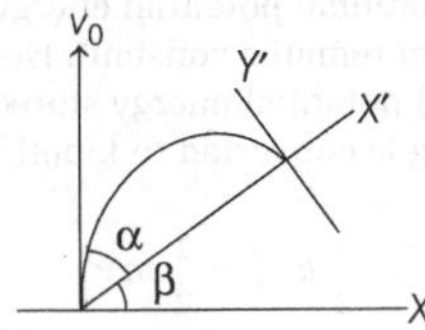

$$v_{0x'} = v_0 \cos\alpha,\ v_{0y'} = v_0 \sin\alpha$$
$$a_{x'} = -g\sin\beta,\ a_{y'} = -g\cos\beta$$
$$\Rightarrow v_{x't} = v_0\cos\alpha - g\sin\beta t$$

At the point of impact $v_{x'} = 0$

$$\Rightarrow t = \frac{v_0 \cos\alpha}{g\sin\beta} \quad \ldots(i)$$

Also, y at this point is zero

$$\Rightarrow v_0 \sin\alpha t - 1/2 g\cos\beta t^2 = 0$$
$$t^2 = \frac{2v_0 \sin\alpha}{g\cos\beta} \quad \ldots(ii)$$

From Eqs. (i) and (ii), we get

$$\tan\alpha = \frac{\cot\beta}{2}$$

$$x = v_0 \cos(\alpha+\beta)t$$
$$= \frac{{v_0}^2}{g}\left[\left(\frac{2}{\sqrt{4+\cot^2\beta}}\right)^2 \cot\beta - \frac{\cot\beta}{\sqrt{4+\cot^2\beta}} \cdot \frac{2}{\sqrt{4+\cot^2\beta}}\right]$$
$$= \frac{{v_0}^2}{g}\frac{2\cot\beta}{(4+\cot^2\beta)}$$
$$\therefore y = y\tan\beta = \frac{{v_0}^2}{g}\frac{2\cot\beta}{4+\cot^2\beta}\cdot\tan\beta$$
$$= \frac{2{v_0}^2}{g(4+\cot^2\beta)}$$

2 Charge of capacitor is the charge on facing surfaces of the plates of capacitor

$$Q = \left(\frac{q_1 - q_2}{2}\right) = \frac{[10-(-2)]}{2} = 6\ \mu\text{C}$$

Potential difference across the capacitor = 12 V

$$\text{So, } C = \frac{Q}{V} = \left[\frac{(6\times10^{-6})}{12}\right] F = 0.5\mu\text{F}$$

3 Given, speed of man (v_m) = 4 kmh^{-1}

Speed of river (v_r) = 3 kmh^{-1}

Width of the river (d) = 1 km

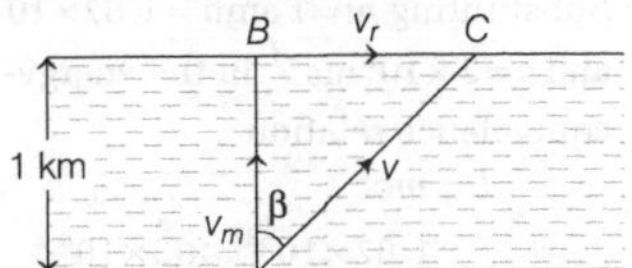

Time taken by the man to cross the river

$$t = \frac{\text{Width of the river}}{\text{Speed of the man}} = \frac{1\text{ km}}{4\text{ Kmh}^{-1}}$$
$$= \frac{1}{4}\text{h} = \frac{1}{4}\times 60 = 15\text{ min}$$

Distance travelled along the river $= v_r \times t$

$$= 3\times\frac{1}{4} = \frac{3}{4}\text{ km}$$
$$= \frac{3000}{4} = 750\text{ m}$$

4 Let particle is projected with speed u, so total time of flight,

$$T = \left(\frac{2u}{g}\right)$$

and $s = 2\times$ maximum height

$$= 2\times\frac{u^2}{2g} = \frac{u^2}{g}$$

If there is no gravity, then $s' = u \times T$

$$= \frac{2u^2}{g} = 2s$$

If gravity is not there, it will never fall back.

5 $N = A - Z = 197 - 79 = 118$

6 Density of every nucleus is same $= 2.29\times10^{17}$ kg m^{-3}

7 In $_3\text{Li}^7$, $Z = 3$; $N = A - Z = 7 - 3 = 4$

$$\therefore \text{Mass of nucleus} = Z m_p + (A-Z)m_n$$
$$= 3\times1.007825 + 4\times1.008665$$
$$= 7.058075\ \text{amu}$$

The actual mass of nucleus is slightly less than this calculated value.

8 $R_1 = \frac{l}{KA}, R_2 = \frac{l}{2KA}, R_3 = \frac{l}{4KA}$

$$R_{eq} = \left(\frac{R_1 R_2}{R_1 + R_2} + R_3\right)$$

$$R_{eq} = \left[\frac{\frac{l}{KA}\cdot\frac{l}{2KA}}{\frac{l}{KA}+\frac{l}{2KA}} + \frac{l}{4KA}\right] = \left[\frac{\frac{l^2}{2K^2A^2}}{\frac{2l+l}{2KA}} + \frac{l}{4KA}\right]$$

$$= \frac{l}{3KA} + \frac{l}{4KA} = \frac{4l+3l}{12KA} = \frac{7l}{12KA}$$

(as rods 1 and 2 are in parallel and equivalent is in series with 3).

9 Free body diagrams of balls A and B,

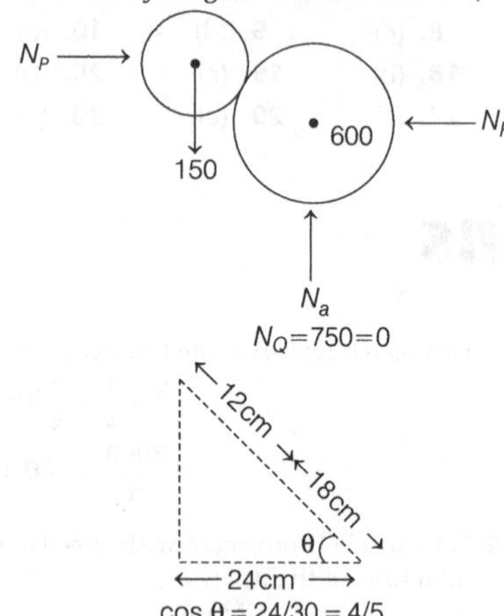

$\cos\theta = 24/30 = 4/5$

Free body diagram of ball B,

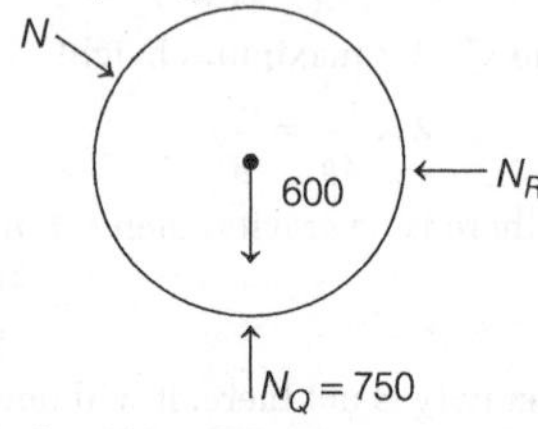

$$N\sin\theta + 600 - 750 = 0$$

$$\frac{3N}{5} = 150$$

$$\Rightarrow \quad N = \frac{150\times 5}{3} = 250$$

$$N\cos\theta - N_R = 0$$

$$N_R = 200 \text{ N}$$

10 $E = \sqrt{\frac{2qV}{M}} \times B$

$$= \sqrt{\frac{2\times 1.60218\times 10^{-19}\times 320}{9.1\times 10^{-31}}} \times 6\times 10^{-5}$$

$$= \sqrt{112.680\times 10^{12}} \times 6\times 10^{-5}$$

$$= 10.61512\times 10^{6}\times 6\times 10^{-5}$$

$$= 63.69072\times 10^{1}$$

$$= 636.9072 \text{ Vm}^{-1} \approx 637 \text{ Vm}^{-1}$$

11 Let a resistance R is connected in series with the battery and wire.

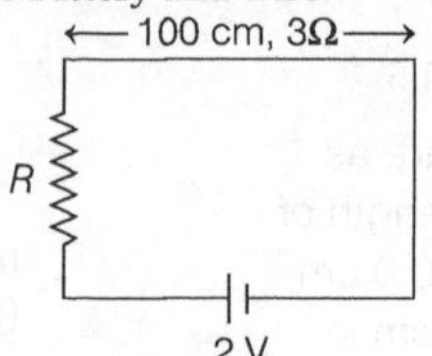

Voltage drop across wire

$= 1\times 10^{-3} \times 100 = 0.1$ V

Let current in circuit is I.

$\therefore \; 0.1 = I\times 3 = 3\times \frac{2}{R+3} \Rightarrow R = 57\;\Omega$

12 The number of beats per second is equal to the difference in frequency between the two sound sources. In this case, because one of the source frequencies is 440 Hz, 4 beats s^{-1} would be heard, if the frequency of the string (the second source) were either 444 Hz or 436 Hz.

13 The work done by gravity is the work done, as if all the mass were concentrated at the centre of mass. The work necessary to lift the object can be thought of as the work done against gravity and is just $W = mgh$, where h is the height through which the centre of mass is raised.

$$W = 180\,(9.8)\,(1.7) = 3.0 \text{ kJ}$$

14 The horizontal range, $R = \frac{u^2 \sin 2\theta}{g}$

Time of flight, $T = \frac{2u\sin\theta}{g}$

Range is maximum, when $\theta = 45°$

So that, $\sin 2\theta = \sin 90° = 1$

$$R_{max} = \frac{u^2}{g}$$

Time of flight is maximum when $\theta = 90°$

So that, $\sin\theta = \sin 90° = 1$

$$T_{max} = \frac{2u}{g}$$

15 Substituting $m = 1$ amu $= 1.67\times 10^{-27}$ kg and $c = 3\times 10^{8}\text{ms}^{-1}$ in the energy-mass equivalence relation

$$E = mc^2$$

$$= 1.67\times 10^{-27} \times (3\times 10^{8})^2$$

$$= 1.67\times 10^{-27} \times 9\times 10^{16} \text{ J}$$

$$= \frac{1.67\times 10^{-27}\times 9\times 10^{16}}{1.6\times 10^{-13}}$$

$(\because 1 \text{ MeV} = 1.6\times 10^{-13} \text{ V})$

$$= 931 \text{ MeV}$$

16 Find acceleration,

$$2ma = mg\sin 53° - \mu\, mg\cos 53° - mg\sin 37° - \mu\, mg\cos 37°$$

Then, use $v^2 = u^2 + 2as$, $v = 0$, $u = 1 \text{ ms}^{-1}$

On solving, we get $s = 0.45$ m

17 The rotational and translational kinetic energy of the ball at the bottom will be changed to gravitational potential energy, when the sphere stops. We therefore write

$$\left(\frac{Mv^2}{2} + \frac{I\omega^2}{2}\right)_{start} = (Mgh)_{end}$$

For a solid sphere, $I = \frac{2\,Mr^2}{5}$

Also, $\omega = \frac{v}{r}$. Then, above equation becomes

$$\frac{1}{2}Mv^2 + \frac{1}{2}\left(\frac{2}{5}Mr^2\right)\left(\frac{v}{r}\right)^2 = Mgh$$

or $$\frac{1}{2}v^2 + \frac{1}{5}v^2 = (9.8)\,h$$

Using $v = 20 \text{ ms}^{-1}$, gives $h = 28.6$ m

18 The initial elastic potential energy of the compressed spring is

$$\text{PE}_s = \frac{1}{2}kx_i^2$$

Because the block is always at the same height above earth's surface, the gravitational potential energy of the system remains constant. Hence, the initial potential energy stored in the spring is converted to kinetic energy at $x = 0$. i.e.

$$\frac{1}{2}kx_i^2 = \frac{1}{2}mv_f^2$$

Solving for v_f gives, $v_f = \sqrt{\frac{k}{m}}\,x_i$

$$= \sqrt{\frac{1.0\times 10^3}{1.6}}\,(2.0\times 10^{-2}) = 0.50 \text{ ms}^{-1}$$

19 The angle between magnetic field and area vector is 90°, so the flux associated with coil is zero. Although magnetic field is changing but flux is remaining constant equal to zero, so emf induced and hence, current in the loop is equal to zero.

20 Resistance of coil $= \frac{V_{DC}}{I_{DC}} = \frac{6}{12} = 0.5\;\Omega$

In an AC circuit, the current is in phase with emf. This means that the net reactance of the circuit is zero. The impedance is equal to the resistance, i.e. $Z = 0.5\;\Omega$

Rms current $= \frac{\text{rms voltage}}{Z}$

$$= \frac{6}{0.5} = 12 \text{ A}$$

21 If the upper wire is to float, it must be in equilibrium under the action of two forces : the force of gravity and magnetic repulsion. The weight per unit length here 1.0×10^{-4} Nm^{-1} must be equal and opposite the magnetic force per unit length. Because the currents are the same, we have

$$\frac{F_1}{l} = \frac{mg}{l} = \frac{\mu_0 I^2}{2\pi d}$$

$$\Rightarrow 1.0 \times 10^{-4} = \frac{(4\pi \times 10^{-7})(I^2)}{(2\pi)(0.10)}$$

We solve for the current to find

$$I = 7.1 \text{ A}$$

22 Because $2nt = \frac{\lambda}{2}$, we have

$$t = \frac{\lambda}{4n} = \frac{602}{(4)(1.33)} = 113 \text{ nm}$$

23 Maximum distance covered by space wave communication

$$= \sqrt{2Rh}$$

$$= \sqrt{2 \times 6.4 \times 10^6 \times 300} = 62 \text{ km}$$

Since, receiver-transmitter distance is 100 km, this is ruled out for signal frequency.

Further f_c for ionospheric propagation is

$$f_c = 9(N_{\max})^{1/2} = 9 \times (10^{12})^{1/2} = 9 \text{ MHz}$$

So, the signal of 100 MHz $(7f_c)$ comes *via* the satellite mode.

24 For $t \le t_1$; $A = 0, B = 0$; Hence $Y = 0$
For t_1 to t_2; $A = 1, B = 0$; Hence $Y = 0$
For t_2 to t_3; $A = 1, B = 1$; Hence $Y = 1$
For t_3 to t_4; $A = 0, B = 1$; Hence $Y = 0$
For t_4 to t_5; $A = 0, B = 0$; Hence $Y = 0$
For t_5 to t_6; $A = 1, B = 0$; Hence $Y = 0$
For $t > t_6$; $A = 0, B = 1$; Hence $Y = 0$

Based on the above, the output waveform for AND gate can be drawn as given below.

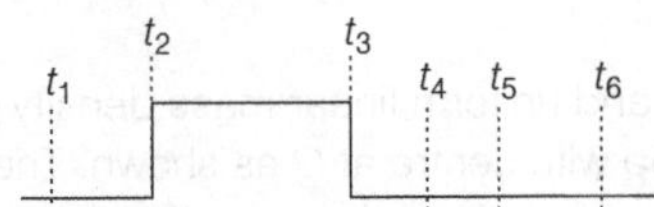

25 The initial energy of the sledge-rider-earth system includes kinetic energy because of the initial speed

$$\frac{1}{2} m v_i^2 + mg y_i = \frac{1}{2} m v_f^2 + mg y_f$$

or $\quad \frac{1}{2} v_i^2 + g y_i = \frac{1}{2} v_f^2 + g y_f$

If we set the origin of our coordinates at the bottom of the incline, the initial and final y-coordinates of the sledge are $y_i = 10.0$ m and $y_f = 0$.

Thus, we get

$$\frac{1}{2} v_i^2 + g y_i = \frac{1}{2} v_f^2 + 0$$

$$v_f^2 = v_i^2 + 2g y_i$$

$$= (5.00)^2 + 2(9.80)(10.0)$$

$$v_f = 14.9 \text{ ms}^{-1}$$

26 The force that produces the centripetal acceleration of the satellite is the gravitational force, so

$$\therefore \quad G\frac{M_e m}{r^2} = \frac{m v^2}{r} \quad \text{...(i)}$$

where, M_e is earth's mass and r is the satellite's distance from the centre of the earth.

Also, we find the speed of the satellite to be

$$v = \frac{d}{T} = \frac{2\pi r}{T} \quad \text{...(ii)}$$

where, T is the orbital period of the satellite.

Solving Eqs. (i) and (ii) simultaneously for r yields,

$$r = \left(\frac{T^2}{4\pi^2} GM_e\right)^{1/3}$$

27 Recall that an isovolumetric process is one that takes place at constant volume. In such a process the work done is equal to zero because there is no change in volume. Thus, the first law of thermodynamics gives

$$\Delta U = Q$$

This indicates that the net energy Q added to the water goes into increasing the internal energy of the water. The net energy added to the water is

$$Q = 10.0 - 2.0 = 8.0 \text{ kJ}$$

Because $Q = mc\,\Delta T$, the temperature increase of the water is

$$\Delta T = \frac{Q}{mc} = \frac{8.0 \times 10^3}{(2.0)(4.186 \times 10^3)}$$

$$= 0.96°\text{C}$$

28 By conservation of momentum, the momentum of the block bullet system just after the interaction is $p = mv$. where, m is the mass of bullet and v is its velocity before striking the block. Hence, the kinetic energy of the system just after the lodging of bullet into the block is

$$K = \frac{p^2}{2(M+m)} = \frac{(mv)^2}{2(M+m)} \quad \text{...(i)}$$

The friction force does work

$$W_f = -f_s = -\mu_k (m+M) gs \quad \text{...(ii)}$$

in stopping the block where s is the distance traversed by block-bullet system on the table top.

From work-energy theorem,

$$\Delta K = W_f$$

$$0 - K = -\mu_k (m+M) gs$$

Substituting values in Eqs.(i) and (ii), we get

$$\therefore \quad s = 0.37 \text{ m}$$

Also the rest mass of photon is zero.

29 The descending part of the rope is in free fall. It has speed $v = \sqrt{2gy}$ at the instant all its points have descended a distance y. The length of the rope which lands on the table during an interval dt following this instant is $v\,dt$. The increment of momentum imparted to the table by this length in coming to rest is $m(v\,dt)v$. Thus, the rate at which momentum is transferred to the table is

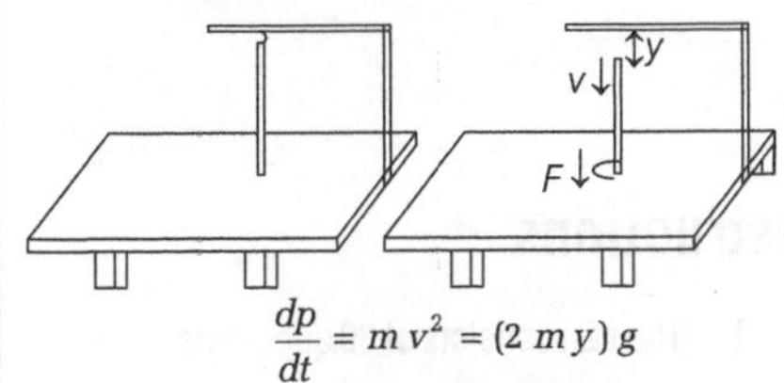

$$\frac{dp}{dt} = mv^2 = (2my)g$$

and this is the force arising from stopping the downward fall of the rope. Since, a length of rope y of the weight $(my)g$, already lies on the tabletop, the total force on the tabletop is $(2my)g + (my)g = (3my)g$, or the weight of a length $3y$ of rope.

So, $\quad k = 3$

30 First we make ray diagrams roughly to scale to see where the image from the first lens falls and how it acts as the object for the second lens. The location of the image formed by the first lens is found *via* the thin lens equation

$$\frac{1}{30.0} + \frac{1}{v_1} = \frac{1}{10.0}, \; v_1 = +15.0 \text{ cm}$$

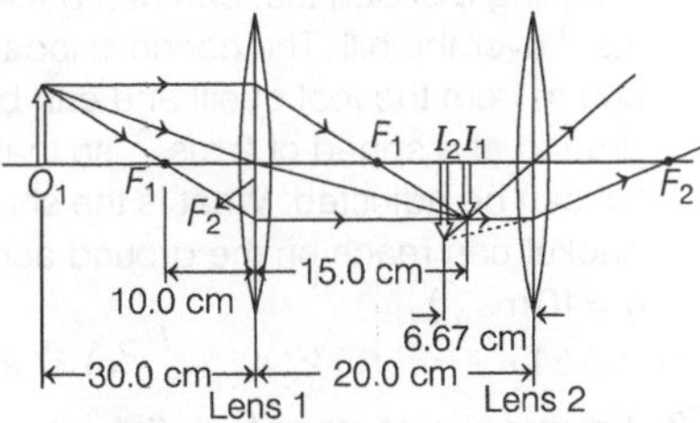

The image formed by this lens becomes the object for the second lens. Thus, the object distance for the second lens is 20.0 cm − 15.0 cm = 5.00 cm. We again apply the thin lens equation to find the location of the final image.

$$\frac{1}{5.00} + \frac{1}{v_2} = \frac{1}{20.0}, \; v_2 = -6.67 \text{ cm}$$

Thus, the final image is 6.67 cm to the left of the second lens.

DAY FOURTY

Mock Test 3

(Based on Complete Syllabus)

Instructions

1. The test consists of 30 questions.
2. Candidates will be awarded marks for correct response of each question. 1/4 (one-fourth) marks will be deducted for indicating incorrect response of each question. No deduction from the total score will be made, if no response is indicated for an item in the answer sheet.
3. There is only one correct response for each question. Filling up more than one response in each question will be treated as wrong response.

1 An organ pipe of length L_0, open at both ends is bound to vibrate in its first harmonic, when sounded with a tuning fork of 480 Hz. What should be the length of a pipe closed at one end, so that it also vibrates in its first harmonic with the same tuning fork?

(a) $L_c = 2L_0$ (b) $L_c = \frac{L_0}{3}$ (c) $L_c = \frac{L_0}{2}$ (d) $L_c = \frac{2L_0}{3}$

2 A hill is 500 m high. Supplies are to be sent across the hill using a canon that can hurl packets at a speed of 125 ms^{-1} over the hill. The canon is located at a distance of 800 m from the foot of hill and can be moved on the ground at a speed of 2 ms^{-1}, so that its distance from the hill can be adjusted. What is the shortest time in which a packet can reach on the ground across the hill? (Take, $g = 10\,ms^{-2}$)

(a) 31 s (b) 27 s (c) 37 s (d) 45 s

3 A motor cyclist starts from the bottom of a slope of angle 45° and travels along the slope to jump clear of the valley AB as shown in figure. The width of the valley is 160 m and the length of the slope is $160\sqrt{2}$ m. The minimum velocity with which he should leave the bottom O, so that he can clear the valley, is (nearest to in ms^{-1})

(a) 50 (b) 56 (c) 60 (d) 70

4 Choose the correct alternative.

(a) Gravitational potential at curvature centre of a thin hemispherical shell of radius R and mass M is equal to GM/R

(b) Gravitational field strength at a point lying on the axis of a thin uniform circular ring of radius R and mass M is equal to $\frac{GM}{(R^2 + x^2)^{3/2}}$, where x is distance of that point from centre of the ring

(c) Newton's law of gravitation for gravitational force between two bodies is applicable only, when bodies have spherically symmetric distribution of mass

(d) None of the above

5 A thin wire of length L and uniform linear mass density ρ is bent into a circular loop with centre at O as shown. The moment of inertia of the loop about the axis XX' is

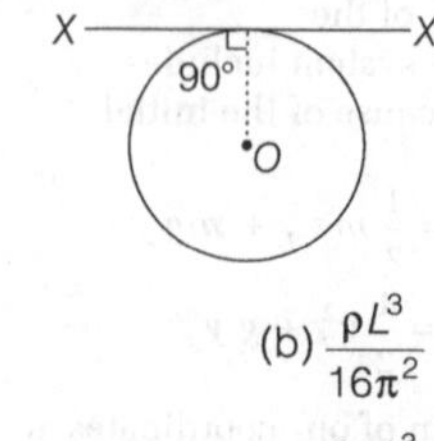

(a) $\frac{\rho L^3}{8\pi^2}$ (b) $\frac{\rho L^3}{16\pi^2}$

(c) $\frac{5\rho L^3}{16\pi^2}$ (d) $\frac{3\rho L^3}{8\pi^2}$

6 If a drop of liquid breaks into smaller droplets, it results in lowering of temperature of the droplets. Let a drop of radius R, break into N small droplets each of radius r. Estimate the temperature in drop.

(a) $\frac{S}{\rho s}\left[\frac{1}{R}\right]$ (b) $\frac{2S}{\rho s}\left[\frac{1}{r}-\frac{1}{R}\right]$

(c) $\frac{3S}{\rho s}\left[\frac{1}{R}-\frac{1}{r}\right]$ (d) $\frac{2S}{\rho s}\left[\frac{1}{R}-\frac{1}{r}\right]$

7 A man beats a drum at a certain distance from a mountain. He slowly increase the rate of beating and finds that the echo is not heard distinctly, when the drum beating is at the rate of 40 per min. He moves by 80 m towards the mountain and finds that the echo is again not heard distinctly, when the rate of beating of the drum is 1 per sec. What is the original distance of the man from the mountain?

(a) 120 m (b) 240 m (c) 270 m (d) 340 m

8 The smiling face in figure consists of three items (i), A thin rod of charge $-3.0\ \mu C$ that forms a full circle of radius 6 cm. (ii) second thin rod of charge $2.0\ \mu C$ that forms a circular arc of radius 4.0 cm, subtending an angle of 20° about the centre of the full circle and (iii) electric dipole with a dipole moment that is $\perp$ to a radial line and has a magnitude 1.28×10^{-21}cm, what is the net electric potential of the centre.

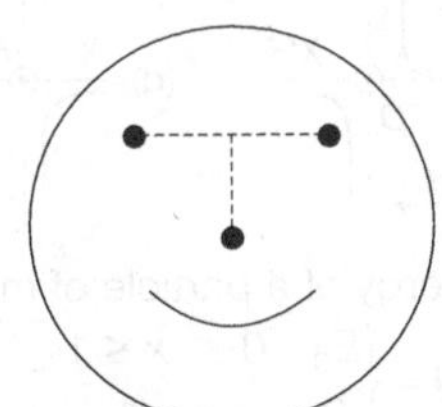

(a) zero (b) 2.3×10^{-3}
(c) 1.28×10^{21} (d) None of these

9 Pressure *versus* temperature graphs of an ideal gas are as shown in figure. Choose the incorrect statement.

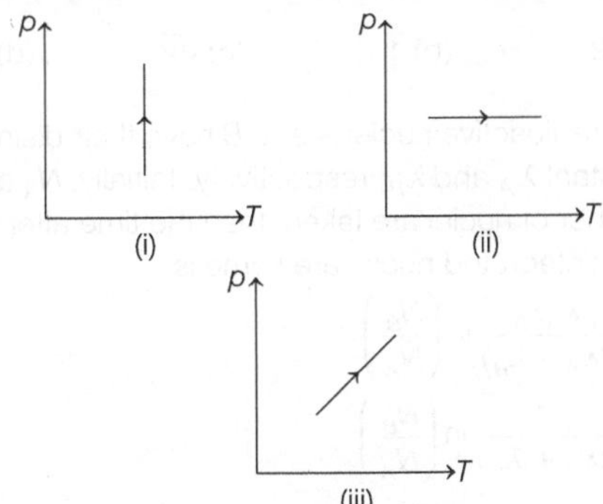

(a) Density of gas is increasing in graph (i)
(b) Density of gas is increasing in graph (ii)
(c) Density of gas is constant in graph (iii)
(d) None of the above

10 Two non-ideal batteries of unequal emf 's are connected in parallel. Consider the following statements.

(A) The equivalent emf is smaller than either of the two emf 's.

(B) The equivalent internal resistance is smaller than either of the two internal resistances.

(a) Both A and B are correct
(b) A is correct but B is incorrect
(c) B is correct but A is incorrect
(d) Both A and B are incorrect

11 A solid conducting sphere of radius a has a net positive charge $2Q$. A conducting spherical shell of inner radius b and outer radius c is concentric with the solid sphere and has a net charge $-Q$.

The surface charge density on the inner and outer surfaces of the spherical shell will be

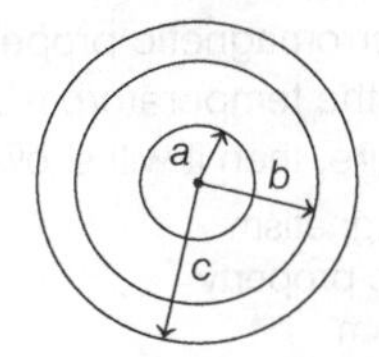

(a) $-\frac{2Q}{4\pi b^2}, \frac{Q}{4\pi c^2}$ (b) $-\frac{Q}{4\pi b^2}, \frac{Q}{4\pi c^2}$

(c) $0, \frac{Q}{4\pi c^2}$ (d) None of these

Direction (Q. Nos. 12-14) *Each of these questions contains two statements : Statement I and Statement II. Each of these questions also has four alternative choices, only one of which is the correct answer. You have to select one of the codes (a), (b), (c), (d) given below.*

(a) Statement I is true, Statement II is true; Statement II is the correct explanation for Statement I
(b) Statement I is true, Statement II is true; Statement II is not the correct explanation for Statement I
(c) Statement I is true; Statement II is false
(d) Statement I is false; Statement II is true

12 **Statement I** Time period of oscillation of two magnets, when like poles are in same direction (in a vibration magnetometer) is smaller, than the period of vibration when like poles are in opposite direction.

Statement II Moment of inertia increases in same position.

13 **Statement I** It is impossible for a ship to use the internal energy of sea water to operate its engine.

Statement II Refrigerator is a type of heat engine.

14 **Statement I** A tennis ball bounces higher on hills than in plains.

Statement II Acceleration due to gravity on the hill is greater than that on the surface of the earth.

15 In the circuit shown, the coil has inductance and resistance. When X is joined to Y, the time constant is τ during growth of current. When the steady state is reached, heat is produced in the coil at a rate P. X is now joined to Z

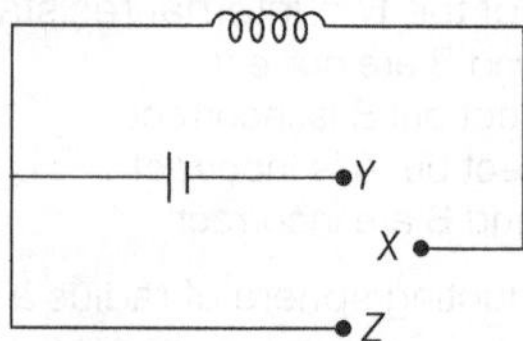

(a) the total heat produced in the coil is $P\tau$
(b) the total heat produced in the coil is $\frac{1}{2}P\tau$
(c) the total heat produced in the coil is $2P\tau$
(d) the data given is not sufficient to reach a conclusion

16 Nickle shows ferromagnetic property at room temperature. If the temperature is increased beyond curie temperature, then it will show
(a) anti-ferromagnetism
(b) no magnetic property
(c) diamagnetism
(d) paramagnetism

17 A glass prism ABC (refractive index 1.5), immersed in water (refractive index 4/3). A ray of light is incident normally on face AB. If it is totally reflected at face AC, then

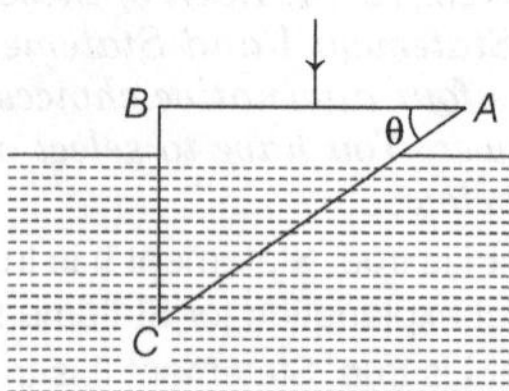

(a) $\sin\theta \geq \frac{8}{9}$ (b) $\sin\theta \geq \frac{2}{3}$
(c) $\sin\theta = \frac{\sqrt{3}}{2}$ (d) $\frac{2}{3} < \sin\theta < \frac{8}{9}$

18 A metal wire of linear mass density of 9.8 gm^{-1} is stretched with a tension of 10 kg-wt between two rigid supports which are 1m apart. The wire passes through the middle points between the poles of a permanent magnet and it vibrates in resonance, when carrying on alternating current of frequency n. The frequency n of the alternating current is
(a) 25 Hz (b) 50 Hz (c) 200 Hz (d) 100 Hz

19 In a given process of an ideal gas, $dW = 0$ and $dQ < 0$. Then, for the gas
(a) the temperature will decrease
(b) the volume will increase
(c) the pressure will remain constant
(d) the temperature will increase

20 The circuit shown in figure, contains a resistance of $R = 6\,\Omega$ connected with a battery of emf 6 V.

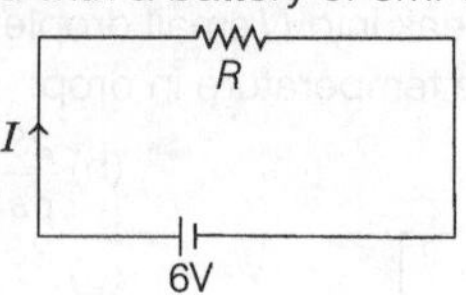

Given, n = number of electrons per volume $= 10^{29}\,/m^3$, length of circuit = 10 cm, cross-section $A = 1\,mm^2$. The energy absorbed by electrons from initial state of no current (ignore thermal motion) to the state of drift velocity is
(a) 1.5×10^{-18} J (b) 3.5×10^{-19} J
(c) 2×10^{-17} J (d) 3×10^{-15} J

21 Two long parallel wires are at a distance $2d$ apart. They carry steady equal currents flowing out of the plane of the paper, as shown. The variation of the magnetic field B along the line XX' is given by

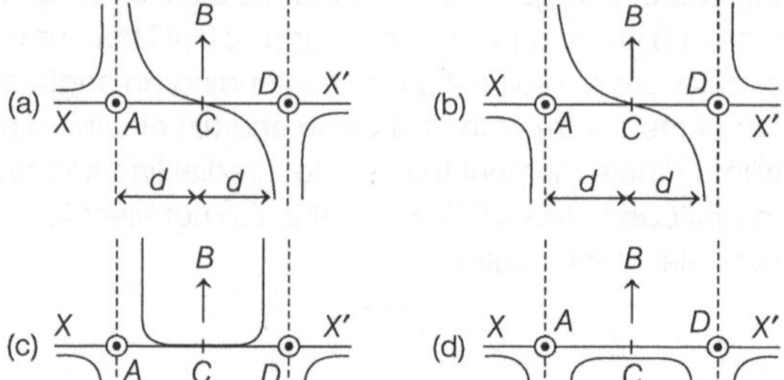

22 The potential energy of a particle of mass m is given by

$$U(x) = \begin{cases} E_0 & 0 \leq x \leq 1 \\ 0 & x > 1 \end{cases}$$

λ_1 and λ_2 are the de-Broglie wavelengths of the particle, when $0 \leq x \leq 1$ and $x > 1$, respectively. If the total energy of particle is $2E_0$, the ratio $\frac{\lambda_1}{\lambda_2}$ will be
(a) 2 (b) 1 (c) $\sqrt{2}$ (d) $\frac{1}{\sqrt{2}}$

23 Two radioactive nuclei A and B have their disintegration constant λ_A and λ_B, respectively. Initially, N_A and N_B number of nuclei are taken, then the time after which their undisintegrated nuclei are same is

(a) $\frac{\lambda_A \lambda_B}{(\lambda_A - \lambda_B)} \ln\left(\frac{N_B}{N_A}\right)$

(b) $\frac{1}{(\lambda_A + \lambda_B)} \ln\left(\frac{N_B}{N_A}\right)$

(c) $\frac{1}{(\lambda_B - \lambda_A)} \ln\left(\frac{N_B}{N_A}\right)$

(d) $\frac{1}{(\lambda_A - \lambda_B)} \ln\left(\frac{N_B}{N_A}\right)$

24 A student constructed a vernier callipers as shown in the figure. He used two identical inclines and tried to measure the length of line PQ. For this instrument determine the least count.

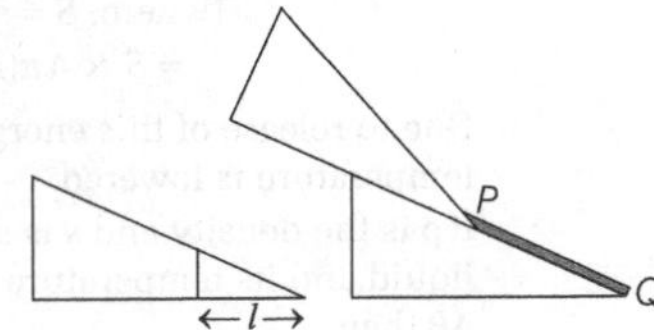

(a) $\frac{l(1-\cos\theta)}{\cos\theta}$ units

(b) $\frac{l}{\cos\theta}$ units

(c) $l(1-\cos\theta)$ units

(d) $\frac{1-\cos\theta}{l}$ units

25 A cubical block of side L rests on a rough horizontal surface with coefficient of friction μ. A horizontal force F is applied on the block as shown in the figure. If the coefficient of friction is sufficiently high, so that the block does not slide before toppling, the minimum force required to topple the block is

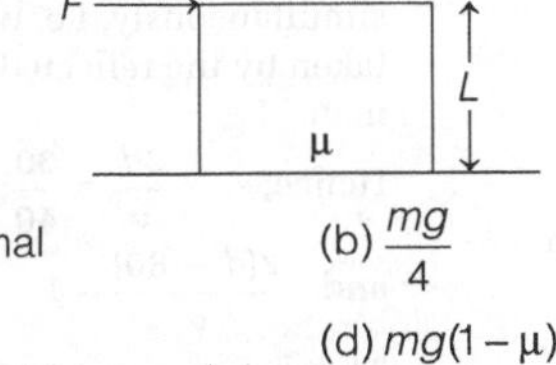

(a) infinitesimal (b) $\frac{mg}{4}$

(c) $\frac{mg}{2}$ (d) $mg(1-\mu)$

26 A double star consists of two stars having masses M and $2M$. The distance between their centres is equal to r. They revolve under their mutual gravitational interaction. Then, which of the following statement(s) is/are correct?

(a) Heavier star revolves in orbit of radius $2r/3$

(b) Both of the stars revolve with the same period which is equal to $\frac{2\pi}{\sqrt{2GM/3}} r^{3/2}$

(c) Kinetic energy of heavier star is twice that of the other star

(d) None of the above

27 The fundamental frequency of a sonometer wire of length l is f_0. A bridge is now introduced at a distance of Δl from the centre of the wire ($\Delta l << l'$). The number of beats heard, if both sides of the bridges are set into vibration in their fundamental modes, are

(a) $\frac{8f_0\,\Delta l}{l}$ (b) $\frac{f_0\,\Delta l}{l}$

(c) $\frac{2f_0\,\Delta l}{l}$ (d) $\frac{4f_0\,\Delta l}{l}$

28 A mason is supplied with bricks by his assistant who is 3 m below him, the assistant is tossing the brick vertically up. The speed of the brick, when it reaches the mason is 2 ms^{-1}. What percentage of energy used up by the servant serves no useful purpose ?

(a) 9.8% (b) 4.9% (c) 5.6% (d) 10%

29 To plot forward characteristic of p-n junction diode, the correct circuit diagram is

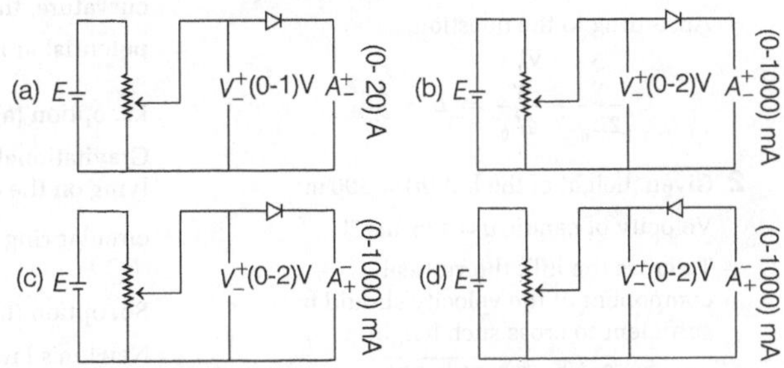

30 A block of mass m moves with a velocity v_o on a smooth horizontal surface. If that passes over a cylinder of radius R and mass m, capable of rotating about its own fixed axis through O the block, while passing over, slips on the cylinder. The slipping stops before, it loses contact. The block then moves on similar smooth horizontal surface with a velocity v. then the velocity v is

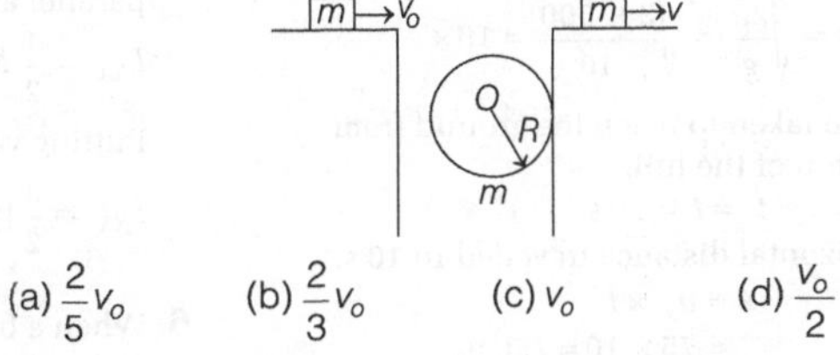

(a) $\frac{2}{5}v_o$ (b) $\frac{2}{3}v_o$ (c) v_o (d) $\frac{v_o}{2}$

ANSWERS

1. (c)	**2.** (d)	**3.** (d)	**4.** (c)	**5.** (d)	**6.** (c)	**7.** (b)	**8.** (a)	**9.** (a)	**10.** (a)
11. (a)	**12.** (c)	**13.** (b)	**14.** (c)	**15.** (b)	**16.** (d)	**17.** (a)	**18.** (b)	**19.** (a)	**20.** (c)
21. (b)	**22.** (c)	**23.** (c)	**24.** (a)	**25.** (c)	**26.** (b)	**27.** (a)	**28.** (c)	**29.** (b)	**30.** (b)

Hints and Explanations

1 The fundamental frequency for an open pipe, $\nu_0 = \frac{\nu}{2L_0}$

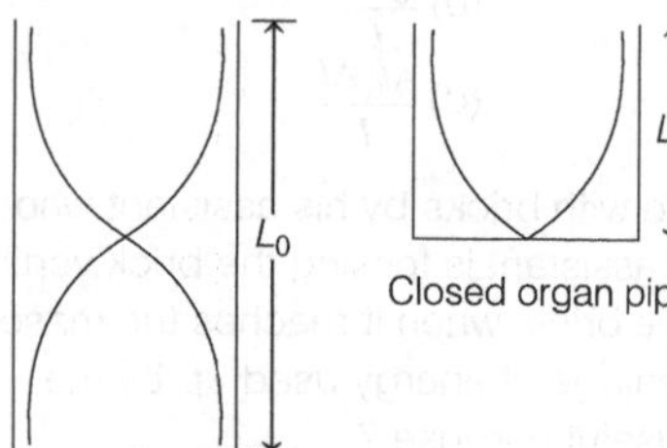

Fundamental frequency for closed pipe,

$$\nu'_0 = \frac{\nu}{4L_c}$$

According to the question,

$$\nu_0 = \nu'_0$$

$$\frac{\nu}{2L_0} = \frac{\nu}{4L_0} \Rightarrow L_c = \frac{L_0}{2}$$

2 Given, height of the hill $(h) = 500$ m

Velocity of canon, $u = 125$ ms^{-1}

To cross the hill, the vertical component of the velocity should be sufficient to cross such height.

$$\therefore u_y \ge \sqrt{2gh} \ge \sqrt{2 \times 10 \times 500}$$

$$\ge 100 \text{ms}^{-1}$$

But $u^2 = u_x^2 + u_y^2$

$\therefore$ Horizontal component of initial velocity,

$$u_x = \sqrt{u^2 - u_y^2} = \sqrt{(125)^2 - (100)^2}$$

$$= 75 \text{ms}^{-1}$$

Time taken to reach the top of the hill,

$$t = \sqrt{\frac{2h}{g}} = \sqrt{\frac{2 \times 500}{10}} = 10 \text{ s}$$

Time taken to reach the ground from the top of the hill,

$$t' = t = 10 \text{ s}$$

Horizontal distance travelled in 10 s,

$$x = u_x \times t$$

$$= 75 \times 10 = 750 \text{ m}$$

$\therefore$ Distance through which canon has to be removed $= 800 - 750 = 50$ m

Speed with which canon can move

$$= 2 \text{ms}^{-1}$$

$\therefore$ Time taken by canon $= \frac{50}{2}$

$$t'' = 25 \text{ s}$$

Hence, total time taken by a packet to reach on the ground $= t'' + t + t'$

$$= 25 + 10 + 10$$

$$= 45 \text{s}$$

3 Velocity to take-off from A to clear the valley is given by

$$R = \frac{u^2}{g} \sin 2\alpha$$

$$\alpha = 45°, u = \sqrt{gR} = 40 \text{ ms}^{-1}$$

Velocity to start from lowest point (due to retardation on inclined plane, $g \sin \alpha$), $v_0^2 = u^2 + 2g \sin \alpha \times s$

$$\therefore \quad v_0 = \sqrt{(40)^2 + 2 \times 10 \times \frac{1}{\sqrt{2}} \times 160 \times \sqrt{2}}$$

$$= \sqrt{4800} \simeq 70 \text{ ms}^{-1}$$

4 Because every element of hemispherical shell is at a distance R from centre of curvature, therefore gravitational potential at its centre $= -\frac{GM}{R}$,

i.e. option (a) is incorrect.

Gravitational field strength at a point, lying on the axis of a thin uniform circular ring of radius R is $\frac{GMx}{(R^2 + x^2)^{3/2}}$

So, option (b) is incorrect.

Newton's law of gravitation is applicable to only those bodies which have spherically symmetric distribution of mass. So, option (c) is correct.

5 Mass of the ring, $M = \rho L$

Let R be the radius of the ring.

Then, $L = 2\pi R$ or $R = \frac{L}{2\pi}$

Moment of inertia about XX' (from parallel axis theorem) will be given by

$$I_{XX'} = \frac{1}{2} MR^2 + MR^2 = \frac{3}{2} MR^2$$

Putting values of M and R,

$$I_{XX'} = \frac{3}{2} (\rho L) \left(\frac{L^2}{4\pi^2} \right) = \frac{3}{8} \frac{\rho L^3}{\pi^2}$$

6 When a big drop of radius R, break into N droplets each of radius r, the volume remains constant.

$\therefore$ Volume of big drop

$$= N \times \text{volume of small drop}$$

$$\frac{4}{3} \pi R^3 = N \times \frac{4}{3} \pi r^3$$

or $\quad R^3 = Nr^3$ or $N = \frac{R^3}{r^3}$

Now, change in surface area

$$= 4\pi R^2 - N 4\pi r^2$$

$$= 4\pi(R^2 - Nr^2)$$

Energy released $= S \times \Delta A$

(where, S = surface tension)

$$= S \times 4\pi(R^2 - Nr^2)$$

Due to release of this energy, the temperature is lowered.

If ρ is the density and s is specific heat of liquid and its temperature is lowered by $\Delta\theta$ then,

Energy released $= ms\Delta\theta$

$$S \times 4\pi(R^2 - Nr^2) = \left(\frac{4\pi}{3} \times R^3 \times \rho \right) s\Delta\theta$$

$$\Delta\theta = \frac{S \times 4\pi(R^2 - Nr^2)}{\frac{4}{3}\pi R^3 \rho \times s}$$

$$= \frac{3S}{\rho s} \left[\frac{R^2}{R^3} - \frac{Nr^2}{R^3} \right]$$

$$= \frac{3S}{\rho s} \left[\frac{1}{R} - \frac{(R^3 / r^3) \times r^2}{R^3} \right]$$

$$= \frac{3S}{\rho s} \left[\frac{1}{R} - \frac{1}{r} \right]$$

7 The echo is not heard distinctly, when the echo and the next beat fall on the ear simultaneously, i.e. time per beat = time taken by the reflected beat to reach the man.

Hence, $\quad \frac{2d}{v} = \frac{60}{40} = \frac{3}{2}$

and $\quad \frac{2(d - 80)}{v} = 1$

This gives, $\quad d = 240$ m

8 The dipole potential $(\theta = 90°)$ is given by the equation,

$$V = \frac{p \cos\theta}{4\pi\varepsilon_0 r^2} = \frac{p \cos 90°}{4\pi\varepsilon_0 r^2} = 0$$

$[\because \cos 90° = 0]$

Also, the potential due to the short arc is $q_1 / 4\pi\varepsilon_0 r_1$ and that caused by the long arc is $q_2 / 4\pi\varepsilon_0 r^2$.

Since, $q_1 = +2\mu$C, $r_1 = 4$cm, $q_2 = -3\mu$C and $r_2 = 6$ cm, the potential of the arcs cancel.

Thus, the result is zero.

9 As, $\rho = \frac{pM}{RT}$

Density ρ remains constant, when p/T or volume remains constant. In graph (i) volume is decreasing, hence density is increasing; while in graphs (ii) and (iii) temperature is increasing, hence, density is decreasing. Note that, volume would have been constant in case the straight line in graph (iii) had passed through origin.

10 In parallel,

$$E_{eq} = \frac{E_1/r_1 + E_2/r_2}{1/r_1 + 1/r_2}$$

$$= E_1\left(\frac{r_2 + (E_2/E_1)\cdot r_1}{r_1 + r_2}\right)$$

$$= E_2\left(\frac{r_1 + (E_1/E_2)\cdot r_2}{r_1 + r_2}\right)$$

Now, if $E_1 = E_2$.

Then, $E_{eq} = E_1$ if $E_2 > E_1$.

Then, $E_{eq} > E_1$ and $E_{eq} < E_2$.

Similarly, if $E_1 > E_2$.

Then, $E_{eq} > E_2$ but $E_{eq} < E_1$.

11 Due to induction, inner surface of spherical shell has charge $-2Q$,

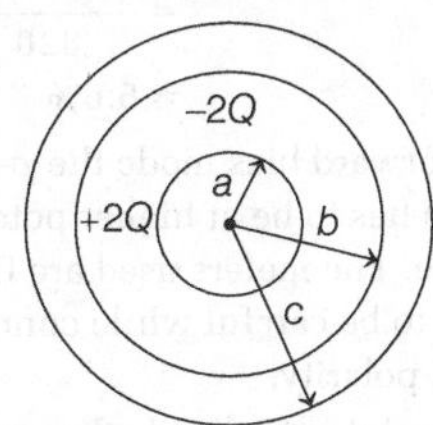

So, surface charge density on inner side

$$\sigma_{inner} = \frac{-2Q}{4\pi b^2}$$

and surface charge density on outer side $\sigma_{outer} = \dfrac{Q}{4\pi c^2}$

12 ***Case* I** When the like poles of two magnets are placed in same direction, then the time period of vibration is expressed as

$$T' = 2\pi\sqrt{\frac{I_1 + I_2}{(M_1 + M_2)B}} \quad \text{...(i)}$$

***Case* II** When the like poles of two magnets are placed in opposite direction, then period of vibration is expressed as

$$T'' = 2\pi\sqrt{\frac{I_1 + I_2}{(M_1 - M_2)B}} \quad \text{...(ii)}$$

It is clear from Eqs. (i) and (ii) that, $T' < T''$.

13 For using the internal energy of sea water to operate the engine of a ship, the internal energy of the sea water has to be converted into mechanical energy. Since, whole of the internal energy cannot be converted into mechanical energy, a part has to be rejected to a colder body (sink). As, no such body is available, the internal energy of the sea water cannot be used to operate the engine of the ship. Note that a refrigerator is a heat engine working in the reverse direction.

14 Suppose that the tennis ball bounces with a velocity u. It will go up, till its velocity becomes zero. If h is the height upto which it rises on the hill, then

$$(0)^2 - u^2 = 2(-g')h$$

where, g' is acceleration due to gravity on the hill

$$\therefore \quad h = \frac{u^2}{2g'}$$

Since, the acceleration due to gravity on the hill (g') is less than that on earth (effect of height), it follows that tennis ball will bounce higher on hills than in plains.

15 As, $P = (I_0)^2 \cdot R$

i.e. $(I_0)^2 = \dfrac{P}{R}$

$$\therefore \quad U = \frac{1}{2}LI_0^2 = \frac{1}{2}(\tau R)\left(\frac{P}{R}\right) = \frac{1}{2}P\tau$$

16 The curie temperature is defined as the temperature beyond which the ferromagnetic material shows paramagnetic behaviour.

17 For total internal reflection at AC-face

$$\sin i \geq \frac{\mu_w}{\mu_g}$$

$$\sin\theta \geq \frac{4}{3\times 1.5}$$

$$\sin\theta \geq \frac{8}{9}$$

18 Since, the tension,

$$T = 10 \text{ kg-wt} = 10\times 9.8 = 98 \text{ N}$$

and $m = 9.8\times 10^{-3}$ kgm^{-1}, $L = 1$m

So, we get

$$n = \frac{1}{2L}\sqrt{\frac{T}{m}} = \frac{1}{2\times 1}\times\sqrt{\frac{98}{9.8\times 10^{-3}}}$$

$$= 50 \text{ Hz}$$

19 From first law of thermodynamics

$$dQ = dU + dW$$

$$dQ = dU \quad (\because dW = 0)$$

Since, $dQ < 0$

Therefore, $dU < 0$

or $U_{final} < U_{initial}$

or temperature will decrease.

20 Given, $V = 6$ V, $R = 6\ \Omega$,

$$A = 1\times 10^{-6} \text{ m}^2$$

and $l = 10$ cm $= 0.1$ m

The current in the circuit,

$$I = \frac{V}{R} = \frac{6}{6} = 1\text{A}$$

Use the relation $I = neAv_d$

Drift velocity of electrons,

$$v_d = \frac{I}{neA}$$

$$= \frac{1}{10^{29}\times 1.6\times 10^{-19}\times 1\times 10^{-6}}$$

$$= \frac{1}{1.6}\times 10^{-4} \text{ms}^{-1}$$

The energy of electrons, (KE) $= \dfrac{1}{2}mv^2$

$= \dfrac{1}{2}\times m_e\times v_d^2\times$ volume $\times$ number of electrons per volume

$$= \frac{1}{2}\times 9.1\times 10^{-31}\times\left(\frac{10^{-4}}{1.6}\right)^2\times A\times l\times n$$

($\because$ Mass of electron $m_e = 9.1\times 10^{-31}$)

$$= \frac{9.1\times 10^{-39}}{2\times 1.6\times 1.6}\times 10^{-6}\times 0.1\times 10^{29}$$

$$= 2\times 10^{-17} \text{ J}$$

21 If the current flows out of the paper, the magnetic field at points to the right of the wire will be upwards and to the left will be downward. Now, magnetic field at C is zero. The field in the region DX' will be upwards (+ve), because all points existing in this region are to the right of both the wires. Similarly, magnetic field in the region AX will be downwards (–ve). The field in the region AC will be upwards (+ve), because points are closer to A as compared to D. Similarly, magnetic field in region DC will be downward (–ve).

22 KE $= 2E_0 - E_0 = E_0$ (for $0\leq x\leq 1$)

So, $\quad \lambda_1 = \dfrac{h}{\sqrt{2mE_0}}$ ...(i)

Again $\quad$ KE $= 2E_0$ (for $x > 1$)

$\therefore \quad \lambda_2 = \dfrac{h}{\sqrt{2m2E_0}}$ ...(ii)

From Eqs. (i) and (ii), we get

$$\frac{\lambda_1}{\lambda_2} = \sqrt{2}$$

23 After disintegration,

$$N_Ae^{-\lambda_A t} = N_Be^{-\lambda_B t} \quad (\text{for } 0\leq x\leq 1)$$

or $\quad e^{(\lambda_B - \lambda_A)t} = \dfrac{N_B}{N_A}$ ...(i)

$$\therefore \quad (\lambda_B - \lambda_A)t = \text{In}\left(\frac{N_B}{N_A}\right)$$

$$\Rightarrow \quad t = \frac{1}{\lambda_B - \lambda_A}\text{In}\left(\frac{N_B}{N_A}\right)$$

24 Let θ be the angle of incline. Here, the incline kept horizontally is working as main scale while the other incline kept on horizontally placed incline is treated as vernier scale.

From the figure, it is clear that,

1 MSD $= \dfrac{l}{\cos\theta}$ unit and 1 VSD $= l$ unit

So, LC of instrument is,

LC = 1MSD – 1 VSD

$$= \left(\frac{l}{\cos\theta} - l\right) = \frac{l(1-\cos\theta)}{\cos\theta} \text{ units}$$

25 At the critical condition, normal reaction N will pass through point P.

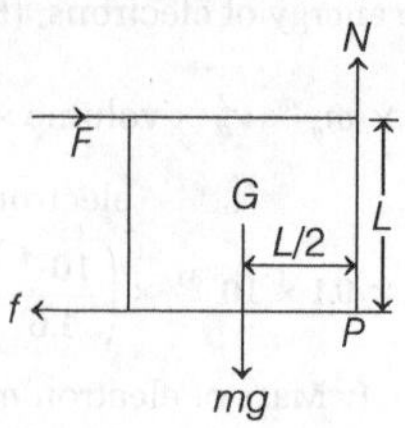

The block will topple when

$$\tau_F > \tau_{mg} \quad \text{or} \quad FL > (mg)\frac{L}{2}$$

$$\therefore \quad F > \frac{mg}{2}$$

Therefore, the minimum force required to topple the block is $F = \frac{mg}{2}$.

26 The centre of mass of the double star system remains stationary and both the stars revolve round in circular orbits, which are concentric with the centre of mass.

The distance of centre of mass from the heavier star $= \frac{Mr + 2M \times 0}{M + 2M} = \frac{r}{3}$

Hence, the heavier star revolves in a circle of radius $\frac{r}{3}$ while the lighter star in a circle of radius $\frac{2r}{3}$.

Reduced mass of the system $= \frac{M \cdot 2M}{M + 2M}$

$$= \frac{2M}{3}$$

Period of revolution of the double star system $= \frac{2\pi}{\sqrt{\frac{2GM}{3}}} r^{3/2}$

where, r is the distance between two stars.

KE of a star $= \frac{1}{2} mv^2$

KE of heavier star

$$E_1 = \frac{1}{2} \times 2M \times \left(\frac{r}{3}\omega\right)^2$$

and that of lighter star

$$E_2 = \frac{1}{2} M \left(\frac{2r}{3}\omega\right)^2$$

So, kinetic energy of lighter star is two times that of heavier star.

27 As, $f_0 = \frac{v}{2l}$

Beat frequency $= f_1 - f_2$

$$= \frac{v}{2\left(\frac{l}{2} - \Delta l\right)} - \frac{v}{2\left(\frac{l}{2} + \Delta l\right)}$$

$$= (f_0 l)\left[\frac{2}{l - 2\Delta l} - \frac{2}{l + 2\Delta l}\right]$$

$$= 2f_0 l\left[\frac{4\Delta l}{l^2}\right] \approx \frac{8f_0 \Delta l}{l}$$

28 Once the bricks leave the assistant's hands the only force that acts on them is gravitational force. Since this produces a constant acceleration $a = -g = -32 \text{ fts}^{-2}$, the kinematic equation, $v^2 = v_0^2 - 2a(x - x_0)$, can be used to describe the motion. The initial velocity v_0 is found by putting known values in the above equation,

$$v_0^2 = 36 + 2 \times 32 \times 10 = 676$$

$$\Rightarrow \quad v_0 = 26 \text{ fts}^{-1}$$

The kinetic energy given to each brick and supplied by the assistant is

$$E_1 = \frac{1}{2} mv_0^2$$

$$= \frac{1}{2} \times m \times 676$$

$$= 338 \text{ m ft}^2\text{s}^{-1}$$

If the brick assistant supplied is only just enough energy to reach the required level and no more, the initial velocity being u, they would have zero velocity at the Mason's hand.

$$\therefore u^2 = 0 + 2g(x - x_0)$$

$$= 2 \times 32 \times 10 = 640$$

$$\Rightarrow u = 8\sqrt{10} \text{ fts}^{-1}$$

KE supplied in this case,

$$E_2 = \frac{1}{2} mu^2 = 320m \text{ ft}^2\text{s}^{-2}$$

$\therefore$ Wasted energy $= E_1 - E_2$

$$\% \text{ waste} = \frac{E_1 - E_2}{E_1} \times 100$$

$$= \frac{338 - 320}{320} \times 100$$

$$= 5.6\%$$

29 For forward bias mode the p-side of diode has to be at higher potential than n-side. The meters used are DC, so we have to be careful while connecting them w.r.t. polarity.

Last point is to decide the range of meters, the range of meters has to be in such a way that we can have the readings which leads to plot on realistic scale. If we take 0-20 A ammeter, then reading we read from this is tending to 0 to 5 divisions which is not fruitful.

30 Using conservation of angular momentum about the axis of cylinder for the (block + cylinder) system

$$mv_o R = mvR + \frac{MR^2\omega}{2}$$

$$\Rightarrow \quad mv_o R = \frac{3}{2} mvR \qquad (\because v = \omega R)$$

(when slipping stops)

$$\Rightarrow \quad v = \frac{2v_o}{3}$$

ONLINE QUESTION PAPERS

JEE Main 2019 (April & January Attempt)

APRIL ATTEMPT

8 April, Shift-I

1 A steel wire having a radius of 2.0 mm, carrying a load of 4 kg, is hanging from a ceiling. Given that $g = 3.1\pi\ \text{ms}^{-2}$, what will be the tensile stress that would be developed in the wire?

(a) $6.2\times10^6\ \text{Nm}^{-2}$ (b) $5.2\times10^6\ \text{Nm}^{-2}$
(c) $3.1\times10^6\ \text{Nm}^{-2}$ (d) $4.8\times10^6\ \text{Nm}^{-2}$

2 If 10^{22} gas molecules each of mass 10^{-26} kg collide with a surface (perpendicular to it) elastically per second over an area 1 m^2 with a speed 10^4 m/s, the pressure exerted by the gas molecules will be of the order of

(a) $10^4\ \text{N/m}^2$ (b) $10^8\ \text{N/m}^2$
(c) $10^3\ \text{N/m}^2$ (d) $10^{16}\ \text{N/m}^2$

3 The bob of a simple pendulum has mass 2g and a charge of 5.0 μC. It is at rest in a uniform horizontal electric field of intensity 2000 V/m. At equilibrium, the angle that the pendulum makes with the vertical is (take $g = 10\ \text{m/s}^2$)

(a) $\tan^{-1}(2.0)$ (b) $\tan^{-1}(0.2)$
(c) $\tan^{-1}(5.0)$ (d) $\tan^{-1}(0.5)$

4 A boy's catapult is made of rubber cord which is 42 cm long, with 6 mm diameter of cross-section and of negligible mass. The boy keeps a stone weighing 0.02 kg on it and stretches the cord by 20 cm by applying a constant force. When released the stone flies off with a velocity of 20 ms^{-1}. Neglect the change in the area of cross-section of the cord while stretched. The Young's modulus of rubber is closest to

(a) $10^6\ \text{Nm}^{-2}$ (b) $10^4\ \text{Nm}^{-2}$
(c) $10^8\ \text{Nm}^{-2}$ (d) $10^3\ \text{Nm}^{-2}$

5 A plane electromagnetic wave travels in free space along the x-direction. The electric field component of the wave at a particular point of space and time is $E = 6\ \text{Vm}^{-1}$ along y-direction. Its corresponding magnetic field component, B would be

(a) 2×10^{-8} T along z - direction
(b) 6×10^{-8} T along x - direction
(c) 6×10^{-8} T along z - direction
(d) 2×10^{-8} T along y - direction

6 Ship A is sailing towards north-east with velocity $\mathbf{v} = 30\hat{\mathbf{i}} + 50\hat{\mathbf{j}}$ km/h, where $\hat{\mathbf{i}}$ points east and $\hat{\mathbf{j}}$ north. Ship B is at a distance of 80 km east and 150 km north of Ship A and is sailing towards west at 10 km/h. A will be at minimum distance from B in

(a) 4.2 h (b) 2.6 h (c) 3.2 h (d) 2.2 h

7 A thin strip 10 cm long is on an U-shaped wire of negligible resistance and it is connected to a spring of spring constant $0.5\ \text{Nm}^{-1}$ (see figure). The assembly is kept in a uniform magnetic field of 0.1 T. If the strip is pulled from its equilibrium position and released, the number of oscillations it performs before its amplitude decreases by a factor of e is N. If the mass of the strip is 50 grams, its resistance 10 Ω and air drag negligible, N will be close to

(a) 1000 (b) 50000 (c) 5000 (d) 10000

8 Four particles A, B, C and D with masses $m_A = m$, $m_B = 2m$, $m_C = 3m$ and $m_D = 4m$ are at the corners of a square. They have accelerations of equal magnitude with directions as shown. The acceleration of the centre of mass of the particles (in ms^{-2}) is

(a) $\frac{a}{5}(\hat{\mathbf{i}} - \hat{\mathbf{j}})$ (b) $a(\hat{\mathbf{i}} + \hat{\mathbf{j}})$
(c) zero (d) $\frac{a}{5}(\hat{\mathbf{i}} + \hat{\mathbf{j}})$

9 A solid conducting sphere, having a charge Q, is surrounded by an uncharged conducting hollow spherical shell. Let the potential difference between the surface of the solid sphere and that of the outer surface of the hollow shell be V. If the shell is now given a charge of $-4Q$, the new potential difference between the same two surfaces is

(a) $-2V$ (b) $2V$
(c) $4V$ (d) V

10 A 20 H inductor coil is connected to a 10 ohm resistance in series as shown in figure. The time at which rate of dissipation of energy (Joule's heat) across resistance is equal to the rate at which magnetic energy is stored in the inductor, is

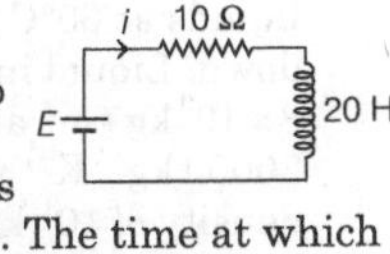

(a) $\frac{2}{\ln 2}$ (b) $\frac{1}{2}\ln 2$ (c) $2\ln 2$ (d) $\ln 2$

11 A thin circular plate of mass M and radius R has its density varying as $\rho(r) = \rho_0 r$ with ρ_0 as constant and r is the distance from its centre. The moment of inertia of the circular plate about an axis perpendicular to the plate and passing through its edge is $I = aMR^2$. The value of the coefficient a is

(a) $\frac{1}{2}$ (b) $\frac{3}{5}$ (c) $\frac{8}{5}$ (d) $\frac{3}{2}$

12 In SI units, the dimensions of $\sqrt{\frac{\varepsilon_0}{\mu_0}}$ is

(a) $\text{A}^{-1}\text{TML}^3$ (b) $\text{AT}^2\text{M}^{-1}\text{L}^{-1}$
(c) $\text{AT}^{-3}\text{ML}^{3/2}$ (d) $\text{A}^2\text{T}^3\text{M}^{-1}\text{L}^{-2}$

13 A thermally insulated vessel contains 150 g of water at 0°C. Then, the air from vessel is pumped out adiabatically. A fraction of water turns into ice and the rest evaporates at 0°C itself. The mass of evaporated water will be closest to (Latent heat of vaporisation of water $= 2.10\times10^6\ \text{J kg}^{-1}$ and latent heat of fusion of water $= 3.36\times10^5\ \text{J kg}^{-1}$)

(a) 150 g (b) 20 g (c) 130 g (d) 35g

14 The reverse breakdown voltage of a Zener diode is 5.6 V in the given circuit. The current I_z through the Zener is

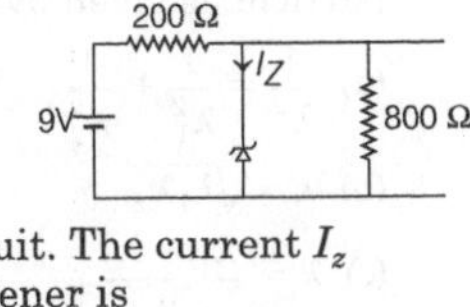

(a) 10 mA (b) 17 mA
(c) 15 mA (d) 7 mA

15 In an interference experiment, the ratio of amplitudes of coherent waves is $\frac{a_1}{a_2} = \frac{1}{3}$. The ratio of maximum and minimum intensities of fringes will be

(a) 2 (b) 18 (c) 4 (d) 9

16 Water from a pipe is coming at a rate of 100 liters per minute. If the radius of the pipe is 5 cm, the Reynolds number for the flow is of the order of (density of water = $1000 kg/m^3$, coefficient of viscosity of water = 1 mPa s)
(a) 10^3 (b) 10^4 (c) 10^2 (d) 10^6

17 Two identical beakers A and B contain equal volumes of two different liquids at 60°C each and left to cool down. Liquid in A has density of $8\times10^2 kg/m^3$ and specific heat of $2000 J kg^{-1}K^{-1}$ while liquid in B has density of $10^3 kg m^{-3}$ and specific heat of $4000 J kg^{-1}K^{-1}$. Which of the following best describes their temperature *versus* time graph schematically? (Assume the emissivity of both the beakers to be the same)

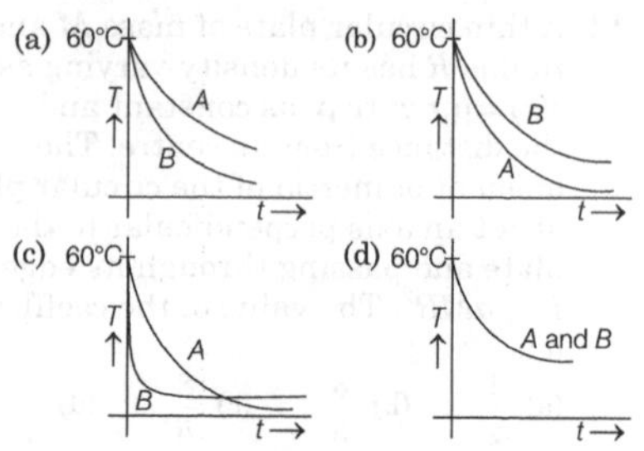

18 Voltage rating of a parallel plate capacitor is 500 V. Its dielectric can withstand a maximum electric field of 10^6 V/m. The plate area is $10^{-4} m^2$. What is the dielectric constant, if the capacitance is 15 pF?
(Take, $\varepsilon_0 = 8.86\times10^{-12} C^2/N\text{-}m^2$)
(a) 3.8 (b) 8.5
(c) 4.5 (d) 6.2

19 Two particles move at right angle to each other. Their de-Broglie wavelengths are λ_1 and λ_2, respectively. The particles suffer perfectly inelastic collision. The de-Broglie wavelength λ of the final particle, is given by
(a) $\frac{1}{\lambda^2} = \frac{1}{\lambda_1^2} + \frac{1}{\lambda_2^2}$
(b) $\lambda = \sqrt{\lambda_1\lambda_2}$
(c) $\lambda = \frac{\lambda_1 + \lambda_2}{2}$
(d) $\frac{2}{\lambda} = \frac{1}{\lambda_1} + \frac{1}{\lambda_2}$

20 A 200 Ω resistor has a certain colour code. If one replaces the red colour by green in the code, the new resistance will be

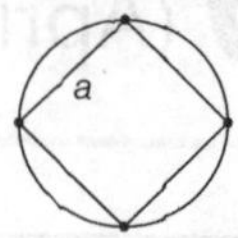

(a) 100 Ω (b) 400 Ω
(c) 300 Ω (d) 500 Ω

21 Four identical particles of mass M are located at the corners of a square of side a. What should be their speed, if each of them revolves under the influence of other's gravitational field in a circular orbit circumscribing the square?
(a) $1.35\sqrt{\frac{GM}{a}}$ (b) $1.16\sqrt{\frac{GM}{a}}$
(c) $1.21\sqrt{\frac{GM}{a}}$ (d) $1.41\sqrt{\frac{GM}{a}}$

22 The wavelength of the carrier waves in a modern optical fibre communication network is close to
(a) 2400 nm (b) 1500 nm
(c) 600 nm (d) 900 nm

23 An upright object is placed at a distance of 40 cm in front of a convergent lens of focal length 20 cm. A convergent mirror of focal length 10 cm is placed at a distance of 60 cm on the other side of the lens. The position and size of the final image will be
(a) 20 cm from the convergent mirror, same size as the object
(b) 40 cm from the convergent mirror, same size as the object
(c) 40 cm from the convergent lens, twice the size of the object
(d) 20 cm from the convergent mirror,twice size of the object

24 A circular coil having N turns and radius r carries a current I. It is held in the XZ-plane in a magnetic field $B\hat{i}$. The torque on the coil due to the magnetic field (in N-m) is
(a) $\frac{Br^2I}{\pi N}$ (b) $B\pi r^2 IN$
(c) $\frac{B\pi r^2 I}{N}$ (d) Zero

25 An alternating voltage $V(t) = 220\sin 100\pi t$ volt is applied to a purely resistive load of 50 Ω. The time taken for the current to rise from half of the peak value to the peak value is
(a) 5 ms (b) 2.2 ms
(c) 7.2 ms (d) 3.3 ms

26 Radiation coming from transitions $n = 2$ to $n = 1$ of hydrogen atoms fall on He^+ ions in $n = 1$ and $n = 2$ states. The possible transition of helium ions as they absorb energy from the radiation is
(a) $n = 2 \to n = 3$ (b) $n = 1 \to n = 4$
(c) $n = 2 \to n = 5$ (d) $n = 2 \to n = 4$

27

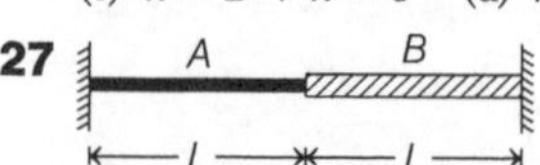

A wire of length $2L$, is made by joining two wires A and B of same length but different radii r and $2r$ and made of the same material. It is vibrating at a frequency such that the joint of the two wires forms a node. If the number of antinodes in wire A is p and that in B is q, then the ratio $p : q$ is
(a) 3 : 5 (b) 4 : 9 (c) 1 : 2 (d) 1 : 4

28 In figure, the optical fibre is $l = 2$ m long and has a diameter of $d = 20\,\mu$m. If a ray of light is incident on one end of the fibre at angle $\theta_1 = 40°$, the number of reflections it makes before emerging from the other end is close to (refractive index of fibre is 1.31 and sin 40° = 0.64)

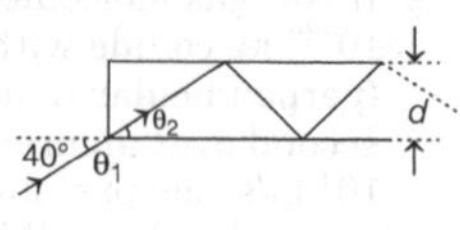

(a) 55000 (b) 66000
(c) 45000 (d) 57000

29 A particle moves in one dimension from rest under the influence of a force that varies with the distance travelled by the particle as shown in the figure. The kinetic energy of the particle after it has travelled 3 m is

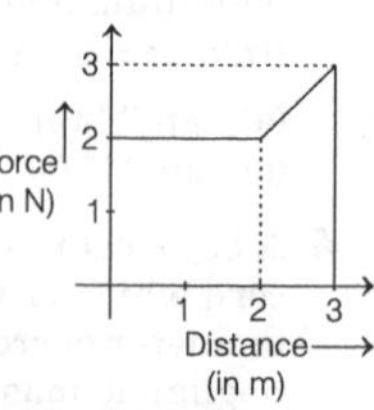

(a) 4 J (b) 2.5 J (c) 6.5 J (d) 5 J

30 For the circuit shown with $R_1 = 1.0\Omega$, $R_2 = 2.0\Omega$, $E_1 = 2$ V and $E_2 = E_3 = 4$ V, the potential difference between the points a and b is approximately (in volt)

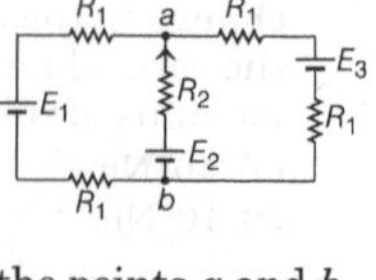

(a) 2.7 (b) 2.3
(c) 3.7 (d) 3.3

ANSWERS

1. (c)	2. (*)	3. (d)	4. (a)	5. (a)	6. (b)	7. (c)	8. (a)	9. (d)	10. (c)
11. (*)	12. (d)	13. (b)	14. (a)	15. (c)	16. (b)	17. (b)	18. (b)	19. (a)	20. (d)
21. (b)	22. (b)	23. (*)	24. (b)	25. (d)	26. (d)	27. (c)	28. (d)	29. (c)	30. (d)

* No option is correct

For Detailed Solutions
Visit : http://tinyurl.com/y277m5o6
Or Scan :

8 April, Shift-II

1 In a simple pendulum, experiment for determination of acceleration due to gravity (g), time taken for 20 oscillations is measured by using a watch of 1 second least count. The mean value of time taken comes out to be 30 s. The length of pendulum is measured by using a meter scale of least count 1 mm and the value obtained 55.0 cm. The percentage error in the determination of g is close to

(a) 0.7% (b) 6.8% (c) 3.5% (d) 0.2%

2 A nucleus A, with a finite de-Broglie wavelength λ_A, undergoes spontaneous fission into two nuclei B and C of equal mass. B flies in the same directions as that of A, while C flies in the opposite direction with a velocity equal to half of that of B. The de-Broglie wavelengths λ_B and λ_C of B and C respectively

(a) $2\lambda_A, \lambda_A$ (b) $\frac{\lambda_A}{2}, \lambda_A$

(c) $\lambda_A, 2\lambda_A$ (d) $\lambda_A, \frac{\lambda_A}{2}$

3 In the figure shown, what is the current (in ampere) drawn from the battery? You are given : $R_1 = 15\,\Omega$, $R_2 = 10\,\Omega$, $R_3 = 20\,\Omega$, $R_4 = 5\,\Omega$, $R_5 = 25\,\Omega$, $R_6 = 30\,\Omega$, $E = 15\,\text{V}$

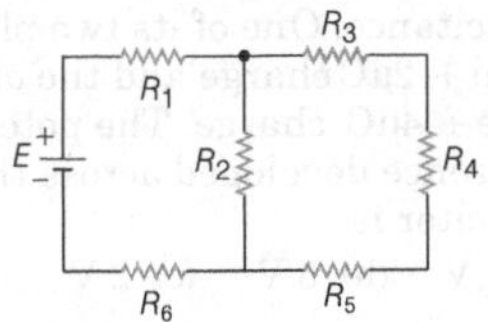

(a) 13/24 (b) 7/18 (c) 20/3 (d) 9/32

4 Two magnetic dipoles X and Y are placed at a separation d, with their axes perpendicular to each other. The dipole moment of Y is twice that of X. A particle of charge q is passing through their mid-point P, at angle $\theta = 45°$ with the horizontal line, as shown in figure. What would be the magnitude of force on the particle at that instant? (d is much larger than the dimensions of the dipole)

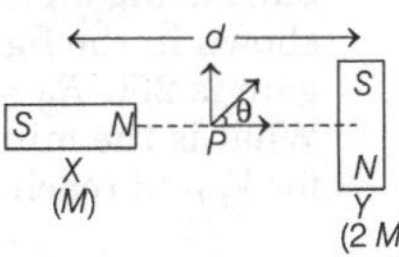

(a) $\left(\frac{\mu_0}{4\pi}\right)\frac{M}{\left(\frac{d}{2}\right)^3} \times qv$ (b) 0

(c) $\sqrt{2}\left(\frac{\mu_0}{4\pi}\right)\frac{M}{\left(\frac{d}{2}\right)^3} \times qv$

(d) $\left(\frac{\mu_0}{4\pi}\right)\frac{2M}{\left(\frac{d}{2}\right)^3} \times qv$

5 A rocket has to be launched from earth in such a way that it never returns. If E is the minimum energy delivered by the rocket launcher, what should be the minimum energy that the launcher should have, if the same rocket is to be launched from the surface of the moon? Assume that the density of the earth and the moon are equal and that the earth's volume is 64 times the volume of the moon.

(a) $\frac{E}{64}$ (b) $\frac{E}{16}$ (c) $\frac{E}{32}$ (d) $\frac{E}{4}$

6 The electric field in a region is given by $\mathbf{E} = (Ax + B)\hat{\mathbf{i}}$, where E is in NC^{-1} and x is in metres. The values of constants are $A = 20$ SI unit and $B = 10$ SI unit. If the potential at $x = 1$ is V_1 and that at $x = -5$ is V_2, then $V_1 - V_2$ is

(a) -48 V (b) -520 V (c) 180 V (d) 320 V

7 Let $|\mathbf{A}_1| = 3$, $|\mathbf{A}_2| = 5$ and $|\mathbf{A}_1 + \mathbf{A}_2| = 5$. The value of $(2\mathbf{A}_1 + 3\mathbf{A}_2)\cdot(3\mathbf{A}_1 - 2\mathbf{A}_2)$ is

(a) -106.5 (b) -112.5
(c) -99.5 (d) -118.5

8 The temperature, at which the root mean square velocity of hydrogen molecules equals their escape velocity from the earth, is closest to :

[Boltzmann constant $k_B = 1.38 \times 10^{-23}$ J/K, Avogadro no. $N_A = 6.02 \times 10^{26}$/kg, Radius of earth $= 6.4 \times 10^6$ m, Gravitational acceleration on earth $= 10\,\text{ms}^{-2}$]

(a) 10^4 K (b) 650 K
(c) 3×10^5 K (d) 800 K

9 Two very long, straight and insulated wires are kept at 90° angle from each other in xy-plane as shown in the fig. These wires carry currents of equal magnitude I, whose directions are shown in the figure. The net magnetic field at point P will be

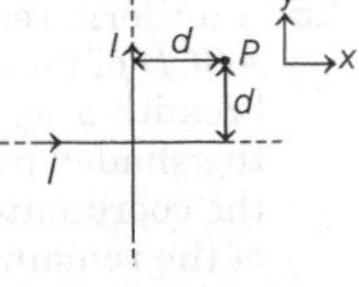

(a) zero (b) $\frac{+\mu_0 I}{\pi d}(\hat{\mathbf{z}})$

(c) $-\frac{\mu_0 I}{2\pi d}(\hat{\mathbf{x}} + \hat{\mathbf{y}})$ (d) $\frac{\mu_0 I}{2\pi d}(\hat{\mathbf{x}} + \hat{\mathbf{y}})$

10 An electric dipole is formed by two equal and opposite charges q with separation d. The charges have same mass m. It is kept in a uniform electric field E. If it is slightly rotated from its equilibrium orientation, then its angular frequency ω is

(a) $\sqrt{\frac{2qE}{md}}$ (b) $2\sqrt{\frac{qE}{md}}$ (c) $\sqrt{\frac{qE}{md}}$ (d) $\sqrt{\frac{qE}{2md}}$

11 A cell of internal resistance r drives current through an external resistance R. The power delivered by the cell to the external resistance will be maximum when

(a) $R = 2r$ (b) $R = r$
(c) $R = 0.001\,r$ (d) $R = 1000\,r$

12 A body of mass m_1 moving with an unknown velocity of $v_1\hat{\mathbf{i}}$, undergoes a collinear collision with a body of mass m_2 moving with a velocity $v_2\hat{\mathbf{i}}$. After collision, m_1 and m_2 move with velocities of $v_3\hat{\mathbf{i}}$ and $v_4\hat{\mathbf{i}}$, respectively.

If $m_2 = 0.5 m_1$ and $v_3 = 0.5\,v_1$, then v_1 is

(a) $v_4 + v_2$ (b) $v_4 - \frac{v_2}{4}$

(c) $v_4 - \frac{v_2}{2}$ (d) $v_4 - v_2$

13 A circuit connected to an AC source of emf $e = e_0 \sin(100t)$ with t in seconds, gives a phase difference of $\frac{\pi}{4}$ between the emf e and current i. Which of the following circuits will exhibit this?

(a) RC circuit with $R = 1\text{k}\Omega$ and $C = 1\,\mu\text{F}$
(b) RL circuit with $R = 1\text{k}\Omega$ and $L = 1\text{mH}$
(c) RC circuit with $R = 1\text{k}\Omega$ and $C = 10\,\mu\text{F}$
(d) RL circuit with $R = 1\text{k}\Omega$ and $L = 10$ mH

14 In the circuit shown, a four-wire potentiometer is made of a 400 cm long wire, which extends between A and B. The resistance per unit length of the potentiometer wire is $r = 0.01\,\Omega$/cm. If an ideal voltmeter is connected as shown with jockey J at 50 cm from end A, the expected reading of the voltmeter will be

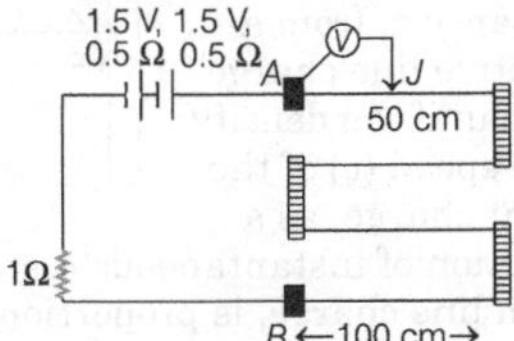

(a) 0.20 V (b) 0.75 V
(c) 0.25 V (d) 0.50 V

15 A solid sphere and solid cylinder of identical radii approach an incline with the same linear velocity (see figure). Both roll without slipping all throughout. The two climb maximum heights h_{sph} and h_{cyl} on the incline. The ratio $\frac{h_{sph}}{h_{cyl}}$ is given by

(a) $\frac{2}{\sqrt{5}}$ (b) $\frac{14}{15}$ (c) 1 (d) $\frac{4}{5}$

16 The given diagram shows four processes, i.e. isochoric, isobaric, isothermal and adiabatic. The correct assignment of the processes, in the same order is given by

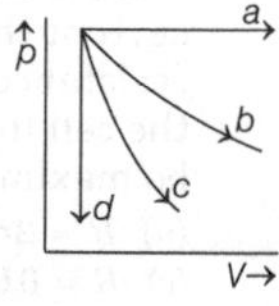

(a) $d\ a\ b\ c$ (b) $a\ d\ b\ c$
(c) $d\ a\ c\ b$ (d) $a\ d\ c\ b$

17 The ratio of mass densities of nuclei of ^{40}Ca and ^{16}O is close to
(a) 5 (b) 2 (c) 0.1 (d) 1

18 A rectangular solid box of length 0.3 m is held horizontally, with one of its sides on the edge of a platform of height 5 m. When released, it slips off the table in a very short time $\tau = 0.01$ s, remaining essentially horizontal. The angle by which it would rotate when it hits the ground will be (in radians) close to
(a) 0.02 (b) 0.3 (c) 0.5 (d) 0.28

19 In a line of sight radio communication, a distance of about 50 km is kept between the transmitting and receiving antennas. If the height of the receiving antenna is 70 m, then the minimum height of the transmitting antenna should be
(Radius of the earth = 6.4×10^6 m)
(a) 20 m (b) 32 m (c) 40 m (d) 51 m

20 A positive point charge is released from rest at a distance r_0 from a positive line charge with uniform density. The speed (v) of the point charge, as a function of instantaneous distance r from line charge, is proportional to

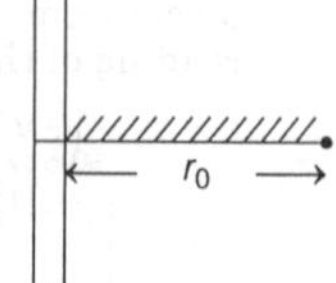

(a) $v \propto \left(\frac{r}{r_0}\right)$ (b) $v \propto e^{+\frac{r}{r_0}}$
(c) $v \propto \ln\left(\frac{r}{r_0}\right)$ (d) $v \propto \sqrt{\ln\left(\frac{r}{r_0}\right)}$

21 A particle starts from origin O from rest and moves with a uniform acceleration along the positive X-axis. Identify all figures that correctly represent the motion qualitatively.
(a = acceleration, v = velocity, x = displacement, t = time)

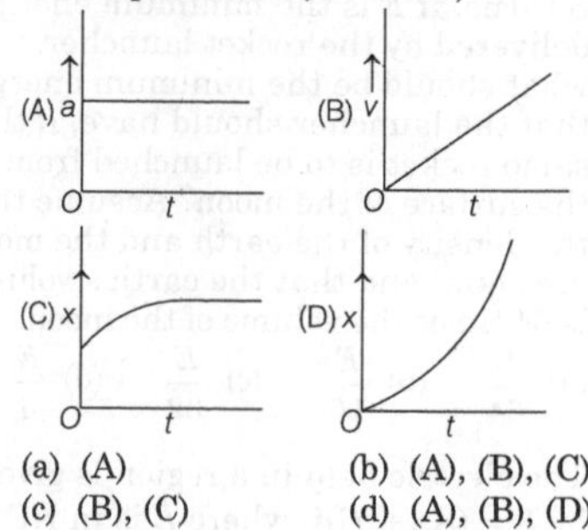

(a) (A) (b) (A), (B), (C)
(c) (B), (C) (d) (A), (B), (D)

22 Calculate the limit of resolution of a telescope objective having a diameter of 200 cm, if it has to detect light of wavelength 500 nm coming from a star.
(a) 610×10^{-9} rad (b) 305×10^{-9} rad
(c) 457.5×10^{-9} rad (d) 152.5×10^{-9} rad

23 If surface tension (S), moment of inertia (I) and Planck's constant (h), were to be taken as the fundamental units, the dimensional formula for linear momentum would be
(a) $S^{1/2}I^{1/2}h^{-1}$ (b) $S^{3/2}I^{1/2}h^{0}$
(c) $S^{1/2}I^{1/2}h^{0}$ (d) $S^{1/2}I^{3/2}h^{-1}$

24 A damped harmonic oscillator has a frequency of 5 oscillations per second. The amplitude drops to half its value for every 10 oscillations. The time it will take to drop to $\frac{1}{1000}$ of the original amplitude is close to
(a) 20 s (b) 50 s
(c) 100 s (d) 10 s

25 A uniform rectangular thin sheet $ABCD$ of mass M has length a and breadth b, as shown in the figure. If the shaded portion $HBGO$ is cut-off, the coordinates of the centre of mass of the remaining portion will be

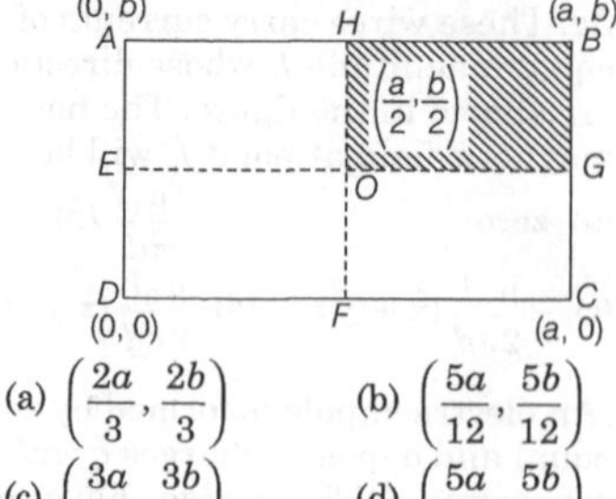

(a) $\left(\frac{2a}{3}, \frac{2b}{3}\right)$ (b) $\left(\frac{5a}{12}, \frac{5b}{12}\right)$
(c) $\left(\frac{3a}{4}, \frac{3b}{4}\right)$ (d) $\left(\frac{5a}{3}, \frac{5b}{3}\right)$

26 A convex lens (of focal length 20 cm) and a concave mirror, having their principal axes along the same lines, are kept 80 cm apart from each other. The concave mirror is to the right of the convex lens. When an object is kept at a distance of 30 cm to the left of the convex lens, its image remains at the same position even if the concave mirror is removed. The maximum distance of the object for which this concave mirror, by itself would produce a virtual image would be
(a) 25 cm (b) 20 cm (c) 10 cm (d) 30 cm

27 The magnetic field of an electromagnetic wave is given by $\mathbf{B} = 1.6 \times 10^{-6} \cos(2 \times 10^7 z + 6 \times 10^{15} t)$ $(2\hat{i} + \hat{j})$ Wbm^{-2}
The associated electric field will be
(a) $\mathbf{E} = 4.8 \times 10^2 \cos(2 \times 10^7 z - 6 \times 10^{15} t)$ $(-2\hat{j} + \hat{i})$ Vm^{-1}
(b) $\mathbf{E} = 4.8 \times 10^2 \cos(2 \times 10^7 z - 6 \times 10^{15} t)$ $(2\hat{j} + \hat{i})$ Vm^{-1}
(c) $\mathbf{E} = 4.8 \times 10^2 \cos(2 \times 10^7 z + 6 \times 10^{15} t)$ $(\hat{i} - 2\hat{j})$ Vm^{-1}
(d) $\mathbf{E} = 4.8 \times 10^2 \cos(2 \times 10^7 z + 6 \times 10^{15} t)$ $(-\hat{i} + 2\hat{j})$ Vm^{-1}

28 A parallel plate capacitor has 1µF capacitance. One of its two plates is given + 2µC charge and the other plate + 4µC charge. The potential difference developed across the capacitor is
(a) 1 V (b) 5 V (c) 2 V (d) 3 V

29 A common emitter amplifier circuit, built using an n-p-n transistor, is shown in the figure. Its DC current gain is 250, $R_C = 1\text{k}\Omega$ and $V_{CC} = 10$ V. What is the minimum base current for V_{CE} to reach saturation?

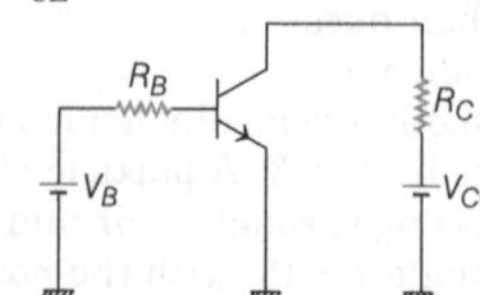

(a) 40 µA (b) 10 µA (c) 100 µA (d) 7 µA

30 Young's moduli of two wires A and B are in the ratio 7 : 4. Wire A is 2 m long and has radius R. Wire B is 1.5 m long and has radius 2 mm. If the two wires stretch by the same length for a given load, then the value of R is close to
(a) 1.3 mm (b) 1.5 mm
(c) 1.9 mm (d) 1.7 mm

ANSWERS

1. (b)	2. (b)	3. (d)	4. (b)	5. (b)	6. (c)	7. (d)	8. (a)	9. (a)	10. (a)
11. (b)	12. (d)	13. (c)	14. (c)	15. (b)	16. (a)	17. (d)	18. (c)	19. (b)	20. (d)
21. (d)	22. (b)	23. (c)	24. (a)	25. (b)	26. (c)	27. (c)	28. (a)	29. (a)	30. (d)

For Detailed Solutions
Visit : http://tinyurl.com/yxjbp89y
Or Scan :

09 April, Shift-I

1 A stationary horizontal disc is free to rotate about its axis. When a torque is applied on it, its kinetic energy as a function of θ, where θ is the angle by which it has rotated, is given as $k\theta^2$. If its moment of inertia is I, then the angular acceleration of the disc is

(a) $\frac{k}{2I}\theta$ (b) $\frac{k}{I}\theta$ (c) $\frac{k}{4I}\theta$ (d) $\frac{2k}{I}\theta$

2 For a given gas at 1 atm pressure, rms speed of the molecules is 200 m/s at 127° C. At 2 atm pressure and at 227° C, the rms speed of the molecules will be

(a) $100\sqrt{5}$ m/s (b) 80 m/s
(c) 100 m/s (d) $80\sqrt{5}$ m/s

3 A wire of resistance R is bent to form a square $ABCD$ as shown in the figure. The effective resistance between E and C is [E is mid-point of arm CD]

(a) $\frac{7}{64}R$ (b) $\frac{3}{4}R$ (c) R (d) $\frac{1}{16}R$

4 A simple pendulum oscillating in air has period T. The bob of the pendulum is completely immersed in a non-viscous liquid. The density of the liquid is $\frac{1}{16}$th of the material of bob. If the bob is inside liquid all the time, its period of oscillation in this liquid is

(a) $2T\sqrt{\frac{1}{10}}$ (b) $2T\sqrt{\frac{1}{14}}$
(c) $4T\sqrt{\frac{1}{14}}$ (d) $4T\sqrt{\frac{1}{15}}$

5 A system of three charges are placed as shown in the figure

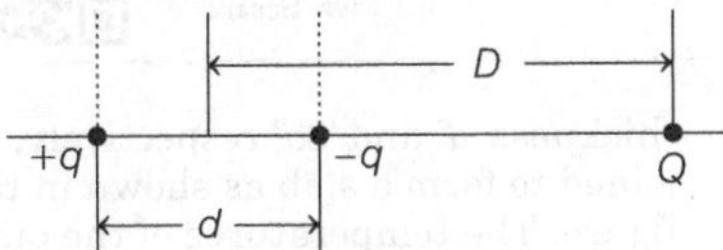

If $D >> d$, the potential energy of the system is best given by

(a) $\frac{1}{4\pi\varepsilon_0}\left[-\frac{q^2}{d}+\frac{2qQd}{D^2}\right]$

(b) $\frac{1}{4\pi\varepsilon_0}\left[+\frac{q^2}{d}+\frac{qQd}{D^2}\right]$

(c) $\frac{1}{4\pi\varepsilon_0}\left[-\frac{q^2}{d}-\frac{qQd}{2D^2}\right]$

(d) $\frac{1}{4\pi\varepsilon_0}\left[-\frac{q^2}{d}-\frac{qQd}{D^2}\right]$

6 A body of mass 2 kg makes an elastic collision with a second body at rest and continues to move in the original direction but with one-fourth of its original speed. What is the mass of the second body ?

(a) 1.5 kg (b) 1.2 kg (c) 1.8 kg (d) 1.0 kg

7 A concave mirror for face viewing has focal length of 0.4 m. The distance at which you hold the mirror from your face in order to see your image upright with a magnification of 5 is

(a) 0.16 m (b) 1.60 m
(c) 0.32 m (d) 0.24 m

8 The stream of a river is flowing with a speed of 2 km/h. A swimmer can swim at a speed of 4 km/h. What should be the direction of the swimmer with respect to the flow of the river to cross the river straight ?

(a) 60° (b) 120° (c) 90° (d) 150°

9 A string is clamped at both the ends and it is vibrating in its 4th harmonic. The equation of the stationary wave is $Y = 0.3\sin(0.157x)\cos(200\pi t)$. The length of the string is (All quantities are in SI units)

(a) 60 m (b) 40 m (c) 80 m (d) 20 m

10 A moving coil galvanometer has resistance 50 Ω and it indicates full deflection at 4 mA current. A voltmeter is made using this galvano- meter and a 5 kΩ resistance. The maximum voltage, that can be measured using this voltmeter, will be close to

(a) 40 V (b) 10 V (c) 15 V (d) 20 V

11 A capacitor with capacitance 5 μF is charged to 5 μC. If the plates are pulled apart to reduce the capacitance to 2 μF, how much work is done?

(a) 6.25×10^{-6} J (b) 2.16×10^{-6} J
(c) 2.55×10^{-6} J (d) 3.75×10^{-6} J

12 In the density measurement of a cube, the mass and edge length are measured as (10.00 ± 0.10) kg and (0.10 ± 0.01) m, respectively. The error in the measurement of density is

(a) 0.01 kg/m^3 (b) 0.10 kg/m^3
(c) 0.07 kg/m^3 (d) 0.31 kg/m^3

13 The total number of turns and cross-section area in a solenoid is fixed. However, its length L is varied by adjusting the separation between windings. The inductance of solenoid will be proportional to

(a) $1/L$ (b) L^2 (c) L (d) $1/L^2$

14 The following bodies are made to roll up (without slipping) the same inclined plane from a horizontal plane: (i) a ring of radius R, (ii) a solid cylinder of radius $R/2$ and (iii) a solid sphere of radius $R/4$. If in each case, the speed of the centre of mass at the bottom of the incline is same, the ratio of the maximum height they climb is

(a) 10 : 15 : 7 (b) 4 : 3 : 2
(c) 14 : 15 : 20 (d) 2 : 3 : 4

15 If M is the mass of water that rises in a capillary tube of radius r, then mass of water which will rise in a capillary tube of radius $2r$ is

(a) $2M$ (b) $4M$ (c) $\frac{M}{2}$ (d) M

16 Following figure shows two processes A and B for a gas. If ΔQ_A and ΔQ_B are the amount of heat absorbed by the system in two cases, and ΔU_A and ΔU_B are changes in internal energies respectively, then

(a) $\Delta Q_A > \Delta Q_B$, $\Delta U_A > \Delta U_B$
(b) $\Delta Q_A < \Delta Q_B$, $\Delta U_A < \Delta U_B$
(c) $\Delta Q_A > \Delta Q_B$, $\Delta U_A = \Delta U_B$
(d) $\Delta Q_A = \Delta Q_B$; $\Delta U_A = \Delta U_B$

17 Determine the charge on the capacitor in the following circuit

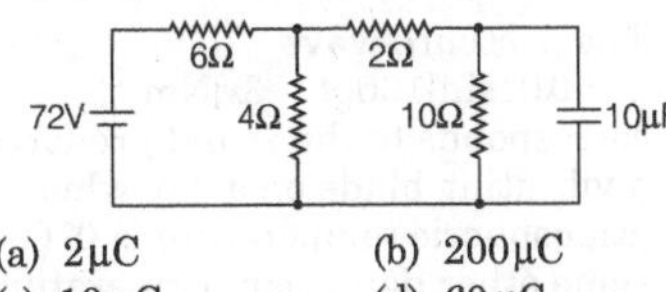

(a) 2μC (b) 200μC
(c) 10μC (d) 60μC

18 A uniform cable of mass M and length L is placed on a horizontal surface such that its $\left(\frac{1}{n}\right)$th part is hanging below the edge of the surface. To lift the hanging part of the cable upto the surface, the work done should be

(a) $\frac{2MgL}{n^2}$ (b) $nMgL$
(c) $\frac{MgL}{n^2}$ (d) $\frac{MgL}{2n^2}$

19 The electric field of light wave is given as

$$E = 10^{-3}\cos\left(\frac{2\pi x}{5\times10^{-7}} - 2\pi\times6\times10^{14}t\right)\hat{x}\ NC^{-1}.$$

This light falls on a metal plate of work function 2eV. The stopping potential of the photoelectrons is

Given, $E \text{ (in eV)} = \frac{12375}{\lambda(\text{in Å})}$

(a) 0.48 V (b) 0.72 V
(c) 2.0 V (d) 2.48 V

20 A ball is thrown vertically up (taken as + Z-axis) from the ground. The correct momentum-height (p-h) diagram is

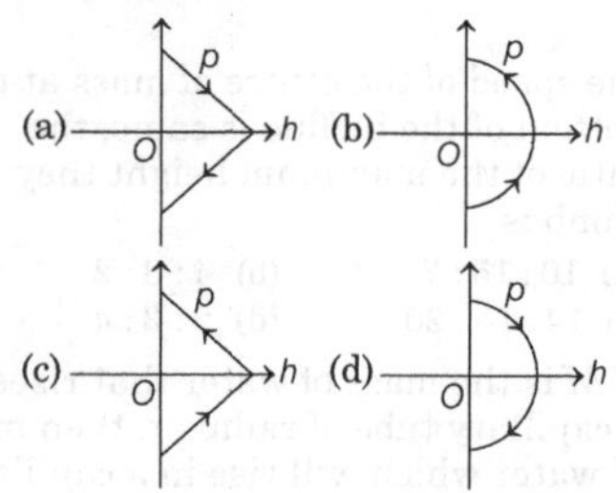

21 A signal $A\cos\omega t$ is transmitted using $v_0 \sin\omega_0 t$ as carrier wave. The correct amplitude modulated (AM) signal is

(a) $(v_0 \sin\omega_0 t + A\cos\omega t$
(b) $(v_0 + A)\cos\omega t \sin\omega_0 t$
(c) $v_0 \sin[\omega_0(1 + 0.01A \sin\omega t)t]$
(d) $v_0 \sin\omega_0 t + \frac{A}{2}\sin(\omega_0 - \omega)t + \frac{A}{2}\sin(\omega_0 + \omega)t$

22 An HCl molecule has rotational, translational and vibrational motions. If the rms velocity of HCl molecules in its gaseous phase is $\bar{v}$, m is its mass and k_B is Boltzmann constant, then its temperature will be

(a) $\frac{m\bar{v}^2}{3k_B}$ (b) $\frac{m\bar{v}^2}{7k_B}$ (c) $\frac{m\bar{v}^2}{5k_B}$ (d) $\frac{m\bar{v}^2}{6k_B}$

23 The pressure wave $p = 0.01\sin[1000t - 3x]\text{Nm}^{-2}$, corresponds to the sound produced by a vibrating blade on a day when atmospheric temperature is 0° C. On some other day when temperature is T, the speed of sound produced by the same blade and at the same frequency is found to be $336\ \text{ms}^{-1}$. Approximate value of T is

(a) 15° C (b) 11° C (c) 12° C (d) 4° C

24 The figure shows a Young's double slit experimental setup. It is observed that when a thin transparent sheet of thickness t and refractive index μ is put in front of one of the slits, the central maximum gets shifted by a distance equal to n fringe widths. If the wavelength of light used is λ, t will be

(a) $\frac{2nD\lambda}{a(\mu - 1)}$ (b) $\frac{2D\lambda}{a(\mu - 1)}$
(c) $\frac{D\lambda}{a(\mu - 1)}$ (d) $\frac{nD\lambda}{a(\mu - 1)}$

25 A rectangular coil (dimension 5 cm × 2.5 cm) with 100 turns, carrying a current of 3A in the clockwise direction, is kept centred at the origin and in the X-Z plane. A magnetic field of 1 T is applied along X-axis. If the coil is tilted through 45° about Z-axis, then the torque on the coil is

(a) 0.27 N-m (b) 0.38 N-m
(c) 0.42 N-m (d) 0.55 N-m

26 A rigid square loop of side a and carrying current I_2 is lying on a horizontal surface near a long current I_1 carrying wire in the same plane as shown in figure. The net force on the loop due to the wire will be

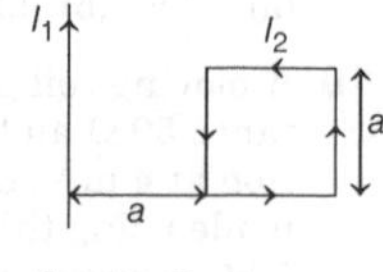

(a) repulsive and equal to $\frac{\mu_0 I_1 I_2}{2\pi}$
(b) attractive and equal to $\frac{\mu_0 I_1 I_2}{3\pi}$
(c) zero
(d) repulsive and equal to $\frac{\mu_0 I_1 I_2}{4\pi}$

27 Taking the wavelength of first Balmer line in hydrogen spectrum ($n = 3$ to $n = 2$) as 660 nm, the wavelength of the 2nd Balmer line ($n = 4$ to $n = 2$) will be

(a) 889.2 nm (b) 388.9 nm
(c) 642.7 nm (d) 488.9 nm

28 An n-p-n transistor is used in common emitter configuration as an amplifier with 1 kΩ load resistance. Signal voltage of 10 mV is applied across the base-emitter. This produces a 3 mA change in the collector current and 15 μA change in the base current of the amplifier. The input resistance and voltage gain are

(a) 0.67 kΩ, 200 (b) 0.33 kΩ, 1.5
(c) 0.67 kΩ, 300 (d) 0.33 kΩ, 300

29 A solid sphere of mass M and radius a is surrounded by a uniform concentric spherical shell of thickness $2a$ and $2M$. The gravitational field at distance $3a$ from the centre will be

(a) $\frac{GM}{9a^2}$ (b) $\frac{2GM}{9a^2}$ (c) $\frac{GM}{3a^2}$ (d) $\frac{2GM}{3a^2}$

30 The magnetic field of a plane electromagnetic wave is given by

$\mathbf{B} = B_0[\cos(kz - \omega t)]\hat{\mathbf{i}} + B_1 \cos(kz + \omega t)\hat{\mathbf{j}}$

where, $B_0 = 3 \times 10^{-5}$ T and $B_1 = 2 \times 10^{-6}$ T. The rms value of the force experienced by a stationary charge $Q = 10^{-4}$ C at $z = 0$ is closest to

(a) 0.1 N (b) 3×10^{-2} N
(c) 0.6 N (d) 0.9 N

ANSWERS

1. (d)	2. (a)	3. (a)	4. (d)	5. (d)	6. (b)	7. (c)	8. (b)	9. (c)	10. (d)
11. (d)	12. (*)	13. (a)	14. (c)	15. (a)	16. (c)	17. (b)	18. (d)	19. (a)	20. (d)
21. (d)	22. (b)	23. (d)	24. (*)	25. (a)	26. (d)	27. (d)	28. (c)	29. (c)	30. (c)

For Detailed Solutions
Visit : http://tinyurl.com/y6x7rxu3
Or Scan :

9 April Shift-II

1 A thin convex lens L (refractive index = 1.5) is placed on a plane mirror M. When a pin is placed at A, such that $OA = 18$ cm, its real inverted image is formed at A itself, as shown in figure. When a liquid of refractive index μ_l is put between the lens and the mirror, the pin has to be moved to A', such that $OA' = 27$ cm, to get its inverted real image at A' itself. The value of μ_l will be

(a) $\sqrt{3}$ (b) $\sqrt{2}$ (c) $\frac{4}{3}$ (d) $\frac{3}{2}$

2 A massless spring ($k = 800$ N/m), attached with a mass (500 g) is completely immersed in 1 kg of water. The spring is stretched by 2 cm and released, so that it starts vibrating. What would be the order of magnitude of the change in the temperature of water when the vibrations stop completely? (Assume that the water container and spring receive negligible heat and specific heat of mass = 400 J/kg K, specific heat of water = 4184 J/kg K)

(a) 10^{-4} K (b) 10^{-3} K
(c) 10^{-1} K (d) 10^{-5} K

3 Two materials having coefficients of thermal conductivity '$3K$' and 'K' and thickness 'd' and '$3d$' respectively, are joined to form a slab as shown in the figure. The temperatures of the outer surfaces are 'θ_2' and 'θ_1' respectively, ($\theta_2 > \theta_1$). The temperature at the interface is

(a) $\frac{\theta_2 + \theta_1}{2}$ (b) $\frac{\theta_1}{3} + \frac{2\theta_2}{3}$
(c) $\frac{\theta_1}{6} + \frac{5\theta_2}{6}$ (d) $\frac{\theta_1}{10} + \frac{9\theta_2}{10}$

4 A test particle is moving in a circular orbit in the gravitational field produced by mass density $\rho(r) = \frac{K}{r^2}$. Identify the correct relation between the radius R of the particle's orbit and its period T

(a) $\frac{T^2}{R^3}$ is a constant

(b) $\frac{T}{R^2}$ is a constant

(c) TR is a constant (d) $\frac{T}{R}$ is a constant

5 A convex lens of focal length 20 cm produces images of the same magnification 2 when an object is kept at two distances x_1 and x_2 $(x_1 > x_2)$ from the lens. The ratio of x_1 and x_2 is

(a) 5 : 3 (b) 2 : 1 (c) 4 : 3 (d) 3 : 1

6 A wedge of mass $M = 4\,m$ lies on a frictionless plane. A particle of mass m aproaches the wedge with speed v. There is no friction between the particle and the plane or between the particle and the wedge. The maximum height climbed by the particle on the wedge is given by

(a) $\frac{2v^2}{7g}$ (b) $\frac{v^2}{g}$ (c) $\frac{2v^2}{5g}$ (d) $\frac{v^2}{2g}$

7 The parallel combination of two air filled parallel plate capacitors of capacitance C and nC is connected to a battery of voltage, V. When the capacitors are fully charged, the battery is removed and after that a dielectric material of dielectric constant K is placed between the two plates of the first capacitor. The new potential difference of the combined system is

(a) $\frac{(n+1)V}{(K+n)}$ (b) $\frac{nV}{K+n}$ (c) V (d) $\frac{V}{K+n}$

8 The area of a square is 5.29 cm^2. The area of 7 such squares taking into account the significant figures is

(a) 37.030 cm^2 (b) 37.0 cm^2
(c) 37.03 cm^2 (d) 37 cm^2

9 The logic gate equivalent to the given logic circuit is

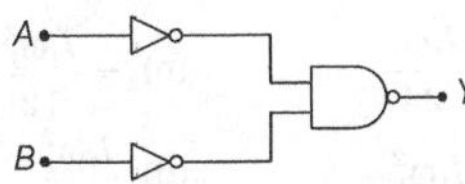

(a) NOR (b) NAND
(c) OR (d) AND

10 The resistance of a galvanometer is 50 ohm and the maximum current which can be passed through it is 0.002 A. What resistance must be connected to it in order to convert it into an ammeter of range 0-0.5 A?

(a) 0.2 ohm (b) 0.5 ohm
(c) 0.002 ohm (d) 0.02 ohm

11 The position vector of particle changes with time according to the relation $\mathbf{r}(t) = 15t^2\hat{\mathbf{i}} + (4 - 20t^2)\hat{\mathbf{j}}$. What is the magnitude of the acceleration (in ms^{-2}) at $t = 1$?

(a) 50 (b) 100 (c) 25 (d) 40

12 A particle P is formed due to a completely inelastic collision of particles x and y having de-Broglie wavelengths λ_x and λ_y, respectively. If x and y were moving in opposite directions, then the de-Broglie wavelength of P is

(a) $\lambda_x - \lambda_y$ (b) $\frac{\lambda_x \lambda_y}{\lambda_x - \lambda_y}$

(c) $\frac{\lambda_x \lambda_y}{\lambda_x + \lambda_y}$ (d) $\lambda_x + \lambda_y$

13 A moving coil galvanometer has a coil with 175 turns and area 1 cm^2. It uses a torsion band of torsion constant 10^{-6} N-m/rad. The coil is placed in a magnetic field B parallel to its plane. The coil deflects by 1° for a current of 1 mA. The value of B (in Tesla) is approximately

(a) 10^{-3} (b) 10^{-4} (c) 10^{-1} (d) 10^{-2}

14 Four point charges $-q, +q, +q$ and $-q$ are placed on Y-axis at $y = -2d$, $y = -d$, $y = +d$ and $y = +2d$, respectively. The magnitude of the electric field E at a point on the X-axis at $x = D$, with $D >> d$, will behave as

(a) $E \propto \frac{1}{D}$ (b) $E \propto \frac{1}{D^3}$

(c) $E \propto \frac{1}{D^2}$ (d) $E \propto \frac{1}{D^4}$

15 The physical sizes of the transmitter and receiver antenna in a communication system are

(a) proportional to carrier frequency
(b) inversely proportional to modulation frequency
(c) independent of both carrier and modulation frequency
(d) inversely proportional to carrier frequency

16 Diameter of the objective lens of a telescope is 250 cm. For light of wavelength 600 nm coming from a distant object, the limit of resolution of the telescope is close to

(a) 3.0×10^{-7} rad (b) 2.0×10^{-7} rad
(c) 1.5×10^{-7} rad (d) 4.5×10^{-7} rad

17 Two cars A and B are moving away from each other in opposite directions. Both the cars are moving with a speed of 20 ms^{-1} with respect to the ground. If an observer in car A detects a frequency 2000 Hz of the sound coming from car B, what is the natural frequency of the sound source in car B? (speed of sound in air = 340 ms^{-1})

(a) 2060 Hz (b) 2250 Hz
(c) 2300 Hz (d) 2150 Hz

18 The position of a particle as a function of time t, is given by

$$x(t) = at + bt^2 - ct^3$$

where a, b and c are constants. When the particle attains zero acceleration, then its velocity will be

(a) $a + \frac{b^2}{2c}$ (b) $a + \frac{b^2}{4c}$

(c) $a + \frac{b^2}{3c}$ (d) $a + \frac{b^2}{c}$

19 In a conductor, if the number of conduction electrons per unit volume is 8.5×10^{28} m^{-3} and mean free time is 25 fs (femto second), it's approximate resistivity is (Take, $m_e = 9.1 \times 10^{-31}$ kg)

(a) $10^{-7}\,\Omega$-m (b) $10^{-5}\,\Omega$-m
(c) $10^{-6}\,\Omega$-m (d) $10^{-8}\,\Omega$-m

20 The specific heats, C_p and C_V of a gas of diatomic molecules, A are given (in units of J mol^{-1} K^{-1}) by 29 and 22, respectively. Another gas of diatomic molecules B, has the corresponding values 30 and 21. If they are treated as ideal gases, then

(a) A has a vibrational mode but B has none
(b) Both A and B have a vibrational mode each
(c) A has one vibrational mode and B has two
(d) A is rigid but B has a vibrational mode

21 50 W/m^2 energy density of sunlight is normally incident on the surface of a solar panel. Some part of incident energy (25%) is reflected from the surface and the rest is absorbed. The force exerted on 1 m^2 surface area will be close to ($c = 3 \times 10^8$ m/s)

(a) 20×10^{-8} N (b) 35×10^{-8} N
(c) 15×10^{-8} N (d) 10×10^{-8} N

22 A string 2.0 m long and fixed at its ends is driven by a 240 Hz vibrator. The string vibrates in its third harmonic mode. The speed of the wave and its fundamental frequency is

(a) 180 m/s, 80 Hz (b) 320 m/s, 80 Hz
(c) 320 m/s, 120 Hz (d) 180 m/s, 120 Hz

23 A thin smooth rod of length L and mass M is rotating freely with angular speed ω_0 about an axis perpendicular to the rod and passing through its centre. Two beads of mass m and negligible size are at the centre of the rod initially. The beads are free to slide along the rod. The angular speed of the system, when the beads reach the opposite ends of the rod, will be

(a) $\frac{M\omega_0}{M+3m}$ (b) $\frac{M\omega_0}{M+m}$
(c) $\frac{M\omega_0}{M+2m}$ (d) $\frac{M\omega_0}{M+6m}$

24 A particle of mass m is moving with speed $2v$ and collides with a mass $2m$ moving with speed v in the same direction. After collision, the first mass is stopped completely while the second one splits into two particles each of mass m, which move at angle 45° with respect to the original direction.The speed of each of the moving particle will be

(a) $\sqrt{2}\,v$ (b) $\frac{v}{\sqrt{2}}$ (c) $\frac{v}{(2\sqrt{2})}$ (d) $2\sqrt{2}\,v$

25 A metal wire of resistance 3 Ω is elongated to make a uniform wire of double its previous length. This new wire is now bent and the ends joined to make a circle. If two points on this circle make an angle 60° at the centre, the equivalent resistance between these two points will be

(a) $\frac{7}{2}\Omega$ (b) $\frac{5}{2}\Omega$ (c) $\frac{12}{5}\Omega$ (d) $\frac{5}{3}\Omega$

26 A very long solenoid of radius R is carrying current $I(t) = kte^{-\alpha t}$ $(k > 0)$, as a function of time $(t \geq 0)$. Counter clockwise current is taken to be positive. A circular conducting coil of radius $2R$ is placed in the equatorial plane of the solenoid and concentric with the solenoid. The current induced in the outer coil is correctly depicted, as a function of time, by

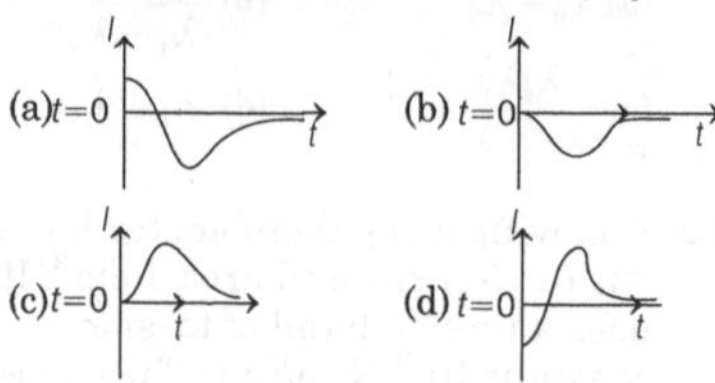

27 A He$^+$ ion is in its first excited state. Its ionisation energy is

(a) 54.40 eV (b) 13.60 eV
(c) 48.36 eV (d) 6.04 eV

28 A wooden block floating in a bucket of water has 4/5 of its volume submerged. When certain amount of an oil is poured into the bucket, it is found that the block is just under the oil surface with half of its volume under water and half in oil. The density of oil relative to that of water is

(a) 0.6 (b) 0.8 (c) 0.7 (d) 0.5

29 Two coils P and Q are separated by some distance. When a current of 3 A flows through coil P, a magnetic flux of 10^{-3} Wb passes through Q. No current is passed through Q. When no current passes through P and a current of 2 A passes through Q, the flux through P is

(a) 6.67×10^{-3} Wb (b) 6.67×10^{-4} Wb
(c) 3.67×10^{-3} Wb (d) 3.67×10^{-4} Wb

30 Moment of inertia of a body about a given axis is 1.5 kg m^2. Initially, the body is at rest. In order to produce a rotational kinetic energy of 1200 J, the angular acceleration of 20 rad/s^2 must be applied about the axis for a duration of

(a) 5 s (b) 2 s (c) 3 s (d) 2.5 s

ANSWERS

1. (c)	2. (d)	3. (d)	4. (d)	5. (d)	6. (c)	7. (a)	8. (c)	9. (c)	10. (a)
11. (a)	12. (b)	13. (a)	14. (d)	15. (d)	16. (a)	17. (b)	18. (c)	19. (d)	20. (a)
21. (a)	22. (b)	23. (d)	24. (d)	25. (d)	26. (d)	27. (b)	28. (a)	29. (b)	30. (b)

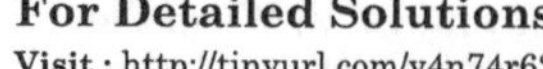
For Detailed Solutions
Visit : http://tinyurl.com/y4n74r62
Or Scan :

10 April Shift-I

1 A uniformly charged ring of radius $3a$ and total charge q is placed in xy-plane centred at origin. A point charge q is moving towards the ring along the Z-axis and has speed v at $z = 4a$. The minimum value of v such that it crosses the origin is

(a) $\sqrt{\frac{2}{m}}\left(\frac{1}{5}\frac{q^2}{4\pi\varepsilon_0 a}\right)^{1/2}$

(b) $\sqrt{\frac{2}{m}}\left(\frac{4}{15}\frac{q^2}{4\pi\varepsilon_0 a}\right)^{1/2}$

(c) $\sqrt{\frac{2}{m}}\left(\frac{1}{15}\frac{q^2}{4\pi\varepsilon_0 a}\right)^{1/2}$

(d) $\sqrt{\frac{2}{m}}\left(\frac{2}{15}\frac{q^2}{4\pi\varepsilon_0 a}\right)^{1/2}$

2 Figure shows charge (q) *versus* voltage (V) graph for series and parallel combination of two given capacitors. The capacitances are

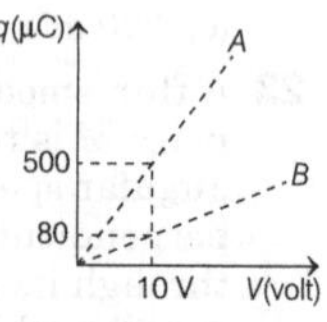

(a) 60 μF and 40 μF (b) 50 μF and 30 μF
(c) 20 μF and 30 μF (d) 40 μF and 10μF

3 Two wires A and B are carrying currents I_1 and I_2 as shown in the figure. The separation between them is d. A third wire C carrying a current I is to be kept parallel to them at a distance x from A such that the net force acting on it is zero. The possible values of x are

(a) $x = \left(\frac{I_2}{I_1+I_2}\right)d$ and $x = \left(\frac{I_2}{I_1-I_2}\right)d$

(b) $x = \left(\frac{I_1}{I_1-I_2}\right)d$ and $x = \left(\frac{I_2}{I_1+I_2}\right)d$

(c) $x = \left(\frac{I_1}{I_1+I_2}\right)d$ and $x = \left(\frac{I_2}{I_1-I_2}\right)d$

(d) $x = \pm\frac{I_1 d}{(I_1-I_2)}$

4 A message signal of frequency 100 MHz and peak voltage 100 V is used to execute amplitude modulation on a carrier wave of frequency 300 GHz and peak voltage 400 V. The modulation index and difference between the two side band frequencies are

(a) 0.25 ; 1×10^8 Hz (b) 4 ; 1×10^8 Hz
(c) 0.25 ; 2×10^8 Hz (d) 4; 2×10^8 Hz

5 Two coaxial discs, having moments of inertia I_1 and $\frac{I_1}{2}$ are rotating with respective angular velocities ω_1 and $\frac{\omega_1}{2}$, about their common axis. They are brought in contact with each other and thereafter they rotate with a common angular velocity. If E_f and E_i are the final and initial total energies, then $(E_f - E_i)$ is

(a) $-\frac{I_1\omega_1^2}{24}$ (b) $-\frac{I_1\omega_1^2}{12}$
(c) $\frac{3}{8}I_1\omega_1^2$ (d) $\frac{I_1\omega_1^2}{6}$

6 An n-p-n transistor operates as a common emitter amplifier, with a power gain of 60 dB. The input circuit resistance is 100 Ω and the output load resistance is 10 kΩ. The common emitter current gain β is

(a) 10^2 (b) 6×10^2 (c) 10^4 (d) 60

7 The displacement of a damped harmonic oscillator is given by $x(t) = e^{-0.1t}\cos(10\pi t + \phi)$. Here, t is in seconds.

The time taken for its amplitude of vibration to drop to half of its initial value is close to

(a) 27 s (b) 13 s (c) 4 s (d) 7 s

8 The ratio of surface tensions of mercury and water is given to be 7.5 while the ratio of their densities is 13.6. Their contact angles with glass are close to 135° and 0°, respectively. It is observed that mercury gets depressed by an amount h in a capillary tube of radius r_1, while water rises by the same amount h in a capillary tube of radius r_2. The ratio (r_1/r_2), is then close to

(a) 3/5 (b) 2/3 (c) 2/5 (d) 4/5

9 A ray of light AO in vacuum is incident on a glass slab at angle 60° and refracted at angle 30° along OB as shown in the figure.

The optical path length of light ray from A to B is

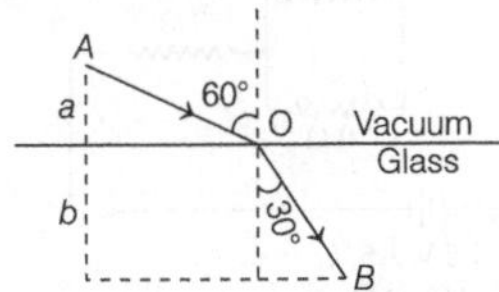

(a) $\frac{2\sqrt{3}}{a} + 2b$ (b) $2a + \frac{2b}{3}$

(c) $2a + 2b$ (d) $2a + \frac{2b}{\sqrt{3}}$

10 n moles of an ideal gas with constant volume heat capacity C_V undergo an isobaric expansion by certain volume. The ratio of the work done in the process, to the heat supplied is

(a) $\frac{4nR}{C_V + nR}$ (b) $\frac{4nR}{C_V - nR}$

(c) $\frac{nR}{C_V - nR}$ (d) $\frac{nR}{C_V + nR}$

11 A thin disc of mass M and radius R has mass per unit area $\sigma(r) = kr^2$, where r is the distance from its centre. Its moment of inertia about an axis going through its centre of mass and perpendicular to its plane is

(a) $\frac{MR^2}{2}$ (b) $\frac{MR^2}{6}$ (c) $\frac{MR^2}{3}$ (d) $\frac{2MR^2}{3}$

12 A particle of mass m is moving along a trajectory given by

$x = x_0 + a\cos\omega_1 t$ and

$y = y_0 + b\sin\omega_2 t$.

The torque acting on the particle about the origin at $t = 0$ is

(a) zero

(b) $m(-x_0 b + y_0 a)\,\omega_1^2\,\hat{k}$

(c) $-m(x_0 b\omega_2^2 - y_0 a\omega_1^2)\,\hat{k}$

(d) $+my_0 a\,\omega_1^2\,\hat{k}$

13 A proton, an electron and a helium nucleus, have the same energy. They are in circular orbits in a plane due to magnetic field perpendicular to the plane. Let r_p, r_e and r_{He} be their respective radii, then

(a) $r_e < r_p = r_{He}$ (b) $r_e > r_p = r_{He}$

(c) $r_e < r_p < r_{He}$ (d) $r_e > r_p > r_{He}$

14 Two radioactive materials A and B have decay constants 10λ and λ, respectively. If initially they have the same number of nuclei, then the ratio of the number of nuclei of A to that of B will be $1/e$ after a time

(a) $\frac{1}{11\lambda}$ (b) $\frac{11}{10\lambda}$ (c) $\frac{1}{9\lambda}$ (d) $\frac{1}{10\lambda}$

15 A 25×10^{-3} m^3 volume cylinder is filled with 1 mole of O_2 gas at room temperature (300 K). The molecular diameter of O_2 and its root mean square speed are found to be 0.3 nm and 200 m/s, respectively. What is the average collision rate (per second) for an O_2 molecule?

(a) $\sim 10^{10}$ (b) $\sim 10^{12}$ (c) $\sim 10^{11}$ (d) $\sim 10^{13}$

16 In an experiment, the resistance of a material is plotted as a function of temperature (in some range). As shown in the figure, it is a straight line.

lnR(T)

1/T²

One may conclude that

(a) $R(T) = R_0 e^{-T^2/T_0^2}$ (b) $R(T) = R_0 e^{T^2/T_0^2}$

(c) $R(T) = R_0 e^{-T_0^2/T^2}$ (d) $R(T) = \frac{R_0}{T^2}$

17 In a photoelectric effect experiment, the threshold wavelength of light is 380 nm. If the wavelength of incident light is 260 nm, the maximum kinetic energy of emitted electrons will be

Given, $E\text{ (in eV)} = \frac{1237}{\lambda\text{(in nm)}}$

(a) 15.1 eV (b) 3.0 eV

(c) 1.5 eV (d) 4.5 eV

18 A moving coil galvanometer allows a full scale current of 10^{-4} A. A series resistance of 2 MΩ is required to convert the above galvanometer into a voltmeter of range 0-5 V. Therefore, the value of shunt resistance required to convert the above galvanometer into an ammeter of range 0.10 mA is

(a) 100 Ω (b) 500 Ω (c) 200 Ω (d) 10 Ω

19 A current of 5 A passes through a copper conductor (resistivity $= 1.7 \times 10^{-8}$ Ω-m) of radius of cross-section 5 mm. Find the mobility of the charges, if their drift velocity is 1.1×10^{-3} m/s.

(a) 1.5 m^2/V-s (b) 1.3 m^2/V-s

(c) 1.0 m^2/V-s (d) 1.8 m^2/V-s

20 In a meter bridge experiment, the circuit diagram and the corresponding observation table are shown in figure

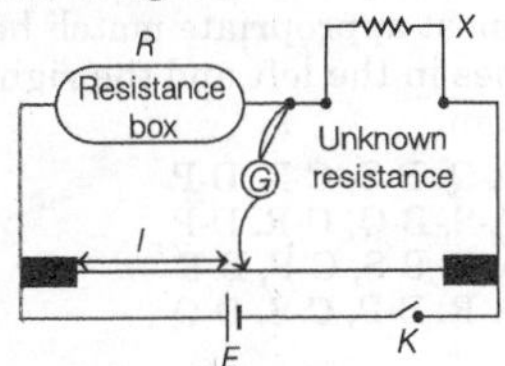

S. No.	R (Ω)	l (cm)
1.	1000	60
2.	100	13
3.	10	1.5
4.	1	1.0

Which of the readings is inconsistent?

(a) 3 (b) 2 (c) 1 (d) 4

21 A ball is thrown upward with an initial velocity v_0 from the surface of the earth. The motion of the ball is affected by a drag force equal to $m\gamma v^2$ (where, m is mass of the ball, v is its instantaneous velocity and γ is a constant). Time taken by the ball to rise to its zenith is

(a) $\frac{1}{\sqrt{2\gamma g}}\tan^{-1}\left(\sqrt{\frac{2\gamma}{g}}\,v_0\right)$

(b) $\frac{1}{\sqrt{\gamma g}}\tan^{-1}\left(\sqrt{\frac{\gamma}{g}}\,v_0\right)$

(c) $\frac{1}{\sqrt{\gamma g}}\sin^{-1}\left(\sqrt{\frac{\gamma}{g}}\,v_0\right)$

(d) $\frac{1}{\sqrt{\gamma g}}\ln\left(1+\sqrt{\frac{\gamma}{g}}\,v_0\right)$

22 One plano-convex and one plano-concave lens of same radius of curvature R but of different materials are joined side by side as shown in the figure. If the refractive index of the material of 1 is μ_1 and that of 2 is μ_2, then the focal length of the combination is

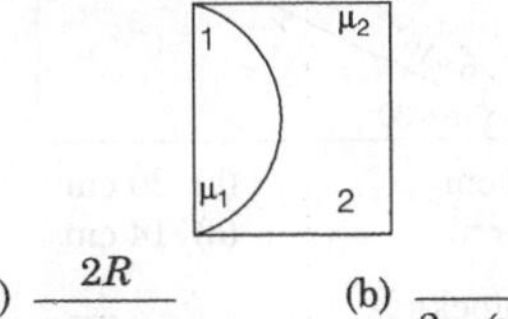

(a) $\frac{2R}{\mu_1 - \mu_2}$ (b) $\frac{R}{2 - (\mu_1 - \mu_2)}$

(c) $\frac{R}{2(\mu_1 - \mu_2)}$ (d) $\frac{R}{\mu_1 - \mu_2}$

23 Given below in the left column are different modes of communication using the kinds of waves given in the right column.

A. Optical fibre communication	P. Ultrasound
B. Radar	Q. Infrared light
C. Sonar	R. Microwaves
D. Mobile phones	S. Radio waves

From the options given below, find the most appropriate match between entries in the left and the right column.

(a) A-Q, B-S, C-R, D-P
(b) A-S, B-Q, C-R, D-P
(c) A-Q, B-S, C-P, D-R
(d) A-R, B-P, C-S, D-Q

24 A stationary source emits sound waves of frequency 500 Hz. Two observers moving along a line passing through the source detect sound to be of frequencies 480 Hz and 530 Hz. Their respective speeds are in ms^{-1}, (Take, speed of sound = 300 m/s)

(a) 12, 16 (b) 12, 18
(c) 16, 14 (d) 8, 18

25 The electric field of a plane electromagnetic wave is given by

$$\mathbf{E} = E_0 \hat{\mathbf{i}} \cos(kz) \cos(\omega t)$$

The corresponding magnetic field **B** is then given by

(a) $\mathbf{B} = \frac{E_0}{c} \hat{\mathbf{j}} \sin(kz) \sin(\omega t)$
(b) $\mathbf{B} = \frac{E_0}{c} \hat{\mathbf{j}} \sin(kz) \cos(\omega t)$
(c) $\mathbf{B} = \frac{E_0}{c} \mathbf{k} \sin(kz) \cos(\omega t)$
(d) $\mathbf{B} = \frac{E_0}{c} \hat{\mathbf{j}} \cos(kz) \sin(\omega t)$

26 A transformer consisting of 300 turns in the primary and 150 turns in the secondary gives output power of 2.2 kW. If the current in the secondary coil is 10 A, then the input voltage and current in the primary coil are

(a) 440 V and 5 A (b) 220 V and 20 A
(c) 220 V and 10 A (d) 440 V and 20 A

27 Two particles of masses M and $2M$, moving as shown, with speeds of 10 m/s and 5 m/s, collide elastically at the origin. After the collision, they move along the indicated directions with speed v_1 and v_2 are nearly

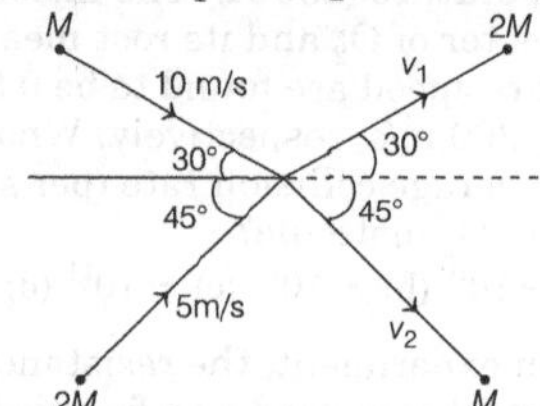

(a) 6.5 m/s and 3.2 m/s
(b) 3.2 m/s and 6.3 m/s
(c) 3.2 m/s and 12.6 m/s
(d) 6.5 m/s and 6.3 m/s

28 The value of acceleration due to gravity at earth's surface is 9.8 ms^{-2}. The altitude above its surface at which the acceleration due to gravity decreases to 4.9 ms^{-2}, is close to (Take, radius of earth = 6.4×10^6 m)

(a) 9.0×10^6 m (b) 2.6×10^6 m
(c) 6.4×10^6 m (d) 1.6×10^6 m

29 A cylinder with fixed capacity of 67.2 L contains helium gas at STP. The amount of heat needed to raise the temperature of the gas by 20°C is [Take, $R = 8.31$ J $mol^{-1}K^{-1}$]

(a) 700 J (b) 748 J
(c) 374 J (d) 350 J

30 In the given circuit, an ideal voltmeter connected across the 10 Ω resistance reads 2 V. The internal resistance r, of each cell is

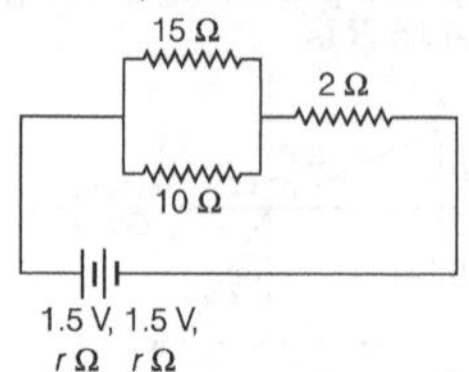

(a) 1.5 Ω (b) 0.5 Ω (c) 1 Ω (d) 0 Ω

ANSWERS

1. (d)	2. (d)	3. (d)	4. (c)	5. (a)	6. (a)	7. (d)	8. (c)	9. (c)	10. (d)
11. (d)	12. (d)	13. (c)	14. (c)	15. (*)	16. (c)	17. (c)	18. (*)	19. (c)	20. (d)
21. (b)	22. (d)	23. (c)	24. (b)	25. (a)	26. (a)	27. (d)	28. (b)	29. (b)	30. (b)

For Detailed Solutions
Visit : http://tinyurl.com/y2kdgeua
Or Scan :

10 April 2019 Shift-II

1 A plane is inclined at an angle $\alpha = 30°$ with respect to the horizontal. A particle is projected with a speed $u = 2\ ms^{-1}$, from the base of the plane, making an angle $\theta = 15°$ with respect to the plane as shown in the figure. The distance from the base, at which the particle hits the plane is close to [Take, $g = 10\ ms^{-2}$]

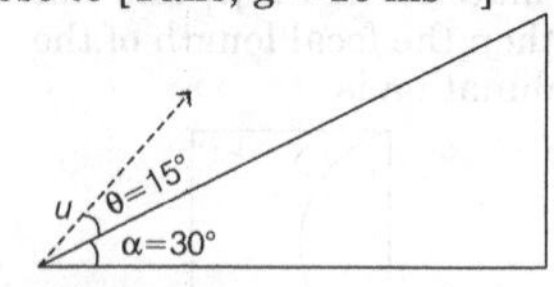

(a) 26 cm (b) 20 cm
(c) 18 cm (d) 14 cm

2 Two blocks A and B of masses $m_A = 1$ kg and $m_B = 3$ kg are kept on the table as shown in figure. The coefficient of friction between A and B is 0.2 and between B and the surface of the table is also 0.2. The maximum force F that can be applied on B horizontally, so that the block A does not slide over the block B is [Take, $g = 10\ m/s^2$]

(a) 12 N (b) 16 N (c) 8 N (d) 40 N

3 The time dependence of the position of a particle of mass $m = 2$ is given by $\mathbf{r}(t) = 2t\hat{\mathbf{i}} - 3t^2\hat{\mathbf{j}}$. Its angular momentum, with respect to the origin, at time $t = 2$ is

(a) $36\ \hat{\mathbf{k}}$ (b) $-34\ (\hat{\mathbf{k}} - \hat{\mathbf{i}})$
(c) $-48\ \hat{\mathbf{k}}$ (d) $48\ (\hat{\mathbf{i}} + \hat{\mathbf{j}})$

4 A spaceship orbits around a planet at a height of 20 km from its surface. Assuming that only gravitational field of the planet acts on the spaceship, what will be the number of complete revolutions made by the spaceship in 24 hours around the planet? [Take, mass of planet = 8×10^{22} kg, radius of planet = 2×10^6 m, gravitational constant $G = 6.67 \times 10^{-11}$ N-m^2/kg^2]

(a) 11 (b) 17 (c) 13 (d) 9

5 In free space, a particle A of charge 1μC is held fixed at a point P. Another particle B of the same charge and mass 4μg is kept at a distance of 1 mm from P. If B is released, then its velocity at a distance of 9 mm from P is [Take, $1/4\pi\varepsilon_0 = 9 \times 10^9$ N-m^2C^{-2}]

(a) 1.5×10^2 m/s (b) 3.0×10^4 m/s
(c) 1.0 m/s (d) 2.0×10^3 m/s

6 A submarine experience a pressure of 5.05×10^6 Pa at a depth of d_1 in a sea. When it goes further to a depth of d_2, it experiences a pressure of 8.08×10^6 Pa, then $d_2 - d_1$ is approximately (density of water = 10^3 kg/m^3 and acceleration due to gravity = 10 ms^{-2})

(a) 500 m (b) 400 m (c) 600 m (d) 300 m

7 Light is incident normally on a completely absorbing surface with an energy flux of 25 W cm^{-2}. If the surface has an area of 25 cm^2, the momentum transferred to the surface in 40 min time duration will be

(a) 3.5×10^{-6} N-s (b) 6.3×10^{-4} N-s
(c) 1.4×10^{-6} N-s (d) 5.0×10^{-3} N-s

8 A square loop is carrying a steady current I and the magnitude of its magnetic dipole moment is m. If this square loop is changed to a circular loop and it carries the same current, the magnitude of the magnetic dipole moment of circular loop (in A-m) will be

(a) $\frac{4m}{\pi}$ (b) $\frac{3m}{\pi}$ (c) $\frac{2m}{\pi}$ (d) $\frac{m}{\pi}$

9 A 2 mW laser operates at a wavelength of 500 nm. The number of photons that will be emitted per second is

[Given, Planck's constant $h = 6.6 \times 10^{-34}$ Js, speed of light $c = 3.0 \times 10^{8}$ m/s]

(a) 1×10^{16} (b) 5×10^{15}
(c) 1.5×10^{16} (d) 2×10^{16}

10 A solid sphere of mass M and radius R is divided into two unequal parts. The first part has a mass of $\frac{7M}{8}$ and is converted into a uniform disc of radius $2R$. The second part is converted into a uniform solid sphere. Let I_1 be the moment of inertia of the disc about its axis and I_2 be the moment of inertia of the new sphere about its axis.

The ratio I_1 / I_2 is given by

(a) 285 (b) 185 (c) 65 (d) 140

11 The correct figure that shows schematically, the wave pattern produced by superposition of two waves of frequencies 9Hz and 11 Hz, is

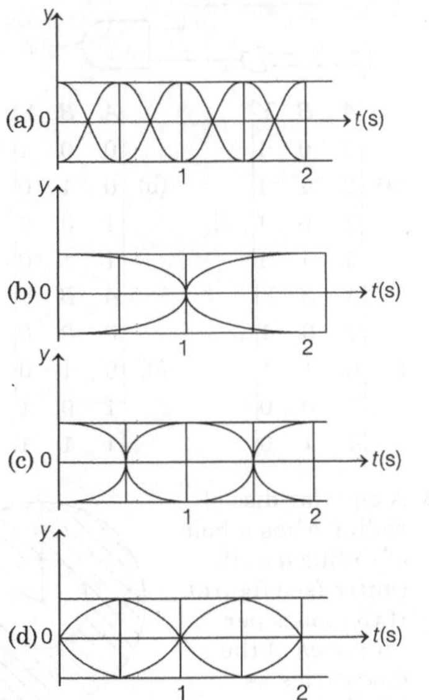

12 A source of sound S is moving with a velocity of 50 m/s towards a stationary observer. The observer measures the frequency of the source as 1000 Hz. What will be the apparent frequency of the source when it is moving away from the observer after crossing him? (Take, velocity of sound in air is 350 m/s)

(a) 807 Hz (b) 1143 Hz
(c) 750 Hz (d) 857 Hz

13 A cubical block of side 0.5 m floats on water with 30% of its volume under water. What is the maximum weight that can be put on the block without fully submerging it under water?

[Take, density of water $= 10^3$ kg/m^3]

(a) 30.1 kg (b) 46.3 kg
(c) 87.5 kg (d) 65.4 kg

14 In the formula $X = 5YZ^2$, X and Z have dimensions of capacitance and magnetic field, respectively. What are the dimensions of Y in SI units?

(a) $[M^{-1}L^{-2}T^{4}A^{2}]$ (b) $[M^{-2}L^{0}T^{-4}A^{-2}]$
(c) $[M^{-3}L^{-2}T^{8}A^{4}]$ (d) $[M^{-2}L^{-2}T^{6}A^{3}]$

15 In a Young's double slit experiment, the ratio of the slit's width is 4 : 1. The ratio of the intensity of maxima to minima, close to the central fringe on the screen, will be

(a) 4 : 1 (b) 25 : 9
(c) 9 : 1 (d) $(\sqrt{3}+1)^4 : 16$

16 The graph shows how the magnification m produced by a thin lens varies with image distance v. What is the focal length of the lens used?

(a) $\frac{b^2c}{a}$ (b) $\frac{b^2}{ac}$ (c) $\frac{a}{c}$ (d) $\frac{b}{c}$

17 A metal coin of mass 5g and radius 1 cm is fixed to a thin stick AB of negligible mass as shown in the figure. The system is initially at rest. The constant torque, that will make the system rotate about AB at 25 rotations per second in 5s, is close to

(a) 4.0×10^{-6} N-m (b) 2.0×10^{-5} N-m
(c) 1.6×10^{-5} N-m (d) 7.9×10^{-6} N-m

18 Space between two concentric conducting spheres of radii a and $b (b > a)$ is filled with a medium of resistivity ρ. The resistance between the two spheres will be

(a) $\frac{\rho}{2\pi}\left(\frac{1}{a}+\frac{1}{b}\right)$ (b) $\frac{\rho}{4\pi}\left(\frac{1}{a}-\frac{1}{b}\right)$
(c) $\frac{\rho}{2\pi}\left(\frac{1}{a}-\frac{1}{b}\right)$ (d) $\frac{\rho}{4\pi}\left(\frac{1}{a}+\frac{1}{b}\right)$

19 A bullet of mass 20 g has an initial speed of 1ms^{-1}, just before it starts penetrating a mud wall of thickness 20 cm. If the wall offers a mean resistance of 2.5×10^{-2}N, the speed of the bullet after emerging from the other side of the wall is close to

(a) 0.3 ms^{-1} (b) 0.4 ms^{-1}
(c) 0.1 ms^{-1} (d) 0.7 ms^{-1}

20 The magnitude of the magnetic field at the centre of an equilateral triangular loop of side 1 m which is carrying a current of 10 A is

[Take, $\mu_0 = 4\pi \times 10^{-7}$ NA^{-2}]

(a) 9 μT (b) 1 μT
(c) 3 μT (d) 18 μT

21 In an experiment, brass and steel wires of length 1 m each with areas of cross-section 1 mm^2 are used. The wires are connected in series and one end of the combined wire is connected to a rigid support and other end is subjected to elongation. The stress requires to produce a net elongation of 0.2 mm is

[Take, the Young's modulus for steel and brass are respectively 120×10^9 N/m^2 and 60×10^9 N/m^2]

(a) 1.2×10^6 N/m^2 (b) 0.2×10^6 N/m^2
(c) 1.8×10^6 N/m^2 (d) 4.0×10^6 N/m^2

22 A coil of self inductance 10 mH and resistance 0.1 Ω is connected through a switch to a battery of internal resistance 0.9 Ω. After the switch is closed, the time taken for the current to attain 80% of the saturation value is [Take, ln 5 = 1.6]

(a) 0.002 s (b) 0.324 s
(c) 0.103 s (d) 0.016 s

23 The figure represents a voltage regulator circuit using a Zener diode. The breakdown voltage of the Zener diode is 6 V and the load resistance is $R_L = 4$ kΩ. The series resistance of the circuit is $R_i = 1$ kΩ. If the battery voltage V_B varies from 8V to 16V, what are the minimum and maximum values of the current through Zener diode?

(a) 1.5 mA, 8.5 mA (b) 1 mA, 8.5 mA
(c) 0.5 mA, 8.5 mA (d) 0.5 mA, 6 mA

24 A simple pendulum of length L is placed between the plates of a parallel plate capacitor having electric field E, as shown in figure. Its bob has mass m and charge q. The time period of the pendulum is given by

(a) $2\pi\sqrt{\dfrac{L}{\sqrt{g^2+\left(\dfrac{qE}{m}\right)^2}}}$ (b) $2\pi\sqrt{\dfrac{L}{\sqrt{g^2-\dfrac{q^2E^2}{m^2}}}}$

(c) $2\pi\sqrt{\dfrac{L}{\left(g+\dfrac{qE}{m}\right)}}$ (d) $2\pi\sqrt{\dfrac{L}{\left(g-\dfrac{qE}{m}\right)}}$

25 When heat Q is supplied to a diatomic gas of rigid molecules, at constant volume, its temperature increases by ΔT. The heat required to produce the same change in temperature, at a constant pressure is

(a) $\dfrac{2}{3}Q$ (b) $\dfrac{5}{3}Q$

(c) $\dfrac{3}{2}Q$ (d) $\dfrac{7}{5}Q$

26 One mole of an ideal gas passes through a process, where pressure and volume obey the relation $p=p_0\left[1-\dfrac{1}{2}\left(\dfrac{V_0}{V}\right)^2\right]$. Here, p_0 and V_0 are constants. Calculate the change in the temperature of the gas, if its volume changes from V_0 to $2V_0$.

(a) $\dfrac{1}{2}\dfrac{p_0V_0}{R}$ (b) $\dfrac{1}{4}\dfrac{p_0V_0}{R}$

(c) $\dfrac{3}{4}\dfrac{p_0V_0}{R}$ (d) $\dfrac{5}{4}\dfrac{p_0V_0}{R}$

27 Water from a tap emerges vertically downwards with an initial speed of $10\ \text{ms}^{-1}$. The cross-sectional area of the tap is $10^{-4}\ \text{m}^2$. Assume that the pressure is constant throughout the stream of water and that the flow is streamlined. The cross-sectional area of the stream, 0.15 m below the tap would be [Take, $g=10\ \text{ms}^{-2}$]

(a) $2\times10^{-5}\ \text{m}^2$
(b) $1\times10^{-5}\ \text{m}^2$
(c) $5\times10^{-4}\ \text{m}^2$
(d) $5\times10^{-5}\ \text{m}^2$

28 In Li^{++}, electron in first Bohr orbit is excited to a level by a radiation of wavelength λ. When the ion gets de-excited to the ground state in all possible ways (including intermediate emissions), a total of six spectral lines are observed. What is the value of λ? [Take, $h=6.63\times10^{-34}$ Js; $c=3\times10^8\ \text{ms}^{-1}$]

(a) 9.4 nm (b) 12.3 nm
(c) 10.8 nm (d) 11.4 nm

29 The elastic limit of brass is 379 MPa. What should be the minimum diameter of a brass rod, if it is to support a 400 N load without exceeding its elastic limit?

(a) 0.90 mm (b) 1.00 mm
(c) 1.16 mm (d) 1.36 mm

30 Two radioactive substances A and B have decay constants 5λ and λ, respectively. At $t=0$, a sample has the same number of the two nuclei. The time taken for the ratio of the number of nuclei to become $\left(\dfrac{1}{e}\right)^2$ will be

(a) $2/\lambda$ (b) $1/2\lambda$ (c) $1/4\lambda$ (d) $1/\lambda$

For Detailed Solutions
Visit : http://tinyurl.com/y67eggxe
Or Scan :

ANSWERS

1. (b)	2. (b)	3. (c)	4. (a)	5. (d)	6. (d)	7. (d)	8. (a)	9. (b)	10. (d)
11. (a)	12. (c)	13. (c)	14. (c)	15. (c)	16. (d)	17. (b)	18. (b)	19. (d)	20. (d)
21. (*)	22. (d)	23. (c)	24. (a)	25. (d)	26. (d)	27. (d)	28. (c)	29. (c)	30. (b)

12 April, Shift-I

1 A shell is fired from a fixed artillery gun with an initial speed u such that it hits the target on the ground at a distance R from it. If t_1 and t_2 are the values of the time taken by it to hit the target in two possible ways, the product t_1t_2 is

(a) $\dfrac{R}{4g}$ (b) $\dfrac{R}{g}$

(c) $\dfrac{R}{2g}$ (d) $\dfrac{2R}{g}$

2 The trajectory of a projectile near the surface of the earth is given as $y=2x-9x^2$.

If it were launched at an angle θ_0 with speed v_0, then (Take, $g=10\ \text{ms}^{-2}$)

(a) $\theta_0=\sin^{-1}\left(\dfrac{1}{\sqrt{5}}\right)$ and $v_0=\dfrac{5}{3}\text{ms}^{-1}$

(b) $\theta_0=\cos^{-1}\left(\dfrac{2}{\sqrt{5}}\right)$ and $v_0=\dfrac{3}{5}\text{ms}^{-1}$

(c) $\theta_0=\cos^{-1}\left(\dfrac{1}{\sqrt{5}}\right)$ and $v_0=\dfrac{5}{3}\text{ms}^{-1}$

(d) $\theta_0=\sin^{-1}\left(\dfrac{2}{\sqrt{5}}\right)$ and $v_0=\dfrac{3}{5}\text{ms}^{-1}$

3 Shown in the figure is a shell made of a conductor. It has inner radius a and outer radius b and carries charge Q. At its centre is a dipole **p** as shown.

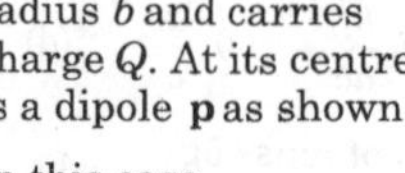

In this case,

(a) surface charge density on the inner surface is uniform and equal to $\dfrac{\left(\dfrac{Q}{2}\right)}{4\pi a^2}$

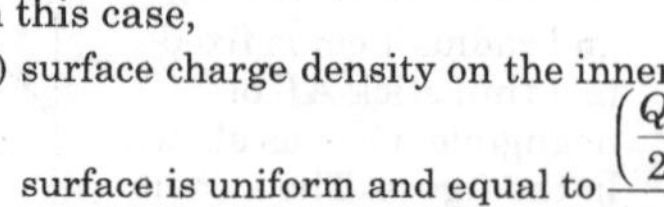

(b) electric field outside the shell is the same as that of a point charge at the centre of the shell

(c) surface charge density on the outer surface depends on $|\mathbf{p}|$

(d) surface charge density on the inner surface of the shell is zero everywhere

4 When M_1 gram of ice at $-10°$C (specific heat $=0.5\ \text{cal g}^{-1}\,°\text{C}^{-1}$) is added to M_2 gram of water at 50°C, finally no ice is left and the water is at 0°C. The value of latent heat of ice, in cal g^{-1} is

(a) $\dfrac{50M_2}{M_1}-5$ (b) $\dfrac{50M_1}{M_2}-50$

(c) $\dfrac{50M_2}{M_1}$ (d) $\dfrac{5M_2}{M_1}-5$

5 The truth table for the circuit given in the figure is

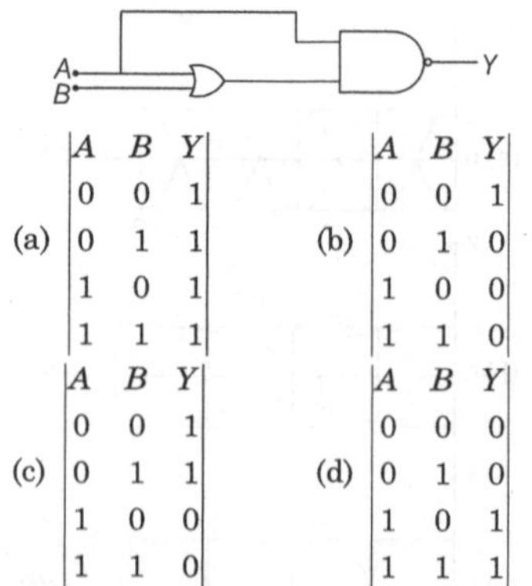

(a)

A	B	Y
0	0	1
0	1	1
1	0	1
1	1	1

(b)

A	B	Y
0	0	1
0	1	0
1	0	0
1	1	0

(c)

A	B	Y
0	0	1
0	1	1
1	0	0
1	1	0

(d)

A	B	Y
0	0	0
0	1	0
1	0	1
1	1	1

6 A circular disc of radius b has a hole of radius a at its centre (see figure). If the mass per unit area of the disc varies as $\left(\dfrac{\sigma_0}{r}\right)$, then the radius of gyration of the disc about its axis passing through the centre is

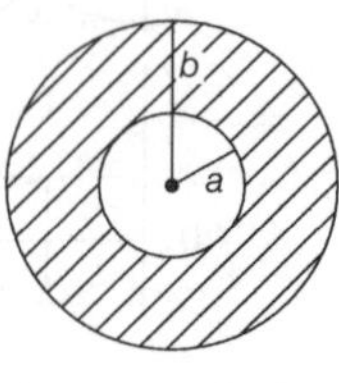

(a) $\sqrt{\dfrac{a^2+b^2+ab}{2}}$ (b) $\dfrac{a+b}{2}$

(c) $\sqrt{\dfrac{a^2+b^2+ab}{3}}$ (d) $\dfrac{a+b}{3}$

7 The stopping potential V_0 (in volt) as a function of frequency (ν) for a sodium emitter, is shown in the figure. The work function of sodium, from the data plotted in the figure, will be (Take, Planck's constant $(h) = 6.63 \times 10^{-34}$ J-s, electron charge, $e = 1.6 \times 10^{-19}$ C]

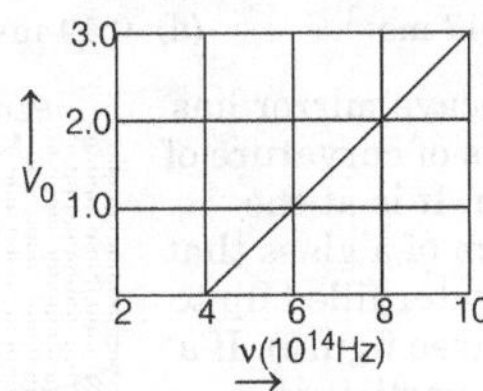

(a) 1.82 eV (b) 1.66 eV
(c) 1.95 eV (d) 2.12 eV

8 A uniform rod of length l is being rotated in a horizontal plane with a constant angular speed about an axis passing through one of its ends. If the tension generated in the rod due to rotation is $T(x)$ at a distance x from the axis, then which of the following graphs depicts it most closely?

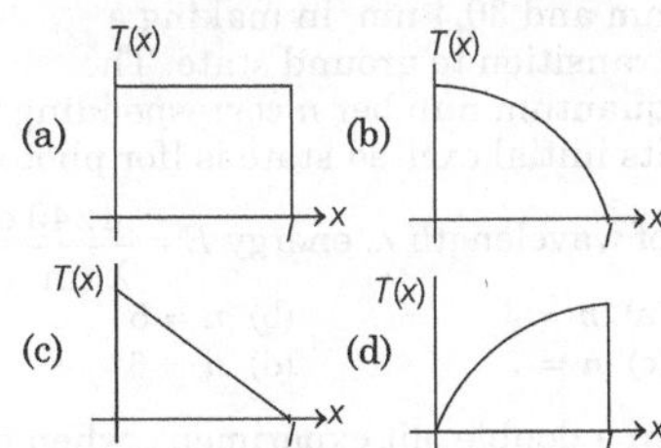

9 To verify Ohm's law, a student connects the voltmeter across the battery as shown in the figure. The measured voltage is plotted as a function of the current and the following graph is obtained

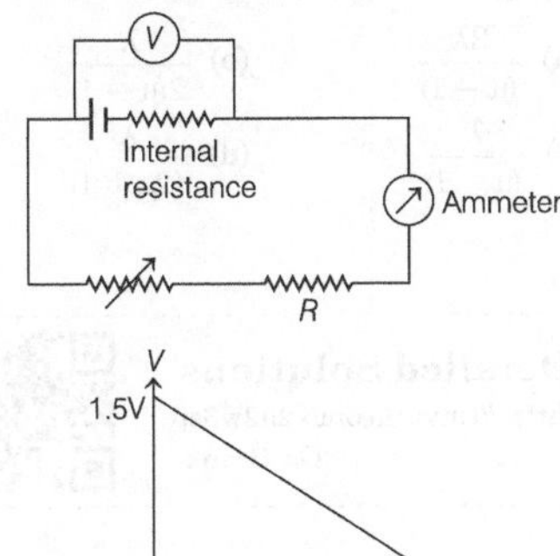

If V_0 is almost zero, then identify the correct statement.

(a) The emf of the battery is 1.5 V and its internal resistance is 1.5 Ω
(b) The value of the resistance R is 1.5 Ω
(c) The potential difference across the battery is 1.5 V when it sends a current of 1000 mA
(d) The emf of the battery is 1.5 V and the value of R is 1.5 Ω

10 A thin ring of 10 cm radius carries a uniformly distributed charge. The ring rotates at a constant angular speed of 40π rad s^{-1} about its axis, perpendicular to its plane. If the magnetic field at its centre is 3.8×10^{-9} T, then the charge carried by the ring is close to ($\mu_0 = 4\pi \times 10^{-7}$ N/A^2).

(a) 2×10^{-6} C (b) 3×10^{-5} C
(c) 4×10^{-5} C (d) 7×10^{-6} C

11 An electromagnetic wave is represented by the electric field $\mathbf{E} = E_0\,\hat{\mathbf{n}}\,\sin[\omega t + (6y - 8z)]$. Taking unit vectors in x, y and z- directions to be $\hat{\mathbf{i}}, \hat{\mathbf{j}}, \hat{\mathbf{k}}$, the direction of propagation $\hat{\mathbf{s}}$, is

(a) $\hat{\mathbf{s}} = \dfrac{3\hat{\mathbf{i}} - 4\hat{\mathbf{j}}}{5}$ (b) $\hat{\mathbf{s}} = \dfrac{-4\hat{\mathbf{k}} + 3\hat{\mathbf{j}}}{5}$

(c) $\hat{\mathbf{s}} = \left(\dfrac{-3\hat{\mathbf{j}} + 4\hat{\mathbf{k}}}{5}\right)$ (d) $\hat{\mathbf{s}} = \dfrac{4\hat{\mathbf{j}} - 3\hat{\mathbf{k}}}{5}$

12 A magnetic compass needle oscillates 30 times per minute at a place, where the dip is 45° and 40 times per minute, where the dip is 30°. If B_1 and B_2 are respectively, the total magnetic field due to the earth at the two places, then the ratio $\dfrac{B_1}{B_2}$ is best given by

(a) 1.8 (b) 0.7
(c) 3.6 (d) 2.2

13 At 40°C, a brass wire of 1 mm radius is hung from the ceiling. A small mass M is hung from the free end of the wire. When the wire is cooled down from 40°C to 20°C, it regains its original length of 0.2 m. The value of M is close to [Coefficient of linear expansion and Young's modulus of brass are 10^{-5}/°C and 10^{11} N/m^2 respectively, $g = 10$ ms^{-2}]

(a) 9 kg (b) 0.5 kg
(c) 1.5 kg (d) 0.9 kg

14 A galvanometer of resistance 100 Ω has 50 divisions on its scale and has sensitivity of 20 μA/division. It is to be converted to a voltmeter with three ranges of 0-2 V, 0-10 V and 0-20 V. The appropriate circuit to do so is

(a)
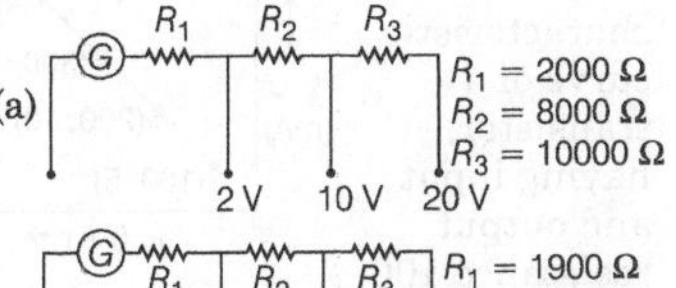

(b)
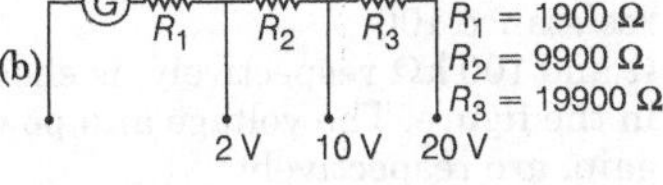

(c)
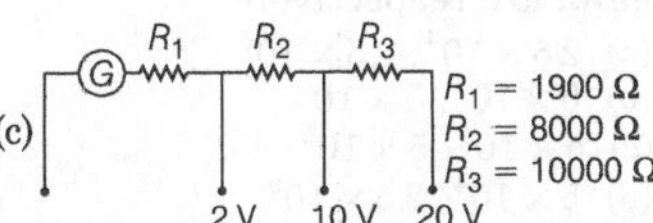

(d)
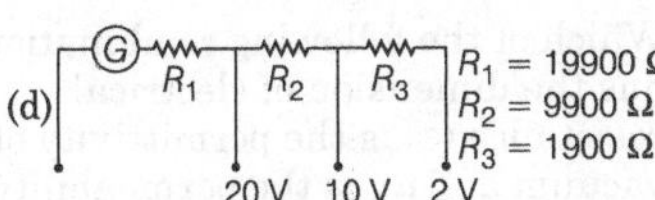

15 A progressive wave travelling along the positive x-direction is represented by $y(x, t) = A \sin(kx - \omega t + \phi)$. Its snapshot at $t = 0$ is given in the figure.

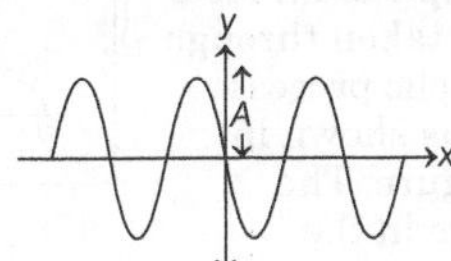

For this wave, the phase ϕ is

(a) $-\dfrac{\pi}{2}$ (b) π (c) 0 (d) $\dfrac{\pi}{2}$

16 The value of numerical aperture of the objective lens of a microscope is 1.25. If light of wavelength 5000Å is used, the minimum separation between two points, to be seen as distinct, will be

(a) 0.24 μm (b) 0.38 μm
(c) 0.12 μm (d) 0.48 μm

17 A point dipole $\mathbf{p} = -p_0\,\hat{\mathbf{x}}$ is kept at the origin. The potential and electric field due to this dipole on the Y-axis at a distance d are, respectively [Take, $V = 0$ at infinity]

(a) $\dfrac{|\mathbf{p}|}{4\pi\varepsilon_0 d^2}, \dfrac{\mathbf{p}}{4\pi\varepsilon_0 d^3}$ (b) $0, \dfrac{-\mathbf{p}}{4\pi\varepsilon_0 d^3}$

(c) $0, \dfrac{\mathbf{p}}{4\pi\varepsilon_0 d^3}$ (d) $\dfrac{|\mathbf{p}|}{4\pi\varepsilon_0 d^2}, \dfrac{-\mathbf{p}}{4\pi\varepsilon_0 d^3}$

18 The resistive network shown below is connected to a DC source of 16 V. The power consumed by the network is 4 W. The value of R is

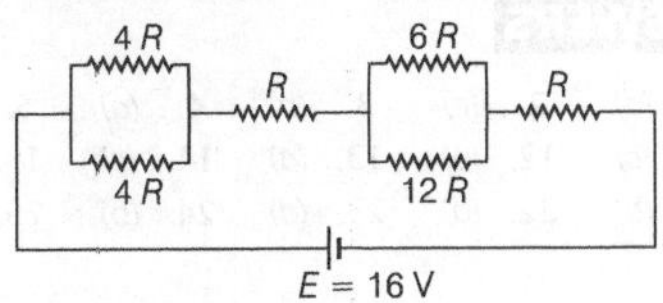

(a) 6 Ω (b) 8 Ω
(c) 1 Ω (d) 16 Ω

19 The transfer characteristic curve of a transistor, having input and output resistance 100 Ω and 100 kΩ respectively, is shown in the figure. The voltage and power gain, are respectively

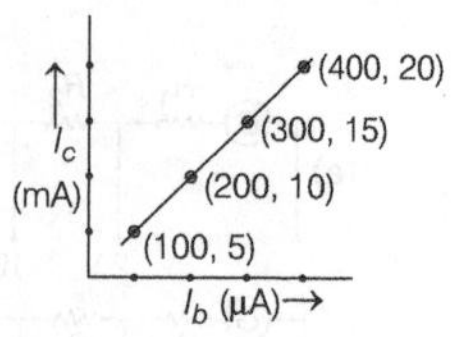

(a) 2.5×10^4, 2.5×10^6
(b) 5×10^4, 5×10^6
(c) 5×10^4, 5×10^5
(d) 5×10^4, 2.5×10^6

20 Which of the following combinations has the dimension of electrical resistance (ε_0 is the permittivity of vacuum and μ_0 is the permeability of vacuum)?

(a) $\sqrt{\frac{\mu_0}{\varepsilon_0}}$ (b) $\frac{\mu_0}{\varepsilon_0}$ (c) $\sqrt{\frac{\varepsilon_0}{\mu_0}}$ (d) $\frac{\varepsilon_0}{\mu_0}$

21 A sample of an ideal gas is taken through the cyclic process *abca* as shown in the figure. The change in the internal energy of the gas along the path *ca* is −180 J. The gas absorbs 250 J of heat along the path *ab* and 60 J along the path *bc*. The work done by the gas along the path *abc* is

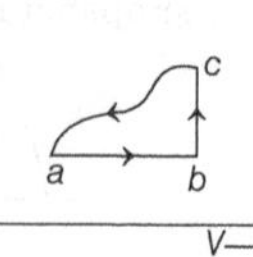

(a) 120 J (b) 130 J (c) 100 J (d) 140 J

22 The figure shows a square loop *L* of side 5 cm which is connected to a network of resistances. The whole setup is moving towards right with a constant speed of 1 cm s^{-1}. At some instant, a part of *L* is in a uniform magnetic field of 1 T, perpendicular to the plane of the loop.
If the resistance of *L* is 1.7 Ω, the current in the loop at that instant will be close to

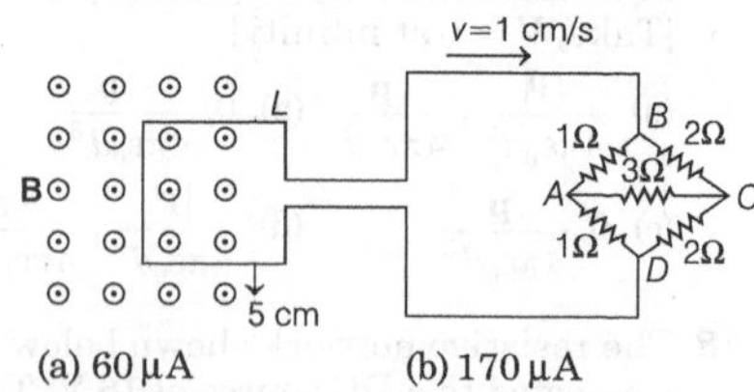

(a) 60 μA (b) 170 μA
(c) 150 μA (d) 115 μA

23 Two identical parallel plate capacitors of capacitance *C* each, have plates of area *A*, separated by a distance *d*. The space between the plates of the two capacitors, is filled with three dielectrics of equal thickness and dielectric constants K_1, K_2 and K_3.

The first capacitor is filled as shown in Fig. I, and the second one is filled as shown in Fig. II. If these two modified capacitors are charged by the same potential *V*, the ratio of the energy stored in the two, would be (E_1 refers to capacitor (I) and E_2 to capacitor (II)) :

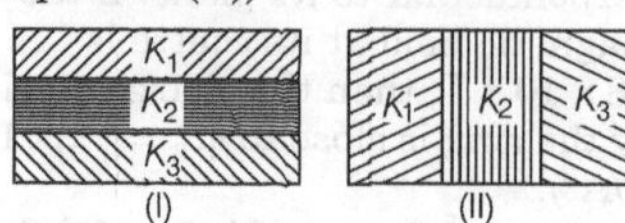

(a) $\frac{E_1}{E_2} = \frac{K_1K_2K_3}{(K_1+K_2+K_3)(K_2K_3+K_3K_1+K_1K_2)}$

(b) $\frac{E_1}{E_2} = \frac{(K_1+K_2+K_3)(K_2K_3+K_3K_1+K_1K_2)}{K_1K_2K_3}$

(c) $\frac{E_1}{E_2} = \frac{9K_1K_2K_3}{(K_1+K_2+K_3)(K_2K_3+K_3K_1+K_1K_2)}$

(d) $\frac{E_1}{E_2} = \frac{(K_1+K_2+K_3)(K_2K_3+K_3K_1+K_1K_2)}{9K_1K_2K_3}$

24 A person of mass *M* is sitting on a swing to length *L* and swinging with an angular amplitude θ_0. If the person stands up when the swing passes through its lowest point, the work done by him, assuming that his centre of mass moves by a distance $l (l << L)$, is close to

(a) $Mgl\,(1-\theta_0^2)$ (b) $Mgl\,(1+\theta_0^2)$
(c) Mgl (d) $Mgl\left(1+\frac{\theta_0^2}{2}\right)$

25 Two moles of helium gas is mixed with three moles of hydrogen molecules (taken to be rigid). What is the molar specific heat of mixture at constant volume?
[Take, $R = 8.3$ J/mol-K]

(a) 19.7 J/mol-K (b) 15.7 J/mol-K
(c) 17.4 J/mol-K (d) 21.6 J/mol-K

26 A submarine *A* travelling at 18 km/h is being chased along the line of its velocity by another submarine *B* travelling at 27 km/h. *B* sends a sonar signal of 500 Hz to detect *A* and receives a reflected sound of frequency ν. The value of ν is close to (Speed of sound in water = 1500 ms^{-1})

(a) 504 Hz (b) 507 Hz
(c) 499 Hz (d) 502 Hz

27 A man (mass = 50 kg) and his son (mass = 20 kg) are standing on a frictionless surface facing each other. The man pushes his son, so that he starts moving at a speed of 0.70 ms^{-1} with respect to the man. The speed of the man with respect to the surface is

(a) 0.28 ms^{-1} (b) 0.20 ms^{-1}
(c) 0.47 ms^{-1} (d) 0.14 ms^{-1}

28 A concave mirror has radius of curvature of 40 cm. It is at the bottom of a glass that has water filled up to 5 cm (see figure). If a small particle is floating on the surface of water, its image as seen, from directly above the glass, is at a distance *d* from the surface of water. The value of *d* is close to [Refractive index of water = 1.33]

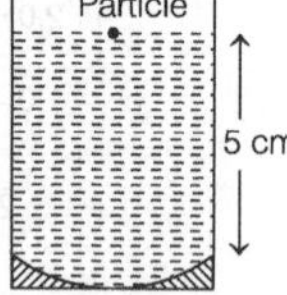

(a) 6.7 cm (b) 13.4 cm
(c) 8.8 cm (d) 11.7 cm

29 An excited He$^+$ ion emits two photons in succession, with wavelengths 108.5 nm and 30.4 nm, in making a transition to ground state. The quantum number *n* corresponding to its initial excited state is [for photon of wavelength λ, energy $E = \frac{1240 \text{ eV}}{\lambda \text{ (in nm)}}$]

(a) $n = 4$ (b) $n = 5$
(c) $n = 7$ (d) $n = 6$

30 In a double slit experiment, when a thin film of thickness *t* having refractive index μ is introduced in front of one of the slits, the maximum at the centre of the fringe pattern shifts by one fringe width. The value of *t* is (λ is the wavelength of the light used)

(a) $\frac{2\lambda}{(\mu-1)}$ (b) $\frac{\lambda}{2(\mu-1)}$
(c) $\frac{\lambda}{(\mu-1)}$ (d) $\frac{\lambda}{(2\mu-1)}$

ANSWERS

1. (d)	2. (c)	3. (b)	4. (a)	5. (c)	6. (c)	7. (b)	8. (b)	9. (a)	10. (b)
11. (c)	12. (b)	13. (a)	14. (c)	15. (b)	16. (a)	17. (b)	18. (b)	19. (d)	20. (a)
21. (b)	22. (b)	23. (d)	24. (b)	25. (c)	26. (d)	27. (b)	28. (c)	29. (b)	30. (c)

For Detailed Solutions
Visit : http://tinyurl.com/y2a2w6sp
Or Scan :

12 April, Shift-II

1 Two particles are projected from the same point with the same speed u such that they have the same range R, but different maximum heights h_1 and h_2. Which of the following is correct?
(a) $R^2 = 4h_1h_2$ (b) $R^2 = 16h_1h_2$
(c) $R^2 = 2h_1h_2$ (d) $R^2 = h_1h_2$

2 In an amplitude modulator circuit, the carrier wave is given by
$C(t) = 4\sin(20000\,\pi t)$ while modulating signal is given by, $m(t) = 2\sin(2000\,\pi t)$. The values of modulation index and lower side band frequency are
(a) 0.5 and 10 kHz (b) 0.4 and 10 kHz
(c) 0.3 and 9 kHz (d) 0.5 and 9 kHz

3 Two sources of sound S_1 and S_2 produce sound waves of same frequency 660 Hz.
A listener is moving from source S_1 towards S_2 with a constant speed u m/s and he hears 10 beats/s. The velocity of sound is 330 m/s. Then, u equal to
(a) 5.5 m/s (b) 15.0 m/s
(c) 2.5 m/s (d) 10.0 m/s

4 Three particles of masses 50 g, 100 g and 150 g are placed at the vertices of an equilateral triangle of side 1 m (as shown in the figure). The (x, y) coordinates of the centre of mass will be

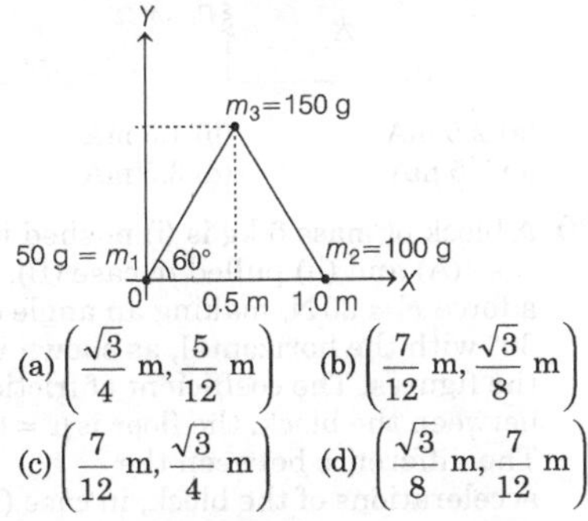

(a) $\left(\frac{\sqrt{3}}{4}\text{ m}, \frac{5}{12}\text{ m}\right)$ (b) $\left(\frac{7}{12}\text{ m}, \frac{\sqrt{3}}{8}\text{ m}\right)$
(c) $\left(\frac{7}{12}\text{ m}, \frac{\sqrt{3}}{4}\text{ m}\right)$ (d) $\left(\frac{\sqrt{3}}{8}\text{ m}, \frac{7}{12}\text{ m}\right)$

5 A Carnot engine has an efficiency of 1/6. When the temperature of the sink is reduced by 62°C, its efficiency is doubled. The temperatures of the source and the sink are respectively,
(a) 62°C, 124°C (b) 99°C, 37°C
(c) 124°C, 62°C (d) 37°C, 99°C

6 A spring whose unstretched length is l has a force constant k. The spring is cut into two pieces of unstretched lengths l_1 and l_2 where, $l_1 = nl_2$ and n is an integer. The ratio k_1 / k_2 of the corresponding force constants k_1 and k_2 will be
(a) n (b) $\frac{1}{n^2}$ (c) $\frac{1}{n}$ (d) n^2

7 A transparent cube of side d, made of a material of refractive index μ_2, is immersed in a liquid of refractive index $\mu_1\,(\mu_1 < \mu_2)$. A ray is incident on the face AB at an angle θ (shown in the figure). Total internal reflection takes place at point E on the face BC.

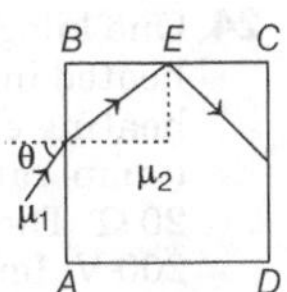

Then, θ must satisfy
(a) $\theta < \sin^{-1}\frac{\mu_1}{\mu_2}$ (b) $\theta > \sin^{-1}\sqrt{\frac{\mu_2^2}{\mu_1^2} - 1}$
(c) $\theta < \sin^{-1}\sqrt{\frac{\mu_2^2}{\mu_1^2} - 1}$ (d) $\theta > \sin^{-1}\frac{\mu_1}{\mu_2}$

8 A tuning fork of frequency 480 Hz is used in an experiment for measuring speed of sound (v) in air by resonance tube method. Resonance is observed to occur at two successive lengths of the air column $l_1 = 30$ cm and $l_2 = 70$ cm. Then, v is equal to
(a) 332 ms^{-1} (b) 384 ms^{-1}
(c) 338 ms^{-1} (d) 379 ms^{-1}

9 The electron in a hydrogen atom first jumps from the third excited state to the second excited state and subsequently to the first excited state. The ratio of the respective wavelengths λ_1 / λ_2 of the photons emitted in this process is
(a) 20/7 (b) 27/5 (c) 7/5 (d) 9/7

10 A diatomic gas with rigid molecules does 10 J of work when expanded at constant pressure. What would be the heat energy absorbed by the gas, in this process?
(a) 25 J (b) 35 J (c) 30 J (d) 40 J

11 Let a total charge $2Q$ be distributed in a sphere of radius R, with the charge density given by $\rho(r) = kr$, where r is the distance from the centre. Two charges A and B, of $-Q$ each, are placed on diametrically opposite points, at equal distance a, from the centre. If A and B do not experience any force, then
(a) $a = 8^{-1/4}R$ (b) $a = \frac{3R}{2^{1/4}}$
(c) $a = 2^{-1/4}R$ (d) $a = R/\sqrt{3}$

12 Consider an electron in a hydrogen atom, revolving in its second excited state (having radius 4.65 Å). The de-Broglie wavelength of this electron is
(a) 3.5 Å (b) 6.6 Å (c) 12.9 Å (d) 9.7 Å

13 A solid sphere of radius R acquires a terminal velocity v_1 when falling (due to gravity) through a viscous fluid having a coefficient of viscosity η. The sphere is broken into 27 identical solid spheres. If each of these spheres acquires a terminal velocity, v_2 when falling through the same fluid, the ratio (v_1 / v_2) equals
(a) 9 (b) 1/27 (c) 1/9 (d) 27

14 A smooth wire of length $2\pi r$ is bent into a circle and kept in a vertical plane. A bead can slide smoothly on the wire. When the circle is rotating with angular speed ω about the vertical diameter AB, as shown in figure, the bead is at rest with respect to the circular ring at position P as shown. Then, the value of ω^2 is equal to
(a) $\frac{\sqrt{3}g}{2r}$ (b) $2g/(r\sqrt{3})$
(c) $(g\sqrt{3})/r$ (d) $2g/r$

15 A particle is moving with speed $v = b\sqrt{x}$ along positive X-axis. Calculate the speed of the particle at time $t = \tau$ (assume that the particle is at origin at $t = 0$).
(a) $\frac{b^2\tau}{4}$ (b) $\frac{b^2\tau}{2}$ (c) $b^2\tau$ (d) $\frac{b^2\tau}{\sqrt{2}}$

16 The ratio of the weights of a body on the earth's surface, so that on the surface of a planet is 9 : 4. The mass of the planet is $\frac{1}{9}$th of that of the earth. If R is the radius of the earth, what is the radius of the planet? (Take, the planets to have the same mass density)
(a) $\frac{R}{3}$ (b) $\frac{R}{4}$ (c) $\frac{R}{9}$ (d) $\frac{R}{2}$

17 A system of three polarisers P_1, P_2, P_3 is set up such that the pass axis of P_3 is crossed with respect to that of P_1. The pass axis of P_2 is inclined at 60° to the pass axis of P_3. When a beam of unpolarised light of intensity I_0 is incident on P_1, the intensity of light transmitted by the three polarisers is I. The ratio (I_0 / I) equals (nearly)
(a) 5.33 (b) 16.00 (c) 10.67 (d) 1.80

18 A uniform cylindrical rod of length L and radius r, is made from a material whose Young's modulus of elasticity equals Y. When this rod is heated by temperature T and simultaneously subjected to a net longitudinal compressional force F, its length remains unchanged. The coefficient of volume expansion of the material of the rod, is (nearly) equal to

(a) $9F/(\pi r^2 YT)$ (b) $6F/(\pi r^2 YT)$
(c) $3F/(\pi r^2 YT)$ (d) $F/(3\pi r^2 YT)$

19 The number density of molecules of a gas depends on their distance r from the origin as, $n(r) = n_0 e^{-\alpha r^4}$. Then, the total number of molecules is proportional to

(a) $n_0\alpha^{-3/4}$ (b) $\sqrt{n_0}\,\alpha^{1/2}$
(c) $n_0\alpha^{1/4}$ (d) $n_0\alpha^{-3}$

20 A small speaker delivers 2 W of audio output. At what distance from the speaker will one detect 120 dB intensity sound? [Take, reference intensity of sound as 10^{-12} W/m^2]

(a) 40 cm (b) 20 cm (c) 10 cm (d) 30 cm

21 Half lives of two radioactive nuclei A and B are 10 minutes and 20 minutes, respectively. If initially a sample has equal number of nuclei, then after 60 minutes, the ratio of decayed numbers of nuclei A and B will be

(a) 3 : 8 (b) 1 : 8 (c) 8 : 1 (d) 9 : 8

22 An electron moving along the X-axis with an initial energy of 100 eV, enters a region of magnetic field $\mathbf{B} = (1.5\times10^{-3}\ \text{T})\,\hat{\mathbf{k}}$ at S (see figure).

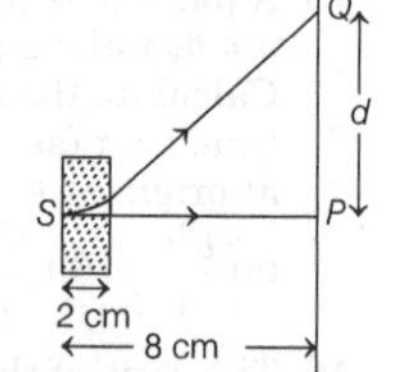

The field extends between $x = 0$ and $x = 2$ cm. The electron is detected at the point Q on a screen placed 8 cm away from the point S. The distance d between P and Q (on the screen) is (Take, electron's charge $= 1.6\times10^{-19}$ C, mass of electron $= 9.1\times10^{-31}$ kg)

(a) 11.65 cm (b) 12.87 cm
(c) 1.22 cm (d) 2.25 cm

23 In the given circuit, the charge on 4 μF capacitor will be

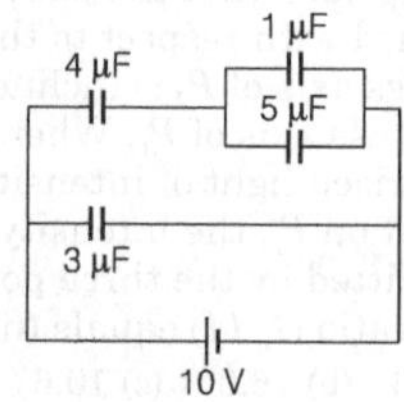

(a) 5.4 μC (b) 9.6 μC
(c) 13.4 μC (d) 24 μC

24 One kilogram of water at 20°C is heated in an electric kettle whose heating element has a mean (temperature averaged) resistance of 20 Ω. The rms voltage in the mains is 200 V. Ignoring heat loss from the kettle, time taken for water to evaporate fully, is close to [Specific heat of water = 4200 J/(kg°C), Latent heat of water = 2260 kJ/kg]

(a) 16 min (b) 22 min
(c) 3 min (d) 10 min

25 A moving coil galvanometer, having a resistance G, produces full scale deflection when a current I_g flows through it. This galvanometer can be converted into (i) an ammeter of range 0 to $I_0\,(I_0 > I_g)$ by connecting a shunt resistance R_A to it and (ii) into a voltmeter of range 0 to $V\ (V = GI_0)$ by connecting a series resistance R_V to it. Then,

(a) $R_A R_V = G^2\left(\dfrac{I_0 - I_g}{I_g}\right)$ and $\dfrac{R_A}{R_V} = \left(\dfrac{I_g}{(I_0 - I_g)}\right)^2$

(b) $R_A R_V = G^2$ and $\dfrac{R_A}{R_V} = \left(\dfrac{I_g}{I_0 - I_g}\right)^2$

(c) $R_A R_V = G^2\left(\dfrac{I_g}{I_0 - I_g}\right)$ and $\dfrac{R_A}{R_V} = \left(\dfrac{I_0 - I_g}{I_g}\right)^2$

(d) $R_A R_V = G^2$ and $\dfrac{R_A}{R_V} = \dfrac{I_g}{(I_0 - I_g)}$

26 Find the magnetic field at point P due to a straight line segment AB of length 6 cm carrying a current of 5 A (See figure). (Take, $\mu_0 = 4\pi\times10^{-7}$ N-A^{-2})

(a) 2.0×10^{-5} T (b) 1.5×10^{-5} T
(c) 3.0×10^{-5} T (d) 2.5×10^{-5} T

27 A plane electromagnetic wave having a frequency ν = 23.9 GHz propagates along the positive z-direction in free space. The peak value of the electric field is 60 V/m. Which among the following is the acceptable magnetic field component in the electromagnetic wave?

(a) $\mathbf{B} = 2\times10^{7}\sin(0.5\times10^3 z + 1.5\times10^{11} t)\,\hat{\mathbf{i}}$
(b) $\mathbf{B} = 2\times10^{-7}\sin(0.5\times10^3 z - 1.5\times10^{11} t)\,\hat{\mathbf{i}}$
(c) $\mathbf{B} = 60\sin(0.5\times10^3 x + 1.5\times10^{11} t)\,\hat{\mathbf{k}}$
(d) $\mathbf{B} = 2\times10^{-7}\sin(1.5\times10^2 x + 0.5\times10^{11} t)\,\hat{\mathbf{j}}$

28 Consider the L-R circuit shown in the figure. If the switch S is closed at $t = 0$, then the amount of charge that passes through the battery between $t = 0$ and $t = \dfrac{L}{R}$ is

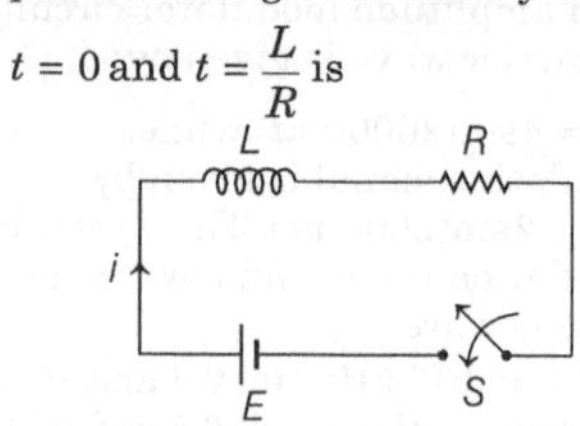

(a) $\dfrac{2.7EL}{R^2}$ (b) $\dfrac{EL}{2.7R^2}$
(c) $\dfrac{7.3EL}{R^2}$ (d) $\dfrac{EL}{7.3R^2}$

29 Figure shows a DC voltage regulator circuit, with a Zener diode of breakdown voltage = 6 V. If the unregulated input voltage varies between 10 V to 16 V, then what is the maximum Zener current?

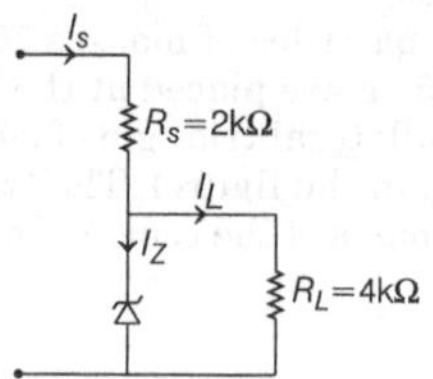

(a) 2.5 mA (b) 1.5 mA
(c) 7.5 mA (d) 3.5 mA

30 A block of mass 5 kg is (i) pushed in case (A) and (ii) pulled in case (B), by a force $F = 20$ N, making an angle of 30° with the horizontal, as shown in the figures. The coefficient of friction between the block, the floor is $\mu = 0.2$. The difference between the accelerations of the block, in case (B) and case (A) will be (Take, $g = 10$ ms^{-2})

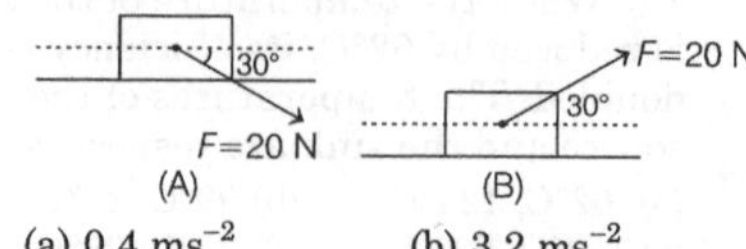

(a) 0.4 ms^{-2} (b) 3.2 ms^{-2}
(c) 0.8 ms^{-2} (d) 0 ms^{-2}

ANSWERS

1. (b)	2. (d)	3. (c)	4. (c)	5. (b)	6. (c)	7. (c)	8. (b)	9. (a)	10. (b)
11. (a)	12. (d)	13. (a)	14. (b)	15. (b)	16. (d)	17. (c)	18. (c)	19. (a)	20. (a)
21. (d)	22. (*)	23. (d)	24. (b)	25. (b)	26. (b)	27. (b)	28. (b)	29. (d)	30. (c)

For Detailed Solutions
Visit : http://tinyurl.com/y36jgozj
Or Scan :

JANUARY ATTEMPT

9 January, Shift-I

1 A bar magnet is demagnetised by inserting it inside a solenoid of length 0.2 m, 100 turns and carrying a current of 5.2 A. The coercivity of the bar magnet is

(a) 1200 A/m (b) 285 A/m
(c) 2600A/m (d) 520A/m

2 A rod of length L at room temperature and uniform area of cross-section A, is made of a metal having coefficient of linear expansion α /°C. It is observed that an external compressive force F, is applied on each of its ends, prevents any change in the length of the rod, when its temperature rises by ΔT K. Young's modulus, Y for this metal is

(a) $\frac{F}{2A\alpha\,\Delta T}$ (b) $\frac{F}{A\alpha(\Delta T - 273)}$
(c) $\frac{2F}{A\alpha\Delta T}$ (d) $\frac{F}{A\alpha\Delta T}$

3 Three charges $+Q$, q, $+Q$ are placed respectively at distance 0, $\frac{d}{2}$ and d from the origin on the X-axis. If the net force experienced by $+Q$ placed at $x = 0$ is zero, then value of q is

(a) $\frac{+Q}{2}$ (b) $\frac{+Q}{4}$
(c) $\frac{-Q}{2}$ (d) $\frac{-Q}{4}$

4 Two masses m and $\frac{m}{2}$ are connected at the two ends of a massless rigid rod of length l. The rod is suspended by a thin wire of torsional constant k at the centre of mass of the rod-mass system (see figure). Because of torsional constant k, the restoring torque is $\tau = k\theta$ for angular displacement θ. If the rod is rotated by θ_0 and released, the tension in it when it passes through its mean position will be

(a) $\frac{2k\theta_0^2}{l}$ (b) $\frac{k\theta_0^2}{l}$ (c) $\frac{3k\theta_0^2}{l}$ (d) $\frac{k\theta_0^2}{2l}$

5 An infinitely long current-carrying wire and a small current-carrying loop are in the plane of the paper as shown. The radius of the loop is a and distance of its centre from the wire is $d (d >> a)$. If the loop applies a force F on the wire, then

(a) $F \propto \left(\frac{a^2}{d^3}\right)$ (b) $F = 0$
(c) $F \propto \left(\frac{a}{d}\right)$ (d) $F \propto \left(\frac{a}{d}\right)^2$

6 A plane electromagnetic wave of frequency 50 MHz travels in free space along the positive x - direction. At a particular point in space and time, $\mathbf{E} = 6.3\,\hat{\mathbf{j}}$ V/m. The corresponding magnetic field $\vec{\mathbf{B}}$, at that point will be

(a) $18.9\times10^{8}\hat{\mathbf{k}}$ T (b) $6.3\times10^{-8}\hat{\mathbf{k}}$ T
(c) $18.9\times10^{-8}\hat{\mathbf{k}}$ T (d) $2.1\times10^{-8}\hat{\mathbf{k}}$ T

7 Drift speed of electrons, when 1.5 A of current flows in a copper wire of cross-section 5 mm^2 is v. If the electron density in copper is 9×10^{28} / m^3, the value of v in mm/s is close to (Take, charge of electron to be $= 1.6\times10^{-19}$ C)

(a) 0.02 (b) 0.2 (c) 2 (d) 3

8 Mobility of electrons in a semiconductor is defined as the ratio of their drift velocity to the applied electric field. If for an n - type semiconductor, the density of electrons is 10^{19}m^{-3} and their mobility is 1.6 m^2 (V-s), then the resistivity of the semiconductor (since, it is an n-type semiconductor contribution of holes is ignored) is close to

(a) 2 Ω-m (b) 0.2 Ω-m
(c) 0.4 Ω-m (d) 4 Ω-m

9 A copper wire is stretched to make it 0.5% longer. The percentage change in its electrical resistance, if its volume remains unchanged is

(a) 2.0% (b) 1.0% (c) 0.5% (d) 2.5%

10 A gas can be taken from A to B *via* two different processes ACB and ADB.

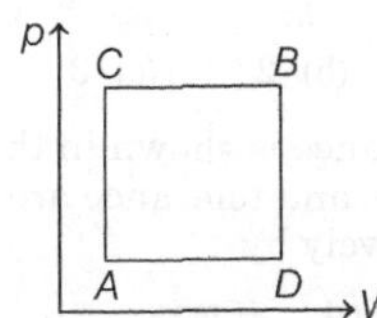

When path ACB is used 60 J of heat flows into the system and 30 J of work is done by the system. If path ADB is used work done by the system is 10 J the heat flow into the system in path ADB is

(a) 80 J (b) 40 J (c) 100 J (d) 20 J

11 Two coherent sources produce waves of different intensities which interfere. After interference, the ratio of the maximum intensity to the minimum intensity is 16. The intensity of the waves are in the ratio

(a) 16 : 9 (b) 5 : 3 (c) 25 : 9 (d) 4 : 1

12 A particle is moving with a velocity $\mathbf{v} = k(y\hat{\mathbf{i}} + x\hat{\mathbf{j}})$, where k is a constant. The general equation for its path is

(a) $y = x^2$ + constant
(b) $y^2 = x$ + constant
(c) xy = constant
(d) $y^2 = x^2$ + constant

13 Temperature difference of 120°C is maintained between two ends of a uniform rod AB of length $2L$. Another bent rod PQ, of same cross-section as AB and length $\frac{3L}{2}$ is connected across AB (see figure). In steady state, temperature difference between P and Q will be close to

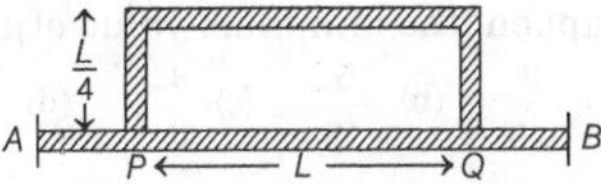

(a) 45° (b) 35°C (c) 75°C (d) 60°C

14 Surface of certain metal is first illuminated with light of wavelength $\lambda_1 = 350$ n-m and then by light of wavelength $\lambda_2 = 540$ n-m. It is found that the maximum speed of the photoelectrons in the two cases differ by a factor of 2. The work function of the metal (in eV) is close to

(energy of photon $= \frac{1240}{\lambda(\text{in n - m})}$ eV)

(a) 5.6 (b) 2.5 (c) 1.8 (d) 1.4

15 A parallel plate capacitor is made of two square plates of side 'a' separated by a distance d (d<<a). The lower triangular portions is filled with a dielectric of dielectric constant k, as shown in the figure. Capacitance of this capacitor is

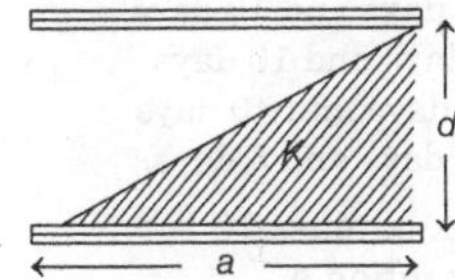

(a) $\frac{K\varepsilon_0 a^2}{d}\ln K$ (b) $\frac{K\varepsilon_0 a^2}{d(K-1)}\ln K$
(c) $\frac{K\varepsilon_0 a^2}{2d(K+1)}$ (d) $\frac{1}{2}\cdot\frac{K\varepsilon_0 a^2}{d}$

16 A current loop, having two circular arcs joined by two radial lines as shown in the figure. It carries a current of 10 A. The magnetic field at point O will be close to

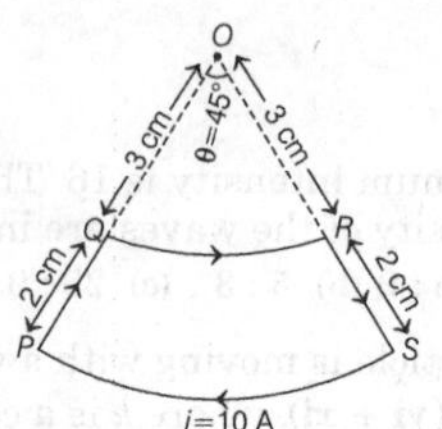

(a) 1.0×10^{-5} T (b) 1.0×10^{-7} T
(c) 1.5×10^{-7} T (d) 1.5×10^{-5} T

17 Consider a tank made of glass (refractive index is 1.5) with a thick bottom. It is filled with a liquid of refractive index μ. A student finds that, irrespective of what the incident angle i (see figure) is for a beam of light entering the liquid, the light reflected from the liquid glass interface is never completely polarised. For this to happen, the minimum value of μ is

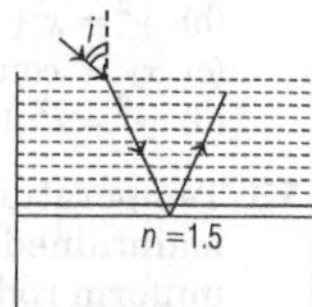

(a) $\frac{3}{\sqrt{5}}$ (b) $\frac{5}{\sqrt{3}}$ (c) $\frac{4}{3}$ (d) $\sqrt{\frac{5}{3}}$

18 A convex lens is put 10 cm from a light source and it makes a sharp image on a screen, kept 10 cm from the lens. Now, a glass block (refractive index is 1.5) of 1.5 cm thickness is placed in contact with the light source. To get the sharp image again, the screen is shifted by a distance d. Then, d is

(a) 0
(b) 1.1 cm away from the lens
(c) 0.55 cm away from the lens
(d) 0.55 cm towards the lens

19 A sample of radioactive material A, that has an activity of 10 mCi (1 Ci $= 3.7 \times 10^{10}$ decays/s) has twice the number of nuclei as another sample of a different radioactive material B which has an activity of 20 mCi. The correct choices for half-lives of A and B would, then be respectively

(a) 20 days and 10 days
(b) 5 days and 10 days
(c) 10 days and 40 days
(d) 20 days and 5 days

20 A block of mass 10 kg is kept on a rough inclined plane as shown in the figure. A force of 3 N is applied on the block. The coefficient of static friction between the plane and the block is 0.6. What should be the minimum value of force F, such that the block does not move downward? (Take, $g = 10\text{ms}^{-2}$)

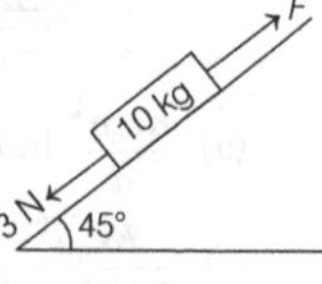

(a) 32 N (b) 25 N (c) 23 N (d) 18 N

21 A heavy ball of mass M is suspended from the ceiling of a car by a light string of mass $m (m << M)$. When the car is at rest, the speed of transverse waves in the string is 60 ms^{-1}. When the car has acceleration a, the wave speed increases to 60.5 ms^{-1}. The value of a, in terms of gravitational acceleration g is closest to

(a) $\frac{g}{20}$ (b) $\frac{g}{5}$ (c) $\frac{g}{30}$ (d) $\frac{g}{10}$

22 An L-shaped object made of thin rods of uniform mass density is suspended with a string as shown in figure. If $AB = BC$ and the angle is made by AB with downward vertical is θ, then

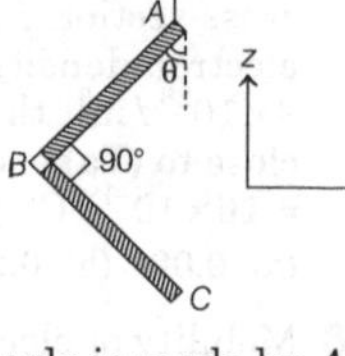

(a) $\tan\theta = \frac{2}{\sqrt{3}}$ (b) $\tan\theta = \frac{1}{2\sqrt{3}}$
(c) $\tan\theta = \frac{1}{2}$ (d) $\tan\theta = \frac{1}{3}$

23 Three blocks A, B and C are lying on a smooth horizontal surface as shown in the figure. A and B have equal masses m while C has mass M. Block A is given an initial speed v towards B due to which it collides with B perfectly inelastically. The combined mass collides with C, also perfectly inelastically $\frac{5}{6}$th of the initial kinetic energy is lost in whole process. What is value of $\frac{M}{m}$?

(a) 4 (b) 2 (c) 3 (d) 5

24 A resistance is shown in the figure. Its value and tolerance are given respectively by

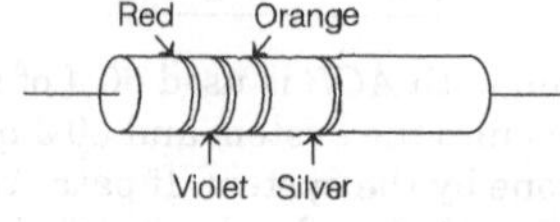

(a) 270 Ω, 5% (b) 27 kΩ, 20%
(c) 27 kΩ, 10% (d) 270 kΩ, 10%

25 When the switch S in the circuit shown is closed, then the value of current i will be

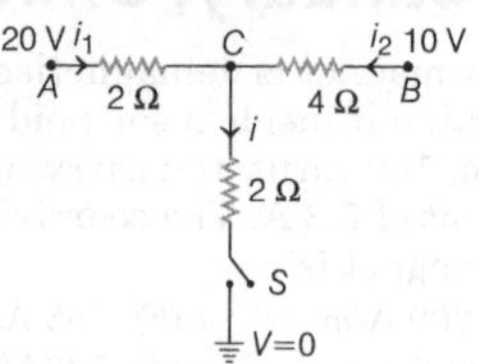

(a) 4A (b) 3A (c) 2A (d) 5A

26 For a uniformly charged ring of radius R, the electric field on its axis has the largest magnitude at a distance h from its centre. Then, value of h is

(a) $\frac{R}{\sqrt{2}}$ (b) $R\sqrt{2}$ (c) R (d) $\frac{R}{\sqrt{5}}$

27 A conducting circular loop is made of a thin wire has area 3.5×10^{-3} m^2 and resistance 10 Ω. It is placed perpendicular to a time dependent magnetic field $B(t) = (0.4\text{T})\sin(0.5\,\pi t)$. The field is uniform in space. Then the net charge flowing through the loop during $t = 0$ s and $t = 10$ ms is close to

(a) 6 mC (b) 21 mC
(c) 7 mC (d) 14 mC

28 A mixture of 2 moles of helium gas (atomic mass = $4u$) and 1 mole of argon gas (atomic mass = $40u$) is kept at 300 K in a container. The ratio of their rms speeds $\left[\frac{v_{\text{rms(helium)}}}{v_{\text{rms(argon)}}}\right]$ is close to

(a) 0.32 (b) 2.24 (c) 3.16 (d) 0.45

29 If the angular momentum of a planet of mass m, moving around sun in a circular orbit is L about the centre of the sun, its areal velocity is

(a) $\frac{4L}{m}$ (b) $\frac{2L}{m}$ (c) $\frac{L}{2m}$ (d) $\frac{L}{m}$

30 A block of mass m lying on a smooth horizontal surface is attached to a spring (of negligible mass) of spring constant k. The other end of the spring is fixed as shown in the figure. The block is initially at rest in its equilibrium position. If now the block is pulled with a constant force F, the maximum speed of the block is

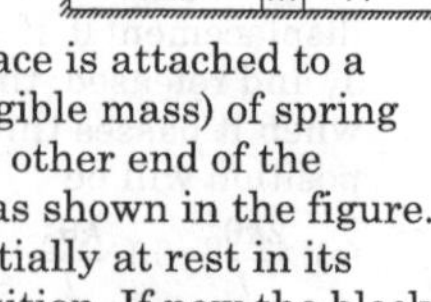

(a) $\frac{\pi F}{\sqrt{mk}}$ (b) $\frac{F}{\sqrt{mk}}$ (c) $\frac{2F}{\sqrt{mk}}$ (d) $\frac{F}{\pi\sqrt{mk}}$

ANSWERS

1. (c)	2. (d)	3. (d)	4. (b)	5. (d)	6. (d)	7. (a)	8. (c)	9. (b)	10. (b)
11. (c)	12. (d)	13. (a)	14. (c)	15. (b)	16. (a)	17. (a)	18. (c)	19. (d)	20. (a)
21. (b)	22. (d)	23. (a)	24. (c)	25. (d)	26. (a)	27. (d)	28. (c)	29. (c)	30. (b)

For Detailed Solutions
Visit : http://tinyurl.com/y4mssqfe *Or* Scan :

9 January, Shift-II

1 In form of G (universal gravitational constant), h (Planck constant) and c (speed of light), the time period will be proportional to

(a) $\sqrt{\frac{Gh}{c^5}}$ (b) $\sqrt{\frac{hc^5}{G}}$ (c) $\sqrt{\frac{c^3}{Gh}}$ (d) $\sqrt{\frac{Gh}{c^3}}$

2 A parallel plate capacitor with square plates is filled with four dielectrics of dielectric constants K_1, K_2, K_3, K_4 arranged as shown in the figure. The effective dielectric constant K will be:

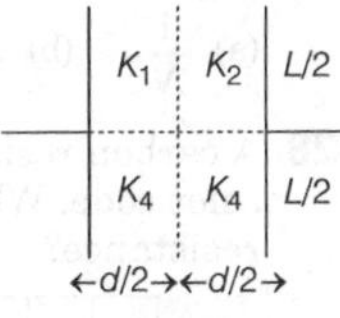

(a) $K = \frac{(K_1 + K_2)(K_3 + K_4)}{2(K_1 + K_2 + K_3 + K_4)}$

(b) $K = \frac{(K_1 + K_2)(K_3 + K_4)}{K_1 + K_2 + K_3 + K_4}$

(c) $K = \frac{(K_1 + K_3)(K_2 + K_4)}{K_1 + K_2 + K_3 + K_4}$

(d) $K = \frac{(K_1 + K_4)(K_2 + K_3)}{2(K_1 + K_2 + K_3 + K_4)}$

3 A series AC circuit containing an inductor (20 mH), a capacitor (120 μF) and a resistor (60 Ω) is driven by an AC source of 24 V/50 Hz. The energy dissipated in the circuit in 60 s is

(a) 3.39×10^3 J (b) 5.65×10^2 J
(c) 2.26×10^3 J (d) 5.17×10^2 J

4 In the given circuit, the internal resistance of the 18 V cell is negligible. If $R_1 = 400\ \Omega$, $R_3 = 100\ \Omega$ and $R_4 = 500\ \Omega$ and the reading of an ideal voltmeter across R_4 is 5 V, then the value of R_2 will be

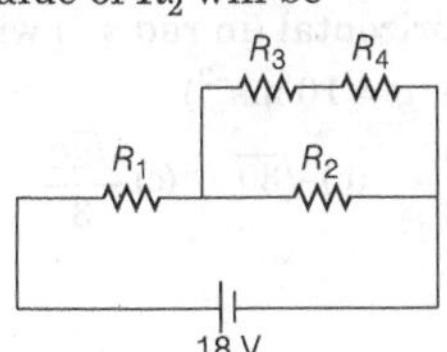

(a) 550 Ω (b) 230 Ω
(c) 300 Ω (d) 450 Ω

5 In a car race on a straight path, car A takes a time t less than car B at the finish and passes finishing point with a speed 'v' more than that of car B. Both the cars start from rest and travel with constant acceleration a_1 and a_2 respectively. Then 'v' is equal to

(a) $\frac{2a_1a_2}{a_1 + a_2}t$ (b) $\sqrt{2a_1a_2}\ t$
(c) $\sqrt{a_1a_2}\ t$ (d) $\frac{a_1 + a_2}{2}t$

6 The energy required to take a satellite to a height 'h' above earth surface (where, radius of earth $= 6.4 \times 10^3$ km) is E_1 and kinetic energy required for the satellite to be in a circular orbit at this height is E_2. The value of h for which E_1 and E_2 are equal is

(a) 3.2×10^3 km (b) 1.28×10^4 km
(c) 6.4×10^3 km (d) 1.6×10^3 km

7 At 0.3V and 0.7 V, the diodes Ge and Si become conductor respectively. In given figure, if ends of diode Ge overturned, the change in potential V_0 will be

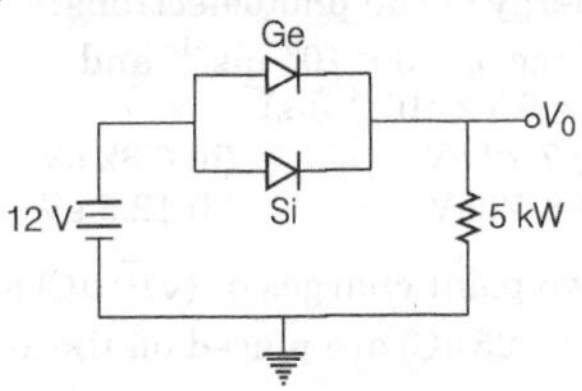

(a) 0.2 V (b) 0.6V (c) 0.4 V (d) 0.8V

8. A particle having the same charge as of electron moves in a circular path of radius 0.5 cm under the influence of a magnetic field of 0.5 T. If an electric field of 100 V/m makes it to move in a straight path, then the mass of the particle is (Take, charge of electron $= 1.6 \times 10^{-19}$ C)

(a) 1.6×10^{-19} kg (b) 1.6×10^{-27} kg
(c) 9.1×10^{-31} kg (d) 2.0×10^{-24} kg

9. A mass of 10 kg is suspended vertically by a rope from the roof. When a horizontal force is applied on the mass, the rope deviated at an angle of 45° at the roof point. If the suspended mass is at equilibrium, the magnitude of the force applied is (Take, $g = 10\ \text{ms}^{-2}$)

(a) 70 N (b) 200 N
(c) 100 N (d) 140 N

10 A 15 g mass of nitrogen gas is enclosed in a vessel at a temperature 27°C. Amount of heat transferred to the gas, so that rms velocity of molecules is doubled is about (Take, $R = 8.3$ J / K- mole)

(a) 10 kJ (b) 0.9 kJ (c) 14 kJ (d) 6 kJ

11 A power transmission line feeds input power at 2300 V to a step-down transformer with its primary windings having 4000 turns. The output power is delivered at 230 V by the transformer. If the current in the primary of the transformer is 5A and its efficiency is 90%, the output current would be

(a) 45 A (b) 50 A (c) 25 A (d) 35 A

12 A rod of mass 'M' and length '$2L$' is suspended at its middle by a wire. It exhibits torsional oscillations. If two masses each of 'm' are attached at distance '$L/2$' from its centre on both sides, it reduces the oscillation frequency by 20%. The value of ratio m/M is close to

(a) 0.57 (b) 0.37 (c) 0.77 (d) 0.17

13 Two Carnot engines A and B are operated in series. The first one, A receives heat at T_1 (= 600 K) and rejects to a reservoir at temperature T_2. The second engine B receives heat rejected by the first engine and in turn rejects to a heat reservoir at T_3 (= 400 K). Calculate the temperature T_2 if the work outputs of the two engines are equal.

(a) 600 K (b) 500 K
(c) 400 K (d) 300 K

14 A musician produce the sound of second harmonics from open end flute of 50 cm. The other person moves toward the musician with speed 10 km/h from the second end of room. If the speed of sound 330 m/s, the frequency heard by running person will be

(a) 666 Hz (b) 500 Hz
(c) 753 Hz (d) 333 Hz

15 The plane mirrors (M_1 and M_2) are inclined to each other such that a ray of light incident on mirror M_1 and parallel to the mirror M_2 is reflected from mirror M_2 parallel to the mirror M_1. The angle between the two mirror is

(a) 45° (b) 75° (c) 90° (d) 60°

16 In free space, the energy of electromagnetic wave in electric field is U_E and in magnetic field is U_B. Then

(a) $U_E = U_B$ (b) $U_E > U_B$
(c) $U_E < U_B$ (d) $U_E = \frac{U_B}{2}$

17 Charge is distributed within a sphere of radius R with a volume charge density $\rho(r) = \frac{A}{r^2}e^{-2r/a}$, where A and a are constants. If Q is the total charge of this charge distribution, the radius R is

(a) $a \log\left(\frac{1}{1 - \frac{Q}{2\pi aA}}\right)$

(b) $a \log\left(1 - \frac{Q}{2\pi aA}\right)$

(c) $\frac{a}{2}\log\left(1-\frac{Q}{2\pi aA}\right)$

(d) $\frac{a}{2}\log\left(\frac{1}{1-\frac{Q}{2\pi aA}}\right)$

18 In a Young's double slit experiment, the slits are placed 0.320 mm apart. Light of wavelength $\lambda = 500$ n-m is incident on the slits. The total number of bright fringes that are observed in the angular range $-30° \leq \theta \leq 30°$ is

(a) 320 (b) 321 (c) 640 (d) 641

19 In communication system, only one percent frequency of signal of wavelength 800 nm can be used as bandwidth. How many channal of 6MHz bandwidth can be broadcast this?

($c = 3\times 10^8$ m/s, $h = 6.6\times 10^{-34}$ J-s)

(a) 3.75×10^6 (b) 3.86×10^6
(c) 6.25×10^5 (d) 4.87×10^5

20 In given time $t = 0$, Activity of two radioactive substances A and B are equal. After time t, the ratio of their activities $\frac{R_B}{R_A}$ decreases according to e^{-3t}. If the half life of A is In 2, the half-life of B will be

(a) 4ln 2 (b) $\frac{\ln 2}{4}$ (c) $\frac{\ln 2}{2}$ (d) 2ln 2

21 In three dimensional system, the position coordinates of a particle (in motion) are given below

$$x = a\cos\omega t$$
$$y = a\sin\omega t$$
$$z = a\omega t$$

The velocity of particle will be

(a) $\sqrt{2}\ a\omega$ (b) $2\ a\omega$ (c) $a\omega$ (d) $\sqrt{3}\ a\omega$

22 A force acts on a 2 kg object, so that its position is given as a function of time as $x = 3t^2 + 5$. What is the work done by this force in first 5 seconds?

(a) 850 J (b) 900 J (c) 950 J (d) 875 J

23 A particle is executing simple harmonic motion (SHM) of amplitude A, along the X-axis, about $x = 0$. when its potential energy (PE) equals kinetic energy (KE), the position of the particle will be

(a) A (b) $\frac{A}{2}$ (c) $\frac{A}{2\sqrt{2}}$ (d) $\frac{A}{\sqrt{2}}$

24 The magnetic field associated with a light wave is given at the origin, by $B = B_0$ [sin $(3.14\times 10^7)\,ct$ + sin $(6.28\times 10^7)ct$]. If this light falls on a silver plate having a work function of 4.7 eV, what will be the maximum kinetic energy of the photoelectrons?

(Take, $c = 3\times 10^8$ ms^{-1} and $h = 6.6\times 10^{-34}$ J-s)

(a) 7.72 eV (b) 6.82 eV
(c) 8.52 eV (d) 12.5 eV

25 Two point charges q_1 ($\sqrt{10}$ μC) and q_2 (− 25 μC) are placed on the x-axis at $x = 1$ m and $x = 4$ m, respectively. The electric field (in V/m) at a point $y = 3$ m on Y-axis is

$$\left(\text{Take, } \frac{1}{4\pi\varepsilon_0} = 9\times 10^9 \text{ N-m}^2\text{C}^{-2}\right)$$

(a) $(63\,\hat{i} - 27\hat{j})\times 10^2$

(b) $(81\,\hat{i} - 81\hat{j})\times 10^2$

(c) $(-81\,\hat{i} + 81\hat{j})\times 10^2$

(d) $(-63\,\hat{i} + 27\hat{j})\times 10^2$

26 The top of a water tank is open to air and its water level is maintained. It is giving out 0.74 m^3 water per minute through a circular opening of 2 cm radius is its wall. The depth of the centre of the opening from the level of water in the tank is close to

(a) 4.8 m (b) 6.0 m
(c) 2.9 m (d) 9.6 m

27 One of the two identical conducting wires of length L is bent in the form of a circular loop and the other one into a circular coil of N identical turns. If the same current is passed in both, the ratio of the magnetic field at the centre of the loop (B_L) to that at the centre of the coil (B_C), i.e. $\frac{B_L}{B_C}$ will be

(a) $\frac{1}{N}$ (b) N (c) $\frac{1}{N^2}$ (d) N^2

28 A carbon resistance has a following color code. What is the value of the resistance?

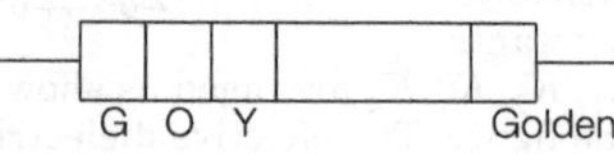

(a) 5.3 MΩ ± 5% (b) 64 kΩ ± 10%
(c) 6.4 MΩ ± 5% (d) 530 kΩ ± 5%

29 The pitch and the number of divisions, on the circular scale for a given screw gauge are 0.5 mm and 100, respectively. When the screw gauge is fully tightened without any object, the zero of its circular scale lies 3 divisions below the mean line. The readings of the main scale and the circular scale for a thin sheet are 5.5 mm and 48 respectively, the thickness of this sheet is

(a) 5.950 mm (b) 5.725 mm
(c) 5.755 mm (d) 5.740 mm

30 A rod of length 50 cm is pivoted at one end. It is raised such that if makes an angle of 30° from the horizontal as shown and released from rest. Its angular speed when it passes through the horizontal (in rad s^{-1}) will be

(Take, $g = 10$ ms^{-2})

(a) $\frac{\sqrt{30}}{2}$ (b) $\sqrt{30}$ (c) $\frac{\sqrt{20}}{3}$ (d) $\sqrt{\frac{30}{2}}$

ANSWERS

1 (a)	2 (*)	3 (d)	4 (c)	5 (c)	6 (a)	7 (c)	8 (d)	9 (c)	10 (a)
11 (a)	12 (b)	13 (b)	14 (a)	15 (d)	16 (a)	17 (d)	18 (d)	19 (c)	20 (b)
21 (a)	22 (b)	23 (d)	24 (a)	25 (a)	26 (a)	27 (c)	28 (d)	29 (c)	30 (b)

* No option is correct

For Detailed Solutions
Visit : http://tinyurl.com/yy9v5wgb
Or Scan :

10 January, Shift-I

1 A potentiometer wire AB having length L and resistance $12r$ is joined to a cell D of EMF ε and internal resistance r. A cell C having emf $\frac{\varepsilon}{2}$ and internal resistance $3r$ is connected. The length AJ at which the galvanometer as shown in figure shows no deflection is

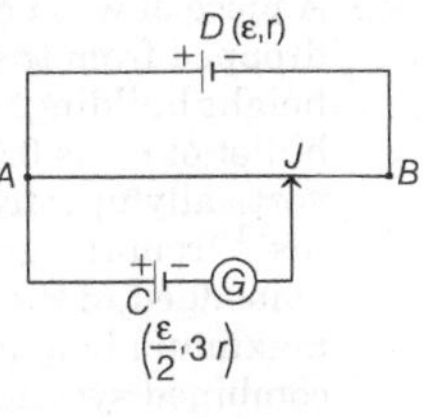

(a) $\frac{5}{12}L$ (b) $\frac{11}{12}L$ (c) $\frac{13}{24}L$ (d) $\frac{11}{24}L$

2 In a Young's double slit experiment with slit separation 0.1 mm, one observes a bright fringe at angle $\frac{1}{40}$ rad by using light of wavelength λ_1. When the light of the wavelength λ_2 is used a bright fringe is seen at the same angle in the same set up. Given that λ_1 and λ_2 are in visible range (380 n-m to 740 n-m), their values are

(a) 380 n-m, 525 n-m
(b) 400 n-m, 500 n-m
(c) 380 n-m, 500 n-m
(d) 625 n-m, 500 n-m

3 An insulating thin rod of length l has a linear charge density $\rho(x) = \rho_0 \frac{x}{l}$ on it. The rod is rotated about an axis passing through the origin ($x = 0$) and perpendicular to the rod. If the rod makes n rotations per second, then the time averaged magnetic moment of the rod is

(a) $n \rho l^3$ (b) $\pi n \rho l^3$
(c) $\frac{\pi}{3} n \rho l^3$ (d) $\frac{\pi}{4} n \rho l^3$

4 A 2 W carbon resistor is color coded with green, black, red and brown respectively. The maximum current which can be passed through this resistor is

(a) 0.4 mA (b) 63 mA
(c) 20 mA (d) 100 mA

5 In the given circuit, the cells have zero internal resistance. The currents (in Ampere) passing through resistances R_1 and R_2 respectively are

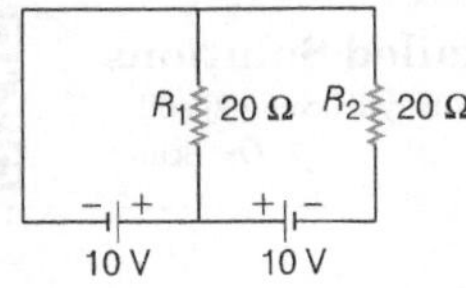

(a) 0.5, 0 (b) 1, 2 (c) 2, 2 (d) 0, 1

6 In an electron microscope, the resolution that can be achieved is of the order of the wavelength of electrons used. To resolve a width of 7.5×10^{-12} m, the minimum electron energy required is close to

(a) 500 keV (b) 1 keV
(c) 100 keV (d) 25 keV

7 To get output '1' at R, for the given logic gate circuit, the input values must be

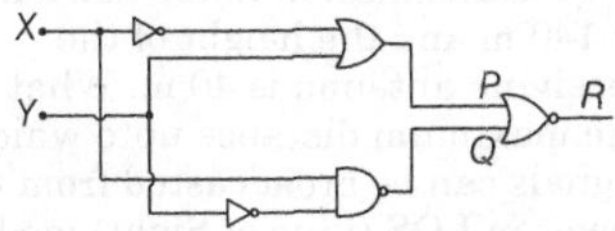

(a) $X = 0, Y = 0$ (b) $X = 1, Y = 0$
(c) $X = 1, Y = 1$ (d) $X = 0, Y = 1$

8 Two guns A and B can fire bullets at speeds 1 km/s and 2 km/s, respectively. From a point on a horizontal ground, they are fired in all possible directions. The ratio of maximum areas covered by the bullets on the ground fired by the two guns is

(a) 1 : 4 (b) 1 : 16 (c) 1 : 8 (d) 1 : 2

9 A satellite is moving with a constant speed v in circular orbit around the earth. An object of mass 'm' is ejected from the satellite such that it just escapes from the gravitational pull of the earth. At the time of ejection, the kinetic energy of the object is

(a) $\frac{3}{2}mv^2$ (b) $2mv^2$ (c) $m\,v^2$ (d) $\frac{1}{2}m\,v^2$

10 A heat source at $T = 10^3$ K is connected to another heat reservoir at $T = 10^2$ K by a copper slab which is 1 m thick. Given that the thermal conductivity of copper is 0.1 $\mathrm{WK^{-1}m^{-1}}$, the energy flux through it in the steady state is

(a) 90 $\mathrm{Wm^{-2}}$ (b) 65 $\mathrm{Wm^{-2}}$
(c) 120 $\mathrm{Wm^{-2}}$ (d) 200 $\mathrm{Wm^{-2}}$

11 A magnet of total magnetic moment $10^{-2}\hat{i}$ A-m^2 is placed in a time varying magnetic field, $B\hat{i}$ (cosωt), where $B = 1$ T and $\omega = 0.125$ rad/s. The work done for reversing the direction of the magnetic moment at $t = 1$ s is

(a) 0.01 J (b) 0.007 J
(c) 0.014 J (d) 0.028 J

12 To mop-clean a floor, a cleaning machine presses a circular mop of radius R vertically down with a total force F and rotates it with a constant angular speed about its axis. If the force F is distributed uniformly over the mop and if coefficient of friction between the mop and the floor is μ, the torque applied by the machine on the mop is

(a) $\frac{2}{3}\mu FR$ (b) $\frac{\mu FR}{6}$ (c) $\frac{\mu FR}{3}$ (d) $\frac{\mu FR}{2}$

13 A homogeneous solid cylindrical roller of radius R and mass m is pulled on a cricket pitch by a horizontal force. Assuming rolling without slipping, angular acceleration of the cylinder is

(a) $\frac{F}{2mR}$ (b) $\frac{2F}{3mR}$
(c) $\frac{3F}{2mR}$ (d) $\frac{F}{3mR}$

14 Using a nuclear counter, the count rate of emitted particles from a radioactive source is measured. At $t = 0$, it was 1600 counts per second and $t = 8$ s, it was 100 counts per second. The count rate observed as counts per second at $t = 6$ s is close to

(a) 400 (b) 200 (c) 150 (d) 360

15 A charge Q is distributed over three concentric spherical shells of radii a, b, c $(a < b < c)$ such that their surface charge densities are equal to one another. The total potential at a point at distance r from their common centre, where $r < a$ would be

(a) $\frac{Q(a^2 + b^2 + c^2)}{4\pi\varepsilon_0(a^3 + b^3 + c^3)}$
(b) $\frac{Q(a + b + c)}{4\pi\varepsilon_0(a^2 + b^2 + c^2)}$
(c) $\frac{Q}{4\pi\varepsilon_0(a + b + c)}$
(d) $\frac{Q}{12\pi\varepsilon_0} \cdot \frac{ab + bc + ca}{abc}$

16 A parallel plate capacitor is of area 6 cm^2 and a separation 3 mm. The gap is filled with three dielectric materials of equal thickness (see figure) with dielectric constants $K_1 = 10$, $K_2 = 12$ and $K_3 = 14$. The dielectric constant of a material which give same capacitance when fully inserted in above capacitor, would be

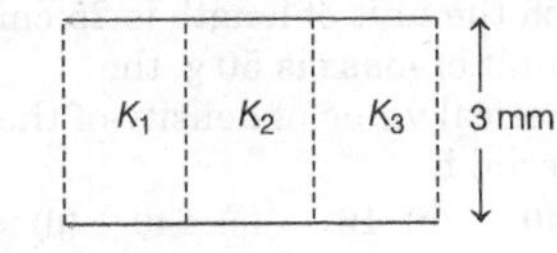

(a) 4 (b) 36 (c) 12 (d) 14

17 A solid metal cube of edge length 2 cm is moving in a positive Y-direction at a constant speed of 6 m/s. There is a uniform magnetic field of 0.1 T in the positive Z-direction. The potential difference between the two faces of the cube perpendicular to the X-axis is

(a) 2 mV (b) 12 mV (c) 6 mV (d) 1 mV

18 In the cube of side 'a' shown in the figure, the vector from the central point of the face $ABOD$ to the central point of the face $BEFO$ will be

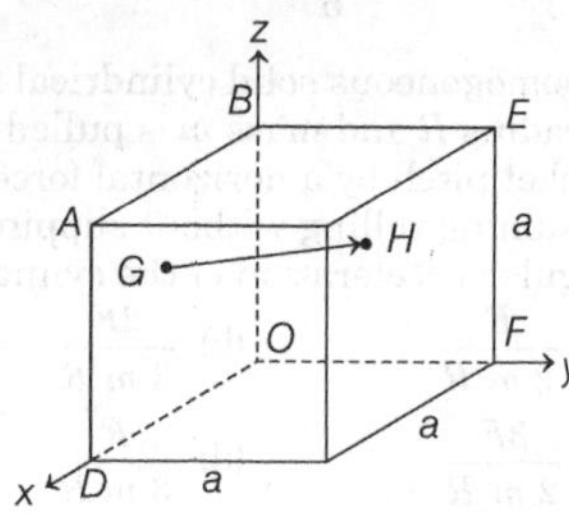

(a) $\frac{1}{2}a(\hat{i}-\hat{k})$ (b) $\frac{1}{2}a(\hat{j}-\hat{i})$
(c) $\frac{1}{2}a(\hat{j}-\hat{k})$ (d) $\frac{1}{2}a(\hat{k}-\hat{i})$

19 If the magnetic field of a plane electromagnetic wave is given by

$$B = 100 \times 10^{-6} \sin\left[2\pi \times 2 \times 10^{15}\left(t - \frac{x}{c}\right)\right]$$

then the maximum electric field associated with it is (Take, the speed of light $= 3 \times 10^8$ m/s)
(a) 6×10^4 N/C (b) 4×10^4 N/C
(c) 3×10^4 N/C (d) 4.5×10^4 N/C

20 A block of mass m is kept on a platform which starts from rest with constant acceleration $\frac{g}{2}$ upwards as shown in figure. Work done by normal reaction on block in time t is

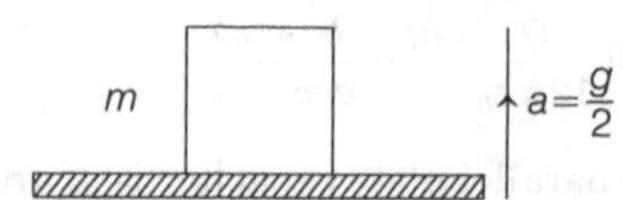

(a) $\frac{mg^2t^2}{8}$ (b) $\frac{3mg^2t^2}{8}$
(c) 0 (d) $-\frac{mg^2t^2}{8}$

21 The density of a material in SI units is 128 kg m^{-3}. In certain units in which the unit of length is 25 cm and the unit of mass is 50 g, the numerical value of density of the material is
(a) 40 (b) 16 (c) 640 (d) 410

22 A train moves towards a stationary observer with speed 34 m/s. The train sounds a whistle and its frequency registered by the observer is f_1. If the speed of the train is reduced to 17 m/s, the frequency registered is f_2. If speed of sound is 340 m/s, then the ratio $\frac{f_1}{f_2}$ is
(a) $\frac{19}{18}$ (b) $\frac{21}{20}$ (c) $\frac{20}{19}$ (d) $\frac{18}{17}$

23 A TV transmission tower has a height of 140 m and the height of the receiving antenna is 40 m. What is the maximum distance upto which signals can be broadcasted from this tower in LOS (Line of Sight) mode? (Take, radius of earth $= 6.4 \times 10^6$ m).
(a) 65 km (b) 80 km
(c) 40 km (d) 48 km

24 Three Carnot engines operate in series between a heat source at a temperature T_1 and a heat sink at temperature T_4 (see figure). There are two other reservoirs at temperatures T_2 and T_3, as shown with $T_1 > T_2 > T_3 > T_4$. The three engines are equally efficient if

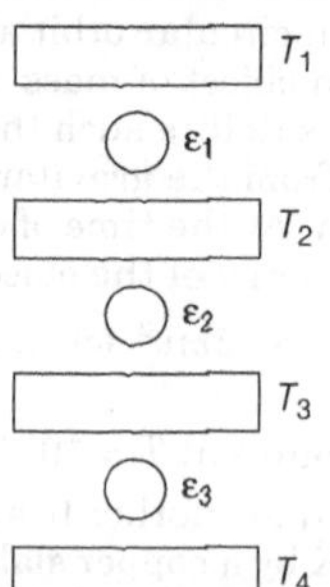

(a) $T_2 = (T_1^3T_4)^{1/4}$; $T_3 = (T_1T_4^3)^{1/4}$
(b) $T_2 = (T_1^2T_4)^{1/3}$; $T_3 = (T_1T_4^2)^{1/3}$
(c) $T_2 = (T_1T_4)^{1/2}$; $T_3 = (T_1^2T_4)^{1/3}$
(d) $T_2 = (T_1T_4^2)^{1/3}$; $T_3 = (T_1^2T_4)^{1/3}$

25 A string of length 1 m and mass 5 g is fixed at both ends. The tension in the string is 8.0 N. The string is set into vibration using an external vibrator of frequency 100 Hz. The separation between successive nodes on the string is close to
(a) 16.6 cm (b) 33.3 cm
(c) 10.0 cm (d) 20.0 cm

26 A piece of wood of mass 0.03 kg is dropped from the top of a 100 m height building. At the same time, a bullet of mass 0.02 kg is fired vertically upward with a velocity 100 ms^{-1} from the ground. The bullet gets embedded in the wood. Then, the maximum height to which the combined system reaches above the top of the building before falling below is (Take, $g = 10$ ms^{-2})
(a) 20 m (b) 30 m (c) 10 m (d) 40 m

27 A uniform metallic wire has a resistance of 18 Ω and is bent into an equilateral triangle. Then, the resistance between any two vertices of the triangle is
(a) 12 Ω (b) 8 Ω (c) 2 Ω (d) 4 Ω

28 Two electric dipoles, A, B with respective dipole moments $\mathbf{d}_A = -4\,qa\,\hat{\mathbf{i}}$ and $\mathbf{d}_B = -2\,qa\,\hat{\mathbf{i}}$ are placed on the X-axis with a separation R, as shown in the figure

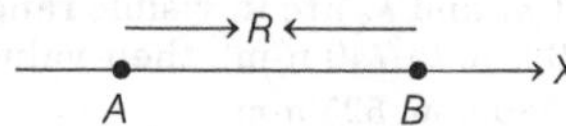

The distance from A at which both of them produce the same potential is
(a) $\frac{\sqrt{2}\,R}{\sqrt{2}+1}$ (b) $\frac{\sqrt{2}\,R}{\sqrt{2}-1}$
(c) $\frac{R}{\sqrt{2}+1}$ (d) $\frac{R}{\sqrt{2}-1}$

29 Water flows into a large tank with flat bottom at the rate of 10^{-4} m^{3}s^{-1}. Water is also leaking out of a hole of area 1 cm^2 at its bottom. If the height of the water in the tank remains steady, then this height is
(a) 4 cm (b) 2.9 cm
(c) 5.1 cm (d) 1.7 cm

30 A plano-convex lens of refractive index μ_1 and focal length f_1 is kept in contact with another plano-concave lens of refractive index μ_2 and focal length f_2. If the radius of curvature of their spherical faces is R each and $f_1 = 2f_2$, then μ_1 and μ_2 are related as
(a) $3\mu_2 - 2\mu_1 = 1$ (b) $2\mu_2 - \mu_1 = 1$
(c) $2\mu_1 - \mu_2 = 1$ (d) $\mu_1 + \mu_2 = 3$

ANSWERS

1 (c)	2 (d)	3 (d)	4 (c)	5 (a)	6 (d)	7 (b)	8 (b)	9 (c)	10 (a)
11 (c)	12 (a)	13 (b)	14 (b)	15 (b)	16 (c)	17 (b)	18 (b)	19 (c)	20 (b)
21 (a)	22 (a)	23 (a)	24 (b)	25 (d)	26 (d)	27 (d)	28 (a)	29 (c)	30 (c)

For Detailed Solutions
Visit : http://tinyurl.com/y2jls3lb
Or Scan :

10 January, Shift-II

1 A particle which is experiencing a force, is given by $\mathbf{F} = 3\hat{\mathbf{i}} - 12\hat{\mathbf{j}}$, undergoes a displacement of $\mathbf{d} = 4\hat{\mathbf{i}}$. If the particle had a kinetic energy of 3 J at the beginning of the displacement, what is its kinetic energy at the end of the displacement ?

(a) 9 J (b) 15J (c) 12 J (d) 10 J

2 An unknown metal of mass 192 g heated to a temperature of 100°C was immersed into a brass calorimeter of mass 128 g containing 240 g of water at a temperature of 8.4°C. Calculate the specific heat of the unknown metal, if water temperature stabilises at 21.5°C. (Take, specific heat of brass is 394 J kg^{-1}K^{-1})

(a) 916 J kg^{-1} K^{-1} (b) 654 J kg^{-1} K^{-1}
(c) 1232 J kg^{-1} K^{-1} (d) 458 J kg^{-1} K^{-1}

3 A current of 2 mA was passed through an unknown resistor which dissipated a power of 4.4 W. Dissipated power when an ideal power supply of 11 V is connected across it is

(a) 11×10^{-4} W (b) 11×10^{-5} W
(c) 11×10^{5} W (d) 11×10^{-3} W

4 A cylindrical plastic bottle of negligible mass is filled with 310 mL of water and left floating in a pond with still water. If pressed downward slightly and released, it starts performing simple harmonic motion at angular frequency ω. If the radius of the bottle is 2.5 cm, then ω is close to (Take, density of water = 10^3 kg/m^3)

(a) 2.50 rad s^{-1} (b) 5.00 rad s^{-1}
(c) 1.25 rad s^{-1} (d) 3.75 rad s^{-1}

5 Two stars of masses 3×10^{31} kg each and at distance 2×10^{11} m rotate in a plane about their common centre of mass O. A meteorite passes through O moving perpendicular to the star's rotation plane. In order to escape from the gravitational field of this double star, the minimum speed that meteorite should have at O is (Take, gravitational constant, $G = 6.67\times 10^{-11}$ N-m^{2}kg^{-2})

(a) 2.8×10^5 m/s (b) 3.8×10^4 m/s
(c) 2.4×10^4 m/s (d) 1.4×10^5 m/s

6 Two identical spherical balls of mass M and radius R each are stuck on two ends of a rod of length $2R$ and mass M (see figure).

The moment of inertia of the system about the axis passing perpendicularly through the centre of the rod is

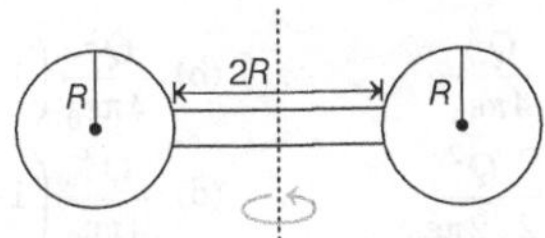

(a) $\frac{137}{15}MR^2$ (b) $\frac{209}{15}MR^2$
(c) $\frac{17}{15}MR^2$ (d) $\frac{152}{15}MR^2$

7 The modulation frequency of an AM radio station is 250 kHz, which is 10% of the carrier wave. If another AM station approaches you for license, what broadcast frequency will you allot ?

(a) 2000 kHz (b) 2250 kHz
(c) 2900 kHz (d) 2750 kHz

8 The actual value of resistance R, shown in the figure is 30 Ω. This is measured in an experiment as shown using the standard formula $R = \frac{V}{I}$, where V and I are the readings of the voltmeter and ammeter, respectively. If the measured value of R is 5% less, then the internal resistance of the voltmeter is

(a) 600 Ω (b) 570 Ω
(c) 350 Ω (d) 35 Ω

9 The eye can be regarded as a single refracting surface. The radius of curvature of this surface is equal to that of cornea (7.8 mm). This surface separates two media of refractive indices 1 and 1.34. Calculate the distance from the refracting surface at which a parallel beam of light will come to focus.

(a) 4.0 cm (b) 2 cm (c) 3.1 cm (d) 1 cm

10 Charges $-q$ and $+q$ located at A and B, respectively, constitute an electric dipole. Distance $AB = 2a$, O is the mid point of the dipole and OP is perpendicular to AB. A charge Q is placed at P, where $OP = y$ and $y >> 2a$. The charge Q experiences an electrostatic force F. If Q is now moved along the equatorial line to P' such that $OP' = \left(\frac{y}{3}\right)$, the force on Q will be close to $\left(\frac{y}{3} >> 2a\right)$

(a) $\frac{F}{3}$ (b) $3F$
(c) $9F$ (d) $27F$

11 A hoop and a solid cylinder of same mass and radius are made of a permanent magnetic material with their magnetic moment parallel to their respective axes. But the magnetic moment of hoop is twice of solid cylinder. They are placed in a uniform magnetic field in such a manner that their magnetic moments make a small angle with the field. If the oscillation periods of hoop and cylinder are T_h and T_c respectively, then

(a) $T_h = 0.5\,T_c$ (b) $T_h = T_c$
(c) $T_h = 2\,T_c$ (d) $T_h = 1.5\,T_c$

12 Half-mole of an ideal monoatomic gas is heated at constant pressure of 1 atm from 20° C to 90° C. Work done by gas is close to (Take, gas constant, $R = 8.31$ J/mol-K)

(a) 291 J (b) 581 J (c) 146 J (d) 73 J

13 A rigid massless rod of length $3l$ has two masses attached at each end as shown in the figure. The rod is pivoted at point P on the horizontal axis (see figure). When released from initial horizontal position, its instantaneous angular acceleration will be

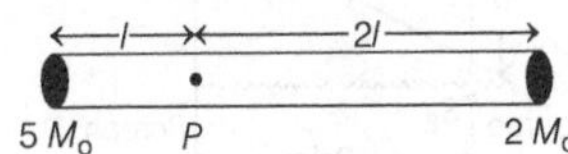

(a) $\frac{g}{13l}$ (b) $\frac{g}{2l}$ (c) $\frac{7g}{3l}$ (d) $\frac{g}{3l}$

14 Consider the nuclear fission

$Ne^{20} \longrightarrow 2He^4 + C^{12}$

Given that the binding energy/nucleon of Ne^{20}, He^4 and C^{12} are respectively, 8.03 MeV, 7.07 MeV and 7.86 MeV, identify the correct statement.

(a) Energy of 3.6 MeV will be released.
(b) Energy of 12.4 MeV will be supplied.
(c) 8.3 MeV energy will be released.
(d) Energy of 11.9 MeV has to be supplied.

15 For the circuit shown below, the current through the Zener diode is

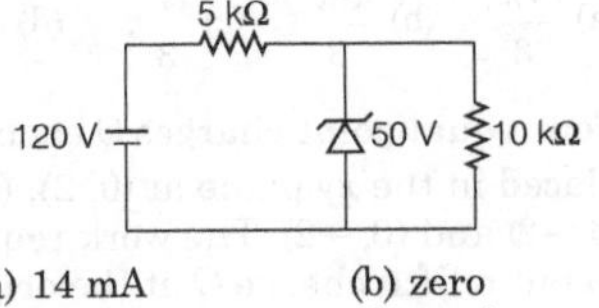

(a) 14 mA (b) zero
(c) 5 mA (d) 9 mA

16 Two forces P and Q of magnitude $2F$ and $3F$, respectively, are at an angle θ with each other. If the force Q is doubled, then their resultant also gets doubled. Then, the angle θ is

(a) 60° (b) 120°
(c) 30° (d) 90°

ONLINE JEE Main 2019

17 A closed organ pipe has a fundamental frequency of 1.5 kHz. The number of overtones that can be distinctly heard by a person with this organ pipe will be (Assume that the highest frequency a person can hear is 20,000 Hz)

(a) 7 (b) 4 (c) 5 (d) 6

18 At some location on earth, the horizontal component of earth's magnetic field is 18×10^{-6} T. At this location, magnetic needle of length 0.12 m and pole strength 1.8 A-m is suspended from its mid point using a thread, it makes 45° angle with horizontal in equilibrium. To keep this needle horizontal, the vertical force that should be applied at one of its ends is

(a) 6.5×10^{-5} N (b) 3.6×10^{-5} N
(c) 1.3×10^{-5} N (d) 1.8×10^{-5} N

19 Consider a Young's double slit experiment as shown in figure.

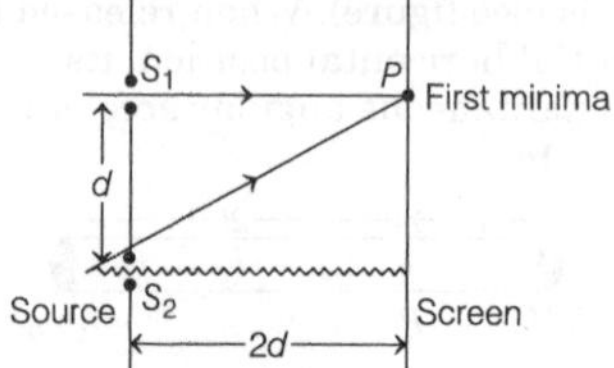

What should be the slit separation d in terms of wavelength λ such that the first minima occurs directly in front of the slit (S_1)?

(a) $\frac{\lambda}{2(5-\sqrt{2})}$ (b) $\frac{\lambda}{(5-\sqrt{2})}$
(c) $\frac{\lambda}{2(\sqrt{5}-2)}$ (d) $\frac{\lambda}{(\sqrt{5}-2)}$

20 A particle executes simple harmonic motion with an amplitude of 5 cm. When the particle is at 4 cm from the mean position, the magnitude of its velocity in SI units is equal to that of its acceleration. Then, its periodic time (in seconds) is

(a) $\frac{4\pi}{3}$ (b) $\frac{8\pi}{3}$ (c) $\frac{7}{3}\pi$ (d) $\frac{3}{8}\pi$

21 Four equal point charges Q each are placed in the xy-plane at (0, 2), (4, 2), (4, –2) and (0, – 2). The work required to put a fifth charge Q at the origin of the coordinate system will be

(a) $\frac{Q^2}{4\pi\varepsilon_0}$ (b) $\frac{Q^2}{4\pi\varepsilon_0}\left(1+\frac{1}{\sqrt{3}}\right)$
(c) $\frac{Q^2}{2\sqrt{2}\pi\varepsilon_0}$ (d) $\frac{Q^2}{4\pi\varepsilon_0}\left(1+\frac{1}{\sqrt{5}}\right)$

22 Two vectors **A** and **B** have equal magnitudes. The magnitude of (**A**+ **B**) is 'n' times the magnitude of (**A** – **B**). The angle between **A** and **B** is

(a) $\sin^{-1}\left(\frac{n^2-1}{n^2+1}\right)$ (b) $\sin^{-1}\left(\frac{n-1}{n+1}\right)$
(c) $\cos^{-1}\left(\frac{n^2-1}{n^2+1}\right)$ (d) $\cos^{-1}\left(\frac{n-1}{n+1}\right)$

23 2 kg of a monoatomic gas is at a pressure of 4×10^4 N/m^2. The density of the gas is 8 kg/m^3. What is the order of energy of the gas due to its thermal motion ?

(a) 10^6 J (b) 10^3 J (c) 10^4 J (d) 10^5 J

24 A parallel plate capacitor having capacitance 12 pF is charged by a battery to a potential difference of 10 V between its plates. The charging battery is now disconnected and a porcelain slab of dielectric constant 6.5 is slipped between the plates. The work done by the capacitor on the slab is

(a) 560 pJ (b) 508 pJ
(c) 692 pJ (d) 600 pJ

25 The diameter and height of a cylinder are measured by a meter scale to be 12.6 ± 0.1 cm and 34.2 ± 0.1 cm, respectively. What will be the value of its volume in appropriate significant figures ?

(a) 4300 ± 80 cm^3
(b) 4260 ± 80 cm^3
(c) 4264.4 ± 81.0 cm^3
(d) 4264 ± 81 cm^3

26 A particle starts from the origin at time $t = 0$ and moves along the positive X-axis. The graph of velocity with respect to time is shown in figure. What is the position of the particle at time $t = 5$ s?

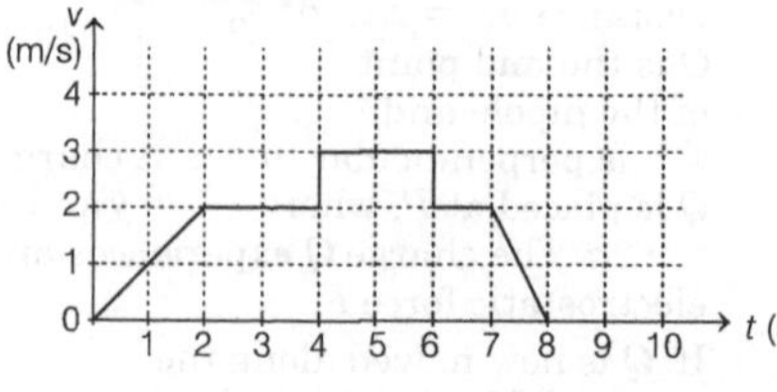

(a) 6 m (b) 3 m (c) 10 m (d) 9 m

27 A metal plate of area 1×10^{-4} m^2 is illuminated by a radiation of intensity 16 m W/m^2. The work function of the metal is 5 eV. The energy of the incident photons is 10 eV and only 10% of it produces photoelectrons. The number of emitted photoelectrons per second and their maximum energy, respectively will be (Take, 1 eV = 1.6×10^{-19} J)

(a) 10^{11} and 5 eV (b) 10^{12} and 5 eV
(c) 10^{10} and 5 eV (d) 10^{14} and 10 eV

28 The self-induced emf of a coil is 25 V. When the current in it is changed at uniform rate from 10 A to 25 A in 1s, the change in the energy of the inductance is

(a) 437.5 J (b) 740 J
(c) 637.5 J (d) 540 J

29 The electric field of a plane polarised electromagnetic wave in free space at time $t = 0$ is given by an expression.

$$\mathbf{E}(x, y) = 10\hat{\mathbf{j}} \cos[(6x + 8z)]$$

The magnetic field $\mathbf{B}(x, z, t)$ is given by (where, c is the velocity of light)

(a) $\frac{1}{c}(6\hat{\mathbf{k}} - 8\hat{\mathbf{i}}) \cos[(6x + 8z + 10ct)]$
(b) $\frac{1}{c}(6\hat{\mathbf{k}} - 8\hat{\mathbf{i}}) \cos[(6x + 8z - 10ct)]$
(c) $\frac{1}{c}(6\hat{\mathbf{k}} + 8\hat{\mathbf{i}}) \cos[(6x - 8z + 10ct)]$
(d) $\frac{1}{c}(6\hat{\mathbf{k}} + 8\hat{\mathbf{i}}) \cos[(6x + 8z - 10ct)]$

30 The Wheatstone bridge shown in figure here, gets balanced when the carbon resistor is used as R_1 has the color code (orange, red, brown). The resistors R_2 and R_4 are 80 Ω and 40 Ω, respectively.

Assuming that the color code for the carbon resistors gives their accurate values, the color code for the carbon resistor is used as R_3 would be

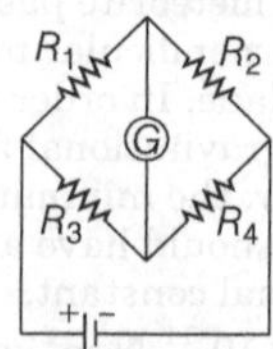

(a) brown, blue, black
(b) brown, blue, brown
(c) grey, black, brown
(d) red, green, brown

ANSWERS

1. (b)	2. (a)	3. (b)	4. (*)	5. (a)	6. (a)	7. (a)	8. (b)	9. (c)	10. (d)
11. (b)	12. (a)	13. (a)	14. (*)	15. (d)	16. (b)	17. (d)	18. (a)	19. (c)	20. (b)
21. (d)	22. (c)	23. (c)	24. (b)	25. (b)	26. (d)	27. (a)	28. (a)	29. (b)	30. (b)

For Detailed Solutions
Visit : http://tinyurl.com/y3zolw34
Or Scan :

11 January, Shift-I

1 A liquid of density ρ is coming out of a hose pipe of radius a with horizontal speed v and hits a mesh. 50% of the liquid passes through the mesh unaffected 25% losses all of its momentum and, 25% comes back with the same speed. The resultant pressure on the mesh will be

(a) ρv^2 (b) $\frac{1}{2}\rho v^2$ (c) $\frac{1}{4}\rho v^2$ (d) $\frac{3}{4}\rho v^2$

2 A particle undergoing simple harmonic motion has time dependent displacement given by $x(t) = A \sin\frac{\pi t}{90}$. The ratio of kinetic to potential energy of this particle at $t = 210$ s will be

(a) 2 (b) 1 (c) $\frac{1}{9}$ (d) 3

3 An electromagnetic wave of intensity 50 Wm^{-2} enters in a medium of refractive index 'n' without any loss. The ratio of the magnitudes of electric fields and the ratio of the magnitudes of magnetic fields of the wave before and after entering into the medium are respectively, given by

(a) $\left(\frac{1}{\sqrt{n}}, \sqrt{n}\right)$ (b) $(\sqrt{n}, \sqrt{n})$

(c) $\left(\frac{1}{\sqrt{n}}, \frac{1}{\sqrt{n}}\right)$ (d) $\left(\sqrt{n}, \frac{1}{\sqrt{n}}\right)$

4 A slab is subjected to two forces $\mathbf{F}_1$ and $\mathbf{F}_2$ of same magnitude F as shown in the figure. Force $\mathbf{F}_2$ is in xy-plane while force $\mathbf{F}_1$ acts along Z-axis at the point $(2\hat{i} + 3\hat{j})$. The moment of these forces about point O will be

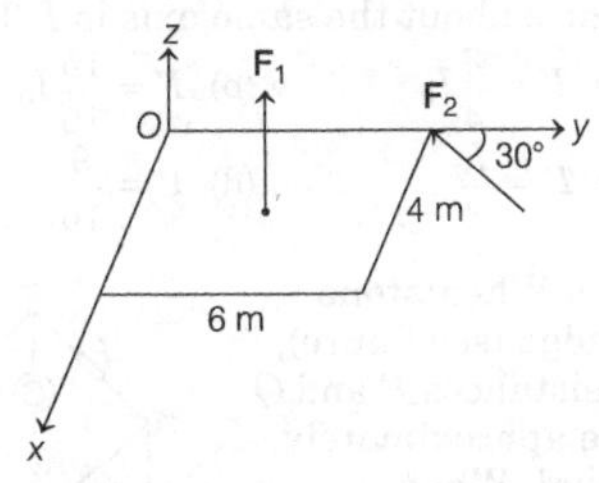

(a) $(3\hat{i} + 2\hat{j} - 3\hat{k})F$ (b) $(3\hat{i} - 2\hat{j} + 3\hat{k})F$

(c) $(3\hat{i} - 2\hat{j} - 3\hat{k})F$ (d) $(3\hat{i} + 2\hat{j} + 3\hat{k})F$

5 There are two long coaxial solenoids of same length l. The inner and outer coils have radii r_1 and r_2 and number of turns per unit length n_1 and n_2, respectively. The ratio of mutual inductance to the self-inductance of the inner coil is

(a) $\frac{n_2}{n_1}\cdot\frac{r_1}{r_2}$ (b) $\frac{n_2}{n_1}\cdot\frac{r_2^2}{r_1^2}$

(c) $\frac{n_2}{n_1}$ (d) $\frac{n_1}{n_2}$

6 In an experiment, electrons are accelerated, from rest by applying a voltage of 500 V. Calculate the radius of the path, if a magnetic field 100 mT is then applied. (Take, charge of the electron $= 1.6 \times 10^{-19}$ C and mass of the electron $= 9.1 \times 10^{-31}$ kg)

(a) 7.5×10^{-2} m (b) 7.5×10^{-4} m

(c) 7.5×10^{-3} m (d) 7.5 m

7 Three charges Q, $+q$ and $+q$ are placed at the vertices of a right angle isosceles triangle as shown below. The net electrostatic energy of the configuration is zero, if the value of Q is

(a) $-2q$ (b) $\frac{-q}{1+\sqrt{2}}$ (c) $+q$ (d) $\frac{-\sqrt{2}q}{\sqrt{2}+1}$

8 An object is at a distance of 20 m from a convex lens of focal length 0.3 m. The lens forms an image of the object. If the object moves away from the lens at a speed of 5 m/s, the speed and direction of the image will be

(a) 3.22×10^{-3} m/s towards the lens

(b) 0.92×10^{-3} m/s away from the lens

(c) 2.26×10^{-3} m/s away from the lens

(d) 1.16×10^{-3} m/s towards the lens

9 The variation of refractive index of a crown glass thin prism with wavelength of the incident light is shown. Which of the following graphs is the correct one, if D_m is the angle of minimum deviation ?

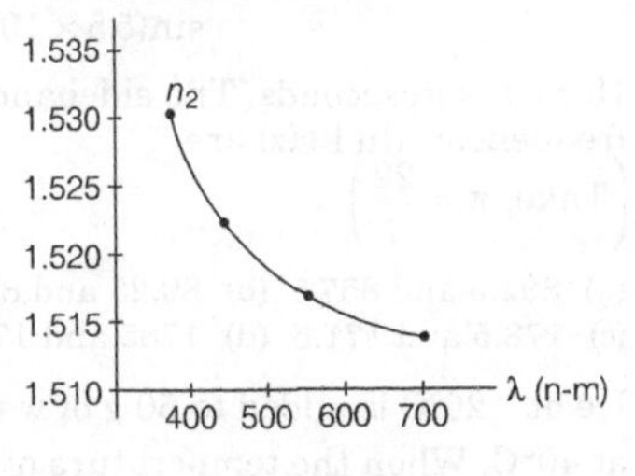

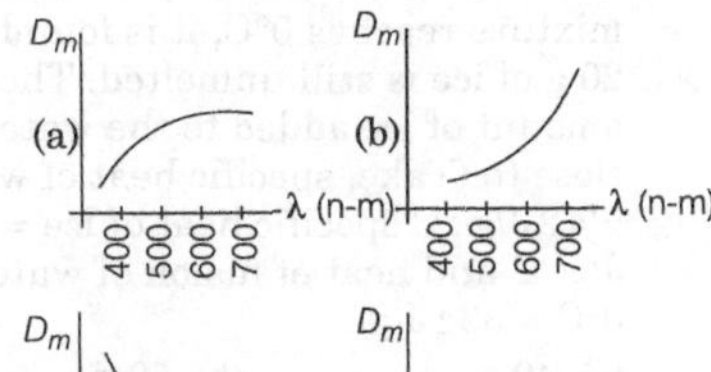

10 In the figure shown below, the charge on the left plate of the 10 μF capacitor is – 30 μC. The charge on the right plate of the 6 μF capacitor is

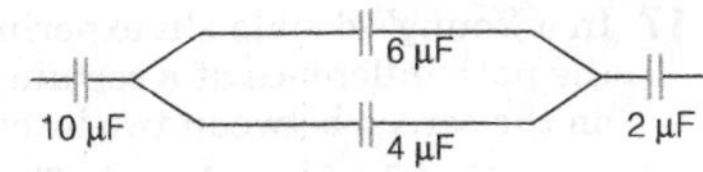

(a) + 12 μC (b) + 18 μC

(c) – 12 μC (d) – 18 μC

11 If the de-Broglie wavelength of an electron is equal to 10^{-3} times, the wavelength of a photon of frequency 6×10^{14} Hz, then the speed of electron is equal to (Take, speed of light $= 3 \times 10^8$ m/s, Planck's constant $= 6.63 \times 10^{-34}$ J-s and mass of electron $= 9.1 \times 10^{-31}$ kg)

(a) 1.45×10^6 m/s (b) 1.8×10^6 m/s

(c) 1.1×10^6 m/s (d) 1.7×10^6 m/s

12 A satellite is revolving in a circular orbit at a height h from the earth surface such that $h << R$, where R is the radius of the earth. Assuming that the effect of earth's atmosphere can be neglected the minimum increase in the speed required so that the satellite could escape from the gravitational field of earth is

(a) $\sqrt{\frac{gR}{2}}$ (b) $\sqrt{gR}$

(c) $\sqrt{2gR}$ (d) $\sqrt{gR}\,(\sqrt{2}-1)$

13 Equation of travelling wave on a stretched string of linear density 5 g/m is $y = 0.03 \sin(450t - 9x)$, where distance and time are measured in SI units. The tension in the string is

(a) 5 N (b) 12.5 N

(c) 7.5 N (d) 10 N

14 A hydrogen atom, initially in the ground state is excited by absorbing a photon of wavelength 980 Å. The radius of the atom in the excited state in terms of Bohr radius a_0 will be (Take $hc = 12500$ eV-Å)

(a) $4a_0$ (b) $9a_0$ (c) $16a_0$ (d) $25a_0$

15 A body is projected at $t = 0$ with a velocity 10 ms^{-1} at an angle of 60° with the horizontal. The radius of curvature of its trajectory at $t = 1$ s is R. Neglecting air resistance and taking acceleration due to gravity $g = 10\ ms^{-2}$, the value of R is

(a) 10.3 m (b) 2.8 m

(c) 5.1 m (d) 2.5 m

16 The force of interaction between two atoms is given by $F = \alpha\beta \exp\left(-\frac{x^2}{\alpha kT}\right)$; where x is the distance, k is the Boltzmann constant and T is temperature and α and β are two constants. The dimension of β is

(a) $[MLT^{-2}]$ (b) $[M^0L^2T^{-4}]$

(c) $[M^2LT^{-4}]$ (d) $[M^2L^2T^{-2}]$

17 In a Young's double slit experiment, the path difference at a certain point on the screen between two interfering waves is $\frac{1}{8}$th of wavelength. The ratio of the intensity at this point to that at the centre of a bright fringe is close to

(a) 0.80 (b) 0.74 (c) 0.94 (d) 0.85

18 A rigid diatomic ideal gas undergoes an adiabatic process at room temperature. The relation between temperature and volume for this process is TV^x = constant, then x is

(a) $\frac{2}{5}$ (b) $\frac{2}{3}$ (c) $\frac{5}{3}$ (d) $\frac{3}{5}$

19 In the given circuit, the current through zener diode is close to

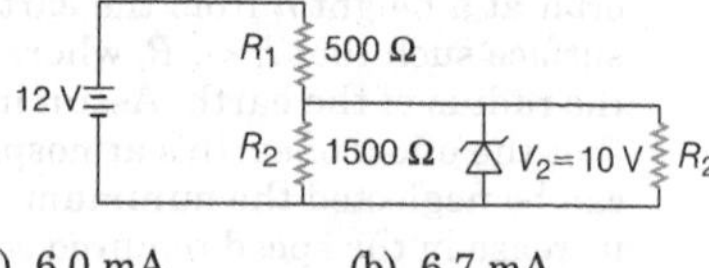

(a) 6.0 mA (b) 6.7 mA
(c) 0 (d) 4.0 mA

20 The resistance of the meter bridge AB in given figure is 4Ω. With a cell of emf ε = 0.5 V and rheostat resistance R_h = 2Ω. The null point is obtained at some point J. When the cell is replaced by another one of emf ε = ε_2, the same null point J is found for R_h = 6Ω. The emf ε_2 is

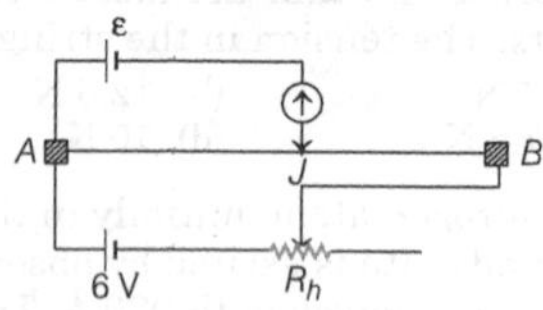

(a) 0.6 V (b) 0.3 V (c) 0.5 V (d) 0.4 V

21 Two equal resistances when connected in series to a battery consume electric power of 60 W. If these resistances are now connected in parallel combination to the same battery, the electric power consumed will be

(a) 60 W (b) 30 W (c) 240 W (d) 120 W

22 In the circuit shown,

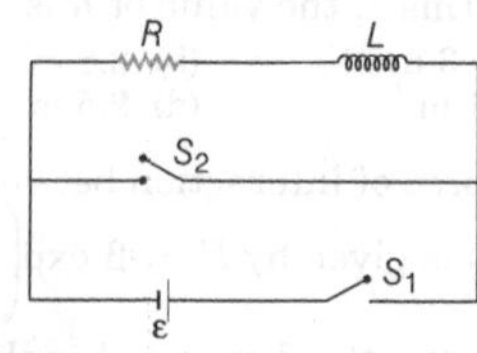

The switch S_1 is closed at time t = 0 and the switch S_2 is kept open. At some later time (t_0), the switch S_1 is opened and S_2 is closed. The behaviour of the current I as a function of time 't' is given by

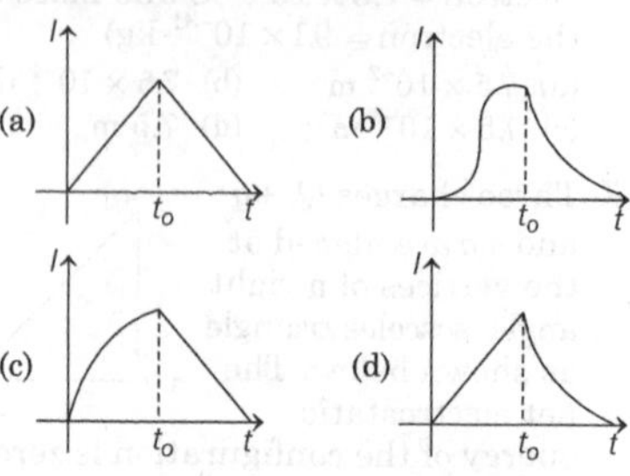

23 The given graph shows variation (with distance r from centre) of

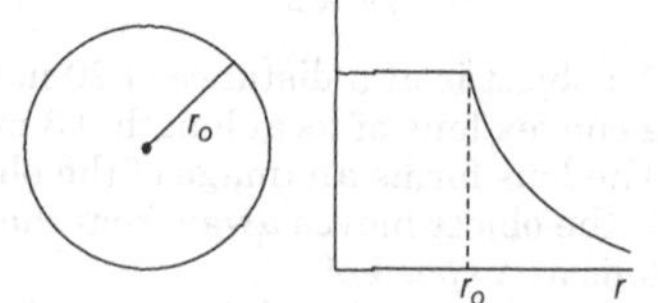

(a) electric field of a uniformly charged spherical shell
(b) potential of a uniformly charged spherical shell
(c) electric field of a uniformly charged sphere
(d) potential of a uniformly charged sphere

24 An amplitude modulates signal is given by

$$v(t) = 10[1 + 0.3\cos(2.2 \times 10^4 t)] \sin(5.5 \times 10^5 t).$$

Here, t is in seconds. The sideband frequencies (in kHz) are $\left(\text{Take, } \pi = \frac{22}{7}\right)$

(a) 892.5 and 857.5 (b) 89.25 and 85.75
(c) 178.5 and 171.5 (d) 1785 and 1715

25 Ice at −20°C is added to 50 g of water at 40°C. When the temperature of the mixture reaches 0°C, it is found that 20 g of ice is still unmelted. The amount of ice added to the water was close to (Take, specific heat of water = 4.2 J/g/°C specific heat of ice = 2.1 J/g/°C and heat of fusion of water at 0°C = 334 J/g)

(a) 40 g (b) 50 g
(c) 60 g (d) 100 g

26 A body of mass 1 kg falls freely from a height of 100 m on a platform of mass 3 kg which is mounted on a spring having spring constant $k = 1.25 \times 10^6$ N/m. The body sticks to the platform and the spring's maximum compression is found to be x. Given that $g = 10\ \text{ms}^{-2}$, the value of x will be close to

(a) 8 cm (b) 4 cm (c) 40 cm (d) 80 cm

27 A gas mixture consists of 3 moles of oxygen and 5 moles of argon at temperature T. Considering only translational and rotational modes, the total internal energy of the system is

(a) 12 RT (b) 15 RT (c) 20 RT (d) 4 RT

28 A particle is moving along a circular path with a constant speed of 10 ms^{-1}. What is the magnitude of the change in velocity of the particle, when it moves through an angle of 60° around the centre of the circle?

(a) $10\sqrt{2}$ m/s (b) 10 m/s
(c) $10\sqrt{3}$ m/s (d) Zero

29 An equilateral triangle ABC is cut from a thin solid sheet of wood. (see figure) D, E and F are the mid points of its sides as shown and G is the centre of the triangle. The moment of inertia of the triangle about an axis passing through G and perpendicular to the plane of the triangle is I_0. If the smaller triangle DEF is removed from ABC, the moment of inertia of the remaining figure about the same axis is I. Then

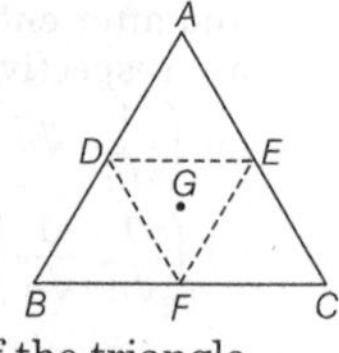

(a) $I' = \frac{3}{4}I_0$ (b) $I' = \frac{15}{16}I_0$
(c) $I' = \frac{I_0}{4}$ (d) $I' = \frac{9}{16}I_0$

30 In a Wheatstone bridge (see figure), resistances P and Q are approximately equal. When R = 400Ω, the bridge is balanced. On interchanging P and Q, the value of R for balance is 405Ω. The value of X is close to

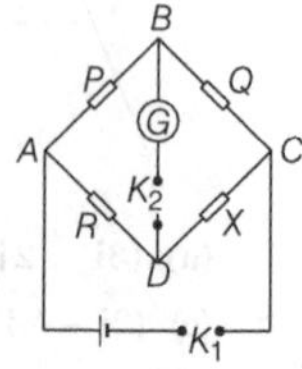

(a) 404.5 Ω (b) 401.5 Ω
(c) 402.5 Ω (d) 403.5 Ω

ANSWERS

1 (d)	2 (*)	3 (d)	4 (b)	5 (c)	6 (b)	7 (d)	8 (d)	9 (c)	10 (b)
11 (a)	12 (d)	13 (b)	14 (c)	15 (b)	16 (c)	17 (d)	18 (a)	19 (c)	20 (b)
21 (c)	22 (b)	23 (b)	24 (b)	25 (a)	26 (*)	27 (b)	28 (b)	29 (b)	30 (c)

For Detailed Solutions
Visit : http://tinyurl.com/y579lef2
Or Scan :

11 January, Shift-II

1 An amplitude modulated signal is plotted below

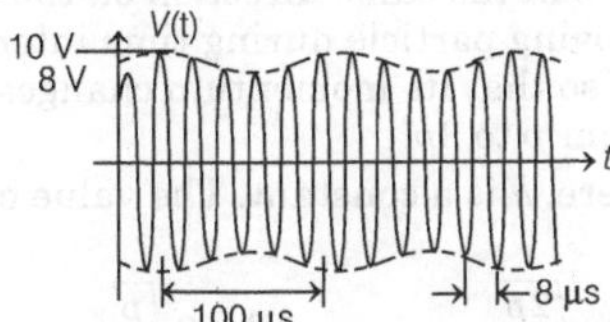

Which one of the following best describes the above signal?

(a) $[1+9\sin(2\pi\times10^4 t)]\sin(2.5\pi\times10^5 t)$V
(b) $[9+\sin(2\pi\times10^4 t)]\sin(2.5\pi\times10^5 t)$V
(c) $[9+\sin(4\pi\times10^4 t)]\sin(5\pi\times10^5 t)$V
(d) $[9+\sin(2.5\pi\times10^5 t)]\sin(2\pi\times10^4 t)$V

2 Two rods A and B of identical dimensions are at temperature 30°C. If A is heated upto 180°C and B upto T°C, then new lengths are the same. If the ratio of the coefficients of linear expansion of A and B is 4 : 3, then the value of T is

(a) 230°C (b) 270°C (c) 200°C (d) 250°C

3 A galvanometer having a resistance of 20 Ω and 30 divisions on both sides has figure of merit 0.005 ampere/division. The resistance that should be connected in series such that it can be used as a voltmeter upto 15 volt is

(a) 100 Ω (b) 125 Ω
(c) 120 Ω (d) 80 Ω

4 A circular disc D_1 of mass M and radius R has two identical discs D_2 and D_3 of the same mass M and radius R attached rigidly at its opposite ends (see figure). The moment of inertia of the system about the axis OO' passing through the centre of D_1, as shown in the figure will be

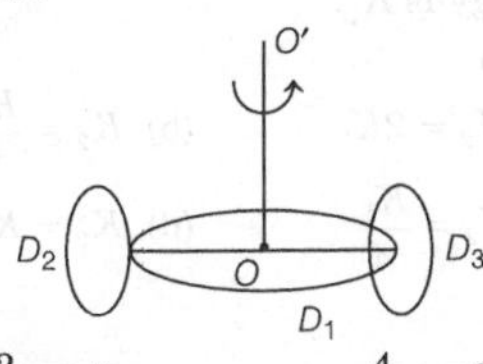

(a) $\frac{2}{3}MR^2$ (b) $\frac{4}{5}MR^2$
(c) $3MR^2$ (d) MR^2

5 The magnitude of torque on a particle of mass 1 kg is 2.5 N-m about the origin. If the force acting on it is 1 N and the distance of the particle from the origin is 5 m, then the angle between the force and the position vector is (in radian)

(a) $\frac{\pi}{8}$ (b) $\frac{\pi}{4}$ (c) $\frac{\pi}{3}$ (d) $\frac{\pi}{6}$

6. The circuit shown below contains two ideal diodes, each with a forward resistance of 50 Ω. If the battery voltage is 6 V, the current through the 100 Ω resistance (in ampere) is

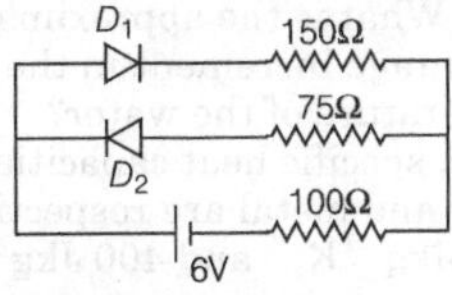

(a) 0.027 (b) 0.020 (c) 0.030 (d) 0.036

7 When 100 g of a liquid A at 100°C is added to 50 g of a liquid B at temperature 75°C, the temperature of the mixture becomes 90°C. The temperature of the mixture, if 100 g of liquid A at 100°C is added to 50 g of liquid B at 50°C will be

(a) 60°C (b) 80°C (c) 70°C (d) 85°C

8 A string is wound around a hollow cylinder of mass 5 kg and radius 0.5 m. If the string is now pulled with a horizontal force of 40 N and the cylinder is rolling without slipping on a horizontal surface (see figure), then the angular acceleration of the cylinder will be (Neglect the mass and thickness of the string)

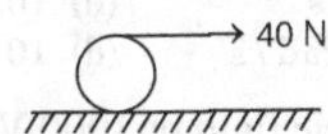

(a) 10 rad/s^2 (b) 16 rad/s^2
(c) 20 rad/s^2 (d) 12 rad/s^2

9 A particle moves from the point $(2.0\hat{i}+4.0\hat{j})$ m at $t=0$ with an initial velocity $(5.0\hat{i}+4.0\hat{j})\text{ ms}^{-1}$. It is acted upon by a constant force which produces a constant acceleration $(4.0\hat{i}+4.0\hat{j})\text{ ms}^{-2}$. What is the distance of the particle from the origin at time 2 s?

(a) 5 m (b) $20\sqrt{2}$ m
(c) $10\sqrt{2}$ m (d) 15 m

10 A monochromatic light is incident at a certain angle on an equilateral triangular prism and suffers minimum deviation. If the refractive index of the material of the prism is $\sqrt{3}$, then the angle of incidence is

(a) 45° (b) 90°
(c) 60° (d) 30°

11 A paramagnetic substance in the form of a cube with sides 1 cm has a magnetic dipole moment of 20×10^{-6} J/T when a magnetic intensity of 60×10^3 A/m is applied. Its magnetic susceptibility is

(a) 3.3×10^{-4} (b) 3.3×10^{-2}
(c) 4.3×10^{-2} (d) 2.3×10^{-2}

12 The mass and the diameter of a planet are three times the respective values for the earth. The period of oscillation of a simple pendulum on the earth is 2 s. The period of oscillation of the same pendulum on the planet would be

(a) $\frac{2}{\sqrt{3}}$ s (b) $\frac{3}{2}$ s (c) $2\sqrt{3}$ s (d) $\frac{\sqrt{3}}{2}$ s

13 A particle of mass m and charge q is in an electric and magnetic field is given by

$$\mathbf{E}=2\hat{i}+3\hat{j},\ \mathbf{B}=4\hat{j}+6\hat{k}.$$

The charged particle is shifted from the origin to the point $P(x=1;\ y=1)$ along a straight path. The magnitude of the total work done is

(a) $(0.35)\,q$ (b) $(0.15)\,q$
(c) $(2.5)\,q$ (d) $5\,q$

14 In a double-slit experiment, green light (5303 Å) falls on a double slit having a separation of 19.44 μ-m and a width of 4.05 μ-m. The number of bright fringes between the first and the second diffraction minima is

(a) 5 (b) 10 (c) 9 (d) 4

15 If speed (V), acceleration (A) and force (F) are considered as fundamental units, the dimension of Young's modulus will be

(a) $[V^{-4}A^{-2}F]$ (b) $[V^{-2}A^2F^2]$
(c) $[V^{-2}A^2F^{-2}]$ (d) $[V^{-4}A^2F]$

16 In a hydrogen like atom, when an electron jumps from the M-shell to the L-shell, the wavelength of emitted radiation is λ. If an electron jumps from N-shell to the L-shell, the wavelength of emitted radiation will be

(a) $\frac{27}{20}\lambda$ (b) $\frac{25}{16}\lambda$ (c) $\frac{20}{27}\lambda$ (d) $\frac{16}{25}\lambda$

17 In the experimental set up of meter bridge shown in the figure, the null point is obtained at a distance of 40 cm from A. If a 10 Ω resistor is connected in series with R_1, the null point shifts by 10 cm. The resistance that should be connected in parallel with (R_1+10) Ω such that the null point shifts back to its initial position is

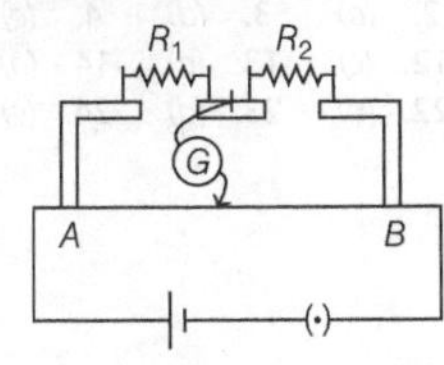

(a) 60 Ω (b) 20 Ω
(c) 30 Ω (d) 40 Ω

18 The region between $y = 0$ and $y = d$ contains a magnetic field $\mathbf{B} = B\hat{\mathbf{k}}$. A particle of mass m and charge q enters the region with a velocity $\mathbf{v} = v\hat{\mathbf{i}}$. If $d = \dfrac{mv}{2qB}$ then the acceleration of the charged particle at the point of its emergence at the other side is

(a) $\dfrac{qvB}{m}\left(\dfrac{\sqrt{3}}{2}\hat{\mathbf{i}} + \dfrac{1}{2}\hat{\mathbf{j}}\right)$

(b) $\dfrac{qvB}{m}\left(\dfrac{1}{2}\hat{\mathbf{i}} - \dfrac{\sqrt{3}}{2}\hat{\mathbf{j}}\right)$

(c) $\dfrac{qvB}{m}\left(\dfrac{-\hat{\mathbf{j}} + \hat{\mathbf{i}}}{\sqrt{2}}\right)$ (d) $\dfrac{qvB}{m}\left(\dfrac{\hat{\mathbf{i}} + \hat{\mathbf{j}}}{\sqrt{2}}\right)$

19 In a process, temperature and volume of one mole of an ideal monoatomic gas are varied according to the relation $VT = k$, where k is a constant. In this process, the temperature of the gas is increased by ΔT. The amount of heat absorbed by gas is (where, R is gas constant)

(a) $\dfrac{1}{2}kR\Delta T$ (b) $\dfrac{2k}{3}\Delta T$

(c) $\dfrac{1}{2}R\Delta T$ (d) $\dfrac{3}{2}R\Delta T$

20 A 27 mW laser beam has a cross-sectional area of 10 mm^2. The magnitude of the maximum electric field in this electromagnetic wave is given by [Take, permittivity of space, $\varepsilon_0 = 9\times10^{-12}$ SI units and speed of light, $c = 3\times10^8$ m/s]

(a) 1 kV/m (b) 0.7 kV/m

(c) 2 kV/m (d) 1.4 kV/m

21 A copper wire is wound on a wooden frame, whose shape is that of an equilateral triangle. If the linear dimension of each side of the frame is increased by a factor of 3, keeping the number of turns of the coil per unit length of the frame the same, then the self-inductance of the coil

(a) increases by a factor of 3

(b) decreases by a factor of $9\sqrt{3}$

(c) increases by a factor of 27

(d) decreases by a factor of 9

22 A metal ball of mass 0.1 kg is heated upto 500°C and dropped into a vessel of heat capacity 800 JK^{-1} and containing 0.5 kg water. The initial temperature of water and vessel is 30°C. What is the approximate percentage increment in the temperature of the water? [Take, specific heat capacities of water and metal are respectively 4200 Jkg^{-1}K^{-1} and 400 Jkg^{-1}K^{-1}]

(a) 25% (b) 15% (c) 30% (d) 20%

23 A thermometer graduated according to a linear scale reads a value x_0, when in contact with boiling water and $x_0/3$, when in contact with ice. What is the temperature of an object in °C, if this thermometer in the contact with the object reads $x_0/2$?

(a) 35 (b) 60 (c) 40 (d) 25

24 A simple pendulum of length 1 m is oscillating with an angular frequency 10 rad/s. The support of the pendulum starts oscillating up and down with a small angular frequency of 1 rad/s and an amplitude of 10^{-2} m. The relative change in the angular frequency of the pendulum is best given by

(a) 1 rad/s (b) 10^{-5} rad/s

(c) 10^{-3} rad/s (d) 10^{-1} rad/s

25 An electric field of 1000 V/m is applied to an electric dipole at angle of 45°. The value of electric dipole moment is 10^{-29} C-m. What is the potential energy of the electric dipole?

(a) -9×10^{-20} J (b) -10×10^{-29} J

(c) -20×10^{-18} J (d) -7×10^{-27} J

26 In the circuit shown, the potential difference between A and B is

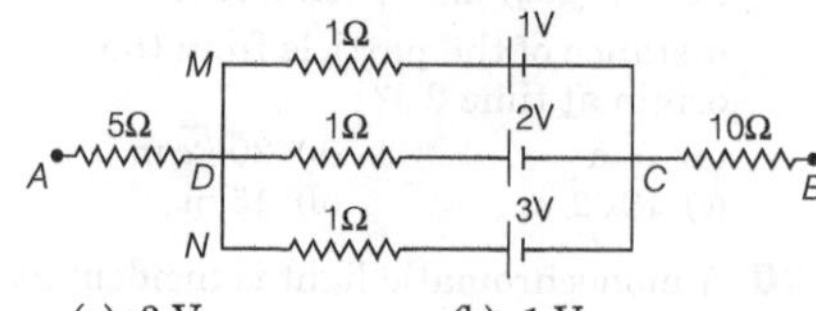

(a) 3 V (b) 1 V

(c) 6 V (d) 2 V

27 A particle of mass m is moving in a straight line with momentum p. Starting at time $t = 0$, a force $F = kt$ acts in the same direction on the moving particle during time interval T, so that its momentum changes from p to $3p$.

Here, k is a constant. The value of T is

(a) $\sqrt{\dfrac{2p}{k}}$ (b) $2\sqrt{\dfrac{p}{k}}$

(c) $\sqrt{\dfrac{2k}{p}}$ (d) $2\sqrt{\dfrac{k}{p}}$

28 Seven capacitors, each of capacitance 2 μF are to be connected in a configuration to obtain an effective capacitance of $\left(\dfrac{6}{13}\right)$ μF. Which of the combinations shown in figures below will achieve the desired value?

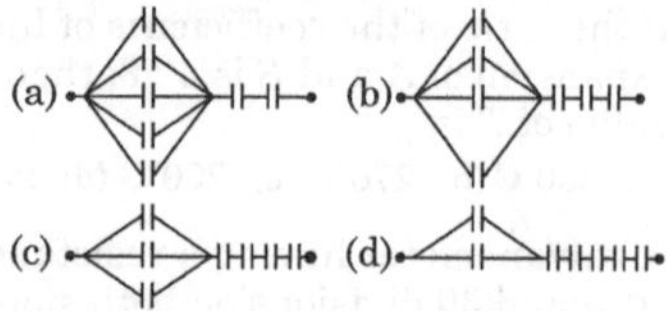

29 In a photoelectric experiment, the wavelength of the light incident on a metal is changed from 300 n-m to 400 n-m. The decrease in the stopping potential is close to $\left(\dfrac{hc}{e} = 1240 \text{ n-mV}\right)$

(a) 0.5 V (b) 2.0 V

(c) 1.5 V (d) 1.0 V

30 A pendulum is executing simple harmonic motion and its maximum kinetic energy is K_1. If the length of the pendulum is doubled and it performs simple harmonic motion with the same amplitude as in the first case, its maximum kinetic energy is K_2.

Then

(a) $K_2 = 2K_1$ (b) $K_2 = \dfrac{K_1}{2}$

(c) $K_2 = \dfrac{K_1}{4}$ (d) $K_2 = K_1$

ANSWERS

1. (b)	2. (a)	3. (d)	4. (c)	5. (d)	6. (b)	7. (b)	8. (b)	9. (b)	10. (c)
11. (a)	12. (c)	13. (d)	14. (a)	15. (d)	16. (c)	17. (a)	18. (*)	19. (c)	20. (d)
21. (c)	22. (d)	23. (d)	24. (c)	25. (d)	26. (d)	27. (b)	28. (c)	29. (d)	30. (b)

For Detailed Solutions

Visit : http://tinyurl.com/y56dxvwz

Or Scan :

12 January, Shift-I

1 For the given cyclic process CAB as shown for a gas, the work done is

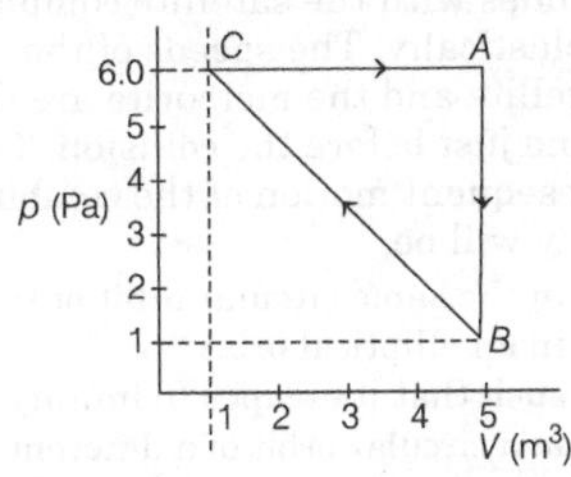

(a) 5 J (b) 10 J (c) 1 J (d) 30 J

2 As shown in the figure, two infinitely long, identical wires are bent by 90° and placed in such a way that the segments LP and QM are along the X-axis, while segments PS and QN are parallel to the Y-axis. If $OP = OQ = 4$ cm and the magnitude of the magnetic field at O is 10^{-4} T and the two wires carry equal currents (see figure), the magnitude of the current in each wire and the direction of the magnetic field at O will be (Take, $\mu_0 = 4\pi \times 10^{-7}$ NA^{-2})

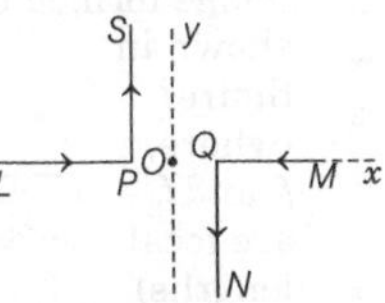

(a) 40 A, perpendicular out of the page
(b) 20 A, perpendicular into the page
(c) 20 A, perpendicular out of the page
(d) 40 A, perpendicular into the page

3 A person standing on an open ground hears the sound of a jet aeroplane, coming from north at an angle 60° with ground level. But he finds the aeroplane right vertically above his position. If v is the speed of sound, then speed of the plane is

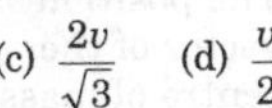

(a) $\frac{\sqrt{3}}{2}v$ (b) v (c) $\frac{2v}{\sqrt{3}}$ (d) $\frac{v}{2}$

4 The galvanometer deflection, when key K_1 is closed but K_2 is open equals θ_0 (see figure). On closing K_2 also and adjusting R_2 to 5Ω, the deflection in galvanometer becomes $\frac{\theta_0}{5}$. The resistance of the galvanometer is given by (neglect the internal resistance of battery) :

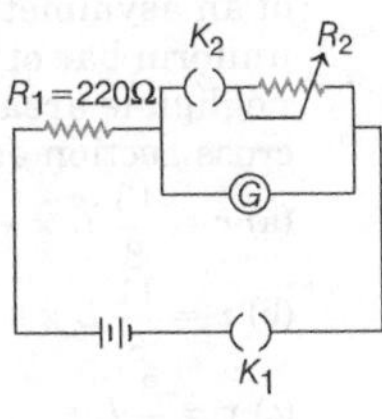

(a) 22Ω (b) 5Ω (c) 25Ω (d) 12Ω

5 A cylinder of radius R is surrounded by a cylindrical shell of inner radius R and outer radius $2R$. The thermal conductivity of the material of the inner cylinder is K_1 and that of the outer cylinder is K_2. Assuming no loss of heat, the effective thermal conductivity of the system for heat flowing along the length of the cylinder is

(a) $\frac{K_1 + K_2}{2}$ (b) $\frac{K_1 + 3K_2}{4}$
(c) $\frac{2K_1 + 3K_2}{5}$ (d) $K_1 + K_2$

6 Two light identical springs of spring constant k are attached horizontally at the two ends of an uniform horizontal rod AB of length l and mass m. The rod is pivoted at its centre 'O' and can rotate freely in horizontal plane. The other ends of the two springs are fixed to rigid supports as shown in figure. The rod is gently pushed through a small angle and released. The frequency of resulting oscillation is

(a) $\frac{1}{2\pi}\sqrt{\frac{2k}{m}}$ (b) $\frac{1}{2\pi}\sqrt{\frac{3k}{m}}$
(c) $\frac{1}{2\pi}\sqrt{\frac{6k}{m}}$ (d) $\frac{1}{2\pi}\sqrt{\frac{k}{m}}$

7 Two electric bulbs rated at 25 W, 220 V and 100 W, 220 V are connected in series across a 220 V voltage source. If the 25 W and 100 W bulbs draw powers P_1 and P_2 respectively, then

(a) $P_1 = 16$ W, $P_2 = 4$W
(b) $P_1 = 4$ W, $P_2 = 16$W
(c) $P_1 = 9$ W, $P_2 = 16$W
(d) $P_1 = 16$ W, $P_2 = 9$W

8 The output of the given logic circuit is

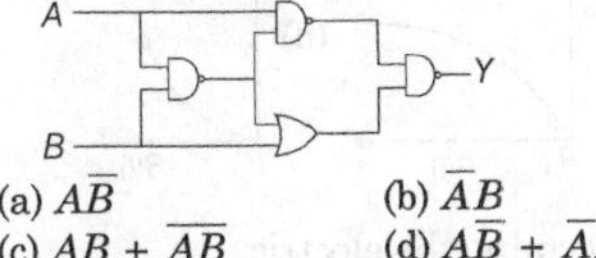

(a) $A\overline{B}$ (b) $\overline{A}B$
(c) $AB + \overline{AB}$ (d) $A\overline{B} + \overline{A}B$

9 The least count of the main scale of a screw gauge is 1 mm. The minimum number of divisions on its circular scale required to measure 5 μm diameter of a wire is

(a) 50 (b) 200
(c) 500 (d) 100

10 A point source of light, S is placed at a distance L in front of the centre of plane mirror of width d which is hanging vertically on a wall. A man walks in front of the mirror along a line parallel to the mirror, at a distance $2L$ as shown below. The distance over which the man can see the image of the light source in the mirror is

(a) $\frac{d}{2}$ (b) d (c) $3d$ (d) $2d$

11 A passenger train of length 60 m travels at a speed of 80 km/hr. Another freight train of length 120 m travels at a speed of 30 km/hr. The ratio of times taken by the passenger train to completely cross the freight train when : (i) they are moving in the same direction and (ii) in the opposite direction is

(a) $\frac{3}{2}$ (b) $\frac{25}{11}$ (c) $\frac{11}{5}$ (d) $\frac{5}{2}$

12 An ideal gas occupies a volume of 2 m^3 at a pressure of 3×10^6 Pa. The energy of the gas is

(a) 6×10^4 J (b) 10^8J
(c) 9×10^6 J (d) 3×10^2 J

13 A simple pendulum is made of a string of length l and a bob of mass m, is released from a small angle θ_0. It strikes a block of mass M, kept on a horizontal surface at its lowest point of oscillations, elastically. It bounces back and goes up to an angle θ_1. Then, M is given by

(a) $m\left(\frac{\theta_0 + \theta_1}{\theta_0 - \theta_1}\right)$ (b) $\frac{m}{2}\left(\frac{\theta_0 - \theta_1}{\theta_0 + \theta_1}\right)$
(c) $m\left(\frac{\theta_0 - \theta_1}{\theta_0 + \theta_1}\right)$ (d) $\frac{m}{2}\left(\frac{\theta_0 + \theta_1}{\theta_0 - \theta_1}\right)$

14 A particle of mass m moves in a circular orbit in a central potential field $U(r) = \frac{1}{2}kr^2$. If Bohr's quantization conditions are applied, radii of possible orbitals and energy levels vary with quantum number n as

(a) $r_n \propto n, E_n \propto n$ (b) $r_n \propto n^2, E_n \propto \frac{1}{n^2}$
(c) $r_n \propto \sqrt{n}, E_n \propto n$ (d) $r_n \propto \sqrt{n}, E_n \propto \frac{1}{n}$

15 A proton and an α-particle (with their masses in the ratio of 1 : 4 and charges in the ratio of 1 : 2) are accelerated from rest through a potential difference V. If a uniform magnetic field B is set up perpendicular to their velocities, the ratio of the radii $r_p : r_\alpha$ of the circular paths described by them will be

(a) $1 : \sqrt{2}$ (b) $1 : \sqrt{3}$ (c) 1 : 3 (d) 1 : 2

16 In the figure shown, a circuit contains two identical resistors with resistance $R = 5\Omega$ and an inductance with $L = 2$ mH. An ideal battery of 15 V is connected in the circuit. What will be the current through the battery long after the switch is closed?

(a) 6 A (b) 3 A
(c) 5.5 A (d) 7.5 A

17 A light wave is incident normally on a glass slab of refractive index 1.5. If 4% of light gets reflected and the amplitude of the electric field of the incident light is 30 V/m, then the amplitude of the electric field for the wave propogating in the glass medium will be

(a) 30 V/m (b) 6 V/m (c) 10 V/m (d) 24 V/m

18 A straight rod of length L extends from $x = a$ to $x = L + a$. The gravitational force it exerts on a point mass m at $x = 0$, if the mass per unit length of the rod is $A + Bx^2$, is given by

(a) $Gm\left[A\left(\frac{1}{a+L} - \frac{1}{a}\right) - BL\right]$

(b) $Gm\left[A\left(\frac{1}{a+L} - \frac{1}{a}\right) + BL\right]$

(c) $Gm\left[A\left(\frac{1}{a} - \frac{1}{a+L}\right) + BL\right]$

(d) $Gm\left[A\left(\frac{1}{a} - \frac{1}{a+L}\right) - BL\right]$

19 A 100 V carrier wave is made to vary between 160 V and 40 V by a modulating signal. What is the modulation index?

(a) 0.4 (b) 0.5 (c) 0.6 (d) 0.3

20 In the figure shown, after the switch 'S' is turned from position 'A' to position 'B', the energy dissipated in the circuit in terms of capacitance 'C' and total charge 'Q' is

(a) $\frac{3}{4}\cdot\frac{Q^2}{C}$ (b) $\frac{5}{8}\cdot\frac{Q^2}{C}$ (c) $\frac{1}{8}\cdot\frac{Q^2}{C}$ (d) $\frac{3}{8}\cdot\frac{Q^2}{C}$

21 An ideal battery of 4 V and resistance R are connected in series in the primary circuit of a potentiometer of length 1 m and resistance 5 Ω. The value of R to give a potential difference of 5 mV across 10 cm of potentiometer wire is

(a) 395 Ω (b) 495 Ω (c) 490 Ω (d) 480 Ω

22 A particle A of mass 'm' and charge 'q' is accelerated by a potential difference of 50 V. Another particle B of mass '$4m$' and charge 'q' is accelerated by a potential difference of 2500 V. The ratio of de-Broglie wavelengths $\frac{\lambda_A}{\lambda_B}$ is close to

(a) 4.47 (b) 10.00
(c) 0.07 (d) 14.14

23 Let the moment of inertia of a hollow cylinder of length 30 cm (inner radius 10 cm and outer radius 20 cm) about its axis be I. The radius of a thin cylinder of the same mass such that its moment of inertia about its axis is also I, is

(a) 16 cm (b) 14 cm (c) 12 cm (d) 18 cm

24 A travelling harmonic wave is represented by the equation $y(x, t) = 10^{-3}\sin(50t + 2x)$, where x and y are in metre and t is in second. Which of the following is a correct statement about the wave?

(a) The wave is propagating along the negative X-axis with speed 25 ms^{-1}.
(b) The wave is propagating along the positive X-axis with speed 25 ms^{-1}.
(c) The wave is propagating along the positive X-axis with speed 100 ms^{-1}.
(d) The wave is propagating along the negative X-axis with speed 100 ms^{-1}.

25 There is uniform spherically symmetric surface charge density at a distance R_0 from the origin. The charge distribution is initially at rest and starts expanding because of mutual repulsion. The figure that represents best the speed $v[R(t)]$ of the distribution as a function of its instantaneous radius $R(t)$ is

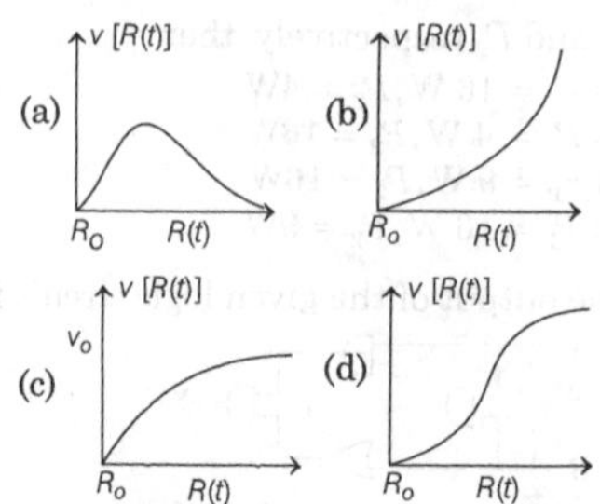

26 Determine the electric dipole moment of the system of three charges, placed on the vertices of an equilateral triangle as shown in the figure.

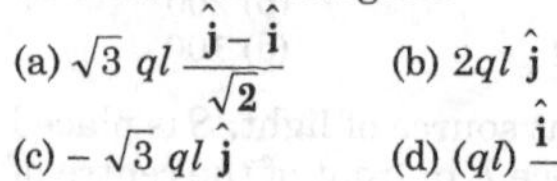

(a) $\sqrt{3}\,ql\,\frac{\hat{j}-\hat{i}}{\sqrt{2}}$ (b) $2ql\,\hat{j}$

(c) $-\sqrt{3}\,ql\,\hat{j}$ (d) $(ql)\,\frac{\hat{i}+\hat{j}}{\sqrt{2}}$

27 A satellite of mass M is in a circular orbit of radius R about the centre of the earth. A meteorite of the same mass falling towards the earth collides with the satellite completely inelastically. The speeds of the satellite and the meteorite are the same just before the collision. The subsequent motion of the combined body will be

(a) in the same circular orbit of radius R
(b) in an elliptical orbit
(c) such that it escapes to infinity
(d) in a circular orbit of a different radius

28 What is the position and nature of image formed by lens combination shown in figure? (where, f_1 and f_2 are focal lengths)

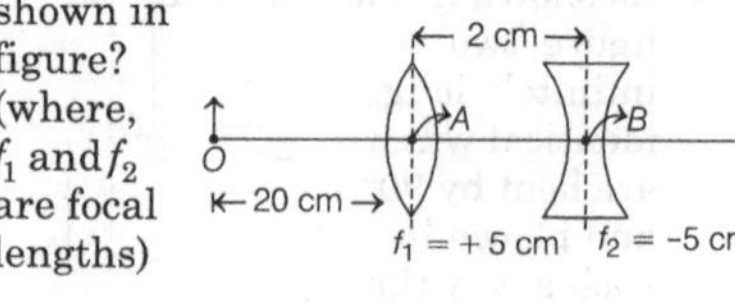

(a) $\frac{20}{3}$ cm from point B at right, real
(b) 70 cm from point B at right, real
(c) 40 cm from point B at right, real
(d) 70 cm from point B at left, virtual

29 In a meter bridge, the wire of length 1m has a non-uniform cross-section such that the variation $\frac{dR}{dl}$ of its resistance R with length l is $\frac{dR}{dl} \propto \frac{1}{\sqrt{l}}$. Two equal resistance are connected as shown in the figure. The galvanometer has zero deflection when the jockey is at point P. What is the length AP?

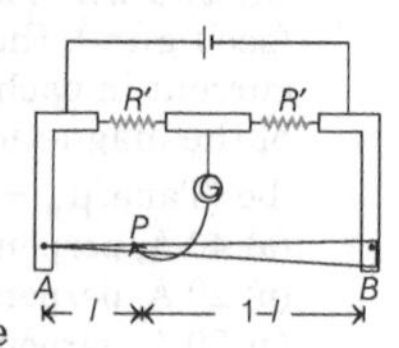

(a) 0.3 m (b) 0.25 m (c) 0.2 m (d) 0.35 m

30 The position vector of the centre of mass $\mathbf{r}_{cm}$ of an asymmetric uniform bar of negligible area of cross-section as shown in figure is

(a) $\mathbf{r} = \frac{13}{8}L\,\hat{\mathbf{x}} + \frac{5}{8}L\,\hat{\mathbf{y}}$

(b) $\mathbf{r} = \frac{11}{8}L\,\hat{\mathbf{x}} + \frac{3}{8}L\,\hat{\mathbf{y}}$

(c) $\mathbf{r} = \frac{3}{8}L\,\hat{\mathbf{x}} + \frac{11}{8}L\,\hat{\mathbf{y}}$

(d) $\mathbf{r} = \frac{5}{8}L\,\hat{\mathbf{x}} + \frac{13}{8}L\,\hat{\mathbf{y}}$

ANSWERS

1. (b)	2. (b)	3. (d)	4. (a)	5. (b)	6. (c)	7. (a)	8. (a)	9. (b)	10. (c)
11. (c)	12. (c)	13. (a)	14. (c)	15. (a)	16. (a)	17. (d)	18. (c)	19. (c)	20. (d)
21. (a)	22. (d)	23. (a)	24. (a)	25. (c)	26. (c)	27. (b)	28. (b)	29. (b)	30. (a)

For Detailed Solutions
Visit : http://tinyurl.com/y5uq4233
Or Scan :

12 January, Shift-II

1 To double the covering range of a TV transmission tower, its height should be multiplied by

(a) $\sqrt{2}$ (b) 4 (c) 2 (d) $\frac{1}{\sqrt{2}}$

2 The charge on a capacitor plate in a circuit as a function of time is shown in the figure.

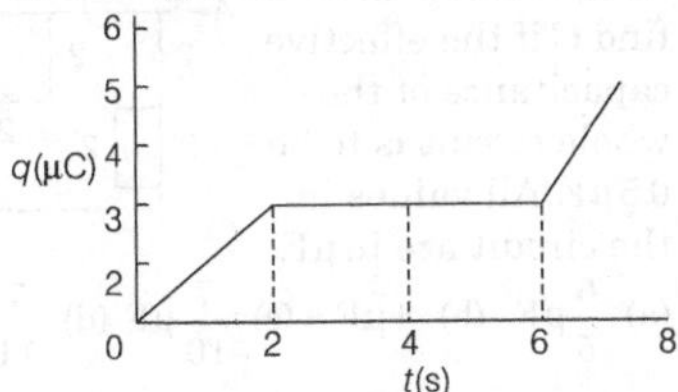

What is the value of current at $t = 4$ s?

(a) 2 μA (b) 1.5 μA (c) Zero (d) 3 μA

3 Two satellites A and B have masses m and $2m$ respectively. A is in a circular orbit of radius R and B is in a circular orbit of radius $2R$ around the earth. The ratio of their kinetic energies, T_A / T_B is

(a) $\frac{1}{2}$ (b) 2 (c) $\sqrt{\frac{1}{2}}$ (d) 1

4 A simple harmonic motion is represented by $y = 5(\sin 3\pi t + \sqrt{3}\cos 3\pi t)$ cm. The amplitude and time period of the motion are

(a) 10 cm, $\frac{3}{2}$ s (b) 5 cm, $\frac{2}{3}$ s

(c) 5 cm, $\frac{3}{2}$ s (d) 10 cm, $\frac{2}{3}$ s

5 In the given circuit diagram, the currents $I_1 = -0.3$ A, $I_4 = 0.8$ A and $I_5 = 0.4$ A, are flowing as shown. The currents I_2, I_3 and I_6 respectively, are

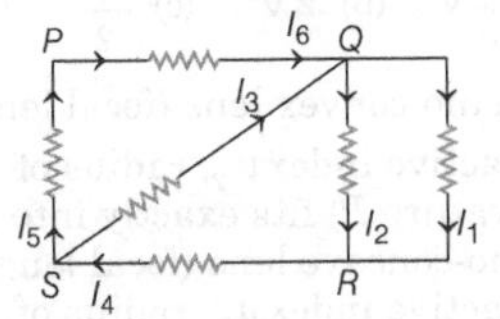

(a) 1.1 A, 0.4 A, 0.4 A
(b) 1.1 A, − 0.4 A, 0.4 A
(c) 0.4 A, 1.1 A, 0.4 A
(d) − 0.4 A, 0.4 A, 1.1 A

6 A load of mass M kg is suspended from a steel wire of length 2 m and radius 1.0 mm in Searle's apparatus experiment. The increase in length produced in the wire is 4.0 mm. Now, the load is fully immersed in a liquid of relative density 2. The relative density of the material of load is 8. The new value of increase in length of the steel wire is

(a) zero (b) 5.0 mm
(c) 4.0 mm (d) 3.0 mm

7 The moment of inertia of a solid sphere, about an axis parallel to its diameter and at a distance of x from it, is '$I(x)$'. Which one of the graphs represents the variation of $I(x)$ with x correctly?

(a) 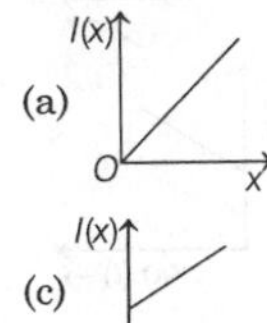 (b) 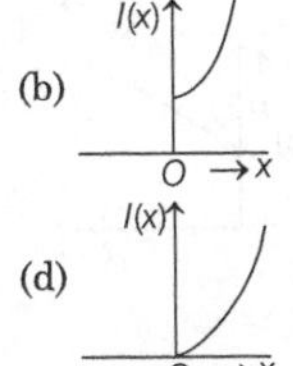

(c) (d)

8 An ideal gas is enclosed in a cylinder at pressure of 2 atm and temperature, 300 K. The mean time between two successive collisions is 6×10^{-8} s. If the pressure is doubled and temperature is increased to 500 K, the mean time between two successive collisions will be close to

(a) 4×10^{-8} s (b) 3×10^{-6} s
(c) 2×10^{-7} s (d) 0.5×10^{-8} s

9 A vertical closed cylinder is separated into two parts by a frictionless piston of mass m and of negligible thickness. The piston is free to move along the length of the cylinder. The length of the cylinder above the piston is l_1 and that below the piston is l_2, such that $l_1 > l_2$. Each part of the cylinder contains n moles of an ideal gas at equal temperature T. If the piston is stationary, its mass m, will be given by (where, R is universal gas constant and g is the acceleration due to gravity)

(a) $\frac{nRT}{g}\left[\frac{l_1 - l_2}{l_1 l_2}\right]$ (b) $\frac{nRT}{g}\left[\frac{1}{l_2} + \frac{1}{l_1}\right]$

(c) $\frac{RT}{g}\left[\frac{2l_1 + l_2}{l_1 l_2}\right]$ (d) $\frac{RT}{ng}\left[\frac{l_1 - 3l_2}{l_1 l_2}\right]$

10 A parallel plate capacitor with plates of area 1 m² each, are at a separation of 0.1 m. If the electric field between the plates is 100 N/C, the magnitude of charge on each plate is

$\left(\text{Take, } \varepsilon_0 = 8.85 \times 10^{-12} \frac{\text{C}^2}{\text{N} - \text{m}^2}\right)$

(a) 9.85×10^{-10} C (b) 8.85×10^{-10} C
(c) 7.85×10^{-10} C (d) 6.85×10^{-10} C

11 A galvanometer whose resistance is 50 Ω, has 25 divisions in it. When a current of 4×10^{-4} A passes through it, its needle (pointer) deflects by one division. To use this galvanometer as a voltmeter of range 2.5 V, it should be connected to a resistance of

(a) 250 Ω (b) 6200 Ω (c) 200 Ω (d) 6250 Ω

12 Formation of real image using a biconvex lens is shown below. If the whole set up is immersed in water without disturbing the object and the screen positions, what will one observe on the screen?

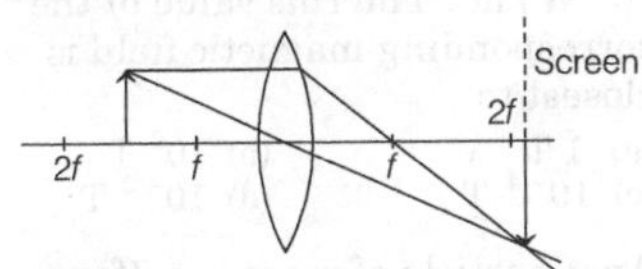

(a) No change
(b) Magnified image
(c) Image disappears
(d) Erect real image

13 Let l, r, c, and v represent inductance, resistance, capacitance and voltage, respectively. The dimension of $\frac{l}{rcv}$ in SI units will be

(a) [LT²] (b) [LTA]
(c) [A⁻¹] (d) [LA⁻²]

14 A particle of mass 20 g is released with an initial velocity 5 m/s along the curve from the point A, as shown in the figure. The point A is at height h from point B. The particle slides along the frictionless surface. When the particle reaches point B, its angular momentum about O will be (Take, $g = 10 \text{ m/s}^2$)

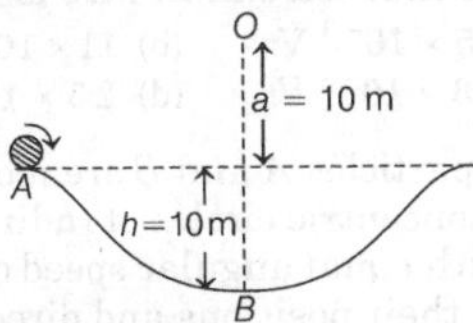

(a) 8 kg - m²/s (b) 3 kg - m²/s
(c) 2 kg - m²/s (d) 6 kg - m²/s

15 In a Frank-Hertz experiment, an electron of energy 5.6 eV passes through mercury vapour and emerges with an energy 0.7 eV. The minimum wavelength of photons emitted by mercury atoms is close to

(a) 250 nm (b) 2020 nm
(c) 1700 nm (d) 220 nm

16 A block kept on a rough inclined plane, as shown in the figure, remains at rest upto a maximum force 2 N down the inclined plane. The maximum external force up the inclined plane that does not move the block is 10 N. The coefficient of static friction between the block and the plane is (Take, $g = 10 \text{ m/s}^2$)

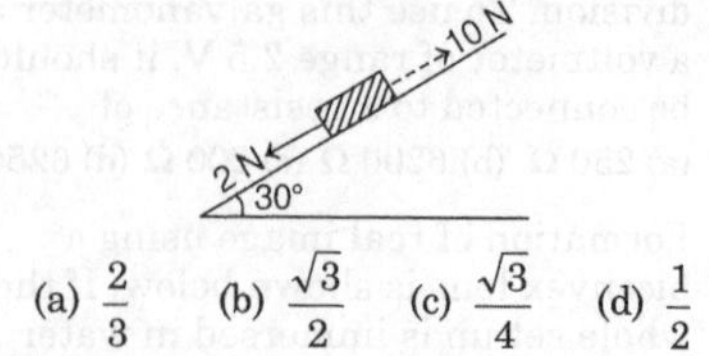

(a) $\frac{2}{3}$ (b) $\frac{\sqrt{3}}{2}$ (c) $\frac{\sqrt{3}}{4}$ (d) $\frac{1}{2}$

17 The mean intensity of radiation on the surface of the sun is about 10^8 W / m^2. The rms value of the corresponding magnetic field is closest to

(a) 1 T (b) 10^2 T
(c) 10^{-4} T (d) 10^{-2} T

18 An α-particle of mass m suffers one-dimensional elastic collision with a nucleus at rest of unknown mass. It is scattered directly backwards losing 64% of its initial kinetic energy. The mass of the nucleus is

(a) 1.5 m (b) 4 m (c) 3.5 m (d) 2 m

19 In a radioactive decay chain, the initial nucleus is $^{232}_{90}$Th. At the end, there are 6 α-particles and 4 β-particles which are emitted. If the end nucleus is $^{A}_{Z}X$, A and Z are given by

(a) $A = 202$; $Z = 80$ (b) $A = 208$; $Z = 82$
(c) $A = 200$; $Z = 81$ (d) $A = 208$; $Z = 80$

20 A 10 m long horizontal wire extends from North-East to South-West. It is falling with a speed of 5.0 ms^{-1} at right angles to the horizontal component of the earth's magnetic field of 0.3×10^{-4} Wb / m^2. The value of the induced emf in wire is

(a) 1.5×10^{-3} V (b) 1.1×10^{-3} V
(c) 0.3×10^{-3} V (d) 2.5×10^{-3} V

21 Two particles A and B are moving on two concentric circles of radii R_1 and R_2 with equal angular speed ω. At $t = 0$, their positions and direction of motion are shown in the figure

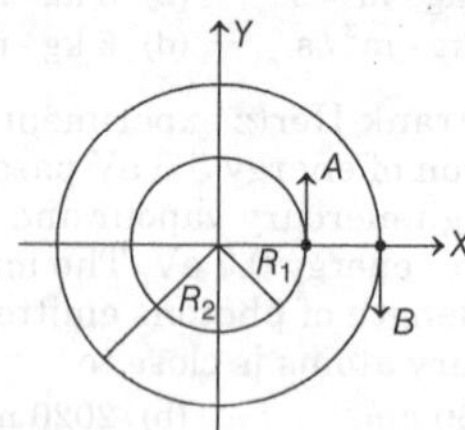

The relative velocity $\mathbf{v}_A - \mathbf{v}_B$ at $t = \frac{\pi}{2\omega}$ is given by

(a) $\omega(R_1 + R_2)\hat{i}$ (b) $-\omega(R_1 + R_2)\hat{i}$
(c) $\omega(R_1 - R_2)\hat{i}$ (d) $\omega(R_2 - R_1)\hat{i}$

22 A soap bubble, blown by a mechanical pump at the mouth of a tube, increases in volume, with time, at a constant rate. The graph that correctly depicts the time dependence of pressure inside the bubble is given by

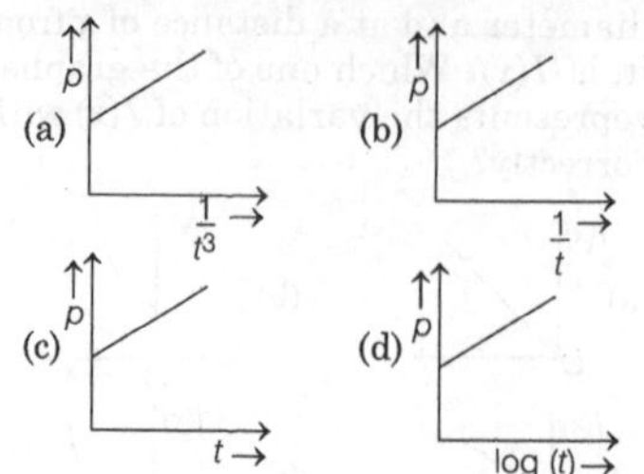

23

In the above circuit, $C = \frac{\sqrt{3}}{2}$ μF, $R_2 = 20\ \Omega$, $L = \frac{\sqrt{3}}{10}$ H and $R_1 = 10\ \Omega$. Current in L - R_1 path is I_1 and in C - R_2 path is I_2. The voltage of AC source is given by $V = 200\sqrt{2}\sin(100t)$ volts. The phase difference between I_1 and I_2 is

(a) 30° (b) 60° (c) 0° (d) 90°

24 A long cylindrical vessel is half-filled with a liquid. When the vessel is rotated about its own vertical axis, the liquid rises up near the wall. If the radius of vessel is 5 cm and its rotational speed is 2 rotations per second, then the difference in the heights between the centre and the sides (in cm) will be

(a) 0.1 (b) 1.2 (c) 0.4 (d) 2.0

25 A paramagnetic material has 10^{28} atoms/m^3. Its magnetic susceptibility at temperature 350 K is 2.8×10^{-4}. Its susceptibility at 300 K is

(a) 3.726×10^{-4} (b) 3.672×10^{-4}
(c) 2.672×10^{-4} (d) 3.267×10^{-4}

26 In the figure, given that V_{BB} supply can vary from 0 to 5.0 V, $V_{CC} = 5$ V, $\beta_{DC} = 200$, $R_B = 100$ k Ω, $R_C = 1$ k Ω and $V_{BE} = 1.0$ V. The minimum base current and the input voltage at which the transistor will go to saturation, will be, respectively

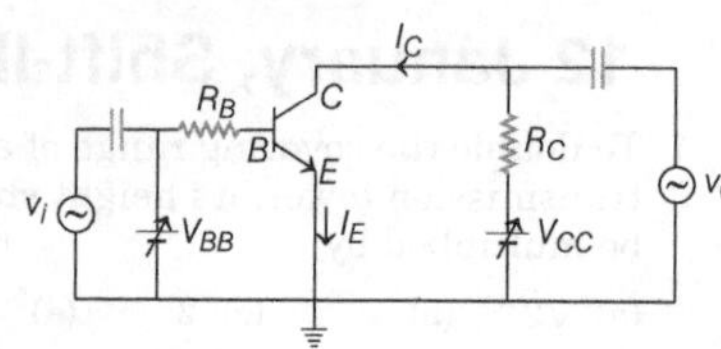

(a) 25 μA and 2.8 V (b) 25 μA and 3.5 V
(c) 20 μA and 3.5 V (d) 20 μA and 2.8 V

27 In the circuit shown, find C if the effective capacitance of the whole circuit is to be 0.5 μF. All values in the circuit are in μF.

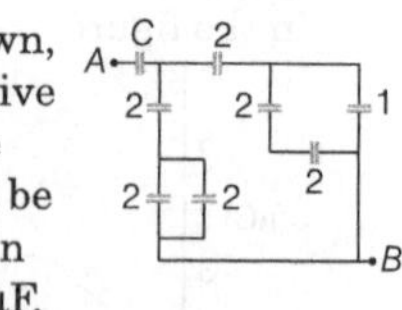

(a) $\frac{6}{5}$ μF (b) 4 μF (c) $\frac{7}{10}$ μF (d) $\frac{7}{11}$ μF

28 A resonance tube is old and has jagged end. It is still used in the laboratory to determine velocity of sound in air. A tuning fork of frequency 512 Hz produces first resonance when the tube is filled with water to a mark 11 cm below a reference mark. near the open end of the tube. The experiment is repeated with another fork of frequency 256 Hz which produces first resonance when water reaches a mark 27 cm below the reference mark. The velocity of sound in air, obtained in the experiment is close to

(a) 328 ms^{-1} (b) 341 ms^{-1}
(c) 322 ms^{-1} (d) 335 ms^{-1}

29 When a certain photosensitive surface is illuminated with monochromatic light of frequency v, the stopping potential for the photocurrent is $-V_0/2$. When the surface is illuminated by monochromatic light of frequency $v/2$, the stopping potential is $-V_0$. The threshold frequency for photoelectric emission is

(a) $\frac{4}{3}v$ (b) $2v$ (c) $\frac{3v}{2}$ (d) $\frac{5v}{3}$

30 A plano-convex lens (focal length f_2, refractive index μ_2, radius of curvature R) fits exactly into a plano-concave lens (focal length f_1, refractive index μ_1, radius of curvature R). Their plane surfaces are parallel to each other. Then, the focal length of the combination will be

(a) $f_1 - f_2$ (b) $\frac{R}{\mu_2 - \mu_1}$
(c) $f_1 + f_2$ (d) $\frac{2f_1 f_2}{f_1 + f_2}$

ANSWERS

1. (b)	2. (c)	3. (d)	4. (d)	5. (a)	6. (d)	7. (b)	8. (a)	9. (a)	10. (b)
11. (c)	12. (c)	13. (c)	14. (d)	15. (a)	16. (b)	17. (c)	18. (b)	19. (b)	20. (a)
21. (d)	22. (b)	23. (a)	24. (d)	25. (d)	26. (b)	27. (d)	28. (a)	29. (c)	30. (b)

For Detailed Solutions
Visit : http://tinyurl.com/y22wuljv
Or Scan :

ONLINE QUESTION PAPERS

JEE Main 2020 (September & January Attempt)

SEPTEMBER ATTEMPT

2 September, Shift-I

1 A gas mixture consists of 3 moles of oxygen and 5 moles of argon at temperature T. Assuming the gases to be ideal and the oxygen bond to be rigid, the total internal energy (in units of RT) of the mixture is

(a) 15 (b) 13
(c) 11 (d) 20

2

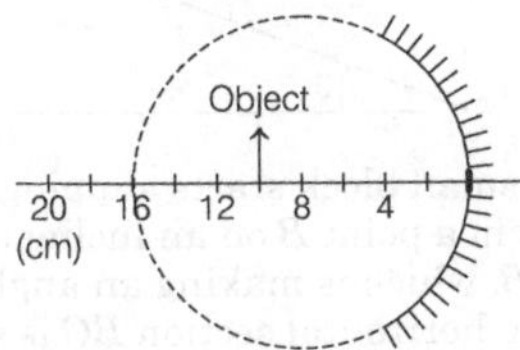

A spherical mirror is obtained as shown in the above figure from a hollow glass sphere. If an object is in front of the mirror, what will be the nature and magnification of the image of the object ? (Figure drawn as schematic and not to scale)

(a) Erect, virtual and unmagnified
(b) Inverted, real and magnified
(c) Erect, virtual and magnified
(d) Inverted, real and unmagnified

3 In a reactor, 2kg of $_{92}U^{235}$ fuel is fully used up in 30 days. The energy released per fission is 200 MeV. Given that, the Avogadro number, $N = 6.023 \times 10^{26}\ K^{-1}\ mol^{-1}$ and $1\ eV = 1.6 \times 10^{-19}$ J. The power output of the reactor is close to

(a) 35 MW (b) 125 MW
(c) 60 MW (d) 54 MW

4 Interference fringes are observed on a screen by illuminating two thin slits 1 mm apart with a light source ($\lambda = 632.8$ nm). The distance between the screen and the slits is 100 cm. If a bright fringe is observed on a screen at a distance of 1.27 mm from the central bright fringe, then the path difference between the waves, which are reaching this point from the slits is close to

(a) 2 nm (b) 1.27 μm
(c) 2.87 nm (d) 2.05 μm

5

0 25 50 75 100
A B
2 m

Shown in the above figure, a rigid and uniform 1 m long rod AB held in horizontal position by two strings tied to its ends and attached to the ceiling. The rod is of mass m and has another weight of mass 2 m hung at a distance of 75 cm from A. The tension in the string at A is

(a) 1 mg (b) 2 mg
(c) 0.75 mg (d) 0.5 mg

6 A uniform cylinder of mass M and radius R is to be pulled over a step of height a ($a < R$) by applying a force F at its centre O perpendicular to the plane through the axes of the cylinder on the edge of the step (see figure). The minimum value of F required is

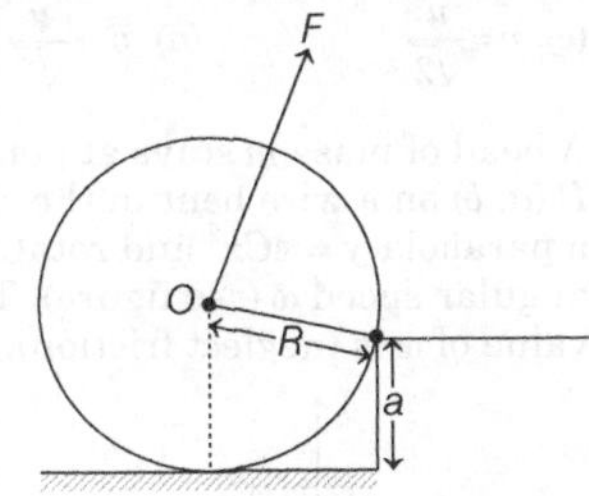

(a) $Mg\dfrac{a}{R}$ (b) $Mg\sqrt{1-\dfrac{a^2}{R^2}}$

(c) $Mg\sqrt{\left(\dfrac{R}{R-a}\right)^2 - 1}$

(d) $Mg\sqrt{1-\left(\dfrac{R-a}{R}\right)^2}$

7 A plane electromagnetic wave, has frequency of 2.0×10^{10} Hz and its energy density is $1.02 \times 10^{-8}\ J/m^3$ in vacuum. The amplitude of the magnetic field of the wave is close to (Take, $\dfrac{1}{4\pi\varepsilon_0} = 9 \times 10^9 \dfrac{Nm^2}{C^2}$ and speed of light $= 3 \times 10^8\ ms^{-1}$)

(a) 190 nT (b) 160 nT
(c) 180 nT (d) 150 nT

8 Magnetic materials used for making permanent magnets (P) and magnets in a transformer (T) have different properties, of the following, which property best matches for the type of magnet required?

(a) T: Large retentivity, small coercivity
(b) P : Large retentivity, large coercivity
(c) P : Small retentivity, large coercivity
(d) T : Large retentivity, large coercivity

9 Trains A and B are running on parallel tracks in the opposite directions with speeds of 36 km/h and 72 km/h, respectively. A person is walking in train A in the opposite direction to its motion with a speed of 1.8 km/h. Speed (in ms^{-1}) of this person as observed from train B will be close to

(Take, the distance between the tracks as negligible)

(a) 28.5 (b) 30.5
(c) 29.5 (d) 31.5

10 The least count of the main scale of a vernier callipers is 1 mm. Its vernier scale is divided into 10th division and coincide with 9th division of the main scale. When jaws are touching each other, the 7th division of vernier scale coincides with a division of main scale and the zero of vernier scale is lying right side of the zero of main scale. When this vernier is used to measure length of a cylinder the zero of the vernier scale lies between 3.1 cm and 3.2 cm and 4th VSD coincides with a main scale division. The length of the cylinder is (VSD is vernier scale division)

(a) 3.2 cm (b) 2.99 cm
(c) 3.07 cm (d) 3.21 cm

11 A charged particle (mass m and charge q) moves along X-axis with velocity v_0. When it passes through the origin it enters a region having uniform electric field $\mathbf{E} = -E\hat{\mathbf{j}}$ which extends upto $x = d$. Equation of path of electron in the region ($x > d$) is

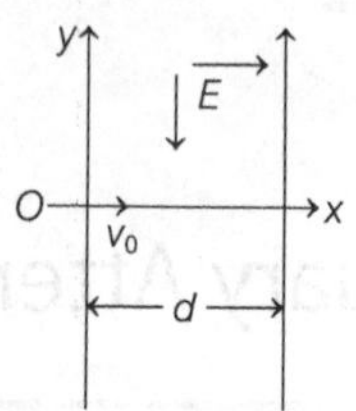

(a) $y = \frac{qEd}{mv_0^2}x$ (b) $y = \frac{qEd}{mv_0^2}(x-d)$

(c) $y = \frac{qEd}{mv_0^2}\left(\frac{d}{2}-x\right)$ (d) $y = \frac{qEd^2}{mv_0^2}x$

12 An amplitude modulated wave is represented by the expression $V_m = 5(1+0.6\cos 6280t)\sin(211\times10^4 t)$ volt. The minimum and maximum amplitudes of the amplitude modulated wave are, respectively

(a) 3 V, 5 V (b) $\frac{5}{2}$ V, 8 V

(c) $\frac{3}{2}$ V, 5 V (d) 5 V, 8 V

13 A beam of protons with speed 4×10^5 ms^{-1} enters a uniform magnetic field of 0.3 T at an angle of 60° to the magnetic field. The pitch of the resulting helical path of protons is close to

(Take, mass of the proton $= 1.67\times10^{-27}$ kg and charge of the proton $= 1.69\times10^{-19}$ C)

(a) 2 cm (b) 4 cm

(c) 5 cm (d) 12 cm

14 The mass density of a spherical galaxy varies as $\frac{K}{r}$ over a large distance r from its centre. In that region, a small star is in a circular orbit of radius R. Then, the period of revolution T depends on R as

(a) $T \propto R$ (b) $T^2 \propto R^3$

(c) $T^2 \propto R$ (d) $T^2 \propto \frac{1}{R^3}$

15 A cylindrical vessel containing a liquid is rotated about its axis, so that the liquid rises at its sides as shown in the figure. The radius of vessel is 5 cm and the angular speed of rotation is ω red s^{-1}. The difference in the height h (in cm) of liquid at the centre of vessel and at the side will be

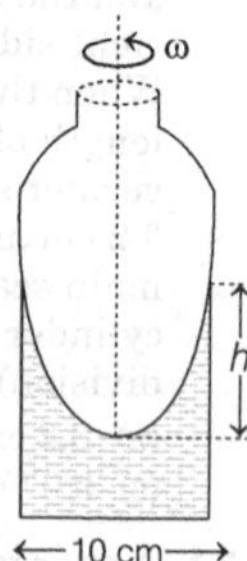

(a) $\frac{25\omega^2}{2g}$ (b) $\frac{5\omega^2}{2g}$

(c) $\frac{2\omega^2}{25g}$ (d) $\frac{2\omega^2}{5g}$

16 Two identical strings X and Z made of same material have tensions T_X and T_Z in them. If their fundamental frequencies are 450 Hz and 300 Hz respectively, then the ratio of T_X/T_Z is

(a) 1.25 (b) 2.25

(c) 0.44 (d) 1.5

17 Consider four conducting materials copper, tungsten, mercury and aluminium with resistivity ρ_C, ρ_T, ρ_M and ρ_A, respectively. Then,

(a) $\rho_A > \rho_T > \rho_C$ (b) $\rho_A > \rho_M > \rho_C$

(c) $\rho_C > \rho_A > \rho_T$ (d) $\rho_M > \rho_A > \rho_C$

18 If speed V, area A and force F are chosen as fundamental units, then the dimensional formula of Young's modulus will be

(a) $[FA^2V^{-3}]$ (b) $[FA^{-1}V^0]$

(c) $[FA^2V^{-2}]$ (d) $[FA^2V^{-1}]$

19 A particle of mass m with an initial velocity $u\hat{i}$ collides perfectly elastically with a mass $3m$ at rest. It moves with a velocity $v\hat{j}$ after collision, then v is given by

(a) $v = \sqrt{\frac{2}{3}}u$ (b) $v = \frac{1}{\sqrt{6}}u$

(c) $v = \frac{u}{\sqrt{2}}$ (d) $v = \frac{u}{\sqrt{3}}$

20 A bead of mass m stays at point $P(a, b)$ on a wire bent in the shape of a parabola $y = 4Cx^2$ and rotating with angular speed ω (see figure). The value of ω is (neglect friction)

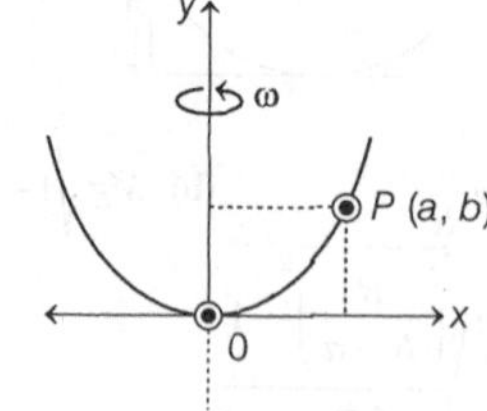

(a) $2\sqrt{2gC}$ (b) $\sqrt{\frac{2g}{C}}$

(c) $2\sqrt{gC}$ (d) $\sqrt{\frac{2gC}{ab}}$

Numerical Type Questions

21 An engine takes in 5 moles of air at 20° C and 1 atm and compresses it adiabatically to 1/10th of the original volume. Assuming air to be a diatomic ideal gas made up of rigid molecules, the change in its internal energy during this process comes out to be XkJ. The value of X to the nearest integer is ....

22 When radiation of wavelength λ is used to illuminate a metallic surface, the stopping potential is V. When the same surface is illuminated with radiation of wavelength 3λ, the stopping potential is $\frac{V}{4}$. If the threshold wavelength for the metallic surface is $n\lambda$, then value of n will be ....

23

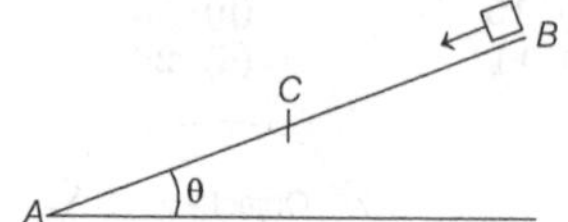

A small block starts slipping down from a point B on an inclined plane AB, which is making an angle θ with the horizontal section BC is smooth and the remaining section CA is rough with a coefficient of friction μ. It is found that the block comes to rest as it reaches the bottom (point A) of the inclined plane. If $BC = 2AC$, the coefficient of friction is given by $\mu = k\tan\theta$. Then, the value of k is........

24 A circular coil of radius 10 cm is placed in a uniform magnetic field of 3.0×10^{-5} T with its plane perpendicular to the field initially. It is rotated at constant angular speed about an axis along the diameter of coil and perpendicular to magnetic field, so that it undergoes half of rotation in 0.2s. The maximum value of emf induced (in μV) in the coil will be close to the integer ....

25 A 5 μF capacitor is charged fully by a 220V supply. It is then disconnected from the supply and is connected in series to another uncharged 2.5 μF capacitor. If the energy change during the charge redistribution is $\frac{X}{100}$ J, then value of X to the nearest integer is ....

ANSWERS

1. (a)	2. (d)	3. (c)	4. (b)	5. (a)	6. (d)	7. (b)	8. (b)	9. (c)	10. (c)
11. (c)	12. (b)	13. (b)	14. (c)	15. (a)	16. (b)	17. (d)	18. (b)	19. (c)	20. (a)
21. (4)	22. (9)	23. (3)	24. (15)	25. (4)					

For Detailed Solutions
Visit : https://bit.ly/3dE3VOs
Or Scan :

2 September, Shift-II

1 A capillary tube made of glass of radius 0.15 mm is dipped vertically in a beaker filled with methylene iodide, which rises to height h in the tube. It is observed that the two tangents drawn from liquid- glass interfaces (from opposite sides of the capillary) make an angle of 60° with one another. Then, h is close to (Given, surface tension = $0.05\,\text{Nm}^{-1}$, density = $667\,\text{kg m}^{-3}$ and $g = 10\,\text{ms}^{-2}$)

(a) 0.049 m (b) 0.087 m
(c) 0.137 m (d) 0.172 m

2 An inductance coil has a reactance of 100 Ω. When an AC signal of frequency 1000 Hz is applied to the coil, the applied voltage leads the current by 45°. The self-inductance of the coil is

(a) 1.1×10^{-2} H (b) 1.1×10^{-1} H
(c) 5.5×10^{-5} H (d) 6.7×10^{-7} H

3 The height h at which the weight of a body will be the same as that at the same depth h from the surface of the earth is (Radius of the earth is R and effect of the rotation of the earth is neglected)

(a) $\frac{\sqrt{5}}{2}(R - R)$ (b) $\frac{R}{2}$
(c) $\frac{\sqrt{5}R - R}{2}$ (d) $\frac{\sqrt{3}R - R}{2}$

4 The figure shows a region of length l with a uniform magnetic field of 0.3 T in it and a proton entering the region with velocity $4 \times 10^5\ \text{ms}^{-1}$ making an angle 60° with the field. If the proton completes 10 revolutions by the time it cross the region shown, l is close to (Take, mass of proton = 1.67×10^{-27} kg, charge of the proton = 1.6×10^{-19} C)

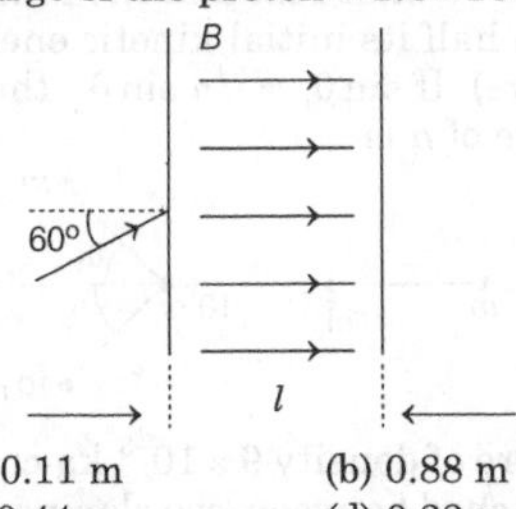

(a) 0.11 m (b) 0.88 m
(c) 0.44 m (d) 0.22 m

5 If momentum P, area A and time T are taken to be the fundamental quantities, then the dimensional formula for energy is

(a) $[P^2AT^{-2}]$ (b) $[PA^{-1}T^{-2}]$
(c) $[PA^{1/2}T^{-1}]$ (d) $[P^{1/2}AT^{-1}]$

6 When the temperature of a metal wire is increased from 0°C to 10°C, its length increases by 0.02%. The percentage change in its mass density will be closest to

(a) 0.06 (b) 2.3
(c) 0.008 (d) 0.8

7 A 10 μF capacitor is fully charged to a potential difference of 50 V. After removing the source voltage it is connected to an uncharged capacitor in parallel. Now, the potential difference across them becomes 20 V. The capacitance of the second capacitor is

(a) 15 μF (b) 30 μF
(c) 20 μF (d) 10 μF

8 In a hydrogen atom, electron makes a transition from $(n + 1)$th level to the nth level. If $n >> 1$, the frequency of radiation emitted is proportional to

(a) $\frac{1}{n}$ (b) $\frac{1}{n^3}$
(c) $\frac{1}{n^2}$ (d) $\frac{1}{n^4}$

9 A heat engine is involved with exchange of heat of 1915 J, – 40 J, + 125 J and – QJ, during one cycle achieving an efficiency of 50.0%. The value of Q is

(a) 640 J (b) 40 J
(c) 980 J (d) 400 J

10 A wire carrying current I is bent in the shape $ABCDEFA$ as shown, where rectangle $ABCDA$ and $ADEFA$ are perpendicular to each other. If the sides of the rectangles are of lengths a and b, then the magnitude and direction of magnetic moment of the loop $ABCDEFA$ is

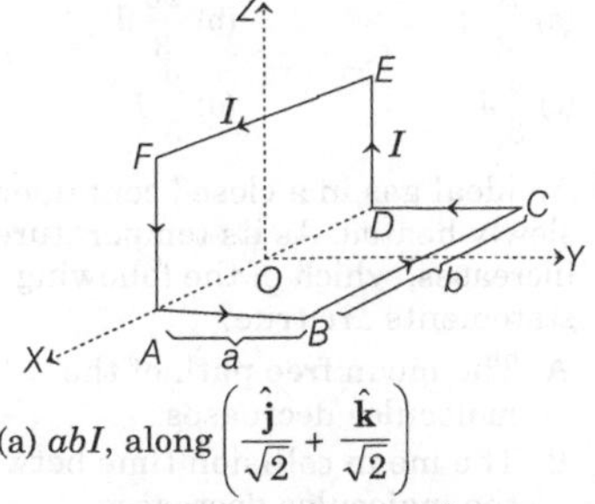

(a) abI, along $\left(\frac{\hat{j}}{\sqrt{2}} + \frac{\hat{k}}{\sqrt{2}}\right)$

(b) $\sqrt{2}\,abI$, along $\left(\frac{\hat{j}}{\sqrt{2}} + \frac{\hat{k}}{\sqrt{2}}\right)$

(c) $\sqrt{2}\,abI$, along $\left(\frac{\hat{j}}{\sqrt{5}} + \frac{2\hat{k}}{\sqrt{5}}\right)$

(d) abI, along $\left(\frac{\hat{j}}{\sqrt{5}} + \frac{2\hat{k}}{\sqrt{5}}\right)$

11 A particle is moving 5 times as fast as an electron. The ratio of the de-Broglie wavelength of the particle to that of the electron is 1.878×10^{-4}. The mass of the particle is close to

(a) 4.8×10^{-27} kg
(b) 9.1×10^{-31} kg
(c) 1.2×10^{-28} kg
(d) 9.7×10^{-28} kg

12 In a plane electromagnetic wave, the directions of electric field and magnetic field are represented by $\hat{k}$ and $2\hat{i} - 2\hat{j}$, respectively. What is the unit vector along direction of propagation of the wave?

(a) $\frac{1}{\sqrt{2}}(\hat{i} + \hat{j})$

(b) $\frac{1}{\sqrt{2}}(\hat{j} + \hat{k})$

(c) $\frac{1}{\sqrt{5}}(\hat{i} + 2\hat{j})$

(d) $\frac{1}{\sqrt{5}}(2\hat{i} + \hat{j})$

13 In the following digital circuit, what will be the output at Z, when the input (A, B) are (1, 0), (0, 0), (1, 1), (0, 1)?

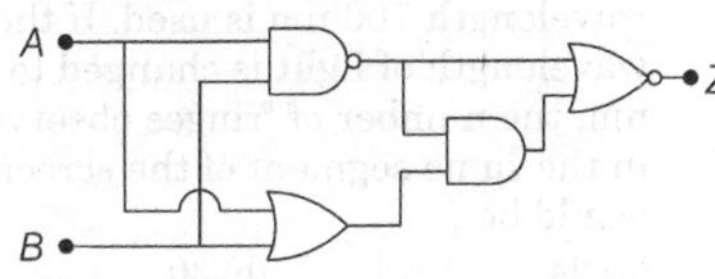

(a) 0, 0, 1, 0
(b) 1, 0, 1, 1
(c) 1, 1, 0, 1
(d) 0, 1, 0, 0

14 A small point mass carrying some positive charge on it, is released from the edge of a table. There is an uniform electric field in this region in the horizontal direction. Which of the following options then correctly describe the trajectory of the mass? (Curves are drawn schematically and are not to scale)

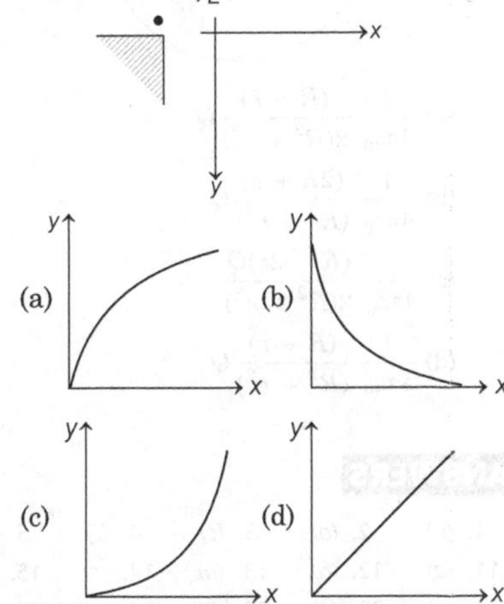

15 A potentiometer wire PQ of 1 m length is connected to a standard cell E_1. Another cell E_2 of emf 1.02 V is connected with a resistance r and switch S (as shown in figure). With switch S open, the null position is obtained at a distance of 49 cm from Q. The potential gradient in the potentiometer wire is

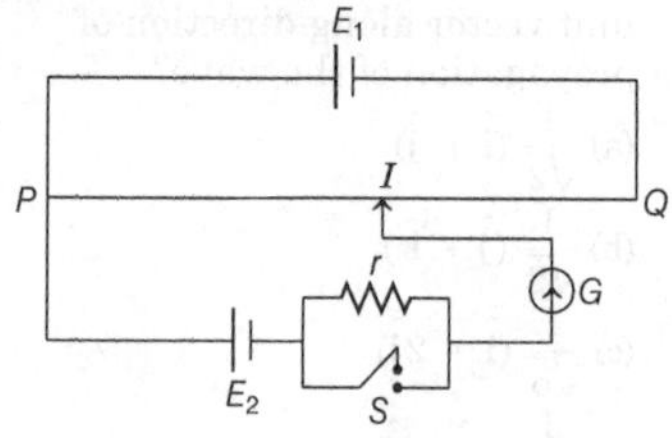

(a) 0.02 V/cm (b) 0.01 V/cm
(c) 0.03 V/cm (d) 0.04 V/cm

16 In a Young's double slit experiment, 16 fringes are observed in a certain segment of the screen when light of wavelength 700 nm is used. If the wavelength of light is changed to 400 nm, the number of fringes observed in the same segment of the screen would be

(a) 24 (b) 30
(c) 18 (d) 28

17 A charge Q is distributed over two concentric conducting thin spherical shells radii r and R $(R > r)$. If the surface charge densities on the two shells are equal, the electric potential at the common centre is

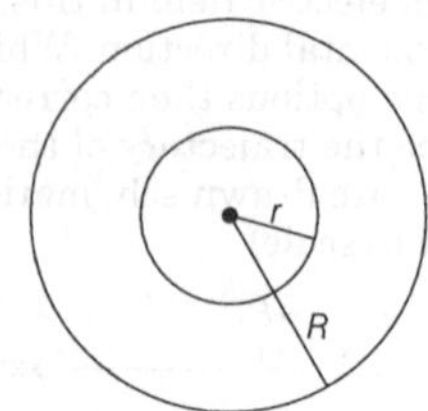

(a) $\frac{1}{4\pi\varepsilon_0}\frac{(R+r)}{2(R^2+r^2)}Q$

(b) $\frac{1}{4\pi\varepsilon_0}\frac{(2R+r)}{(R^2+r^2)}Q$

(c) $\frac{1}{4\pi\varepsilon_0}\frac{(R+2r)Q}{2(R^2+r^2)}$

(d) $\frac{1}{4\pi\varepsilon_0}\frac{(R+r)}{(R^2+r^2)}Q$

18 The displacement-time graph of a particle executing SHM is given in figure (sketch is schematic and not to scale)

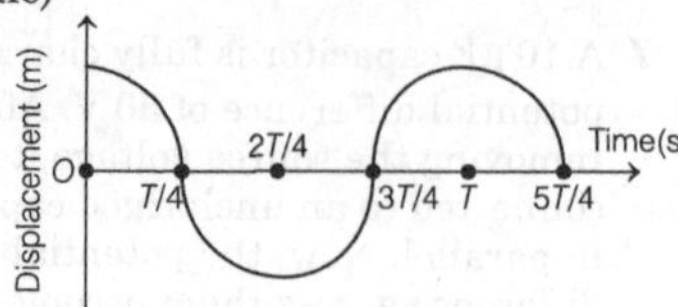

Which of the following statement(s) is/are true for this motion?

A. The force is zero at $t = \frac{3T}{4}$.
B. The acceleration is maximum at $t = T$.
C. The speed is maximum at $t = \frac{T}{4}$.
D. The potential energy is equal to kinetic energy of the oscillation at $t = \frac{T}{2}$.

(a) A, B and D (b) B, C and D
(c) A, B and C (d) A and D

19 Two uniform circular discs are rotating independently in the same direction around their common axis passing through their centres. The moment of inertia and angular velocity of the first disc are 0.1 kg-m^2 and 10 rad s^{-1} respectively, while those for the second one are 0.2 kg-m^2 and 5 rad s^{-1}, respectively. At some instant they get stuck together and start rotating as a single system about their common axis with some angular speed. The kinetic energy of the combined system is

(a) $\frac{10}{3}$ J (b) $\frac{20}{3}$ J
(c) $\frac{5}{3}$ J (d) $\frac{2}{3}$ J

20 An ideal gas in a closed container is slowly heated. As its temperature increases, which of the following statements are true?

A. The mean free path of the molecules decreases.
B. The mean collision time between the molecules decreases.
C. The mean free path remains unchanged.
D. The mean collision time remains unchanged.

(a) B and C (b) A and B
(c) C and D (d) A and D

Numerical Type Questions

21 An ideal cell of emf 10 V is connected in circuit shown in figure. Each resistance is 2 Ω. The potential difference (in V) across the capacitor when it is fully charged is ....

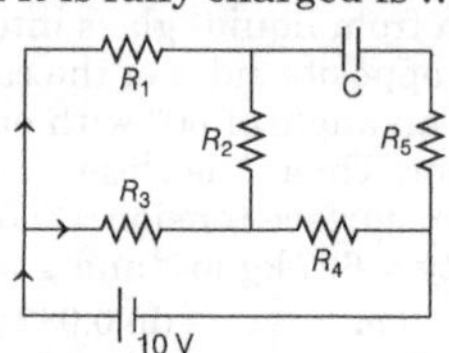

22 A square shaped hole of side $l = \frac{a}{2}$ is curved out at a distance $d = \frac{a}{2}$ from the centre O of a uniform circular disc of radius a. If the distance of the centre of mass of the remaining portion from O is $-\frac{a}{X}$, value of X (to the nearest integer) is ....

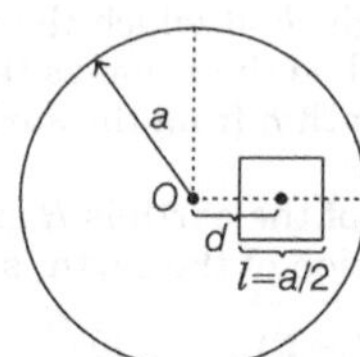

23 A light ray enters a solid glass sphere of refractive index $\mu = \sqrt{3}$ at an angle of incidence 60°. The ray is both reflected and refracted at the farther surface of the sphere. The angle (in degree) between the reflected and refracted rays at this surface is .... .

24 A particle of mass m is moving along the X-axis with initial velocity $u\hat{i}$. It collides elastically with a particle of mass 10 m at rest and then moves with half its initial kinetic energy (see figure). If $\sin\theta_1 = \sqrt{n}\sin\theta_2$, then value of n is ....

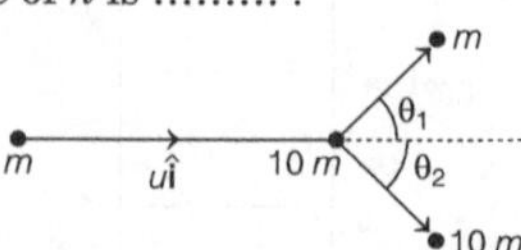

25 A wire of density 9×10^{-3} kg-cm^{-3} is stretched between two clamps 1 m apart. The resulting strain in the wire is 4.9×10^{-4}. The lowest frequency of the transverse vibrations in the wire is (Young's modulus of wire, $Y = 9 \times 10^{10}$ Nm^{-2}), (to the nearest integer) ....

ANSWERS

1. (b)	2. (a)	3. (c)	4. (c)	5. (c)	6. (a)	7. (a)	8. (b)	9. (c)	10. (b)
11. (d)	12. (a)	13. (a)	14. (d)	15. (a)	16. (d)	17. (d)	18. (c)	19. (b)	20. (a)
21. (8)	22. (23)	23. (90)	24. (10)	25. (35)					

For Detailed Solutions
Visit : https://bit.ly/31hyvZe
Or Scan :

3 September, Shift-I

1 In a radioactive material, fraction of active material remaining after time t is 9/16. The fraction that was remaining after time $t/2$ is

(a) $\frac{4}{5}$ (b) $\frac{3}{5}$
(c) $\frac{3}{4}$ (d) $\frac{7}{8}$

2 An elliptical loop having resistance R, of semi-major axis a and semi-minor axis b is placed in a magnetic field as shown in the figure. If the loop is rotated about the X-axis with angular frequency ω, then the average power loss in the loop due to joule's heating is

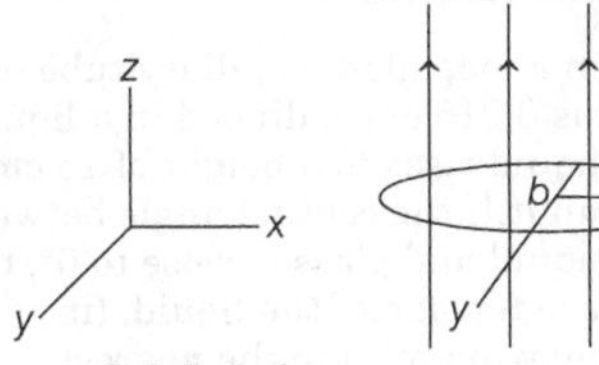

(a) $\frac{\pi^2 a^2 b^2 B^2 \omega^2}{2R}$ (b) zero
(c) $\frac{\pi a b B \omega}{R}$ (d) $\frac{\pi^2 a^2 b^2 B^2 \omega^2}{R}$

3

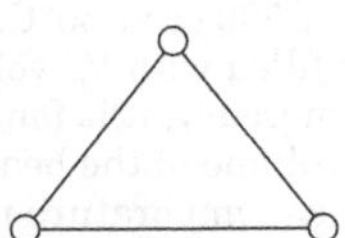

Consider a gas of triatomic molecules. The molecules are assumed to be triangular and made of massless rigid rods whose vertices are occupied by atoms. The internal energy of a mole of the gas at temperature T is

(a) $\frac{5}{2}RT$ (b) $\frac{3}{2}RT$
(c) $\frac{9}{2}RT$ (d) $3\,RT$

4 Two isolated conducting spheres S_1 and S_2 of radii $\frac{2}{3}R$ and $\frac{1}{3}R$, have charges $12\,\mu C$ and $-3\mu C$ respectively, and are at a large distance from each other. They are now connected by a conducting wire. A long time after this is done, the charges on S_1 and S_2 respectively, are

(a) $4.5\,\mu C$ on both
(b) $+4.5\,\mu C$ and $-4.5\mu C$
(c) $3\,\mu C$ and $6\,\mu C$
(d) $6\,\mu C$ and $3\,\mu C$

5 Moment of inertia of a cylinder of mass M, length L and radius R about an axis passing through its centre and perpendicular to the axis of the cylinder is $I = M\left(\frac{R^2}{4} + \frac{L^2}{12}\right)$. If such a cylinder is to be made for a given mass of a material. To have minimum possible moment of inertia, the ratio L/R for cylinder is

(a) $\frac{2}{3}$ (b) $\frac{3}{2}$ (c) $\sqrt{\frac{3}{2}}$ (d) $\sqrt{\frac{2}{3}}$

6 A torch battery of length l is to be made up of a thin cylindrical bar of radius a and a concentric thin cylindrical shell of radius b is filled in between with an electrolyte of resistivity ρ (see figure). If the battery is connected to a resistance R, the maximum joule's heating in R will takes place for

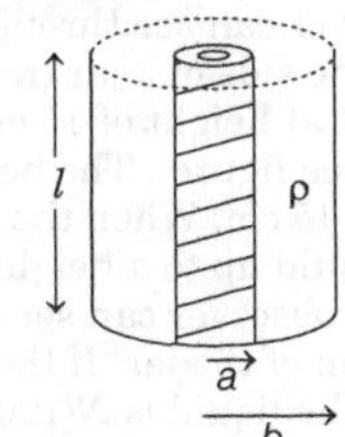

(a) $R = \frac{\rho}{2\pi l}\left(\frac{b}{a}\right)$ (b) $R = \frac{\rho}{2\pi l}\ln\left(\frac{b}{a}\right)$
(c) $R = \frac{\rho}{\pi l}\ln\left(\frac{b}{a}\right)$ (d) $R = \frac{2\rho}{\pi l}\ln\left(\frac{b}{a}\right)$

7 Magnitude of magnetic field (in SI unit) at the centre of a hexagonal shaped coil of side 10 cm, 50 turns and carrying current I ampere in units of $\frac{\mu_0 I}{\pi}$ is

(a) $250\sqrt{3}$ (b) $50\sqrt{3}$
(c) $500\sqrt{3}$ (d) $5\sqrt{3}$

8 A balloon filled with helium (32°C and 1.7 atm) bursts. Immediately afterwards; the expansion of helium can be considered as

(a) irreversible isothermal
(b) irreversible adiabatic
(c) reversible adiabatic
(d) reversible isothermal

9 A 750 Hz, 20 V (rms) source is connected to a resistance of 100 Ω, an inductance of 0.1803 H and a capacitance of 10 μF all in series combination. The time in which the resistance (heat capacity 2 J/°C) will get heated by 10°C is close to. (Assume no loss of heat to the surroundings)

(a) 418 s (b) 245 s (c) 365 s (d) 348 s

10 The magnetic field of a plane electromagnetic wave is $\mathbf{B} = 3\times10^{-8} \sin[200\pi(y+ct)]\hat{\mathbf{i}}$ T, where $c = 3\times10^8$ ms^{-1} is the speed of light. The corresponding electric field is

(a) $\mathbf{E} = 9\sin[200\pi(y+ct)]\hat{\mathbf{k}}$ V/m
(b) $\mathbf{E} = -10^{-6}\sin[200\pi(y+ct)]\hat{\mathbf{k}}$ V/m
(c) $\mathbf{E} = 3\times10^{-8}\sin[200\pi(y+ct)]\hat{\mathbf{k}}$ V/m
(d) $\mathbf{E} = -9\sin[200\pi(y+ct)]\hat{\mathbf{k}}$ V/m

11 A charged particle carrying charge 1 μC is moving with velocity $(2\hat{i} + 3\hat{j} + 4\hat{k})$ ms^{-1}. If an external magnetic field of $(5\hat{i} + 3\hat{j} - 6\hat{k})\times10^{-3}$ T exists in the region, where the particle is moving, then the force on the particle is $\mathbf{F}\times10^{-9}$ N. The vector $\mathbf{F}$ is

(a) $-0.30\hat{i} + 0.32\hat{j} - 0.09\hat{k}$
(b) $-30\hat{i} + 32\hat{j} - 9\hat{k}$
(c) $-300\hat{i} + 320\hat{j} - 90\hat{k}$
(d) $-3.0\hat{i} + 3.2\hat{j} - 0.9\hat{k}$

12 In the circuit shown in the figure, the total charge is 750 μC and the voltage across capacitor C_2 is 20 V. Then, the charge on capacitor C_2 is

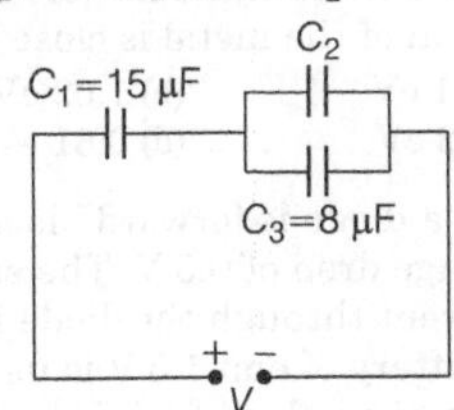

(a) 450 μC (b) 590 μC
(c) 160 μC (d) 650 μC

13 In Young's double slit experiment, light of 500 nm is used to produce an interference pattern. When the distance between the slits is 0.05 mm, the angular width (in degree) of the fringes formed on the distance screen is close to

(a) 0.17° (b) 0.57°
(c) 1.7° (d) 0.07°

14 A block of mass $m = 1$ kg slides with velocity $v = 6$ m/s on a frictionless horizontal surface and collides with a uniform vertical rod and sticks to it as shown. The rod is pivoted about O and swings as a result of the collision, making angle θ before momentarily coming to rest. If the rod has mass $M = 2$ kg and length $l = 1$ m, then the value of θ is approximately (Take, $g = 10$ m/s^2)

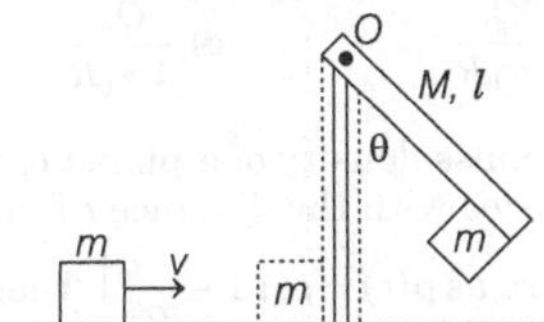

15 Using screw gauge of pitch 0.1 cm and 50 divisions on its circular scale, the thickness of an object is measured. It should correctly be recorded as
(a) 2.121 cm (b) 2.124 cm
(c) 2.125 cm (d) 2.123 cm

16 A uniform thin rope of length 12 m and mass 6 kg hangs vertically from a rigid support and a block of mass 2 kg is attached to its free end. A transverse short wavetrain of wavelength 6 cm is produced at the lower end of the rope. What is the wavelength of the wavetrain (in cm) when it reaches the top of the rope?
(a) 3 (b) 6
(c) 12 (d) 9

17 When the wavelength of radiation falling on a metal is changed from 500 nm to 200 nm, the maximum kinetic energy of the photoelectrons becomes three times larger. The work function of the metal is close to
(a) 0.81 eV (b) 1.02 eV
(c) 0.52 eV (d) 0.61 eV

18 When a diode is forward biased, it has a voltage drop of 0.5 V. The safe limit of current through the diode is 10 mA. If a battery of emf 1.5 V is used in the circuit, the value of minimum resistance to be connected in series with the diode, so that the current does not exceed the safe limit is
(a) 300 Ω (b) 50 Ω
(c) 100 Ω (d) 200 Ω

19 Pressure inside two soap bubbles are 1.01 atm and 1.02 atm, respectively. The ratio of their volume is
(a) 4 : 1 (b) 0.8 : 1 (c) 8 : 1 (d) 2 : 1

20 A satellite is moving in a low nearly circular orbit around the earth. Its radius is roughly equal to that of the earth's radius R_e. By firing rockets attached to it, its speed is instantaneously increased in the direction of its motion, so that it becomes $\sqrt{3/2}$ times larger. Due to this, the farthest distance from the centre of the earth that the satellite reaches is R. Value of R is
(a) $4R_e$ (b) $2.5R_e$ (c) $3R_e$ (d) $2R_e$

Numerical Type Questions

21 An observer can see through a small hole on the side of a jar (radius 15 cm) at a point at height of 15 cm from the bottom (see figure). The hole is at a height of 45 cm. When the jar is filled with a liquid up to a height of 30 cm, the same observer can see the edge at the bottom of the jar. If the refractive index of the liquid is $N/100$, where N is an integer, the value of N is ....

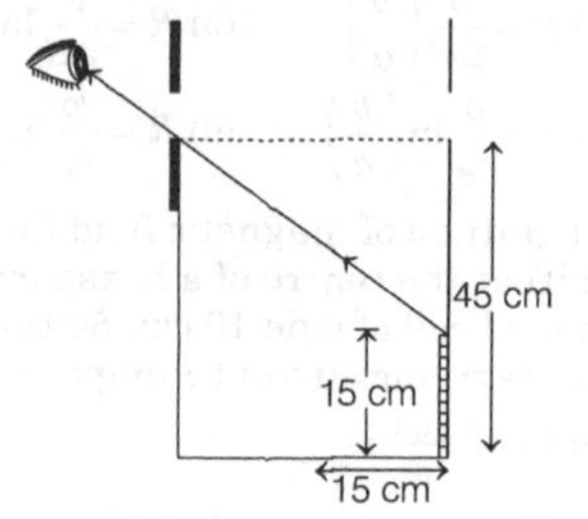

22 A cricket ball of mass 0.15 kg is thrown vertically up by a bowling machine, so that it rises to a maximum height of 20 m after leaving the machine. If the part pushing the ball applies a constant force F on the ball and moves horizontally a distance of 0.2 m, while launching the ball, the value of F (in N) is (Take, $g = 10\,\text{ms}^{-2}$) ....

23 A person of 80 kg mass is standing on the rim of a circular platform of mass 200 kg rotating about its axis at 5 rpm. The person now starts moving towards the centre of the platform. What will be the rotational speed (in rpm) of the platform when the person reaches its centre?

24 When a long glass capillary tube of radius 0.015 cm is dipped in a liquid, the liquid rises to a height of 15 cm within it.If the contact angle between the liquid and glass is close to 0°, the surface tension of the liquid, (in millinewton m^{-1}) to the nearest integer is
(Take, $\rho_{(\text{liquid})} = 900\,\text{kgm}^{-3}$, $g = 10\,\text{ms}^{-2}$) ....

25 A bakelite beaker has volume capacity of 500 cc at 30°C. When it is partially filled with V_m volume (at 30°C) of mercury, it is found that the unfilled volume of the beaker remains constant as temperature is varied. If $\gamma_{(\text{beaker})} = 6 \times 10^{-6}\ ^\circ\text{C}^{-1}$ and $\gamma_{(\text{mercury})} = 1.5 \times 10^{-4}\ ^\circ\text{C}^{-1}$, where γ is the coefficient of volume expansion, then V_m (in cc) is close to ....

ANSWERS

1. (c)	2. (a)	3. (d)	4. (d)	5. (c)	6. (d)	7. (c)	8. (b)	9. (d)	10. (d)
11. (b)	12. (b)	13. (b)	14. (a)	15. (b)	16. (c)	17. (d)	18. (c)	19. (c)	20. (d)
21. (158)	22. (150)	23. (9)	24. (101)	25. (20)					

For Detailed Solutions
Visit : https://bit.ly/34bgKNd
Or Scan :

3 September, Shift-II

1 Concentric metallic hollow spheres of radii R and $4R$ hold charges Q_1 and Q_2, respectively. Given that, surface charge densities of the concentric spheres are equal. The potential difference $V(R) - V(4R)$ is
(a) $\frac{3Q_2}{4\pi\varepsilon_0 R}$ (b) $\frac{3Q_1}{16\pi\varepsilon_0 R}$
(c) $\frac{3Q_1}{4\pi\varepsilon_0 R}$ (d) $\frac{Q_2}{4\pi\varepsilon_0 R}$

2 The mass density of a planet of radius R varies with the distance r from its centre as $\rho(r) = \rho_0\left(1 - \frac{r^2}{R^2}\right)$. Then, the gravitational field is maximum at
(a) $r = \sqrt{\frac{3}{4}}R$ (b) $r = R$
(c) $r = \sqrt{\frac{5}{9}}R$ (d) $r = \frac{1}{\sqrt{3}}R$

3 A particle is moving unidirectionally on a horizontal plane under the action of a constant power supplying energy source. The displacement (s)-time (t) graph that describes the motion of the particle is (graphs are drawn schematically and are not to scale)

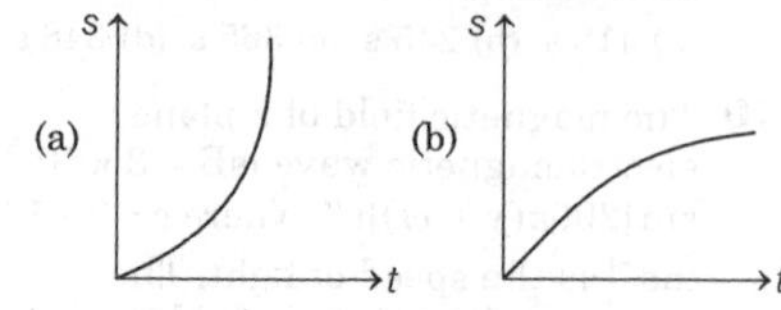

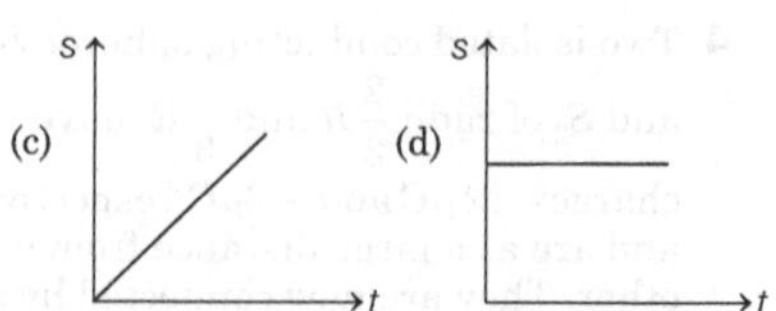

4 Two resistors 400 Ω and 800 Ω are connected in series across a 6 V battery. The potential difference measured by a voltmeter of 10 kΩ across 400 Ω resistor is close to
(a) 1.95 V
(b) 2 V
(c) 1.8 V
(d) 2.05 V

5 A perfectly diamagnetic sphere has a small spherical cavity at its centre, which is filled with a paramagnetic substance. The whole system is placed in a uniform magnetic field **B**. Then, the field inside the paramagnetic substance is

(a) much larger than $|\mathbf{B}|$ and parallel to **B**
(b) **B**
(c) zero
(d) much larger than $|\mathbf{B}|$ but opposite to **B**

6 Two light waves having the same wavelength λ in vacuum are in phase initially. Then, the first wave travels a path L_1 through a medium of refractive index n_1 while the second wave travels a path of length L_2 through a medium of refractive index n_2. After this, the phase difference between the two waves is

(a) $\frac{2\pi}{\lambda}(n_1L_1 - n_2L_2)$ (b) $\frac{2\pi}{\lambda}\left(\frac{L_1}{n_1} - \frac{L_1}{n_2}\right)$
(c) $\frac{2\pi}{\lambda}\left(\frac{L_2}{n_1} - \frac{L_1}{n_2}\right)$ (d) $\frac{2\pi}{\lambda}(n_2L_1 - n_1L_2)$

7 A block of mass m attached to a massless spring is performing oscillatory motion of amplitude A on a frictionless horizontal plane. If half of the mass of the block breaks off when it is passing through its equilibrium point, the amplitude of oscillation for the remaining system becomes fA. The value of f is

(a) $\sqrt{2}$ (b) $\frac{1}{2}$ (c) $\frac{1}{\sqrt{2}}$ (d) 1

8 A calorimeter of water equivalent 20 g contains 180 g of water at 25°C. 'm' grams of steam at 100°C is mixed in it till the temperature of the mixture is 31°C. The value of m is close to (Take, latent heat of water = $540\ \text{cal g}^{-1}$, specific heat of water = $1\ \text{cal g}^{-1}\ {}^\circ\text{C}^{-1}$)

(a) 2 g (b) 3.2 g
(c) 2.6 g (d) 4 g

9 Hydrogen ion and singly ionised helium atom are accelerated from rest, through the same potential difference. The ratio of final speeds of hydrogen and helium ions is close to

(a) 1 : 2 (b) 10 : 7
(c) 2 : 1 (d) 5 : 7

10 A metallic sphere cools from 50°C to 40°C in 300 s. If atmospheric temperature around is 20°C, then the sphere's temperature after the next 5 min will be close to

(a) 31°C (b) 35°C
(c) 28°C (d) 33°C

11 To raise the temperature of a certain mass of gas by 50°C at a constant pressure, 160 cal of heat is required. When the same mass of gas is cooled by 100°C at constant volume, 240 cal of heat is released. How many degrees of freedom does each molecule of this gas have (assume gas to be ideal)?

(a) 5 (b) 7 (c) 6 (d) 3

12 Two sources of light emit X-rays of wavelength 1 nm and visible light of wavelength 500 nm, respectively. Both the sources emit light of the same power 200 W. The ratio of the number density of photons of X-rays to the number density of photons of the visible light of the given wavelengths is

(a) $\frac{1}{250}$ (b) 500
(c) 250 (d) $\frac{1}{500}$

13 The electric field of a plane electromagnetic wave propagating along the x-direction in vacuum is $\mathbf{E} = E_0\hat{\mathbf{j}}\cos(\omega t - kx)$. The magnetic field **B**, at the moment $t = 0$ is

(a) $\mathbf{B} = E_0\sqrt{\mu_0\varepsilon_0}\cos(kx)\,\hat{\mathbf{j}}$
(b) $\mathbf{B} = \frac{E_0}{\sqrt{\mu_0\varepsilon_0}}\cos(kx)\hat{\mathbf{j}}$
(c) $\mathbf{B} = \frac{E_0}{\sqrt{\mu_0\varepsilon_0}}\cos(kx)\hat{\mathbf{k}}$
(d) $\mathbf{B} = E_0\sqrt{\mu_0\varepsilon_0}\cos(kx)\hat{\mathbf{k}}$

14

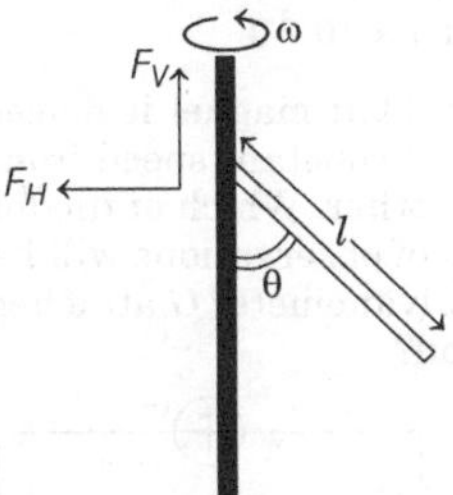

A uniform rod of length l is pivoted at one of its ends on a vertical shaft of negligible radius. When the shaft rotates at angular speed ω, the rod makes an angle θ with it (see figure). To find θ, equate the rate of change of angular momentum (direction going into the paper) $\frac{ml^2}{12}\omega^2\sin\theta\cos\theta$ about the centre of mass to the torque provided by the horizontal and vertical forces F_H and F_V about the centre of mass. The value of θ is then such that

(a) $\cos\theta = \frac{g}{l\omega^2}$ (b) $\cos\theta = \frac{g}{2l\omega^2}$
(c) $\cos\theta = \frac{2g}{3l\omega^2}$ (d) $\cos\theta = \frac{3g}{2l\omega^2}$

15 A block of mass 1.9 kg is at rest at the edge of a table of height 1 m. A bullet of mass 0.1 kg collides with the block and sticks to it. If the velocity of the bullet is 20 m/s in the horizontal direction just before the collision, then the kinetic energy just before the combined system strikes the floor, is (Take, $g = 10\ \text{m/s}^2$ and assume there is no rotational motion and loss of energy after the collision is negligible)

(a) 20 J (b) 19 J (c) 21 J (d) 23 J

16 Amount of solar energy received on the earth's surface per unit area per unit time is defined as solar constant. Dimensional formula of solar constant is

(a) $[\text{MLT}^{-2}]$ (b) $[\text{ML}^0\text{T}^{-3}]$
(c) $[\text{M}^2\text{L}^0\text{T}^{-1}]$ (d) $[\text{ML}^2\text{T}^{-2}]$

17 If a semiconductor photodiode can detect a photon with a maximum wavelength of 400 nm, then its band gap energy is (Take, Planck's constant, $h = 6.63 \times 10^{-34}$ J-s and speed of light, $c = 3 \times 10^8$ m/s)

(a) 1.5 eV (b) 1.1 eV
(c) 3.1 eV (d) 2.0 eV

18 A uniform magnetic field B exists in a direction perpendicular to the plane of a square loop made of a metal wire. The wire has a diameter of 4 mm and a total length of 30 cm. The magnetic field changes with time at a steady rate $dB/dt = 0.032\ \text{Ts}^{-1}$. The induced current in the loop is close to (Take, resistivity of the metal wire $= 1.23 \times 10^{-8}\ \Omega\text{m}$)

(a) 0.61 A (b) 0.43 A
(c) 0.53 A (d) 0.34 A

19 Which of the following will not be observed when a multimeter (operating in resistance measuring mode) probes connected across a component, are just reversed?

(a) Multimeter shows no deflection in both cases, i.e. before and after reversing the probes if the chosen component is metal wire.
(b) Multimeter shows an equal deflection in both cases, i.e. before and after reversing the probes if the chosen component is resistor.
(c) Multimeter shows no deflection in both cases, i.e. before and after reversing the probes if the chosen component is capacitor.
(d) Multimeter shows a deflection, accompanied by a splash of light out of connected component in one direction and no deflection on reversing the probes if the chosen component is LED.

20 The radius R of a nucleus of mass number A can be estimated by the formula $R = (1.3 \times 10^{-15}) A^{1/3}$ m. It follows that the mass density of a nucleus is of the order of ($M_{proton} \cong M_{neutron} \simeq 1.67 \times 10^{-27}$ kg)

(a) 10^{17} kg m^{-3} (b) 10^{24} kg m^{-3}
(c) 10^{10} kg m^{-3} (d) 10^{3}kg m^{-3}

Numerical Type Questions

21 A galvanometer coil has 500 turns and each turn has an average area of 3×10^{-4} m^2. If a torque of 1.5 N-m is required to keep this coil parallel to a magnetic field when a current of 0.5 A is flowing through it, the strength of the field (in T) is ....

22 A block starts moving up an inclined plane of inclination 30° with an initial velocity of v_0. It comes back to its initial position with velocity $v_0/2$. The value of the coefficient of kinetic friction between the block and the inclined plane is close to $I/1000$, the nearest integer to I is ....

23 If minimum possible work is done by a refrigerator in converting 100 g of water at 0°C to ice, how much heat (in cal) is released to the surroundings at temperature 27°C to the nearest integer ...... ?
(Take, latent heat of ice = 80 cal/g)

24 A massless equilateral triangle EFG of side a (as shown in figure) has three particles of mass m situated at its vertices. The moment of inertia of the system about the line EX perpendicular to EG in the plane of EFG is $\frac{N}{20} ma^2$, where N is an integer. The value of N is ....

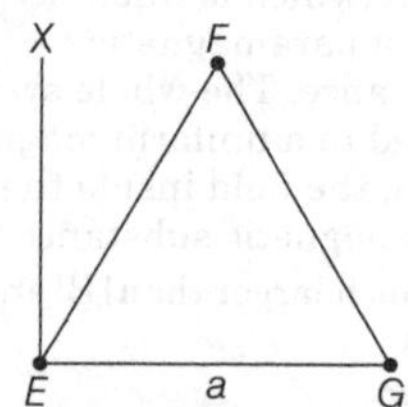

25 When an object is kept at a distance of 30 cm from a concave mirror, the image is formed at a distance of 10 cm from the mirror. If the object is moved with a speed of 9 cms^{-1}, the speed (in cms^{-1}) with which image moves at that instant is ....

ANSWERS

1. (b)	2. (c)	3. (a)	4. (a)	5. (c)	6. (a)	7. (c)	8. (a)	9. (c)	10. (d)
11. (c)	12. (b)	13. (d)	14. (d)	15. (c)	16. (b)	17. (c)	18. (a)	19. (c)	20. (a)
21. (20)	22. (346)	23. (8791)	24. (25)	25. (1)					

For Detailed Solutions
Visit : https://bit.ly/3lYCqlE
Or Scan :

4 September, Shift-I

1 A beam of plane polarised light of large cross-sectional area and uniform intensity of 3.3 Wm^{-2} falls normally on a polariser, which rotates about its axis with an angular speed of 31.4 rad/s. The energy of light passing through the polariser per revolution is close to
(Take, cross-sectional area = 3×10^{-4} m^2)

(a) 1.5×10^{-4} J (b) 5.0×10^{-4} J
(c) 1.0×10^{-4} J (d) 1.0×10^{-5} J

2 Two point charges $4q$ and $-q$ are fixed on the X-axis at $x = \frac{-d}{2}$ and $x = \frac{d}{2}$, respectively. If a third point charge q is taken from the origin to $x = d$ along the semi-circle as shown in the figure, the energy of the charge will

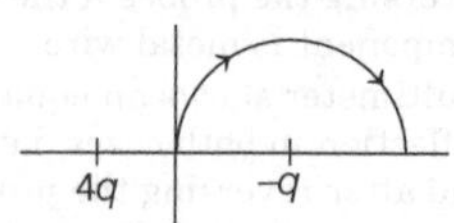

(a) increase by $\frac{2q^2}{3\pi\varepsilon_0 d}$

(b) decrease by $\frac{4q^2}{3\pi\varepsilon_0 d}$

(c) increase by $\frac{3q^2}{4\pi\varepsilon_0 d}$

(d) decrease by $\frac{q^2}{4\pi\varepsilon_0 d}$

3 A small bar magnet placed with its axis at 30° with an external field of 0.06 T experiences a torque of 0.018 Nm. The minimum work required to rotate it from its stable to unstable equilibrium position is

(a) 9.2×10^{-3} J
(b) 7.2×10^{-2} J
(c) 6.4×10^{-2} J
(d) 11.7×10^{-3} J

4 A small bar magnet is moved through a coil at constant speed from one end to the other. Which of the following series of observations will be seen on the galvanometer G attached across the coil?

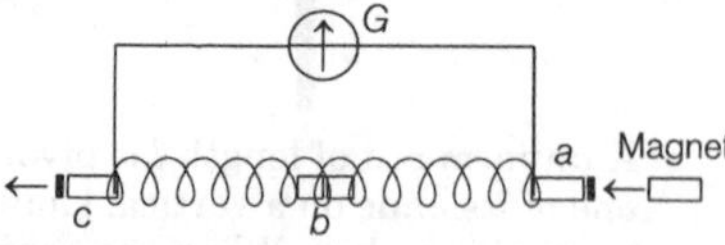

Three positions shown describe: (i) the magnet's entery (ii) magnet is completely inside and (iii) magnet's exit.

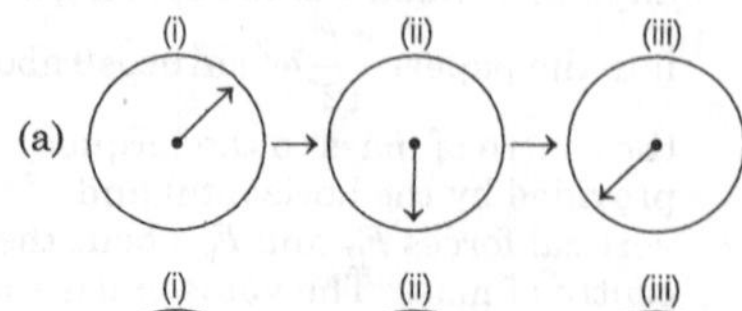

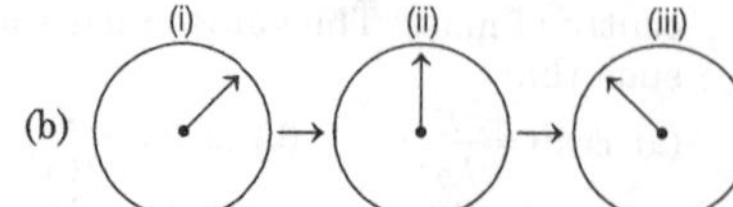

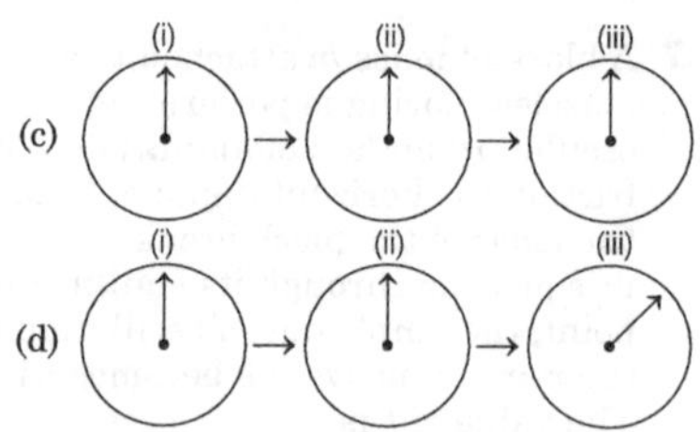

5 Two charged thin infinite plane sheets of uniform surface charge densities σ_+ and σ_-, where $|\sigma_+| > |\sigma_-|$, intersect at right angle. Which of the following best represents the electric field lines for this system?

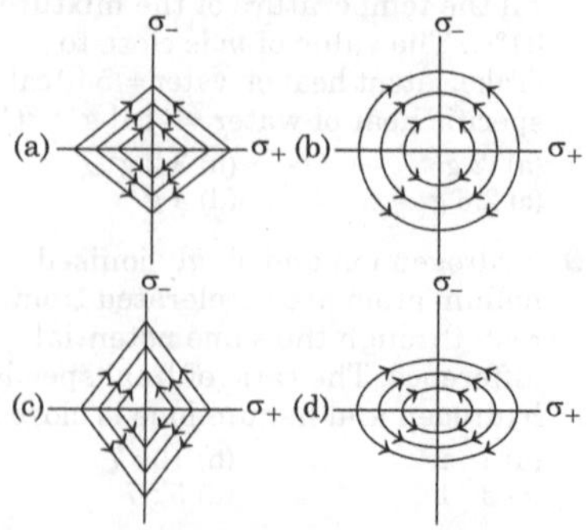

6 An air bubble of radius 1 cm in water has an upward acceleration 9.8 cm s^{-2}. The density of water is 1 g/cm^3 and water offers negligible drag force on the bubble. The mass of the bubble is (Take, g = 980 cm/s^2)

(a) 1.52 g (b) 4.51 g
(c) 4.15 g (d) 3.15 g

7 Take the breakdown voltage of the zener diode used in the given circuit as 6V. For the input voltage shown in figure below, the time variation of the output voltage is (Graphs are drawn schematically and on not to scale)

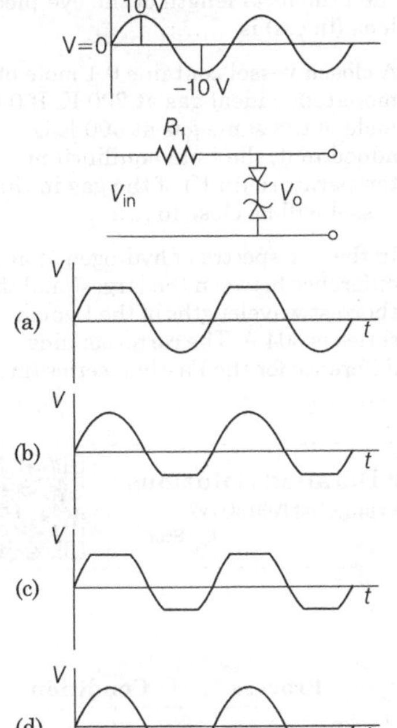

8 Particle A of mass $m_A = \frac{m}{2}$ moving along the X-axis with velocity v_0 collides elastically with another particle B at rest having mass $m_B = \frac{m}{3}$. If both particles move along the X-axis after the collision, the change $\Delta\lambda$ in de-Broglie wavelength of particle A, in terms of its de-Broglie wavelength λ_0 before collision is

(a) $\Delta\lambda = \frac{5}{2}\lambda_0$ (b) $\Delta\lambda = 4\lambda_0$

(c) $\Delta\lambda = 2\lambda_0$ (d) $\Delta\lambda = \frac{3}{2}\lambda_0$

9 Dimensional formula for thermal conductivity is
(Here, K denotes the temperatuere)

(a) $[MLT^{-2}K]$ (b) $[MLT^{-2}K^{-2}]$
(c) $[MLT^{-3}K^{-1}]$ (d) $[MLT^{-3}K]$

10 A tennis ball is released from a height h and after freely falling on a wooden floor, it rebounds and reaches height $\frac{h}{2}$. The velocity *versus* height of the ball during its motion may be represented graphically by
(Graphs are drawn schematically and on not to scale)

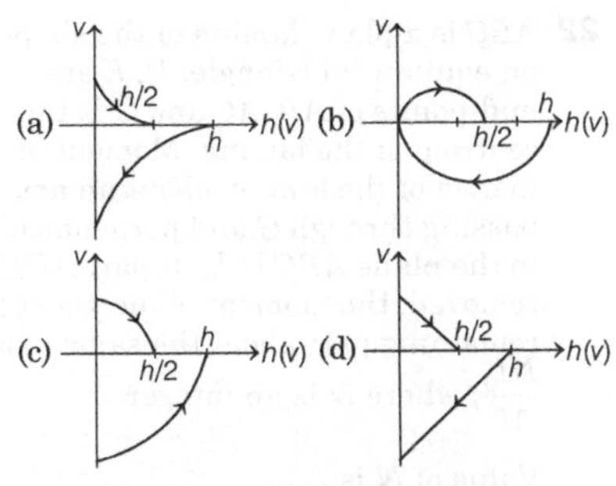

11 For a transverse wave travelling along a straight line, the distance between two peaks (crests) is 5 m, while the distance between one crest and one trough is 1.5 m. The possible wavelengths (in metre) of the waves are

(a) 1, 2, 3,...... (b) $\frac{1}{1}, \frac{1}{3}, \frac{1}{5}$,.....

(c) 1, 3, 5,..... (d) $\frac{1}{2}, \frac{1}{4}, \frac{1}{6}$, ...

12 Blocks of masses m, $2m$, $4m$ and $8m$ are arranged in a line on a frictionless floor. Another block of mass m, moving with speed v along the same line (see figure) collides with mass m in perfectly inelastic manner. All the subsequent collisions are also perfectly inelastic. By the time, the last block of mass $8m$ starts moving, the total energy loss is p% of the original energy. Value of p is close to

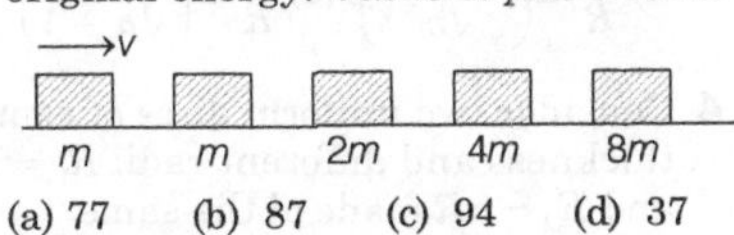

(a) 77 (b) 87 (c) 94 (d) 37

13 Choose the correct option relating wavelengths of different parts of electromagnetic wave spectrum

(a) $\lambda_{\text{X-rays}} < \lambda_{\text{microwaves}} < \lambda_{\text{radio waves}} < \lambda_{\text{visible}}$
(b) $\lambda_{\text{visible}} > \lambda_{\text{X-rays}} > \lambda_{\text{radio waves}} > \lambda_{\text{microwaves}}$
(c) $\lambda_{\text{radio waves}} > \lambda_{\text{microwaves}} > \lambda_{\text{visible}} > \lambda_{\text{X-rays}}$
(d) $\lambda_{\text{visible}} < \lambda_{\text{microwaves}} < \lambda_{\text{radio waves}} < \lambda_{\text{X-rays}}$

14 A wire A, bent in the shape of an arc of a circle, carrying a current of 2 A and having radius 2 cm and another wire B, also bent in the shape of arc of a circle, carrying a current of 3 A and having radius of 4 cm, are placed as shown in the figure. The ratio of the magnetic fields due to the wires A and B at the common centre O is

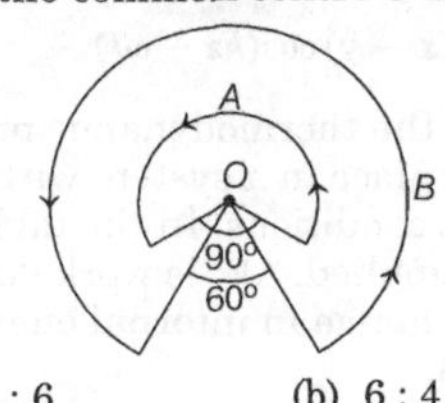

(a) 4 : 6 (b) 6 : 4
(c) 6 : 5 (d) 2 : 5

15 On the X-axis and at a distance x from the origin, the gravitational field due to a mass distribution is given by $\frac{Ax}{(x^2 + a^2)^{3/2}}$ in the x-direction. The magnitude of gravitational potential on the X-axis at a distance x, taking its value to be zero at infinity, is

(a) $\frac{A}{(x^2 + a^2)^{1/2}}$ (b) $A(x^2 + a^2)^{3/2}$

(c) $\frac{A}{(x^2 + a^2)^{3/2}}$ (d) $A(x^2 + a^2)^{1/2}$

16 A battery of 3.0 V is connected to a resistor dissipating 0.5 W of power. If the terminal voltage of the battery is 2.5 V, the power dissipated within the internal resistance is

(a) 0.072 W (b) 0.125 W
(c) 0.50 W (d) 0.10 W

17 Match the C_p/C_V ratio for ideal gases with different type of molecules:

	Molecule type	C_p/C_V
(A)	Monatomic molecules	I. 7/5
(B)	Diatomic rigid molecules	II. 9/7
(C)	Diatomic non-rigid molecules	III. 4/3
(D)	Triatomic rigid molecules	IV. 5/3

	A	B	C	D
(a)	IV	I	II	III
(b)	III	IV	II	I
(c)	II	III	I	IV
(d)	IV	II	I	III

18 The following figure shows few data points in a photoelectric effect experiment for a certain metal. The minimum energy for ejection of electron from its surface is (Take, Planck's constant, $h = 6.62 \times 10^{-34}$ J-s)

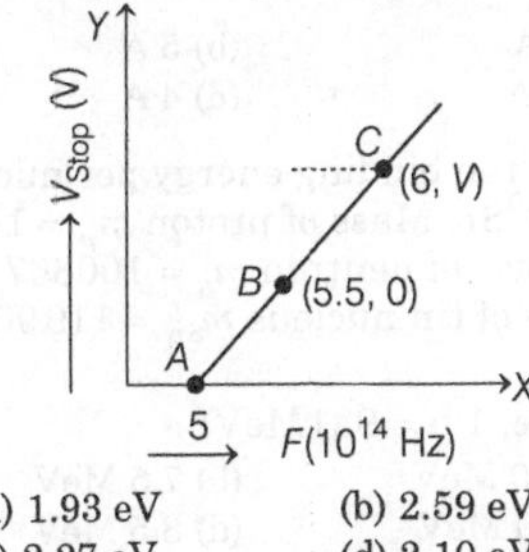

(a) 1.93 eV (b) 2.59 eV
(c) 2.27 eV (d) 2.10 eV

19 The specific heat of water = 4200 J $kg^{-1}K^{-1}$ and the latent heat of ice $= 3.4 \times 10^5$ J kg^{-1}. 100 g of ice at 0°C is placed in 200 g of water at 25°C. The amount of ice that will melt as the temperature of water reaches 0°C is close to (in grams)

(a) 63.8 (b) 69.3 (c) 64.6 (d) 61.7

20 Starting from the origin at time $t = 0$, with initial velocity $5\hat{j}\ ms^{-1}$, a particle moves in the xy-plane with a constant acceleration of $(10\hat{i} + 4\hat{j})\ ms^{-2}$. At time t, its coordinates are (20 m, y_0 m). The values of t and y_0 respectively, are
(a) 2 s and 18 m (b) 5 s and 25 m
(c) 2 s and 24 m (d) 4 s and 52 m

Numerical Type Questions

21 A circular disc of mass M and radius R is rotating about its axis with angular speed ω_1. If another stationary disc having radius $R/2$ and same mass M is dropped co-axially on to the rotating disc. Gradually, both discs attain constant angular speed ω_2. The energy lost in the process is $p\%$ of the initial energy. Value of p is ....

22 ABC is a plane lamina of the shape of an equilateral triangle. D, E are mid-points of AB, AC and G is the centroid of the lamina. Moment of inertia of the lamina about an axis passing through G and perpendicular to the plane ABC is I_0. If part ADE is removed, the moment of inertia of the remaining part about the same axis is $\frac{NI_0}{16}$, where N is an integer.

Value of N is ....

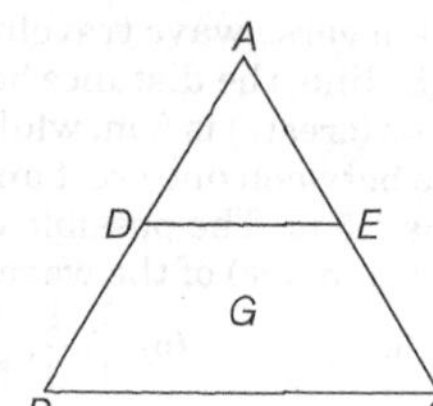

23 In a compound microscope, the magnified virtual image is formed at a distance of 25 cm from the eye-piece. The focal length of its objective lens is 1 cm. If the magnification is 100 and the tube length of the microscope is 20 cm, then the focal length of the eye-piece lens (in cm) is ....

24 A closed vessel contains 0.1 mole of a monatomic ideal gas at 200 K. If 0.05 mole of the same gas at 400 K is added to it, the final equilibrium temperature (in K) of the gas in the vessel will be close to ....

25 In the line spectra of hydrogen atom, difference between the largest and the shortest wavelengths of the Lyman series is 304 Å. The corresponding difference for the Paschan series (in Å) is ....

ANSWERS

1. (c)	2. (b)	3. (b)	4. (b)	5. (c)	6. (c)	7. (c)	8. (b)	9. (c)	10. (c)
11. (b)	12. (c)	13. (c)	14. (c)	15. (a)	16. (d)	17. (a)	18. (c)	19. (d)	20. (a)
21. (20)	22. (11)	23. (6)	24. (266)	25. (10553)					

For Detailed Solutions
Visit : https://bit.ly/31nOYv7
Or Scan :

4 September, Shift-II

1 The value of current i_1 flowing from A to C in the circuit diagram is

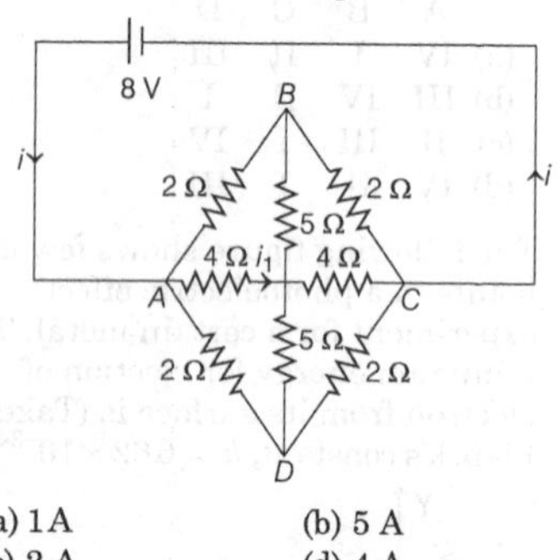

(a) 1 A (b) 5 A
(c) 2 A (d) 4 A

2 Find the binding energy per nucleon for $^{120}_{50}Sn$. Mass of proton $m_p = 1.00783$ u, mass of neutron $m_n = 1.00867$ u and mass of tin nucleus $m_{Sn} = 119.902199$ u.
(Take, 1 u = 931MeV)
(a) 9.0 MeV (b) 7.5 MeV
(c) 8.0 MeV (d) 8.5 MeV

3 A series L-R circuit is connected to a battery of emf V. If the circuit is switched ON at $t = 0$, then the time at which the energy stored in the inductor reaches $\left(\frac{1}{n}\right)$ times of its maximum value, is

(a) $\frac{L}{R}\ln\left(\frac{\sqrt{n}+1}{\sqrt{n}-1}\right)$ (b) $\frac{L}{R}\ln\left(\frac{\sqrt{n}}{\sqrt{n}+1}\right)$

(c) $\frac{L}{R}\ln\left(\frac{\sqrt{n}-1}{\sqrt{n}}\right)$ (d) $\frac{L}{R}\ln\left(\frac{\sqrt{n}}{\sqrt{n}-1}\right)$

4 Consider two uniform discs of same thickness and different radii $R_1 = R$ and $R_2 = \alpha R$ made of the same material. If the ratio of their moments of inertia I_1 and I_2 respectively, about their axes is $I_1 : I_2 = 1 : 16$, then the value of α is
(a) 2 (b) $2\sqrt{2}$
(c) 4 (d) $\sqrt{2}$

5 The electric field of a plane electromagnetic wave is given by

$$\mathbf{E} = E_0(\hat{\mathbf{x}} + \hat{\mathbf{y}})\sin(kz - \omega t)$$

Its magnetic field will be given by

(a) $\frac{E_0}{c}(\hat{\mathbf{x}} + \hat{\mathbf{y}})\sin(kz - \omega t)$

(b) $\frac{E_0}{c}(-\hat{\mathbf{x}} + \hat{\mathbf{y}})\sin(kz - \omega t)$

(c) $\frac{E_0}{c}(\hat{\mathbf{x}} - \hat{\mathbf{y}})\sin(kz - \omega t)$

(d) $\frac{E_0}{c}(\hat{\mathbf{x}} - \hat{\mathbf{y}})\cos(kz - \omega t)$

6 Match the thermodynamic processes taking place in a system with the correct conditions. In the table, ΔQ is heat supplied, ΔW is work done and ΔU is change in internal energy of the system.

Process	Condition
(I) Adiabatic	(A) $\Delta W = 0$
(II) Isothermal	(B) $\Delta Q = 0$
(III) Isochoric	(C) $\Delta U \neq 0, \Delta W \neq 0$ and $\Delta Q \neq 0$
(IV) Isobaric	(D) $\Delta U = 0$

	I	II	III	IV
(a)	B	A	D	C
(b)	A	A	B	C
(c)	B	D	A	C
(d)	A	B	D	D

7

For a uniform rectangular sheet, shown in the above figure, the ratio of moments of inertia about the axes perpendicular to the sheet and passing through O (the centre of mass) and O' (corner point) is
(a) 1/8 (b) 2/3
(c) 1/4 (d) 1/2

8 A paramagnetic sample shows a net magnetisation of 6 A/m, when it is placed in an external magnetic field of 0.4 T at a temperature of 4 K. When the sample is placed in an external magnetic field of 0.3 T at a temperature of 24 K, then the magnetisation will be

(a) 4 A/m (b) 0.75 A/m
(c) 1 A/m (d) 2.25 A/m

9 A circular coil has moment of inertia 0.8 kg m^2 around any diameter and is carrying current to produce a magnetic moment of 20 Am^2. The coil is kept initially in a vertical position and it can rotate freely around a horizontal diameter. When a uniform magnetic field of 4 T is applied along the vertical, it starts rotating around its horizontal diameter. The angular speed of the coil, acquired after rotating by 60°, will be

(a) 10 rad s^{-1} (b) 20π rad s^{-1}
(c) 10π rad s^{-1} (d) 20 rad s^{-1}

10 A particle of charge q and mass m is subjected to an electric field $E = E_0(1 - ax^2)$ in the x-direction, where a and E_0 are constants. Initially, the particle was at rest at $x = 0$. Other than the initial position, the kinetic energy of the particle becomes zero when the distance of the particle from the origin is

(a) $\sqrt{\frac{1}{a}}$ (b) a
(c) $\sqrt{\frac{3}{a}}$ (d) $\sqrt{\frac{2}{a}}$

11 A capacitor C is fully charged with voltage V_0. After disconnecting the voltage source, it is connected in parallel with another uncharged capacitor of capacitance $\frac{C}{2}$. The energy loss in the process after the charge is distributed between the two capacitors is

(a) $\frac{1}{3}CV_0^2$ (b) $\frac{1}{6}CV_0^2$
(c) $\frac{1}{2}CV_0^2$ (d) $\frac{1}{4}CV_0^2$

12 A small ball of mass m is thrown upward with velocity u from the ground. The ball experiences a resistive force mkv^2, where v is its speed. The maximum height attained by the ball is

(a) $\frac{1}{k}\tan^{-1}\left(\frac{ku^2}{2g}\right)$ (b) $\frac{1}{2k}\ln\left(1+\frac{ku^2}{g}\right)$
(c) $\frac{1}{k}\ln\left(1+\frac{ku^2}{2g}\right)$ (d) $\frac{1}{2k}\tan^{-1}\left(\frac{ku^2}{g}\right)$

13 A body is moving in a low circular orbit about a planet of mass M and radius R. The radius of the orbit can be taken to be R itself. Then, the ratio of the speed of this body in the orbit to the escape velocity from the planet is

(a) 1 (b) $\frac{1}{\sqrt{2}}$ (c) $\sqrt{2}$ (d) 2

14 Two identical cylindrical vessels are kept on the ground and each contains the same liquid of density d. The area of the base of both vessels is S but the height of liquid in one vessel is x_1 and in the other, x_2. When both cylinders are connected through a pipe of negligible volume very close to the bottom, the liquid flows from one vessel to the other until it comes to equilibrium at a new height. The change in energy of the system in the process is

(a) $gdS(x_2^2 + x_1^2)$ (b) $\frac{3}{4}gdS(x_2 - x_1)^2$
(c) $gdS(x_2 + x_1)^2$ (d) $\frac{1}{4}gdS(x_2 - x_1)^2$

15 The driver of a bus approaching a big wall notices that the frequency of his bus's horn changes from 420 Hz to 490 Hz, when he hears it after it gets reflected from the wall. Find the speed of the bus if speed of the sound is 330 ms^{-1}.

(a) 81 kmh^{-1} (b) 91 kmh^{-1}
(c) 71 kmh^{-1} (d) 61 kmh^{-1}

16 A person pushes a box on a rough horizontal platform surface. He applies a force of 200 N over a distance of 15 m. Thereafter, he gets progressively tired and his applied force reduces linearly with distance to 100 N. The total distance through which the box has been moved is 30 m. What is the work done by the person during the total movement of the box?

(a) 5250 J (b) 2780 J
(c) 3280 J (d) 5690 J

17 In a photoelectric effect experiment, the graph of stopping potential V *versus* reciprocal of wavelength ($1/\lambda$) obtained is shown in the figure.

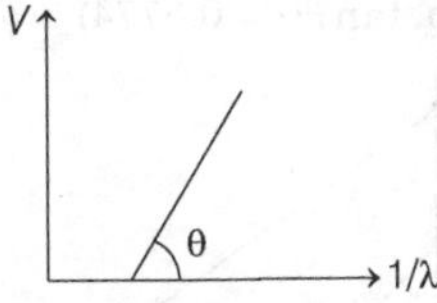

As the intensity of incident radiation is increased,

(a) graph does not change
(b) straight line shifts to left
(c) slope of the straight line get more steep
(d) straight line shifts to right

18 A cube of metal is subjected to a hydrostatic pressure of 4 GPa. The percentage change in the length of the side of the cube is close to (Take bulk modulus of metal, $B = 8 \times 10^{10}$ Pa)

(a) 1.67 (b) 0.6 (c) 20 (d) 5

19 Identify the operation performed by the circuit given below.

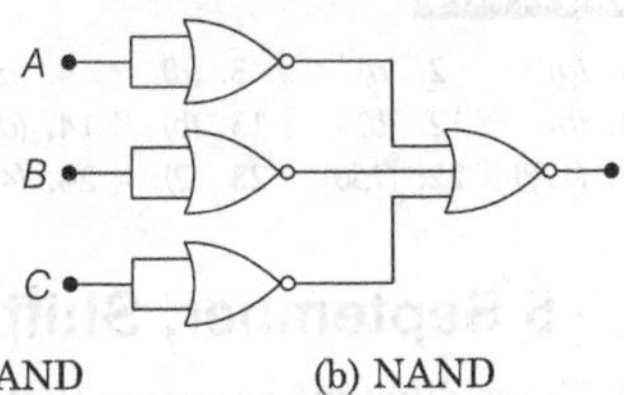

(a) AND (b) NAND
(c) NOT (d) OR

20 A quantity x is given by $x = \frac{IFv^2}{WL^4}$ where, I is moment of inertia, F is force, v is work and L is length. The dimensional formula for x is same as that of

(a) Planck's constant
(b) force constant
(c) coefficient of viscosity
(d) energy density

Numerical Type Questions

21 Orange light of wavelength 6000×10^{-10} m illuminates a single slit of width 0.6×10^{-4} m. The maximum possible number of diffraction minima produced on both sides of the central maximum is ....

22 The change in the magnitude of the volume of an ideal gas when a small additional pressure Δp is applied at a constant temperature, is the same as the change when the temperature is reduced by a small quantity ΔT at constant pressure. The initial temperature and pressure of the gas are 300 K and 2 atm, respectively. If $|\Delta T| = C|\Delta p|$, then value of C (in K/atm) is ....

23

Four resistances $40\,\Omega$, $60\,\Omega$, $90\,\Omega$ and $110\,\Omega$ make the arms of a quadrilateral $ABCD$. Across AC is a battery of emf 40 V and internal resistance negligible. The potential difference across BD (in volt) is ....

24 The distance between an object and a screen is 100 cm. A lens can produce real image of the object on the screen for two different positions between the screen and the object. The distance between these two positions is 40 cm. If the power of the lens is close to $(N/100)$ D, where N is an integer. Then, the value of N is ....

25 The speed *versus* time graph for a particle is shown in the figure. The distance travelled (in metre) by the particle during the time interval $t = 0$ s to $t = 5$ s will be ....

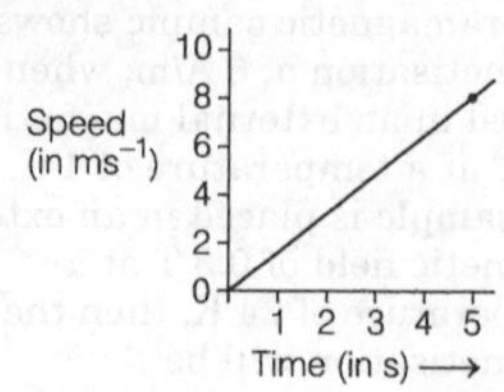

ANSWERS

1. (a)	2. (d)	3. (d)	4. (a)	5. (b)	6. (c)	7. (c)	8. (b)	9. (d)	10. (c)
11. (b)	12. (b)	13. (b)	14. (d)	15. (b)	16. (a)	17. (a)	18. (a)	19. (a)	20. (d)
21. (198)	22. (150)	23. (2)	24. (476)	25. (20)					

For Detailed Solutions
Visit : https://bit.ly/3dFmba9
Or Scan :

5 September, Shift-I

1 Three different processes that can occur in an ideal monoatomic gas are shown in the p *versus* V diagram. The paths are labelled as A → B, A → C and A → D. The change in internal energies during these process are taken as E_{AB}, E_{AC} and E_{AD} and the work done as W_{AB}, W_{AC} and W_{AD}. The correct relation between these parameters are

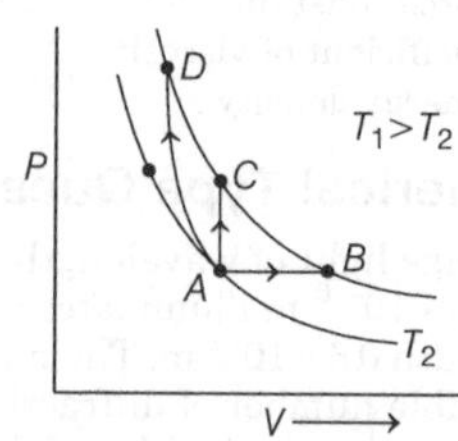

(a) $E_{AB} = E_{AC} < E_{AD}$, $W_{AB} > 0$, $W_{AC} = 0$, $W_{AD} < 0$
(b) $E_{AB} = E_{AC} = E_{AD}$, $W_{AB} > 0$, $W_{AC} = 0$, $W_{AD} > 0$
(c) $E_{AB} < E_{AC} < E_{AD}$, $W_{AB} > 0$, $W_{AC} > W_{AD}$
(d) $E_{AB} > E_{AC} > E_{AD}$, $W_{AB} < W_{AC} < W_{AD}$

2 An electrical power line, having a total resistance of 2Ω, delivers 1 kW at 220 V. The efficiency of the transmission line is approximately
(a) 91% (b) 85% (c) 96% (d) 72%

3 Number of molecules in a volume of 4 cm^3 of a perfect monoatomic gas at some temperature T and at a pressure of 2 cm of mercury is close to (Given, mean kinetic energy of a molecule at T is 4×10^{-14} erg, $g = 980$ cm/s^2, density of mercury = 13.6 g/cm^3)
(a) 5.8×10^{16} (b) 4.0×10^{16}
(c) 4.0×10^{18} (d) 5.8×10^{18}

4 For a concave lens of focal length f, the relation between object and image distances u and v, respectively, from its pole can best be represented by ($u = v$ is the reference line)

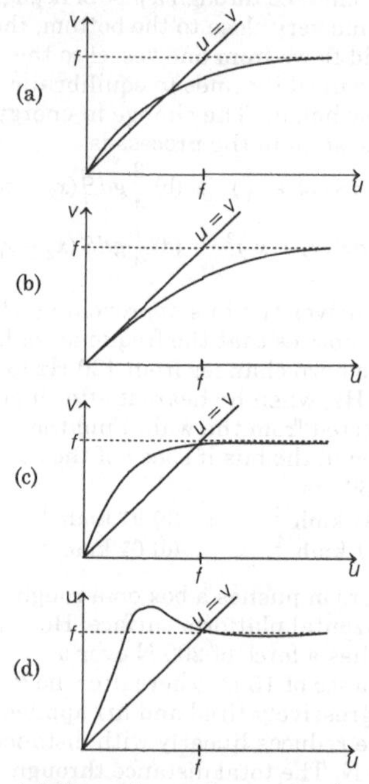

5 A balloon is moving up in air vertically above a point A on the ground. When it is at a height h_1, a girl standing at a distance d (point B) from A (see figure) sees it at an angle 45° with respect to the vertical. When the balloon climbs up a further height h_2, it is seen at an angle 60° with respect to the vertical if the girl moves further by a distance 2.464 d (point C). Then, the height h_2 is (Given, tan 30°= 0.5774)

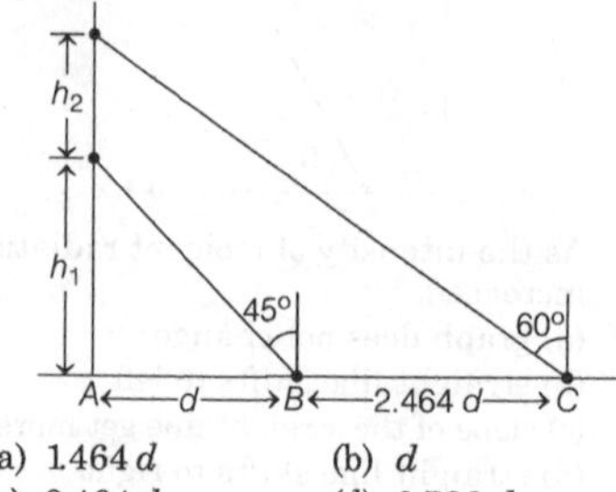

(a) 1.464 d (b) d
(c) 0.464 d (d) 0.732 d

6 A bullet of mass 5g, travelling with a speed of 210 m/s, strikes a fixed wooden target. One-half of its kinetic energy is converted into heat in the bullet while the other half is converted into heat in the wood. The rise of temperature of the bullet, if the specific heat of its material is 0.030 cal g^{-1} °C^{-1}, is close to
(a) 87.5°C (b) 83.3°C
(c) 119.2°C (d) 38.4°C

7 A physical quantity z depends on four observables a, b, c and d, as $z = \frac{a^2 b^{2/3}}{\sqrt{c} d^3}$.
The percentages of error in the measurement of a, b, c and d are 2%, 1.5%, 4% and 2.5% respectively.
The percentage of error in z is
(a) 13.5% (b) 16.5% (c) 14.5% (d) 12.25%

8 An electron is constrained to move along the Y-axis with a speed of 0.1c (c is the speed of light) in the presence of electromagnetic wave, whose electric field is
$\mathbf{E} = 30\hat{j} \sin(1.5 \times 10^7 t - 5 \times 10^{-2} x)$ V/m.
The maximum magnetic force experienced by the electron will be (Given, $c = 3 \times 10^8$ ms^{-1} and electron charge = 1.6×10^{-19} C)
(a) 2.4×10^{-18} N (b) 4.8×10^{-19} N
(c) 1.6×10^{-19} N (d) 3.2×10^{-18} N

9 A galvanometer of resistance G is converted into a voltmeter of range 0-1V by connecting a resistance R_1 in series with it. The additional resistance that should be connected in series with R_1 to increase the range of the voltmeter to 0-2V will be
(a) $R_1 + G$ (b) R_1 (c) G (d) $R_1 - G$

10 A wheel is rotating freely with an angular speed ω on a shaft. The moment of inertia of the wheel is I and the moment of inertia of the shaft is negligible. Another wheel of moment of inertia $3I$ initially at rest is suddenly coupled to the same shaft. The resultant fractional loss in the kinetic energy of the system is
(a) 3/4 (b) 5/6 (c) 0 (d) 1/4

11 Two capacitors of capacitances C and $2C$ are charged to potential differences V and $2V$ respectively. These are then connected in parallel in such a manner that the positive terminal of one is connected to the negative terminal of the other. The final energy of this configuration is

(a) $\frac{9}{2}CV^2$ (b) $\frac{3}{2}CV^2$ (c) $\frac{25}{6}CV^2$ (d) zero

12 The value of the acceleration due to gravity is g_1 at a height $h = R/2$ (where, R = radius of the earth) from the surface of the earth. It is again equal to g_1 at a depth d below the surface of the earth. The ratio (d/R) equals

(a) $\frac{7}{9}$ (b) $\frac{1}{3}$ (c) $\frac{4}{9}$ (d) $\frac{5}{9}$

13 A helicopter rises from rest on the ground vertically upwards with a constant acceleration g. A food packet is dropped from the helicopter when it is at a height h. The time taken by the packet to reach the ground is close to (Here, g is the acceleration due to gravity).

(a) $t = \frac{2}{3}\sqrt{\frac{h}{g}}$ (b) $t = 1.8\sqrt{\frac{h}{g}}$

(c) $t = \sqrt{\frac{2h}{3g}}$ (d) $t = 3.4\sqrt{\frac{h}{g}}$

14 A square loop of side $2a$ and carrying current I is kept in xy-plane with its centre at origin. A long wire carrying the same current I is placed parallel to the Z-axis and passing through the point $(0, b, 0)$, $(b >> a)$. The magnitude of the torque on the loop about Z-axis is given by

(a) $\frac{2\mu_0 I^2 a^3}{\pi b^2}$ (b) $\frac{2\mu_0 I^2 a^2}{\pi b}$

(c) $\frac{\mu_0 I^2 a^3}{2\pi b^2}$ (d) $\frac{\mu_0 I^2 a^2}{2\pi b}$

15 Activities of three radioactive substances A, B and C are represented by the curves A, B and C in the figure. Then, their half-lives $T_{1/2}(A):T_{1/2}(B):T_{1/2}(C)$ are in the ratio

(a) 4 : 3 : 1 (b) 3 : 2 : 1
(c) 2 : 1 : 1 (d) 2 : 1 : 3

16 In a resonance tube experiment, when the tube is filled with water up to a height of 17.0 cm from bottom, it resonates with a given tuning fork. When the water level is raised, the next resonance with the same tuning fork occurs at a height of 24.5 cm. If the velocity of sound in air is 330 m/s, the tuning fork frequency is

(a) 1100 Hz (b) 3300 Hz
(c) 2200 Hz (d) 550 Hz

17 Assume that the displacement of air is proportional to the pressure difference Δp created by a sound wave. Displacement further depends on the speed of sound v, density of air ρ and the frequency f. If $\Delta p \sim 10$Pa, $v \sim 300$ m/s, $\rho \sim 1$ kg/m^3 and $f \sim 1000$ Hz, then s will be of the order of (Take the multiplicative constant to be 1.)

(a) 10 mm (b) 1 mm
(c) $\frac{1}{10}$ mm (d) $\frac{3}{100}$ mm

18 With increasing biasing voltage of a photodiode, the photocurrent magnitude

(a) increases initially and after attaining certain value, it decreases
(b) increases initially and saturates finally
(c) remains constant
(d) increases linearly

19 A hollow spherical shell at outer radius R floats just submerged under the water surface. The inner radius of the shell is r. If the specific gravity of the shell material is $\frac{27}{8}$ with respect to water, the value of r is

(a) $\frac{8}{9}R$ (b) $\frac{1}{3}R$ (c) $\frac{2}{3}R$ (d) $\frac{4}{9}R$

20 A solid sphere of radius R carries a charge $Q + q$ distributed uniformly over its volume. A very small point-like piece of it of mass m gets detached from the bottom of the sphere and falls down vertically under gravity. This piece carries charge q. If it acquires a speed v when it has fallen through a vertical height y (see figure), then (Assume the remaining portion to be spherical.)

(a) $v^2 = y\left[\frac{qQ}{4\pi\varepsilon_0 R^2 ym} + g\right]$

(b) $v^2 = 2y\left[\frac{qQ}{4\pi\varepsilon_0 R(R+y)m} + g\right]$

(c) $v^2 = y\left[\frac{qQ}{4\pi\varepsilon_0 R(R+y)m} + g\right]$

(d) $v^2 = 2y\left[\frac{qQR}{4\pi\varepsilon_0 (R+y)^3 m} + g\right]$

Numerical Type Questions

21 A beam of electrons of energy E scatters from a target having atomic spacing of 1Å. The first maximum intensity occurs at $\theta = 60°$. Then, E (in eV) is.... (Given, Planck's constant, $h = 6.64 \times 10^{-34}$ Js, 1 eV $= 1.6 \times 10^{-19}$ J and electron mass, $m = 9.1 \times 10^{-31}$ kg

22 A compound microscope consists of an objective lens of focal length 1 cm and an eye piece of focal length 5 cm with a separation of 10 cm. The distance between an object and the objective lens, at which the strain on the eye is minimum is $\frac{n}{40}$ cm. The value of n is... .

23 A particle of mass 200 MeV/c^2 collides with a hydrogen atom at rest. Soon after the collision, the particle comes to rest and the atom recoils and goes to its first excited state. The initial kinetic energy of the particle (in eV) is $N/4$. The value of N is .... (Given, the mass of the hydrogen atom to be 1 GeV/c^2)

24 Two concentric circular coils C_1 and C_2 are placed in the xy-plane. C_1 has 500 turns and radius of 1 cm. C_2 has 200 turns and radius of 20 cm. C_2 carries a time dependent current $I(t) = (5t^2 - 2t + 3)$A, where t is in second. The emf induced in C_1 (in mV), at the instant $t = 1$ s is $\frac{4}{x}$. The value of x is ....

25 A force $\mathbf{F} = (\hat{i} + 2\hat{j} + 3\hat{k})$ N acts at a point $(4\hat{i} + 3\hat{j} - \hat{k})$m. Then, the magnitude of torque about the point $(\hat{i} + 2\hat{j} + \hat{k})$m will be $\sqrt{x}$ N-m. The value of x is ......

ANSWERS

1. (*)	2. (c)	3. (c)	4. (b)	5. (b)	6. (a)	7. (c)	8. (b)	9. (a)	10. (a)
11. (b)	12. (d)	13. (d)	14. (b)	15. (d)	16. (c)	17. (d)	18. (b)	19. (a)	20. (b)
21. (50)	22. (50)	23. (51)	24. (5)	25. (195)					

For Detailed Solutions
Visit : https://bit.ly/2FENPaG
Or Scan :

5 September, Shift-II

1 A spaceship in space sweeps stationary interplanetary dust. As a result, its mass increases at a rate $\frac{dM(t)}{dt} = bv^2(t)$, where $v(t)$ is its instantaneous velocity. The instantaneous acceleration of the satellite is

(a) $-bv^3(t)$ (b) $-\frac{bv^3}{M(t)}$

(c) $-\frac{2bv^3}{M(t)}$ (d) $-\frac{bv^3}{2M(t)}$

2 Ten charges are placed on the circumference of a circle of radius R with constant angular separation between successive charges. Alternate charges 1, 3, 5, 7, 9 have charge $+q$ each, while 2, 4, 6, 8, 10 have charge $-q$ each. The potential V and the electric field E at the centre of the circle respectively, are (Take, $V = 0$ at infinity)

(a) $V = \frac{10q}{4\pi\varepsilon_0 R}$; $E = 0$

(b) $V = 0$; $E = \frac{10q}{4\pi\varepsilon_0 R^2}$

(c) $V = 0$; $E = 0$

(d) $V = \frac{10q}{4\pi\varepsilon_0 R}$; $E = \frac{10q}{4\pi\varepsilon_0 R^2}$

3 An infinitely long straight wire carrying current I, one side opened rectangular loop and a conductor C with a sliding connector are located in the same plane, as shown in the figure. The connector has length l and resistance R. It slides to the right with a velocity v. The resistance of the conductor and the self-inductance of the loop are negligible. The induced current in the loop, as a function of separation r between the connector and the straight wire is

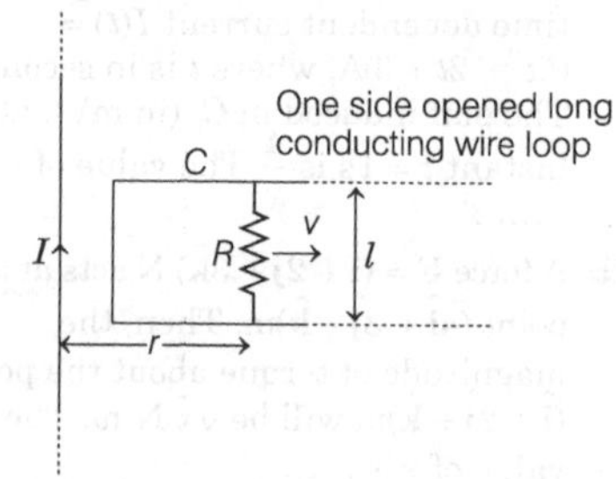

(a) $\frac{\mu_0}{4\pi}\frac{Ivl}{Rr}$ (b) $\frac{\mu_0}{\pi}\frac{Ivl}{Rr}$

(c) $\frac{2\mu_0}{\pi}\frac{Ivl}{Rr}$ (d) $\frac{\mu_0}{2\pi}\frac{Ivl}{Rr}$

4 In the circuit shown, charge on the 5μF capacitor is

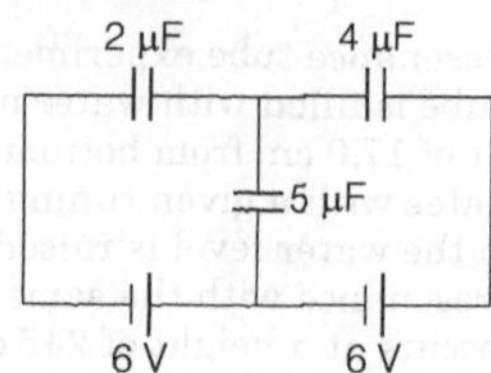

(a) 18.00 μC (b) 10.90 μC
(c) 16.36 μC (d) 5.45 μC

5 A galvanometer is used in laboratory for detecting the null point in electrical experiments. If on passing a current of 6 mA, it produces a deflection of 2°, its figure of merit is close to

(a) 333° A/div (b) 6×10^{-3} A/div
(c) 666° A/div (d) 3×10^{-3} A/div

6 The correct match between the entries in table are :

Column I (Radiation)	Column II (Wavelength)
(A) Microwaves	(I) 100 m
(B) γ-rays	(II) 10^{-15} m
(C) AM radiowaves	(III) 10^{-10} m
(D) X-rays	(IV) 10^{-3} m

	A	B	C	D
(a)	II	I	IV	III
(b)	I	III	IV	II
(c)	III	II	I	IV
(d)	IV	II	I	III

7 Two coherent sources of sound S_1 and S_2, produce sound waves of the same wavelength $\lambda = 1$ m, in phase. S_1 and S_2 are placed 1.5 m apart (see figure). A listener, located at L, directly in front of S_2 finds that the intensity is at a minimum when he is 2 m away from S_2. The listener moves away from S_1, keeping his distance from S_2 fixed. The adjacent maximum of intensity is observed when the listener is at a distance d from S_1. Then, d is

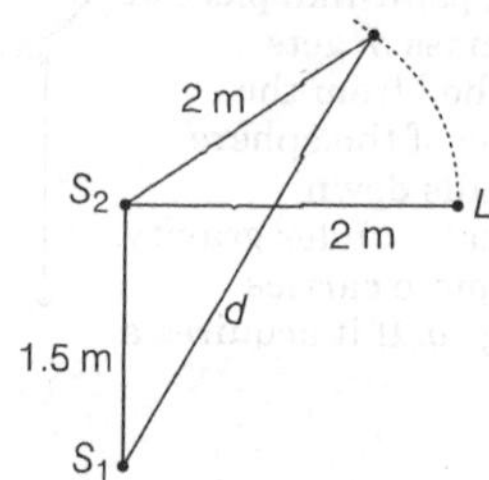

(a) 12 m (b) 5 m
(c) 2 m (d) 3 m

8 Two different wires having lengths L_1 and L_2 and respective temperature coefficients of linear expansion α_1 and α_2, are joined end-to-end. Then the effective temperature coefficient of linear expansion is

(a) $\frac{\alpha_1 L_1 + \alpha_2 L_2}{L_1 + L_2}$

(b) $2\sqrt{\alpha_1 \alpha_2}$

(c) $\frac{\alpha_1 + \alpha_2}{2}$

(d) $4\frac{\alpha_1 \alpha_2}{\alpha_1 + \alpha_2}\frac{L_2 L_1}{(L_2 + L_1)^2}$

9 The v-t graph of a body in a straight line motion is shown in the figure. The point S is at 4.333 s. The total distance covered by the body in 6 s is

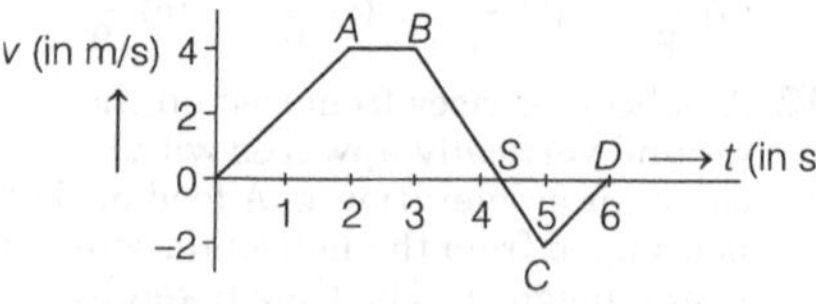

(a) $\frac{37}{3}$ m (b) 12 m

(c) 11 m (d) $\frac{49}{4}$ m

10 In an experiment to verify Stoke's law, a small spherical ball of radius r and density ρ falls under gravity through a distance h in air before entering a tank of water. If the terminal velocity of the ball inside water is same as its velocity just before entering the water surface, then the value of h is proportional to (Ignore viscosity of air)

(a) r^4 (b) r
(c) r^3 (d) r^2

11 A parallel plate capacitor has plate of length l, width w and separation of plates is d. It is connected to a battery of emf V. A dielectric slab of the same thickness d and of dielectric constant $k = 4$ is being inserted between the plates of the capacitor. At what length of the slab inside plates, will the energy stored in the capacitor be two times the initial energy stored?

(a) $\frac{2l}{3}$ (b) $\frac{l}{3}$ (c) $\frac{l}{4}$ (d) $\frac{l}{2}$

12 A driver in a car, approaching a vertical wall notices that the frequency of his car horn, has changed from 440 Hz to 480 Hz, when it gets reflected from the wall. If the speed of sound in air is 345 m/s, then the speed of the car is

(a) 54 km/h (b) 36 km/h
(c) 18 km/h (d) 24 km/h

13 A ring is hung on a nail. It can oscillate without slipping or sliding

(i) in its plane with a time period T_1 and

(ii) back and forth in a direction perpendicular to its plane, with a period T_2.

The ratio $\frac{T_1}{T_2}$ will be

(a) $\frac{2}{\sqrt{3}}$ (b) $\frac{2}{3}$

(c) $\frac{3}{\sqrt{2}}$ (d) $\frac{\sqrt{2}}{3}$

14 Two zener diodes (A and B) having breakdown voltages of 6 V and 4 V respectively, are connected as shown in the circuit below. The output voltage V_{out} variation with input voltage linearly increasing with time, is given by (V_{in} = 0 V at t = 0) (figures are qualitative)

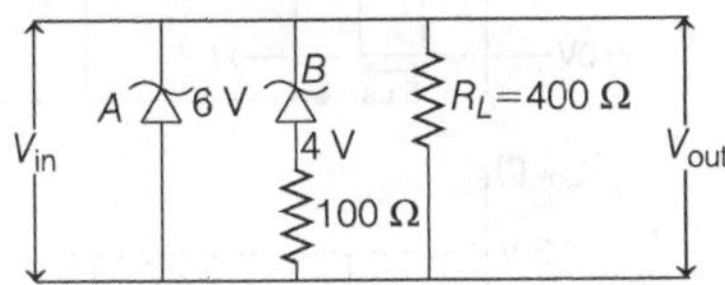

(a)

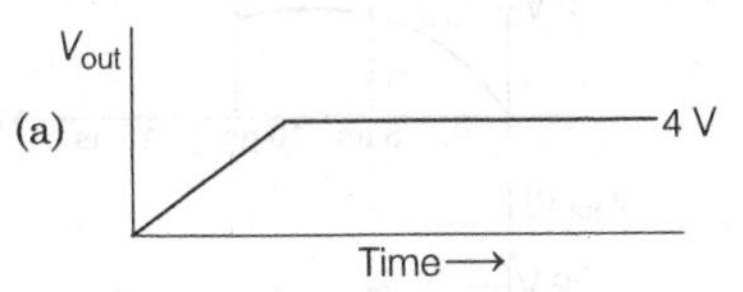

(b)

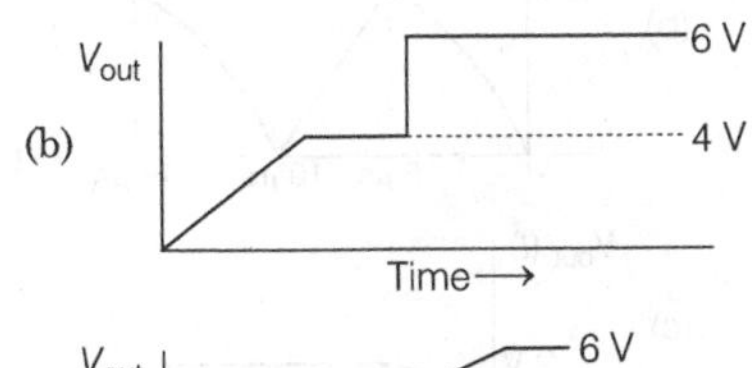

(c)

(d)

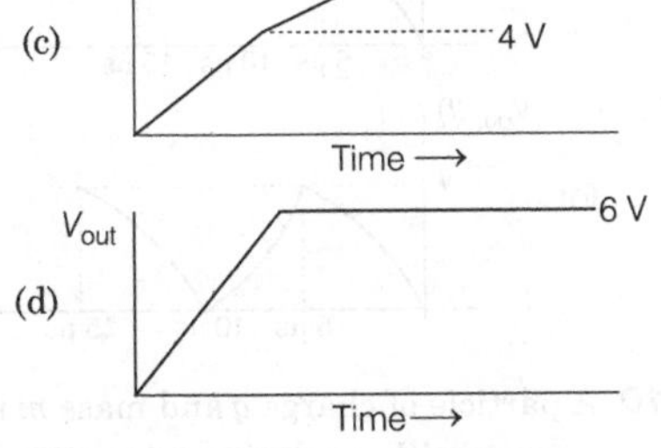

15 A radioactive nucleus decays by two different processes. The half-life for the first process is 10 s and that for the second is 100 s. The effective half-life of the nucleus is close to

(a) 9 s (b) 6 s

(c) 55 s (d) 12 s

16 The quantities $x = \frac{1}{\sqrt{\mu_0 \varepsilon_0}}$, $y = \frac{E}{B}$ and $z = \frac{l}{CR}$ are defined, where C is capacitance, R is resistance, l is length, E is electric field, B is magnetic field, ε_0 is free space permittivity and μ_0 is permeability, respectively. Then,

(a) x, y and z have the same dimension

(b) Only x and z have the same dimension

(c) Only x and y have the same dimension

(d) Only y and z have the same dimension

17 In an adiabatic process, the density of a diatomic gas becomes 32 times of its initial value. The final pressure of the gas is found to be n times the initial pressure. The value of n is

(a) 32 (b) 326

(c) 128 (d) $\frac{1}{32}$

18 The acceleration due to gravity on the earth's surface at the poles is g and angular velocity of the earth about the axis passing through the pole is ω. An object is weighed at the equator and at a height h above the poles by using a spring balance. If the weights are found to be same, then h is ($h << R$, where R is the radius of the earth)

(a) $\frac{R^2\omega^2}{2g}$ (b) $\frac{R^2\omega^2}{g}$

(c) $\frac{R^2\omega^2}{4g}$ (d) $\frac{R^2\omega^2}{8g}$

19 In the given circuit, currents in different branches and value of one resistor are shown. Then, potential at point B with respect to the point A is

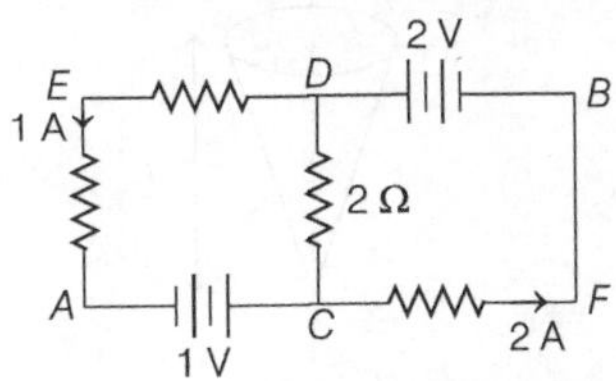

(a) +2 V (b) −2 V (c) −1 V (d) +1 V

20 An iron rod of volume 10^{-3} m^3 and relative permeability 1000 is placed as core in a solenoid with 10 turns/cm. If a current of 0.5 A is passed through the solenoid, then the magnetic moment of the rod will be

(a) 50×10^2 Am^2

(b) 5×10^2 Am^2

(c) 500×10^2 Am^2

(d) 0.5×10^2 Am^2

Numerical Type Questions

21 A body of mass 2 kg is driven by an engine delivering a constant power of 1 J/s. The body starts from rest and moves in a straight line. After 9 s, the body has moved a distance (in m) ....

22 Nitrogen gas is at 300°C temperature. The temperature (in K) at which the rms speed of a H_2 molecule would be equal to the rms speed of a N_2 molecule is ....

(Molar mass of N_2 gas is 28 g.)

23 A prism of angle A = 1° has a refractive index μ = 1.5. A good estimate for the minimum angle of deviation (in degree) is close to N/10. Value of N is ....

24 The surface of a metal is illuminated alternately with photons of energies E_1 = 4 eV and E_2 = 2.5 eV respectively. The ratio of maximum speeds of the photoelectrons emitted in the two cases is 2. The work function of the metal (in eV) is ....

25 A thin rod of mass 0.9 kg and length 1 m is suspended at rest from one end, so that it can freely oscillate in the vertical plane.

A particle of mass 0.1 kg moving in a straight line with velocity 80 m/s hits the rod at its bottom-most point and sticks to it (see figure). The angular speed (in rad/s) of the rod immediately after the collision will be ....

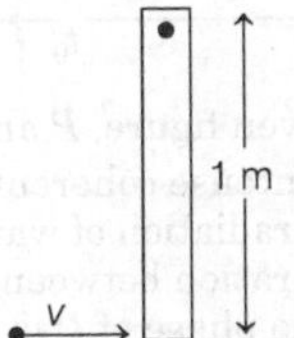

ANSWERS

1. (b)	2. (c)	3. (d)	4. (c)	5. (d)	6. (d)	7. (d)	8. (a)	9. (a)	10. (a)
11. (b)	12. (a)	13. (a)	14. (c)	15. (a)	16. (a)	17. (c)	18. (a)	19. (d)	20. (b)
21. (18)	22. (41)	23. (5)	24. (2)	25. (20)					

For Detailed Solutions

Visit : https://bit.ly/3o6TQOY

Or Scan :

6 September, Shift-I

1 A screw gauge has 50 divisions on its circular scale. The circular scale is 4 units ahead of the pitch scale marking, prior to use. Upon one complete rotation of the circular scale, a displacement of 0.5 mm is noticed on the pitch scale. The nature of zero error involved and the least count of the screw gauge, are respectively

(a) negative, 2 μm (b) positive, 10 μm
(c) positive, 0.1 mm (d) positive, 0.1 μm

2 A sound source S is moving along a straight track with speed v and is emitting sound of frequency ν_o (see figure). An observer is standing at a finite distance, at the point O, from the track. The time variation of frequency heard by the observer is best represented by

(Here, t_0 represents the instant when the distance between the source and observer is minimum.)

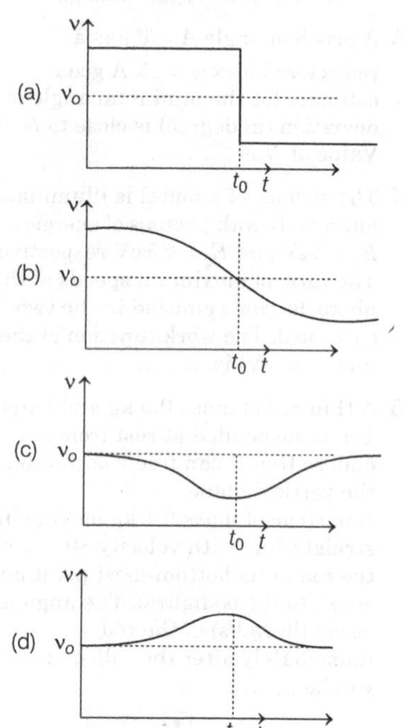

3 In the given figure, P and Q are two equally intense coherent sources emitting radiation of wavelength 20 m. The separation between P and Q is 5 m and the phase of P is ahead of that of Q by 90°. A, B and C are three distinct points of observation, each equidistant from the mid-point of PQ. The intensities of radiation at A, B and C will be in the ratio

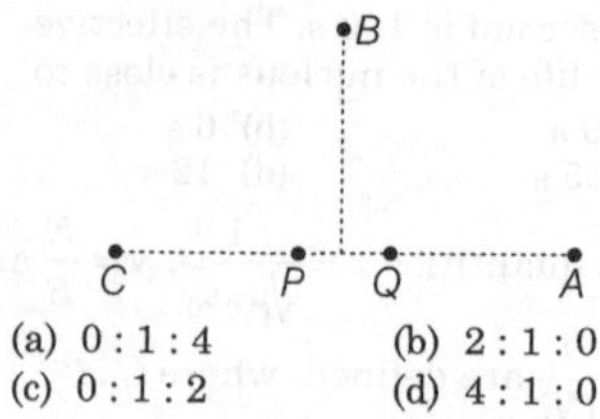

(a) 0 : 1 : 4 (b) 2 : 1 : 0
(c) 0 : 1 : 2 (d) 4 : 1 : 0

4 If the potential energy between two molecules is given by $U = -\frac{A}{r^6} + \frac{B}{r^{12}}$, then at equilibrium, separation between molecules and the potential energy are

(a) $\left(\frac{B}{2A}\right)^{1/6}, -\frac{A^2}{2B}$ (b) $\left(\frac{B}{A}\right)^{1/6}, 0$
(c) $\left(\frac{2B}{A}\right)^{1/6}, -\frac{A^2}{4B}$ (d) $\left(\frac{2B}{A}\right)^{1/6}, -\frac{A^2}{2B}$

5 An AC circuit has $R = 100\,\Omega$, $C = 2\,\mu F$ and $L = 80$ mH connected in series. The quality factor of the circuit is

(a) 2 (b) 0.5 (c) 20 (d) 400

6 Charges Q_1 and Q_2 are at points A and B of a right angle triangle OAB (see figure). The resultant electric field at point O is perpendicular to the hypotenuse, then Q_1/Q_2 is proportional to

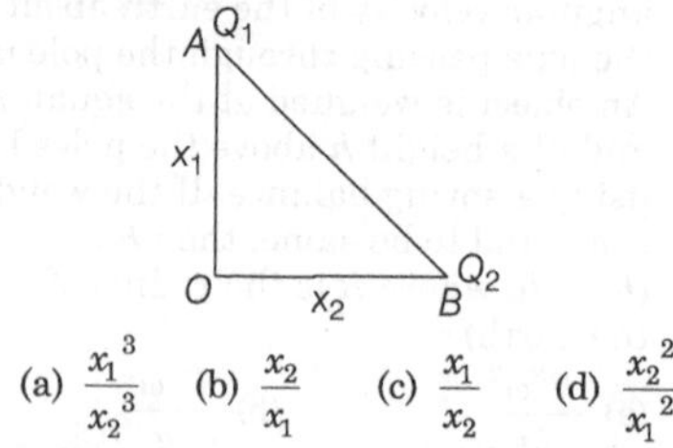

(a) $\frac{x_1^3}{x_2^3}$ (b) $\frac{x_2}{x_1}$ (c) $\frac{x_1}{x_2}$ (d) $\frac{x_2^2}{x_1^2}$

7 Shown in the figure is a hollow icecream cone (it is open at the top). If its mass is M, radius of its top R and height H, then its moment of inertia about its axis is

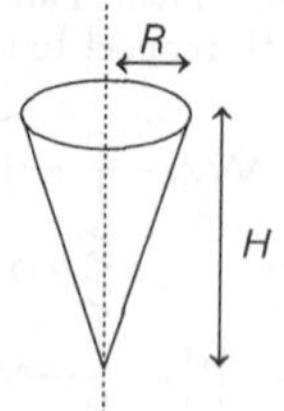

(a) $\frac{MR^2}{2}$ (b) $\frac{M(R^2 + H^2)}{4}$
(c) $\frac{MH^2}{3}$ (d) $\frac{MR^2}{3}$

8 Four point masses, each of mass m are fixed at the corners of a square of side l. The square is rotating with angular frequency ω, about an axis passing through one of the corners of the square and parallel to its diagonal, as shown in the figure. The angular momentum of the square about this axis is

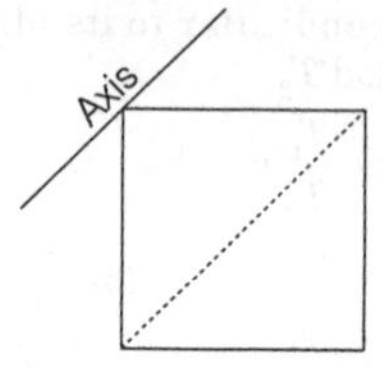

(a) $ml^2\omega$ (b) $4\,ml^2\omega$
(c) $3\,ml^2\omega$ (d) $2\,ml^2\omega$

9 For the given input voltage waveform $V_{in}(t)$, the output voltage waveform $V_{out}(t)$, across the capacitor is correctly depicted by

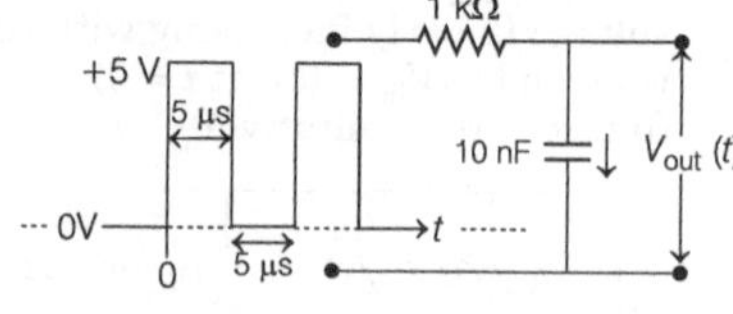

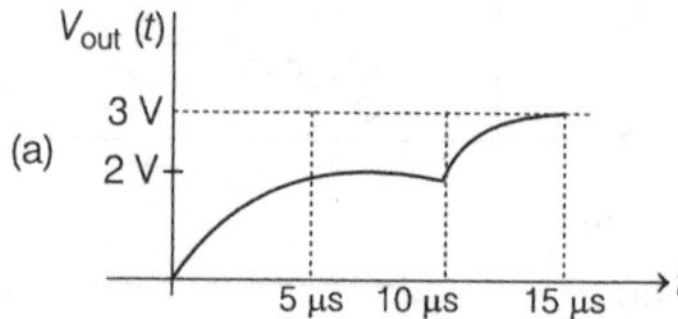

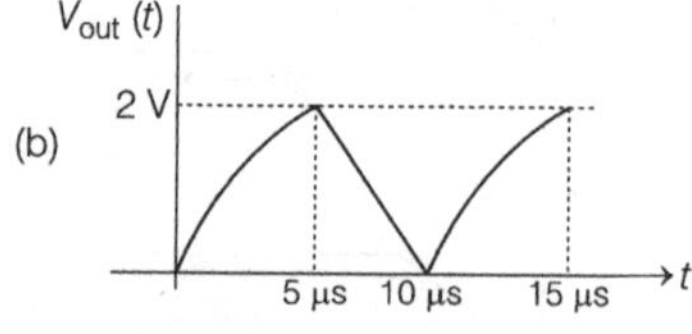

(c)

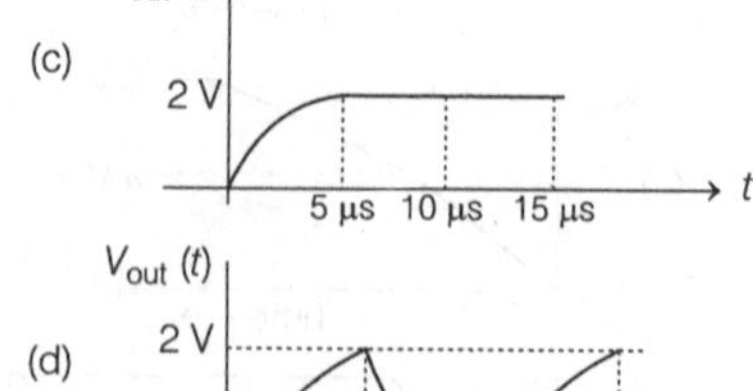

10 A particle of charge q and mass m is moving with a velocity $-v\hat{\mathbf{i}} (v \neq 0)$ towards a large screen placed in the yz-plane at a distance d. If there is a magnetic field $\mathbf{B} = B_0\hat{\mathbf{k}}$, the minimum value of v for which the particle will not hit the screen is

(a) $\frac{qdB_0}{3m}$ (b) $\frac{2qdB_0}{m}$
(c) $\frac{qdB_0}{m}$ (d) $\frac{qdB_0}{2m}$

11 An insect is at the bottom of a hemispherical ditch of radius 1 m. It crawls up the ditch but starts slipping after it is at height h from the bottom. If the coefficient of friction between the ground and the insect is 0.75, then h is (Take, $g = 10 \text{ ms}^{-2}$)

(a) 0.20 m (b) 0.45 m
(c) 0.60 m (d) 0.80 m

12 A satellite is in an elliptical orbit around a planet P. It is observed that the velocity of the satellite when it is farthest from the planet is 6 times less than that when it is closest to the planet. The ratio of distances between the satellite and the planet at closest and farthest points is

(a) 1 : 6 (b) 1 : 3
(c) 1 : 2 (d) 3 : 4

13 An electron, a doubly ionised helium ion (He^{++}) and a proton are having the same kinetic energy. The relation between their respective de-Broglie wavelengths λ_e, $\lambda_{He^{++}}$ and λ_p is

(a) $\lambda_e > \lambda_{He^{++}} > \lambda_p$ (b) $\lambda_e < \lambda_{He^{++}} = \lambda_p$
(c) $\lambda_e > \lambda_p > \lambda_{He^{++}}$ (d) $\lambda_e < \lambda_p < \lambda_{He^{++}}$

14 A clock has a continuously moving second's hand of 0.1 m length. The average acceleration of the tip of the hand (in ms^{-2}) is of the order of

(a) 10^{-3} (b) 10^{-4} (c) 10^{-2} (d) 10^{-1}

15 You are given that mass of ${}^7_3\text{Li} = 7.0160\text{u}$, mass of ${}^4_2\text{He} = 4.0026\text{u}$ and mass of ${}^1_1\text{H} = 1.0079\text{u}$.

When 20g of ${}^7_3\text{Li}$ is converted into ${}^4_2\text{He}$ by proton capture, the energy liberated (in kWh), is
[Mass of nucleon = 1 GeV/c^2]

(a) 4.5×10^5 (b) 8×10^6
(c) 6.82×10^5 (d) 1.33×10^6

16 A point like object is placed at a distance of 1 m in front of a convex lens of focal length 0.5 m. A plane mirror is placed at a distance of 2 m behind the lens. The position and nature of the final image formed by the system is

(a) 2.6 m from the mirror, real
(b) 1 m from the mirror, virtual
(c) 1 m from the mirror, real
(d) 2.6 m from the mirror, virtual

17 Identify the correct output signal Y in the given combination of gates (as shown) for the given inputs A and B.

18 Molecules of an ideal gas are known to have three translational degrees of freedom and two rotational degrees of freedom. The gas is maintained at a temperature of T.

The total internal energy U of a mole of this gas, and the value of $\gamma \left(= \frac{C_p}{C_V} \right)$ are given respectively, by

(a) $U = \frac{5}{2}RT$ and $\gamma = \frac{6}{5}$
(b) $U = 5RT$ and $\gamma = \frac{7}{5}$
(c) $U = \frac{5}{2}RT$ and $\gamma = \frac{7}{5}$
(d) $U = 5RT$ and $\gamma = \frac{6}{5}$

19 An object of mass m is suspended at the end of a massless wire of length L and area of cross-section A. Young modulus of the material of the wire is Y. If the mass is pulled down slightly its frequency of oscillation along the vertical direction is

(a) $f = \frac{1}{2\pi}\sqrt{\frac{mL}{YA}}$ (b) $f = \frac{1}{2\pi}\sqrt{\frac{YA}{mL}}$
(c) $f = \frac{1}{2\pi}\sqrt{\frac{mA}{YL}}$ (d) $f = \frac{1}{2\pi}\sqrt{\frac{YL}{mA}}$

20 An electron is moving along $+x$-direction with a velocity of $6 \times 10^6 \text{ ms}^{-1}$. It enters a region of uniform electric field of 300 V/cm pointing along $+y$-direction. The magnitude and direction of the magnetic field set up in this region such that the electron keeps moving along the x-direction will be

(a) 3×10^{-4} T, along $+z$-direction
(b) 5×10^{-3} T, along $-z$-direction
(c) 5×10^{-3} T, along $+z$-direction
(d) 3×10^{-4} T, along $-z$-direction

Numerical Type Questions

21 The density of a solid metal sphere is determined by measuring its mass and its diameter. The maximum error in the density of the sphere is $\left(\frac{x}{100} \right)$%. If the relative errors in measuring the mass and the diameter are 6.0% and 1.5% respectively, the value of x is ....

22 Two bodies of the same mass are moving with the same speed, but in different directions in a plane. They have a completely inelastic collision and move together thereafter with a final speed which is half of their initial speed. The angle between the initial velocities of the two bodies (in degree) is ....

23 Suppose that intensity of a laser is $\left(\frac{315}{\pi} \right) \text{W/m}^2$. The rms electric field (in V/m) associated with this source is close to the nearest integer is .... (Take, $\varepsilon_0 = 8.86 \times 10^{-12} \text{ C}^2\text{Nm}^{-2}$ and $c = 3 \times 10^8 \text{ ms}^{-1}$)

24 Initially a gas of diatomic molecules is contained in a cylinder of volume V_1 at a pressure p_1 and temperature 250 K. Assuming that 25% of the molecules get dissociated causing a change in number of moles. The pressure of the resulting gas at temperature 2000 K, when contained in a volume $2V_1$ is given by p_2. The ratio p_2/p_1 is ....

25 A part of a complete circuit is shown in the figure. At some instant, the value of current I is 1 A and it is decreasing at the rate of 10^2 As^{-1}. The value of the potential difference $V_P - V_Q$ (in volt) at that instant, is ....

ANSWERS

1. (b)	2. (b)	3. (b)	4. (c)	5. (a)	6. (c)	7. (a)	8. (c)	9. (a)	10. (c)
11. (a)	12. (a)	13. (c)	14. (a)	15. (d)	16. (a)	17. (b)	18. (c)	19. (b)	20. (c)
21. (1050)	22. (120)	23. (194)	24. (5)	25. (33)					

For Detailed Solutions
Visit : https://bit.ly/3kbXJ2P
Or Scan :

6 September, Shift-II

1

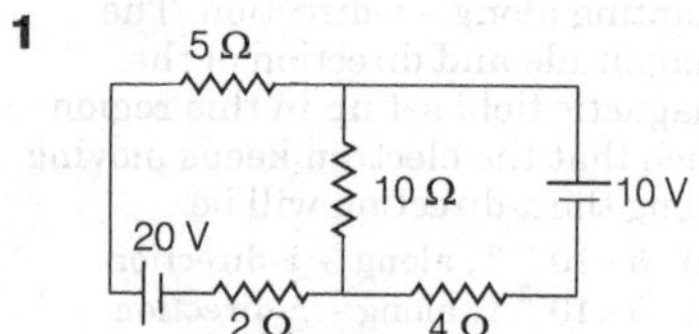

In above figure shown, the current in the 10 V battery is close to

(a) 0.71 A from positive to negative terminal
(b) 0.42 A from positive to negative terminal
(c) 0.21 A from positive to negative terminal
(d) 0.36 A from negative to positive terminal

2 A charged particle going around in a circle can be considered to be a current loop. A particle of mass m carrying charge q is moving in a plane with speed v under the influence of magnetic field $\mathbf{B}$. The magnetic moment of this moving particle

(a) $\frac{mv^2\mathbf{B}}{2B^2}$ (b) $-\frac{mv^2\mathbf{B}}{2\pi B^2}$
(c) $-\frac{mv^2\mathbf{B}}{B^2}$ (d) $-\frac{mv^2\mathbf{B}}{2B^2}$

3 Three rods of identical cross-section and lengths are made of three different materials of thermal conductivity K_1, K_2 and K_3, respectively. They are joined together at their ends to make a long rod (see figure). One end of the long rod is maintained at 100°C and the other at 0°C (see figure). If the joints of the rod are at 70°C and 20°C in steady state and there is no loss of energy from the surface of the rod, the correct relationship between K_1, K_2 and K_3 is

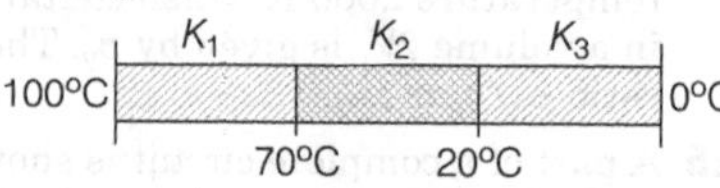

(a) $K_1 : K_3 = 2:3; K_2 : K_3 = 2:5$
(b) $K_1 < K_2 < K_3$
(c) $K_1 : K_2 = 5:2; K_1 : K_3 = 3:5$
(d) $K_1 > K_2 > K_3$

4 Two identical electric point dipoles have dipole moments $\mathbf{p}_1 = p\hat{\mathbf{i}}$ and $\mathbf{p}_2 = -p\hat{\mathbf{i}}$ are held on the X-axis at distance a from each other. When released, they move along the X-axis with the direction of their dipole moments remaining unchanged. If the mass of each dipole is m, their speed when they are infinitely far apart is

(a) $\frac{p}{a}\sqrt{\frac{1}{\pi\varepsilon_0 ma}}$ (b) $\frac{p}{a}\sqrt{\frac{1}{2\pi\varepsilon_0 ma}}$
(c) $\frac{p}{a}\sqrt{\frac{2}{\pi\varepsilon_0 ma}}$ (d) $\frac{p}{a}\sqrt{\frac{3}{2\pi\varepsilon_0 ma}}$

5 For a plane electromagnetic wave, the magnetic field at a point x and time t is

$$\mathbf{B}(x, t) = [1.2\times10^{-7}\sin(0.5\times10^3 x + 1.5\times10^{11} t)\hat{\mathbf{k}}]\text{ T}$$

The instantaneous electric field $\mathbf{E}$ corresponding to $\mathbf{B}$ is
(Given, speed of light, $c = 3\times10^8$ m/s)

(a) $\mathbf{E}(x, t) = [-36\sin(0.5\times10^3 x + 1.5\times10^{11} t)\hat{\mathbf{j}}]\frac{\text{V}}{\text{m}}$
(b) $\mathbf{E}(x, t) = [36\sin(1\times10^3 x + 0.5\times10^{11} t)\hat{\mathbf{j}}]\frac{\text{V}}{\text{m}}$
(c) $\mathbf{E}(x, t) = [36\sin(0.5\times10^3 x + 1.5\times10^{11} t)\hat{\mathbf{k}}]\frac{\text{V}}{\text{m}}$
(d) $\mathbf{E}(x, t) = [36\sin(1\times10^3 x + 1.5\times10^{11} t)\hat{\mathbf{j}}]\frac{\text{V}}{\text{m}}$

6 Two planets have masses M and $16\,M$ and their radii are a and $2a$, respectively. The separation between the centres of the planets is $10a$. A body of mass m is fired from the surface of the larger planet towards the smaller planet along the line joining their centres. For the body to be able to reach at the surface of smaller planet, the minimum firing speed needed is

(a) $2\sqrt{\frac{GM}{a}}$ (b) $4\sqrt{\frac{GM}{a}}$
(c) $\sqrt{\frac{GM^2}{ma}}$ (d) $\frac{3}{2}\sqrt{\frac{5GM}{a}}$

7 A particle moving in the xy-plane experiences a velocity dependent force

$\mathbf{F} = k(v_y\hat{\mathbf{i}} + v_x\hat{\mathbf{j}})$, where v_x and v_y are the x and y components of its velocity $\mathbf{v}$.

If $\mathbf{a}$ is the acceleration of the particle, then which of the following statements is true for the particle?

(a) Quantity $\mathbf{v}\times\mathbf{a}$ is constant in time.
(b) $\mathbf{F}$ arises due to a magnetic field.
(c) Kinetic energy of particle is constant in time.
(d) Quantity $\mathbf{v}\cdot\mathbf{a}$ is constant in time.

8 Particle A of mass m_1 moving with velocity $(\sqrt{3}\hat{\mathbf{i}} + \hat{\mathbf{j}})\text{ms}^{-1}$ collides with another particle B of mass m_2 which is at rest initially. Let $\mathbf{v}_1$ and $\mathbf{v}_2$ be the velocities of particles A and B after collision, respectively. If $m_1 = 2m_2$ and after collision $\mathbf{v}_1 = (\hat{\mathbf{i}} + \sqrt{3}\hat{\mathbf{j}})\text{ ms}^{-1}$, then the angle between $\mathbf{v}_1$ and $\mathbf{v}_2$ is

(a) 15° (b) 60°
(c) −45° (d) 105°

9 When a car is at rest, its driver sees rain drops falling on it vertically. When driving the car with speed v, he sees that rain drops are coming at an angle 60° from the horizontal. On further increasing the speed of the car to $(1+\beta)v$, this angle changes to 45°. The value of β is close to

(a) 0.50 (b) 0.41 (c) 0.37 (d) 0.73

10 Given the masses of various atomic particles $m_p = 1.0072$ u, $m_n = 1.0087$ u, $m_e = 0.000548$ u, $m_{\bar{\nu}} = 0$, $m_d = 2.0141$u, where p = proton, n = neutron, e = electron $\bar{\nu}$ = antineutrino and d = deuteron. Which of the following process is allowed by momentum and energy conservation?

(a) $n + n \to$ deuterium atom (electron bound to the nucleus)
(b) $p \to n + e^+ + \bar{\nu}$
(c) $n + p \to d + \bar{\nu}$
(d) $e^+ + e^- \to \gamma$

11 A circuit to verify Ohm's law uses ammeter and voltmeter in series or parallel connected correctly to the resistor. In the circuit,

(a) ammeter is always used in parallel and voltmeter in series
(b) Both ammeter and voltmeter must be connected in parallel
(c) ammeter is always connected in series and voltmeter in parallel
(d) Both ammeter and voltmeter must be connected in series

12 Consider the force F on a charge q due to a uniformly charged spherical shell of radius R carrying charge Q distributed uniformly over it. Which one of the following statement is true for F, if q is placed at distance r from the centre of the shell ?

(a) $F = \frac{1}{4\pi\varepsilon_0}\frac{Qq}{R^2}$ (for $r < R$)
(b) $\frac{1}{4\pi\varepsilon_0}\frac{Qq}{R^2} > F > 0$ (for $r < R$)
(c) $F = \frac{1}{4\pi\varepsilon_0}\frac{Qq}{r^2}$ (for $r > R$)
(d) $F = \frac{1}{4\pi\varepsilon_0}\frac{Qq}{R^2}$ (for all r)

13 A student measuring the diameter of a pencil of circular cross-section with the help of a vernier scale records the following four readings; 5.50 mm, 5.55 mm, 5.45 mm and 5.65 mm. The average of these four readings is 5.5375 mm and the standard deviation of the data is 0.07395 mm. The average diameter of the pencil should be therefore recorded as

(a) (5.5375 ± 0.0739) mm
(b) (5.5375 ± 0.0740) mm
(c) (5.538 ± 0.074) mm
(d) (5.54 ± 0.07) mm

14 A double convex lens has power P and same radii of curvature R of both the surfaces. The radius of curvature of a surface of a plano-convex lens made of the same material with power $1.5\,P$ is

(a) $2R$ (b) $\frac{R}{2}$ (c) $\frac{3R}{2}$ (d) $\frac{R}{3}$

15 A square loop of side $2a$ and carrying current I is kept in xz-plane with its centre at origin. A long wire carrying the same current I is placed parallel to Z-axis and passing through point $(0, b, 0)$, $(b >> a)$. The magnitude of torque on the loop about Z-axis will be

(a) $\frac{2\mu_0 I^2 a^2}{\pi b}$ (b) $\frac{2\mu_0 I^2 a^2 b}{\pi(a^2 + b^2)}$

(c) $\frac{\mu_0 I^2 a^2 b}{2\pi(a^2 + b^2)}$ (d) $\frac{\mu_0 I^2 a^2}{2\pi b}$

16 A fluid is flowing through a horizontal pipe of varying cross-section with speed v ms^{-1} at a point where the pressure is p Pa. At another point, where pressure is $\frac{p}{2}$ Pa its speed is V ms^{-1}. If the density of the fluid is ρ kg m^{-3} and the flow is streamline, then V is equal to

(a) $\sqrt{\frac{p}{\rho} + v}$ (b) $\sqrt{\frac{2p}{\rho} + v^2}$

(c) $\sqrt{\frac{p}{2\rho} + v^2}$ (d) $\sqrt{\frac{p}{\rho} + v^2}$

17 When a particle of mass m is attached to a vertical spring of spring constant k and released, its motion is described by $y(t) = y_0 \sin^2 \omega t$, where y is measured from the lower end of unstretched spring. Then ω is

(a) $\frac{1}{2}\sqrt{\frac{g}{y_0}}$ (b) $\sqrt{\frac{g}{y_0}}$

(c) $\sqrt{\frac{g}{2y_0}}$ (d) $\sqrt{\frac{2g}{y_0}}$

18 In a dilute gas at pressure p and temperature T, the mean time between successive collisions of a molecule varies with T as

(a) T (b) $\frac{1}{\sqrt{T}}$

(c) $\frac{1}{T}$ (d) $\sqrt{T}$

19 Assuming the nitrogen molecule is moving with rms velocity at 400 K, the de-Broglie wavelength of nitrogen molecule is close to
(Given, weight of N_2 molecule $= 4.64 \times 10^{-26}$ kg, Boltzmann constant $= 1.38 \times 10^{-23}$ J/K and Planck's constant $= 6.63 \times 10^{-34}$ J-s)

(a) 0.24Å (b) 0.20Å
(c) 0.34Å (d) 0.44Å

20 The linear mass density of a thin rod AB of length L varies from A to B as $\lambda(x) = \lambda_0\left(1 + \frac{x}{L}\right)$, where x is the distance from A. If M is mass of the rod, then its moment of inertia about an axis passing through A and perpendicular to the rod is

(a) $\frac{5}{12}ML^2$ (b) $\frac{7}{18}ML^2$

(c) $\frac{2}{5}ML^2$ (d) $\frac{3}{7}ML^2$

Numerical Type Questions

21 The output characteristics of a transistor is shown in the figure.

When V_{CE} is 10V and $I_C = 4.0$ mA, then value of β_{ac} is

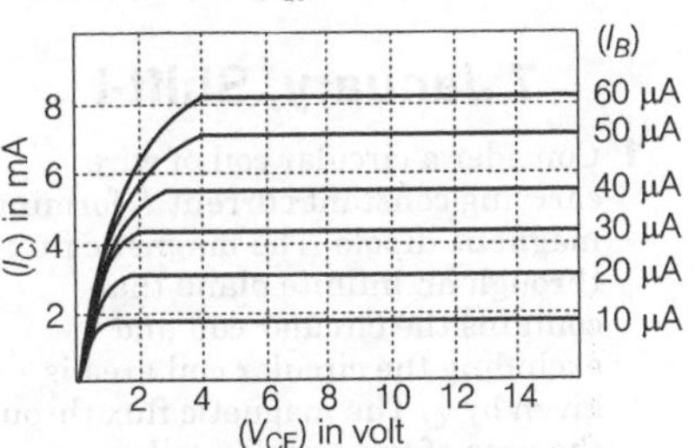

22 The centre of mass of a solid hemisphere of radius 8 cm is x cm from the centre of the flat surface. Then, value of x is ....

23 An engine operates by taking a monatomic ideal gas through the cycle shown in the figure. The percentage efficiency of the engine is close to ....

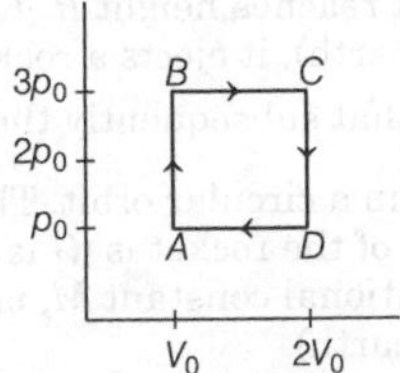

24 A Young's double slit experiment is performed using monochromatic light of wavelength λ. The intensity of light at a point on the screen, where the path difference is λ, is K units. The intensity of light at a point where the path difference is $\frac{\lambda}{6}$ is given by $\frac{nK}{12}$, where n is an integer. The value of n is ....

25 In a series L-R circuit, power of 400W is dissipated from a source of 250 V, 50 Hz. The power factor of the circuit is 0.8. In order to bring the power factor to unity, a capacitor of value C is added in series to the L and R. Taking the value of C as $\left(\frac{n}{3\pi}\right)$ μF, then value of n is ....

ANSWERS

1. (c)	2. (d)	3. (a)	4. (b)	5. (a)	6. (d)	7. (a)	8. (d)	9. (d)	10. (c)
11. (c)	12. (c)	13. (d)	14. (d)	15. (a)	16. (d)	17. (c)	18. (b)	19. (a)	20. (b)
21. (150)	22. (3)	23. (19)	24. (9)	25. (400)					

For Detailed Solutions
Visit : https://bit.ly/31lbmFq
Or Scan :

JANUARY ATTEMPT

7 January, Shift-I

1 Consider a circular coil of wire carrying constant current I, forming a magnetic dipole. The magnetic flux through an infinite plane that contains the circular coil and excluding the circular coil area is given by ϕ_i. The magnetic flux through the area of the circular coil area is given by ϕ_o. Which of the following option is correct ?

(a) $\phi_i > \phi_o$
(b) $\phi_i < \phi_o$
(c) $\phi_i = \phi_o$
(d) $\phi_i = -\phi_o$

2 A satellite of mass m is launched vertically upwards with an initial speed u from the surface of the earth. After it reaches height R (R = radius of the earth), it ejects a rocket of mass $\frac{m}{10}$, so that subsequently the satellite moves in a circular orbit. The kinetic energy of the rocket is (G is the gravitational constant M, is the mass of the earth)

(a) $\frac{3m}{8}\left(u + \sqrt{\frac{5\,GM}{6R}}\right)^2$
(b) $\frac{m}{20}\left(u - \sqrt{\frac{2GM}{3R}}\right)^2$
(c) $\frac{m}{20}\left(u^2 + \frac{113}{200}\frac{GM}{R}\right)$
(d) $5m\left(u^2 - \frac{119}{200}\frac{GM}{R}\right)$

3 If the magnetic field in a plane electromagnetic wave is given by $\mathbf{B} = 3\times10^{-8}\sin(1.6\times10^3 x + 48\times10^{10} t)\hat{\mathbf{j}}$T, then what will be expression for electric field?

(a) $\mathbf{E} = 60\sin(1.6\times10^3 x + 48\times10^{10} t)\,\hat{\mathbf{k}}$ V/m
(b) $\mathbf{E} = 3\times10^{-8}\sin(1.6\times10^3 x + 48\times10^{10} t)\,\hat{\mathbf{i}}$ V/m
(c) $\mathbf{E} = 3\times10^{-8}\sin(1.6\times10^3 x + 48\times10^{10} t)\,\hat{\mathbf{j}}$ V/m
(d) $\mathbf{E} = 9\sin(1.6\times10^3 x + 48\times10^{10} t)\,\hat{\mathbf{k}}$ V/m

4 Visible light of wavelength 6000×10^{-8} cm falls normally on a single slit and produces a diffraction pattern. It is found that the second diffraction minimum is at 60° from the central maximum. If the first minimum is produced at θ_1, then θ_1 is close to

(a) 25° (b) 30°
(c) 20° (d) 45°

5 An L-C-R circuit behaves like a damped harmonic oscillator. Comparing it with a physical spring-mass damped oscillator having damping constant b, the correct equivalence would be

(a) $L \leftrightarrow \frac{1}{b}, C \leftrightarrow \frac{1}{m}, R \leftrightarrow \frac{1}{k}$
(b) $L \leftrightarrow m, C \leftrightarrow \frac{1}{k}, R \leftrightarrow b$
(c) $L \leftrightarrow m, C \leftrightarrow k, R \leftrightarrow b$
(d) $L \leftrightarrow k, C \leftrightarrow b, R \leftrightarrow m$

6 As shown in the figure, a bob of mass m is tied by a massless string whose other end portion is wound on a flywheel (disc) of radius r and mass m. When released from rest the bob starts falling vertically. When it has covered a distance of h, the angular speed of the wheel will be

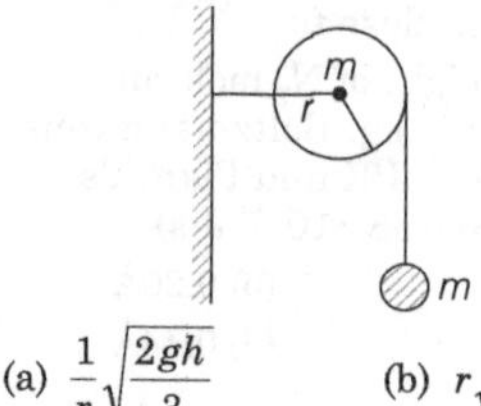

(a) $\frac{1}{r}\sqrt{\frac{2gh}{3}}$ (b) $r\sqrt{\frac{3}{4gh}}$
(c) $\frac{1}{r}\sqrt{\frac{4gh}{3}}$ (d) $r\sqrt{\frac{3}{2gh}}$

7 Speed of a transverse wave on a straight wire (mass 6.0 g, length 60 cm and area of cross-section 1.0 mm^2) is 90 ms^{-1}. If the Young's modulus of wire is 16×10^{11} Nm^{-2}, the extension of wire over its natural length is

(a) 0.01 mm (b) 0.04 mm
(c) 0.03 mm (d) 0.02 mm

8 Two infinite planes each with uniform surface charged density $+\sigma$ are kept in such a way that the angle between them is 30°. The electric field in the region shown between them is given by

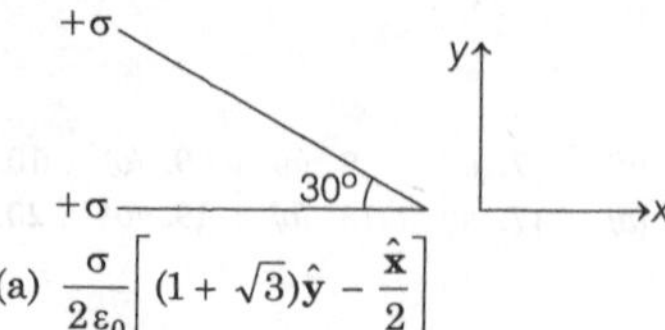

(a) $\frac{\sigma}{2\varepsilon_0}\left[(1+\sqrt{3})\hat{\mathbf{y}} - \frac{\hat{\mathbf{x}}}{2}\right]$
(b) $\frac{\sigma}{2\varepsilon_0}\left[\left(1-\frac{\sqrt{3}}{2}\right)\hat{\mathbf{y}} - \frac{\hat{\mathbf{x}}}{2}\right]$
(c) $\frac{\sigma}{2\varepsilon_0}\left[(1+\sqrt{3})\hat{\mathbf{y}} + \frac{\hat{\mathbf{x}}}{2}\right]$
(d) $\frac{\sigma}{\varepsilon_0}\left[\left(1+\frac{\sqrt{3}}{2}\right)\hat{\mathbf{y}} + \frac{\hat{\mathbf{x}}}{2}\right]$

9 The current I_1 (in ampere) flowing through 1Ω resistor in the following circuit is

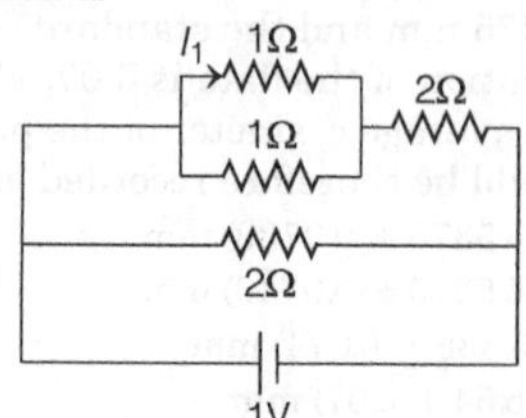

(a) 0.25 (b) 0.4
(c) 0.5 (d) 0.2

10 Which of the following gives a reversible operation?

(a) (b)

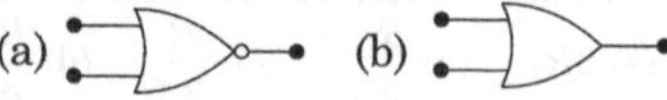

(c) (d)

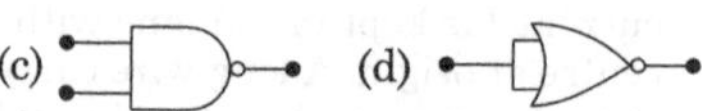

11 Two moles of an ideal gas with $\frac{C_p}{C_V} = \frac{5}{3}$ are mixed with 3 mol of another ideal gas with $\frac{C_p}{C_V} = \frac{4}{3}$. The value of $\frac{C_p}{C_V}$ for the mixture is

(a) 1.42 (b) 1.47
(c) 1.50 (d) 1.45

12 The time period of revolution of electron in its ground state orbit in a hydrogen atom is 1.6×10^{-16} s. The frequency of revolution of the electron in its first excited state (in Hz) is

(a) 1.6×10^{14} (b) 5.6×10^{12}
(c) 6.2×10^{15} (d) 7.8×10^{14}

13 A parallel plate capacitor has plates of area A separated by distance d between them. It is filled with a dielectric which has a dielectric constant that varies as $k(x) = K(1+\alpha x)$, where x is the distance measured from one of the plates. If $(\alpha d) << 1$, the total capacitance of the system is best given by the expression

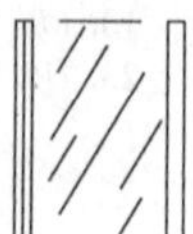

(a) $\frac{AK\varepsilon_0}{d}(1+\alpha d)$
(b) $\frac{A\varepsilon_0 K}{d}\left(1+\left(\frac{\alpha d}{2}\right)^2\right)$

(c) $\frac{AK\varepsilon_0}{d}\left(1+\frac{\alpha d}{2}\right)$

(d) $\frac{A\varepsilon_0 K}{d}\left(1+\frac{\alpha^2 d^2}{2}\right)$

14 If we need a magnification of 375 from a compound microscope of tube length 150 mm and an objective of focal length 5 mm, the focal length of the eyepiece should be close to

(a) 22 mm (b) 2 mm
(c) 12 mm (d) 33 mm

15 A litre of dry air at STP expands adiabatically to a volume of 3 L. If $\gamma = 1.40$, the work done by air is ($3^{1.4} = 4.6555$)
[Take, air to be an ideal gas]

(a) 100.8 J (b) 90.5 J
(c) 48 J (d) 60.7 J

16 A 60 HP electric motor lifts an elevator having a maximum total load capacity of 2000 kg. If the frictional force on the elevator is 4000 N, the speed of the elevator at full load is close to
(Take, 1 HP = 746 W, $g = 10\,\text{ms}^{-2}$)

(a) $2.0\ \text{ms}^{-1}$ (b) $1.5\ \text{ms}^{-1}$
(c) $1.9\ \text{ms}^{-1}$ (d) $1.7\ \text{ms}^{-1}$

17 A polariser-analyser set is adjusted such that the intensity of light coming out of the analyser is just 10% of the original intensity. Assuming that the polariser-analyser set does not absorb any light, the angle by which the analyser need to be rotated further to reduce the output intensity to be zero, is

(a) 71.6° (b) 90°
(c) 45° (d) 18.4°

18 A long solenoid of radius R carries a time (t)-dependent current $I(t) = I_0 t(1-t)$. A ring of radius $2R$ is placed coaxially near its middle. During the time interval $0 \le t \le 1$, the induced current (I_R) and the induced EMF(V_R) in the ring change as ;

(a) Direction of I_R remains unchanged and V_R is maximum at $t = 0.5$s
(b) At $t = 0.25$s direction of I_R reverses and V_R is maximum
(c) Direction of I_R remains unchanged and V_R is zero at $t = 0.25$s
(d) At $t = 0.5$ direction of I_R reverses and V_R is zero

19 Three point particles of masses 1.0 kg, 1.5 kg and 2.5 kg are placed at three corners of a right angle triangle of sides 4.0 cm, 3.0 cm and 5.0 cm as shown in the figure. The centre of mass of the system is at a point

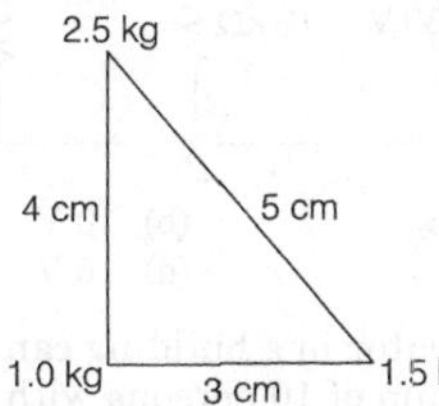

(a) 2.0 cm right and 0.9 cm above 1 kg mass
(b) 0.6 cm right and 2.0 cm above 1 kg mass
(c) 1.5 cm right and 1.2 cm above 1 kg mass
(d) 0.9 cm right and 2.0 cm above 1 kg mass

20 The radius of gyration of a uniform rod of length l, about an axis passing through a point $\frac{l}{4}$ away from the centre of the rod and perpendicular to it, is

(a) $\frac{1}{8}l$ (b) $\sqrt{\frac{3}{8}}\,l$
(c) $\sqrt{\frac{7}{48}}\,l$ (d) $\frac{1}{4}l$

Numerical Type Questions

21 A loop $ABCDEFA$ of straight edges has six corner points $A(0, 0, 0)$, $B(5, 0, 0)$, $C(5, 5, 0)$, $D(0, 5, 0)$, $E(0, 5, 5)$ and $F(0, 0, 5)$. The magnetic field in this region is $\mathbf{B} = (3\hat{i} + 4\hat{k})$T. The quantity of flux through the loop $ABCDEFA$ (in Wb) is ....

22 A beam of electromagnetic radiation of intensity $6.4 \times 10^{-5}\ \text{W/cm}^2$ is comprised of wavelength $\lambda = 310$ nm. It falls normally on a metal (work function $\phi = 2$ eV) of surface area of 1 cm^2. If one in 10^3 photons ejects an electron, total number of electrons ejected in 1 s is 10^x.
($hc = 1240$ eVnm, 1 eV $= 1.6 \times 10^{-19}$ J), then x is ....

23 A Carnot engine operates between two reservoirs of temperatures 900 K and 300 K. The engine performs 1200 J of work per cycle. The heat energy (in J) delivered by the engine to the low temperature reservoir in a cycle, is ....

24 A non-isotropic solid metal cube has coefficients of linear expansion as : 5×10^{-5}/°C along the X-axis and 5×10^{-6}/°C along the Y and the Z-axis. If the coefficient of volume expansion of the solid is $C \times 10^{-6}$/°C, then the value of C is ....

25 A particle ($m = 1$kg) slides down a frictionless track (AOC) starting from rest at a point A (height 2 m). After reaching C, the particle continues to move freely in air as a projectile. When it reaching its highest point P (height 1 m), the kinetic energy of the particle (in J) is : (Figure drawn is schematic and not to scale; take $g = 10\,\text{ms}^{-2}$) ....

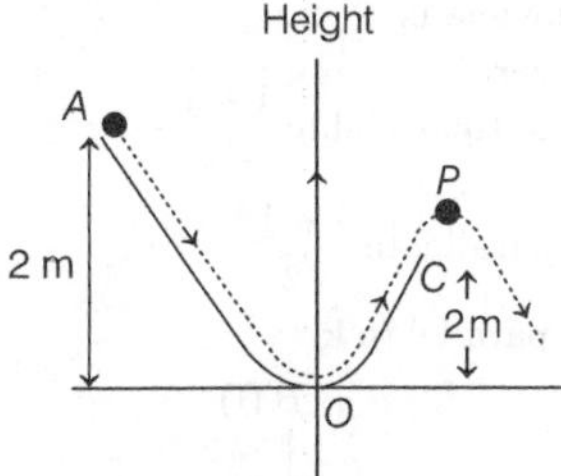

ANSWERS

1. (d)	2. (d)	3. (d)	4. (a)	5. (b)	6. (c)	7. (c)	8. (b)	9. (d)	10. (d)
11. (a)	12. (d)	13. (c)	14. (a)	15. (b)	16. (c)	17. (d)	18. (d)	19. (d)	20. (c)
21. (175)	22. (11)	23. (600)	24. (60)	25. (10)					

For Detailed Solutions
Visit : https://bit.ly/35770Tm
Or Scan :

7 January, Shift-II

1 A thin lens made of glass (refractive index = 1.5) of focal length $f = 16$ cm in immersed in a liquid of refractive index 1.42. If its focal length in liquid is f_1, then the ratio f_1/f is closest to the integer

(a) 1 (b) 5
(c) 9 (d) 17

2 A mass of 10 kg is suspended by a rope of length 4 m, from the ceiling. A force F is applied horizontally at the mid-point of the rope such that the top half of the rope makes an angle of 45° with the vertical. Then, F equals (Take, $g = 10\,\text{ms}^{-2}$ and the rope to be massless)

(a) 75 N (b) 70 N
(c) 100 N (d) 90 N

3 A box weighs 196 N on a spring balance at the north pole. Its weight recorded on the same balance, if it is shifted to the equator is close to (Take, $g = 10\,\text{ms}^{-2}$ at the north pole and the radius of the earth = 6400 km)

(a) 195.66 N (b) 195.32 N
(c) 194.66 N (d) 194.32 N

4 The electric field of a plane electromagnetic wave is given by

$$\mathbf{E} = E_0 \frac{\hat{\mathbf{i}} + \hat{\mathbf{j}}}{\sqrt{2}} \cos(kz + \omega t)$$

At $t = 0$, a positively charged particle is at the point $(x, y, z) = \left(0, 0, \frac{\pi}{k}\right)$.

If its instantaneous velocity at $(t = 0)$ is $v_0 \hat{\mathbf{k}}$, the force acting on it due to the wave is

(a) zero
(b) antiparallel to $\frac{\hat{\mathbf{i}} + \hat{\mathbf{j}}}{\sqrt{2}}$
(c) parallel to $\frac{\hat{\mathbf{i}} + \hat{\mathbf{j}}}{\sqrt{2}}$
(d) parallel to $\hat{\mathbf{k}}$

5

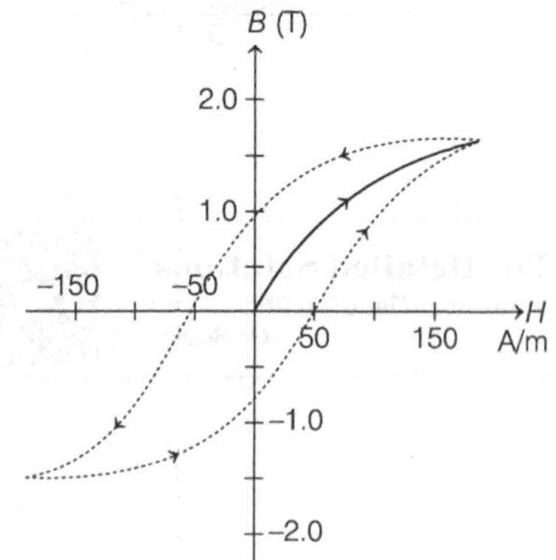

The figure gives experimentally measured B *versus* H variation in a ferromagnetic material. The retentivity, coercivity and saturation respectively of the material are

(a) 1.0 T, 50 A/m and 1.5 T
(b) 150 A/m, 1.0 T and 1.5 T
(c) 1.5 T, 50 A/m and 1.0 T
(d) 1.5 T, 50 A/m and 1.0 T

6 In the figure, potential difference between A and B is

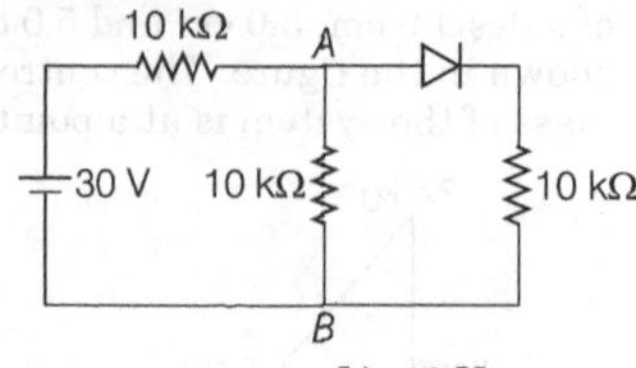

(a) zero (b) 10 V
(c) 5 V (d) 15 V

7 An elevator in a building can carry a maximum of 10 persons with the average mass of each person being 68 kg. The mass of the elevator itself is 920 kg and it moves with a constant speed of 3 m/s. The frictional force opposing the motion is 6000 N. If the elevator is moving up with its full capacity, the power delivered by the motor to the elevator ($g = 10\,\text{m/s}^2$) must be at least

(a) 62360 W (b) 48000 W
(c) 56300 W (d) 66000 W

8 A stationary observer receives sound from two identical tuning forks, one of which approaches and the other one recedes with the same speed (much less than the speed of sound). The observer hears 2 beats/s. The oscillation frequency of each tuning fork is $\nu_0 = 1400$ Hz and the velocity of sound in air is 350 m/s. The speed of each tuning fork is close to

(a) $\frac{1}{8}$ m/s (b) $\frac{1}{2}$ m/s (c) 1 m/s (d) $\frac{1}{4}$ m/s

9 An electron (of mass m) and a photon have the same energy E in the range of a few electron volt. The ratio of the de Broglie wavelength associated with the electron and the wavelength of the photon is (c = speed of light in vacuum)

(a) $\left(\frac{E}{2m}\right)^{1/2}$ (b) $\frac{1}{c}\left(\frac{2E}{m}\right)^{1/2}$
(c) $c\,(2mE)^{1/2}$ (d) $\frac{1}{c}\left(\frac{E}{2m}\right)^{1/2}$

10 In a building, there are 15 bulbs of 45 W, 15 bulbs of 100 W, 15 small fans of 10 W and 2 heaters of 1 kW. The voltage of electric main is 220 V. The minimum fuse capacity (rated value) of the building will be

(a) 25 A (b) 10 A (c) 20 A (d) 15 A

11 Two ideal Carnot engines operate in cascade (all heat given up by one engine is used by the other engine to produce work) between temperatures T_1 and T_2. The temperature of the hot reservoir of the first engine is T_1 and the temperature of the cold reservoir of the second engine is T_2. T is temperature of the sink of first engine which is also the source for the second engine. How is T related to T_1 and T_2, if both the engines perform equal amount of work?

(a) $T = 0$ (b) $T = \frac{2T_1T_2}{T_1 + T_2}$
(c) $T = \sqrt{T_1 T_2}$ (d) $T = \frac{T_1 + T_2}{2}$

12 In a Young's double slit experiment, the separation between the slits is 0.15 mm. In the experiment, a source of light of wavelength 589 nm is used and the interference pattern is observed on a screen kept 1.5 m away. The separation between the successive bright fringes on the screen is

(a) 4.9 mm (b) 3.9 mm
(c) 6.9 mm (d) 5.9 mm

13 The dimension of $\frac{B^2}{2\mu_0}$, where B is magnetic field and μ_0 is the magnetic permeability of vacuum, is

(a) $[\text{ML}^{-1}\text{T}^{-2}]$ (b) $[\text{MLT}^{-2}]$
(c) $[\text{ML}^2\text{T}^{-1}]$ (d) $[\text{ML}^2\text{T}^{-2}]$

14 Mass per unit area of a circular disc of radius a depends on the distance r from its centre as $\sigma(r) = A + Br$. The moment of inertia of the disc about the axis, perpendicular to the plane and passing through its centre is

(a) $2\pi a^4\left(\frac{A}{4} + \frac{B}{5}\right)$
(b) $2\pi a^4\left(\frac{aA}{4} + \frac{B}{5}\right)$
(c) $\pi a^4\left(\frac{A}{4} + \frac{aB}{5}\right)$
(d) $2\pi a^4\left(\frac{A}{4} + \frac{aB}{5}\right)$

15 Under an adiabatic process, the volume of an ideal gas gets doubled. Consequently, the mean collision time between the gas molecule changes from τ_1 to τ_2. If $\frac{C_p}{C_V} = \gamma$ for this gas, then a good estimate for $\frac{\tau_2}{\tau_1}$ is given by

(a) $\left(\frac{1}{2}\right)^{\frac{\gamma+1}{2}}$ (b) $\frac{1}{2}$
(c) 2 (d) $\left(\frac{1}{2}\right)^{\gamma}$

16 A planar loop of wire rotates in a uniform magnetic field. Initially at $t = 0$, the plane of the loop is perpendicular to the magnetic field. If it rotates with a period of 10s about an axis in its plane, then the magnitude of induced emf will be maximum and minimum respectively at

(a) 2.5 s and 7.5 s (b) 2.5 s and 5.0 s
(c) 5.0 s and 10.0 s (d) 5.0 s and 7.5 s

17 A particle of mass m and charge q has an initial velocity $\mathbf{v} = v_0\hat{\mathbf{j}}$. If an electric field $\mathbf{E} = E_0\hat{\mathbf{i}}$ and magnetic field $\mathbf{B} = B_0\hat{\mathbf{i}}$ act on the particle, its speed will double after a time

(a) $\frac{3mv_0}{qE_0}$ (b) $\frac{\sqrt{3}mv_0}{qE_0}$
(c) $\frac{2mv_0}{qE_0}$ (d) $\frac{\sqrt{2}mv_0}{qE_0}$

18 The activity of a radioactive sample falls from 700s^{-1} to 500s^{-1} in 30 min. Its half-life is close to

(a) 62 min (b) 66 min
(c) 72 min (d) 52 min

19 An ideal fluid flows (laminar flow) through a pipe of non-uniform diameter. The maximum and minimum diameters of the pipes are 6.4 cm and 4.8 cm, respectively. The ratio of the minimum and the maximum velocities of fluid in this pipe is

(a) $\frac{9}{16}$ (b) $\frac{81}{256}$
(c) $\frac{\sqrt{3}}{2}$ (d) $\frac{3}{4}$

20 An emf of 20 V is applied at time $t = 0$ to a circuit containing in series 10 mH inductor and 5Ω resistor. The ratio of the currents at time $t = \infty$ and at $t = 40$ s is close to (Take, $e^2 = 7.389$)

(a) 1.06 (b) 1.15
(c) 1.46 (d) 0.84

Numerical Type Questions

21 A 60 pF capacitor is fully charged by a 20 V supply. It is then disconnected from the supply and is connected to another uncharged 60 pF capacitor in parallel. The electrostatic energy that is lost in this process by the time, the charge is redistributed between them is (in nJ) ....

22 The balancing length for a cell is 560 cm in a potentiometer experiment. When an external resistance of 10Ω is connected in parallel to the cell, the balancing length changes by 60 cm. If the internal resistance of the cell is $\frac{N}{10}$Ω, where N is an integer, then value of N is ....

23 The sum of two forces **P** and **Q** is **R** such that $|\mathbf{R}| = |\mathbf{P}|$. The angle θ (in degrees) that the resultant of 2**P** and **Q** will make with **Q** is, ...

24 Consider a uniform cubical box of side a on a rough floor that is to be moved by applying minimum possible force F at a point b above its centre of mass (see figure). If the coefficient of friction is $\mu = 0.4$, the maximum possible value of $100 \times \frac{b}{a}$ for box not to topple before moving is ....

25 M grams of steam at 100°C is mixed with 200 g of ice at its melting point in a thermally insulated container. If it produces liquid water at 40°C [heat of vaporisation of water is 540 cal/g and heat of fusion of ice is 80 cal/g], the value of M is ....

ANSWERS

1. (c)	2. (c)	3. (b)	4. (b)	5. (a)	6. (b)	7. (d)	8. (d)	9. (d)	10. (c)
11. (d)	12. (d)	13. (a)	14. (d)	15. (a)	16. (b)	17. (b)	18. (a)	19. (a)	20. (a)
21. (6)	22. (12)	23. (90)	24. (75)	25. (40)					

For Detailed Solutions
Visit : https://bit.ly/3kboTqx
Or Scan :

8 January, Shift-I

1 A leak proof cylinder of length 1 m, made of a metal which has very low coefficient of expansion is floating vertically in water at 0 °C such that its height above the water surface is 20 cm. When the temperature of water is increased to 4°C, the height of the cylinder above the water surface becomes 21 cm. The density of water at $T = 4$°C, relative to the density at $T = 0$°C is close to

(a) 1.26 (b) 1.03 (c) 1.01 (d) 1.04

2 The magnifying power of a telescope with tube length 60 cm is 5. What is the focal length of its eyepiece?

(a) 10 cm (b) 20 cm
(c) 30 cm (d) 40 cm

3 The plot that depicts the behaviour of the mean free time τ (time between two successive collisions) for the molecules of an ideal gas, as a function of temperature (T), qualitatively is (graphs are schematic and not drawn to scale)

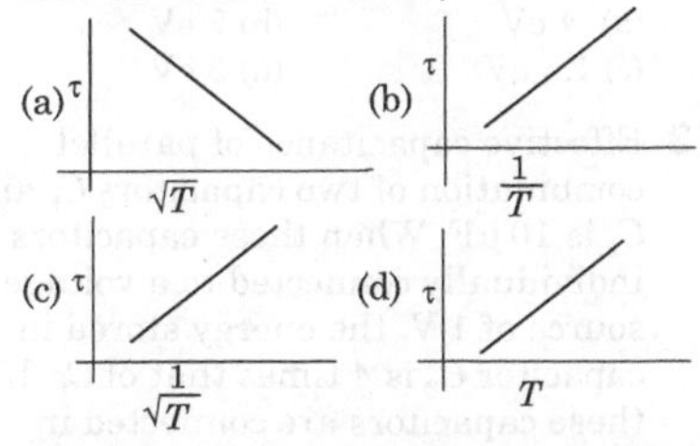

4 Consider two solid spheres of radii $R_1 = 1$ m, $R_2 = 2$ m and masses M_1 and M_2, respectively. The gravitational field due to sphere 1 and 2 are shown. The value of $\frac{M_1}{M_2}$ is

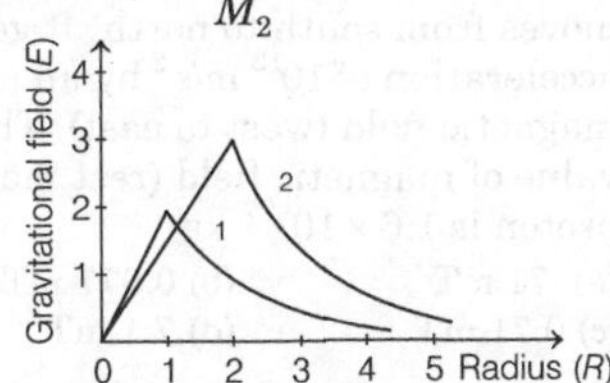

(a) $\frac{1}{3}$ (b) $\frac{1}{2}$ (c) $\frac{1}{6}$ (d) $\frac{2}{3}$

5 A particle of mass m is fixed to one end of a light spring having force constant k and unstretched length l. The other end is fixed. The system is given an angular speed ω about the fixed end of the spring such that it rotates in a circle in gravity free space. Then, the stretch in the spring is

(a) $\frac{ml\omega^2}{k + m\omega}$ (b) $\frac{ml\omega^2}{k + m\omega^2}$
(c) $\frac{ml\omega^2}{k - m\omega^2}$ (d) $\frac{ml\omega^2}{k - m\omega}$

6 A thermodynamic cycle $xyzx$ is shown on a V - T diagram.

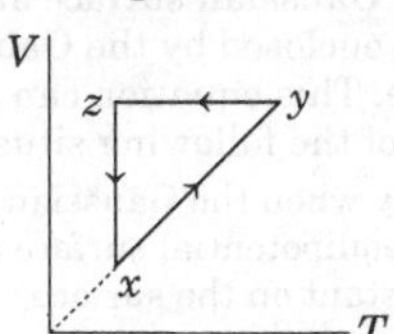

ONLINE JEE Main 2020

The p-V diagram that best describes this cycle is (diagrams are schematic and not to scale)

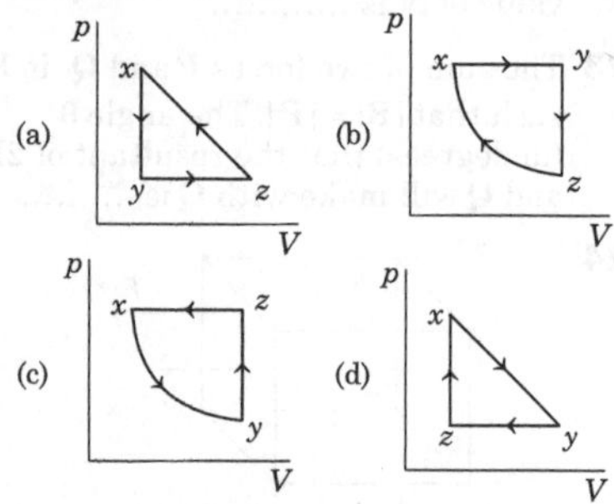

7 Consider a solid sphere of radius R and mass density $\rho(r) = \rho_0\left(1 - \frac{r^2}{R^2}\right)$, $0 < r \le R$. The minimum density of a liquid in which it will float is

(a) $\frac{\rho_0}{3}$ (b) $\frac{2\rho_0}{5}$
(c) $\frac{2\rho_0}{3}$ (d) $\frac{\rho_0}{5}$

8 Three charged particles A, B and C with charges $-4q$, $2q$ and $-2q$ are present on the circumference of a circle of radius d. The charged particles A, C and centre O of the circle formed an equilateral triangle as shown in figure. Electric field at O along x-direction is

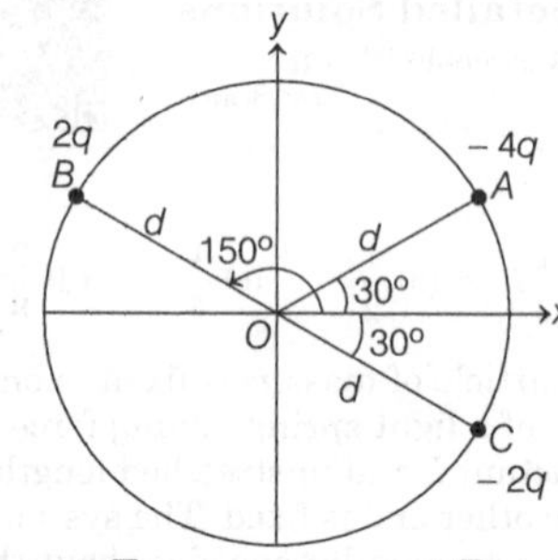

(a) $\frac{2\sqrt{3}q}{\pi\varepsilon_0 d^2}$ (b) $\frac{\sqrt{3}q}{4\pi\varepsilon_0 d^2}$
(c) $\frac{3\sqrt{3}q}{4\pi\varepsilon_0 d^2}$ (d) $\frac{\sqrt{3}q}{\pi\varepsilon_0 d^2}$

9 In finding the electric field using Gauss law the formula $|\mathbf{E}| = \frac{q_{\text{enc}}}{\varepsilon_0 |A|}$ is applicable. In the formula, ε_0 is permittivity of free space, A is the area of Gaussian surface and q_{enc} is charge enclosed by the Gaussian surface. This equation can be used in which of the following situation?

(a) Only when the Gaussian surface is an equipotential surface and $|\mathbf{E}|$ is constant on the surface.
(b) Only when the Gaussian surface is an equipotential surface.
(c) For any choice of Gaussian surface.
(d) Only when $|\mathbf{E}|$ = constant on the surface.

10 At time $t = 0$ magnetic field of 1000 gauss is passing perpendicularly through the area defined by the closed loop shown in the figure. If the magnetic field reduces linearly to 500 gauss, in the next 5 s, then induced emf in the loop is

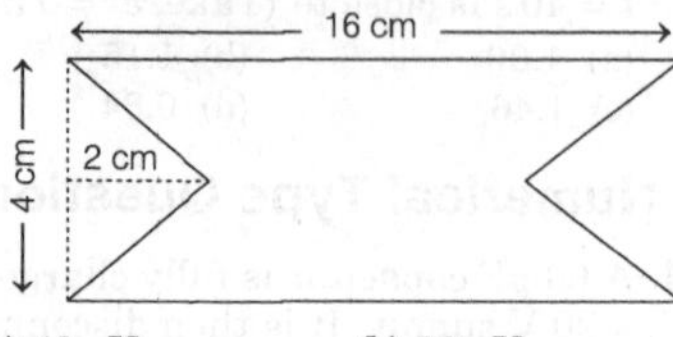

(a) 48 μV (b) 28 μV
(c) 56 μV (d) 36 μV

11 The dimension of stopping potential V_0 in photoelectric effect in units of Planck's constant h, speed of light c and gravitational constant G and ampere A is

(a) $h^{-2/3}c^{-1/3}G^{4/3}A^{-1}$ (b) $h^{1/3}G^{2/3}c^{1/3}A^{-1}$
(c) $h^2G^{3/2}c^{1/3}A^{-1}$ (d) $h^{2/3}c^{5/3}G^{1/3}A^{-1}$

12 When photon of energy 4.0 eV strikes the surface of a metal A, the ejected photoelectrons have maximum kinetic energy T_A eV and de-Broglie wavelength λ_A. The maximum kinetic energy of photoelectrons liberated from another metal B by photon of energy 4.50 eV is $T_B = (T_A - 1.5)$ eV. If the de-Broglie wavelength of these photoelectrons $\lambda_B = 2\lambda_A$, then the work function of metal B is

(a) 4 eV (b) 2 eV
(c) 1.5 eV (d) 3 eV

13 Effective capacitance of parallel combination of two capacitors C_1 and C_2 is 10 μF. When these capacitors are individually connected to a voltage source of 1 V, the energy stored in the capacitor C_2 is 4 times that of C_1. If these capacitors are connected in series, then their effective capacitance will be

(a) 4.2 μF (b) 3.2 μF
(c) 1.6 μF (d) 8.4 μF

14 Proton with kinetic energy of 1 MeV moves from south to north. It gets an acceleration of 10^{12} m/s^2 by an applied magnetic field (west to east). The value of magnetic field (rest mass of proton is 1.6×10^{-27} kg)

(a) 71 mT (b) 0.071 mT
(c) 0.71 mT (d) 7.1 mT

15 Consider a uniform rod of mass $M = 4m$ and length l pivoted about its centre. A mass m moving with velocity v making angle $\theta = \frac{\pi}{4}$ to the rod's long axis collides with one end of the rod and sticks to it. The angular speed of the rod-mass system just after the collision is

(a) $\frac{3\sqrt{2}}{7}\frac{v}{l}$ (b) $\frac{4}{7}\frac{v}{l}$
(c) $\frac{3}{7}\frac{v}{l}$ (d) $\frac{3}{7\sqrt{2}}\frac{v}{l}$

16 The critical angle of a medium for a specific wavelength, if the medium has relative permittivity 3 and relative permeability $\frac{4}{3}$ for this wavelength, will be

(a) 45° (b) 60°
(c) 15° (d) 30°

17 Boolean relation at the output stage Y for the following circuit is

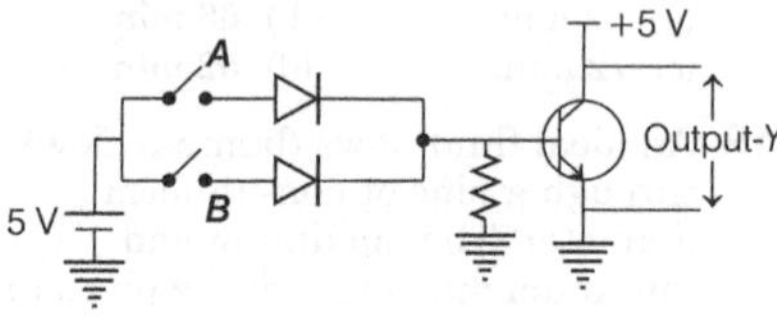

(a) $A \cdot B$ (b) $A + B$
(c) $\overline{A} + \overline{B}$ (d) $\overline{A} \cdot \overline{B}$

18 The length of a potentiometer wire is 1200 cm and it carries a current of 60 mA. For a cell of emf 5 V and internal resistance of 20 Ω, the null point on it is found to be at 1000 cm. The resistance of whole wire is

(a) 80 Ω (b) 100 Ω
(c) 60 Ω (d) 120 Ω

19 The graph which depicts the results of Rutherford gold foil experiment with α-particles is
θ : Scattering angle
Y : Number of scattered α-particles detected
(plots are schematic and not to scale)

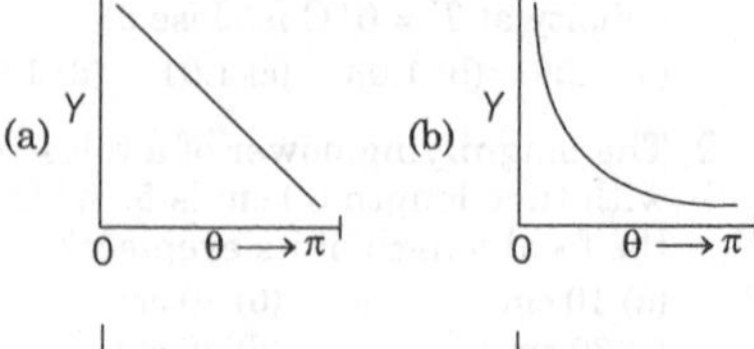

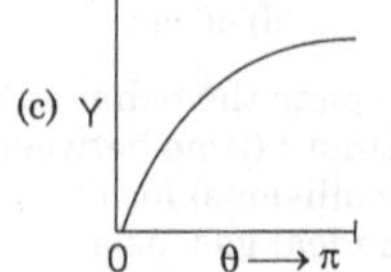

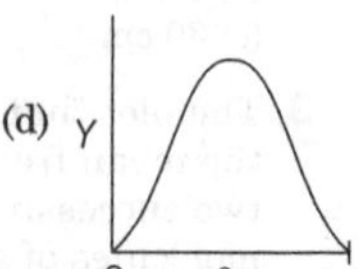

20 The coordinates of centre of mass of a uniform flag shaped lamina (thin flat plate) of mass 4 kg.
(The coordinates of the same are shown in figure) are

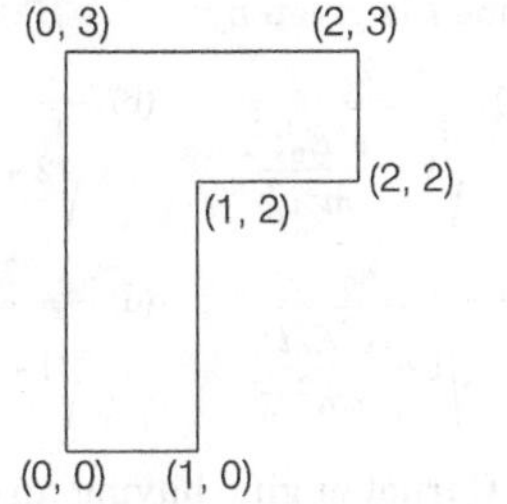

(a) (1.25 m, 1.50 m)
(b) (1 m, 1.75 m)
(c) (0.75 m, 0.75 m)
(d) (0.75 m, 1.75 m)

Numerical Type Questions

21 Four resistances of $15\,\Omega$, $12\,\Omega$, $4\,\Omega$ and $10\,\Omega$ respectively in cyclic order to form Wheatstone's network. The resistance that is to be connected in parallel with the resistance of $10\,\Omega$ to balance the network is .... Ω.

22 A one metre long (both ends open) organ pipe is kept in a gas that has double the density of air at STP. Assuming the speed of sound in air at STP is 300 m/s, the frequency difference between the fundamental and second harmonic of this pipe is ......Hz.

23 A particle is moving along the x-axis with its coordinate with the t given by $x(t) = 10 + 8t - 3t^2$. Another particle is moving along the y-axis with its coordinate as a function of time given by $y(t) = 5 - 8t^3$. At $t = 1$ s, the speed of the second particle as measured in the frame of the first particle is given as $\sqrt{v}$. Then v (in m/s) is ....

24 A point object in air is in front of the curved surface of a *plano-convex* lens. The radius of curvature of the curved surface is 30 cm and the refractive index of the lens material is 1.5, then the focal length of the lens (in cm) is ....

25 A body A, of mass $m = 0.1$ kg has an initial velocity of $3\hat{i}$ ms^{-1}. It collides elastically with another body, B of the same mass which has an initial velocity of $5\hat{j}$ ms^{-1}. After collision, A moves with a velocity $\mathbf{v} = 4(\hat{i} + \hat{j})$. The energy of B after collision is written as $\frac{x}{10}$ J. The value of x is ....

ANSWERS

1. (c)	2. (a)	3. (c)	4. (c)	5. (c)	6. (b)	7. (b)	8. (d)	9. (a)	10. (c)
11. (*)	12. (a)	13. (c)	14. (c)	15. (a)	16. (d)	17. (d)	18. (b)	19. (b)	20. (d)
21. (10)	22. (106.06)	23. (580)	24. (60)	25. (1)					

For Detailed Solutions
Visit : https://bit.ly/3o4skkW
Or Scan :

8 January, Shift-II

1 As shown in the figure, a battery of emf ε is connected to an inductor L and resistance R in series. The switch is closed at $t = 0$. The total charge that flows from the battery, between $t = 0$ and $t = t_c$ (t_c is the time constant of the circuit) is

(a) $\frac{\varepsilon L}{R^2}\left(1 - \frac{1}{e}\right)$ (b) $\frac{\varepsilon L}{R^2}$
(c) $\frac{\varepsilon R}{eL^2}$ (d) $\frac{\varepsilon L}{eR^2}$

2 A simple pendulum is being used to determine the value of gravitational acceleration g at a certain place. The length of the pendulum is 25.0 cm and a stop watch with 1s resolution measures the time taken for 40 oscillations to be 50 s. The accuracy in g is
(a) 2.40% (b) 5.40%
(c) 4.40% (d) 3.40%

3 A particle of mass m is dropped from a height h above the ground. At the same time another particle of the same mass is thrown vertically upwards from the ground with a speed of $\sqrt{2gh}$. If they collide head-on completely inelastically, then the time taken for the combined mass to reach the ground, in units of $\sqrt{\frac{h}{g}}$ is

(a) $\sqrt{\frac{1}{2}}$ (b) $\frac{1}{2}$ (c) $\sqrt{\frac{3}{2}}$ (d) $\sqrt{\frac{3}{4}}$

4

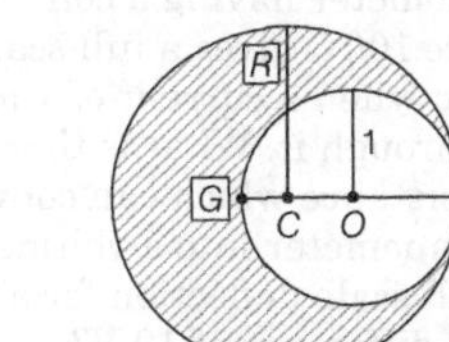

As shown in figure, when a spherical cavity (centred at O) of radius 1 is cut out of a uniform sphere of radius R (centred at C), the centre of mass of remaining (shaded) part of sphere is at G, i.e. on the surface of the cavity. R can be determined by the equation
(a) $(R^2 + R + 1)(2 - R) = 1$
(b) $(R^2 + R - 1)(2 - R) = 1$
(c) $(R^2 - R - 1)(2 - R) = 1$
(d) $(R^2 - R + 1)(2 - R) = 1$

5 A plane electromagnetic wave of frequency 25 GHz is propagating in vacuum along the z-direction. At a particular point in space and time, the magnetic field is given by $\mathbf{B} = 5 \times 10^{-8}\hat{j}$ T. The corresponding electric field $\mathbf{E}$ is
(speed of light, $c = 3 \times 10^8$ ms^{-1})

(a) $-1.66 \times 10^{-16}\hat{i}$ V/m
(b) $1.66 \times 10^{-16}\hat{i}$ V/m
(c) $-15\hat{i}$ V/m
(d) $15\hat{i}$ V/m

6 A particle of mass m and charge q is released from rest in a uniform electric field. If there is no other force on the particle, the dependence of its speed v on the distance x travelled by it is correctly given by (graphs are schematic and not drawn to scale)

(a)
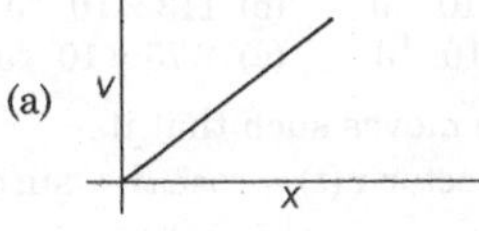

(b)
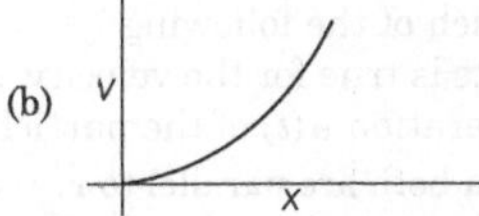

(c)
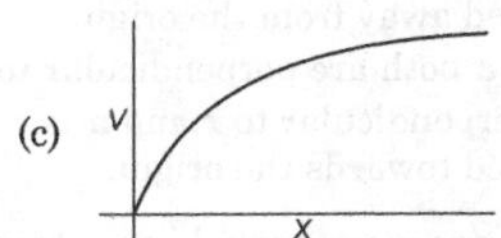

(d)
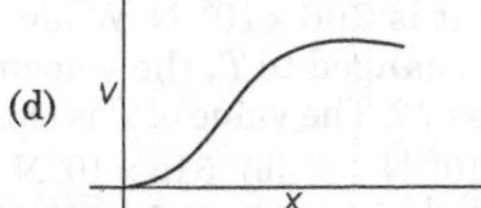

7 An object is gradually moving away from the focal point of a concave mirror along the axis of the mirror. The graphical representation of the magnitude of linear magnification (m) *versus* distance of the object from the mirror (x) is correctly given by (graphs are drawn schematically and are not to scale)

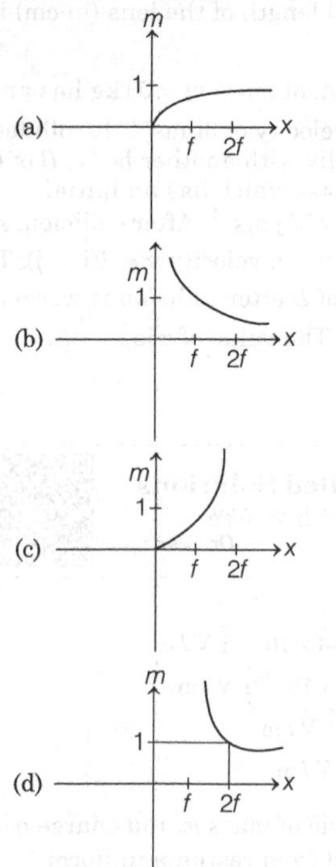

8 A uniform sphere of mass 500 g rolls without slipping on a plane horizontal surface with its centre moving at a speed of 5.00 cm/s. Its kinetic energy is

(a) 6.25×10^{-4} J (b) 1.13×10^{-3} J
(c) 8.75×10^{-4} J (d) 8.75×10^{-3} J

9 A particle moves such that its position vector $\mathbf{r}(t) = \cos\omega t\hat{\mathbf{i}} + \sin\omega t\hat{\mathbf{j}}$, where ω is a constant and t is time. Then, which of the following statements is true for the velocity $\mathbf{v}(t)$ and acceleration $\mathbf{a}(t)$ of the particle?

(a) $\mathbf{v}$ and $\mathbf{a}$ both are parallel to $\mathbf{r}$.
(b) $\mathbf{v}$ is perpendicular to $\mathbf{r}$ and $\mathbf{a}$ is directed away from the origin.
(c) $\mathbf{v}$ and $\mathbf{a}$ both are perpendicular to $\mathbf{r}$.
(d) $\mathbf{v}$ is perpendicular to $\mathbf{r}$ and $\mathbf{a}$ is directed towards the origin.

10 A transverse wave travels on a taut steel wire with a velocity of v when tension in it is 2.06×10^4 N. When the tension is changed to T, the velocity changed to $v/2$. The value of T is close to

(a) 10.2×10^2 N (b) 5.15×10^3 N
(c) 2.50×10^4 N (d) 30.5×10^4 N

11 A very long wire $ABDMNDC$ is shown in figure carrying current I. AB and BC parts are straight, long and at right angle.
At D wire forms a circular turn $DMND$ of radius R. AB, BC parts are tangential to circular turn at N and D. Magnetic field at the centre of circle is

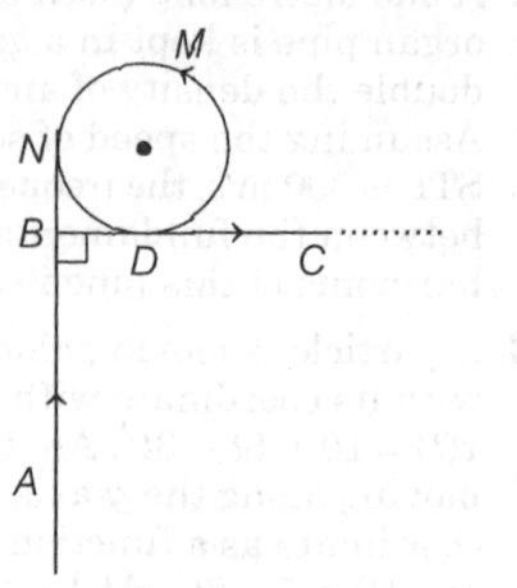

(a) $\frac{\mu_0 I}{2\pi R}\left(\pi + \frac{1}{\sqrt{2}}\right)$ (b) $\frac{\mu_0 I}{2R}$
(c) $\frac{\mu_0 I}{2\pi R}(\pi + 1)$ (d) $\frac{\mu_0 I}{2\pi R}\left(\pi - \frac{1}{\sqrt{2}}\right)$

12 In a double-slit experiment, at a certain point on the screen the path difference between the two interfering waves is $\frac{1}{8}$ th of a wavelength. The ratio of the intensity of light at that point to that at the centre of a bright fringe is

(a) 0.568 (b) 0.853 (c) 0.760
(d) 0.672

13 A galvanometer having a coil resistance 100 Ω gives a full scale deflection when a current of 1 mA is passed through it. What is the value of the resistance which can convert this galvanometer into a voltmeter giving full scale deflection for a potential difference of 10 V?

(a) 8.9 kΩ (b) 10 kΩ
(c) 9.9 kΩ (d) 7.9 kΩ

14 A capacitor is made of two square plates each of side a making a very small angle α between them, as shown in figure. The capacitance will be close to

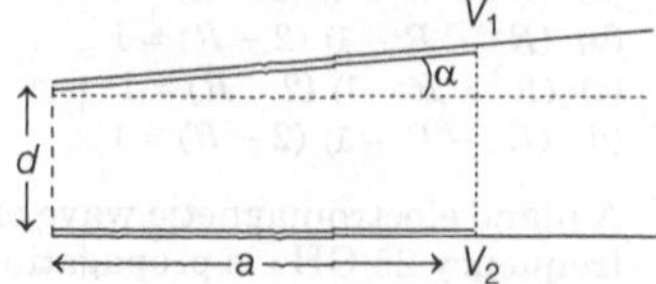

(a) $\frac{\varepsilon_0 a^2}{d}\left(1 - \frac{\alpha a}{4d}\right)$ (b) $\frac{\varepsilon_0 a^2}{d}\left(1 + \frac{\alpha a}{d}\right)$
(c) $\frac{\varepsilon_0 a^2}{d}\left(1 - \frac{\alpha a}{2d}\right)$ (d) $\frac{\varepsilon_0 a^2}{d}\left(1 - \frac{3\alpha a}{2d}\right)$

15 An electron (mass m) with initial velocity $\mathbf{v} = v_0\hat{\mathbf{i}} + v_0\hat{\mathbf{j}}$ is in an electric field $\mathbf{E} = -E_0\hat{\mathbf{k}}$. If λ_0 is initial de-Broglie wavelength of electron, then its de Broglie wavelength at time t is given by

(a) $\frac{\lambda_0}{\sqrt{1 + \frac{e^2E_0^2t^2}{m^2v_0^2}}}$ (b) $\frac{\lambda_0}{\sqrt{2 + \frac{e^2E_0^2t^2}{m^2v_0^2}}}$
(c) $\frac{\lambda_0}{\sqrt{1 + \frac{e^2E_0^2t^2}{2m^2v_0^2}}}$ (d) $\frac{\lambda_0\sqrt{2}}{\sqrt{1 + \frac{e^2E_0^2t^2}{m^2v_0^2}}}$

16 A Carnot engine having an efficiency of $\frac{1}{10}$ is being used as a refrigerator. If the work done on the refrigerator is 10 J, then the amount of heat absorbed from the reservoir at lower temperature is:

(a) 99 J (b) 100 J (c) 90 J
(d) 1 J

17 In the given circuit, value of Y is

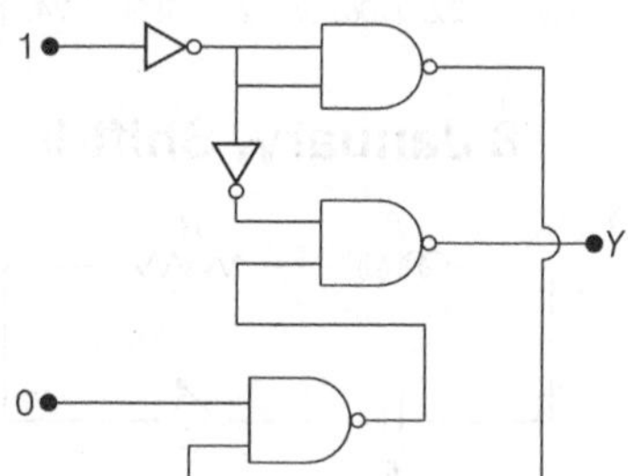

(a) toggles between 0 and 1
(b) 1
(c) 0
(d) will not execute

18

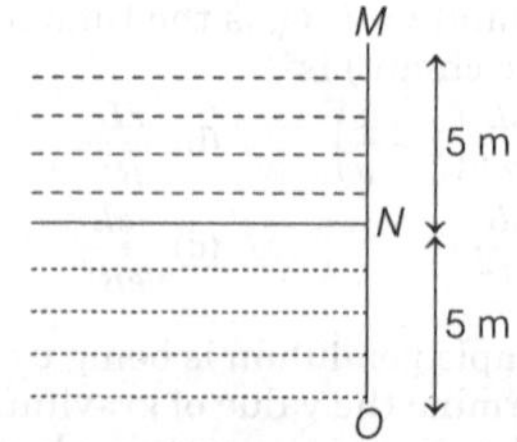

Two liquids of densities ρ_1 and $\rho_2(\rho_2 = 2\rho_1)$ are filled up behind a square wall of side 10 m as shown in figure. Each liquid has a height of 5 m. The ratio of the forces due to these liquids exerted on upper part MN to that at the lower part NO is (assume that the liquids are not mixing)

(a) 1/2 (b) 2/3
(c) 1/4 (d) 1/3

19 Consider a mixture of n moles of helium gas and $2n$ moles of oxygen gas (molecules taken to be rigid) as an ideal gas. Its C_p / C_V value will be

(a) 40/27 (b) 23/15
(c) 19/13 (d) 67/45

20 Consider two charged metallic spheres S_1 and S_2 of radii R_1 and R_2, respectively. The electric fields E_1 (on S_1) and E_2 (on S_2) on their surfaces are such that $E_1 / E_2 = R_1 / R_2$. Then the ratio V_1(on S_1)/V_2 (on S_2) of the electrostatic potentials on each sphere is

(a) $\left(\frac{R_1}{R_2}\right)^3$ (b) R_2 / R_1
(c) R_1 / R_2 (d) $(R_1 / R_2)^2$

Numerical Type Questions

21 The series combination of two batteries, both of the same emf 10 V, but different internal resistance of $20\,\Omega$ and $5\,\Omega$, is connected to the parallel combination of two resistors $30\,\Omega$ and $R\,\Omega$. The voltage difference across the battery of internal resistance $20\,\Omega$ is zero, the value of R(in Ω) is ....

22 Three containers C_1, C_2 and C_3 have water at different temperatures. The table below shows the final temperature T when different amounts of water (given in liters) are taken from each container and mixed (assume no loss of heat during the process)

C_1	C_2	C_3	T
$1l$	$2l$	–	60°C
–	$1l$	$2l$	30°C
$2l$	–	$1l$	60°C
$1l$	$1l$	$1l$	θ

The value of θ (in °C to the nearest integer) is ....

23 A ball is dropped from the top of a 100 m high tower on a planet. In the last $\frac{1}{2}$ s before hitting the ground, it covers a distance of 19 m. Acceleration due to gravity (in ms^{-2}) near the surface on that planet is ....

24 An asteroid is moving directly towards the centre of the earth. When at a distance of $10R$ (R is the radius of the earth) from the earth's centre, it has a speed of 12 km/s. Neglecting the effect of earth's atmosphere, what will be the speed of the asteroid when it hits the surface of the earth (escape velocity from the earth is 11.2 km/s)? Give your answer to the nearest integer in km/s ....

25 The first member of the Balmer series of hydrogen atom has a wavelength of 6561 Å. The wavelength of the second member of the Balmer series (in nm) is ..... .

ANSWERS

1. (d)	2. (c)	3. (c)	4. (a)	5. (d)	6. (c)	7. (d)	8. (c)	9. (d)	10. (b)
11. (a)	12. (b)	13. (c)	14. (c)	15. (c)	16. (c)	17. (c)	18. (c)	19. (c)	20. (d)
21. (30)	22. (50)	23. (8)	24. (16)	25. (486)					

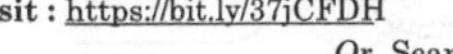

For Detailed Solutions
Visit : https://bit.ly/37jCFDH
Or Scan :

9 January, Shift-I

1 An electric dipole of moment $\mathbf{p} = (-\hat{\mathbf{i}} - 3\hat{\mathbf{j}} + 2\hat{\mathbf{k}}) \times 10^{-29}$ C-m is at the origin (0, 0, 0). The electric field due to this dipole at $\mathbf{r} = +\hat{\mathbf{i}} + 3\hat{\mathbf{j}} + 5\hat{\mathbf{k}}$ (note that $\mathbf{r} \cdot \mathbf{p} = 0$) is parallel to

(a) $(+\hat{\mathbf{i}} - 3\hat{\mathbf{j}} - 2\hat{\mathbf{k}})$ (b) $(-\hat{\mathbf{i}} - 3\hat{\mathbf{j}} + 2\hat{\mathbf{k}})$
(c) $(+\hat{\mathbf{i}} + 3\hat{\mathbf{j}} - 2\hat{\mathbf{k}})$ (d) $(-\hat{\mathbf{i}} + 3\hat{\mathbf{j}} - 2\hat{\mathbf{k}})$

2 Consider two ideal diatomic gases A and B at some temperature T. Molecules of the gas A are rigid and have a mass m. Molecules of the gas B have an additional vibrational mode and have a mass $\frac{m}{4}$. The ratio of the specific heats (C_V^A and C_V^B) of gas A and B respectively is

(a) 5 : 9 (b) 7 : 9
(c) 3 : 5 (d) 5 : 7

3 Two particles of equal mass m have respective initial velocities $u\hat{\mathbf{i}}$ and $u\left(\frac{\hat{\mathbf{i}} + \hat{\mathbf{j}}}{2}\right)$. They collide completely inelastically. The energy lost in the process is

(a) $\frac{3}{4}mu^2$ (b) $\sqrt{\frac{2}{3}}mu^2$
(c) $\frac{1}{3}mu^2$ (d) $\frac{1}{8}mu^2$

4 A charged particle of mass m and charge q moving under the influence of uniform electric field $E\hat{\mathbf{i}}$ and a uniform magnetic field $B\hat{\mathbf{k}}$ follows a trajectory from point P to Q as shown in the figure. The velocities at P and Q are respectively, $v\hat{\mathbf{i}}$ and $-2v\hat{\mathbf{j}}$. Then, which of the following statements (A, B, C, D) are the correct? (Trajectory shown is schematic and not to scale)

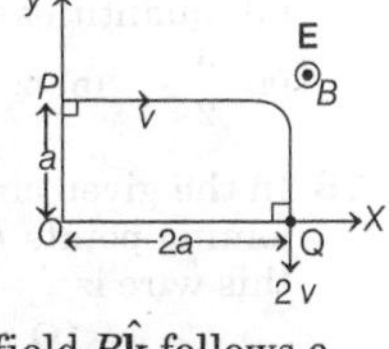

1. $E = \frac{3}{4}\left(\frac{mv^2}{qa}\right)$.
2. Rate of work done by the electric field at P is $\frac{3}{4}\left(\frac{mv^3}{a}\right)$.
3. Rate of work done by both the fields at Q is zero.
4. The difference between the magnitude of angular momentum of the particle at P and Q is $2\,mav$.

(a) (1), (3),(4) (b) (1), (2), (3)
(c) (1), (2), (3), (4) (d) (2), (3), (4)

5 A particle moving with kinetic energy E has de Broglie wavelength λ. If energy ΔE is added to its energy, the wavelength become $\lambda / 2$. Value of ΔE is

(a) $2E$ (b) $4E$ (c) $3E$ (d) E

6 A vessel of depth $2h$ is half filled with a liquid of refractive index $2\sqrt{2}$ and the upper half with another liquid of refractive index $\sqrt{2}$. The liquids are immiscible. The apparent depth of the inner surface of bottom of vessel will be

(a) $\frac{h}{\sqrt{2}}$ (b) $\frac{3}{4}h\sqrt{2}$ (c) $\frac{h}{3\sqrt{2}}$ (d) $\frac{h}{2(\sqrt{2}+1)}$

7 Radiation with wavelength 6561Å falls on a metal surface to produce photoelectrons. The electrons are made to enter a uniform magnetic field of 3×10^{-4}T. If the radius of the largest circular path followed by the electrons is 10 mm, the work function of the metal is close to

(a) 0.8 eV (b) 1.1 eV
(c) 1.8 eV (d) 1.6 eV

8 Consider a sphere of radius R which carries a uniform charge density ρ. If a sphere of radius $\frac{R}{2}$ is carved out of it, as shown in the figure the ratio $\frac{|\mathbf{E}_A|}{|\mathbf{E}_B|}$ of magnitude of electric field $\mathbf{E}_A$ and $\mathbf{E}_B$ respectively, at points A and B due to the remaining portion is

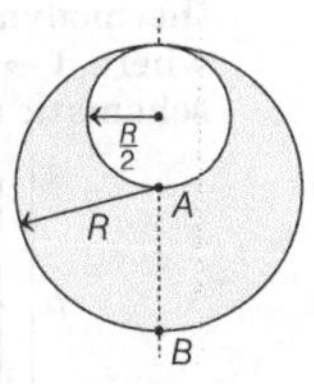

(a) $\frac{21}{34}$ (b) $\frac{18}{54}$ (c) $\frac{17}{54}$ (d) $\frac{18}{34}$

9 Three solid sphere each of mass m and diameter d are stuck together such that the lines connecting the centres form an equilateral triangle of side of length d. The ratio I_0 / I_A of moment of inertia I_0 of the system about an axis passing the centroid and about centre of any of the spheres I_A and perpendicular to the plane of the triangle is

(a) $\frac{15}{13}$ (b) $\frac{13}{15}$ (c) $\frac{13}{23}$ (d) $\frac{23}{13}$

10 Water flows in a horizontal tube (see figure). The pressure of water changes by 700 Nm^{-2} between A and B, where the area of cross-section are 40 cm^2 and 20 cm^2, respectively. Find the rate of flow of water through the tube. (Take, density of water = 1000 kgm^{-3})

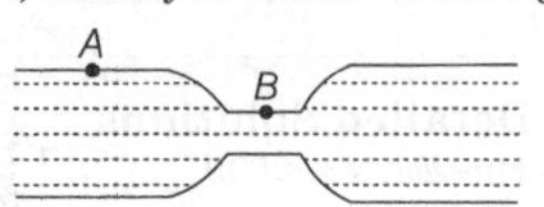

(a) 3020 cm^3/s (b) 2420 cm^3/s
(c) 2720 cm^3/s (d) 1810 cm^3/s

11 Three harmonic waves having equal frequency ν and same intensity I_0, have phase angles 0, $\frac{\pi}{4}$ and $-\frac{\pi}{4}$, respectively. When they are superimposed, the intensity of the resultant wave is close to

(a) $0.2I_0$ (b) I_0 (c) $3I_0$ (d) $5.8I_0$

12 The aperture of a telescope is 5m. The separation between the moon and the earth is 4×10^5 km. With light of wavelength of 5500 Å, the minimum separation between objects on the surface of moon, so that they are just resolved, is close to

(a) 600 m (b) 20 m (c) 60 m (d) 200 m

13 Which of the following is an equivalent cyclic process corresponding to the thermodynamic cyclic given in the fig.? where, $1\rightarrow2$ is adiabatic. (Graphs are schematic and are not to scale)

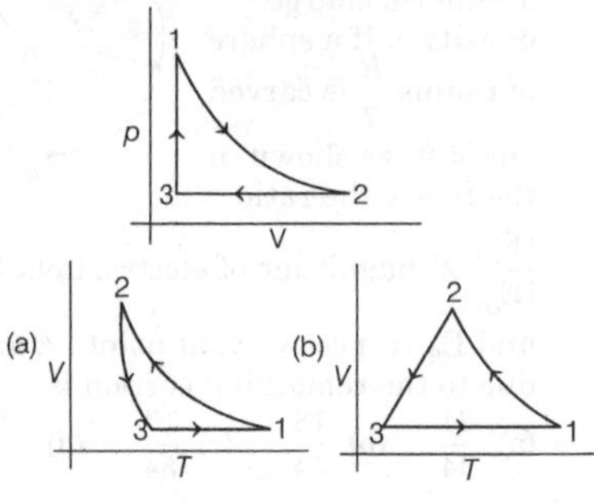

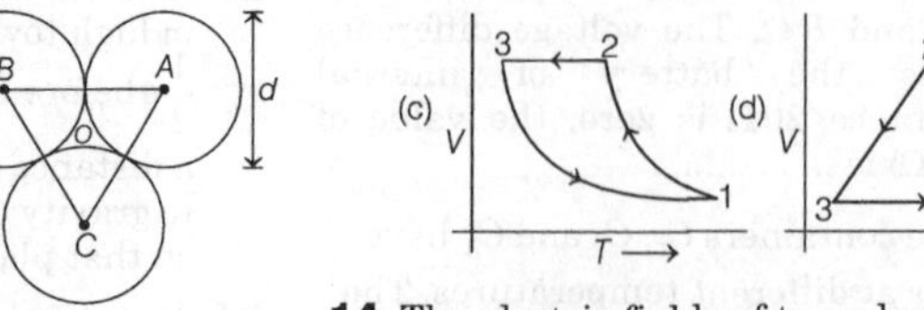

14 The electric fields of two plane electromagnetic plane waves in vacuum are given by

$$\mathbf{E}_1 = E_0\hat{\mathbf{j}}\cos(\omega t - kx)$$

and $\mathbf{E}_2 = E_0\hat{\mathbf{k}}\cos(\omega t - ky)$

At $t = 0$, a particle of charge q is at origin with a velocity $\mathbf{v} = 0.8c\hat{\mathbf{j}}$ (c is the speed of light in vacuum). The instantaneous force experienced by the particle is

(a) $E_0q\,(0.8\hat{\mathbf{i}} - \hat{\mathbf{j}} + 0.4\hat{\mathbf{k}})$
(b) $E_0q\,(0.4\hat{\mathbf{i}} - 3\hat{\mathbf{j}} + 0.8\hat{\mathbf{k}})$
(c) $E_0q(-0.8\hat{\mathbf{i}} + \hat{\mathbf{j}} + \hat{\mathbf{k}})$
(d) $E_0q\,(0.8\hat{\mathbf{i}} + \hat{\mathbf{j}} + 0.2\hat{\mathbf{k}})$

15 Consider a force $\mathbf{F} = -x\hat{\mathbf{i}} + y\hat{\mathbf{j}}$. The work done by this force in moving a particle from point $A(1, 0)$ to $B(0, 1)$ along the line segment is (all quantities are in SI units)

(a) $\frac{3}{2}$ (b) 2 (c) 1 (d) $\frac{1}{2}$

16 In the given circuit diagram, a wire is joining points B and D. The current in this wire is

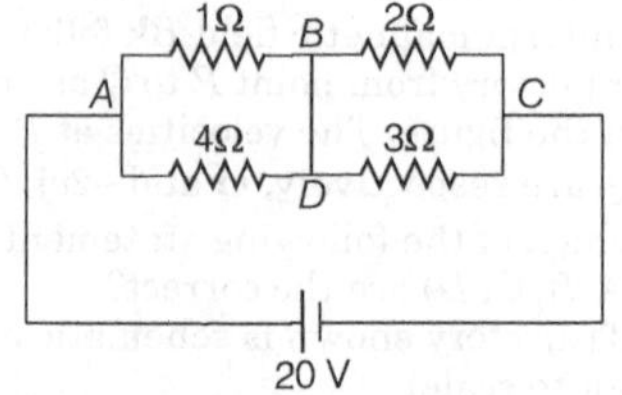

(a) 0.4 A (b) zero (c) 2A (d) 4A

17 A long, straight wire of radius a carries a current distributed uniformly over its cross-section. The ratio of the magnetic fields due to the wire at distance $\frac{a}{3}$ and $2a$ respectively, from the axis of the wire is

(a) $\frac{3}{2}$ (b) 2 (c) $\frac{2}{3}$ (d) $\frac{1}{2}$

18 A quantity f is given by $f = \sqrt{\frac{hc^5}{G}}$, where c is speed of light, G universal gravitational constant and h is the Planck's constant. Dimension of f is that of

(a) area (b) volume
(c) momentum (d) energy

19 If the screw on a screw gauge is given six rotations, it moves by 3 mm on the main scale. If there are 50 divisions on the circular scale, the least count of the screw gauge is

(a) 0.001 cm (b) 0.01 cm
(c) 0.02 cm (d) 0.001 cm

20 A body A of mass m is moving in a circular orbit of radius R about a planet. Another body B of mass $m/2$ collides with A with a velocity which is half ($\mathbf{v}/2$), the instantaneous velocity $\mathbf{v}$ of A. The collision is completely inelastic. Then, the combined body

(a) escapes from the planet's gravitational field
(b) starts moving in an elliptical orbit around the planet
(c) falls vertically downward towards the planet
(d) continues to move in a circular orbit

Numerical Type Questions

21 Both the diodes used in the circuit shown are assumed to be ideal and have negligible resistance when these are forward biased. Built in potential in each diode is 0.7V. For the input voltages shown in the figure, the voltage (in volts) at point A is ....

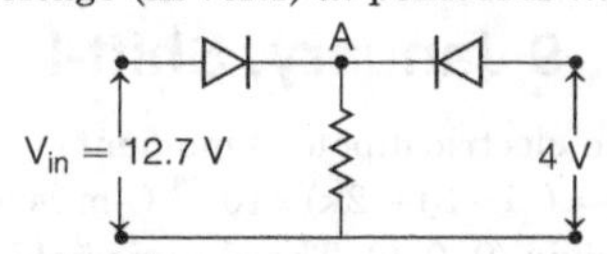

22 The distance x covered by a particle in one dimensional motion varies with time t as $x^2 = at^2 + 2bt + c$. If the acceleration of the particle depends on x as x^{-n}, where n is an integer, the value of n is ....

23 One end of a straight uniform 1 m long bar is pivoted on horizontal table. It is released from rest when it makes an angle 30° from the horizontal (see figure). Its angular speed when its hits the table is given as $\sqrt{n}\ \text{s}^{-1}$, where n is an integer. The value of n is ....

24 A body of mass $m = 10$ kg is attached to one end of a wire of length 0.3 m. The maximum angular speed (in rad s^{-1}) with which it can be rotated about its other end in space station is (breaking stress of wire = 4.8×10^7 Nm^{-2} and area of cross-section of the wire $=10^{-2}$ cm^2) is

25 In a fluorescent lamp choke (a small transformer) 100 V of reverse voltage is produced when the choke current changes uniformly from 0.25A to 0 in a duration of 0.025 ms.The self-inductance of the choke (in mH) is estimated to be ....

ANSWERS

1. (c)	2. (d)	3. (d)	4. (b)	5. (c)	6. (b)	7. (b)	8. (d)	9. (c)	10. (c)
11. (d)	12. (c)	13. (b)	14. (d)	15. (c)	16. (c)	17. (c)	18. (d)	19. (a)	20. (b)
21. (12)	22. (3)	23. ($\sqrt{15}$)	24. (4)	25. (10)					

For Detailed Solutions
Visit : https://bit.ly/2HeKKPg
Or Scan :

9 January, Shift-II

1 A rod of length L has non-uniform linear mass density given by $\rho(x) = a + b\left(\frac{x}{L}\right)^2$, where a and b are constants and $0 \le x \le L$. The value of x for the centre of mass of the rod is at

(a) $\frac{3}{4}\left(\frac{2a+b}{3a+b}\right)L$ (b) $\frac{4}{3}\left(\frac{a+b}{2a+3b}\right)L$

(c) $\frac{3}{2}\left(\frac{a+b}{2a+b}\right)L$ (d) $\frac{3}{2}\left(\frac{2a+b}{3a+b}\right)L$

2 Planet A has mass M and radius R. Planet B has half the mass and half the radius of planet A. If the escape velocities from the planets A and B are v_A and v_B respectively, then $\frac{v_A}{v_B} = \frac{n}{4}$. The value of n is

(a) 1 (b) 2 (c) 3 (d) 4

3 Two steel wires having same length are suspended from a ceiling under the same load. If the ratio of their energy stored per unit volume is 1 : 4, the ratio of their diameters is

(a) $\sqrt{2}:1$ (b) $1:\sqrt{2}$
(c) 2 : 1 (d) 1 : 2

4 The energy required to ionise a hydrogen like ion in its ground state is 9 Rydbergs. What is the wavelength of the radiation emitted when the electron in this ion jumps from the second excited state to the ground state ?

(a) 8.6 nm (b) 24.2 nm
(c) 11.4 nm (d) 35.8 nm

5 For the four sets of three measured physical quantities as given below. Which of the following options is correct?

(i) $A_1 = 24.36$, $B_1 = 0.0724$, $C_1 = 256.2$
(ii) $A_2 = 24.44$, $B_2 = 16.082$, $C_2 = 240.2$
(iii) $A_3 = 25.2$, $B_3 = 19.2812$, $C_3 = 236.183$
(iv) $A_4 = 25$, $B_4 = 236.191$, $C_4 = 19.5$

(a) $A_1 + B_1 + C_1 < A_3 + B_3 + C_3 < A_2 + B_2 + C_2 < A_4 + B_4 + C_4$
(b) $A_4 + B_4 + C_4 < A_1 + B_1 + C_1 = A_2 + B_2 + C_2 = A_3 + B_3 + C_3$
(c) $A_4 + B_4 + C_4 < A_1 + B_1 + C_1 = A_3 + B_3 + C_3 < A_2 + B_2 + C_2$
(d) $A_1 + B_1 + C_1 = A_2 + B_2 + C_2 = A_3 + B_3 + C_3 = A_4 + B_4 + C_4$

6 A particle of mass m is projected with a speed u from the ground at an angle $\theta = \frac{\pi}{3}$ w.r.t. horizontal (X-axis). When it has reached its maximum height, it collides completely inelastically with another particle of the same mass and velocity $u\hat{i}$. The horizontal distance covered by the combined mass before reaching the ground is

(a) $\frac{3\sqrt{3}}{8}\frac{u^2}{g}$ (b) $\frac{3\sqrt{2}}{4}\frac{u^2}{g}$

(c) $\frac{5}{8}\frac{u^2}{g}$ (d) $2\sqrt{2}\frac{u^2}{g}$

7 The current i in the network is

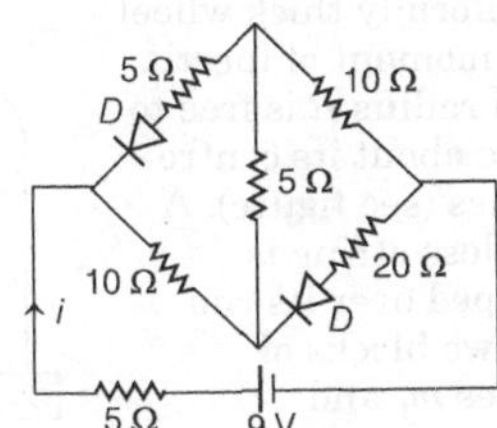

(a) 0.2A (b) 0 A
(c) 0.6 A (d) 0.3 A

8 A wire of length L and mass per unit length 6.0×10^{-3} kgm^{-1} is put under tension of 540 N. Two consecutive frequencies that it resonates at are : 420 Hz and 490 Hz. Then, L in metres is

(a) 8.1 m (b) 2.1 m
(c) 5.1 m (d) 1.1 m

9 Two identical capacitors A and B, charged to the same potential 5V are connected in two different circuits as shown below at time $t = 0$. If the charge on capacitors A and B at time $t = CR$ is Q_A and Q_B respectively, then (here e is the base of natural logarithm)

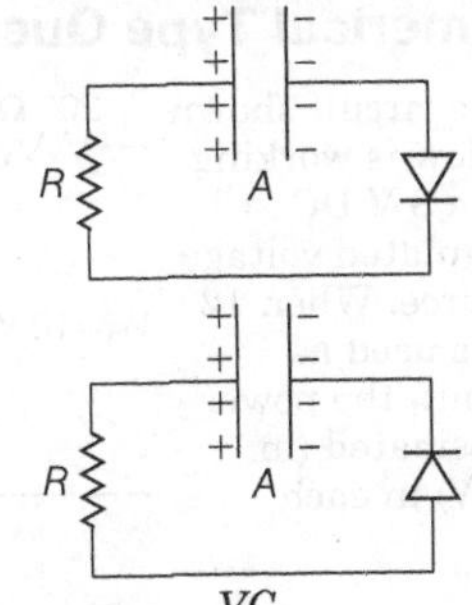

(a) $Q_A = VC$, $Q_B = \frac{VC}{e}$
(b) $Q_A = \frac{VC}{e}$, $Q_B = \frac{CV}{2}$
(c) $Q_A = \frac{CV}{2}$, $Q_B = \frac{VC}{e}$
(d) $Q_A = VC$, $Q_B = CV$

10 Two gases-argon (atomic radius 0.07 nm, atomic weight 40) and xenon (atomic radius 0.1 nm, atomic weight 140) have the same number density and are at the same temperature. The ratio of their respective mean free times is closest to

(a) 4.67 (b) 2.3
(c) 3.67 (d) 1.83

11 An electron of mass m and magnitude of charge $|e|$ initially at rest gets accelerated by a constant electric field **E**. The rate of change of de Broglie wavelength of this electron at time t ignoring relativistic effects is

(a) $\frac{h}{|e|\mathbf{E}\sqrt{t}}$ (b) $-\frac{h}{|e|\mathbf{E}t}$

(c) $\frac{-h}{|e|\mathbf{E}t^2}$ (d) $\frac{|e|\mathbf{E}t}{h}$

12 A small spherical droplet of density d is floating exactly half immersed in a liquid of density ρ and surface tension T. The radius of the droplet is (take note that the surface tension applies an upward force on the droplet)

(a) $r = \sqrt{\frac{3T}{(2d-\rho)g}}$ (b) $r = \sqrt{\frac{T}{(d-\rho)g}}$

(c) $r = \sqrt{\frac{T}{(d+\rho)g}}$ (d) $r = \sqrt{\frac{2T}{3(d+\rho)g}}$

13 In L-C circuit, the inductance $L = 40$ mH and capacitance $C = 100\mu$F. If a voltage $V(t) = 10\sin(314t)$ is applied to the circuit, the current in the circuit is given as

(a) $0.52\cos 314t$ (b) $0.52\sin 314t$
(c) $10\cos 314t$ (d) $5.2\cos 314t$

14 A plane electromagnetic wave is propagating along the direction $\frac{\hat{i}+\hat{j}}{\sqrt{2}}$ with its polarisation along the direction $\hat{k}$. The correct form of the magnetic field of the wave would be (here B_0 is an appropriate constant)

(a) $B_0\hat{k}\cos\left(\omega t - k\frac{\hat{i}+\hat{j}}{\sqrt{2}}\right)$

(b) $B_0\frac{\hat{j}-\hat{i}}{\sqrt{2}}\cos\left(\omega t + k\frac{\hat{i}+\hat{j}}{\sqrt{2}}\right)$

(c) $B_0\frac{\hat{i}-\hat{j}}{\sqrt{2}}\cos\left(\omega t - k\frac{\hat{i}+\hat{j}}{\sqrt{2}}\right)$

(d) $B_0\frac{\hat{i}+\hat{j}}{\sqrt{2}}\cos\left(\omega t - k\frac{\hat{i}+\hat{j}}{\sqrt{2}}\right)$

15 An electron gun is placed inside a long solenoid of radius R on its axis. The solenoid has n turns/length and carries a current I. The electron gun shoots an electron along the radius of the solenoid with speed v. If the electron does not hit the surface of the solenoid, maximum possible value of v is (all symbols have their standard meaning)

(a) $\frac{2e\mu_0 nIR}{m}$ (b) $\frac{e\mu_0 nIR}{4m}$

(c) $\frac{e\mu_0 nIR}{2m}$ (d) $\frac{e\mu_0 nIR}{m}$

16 A small circular loop of conducting wire has radius a and carries current i. It is placed in a uniform magnetic field B perpendicular to its plane such that when rotated slightly about its diameter and released, it starts performing simple harmonic motion of time period T. If the mass of the loop is m, then

(a) $T = \sqrt{\frac{\pi m}{iB}}$ (b) $T = \sqrt{\frac{2m}{iB}}$

(c) $T = \sqrt{\frac{2\pi m}{iB}}$ (d) $T = \sqrt{\frac{\pi m}{2iB}}$

17 There is a small source of light at some depth below the surface of water (refractive index $= \frac{4}{3}$) in a tank of large cross-sectional surface area. Neglecting any reflection from the bottom and absorption by water, percentage of light that emerges out of surface is (nearly)

[Use the fact that surface area of a spherical cap of height h and radius of curvature R is $2\pi Rh$]

(a) 34% (b) 50%

(c) 17% (d) 21%

18 A spring mass system (mass m, spring constant k and natural length l) rests in equilibrium on horizontal disc. The free end of the spring is fixed at the centre of the disc. If the disc together with spring mass system, rotates about it 's axis with an angular velocity ω, ($k >> m\omega^2$), the relative change in the length of the spring is best given by the option

(a) $\frac{m\omega^2}{3k}$ (b) $\frac{2m\omega^2}{k}$

(c) $\sqrt{\frac{2}{3}}\left(\frac{m\omega^2}{k}\right)$ (d) $\frac{m\omega^2}{k}$

19 A particle starts from the origin at $t = 0$ with an initial velocity of $3.0\hat{i}$ m/s and moves in the xy-plane with a constant acceleration $(6.0\hat{i} + 4.0\hat{j})\text{m/s}^2$. The x-coordinate of the particle at the instant when its y-coordinate is 32 m is D metres. The value of D is

(a) 50 (b) 60

(c) 40 (d) 32

20 A uniformly thick wheel with moment of inertia I and radius R is free to rotate about its centre of mass (see figure). A massless string is wrapped over its rim and two blocks of masses m_1 and $m_2(m_1 > m_2)$ are attached to the ends of the string. The system is released from rest. The angular speed of the wheel when m_1 descents by a distance h is

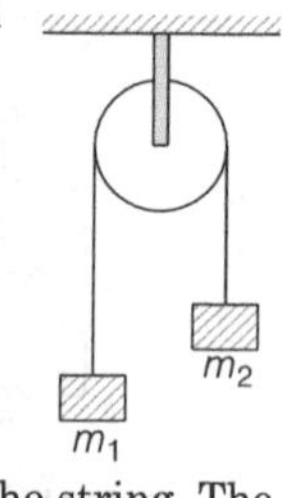

(a) $\left[\frac{2(m_1 + m_2)gh}{(m_1 + m_2)R^2 + 1}\right]^{\frac{1}{2}}$

(b) $\left[\frac{m_1 + m_2}{(m_1 + m_2)R^2 + 1}\right]^{\frac{1}{2}} gh$

(c) $\left[\frac{m_1 - m_2}{(m_1 + m_2)R^2 + 1}\right]^{\frac{1}{2}} gh$

(d) $\left[\frac{2(m_1 - m_2)gh}{(m_1 + m_2)R^2 + 1}\right]^{\frac{1}{2}}$

Numerical Type Questions

21 The circuit shown below is working as a 8 V DC regulated voltage source. When 12 V is used as input, the power dissipated (in mW) in each diode is; (considering both Zener diodes are identical) ....

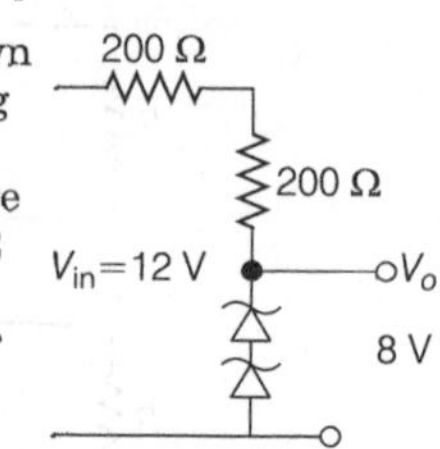

22 In a meter bridge experiment, S is a standard resistance, R is a resistance wire. It is found that balancing length is $l = 25$ cm. If R is replaced by a wire of half length and half diameter that of R of same material, then the balancing distance l'(in cm) will now be ....

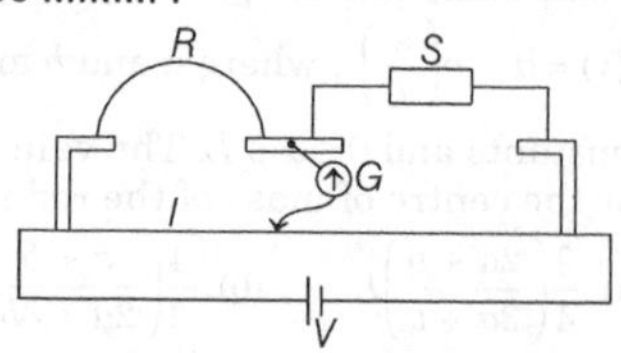

23 Starting at temperature 300 K, one mole of an ideal diatomic gas ($\gamma = 1.4$) is first compressed adiabatically from volume V_1 to $V_2 = \frac{V_1}{16}$. It is then allowed to expand isobarically to volume $2V_2$. If all the processes are the quasi-static, then the final temperature of the gas (in °K) is (to the nearest integer) ....

24 An electric field $\mathbf{E} = 4x\hat{i} - (y^2 + 1)\hat{j}$ N/C passes through the box shown in figure. The flux of the electric field through surfaces $ABCD$ and $BCGF$ are marked as ϕ_I and ϕ_{II}, respectively. The difference between $(\phi_I - \phi_{II})$ is (in N-m²/C) ....

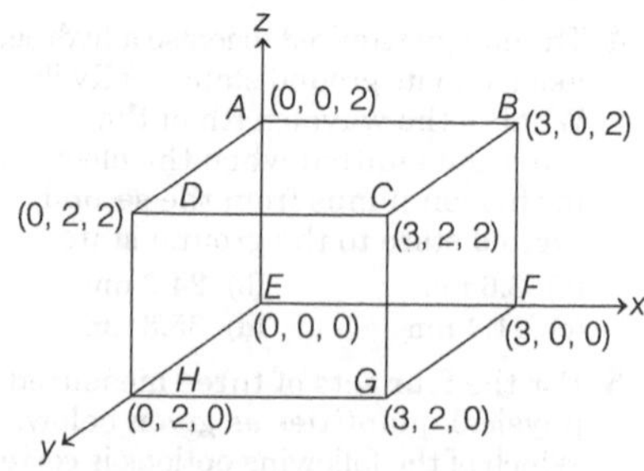

25 In a Young's double slit experiment, 15 fringes are observed on a small portion of the screen when light of wavelength 500 nm is used. Ten fringes are observed on the same section of the screen when another light source of wavelength λ is used. Then, the value of λ is (in nm) ....

ANSWERS

1. (a)	2. (d)	3. (a)	4. (c)	5. (*)	6. (a)	7. (d)	8. (b)	9. (a)	10. (*)
11. (c)	12. (a)	13. (a)	14. (c)	15. (c)	16. (c)	17. (c)	18. (d)	19. (b)	20. (d)
21. (40)	22. (40)	23. (1819)	24. (–48)	25. (750)					

For Detailed Solutions

Visit : https://bit.ly/2T6a5h1

Or Scan :